Metrology for Inclusive Growth of India

Dinesh K. Aswal
Editor

Metrology for Inclusive Growth of India

Springer

Editor
Dinesh K. Aswal
CSIR-National Physical Laboratory
New Delhi, India

ISBN 978-981-15-8871-6 ISBN 978-981-15-8872-3 (eBook)
https://doi.org/10.1007/978-981-15-8872-3

This Springer imprint is published by the registered company Springer Nature Singapore Pte Ltd.
The registered company address is: 152 Beach Road, #21-01/04 Gateway East, Singapore 189721, Singapore

Dedicated to

75 Glorious Years of CSIR-National Physical Laboratory in the Service of India

75th Foundation Day

4th January 2021
Platinum Jubilee

Message of Honorable PM

सत्यमेव जयते

प्रधान मंत्री

Prime Minister

MESSAGE

I am pleased to learn that Council of Scientific and Industrial Research-National Physical Laboratory (CSIR-NPL) is celebrating its 75th Foundation Day on January 04, 2021. It is a memorable occasion for everyone associated with the institution.

India is a youthful nation with youthful dreams and aspirations. With holistic efforts, we are creating the right ecosystems where energy, strong intent and good ideas of our youth power the vision of self-reliant India. Role of research, innovation and enterprise is vital to further this vision.

We have no dearth of talent, hard working workforce or purpose. This is evident from India's rapid emergence in the field of science and technology in the last few decades. We need to take leverage of our strong scientific research ecosystem to find solutions that help enhance the quality of life of common citizens.

On the occasion of the Platinum Jubilee celebration of CSIR-NPL, different technology-based initiatives will showcase the scientific progress made by the nation. These achievements will further the spirit of *Aatmanirbhar Bharat* in every sector of activity.

The book titled 'Metrology for Inclusive Growth of India' will reflect the contribution made by the institution. It will also serve as a reference guide for the scientists, researchers and science students in the times to come.

My greetings and best wishes to all the members of CSIR-NPL for the 75th Foundation Day celebrations. May the occasion strengthen scientific temper in younger generations and inspire everyone to serve the nation with renewed dedication and commitment.

(Narendra Modi)

New Delhi
भाद्रपद 10, शक संवत, 1942
1st September, 2020

Dr. Harsh Vardhan
Union Minister for Health & Family Welfare,
Science & Technology and Earth Sciences
348, A-Wing, Nirman Bhawan
New Delhi – 110011

Foreword

It indeed is a matter of great pleasure that CSIR-The National Physical Laboratory (CSIR-NPL), one of the premier laboratories of Council of Scientific and Industrial Research (CSIR), is celebrating its 75th Foundation Day on 4 January 2021 and bringing a book on *Metrology for Inclusive Growth of India* through Springer Nature to commemorate Platinum Jubilee Foundation Day. CSIR-NPL is the National Measurement Institute (NMI) of the country by an Act of Parliament and represents India as a permanent member in International Bureau of Weights and Measure (BIPM), France.

This book is an excellent initiative by Team CSIR-NPL which elaborates the role of metrology in the governance for ensuring the conformity assessment along with the gap areas and future directions in order to make India as knowledge and economic power. It presents high technicality of national/primary measurement standards of India in the areas of physico-mechanical, electrical and electronics, Indian Standard Time (IST), electromagnetic, environment, biomedical, materials and Bhartiya Nirdeshak Dravyas (BND®). The dissemination of measurement traceability from primary/national standards of CSIR-NPL to various stakeholders across the country through national legal metrology framework and accreditation process is elaborated.

The book has been written using a motivational approach and impresses upon the fact that the measurement capabilities of India are at par with international standards. Adoption of these measurement standards by all the stakeholders will ensure a self-reliant India. The book brilliantly showcases the contributions made by CSIR-NPL during its glorious 75 years, including discovery of indelible ink, preservation of the "Original Constitution of India" kept at Parliament Library, New Delhi, dissemination of Indian Standard Time (IST) with few nanoseconds for a safe digital India, supporting micro, small and medium enterprises (MSMEs) through calibrations and technology transfers, use of BND® by testing and calibration laboratories for quality assurance, India's first national environmental standard laboratory for certification of ambient air and industrial emission measuring equipment, calibration facilities for renewable energy and biomedical equipment, etc.

The uniqueness of the book is that it connects the metrology, in association with accreditation and standards, to the science and technology, government and regulators, civil society and media and the industries. I am sure that this book will be appreciated by science and engineering graduates and postgraduates, researchers, policy-makers and entrepreneurs. On the occasion of its Platinum Jubilee Foundation Day, I congratulate all the members of CSIR-NPL. I am sure the celebration of this day will inspire each individual working with CSIR-NPL and all those working with testing and calibration laboratories across the country. I also extend my best wishes for publication and wide circulation of the book.

New Delhi, India

Shekhar C. Mande
Secretary, DSIR, and DG CSIR

Preface

Metrology—the science of measurements, embracing both experimental and theoretical determinations at any level of uncertainty—plays a key role in growth of national economy as well as in improving the quality of life of citizens. Accurate and reliable measurements are essential to ensure quality of food and products, defect-free and efficient manufacturing, affordable healthcare—both diagnosis and therapy, high impact science and technology, sustainable energy, clean environment, climate change studies, international trade without technical barrier, navigation and telecommunications, cyber-physical systems, cybersecurity, equality and trust in society, policy-making, etc.

For India, CSIR-National Physical Laboratory (CSIR-NPL), established on January 4, 1947 and mandated as the National Metrology Institute (NMI), ensures that the primary/national measurement standards of the country not only have the international equivalences concluded at International Bureau of Weights and Measures (BIPM), France, but also these measurement capabilities get disseminated to all the stakeholders across the country. To mark the 75th Foundation Day, this book is edited to showcase the glorious metrology contributions made by CSIR-NPL for inclusive growth of India. The book has 21 Chapters—all written by the scientists of CSIR-NPL having vast domain expertise in metrology.

The introductory Chapter 1 reviews various policy-making innovation models for creation of knowledge economy and knowledge society. In the Aswal model, the metrology fortifies strong interactions among government –academia–industry–society, which in turn, contribute to inclusive growth of the country and improves economy and quality of life. Chapters 2 and 3 describe the evolution of defining 7 basic SI units of measurements from artefacts to recent revisions based on fundamental constants of nature. The international metrology network to harmonize the measurement standards across the globe as well as their dissemination are described. The role of CSIR-NPL in the quality-infrastructure (metrology, accreditation and standards) of India is presented in a lucid way. Chapters 4 and 5 deal with the significance of "second"—the unit of time, its generation though atomic clock ensembles and application in navigation, telecommunications, internet, power grid, banking, digital governance, transport, finance, defence, cyber-physical systems,

5G technologies, artificial intelligence, internet of things etc. These chapters delightfully explain how CSIR-NPL, the timekeeper of nation, is realizing the Indian Standard Time (IST) with an accuracy of few nanoseconds and contributing towards the growth of a safe digital India. Chapters 6–9 elaborate on various national/primary standards on physico-mechanical metrology related to mass, length, thermal, optical, acoustic, vacuum, pressure, force and hardness, ultrasound, and fluid flow. These chapters describe the role of these parameters in the growth of industrial and strategic in the country. Chapter 10 is dedicated to the electrical and electronics (primary Josephson voltage standard (JVS), quantum Hall resistance (QHR) and derived parameters related to DC/AC/RF voltage, current, power, energy, etc.). Chapters 11 and 12 presents the electromagnetic standards (measurement facilities for 1 Hz to 110 GHz of electromagnetic spectrum, microwave power, electric-field, specific absorption rate, phasor measurement unit, magnetic field measurement parameters, etc.) for upcoming smart technologies. Chapters 13 and 14 are dedicated to the metrology for the environment and provide an in-depth assessment of environmental governance in India, including needs of environmental metrology, ideal quality infrastructure, interface among stakeholders, role of CSIR-NPL, existing national facilities traceability to SI units, upcoming national services/facilities, etc. The glimpses of the environmental regulatory guidelines of India which focus on quality and transparency of measurement data are also discussed. Chapter 15 describes biomedical metrology activities of CSIR-NPL in the context of new medical device rules notified by the government of India. The establishment of medical device calibration facilities and the development of diagnostic devices are presented. Chapters 16 and 17 deal with the materials metrology activities of CSIR-NPL as well as calibration and testing facilities for materials and devices. The development of advanced materials for their applications in strategic, energy and societal applications are described. Chapter 18 and Chapter 19 are dedicated to the Bharatiya Nirdeshak Dravyas (BND®), the trademarked Indian certified refence material. These chapters elaborate on the fundamental of reference materials to their production and marketing. The tremendous efforts of CSIR-NPL on the production of BNDs in association with various reference materials producers in the areas viz. water, cement, high purity precious metals, petroleum, metals and alloys, food, pesticides, calibration of analytical equipment, Newtonian fluids, etc. are illustrated. Chapter 20 presents the role of human resources in metrology for effective implementation of national quality infrastructure and its socioeconomic impact. Chapter 21 summarises the contributions of CSIR-NPL in supporting government, industry, academia and society for the growth of nation.

All the chapters put together constitute a resource material on understanding the role of CSIR-NPL for inclusive growth of India. The data on metrology, standards and accreditation related to national quality infrastructure are uniquely presented under a single window. All the chapters have data presented in the form of tables, block diagrams, photographs, layouts and line diagrams, which make them interesting for readers without compromising the technical richness of the contents. Therefore, this book will emerge as an important reference book for science and engineering graduates and post graduates, researchers, scientists, regulators, policymakers and entrepreneurs. All the chapters put together constitute a resource material on understanding the role of CSIR-NPL for inclusive growth of India. The data on metrology, standards and accreditation related to national quality infrastructure are uniquely presented under a single window. All the chapters have data presented in the form of tables, block diagrams, photographs, layouts and line diagrams, which make them interesting for readers without compromising the technical richness of the contents. Therefore, this book will emerge as an important reference book for science and engineering graduates and post graduates, researchers, scientists, regulators, policymakers and entrepreneurs.

I feel fortunate being the director of CSIR-NPL on its platinum jubilee foundation day celebration. I sincerely acknowledge the enormous contributions made by former directors namely, Dr. K S Krishnan (FRS), Prof. P K Kichlu, Prof. A R Verma, Dr. A P Mitra (FRS), Prof. S K Joshi, Prof. E S Raja Gopal, Prof. A K Raychaudhuri, Dr. Krishan Lal, Prof. Vikram Kumar and Prof. R.C. Budhani. I also thank Dr. Shekhar C Mande, Director General, CSIR and Prof. Arun K Grover, Chairman, Research Council, CSIR-NPL, for their consistent support and encouragements. Finally, my deepest appreciation to all the contributing authors for writing the chapters in the shortest possible time, and expressly to Dr. Sushil Kumar, for helping me in editing the book.

New Delhi, India
August 5, 2020

Dinesh K. Aswal
(Editor)

Contents

About the Editor

Dr. Dinesh K. Aswal is present Director of CSIR-National Physical Laboratory (CSIR-NPL), maintaining lien with his position as "Outstanding Scientist" at Bhabha Atomic Research Centre (BARC), Department of Atomic Energy (DAE), Government of India, Mumbai. Dr. Aswal, during 2012–2015, has also served as Secretary, Atomic Energy Education Society (AEES), Mumbai, which is an autonomous institution under DAE. He has also held additional charges of Director, Central Electronics Engineering Research Institute (CEERI, Pilani, 2019–2020); Director, National Institute of Science, Technology and Development Studies (NISTADS, New Delhi, 2019), and Chairman, National Accreditation Board for Testing and Calibration Laboratories (NABL, India, 2019–20).

His Aswal Model for Inclusive Growth i.e. the national metrology infrastructure of international equivalence is indispensable for establishing synergy among government, academia, industries and civil society, is a key enabler for realizing Aatmanirbhar Bharat. A convincing representative of India, he advocates for national interests in the meetings of International Bureau of Weights and Measures (BIPM) and Asia Pacific Metrology Program (APMP) and ensures that the metrology capabilities of India are recognized internationally, which are so important for international trade without technical barrier, manufacturing of quality products in the country, environment and climate change, sustainable energy, affordable health, cyber-security, and quality of life.

Dr. Aswal was born on June 30, 1965 in a village of Uttarakhand state. After completion of M.Sc. in Physics (Gold Medalist) from Garhwal University, he joined Bhabha Atomic Research Center (BARC) as a scientist in 1986 where he rose to a position of outstanding scientist and was deputed as Director CSIR-NPL in 2015. His visionary scientific leadership has put India several foot forward in realizing the measurement capabilities at par with the developed nations. He implemented redefined international system of units (the SI) based on the fundamental and atomic constants in the country on May 20, 2019. He architected the "National Atomic Clock Ensemble" to generate Indian Standard Time (IST™) that registered an accuracy of 2.8 nanosecond at BIPM, which is one of the best in the world. To ensure a safe digital India through accurate and reliable timestamps, he launched "One Nation One Time" mission in which all device clocks in the country need to be synchronized with IST™. He introduced an ambitious national program on indigenous production of Certified Reference Materials (CRM) under the trademark of Bhartiya Nirdeshak Dravyas (BND®), and got produced them through various reference material producers for cement, petroleum, high purity gold, minerals, water, food, pesticides, gases, etc. He is brainchild of India's first certification scheme (NPLICS™) for the ambient air and industrial emission monitoring equipment, which is Gazette notified by the government of India. Other national metrology programs originated by him include calibration for biomedical devices, light-efficiency measurements, efficiency of solar cells and modules, strategic materials, etc., which are reaping huge benefits to the country through quality control in indigenous products.

He has contributed in the indigenous development of several key strategic and societal technologies, including hydrogen-sulphide gas sensors for heavy water plants, thermoelectric power-generators for conversion of waste heat into electricity, recycling of plastic waste into structural tiles, $PM_{2.5}$ and PM_{10} measuring devices, new security ink for the Indian currency notes, magnetron microwave devices, ultraviolet disinfection caskets, etc.

He also made enormous contributions in science education and dissemination. In fighting humongous situation arisen due to COVID-19 in the year 2020, as a Chairman of NABL he steered accreditation of more than 500 RT-PCR testing labs across the country (in 26 states) with a testing capacity of more than 2 lakhs samples per day. As a Secretary of AEES, he made a policy for admission of students belonging to poor backgrounds in the atomic energy schools (spread across 10 states). In 2016, he organized a mega India International Science Festival for the masses at CSIR-NPL. His efforts in providing metrological traceability to testing and calibration laboratories for their accurate and reliable results are benefitting small and medium industries in quality control.

His fundamental scientific contributions in the field of gas sensors, thermoelectricity, superconductivity, molecular electronics and metrology are internationally recognized. He had visiting professor/scientist positions at several international institutes/universities viz. Institut d'Electronique de Microelectronique et de Nanotechnologie, Lille (France), Sizuoka University (Japan), Commissariat à l'Energie Atomique, Saclay (France), Weizmann Institute of Science (Israel), University of Yamanashi (Japan), University of Paris VII (France), Karlsruhe institute of Technology (Germany), and the University of South Florida (USA). He has over 500 publications to his credit in national and international conferences/journals, filed 6 patents, secured 4 trademarks for CSIR-NPL and edited 5 books.

He is a recipient of several national and international recognitions including Fellow of International Association of Advanced Materials, FIAAM, Sweden (2020); Fellow of National Academy of Sciences, India (FNASc), (2019); Academician, Asia Pacific Academy of Materials (APAM), Singapore (2019);

Science and Technology Excellence Award, Government of Uttarakhand (2017); Distinguished Faculty Award of Homi Bhabha National Institute (2015); Materials Research Society of India Medal (2010); DAE-SRC Outstanding Research Investigator Award (2008); Homi Bhabha Science and Technology Award, DAE, (2007), etc.

Chapter 1
Introduction: Metrology for All People for All Time

Dinesh K. Aswal

Abstract In this introductory chapter, a brief account of international metrology program for ensuring a single, coherent system of measurements traceable to SI units throughout the world is presented. The making of National Physical Laboratory (CSIR-NPL) as the National Metrology Institute of India is highlighted. Various policy-making innovation models for the creation of knowledge economy and knowledge society are reviewed. A new and innovative model, i.e., *Aswal model* for inclusive growth is introduced. In this model, metrology fortifies strong interactions among the four helices of Quadruple Helix (QH) model, i.e., government, academia, industry, and society, to ensure the inclusive and sustainable growth of a nation. The implications of this model on various sectors contributing to national economy and quality of life are discussed, which include equality and trust in society, affordable health, high-quality science and technology, sustainable energy, clean environment, and climate change, international trade, cyber-physical systems, cybersecurity, etc. Finally, the roles of CSIR-NPL in nation-building are highlighted.

Measurement is the first step that leads to control and eventually to improvement. If you can't measure something, you can't understand it. If you can't understand it, you can't control it. If you can't control it, you can't improve it.

– H. James Harrington

D. K. Aswal (✉)
CSIR-National Physical Laboratory, Dr. K.S. Krishnan Marg, New Delhi 110012, India
e-mail: dkaswal.npl@nic.in

D. K. Aswal (ed.), *Metrology for Inclusive Growth of India*,
https://doi.org/10.1007/978-981-15-8872-3_1

1.1 From a Landmark Initiative "Metre Convention" to a Milestone Decision "Revision of SI Units"

For all people for all time was the motive of a landmark initiative Metre Convention (Convention du Mètre), an international diplomatic treaty, for establishing a rational and universal set of measurement units, which was signed in Paris on May 20, 1875, by seventeen nations [1]. The Metre Convention led the worldwide focus on Metrology—the science of measurement—that resulted in an international structure, as depicted in Fig. 1.1, to ensure that all member states have identical capabilities on measurements [2]. This treaty founded the International Bureau of Weights and Measures (Bureau International des Poids et Mesures, BIPM) at Pavillon de Breteuil in Sèvres, France. BIPM is an intergovernmental organization under the authority of the General Conference on Weights and Measures (Conférence Générale des Poids et Mesures, CGPM) and under the supervision of the International Committee for Weights and Measures (Comité international des Poids et Mesures, CIPM). At present, CGPM has 62 Member States and observers from 40 Associate states and economies. The Republic of India signed the diplomatic treaty and became the Member State of CGPM in 1957 with CSIR-National Physical Laboratory (CSIR-NPL), New Delhi as its National Metrology Institute (NMI). CGPM meets once in every four years and has the responsibility to oversee the systems of measurement and adopt new definitions based on the recommendations of CIPM, which is supported by ten consultative committees related to Acoustics, Ultrasound, and Vibration; Electricity and Magnetism; Length; Mass and Related Quantities; Photometry and Radiometry; Amount of Substance: Metrology in Chemistry and Biology; Ionizing Radiation; Thermometry, Time and Frequency' and Units. BIPM coordinates international metrology and the development of the measurement system in association with the NMIs of the Member States and the Associates States. The

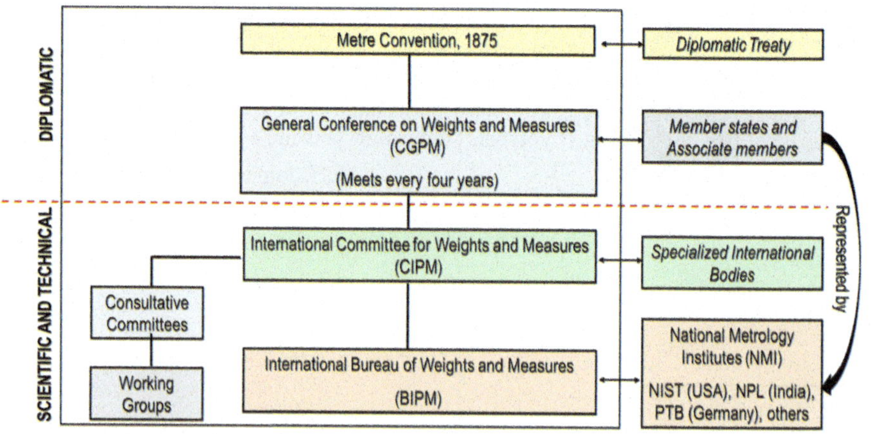

Fig. 1.1 International agreement on metrology for ensuring a single, coherent system of measurements traceable to SI units throughout the world

directors of NMIs meet annually at BIPM, France, to discuss the scientific and technical aspects of the International System of Units (the SI units) and their applications. BIPM, therefore, provides the basis for a single, coherent system of measurements traceable to SI units throughout the world.

On 16 November 2018, during the 26th CGPM meeting held at Versailles, France, 60 member states voted for a landmark revision of the SI units on the basis of fundamental constants that describe the natural world [3]. These seven SI units are: (i) second (*s*): caesium hyperfine frequency ($\Delta\nu_{Cs}$), (ii) meter (m): the speed of light in vacuum (*c*), (iii) kilogram (kg): the Planck constant (*h*), (iv) Ampere (A): the elementary charge (*e*), (v) Temperature (T): the Boltzmann constant (*k*), (vi) mole (M): the Avogadro constant (N_A), and (vii) candela (cd): the luminous efficacy of a defined visible radiation (K_{cd}). On behalf of India, a team lead by Dr. D.K. Aswal, Director CSIR-NPL, voted for the revision of the SI units (see Fig. 1.2a). The values of these fundamental constants have been fixed without any uncertainty, and therefore, the SI units are the apex measurement standards across the globe for any kind of measurements. To highlight the significance of this revision, a wall of SI units was constructed at CSIR-NPL campus, as shown in Fig. 1.2b, in which the left and right wings of the butterflies show a correlation between the SI unit and the associated fundamental constant. On May 20, 2019, CSIR-NPL implemented the revised SI units in India. The detailed explanation of revised SI units, the calibration and measurement capabilities (CMCs), and the role of regional metrology organization, e.g., Asia Pacific Metrology Program (APMP) are presented in Chap. 2.

1.2 National Physical Laboratory—The NMI of India

1.2.1 A Brief Historical Account

On September 26, 1942, the Council of Scientific and Industrial Research (CSIR), India was established under the leadership of Sir Shanti Swarup Bhatnagar. In 1943 the governing body of CSIR decided to establish five national laboratories—National Physical Laboratory, National Chemical Laboratory, National Metallurgical Laboratory, Fuel Research Station, and Glass and Ceramics Research Institute. The foundation for the National Physical Laboratory was laid on 4 January 1947, eight months before India became independent, by Pandit Jawaharlal Nehru, the first Prime Minister of India in the presence of Sir Charles Darwin, Director NPL, UK. The CSIR-NPL was expected to fill a wide gap in the scientific and industrial organizations of India. The account of the foundation of CSIR-NPL was published in *Nature, February 8, 1947* issue [4]. At that point of time, India neither had a well-equipped laboratory to take up the measurement standards work nor had the statutory acceptance from BIPM for the mass and length standards. In addition, the country also did not have the capability to disseminate the length and mass standards with scientific precession for the use of industry. Therefore, the core mandate assigned to

Fig. 1.2 **a** Photograph after India voting for the revised SI units at 26th CGPM meeting, Versailles, France, on 16 November 2018 (Center: Dr. D. K. Aswal, Director, NPL. Right: A. K. Srivastava, Secretary, Ministry of Consumer Affairs, Govt. of India; Left: Dr. T. D. Senguttuvan, NPL). **b** Wall of SI units at the campus of CSIR-National Physical Laboratory, New Delhi, depicting the relationship between the SI units of measurement and fundamental constants (conceptualized by Dr. D. K. Aswal and Dr. V. N. Ojha in 2016)

CSIR-NPL was to maintain the fundamental and derived standards and to undertake research for achieving a high degree of accuracy in the measurements through regular inter-comparisons with standard laboratories of the world.

The responsibility assigned to CSIR-NPL clearly indicated that it would in future represent India in the BIPM and act as NMI of the country, and therefore, CSIR-NPL was made a lab of unique importance to the Nation. The CSIR-NPL was formally opened by former Deputy Prime Minister Sardar Vallabhbhai Patel on 21 January 1950, and an account of this ceremony was published in *Nature March 24, 1951* issue [5]. The inaugural ceremony of CSIR-NPL was attended by a galaxy of eminent people from many countries and a photograph of this obtained from the album of CSIR-NPL is reproduced in Fig. 1.3. The prominent dignitaries present during the inauguration were Mr. C. Rajagopalachari, Pandit Jawaharlal Nehru, Dr. Shyama Prasad Mookerjee (at that time Minister of Industries and President of CSIR), Sir Shanti Swarup Bhatnagar (Director of CSIR), Dr. K. S. Krishnan (Director, NPL), Dr. K. N. Mathur (Assistant Director, Planning, NPL), Dr. Homi Jahangir Bhabha (Director, Tata Institute of Fundamental Research), Sir Robert Robinson (President of Royal Society), Prof. P. V. Auger (Director, Natural Sciences, UNESCO), Prof. W. D. Englehardt, Prof. D. E. H. Rydbeck (Chalmers Institute of Technology, Sweden), Prof. J. D. Bernal and Dr. E. U. Condon (National Bureau of Standards, USA now National Institute of Standards and Technology, NIST). It is apparent that the world had very high expectations from the CSIR-NPL towards the growth of India.

CSIR-NPL is the founding member of Asia Pacific Metrology Programme (APMP)—the associations of national metrology institutes in the Asia region—which is one of the six Regional metrology organizations (RMOs) of the world. RMOs function within the framework of CIPM Mutual Recognition Arrangement (MRA). Other 5 RMOs are: Intra-Africa Metrology System (AFRIMETS), Euro-Asian Cooperation of National Metrological Institutions (COOMET), European Association of National Metrology Institutes (EURAMET), Gulf Association for Metrology (GULFMET) and Inter-American Metrology System (SIM). The RMOs play an important role in establishing mutual confidence in the validity of calibration and measurement certificates (CMCs) issued by NMIs by carrying out key and supplementary comparisons and other related activities.

1.2.2 Weights and Measures Act of India

The "Standards of Weights and Measures Act" was enacted by the Government of India for the first time in 1956 to ensure that each of its citizens has an access to uniform standards of weights and measures that are traceable to the SI units. As mentioned earlier, India became a Member State of CGPM and BIPM in 1957. Depending upon the developments on SI units at CGPM, the Rules under the Weights and Measures Act are revised from time to time [6]. The Legal Metrology (National Standards) Rules, 2011 under Chapter III states: "*The work relating to the realization, establishment, custody, maintenance, determination, reproduction and updating of national standards of weights and measures shall, on the commencement of these rules, be the responsibility of the National Physical Laboratory.*" Therefore, this is the core mandate of CSIR-NPL.

The Legal Metrology under the Ministry of Consumer Affairs has established 5 Regional Reference Standard Laboratories (RRSLs) at Ahmedabad, Bangalore,

Fig. 1.3 A photograph of the opening ceremony of NPL-India on January 21, 1950 (obtained from the album of NPL). From left to right: Mr. G. M. Bhuta, Dr. K. N. Mathur, Prof. O. E. H. Rydbeck, Prof. H. J. Bhabha, Prof. P. Auger, Dr. E. U. Condon, Dr. S. P. Mookerjee, Pandit Jawahar Lal Nehru, Sri C. Rajgopalachari, Sardar Vallabhbhai Patel, Dr. S. S. Bhatnagar, Dr. K. S. Krishnan, Sir Robert Robinson, Lady Robinson, Prof. W. A. Englehardt, and Prof. J. D. Bernal

Bhubaneswar, Faridabad, and Guwahati. These RRSLs work as a central agency between CSIR-NPL, the NMI of the country, and the state government laboratories for ensuring the metrological traceability to the SI units. The responsibilities of RRSLs also include testing of models of weights and measures, verification of secondary standards of state government, calibration of sophisticated weighing and measuring instruments, consumer awareness program, etc. In order to disseminate the mass and length standards to the actual users, the secondary and working standards are fabricated by the India Government Mint, Mumbai. Legal Metrology has the mandate to establish and enforce standards of weights and measures, regulate trade and commerce in weights, measures, and other goods which are sold or distributed by weights, measures or numbers.

1.3 Knowledge Creation Models for Inclusive Growth

Inclusive growth as defined by *Organization for Economic Co-operation and Development* (OECD) is the economic growth that is distributed fairly across society and creates opportunities for all [7]. The inclusive growth denotes that every member of the society should have a chance to participate in the development and get benefited from its economic outcome. The implementation of conditions required for inclusive growth must ensure the compliance of 3 dimensions of sustainability, i.e., economic, environmental, and social. In 2015, the United Nation along with its 195 member states agreed upon a global movement on the Sustainable Development Goals (SDG), as depicted in Fig. 1.4, which will improve the quality of lives of the people [8]. These goals need to be accomplished by the year 2030 through the involvement of government, business, media, academia, science and technology institutions, and civil society.

Fig. 1.4 The sustainable development goals (SDGs) of the United Nation to be implemented by the year 2030 for the improved lives of the people across the globe

Unfortunately, in December 2019 a new coronavirus (nCov19) pandemic has spread globally, which so far has infected millions of people and killed more than five hundred thousand. The positive side is that majority of the people are getting recovered [9]. The nCov19 induced disease, known as COVID-19, mainly gets transmitted through droplets released by an infected person during his exhales, coughs, or sneezes. In the absence of vaccines and medicines, the only way to control the spread of pandemic is to minimize person-to-person contacts. Therefore, most governments imposed upon nation-wide lock-down for months together, e.g., India imposed the lock-down during March-May, 2020. However, these lock-downs halted economic activities worldwide. Consequently, many people lost employment, which also incited them to return to their native places [10]. Therefore, governments are thinking a fresh to revive the economy for achieving set targets and accomplishments of SDGs.

In June 2020, the Government of India launched *AtmaNirbhar Bharat* (self-reliant India) mission to enable the resurgence in the Indian economy [11]. *AtmaNirbhar Bharat* focuses on the following five pillars: economy, infrastructure, system, vibrant demography, and demand. *AtmaNirbhar Bharat* envisions the upliftment of (i) businesses including micro, small and medium enterprises (MSMEs)—the backbone of manufacturing and services in India, (ii) poor people, including migrants and farmers, (iii) agriculture—enhanced production, (iv) new horizons of growth, including Industry 4.0, and (v) government reforms and enablers. *AtmaNirbhar Bharat* program in a way promotes the glocalization model, i.e., a concept that combines "globalization" and "localization". The significance of the glocalization lies in the development of products that not only caters to the needs of high-end consumers and foreign markets but also are suitable for consumer segments of low-end and middle economy class [12]. Glocalization can only be empowered by the 'knowledge economy'—a concept that was coined during the 1960s, which epitomizes production of new knowledge and their utilization for economic growth [13, 14]. Knowledge economy is a shift from earlier growth models, which depended upon the labor force and utilization of natural resources. Therefore, in order to make glocalization a successful model for inclusive growth, there is a need to envisage and develop policies that enable creation of new and usable knowledge. Several models of knowledge creation activities exist in the literature, which is summarized in Fig. 1.5. A brief overview of these models is described below.

1.3.1 Mode 1, Mode 2, and Mode 3 Models

One of the earliest knowledge creation models is known as Mode 1, which was conceptualized by Gibbons et al. [15]. In this model, the university research is only to gain fundamental knowledge and may not contribute to the industry or economy. The validation of knowledge is mainly done through peer-reviewed research publications. In Mode 2, the research is focused, interdisciplinary, and context-driven to enable knowledge for application, quality assurance, and economic and social accountability

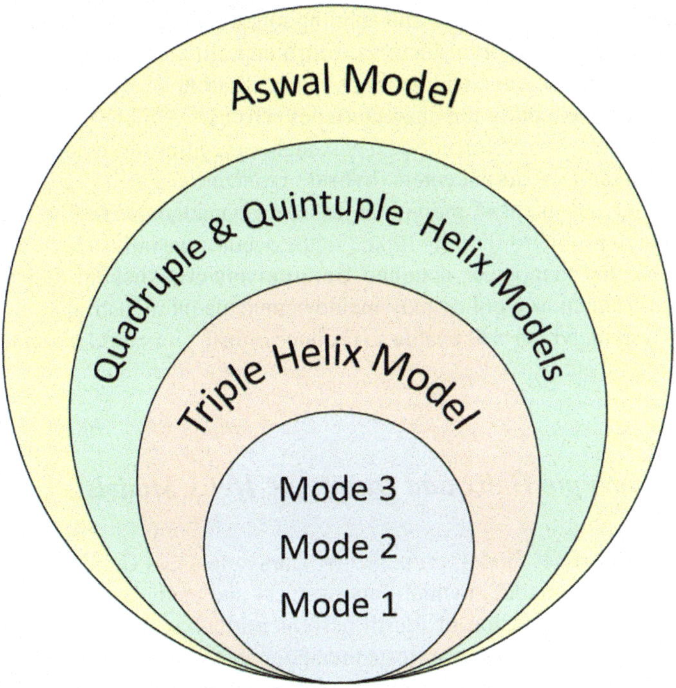

Fig. 1.5 Various models of knowledge creation activities for knowledge economy

[15]. Mode 3 model emphasizes the co-evolution of Mode 1 and Mode 2 innovation modes. The basic research (Mode 1) and the context-driven research (Mode 2) modes are envisioned to get link together for creation of interdisciplinary and transdisciplinary knowledge and innovation, which can be utilized for the advancement of societies, economic growth, and quality of life improvement [16].

1.3.2 Triple Helix Model

The Triple Helix (TH) model of Etzkowitz and Leydesdorff is based on the university-industry-government interactions for innovation and economic development [17–19]. In this model, each sector namely university, industry, and government is represented by helices. The trilateral interactions amongst these helices are very complex in nature and the mutual bilateral interactions play important roles. In TH innovation model, universities produce knowledge through basic and applied sciences, industries produce products based on the knowledge and innovations, and the government makes regulations for marketing. In addition, the government plays a crucial role in formulating appropriate policies and funding to support academia as well as industries. Industries may support academia through appropriate funding for further innovation. In addition, prolonged interactions can facilitate helices to change their partial characteristics by adopting the characteristics of others, which in turn, may lead to the creation of new hybrid institutions, e.g., technology transfer mechanisms, setting of science parks, incubation centers, hybrid organizations, etc. The TH model also helps in the advancement of individual helix, e.g., academia can find new context-driven research problems directly linked to the need of the industries; industries can utilize the skilled manpower obtained from universities to setup their own R&D centers; government as a policy may include curricula on industrial approach in the education system, which will enable to produce industry-oriented manpower for the national growth, etc.

1.3.3 Quadruple Helix and Quintuple Helix Models

Quadruple Helix (QH) model, proposed by Carayannis and Campbell, is an extension of Triple Helix model in which 'civil society and media' is added as the fourth helix [19, 20]. The inclusion of fourth helix is critical as the innovation approach becomes the user-driven and also tests the applicability of scientific knowledge for its robustness in the society. The QH model provides a broader picture of strong interactions amongst the helices that would assist in co-evolution, co-specialization, and co-development leading to innovations, e.g., technology, social reforms, products, services, etc. The QH model deals with the dynamics and evolution of these four helices and their impact on the economy and economic growth, unemployment, healthcare needs, environment, regulatory systems, and quality of life. The interactions amongst government-academia-industries-society though very complex in nature but need to be enhanced for the inclusive growth of a country. The QH model encourages networking of democracy, glocalization, knowledge economy, and knowledge society. Adding "natural environment" as fifth helix becomes the Quintuple Helix model, and therefore, deals with socio-ecological prospects for the sustainable development of knowledge-based society and economy, which also addresses climate change [20–22].

1.3.4 Aswal Model of Inclusive Growth

Though TH and QH models have been widely used in policy-making by developed nations, their utilities in developing nations are limited as they have the following major drawbacks.

(i) These models do not adequately elaborate essential infrastructure needed for their implementation, e.g., sophisticated equipment for conducting fundamental studies, the manufacturing and quality-assurance facilities with industries, access to information and communication technologies (ICT) for fast access to information, skilled human resources, etc. These models probably assume that such basic requirements are available naturally, which of course is true for the developed world and India. However, implementation of these models in lesser developing nations can fail if prior critical analyses of the essential requirements are not done.

(ii) The co-evolution, co-specialization, and co-development amongst the helices for creation of knowledge-intensive activities are assumed to occur naturally. The models do not divulge essential underlying scientific/technological mechanisms required for the facilitation of strong interactions amongst the helices. Therefore, utilizing these innovation models in policymaking without any understanding interaction mechanisms may not yield fruitful and desired outcomes.

To address the issues raised above amicably, recently Aswal has analyzed the QH model, particularly to understand the interaction mechanisms amongst the helices, and proposed that metrology is at core of such interactions [23]. The Aswal model of inclusive growth, as shown in Fig. 1.6, illustrates broad roles and responsibilities of four helices namely, government, academia, industries, and society as well as their bindings through national quality-infrastructure (QI). A formal definition of the QI,

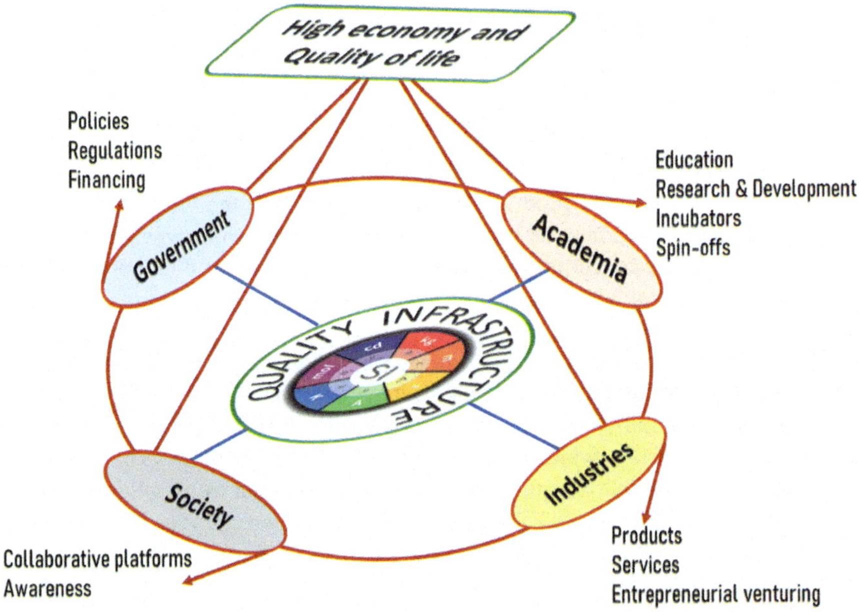

Fig. 1.6 The Aswal model of inclusive growth [23]. The trustworthy data measured with SI traceability from the national quality-infrastructure facilitates strong interactions amongst government, academia, industry, and society, which is essential for the national economy and quality of life

a brief overview of four helices is described later. A robust QI of a country ensures that all experimentally measured data are metrologically traceable to SI units, and therefore, have international equivalence. Such trustworthy measurement data are essential for the growth of individual helices as well as facilitating inherent trust for the collaboration amongst them in all directions, i.e., sideways, bottom-up, top-down, and crisscross, which in turn create knowledge and innovation for yielding high economy and improved quality of life. In addition, Aswal model envisages the requirement of qualified metrologists in each of the helices to ensure accuracies in the measurements along with associated uncertainties. The roles of government and society are paramount in sensitizing and formulation of industrial and science & technology policies that ensure sustainability through circular economy (a technology-enabled closed-loop system in which the creation of waste and pollution is minimized by recycling/reuse) as well as environmental issues and climate change.

A formal definition of QI has been proposed by the United Nations Industrial Development Organization (UNIDO), which states [24]: "*The system comprising the organizations (public and private) together with the policies, relevant legal and regulatory framework, and practices needed to support and enhance the quality, safety and environmental soundness of goods, services and processes. The quality infrastructure is required for the effective operation of domestic markets, and its international recognition is important to enable access to foreign markets. It is a critical element in promoting and sustaining economic development, as well as environmental and social wellbeing. It relies on metrology, standardization, accreditation, conformity assessment, and market surveillance.*"

In a country, the national quality policy ensures that institutions of the quality infrastructure, i.e., metrology, accreditation, and standards are well established. These QI institutions ensure the conformity assessment through calibration, testing, certification, and inspection. The QI institutions in developed nations, e.g., USA, are very strong, and therefore, have robust S&T and industrial policies, making them the highest gross domestic product (GDP) value nation in the world [25]. UNIDO supports the developing nations for their industrial growth and advocates for setting up a strong QI for the conformity assessment [26, 27]. If a nation does not have robust QI that ensures metrological traceability to SI units, a mistrust amongst government, academia, industries, and society may occur, which is illustrated here by citing a simple example. In a metro city, if the society feels that the pollution level is consistently high and posing risk to the health of its citizens and suppose the government data suggest that pollution is under control. To address this conflict, civil society may approach a court of law, which may order concerned regulator/government departments to provide reliable air-quality data and emission data from polluting sources (e.g., industries, vehicles, etc.). In order to obtain trustworthy data, the concerned regulator/government department will ensure that all used monitoring equipment is calibrated correctly. Therefore, equipment manufactures would be asked to provide certification of calibration from an accredited calibration laboratory. If the calibration laboratory finds that the equipment is not designed to measure the values correctly, the manufacturer (industry) may approach academia or scientific organization to help them to improve their designs. To verify the performance of design, the academia of course must have an access to a calibration laboratory. If measurements of all the concerned laboratories are not metrologically traceable to SI units, their results may vary significantly owing to biases and uncertainties in the measurements, as discussed in the next section. Thus, the absence of metrologically traceable QI in a country, varying and unreliable data not only creates mistrust amongst stakeholders but also creates a negative impact on the economy and quality of life. On the other hand, if a country has a strong QI, the question of suspicion does not arise as all the helices utilize accurate and precise measurement data traceable to SI units.

An internationally recognized QI in a country facilitates the growth of individual helices in the Aswal model. Government and regulators are benefited from having trustworthy data inputs for the formulation of national policies related to science and technology, industrial development, international trade, etc. as well as their implementations. The academia and S&T organizations have the advantages of getting appropriately calibrated equipment, which enables them to get high-quality scientific data for research as unwanted data can easily be disregarded. The original data contribute to discovering new fundamental phenomena in sciences as well as innovative technologies. The accurate and precise measurements allow industries to produce internationally competitive products according to trade standards. QI also satisfies the consumers as they can trust the quality of products and services manufactured and/or imported in the country.

1.3.4.1 The Four Helices of India

The four helices of India namely government, academia, industry, and society are so massive that it is next to impossible to provide a detailed description of them, and therefore, only a brief sketch is presented here.

Government

The Government of India is constitutionally a democratic republic and has a union government of 28 states and 8 union territories. The union has the legislative, executive, and judicial authority. The government creates steering mechanisms though the formation of the policies and regulations as well as financing for the country's growth in the right direction. The present Government has launched several new schemes for the development of the country and to enhance the quality of life of its citizens. Some of the schemes are Skill India Mission; Make in India; Digital India Mission; Start-up India; Stand Up India; Pradhan Mantri Kisan Samman Nidhi (PM-KISAN) Scheme; PM Garib Kalyan Yojana (PMGKY); Pradhan Mantri Kisan Pension Yojana; Jal Shakti Yojana; Jan Dhan Yojana; Swachh Bharat Mission; Sansad Adarsh Gram Yojana; Beti Bachao Beti Padhao; Hridaya Plan; PM Mudra Yojna; Ujala Yojna; Prime Minister Jeevan Jyoti Bima Yojana; Atal Pension Yojana; Pradhan Mantri Suraksha Bima Yojana; Gold Monetization Scheme; Setu Bhartam Yojana; Atal Bhujal Yojana (ABY); National Mission for Clean Ganga (NMCG);

Table 1.1 A partial list of regulatory bodies in India [23]

Category	Authority
Agricultural and processed food products	Agricultural and Processed Food Products Export Development Authority (APEDA)
Digital signatures	Controller of Certifying Authorities (CCA)
Drugs and medical devices	Drugs Controller General of India (DGCI)
Electronics and IT	Standardization Testing and Quality Certification (STQC)
Energy efficiency	Bureau of Energy Efficiency (BEE)
Environmental pollution	Central Pollution Control Board (CPCB)
Export certification	Export Inspection Council (EIC)
Food safety	Food Safety and Standards Authority of India (FSSAI)
Forensic	Central Forensic Science Laboratory (CFSL)
Industrial policy	Department of Industrial Policy and Promotion (DIPP)
Petroleum products	Petroleum & Natural Gas Regulatory Board (PNGRB)
Radiation safety	Atomic Energy Regulatory Board (AERB)
Renewable energy	Ministry of New and Renewable Energy (MNRE)
Securities market	Securities and Exchange Board of India (SEBI)
Telcom regulatory	Telcom Regulatory Authority of India (TRAI)
Weights and measures	Legal Metrology

Aarogya Setu; Ayushman Bharat; Unified Mobile Application for New-age Governance (UMANG); Pilgrimage Rejuvenation And Spirituality Augmentation Drive (PRASAD) Scheme, Prime Minister's Citizen Assistance and Relief in Emergency Situation (PM CARES); etc. As mentioned earlier, to counter the economic slowdown due to COVID-19 pandemic, government of India has made path-breaking reforms and pumped in >265 billion US dollars as stimulus package, which is equivalent to ~10% of India's GDP. Honorable Prime Minister gave a clarion call for *AtmaNirbhar Bharat* or Self-Reliant India Movement and outlined five pillars—economy, infrastructure, system, vibrant demography, and demand [28]. In order to implement these schemes, various government departments/regulators related to manufacturing, food, energy, health, environment, electronics, export, etc. are listed in Table 1.1, and all of them need national quality-infrastructure for conformity assessment.

The Academia—Universities, Science and Technology Institutions

The academia—universities and science and technology institutions, have the responsibility for the education, research and development, innovations, incubation, and spin-offs. India has a strong network of universities and S&T institutions. University Grant Commission (UGC) has recognized 825 universities belonging to central, state, deemed, and private [29]. UGC has also established 15 professional councils. In addition, the country has many other central institutions including, Indian Institute of Sciences (IISc), Indian Institutes of Technology (IITs), National Institutes of Technology (NITs), Indian Institutes of Management (IIMs), Indian Institutes of Information Technology (IIITs), School of Planning and Architecture (SPAs), National Institute of Science Education and Research (NISER), Indian Institutes of Science Education and Research (IISERs), All India Institute of Medical Sciences (AIIMS), etc. India also has many specialized departments and organization for R&D, such as Department of Atomic Energy (DAE), Indian Space Research Organization (ISRO), Defense Research and Development Organization (DRDO), Indian Council of Medical Research (ICMR), Council of Scientific and Industrial Research (CSIR), Indian Council of Agriculture Research (ICAR), etc.

Industries

The industries manufacture the products and related services and are major contributors to the economic growth of a country. India has a strong industry base almost in all sectors [30] including agriculture; automobiles and auto components; aviation and aeronautics; banking, finance and insurance; cement and building materials; engineering and manufacturing' gems and jewelry; healthcare; information technology; media and entertainment; metals, metallurgy and mining; oil and natural gas; pharmaceuticals; power, railways; renewable energy; steel; telecommunication; textiles; tourism and hospitality, etc.

The Civil Society and Media

The civil society and media facilitate the integration of local cultural values to the innovation and provide a platform to generate the reputation and acceptability of the products. India has a current population of 1.32 billion, which is likely to reach 1.54 billion by 2035 [31]. India with such a large population is also known to be a country of magnificent paradoxes, e.g., very poor and very rich people, extreme cold and extremely hot weather conditions, traditional knowledge and modern technologies, demographics and democracy, culture and modernity, etc. Therefore, civil society and media will have to play a dominant role in the realization of Aswal model for inclusive growth.

1.3.4.2 Quality-Infrastructure of India

Figure 1.7 summarizes India's QI institutions namely, metrology, accreditation, and standards as well as their international affiliations. The scientific metrology is the

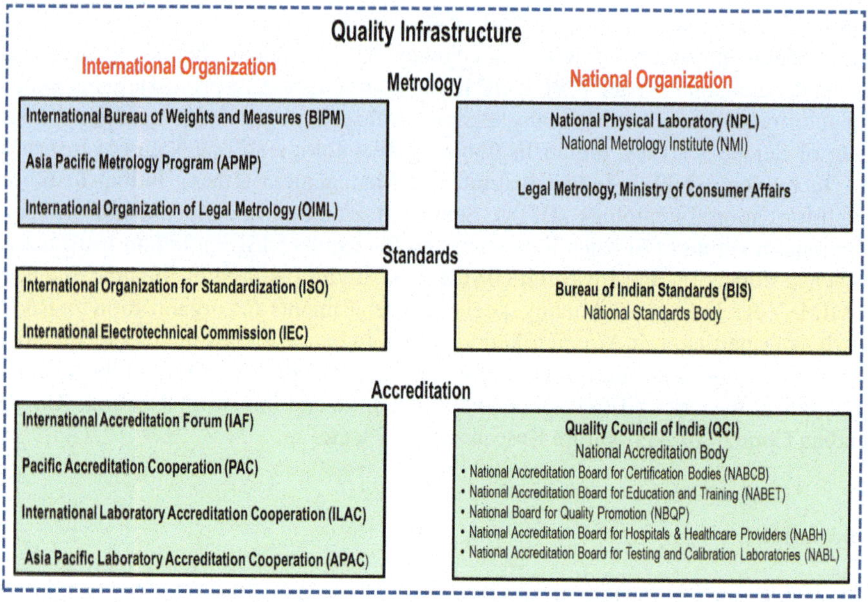

Fig. 1.7 The Quality infrastructure of India and its international recognition

responsibility of CSIR-NPL [32] along with the Bhabha Atomic Research Centre (BARC) [33], the designated institute for the ionization radiation. The legal and enforcement of metrology is the responsibility of Legal Metrology, Ministry of Consumer Affairs, Government of India. Various boards under the Quality Council of India (QCI) [34] namely NABCB, NABET, NBQP, NABH, and NABL look after the accreditation and certification. Indian Standards are developed by the Bureau of Indian Standards (BIS) [35]. Regardless of having international affiliations of QI institutions, the QI strength of countries can be widely different. The QI strength of a country depends upon (i) overall international capabilities of each QI institution and (ii) the coordination amongst the QI institutions within the country and their effective dissemination to various stakeholders across the country. As discussed in Aswal model, the responsibilities of QI institutions are predominantly high in realizing the national schemes of government for the economic growth and improvement of quality of life.

One of the major roles of QI is to ensure that the measurement results are accurate and precise. The reliability of test/experiment results is important for decision making in both scientific research and routine daily life with different professions. For example, doctors use the patient's blood test values for the presence/absence of a particular disease, environmental engineers test samples of drinking water for the presence of toxic elements and harmful bacteria and determine whether their concentration is within the safety limits, import agencies may test the items for the harmful contaminates, various regulators as listed in Table 1.1 may seek tests results for the compliance of technical regulations, so on. As shown in Fig. 1.8, the measured

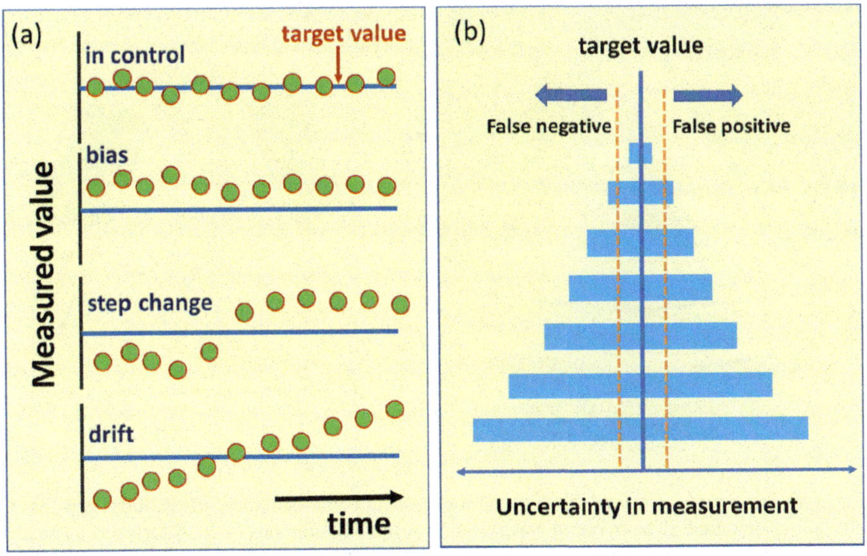

Fig. 1.8 **a** A schematic representing that measured value should be in control as a function of time and close to the target value. However, if measuring equipment is not calibrated appropriately, the measured values can exhibit bias, step-change, or drift. **b** The bias and large uncertainty in the measurements can lead to false positive or false negative results, which can have serious implications

value from a test may deviate from the actual value due to various sources of errors. In addition, if the measuring equipment is not appropriately calibrated or the standard or control samples are not used during the measurements, the measured value may exhibit bias or drift/change with time. A bias and/or large uncertainty in the measurement values can lead to a big problem where test results are used as binary classification, i.e., either positive or negative. One can end up getting false positive, i.e., test result wrongly shows presence of a disease but in reality, it is not present, or false-negative, i.e., test result wrongly indicates absence of the disease but in reality, it is present. Both false negative and false-positive results are not only bad for patient but can also create problems in the society. For example, if the Covid-19 test is a false negative, the patient unknowingly will keep spreading the disease in society. On the other hand, if the Covid-19 test is false positive, the patient will have to undergo unnecessary trauma of quarantine and treatment. Therefore, all the experimentally measured values should be close to the true value with minimum uncertainty, and this can be ensured if the country has a strong QI.

Metrology in India

The responsibility of CSIR-NPL, in collaboration with NABL and BIS, is to ensure that all the testing labs across the country (which are estimated to be >400,000 in numbers) produce accurate and precise measurement values. A measurement is valid

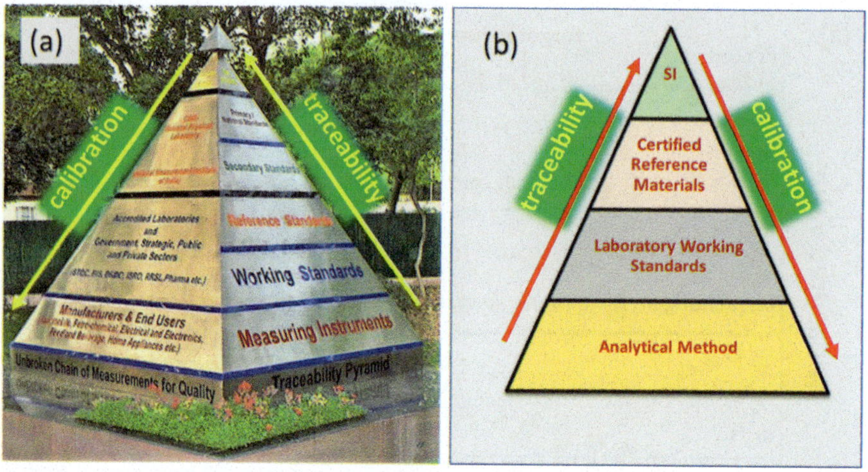

Fig. 1.9 Two routes of establishing an unbroken chain of measurement traceability: **a** using the traceability pyramid. (The pyramid was installed at the NPL campus in 2016 for mass awareness based on the concepts of Dr. D. K. Aswal and Dr. V. N. Ojha). **b** using the certified reference material

only if: (i) the measurement has demonstrated an unbroken chain of traceability to the SI unit though the primary/national measurements available at the NMI, and (ii) the measurement uncertainty has been estimated in each of the transfer chains. In the absence of either of the two, the measurement would be purely indicative with no legal validity, i.e., would not be accepted in case of any dispute. The measurement traceability, as shown in Fig. 1.9, can be obtained through two routes: (i) calibrating the measuring instrument through working standards, reference standard, secondary/primary standards to the SI units, as depicted in the photograph of a Traceability Pyramid installed at the NPL campus; and (ii) using the certified reference materials (CRM), which are trademarked as Bhartiya Nirdeshak Dravyas (BND®) by CSIR-NPL.

CSIR-NPL maintains a quality system in accordance with ISO 17025:2017 and ISO 17034, and has the following responsibilities:

(i) To actively participate in technical meetings of BIPM and APMP related to the development of metrology. RMOs like APMP, in association with BIPM, homogenize measurement capabilities across the world through NMIs.
(ii) Realization of primary/national measurement standards and establish the calibration and measurement capabilities (CMCs). The joint committee of the regional organizations and the BIPM (JCRB) analyses the CMCs of NMIs and the information is uploaded on the website of Key Comparison Database of BIPM (KCDB-BIPM) [36]. CSIR-NPL has 236 CMCs and 130 key/supplementary comparisons in metrology areas related to physicomechanical, electrical and electronics, time and frequency, chemistry, etc. In addition, BARC has 52 key/supplementary comparisons in the area of ionization radiation. A detailed account of these is presented in the subsequent Chapters.

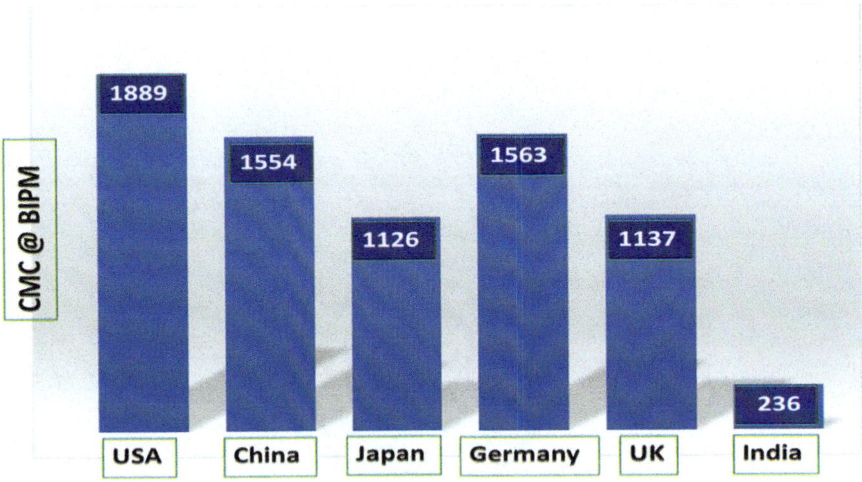

Fig. 1.10 The calibration and measurement capabilities of the leading economies at KCDB-BIPM

However, as shown in Fig. 1.10, the CMCs of India are far low as compared to NMIs of developed nations viz. USA (1889), China (1554), Japan (1126), Germany (1563), UK (1136). Large number of CMCs indicate that country has measurement capabilities of international equivalence in several frontiers of science and technology. Thus, both CSIR-NPL and BARC have enough scope to enhance their CMCs at the KCDB.

(iii) Dissemination of measurement traceability to all the stakeholders in the country using calibration of measurement equipment and/or CRMs or BNDs. Over the past 7 decades, CSIR-NPL has carried out dissemination of primary/national measurement facilities to several thousands of industries and strategic sectors. CSIR-NPL worked as "Growth Engine" of the country as these calibration services facilitated Indian industries to enhance the quality of their products.

In 2016, based on the need of the country, CSIR-NPL reorganized itself to take up the metrology in areas of Indian Standard Time (IST), physico-mechanical, electrical and electronics, environment, energy, biomedical, quantum standards, materials, etc., as depicted in Fig. 1.11. Of course, the setting of new primary/national standards require high capital investment, trained manpower, and international intercomparison. CSIR-NPL has taken up setting up of primary standards in the areas of air quality and emission monitoring, solar cell efficiency, biomedical devices, etc. CSIR-NPL produces the CRMs under the trademark of Bhartiya Nirdeshak Dravya (BND®) and has a national mechanism for the production of BNDs in collaboration with reference material producers in the country.

Legal metrology through appropriate law enforcement transfers the benefits of metrology to common citizens of the country by ensuring fair trade through correct measurements. Legal Metrology ensures this by Type approval of measuring equipment; calibration, verification, and market surveillance of measuring equipment; and

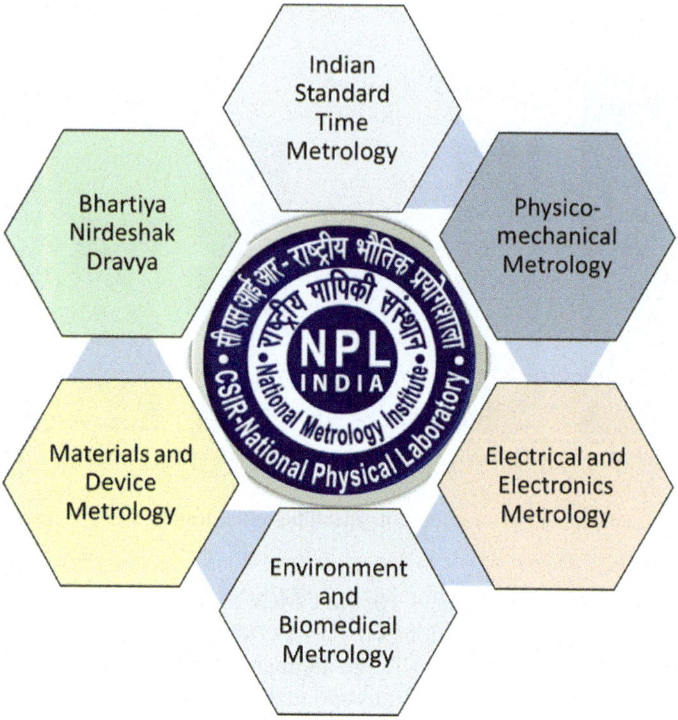

Fig. 1.11 Various metrology activities at the National Physical Laboratory, New Delhi—the National Metrology Institute (NMI) of India

controls on pre-packaged goods. Legal metrology is an integral part of the technical regulation regime and complies with the requirements of the Agreement on Technical Barriers to Trade (TBT Agreement) of the World Trade Organization (WTO) [37]. Legal metrology also defines the roles and responsibilities of central and state governments in the implementation of the Weights and Measure Act of the government. The central government deals with matters related to the implementation of International Organization of Legal Metrology (OIML) recommendations [38], formulation of national policy and laws on weights and measures, technical regulations, training, and setting up of precision laboratory. State and union territory governments have the responsibility for the enforcement of the laws framed by the central government through their offices of the Controller of Legal Metrology (Weights and Measures). The nationwide network of the legal metrology laboratories is responsible for the calibration and testing of equipment of weights and measures, ensuring the measurement traceability is available to end-users. Legal metrology has initiated a nationwide program on dissemination of Indian Standard Time (IST) to common citizens.

Standardization in India

Bureau of Indian Standards (BIS) is an exclusive standardization body of India for development of Indian Standards (IS). BIS is a member International Organization for Standardization (ISO) [39] and International Electrotechnical Commission (IEC) [40]. Standards offer numerous economic benefits to industries: (i) assist in the optimization of internal operations, which may lead to reduced time of operation, enhanced productivity, reduction procurement costs, and reduction in waste; (ii) support in the utilization of innovative processes for scalable manufacturing of new products; and (iii) facilitate in identifying new domestic and export markets. More than 20,000 Indian standards have been developed so far by 344 technical committees belonging to 16 Division Councils of BIS. Many technical committees act as mirror committees at ISO and IEC. Specialized Standards are also developed by many regulators, as listed in Table 1.1, and often these standards are adopted as IS by BIS. Indian standards are widely used for ensuring necessary characteristics of products and services, which include quality, reliability, safety, efficiency, environment-friendly, and interchangeability. BIS also has schemes of conformity assessment, e.g., ISI certification for products, hall-marking of gold jewelry, and foreign manufacturers' certification scheme (FMCS), etc.

The most widely adopted ISO standards are: (i) quality management systems (ISO 9001), (ii) general requirements for the competence of testing and calibration laboratories (ISO/IEC 17025:2017), (iii) environmental management systems (ISO 14001), (iv) conformity assessment—general requirements for proficiency testing (ISO/IEC 17043:2010), (v) general requirements for the competence of reference material producers (ISO 17034:2016), (vi) information security management systems (ISO/IEC 27001:2005) and (vii) risk management (ISO 31000:2009).

Accreditation in India

National Accreditation Board for Testing and Calibration Laboratories (NABL) [41] undertakes the assessment and accreditation of testing and calibration laboratories in India. NABL is a signatory of Mutual Recognition Arrangement (MRA) of International Laboratory Accreditation Co-operation (ILAC) [42] and Asia Pacific Laboratory Accreditation Co-operation (APAC) [43]. NABL does accreditation of conformity assessment bodies (CAB) including calibration and testing laboratories (ISO17025), proficiency testing provider (PTP) using ISO 17043, and reference material producers (RMP) using ISO 17034. The accreditation process ensures that CAB meets metrological traceability to SI units through the primary/national measurement standards of the NMI, as depicted in Fig. 1.9. Thus, NABL accreditation of testing laboratories produces test results of an internationally acceptable level of competence.

NABL so far has granted >5600 accreditations to calibration laboratories, testing laboratories, medical laboratories, proficiency testing providers and reference material producers. The scope of NABL accreditation are:

(i) calibration laboratories: mechanical, thermal, electro-technical, optical, fluid flow, radiological and medical devices;
(ii) testing laboratories: chemical, electrical, electronics, biological, fluid flow, mechanical, non-destructive testing (NDT), photometry, radiological, forensic, diagnostic radiology QA testing and software and IT system;
(iii) medical laboratories: clinical biochemistry, hematology and immunohematology, clinical pathology, histopathology, microbiology, and infectious disease serology, genetics, cytopathology, flow cytometry, nuclear medicine (in vitro tests only). In the era of nCov19 pandemic, through on-line assessment procedure, NABL accredited >300 medical testing laboratories for RT-PCR RNA Virus/COVID-19, enhancing the capabilities of the country for timely testing of large number of patients;
(iv) reference material producers: physical properties, chemical composition, engineering properties, biological and clinical properties, and miscellaneous properties.

NABL accreditation has been mandated by the majority of regulators listed in Table 1.1 for ensuring conformity assessment of their schemes/regulation, and therefore, provides an advantage to industries in getting their products to get tested from NABL accredited laboratories. Similarly, reference materials producers have distinct advantages of getting accredited by NABL to ensures metrological traceability is from CSIR-NPL. In addition, they can produce and market their certified reference materials under the trademark of BND®, which has visibility in the international market.

The number of testing/calibration laboratories in India is >40,000 and only 5600 laboratories have been accredited so far by NABL. Therefore, in order to provide accreditation to such a large number of laboratories while ensuring the metrological traceability to SI units, NABL needs to create two tiers of laboratories. Tier-1 calibration laboratories: these would be equivalent to those of regional reference laboratories of Legal Metrology and obtain measurement traceability directly from the CSIR-NPL. Therefore, these Tier-1 laboratories are expected to have much lower measurement uncertainty. Laboratories falling under government and/or private sectors having essential infrastructure can be assigned Tier-1 grade after the approval from CSIR-NPL. Tier 2 calibration laboratories: rest of the laboratories would be accredited under Tier-2 and they will obtain measurement traceability from Tier-1 laboratories. Tier-2 laboratories will disseminate measurement traceability to testing laboratories across the country.

The conformity assessment bodies (CABs) not requiring the metrology traceability are accredited by other boards of QCI, namely the National Accreditation Board for Certification Bodies (NABCB) [44], National Accreditation Board for Hospitals and Healthcare Providers (NABH) [45], and National Accreditation Board for Education and Training (NABET) [46]. The NABCB operates under the framework of ISO 17011 and accredits CABs for inspection bodies (e.g., BIS for product certification), medical devices, management systems for food safety, environmental monitoring, information security, health safety, occupational hazards, energy, etc. NABET provides accreditation in the management of education and training and NABH in management of hospitals. Both NABET and NABH function under self-regulation.

1.4 Implications of Aswal Model

As discussed in the previous section, India has a robust and effective system of government, academia, industries, and civil society. In addition, the quality-infrastructure in India is reasonably well established. In order to ensure the *AtmaNirbhar Bharat* with a goal of $ 5 trillion economies by 2025 [47], implementation of Aswal model of inclusive growth would be highly beneficial. In this model, metrology is at the core of the national quality infrastructure, which is a must for all the helices for creating knowledge and innovation for the knowledge economy, sustainable development, and knowledge society. As depicted in Fig. 1.12, the Aswal model has an enormous impact, which is discussed below.

1.4.1 Technology Development and Industrialization

One of the effective methods to enhance the national economy and eliminate poverty is to undergo sustainable industrialization with a focus to cater needs of local market as well as for export. UNIDO measures the industrialization level of a country using an indicator known as manufacturing value added (MVA) per capita, which is a difference between the net manufacturing output and intermediate inputs. Figure 1.13 shows the MVA per capita of leading economies of the world for the years 2014-2019

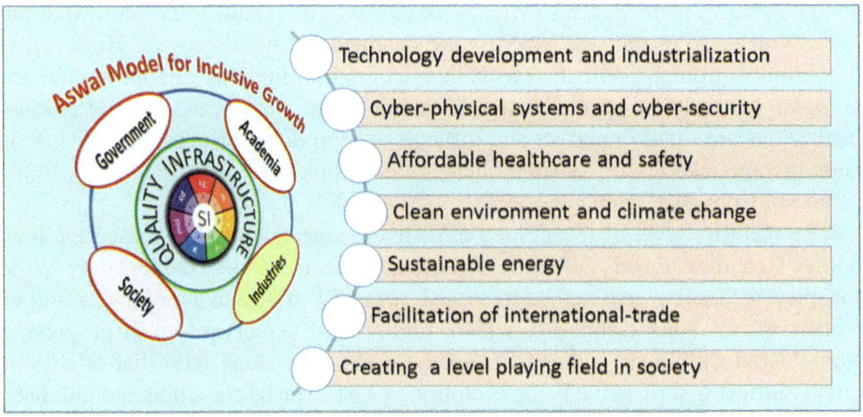

Fig. 1.12 The applicability of the Aswal model for the all-round growth of a nation

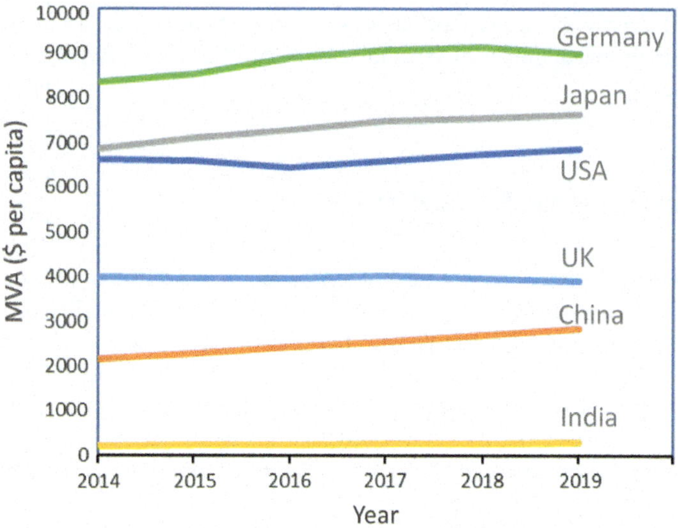

Fig. 1.13 Manufacturing value added (MVA) per capita (in US dollars) of the leading economies of the world for the years 2014–2019 for various countries (data source Ref. [48])

[48]. It is evident that India has to work very hard on industrialization to catch up the developed world.

In India, the *Micro, Small and Medium Enterprises* (MSMEs) play a significant role in economic growth by providing the solutions to the local needs, develop products and services at reasonable prices, provide large employment opportunities (second after the agriculture sector) to both rural and urban population, and act as nucleation for entrepreneurship and innovation. Under *AtmaNirbhar Bharat* program, the government has taken a pathbreaking step not only to incentivize MSMEs but also made fundamental changes the way MSMEs are classified, i.e., annual turnover up to 5 crores (Micro), 10 crores (small), and 250 crores (Medium) [49]. With these incentives, MSMEs are destined for spiral growth. However, on the technical front, MSMEs face a challenge of non-availability of appropriate technologies as the know-how of technologies obtained from academia and research institutions are often imperfect and difficult to reproduce for mass manufacturing. Here the applicability of Aswal model for development of quality products from a technology obtained from academia is highlighted.

The maturity level of technology at different stages is often assessed by Technology Readiness Levels (TRL) measurement system, initially developed by NASA for space technology and ranked from TRL-1 to TRL-9, which later on was adopted worldwide for other technologies [50]. Figure 1.14 depicts a typical progression from TRL-1 to TRL-9, roles of involved stakeholders along with that of national quality infrastructure. Broadly the evolution of TRLs can be classified into four parts, i.e., concept (TRL-1 and TRL-2), development (TRL-3 and TRL-4), manufacturing (TRL-5 and TRL-6), and trading (TRL-8 and TRL-9).

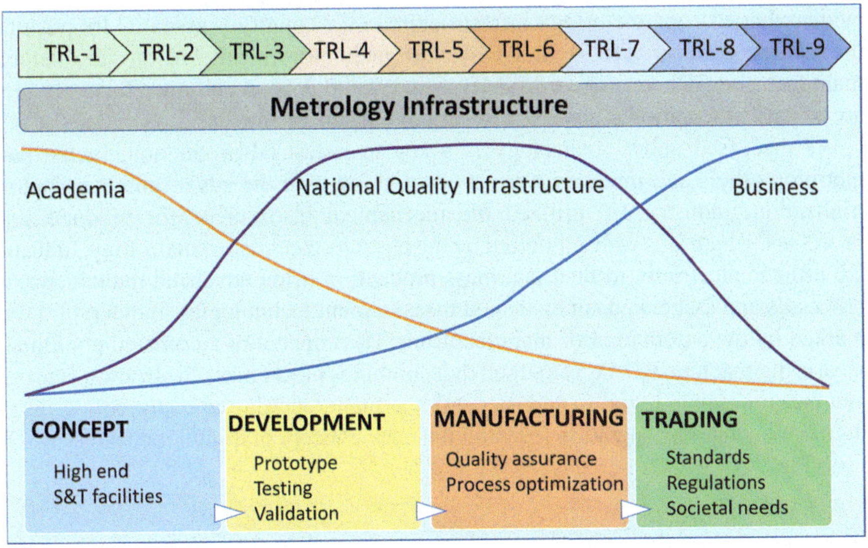

Fig. 1.14 A typical progression of technology from TRL-1 to TRL-9, concerned stakeholders, and their responsibilities. It is noted that a strong national metrology infrastructure is a must for all the TRLs

Usually, the TRL-1 and TRL-2 are taken up by academia and evaluated for concepts based on scientific research that can be translated into product development. The experimental proof for the technology has also been established to a certain extent. TRL-1 and TRL-2 require laboratories with S&T facilities to have creditable metrology to ensure that experimental results are genuine enough for the analyses for ensuring that if concept can be taken further or not for the development.

TRL-3 and TRL-4 belong to the development stage of the product. TRL-3 is assigned when the design and development of the product begin and a proof-of-concept is established. TRL-4 is the stage when the creditable level of the prototype has been built. All of the components and tested and validated independently, which requires a strong quality infrastructure in the country. TRL-5 is the rigorous testing and validation of the prototypes developed at TRL-4 under the simulated environmental conditions equivalent to those where the product is eventually going to be deployed. The technology advances to TRL-6, once the conformity assessment of the product developed at TRL-5 is complete. At TRL-6 the product undergoes for manufacturing and undergoes process optimization and quality assurance, which also demand a strong quality infrastructure. TRL-7 is a further extension of TRL-6 and requires the testing of the product under the actual environmental conditions. TRL-8 and TRL-9 are making the product ready for the trading. At TRL-8, the standard and regulatory compliance of the technology are ensured. In order to issue a certificate of regulatory compliance of the product, the regulators, as discussed earlier, would require the quality infrastructure for obtaining trustworthy test results. In TRL-9 the societal and market needs are assessed and technology is launched in the market. It is evident that a strong metrology infrastructure in the country is essential for pushing the technology from TRL-1 to TRL-9 with technologies suitable for zero defects manufacturing. The growth of MSME sector would depend on how effectively they are utilizing the national quality infrastructure of the country.

As discussed above, metrology is a key to industrial production. In the past metrology tools and methods were evolved along with the advancement of industrialization. Industry 1.0 utilized the mechanical instruments for products and processes, which was well supported by the physico-mechanical metrology. Industry 2.0 utilized electricity to facilitate mass production using advanced manufacturing processes and associated sophisticated measurement technologies. Industry 3.0 was marked by the automation of manufacturing. The concept of automated coordinate measuring machine (CMM) emerged that combines mechanics, electronics, sensors, and programmable logic. In Industry 4.0, the manufacturing industry will undergo digitization and virtualization [51] and this new concept of digital metrology is still evolving.

1.4.2 Cyber-Physical Systems and Cyber Security

The Industry 4.0 paradigm is based on the intuitive interactions between the machines and humans, which will be governed by the advancements in the cyber-physical systems (CPS). In CPS, through integrated physics and logic, deep intertwining amongst the analog, digital, physical, network, and human components are engineered to obtain desired functions, i.e., automated and self-organizing physical processes in real-time [51]. Therefore, new industries will have fully automatic (auto-configuring, auto-adjusting, and auto-optimizing) production environment, leading to greater quickness, flexibility, and cost-effectiveness. Therefore, these manufacturing plants would require highly skilled workers, who will have to work from the central control room and interacting directly with synchronized intelligent machines. The application of CPS includes autonomous automobiles, smart grid, manufacturing, robotics systems, smart cities, healthcare, aerospace, chemical processes, civil infrastructure, energy, entertainment, consumer appliances, emergency response, traffic flow management, indoor automatic agriculture, etc., which are going to improve the quality of life.

Metrology remains the backbone of the "autonomy" in the CPS technologies [52]. In CPS, the decision-making is to be carried out by the system itself based on the received measurement data. Therefore, in CPS, first task is to measure the "intelligence" of a system, and for this metrology plays a predominant role in identifying indicators of intelligence and their measurement. CPS requires a constant collection of measurement information using calibrated sensors and actuators. The real-time application of CPS can only be ensured if all of its components—physical and software, are synchronized to Universal Coordinated Time (UTC) through the official/legal time of the respective country, e.g., Indian Standard Time generated by

CSIR-NPL, and this is discussed in detail in Chaps. 4 and 5. Therefore, management of complex but critical time synchronization for achieving the real-time CPS is another key component of time and frequency metrology. Time metrology also ensures secure and trusted timestamping which once recorded cannot even be modified by its owner, and therefore, is essential for safe national digital services, e.g., electronic signatures, electronic seals, electronic delivery services, electronic time stamps, and website authentication [53]. Electronic timestamp provides the guarantee of accuracy of bound date and time to the data. The metrology of CPS is still evolving as is complex in nature due to various factors: (i) a large number of sensors are often used for measurements and most of them have different functional output; (ii) drift in the characteristics of sensors and other electronic components due to changes in environmental conditions; (iii) all components need to be synchronized in time for obtaining real-time processing of measurements; (iv) software of device components are usually different (different manufacturers use different data transfer protocols), which cause problems in data transfers without errors, etc. However, it is expected that in the very near future, metrology for Industry 4.0 would be ready.

1.4.3 *Affordable Healthcare and Safety*

A strong healthcare system is important keeping in view population growth as well as the aging population. Affordable healthcare requires accurate physical, chemical, and biological measurements in the everyday life of people as they are fundamental in the prevention, diagnosis, and treatment of disease [54]. Various aspects of a patient's health can be diagnosed correctly if diagnostic medical equipment (e.g. clinical thermometer for body temperature; stethoscopes for the sounds originating from hearts, lungs, etc.; electrocardiographs for measuring the electrical activity of the heart; ophthalmoscopes to see into the fundus of a patient's eye; medical X-rays; magnetic resonance imaging (MRI); computed tomography (CT) scan; and the list is very long) are calibrated to ensure the metrological traceability. Once a correct diagnosis is made, the doctor can easily recommend an appropriate treatment plan. Even during the treatment, the quality of medicine as well as safe and effective delivery of therapies need to be ensured. Therefore, there is a growing interest in ensuring the metrological traceability of the medical devices as well as in regulation on conformity assessment of medical devices for their measuring functions. The innovation in the diagnosis and treatment of diseases is essential for the good health of humans and enhancing their life expectancy. Metrology ensures accurate measurements to assess the performance of diagnostic methods and therapies, and ensure the effective treatment of patients. In order to understand therapy-induced changes and their subsequent control for treatment of a patient, accurate measurements on physiological variables are essential. Wrong measurements can endanger patient safety and put critically ill patients in deep troubles.

Use of appropriately calibrated medical devices will facilitate affordable healthcare owing to the following advantages: (i) the false positive or false negative results are minimized; (ii) unnecessary harassment to the patient for repeated and unnecessary associated tests; (iii) right treatment and therefore less medicinal side effects to the patient; and (iv) reduction in the treatment cost. This also reduces the burden in government hospitals, where the healthcare facilities are limited. CSIR-NPL is playing a key role in the calibration of biomedical equipment, which is elaborated in Chap. 15.

1.4.4 Clean Environment and Climate Change

A delicate balance between the protection of the environment to keep the air clean enough for breathing and sustainable industrial growth for the economy and job creation with environmental legislation is a challenging task. This balance can be achieved through the application of metrology. Measurements of ambient air quality parameters (i.e., particulate matters, CO, CO_2, ozone, SO_2, etc.) and emission from the industries with part per billion levels of accuracies are crucial for managing and protecting the environment. A terrestrial network of pollution measuring equipment is used for monitoring air quality, while climate change studies are often carried on data collected from satellite-borne measuring devices. Analyses of such robust and trustworthy environmental data obtained from metrologically traceable equipment provide useful information about climate change or pollution emission trends, which can be utilized for appropriate remedial actions. To ensure the generation of reliable data, most nations enforce mandatory certification of pollution measuring equipment. In India, CSIR-NPL is also working on a national certification scheme in association with *Central Pollution Control Board* (CPCB) and Ministry of Environment and Climate Change. Accurate environmental data is also important for the formulation of regulations that can be enforced effectively to ensure a clean environment, and this is elaborated in Chaps. 13 and 14. As environmental monitoring is a complex issue, scientists from the metrology institute, academia, and industry need to work in collaboration for improved measurement of hazardous emissions to ensure the quality of life and quantification of parameters that influence climate. Moreover, after the Climate Convention at Paris in 2015, most of the countries have agreed to align their national environmental regulations with international agreements [55]. Therefore, to obtain reliable data for climate goals, harmonized measuring system traceable to SI units are essential [56].

1.4.5 Sustainable Energy

In order to reduce carbon-dioxide emissions, the majority of the nation are focusing on the production of their energy needs from renewable sources, e.g., solar, wind, biomass, geothermal, and hydropower. However, to ensure secure and sustainable energy resources, a national strategy needs to be evolved that would address all aspects of the energy, including production of energy from both non-renewable and renewable, supply chain, reduction in greenhouse gas emission, ensuring the stability of power grid and to introduce methods for reducing the energy consumption, e.g., by utilizing efficient devices. In addition, research and development should continue for the development of new technologies for energy production as well as new methodologies to optimize the operation of conventional and nuclear power plants in most efficient manner. All these require a strong metrology infrastructure to ensure better quantification of measurement parameters across the full energy chain, i.e., generation, transmission, and consumption [57]. The measurement capabilities enable regulators to perform the tests for ensuring the conformity assessment of equipment and appliances for their energy efficiency. CSIR-NPL has established calibration facilities for the measurement of the efficiency of organic solar cells and details of this are presented in Chap. 16.

1.4.6 Facilitation of International Trade

World Trade Organization (WTO) keeps making efforts in removing technical barriers to ensure free international trade and promote free competition worldwide. Most nations adopt their standards and regulations on measuring devices to protect both manufacturers and customers in their respective countries. However, if these standards and regulations are not backed by a functioning measurement system traceable to SI units, their noncompliance can lead to technical barriers to trade. In modern time, it is normal that various components of "a system device" are manufactured in different countries, and therefore, the compatibility of components, in any case, is required even though their trade are coming under non-regulated areas of trade. Thus, in order to produce goods of high quality, nations must ensure that their measurement system is on par with an international one. As has been discussed earlier, CIPM MRA and the OIML Mutual Acceptance Arrangement (MAA) facilitate the promotion of setting up of uniform international measurement system to ensure consistency in testing for global trade. In order to eliminate the technical barriers to trade (TBT), WTO introduced TBT Agreement in 1995, which sets out a code of good practice for both governments and non-governmental or industry bodies to prepare, adopt and apply voluntary standards [58]. Removal of technical barriers to trade contribute to several benefits: (i) the time and cost of transactions are reduced; (ii) the possibilities of international trade disputes reduces; (iii) low-quality imports into the country can be minimized; and (iv) the possibility of export rejection is minimized. Therefore, a strong quality infrastructure of a country is the backbone for improving the quality of the indigenously developed products and enhance their competitiveness for global trade. This in turn builds the confidence of investors, particularly in developing countries like India, where efforts are being made to attract foreign direct investment for economic growth. A poor conformity assessment service that is not traceable to SI units can allow entry of imports that are sub-standard and toxic in nature. There are numerous cases where poor-quality import items could enter in the country, e.g., steel, toys, tyres, light-emitting diodes (LEDs), electronic items, solar panels, food items, etc. [59–64]. These poor-quality imports having toxicity not only make their disposal an economic burden but also pose health hazards. Similarly, the export consignments get rejected if the quality does not pass through the conformity assessment of the importing country [65–67]. Therefore, the country must have a strong quality-infrastructure for international trade to avoid huge economic losses. As discussed above, BNDs are utilized to calibrate the analytical equipment of testing laboratory, and the details of the BND program (production to marketing) are discussed in Chaps. 18 and 19. In fact, the theme for World Metrology Day 2020 was "Measurements for Global Trade", which was chosen to create awareness of: (i) importance of measurement in enabling a fair global trade. (ii) measurements ensure that products meet standards and regulations, and (iii) measurements safeguard the quality expectations of customers.

1.4.7 Creating a Level Playing Field in Society

Accurate measurements in day to day activities, safeguarded through national quality-infrastructure, generate a state of affairs in the society in which every citizen gets a fair deal. This can be achieved if all the regulators (as mentioned in Table 1.1) in their domain ensure that all the measuring instruments, e.g., electronic instruments, weighbridges, petrol pumps, water meter, sphygmomanometer, clinical thermometer, medical equipment, scientific instruments, etc. are calibrated periodically. The accurate measurement values of these instruments are used by traders, industries, hospitals, academia, and other organizations are essential to ensure that common citizens are not cheated, e.g., the farmers producing the food commodities, milk, etc. get the values worth their products, the consumer gets the right amount of the quantity for their money, the trucks carry the right kind of load to avoid damage to the roads, the doctors measure correct values of body parameters for medical diagnosis of poorest of the people enabling to save their hard-earned money, customers get the right quantity in the prepacked food items, etc. Measurements traceable to SI units create trust in the society and benefit one and all. The significance of accurate measurements in a developed society is evident when a Japanese rail company expressed regret after a train left a station 25 s early in May 2018 and the second similar incidence happened again after 8 months. The society felt that the mistake is truly inexcusable and customers are faced with slipping standards of the Japanese rail company [68].

1.5 Digital SI units

As highlighted earlier, Industry 4.0 will require the new digital metrology. This implies that measured values obtained from different sub-components are converted into reliable machine-readable digital data. High-quality digital data as well as their exchange also are the foundation for reaping benefits of artificial intelligence and machine-learning applications. To establish the concepts of digital metrology, the International Committee for Weights and Measures has already set up a Task Force on "Digital SI" (D-SI) and authorized to develop a world-wide harmonized, unambiguous, and secure basic standard for measurement data exchange, i.e., transformation of the SI units into a meta-data format for digitized world [69]. The meta-data format of metrology would require complete coverage of the information, i.e., the numerical value of a measured quantity, assigned SI unit, measurement uncertainty, timestamp, etc. It is expected that D-SI will be established in the coming years and will give a new fillip to cyber-secure digital science and technology.

1.6 CSIR-NPL in Service of Nation

CSIR-NPL through its primary/national calibration facilities supports thousands of quality-assurance testing and calibration laboratories across the country, which belong to the government, strategic sector, industries, etc. Detailed analysis of these services is presented in Chap. 21. In addition, CSIR-NPL also supports NMIs of the South Asian Association for Regional Cooperation (SAARC) nations. CSIR-NPL is the "Timekeeper of the Nation" and generates Indian Standard Time (IST) traceable to Coordinated Universal Time (UTC). CSIR-NPL disseminates IST to various sectors across the country. Detailed discussion on these contributions is presented in the subsequent chapters of this book.

CSIR-NPL also develops technologies and provides consultancies to industries and the strategic sector. One of the earliest technologies developed by CSIR-NPL was the Indelible-ink—the mark of democracy, which was transferred to Mysore Varnish and Paints Ltd. (MVPL) for mass production [72]. During elections, the purple ink is applied to the forefinger of voters, as shown in Fig. 1.15a, to prevent electoral fraud, e.g., multiple voting and malpractices. MVPL exports the indelible ink to several countries like Thailand, Nigeria, Singapore, South Africa, Malaysia, etc. CSIR-NPL has developed technologies and provided several consultancy services in the areas of strategic sectors, societal, environmental, energy, biomedical, and n-Cov19 related, e.g., ultraviolet disinfectant products, contactless thermometers, calibration services to ventilators and personal protection equipment, etc.

Another great contribution of CSIR-NPL is the preservation and maintenance of the "Original Copy of the Constitution of India" which is kept at Parliament Library, New Delhi, see Fig. 1.15(b) [73]. For this CSIR-NPL designed and developed two hermetically sealed glass cases in association with Getty Conservation Institute, USA. On 26 November 1949, the Constituent Assembly adopted "Constitution of India" written under the chairmanship of B. R. Ambedkar, which came into force on 26 January 1950. The original copy of the constitution of India has two following versions.

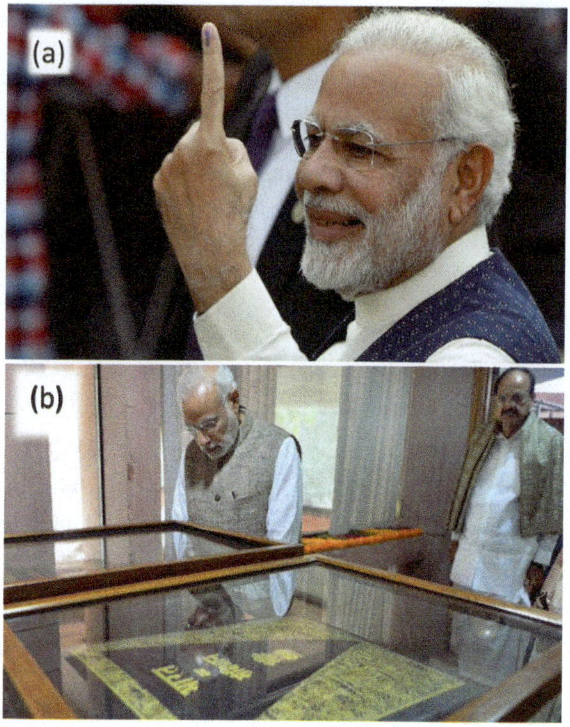

Fig. 1.15 Shri Narendra Modi, Prime Minister of India, **a** displaying the mark of indelible-ink developed by CSIR-NPL and manufactured by Mysore Paints and Varnish Ltd. *Photo courtesy* Ref. [70], **b** paying respect to the "Original copy of the Constitution of India" (Hindi and English versions) at Parliament Library, preserved in hermetically sealed cases developed by CSIR-NPL and Getty Conservation Institute, USA. *Photo courtesy* Ref. [71]

(i) English version: Calligraphed by Prem Behari Narain Raizada. It consists of 221 pages of hand-made parchment paper of size 45.7 cm × 58.4 cm and weighs ~13 kg. The decoration of these sheets was done by Nand Lal Bose, which depicts a journey from the Mohen-jo-Daro and Vedic periods to the Indian freedom movement.

(ii) Hindi version: comprises 252 calligraphed sheets and weighs ~14 kg, and was calligraphed by Basantrao Vaidya almost similar to the English version.

Both documents are bound in first-class Morocco leather embossed in gold. These original calligraphed copies of the 'Constitution of India' are kept in the Parliament Library, and have great autographic and historical value as they contain the signatures of the founding fathers of the Constitution. The preservation of these prestigious documents is being done by the scientists of CSIR-NPL.

CSIR-NPL also provides training programs in the areas of metrology and quality assurance for the personnel of industries and regulators. In addition, CSIR-NPL has a diploma program on "*Precision Measurement and Quality Control,* (PMQC)" and offers Ph.D. programs on metrology and advance topics of research. The details on various training programs of CSIR-NPL on skill development are presented in Chap. 20.

1.7 Summary

A hard work on world metrology for more than 140 years has resulted in SI units that are defined in terms of fundamental constants of nature. The advantages of revised SI units are that they are sustainable infinitely and would drive innovation for future technologies. The D-SI is expected to be established soon, which will benefit the world economy through secure digitalization. The CSIR-NPL, as the NMI of India, not only has been participating and contributing to the BIPM and APMP but also realized many CMCs for the country. The metrological trackabilities' are being disseminated to various stakeholders across the county either directly or through Legal Metrology and NABL accredited laboratories. Analyses of various knowledge creation models, e.g., Quadruple and Quintuple Helix models indicated that these models are inadequate to be implemented in the developing countries as they lack in explaining the underlying binding forces among the four helices namely government, industries, academia, and society. The Aswal model of inclusive growth suggests that metrology ensured through national quality infrastructure fortifies interactions amongst the four helices. An analysis of national quality infrastructure in India revealed that it requires to be strengthened to realize the goals of self-reliance. The metrology plays a key role in improvising the technology development and industrialization, cyber-physical systems and cybersecurity, affordable healthcare and safety, clean environment and climate change, sustainable energy, facilitation of international trade, and creation of a level playing field in the society.

Acknowledgements The author acknowledges Prof. Arun K. Grover, Chairman, Research Council, CSIR-NPL for his valuable guidance. Technical discussions with ex-colleagues namely, Dr. R. K. Kotnala, Dr. V. N. Ojha, Dr. Ranjana Mehrotra, Dr. Ajay Dhar, Dr. S. K. Dhawan and Dr. Suresh Chand, are acknowledged, which resulted in reshaping the activities of CSIR-NPL during 2015–2016. The support of entire staff of CSIR-NPL is highly appreciated.

References

1. BIPM—History of the SI. https://www.bipm.org/en/measurement-units/history-si/. Accessed on 30 June 2020
2. D.K. Aswal, The SI redefinition to come into force from 20 May 2019. Curr Sci **113**, 2026 (2017)
3. 26th meeting of the CGPM: 13–16 November 2018, https://www.bipm.org/en/cgpm-2018/. Accessed on 30 June 2020
4. K.N. Mathur, The national physical laboratory of India. Nature **159**, 184 (1947)
5. National Physical Laboratory of India, Nature **167**, 468–470 (1951)
6. Legal Metrology, Department of Consumers Affairs, New Delhi, https://consumeraffairs.nic.in/organisation-and-units/division/legal-metrology/overview. Accessed on 30 June 2020
7. OECD, Inclusive growth. http://www.oecd.org/inclusive-growth/#introduction. Accessed on 30 June 2020
8. United Nation, About the sustainable development goals. https://www.un.org/sustainabledevelopment/sustainable-development-goals/. Accessed on 30 June 2020

9. World Health Organization, Rolling updates on coronavirus disease (COVID-19). https://www.who.int/emergencies/diseases/novel-coronavirus-2019/events-as-they-happen, Accessed on 26 June 2020
10. Business Today, Coronavirus: Lockdown in India has affected 40 million migrants, says World Bank, April 23, 2020. https://www.businesstoday.in/current/economy-politics/coronavirus-lockdown-in-india-has-affected-40-million-migrants-says-world-bank/story/401762.html. Accessed on 30 June 2020
11. National Portal of India, Building Atmanirbhar Bharat & Overcoming COVID-19. https://www.india.gov.in/spotlight/building-atmanirbhar-bharat-overcoming-covid-19. Accessed on 30 June 2020
12. V. Roudometof, *Glocalization: a critical introduction* (Routledge Taylor & Francis, London, 2016)
13. OCED, Global forum on the knowledge economy. http://www.oecd.org/innovation/inno/globalforumontheknowledgeeconomy.htm. Accessed on 30 June 2020
14. D. Rooney, G. Hearn, A. Ninan (eds.), *Handbook on the knowledge economy* (Edward Elgar Publishing Ltd., UK, 2005)
15. M. Gibbons, C. Limoges, H. Nowotny, S. Schwartzman, P. Scott, M. Trow (eds.), *The New Production of Knowledge: The Dynamics of Science And Research in Contemporary Societies* (SAGE, London, 1994)
16. E.G. Carayannis, D.F.J. Campbell, *Knowledge Creation, Diffusion, and Use in Innovation Networks and Knowledge Clusters: A Comparative Systems Approach Across the United States* (Europe and Asia, Westporti, Praeger, 2006)
17. H. Etzkowitz, L. Leydesdorff, The dynamics of innovation: from National Systems and "Mode 2" to a Triple Helix of university-industry-government relations. Res. Policy **29**, 109 (2000)
18. H. Etzkowitz, L. Leydesdorff, The triple Helix—University-Industry-Government Relations: a laboratory for knowledge-based economic development. EASST Review **14**, 14–19 (1995)
19. E.G. Carayannis, D.F.J. Campbell, Mode 3 and quadruple helix: toward a 21st century fractal innovation ecosystem. Int. J. Technol. Manage. **46**, 201 (2009)
20. E.G. Carayannis, D.F.J. Campbell, Triple Helix, Quadruple Helix and Quintuple Helix and how do knowledge, innovation and the environment relate to each other? A proposed framework for a trans-disciplinary analysis of sustainable development and social ecology. Int. J. Soc. Ecol. Sustain. Dev **1**, 41–69 (2010)
21. E.G. Carayannis, D.F.J. Campbell, Open innovation diplomacy and a 21st century fractal research, education and innovation (FREIE) ecosystem: building on the Quadruple and Quintuple Helix innovation concepts and the "Mode 3" knowledge production system. J. Knowl. Econ. **2**, 327–372 (2011)
22. E.G. Carayannis, T.D. Barth, D.F.J. Campbell, The Quintuple Helix innovation model: global warming as a challenge and driver for innovation. J. Innov. Entrep. **1**, 1 (2012)
23. D.K. Aswal, Quality infrastructure of india and its importance for inclusive national growth. MAPAN-J. Metrol. Soc. India **35**, 139 (2020)
24. Quality Infrastructure: UNIDO's unique approach, 2018, https://www.unido.org/sites/default/files/files/2018-.08/UNIDO_QI_CASE_FINAL_ONLINE_2.pdf. Accessed on 30 June 2020
25. Quality infrastructure, United State of America. https://www.bipm.org/en/about-us/member-states/us/links.html. Accessed on 30 June 2020
26. Quality infrastructure-building trust for trade, The United Nations Industrial Development Organization (UNIDO), 2016. https://www.unido.org/sites/default/files/2016-05/UNIDO_Quality_system_0.pdf. Accessed on 30 June 2020
27. Ulrich Harmes-Liedtke, Juan José Oteiza Di Matteo, Measurement of Quality Infrastructure, published by Physikalisch-Technische Bundesanstalt, Germany, 2011. https://www.ptb.de/cms/fileadmin/internet/fachabteilungen/abteilung_9/9.3_internationale_zusammenarbeit/q5_publikationen/305_Discussion_5_Measurement_QI/PTB_Q5_Discussion5_Measurement_QI_EN.pdf. Accessed on 30 June 2020
28. Major announcements and policy reforms under Aatma Nirbhar Bharat Abhiyan. https://blog.mygov.in/major-announcements-and-policy-reforms-under-aatma-nirbhar-bharat-abhiyan/. Accessed on 30 June 2020

29. University Grant Commission. https://www.ugc.ac.in/. Accessed on 30 June 2020
30. India Brand Equity Foundation. https://www.ibef.org/industry.aspx. Accessed on 30 June 2020
31. Population of India, 2020. https://www.livepopulation.com/country/india.html. Accessed on 30 June 2020
32. National Physical Laboratory, New Delhi, http://www.nplindia.in/. Accessed on 30 June 2020
33. Bhabha Atomic Research Centre (BARC), Trombay, Mumbai, http://www.barc.gov.in/. Accessed on 30 June 2020
34. Quality Council of India, New Delhi, https://www.qcin.org/. Accessed on 30 June 2020
35. Bureau of Indian Standards, Retrieved from https://bis.gov.in/. Accessed on 30 June 2020
36. Calibration and measurement capabilities (CMCs), International bureau of weights and measures (BIPM), https://www.bipm.org/utils/common/pdf/KCDB/KCDB_CMCs.pdf. Accessed on 30 June 2020
37. WTO, Technical barriers to trade. https://www.wto.org/english/tratop_e/tbt_e/tbt_e.htm. Accessed on 30 June 2020
38. International Organization of Legal Metrology. https://www.oiml.org/en. Accessed on 30 June 2020
39. International Organization for Standardization. https://www.iso.org/home.html. Accessed on 30 June 2020
40. International Electrotechnical Commission. https://www.iec.ch/. Accessed on 30 June 2020
41. National Accreditation Board for Testing and Calibration Laboratories (NABL). http://www.nabl-india.org/. Accessed on 30 June 2020
42. International Laboratory Accreditation Co-operation (ILAC). https://ilac.org/. Accessed on 30 June 2020
43. Asia Pacific Laboratory Accreditation Co-operation (APLC). https://aplac.org/. Accessed on 30 June 2020
44. National Accreditation Board for Certification Bodies (NABCB). http://nabcb.qci.org.in/. Accessed on 30 June 2020
45. National Accreditation Board for Hospitals and Healthcare Providers (NABH), https://www.nabh.co/. Accessed on 30 June 2020
46. National Accreditation Board for Education & Training (NABET), http://nabet.qci.org.in/. Accessed on 30 June 2020
47. From fifth-largest to $5-trillion economy, 2020. https://economictimes.indiatimes.com/blogs/et-editorials/from-fifth-largest-to-5-trillion-economy/. Accessed on 30 June 2020
48. Monitoring the SDGs, UNIDO, https://stat.unido.org/SDG/DEU. Accessed on 1 July 2020
49. What's MSME. https://msme.gov.in/know-about-msme. Accessed on 1 July 2020
50. Technology Readiness Level Definitions, NASA. https://www.nasa.gov/pdf/458490main_TRL_Definitions.pdf. Accessed on 1 July 2020
51. E.G. Popkova, Y.V. Ragulina, A.V. Bogoviz (eds.), *Industry 4.0: Industrial Revolution of the 21st Century* (Springer Nature, Berlin, 2019)
52. R. Niyazbekova, M. Konkanov, Metrology aspects in cyber-physical systems. IOP Conf. Series: J. Phys.: Conf. Ser. **1065**, 072023 (2018)
53. Management Association, Information Resources USA (ed.), *Cyber Security and Threats: Concepts, Methodologies, Tools, and Applications*, (IGI Global, 2018)
54. J.O. Westgard, D. Armbruster, S. Westgard, *Risk, Error and Uncertainty: Laboratory Quality Management in the Age of Metrology* (Elsevier, Amsterdam, 2017)
55. UN Climate Change Conference Paris 2015. https://www.un.org/sustainabledevelopment/cop21/. Accessed on 2 July 2020
56. Measurement Challenges for Global Observation Systems for Climate Change Monitoring Traceability, Stability and Uncertainty, Report on the WMO-BIPM workshop, World Meteorological Organization, 2010. https://www.bipm.org/utils/common/pdf/rapportBIPM/2010/08.pdf. Accessed on 1 July 2020
57. Towards Secure, Sustainable Energy. https://www.euramet.org/metrology-for-societys-challenges/metrology-for-energy/. Accessed on 1 July 2020

58. Technical barriers to trade. https://www.wto.org/english/tratop_e/tbt_e/tbt_e.htm. Accessed on 1 July 2020
59. Cheap imports and quality-killing cancer, The Statesman, February 2, 2018. https://www.thestatesman.com/opinion/cheap-imports-quality-killing-cancer-1502577960.html. Accessed on 1 July 2020
60. India's booming solar sector has one major flaw: poor quality, Quartz India, August 2, 2018, https://qz.com/india/1345508/poor-quality-solar-panels-may-ruin-indias-renewable-energy-boom/. Accessed on 1 July 2020
61. Eight Chinese pharma cos to be blacklisted for quality issues, may lead to shortage of medicines, Firstpost, January 8, 2018, https://www.firstpost.com/india/centre-issues-showcause-notices-to-8-chinese-pharmaceutical-companies-for-poor-quality-raw-materials-4291337.html. Accessed on 1 July 2020
62. Low quality imports harm India's LED market, Lux, August 5, 2014, http://luxreview.com/article/2014/08/low-quality-imports-harm-india-s-led-market. Accessed on 1 July 2020
63. Battery making firms wants government to prevent poor quality imports from China, DNA, February 1, 2016, https://www.dnaindia.com/business/report-battery-making-firms-wants-government-to-prevent-poor-quality-imports-from-china-2172441. Accessed on 1 July 2020
64. Chinese imports hurt Eveready, The Telegraph, May 7, 2016, https://www.telegraphindia.com/1160507/jsp/business/story_84296.jsp. Accessed on 1 July 2020
65. Why so many Indian food product exports get rejected, Rediff.com, September 13, 2016. http://www.rediff.com/money/report/why-so-many-indian-food-product-exports-get-rejected/20160913.htm. Accessed on 1 July 2020
66. 13,334 'made-in-India' items rejected by US-FDA in 6 yrs, Times of India, February 11, 2016. https://timesofindia.indiatimes.com/india/13334-made-in-India-items-rejected-by-US-FDA-in-6-yrs-1753-less-than-China/articleshow/50909128.cms. Accessed on 1 July 2020
67. Food industry must focus on quality to avoid rejection in export mkt: Govt, India Today, June 19, 2018. https://www.indiatoday.in/pti-feed/story/food-industry-must-focus-on-quality-to-avoid-rejection-in-export-mkt-govt-1264269-2018-06-19. Accessed on 1 July 2020
68. Japanese train departs 25 seconds early—again. https://www.bbc.com/news/world-asia-44149791. Accessed on 1 July 2020
69. The International System of Units (SI) in Digital Communication, https://www.bipm.org/en/conference-centre/bipm-workshops/digital-si/. Accessed on 1 August 2020.
70. https://www.deccanherald.com/national/voter-id-more-powerful-than-ied-modi-after-voting-730198.html. Accessed on 1 Aug 2020.
71. https://indianexpress.com/article/opinion/columns/preamble-to-a-2015-debate/. Accessed on 1 Aug 2020.
72. Indelible ink. https://mysorepaints.com/product/indelible-ink/. Accessed on 1 July 2020
73. D.K. Aswal, Constitution of India: preservation of original. Curr. Sci. **115**, 788 (2018)

Chapter 2
International Harmonization of Measurements

Part I: International Measurement System

Sanjay Yadav, Goutam Mandal, Dilip D. Shivagan, Parag Sharma, Afaqul Zafer, and Dinesh K. Aswal

Abstract The ***International System of Units (SI)*** of physical quantities is very important and necessary toolkit for any scientific development. The existing system adopted in 1960 by CGPM, France, is the globally well accepted, up-to-date, modern, coherent, having official status and even contemporary arrangement or form of the metric system starting with seven base units. It is almost dominant and globally accepted modern metric system of measurements used in science, engineering, technology, international trade and commerce. The system also embodied and agrees with limitless additional units, called derived units and special units, identified by special names and symbols which are always epitomized through seven base units. The SI is always envisioned to be an evolving system, wherein units, prefixes and definitions are created, modified and adapted as a result of international agreement arrived on the basis of advancement of measurement technologies and improvement in precision in measurements. As science evolves, advance technologies are developed which demand even more precise and accurate measurements. Due to advent of new technologies, the improvement in the accuracy of the products is carried out through development of new and improved measurement techniques, measurement standards and their definitions. Daily a number of precision measurements

S. Yadav (✉) · G. Mandal · D. D. Shivagan · P. Sharma · A. Zafer · D. K. Aswal
CSIR-National Physical Laboratory, Dr. K.S. Krishnan Marg, New Delhi 110012, India
e-mail: syadav@nplindia.org

G. Mandal
e-mail: goutam@nplindia.org

D. D. Shivagan
e-mail: shivagand@nplindia.org

P. Sharma
e-mail: sharmap2@nplindia.org

A. Zafer
e-mail: afaqul.zafer@nplindia.org

D. K. Aswal
e-mail: dkaswal@nplindia.org

D. K. Aswal (ed.), *Metrology for Inclusive Growth of India*,
https://doi.org/10.1007/978-981-15-8872-3_2

are performed which make our lives easier, comfortable, healthy and safer. In this chapter, authors have attempted to explain the historical aspects of metrology; international system of measurements, various international metrological organizations, for example, BIPM, NMIs, Metre Convention, MRA, etc. A brief overview of the importance and significance of the SI units; fundamental constants of nature (FCN) and their relevance in SI, historical aspects and evolution of SI units; redefinition of SI units in terms of FCN and current status and implication of redefined SI units, derived and other non-metric units, has also been presented. The whole contents are presented in a unique way to include scientific and analyzed data in tables, graphs, diagrams and flow charts to make the reading interesting for the readers.

"I often say that you can measure what you are speaking about, and express it in numbers, you know something about it; but when you cannot measure it, when you cannot express it in numbers, your knowledge is of a meagre and unsatisfactory kind; it may be the beginning of knowledge, but you have scarcely, in your thoughts, advanced to the stage of science, whatever the matter may be."

– Lord Kelvin (1824-1907)

2.1 Introduction

It is well-known fact that metrology drives the growth engine of the any economy. No country can grow at its desired pace if metrology is ignored or not given due importance. The foundation of the metrology lies in the realization of SI units and a globally accepted uniform system of measurement. The current *International System of Units (SI)* is a very strong, universal, internationally well accepted, uniform, modern and lucid system, having official status. The existing arrangement or form of the metric system starts with realization of seven base units, which are now connected with nature's physical constants. The redefinitions of the seven base units are recently adopted and implemented. The system also embodied and agrees with limitless additional units, called derived units and special units, identified by special names and symbols. In this context, it is appropriate to review and describe the modern SI system, metrology, its significance, relevance and historical aspects.

2.2 Metrology and Measurement Systems

2.2.1 Introduction

Metrology plays a central role in scientific discoveries and innovation, industrial manufacturing and international trade, in improving the quality of life and in protecting the global environment. It not only provides the basis for the development of science and technology but also plays a vital role in our daily life. The economy of any state depends upon the manufacturing capability and global business which depends on international comparability of the manufactured products. Almost all aspects of the life including health, food, daily necessities, transportation, communication, business, economy, defence the measurements play key role. For global relevance of any society, international comparability of measurements is necessary. Incorrect measurements lead to wrong decisions, which can have serious consequences. Metrology helps to take logical judgments on the basis of uncertainty with the measurements of results and process. It also enhances confidence in the measurements and science behind it. Without the correct measurements, the ability to produce quality goods and fulfil consumer requirements would not be possible.

It is clear now that accurate and precise measurements are essentially required for better quality and confidence in the product. For ascertaining the accuracy and precision of a measurement, a calibrated reference standard is a must. For this, the measurement standards are developed, which not just only provide the reference for the measurement value but also provide the quality of measurements by stating the uncertainty related to the measurement. A better uncertainty is a proof, mark or evidence of the better quality of a measurement.

A user of any instruments should compare (calibrate) its instrument's measurement results with its respective measurement standard's results in order to judge the quality of its measurement by determining its uncertainty and hence the performance of the instrument. Now, this calibrated instrument can be used to calibrate another similar instrument or product in order to determine its quality but the only condition is that this chain of determining uncertainty should not be broken and ultimately traced to the national or international measurement standard. This unbroken chain of documented calibration with measurement uncertainty is known as *metrological traceability*. So any instrument used for measurement or producing final product/services should be traceable to measurement standard in order to obtain the required quality. If for a parameter, a measurement standard is not present, then certified reference material can be used as a reference for true measurement value. The national standards and certified reference materials are maintained by the *National Metrology Institute (NMI)* of the country.

Admittedly, metrology may not look very popular and crowd preferment but still it attracts the attention of experts and touches all walks of human life in almost all the socio, economic and industrial sectors. It plays a pivotal role as growth engine of industry for any country. It is an enabling tool for the development of technology, enhancing world-wide competence and building confidence. With the advent of better quality products through metrological advancement, our industries can compete globally, remove barriers and constraints in trade and commerce, boosting product demand and exports. This translates into growth for industries through rapid industrialization, economic growth and societal upliftment.

As an example, some of the metrological applications are described here. There are three main facets in *aerospace and automotive industry* for invention, creation, competitiveness and safety. Starting from a bike, cars to airplane, all rely upon metrology to keep their systems within certain quality parameters. In *automation and manufacturing sector*, nowadays, mostly manufacturing processing in plants is automated for control, monitoring, recording, production, quality checks by machines. Metrology plays a key role, in *energy and petrochemical* to cater the ever-increasing demand for more and more energy and petroleum products for keep on running the machines and the plants. In case of the *marine industry*, metrological instruments are used inside or outside of ships, tankers, vessels and much more. These are only few examples enough to signify the importance of metrology. It helps to create safe *environments and processes*, promotes innovation and increases knowledge and development.

As per *Vocabulary of Metrology (VIM)* document, metrology is defined as science of measurements and its applications. It includes all theoretical and practical aspects of measurement, whatever the measurement uncertainty and field of application [1]. It is often classified into *scientific or fundamental metrology, industrial or applied metrology* and *legal metrology*.

Scientific metrology is carried out mainly for compatibility of the measurements through research and development in the field of measurement standards, their maintenance and dissemination, is normally practiced at '*International Bureau of Weights and Measures (BIPM)*' and its associated organizations, specially NMI of the country. *Industrial metrology* is practiced for utility of measurements, including theoretical and practical, in manufacturing processes, ensuring the quality and accuracy of manufactured parts/components/products as well correctness of measuring instruments through test and calibration. Precise and accurate measurements improve the quality of the product, and therefore, support manufacturing and trade. *Legal Metrology* is performed for legitimacy of measurements with the statutory requirements concerning measurements as well as the law enforcement on measurements. These statutory requirements arise from a need for the protection of public health and safety, consumers rights, enabling taxation, the environment and fair trade are performed by competent bodies. In nutshell, legal metrology is responsible for improvement in quality of life. Figure 2.1 shows a simple diagram depicting conceptual classifications, specific purpose and applications of metrology in all the above three categories. For further details, please also refer Fig. 2.2.

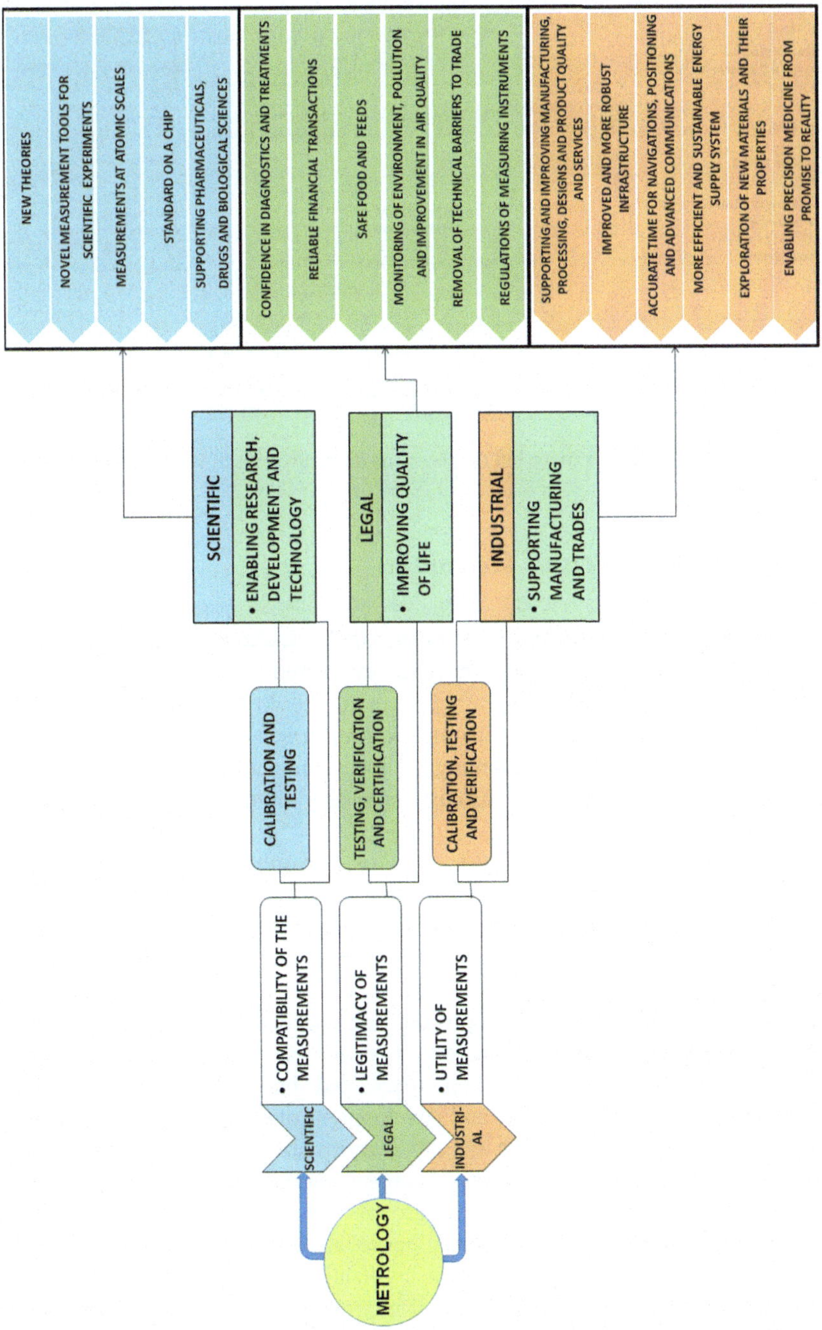

Fig. 2.1 Conceptual classifications and applications of scientific, legal and industrial metrology

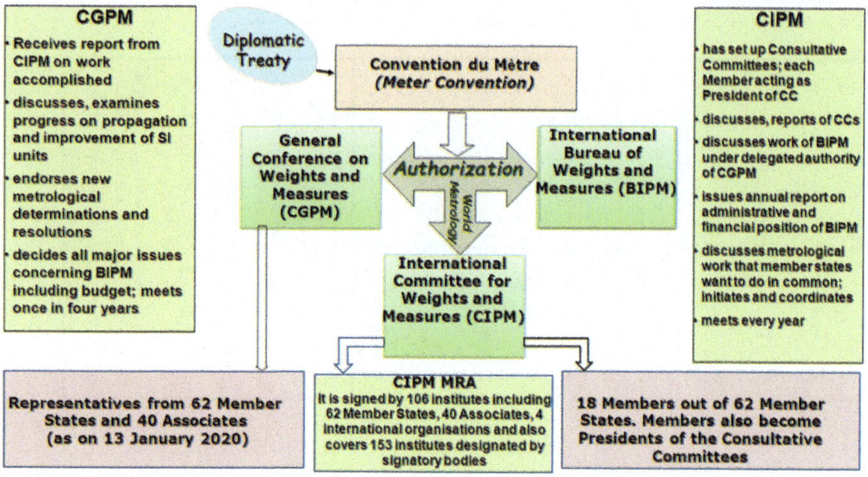

Fig. 2.2 Pictorial view of the hierarchy, interlinkages and main functions of BIPM, CGPM, CIPM and CIPM-MRA

2.2.2 *Historical Aspects of Metrology*

As a preliminary introductory point of view for the readers, the *property* of a phenomenon, body or substance, which can be expressed in terms of numbers, is defined as quantity, and the process of assigning a number to that property is termed as *measurement*, which facilitate the quantitative comparison of that property. Quantitative '*estimate*' is the value representing the measurement result. As a matter of fact, lot of science is involved in metrology including almost all the branches of science [1–3].

Life is about deal with physical world around us. Not only for human beings, but animals also, consciously and unconsciously, had been trying to quantify different things, such as food, distance to jump, for surviving. Even some of the animals use to mark their territory in forest, i.e. some sought of dimensions. Preparation of different type of nests by various types of birds, honeycomb are beautiful examples of quantification and measurements. With the progress of society, the demand of accurate measurements also increases side by side.

We live in the physical world, and feel 'space' and 'materials' around us. With the beginning of human civilization, and development of agriculture, the need of measurement of distribution of crops and the quantity of food were realized. After the age of nomadic groupings, when the humans started settling, measurements became important in managing population growth and agricultural food. Initially, measurements were based on natural things; hence, accuracy was not ascertained.

In the *most ancient age*, the great historic civilizations, China, India, Egypt, and Mesopotamia, were having the knowledge of metrology, since long back. In 1600 BC, using decimal metric system, China made archaeological discoveries, and the entire country was unified by spreading a uniform unit system. Egyptian pyramids are the witnesses of accurate dimensional metrology [4].

In the *middle ages*, to express their power, the different political authorities used to define units by their own. Hence, a large number of different units were defined differently for different goods, e.g. the weight pound was different for wheat and barley. In the nearby places of Paris, about forty-eight definitions for the measurement of surfaces existed during 1780 and long after that. Hence, the transactions for the unavoidable conversions required many cumbersome calculations [4].

For the social and ethical purposes, the field of measurement was intervened in the *French Revolution*. The need of uniform and simplified unit system was emphasized. For the simplifications in measurement and calculations, the use of decimal system was started. These simplifications resulted easier exchanges and hence increase in general welfare. The proposed reforms were adopted by other countries also. However, the goal to implement the revolutionary calendar, weeks of 10 days, hours of 100 min, and minutes of 100 s, could not be succeeded [4].

The development of scientific discoveries, the intensification of exchanges and the settlement of international institutions have been made during seventeenth to twenty-first century.

2.2.3 International Measurement System (IMS)

Due to regional and cultural variations in the different parts of the world, the systems of measurements were not consistent. For the progress of the society, the exchange and distribution of the knowledge is necessary. Due to lack of uniform measurement system and standards, the progress in scientific manner was not possible. The discovery of basic scientific laws and principles, such as electricity, magnetism and thermal physics, motivated to apply uniform standard of measurements around the world.

2.2.3.1 Introduction and Structure

The international system of measurement comprises of the globally accepted existing SI system and related activities and their operational mechanism through various intergovernmental and other international organizations on mutually agreed terms and conditions, Mutual Recognition Arrangements (MRAs) etc. A brief description of such operational mechanism is discussed in succeeding sections.

2.2.3.2 International Organizations and Their Operational Mechanism for IMS

Metre Convention and BIPM

After realizing the utmost need of global agreement on matters related to metrology, an international diplomatic treaty '*Metre Convention*' was established. The authorized government representatives of 17 countries signed the Convention on *20 May 1875* in Paris. The Convention set the framework for global collaboration in the science of measurement in its industrial, commercial and societal applications. In this treaty, an intergovernmental organization, 'BIPM' was established in Paris, France, through which *Member States* act together on matters related to measurement science and measurement standards. Although, at the time of signing *Metre Convention*, only 17 Member States participated; however, as of early 2020, the 62 Member States and 40 Associate States and Economies exist [5].

World Metrology Day

In view of ever-increasing relevance of global uniformity of measurements, the day 20 May 1875, on which '*Metre Convention*' was signed becomes very important. The *World Metrology Day (WMD)* is an annual celebration of the signature of the *Metre Convention* on 20 May 1875. The WMD was realized and initiated jointly by the BIPM and the International Organization of Legal Metrology (OIML). Every year, a theme is chosen to create awareness of the important role of measurements on different aspects of life, and a WMD Poster is releazed on that theme [6].

As stated very clearly on the website of the BIPM itself [7], the vision of the BIPM is 'to be universally recognized as the world focus for the international system of measurement'. With the mission, for the promotion and the advancement of global comparability of measurements, BIPM works with the RMOs and Member State's NMIs. This directly impacts the scientific discoveries, technological advancements, global trade and of course quality of life.

CGPM, CIPM and CIPM Mutual Recognition Arrangement (CIPM MRA)

BIPM operates under the direction and supervision of CIPM, which itself is under the authority of a '*General Conference on Weights and Measures (CGPM)*'. The CIPM consists of 18 members of different nationality. The appointing authority of the presidents of different *Consultative Committees (CCs)* is CIPM. The CIPM meetings are organized annually (since 2011 in two sessions per year). In the meetings, the reports submitted and presented by CCs are discussed. BIPM publishes the reports of the CGPM, CIPM and CCs meetings. The CIPM is mandated for the promotion of global uniformity in measurement units. CIPM discusses the work done by BIPM, and collaborative proposals from NMIs related to metrology research and activities, and after review make recommendations to CGPM.

CGPM, a supreme authority for all actions related to global metrology, is an intergovernmental conference participated by official delegates of Member States. Full Member States participate in all the BIPM activities, whereas Associate Member States are allowed to participate in the CIPM MRA programme only. Associate members have observer status at the CGPM.

Representing the global metrology community, BIPM is dedicated towards maximizing the impact of metrology on every important aspect of life, and developing opportunities for the application of metrology to global challenges. It makes liaison with other relevant intergovernmental organizations and international bodies and also provides an international platform for collaborative work related to metrology for Member States. It coordinates international comparisons of national measurement standards and establishes and maintains appropriate reference standards. The reference standards thus established are used for *International Key Comparisons* and also to provide calibrations for Member State's NMIs. As a coordinator of the worldwide measurement system, BIPM coordinates activities among the NMIs of Member States and the RMOs. It technically supports the CIPM MRA and the infrastructure for the development and promotion of the SI. To increase the effectiveness of the emerging metrology systems in the metrology community, BIPM also runs Capacity Building and Knowledge Transfer Programme [7].

A brief summary of the hierarchy, interlinkages and main functions of the diplomatic treaty, BIPM, CGPM, CIPM and CIPM MRA are shown in Fig. 2.2 and hierarchy of scientific, legal and industrial measurements performed world over in Fig. 2.3.

The CIPM, consisted of eminent metrologist and scientists of international community, is appointed by CGPM. It is responsible for the preparation and execution of the decisions made in CGPM. The reports submitted by CIPM to CGPM are passed to the governments and NMIs, the inputs received are examined and the proposals appropriate and related to changes to the SI units from the CIPM are approved. CIPM is also responsible for the supervision of the BIPM. Based on various core areas of metrology identified, the following 10 CCs are set up to assist CIPM [8]

CCAUV	*Consultative Committee for Acoustics, Ultrasound and Vibration*
CCEM	*Consultative Committee for Electricity and Magnetism*
CCL	*Consultative Committee for Length*
CCM	*Consultative Committee for Mass and Related Quantities*
CCPR	*Consultative Committee for Photometry and Radiometry*
CCQM	*Consultative Committee for Amount of Substance: Metrology in Chemistry and Biology*
CCRI	*Consultative Committee for Ionizing Radiation*
CCT	*Consultative Committee for Thermometry*
CCTF	*Consultative Committee for Time and Frequency*
CCU	*Consultative Committee for Units*

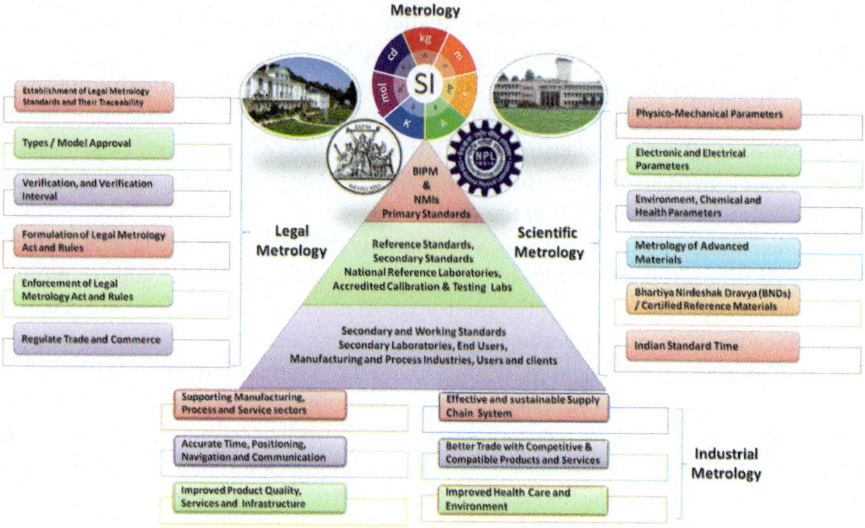

Fig. 2.3 Pictorial view of the hierarchy of scientific, legal and industrial measurements performed world over

The main motive of CCs is establishment of metrological traceability for related parameters to the SI, and hence, global comparability of measurements. Where traceability to SI is not possible; CCs make efforts for the establishment of internationally agreed reference standards. Hence, CCs play key role in implementation of CIPM MRA, and its maintenance, which contributes in establishment of globally recognized metrological standards, methods and measurement systems.

The CCs in their respective areas discuss all the matters that make impact on metrology and other scientific activities organized by BIPM and advise to CIPM accordingly. CCs also review the calibration and measurement services of BIPM including uncertainties in measurements and advice CIPM accordingly. CCs are the platforms for discussion among members and observers, regarding metrology and collaboration.

The President of the respective committee, in general, chairs respective CC meetings, also usually works as a member of the CIPM. NMIs from the Member States that are active in the field, but lack the expertise to become members, are able to attend the CC meetings as observers. The CCU advises on the matters related to the development of the SI. It works in association with other relevant international bodies, e.g. *International Organization for Standardization (ISO), International Astronomical Union (IAU), International Union of Pure and Applied Chemistry (IUPAC) and International Union of Pure and Applied Physics (IUPAP)*. The SI brochure, which contains formal definitions of the SI units, is produced by the CCU in conjunction with a number of other international organizations. The most recent 9th edition of SI brochure was published in 2019 [9].

In general, all the member countries/economies, participating in international metrology programmes, designate a Scientific Institute at National Level, as NMI, e.g. *National Institute of Standards and Technology (NIST) in USA, National Physical Laboratory (NPL) in UK, The Physikalisch-Technische Bundesanstalt (PTB) in Germany*, etc. The NMIs are the custodian of national level primary and secondary standards, and responsible to disseminate the metrological traceability of SI units to respective country. The directors of the NMIs participate in the meetings organized by BIPM, in which the global concerns related to metrology are discussed. The government representatives of Member State's government representatives can also participate in these meetings.

For demonstrating the international acceptability, equivalence of the measurement standards and the calibration and measurement certificates issued by NMIs, CIPM MRA was established. It was signed by the representatives of one hundred and six institutes of sixty-two numbers of Member States, forty numbers of associate members of CGPM and four numbers of international organizations. It also covers one hundred and fifty-three numbers of institutes designated by members and other signatories [10].

CMCs, Key Comparisons, Peer Review, Global Equivalence, RMOs and KCDB

The CIPM MRA results are globally recognized as *Calibration and Measurement Capabilities (CMCs)* of signatory and other participating members. Approved CMCs are publicly available in the *Key Comparison Database (KCDB)* on the BIPM website. CMCs are stated in terms of a measurand and its uncertainty, and may include advice about the instrumentation used. The operation of an appropriate and approved quality management system; participation in reviewed and approved intercomparisons; and international *Peer Review* of claimed measurement capabilities, are necessary for approval and publication of CMCs [11].

To facilitate the ease of participation of different countries in the international metrological activities organized by BIPM, and representation in CCs, RMOs were established [12]. RMOs work within the framework of the CIPM MRA.

For getting set of results for measurement of any parameter for CIPM MRA, key comparisons are conducted by CCs, BIPM and RMOs. The results of these key comparisons are published and maintained in the KCDB. Only NMIs or *Designated Institutes (DIs)* of Member States, who have highest level of technical competency, participate in key comparisons. These comparisons provide '*the reference value*' for the quantity for which comparison was conducted. Whereas RMO key comparisons are organized at regional level, and metrology laboratories of Member States and also associates can participate in the comparisons. The comparisons organized by RMOs provide complementary information, however, with no change in the reference value. The development of the CMCs depends on the results of key comparisons.

CMC is available to customers under normal conditions; (i) as published in the BIPM KCDB of the CIPM MRA; and (ii) as described in the laboratory's scope,

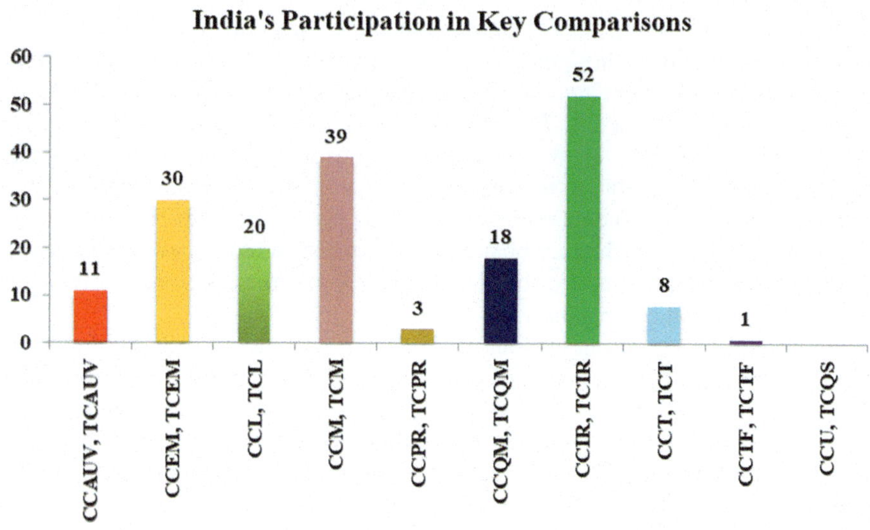

Fig. 2.4 Successful Indian participation in key comparison exercises. CCU and TCQS are shown only as representation of the CC and TC. However, these committees do not participate in any key comparison, and accordingly, no key comparison is shown against their names

duly implemented through the *Quality Management System (QMS)* as per ISO/IEC 17025 standard, and approved through a 'Peer Review' process in the laboratory's scope through the accreditation granted by a accreditation body as signatory to the ILAC MRA. The CMCs, developed by the participants, are submitted through their RMOs for Peer Review. Once the process of Peer Review is successfully completed, the RMOs submit the CMCs for second stage interregional review process. After completion of both the review processes successfully, the CMCs are published in KCDB, through *Joint Committee of the RMOs and the BIPM (JCRB)*. The CSIR-NPL, being NMI of India and *Bhabha Atomic Research Centre (BARC)* as DI for ionization radiation, have successfully participated in 182 key comparison exercises as shown in Fig. 2.4.

After approval and publication of CMCs in KCDB, the metrology institutes can use CIPM MRA logo in the calibration and measurement certificates, which make the certificate recognition, by all other signatories of the CIPM MRA. To conduct intercomparisons within, the region is the major task of RMOs. Before publishing approved CMCs in the KCDB, RMOs, through JCRB, carry out an interregional review of declared measurement capabilities. RMOs also make suggestions to the CIPM regarding policy of operation of the CIPM MRA. After rigorous exercises of key comparisons and Peer Review processes, India has published total of 236 CMCs in the KCDB over the years as shown in Fig. 2.5a for parameter wise and Fig. 2.5b under various CCs and *Technical Committees (TCs)* of RMOs.

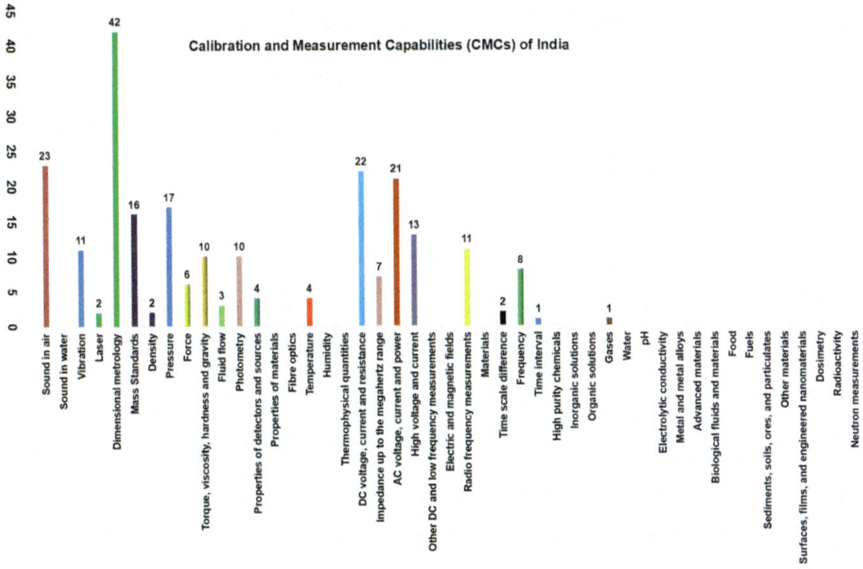

(a)

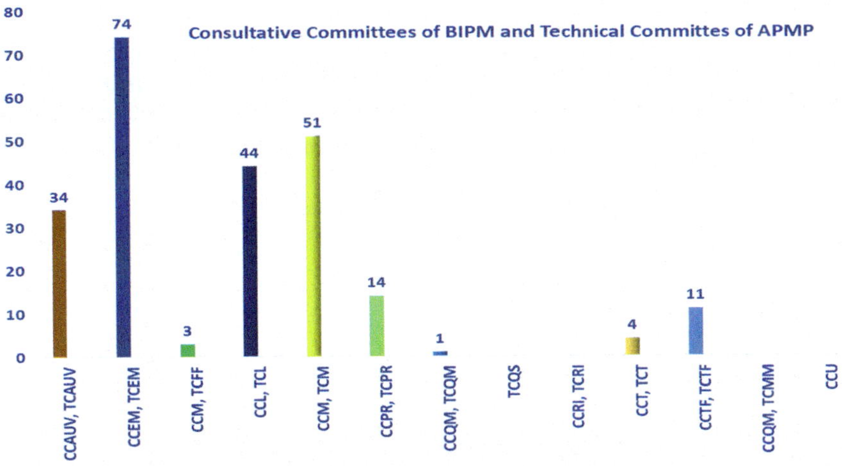

(b)

Fig. 2.5 Indian CMCs published in KCDB, **a** parameter wise and **b** as per organization by CCs and TCs. CCU and TCQS are shown only as representation of the CC and TC. However, no CMC is published by these committees, and accordingly, no CMC is shown against their names

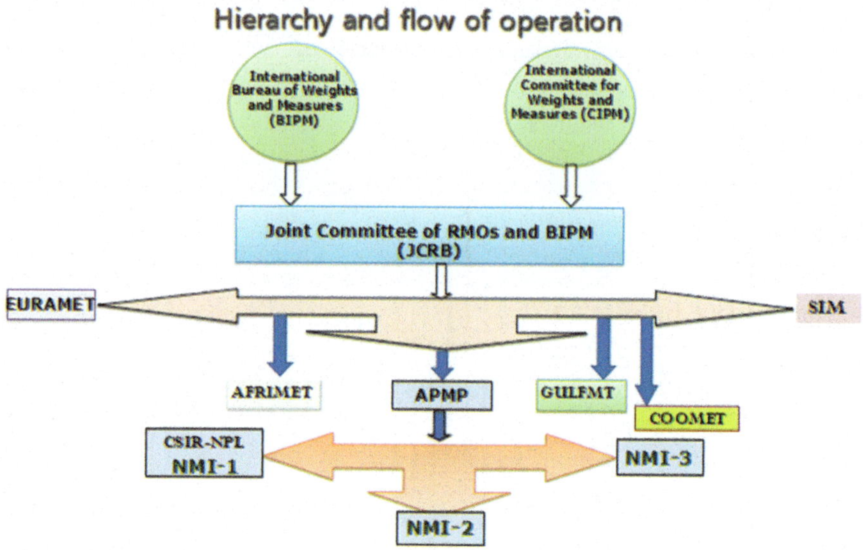

Fig. 2.6 A flow diagram showing linkages and operations between RMOs, NMIs, BIPM and CIPM

Making proposals to the CCs regarding key comparisons; carrying out the RMO key comparisons, participation in the JCRB; carrying out supplementary comparisons, supporting mutual confidence in the measurement capabilities of participating institutes are the key responsibilities of RMOs. The hierarchy and flow of operations of RMOs with NMIs, CIPM and BIPM are depicted in Fig. 2.6. Currently, following six RMOs are recognized within the framework of the CIPM MRA:

Intra-Africa Metrology System (AFRIMETS);

Asia Pacific Metrology Programme (APMP);

Euro-Asian Cooperation of National Metrological Institutions (COOMET);

European Association of National Metrology Institutes (EURAMET);

Gulf Association for Metrology (GULFMET);

Inter-American Metrology System (SIM).

For particular tasks of common interest of BIPM and other international organizations, a number of Joint Committees were formed [13]: (i) *JCGM: Joint Committee for Guides in Metrology*, for maintaining and promoting the use of the *'Guide to the Expression of Uncertainty in Measurement (GUM)'* and the 'International Vocabulary of Basic and General Terms in Metrology (VIM)'. (ii) *JCRB: Joint Committee of the Regional Metrology Organizations and the BIPM*, for coordinating the activities among the RMOs in establishing confidence for the mutual recognition of certificates of measurements in the purview of CIPM MRA. It also makes policy suggestions to the RMOs, analyze the application of RMOs for CIPM MRA, analyze the proposals of each RMO for CMCs of member NMIs and report to the CIPM, facilitate appropriate interregional supplementary comparisons, and finally write annual report on the activities of the Joint Committee to the CIPM and to the signatories of the CIPM MRA. (iii) *JCTLM: Joint Committee for Traceability in Laboratory Medicine*, for responding to the need of establishing lists of available higher-order reference materials, available higher-order reference measurement procedures and reference measurement laboratories for laboratory medicine. (iv) *JCTLM Executive Committee, JCTLM Database WG for Reference Materials, Procedures and Measurement Laboratories, JCTLM WG on Traceability for Education and Promotion on Traceability in Laboratory Medicine.*

2.3 International System of Units (SI)

2.3.1 Introduction of SI Units

The International System of Units (Système international in french) abbreviated as SI is the system of measurement which has an official status, and adopted by almost all the countries. SI is formed of coherent system of seven base units, and twenty-two derived units represented in the form of multiplication of the power of the base units and given special names and symbols. These base and derived units form as much as other derived units to express large numbers of quantities. To express multiples, sub-multiples of the units, twenty prefixes to the unit names and symbols are used. Based on the need of the time and the scientific and technological advancements, SI is continuously evolving system, in which the definitions are also modified through international consensus and agreement. Since the adoption of the system, the efforts are being made to make the system independent from the effect of physical factors and time, and hence, the base units are being defined in terms of universal or fundamental constants rather than artefact. In 2019, all the SI units have been defined by fixing the numerical values for seven defining constants when expressed in terms of their SI units [14]. The current section of the chapter comprises of the historical evaluation and development of the SI system, current status, fundamental constants, derived units, dissemination of the units at national level.

2.3.2 Historical Aspects and Evaluation of SI Units

The most necessary measurement in life is '*length*'. Initially, the units of length or distance were defined in reference to King's thumb and foot and pace, as 'inch', 'foot' and 'yard', respectively. The mile was defined as 'thousand paces'. The unit 'cubit' was defined as length of forearm between elbow to tip of middle finger. However, due to inconsistency in the body parts of the kings, more consistent standardization was needed. The standardization of the 'cubit' was done in a royal master cubit made of black marble (~52 cm). The standard cubit was divided into 28 digits (about finger width), and further fractions resulted the smallest unit of millimetre (mm) [15].

Another next most important quantity in human life is '*weight*'. Due to wide variations in the weight of humans, it was not practical to define weight in reference to human body. Initially, weight of standard size of grains was considered as unit. For example, 'carat' (quirat in Arab) was the seed of the coral tree, which has now been standardized as 0.2 g. The lump of metal corresponding to given number of grains had been considered as standard [16].

One of the related quantity '*volume*' was standardized using baskets, sacks or pottery jars of consistent sizes. The counts of seed of plants filled in certain capacity vessels were used for measurement of volume. The unit used for volume in ancient Egyptians time was 'hen' nearly equal to present day 477 cm^3 [16].

For understanding dynamic process and variations, '*time*' was defined. Initially, the units of time were defined in terms of solar cycle, e.g. day, week, month and year. In the day, the time was divided with respect to position of the Sun. Further, the concept of hours, minutes and seconds was developed [16].

With the progress in human civilization, as per need, the relatively consistent devices were developed for the measurements of above mentioned quantities. The relatively more consistent measurement of time was done by examining the position of the shadow of vertical stick of given length. However, due to variations in the position of sun in different seasons, this technique needed improvements. The flow of sand or fluid from pan of given volume through hole of given dimension was another device to measure time in more consistent manner [16].

In the first-century AD, a technique for the surveying of land using plotting of relative position of features in a landscape was developed. Concept of angles was used in such measurements. The technique was further improved for the measurement of angle of stars.

In fourteenth century, concept of clock dividing the day in forenoon and after noon having twelve hours in each segment was introduced. In fourteenth—sixteenth century, the hour was further subdivided in terms of minutes and seconds [16].

In seventeen century, the concept of atmospheric pressure and barometer for pressure measurement was discovered. After the need of quantification of the warmness, and conceptualization of the physical quantity '*temperature*', efforts towards measurement of 'temperature' were started. In 1650, Gabriel Daniel Fahrenheit designed an alcohol-filled glass thermometer. Further, the performance of liquid in glass thermometer was improved by filling the Hg in place of alcohol. Further, the freezing point and boiling point of water were used for scaling of the thermometer [16].

At the time of the French Revolution, the metric system was created, and further on 22 June 1799, the two platinum standards for the unit of length and mass were made, which laid down the foundation of present SI system [17].

In 1832, using the metric system, for the first time, Gauss did absolute measurements of the Earth's magnetic field in terms of a decimal system based on the three units millimetre, gram and second. A further electrical phenomenon was also added in these measurements. In 1860, with the base and derived units, the need of the coherent system of units was formulated by Maxwell and Thomson. Both of them worked through the British Association for the Advancement of Science (BAAS). In 1874, the centimetre, gram, second (CGS) system was introduced by the BAAS. In the CGS system, the prefix for representing the decimal and multiples from micro to mega were also introduced. In 1880s, a mutually coherent set of units ohm, volt and ampere (for electrical resistance, electromotive force and electric current, respectively) was approved by the BAAS and the International Electrical Congress [predecessor of the International Electrotechnical Commission (IEC)] [17].

After the Metre Convention, the new international prototypes for the 'metre' and the 'kilogram' were established in the 1st CGPM in 1889. Together with the astronomical second, these units created the metre, kilogram, second (MKS) system. By adding a fourth unit of an electrical nature 'ampere' in 1901, a coherent four-dimensional system was formed, which opened the path to a number of new developments [17].

In the 6th CGPM held in 1921, the Metre Convention was revised, and the four-dimensional MKSA system based on metre, kilogram, second and ampere, was approved and adopted in 1946. In 1948, the 9th CGPM instructed the CIPM to get the opinion and suggestions of scientific, technical, academic stakeholder for establishment of a practical system of units useful for all the signatories of Metre Convention, and to study about establishment of a complete set of rules for units of measurement. The 'general principles for the writing of unit symbols' were also laid down by CGPM. In addition, some coherent derived units were assigned special names [17].

In the 10th CGPM (1954), the kelvin and the candela, as base units for thermodynamic temperature and luminous intensity, respectively, were also incorporated in base units. The name International System of Units was abbreviation as SI, in 11th CGPM (1960). In 1971, in the 14th CGPM, the seventh base unit, the mole, was adopted for the quantity amount of substance [17].

From then, the extraordinary advances in SI units have been started. Full efforts to define SI units in terms of invariant quantities such as the fundamental constants of physics and the properties of atoms were started. In 2011, in the 24th CGPM, the principles of a new definition of the SI base unites were adopted. A set of seven constants was identified as references for the definitions. At that time, the experimental determine of the values of these constants in terms of base units was not completely consistent. Till 2018, it was achieved, and in the 26th CGPM (2018), the new definitions of the SI were approved and adopted [17].

2.3.2.1 Unit of Length, Metre

In 8 May 1790, it was decided by the French National Assembly that the units of the physical quantities should be defined in terms of basic physical principles, and the definition of the metre was proposed as the length of a pendulum with a half-period of one second. In 30 March 1791, the French Academy of Sciences proposed a new definition for the metre as one ten millionth of the length of a great circle quadrant along the Earth's meridian through Paris, that is the distance from the equator to the north pole along that quadrant, which was accepted by the French National Assembly. In 1795, a length bar made of brass was considered as standard for metre. In 1799, the platinum metre bar was deposited in the National Archives, as the final standard.

After the Metre Convention, in 1889, the 1st CGPM defined 'the metre as the distance between two lines on a standard bar of an alloy of platinum with 10% iridium, measured at the melting point of ice'. In 1927, 7th CGPM redefined the metre as the distance, at 0 °C (273 K), between the axis of the two central lines marked on the prototype bar of platinum–iridium, this bar being subject to one standard atmosphere of pressure and supported on two cylinders of at least 10 mm (1 cm) diameter, symmetrically placed in the same horizontal plane at a distance of 571 mm (57.1 cm) from each other. To avoid artefact-based realization of unit, in 1960, the 11th CGPM redefined the metre in terms of a certain number of wavelengths of a certain emission line of krypton-86. The metre is 1650763.73 wavelengths in a vacuum of the radiation corresponding to the transition between the $2p^{10}$ and $5d^5$ quantum levels of the krypton-86 atom. In 1983, the 17th CGPM defined the metre as the length of the path travelled by light in a vacuum during a time interval of 1/299,792,458 of a second. The definition was based on fundamental constant, velocity of light in vacuum '*c*'. In order to harmonize with the definitions of the other units, and to make clear its dependence on the fixed numerical value of the speed of light, *c*, the wording of the definition was changed in the 26th CGPM (2018) [17].

2.3.2.2 Unit of Mass, Kilogram

In 1793, the kilogram was defined as the mass of 1 litre (dm^3) of water, which was determined to be 18841 grains. In 1795, the 1000th part of kilogram, the unit gram was provisionally defined as the mass of one cubic centimetre of water at the melting point of ice. In 1799, an artefact prototype for kilogram was manufactured. After signing the 'Metre Convention', in 1879, the IPK was prepared, which was adopted in 1889. It had a mass equal to the mass of 1 dm^3 of water under atmospheric pressure and at the temperature of its maximum density, which is approximately 4 °C. In order to assure universality of the unit of mass, a new definition for the kilogram based on the value of a fundamental Planck' constant was adopted by the 26th CGPM in 2018 [17].

2.3.2.3 Unit of Time, Second

Before 1960, the unit of time, the second, was defined as the fraction 1/86,400 of the 'mean solar day'. However, rotation of the Earth measured to be irregular. Hence, to improve the definition, in 1960, the 11th CGPM adopted a definition based on the tropical year 1900. However, experimental studies on atomic and molecular transitions proved that the atomic frequency standard can be realized and reproduced much more accurately. Considering that the 13th CGPM has chosen a new definition of the second in 1967 terms of the frequency of the ground-state hyperfine transition in the caesium 133 atom. In 2018, 26th CGPM approved a revised and more precise statement of the same definition, in terms of a fixed numerical value of the unperturbed ground-state hyperfine transition frequency of the caesium 133 atom, $\Delta\nu_{Cs}$ [17].

2.3.2.4 Unit of Electric Current, Ampere

In 1893, the units for the electrical parameters current and resistance were introduced by the International Electrical Congress, and the definitions of the 'international ampere' and 'international ohm' were introduced in 1908. In 1948, the 9th CGPM approved the definition of the ampere, the force between parallel wires carrying an electric current and it had the effect of fixing the numerical value of the vacuum magnetic permeability μ_0. The numerical value of the vacuum electricpermittivityε_0 then became fixed as a consequence of the new definition of the metre adopted in 1983. After facing difficulties in the realization of the ampere, and subsequent advancement in electrical quantum standards of voltage and resistance, based on Josephson and quantum Hall effects, which are linked through Ohm's law, the ampere was realized in terms of volt and ohm. The volt and the ohm are realized using combinations of the Planck constant (h) and elementary charge (e). As a consequence, it became natural not only to fix the numerical value of h to redefine the kilogram, but also to fix the numerical value of e to redefine the ampere in order to bring the practical quantum electrical standards into exact agreement with the SI. In 2018, 26th CGPM adopted the definition of ampere based on a fixed numerical value for the elementary charge, e [17].

2.3.2.5 Unit of Thermodynamic Temperature, Kelvin

In 1954, 10th CGPM approved the triple point of water (TPW) as a fundamental fixed point by assigning it the temperature 273.16 K, for defining the 'kelvin', the unit of thermodynamic temperature. In 1967-68, the 13th CGPM adopted the name 'kelvin', symbol K, instead of 'degree kelvin', symbol °K. The practical realization required a sample of pure water of well-defined isotopic composition that is material and process dependent. The technological advancements in the field of thermometry motivated to realize the 'kelvin' in terms of Boltzmann constant k. In 2018, the 26th

CGPM approved the definition of unit kelvin by fixing the numerical value of the k [17].

2.3.2.6 Unit of Luminous Intensity, Candela

Prior to 1948, the units of luminous intensity were based on flame of specific candle or illumination of incandescent filament standards. Further, the unit of luminous intensity 'new candle' was defined in reference to the luminance of a Planckian radiator (a black body) at the temperature of freezing platinum. In 1948, 9th CGPM adopted the name 'the candela', symbol 'cd' for the units of luminous intensity. In 1954, it was established as a base unit. In 1967, the 13th CGPM amended this definition. After technological advancements in the field of detector-based absolute radiometry, in 1979, the candela was redefined in reference to the human eye response. In 2018, 26th CGPM adopted the definition of candela based on fixed numerical value of the luminous efficacy of monochromatic radiation of frequency 540×10^{12} Hz, K_{cd} [17].

2.3.2.7 Unit of Amount of Substance, Mole

The quantity 'amount of substance' (n) is used to specify the amount of chemical elements or compounds. This is defined to be proportional to the number of specified elementary entities N in a sample, and the proportionality constant is a universal constant for all entities. The proportionality constant is the reciprocal of the Avogadro constant N_A, so that $n = N/N_A$. The unit of amount of substance is called the 'mole'. Based on the proposals of the IUPAP, IUPAC and ISO, in 1969, the CIPM confirmed the definition of mol, by specifying that the molar mass of C-12 as exactly 0.012 kg/mol. The definition of the mole was dependent on the artefact definition of the kilogram. The N_A defined here was equal to the number of atoms in12 grams of C-12, and the precise knowledge of N_A was not required in the realization of 'mol' [17].

After continuous technological advancements during decades, the N_A is now known with fairly high precision and can be used to define mole. Hence, mole can be defined by specifying the exact number of entities in one mole, i.e. by fixing the numerical value of the N_A. In result, the definition of mole and the N_A need not to depend on definition of the 'kilogram'. Thus, now the difference between fundamentally distinct quantities 'mass' and 'amount of substance' can be emphasized.

In 2018, 26th CGPM adopted the definition of the mole based on a fixed numerical value for the Avogadro constant, N_A [17].

2.3.3 *Fundamental Constants of Nature (FCN)*

To achieve the universal consistency in SI, in 2011, it was proposed to define SI fully in terms of natural constants. In 2017, the updated values of the constants were published. In November 2018, members of sixty nations unanimously voted for the change of SI in terms of fundamental/natural constants. On 20 May 2019, at the occasion of the '*World Metrology Day*', the revised SI was implemented. Since May 2019, all of the SI base units have been implemented in terms of physical constants.

A physical quantity, having constant value irrespective of time and space, is called fundamental constant of nature or universal constant. The speed of light in vacuum '*c*', the gravitational constant '*G*', the Planck constant '*h*', the electric constant 'ε_0', and the elementary charge '*e*' are some of the well-known examples of physical constants. The term 'physical constant' does not refer to a numerical value, whereas only physical quantity. Numerical values of the dimensional physical constants are dependent upon the system of units used to express. The physical quantities regarded as immutable and non-derivable from more fundamental principles are termed as 'fundamental physical constant', e.g. *c* and *G*. Fixed quantities of any desired dimension can be defined by combining dimensional universal physical constants [18].

The revised SI rests on a foundation of seven values, known as the constants. The values of the constants are the same everywhere in the universe. In the revised SI, these constants completely define the seven base SI units, from the second to the candela [14].

As a result, the constants: the speed of light in vacuum (*c*), the Planck constant (*h*), the elementary charge (*e*), the Avogadro constant (N_A), and the Boltzmann constant (*k*), have known exact numerical values when expressed in SI units.

The *c*, *h* and *e* are fundamental constants. The speed of light in vacuum '*c*' is the limit of speed in the universe and nothing can travel faster than light. The Planck constant '*h*' is one of the fundamental constants of quantum physics, which establishes the relation between energy and frequency of light quanta. This also limits the measurement accuracy of two complementary physical quantities simultaneously by Heisenberg uncertainty principle. The elementary charge '*e*' is the amount of charge in an electron.

The N_A and *k* are the technical constants. They give a proportionality factor for defining the units used. The *k* relates energy to its temperature, and appeared along with the absolute temperature in almost all the physical laws and principles. The N_A defines the number of particles in a mole.

The time is defined in terms of the hyperfine transition frequency of caesium 133 ($\Delta\nu_{Cs}$). Caesium 133 is a metal atom. The energy of Cs outermost electron can be controlled with microwave radiation. The frequency of microwave radiation that causes this electron to jump between two closely spaced low-energy states is known as the hyperfine transition frequency.

The *candela* is defined in terms of the luminous efficacy of monochromatic radiation of frequency 540×10^{12} Hz, denoted as K_{cd}, which is related to human eye

Table 2.1 Information about defining constants [14]

Defining constant	Numerical value
Hyperfine transition frequency of caesium 133 ($\Delta\nu_{Cs}$)	9 192 631 770 Hz
Speed of light in vacuum (c)	299 792 458 m s^{-1}
Planck constant (h)	$6.626\ 070\ 15 \times 10^{-34}$ J s
Elementary charge (e)	$1.602\ 176\ 634 \times 10^{-19}$ C
Boltzmann constant (k_B)	$1.380\ 649 \times 10^{-23}$ J K^{-1}
Avogadro constant (N_A)	$6.022\ 140\ 76 \times 10^{23}$ mol^{-1}
luminous efficacy of monochromatic radiation of frequency 540×10^{12} Hz (K_{cd})	683 lm W^{-1}

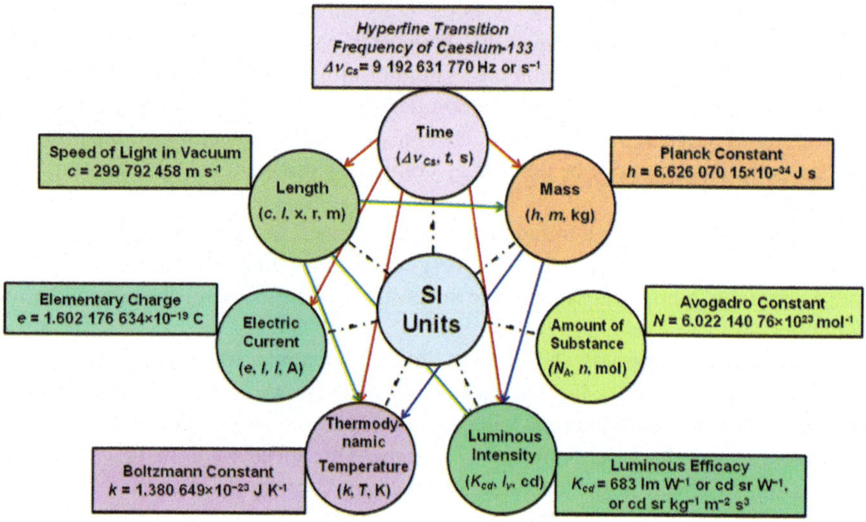

Fig. 2.7 Seven SI base units, relation with each other and their defining constants

response. 'luminous efficacy' is the total amount of visible light (sensed by human eye) produced by a light source using a certain amount of power. Information about the defining constants is included in Table 2.1 and about seven base units in Fig. 2.7.

2.3.4 Redefinition of SI Units in Terms of FCN and Current Status

The redefinition of SI based units in terms of fixed numerical values of the defining constants, effective from 20 May 2019, reproduced in Table 2.2, as published in Resolution 1 of 26th CGPM [19]:

Table 2.2 Redefinition of seven SI base units and their expressions [20]

SI base unit and symbol	Definition
second, s	The **second**, symbol **s**, is the SI unit of time. It is defined by taking the fixed numerical value of the caesium frequency $\boldsymbol{\Delta\nu_{Cs}}$, the unperturbed **ground-state hyperfine transition frequency of the caesium-133 atom**, to be 9 192 631 770 when expressed in the unit Hz, which is equal to $s^{-1.}$
	This definition implies the exact relation $\Delta\nu_{Cs} = 9\,192\,631\,770$ Hz. Inverting this relation gives an expression for the unit second in terms of the defining constant $\Delta\nu_{Cs}$: $1\text{Hz} = \frac{\Delta\nu_{Cs}}{9\,192\,631\,770}$ or $1\text{s} = \frac{9\,192\,631\,770}{\Delta\nu_{Cs}}$ The effect of this definition is that the second is equal to the duration of 9 192 631 770 periods of the radiation corresponding to the transition between the two hyperfine levels of the unperturbed ground state of the ^{133}Cs atom
metre, m	The **metre**, symbol **m**, is the SI unit of length. It is defined by taking the fixed numerical value of the **speed of light in vacuum** $\boldsymbol{c}$ to be 299 792 458 when expressed in the unit ms^{-1}, where the second is defined in terms of $\Delta\nu_{Cs}$
	This definition implies the exact relation $c = 299\,792\,458$ m $s^{-1.}$ Inverting this relation gives an exact expression for the metre in terms of the defining constants c and $\Delta\nu_{Cs}$: $1\text{m} = \left(\frac{c}{299\,792\,458}\right)\text{s} = \left(\frac{9\,192\,631\,770}{299\,792\,458}\right)\frac{c}{\Delta\nu_{Cs}} \approx 30.663\,319\frac{c}{\Delta\nu_{Cs}}$ The effect of this definition is that one metre is the length of the path travelled by light in vacuum during a time interval with duration of 1/299 792 458 of a second
kilogram, kg	The **kilogram**, symbol **kg**, is the SI unit of mass. It is defined by taking the fixed numerical value of the **Planck constant** $\boldsymbol{h}$ to be $6.626\,070\,15 \times 10^{-34}$ when expressed in the unit J s, which is equal to kg m^2 s^{-1}, where the metre and the second are defined in terms of c and $\Delta\nu_{Cs}$
	This definition implies the exact relation $h = 6.626\,070\,15 \times 10^{-34}$ kg m^2 s^{-1}. Inverting this relation gives an exact expression for the kilogram in terms of the three defining constants h, $\Delta\nu_{Cs}$ and c: $1\text{ kg} = \left(\frac{h}{6.626\,070\,15 \times 10^{-34}}\right)\text{m}^{-2}\text{ s}$ which is equal to: $1\text{ kg} = \frac{(299\,792\,458)^2}{(6.626\,070\,15 \times 10^{-34})(9\,192\,631\,770)}\frac{h\Delta\nu_{Cs}}{c^2} \approx 1.475\,521\,4 \times 10^{40}\frac{h\Delta\nu_{Cs}}{c^2}$ The effect of this definition is to define the unit kg m^2 s^{-1} (the unit of both the physical quantities action and angular momentum). Together with the definitions of the second and the metre, this leads to a definition of the unit of mass expressed in terms of the Planck constant h
ampere, A	The **ampere**, symbol **A**, is the SI unit of electric current. It is defined by taking the fixed numerical value of **the elementary charge** $\boldsymbol{e}$ to be $1.602\,176\,634 \times 10^{-19}$ when expressed in the unit C, which is equal to A s, where the second is defined in terms of capital $\Delta\nu_{Cs}$

(continued)

Table 2.2 (continued)

SI base unit and symbol	Definition
	This definition implies the exact relation $e = 1.602\ 176\ 634 \times 10^{-19}$ A s. Inverting this relation gives an exact expression for the unit ampere in terms of the defining constants e and $\Delta\nu_{Cs}$: $1\ \text{A} = \left(\frac{e}{1.602\,176\,634 \times 10^{-19}}\right)\text{s}^{-1}$ which is equal to: $1\ \text{A} = \frac{1}{(9\,192\,631\,770)(1.602\,176\,634 \times 10^{-19})}\Delta\nu_{Cs}e \approx 6.789\,6868 \times 10^{8}\Delta\nu_{Cs}e$ The effect of this definition is that one ampere is the electric current corresponding to the flow of 1/(1.602 176 634 × 10^{-19}) elementary charges per second.
kelvin, K	The **kelvin**, symbol **K**, is the SI unit of thermodynamic temperature. It is defined by taking the fixed numerical value of the **Boltzmann constant *k*** to be 1.380 649 × 10^{-23} when expressed in the unit J K^{-1}, which is equal to kg m^2 s^{-2} K^{-1}, where the kilogram, metre and second are defined in terms of *h*, *c* and $\Delta\nu_{Cs}$
	This definition implies the exact relation $k = 1.380\ 649 \times 10^{-23}$ kg m^2 s^{-2} K^{-1}. Inverting this relation gives an exact expression for the kelvin in terms of the defining constants *k, h* and $\Delta\nu_{Cs}$: $1\ \text{K} = \left(\frac{1.380\,649}{k}\right)10^{-23}\text{kg m}^2\text{s}^{-2}$ which is equal to: $1\text{K} = \frac{(1.380\,649 \times 10^{-23})}{(6.626\,070\,15 \times 10^{-34})(9\,192\,631\,770)}\frac{\Delta\nu_{Cs}h}{k} \approx 2.266\,6653\frac{\Delta\nu_{Cs}h}{k}$ The effect of this definition is that one kelvin is equal to the change of thermodynamic temperature that results in a change of thermal energy *k* T by 1.380 649 × 10^{-23} J
mole, mol	The **mole**, symbol **mol**, is the SI unit of amount of substance. One mole contains exactly 6.022 140 76 × 10^{23} elementary entities. This number is the fixed numerical value of the **Avogadro constant,** N_A, when expressed in the unit mol^{-1} and is called the Avogadro number The amount of substance, symbol n, of a system is a measure of the number of specified elementary entities. An elementary entity may be an atom, a molecule, an ion, an electron, any other particle or specified group of particles
	This definition implies the exact relation $N_A = 6.022\ 140\ 76 \times 10^{23}$ mol^{-1}. Inverting this relation gives an exact expression for the mole in terms of the defining constant N_A: $1\ \text{mol} = \left(\frac{6.022\,140\,76 \times 10^{23}}{N_A}\right)$ The effect of this definition is that the mole is the amount of substance of a system that contains 6.022 140 76 × 10^{23} specified elementary entities
candela, cd	The **candela**, symbol **cd**, is the SI unit of luminous intensity in a given direction. It is defined by taking the fixed numerical value of the **luminous efficacy** of monochromatic radiation of frequency 540 × 10^{12} Hz, K_{cd}, to be 683 when expressed in the unit lm W^{-1}, which is equal to cd sr W^{-1}, or cd sr kg^{-1} m^{-2} s^3, where the kilogram, metre and second are defined in terms of *h,c* and $\Delta\nu_{Cs}$

(continued)

Table 2.2 (continued)

SI base unit and symbol	Definition
	This definition implies the exact relation $K_{cd} = 683$ cd sr kg^{-1} m^{-2} s^{3} for monochromatic radiation of frequency $\nu = 540 \times 10^{12}$ Hz. Inverting this relation gives an exact expression for the candela in terms of the defining constants K_{cd}, h and $\Delta\nu_{Cs}$: $1\text{cd} = \left(\frac{K_{cd}}{683}\right)\text{kg m}^2\text{s}^{-3}\text{sr}^{-1}$ Which is equal to $1\text{ cd} = \frac{1}{(6.626\,070\,15 \times 10^{-34})(9\,192\,631\,770)^2 683}(\Delta\nu_{Cs})^2 h K_{cd} \approx 2.614\,8305 \times 10^{10}(\Delta\nu_{Cs})^2 h K_{cd}$ The effect of this definition is that one candela is the luminous intensity, in a given direction, of a source that emits monochromatic radiation of frequency 540×10^{12} Hz and has a radiant intensity in that direction of (1/683) W/sr

A unique way is proposed to represent the redefinitions in a common format as written in first row of the (Table 2.3), in which after filling the respective information of different SI units given in the seven rows can help in understanding and remember the definitions of seven base units. As definition of 'mol' is slightly different, which is written separately.

After the linkages with the other defined units like second, metre, kilogram, these can be represented to express the dimensional quantities. The metre–kilogram–second and ampere as the core units and the candela–kelvin–mole as the non-core SI units, as depicted in Fig. 2.8 [21]. This effort results into the simpler and more fundamental representation of the entire SI. They are now artefact free definitions, particularly the international prototype of kilogram for mass and isotopic composition dependent triple point of water for kelvin.

2.3.5 *Implication of Redefined SI Units*

In the Sect. 2.3.3, we have described the fundamental constants of nature (FCN) and how they are used to redefine the SI base units. The unanimous historic decision was made in 26th CGPM on the redefinition of four of the base units of; kilogram, kelvin, ampere and mole by fixing the numerical values of Planck constant, Boltzmann constant, elementary charge and Avogadro constant, respectively. These fundamental constants are inherently stable. Therefore, FCNs have been chosen so that the revised definitions need not be modified again in foreseeable time to accommodate future improvements in the technologies used to realize the base units.

Table 2.3 The harmonized redefinitions of the base units are attempted to be represented in a common format, as given in the first row of the table. The definition of any base unit can be written by filling the blanks by the information given in front of the base unit in the respective row. As definition of 'mol' is slightly different, which is written separately [20]

The … (SI base unit name), symbol … (symbol of SI base unit name), is the SI unit of … (name of base quantity)
It is defined by taking the fixed numerical value of the … (name and symbol and the description of defining constant), to be … (value of the defining constant)
when expressed in the unit … (unit of the defining constant), which is equal to … (unit of defining constant in terms of SI units), where …(relevant SI base units with respective defining constants for representing respective base unit)

SI base unit name	SI base unit symbol	Name of SI base quantity	Name and symbol, and description of defining constant	Value of the defining constant	Unit of the defining constant	In terms of SI base units	Relevant SI base units with respective defining constants for representing respective base unit
second	s	time	caesium frequency $\Delta\nu_{Cs}$, the unperturbed ground-state hyperfine transition frequency of the caesium 133 atom	9 192 631 770	Hz	s^{-1}	
metre	m	length	speed of light in vacuum, c	299 792 458	$m\ s^{-1}$		the second is defined in terms of the caesium frequency $\Delta\nu_{Cs}$

(continued)

Table 2.3 (continued)

The … (SI base unit name), symbol … (symbol of SI base unit name), is the SI unit of … (name of base quantity)
It is defined by taking the fixed numerical value of the … (name and symbol and the description of defining constant), to be … (value of the defining constant)
when expressed in the unit … (unit of the defining constant), which is equal to … (unit of defining constant in terms of SI units), where …(relevant SI base units with respective defining constants for representing respective base unit)

SI base unit name	SI base unit symbol	Name of SI base quantity	Name and symbol, and description of defining constant	Value of the defining constant	Unit of the defining constant	In terms of SI base units	Relevant SI base units with respective defining constants for representing respective base unit
kilogram	kg	mass	Planck constant, h	$6.626\ 070\ 15 \times 10^{-34}$	J s	kg m^2 s^{-1}	the metre and the second are defined in terms of c and $\Delta\nu_{Cs}$

(continued)

Table 2.3 (continued)

The … (SI base unit name), symbol … (symbol of SI base unit name), is the SI unit of … (name of base quantity)
It is defined by taking the fixed numerical value of the … (name and symbol and the description of defining constant), to be … (value of the defining constant)
when expressed in the unit … (unit of the defining constant), which is equal to … (unit of defining constant in terms of SI units), where …(relevant SI base units with respective defining constants for representing respective base unit)

SI base unit name	SI base unit symbol	Name of SI base quantity	Name and symbol, and description of defining constant	Value of the defining constant	Unit of the defining constant	In terms of SI base units	Relevant SI base units with respective defining constants for representing respective base unit
ampere	A	electric current	elementary charge, e	$1.602\ 176\ 634 \times 10^{-19}$	C	A s	the second is defined in terms of the caesium frequency $\Delta\nu_{Cs}$

(continued)

Table 2.3 (continued)

The … (SI base unit name), symbol … (symbol of SI base unit name), is the SI unit of … (name of base quantity) It is defined by taking the fixed numerical value of the … (name and symbol and the description of defining constant), to be … (value of the defining constant) when expressed in the unit … (unit of the defining constant), which is equal to … (unit of defining constant in terms of SI units), where …(relevant SI base units with respective defining constants for representing respective base unit)							
SI base unit name	SI base unit symbol	Name of SI base quantity	Name and symbol, and description of defining constant	Value of the defining constant	Unit of the defining constant	In terms of SI base units	Relevant SI base units with respective defining constants for representing respective base unit
kelvin	K	thermodynamic temperature.	Boltzmann constant, k	$1.380\,649 \times 10^{-23}$	$\mathrm{J\,K^{-1}}$	$\mathrm{kg\,m^2\,s^{-2}\,K^{-1}}$	the kilogram, metre and second are defined in terms of h, c and $\Delta\nu_{Cs}$
candela	cd	luminous intensity in a given direction	luminous efficacy of monochromatic radiation of frequency 540×10^{12} Hz, K_{cd}	683	$\mathrm{lm\,W^{-1}}$	$\mathrm{cd\,sr\,W^{-1}}$, or $\mathrm{cd\,sr\,kg^{-1}\,m^{-2}\,s^3}$	the kilogram, metre and second are defined in terms of h, c and $\Delta\nu_{Cs}$

(continued)

Table 2.3 (continued)

The … (SI base unit name), symbol … (symbol of SI base unit name), is the SI unit of … (name of base quantity)
It is defined by taking the fixed numerical value of the … (name and symbol and the description of defining constant), to be … (value of the defining constant)
when expressed in the unit … (unit of the defining constant), which is equal to … (unit of defining constant in terms of SI units), where …(relevant SI base units with respective defining constants for representing respective base unit)

SI base unit name	SI base unit symbol	Name of SI base quantity	Name and symbol, and description of defining constant	Value of the defining constant	Unit of the defining constant	In terms of SI base units	Relevant SI base units with respective defining constants for representing respective base unit
mole	mol	amount of substance	One mole contains exactly 6.022 140 76 × 10^{23} elementary entities. This number is the fixed numerical value of the Avogadro constant, N_A, when expressed in the unit mol^{-1} and is called the Avogadro number The amount of substance, symbol *n*, of a system is a measure of the number of specified elementary entities. An elementary entity may be an atom, a molecule, an ion, an electron, any other particle or specified group of particles.				

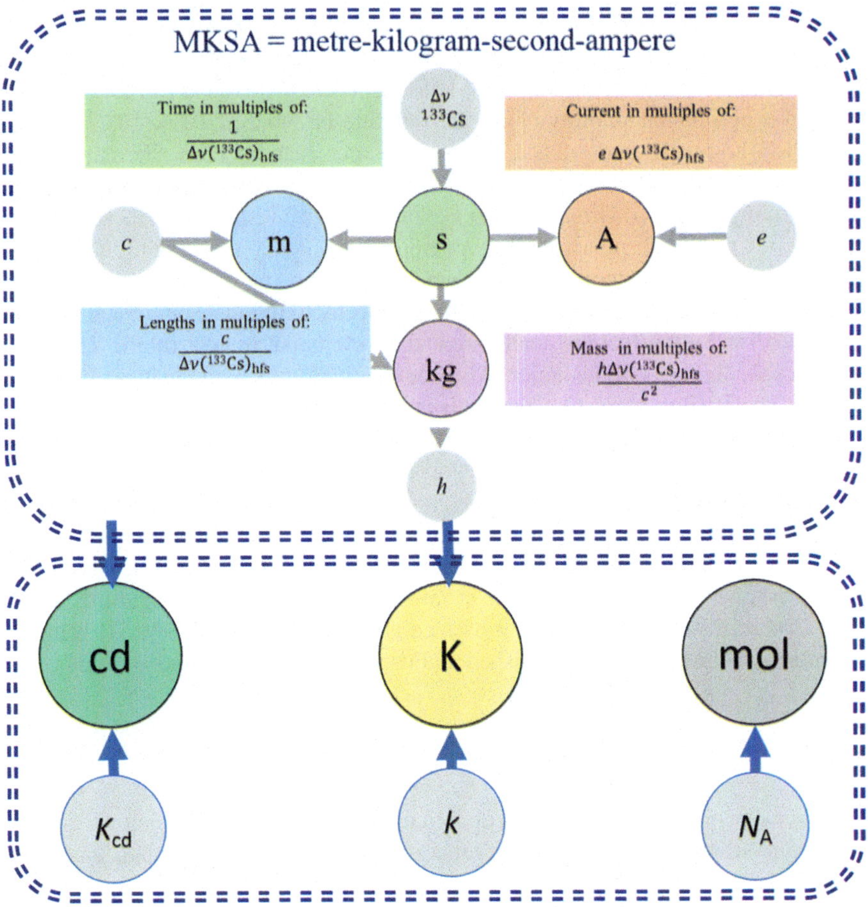

Fig. 2.8 Illustrating the SI base units in core and non-core unit groups depending on their usages in expressing the dimensionality of quantities for other SI and derived units [21]

2.3.5.1 The Robustness of Fundamental Constants Used for New SI

It is the understanding and the belief of the scientific community that the fundamental constants such as c, h and e have not changed for a long time. The search for the variation of fundamental constants is always motivated by the theories unifying the fundamental interactions such as fine structure constant (α) and the proton-to-electron mass ratio (μ).

$$\alpha = \frac{1}{4\pi\varepsilon_0}\frac{e^2}{hc} \tag{2.1}$$

where the stability of α can be measured using the high precision atomic clocks of relative uncertainties of 10^{-16} and even below in terms of frequency ratio. Four fundamental constants related to four SI base units are involved in Eq. 2.1. The 2014 tests on experimental stability of α were found to be $-0.7 \pm 2.1 \times 10^{-17}$ per year for NPL(UK) results and $-2.0 \pm 2.0 \times 10^{-17}$ per year for PTB, Germany results [22, 23], respectively.

The future progress in time will see even more robust consistency in these measurements with the direct measurements of optical frequency ratios than the caesium atomic clock. This suggests that the α is one million times more stable than the kilogram mass. Therefore, redefinition of SI units by fixing their numerical values is robust and will not affect any technological development in near future. This motivated for the linkages of the major of the redefined SI units with the unit of time, the second. The revolutionary advancements in atomic physics and quantum metrology, in past 50 years, have enabled the definitions of the second and metre, and the practical realization of electrical units with unprecedented accuracy. Here, the definitions of the SI units of the second, the metre and the candela are also rephrased in 2018, as presented in the following section.

Before fixing the values of the fundamental constants, the well-defined conditions of concerned CCs and CCU have been satisfied by the results obtained till 1st July 2017. These results were then summarized and the CODATA 2017 [24] values of fundamental constants were published and used for the redefinitions of units.

2.3.5.2 The Implementation of Redefined SI Based Units

The traceable measurements are the quantitative comparison of an unknown quantity with standard quantity. Historically, most of the definitions of standard units were also part of its realization and dissemination. Traditionally, it includes some part of their preferred mode of realizations. If we separate the definition from realization, we allow realization to evolve over decades in the ways we cannot anticipate. Say, for example, the IPK was part of the realization and dissemination of mass. The unit definition of ampere was abstract or idealized, where magnetic effect of current was a measure. However, the ampere was being realized using the Quantum Hall Resistance and Josephson voltage standards.

In the new SI, the realizations are separated conceptually from the definitions so that the units based on fundamental constants can be realized independently at any place and at any time. Mise-in-pratique (MeP) [25] for each of the redefined SI unit has been given with the available methods of primary standards for the realization and dissemination. There is no single preferred method. In addition to above, with the development of science and technology, several new and advantageous realization methods shall be developed and introduced in future, without any need to redefine the unit. It can be clearly visualized from the historical developments happened with the definition of SI metre, it transformed from Pt–Ir artefact to atomic reference transition to present definition of fixed numerical value of the speed of light, led to

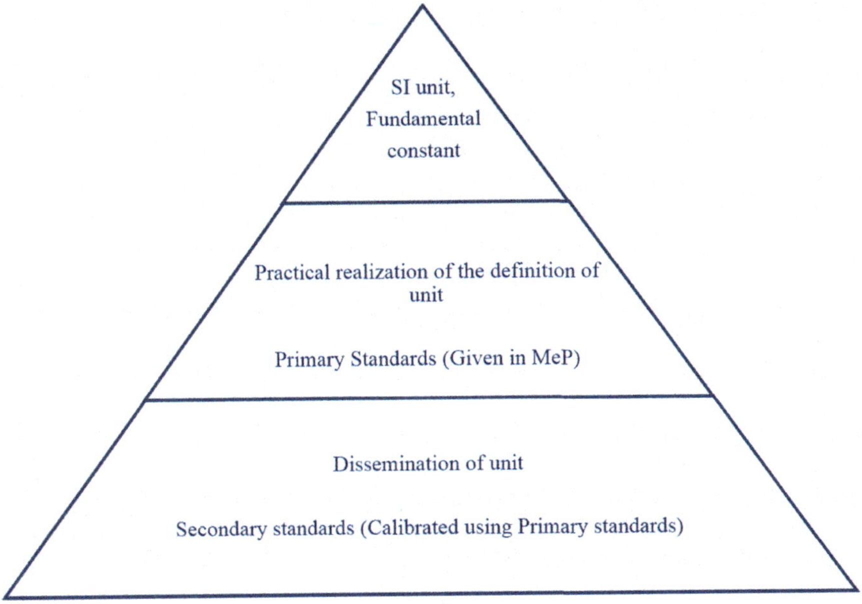

Fig. 2.9 Illustration of the traceability chain from the definition of SI unit

the decision to define all units by using defining constants. Figure 2.9 illustrates the traceability chain from the definition of SI unit

2.3.5.3 The Implications of Redefined SI Based Units

As mentioned earlier, the revised definitions are based on seven fundamental physical constants (e.g. the Planck constant, the Boltzmann constant, the elementary charge and the Avogadro constant) and are therefore inherently stable. Most common users outside the NMIs will not notice the change, and implications will be evolved through the process of realizations and dissemination over the coming years. The relevant mise-en-pratique has been published describing the methods for the practical realizations of the new definitions at NMIs. Some information about how these changes might affect the different areas of measurement and the implications of redefinitions, is given below, reproduced from the 'practical realizations of the definitions of some important units' BIPM website and published literature [25–29].

The Kilogram Defined in Terms of the Planck Constant (*h*)

- The **kilogram** is now defined in terms of the Planck constant (*h*), guaranteeing long-term stability of the SI mass scale. The *h* has an unprecedented realization uncertainty of 9.1×10^{-9} by Kibble balance at NRC Canada, before fixing its value for the redefinition.

- The kilogram can be realized by any suitable primary method, as described in MeP, such as the *Kibble (watt) balance* or the *X-ray crystal density method (XRCD)* method used for Avogadro constant, where h is calculated using the Bohr model of the hydrogen atom.
- IPK at the time of redefinition is 1 kg exactly; however, NMIs need not have to necessarily get the traceability from BIPM, if the h is realized with above primary methods. The realization uncertainty of h about $\pm 10\ \mu g$ will be added to the uncertainty of the IPK and correspondingly at the CMC of copy of IPK at respective NMIs [30].
- The substitution values of mass realized by primary methods, such as few mg, µg can be set as primary, instead of 1 kg.
- Users can also continue to obtain traceability to the SI from the same sources used as before through the BIPM, national metrology institutes and accredited laboratories. International comparisons will ensure their consistency.
- There was no change in the SI kilogram at the time of redefinition, to maintain the consistency to previous measurements.
- The uncertainties given by NMIs to their calibration customers will be mostly unaffected. The traceability chart for the redefined kilogram is shown in Fig. 2.10.

The Ampere Defined in Terms of Elementary Charge (e)

- The ampere and other electrical units are practically realized at the highest metrological level and consistent with the definitions.
- The determination of elementary charge was calculated from experimental values of fine structure constant α, h and c and μ_0. The *CODATA* value thus fixed has the relative standard uncertainty of 5.2×10^{-9}.
- As per MeP, the unit ampere can be realized in three ways; (i) by using *Ohm's law*, the unit relation $A = V/\Omega$, using the SI derived units the volt V and the ohm Ω, based on the Josephson and quantum Hall effects, respectively; (ii) by using a *single electron transport (SET)* or similar device, the unit relation $A = C/s$, the value of e given in the definition of the ampere and realized second s; and (iii) by using the relation $I = C \cdot dU/dt$, the unit relation $A = F \cdot V/s$, with volt V and the farad F and of the SI base unit second s.
- With the reassessment of h and e, the effect on the practical realization of volt and resistance, involving Josephson constant and von Klitzing constants, associated relative uncertainty would be approximately of the order of 19×10^{-9} and 22×10^{-9}, respectively. These changes would be visible to NMI level users and would be most unaffected for majority of users.
- SET has higher uncertainty now, but it is a most appropriate method for the realization of e, and expected to improve with the advent of technology.
- The revised SI has resulted in small changes to all other disseminated electrical units, say because of the resultant adjustments in K_J and R_K, because of the fixing

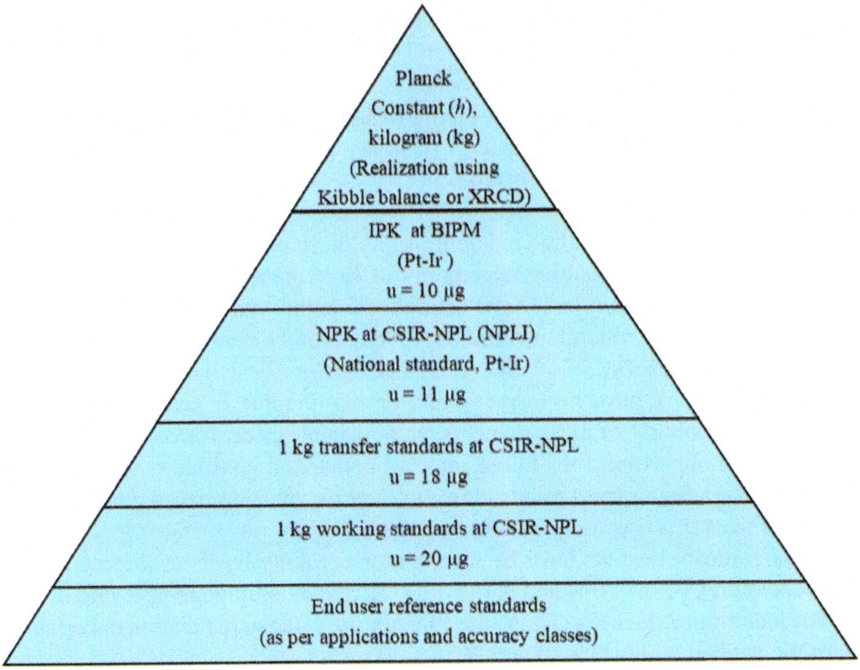

Fig. 2.10 The traceability chart for the mass realization after the implementation of redefined kilogram [26, 27]

of the values of e and h. Therefore, metrologists working at the highest precision level may need to do these slight adjustments of their standards and corresponding measurement uncertainty budgets.

- Any step changes in electrical units will be hidden by the drift or instabilities of the widely used secondary standards.
- The revised definition based on e has a consequence of the conversion factor between SI and electrostatic cgs system, e.g. fine structure constant.

The Kelvin Defined in Terms of Boltzmann Constant (k)

- The **kelvin** has been redefined based on Boltzmann constant, which relate the thermodynamic temperature to the energy, the coherent unit Jule. The material-dependent definition based on water triple point (TPW) has been abrogated, which has isotopic composition dependency in water samples.
- The value of the Boltzmann constant has been fixed with relative standard uncertainty of 3.7×10^{-7}. The primary thermometry methods such as *acoustic gas thermometry (AGT), dielectric constant gas thermometry (DCGT), Johnson noise*

thermometry (JNT), Doppler broadening thermometry and radiometric thermometry methods are successfully used and suggested in MeP-K for the realization and dissemination of redefined kelvin.

- Direct practical measurement of thermodynamic temperature is now possible at any point in the temperature scale. No primary method or the interpolation device is preferred in the direct realization.
- The thermodynamic temperature of TPW was exact to 273.16 K at the time of redefinition, However, after redefinition, its thermodynamic temperature T_{TPW} will have additional relative uncertainty of Boltzmann constant, about ±0.1 K [31]. TPW will remain as ITS-90 define fixed point without any additional realization uncertainty due to redefinition, to maintain the continuity of the previous measurements on T_{90}.
- Redefinition will have no instant effect on the ongoing measurement practices and the traceability of measurements, no noticeable effect for common users.
- It gives an opportunity for future improvements, and possible new temperature scale with defining fixed points closer to thermodynamic temperatures, after the assessments of T-T_{90} temperatures.
- The significant benefits have been seen in uncertainty levels of thermodynamic temperatures below 20 K and above 1300 K, and at extreme temperatures.
- The redefinition does not change the temperature value or realization uncertainties of the defined scales ITS-90 and PLTS-2000.

The traceability chart for the realization of Boltzmann constant based kelvin is given in Fig. 2.11.

The Mole Defined in Terms of Avogadro Constant (N_A)

- The **mole** has been redefined with respect to a specified number of entities (typically atoms or molecules) and no longer depends on the unit of mass, the kilogram.
- The value of Avogadro constant N_A adjusted by CODATA-2017 is $N_A = 6.022140758 \times 10^{23}$ mol^{-1} with a relative uncertainty of 2.0×10^{-8}.
- Practical realization of the definition of the mole with the smallest uncertainty was achieved in an International Avogadro Coordination experiment involving the determination of the number of ^{28}Si atoms (N) in a single crystal of Si. Both the constant N_A and h are realized by this method.
- Traceability to the mole can be established with all previously applicable methods with tables of atomic weights and the molar mass constant (Mu).
- Atomic weights are unaffected by the redefinition and Mu will remain 1 g/mol. The uncertainty is so small that the revised definition of the mole does not require any change to the common practice.
- The realization of mole in chemical entities is carried out using various primary methods. This includes gravimetric, use of the ideal gas law and electrolysis,

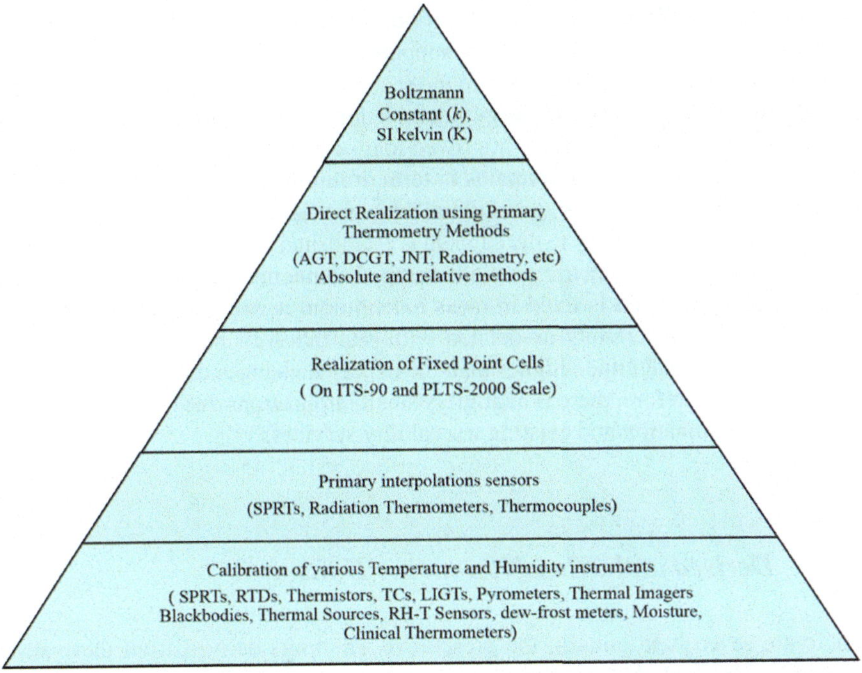

Fig. 2.11 The traceability chain of the realization of Boltzmann constant-based new kelvin

for the measurement of amount of substance concentration (mol/m^3), substance content (mol/kg) and substance fraction (mol/mol).

- The new definition eliminates any linkages between mass and amount of substance,
- The major implication of redefinition is that the mole is being used in the evolving areas of ultra-low chemical and biological quantities.
- It is stated here that there will not be any impact on the second, the metre and the candela because of the redefinitions of the kilogram, ampere, kelvin and mole.

The Implications on Second, Metre and Candela

- The definition of **second** continues to be in terms of the hyperfine transition frequency of the caesium 133 atom. The rewording has no effect on the traceability chain of second. In general, the time and frequency metrology is not impacted by any other redefinitions, instead major of the other units are linked with the second. Research on development on optical clocks is rigorously going on in the leading NMIs.
- The definition of **metre** remains in terms of the speed of light, one of the fundamental constants of physics. The MeP prepared for the implementation of new SI gave (i) direct measurement of light travelling time (time of flight) or (ii) indirect

measurement of light travelling time (optical interferometer) as the primary standard methods. Also the traceability support is required for the emerging areas of dimensional nanometrology, where the secondary method of realization of metre are used. Dimensional metrology practice does not need to be modified in any way and will benefit from the improved long-term stability of the system.

- The definition of **candela** remains in term of luminous efficacy of radiation at a specified frequency, and K_{cd}, is still linked to watt as being a technical constant for photometry. Traceability to the candela is established with the same measurement uncertainty via radiometric methods using absolute measurements by detectors. Since the value of h is fixed in mass redefinition, it will have certain effect on candela. The uncertainty associated with realization of candela is about 10^{-4} [32], order of magnitude higher than the expected changes due to redefinition of kilogram. Therefore, there is negligibly small implications due to the redefinition, on its dissemination and existing traceability services.

2.3.6 *Derived and Other Non-metric Units*

There are seven base units in the present SI. The units derived from these seven SI base units are called derived units, as specified by the SI. The base units are either dimensionless or can be expressed as a product of one or more of the base units, possibly also scaled by an appropriate power of exponentiation. Since the seven base units are redefined based on fundamental constants of nature, by fixing their numerical values, after the rigorous research work in last two decades, it is expected that there would be implications on some of the derived units. Therefore, CIPM advised SI to come up with the updated SI and the 9th edition is published in 2019 [9]. In this section, the descriptions on derived units, decimal multiplexes and sub-multiples and non-SI units are presented.

2.3.6.1 Derived Units

The implications of revised SI base units will be reflected in the derived units based on them. The units of the seven fundamental constants include both base and derived units. As the unit of mass, kilogram is redefined in terms of h, the realization uncertainty added to IPK or the NPK, about 11 μg is so small that the corresponding derived quantities of mass, such as density (kg m^{-3}), force (kg m s^{-2} which is 1 N or 1 N), pressure (kg m^{-1} s^{-2}, which is 1 Pa), will have noticeable change in their realization process or CMCs. Some of these derived units, such as pascal, are being realized using the similar primary method used to realize the fundamental constants. Any possible changes in the realization uncertainties of derived units would be evolved over the time. Derived units are the products of powers of the base units. The coherent derived units are the ones, where the numerical factor of this product is one, such as radian for plane angle. Some of the coherent derived units in SI system have special names given to them, such as Newton, pascal, farad. The list of 22 SI units with special names is given in Table 2.4. Seven defining constants can directly construct the set of seven base units and 22 SI units with special names. These 29 units are used in combination to make other SI units.

The series of prefixes in the form of decimal multiples and sub-multiples suitable for expressing the coherent SI units has been approved by the CGPM, to express larger or smaller quantities. However, when prefixes are used with SI units, it is not coherent. Prefixes can be applied to all 29 SI units of special names, except the base unit kilogram.

Table 2.4 The 22 SI units with special names and symbols [9]

Derived quantity	Special name of unit	Unit expressed in terms of base units	Unit expressed in terms of other SI units
Plane angle	radian	rad = m/m	
Solid angle	steradian	sr = m^2/m^2	
Frequency	hertz	Hz = s^{-1}	
Force	newton	N = kg m s^{-2}	
Pressure, stress	pascal	Pa = kg m^{-1} s^{-2}	
Energy, work, amount of heat	joule	J = kg m^2 s^{-2}	N m
Power, radiant flux	watt	W = kg m^2 s^{-3}	J/s
Electric charge	coulomb	C = A s	
Electric potential difference	volt	V = kg m^2 s^{-3} A^{-1}	W/A
Capacitance	farad	F = kg^{-1} m^{-2} s^4 A^2	C/V
Electric resistance	ohm	Ω = kg m^2 s^{-3} A^{-2}	V/A
Electric conductance	siemens	S = kg^{-1} m^{-2} s^3 A^2	A/V
Magnetic flux	weber	Wb = kg m^2 s^{-2} A^{-1}	V s
Magnetic flux density	tesla	T = kg s^{-2} A^{-1}	Wb/m^2
Inductance	henry	H = kg m^2 s^{-2} A^{-2}	Wb/A
Celsius temperature	degree Celsius	°C = K	
Luminous flux	lumen	lm = cd sr	cd sr
Illuminance	lux	lx = cd sr m^{-2}	lm/m^2
Activity referred to a radionuclide	becquerel	Bq = s^{-1}	
Absorbed dose, kerma	grey	Gy = m^2 s^{-2}	J/kg
Dose equivalent	sievert	Sv = m^2 s^{-2}	J/kg
Catalytic activity	katal	kat = mol s^{-1}	

Table 2.5 Examples of coherent derived units in the SI expressed in terms of base units [9]

Derived quantity	Typical symbol of quantity	Derived unit expressed in terms of base units
Area	A	m^2
Volume	V	m^3
speed, velocity	V	$m\ s^{-1}$
Acceleration	A	$m\ s^{-2}$
Wavenumber	Σ	m^{-1}
density, mass density	ρ	$kg\ m^{-3}$
surface density	ρ_A	$kg\ m^{-2}$
specific volume	v	$m^3\ kg^{-1}$
current density	j	$A\ m^{-2}$
magnetic field strength	H	$A\ m^{-1}$
amount of substance concentration	c	$mol\ m^{-3}$
mass concentration	ρ, γ	$kg\ m^{-3}$
luminance	L_v	$cd\ m^{-2}$

As described, the units of other derived quantities can be expressed in combination of the set of the seven base units and 22 derived units. Being large in number, it is impractical to give here a complete list of derived quantities and units. Some of the quantities and their coherent derived units, given in terms of base units, are listed in Table 2.5.

In addition, Table 2.6 lists the coherent derived units such as dynamic viscosity, entropy, energy density whose names and symbols also include derived units.

Each physical quantity is represented by only one coherent SI unit, irrespective of different form or special names and symbol. However, several different quantities shares the same SI unit, such as quantity heat capacity or the quantity entropy, the SI unit remains J/K. It is also true for the base quantity electric current and the derived quantity magnetomotive force the SI unit is the ampere. Therefore, unit alone can not specify the physical quantity. The same fact applies to the measuring instruments, read-out unit and quantity needs to be indicated clearly.

The ionizing radiation is quantized in terms of the SI unit becquerel rather than the reciprocal second is used. For the quantities absorbed dose and dose equivalent, the SI units grey and sievert are used and not joule per kilogram. Some of the special names are specifically added to avoid misinterpretations and hence the dangers to human life and health, such as becquerel, sievert and grey to avoid the mistakes. Also, some terms should be carefully used, such as temperatures or temperature differences, temperature difference of 1 K is equal to 1 °C, absolute values of temperature not coherent.

Table 2.6 Examples of SI coherent derived units whose names and symbols include SI coherent derived units with special names and symbols [9]

Derived quantity	Name of coherent derived unit	Symbol	Derived unit expressed in terms of base units
Dynamic viscosity	pascal second	Pa s	kg m^{-1} s^{-1}
Moment of force	newton metre	N m	kg m^2 s^{-2}
Surface tension	newton per metre	N m^{-1}	kg s^{-2}
Angular velocity, angular frequency	radian per second	rad s^{-1}	s^{-1}
Angular acceleration	radian per second squared	rad/s^2	s^{-2}
Heat flux density, irradiance	watt per square metre	W/m^2	kg s^{-3}
Heat capacity, entropy	joule per kelvin	J K^{-1}	kg m^2 s^{-2} K^{-1}
Specific heat capacity,	joule per kilogram kelvin	J K^{-1} kg^{-1}	m^2 s^{-2} K^{-1}
Specific entropy, specific energy	joule per kilogram	J kg^{-1}	m^2 s^{-2}
Thermal conductivity	watt per metre kelvin	W m^{-1} K^{-1}	kg m s^{-3} K^{-1}
Energy density	joule per cubic metre	J m^{-3}	kg m^{-1} s^{-2}
Electric field strength	volt per metre	V m^{-1}	kg m s^{-3} A^{-1}
Electric charge density	coulomb per cubic metre	C m^{-3}	A s m^{-3}
Surface charge density	coulomb per square metre	C m^{-2}	A s m^{-2}
Electric flux density, electric displacement	coulomb per square metre	C m^{-2}	A s m^{-2}
Permittivity	farad per metre	F m^{-1}	kg^{-1} m^{-3} s^4 A^2
Permeability	henry per metre	H m^{-1}	kg m s^{-2} A^{-2}
Molar energy	joule per mole	J mol^{-1}	kg m^2 s^{-2} mol^{-1}
Molar entropy, molar heat capacity	joule per mole kelvin	J K^{-1} mol^{-1}	kg m^2 s^{-2} mol^{-1} K^{-1}
Exposure (x- and γ-rays)	coulomb per kilogram	C kg^{-1}	A s kg^{-1}
Absorbed dose rate	grey per second	Gy s^{-1}	m^2 s^{-3}
Radiant intensity	watt per steradian	W sr^{-1}	kg m^2 s^{-3}
Radiance	watt per square metre steradian	W sr^{-1} m^{-2}	kg s^{-3}
Catalytic activity concentration	katal per cubic metre	kat m^{-3}	mol s^{-1} m^{-3}

Table 2.7 SI prefixes [9]

Factor	Name	Symbol	Factor	Name	Symbol
10^1	deca	da	10^{-1}	deci	D
10^2	hecto	h	10^{-2}	centi	C
10^3	kilo	k	10^{-3}	milli	M
10^6	mega	M	10^{-6}	micro	μ
10^9	giga	G	10^{-9}	nano	N
10^{12}	tera	T	10^{-12}	pico	P
10^{15}	peta	P	10^{-15}	femto	F
10^{18}	exa	E	10^{-18}	atto	A
10^{21}	zetta	Z	10^{-21}	zepto	Z
10^{24}	yotta	Y	10^{-24}	yocto	Y

2.3.6.2 Decimal Multiples and Sub-multiples of SI Units

The decimal multiples and sub-multiples make the representation of larges and smaller quantities easy, ranging from 10^{24} to 10^{-24} provided for use with the SI units and are presented in Table 2.7.

2.3.6.3 Non-SI Units that Are Accepted for Use with the SI

In the measurement systems, there are few non-SI units extensively used in daily practice, accepted by CIPM to use along with the SI. These non-SI units are listed in Table 2.8. However, whenever these units are used, some advantages of the SI are lost. The SI prefixes cannot be used with the non-SI units of time.

2.4 Conclusions and Future Perspectives

The existing SI system of units is the official system, adopted and accepted by most of the nations. It is a universal coherent system of seven base units and several other derived and conventionally accepted units. Recently, the metrology has undergone a sea change when in November 2018, world metrology leaders of 60 countries, all together voted for the adoption of redefined SI units system based on *fundamental constants of nature (FCN)*, which later on has been implemented all around world from 20 May 2019, the *World Metrology Day*. This has replaced the earlier system, which was mix of relation of some of the units with FCN and other based on artefact. As a matter of fundamentally importance, such novel innovations for next generation, useful for stakeholders, has been discussed in this chapter in a concise and articulate manner with full of resource data and pictorial depictions to clear concepts.

Table 2.8 Non-SI units accepted for use with the SI Units [9]

Quantity	Name of unit	Symbol for unit value in SI units	Quantity
Time	Minute	min	1 min = 60 s
	Hour	h	1 h = 60 min = 3600 s
	Day	d	1 d = 24 h = 86,400 s
Length	Astronomical unit	au	1 au = 149,597,870,700 m
Plane and	Degree	o	$1^{o} = (\pi/180)$ rad
Phase angle	Minute	′	$1' = (1/60)^{o} = (\pi/10{,}800)$ rad
	Second	″	$1'' = (1/60)' = (\pi/648{,}000)$ rad
Area	Hectare	ha	1 ha = 1 $hm^2 = 10^4$ m^2
Volume	Litre	l, L	1 l = 1 L = 1 $dm^3 = 10^3$ $cm^3 = 10^{-3}$ m^3
Mass	Tonne	t	1 t = 10^3 kg
	Dalton	Da	1 Da = 1.660 539 066 60 (50) × 10^{-27} kg
Energy	Electronvolt	eV	1 eV = 1.602 176 634 × 10^{-19} J
Logarithmic	Neper	Np	
Ratio quantities	Bel Decibel	B dB	

The authors have tried to emphasize on importance of metrology and quality infrastructure required for a country. It is very important to know the right quantity with sufficient accuracy (uncertainty) to take the right decision in respective areas of applications like health, education, safety, economy, trade, etc. Metrology covers a vast area, since, all of us are surrounded by metrological applications in one way or other. Metrology has been existing since human civilization. Initially, though there was bit uniform measurement system applicable within small community only, but not at a large scale or community. In some of the cases, the measurement system was not even consistent and uniform within the country. This restricted the individual country to become self-reliant and fulfil all the needs of the citizen, because import–export as an essential part of the trade becomes difficult for them due to inconsistent measurement system. Hence, uniform measurement system is also essentially required for barrier free trade. Both CIPM MRA and ILAC MRA have fulfilled the requirements of WTO for barrier free trade between member nations. All the elements related to consistent and uniform measurement system are described.

As per applications, metrological activities are classified into scientific or fundamental metrology, industrial or applied metrology and legal metrology, but all are interrelated and their domains are well defined and discussed. For improving quality of products and human lives, the four pillars for establishing quality infrastructure, i.e.

custodian of national standards (NMI), documentary standards body, legal metrology department and accreditation body are described in details in chap. 3. India though has developed a good quality infrastructure for implementation of measurement system wherein four pillars, i.e. CSIR-NPL, BIS, LMD and NABL work as NMI, documentary standards body, legal metrology department and national accreditation body, respectively. All these organizations represent respective international metrological organizations, i.e. BIPM, ISO & IEC, OIML and ILAC, respectively, and doing their best for harmonization of measurement system and global acceptance of Indian products.

While going through the contents of the Chap. 1, the readers can easily understand the close relationship and interactions among government, academia, industry and civil society, through strong national quality infrastructure with metrology being an essential element. The SI system is a pivot around which modern metrology evolves. It signifies and disseminates the contemporary information about the SI system and its close relationship with measurement techniques, standards, measurement methods, industrial and technological developments. Special features discussed are the quality, metrology concept, its constituents, national and international measurement systems responsible for global compatibility, harmonization and acceptability of measurement standards, results and certification systems to remove trade barriers and improve the quality of life, environment and health. Such advanced information and sophisticated knowledge in the areas of high precision metrology, related research, innovation and resultant developments, specially manufacturing, health and environment exhibit close interaction to the modern methods of quality products, processes and services.

CSIR-NPL maintains 'national standards' of all the parameters except ionizing radiation. As per requirements of CIPM MRA, it maintains quality system based on ISO/IEC 17025 and ISO 17034 and periodically participates in international intercomparisons to maintain degree of equivalence in measurement system. At present, CSIR-NPL has registered total 236 CMCs (235 in physico-mechanical, electrical and electronic parameters and one in chemical parameter) which are available in 'CMCs' of BIPM KCDB. CSIR-NPL is moving forward to increase the capabilities in terms of addition of new parameters, enhancement of range, improvement of uncertainty, etc. to support industrial growth in the country.

CSIR-NPL not only maintains the national measurement standards but also disseminate the same to the various users, accredited laboratories, industries, etc. As a core pillar of NQI, it also supports the other three main pillars, i.e. NABL, LMD and BIS those in various capacities. Details of dissemination of measurement system are described in Chap. 3.

Acknowledgements Authors wish to express their whole hearted thanks to all the authors, organizations, members and students of CSIR-NPL past and present, whose works and activities are accommodated in this chapter. Authors also place on record our thanks for all even if the works of some of the organizations are not properly cited, which may be unintentional.

References

1. JCGM 200:2012, International vocabulary of metrology—basic and general concepts and associated terms (VIM), 2012
2. D.K. Aswal, Quality infrastructure of India and its importance for inclusive national growth. Mapan J. Metrol. Soc. India **35**(2), 139–150 (2020). https://doi.org/10.1007/s12647-020-00376-3
3. P. Tavella, M.J.T. Milton, M. Inguscio, N. De Leo, *Metrology: From Physics Fundamentals to Quality of Life*. IOS Press, Proceedings of the International School of Physics "Enrico Fermi," 2017
4. J. Fanton, A brief history of metrology: past, present, and future. Int. J. Metrol. Qual. Eng. **10**(5), 1–8 (2019)
5. BIPM Meter Convention Webpage. https://www.bipm.org/en/worldwide-metrology/metre-convention/. Accessed 6 Jul 2020
6. World Metrology Day webpage. http://www.worldmetrologyday.org/. Accessed 6 Jul 2020
7. The role and objectives of the BIPM webpage. https://www.bipm.org/en/about-us/role.html. Accessed 6 Jul 2020
8. Consultative Committees of the CIPM. https://www.bipm.org/en/committees/cc/. Accessed 6 Jul 2020
9. BIPM SI-Brochure webpage. https://www.bipm.org/utils/common/pdf/si-brochure/SI-Brochure-9.pdf. Accessed 6 Jul 2020
10. International equivalence of measurements: the CIPM MRA webpage. https://www.bipm.org/en/cipm-mra/. Accessed 06 Jul 2020
11. BIPM KCDB webpage. https://www.bipm.org/kcdb/. Accessed 6 Jul 2020
12. Regional Metrology Organizations (RMOs) recognized within the framework of the CIPM MRA webpage. https://www.bipm.org/en/worldwide-metrology/regional/. Accessed 6 Jul 2020
13. Joint Committees of the BIPM and other international organizations, BIPM Webpage. https://www.bipm.org/en/committees/jc/. Accessed 7 Jul 2020
14. The International System of Units (SI) BIPM Webpage. https://www.bipm.org/en/measurement-units/. Accessed 7 Jul 2020
15. The Measure of All Things: A Brief History of Metrology Webpage. https://www.azom.com/article.aspx?ArticleID=12035. Accessed 6 Jul 2020
16. History of Measurement Webpage. http://www.historyworld.net/wrldhis/plaintexthistories.asp?historyid=ac07. Accessed 7 Jul 2020
17. History of the SI BIPM Webpage. https://www.bipm.org/en/measurement-units/history-si/. Accessed 6 Jul 2020
18. Physical Constant Wikipedia, webpage. https://en.wikipedia.org/wiki/Physical_constant. Accessed 18 Jul 2020
19. Resolution 1 of the 26th CGPM (2018), BIPM Webpage. https://www.bipm.org/en/CGPM/db/26/1/. Accessed 2 Jul 2020
20. Official Website of BIPM. https://www.bipm.org/en/measurement-units/. Accessed 9 May 2019
21. M. de Podesta, The impact of the kelvin redefinition and recent primary thermometry on temperature measurements for meteorology and climatology. https://www.wmo.int/pages/prog/www/IMOP/publications/IOM-125_TECO_2016/Session_1/O1(1)_dePodesta.pdf. Accessed 18 Jul 2020
22. R.M. Godun et al., Frequency ratio of two optical clock transitions in 171 Yb^+ and constraints on the time variation of fundamental constants. Phys. Rev. Lett. **210801**(November), 1–5 (2014). https://doi.org/10.1103/PhysRevLett.113.210801
23. N. Huntemann, B. Lipphardt, C. Tamm, V. Gerginov, S. Weyers, E. Peik, Improved limit on a temporal variation of m_p/m_e from comparisons of Yb^+ and Cs atomic clocks. Phys. Rev. Lett. **210802**(November), 1–5 (2014). https://doi.org/10.1103/PhysRevLett.113.210802

24. P.J. Mohr, D.B. Newell, B.N. Taylor, E. Tiesinga, Data and analysis for the CODATA 2017 special fundamental constants adjustment*. Metrologia **55**, 125–146 (2018)
25. Practical realizations of the definitions of some important units, BIPM Webpage. https://www.bipm.org/en/publications/mises-en-pratique/. Accessed 16 Jul 2020
26. Information for users about the redefinition of the SI Webpage. https://www.bipm.org/utils/common/pdf/SI-statement.pdf. Accessed 16 Jul 2020
27. S. Yadav, D.K. Aswal, Redefined SI units and their implications. Mapan - J. Metrol. Soc. India **35**(March), 1–9 (2020). https://doi.org/10.1007/s12647-020-00369-2
28. M. Stock, R. Davis, E. De Mirandés, M.J.T. Milton, Corrigendum: the revision of the SI—the result of three decades of progress in metrology (Metrologia (2019) 56 (022001). Metrologia, **56**(4), 2019. https://doi.org/10.1088/1681-7575/ab0013, https://doi.org/10.1088/1681-7575/ab28a8
29. K. Blaum, D. Budker, A. Surzhykov, J. Ullrich, The revised SI: fundamental constants, basic physics and units. Ann. Phys. **531**(5), 12–13 (2019). https://doi.org/10.1002/andp.201900148
30. CCM detailed note on the dissemination process after the redefinition of the kilogram, webpage. https://www.bipm.org/cc/CCM/Allowed/17/06B_CCM-DetailedNote_Dissemination-after-redefinition.pdf. Accessed 18 Jul 2020
31. J. Fischer, The Boltzmann constant for the definition and realization of the Kelvin. Ann. Phys. **531**(5), 1–12 (2019). https://doi.org/10.1002/andp.201800304
32. A. Sperling, S. Kück, The SI Unit Candela. Ann. Phys. **531**(5), 1–6 (2019). https://doi.org/10.1002/andp.201800305

Chapter 3
International Harmonization of Measurements

Part II: International and National Dissemination Mechanisms

Sanjay Yadav, Goutam Mandal, Dilip D. Shivagan, Parag Sharma, Afaqul Zafer, and Dinesh K. Aswal

Abstract The foundation of the metrology lies in the realization of SI units and a globally accepted uniform system of measurement. In this context, it is appropriate to review and describe the international and national measurement systems; various responsible international and Indian metrological organizations for scientific, legal and industrial metrology. In this Chapter, authors have attempted to describe the brief overview of international trade and related organizations, for example, WTO, ISO, etc.; national measurement system, its structure, responsible organizations like CSIR-NPL for scientific and industrial metrology, BARC for ionizing radiation, Legal Metrology Department, Govt. of India for legal metrology and BIS for documentary standards and product specifications. Once SI units are established, the next goal of the NMIs is to disseminate the SI units to users and client for barrierless international trade, scientific and engineering developments and academic activities. This is to supplement, strengthen and harmonize the national quality infrastructure as proposed in Chap. 1 in a new era of redefined SI units, described in Chap. 2. Accordingly this important aspect of dissemination of SI is also included highlighting

---

S. Yadav (✉) · G. Mandal · D. D. Shivagan · P. Sharma · A. Zafer · D. K. Aswal
CSIR-National Physical Laboratory, Dr. K.S. Krishnan Marg, New Delhi 110012, India
e-mail: syadav@nplindia.org

G. Mandal
e-mail: goutam@nplindia.org

D. D. Shivagan
e-mail: shivagand@nplindia.org

P. Sharma
e-mail: sharmap2@nplindia.org

A. Zafer
e-mail: afaqul.zafer@nplindia.org

D. K. Aswal
e-mail: dkaswal@nplindia.org

D. K. Aswal (ed.), *Metrology for Inclusive Growth of India*,
https://doi.org/10.1007/978-981-15-8872-3_3

National Policy on Metrological Traceability developed by CSIR-NPL; dissemination through calibration, testing and verification services; dissemination services by NMIs, NABL accredited laboratories, Legal Metrology Department, BIS and its authorized laboratories; BARC and its authorized laboratories; DRDO's laboratories and other organization; dissemination through training and academic activities; proposed changes in metrology courses of engineering and NCERT books and finally conclusions and future perspectives. A new concept of '*Metrology Cycle*' is introduced for the first time to accommodate and explain metrology in a cyclic single window.

The same applied to the concept of force as does to any other physical concept: Verbal definitions are meaningless; real definitions are given through measuring process.

– Arnold Sommerfeld (1868-1951)

3.1 Introduction

For the establishment of a uniform, coherent and effective metrological system in the country, a strong, robust, cohesive and effective *national quality infrastructure (NQI)* is essentially required and needs to be established and put in force in accordance with globally agreed practices. An NQI is a tool for competitiveness, trade, and social well-being. The three major pillars of NQI are namely metrology; standardization; accreditation and conformity assessment. A detailed description of the NQI is given in the introductory Chapter of this book and article published elsewhere by one of the authors [1].

As NQI is comprised of several components that are closely related and form a network, whose logical links are connected in a technical hierarchy. This national network is linked to the relevant international systems. Having established the international link, the commodity trading and exchange of services, globally are easy, way forward and possible. The NQI is closely linked with international organizations. In the case of NQI, the linkages based on countries' needs are established between national institutes and *International Bureau of Weights and Measure (BIPM); Metre Convention, International Committee for Weights and Measure (CIPM), International Organisation of Legal Metrology (OIML), Regional Metrology Organizations (RMOs), International Organisation for Standardization (ISO). World Trade Organization (WTO), International Laboratory Accreditation Cooperation (ILAC), International Accreditation Forum (IAF), Regional Conformity Boards (RCBs),* to name few. These organizations are described briefly in the latter part of the chapter.

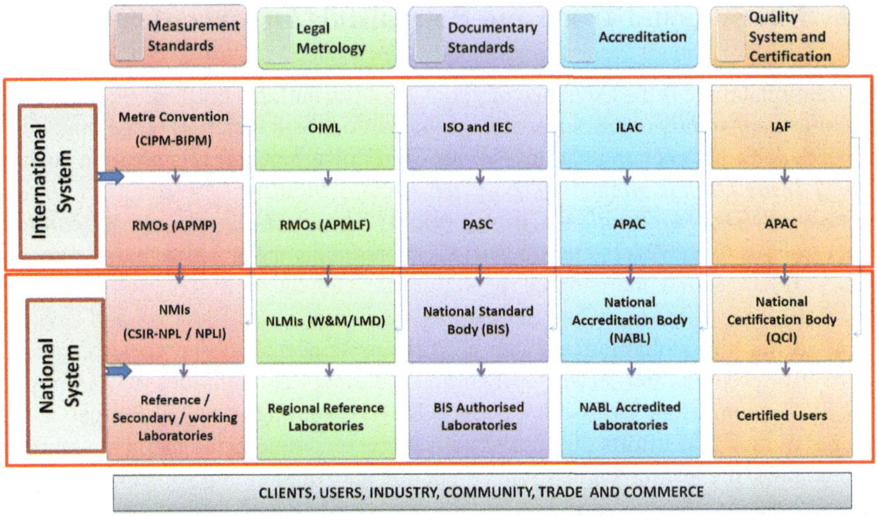

Fig. 3.1 A brief depiction of national quality infrastructure (NQI) with special reference to India

The strong NQI encompasses both international and national measurement systems and is shown in Fig. 3.1. The different constituents of NQI will be described in the succeeding sections of the chapter.

The strong NQI is therefore essentially required for achieving the following objectives:

(i) Ensuring that all producers and manufacturers, including farmers, receive the correct payments for their produces and all consumers get the correct amount of goods.
(ii) To develop and deploy appropriate metrological methods for industry, support product innovation, process improvement and quality assurance.
(iii) Ensuring that components and finished products meet regulatory requirements, documentary standards and specifications.
(iv) Consumer and industrial quality expectations are met, including product value/price and reliability.
(v) The control of pre-packed goods/commodities helps to reduce fraud.
(vi) Correct measurements of raw materials exported in bulk enable them to pay the correct price. It also helps governments to collect correct taxes on the exports.
(vii) The metrological controls also help to improve the economic conditions for all concerned and assist in poverty reduction.
(viii) Global acceptability of measurements and test results removes technical barriers to trade.
(ix) It facilitates international trade and contributes to inclusive economic growth, access to opportunities for *Micro, Small and Medium Enterprises (MSMEs)*, and a level playing field for developing economies.

3.2 International Trade and Responsible Organizations

International trade is the exchange of goods between national borders. World boundaries have apparently come closer after globalization, with various linkages and interdependencies between the nations. Various intergovernmental organization such as *World Trade Organisation (WTO), United Nations Industrial Development Organisation (UNIDO), International Atomic Energy Agency (IAEA), World Meteorological Organisation (WMO),* etc. are working on reducing the technical barriers for the trade, exchanges and international regulations through the tools of metrology, quality assurance and negotiations by safeguarding the human health, ecology and environment so that the global citizens can get the ample access to the world resources, innovative products and services. NMIs and metrology communities play an important role in conformity assessment bodies (CAB) obtaining technical competence in their activities. The efforts of these organizations recommend the member countries to recognize the testing, calibration, inspection and validation results of each other, thereby reducing the technical barriers to the trade and instill the harmonization and transparency.

3.2.1 The World Trade Organization (WTO)

WTO [2] is the only global international organization that deals with the rules of trade between member nations. It is situated in Geneva, Switzerland and commenced from 1st January 1995 with 123 member nations, signed under the *Marrakesh Agreement* on 15th April 1994, replacing the *General Agreement on Tariffs and Trade (GATT)*, which was started in 1948. India has been a WTO member since 1st January 1995 and a member of GATT since 8th July 1948. The representatives of the Ministry of Commerce and Industry of India represent India in WTO. WTO prepares various agreements, which are negotiated and signed by most of the trading nations and endorsed by their parliaments. The overall goal is to ensure that trade flows as smoothly, predictably and freely as possible. Good measurement allows the country to remain competitive, trade throughout the world and improve quality of life.

WTO's agreement on *Technical Barriers to Trade (TBT)* [3] is an apex document that ensures the trading nations that the technical regulations, standards, and conformity assessment procedures do not discriminate or create unnecessary obstacles to trade. WTO-TBT Agreement recommends members to enter into negotiations for the mutual recognition of results of each other's conformity assessment, as well as to permit the participation of conformity assessment bodies located in other member economies. It also provides the members with the right to implement measures to protect human health, safety and environment. The broad mechanism depicting the process of TBT Agreement is shown in Fig. 3.2.

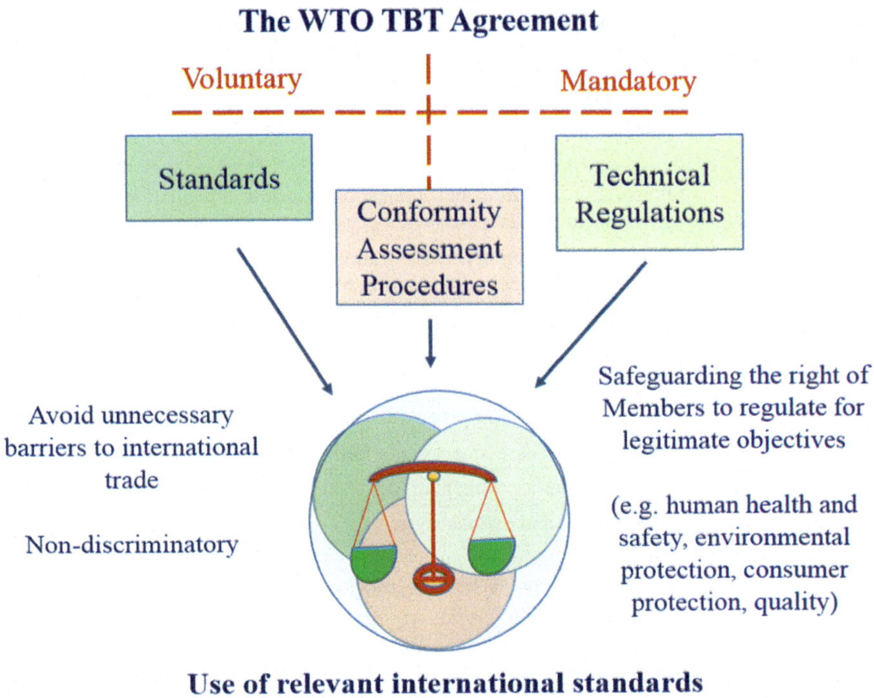

Fig. 3.2 Depicting the mechanism of the technical barrier to trade agreement

BIPM holds an ad-hoc observer status at the WTO's-TBT Committee. The BIPM and the WTO cooperate with each other to ensure effective dissemination of information on the importance of metrology and the quality infrastructure, as one of its key components among trade regulators and international/intergovernmental organizations. The essential components and articles involved in the TBT agreement are presented in Table 3.1.

3.2.2 The United Nations Industrial Development Organization (UNIDO)

The UNIDO [4] is a specialized agency of the *United Nations* established in 1966 by the *UN General Assembly* to promote and accelerate the economic and industrial development in developing countries. It is headquartered at the UN Office in Vienna, Austria, with 170 member states, which sets the organization's policies, programs, and principles through the biannual General Conferences. UNIDO focuses on creating shared prosperity, advancing economic competitiveness, safeguarding the environment, strengthening knowledge and institutions, cross-cutting services.

Table 3.1 Different provisions in agreement on technical barrier to trade [3]

WTO agreement on technical barriers to trade	
Articles/Annex	Provisions
Article 1	General provisions
Article 2	*Technical Regulations and Standards* Preparation, adoption and application of technical regulations by central government bodies
Article 3	Preparation, adoption and application of technical regulations by local government bodies and non-governmental bodies
Article 4	Preparation, adoption and application of standards
Article 5	*Conformity with Technical Regulations and Standards* Procedures for assessment of conformity by central government bodies
Article 6	Recognition of conformity assessment by central government bodies
Article 7	Procedures for assessment of conformity by local government bodies
Article 8	Procedures for assessment of conformity by non-governmental bodies
Article 9	International and regional systems
Article 10	*Information and Assistance* Information about technical regulations, standards and conformity assessment procedures
Article 11	Technical assistance to other members
Article 12	Special and differential treatment of developing country members
Article 13	*Institutions, Consultation and Dispute Settlement* The committee on technical barriers to trade
Article 14	Consultation and dispute settlement
Article 15	Final provisions
Annex 1	Terms and their definitions for the purpose of this agreement
Annex 2	Technical expert groups
Annex 3	Code of good practice for the preparation, adoption and application of standards

UNIDO works to improve the quality of life of the world's poor by managing the global resources and has expertise in the three key areas; (i) poverty reduction through productive activities, (ii) trade capacity-building and (iii) energy and environment.

UNIDO is one of the largest providers of trade-related development services, offering focused and neutral advice and technical cooperation in the areas of competitiveness, industrial modernization and upgrading, compliance with international trade standards, testing methods and metrology. It is also the leading provider of services for improved industrial energy efficiency and promotion of renewable sources of energy for "*Inclusive and Sustainable Industrial Development (ISID)*" goal. These goals are achieved through technical cooperation, policy advice, analysis and research, the development of uniform standards and quality control, and partnerships for knowledge transfer, networking and industrial cooperation.

The Country Service Framework for India as a dynamic, flexible and strategic tool for greater coherence and the developmental impact was developed by UNIDO in close consultation with the Ministry of Commerce and Industry. The special report on 50 Years of UNIDO-India Cooperation 1966–2016, describes various UNIDO projects carried out in India [5]. UNIDO also associates with other Indian ministries, including the Ministry of Environment, Forest and Climate Change (MoEF & CC), Ministry of Urban Development (MoUD), Ministry of New and Renewable Energy (MNRE), Bureau of Energy Efficiency (BEE), Ministry of Micro, Small and Medium Enterprises (MoMSME), Ministry of Heavy Industries (MoHI), Ministry of Health and Family Welfare (MoHFW), Ministry of Earth Sciences (MoES), Ministry of Chemicals and Petrochemicals, Ministry of Food Processing, among others.

3.2.3 The International Atomic Energy Agency (IAEA)

IAEA [6] is the world's central intergovernmental forum for scientific and technical co-operation in the nuclear field. It was established as an autonomous organization of the United Nations on 29 July 1957, with its headquarters in Vienna, Austria. IAEA reports to both the United Nations General Assembly and Security Council. It works for the safe, secure and peaceful uses of nuclear science and technology, contributing to international peace and security, including radiation protection, and the United Nations' *Sustainable Development Goals (SDGs).* The IAEA and its former Director-General, Mohamed ElBaradei, were jointly awarded the Nobel Peace Prize on 7 October 2005.

Radioactivity and radioactive substances are a phenomenon and natural sources, have numerous beneficial applications, such as power generation, uses in medicine, industry and agriculture. The radiation risks that may arise to the workers, public and environment have to be seriously assessed and controlled. Therefore, activities such as the production, transport and use of radioactive materials, operation of nuclear installations, medical and agricultural uses of radiation, management of radioactive waste must be subjected to standards of safety. The IAEA commenced its safety standards programme in 1958. Although it is a nuclear nation's responsibility to regulate safety, the radiation risks may go beyond national borders. Therefore, international cooperation serves to promote and enhance global safety by sharing experiences on accidents, hazards, emergency response mechanisms, and harmful consequences. Hence, these safety standards are mandatory obligations to member nations. The various safety standards governed by IAEA right from site selection, operations and decommissioning of nuclear power plants, medical, agricultural, research installations, etc., can be found [7]. The structure of IAEA safety standards is described in Fig. 3.3, mentioning general and specific safety considerations made during the formulations.

India is a founding member of IAEA. *Atomic Energy Commission* and Department of Atomic Energy, responsible for nuclear research and development, represents India

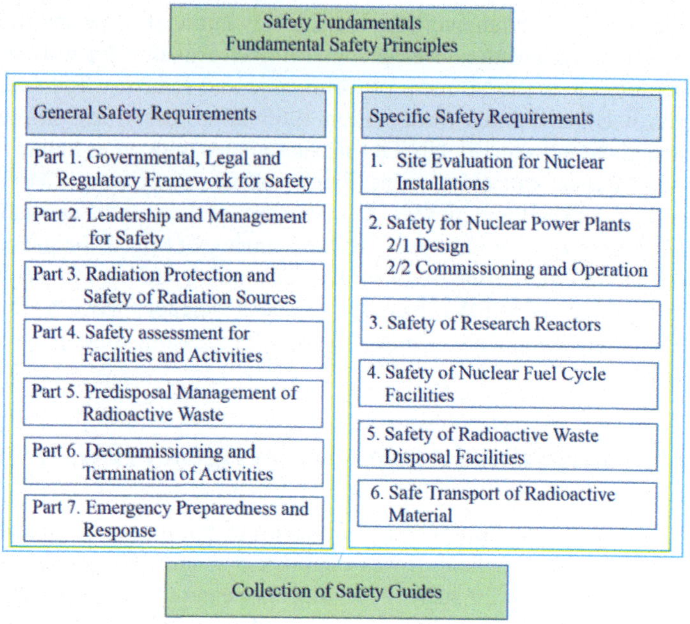

Fig. 3.3 The long term structure of the IAEA safety standards [7]

in IAEA. BARC is the Designated Institute (DI) for the Ionizing Radiation parameter in BIPM activities of international linkages to the standards of measurements. India has joined the IAEA's *Response and Assistance Network (RANET)*, a group of states which offer assistance to mitigate the consequences of nuclear or radiological emergencies.

3.2.4 World Meteorological Organization (WMO)

WMO is an intergovernmental organization with a membership of 193 Member States and Territories [8]. It is a specialized agency of the United Nations formally became operational as WMO on 17 March 1951, with its headquarters in Geneva, Switzerland. WMO is devoted to Earth's atmosphere, interaction with the land and oceans, the weather and climate, distribution of water resources. WMO manages this through international cooperation and coordination with member nations. WMO mandate is in the areas of weather, climate, water and environment. WMO programmes assist and promote the following activities by establishing a facility or creating networks [8]:

- observational stations for the weather, climate and water-related data;

- data management centres and telecommunication systems for the provision and rapid exchange of weather, climate and water-related data;
- standards for observation and monitoring for uniformity in the practices and procedures adopted worldwide to ascertain the homogeneity of data and statistics;
- extension of activities and services of operational meteorology and hydrology to aviation, transport (air, land and maritime), agriculture, water resource management, etc.;
- harmonization of activities in hydrology and cooperation between National Meteorological and Hydrological Services.
- to coordinate the research and training in meteorology and synergetic fields.

For the fields of meteorology, hydrology, climatology and related environmental disciplines, WMO prepares international standards. WMO Technical Regulations, consist of standard and recommended practices and procedures, were adopted by *World Meteorological Congress (WMC)* for universal application by all Members. Because of these regulatory provisions, a variety of users in each WMO Member State and Member Territory make it possible to operate the 24/7/365 global systems for observations, data exchange and management, forecasting and provision of authoritative scientific assessments and standardized service products. WMO prepares the Technical Regulations based on requirements and to bring efficiency, interoperability and to support policy and decision-making in many areas of disaster risk management, agriculture, aviation, shipping, water management, health, etc. These standards are based on robust metrological practices founded on BIPM, ISO, IEC, etc. programmes.

These WMO standards on various parameters are published in the form of basic documents, Technical regulations, Manuals, Guides and Agreements [9]. Various WMO projects on aeronautical, agricultural, flood management, etc. can be found in detail in Ref. [10]. The *India Meteorological Department (IMD)* is an agency of the Ministry of Earth Sciences represents the WMO activities. IMD collaborates with ISRO, *Tata Energy Research Institute (TERI), Indian Institute of Tropical Meteorology (IITM), National Centre for Medium Range Weather Forecasting (NCMRWF)* and the *National Institute of Ocean Technology (NIOT).*

3.2.5 International Organisation for Standardization (ISO)

ISO is an independent, non-governmental international organization with a membership of 164 national standards bodies [11]. ISO brings the experts together to develop market-relevant International Standards to support innovation and provide solutions to international challenges. Its Central Secretariat is based in Geneva, Switzerland. An International Standard is a document containing practical information and best practice. It often describes an agreed way of doing something or a solution to a global problem. Standards give world-class specifications for products, services and systems, to ensure quality, safety and efficiency. They are helpful in international

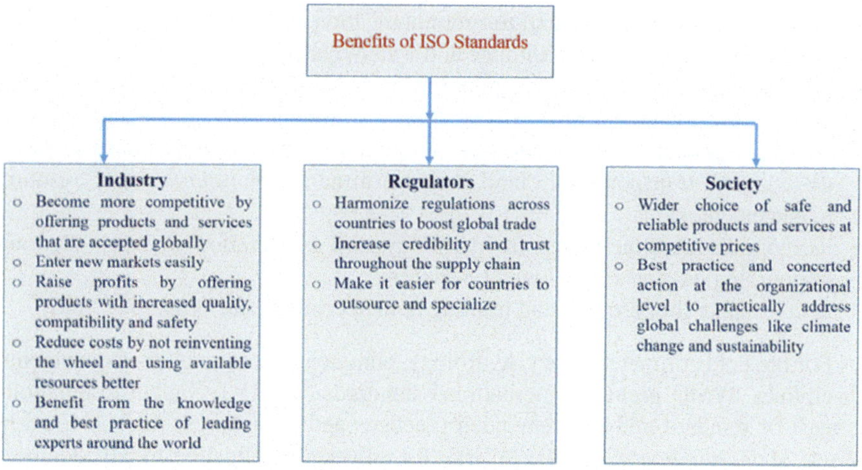

Fig. 3.4 Benefits of various ISO standards to Industry, Regulators and Society

trade. ISO has published more than 23,244 International Standards and related documents for all industries, from technology, to food and safety, to agriculture and healthcare. There are about 787 technical committees and subcommittees to take care of the standards development. ISO International Standards impact everyone, everywhere. ISO works closely with the *International Electrotechnical Commission (IEC) and the International Telecommunication Union (ITU)*. ISO also has a strategic partnership with the WTO to promote free and fair trade. The benefits of the ISO standards are depicted in Fig. 3.4.

ISO develops the documentary standards related to almost all the fields.

- **Quality management standards** to help work more efficiently and reduce product failures.
- **Environmental management standards** to help reduce environmental impacts, reduce waste, and be more sustainable.
- **Health and safety standards** to help reduce accidents in the workplace.
- **Energy management standards** to help to reduce energy consumption.
- **Food safety standards** to help prevent food from being contaminated.
- **IT security standards** to help keep sensitive information secure.

Committee on Conformity Assessment (CASCO) is the ISO committee that publishes standards related to conformity assessment. *Bureau of Indian Standards (BIS)* of Ministry of Consumer Affairs, Food and Public Distribution is a member of ISO and publishes the relevant Indian Standards.

3.2.6 International Electrotechnical Commission (IEC)

IEC [12], founded in 1906 with its central office in Geneva, Switzerland. IEC prepares and publishes International Standards for all electrical, electronic and related technologies, broadly called electrotechnical. The experts from industry, commerce, test and research labs, government, academia and consumer groups participate in IEC Standardization work. Millions of electronic devices rely on IEC International Standards and Conformity Assessment Systems to perform, fit for the purpose and work safely together.

Joint committees ensure that International Standards combine all relevant knowledge of experts working in related areas. IEC is one of the three global sister organizations (IEC, ISO, ITU) that develops major International Standards for the world. IEC cooperates with ISO or ITU to ensure that the International Standards fit together for the purpose.

IEC operates four global conformity assessment (CA) Systems, IECEE, IECEx, IECQ and IECRE, each with participating testing laboratories and certification bodies from all over the world. Harmonized and documented operational and testing procedures combined with the common interpretation of the international standards, then using peer assessment to ensure that the harmonized procedures and common interpretation are applied, ensures consistent and reproducible results from anywhere in the world.

BIS is a member of ISO and through the *Indian National Committee (INC)* is also a member of IEC.

3.2.7 International Organization of Legal Metrology (OIML)

OIML [13] is an intergovernmental treaty organization. Legal metrology is the application of legal requirements to measurements and measuring instruments. Mandate of OIML is:

- Development of model regulations, standards, and documents related to legal metrology authorities and industry, to provide mutual recognition to reduce trade barriers and global market costs,
- to represent legal metrology community in other international organizations and forums related to metrology, standardization certification, testing, and accreditation,
- to promote the exchange of knowledge, information and competencies in worldwide legal metrology community,
- to cooperate to raise awareness about legal metrology infrastructure for modern economy.

OIML publications are in the following categories [14];

(i) International Recommendations—intended as model regulations for various measuring instruments, so that all OIML Member States implement them
(ii) International Documents—intended for information and guidance purposes
(iii) Other—Basic Publications, Vocabularies, Guides, and Expert Reports.

OIML has developed its own *Certification System (OIML-CS)*. It also works as an "International Standard-Setting Body" in the sense of the WTO-TBT. OIML-CS participants utilize OIML Certificates and/or OIML type evaluation reports from those participants that are entitled to issue OIML Certificates. All signatories of the TBT Agreement will apply OIML publications while formulating, technical regulations, as per Article 2.4 of TBT Agreement. There are 16 different TCs on various subjects of its functions.

Department of Consumer Affairs of India participates in the OIML activities and implements the regulations and enforcement.

3.2.8 International Laboratory Accreditation Cooperation (ILAC)

ILAC [15] is an international organization for accreditation bodies that are working as per ISO/IEC 17011. It does the accreditation of CABs, such as calibration laboratories (using ISO/IEC 17025), testing laboratories (using ISO/IEC 17025), medical testing laboratories (using ISO 15189), proficiency testing providers (using ISO/IEC 17043) and inspection bodies (using ISO/IEC 17020). In 2000, *ILAC Mutual Recognition Arrangement (ILAC MRA)* was signed by the 36 ILAC's full members, from 28 economies consisting of laboratory accreditation bodies, effective from 31 January 2002. In 2012 it included the accreditation of inspection bodies and in 2019 it was extended to proficiency testing providers, and in 2020 for reference material producer's accreditation.

ILAC manages the international arrangements in the fields of calibration, testing, medical testing, inspection and proficiency testing provider's accreditation. IAF is working in management systems, services, products, personnel, etc. of conformity assessment. ILAC and IAF together coordinate to enhance the accreditation and conformity assessment activity globally. ILAC works in coordination with the regional accreditation bodies, notably EA in Europe, IAAC in America, APAC in the Asia-Pacific, ARAC in the Arab region, AFRAC in Africa, and SADCA in Southern Africa.

Two organizations of *Quality Council of India (QCI)* are signatory to ILAC-MRA, *National Accreditation Board for Certification Bodies (NABCB)* for Inspection: ISO/IEC 17020 and *National Accreditation Board for Testing & Calibration Laboratories (NABL)* for Calibration: ISO/IEC 17025, Testing: ISO/IEC 17025, Medical Testing: ISO 15189, Proficiency Testing Providers: ISO/IEC 17043.

3.2.9 International Accreditation Forum (IAF)

IAF [16] is the world association of Conformity Assessment Accreditation Bodies. It also involves other bodies working in conformity assessment towards management systems, services, products, personnel, etc. IAFs main function is to develop a unified conformity assessment programme to reduce risk in business. Accreditation bodies are established in many countries with the primary purpose of confirming that CABs are subject to lapse by an authoritative body. IAF works on two fronts, (i) it ensures that the members of the accreditation body only accreditate competent bodies and are not subject to conflicts of interest, (ii) to establish *Multilateral Recognition Arrangements (MLA)* as tool of mutual recognition arrangements, between its accreditation body members which reduce risk to business and its customers, (iii) ensures that an accredited certificate is recognized and accepted worldwide to eliminate technical barriers to trade.

IAF aims for single effective system to enable companies with an accredited conformity assessment certificate in one part of the world recognized elsewhere in the world—*Certified once—accepted everywhere.* Two QCI bodies from India have signed the IAF-MLA, NABCB for Product Certification—ISO/IEC 17065, Management Systems Certification—ISO/IEC 17021-1 and Certification of Persons—ISO/IEC 17024; and National Accreditation Board for Education and Training *(NABET).*

3.2.10 Asia Pacific Metrology Programme (APMP)

APMP [17] is a group of NMIs from the Asia-Pacific region. It aims to improve metrological capabilities in the region by sharing the expertise and technical services among Member NMIs.

Specific objectives of APMP are as follows:

- To develop collaboration between Members in measurement standards.
- To advance the level of metrology, and to gain international recognition for Members, by conducting the key and supplementary comparisons.
- To encourage Members to establish traceability by means of calibration services.
- To ensure that measurement standards in APMP region are traceable to standards maintained or coordinated by BIPM in the realization of the international system (SI) of units, and to reinforce the objectives of the Metric Treaty.

APMP has been operating in the Asia-Pacific since its inception as a *Commonwealth Science Council* initiative in 1977 and broadened to the present title 'APMP' in 1980. As such, it is the oldest continually operating metrological grouping in the world. APMP operates under the framework of *APMP Memorandum of Understanding (MoU)*, latest in December 2009.

APMP was re-arranged, in 1998, with the Committee Meeting re-named as *General Assembly (GA)*, an *Executive Committee (EC)* established to support Chairperson, and TCs established for each quantity. APMP has 12 TCs for coordinating technical activities in each field covered by CCs of CIPM, and one *Developing Economies' Committee (DEC)* for initiating and coordinating work programs to address the needs of APMP member *Developing Economy NMIs (DENs)*. APMP has established *5 Focus Groups* (for *Climate Change and Clean Air, Clean Water, Energy Efficiency, Food Safety, and Medical Metrology*) to provide measurement support to stakeholders of identified priority fields. APMP has 45 full members (27 economies) and 14 associate members (13 economies). Out of them, 21 are Member States of the BIPM and 9 Associates of the CGPM, 30 Signatories of the CIPM MRA.

CSIR-NPL, is the founder member of APMP as NMI of India. It is a signatory to MoU and represents India in these committees and Focus groups. BARC (DI) represents in the Ionizing Radiation TC.

3.2.11 The Asia-Pacific Legal Metrology Forum (APLMF)

APLMF [18] is a group of legal metrology authorities. The objective APLMF is the development and promotion of legal metrology activities for open and free trade in the region. It brings the harmonization by removing the technical and administrative barriers to trade in the field of legal metrology. It works closely with OIML and promotes communication and interaction among the legal metrology organizations for the harmonization in Asia Pacific region.

Asia-Pacific Economic Cooperation (APEC) recognizes the APLMF, APMP and APLAC, as the Specialist Regional Bodies in their fields to help APEC Subcommittee on Standards and Conformance to meet the aims of removing technical barriers to trade in the region.

3.2.12 Asia Pacific Accreditation Cooperation (APAC)

APAC [19], established recently in 2019 is the amalgamation of the *Asia Pacific Laboratory Accreditation Cooperation (APLAC)* and the *Pacific Accreditation Cooperation (PAC)* (Fig. 3.1). APAC's objectives are to manage and expand a MRA within accreditation bodies in Asia Pacific region, and to ensure that, any Conformity results are produced by conformity assessment bodies (CABs) that are accredited by one APAC MRA signatory may be accepted by all the other APAC MRA signatories. The MRA facilitates the acceptance of test reports, test certificates, inspection reports, and certification across the region and with other regions worldwide. This avoids the repetition of testing, inspection or certification thereby saves time and money, enhances economic efficiency and ease of international trade.

APAC's members are accreditation bodies and other organizations that are interested in accredited conformity assessment results. It is recognized by APEC, as one of four APEC Specialist Regional Bodies (SRBs) and supports the APEC Sub-Committee work on Standards and Conformity.

NABCB and NABL are the full members of APAC from India.

3.2.13 *South Asian Regional Standards Organization (SARSO)*

The *South Asian Regional Standards Organization (SARSO)* [20] is a Specialized Body of *South Asian Association for Regional Cooperation (SAARC)* and *Regional Standards Organization of South Asia.* SARSO was established on 25 August 2011 after the ratification of the Agreement on the establishment of SARSO by all Member States of SAARC. SARSO has its Secretariat in Dhaka, Bangladesh and commenced its operations with effect from 03 April 2014.

SARSO has been established to achieve and enhance coordination and cooperation among SAARC member countries in the fields of standardization and conformity assessment and is aimed to develop harmonized standards for the region to facilitate intra-regional trade and to have access in the global market. The following are the member bodies of SARSO;

(i) Afghanistan National Standards Authority (ANSA)
(ii) Bangladesh Standards and Testing Institution (BSTI)
(iii) Bhutan Standards Bureau (BSB)
(iv) Bureau of Indian Standards (BIS)
(v) Ministry of Economic Development, Maldives (MoED)
(vi) Nepal Bureau of Standards and Metrology (NBSM)
(vii) Pakistan Standards and Quality Control Authority (PSQCA)
(viii) Sri Lanka Standards Institution (SLSI).

SARSO has published some SAARC Standards [20].

3.3 Metrology and Quality Assurance Organizations of India

After understanding the operational mechanism of international metrology system in Chap. 2 and international organizations responsible of international trade, it is imperious to describe the Indian premier organizations responsible for strong national quality infrastructure (Described in Chap. 1) here in this section. The major pillars organizations are:

3.3.1 CSIR-National Physical Laboratory (CSIR-NPL)/NMI of India

Among the mainstays of the national measurement system, CSIR-NPL plays a vital role as an NMI of India. An NMI is a globally recognized body which normally has uniquely provided status by legal act of country, carrying the responsibility to facilitate recognition of national measurement standards and CMCs at international platform along with disseminating of the international system of units nationally. In India, this responsibility is given to CSIR-NPL, which is recognized as NMI of India by the Legal act of Indian constitution. CSIR-NPL provides measurement standards, testing and calibration facilities to India. It also maintains the necessary metrology infrastructure, which represents the position of India's capabilities in measurement internationally. In India, the unbroken chain for metrological traceability for any measurement is originated from CSIR-NPL. It actively influences the other pillars of national measurement systems and is passively involved in the quality management of products and services in India.

CSIR-NPL is mandated to be India's NMI by an act of Parliament (The Gazette of India) and also legal metrology rules associating with it. Being the custodian of the National Standards of Measurements, CSIR-NPL is responsible for the realization, establishment, up-gradation, maintenance and dissemination of standards at par to international level, providing metrological traceability across the country.

The mandate of CSIR-NPL is to ensure the realizations of SI units as national standards and to transfer the values of the standards to the NQI through calibrations, testing and other measurement related services. The customers who are the direct beneficiaries of CSIR-NPL make their measurement results traceable to those of CSIR-NPL, and hence to the SI units. These companies, in turn, use their standards to provide measurement services to their customers hence assuring quality in the measurements across the country. CSIR-NPL is also timekeeper of India [1].

The missions (Fig. 3.5) of the CSIR-NPL are [1];

- Developing India's measurement standards that are internationally accepted and disseminating the measurement capabilities to industry, government, strategic and academia that underpin India's prosperity and quality of life;
- Conducting multidisciplinary R&D with a mission to establish the futuristic quantum standards and upcoming technologies so that India remains at par with international measurement laboratories
- Developing sophisticated analytical equipment (i.e. import substitutes) under "Make in India" programme to cater to the ever-increasing demands of emerging India; and
- Training of young scientists and industry personnel in the areas of measurements under "Skill India" programme.

Fig. 3.5 CSIR-NPL mission overview

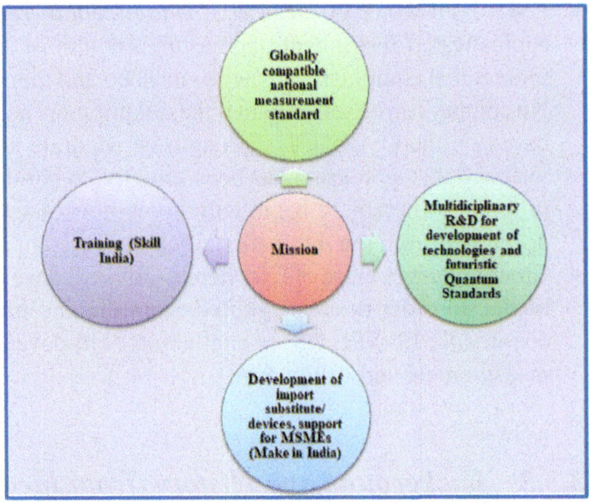

The visions (Fig. 3.6) of the CSIR-NPL are [1];

- To be "*Giant in measurements*" to support the needs of the country by enhancing and expanding the existing apex calibration capabilities to "*Solar Cells*" and "*Biomedical Equipment*" that currently do not exist in the country. New calibration capabilities for environmental gases will be set up. Clean environment demands accurate measurements of pollutants (toxic gases and matter particulate), and as of now, country does not have any "Apex calibration laboratory" for these pollutants. In addition, R&D Centre for Quantum Standards will be set up for game-changing applications ranging from faster communications to navigation, which will be key to improve lives and creation of new businesses.

Fig. 3.6 CSIR-NPL vision overview

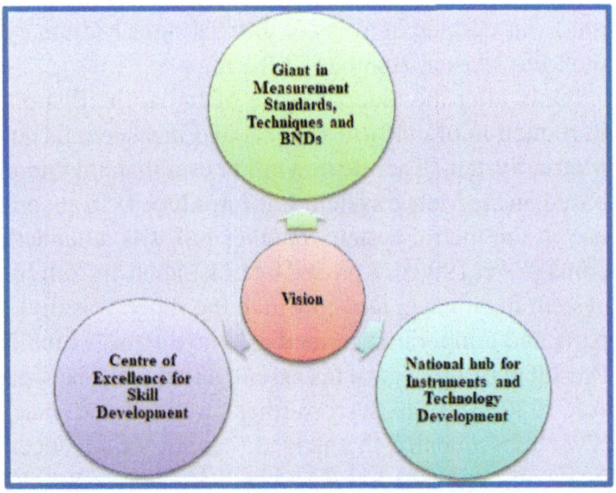

- CSIR-NPL would be creating a "*National hub*" for the development of calibrated sophisticated instruments. India imports almost all sophisticated analytical and biomedical equipment to cater to its need and this requirement is ever increasing. No country can be self-reliant if these equipment are not made in-house. Therefore, based on the CSIR-NPL's strength on accurate and precise measurements, it is visioned that a program has been initiated to build these equipment under "Make in India", whereby to begin with know-how of some of the equipment would be developed from the internationally branded companies.
- Creation of a "Centre of Excellence" for skill development in the area of measurements. In order to create skilled scientific and industry-ready manpower in the country, CSIR-NPL will strengthen its skill development program in the area of measurement and calibration.

3.3.2 Background and History (from Idea to the Institutionalization of CSIR-NPL)

In India, the process of unification and standardization of units of measurement can be traced back from the age of the British period. The metric change in India started with the implementation of the Weights and Measures Act (Act XI of 1870) which promoted the 'seer' (1 seer = 80 tolas) as equal in weight to the kilogram and metre as the primary standard of length. In the next revision of this act, the British system was thoroughly implemented and known as a new Act (ACT XXXI of 1871) [22]. This act was based on the idea and principle that as India is dependent on England, for convenience of foreign trade, the English System should strictly be followed. However, the Metre Convention on 20th May 1875 brought a new paradigm shift in metrological ideology with the adoption of the Metric System. So, in 1939 Central Legislature passed the Standards of Weights Act (Act IX of 1939) which was enforced on whole British India. According to this Act, the standard 'grain' (180 grain = 1 tola) was defined in terms of the Platinum Iridium cylinder placed in the custody of the Mint Master, Bombay [22].

Only after the independence of India, the Planning Commission considered the introduction of uniform weights and measures throughout the country based on the Metric System. Therefore, a bill to establish the Standards of Weights and Measures based on the Metric system was introduced. To accompaniment this bill and effective use of the metric system, another bill was amended. The bill to amend the Indian coinage Act 1906 was passed by Lok Sabha on 26th July 1955 introducing the decimal system in coinage. Due to which the rupee was divided into 100 parts instead of 64 parts and temporarily termed as 'Naya Paisa'. Finally, on 28th December 1956, the President of India gave his assent, and the bill became law as Weights and Measures Act 1956. The acts were further modified and enacted as Act 1976 and Act 1985 [22]. Subsequently, in changed scenario of advancement in technologies, the act is made and published as *The Legal Metrology Act 2009 and associated rules 2011*, which were enforced from April 2011. More details are given in Sect. 3.3.4. For effective implementation of the Act, the Government of India also provided the necessary infrastructure by reinforcing and establishing the vital organization with their suitable roles and objective which are CSIR-NPL, Indian Standards Institute (ISI) and Directorate of Weights and Measures (DWM).

In 1945, the mandate of the CSIR-NPL was conceptualized as the establishment and dissemination of standards of physical measurements, and initially, 3 basic parameters mass, length and time were focused as a priority. The foundation stone of the CSIR-NPL was laid down on 4th January 1947 by the Honourable Pandit Jawaharlal Nehru. After establishing the basic facilities for the measurements related to these parameters, in parallel to the basic research and developmental work, the testing and calibration services were also started in October 1953. Further, the calibration and testing work in the area of electrical and optical was also initiated, after establishing the related facilities such as standard cells and resistance coils, lamps, etc.

To solidify the adoption and acceptation of SI units within the international metrological community, in 1957, India signed the Metre Convention Treaty and became a member of the CGPM. In 1957, the Government of India signed the Metre Convention, and became a member of CGPM. In January 1963, Nobel Laureate P. M. S. Blackett, after a thorough review of the laboratory activities, recommended: "The major role of the NPL over the next decade should be in the fields of standards and testing commercially manufactured instruments, equipment and apparatus. He pointed that at present the maintenance of the standards and the testing in the Indian NPL is spread over the several divisions namely Weights & Measures, Acoustics, Heat & Power, Electricity, Electronics, Applied Mechanics and Optics. He suggested that it would be advantageous if all groups work together in one powerful division for the maintenance of standards and for routine testing of instruments as laboratory under a single Director, preferably with no other responsibilities" [22].

Hence, efforts towards establishment of new measurement standards and achieving the requisite accuracy and precision in the measurements, are the continued activity of CSIR-NPL. In the 1970s, with the support of Department of Electronics, the primary facilities for electrical and electronic measurements were established. Since 1970, collaborative initiatives with USA, NPL UK and PTB Germany have also been undertaken, in the areas of mass metrology, length metrology, time and frequency metrology, DC metrology, pressure and vacuum metrology, optical radiation metrology and surface analytical techniques. In which, PTB Germany's (NMI of Germany) contributions deserve special mention, in the upliftment of metrology programs at CSIR-NPL.

In the following years, in the fifth five-year plan, a core committee was appointed by the Department of Science and Technology to examine and evaluate the existing infrastructure of calibration and testing facilities. At that time (the year 1975), with the starting of the first test facilities, the *Alipore Test House (National Test House)* under the Department of Supply, large numbers of test facilities were established which were under different departments and autonomous bodies. For better accuracy and quality of measurement, all these bodies had to be metrologically traced back to CSIR-NPL. With technological advancement, many changes were brought forth in the Standard of Weights and Measures Act till the acceptation of new definitions of four SI units in 2019. These subsequent changes were enacted by CSIR-NPL in India in order to not lack behind the international metrological community.

A pictorial diagram representing the main historical events related to the Weights and Measure Act and that of CSIR-NPL are shown in Fig. 3.7 and the major services the CSIR-NPL provides to its customers are shown in Fig. 3.8.

With the signing of CIPM MRA in November 1999, the CSIR-NPL initiated the implementation of QMS based on the international standard ISO/IEC 17025:2017,

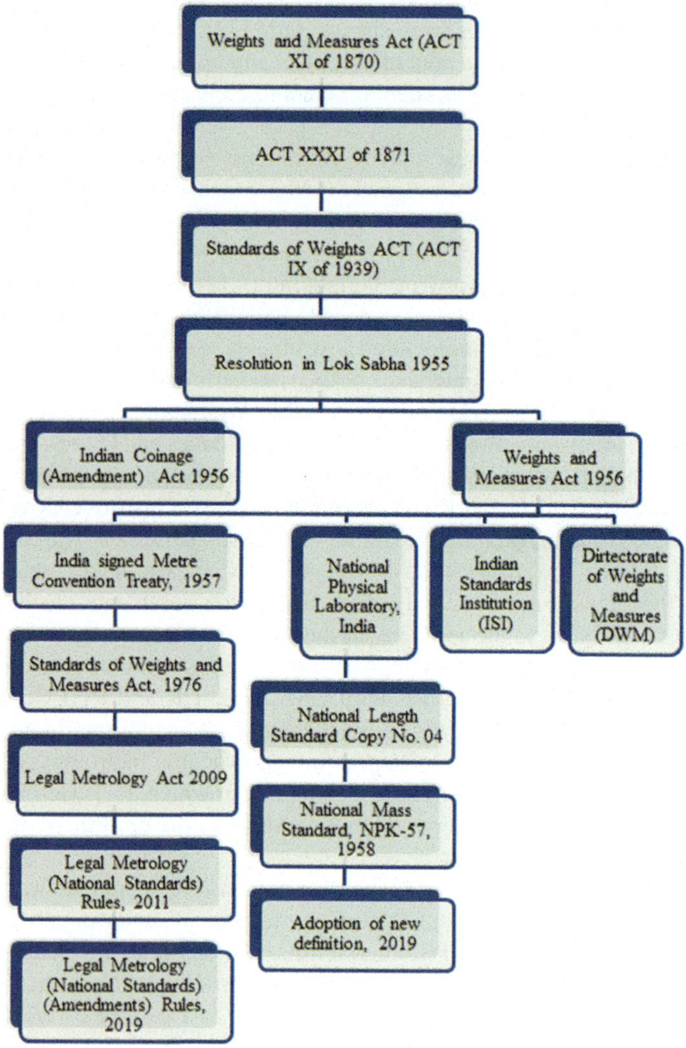

Fig. 3.7 Evolution of CSIR-NPL with the historical development

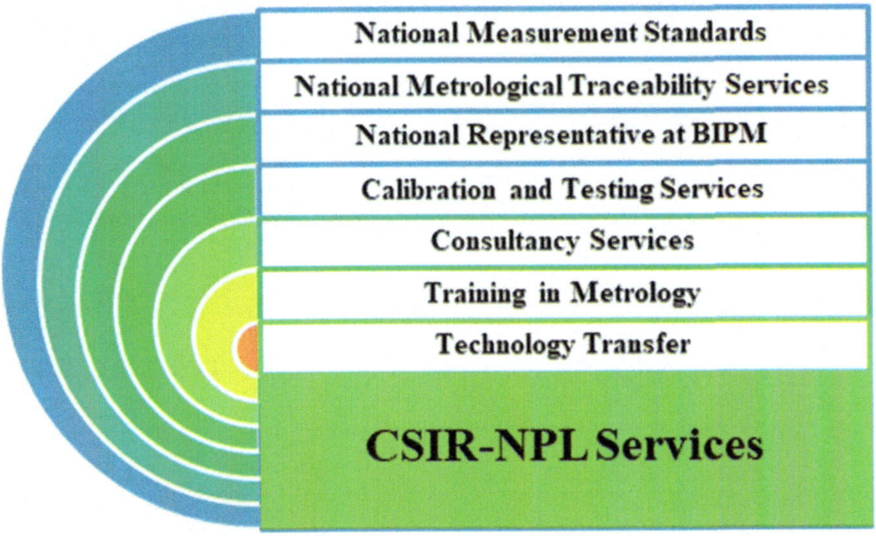

Fig. 3.8 Services provided by CSIR-NPLCSIR-NPL services

in its metrological activities. All the testing and calibration areas of CSIR-NPL are periodically audited and peer-reviewed to confirm the continued compliance with international standards. To maintain the quality, CSIR-NPL adopted the Quality Policy with the main features shown in Fig. 3.9.

A brief description of the establishment and realization of SI base units at CSIR-NPL is summarized in this section. After signing the Meter Convention, Copy No. 57 of the *International Prototype Kilogram (IPK)* as *National Prototype Kilogram (NPK-57)*, shown in Fig. 3.10 was given by BIPM to India in 1958. NPK is periodically calibrated at BIPM, France, and is used as a primary standard of mass at CSIR-NPL [23]. However, to make pace with international developments and to realize 'kilogram' in terms of Planck constant, efforts are being made towards the establishment of the 'Watt Balance' based primary standard for mass.

Since 1957, 'metre' is also being realized and maintained at the CSIR-NPL. The platinum-iridium (Pt–Ir) metre bar (copy No. 4 of International Prototype, Fig. **3.**11) was provided by BIPM to CSIR-NPL to realize the unit 'meter'. After the revision of the definition of 'meter', the orange wavelength of 86Kr lamp was used to realize the unit till 1983. To make pace with the global developments in length metrology, CSIR-NPL also developed the iodine ($^{127}I_2$) frequency stabilized He–Ne laser, and realized 'meter' using it in 1978, and participated in the intercomparison in 1982. In 1983, the 'meter' was redefined in terms of distance travelled by light in vacuum. As per recommendations of the CIPM, CSIR-NPL India used the developed laser for realization of 'meter', which was later on replaced with the commercial I_2 frequency stabilized laser [23]. Efforts are being made for the establishment of frequency comb based primary length standard. For more details, readers can refer Chap. 7.

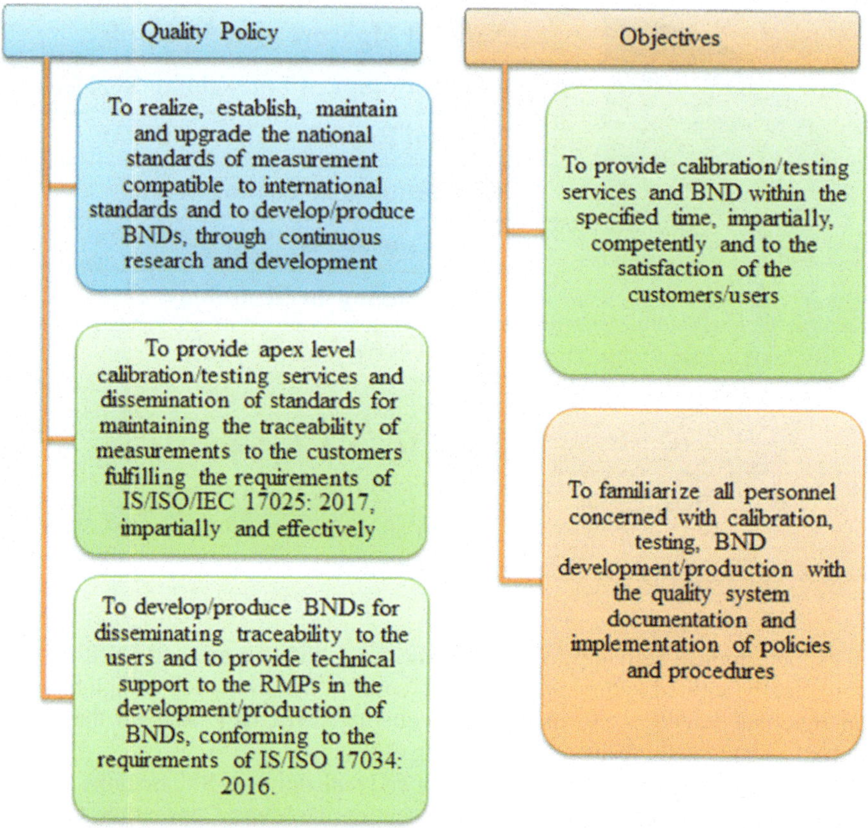

Fig. 3.9 Quality policy of CSIR-NPL

CSIR-NPL has also been instrumental in establishment of time standards for India, since 1950. Initially, the *Indian Standard Time (IST)* was realized using a set of three Essen Ring quartz crystal oscillators. After re-defining 'second' in terms of microwave frequency in 1967, the caesium atomic clocks are being used as a primary standard for time since 1972 [23]. For achieving higher accuracy, a laser-cooled caesium frequency standard was also realized in 2011, and also inter-compared internationally. Further, to achieve the next level of accuracy, efforts are being made to realize optical frequency standard based on a single trapped Ytterbium ion. Detailed description is given Chaps. 4 and 5.

After initiating the electrical metrology activity in 1962, the voltage and resistance standards were established in the form of a bank of 10 Weston Cadmium cells at 2.019 V level, and a bank of six resistances of 1 Ohm (Ω), respectively, and internationally intercompared in 1964. In 1967–66, with the objective of raising the strength, twelve 1 Ω standard resistances were also fabricated. However, after advancement in quantum electronics and development of quantum standards of voltage and current,

Fig. 3.10 National prototype of the kilogram (NPK-57) at CSIR-NPL, India

internationally, the SI base unit of current is realized from 'voltage' and 'resistance' standards [23].

After realization of first voltage standard using AC Josephson Effect in 1970, CSIR-NPL also started R&D to establish quantum voltage standard in 1973–74. The primary Josephson voltage standard was realized at CSIR-NPL in 1985. After adopting the Quantum Hall Effect (QHE) for realization of resistance standard internationally, the primary Quantum Hall Effect (QHE) was established at CSIR-NPL in 2003 [23]. Further details can be seen in Chap. 10.

In 1954, the unit of thermodynamic temperature, the kelvin, was defined based on triple point of water, T(TPW), as a fundamental fixed point and assigned the temperature of 273.16 K. The definition of kelvin was thus adopted in 1967. In the '70s, TPW cell No. 31 was given by BIPM to CSIR-NPL, which was used for practical realizations. Meanwhile, in '90s CSIR-NPL Temperature group has successfully developed the in-house TPW cell with triple distilled de-ionized water and realized (Fig. 3.12). Few copies of the CSIR-NPL' developed TPW cells are also distributed to the Indian users. CSIR-NPL developed cell is the basis of CSIR-NPL CMCs and was used in 2005 and 2010 peer-review of Temperature and Humidity Metrology activity. Recently, in 2016, CSIR-NPL has also established the ensemble of 3 commercially

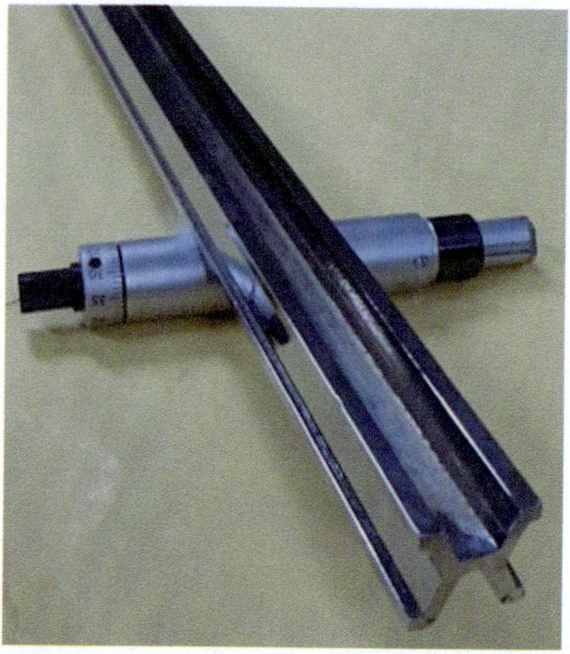

Fig. 3.11 The international prototype of the Pt–Ir metre bar (copy No. 4) at CSIR-NPL

Fig. 3.12 Triple point of water cell No. 31 at CSIR-NPL

available cells with recommended impurity and isotopic composition traceable to Vienna Standard Mean Ocean Water (VSMOW). These TPW cells will be used for the calibration of SPRTs in the Boltzmann Constant based new kelvin realization at CSIR-NPL (Also see Chap. 8).

The unit of luminous intensity, the candela was identified as a SI base unit in 1954. Initially, at CSIR-NPL, the units related to photometry and radiometry were realized using the lamps calibrated at BIPM, NPL UK and PTB Germany [23]. However, in collaboration with PTB Germany, the goniophotometer, intensity bench, and spectroradiometer were established, and primary scale of luminous flux was established. In 2007, the source-based primary standard of optical radiation was established, and in 2012, the detector-based primary standard of optical radiation in the form of a cryogenic radiometer was established. To make the pace with the international developments in quantum radiometry, basic research in the few photon measurements was also initiated. Further details are available in Chap. 8.

3.3.3 Bhabha Atomic Research Centre as Designated Institute of India

CSIR-NPL, as a NMI is responsible for maintaining the national measurement standards for India. However, as the importance of metrology and its influence grows in different fields like medicine, chemistry, food sector, strategic and defense sector, it is not possible to provide all the services from a single institute. In the general case, Governments, an appropriate authority, or in some cases the NMI itself when it is authorized to do so, have therefore identified and appointed specialist institutes to deal with metrological responsibilities in many of these new areas. These are known as Designated Institute (DI). That is why BIPM also recognizes Designated Institute (DI) status to institute other than NMI.

In India, CSIR-NPL provides national standards and dissemination services in almost all the parameters except the ionization radiation which is maintained by BARC. CSIR-NPL, under the authority of Legal Metrology Act and as per CIPM MRA, had recommended BIPM to approve BARC to be the DI for the ionizing radiation during 2003. BARC now acts as the DI of India for the ionizing radiation and has the responsibility for the establishment, maintenance, dissemination of ionizing radiation in India. BARC has participated in 52 intercomparison exercises and has now initiated necessary actions to go for the approval and registration of CMCs in ionizing radiation in close association of CSIR-NPL.

3.3.4 Department of Consumer Affairs, Government of India (Legal Metrology)

The importance of measurement activities was felt significantly around the globe during the mid of the twentieth century (after second world war). Government around the globe converge together and a treaty was signed amongst member countries and the OIML was formed in 1955. The aims and objectives of this organization were to ensure an appropriate level of credibility of measurement results in the national regulation of measurement. India is one of the member countries which is a signatory to OIML.

In India, the Department of Consumer Affairs is one of the separate wings under the Ministry of Consumer Affairs, Food & Civil Supplies, Government of India to enforce the Legal Metrology Act and subsequent Rules. The Director, Department of Consumer Affairs, at the centre who delegates his powers to the Controllers of the states for effective enforcement of the provisions of Legal Metrology Act and Rules.

Every Indian is a consumer whether they are buying foods, vegetables, clothes, etc. for their basic need or fuel for an automobile, etc. When people buy something and pay for the same, they should feel confident that they are getting the right quantity for which they are paying because businesses and retailers are required to measure their products and services accurately. At present, all progressive countries have a governmental organization responsible for monitoring accuracy in measurement (right quantity), integrity of measurement results to the nation's economic growth and prosperity. India is not exceptional.

With a view to providing a coherent system and uniform standards of weights and measures, the first Act named "*The Standards of Weight and Measurement Act, 1956*" was approved based on a system of units and metric systems recognized by CGPM. The Central Government constituted a committee to consider the amendment required to be made in Act, 1956 on the recommendations of International organizations, resulting in which "*The Standards of Weights and Measures Act, 1976*" came into existence. The parliament further enacted "*The Standards of Weights & Measures (Enforcement) Act, 1985*" thereafter.

In view of the fast advancement of science and technology and economic globalization, there has been enormous improvement in weighing and measurement techniques and has extended the scope of weights and measures. With the view to establish the standards of weights and measures, regulate trade and commerce in weights and measures and other goods which are sold or distributed by weight, measure or quantity and for matter connected therewith or incidental thereto, the *Legal Metrology Act, 2009 was enforced on 1st April 2011*. The Legal Metrology Act, 2009 is a single Act covering the earlier both the provisions of the *Standards of Weights and Measures Act, 1976* and *Standards of Weights and Measures (Enforcement) Act, 1985*.

The responsibility related to legal metrology is shared between the Central Government and the States Government. Matters of national policy and other related activities such as uniform laws on weights and measures, technical regulations, training, precision laboratory facilities and implementation of the international recommendations are the concern of the Central Government. The State Governments and Union Territory Administrations are responsible for day-to-day activities related to the enforcement of the laws. Enforcement of standards and Weights and Measures Act is entrusted with the Directorate of Legal Metrology in each state which is essentially an Organisation comprising Controller of Legal Metrology at the state level, Assistant Controller of Legal Metrology at the district level and Inspectors of Legal Metrology at the field level.

The following seven rules have been framed under the Legal Metrology Act, 2009. The glimpses of applications of the rules are mentioned as follows:

3.3.4.1 The Legal Metrology (General) Rules, 2011

Specifications for weighing and measuring instruments have been prescribed in the Legal Metrology (General) Rules, 2011 which include around 40 types of weighing and measuring instruments such as electronic weighing instruments, weighbridges, petrol pumps, water meter, sphygmomanometer, clinical thermometer, etc. These weighing and measuring instruments are used by industries, traders, hospitals and various government and non-government organizations for the weighing and measurement purpose and the end results of the weighing and measuring are directly for the benefit of the common people. These weighing and measuring instruments are periodically verified by the State Government officers using the Standard Weights and Measures and the procedure prescribed in the Rules.

3.3.4.2 The Legal Metrology (Packaged Commodities) Rules, 2011

'Pre-packaged commodity' is defined under the Act as, 'a commodity which without the purchaser being present is placed in a package of whatever nature, whether sealed or not so that the product contained therein has a pre-determined quantity'. As per the Legal Metrology (Packaged Commodities) Rules, 2011 certain mandatory declarations have to be made on every package, which are;

(i) Name and address of the manufacturer/packer/importer;
(ii) Country of origin in case of imported packages;
(iii) Common or generic name of the commodity contained in the package;
(iv) Net quantity, in terms of standard unit of weight or measure or in number;
(v) Month and year of manufacture/pack/import;
(vi) Retail sale price in the form of Maximum Retail Price (MRP) Rs... Inclusive of all taxes;
(vii) Consumer care details.

Besides the above, the Government has made it mandatory to pack 19 commodities in the prescribed sizes in the interest of common consumers.

3.3.4.3 The Legal Metrology (Approval of Models) Rules, 2011

Manufacturers/Importers of Weighing and Measuring equipment, which is prescribed Under the Legal Metrology Act, 2009 and rules made thereunder, are required to take the approval of the Government of India before manufacturing/import.

Some of the equipment like cast iron, brass, bullion, or carat weight or any beam scale, length measures (not being measuring tapes) which are ordinarily used in retail trade for measuring textiles or timber, capacity measures, not exceeding twenty litre incapacity are not required to obtain Model approval.

3.3.4.4 The Legal Metrology (National Standards) Rules, 2011

(i) Under the Rules there is provision of National Prototypes/various standards that are kept at National Physical Laboratory.
(ii) Reference Standards of weights and measures are kept at Regional Reference Standard Laboratories (RRSLs) at Ahmedabad, Bangalore, Bhubaneswar, Faridabad and Guwahati.
(iii) Reference Standards are used for the verification of Secondary Standards Weights & Measures which are part of State Government Laboratories.
(iv) The Working Standard Weights & Measures are available at the District levels which are used for the verification of any Weight & Measures used by traders and manufacturers for the transaction and protection purposes. The Working Standard Weights & Measures are verified by the Secondary Standard Weights & Measures.

3.3.4.5 The Legal Metrology (Numeration) Rules, 2011

Under these rules, the provision is made for making Numeration and the Manner in which numbers shall be written.

3.3.4.6 Indian Institute of Legal Metrology Rules, 2011

Indian Institute of Legal Metrology (IILM), Ranchi is the national training institute for providing basic training in the field of Legal Metrology to the Legal Metrology Officers of States/UTs/Union of India, which are compulsory, under the administrative control of this Department. It also provides training to the foreign participants of the neighbouring/developing countries. Under these rules provisions regarding courses to be imparted at the Institute, Obligatory functions of the Institute, Qualification of persons to be eligible for admission in the Institute are prescribed. Apart from this, it organizes various refresher courses and seminars on Legal Metrology on a regular basis. In a calendar year, approximately 200 officials of Legal Metrology of different States and UTs are being trained by IILM.

3.3.4.7 The Legal Metrology (Government Approved Test Centre) Rules, 2013

The Government Approved Test Centre (GATC) Rules are framed for approval of GATCs established by the private persons for the verification of some of the weights and measures, in addition to verification done by the State Government Officers. The weights and measures prescribed under these rules for the verification by a GATC are (i) Water meter, (ii) Sphygmomanometer, (iii) Clinical Thermometer, (iv) Automatic Rail Weighbridges, (v) Tape Measures, (vi) Non-automatic weighing instrument of Accuracy Class-IIII/Class-III (up to 150 kg), (vii) Load cell, (viii) Beam Scale, (ix) Counter Machine and (x) Weights of all Categories.

3.3.5 Quality Council of India (QCI)

The QCI is a revolutionary experiment of the Government of India in setting up organizations in partnership with the Indian industry. Since 1992, an essential requirement had been felt for the establishment of an accreditation body in the country for a globally accepted mechanism for recognition of results of conformity assessment bodies. In 1996, the then Department of Industries (Department of Industrial Policy and Promotion) coordinated the activities and submitted recommendations to the Cabinet. Key recommendations included the need for establishment of an organization jointly by the Government and the industry and it to be self-sustaining and be away from the government. Accepting the recommendations, the Cabinet Committee decided to set up QCI as a non-profit autonomous society registered under Societies Registration Act, 1860 to establish an accreditation structure and to spread quality movement in India by undertaking a National Quality Campaign.

3.3.5.1 Mission of QCI

The Mission of QCI is to lead a nationwide quality movement in India by involving all stakeholders for emphasis on adherence to quality standards in all spheres of activities primarily for promoting and protecting the interests of the nation and its citizens.

3.3.5.2 Main Objectives

To achieve the Mission of QCI by playing a pivotal role in propagating, adoption and adherence to quality standards in all important spheres of activities including education, health-care, environment protection, governance, social sectors, infrastructure sector and such other areas of organized activities that have significant bearing in improving the quality of life and well being of the citizens of India and without restricting its generality shall inter-alia include:

(a) To lead nationwide quality movement in the country through National Quality Campaign aimed at creating awareness amongst citizens, empowering them to demand quality in all spheres of activities, and promoting and protecting their well being by encouraging manufacturers and suppliers of goods and service providers for adoption of and adherence to quality standards and tools.
(b) To develop apropos capacities at the level of Governments, Institutions and enterprises for implementing and institutionalizing continuous quality improvement.
(c) To develop, establish and operate National Accreditation programmes in accordance with the relevant international standards and guides for the conformity assessment bodies certifying products, personnel, management systems, carrying out inspection and for the laboratories undertaking testing and calibration and such other areas of organized activities that have a significant bearing in improving the quality of life and well being of the citizens of India.
(d) To develop, establish and operate National Accreditation Programmes for various service sectors such as education, healthcare, environment protection, governance, social sectors, infrastructure sector, vocational training, etc., to site a few, as may be required, based on national/international standards and guidelines and where such standards are not available, to develop accreditation standards to support accreditation programs.
(e) To build capacities in the areas of regulation, conformity assessment and accreditation to overcome constraints of Technical Barriers to Trade (TBT) and Sanitary and Phytosanitary Measures (SPS) of the World Trade Organization (WTO) through agreement.
(f) To encourage development and application of third party assessment model for use in government, regulators, organizations and society.
(g) To promote quality competitiveness of India's enterprises especially MSMEs through the adoption of and adherence to quality management standards and quality tools.
(h) Promoting the establishment of quality improvement and benchmarking centre, as a repository of best international/national practices and their dissemination among the industry in all the sectors.

    (i) To encourage industrial/applied research and development in the field of quality and dissemination of its result in relevant publication including and trade journals;
    (ii) To build capacities including development of appropriate quality accreditation mechanism for other emerging areas that have a significant bearing in improving the quality of life and well being of the citizens of India such as food sector, oil and gas, forestry, agriculture/animal husbandry, warehouse, pharmacy, etc. or as the need may arise from time to time.

(i) To facilitate effective functioning of a National Information & Enquiry Services on standards and quality including an appeal mechanism to deal with unresolved complaints.
(j) To develop and operate an appeal mechanism to deal with unresolved complaints.

3.3.6 National Accreditation Board for Testing and Calibration Laboratories (NABL)

Over the years, a large infrastructure was established in India for testing and calibration. In the 1980s, about 130 calibration laboratories and 300 testing laboratories were functioning under various agencies such as Central and State government departments, defence establishments, R&D laboratories, academic institutions, public and private sector industries, etc. To ensure the validity of calibration/test results getting from various laboratories and to establish the traceability of all measurement results to the national standards, the Government of India approved a scheme entitled "*National Coordination of Testing and Calibration Facilities (NCTCF)*". The National Measurement System which was in operation informally, for the last couple of decades, was formalized by NCTCF under Department of Science & Technology (DST). NCTCF, with the cooperation of India's NMI, i.e. CSIR-NPL, started providing accreditation to calibration laboratories in 1982. In 1993, NCTCF was renamed as "*National Accreditation Board for Testing and Calibration Laboratories (NABL)*".

Thereafter, NABL was registered as a society under the Societies Registration Act, 1860 on 12th August 1998 with the objective to promote, coordinate, guide, implement and maintain an accreditation system for laboratories. NABL operated as an autonomous body under the aegis of the DST and started its operations from CSIR-NPL.

In the year 2016, in pursuance of cabinet decision (February 1996), NABL along with the support mechanism existing under the DST, Ministry of Science and Technology was transferred to the Department of Industrial Policy and Promotion (DIPP), Ministry of Commerce and Industry and subsequently transferred to QCI as one of its Boards.

NABL has been established with the objective of providing accreditation to CAB under Government, Industry Associations and Industry with a scheme of which involves third-party assessment of the technical competence of testing including medical and calibration laboratories, *Proficiency Testing Providers (PTPs)* and *Reference Material Producers (RMPs)*.

The laboratory accreditation services to testing and calibration laboratories are provided in accordance with ISO/IEC 17025: 2017, 'General Requirements for the Competence of Testing and Calibration Laboratories' and ISO 15189: 2012, 'Medical laboratories—Requirements for quality and competence'. The accreditation to PTPs are based on ISO/IEC 17043: 2010, "Conformity assessment—General requirements for proficiency testing" and to RMPs based on ISO 17034: 2016, "General requirements for the competence of reference material producers". The fields, disciplines and groups for which the accreditation services are offered are listed in 'Scope of NABL Accreditation'.

NABL has established its accreditation system in accordance with ISO/IEC 17011: 2017 'Conformity Assessment—Requirements for Accreditation bodies accrediting conformity assessment bodies'. NABL accreditation system also fulfils the requirements of ILAC MRA of which NABL is a member.

NABL vision and mission are "To be the world's leading accreditation body and to enhance stakeholders' confidence in its services and "To strengthen the accreditation system accepted across the globe by providing high quality, value-driven services, fostering APAC/ILAC MRA, empanelling competent assessors, creating awareness among the stakeholders, initiating new programs supporting accreditation activities and pursuing organizational excellence', respectively. Formal recognition of competence of a laboratory by an Accreditation body in accordance with international criteria has various advantages:

- Increased confidence in testing/calibration reports issued by the accredited laboratory.
- Better control of laboratory operations and feedback to laboratories as to whether they have sound Quality Assurance System and are technically competent.
- Potential increase in business due to enhanced customers' confidence and satisfaction.
- Customers can search and identify NABL accredited laboratories for their specific requirements from the NABL website or Directory of Accredited Laboratories.
- Users of accredited laboratories get benefit from greater access for their products, in both domestic and international markets.
- Savings of time and money due to reduction or elimination of the need for retesting/recalibration of products.

3.3.7 Bureau of Indian Standards (BIS) (Documentary Standardisation and Product Specifications)

BIS is the National Standard Body of India functions under the *BIS Act 2016* for the harmonious development of the activities of standardization, marking and quality certification of goods and for matters connected therewith or incidental thereto. BIS has been providing traceability and tangibility benefits to the national economy in a number of ways—providing safe reliable quality goods; minimizing health hazards to consumers; promoting exports and imports substitute; control over proliferation of varieties, etc. through standardization, certification and testing.

During the British period in India, when the country was facing with the massive task of building up the industrial infrastructure, the Institution of Engineers (India) prepared the first draft of the constitution of an institute which could take up the task of formulation of National Standards. This led to the Department of Industries and Supplies issuing a memorandum on 3rd September 1946, formally announced the setting of an organization named "*Indian Standards Institution (ISI)*". ISI came into existence on 6th January 1947.

In the beginning, the organization concentrated on standardization activity. To provide the advantages of standardization to common consumers, the ISI started operating the Certification Marks Scheme under the Indian Standards Institution (Certification Marks) Act, 1952. The scheme, which was formally launched by ISI in 1955–56, enabled it to grant licences to manufacturers produced goods in conformity with Indian Standards and to apply ISI Mark on their products. To meet the requirements of the Certification Marks scheme, the nucleus of a laboratory was started in 1963. While the product certification was being operated under the *Indian Standards Institution (Certification Marks) Act, 1952*, the formulation of standards and other related work was not governed by any legislation. A Bill with this objective was therefore introduced in the Parliament on 26th November 1986. BIS came into existence, through an act of parliament dated 26th November 1986 and on 1st April 1987, with a broadened scope and more responsibilities such as the staff, assets, liabilities and functions of erstwhile ISI. Through this changeover, the government envisaged building a climate for quality culture, awareness and greater participation of consumers in formulation and implementation of national standards. Keeping in view, the interest of consumers as well as the industry, BIS is involved in various activities as given below:

- Standards Formulation
- Product Certification Scheme
- Compulsory Registration Scheme
- Foreign Manufacturers Certification Scheme
- Hall Marking Scheme
- Laboratory Services
- Laboratory Recognition Scheme
- Sale of Indian Standards
- Consumer Affairs Activities
- Promotional Activities
- Training Services, National and International level
- Information Services.

3.4 Dissemination of SI Units in India

3.4.1 Introduction

In this section, the dissemination of SI units in India is explained thoroughly. The main focal points included are national metrology policy of India, dissemination through calibration, testing and various verification services, along with the proposed changes in academic syllabus involving SI units.

3.4.2 National Metrology Policy of India

The main purpose of the CSIR-NPL Policy on Metrological Traceability is to state the CSIR-NPL role in the establishment of unbroken chain of metrological traceability. The Policy defines the metrological traceability used by CSIR-NPL and its role along with its beneficiaries in achieving traceability of measurement results for measurements performed within the laboratory and for the services being rendered to CSIR-NPL clients. CSIR-NPL is also the signatory of CIPM-MRA through which CSIR-NPL demonstrates the international equivalence of its measurement standards and the calibration and international acceptability of measurement certificates it issues.

In addition to the above, '*CSIR-NPL Policy on Metrological Traceability*' assists its beneficiaries in the establishment of traceability of their measurement results and to assess the claims of traceability made by others. This is done directly through the apex level calibration services provided by CSIR-NPL, through the provision of CSIR-NPL measurement-related products and technologies, certified reference materials (CRM), proficiency testing, consultancy services and collaboration with organizations, through the development and dissemination of technical information on traceability and through conducting industrial training programs, skill development programs on metrology and coordinated outreach programs. policy document is intended to communicate the role of CSIR-NPL in metrological traceability for the overall growth of the nation and to focus its efforts to the metrology related training and activities. This policy, along with supplementary documents and publications, are aimed to assist CSIR-NPL staff and beneficiaries in their understanding, interpretation, and implementation of the Policy, and to serve as a resource for the customers.

The CSIR-NPL has documented this Policy on Metrological Traceability which stipulates its role as NMI in India in establishing, maintaining, upgrading and disseminating national standards of measurements through an unbroken chain of traceability to SI units. The policy and supported reference documents are for intended use within CSIR-NPL and also to serve its clients/customers. The policy document addresses many of the frequently asked queries and doubts raised by customers.

3.4.2.1 Terms and Definitions

For the sake of clarity. the main terms and definitions, some of them though already discussed above, referred and used in National Metrology Policy are presented in Table 3.2.

3.4.2.2 Statement of Policy

In order to achieve its mandate and to establish the quality in measurement while ensuring that the use of CSIR-NPL name, products, and services is within the ambit of its authority and responsibility, the metrological policy statement of CSIR-NPL stipulates that, CSIR-NPL;

Table 3.2 Terms used in the CSIR-NPL Policy document and their related definitions

Term	Definition
Act of parliament	The Legal Metrology (National Standards) Rules (2011), Chapter III 'National Standard', section 23(1) states that "the work relating to the realization, establishment, custody, maintenance, determination, reproduction and updating of national standards of weights and measures shall, on the commencement of these rules, be the responsibility of the National Physical Laboratory" [24]
Beneficiaries/clients/customers	These terms are used in the policy to indicate industries, laboratories and academic institutes or any other user which produce or utilize the products or services of CSIR-NPL with traceable measurements
BARC	BARC is India's premier nuclear research organization which also maintains the national standards for nuclear radiation as Designated Institute [25]
Bhartiya Nirdeshak Dravya (BND)	A reference material produced by CSIR-NPL or a commercial supplier with a well-defined traceability to CSIR-NPL established via specific criteria and protocols defined and documented by CSIR-NPL and tailored to meet the needs of the beneficiaries/clients/customers [26]
Bureau of Indian Standards (BIS)	BIS is the National Standard Body of India established for the harmonious development of the activities of standardization, marking and quality certification of goods and for matters connected there with or incidental thereto [27]
Calibration	Set of operations that establish, under specified conditions, the relationship between values of quantities indicated by a measuring instrument or measuring system, or values represented by a material measure or reference material, and the corresponding values realized by standards. (International vocabulary of metrology—Basic and general concepts and associated terms [21], definition 2.39)

(continued)

Table 3.2 (continued)

Term	Definition
Calibration and measurement capabilities (CMC)	In the context of the CIPM MRA and ILAC Arrangement, and in relation to the CIPM-ILAC Common Statement, the following shared definition is agreed upon: A CMC is a calibration and measurement capability available to customers under normal conditions: (a) as published in the BIPM key comparison database (KCDB) of the CIPM MRA; or (b) as described in the laboratory's scope of accreditation granted by a signatory to the ILAC Arrangement [28]
Certificates	It is a documented result of a calibration. Sometimes it is called a calibration certificate or a calibration report. In case of certified reference material, this certificate can also refer to the document stating one or more property values and their uncertainties, and confirming that the necessary procedures have been carried out to ensure their validity and traceability [21, 29, 30]
Certified reference material (CRM)	A reference material (RM) characterized by a metrological valid procedure for one or more specified properties, accompanied by a certificate that provides the value of the specified property, its associated uncertainty, and a statement of metrological traceability. The concept of value includes qualitative attributes such as identity or sequence. Uncertainties for such attributes may be expressed as probabilities [30]
International key comparisons	As per CIPM-MRA, "a key comparison is one of the set of comparisons selected by a Consultative Committee to test the principal techniques and methods in the field. Key comparisons may include comparisons of representations of multiples and sub-multiples of SI base and derived units as well as comparisons of artefacts" [31]
Legal Metrology Department	Legal metrology Department, earlier called Weights and Measures Department is a part of Ministry of Consumer Affairs, Govt. of India. The department works in relation to the mandatory technical and legal requirements which have the objects of ensuring public guarantee from the point of view of security and accuracy of the weights and measurements [32]

(continued)

Table 3.2 (continued)

Term	Definition
Measurement result	According to the VIM: set of quantity values being attributed to a measure and together with any other available relevant information, generally expressed as a single measured quantity value and a measurement uncertainty. The concept of traceability covers both measurement results of quantity values and of nominal property values such as chemical identity and sequence ([21], definition 2.9)
Measurement uncertainty	Non-negative parameter characterizing the dispersion of the quantity values being attributed to a measurand, based on the information used [21] (definition 2.26 (3.9))
National Metrology Institute (NMI)	NMI of a country realize SI units, establish, maintain, upgrade and disseminate national standards of measurements and conducts research and developmental activities in the field of scientific metrology. An NMI also provides metrological traceability to international standards. As per CIPM-MRA, CSIR-NPL is a signatory NMI of India [33]
National or international standards	Measurement standards (national) recognized by national authority to serve in the country; measurement standards (international) recognized by signatories to an international agreement and intended to serve worldwide [21]
Proficiency testing	As per ISO/IEC 17025: 2017 "evaluation of participant performance against pre-established criteria by means of inter-laboratory comparisons" (definition 3.5) [34]
Quality Council of India (QCI)	It is an accreditation body in the country to establish internationally acceptable mechanism for recognition of conformity assessment of results [35]
Quality infrastructure	The system comprising the organizations (public and private) together with the policies, relevant legal and regulatory framework, and practices needed to support and enhance the quality, safety and environmental soundness of goods, services and processes [36]
RMP outreach program	Here meaning a CSIR-NPL program or collaboration, liasoning with external reference material producing groups [26]

(continued)

Table 3.2 (continued)

Term	Definition
Unbroken chain of calibrations	Here it means that the traceability is obtained with successively linked chain of calibrations each with documented uncertainty and finally linked to SI units with the help of national or international standards

- Accepts the definition of metrological traceability provided in the International vocabulary of metrology—Basic and general concepts and associated terms (VIM): "property of a measurement result whereby the result can be related to a reference through a documented unbroken chain of calibrations, each contributing to the measurement uncertainty." [21], for own use and endorses for use by others;
- Follows the Quality System as per IS/ISO/IEC 17025:2005 or any updated version, consciously and effectively;
- Establishes metrological traceability to SI units by providing apex level calibration, certification and dissemination of standards compatible with the international standards through the continuous research and development;
- Asserts, in cases where the metrological traceability to SI units is not possible directly, it establishes the metrological traceability by using the certified reference materials, e.g. Bhartiya Nirdeshak Dravya (BND) developed in-house or provided by the accredited reference material producers through RMP outreach program;
- Takes the responsibility of metrological traceability of measurement results certified by CSIR-NPL, while assessing the validity of such a claim is the responsibility of the user of that result;
- Confirms, especially in the conflict of interest regarding the claim that CSIR-NPL does not define, state, guarantee, or certify metrological traceability of the results of measurements except those that certified by CSIR-NPL itself, either directly or through official collaborations;
- Cooperates on formulation of standard definitions, its clarifications, and recommended procedure with national and international organizations bearing the responsibility for defining, stating, assuring, or certifying metrological traceability; and
- Prepares and disseminates technical information on metrological traceability through industrial training programs, skill development programs on metrology and coordinated outreach programs.

A lot of frequent questions and queries are received from the stakeholders about the metrological activities of CSIR-NPL. Table 3.3. Frequently asked questions related to metrology is aimed to address these Frequently Asked Questions (FAQs).

Table 3.3 Frequently asked questions related to metrology

Q	**What is metrological traceability?**
A	The definition of traceability provided in the International vocabulary of metrology—basic and general concepts and associated terms (VIM): "property of a measurement result whereby the result can be related to a reference through a documented unbroken chain of calibrations, each contributing to the measurement uncertainty." [21], for own use and endorses for use by others. As clear from the definition, the traceability is a property of the measurement result, not of an instrument, lab or report
Q	**What is CSIR-NPL?**
A	CSIR-National Physical Laboratory (CSIR-NPL) is mandated to be India's NMI by the act of parliament (gazette of India) and also legal metrology rules associating with it. Being the custodian of the National Standards of Measurements, CSIR-NPL is responsible for the realization, establishment, up-gradation, maintenance and dissemination of standards at par to international level, providing metrological traceability across the country The mandate of CSIR-NPL is to ensure the realizations of SI units as national standards and to transfer the values of the standards to the national measurement system through calibrations, testing and other measurement related services. The customers who are the direct beneficiaries of CSIR-NPL make their measurements results traceable to those of CSIR-NPL, and hence to the SI units. These companies, in turn, use their standards to provide measurement services to their customers hence assuring quality in the measurements across the country
Q	**What is the CIPM MRA and what is its importance in metrological traceability?**
A	The CIPM Mutual Recognition Arrangement (CIPM MRA) is the framework through which NMI's demonstrate the international equivalence of their measurement standards and the calibration and measurement certificates they issue. The mutual recognition of calibration and measurement certificates requires that each NMI participate in the activities of the BIPM (including key measurement comparisons) and have a suitable way of assuring quality in the results of its measurement services. The results of the key comparisons and specific statements of the CMCs of the signatory NMI are entered in CIPM MRA Key Comparison Data Base (KCDB) [33]
Q	**I want my measurement results to be traceable to SI units through CSIR-NPL. What do I have to do?**
A	To achieve traceability of measurement results to SI units through standards maintained by CSIR-NPL, you need to trace your measurement results through an unbroken chain of calibrations, including determining the uncertainties at each step, to CSIR-NPL standards as the specified references. One can also obtain traceability directly from CSIR-NPL or can obtain traceability from laboratories/industries which have their measurement results traceable to CSIR-NPL through unbroken chain of calibrations with documented uncertainty calculation at each step
Q	**How does CSIR-NPL establish the traceability of its own measurement results?**
A	CSIR-NPL establishes its traceability following the definition of the metrological traceability as per International vocabulary of metrology—Basic and general concepts and associated terms (VIM). CSIR-NPL realizes the SI units of measurements through internationally accepted procedure and perform international intercomparisons with other NMIs of the world to authenticate its results. CSIR-NPL maintains the national measurement standards of India

(continued)

Table 3.3 (continued)

Q	**Is CSIR-NPL NABL (National accreditation board for Testing and Calibration Laboratories) accredited or not?**
A	Most of the NMIs, CSIR-NPL in case of India, are internationally peer reviewed and are signatories of the CIPM MRA. Hence such NMIs, (CSIR-NPL in case of India) do not require accreditation from national bodies like NABL. For details please refer ISO/IEC 17025:2017, A.3.1 [34]
Q	**Does CSIR-NPL provide Certified Reference Materials?**
A	Yes. The Certified Reference Materials provided by CSIR-NPL are called BhartiyaNirdeshakDravya (BND). BND is a registered trademark name of CSIR-NPL. CSIR-NPL's BND programme is introduced very recently. Although, it has started Certified Indian Reference Material program way back in 1986. CSIR-NPL has released several BNDs recently and Govt of India has supported CSIR-NPL to strengthen its BND program by developing BNDs in the area of AYUSH, Materials, Nanotechnology, Medicine, Food and Agriculture and Biologics [26]
Q	**If I purchase a BND, does that make my measurement results traceable to CSIR-NPL?**
A	Merely purchasing BND does not automatically make the customer's measurement results traceable to reference standards developed and maintained by CSIR-NPL. A claim of traceability to reference standards developed and maintained by CSIR-NPL can be asserted by proper use of appropriate BNDs and reference to the certified values and uncertainties provided by a CSIR-NPL Certificate along with the purchased BND
Q	**Can CSIR-NPL provide me certificate with validity of four years or more?**
A	We provide calibration certificates with next recommended date based on the long research experience, trends of data, types and handling of the particular type of instruments. However, the decision of the validity of the calibration intervals lies with the users and which can be ascertained based on the guidelines given in OIML D10 (2007)
Q	**Who is responsible for establishing the traceability of measurement results?**
A	The responsibility for establishing the traceability of measurement result is with the provider of the result. If CSIR-NPL is providing the measurement results then CSIR-NPL takes the responsibility for the measurement traceability. Same is the case with other organizations
Q	**Is there a need to re-examine the metrological traceability periodically?**
A	Metrological traceability depends upon several factors, e.g. the measurement requirements, the needs of the clients, the dependability of the equipment and standards, the environmental effects, etc. So there is a need to re-examine the metrological traceability on a regular basis to ensure its integrity and continuity
Q	**Does CSIR-NPL provide the metrological traceability for the ionizing radiation?**
A	No, as per CIPM-MRA, BARC, Mumbai, a designated institute, is responsible for providing the metrological traceability for ionizing radiation parameters in India. CSIR-NPL, New Delhi, being NMI of India is responsible for providing metrological traceability for all other parameters
Q	**Does CSIR-NPL provide the metrological traceability for the health instruments/equipment?**
A	Yes, CSIR-NPL provides metrological traceability to health instruments. The CSIR-NPL has further strengthened its metrology programme for proving metrological traceability to wide range of medical instruments/equipment

(continued)

Table 3.3 (continued)

Q	**Does CSIR-NPL provide any training/certification course in the field of metrology and metrological traceability?**
A	Yes, CSIR-NPL provides training in metrology on regular basis. CSIR-NPL has formulated many training programs/courses in different measurement parameters like mass, length, temperature etc. for industries, public and private organizations and research laboratories. The annual training calendar is uploaded/available on CSIR-NPL website (www.nplindia.org). CSIR-NPL also provides one year certification course on 'Precision Measurements and Quality Control (PMQC)'. There is also a provision for short term training programs especially for post-graduate students. CSIR-NPL is also planning to launch e-Learning modules for such training courses

3.4.3 Dissemination Through Calibration, Testing and Verification Services

3.4.3.1 NMI Services

As per Legal Metrology (National Standards) Rules, 2011, CSIR-NPL is the custodian of "National Standards" with a responsibility of realization, establishment, upgradation, maintenance and dissemination of standards at par to the international level through R&D and latest technology. CSIR-NPL is not only complying with its primary mandate but also provides apex level calibration and testing services on various parameters to the various industries, strategic sectors, user organizations, etc. to excel in their endeavours and to support the economic growth of the country. These include Mass, Volume, Density, Viscosity, Length, Dimension, Nanometrology, Temperature, Humidity, Optical Radiation, Force, Hardness, Torque, Pressure, Vacuum, Ultrasonic, Acoustics, Vibration, Fluid Flow, DC Metrology, AC High Voltage, AC High Current, AC Power, AC Energy, Time, Frequency, LF & HF Impedance, Voltage & Current, Microwave, Magnetic, Ozone and other parameters. It has been working on the development of primary gas reference standards and currently focusing on providing the greenhouse gas (e.g. CH_4 and CO_2) and pollution gas (e.g. SO_2, NO, CO) standards to various stakeholders which include air quality monitoring stations being run by different pollution boards.

CSIR-NPL has started a program for the development/production of Bharatiya Nirdeshak Dravya, BND® (Indian Reference Material) both in-house as well as supporting the RMPs. More than 80 BNDs are already released in the market for import substitution. The Government of India has supported CSIR-NPL to strengthen its BND program by developing BNDs in the area of AYUSH, Materials, Nanotechnology, Medicine, Food & Agriculture and Biologics.

CSIR-NPL not only supports Indian industries and organizations but also to the SAARC and other Nations.

Each parameter is described in detail in the respective chapters of this book.

3.4.3.2 Services Through NABL Accredited Laboratories

Accreditation is the third party attestation related to a CAB conveying the formal demonstration of its technical competence to carry out specific conformity assessment tasks. CAB is a body that includes calibration laboratory, testing laboratory, medical laboratory, PTPs and RMPs.

The liberalization of trade and industry policies of the Government of India have created quality awareness in domestic trade and provided greater thrust for export. As a result, calibration and testing centers and laboratories have to demonstrate technical competence at an international level.

Laboratory accreditation is a procedure by which an authorized body gives formal recognition of technical competence for specific calibrations/tests/measurements, based on third party assessment and following international standards and guidelines in the respective area. The various current NABL accreditation services provided, are shown in Fig. 3.13 as per scheduled fields and disciplines.

3.4.3.3 Services Through Legal Metrology Department

National Prototypes

(i) The Central Government shall, for the purpose of deriving the value of kilogram, cause to be prepared a national prototype of the kilogram and shall cause its accuracy to be certified by the International Bureau of Weights and Measures equivalent to the international prototype of kilogram and shall thereupon deposit the same in the custody of the National Physical Laboratory, New Delhi.
(ii) The Central Government shall, for the purpose of deriving the value of metre, cause to be prepared a national prototype of the metre and shall cause its accuracy to be certified by the International Bureau of Weights and Measures and shall thereupon deposit the same in the custody of the National Physical Laboratory, New Delhi.

Custody, Maintenance, Etc. of National Standards of Weights and Measures

(i) The work relating to the realization, establishment, custody, maintenance, determination, reproduction and updating of national standards of weights and measures shall, on the commencement of these rules, be the responsibility of the National Physical Laboratory.
(ii) The Central Government may call for such reports from, or issue such directions to, the National Physical Laboratory as it may think fit, in relation to all or any of the matters specified in sub-rule (1).

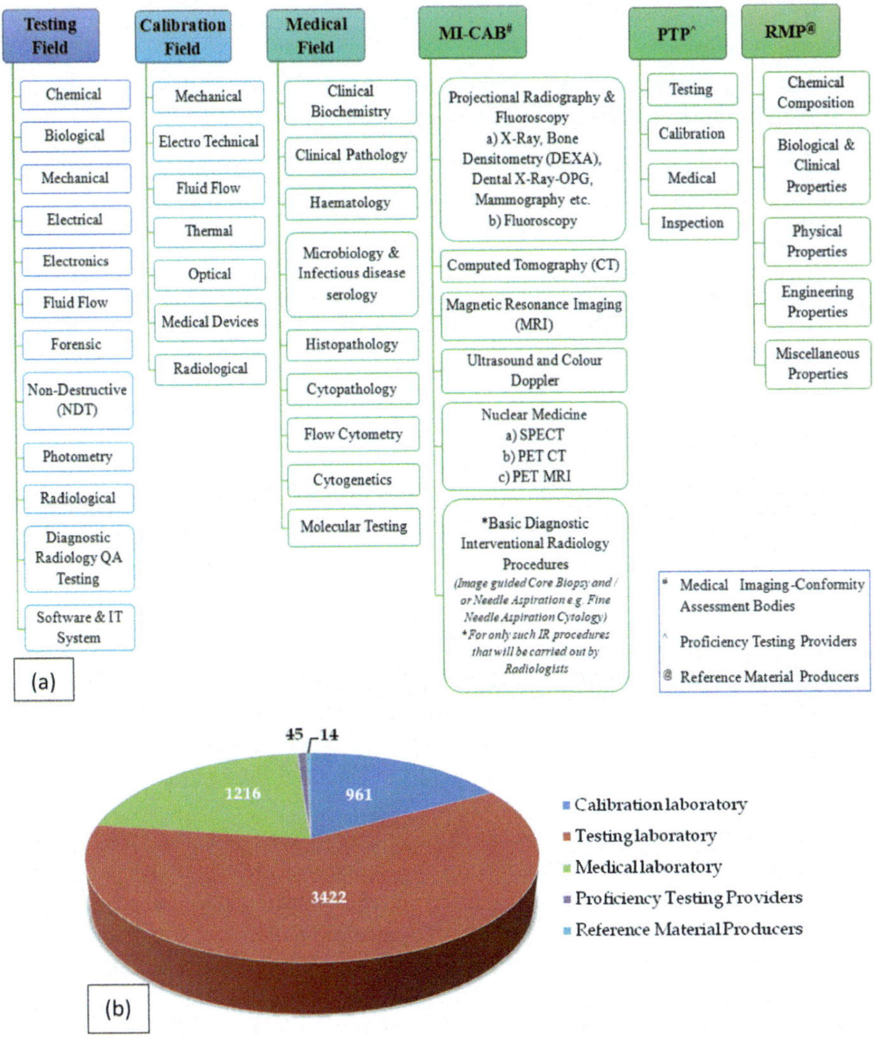

Fig. 3.13 **a** Fields and disciplines in which NABL accreditation is provided. **b** Number of accredited laboratories (as on 22.06.2020)

Realization and Establishment of the National Standards of Weights and Measures Based on SI Units

(i) The National Physical Laboratory shall discharge the responsibility of realizing and establishing the national standards of weights and measures on the basis of recommendations made from time to time, by the General Conference on Weights and Measures or the International Organisation of Legal Metrology, as the case may be.

(ii) The standards of weights and measures, so realized and established, shall be self-consistent.
(iii) For the purpose of establishing the national standards for the base units other than of mass, the National Physical Laboratory shall;

   (a) prepare or cause to be prepared such objects or equipments, or reproduce such phenomena, or both, as may be necessary for the purpose; and
   (b) determine or cause to be determined the value of the national standards as recommended by the General Conference on Weights and Measures and intercompare them, or cause to be inter compared, with the corresponding international standards.

(iv) For the purpose of deriving the value of the kilogram, the National Physical Laboratory shall arrange the periodical determination of the value of the national prototype of the kilogram and the value of which is so determined, shall be the national standards of mass.
(v) For the purpose of establishing the national standards for the derived and supplementary units the National Physical Laboratory shall prepare such standards, or objects or equipment, or both and determine periodically their value and accuracy in relation to the national standards of base units.

Custody and Maintenance of Prototype Standards

(i) The national prototype of the kilogram and other standards, equipments and objects shall remain in the custody of the National Physical Laboratory, New Delhi.
(ii) The national prototype of the kilogram and every other national standard, standard equipments and objects shall be maintained and realized periodically in accordance with such instructions as the General Conference on Weights and Measures on the International Organisation of Legal Metrology or any organization constituted by either of them may issue from time to time.
(iii) Where no instructions have been issued by the International Organisation referred to in sub-rule (2), any Consultative Committee constituted may compile instructions for the proper maintenance of national prototype, national standards, standards equipments and objects.
(iv) The National Physical Laboratory shall arrange, where necessary, to have the national prototype and national standards of physical measurements realized and established in accordance with the recommendations of the General Conference on Weights and Measures and to get them calibrated on inter compared with reference to the appropriate international standards of physical measurements at a periodical interval of not more than ten years.
(v) The value of the national prototype and other national standards shall be the value determined by the National Physical Laboratory or assigned by the National Physical Laboratory on the basis of the technical information provided by the International Bureau of Weights and Measures and the National Physical

Laboratory shall publish such values periodically but in any case at least once in every five years.

(vi) The value determined in accordance with sub-rule (5) shall be deemed to represent the higher obtainable accuracy of such value in the country.

Period and Manner of Verification of Reference, Secondary and Working Standards

(i) Every reference standard shall be verified and certified in terms of the National Standards by the National Physical Laboratory, at an interval not exceeding three years.
(ii) Provided that in the case of length measures such interval shall not exceed five years.
(iii) Every secondary standard shall be verified against the appropriate reference standard by the Reference Standard Laboratory, at an interval not exceeding two years.
(iv) Every working standard shall be verified against the appropriate secondary standard, by any of the laboratories where secondary standards are maintained, at an interval not exceeding one year.

Maintenance of Reference Secondary and Working Standards

Every reference standard, every secondary standard and every working standard, irrespective of the place where they are kept, shall be maintained as far as practicable in accordance with the guidelines issued by the National Physical Laboratory from time to time.

Regional Reference Standard Laboratories (RRSLs)

RRSLs are set up to meet the Legal Metrology requirements of the State Governments, Industries and Consumers in the country. There are 5 RRSLs, situated at Ahmedabad, Bangalore, Bhubaneswar, Faridabad and Guwahati. The RRSLs serve as a link between the CSIR-National Physical Laboratory and the States Weights and Measure laboratories to ensure the correct weighment and measurement in trade and transaction. Two new RRSLs are being established at Nagpur and Varanasi. All the 5 established RRSLs mentioned above are accredited by NABL.

These laboratories are responsible for verification of secondary standards of State Government, testing of models of weights and measures, calibration of sophisticated weighing and measuring instruments and organizing of consumer awareness program, etc.

3.4.3.4 Services Through BIS and Its Authorized Laboratories

Product Certification Overview

The BIS Act 2016, Rules and Regulations are framed to undertake conformity assessment of products, services, systems and processes. Under the product certification scheme—I, BIS grants license to use the standard mark or grants a certificate of conformity as per conformity assessment schemes given in *BIS (Conformity Assessment) Regulations, 2018* which are based on principles laid down in IS/ISO/IEC 17067: 2013.

The Product Certification Schemes laid down are for providing third party assurance of quality, safety and reliability of products to the customer. Presence of BIS certification mark, known as Standard Mark, on a product is an assurance of conformity to the specifications. The manufacturer is permitted to self certify the licensed products after ascertaining its conformity to the standard. Through its surveillance operations, BIS maintains a close vigil on the quality of certified goods. The conformity is ensured by regular surveillance of the licensee's performance by surprise inspections and testing of samples, drawn both from the market/factory.

The certification scheme operates through a network of 32 Branch Offices set up in State's capitals or major industrial towns and 5 Regional Offices coordinating the work of the Branch offices. Although the scheme itself is voluntary in nature, the Government of India, on considerations of public health and safety, security, infrastructure requirements and mass consumption has enforced compulsory certification on various products through orders issued from time to time under various acts.

In order to ensure enhanced consumer safety and compliance to statutory provisions, some products like gas cylinders, regulators and valves, the BIS certification scheme requires that each lot or batch needs to be inspected by BIS certification officers before the release of the product.

Systems Certification

Management Systems Certification Scheme (MSCS)

- MSCS of the BIS is operating since 1992.
- Initially, BIS started the scheme with QMS Certification (IS/ISO 9001) and over the years it has gradually expanded its activities to various other Management Systems.
- NABCB, QCI has accredited QMS Certification Schemes and Environmental Management System Certification Schemes of BIS.
- The Management Systems Certification Department at Head Quarter, New Delhi is the policy making department that also coordinates the Systems Certification activities from the country in a uniform and impartial manner as per ISO/IEC 17022.

- The Management Systems Certification activities are die-centralized to its 5 regional offices: the central region located in Delhi, the eastern region located in Kolkata, the northern region located in Chandigarh, the southern region located in Chennai and the western region located in Mumbai. The Deputy Director General (Region) is the granting, renewing and cancellation authority of the licence and this activity is coordinated by the Management Systems Certification Officer of the region.

Features of Management Systems Certification Scheme (FMSCS)

- BIS is the only organization in India that operates Management Systems Certification Scheme under an Act of Parliament.
- BIS represents the country in the international forum on standardization including ISO and IEC as National Standards Body (NSB) of India.
- Being member of ISO, BIS takes part in the deliberations of the TCs such as ISO/TC 176 and ISO/TC 207 which are responsible for the formulation of ISO 9000 series of standards and 14,000 series.
- ISO does not operate any certification scheme. However, it has compiled a directory of certification bodies world over and as expected, the name of BIS is included in the directory.
- BIS has large number of auditors and experts throughout the country leading to more economic operations of the certifications scheme.
- BIS has a wide network of regional offices and branch offices for providing timely and efficient and economical services.
- BIS is a non-profit organization and hence certification services are very cost-effective.

3.4.3.5 Services Through BARC Authorized Laboratory

BARC was conceived for the Nuclear Program in India by the initiatives taken by none other than Dr. Homi Jehangir Bhabha. In Dr. Bhabha's own words "*When Nuclear Energy has been successfully applied for power production in, say a couple of decades from now, India will not have to look abroad for its experts but will find them ready at hand*". In order to establish the self-reliance in all the fields related to nuclear science and engineering, BARC was founded with the base as *Atomic Energy Establishment, Trombay (AEET*. It also caters to the manpower needed for expanding the atomic energy research program. BARC is the base for the many nuclear R&D institutes like IGCAR, VECC, RRCAT, etc. with the aim to provide advancement in nuclear power production and related technologies. It provides pioneer research for many industries like NPCIL, ECIL, NFC, etc. BARC contains many research reactors like APSARA, ZERLINA, CIRUS reactor, DHRUVA, KAMINI, etc. [25]. BARC forms a very important part for national traceability as a DI for the ionization radiation. It provides the metrological services related to ionization parameters to the Indian industries and R&D institutes (Fig. 3.14).

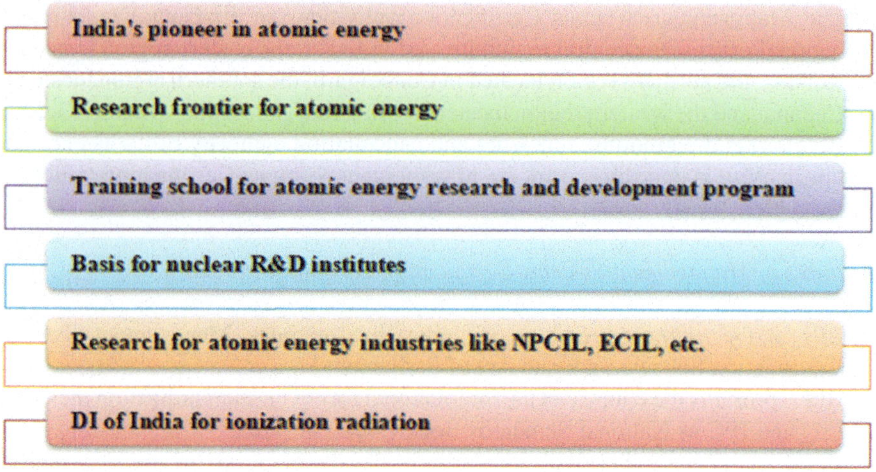

Fig. 3.14 BARC services

3.4.3.6 Services Through DRDO's Laboratory

As an agency of the Government of India, The *Defence Research and Development Organisation (DRDO)* has the responsibility for the military's research and development which has its headquarter in Delhi, India, and is under the administrative control of the Ministry of Defence, Government of India. By the merger of the Technical Development Establishment and the Directorate of Technical Development and Production of the Indian Ordnance Factories with the Defence Science Organisation, it was formed in 1958. It has a wide network for engaging and development of defense technologies consisting of 52 laboratories which encompass various fields, like aeronautics, land combat engineering, life sciences, materials, armaments, electronics, missiles, and naval systems (Fig. 3.15), DRDO is India's largest and most diverse research organization. It has a very wide organizational structure, comprising of many laboratories, institutes, etc. as shown in Table 3.4 [37].

DRDO has responsibilities for providing the services related to defense and its technologies to the nation. DRDO labs use various equipment under different clusters that are calibrated by CSIR-NPL to provide traceability continuously. Therefore it forms an inseparable part of the traceability chain for quality management of products by providing indirect metrological traceability to the nation in defense-related areas and technological fields.

3.4.3.7 Services Through Other Organizations

Apart from the above mentioned important organizations, there are several institutes which are part of the metrological traceability directly or indirectly. The main aim

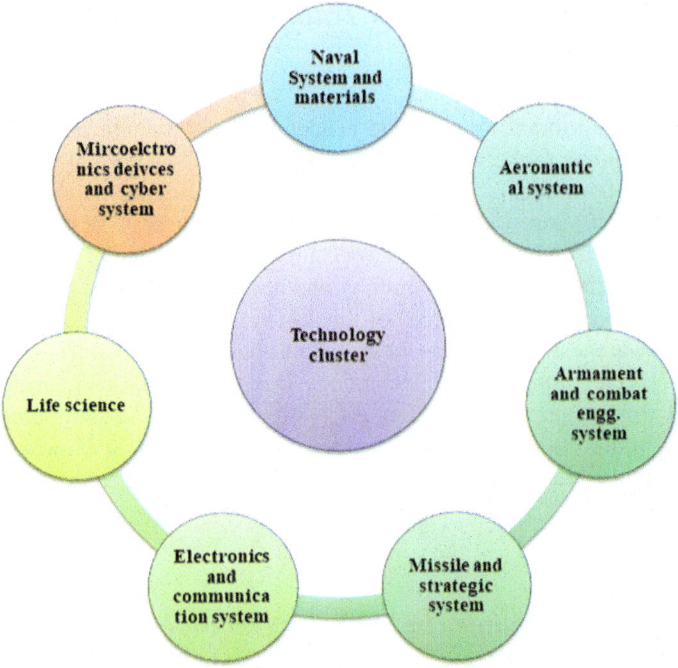

Fig. 3.15 Technology cluster at DRDO

Table 3.4 DRDO's different institutes and their corresponding numbers

Institute	Numbers
Laboratories	52
HR Institutes	3
Other institutes	8
Centre of excellence	10

is to maintain the product quality and removal of technical barriers for international and national trade of goods and services. These organizations include but are not limited to IAF, ITU, IEC, ILAC, ISO, OIML, *International Trade Centre (ITC), United Nations Economic Cooperation for Europe (UNECE), World Bank Group (WBG), WTO, UNIDO,* etc. There are many other Indian organizations, described in Chap. 1, which are involved in metrological activities and mandatory or voluntary regulations and standardization.

3.4.4 Dissemination Through Training and Academic Activities

CSIR-NPL has a vibrant training and academic dissemination programmes of its metrology activities to the users of India, SAARC and visitors of other countries. Normally 25–30 training programmes on various metrological parameters are organized, imparting training to 80–100 participants, annually at CSIR-NPL. The annual training calender is published on NPL website at the beginning of the financial year. A detailed description of various training and academic activities being undertaken are reported separately in Chap. 20 of this book. CSIR-NPL also disseminates the metrology awareness programme in collaborations with the *Metrology Society of India (MSI),* celebrating WMD, Technology Day, Open Day, and various Industry Meets and workshops. CSIR-NPL in association with MSI also organizes its flagship annual conference on Advances in Metrology (AdMet) series, which is widely attended by more than 300 participants from Academia and Industry, calibration, and testing laboratories.

3.4.5 Proposed Changes in Metrology Courses of Engineering

In engineering courses like Mechanical Engineering, the most important aspect of the curriculum is the practical and experiments carried in order to inculcate technological know-how and realization of the technical principles amongst the students. The basic fundamental steps in these experiments include postulate a theory, identify the main aim or objective, prepare a mathematical model, perform predefined sets of measurements within control environment, observe the outputs, analysis the output data and conclude results. If the output of the measurements is incorrect then it will lead to a wrong conclusion and whole experiment process will be rendered useless.

Accurate and precise measurements are the backbone of reaching a conclusive idea after an experiment. In engineering courses theory behind a concept, mathematical model and instruments used, all are emphasized, only leaving behind the importance of correct measurements and how to obtained measurement with quality. These are the hindrances to the growth of the innovative science and technology of the country. Even if some topics on metrology are covered, these are not adequate and are also not directed towards the quality-oriented measurements. Therefore, some revisions in engineering courses regarding metrology were proposed by CSIR-NPL to infuse the correct insights in the future torchbearers of technology of the country.

The existing course on metrology in the engineering syllabus mainly emphasizes the measurement of errors, dimensional metrology, transducers, limits, fits and tolerances, etc. In some universities, the topic metrology is covered as one or two Chapters in the courses like Manufacturing Technology, Instrumentation and Control, Measurement and Instrumentation, etc. However, the comprehensive description about the SI units derived units, role of NMI of the country, realization of measurement standards for the SI units and derived units, etc. is completely missing. As a general conclusion on the education of metrology and measurement science and theory in the bachelor in engineering programmes of the Indian universities, the introduction to the modern metrology in the frame of measurement science and knowledge of the NMI and National Measurement Standards, will offer the students a sound basis for further continuing studies in a Master programme, Research (e.g. Ph.D.), and shall equip them to face the challenges of engineering industries in Indian perspectives.

Based on exhaustive review of the engineering courses in metrology, the following recommendations were proposed for the metrology courses to educate the students with the latest technological know-how and make them more competent to face the growing challenges in the Indian industry:

- The Metrology Courses should specifically cover the metrological aspects on SI units, Derived units, Realization of Measurement Standards, Measurement Traceability, Accuracy and Precision, Introduction to international and national measurement system including standards, accreditation, legal metrology, Introduction to metrology: Definition, types, need of inspection, terminologies, methods of measurement, selection of instruments, measurement errors, units, measurements standards, calibration, statistical concepts in metrology, Basics of evaluation of uncertainty in measurements, measurement system analysis [38].
- The course work should provide at least introductory knowledge about the realization of standards and the measurement process and related instrumentation for Physico-mechanical parameters: Mass, Length, Temperature, Optical Radiation, Force and Hardness, Pressure, Vacuum, Acoustics and Vibration, Fluid Flow and Electrical and Electronic parameters: LF, HF Impedance and DC Metrology, LF, HF Voltage, Current & Microwave Metrology, AC High Voltage & Current Metrology and AC Power & Energy Metrology.
- The practical demonstration of the realization of measurement standards or some of the instruments like Verneir Calliper, Micrometer, Weighing balances, Densitometer, Load cells, Torque transducers, Thermocouples, Hygrometers, Sound level and Vibration meters, Instrumentation for measurement of illuminances, luminous intensity, capacitance, inductance, current, voltage, magnetic field, microwave, AC power, etc. shall be very helpful in nurturing the students with the technological know-how and train them to become professional in precision measurements and quality controls.
- The metrology course should also focus on the redefinition of four recently modified SI base units which are namely the kilogram, the kelvin, the ampere, and the mole. these are currently redefined in terms of constants; the new definitions are based on fixed numerical values of the Planck constant (h), the elementary charge (e), the Boltzmann constant (k), and the Avogadro constant (N_A), respectively.
- Also, the course work should cover the attributes of QMS as per ISO 17025 [34] and ISO 9001 so as to impart education about the standardization of products and enhancement of quality of manufacturing in industrial units in the Indian scenario.

- Furthermore, the students undergoing graduate and post-graduate in Engineering can explore the possibility of practical training at Regional Metrology Laboratories or at CSIR-NPL, New Delhi so as to impart training and hands-on experience on the various aspects of metrology and quality management system to engineering students. Also, the students can pursue their masters' or doctoral work at CSIR-NPL, New Delhi for enhancing their expertise in this area.

Accurate and precise measurements are essential to driving the growth engines of Indian and prompts innovations, which in turn, would save precious lives, resources and time. Today due to increasing pressure of demand of high-quality products as well as their global acceptability at the same time, Industries and production units are standardizing their products with ISO/IEC conformity assessments following several International Standards of quality control, namely; ISO/IEC 17025: 2017 (General Requirements for the Competence of Testing and Calibration Laboratories), ISO/IEC 17043: 2010 (Conformity assessment—General requirements for Proficiency testing), ISO 15189: 2012 (Medical laboratories—Requirements for quality and competence), ISO17034: 2016 (General requirements for the competence of reference material producers), etc. to overcome technical trade barriers, compete globally and also to cater international needs making their brand recognized internationally.

During interactions with Industry, it is emerged that knowledge of precision measurements and quality control is generally, available with senior management and Technical persons. However, the people at shop floor level who are involved in the day-to-day production and measurement do not have adequate knowledge of the subject. In today's competitive world, in order to maintain the quality of products and reduce the rejection ratio, it is essentially required that the staff actually working at shop floor level must have training exposure.

The young minds trained in this discipline would have opportunities to become professional in precision measurements and quality controls to serve industries, accredited testing and calibration laboratories, MSMEs, strategic sectors, and other related areas. They will have the unique advantage of getting exposure and hands-on training on best-measuring equipment through this course. The demand for skilled manpower in precision measurements in the manufacturing sector and quality sectors identified above will be increasing day by day. This way the trained manpower in precision measurements through the revised curriculum course would have ample job opportunity.

All the engineering institutes in India use different tools and techniques to teach students and test their skills. However, for years, many of them have neglected to revise their curricula to keep up with the rapid strides in technological advancement. Engineering courses in India are set to receive a syllabi revision that will make them contemporary enough to fulfill industry requirements. Thus, it is the need of the hour to encourage the amendments in the metrology course with the suggested recommendations so that the skilled manpower is generated which can take the growing technological challenges in the Indian industry.

3.4.6 Proposed Changes in NCERT Books

Keeping in view of the importance and recognition of newly defined SI units, CSIR-NPL has also submitted its recommendations to *National Council of Educational Research and Training (NCERT), New Delhi*, the central organization for qualitative improvement in school education system, to revise its textbooks adding new changes to propagate the globally accepted new definition of SI units among the students. The recommendations submitted are summarized in Table 3.5 for modifications in the textbooks of Physics of NCERT to encompass the contemporary revision in SI system of units and new definitions, specially to be incorporated in the NCERT textbook of Physics Part I, a Text book for Class XI (First Ed., Reprinted January 2011).

In addition to this, global metrology community also agreed upon to adopt the 'concept of uncertainty in measurement' to specify ambiguity in measurement results, rather than 'errors in the measurement'. Accordingly, it was suggested to include uncertainty determination procedure, '*Guide to the expression of uncertainty in measurement*', along with giving emphasis on errors in measurement in the text book.

3.5 A Novel Concept of Metrology Cycle

As described in Aswal model in Chap. 1 of this book and also published in [1], *metrology* is one of the most prominent and essential components of the NQI, apart from the standardization and the certification, and achieving SDGs of the UNIDO [3]. Taking clues from the model, a novel conceptual model is developed which is based on the cyclic phenomenon of the metrology, proposed to be called '*Metrology Cycle*', depicted in Fig. 3.16.

The complete metrology cycle comprises of all the components, be it scientific, legal or industrial metrology, for compatibility, legitimacy and utility of measurements, respectively. All such measurements together play a central role in driving the growth engine for any country. It addresses the central query, which often is raised by many curious minds: metrology for whom, why and how? The answer lies in the proposed *Metrology Cycle*. The straightforward answer is metrology, which is for human beings and the surrounding environment. It is performed for their improved life, trade and environment. However, the answer to 'how' is a bit tricky and is easily explained through the above conceptual diagram. The existence of human beings creates the needs, maybe food, shelter, education, health, finance, etc. These needs are qualitative and quantitative, both co-exist. Both will remain together forever. For example, qualitative requirements are fulfilled through assessment metrology, a bucket of water for volume, the amount of grain equivalent to apple, forecasting of weather looking at air direction and movement of clouds, etc., our traditional ways

Table 3.5 Proposed modifications or addition in NCERT book

S. No.	Existing content	Proposed content/comments
1	The SI, with standard scheme of symbols, units and abbreviations, was developed and recommended by General Conference on Weights and Measures in 1971 for international usage in scientific, technical, industrial and commercial work	The SI, with standard scheme of symbols, units and abbreviations, was developed and recommended by General Conference on Weights and Measures in 1971 for international usage in scientific, technical, industrial and commercial work However, very recently BIPM, France organized 26th meeting of General Conference on Weights and Measures (CGPM) on 16th November, 2018, and CSIR-NPL, India along with global metrology community, agreed upon to accept the landmark change for introducing new SI system based upon redefinition of four.SI base units, namely the kilogram, ampere, kelvin and mole
2	The kilogram is equal to the mass of the international prototype of the kilogram (a platinum-iridium alloy cylinder) kept at international Bureau of Weights and Measures, at Sevres, near Paris, France. (1889)	Definition: The kilogram, symbol kg, is the SI unit of mass. It is defined by taking fixed numerical value of the Planck constant h to be $6.626\ 070\ 15 \times 10^{-34}$ when expressed in the unit J-s, which is equal to kg $m^2 s^{-1}$, where the meter and the second are defined in terms of c and $\Delta \nu_{Cs}$
3	The ampere is that constant current which, if maintained in two straight parallel conductors of infinite length, of negligible circular cross-section, and placed 1 m apart in vacuum, would produce between these conductors a force equal to 2×10^{-7} N per meter of length (1948)	Definition: The ampere, symbol A, is the SI unit of electric current. It is defined by taking the fixed numerical value of the elementary charge e to be $1.602\ 176\ 634 \times 10^{-19}$ when expressed in the unit C, which is equal to As, where the second is defined in terms of $\Delta \nu_{Cs}$
4	The kelvin, is the fraction 1/273.16 of the thermodynamic temperature of the triple point of water (1967)	Definition: The Kelvin, symbol K, is the SI unit of thermodynamic temperature. It is defined by taking the fixed numerical value of the Boltzmann constant k to be $1.380\ 649 \times 10^{-23}$ when expressed in the unit J K^{-1}, which is equal to kg $m^2 s^{-2} K^{-1}$, where the kilogram, meter and the second are defined in terms of h, c and $\Delta \nu_{Cs}$

(continued)

Table 3.5 (continued)

S. No.	Existing content	Proposed content/comments
5	The mole is the amount of substance of a system, which contains as many elementary entities as there are atoms in 0.012 kg of carbon—12. (1971)	Definition: The mole, symbol mol, is the SI unit of amount of substance. One mole contains exactly $6.022\ 140\ 76 \times 10^{23}$ elementary entities. This number is the fixed numerical value of the Avogadro constant, N_A, when expressed in the unit mol^{-1} and is called the Avogadro number The amount of substance, symbol n, of a system is a measure of the number of specified elementary entities. An elementary entity may be an atom, a molecule, an ion, an electron, any other particle or specified group of particles
6	Mass is a basic property of matter. It does not depend on the temperature, pressure or location of the object in space. The SI unit of mass is kilogram (kg). The prototypes of the International standard kilogram supplied by the International Bureau of Weights and Measures (BIPM) are available in many other laboratories of different countries. In India, this is available at the National Physical Laboratory (CSIR-NPL), New Delhi	It is defined by taking fixed numerical value of the Planck constant h to be $6.626\ 070\ 15 \times 10^{-34}$ when expressed in the unit J-s, which is equal to kg m^2s^{-1}, where the meter and the second are defined in terms of c and $\Delta\nu_{Cs}$
7	In India, this is available at the National Physical Laboratory (CSIR-NPL), New Delhi	In India, this is available at the National Physical Laboratory (CSIR-NPL), New Delhi, which is recently redefined by taking fixed numerical value of the Planck constant h to be $6.626\ 070\ 15 \times 10^{-34}$
8	The efficient caesium atomic clocks are so accurate that they impart the uncertainty in time realization as 1×10^{-13}, i.e. 1 part in 10^{-13}. This implies that the uncertainty gained over time by such a device is less than 1 part in 10^{-13}; they lose or gain no more than 3 μs in one year	The efficient caesium atomic clocks are so accurate that they impart the uncertainty in time realization as 1×10^{-15}, i.e. 1 part in 10^{15}. This implies that the uncertainty gained over time by such a device is less than 1 part in 10^{-15}; they lose or gain no more than 32 ns in one year
9	Measurement is the foundation of all experimental science and technology. The result of every measurement by any measuring instrument contains some uncertainty. This uncertainty is called error	This whole section has to be revised in context with 'Guide to the expression of uncertainty in measurement' [39]
10	Any other section of the book that might be revised in line with the current redefinition of SI units	In the context of redefinition of SI units based upon fundamental constants of nature and [21, 40–43]

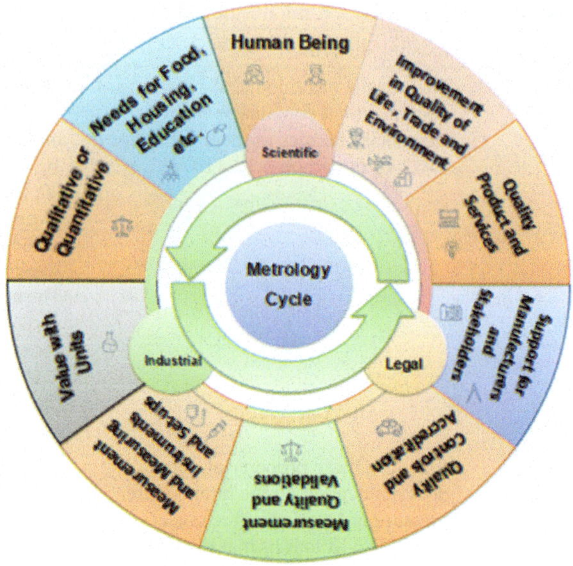

Fig. 3.16 Concept of metrology cycle

of assessment metrology. These methods are still in use in our countryside in-spite of adopting the new SI system.

When the requirement is quantified or made quantitative, it is addressed through 'quantity' in terms of 'value' and 'units'. The exact quantity can only be judged performing 'measurements' using a 'measurement set-up', comprising of one or more 'measuring instruments/devices/sensors' by trained/skilled operator(s). In order to judge the operational integrity of the measurements performed, 'quality control and validation' is essential. Operational integrity of the instruments is judged through a process 'calibration' and established measurement 'traceability'.

For the recognition of the procedural methods, actions or processes for performing measurements for having a particular status or being qualified to perform a particular activity is done through implementation of QMS and approval by BIPM through '*Peer Review*', specially for NMIs and DIs or 'accreditation' by '*National Certification Body (NCB)*'. However, for the "Peer Review' or accreditation process, additional requirement is successful participation in 'Key Comparisons' and Inter Laboratory Comparisons (ILCs)/Proficiency Testing (PTs) exercises, respectively. The Peer Review validates the global compatibility of the measurement standards, acceptability of measurement results and certificates issued and its outcome is published as CMC in KCDB of BIPM while outcome of the successful accreditation process is 'Accreditation Certificate' issued/granted to the applicants.

This overall process supports the manufacturers, MSMEs, users and stakeholders to produce quality products and services as well as improvement in the existing products, measurement results and services. Finally, the metrology cycle leads to improvement of quality of life, health, trade and environment of the human being. Thus the metrology cycle which starts from the needs of human beings comes to improved quality of life of the human being. Thereafter, the next cycle of the whole process starts with much improved NQI, financial status, academic and educational upliftment and this process goes on, new requirements come, new methodologies, technologies and systems are adopted and implemented.

In conclusion, the metrology cycle is an enabling tool for the R&D, innovation, development of technology, promotes global competence and confidence. The metrological advancements and advent of better quality products and services enable our industries to compete globally, remove trade impediments and constraints, and achieve greater demands for their products. This ultimately translates into growth for industries, specially when they are aiming for '*AtmaNirbhar Bharat*' through '*Make in India*' initiative, raising voice for '*Vocal for Local*', rapid industrialization, economic growth and societal upliftment.

3.6 Conclusions and Future Perspectives

In Chap. 2, the international measurement system and its progress with time, i.e. scientific development, harmonization, realization, etc. in measurement system have been described. On the basis of such advanced metrolological knowledge, essential and indispensable know-how is developed as the next-generation manufacturing technology. This is of utmost importance in the present time of worldwide international competition in industry and production engineering, specially for India wherein industries are aiming for import substitute technologies, as per Indian Government initiative of '*AtmaNirbhar Bharat*' or '*Self Reliant India*'. The metrology thus has to play a big role and certainly would lead the momentum for 'Make in India' products. The advance metrological information also helps to clear the confusion, doubts and reliability of Indian products and services and the stakeholders would be much more confident to voice for another govt. initiative '*Vocal for Local*'.

As per the mandate, CSIR-NPL maintains the "National Standards" of various parameters and disseminate the same to the strategic sectors, accredited laboratories, industries, etc. Although, presently CSIR-NPL's total of 236 registered CMCs are not adequate to serve the country's needs as India is a big, diversified and rapidly developing country. Further, the 236 CMCs are not sufficient in number as compared to other leading NMIs in the world. To improve the quality infrastructure of the country, CSIR-NPL should enhance CMCs many folds in terms of addition of new parameters and/or in terms of an increase in range of existing parameters. For this enhancement, CSIR-NPL needs sustained support, be it administrative, scientific, or financial for maintaining existing national standards as well as up-gradating them time to time at par to the other NMIs. It is well-known fact now that those nations which have developed more CMCs, their industries have grown much faster which ultimately leads to the overall growth of the country.

Apart from maintaining and disseminating the national standards of various parameters to the stakeholders, CSIR-NPL also disseminates knowledge through various training programmes, technologies transfer, technical/consultancy services, international/national conferences, etc. To provide trained manpower in metrology to the laboratories, industries; CSIR-NPL has recently started one year Post Graduate Diploma (PGD) course on "*Precision Measurement and Quality Control (PMQC)*". It also offers Ph.D. through AcSIR in various physical and engineering areas under the supervision of highly qualified and experienced scientists. After the implementation of redefinition of SI base units on 20th May 2019, CSIR-NPL had taken initiative as NMI and advised various academic organizations such as UGC, AICTE, NCERT, etc. to update the textbooks incorporating new definitions of SI base units based on physical constants to disseminate the updated knowledge at school, college and university levels. As a result, some of the academic organizations have already amended the same in the textbooks.

In India, NABL accreditation is provided to calibration, testing and medical laboratories on a voluntary basis. As per liberal estimate, India has a huge base of more than 4 lakh laboratories, both in organized and unorganized sectors, including testing, calibration, medical, pathological laboratories and radiology centres, but most of them are non-accredited. As a result, most of the cases, results of one laboratory do not match with other one which leads to wrong diagnosis and wrong treatment. Therefore, it would not be a bad idea if NABL accreditation could be made mandatory, rather than voluntary, for all the laboratories to improve the quality of services. At the same time, NABL needs support to create infrastructure to cater to this huge demand like adequate number of experts in respective fields (trained assessors), experts in accreditation boards, accreditation officers, data management systems, etc.

BIS is responsible for the preparation of documentary standards and more than 19,500 standards have been formulated so far. But, most of the Indian standards (ISs) are very old and based on specifications for manufacturers. There are very few ISs based on measurements (testing/calibration of measuring instruments and final products and its associated uncertainty). Now BIS is trying to align ISs with ISO and/or IEC published standards. On the other hand, in India, it is not yet mandatory to make any product following the respective IS. It would again be a good initiative if it is made mandatory for the manufacturers to make products based on ISs only and measurements should trace back to national standards maintained at CSIR-NPL for global acceptability of the Indian products and reduction of rejection rate in export.

As per BIS report, the annual consumption of gold is over 800 tonnes presently. Out of that, about 80% is for jewellery fabrication (mainly 22-carat purity) for domestic demand, 15% for investor demand, and 5% for industrial use. To assure the purity of precious metals, many countries have hallmarking scheme. In India also BIS has started hallmarking scheme for two precious metals namely gold and silver, but at present, hallmarking is optional. Because of that, customers are cheated by jewellers and Govt. also loses huge revenue. In India, hallmarking should be mandatory for fair transactions. To make hallmarking mandatory in India, number of recognized assaying and hallmarking centre also to be increased significantly. There will also be a huge demand for CRM of gold and silver in assaying and hallmarking centre. CSIR-NPL and India Govt. Mint, Mumbai have jointly developed BND® (CRM) of gold to cater to this demand.

LMD ensures the right quantity of any commodity for which consumers paid. LMD prepares and fulfils the requirements of Weights and Measures Act, 2009 and subsequent Legal Metrology Rules such as packaged commodities, approval of models, numeration, etc. Existing infrastructure is not enough to ensure the fulfilment of legal metrology rules for each and every product. LMD also needs improvement in infrastructure like adequate number of laboratories, manpower, etc.

However, all the Indian organizations are doing their best in their respective fields but still there is a big scope for improvement in coordination and relationship by increasing interaction among them. In some developed countries, national standards (scientific metrology) and legal metrology are looked after by the same organization, i.e. NMI for better synchronization.

Authors have expressed their views based on available data, global scenarios and years of experiences.

Acknowledgements Authors wish to express their whole hearted thanks to all the authors, organizations, members and students of CSIR-NPL past and present, whose works and activities are accommodated in this Chapter. Authors also place on record their thanks for all, even if the works of some of the organizations are not properly cited, which may be unintentional.

References

1. D.K. Aswal, Quality infrastructure of India and its importance for inclusive national growth. Mapan—J. Metrol. Soc. India **35**(2), 139–150 (2020). https://doi.org/10.1007/s12647-020-00376-3
2. World Trade Organization. https://www.wto.org/index.htm. Accessed 06 Jul 2020
3. Technical Barrier to Trade (TBT) Agreement Webpage. https://www.wto.org/english/docs_e/legal_e/17-tbt_e.htm. Accessed 06 Jul 2020
4. United Nations Industrial Development Organization. https://www.unido.org/who-we-are/unido-brief. Accessed 07 Jul 2020
5. Marking the 50th Anniversary of UNIDO, UNIDO-India cooperation, Vienna, 2016 Webpage. https://www.unido.org/sites/default/files/2017-02/MARKING_50_Years_UNIDO-wrap-up_brochure_0.pdf. Accessed 06 Jul 2020
6. The International Atomic Energy Agency (IAEA) Webpage. https://www.iaea.org/. Accessed 07 Jul 2020
7. IAEA Safety Standards Webpage. https://www.iaea.org/resources/safety-standards. Accessed 07 Jul 2020
8. World Meteorological Organization (WMO) Webpage. https://public.wmo.int/en. Accessed 07 Jul 2020
9. WMO standards on various parameters webpage. https://public.wmo.int/en/resources/standards-technical-regulations. Accessed 06 Jul 2020
10. Various WMO projects on Aeronautical, Agricultural, flood managements, etc. webpage. https://public.wmo.int/en/projects. Accessed 06 Jul 2020
11. The International Organization for Standardization, ISO webpage. https://www.iso.org/home.html. Accessed 17 Jul 2020
12. International Electrotechnical Commission (IEC) Webpage. www.iec.ch. Accessed 06 Jul 2020
13. International Organization of Legal Metrology (OIML) Webpage. https://www.oiml.org/en. Accessed 06 Jul 2020

14. OIML Publications Webpage. https://www.oiml.org/en/publications/introduction. Accessed 07 Jul 2020
15. International Laboratory Accreditation Cooperation (ILAC) Webpage. https://ilac.org/. Accessed 01 Jul 2020
16. International Accreditation Forum (IAF) Webpage. https://www.iaf.nu/. Accessed 01 Jul 2020
17. Asia Pacific Metrology Programme (APMP) webpage. http://www.apmpweb.org/. Accessed 01 Jul 2020
18. The Asia-Pacific Legal Metrology Forum (APLMF) webpage. https://www.aplmf.org/. Accessed 01 Jul 2020
19. Asia Pacific Accreditation Cooperation (APAC) webpage. www.apac-accreditation.org. Accessed 01 Jul 2020
20. SARSO published standards webpage. https://sarso.org/published-saarc-standards/. Accessed 01 Jul 2020
21. JCGM 200:2012, International vocabulary of metrology – Basic and general concepts and associated terms (VIM), 2012.
22. A.R. Verma, *National Measurements and Standards* (Dev Press Photocomp, 1987)
23. CSIR-NPL Diamond Jubilee Volume, CSIR-NPL Webpage. http://www.nplindia.in/data/npl djvsec.pdf. Accessed 07 Jul 2020
24. The Legal Metrology (National Standards) Rules, 2011. https://consumeraffairs.nic.in/acts-and-rules/legal-metrology/the-legal-metrology-act-2009. Accessed 02 Jul 2020
25. Official Website of Bhabha Atomic Research Centre. http://www.barc.gov.in/. Accessed 12 Apr 2020
26. Bhartiya Nirdeshak Dravya. http://www.nplindia.in/bhartiya-nirdeshak-dravya-bnd-indian-ref erence-materials. Accessed 28 Feb 2019
27. Official Wesbsite for BIS. https://bis.gov.in. Accessed 04 Mar 2019
28. Calibration and Measurement Capabilities (CMCs), A paper by the joint BIPM/ILAC WG (2007), **7**(May), 1–6 (2007)
29. ISO Guide 31: Reference Materials, Contents of Certificates and Labels (2000)
30. ISO Guide 30:1992/Amd 1:2008, Terms and definitions used in connection with reference materials (2008)
31. Measurement comparisons in the CIPM MRA, CIPM MRA-D-05 Version 1.6. https://www.bipm.org/utils/common/documents/CIPM-MRA/CIPM-MRA-D-05.pdf. Accessed 12 Apr 2019
32. Department of Consumer Affairs. https://consumeraffairs.nic.in/organisation-and-units/div ision/legal-metrology/weights-and-measures. Accessed 12 Apr 2019
33. CIPM MRA Website link. https://www.bipm.org/en/cipm-mra/. Accessed 29 Mar 2019
34. ISO/IEC 17025, General requirements for the competence of testing and calibration laboratories (2017)
35. Official website of Quality Council of India. https://www.qcin.org/index.php. Accessed 12 Apr 2019
36. Quality Infrastructure's Definition adopted in June 2017 by INetQI (then DCMAS Network: BIPM, I A F, IEC, ILAC, ISO, ITC, ITU, OIML, UNECE and UNIDO) + the World Bank.
37. Official Website of DRDO. https://www.drdo.gov.in/home. Accessed 02 Jul 2020
38. BIPM et al., Evaluation of measurement data—Guide to the expression of uncertainty in measurement (2008)
39. I.O., BIPM, IEC, IFCC, ISO, IUPAC, Evaluation of measurement data—Guide to the expression of uncertainty of measurement (2008). https://doi.org/10.1373/clinchem.2003.030528
40. Draft Resolution A on the revision of the International System of Units (SI), BIPM webpage. https://www.bipm.org/utils/en/pdf/CGPM/Draft-Resolution-A-EN.pdf. Accessed 02 Jul 2020
41. Resolution 1 of the 26th CGPM (2018), BIPM Webpage. https://www.bipm.org/en/CGPM/db/26/1/. Accessed 02 Jul 2020
42. Unit of mass (kilogram) BIPM Webpage. https://www.bipm.org/en/CGPM/db/24/1/. Accessed 02 Jul 2020

43. Resolution 1 of the 24th CGPM (2011), On the possible future revision of the International System of Units, the SI Webpage. https://www.bipm.org/en/CGPM/db/24/1/. Accessed 02 Jul 2020

Chapter 4
Time and Frequency Metrology

Part I: Realization and Dissemination of Indian Standard Time

Bharath Vattikonda, Manoj Das, Trilok Bhardwaj, Subhasis Panja, Poonam Arora, Anurag Gupta, and D. K. Aswal

Abstract Time is one of the seven base physical quantities in the SI system. In the modern world, time not only forms an essential part of day-to-day activities, but many sectorial applications related to navigation, telecommunications, and internet, power grid, banking, digital governance, transport, finance, defense, upcoming technologies of 5G, artificial intelligence, and internet of things rely on time synchronization with second to nanosecond accuracy level. Any level of time synchronization is a continuous process and requires a common national reference time that is part of the unbroken chain of international SI traceability. In many developed nations, the National Measurement Institute (NMI) provides such a common national reference time. CSIR-NPL, the NMI of India, realizes the Indian Standard Time (IST) with a timescale system consisting of bank of atomic clocks and other precision measurement instruments. Highly precise satellite links are used to keep IST traceable to UTC (coordinated universal time) within 3 ns uncertainty. Over the years, CSIR-NPL has gained knowledge and knowhow to disseminate time with various time transfer

B. Vattikonda · M. Das · T. Bhardwaj · S. Panja · P. Arora · A. Gupta (✉) · D. K. Aswal
CSIR-National Physical Laboratory, Dr. K. S. Krishnan Marg, New Delhi 110012, India
e-mail: anurag@nplindia.org

B. Vattikonda
e-mail: bharath.v@nplindia.org

M. Das
e-mail: manoj.das@nplindia.org

T. Bhardwaj
e-mail: trilok@nplindia.org

S. Panja
e-mail: panjas@nplindia.org

P. Arora
e-mail: arorap@nplindia.org

D. K. Aswal
e-mail: dkaswal.npl@nic.in

D. K. Aswal (ed.), *Metrology for Inclusive Growth of India*,
https://doi.org/10.1007/978-981-15-8872-3_4

technologies. Currently, it possesses eleven calibration and measurement capabilities related to time and frequency parameters, at par with international standards. In addition, rigorous R&D activities are pursued on time transfer techniques and advanced atomic clocks like Cesium fountain and single ion trap based optical clock for more accurate and highly stable measurements of time and frequency. Metrological services in time and frequency domain viz. calibration, traceability, training, and indigenously developed products are provided to various aforementioned sectors and industries. For Pan-India time synchronization, two national metrological programs are being carried out to provide IST traceability to ISRO's NavIC (Navigation with Indian Constellation) system and secondary timing centers at five locations for the ministry of consumer affairs (MoCA).

A careful analysis of the process of observation in atomic physics has shown that the subatomic particles have no meaning as isolated entities, but can only be understood as interconnections between the preparation of an experiment and the subsequent measurement.

– Erwin Schrodinger (1887-1961)

4.1 Introduction

Second is one of the SI units which is measured with the highest precision among other SI units and is required to realize five other SI units namely Meter, Kilogram, Ampere, Candela, and Kelvin. It is not possible to measure the exact time but time interval can be measured with highest precision than any other unit. Frequency, with unit 'Hz', is a derived quantity that is reciprocal of the time interval/period. The main components of any clock are the oscillator which is a frequency source generating stable periodic oscillations, a counter which counts the oscillations, and a display that shows and advances the time based on the increments in the counter. Continuous research efforts in the field of timekeeping show that there is much more to simply keep time with a clock. The chapter begins with the history of clocks from early ages till modern times.

Over the last six decades, timekeeping has shifted from mechanical clocks to atomic clocks and the timekeeping accuracies have improved by almost eight orders (few μs to ps) of magnitude. One may wonder about such a phenomenal growth in timekeeping accuracies when ms-level accuracy is sufficient for day-to-day activities. Rapidly growing technological applications like communication networks, satellite navigation, deep-space exploration, digital financial markets, etc. require ns-level or better time synchronization. Global timekeeping and time synchronization with nanosecond accuracy is maintained by BIPM which collects data from about 500 atomic clocks distributed over national metrology institutes (NMIs) and designated institutes (DIs) in more than 70 countries including India.

CSIR-NPL is the NMI of India and authorized (by an act of Parliament) to realize and maintain the **Indian Standard Time (IST™)**, which is kept traceable to coordinated universal time (UTC; international reference time) with an uncertainty of 3 ns. Dissemination of IST across Indian subcontinent is done through calibration, network time service, and satellite links. R&D is also pursued in the area of advanced atomic clocks and various time transfer techniques. The details on the realization of IST, its dissemination through various techniques, CMCs, and R&D efforts in T&F metrology at CSIR-NPL have been summarized in Sect. 4.3.

The present contribution of CSIR-NPL in supporting the industry and various sectors through calibration, traceability, training, consultancy services, and product development, in the domain of Time and Frequency Metrology, have been discussed in Sect. 4.4. Special emphasis has been given to discuss the national level time synchronization programs taken up by CSIR-NPL.

A detailed roadmap, required for the legalization and implementation of IST in India, will be dealt with separately in Chap. 5, which is Part 2 of Time and Frequency Metrology. All the efforts towards pan-India IST synchronization, legalization, and implementation of IST will lead to significant improvement in the efficiency and reliability of various sectors as well as a measurable economic impact.

4.2 Atomic Clocks and Timekeeping

Energy difference between designated levels of an atom and the corresponding frequency provides the precise reference for making a highly accurate clock. Timekeeping and measurement of time intervals have been explored since early civilization till date. Especially, since the middle of the last century till now, there have been tremendous research and development efforts in the field of atomic timekeeping which lead to making time the most accurately measured physical quantity. This chapter covers the history of timekeeping with celestial objects in early times till present-day atomic era of timekeeping. A detailed discussion on worldwide time synchronization with the contribution of atomic clock data from various NMIs located in different countries has been included.

4.2.1 History of Time Measurement

Time interval, one of the seven fundamental physical quantities in the International System of Units (SI), is the quantity that can be measured with highest accuracy (part in 10^{16} and still improving) and best resolution [1]. The accuracy of timekeeping

Fig. 4.1 Basic elements of any clock

with modern atomic clocks is such that these will not lose or gain a second over age of the universe. Another way to appreciate the impact of improving timekeeping accuracies is that the distance between the earth and the moon can be measured with an uncertainty of few cm with the help of atomic clocks. For understanding the basic concepts about time and frequency metrology, the readers are referred to [2].

Precise time measurement is essential for a number of systems which include navigation, telecommunications and internet, power grid, banking, digital governance, transport, finance, defence, cyber-physical systems, and upcoming technologies of 5G, artificial intelligence, and internet of things, etc. Precise time with ns accuracy is also required for high-end research in science, namely, measurement of fundamental physical constants, detection of the gravitational waves, geodesy, deep space navigation, radio telescopes, etc. Detailed analysis of time synchronization requirements in various sectors is presented in Chap. 5, Sect. 5.2.2. The present section is dedicated to the evolution of timekeeping over the last few centuries.

Basic constituent elements of any clock to measure the passage of time, as shown in Fig. 4.1, are an oscillator (a device or process to designate equal increments of time) and a counter (for counting increments of time). An oscillator produces highly stable periodic oscillations and a counter is used to count the number of oscillations. In essence, the quality and accuracy of timekeeping with a clock is decided by the quality of its oscillator. The quality of an oscillator depends on its degree of stability and accuracy and its capacity to remain immune to environmental changes. Any instability and change in the frequency of the oscillator reflect in a corresponding instability and deviation in the timekeeping accuracy of the clock. Stability of an oscillator indicates how well it can produce the same frequency over a given period of time and is quantified by Allan deviation. Accuracy indicates how well set or how close is the frequency to the true value. Accuracy is quantified with a dimensionless quantity: ($\Delta f/f$) in the frequency domain and ($\Delta t/T$) in the time domain; where f is the nominal frequency, Δf is the difference between the measured and the nominal frequency, T is duration of the measurement and Δt is the measured phase/time difference.

4.2.1.1 Pre-atomic Era

In early clocks till the end of the middle ages, most of the clock designs, e.g., stone structures like Egyptian obelisks, Stonehenge, sundials were based on great movement great movements of the celestial objects like the Sun, Moon, and stars. Among other notable early clock designs were oil lamps, sand glasses, water flow towers, marked candles, etc. [3]. In India, Jantar Mantar at New Delhi and Konark temple in Odisha depict the use of sundial for time measurement. With the progress in civilization, a variety of modifications were made in clock designs, but a major breakthrough in the history of timekeeping came with the discovery of the process of oscillation by Italian researcher Galileo Galilei. There were mechanical clocks (based on balanced wheel, pendulum, etc. as oscillators), quartz clocks (based on periodically vibrating crystal as an oscillator), and finally atomic clocks (based on oscillating electromagnetic waves) were discovered [4].

In the pre-atomic era, timekeeping and standard frequency measurements were mostly based on mechanical and electrical oscillators. Bases on the principles of oscillations, discovered by Galileo Galilei, Dutch physicist Christian Huygens made the first practical pendulum clock in 1656. The daily error of this clock was of the order of 10 s, or a relative error of only 10^{-4}. After pendulum clocks, mechanical clocks were made in the eighteenth century. Harrison made a marine chronometer for the sailors and it was made with a spring and balance wheel escapement. Over the next century, free pendulum clocks were developed by Riefler and Rudd. But, Shrott's free pendulum clock became the best timekeeper in the nineteenth century. In the beginning of the twentieth century, free pendulum clocks were overtaken by Quartz crystal oscillators. Quartz crystal exhibits a unique phenomenon of piezoelectricity which was discovered by P. Curie in 1880. Quartz crystal resonates at a nearly constant frequency (nominally 32.768 kHz) when an electric current was applied. In 1929, American scientist Warren A. Marrison made the first quartz-crystal clock and it brought an improvement in timekeeping accuracy by an order of magnitude. A chronological advancement in clocks and improvement in timekeeping accuracies has been depicted in Fig. 4.2.

However, quartz oscillators are not suitable for making ideal frequency standards because their resonance frequency varies with the size and shape of the crystal which is man-made and no two crystals can exactly identical. Furthermore, the frequency of a quartz oscillator changes slowly over time due to aging, changes more rapidly due to the effects of environmental factors, such as temperature, humidity, pressure, and vibration [5]. Due to aging and environmental factors, the relative uncertainty of a quartz crystal oscillator cannot be improved beyond 10^{-9}. However, it is still widely used in everyday applications where such high accuracy is not needed like wristwatches, wall clocks, etc. Some better versions of the quartz clock such as oven-controlled crystal oscillator (OCXO) and temperature-compensated crystal oscillator (TCXO) are used for various precision test and measurement purposes in laboratories. The quest for finding oscillators with accuracy better than the quartz oscillator led to the development of atomic oscillators.

Atoms are the fundamental units of matter and have resonances that are stable over time and space. Each atom has designated energy levels with the fixed energy difference between the levels. A unique frequency corresponds to the energy difference and atom absorbs and emits electromagnetic radiation at its own 'characteristic transition frequency' also known as resonance frequency. It is this characteristic resonance frequency which is counted for the measurement of time interval in an atomic clock. Any unperturbed atomic transition is identical from atom to atom and thus, a

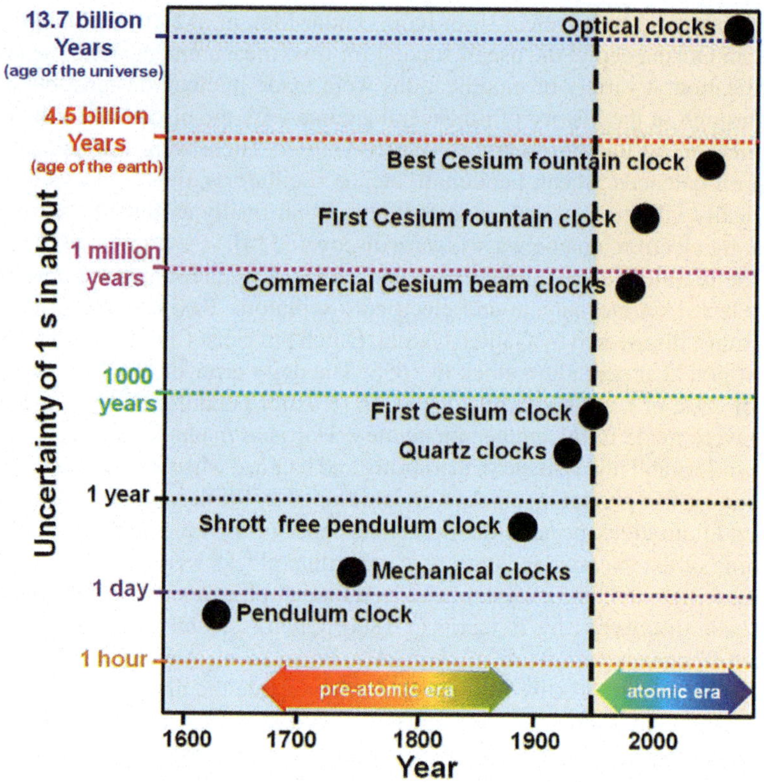

Fig. 4.2 Improvement in timekeeping accuracies over the last few centuries

group of atomic oscillators generates the same frequency, unlike that of a group of quartz oscillators. Also, unlike all electrical or mechanical resonators, atoms do not change their properties with time and position.

4.2.1.2 Atomic Clocks

Atomic clocks are based on the transition frequency between energy levels in a quantum system, which can be an atom, an ion, or a molecule. Due to the unmatched accuracy of timekeeping, the atomic clocks are the best means of measuring time interval and standard frequency. In an atomic clock, a suitably selected atomic species is interrogated with an external oscillator called local oscillator (LO). Error in the LO frequency is generated using a detection scheme and a feedback/servo loop periodically corrects the LO frequency until it becomes equal to the atomic resonance frequency. LO output locked to the atomic transition embodies the stability of the atomic resonance. Thus, a frequency-divided output of the LO provided the accurate clock output or ticks as shown in the schematic of the atomic clock in Fig. 4.3. A

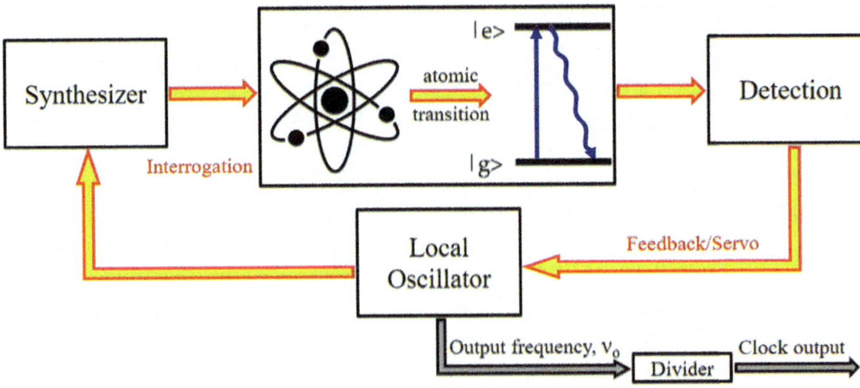

Fig. 4.3 Schematic of a basic atomic clock

comparison of the different frequency standards is given in [6]. For the comprehensive knowledge in the field of atomic clocks, the readers are referred to the two-volume book on atomic clocks and atomic frequency standards by Vanier and Audoin [7].

With the tremendous improvement in timekeeping accuracy obtained using an atomic standard of time, the existing astronomical definition of time (based on earth's rotation) was changed to a very precise definition of the unit of time which is still valid. The SI second was redefined in the 13th General Conference of Weights and Measures (CGPM) (1967/68, Resolution 1; CR, 103 and Metrologia, 1968, 4, 43) as: The second is the duration of 9 192 631 770 periods of the radiation corresponding to the transition between the two hyperfine levels of the ground state of the cesium 133 atoms. In the CIPM meeting held in 1997, it was affirmed that this definition refers to a Cesium atom at rest at a temperature of 0 K.

There are many reasons for selecting Cesium to define the SI second. The frequency corresponding to the hyperfine transition that defines the second is relatively high and with a suitable design, the linewidth can be made quite narrow. The transition frequency lies in the microwave region and could be manipulated and measured using existing standard microwave techniques and circuits [8]. Additionally, Cs-133 is the only stable isotope which ensures that all atoms in the beam have the same hyperfine structure. The hyperfine resonance frequency of 9.19 GHz is relatively high compared to that of the other alkali atoms [9]. Moreover, Cesium atoms are relatively heavy (133 amu) and move at a relatively slow speed of about 130 m/s at room temperature allowing cesium atoms to stay for a longer duration in the interaction zone. Hydrogen atoms, also used for making atomic frequency standards, travel at a speed of about 1600 m/s at room temperature [10]. Cesium is readily available and non-poisonous too. For the standards based on laser-cooled Cesium atoms known as Cesium fountain clocks, Cesium has a rather simple level scheme, can be easily manipulated with laser beams, and the 852-nm wavelength of its D2 line falls into the range where high power, narrow linewidth laser diodes, and efficient silicon photodiodes are easily available [11]. Cesium is chosen for defining

SI second, transitions in rubidium and hydrogen are also commonly used as reference for secondary representation of the SI second. Rubidium clocks are mostly used for satellite payloads while hydrogen masers (active as well as passive) are used for precision timekeeping along with commercial Cs clocks.

Maxwell, Thomson, and Tait proposed the idea of using atomic resonance to measure time in 1873 [12]. Many years later, in 1945, Rabi suggested that atomic beam magnetic resonance can be used as the basis of a clock [13]. The separated oscillatory field method, invented for measuring the resonance of hydrogen molecular beam Ramsey and Silsbee in 1947, improved the accuracy of the transition frequency determination [14]. The atomic timekeeping began in 1949 with the design of the first atomic clock. This clock was based on resonance in the ammonia molecule and was constructed at the National Institute of Standards and Technology, USA. The ammonia-based standard was quickly replaced by the cesium-beam frequency standard [15].

First-ever Cs atomic clock based on the hyperfine transition of the Cesium-133 atom was built by Louis Essen and JVL Parry at National Physical Laboratory (NPL), the UK in 1955 and it had a relative uncertainty of 1×10^{-9} [16]. The NPL clock became the world's first cesium standard which was used for the calibration of secondary working frequency standards [17]. This design of the Cesium clock is also known as the thermal beam clock. It consisted of an oven to produce beam of Cs atoms, state selecting magnets, microwave cavity for interaction with atoms, and state detection at the end. It was operated in a "closed-loop" mode with an external microwave interrogation oscillator locked to the cesium hyperfine transition frequency using feedback and the servo lock mechanism. The details on different types of atomic standards are given in [18, 19]. After the realization of first thermal beam Cesium clock at NPL UK, several attempts were made in the design of the clock to improve the accuracy for the next four decades. First commercial Cs clock named "Atomichron" was manufactured in 1956 [20]. Hewlett-Packard (HP) also started making the commercial Cs clocks by 1964 [21]. The improved version of commercial Cs clock HP 5071 with accuracy of 1 in 10^{12} was later manufactured by Agilent and subsequently with improved accuracies by Symmetricom Inc., Microsemi, and now Microchip. Research efforts at many laboratories around the world resulted in the successful commercialization of Cs clocks and in consequent continuous improvisation in the clock design to increase the accuracy. NIST, USA developed six different versions of thermal beam clocks (named NBS-1 to NBS-6) with longer cavity lengths and velocity selective atomic selection which resulted in successively increasing accuracy. NIST used the novel technique of optical pumping in NIST-7 which reached an accuracy of 5×10^{-15} [22]. NIST-7, the last Cesium beam standard developed at NIST, served as the primary standard till 1998 when it was replaced by the first fountain clock developed at NIST. Apart from NIST, high-accuracy thermal beam clocks were also developed at leading NMIs in Germany, France, UK, Canada, etc. before Cs fountain took over as the primary frequency standard in the late 1990s.

4.2.1.3 Atomic Fountain Clocks

Laser cooling and trapping techniques were demonstrated and explained by Chu, Cohen-Tannoudji, and Phillips in the 1990s [23–25]. The Doppler Effect, which limited the performance of the thermal beam clocks, could be significantly reduced by using laser cooling of atoms. Jerrold Zacharias was the first to propose the concept of a cesium fountain 1950s, much before laser cooling techniques were invented [26]. He proposed to build a vertical cesium beam clock with one Ramsey interaction zone. As per his design, slow atoms from the cesium oven would traverse the microwave interaction zone on the way upward and traverse the same microwave interaction zone while falling downward under the influence of gravity. He suggested that dual interactions with the microwaves would reproduce Ramsey's separated oscillatory field method, and interaction times can approach one second (instead of the 10 ms typical of thermal beam clocks up to that time) if the Cesium beam travels a meter upward on ballistic flight. IN late 1980s, Steven Chu and coworkers at Stanford experimentally implemented Zacharias' idea by incorporating a laser cooling technique for cooling Cesium atoms in the atomic clock [27]. Finally, the first Cesium atomic fountain primary frequency standard, initially proposed by, was developed and operated in 1995 at LNE-SYRTE in Paris [28]. With this fountain clock, a fractional frequency uncertainty of 10^{-15} was obtained. Implementation of the fountain concept increased the interrogation time of atoms with microwaves which was limiting the accuracy in thermal beam clocks. Since 1995, the Cesium fountain primary frequency standards have been developed at about ten NMIs around the world. With continuous research and design improvisations, fountain accuracy has been improved by about an order of magnitude and is mainly limited by the quantum projection noise.

Apart from the Cesium fountain, Rubidium-based fountains have also been realized in France and Russia and they contribute to TAI as a secondary representation of SI second. The international atomic time (TAI) is a weighted average of time kept by atomic clocks in more than 70 national laboratories and designated institutes around the world. Cesium fountains are the primary standards and provide the most accurate measurement of SI second. Due to this reason, they get the highest weight for the calibration of TAI. In order to contribute to the maintenance of the global timescale with the best possible accuracy, it is essential to increase the number of national primary frequency standards contributing to TAI. Presently, around ten Cs atomic fountain clocks contribute to the international atomic time at a given time [29]. Worldwide, there are only 10 countries to date that has indigenously developed Cesium fountain clocks and sent their data for the calibration of TAI. These countries are France, the USA, UK, Germany, Italy, Japan, India, Russia, China, and Canada whereas South Korea and Mexico are still developing their Cs fountains [30–43]. India's first-ever Cesium fountain primary frequency standard, NPLI-CsF1 was completely designed and developed at CSIR-NPL. It was successfully operated, evaluated, and was approved as a primary frequency standard by BIPM. Details on the development of NPLI-CsF1 and NPLI-CsF2 are given in the Sect. 4.3.5.1.

4.2.1.4 Future Frequency Standards

Present primary atomic frequency standard is based on a microwave transition in Cesium atom. It is likely to be replaced with optical clocks based on a forbidden transition in the optical frequency region. With similar linewidth as for microwave frequency standards, optical clocks have much higher resonance frequency (10^{15}) and thus can be accurate to 10^{-18} level. Optical atomic clocks based on single trapped ion or neutral atoms trapped in optical lattices are being developed at many NMIs around the world and they have been outperforming the best Cesium clocks in terms of accuracy as well as stability [44–47]. These developments will eventually lead to the redefinition of SI second on the basis of optical clocks. This redefinition may happen as early as 2026. As optical frequency standards will replace the Cs fountain as primary frequency standards in a few years, India's first-ever optical clock is being developed at CSIR-NPL. With the ongoing efforts on the development of an optical frequency standard based on single trapped Ytterbium ion, CSIR-NPL is gearing up for the proposed redefinition of SI second [48, 49]. More details on the development of the single trapped Ytterbium ion optical clock at NPL are given in Sect. 4.3.5.2.

4.2.2 Global Timekeeping

4.2.2.1 International Atomic Time (TAI)

BIPM along with the support of the metrology community of the world maintains international timescales. International Atomic Time (TAI) and Coordinated Universal Time (UTC). TAI is retrospectively computed at BIPM from the data from more than 450 atomic clocks distributed over more than 70 National Metrology Institutes (NMIs) and designated institutes (DIs). BIPM uses an algorithm to compute the weighted average of the data provided by the participating institutes to produce free-atomic timescale known as EAL. This timescale is then compared with data from the primary frequency standards and a correction is applied to keep it as close as possible to SI second realized with primary standards. The resulting corrected timescale is TAI. TAI is a uniform and stable scale which does not keep in step with the slightly irregular rotation of the Earth. The timescale with astronomical definition based on rotation of earth is known as UT1.

4.2.2.2 Coordinated Universal Time (UTC)

UTC is derived from TAI by adding an integral number of seconds that compensate for the irregular rate of rotation of the Earth. Most of the countries have a National Measurement Institute (NMI) or a Designated Institute (DI) for keeping national time through atomic clocks and to keep it traceable to UTC. These NMIs or DIs, including NMI of India (CSIR-NPL) contribute to UTC by providing their atomic

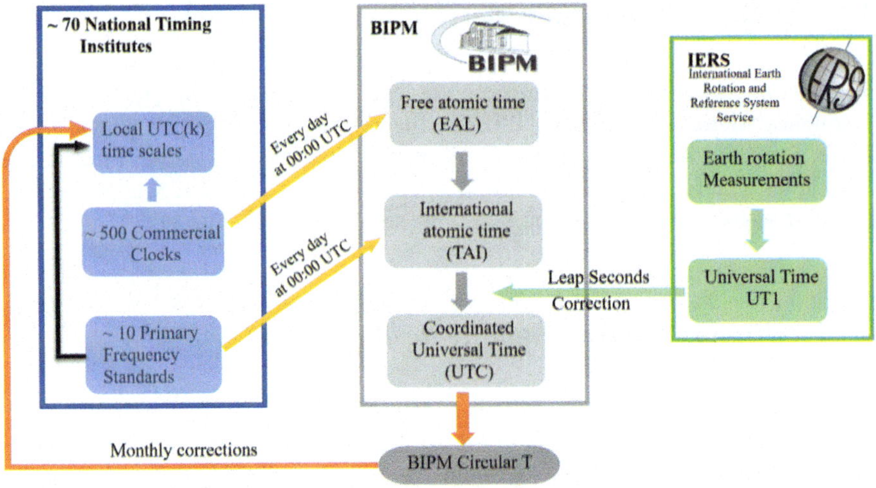

Fig. 4.4 Global timekeeping at BIPM with atomic clock data from various countries and leap second corrections provided by IERS

clock data to BIPM on a daily basis. As shown in Fig. 4.4, BIPM collects the data sent by almost 70 countries, applies rigorous statistical tools (by assigning rate and weight to each contributing atomic clock), and generates a universally synchronized timescale named UTC. UTC is identical with TAI except that from time to time a leap second is added to ensure that the difference between UTC and UT1 remains within 0.9 s. The dates of application of the leap second are decided by the International Earth Rotation Service (IERS). The local realization of UTC at laboratory 'k' is named UTC(k). UTC (NPLI) is the local realization of UTC at CSIR-NPL. On a monthly basis, BIPM issues the scientific bulletin 'Circular T' with deviation and uncertainties of UTC-UTC(k). This data is used by NMIs and DIs for correcting their local timescales.

After the implementation of CIPM MRA in 1999, CCTF, in 2006 assigned computation of UTC as a unique key comparison in time and frequency field. The name of the key comparison is "CCTF-k001.UTC" with UTC as key comparison reference value [50]. Each month, after the publication of BIPM Circular T, the degrees of equivalence of participants 'k' i.e. [UTC-UTC(k)] are published in KCDB (Key Comparison Data Base).

Post realization of UTC (k), standard time zone difference (on the basis of longitude) is either added or subtracted in order to derive the local time at a certain place. For example, time zone for India is UTC + 05:30. Indian standard time (IST) is thus derived by adding 05:30 h. to the local realization of UTC at CSIR-NPL known as UTC(NPLI) i.e. IST = UTC(NPLI) + 05:30 h. More details on the realization of IST are covered in Sect. 4.3.1.

4.3 Time and Frequency Metrology at CSIR–NPL

This section describes the primary activities carried under the umbrella of Time and Frequency Metrology at CSIR-NPL. Detailed description has been given on how IST is generated with a primary timescale and disseminated via various time transfer techniques. Calibration procedures and activities are discussed. A brief description of major ongoing R&D activities is also presented in this section.

4.3.1 Primary Standard and Realization of IST

As discussed in Sect. 4.2, the SI unit of "time" *i.e.,* "second" as defined in 1967 as "One second is the time required for 9 192 631 770 oscillations between doubly splitted hyperfine ground states of Cesium atoms (^{133}Cs)" [51, 52]. After the invention of the laser cooling technology for the atoms, several dominant systematic effects were further reduced. Hence in the year 1997, an amendment was introduced to the SI definition of second as "One second is the time required for 9 192 631 770 oscillations between doubly splitted hyperfine ground states of ^{133}Cs atoms *at rest at 0 K temperature*" [53]. Commercial Cesium clocks realize the SI second and are most widely used for precise timekeeping around the world. However, after the advent of laser cooling, SI second was realized more accurately using Cesium fountain based on laser-cooled Cesium atoms. Thus, Cesium fountain is recognized as the primary standard for the realization of SI second. Unlike, commercial Cesium clocks, Cesium fountain cannot be operated continuously and hence is not used for regular timekeeping. Cesium fountain data is however most accurate and is used for the calibration of TAI as depicted in Fig. 4.5. The frequency accuracy of TAI is improved by comparing its frequency with the primary frequency standard (PFS), and by applying frequency corrections if necessary. There are only about 15 Cesium fountains, developed and maintained in about 10 NMIs, have been contributing frequency measurements since 2007 [54].

International Bureau of Weights and Measures (BIPM), with the support of the timing community across the globe, maintain the international time scales *i.e.,* International Atomic Time (TAI) and Coordinated Universal Time (UTC) [55, 56]. TAI is an integrated time scale based on the weighted average of the contributing atomic clocks. An independent timescale is also realizing time on the basis of the rotation of earth and known as UT1. When the time difference between UT1 and TAI is more than 0.9 s, an additional second, i.e. leap second is added to TAI to compensate that time difference between the two-time scales. As already discussed, the reliability of TAI and UTC is ensured by about 500 commercial clocks (mainly cesium clocks and active and passive hydrogen masers) operated in NMIs and DIs in about 70 countries of the world.

CSIR-NPL, the National Measurement Institute (NMI) of India, is the custodian of national standards of measurements in India through the act of Parliament (Govt.

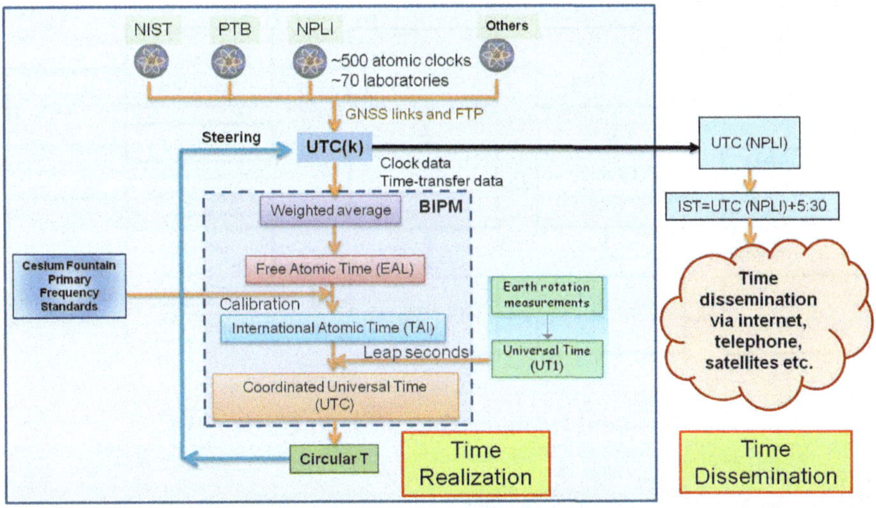

Fig. 4.5 Overview of realization of UTC at BIPM and UTC(k) by NMIs along with dissemination of IST realized at CSIR-NPL

of India Gazette No 589). The mandate of the institute is establishment, maintenance, and dissemination of the units of physical measurements based on the International System of Units (SI units). As per the mandate, the SI unit of time, i.e., second is realized, maintained, and disseminated to the entire nation as Indian Standard Time (IST) at CSIR-NPL. Due to this reason, CSIR-NPL, internationally known as NPLI, is also known as the timekeeper of the nation.

Based on the longitudinal offset from the prime meridian (0° longitude) passing through Greenwich, London, United Kingdom, the globe is divided into time zones. A time zone is a region of the globe that observes a uniform standard time for legal, commercial, and social purposes. Time zones tend to follow the boundaries of countries and their subdivisions because it is convenient for areas in close commercial or other communication to keep the same time. The local time within a time zone is defined by its offset (difference) from UTC. The offset is expressed as either UTC- or UTC + and the number of hours and minutes. Indian Standard Time (IST) is realized by adding ***5.30 h*** to UTC which corresponds to a geographical location at longitude 82.5° E which passes through Mirzapur, near Allahabad in Uttar Pradesh [57]. CSIR-NPL realizes UTC as UTC(NPLI) and thus the national time is UTC(NPLI) + 05:30 which is also known as IST.

CSIR-NPL realizes IST with a state-of-the-art timescale setup namely "primary timescale" consisting of an ensemble of atomic clocks (Cesium clocks and Hydrogen masers), a precision measurement system, and ultra-stable satellite-based traceability links. The cesium clocks provide absolute atomic reference of the time which has exceptional long-term stability, whereas the hydrogen maser has ultimate short-term stability. UTC(NPLI), the local representation of UTC, is realized as the steered output of an Active Hydrogen Maser (AHM), as shown in Fig. 4.6. However,

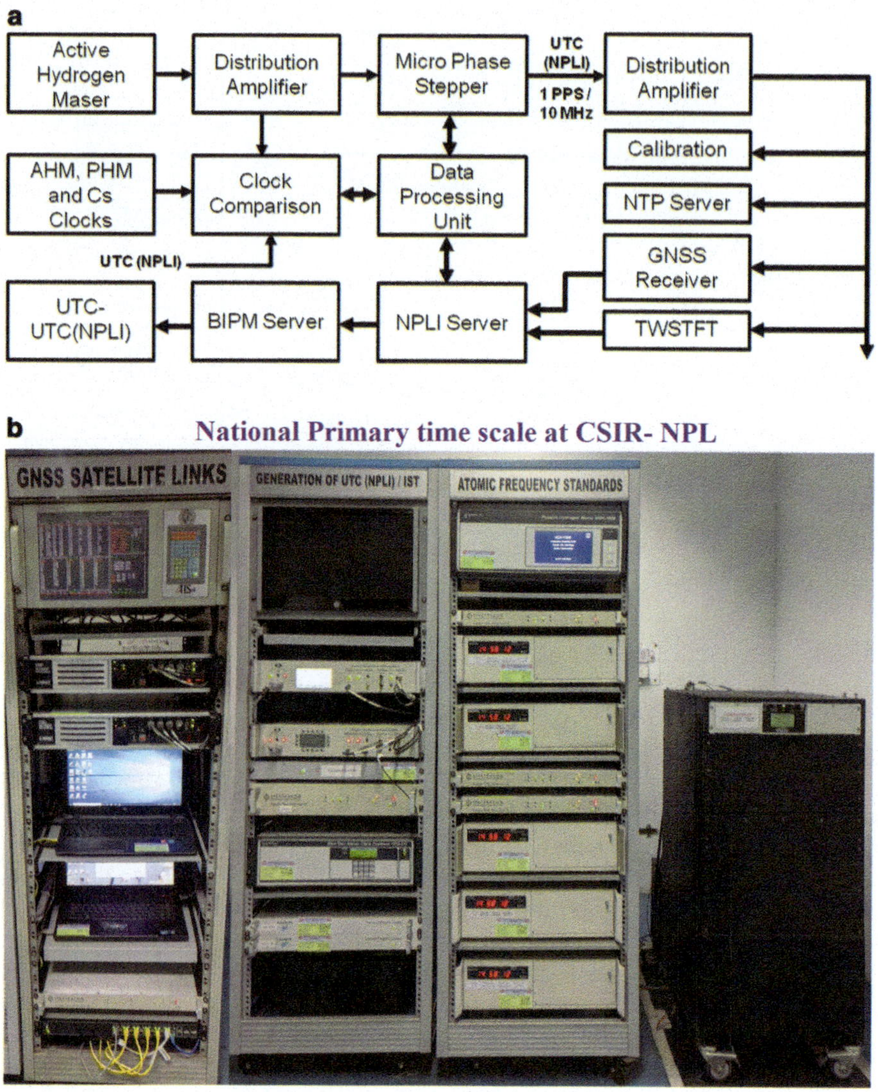

Fig. 4.6 (a) Schematic of the primary timescale and traceability link with BIPM (b) Primary time scale at CSIR- NPL

the timescale ensemble has five high-performance Cesium clocks and one passive Hydrogen maser (PHM) as well. All Cs and PHM clock output frequencies and steered output from microphase stepper are connected to a clock comparison unit which enables time-based switching of measurement of a pair of clocks through a frequency/phase comparator or a time interval counter.

An indigenous software algorithm has been developed for analyzing performances of each of the contributing clocks. The algorithm computes frequency correction based on the weighted average of all contributing clocks in the timescale ensemble. The weights and rates of the clocks used in the algorithm are the same which are assigned by BIPM in Circular T. Necessary steering/correction is introduced at regular intervals in the microphase stepper (also known as phase and frequency offset generator) to steer the incoming frequency in order to compensate the frequency drift and to keep track of the UTC with minimum possible offset. UTC(NPLI) is the output of the microphase stepper which is AHM output corrected with a timescale algorithm through data processing unit. NPLI has also indigenously developed India's first Cesium fountain primary frequency standard NPLI-CsF1 and it also contributed to the calibration of TAI for few months. A primary standard realizes the SI second most accurately and gets highest weight by BIPM as well. However, it is not operational continuously. If it operates continuously, timescale can be steered with Cesium fountain data which will keep the timescale extremely accurate and stable. More details about the NPLI-CsF1 are given in Sect. 4.3.5.1.

Inter-comparison between the UTC (NPLI) and other international time scales UTC(*k*) is performed through a common view of global navigational satellite systems (CVGNSS) method and/or through a two-way satellite time and frequency transfer (TWSTFT) link. UTC(NPLI) is obtained in the form of signals of 1 PPS and 5/10 MHz frequency. These signals are fed to GNSS receivers and TWSTFT system. GNSS receivers log the time difference [UTC(NPLI)- GNSS] in the form of RINEX (Receiver independent exchange format) and CGGTTS (Common GNSS Generic Time Transfer Standard) file formats as per the guidelines of BIPM. TWSTFT system logs the time difference between UTC(NPLI) and the other reference station and is logged in ITU (International Telecommunication Union) format [58]. These files along with clock comparison data are stored in NPLI server and are uploaded on the BIPM server on a daily basis.

The performance of the time scale is published in the bulletin of the BIPM available online and called Circular-T (every month) and UTC rapid (every week). The traceability of the timescale with respect to UTC maintained by BIPM is at the level of few nanoseconds (ns). As an example, performance of UTC (NPLI) with respect to UTC over last seven years is shown in Fig. 4.7.

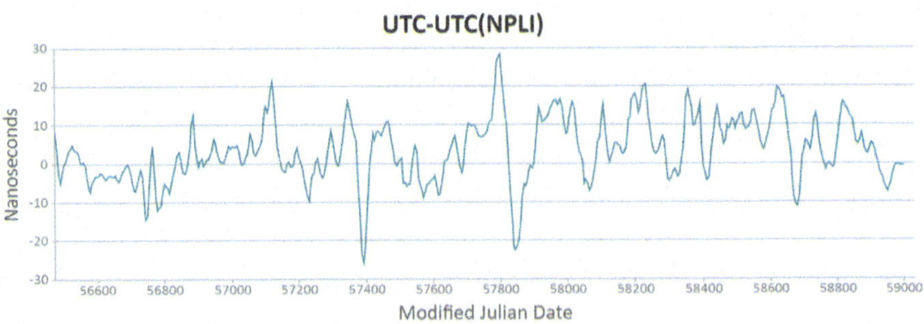

Fig. 4.7 UTC–UTC(NPLI) last seven years (July 1, 2013–July 1, 2020)

The primary timescale is maintained with utmost care as it has to run continuously to keep the traceability with UTC. The timescale is protected with environment control system, dedicated and secure internet system, multi-layer power backup system, and authorized access system. In addition, the timescale is regularly upgraded with new clocks, GNSS receivers, and other precision measurement instruments. Recently, two new GNSS timing receivers have been installed and have been calibrated using the traveling GNSS calibrator from Group-1 laboratory, i.e., NICT, Japan, and the internal delays were calculated with respect to NICT G1. With these efforts, the traceability link to UTC was calibrated and the associated systematic uncertainty improved to ±2.8 ns with effect from October 2018 [59].

In addition, a new time scale setup has been created which acts as a backup time scale. The new setup is analogues to the existing setup consisting of ensemble of number of Cesium clocks, distribution amplifiers, phase steppers, and a pair of hydrogen masers. The antennas of GNSS timing receivers and NTP servers are mounted on high rise antenna structure created at the top of the building to allow clear visibility to satellites. The existing time scale and backup time scale at the CSIR-NPL are interlinked via RF link and fiber optic connections. The dual-time scale setup within the CSIR-NPL will help to provide redundancy and uninterrupted IST to the nation in case of any failure in one of them.

4.3.2 Time Synchronization Technologies and IST Traceability

The importance of maintaining accurate and uniform time as well as its precise measurement has already been described in the above sections. Keeping in view of the above facts CSIR-NPL is providing pan-India IST traceability or disseminating IST across the country that it can be used as a reference and each clock of the country can be synchronized to IST. Considering the requirement and criticality of its application, CSIR-NPL provides IST traceability services or time transfer links with varying accuracies from milliseconds to nanoseconds.

There are several advanced ways for establishing the IST traceability links as for example, Network Time Protocol (NTP) and Precision Time Protocol (PTP) are commonly used for time synchronization in internet protocol (IP) networks. NTP is usually being used through Wide Area Network (WAN), whereas PTP is mostly useful in Local Area Network (LAN). A secure and accurate time transfer link can also be established utilizing public telephone network which is commonly known as Teleclock or FonOclock. There are two other space communication-based methods for most accurate dissemination of time through (i) Common view or All in view of Global Navigational satellites and (ii) Two Way link of geostationary satellites. Here below, a brief description about the NTP, PTP, telephone time services, all in view or common view of a global navigational satellite system (CVGNSS) and two-way satellite time and frequency transfer (TWSTFT) link have been presented

4.3.2.1 Network Time Protocol (NTP)

NTP is a time transfer protocol based on data packet transfer using the internet or ethernet [60]. The protocol is originally developed at University Delaware and it has four different versions based on different Request for Comments (RFC) documents, e.g., RFC-1059 (NTPv1.0), RFC-1119 (NTPv2.0), RFC-1305(NTPv3.0), and RFC-5905 (NTPv4.0). NTP uses a reliable time source, *i.e.,* UTC (in India, UTC(NPLI)as reference for precise synchronization of servers and network devices. In a network, it utilizes UDP (User Datagram Protocol) on port 123 and estimates three key variables in the relationship between a user and a timeserver (i) network delay, (ii) dispersion of time, which is a measure of maximum clock error between two hosts, and (iii) clock offset, which is the amount of correction that shall be applied to the client clock in order to synchronize it with UTC. NTP applies the estimates of these three variables to a client clock in order to synchronize it with UTC and the goal is twofold (i) maintaining the absolute time as close to UTC as possible and (ii) synchronizing the system time of all machines or devices within a network, which are linked to the client clock. The latest version of NTP (NTPv4) can maintain time with an uncertainty of less than 50 ms on WAN. The main sources of uncertainty in the performance of NTP are the random delays in WAN, caused by intermediate switches and routers in network paths. NTP time distribution uses a stratum hierarchy, where the reference time *i.e.,* UTC (NPLI) is defined as the stratum 0 server, and the followers downwards are defined as stratum 1, stratum 2, stratum 3, up to maximum depending on the network architecture. In order to achieve redundancy and reliability, servers and clients access multiple sources over diverse network paths. NTP contacts the servers only as frequently as needed, between once a minute to about once in 17 min (as the polling interval in NTP varies from 64 s to1024 s), so that the clock will stay as close as possible to UTC (NPLI) between corrections. The time accuracy of an NTP cannot be defined precisely as it cannot account for network delays due to number of router hops between the server and client NTP server, Network congestion, outage, routing changes, etc. Figure 4.8 shows a screen shot of the web time display from http://www.nplindia.in/ of NTP service which shows the difference between computer time and IST, it also shows the steps to be followed to synchronize a computer time permanently to IST.

In the absence of hardwire time stamping and presence of network elements, variable latencies limit the accuracy achieved using NTP. Figure 4.8 shows a typical architecture where a client synchronizes its time with NTP server. At present CSIR-NPL is broadcasting the NTP time service within 100 ms accuracy across the country. For this, users need to send a query at CSIR-NPL's time domain "time.nplindia.org" from their computer and it will get synchronized to IST. At present multiple numbers of NTP servers are installed and connected to the web domain of CSIR-NPL. The pool of NTP servers located at CSIR-NPL premises is capable of providing service to about 100,000 clients per second and further work in progress towards strengthening its capacity in order to address more numbers of client at any given time.

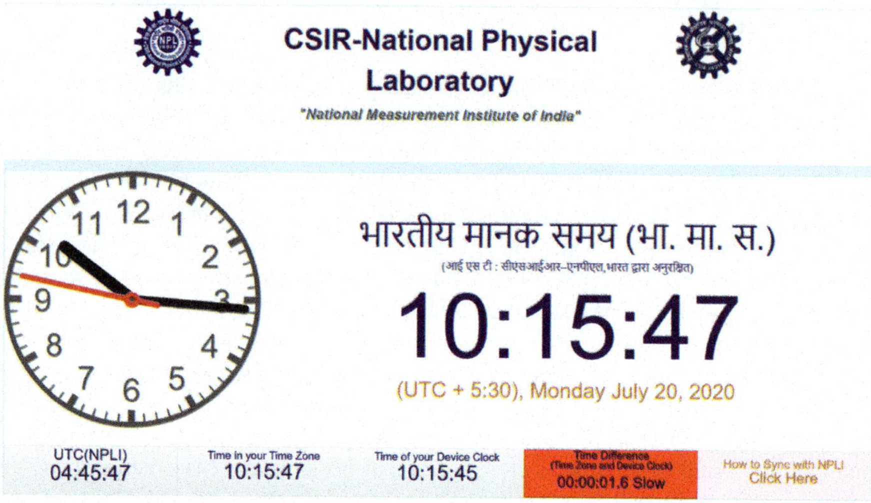

Fig. 4.8 Web time display of NTP service that shows the difference between a computer time and IST, and the steps to be followed to synchronize a computer time permanently to IST

4.3.2.2 Precision Time Protocol (PTP)

The Precision Time Protocol (PTP) is a time transfer technique based on data packets (carrying the time information) transfer in dedicated internet/ethernet networks. The protocol is defined by IEEE 1588 [61] and it enables accurate synchronization of clocks or devices (referred to as slaves) to a master server (referred to as grandmaster). The PTP synchronization mechanism involves the communication of time-stamped data packets between the grandmaster and the slave. The process allows the slave to not only compensate for its offset with the grandmaster but also the propagation delay of the network. All the elements in the network should be PTP compliant-which means the latency time (time that the PTP packets spend in those elements like routers, etc.) for the PTP packets are deterministic and can be accounted for when estimating the final correction of the slave with respect to the master clock.

Figure 4.9 depicts a typical PTP time synchronization architecture where a master (PTP Grandmaster) after getting synchronized with IST (usually 10 MHz for synchronization signal, and GNSS/IRIG for Time of Day (ToD)), transmits PTP time-stamped data packets through a dedicated network (dark fiber or network where

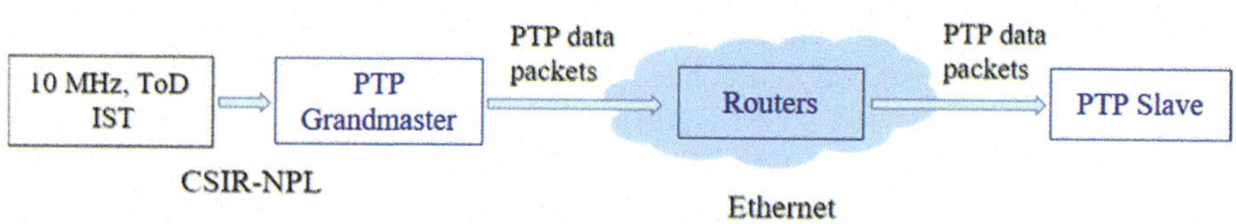

Fig. 4.9 Schematic of PTP time synchronization architecture

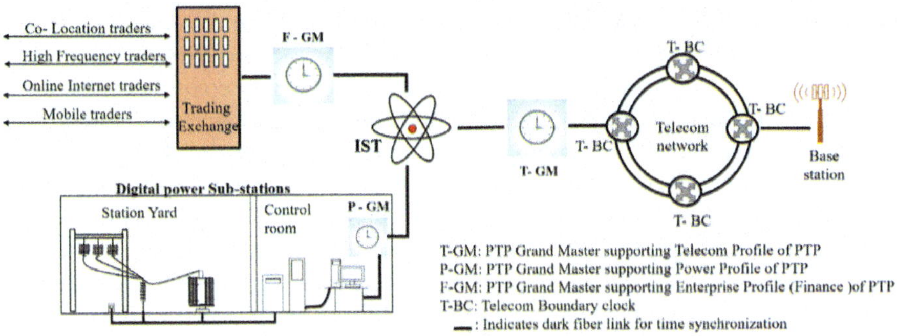

Fig. 4.10 Schematic for various applications using PTP based time synchronization

all the elements are PTP compatible) to client device (PTP slave). The frequency of communication between the master and slave is many times faster than in the case of NTP which results in the realization of accuracy in tens of ns level.

In PTP, apart from the offset compensation between the slave and the master, there is a requirement to measure the propagation delay added in the network. For symmetric networks (where the forward and backward propagation delays are the same), it is straightforward to calculate such delays and in asymmetric networks, advanced algorithms are used to find such delays with uncertainties.

The PTP based time dissemination will provide the redundancy required for critical applications which require time accuracy with respect to IST in tens of ns scale. Figure 4.10 shows the schematic representation for various applications using PTP based time synchronization.

Work is in progress at CSIR-NPL for establishing IST dissemination services utilizing PTP Grandmasters. These grandmasters at CSIR-NPL, synchronized with IST will be linked to client devices (PTP slaves) through dedicated ethernet lines, usually an optical fiber link.

4.3.2.3 Telephone Time Dissemination

Since long, time dissemination over telephone lines has been a very effective means of time transfer. Time code is transferred from a transmitter to a receiver over public switched telephone network (PSTN) as shown in Fig. 4.11. In 1998, CSIR-NPL launched its first telephone time dissemination service known as Teleclock. Teleclock was commercialized and provided IST time synchronization with few seconds accuracy over many years. Recently, the same concept has been upgraded with new

Fig. 4.11 Schematic of time dissemination over PSTN

hardware, software, and algorithm. This new telephone time dissemination system has been named Fonoclock and is capable of providing time synchronization with an uncertainty of ±10 ms [62]. The improvement in uncertainty comes from the use of communication delay compensation scheme in the algorithm where on establishing the call between the transmitter and receiver, the transmitter estimates the round-trip delay and compensates it in the transmitted time. Apart from better synchronization accuracy, Fonoclock provides complete Time-of-the-year information and the receivers are also equipped to provide 1 PPS signal for further synchronization. The technology has been completely tested and demonstrated within CSIR-NPL and few remote locations as well. The technology is ready for commercialization and more on it is discussed in Sect. 4.4.2. Final aim is to transfer the know-how to industry for compact and affordable receivers for accessing IST using a commonly available telephone line. All users who need time synchronization to IST with millisecond accuracy especially defence, telecommunication, banking, educational sectors, etc. will benefit from this technology.

Four Fonoclock transmitters have been developed, installed, and attached to dedicated phone lines and are synchronized to the IST. The users can access IST by connecting the indigenously developed receivers to a telephone line. The receiver is programmed to dial the transmitter number. On establishing the call, the data channel is attached, and delay compensated IST time code is sent to the receiver from the transmitter. The receiver decodes the time information and displays it in the display attached to the receiver. The performance of the system is evaluated using a time interval counter, as shown in Fig. 4.12. The synchronization accuracy of the system has been measured to be ±10 ms.

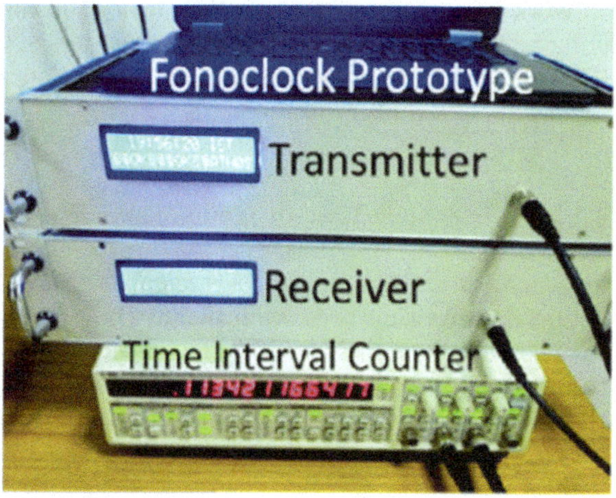

Fig. 4.12 Prototype of indigenously developed Fonoclock transmitter and receiver under test

4.3.2.4 Common View and All in View Global Navigational Satellite System (CVGNSS & AVGNSS)

The CVGNSS method is simple in principle but very effective and it was proposed in the 1980s by Allan and Weiss relies on reception of the same emitted signal by several receivers [63, 64]. When the two clocks are located at two different positions, their time difference Δt can still be estimated by comparing their time with a common clock, which is visible to both the clock. In another word, both clocks can simultaneously record their local time for a signal which is visible to both of them, and comparing those time, they can estimate their time differences. The time signal embedded in a Global Navigational Satellite System (GNSS) signal mostly transmitted through a frequency rang within 1.1–1.7 GHz, those are less sensitive to propagation effects. It provides wide visibility along with good signal-to-noise ratio and because of all those reasons GNSS signals are popularly used as the sources for a common view of clocks. At present several GNSS constellation are available (GPS by US, GLONASS by Russia, GALLILEO by European consortium) and all of them are fully deployed and has global or local coverage with high reliability. Indian space research organization (ISRO) also introduced a regional navigational satellite constellation (NavIC), which is visible from all parts of India and also from some parts of the sub-continent. A pictorial description of the CVGNSS system and its working principle has been shown in Fig. 4.13, where clock 1 and clock 2 needs to be synchronized with each other. They share with each other the GNSS clock data for further processing and the consumer clock steers its clock towards minimization of the time difference with the reference clock.

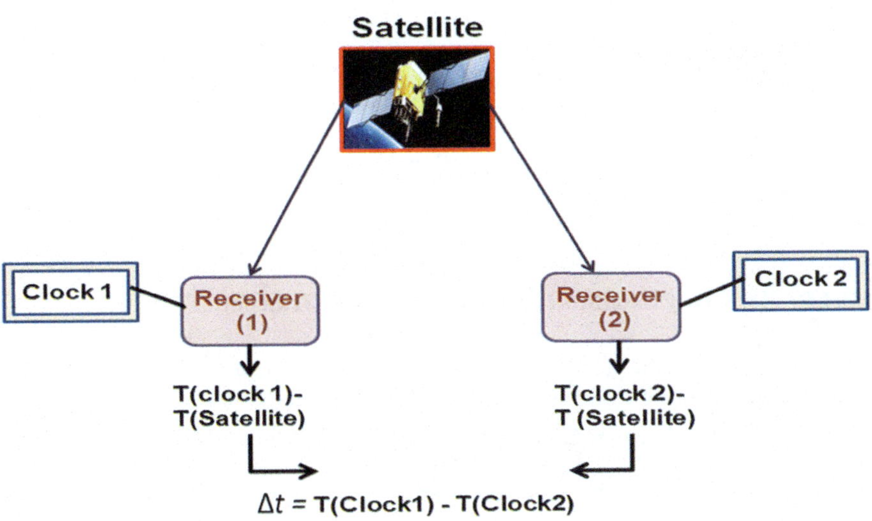

Fig. 4.13 Pictorial representation of CVGNSS method

The use of the global navigational satellites introduced a major improvement in time synchronization among the clocks. In case the clocks are located very far and there are no common navigational satellites visible to both of them at any particular time, clocks times are compared with the time of all visible satellites and this is known as all in view (AVGNSS) method instead of common view method. AVGNSS enables inter-continent clock comparison. Both CVGNSS and AVGNSS time transfer are a one-way method where the signal being emitted by a satellite is received by specific equipment installed in a laboratory. Estimating all systematic uncertainties, accurate time synchronization can easily be achieved by those methods. Now a day's most of the GNSS based systems are equipped with multi-channel and multi-frequency receiver instead of the single-channel and single frequency C/A receiver. The advantage of the multi-channel receiver is it allows simultaneous receiving of signals from all the satellites within field of view. For far distance intercomparison among clocks via GNSS link, the inhomogeneous ionosphere delays cause the most significant error in time comparison. Dual-frequency receivers permit the estimate of the ionospheric delay, thus improving the accuracy of time transfer. Such data, known as GPS P3, which allows clock comparisons within a few nanoseconds' statistical uncertainties. Other than ionospheric delay, tropospheric delays, environmental conditions like temperature and humidity and multipath reflection of the signal limits the performance of the AVGNSS or CVGNSS time transfer method. The multipath reflection can be avoided with specially designed placed with clear view of sky that no reflected signal will reach to it from nearby tall buildings, towers, or trees like things. Post-processing of the CVGNSS data by incorporating satellite ephemeris, known as the Precise Point Positioning (GPS PPP) method, can be utilized for achieving further accuracy for clock comparison within 1 ns.

At present, the traceability of UTC(NPLI) to UTC is achieved through AV/CVGNSS method and the current uncertainty of that link is about ±3 ns. In addition, CSIR-NPL also providing time traceability utilizing CVGNNS methods to the IRNSS timescales, i.e., IRNWT-1 and IRNWT-2 located at Bengaluru and Lucknow, respectively. For precise time synchronization (within 10 ns) or time (IST) traceability, CVGNSS is a unique method that can be implanted anywhere against the requirement.

4.3.2.5 Two-Way Satellite Time and Frequency Transfer (TWSTFT)

Independent to GNSS, the TWSTF T [65] technique utilizes a telecommunications geostationary satellite to compare clocks located in two different positions *i.e.*, at receiving and emitting stations. A pair of clocks located at a far distance can simultaneously be compared using a transponder of a geostationary satellite, which is accessible from both the end. The main advantage of the two-way methods is, most of the systematic errors are common at both ends as they receive/transmit signals following the same path and they cancel each other. One of the constraints of this technique is that the simultaneous measurement should be done at both ends and necessary arrangements like the hiring of transponder time and infrastructural development

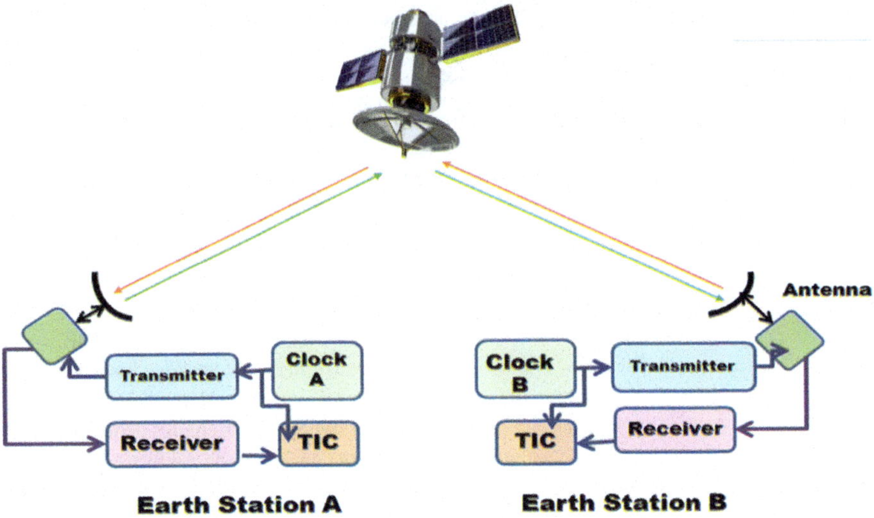

Fig. 4.14 Pictorial representation of TWSTFT method

should be made. With the installation of automated stations in most laboratories, the TWSTFT link observations among the stations can be made at regular intervals, with the consequence of achieving a statistical uncertainty below 1 ns. A schematic diagram for TWSTFT has been shown in Fig. 4.14.

CSIR-NPL established an international TWSTFT satellite link in order to achieve UTC traceability to its locally realized time, i.e., UTC(NPLI). CSIR-NPL also providing IST traceability to IRNSS timescale, i.e., IRNWT-1 and IRNWT-2 located at Bengaluru and Lucknow, respectively, through a dedicated TWSTFT link. TWSTFT is one of best of option for providing time and frequency traceability as it is capable of providing sub-nanosecond accuracy but its an expensive method which require lot infrastructural support, special equipment and rent of the satellite link.

Different methods adopted at CSIR-NPL for providing time and frequency traceability and their technical detail have already been described. Table 4.1 summarizes the mechanism, accuracies, and limitations of each of the techniques.

4.3.3 Calibration and Measurement Capabilities

As the NMI of the country, CSIR-NPL (NPLI) realizes and maintains the UTC(NPLI) & IST and also disseminates it across the country. It also carries other important metrological activities like calibration of the precision reference standards like Cesium and Rubidium clocks, Time interval counters (TIC), Frequency Counters, Stopwatches, volt-second generators, etc. CSIR-NPL has in place a recognized Quality Management System (encompassing quality system, technical capabilities,

Table 4.1 Time transfer techniques with their accuracies, mechanism and limitations

Time transfer technology	Time accuracy	Time transfer mechanism	Limitations
NTP	<100 ms	Time transfer through data packets in public internet	Only millisecond accuracy; no PPS or 10 MHz output; can be used only for internet packets; depends on internet connection and traffic
FonOclock	≤10 ms	Time transfer through public telephone lines (PSTN)	Only millisecond accuracy; only through telephone lines; PSTN is degrading technology
PTP	≤10 μs	Time transfer through data packets in dedicated LAN	Requires dedicated LAN, optical fibers and network elements; not more than 200 km; performance limited by the latency generated from active components used for establishing link
CVGNSS/AVGNSS	≤10 ns	Time transfer using microwaves signals emitted by navigational satellites	Common view data should be processed to provide traceability to IST
TWSTFT	<1 ns	Time transfer through microwaves via leased geosynchronous satellite link	Very costly method; requires huge infrastructural setup; expensive equipment and lease of satellite link

and administrative supports) and is able to demonstrate technical competence at the international level. The Quality Management System has been implemented based on IS/ISO/IEC 17025: 2017 and/or IS/ISO 17034: 2016 in respective Metrological Activities.

Following the "Certificate in Investment Performance Measurement- Mutual Recognition Arrangement" (CIPM-MRA) every metrological activity at CSIR-NPL, *i.e.,* measurement standards, calibration, and testing facilities are peer-reviewed by international experts appointed through Asia Pacific Metrology Program (APMP). Every service/parameter offered by CSIR-NPL, covered by the CIPM MRA, is listed in the BIPM key comparison database (BIPM-KCDB) under Calibration and Measurement Capabilities "CMCs", which the details on the range and measurement uncertainty for each listed service/parameter [66].

A CMC is a calibration and measurement capability offered by NMIs or designated Institutes (DIs) under the CIPM MRA. All the CMCs under CIPM MRA are listed on BIPM key comparison database (KCDB). Under a CMC, the measurement or calibration should be:

- Performed according to a documented procedure and have an established uncertainty budget under the management system of the NMI or the accredited laboratory;
- Performed on a regular basis (including on-demand or scheduled for convenience at specific times in the year); and
- Available to all clients.

At present CSIR-NPL possesses 11 CMCs related to Time and Frequency parameters and those CMCs along with their measurement techniques and associated uncertainties have been listed in Table 4.2.

Calibration or traceability to UTC(NPLI)/IST for any device helps by setting its defined parameters as close to its nominal value as possible. The detail method of realization and maintenance of UTC(NPLI) or IST has been described earlier and in the following section, a brief description of the procedure for providing traceability or calibration of the oscillators, instruments, or timing devices have been presented. At present four different methods, namely (i) Direct Frequency Measurement Methods and (ii) Time Interval measurement method (iii) frequency calibration, and (iv) Epoch calibration are followed for the calibration of atomic frequency standards, frequency sources, clock, timer, stopwatch or any other timing devices.

4.3.3.1 Direct Frequency Measurement Method

The schematic diagram of the experimental set up for direct frequency measure method has been depicted in Fig. 4.15. The Universal counter uses the 10 MHz output signal from UTC(NPLI) as reference to locking its internal oscillator and the output of the Device Under Calibration (DUC) is measured using this frequency counter. The counter gate-time is adjusted to 10 s and all of its readings are either manually noted down at intervals of 30 s or can be stored on a computer.

4.3.3.2 Time Interval Measurement

The schematic diagram of the experimental set up for time interval measurement method has been depicted in Fig. 4.16. The time interval of the DUC or measured by the DUC is compared with that of standard reference time source i.e., 1PPS output of the DUC is compared with the 1PPS output of UTC (NPLI). The time-base of the equipment i.e., a time interval counter (TIC) is locked to the reference 10 MHz of UTC(NPLI). For some of the DUC where no direct 1 PPS output is available, a frequency divider is used for generating 1PPS signal out of the DUC. The mean of the measured values against the standard time interval used indicates the accuracy of the DUC.

Table 4.2 List of CMCs and their associated uncertainties

S. No.	Quantity	Instrument or artefact	Instrument type or method	Values, units	Uncertainties
1	Frequency	General frequency source	Time interval measurements	1, Hz	9.0E−14 Hz/Hz
2	Frequency	General frequency source	Time interval measurements	0.1, MHz	9.0E−14 Hz/Hz
3	Frequency	General frequency source	Time interval measurements	1, MHz	9.0E−14 Hz/Hz
4	Frequency	General frequency source	Time interval measurements	5, MHz	9.0E−14 Hz/Hz
5	Frequency	General frequency source	Time interval measurements	10, MHz	9.0E−14 Hz/Hz
6	Frequency	General frequency source	Direct frequency measurements	1, MHz	6.4E−12 Hz/Hz
7	Frequency	General frequency source	Direct frequency measurements	5, MHz	6.4E−12 Hz/Hz
8	Frequency	General frequency source	Direct frequency measurements	10, MHz	6.4E−12 Hz/Hz
9	Time scale difference	Local clock versus UTC(NPLI)	Time interval measurements	−0.5 to 0.5, s	2 ns
10	Time scale difference	Local clock versus UTC	Time interval measurements	−0.5 to 0.5, s	38 ns
11	Time interval	Time difference meter	Time interval measurements	1 to 1740, s	1.4E+02 ms

Uncertainty of type A for both measurement type is obtained on the basis of the data taken using the DUC. Modified Allan deviation (2 samples) is used to measure the stability and statistical uncertainty. Factors contributing to the systematic uncertainty (Type B) are from the calibration certificate of the measuring instruments, temperature fluctuations of the measurement system, etc., For timing devices, the resolution from the specification is used for calculating the uncertainty of type B, *i.e.*, the uncertainty due to finite resolution of the DUC. The combined uncertainty is calculated by adding the individual sources of uncertainty in quadrature as per Law of Propagation of Uncertainty (LPU) considering the input quantities are independent.

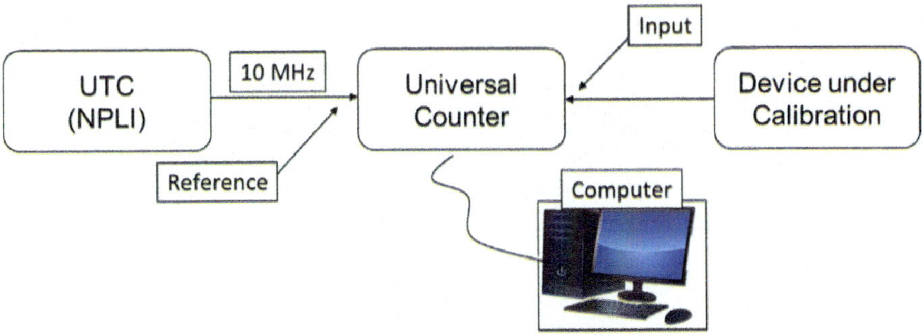

Fig. 4.15 Schematic of the experimental setup for direct frequency measurement method

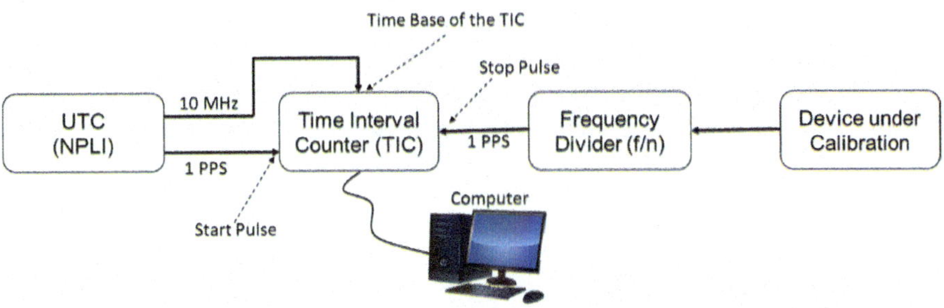

Fig. 4.16 Schematic of the experimental setup for time interval measurement method

Expanded uncertainty is defined as the combined uncertainty multiplied by a coverage factor k = 2 considering a normal distribution with 95% confidence of interval.

4.3.3.3 Frequency Calibration

Any frequency source can be calibrated by either measuring its frequency directly or by generating a 1 PPS from it and comparing it with the 1 PPS from the reference standard UTC(NPLI) in the time interval mode. In the direct frequency measurement technique similar experimental setup as shown in Fig. 4.15 is used where the universal frequency counter counts the number of events per unit time interval through a timed gate. The time-base of a frequency counter controls timing operation in the system, it is usually a stable quartz crystal oscillator. This oscillator is used to provide the gate time and the accuracy or the time-base determines the accuracy of the gate width and hence the accuracy of the measurement. The input gate opens for the selected gate-time [10 s in our case], during which the input transitions are counted. After stopping the count, the registered count will be a decimal fraction, and ten times the measured frequency in hertz and establishing the correct value is only a matter of placing the decimal point at a proper place [all this is automatically taken care of by the counter]. Errors of time base frequency have a direct influence on the result. Therefore, the more expensive counters usually include a high-quality oven-controlled crystal oscillator with low aging. As the atomic frequency standard is available in NPL, it is always advantageous to use its output frequency to lock the time-base of the counter. This allows high precision measurement to be made with the counter irrespective of the quality of its time base frequency. Frequency calibration is also done by comparing one PPS signal of UTC (NPLI) and the DUT as shown in Fig. 3.16. Measuring the time difference between the two 1 PPS signal, the frequency offset of the DUT can easily be estimated.

4.3.3.4 Epoch Calibration

Calibration of time (epoch) of any clock with respect to IST, realized through UTC(NPLI), can be done by comparing their times utilizing a similar experimental set up as described in Fig. 4.16. The steered 1PPS from the Primary Frequency Standard, i.e., UTC(NPLI) is used as a start pulse and that of the DUC is used as a stop pulse in the Time Interval Counter. The time-base of the counter is also locked to the UTC(NPLI). Taking the statistical average of the time difference between 1PPS pulses over a very long period, the epoch of the clock can be calibrated.

4.3.4 Traceability to Other SI Units

CSIR-NPL realizes, develops, maintains, and disseminates reference standards for the seven SI base units meter, kilogram, ampere, Candela, Kelvin, time, and mole. Such standards provide traceability of each measurement to the SI thus ensuring reliability, consistency, and equivalence at the international level. SI meter, SI kilogram, SI ampere, SI Candela, and SI Kelvin depend on the SI unit of time. Out of six other base units, five of them depend on SI second for their realization. Apart from the base units, many derived units also need traceability to SI second for their realization. Considering the Critical requirement of the SI second for realizing other SI units, time and frequency traceability services are being provided through co-axial RF cable for other metrological activities within CSIR-NPL. At present, 10 MHz signals from UTC(NPLI)/IST are being provided for realizing the standards of the length and dimension, microwaves, LF & HF current and voltage, and Josephson voltage. For the realization of a new definition of mass, temperature, and luminous intensity, time and frequency traceability will be provided as well.

4.3.5 R&D Activities

CSIR-NPL is engaged in extensive R&D about developing the most advanced frequency standards or atomic clocks like (a) Cesium fountains, (b) Ytterbium ion optical clock. CSIR-NPL is also working towards developing optical fiber-based ultraprecise time and frequency transfer links utilizing White Rabbit Network. A brief description of each of such R&D activities are given below.

4.3.5.1 Cesium Fountain Clocks

As already discussed, the transition frequency between two hyperfine levels (i.e. |F = 4|mf = 0 and |F = 3|mf = 0) of natural Cs-133 atoms is used to define SI second. This definition of SI second is realized in practice using Cs atomic fountain primary frequency standard (PFS) (see Sect. 4.2.1.3). International atomic timescale (TAI) is a weighted average of time kept by atomic clocks all over the world. The increasing number of operational fountain frequency standards has contributed to the maintenance of TAI at an unprecedented level of accuracy, with fractional uncertainty of the time interval below 10^{-15} (such atomic clock when operated continuously will neither loose nor gain a second in more than a million years). Cs fountain clock is not commercially available technology and has been developed indigenously only at a few NMIs. Worldwide, leading National Metrology Institutes (NMI) in the USA, UK, France, Germany, Italy, Japan, India, Russia, China, and recently Canada are the only ones to have successfully developed their Cs fountain primary standards. Achieving this level of accuracy requires a remarkable combination of technological innovations in precision lasers, vacuum technology, magnetic shielding, and advanced optical, electronic, and mechanical systems besides control over the atomic and optical environment.

In case of a Cesium beam clock or commercial Cesium clock, the accuracy is limited due to the fast-moving hot Cesium atoms and hence interaction time of few ms between the atoms and the microwaves. In a Cesium fountain clock, laser-cooling (to few μK) of Cesium atoms is done for slowing down (to a speed of few cm/s) the fast-moving atoms in order to increase the interaction time between the atoms and the microwaves. As per the Heisenberg uncertainty principle, longer interaction time leads to a more accurate determination of frequency. In a Cesium fountain, Cesium atoms are first cooled in the intersection volume of six mutually orthogonal laser beams and then tossed up with the frequency detuning between laser beams leading to a push in the vertically upward direction. The cooled atoms travel through the microwave cavity on the way up, achieve the toss height (typically 65 cm above trap centre in case of NPLI-CsF1, generally depends on the launch velocity) before the gravity pulls them down. The atoms make the second pass through the microwave cavity on the way down and are then interrogated for their state with laser-induced fluorescence in the detection region. Typically, the interaction time is of the order of half a second which is about two orders of magnitude better than commercial Cesium clock. Therefore, the relative frequency uncertainty of a fountain clock is at least two orders better than a commercial Cesium beam clock.

At CSIR-NPL (NPLI), India's first-ever Cesium fountain primary frequency standard was designed, developed, and assembled indigenously, see Fig. 4.17. The best measured fractional frequency uncertainty of the fountain is 2.5×10^{-15}. The

Fig. 4.17 Cesium fountain clocks at CSIR-NPL

fountain has three major parts: physics package, optical set-up and electronics, and microwave controls. Physics package has the ultra-high vacuum chamber, Cesium source, microwave cavities, vacuum pumps, etc. Atoms are cooled, trapped, launched, and interrogated in the physics package. The optical set-up has lasers and other optical devices needed for generating frequency and intensity-controlled laser beams (6 for cooling and 2 for detection). The laser beams are transported to the physics package via optical fibres where these are expanded using beam expanders mounted on the vacuum chamber. Electronics and microwave controls are critical for the fountain operation. Three main components are precision microwave synthesizer, fountain sequence controller, and data-acquisition system. All these are required for automated operation, control, and evaluation of the fountain.

The practical implementation scheme and technical realization differ from one fountain clock to another; however, the basic principle of fountain operation remains the same. Specific to the development of NPLI-CsF1, (0, 0, 1) geometry is used for cooling, trapping, and launching atoms in a magneto-optical trap (MOT). (0, 0, 1) geometry means that out of the six laser beams used for cooling the atoms, four are in the horizontal plane and two are in the vertical direction. In an MOT, magnetic field is used for confinement/trapping of the laser-cooled atoms. In one fountain cycle, after loading and cooling in MOT, atoms are further cooled in optical molasses and launched in moving molasses using the vertical laser beams. Just after launch, atoms are cooled to few μK temperatures using polarization gradient cooling.

During the ballistic flight, atoms pass through a microwave cavity twice. This method of interrogation with dual interaction with microwaves (on the way up and down) is known as Ramsey's method of separated oscillatory field. Ramsey interaction method gives fringes in the temporal domain and is analogous to Young's double-slit experiment which gives fringes in the spatial domain. One method provides better spatial localization and the other provides better frequency localization. After the Ramsey interaction, the state-interrogation of atoms is done using laser-induced fluorescence.

For detailed frequency evaluation of NPLI-CsF1, the fountain frequency is compared with the Hydrogen Maser (H-Maser). The H-Maser is a part of the timescale UTC (NPLI) which is traceable to UTC via satellite time transfer links and hence fountain is evaluated with respect to the UTC (see Sect. 4.3.1). During each evaluation (lasting 10–20 days), the fountain is operated continuously and all major systematic shifts due to blackbody radiation (BBR), second order Zeeman, gravitational potential, and collision shift are measured and corrected.

With the Cs fountain becoming fully operational in 2012 at CSIR-NPL, India joined the elite club of only 6 (in 2012) developed nations in the whole world to have an ultra-precise primary standard of time and frequency. NPLI-CsF1 was then operated and evaluated several times for measuring its frequency stability and accuracy. In 2013, it was successfully compared with fountains from at Physikalisch-Technische Bundesanstalt (PTB), Germany, Scientific Research Institute of Physical-Technical, and Radio Technical Measurements (VNIIFTRI-SU), Russia, and National Institute of Metrology (NIM), China [67–69]. All the frequency differences between the fountains agreed within the 1-sigma uncertainty in the low 10^{-15} level [70]. The fountain data also contributed to international atomic time (TAI) for a couple of months and was published in Circular T as well. The Allan deviation of the frequency difference between the fountain and the H-Maser showed that fountain frequency was stable to a few parts in 10^{15} at less than one day of averaging. NPLI-CsF1 was approved as a primary frequency standard (PFS) in 2015 by CCTF (consultative committee for Time ad Frequency) at BIPM. For further details on NPLI-CsF1, readers are referred to [71–77].

With the motivation to have a better and stable fountain with improvised design features, a second-generation Cs fountain is currently under development at CSIR-NPL, see Fig. 4.17. Some new design features have been added to the second fountain which enables us to carefully investigate the systematic errors in order to enhance the accuracy of NPLI-CsF2 to a few parts in 10^{16}.

In NPLI-CsF2, optical pumping will be used to increase the S/N ratio which has been a major limitation in NPLI-CsF1. The cavity design has been improved to reduce distributed cavity phase shifts. The (1,1,1) geometry has been chosen for cooling, trapping, and launching instead of (0, 0, 1) geometry used in the first fountain. In (1, 1, 1) geometry, none of the six cooling beams is along the direction of flight of the cold atoms and thus will reduce light shift (occurs due to the presence of resonant light during microwave interrogation). Another change is that optical molasses will be used for cooling the atoms instead of MOT. This will help in the reduction of frequency shifts resulting from pulsating high power magnetic fields (used for MOT) in the cooling region. With all these enhanced design features, NPLI-CsF2 is expected to have better stability and accuracy over NPLI-CsF1.

In the process to develop the second fountain, various sub-systems are being indigenously developed. On the vacuum part, various sub-assemblies like cooling chamber, detection chamber, flight tube, etc. have been individually assembled and vacuum tested. Most of the electronic subsystems have also been developed and tested. The optical set-up is being laid out for this fountain. The assembly will then be done for the Physics package followed by complete automation. The fountain is expected to be operational in a few years.

Although optical clocks are showing better accuracy (see Sect. 4.3.5.2), Cesium fountain atomic clock still holds its significance as a primary frequency standard (PFS). Off late, many National Metrology Institutes (NMIs), such PTB (Germany) and LNE-SYRTE (France), have started steering their timescales with Cesium fountain data in order to achieve extremely stable timescales [78, 79]. Thus, in order to improve the accuracy and stability of the IST timescale, regular operation of either one or both fountains will be crucial in the coming years.

4.3.5.2 Ytterbium Ion Optical Clock

A clock with better accuracy or stability is an essential requirement for many of the applied research as well as to address many of the advanced physics problems related to communications, surveillances, testing temporal constancy of fine structure constant, or testing electron-proton mass ratio. For precision experiments like detecting gravitational waves or testing general relativity require a clock have stability $\sim 10^{-18}$ or better [80, 81]. The uncertainty of a clock in terms of Allan deviation decreases with the increase of operational frequency of the clock, so an optical clock of operational frequency of several hundred of THz is capable of providing orders of magnitude higher stability than the microwave Cs atomic clocks. It is also expected that in near future the SI definition of time, i.e., second will redefined on the basis of optical transition or through an optical clock. There are mainly two types of optical clocks based on the interrogations of ultra-narrow transitions of neutral atoms or atomic ions. The neutral atoms are confined within optical lattices tuned to so-called "magic wavelength" by inducing a dipole moment in the atom and exerting a force on this dipole through a laser field gradient [82, 83]. On the other hand, ions are confined within radiofrequency ion trap and in order to avoid any perturbation of the ionic clock transition due to Columbic interaction among the charged particles, only a single ion is preferred to be interrogated in the field-free region of the trap centre [84, 85].

Ytterbium ion is a potential candidate for optical frequency standards as it has recently achieved an accuracy 3 part in 10^{18} demonstrated by PTB Germany [86]. The international committee for weights and measurements (CIPM) has endorsed the clock transitions in Ytterbium-ion as a secondary frequency standard. This may be considered for redefinition of the next generation SI second in the optical frequency domain. Keeping birds' eye on that we have started developing the state-of-the-art Ytterbium ion optical frequency standard which will keep us in the frontier of this field. Our focus is to build an optical frequency standard of accuracy ~10^{-18} so that we can indulge ourselves in the frontline experiments for the first time in the country. Singly charged ytterbium ($^{171}Yb^{+}$) has two dipole-forbidden transitions from the ground state ($^{2}S_{1/2} - {}^{2}D_{3/2}$ at ~ 436 nm and $^{2}S_{1/2} - {}^{2}F_{7/2}$ at ~467 nm) and those are attractive as reference transitions for optical frequency standards [87–89]. Transition to metastable states, particularly $^{2}S_{1/2} - {}^{2}F_{7/2}$ electric octupole transition in $^{171}Yb^{+}$ is of great interest since the excited ($^{2}F_{7/2}$) state is extremely long-lived and the lifetime estimated to be 6–10 years [89]. Because of the extremely long lifetime the transition has an ultra-narrow transition linewidth on the order of ~nHz, which provides an excellent basis for a highly accurate atomic clock transition. At CSIR-NPL, we are building an optical frequency standard based on interrogation of the octuplet transition at wavelength 467 nm of a single $^{171}Yb^{+}$ [90–97].

The single-ion will be trapped and laser-cooled within a specially designed Paul trap [94, 95]. In order to achieve three-dimensional confinement of the ion, an oscillating radio frequency along with Dc voltage are applied to the trap electrodes and the specially designed Paul trap will produce a point like field-free region and eventually trap a single ion. The schematic of the experimental set up for single trapped ion-based optical clock has been depicted in Fig. 4.18. The entire experiment will be carried out at ultra-high vacuum (UHV) conditions for avoiding any collision between the trapped ion s and residual gases. Photoionization of ytterbium atoms

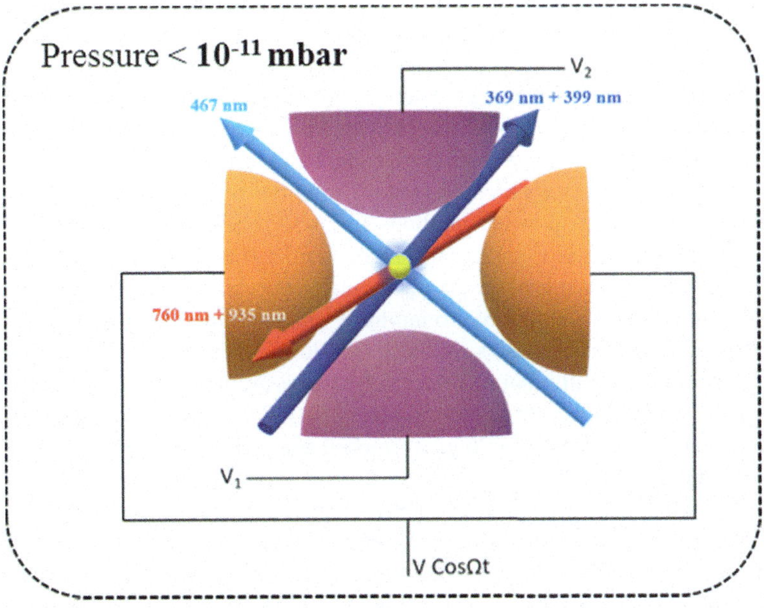

Fig. 4.18 Schematic representation of an optical clock based on single trapped and laser cooled ^{171}Yb ion

will be achieved via two-step resonant process, first exciting an electron from the $^1S_0$ level to the $^1P_1$ level of the neutral Yb with a 399 nm laser and then another radiation at 369.5 nm will supply energy to bring it to the continuum and ionizes the atom into a $^{171}Yb^+$ ion [98]. After production, the ions will be held in a well-characterized harmonic potential within the endcap trap and the whole system is kept inside an ultrahigh vacuum chamber. A high voltage rf will be delivered to the trap electrode for successful trapping of single ion [92, 93]. Narrow bandwidth will help to get rid of multiple frequencies in the trapping oscillation, as multiple frequencies will cause unwanted heating and very much unstable trapping conditions. Confined ions will be Doppler cooled on the $^2S_{1/2} - {}^2P_{1/2}$ transition with 369.5 nm radiation. The state of the ion can be measured using Dehmelt's electron shelving technique by detecting fluorescence from the ions [99, 100]. Single ion trapping will be achieved by optimizing the trapping potential against the calibrated fluorescence intensity of the trapped ions. A specially designed ultra-narrow linewidth laser will be locked to that desired atomic transition of the single trapped ytterbium ion and will be used as the reference frequency for the optical clock. Single trapped ion-based optical clocks are capable of providing excellent long-term stability since the ions are free from interatomic and Columbic interaction and in that way can have a very long coherence time [101].

4.3.5.3 Optical Fiber Link for Precise Time and Frequency Transfer

Timekeeping as well as accurate time transfer is very crucial for timing laboratories. Coaxial RF cables are commonly used to transfer time and frequency signals from the central location (time scale) to different places including other metrological activities within the premises. But, there are some factors such as attenuation of signal, undeterministic delays, and noises which makes coaxial RF cables less reliable for accurate time and frequency transfer or ultra-stable clock comparison. Time transfer and comparison among remotely located atomic clocks had been started utilizing portable clock. After that satellite-based methods like CVGNSS and TWSTFT have been utilized for long-distance clock comparison. Satellite-based methods are capable of transferring time with sub-nanosecond accuracy but, these methods are expensive as well as some extent vulnerable to jamming and spoofing of satellite signals. The time realized through timescales has very high level of accuracy as well as stability and it compels us to think about a medium that can transfer the signal within sub-nanosecond or hundreds of picosecond accuracy. Over the last two decades, optical fiber links have been studied with great importance and found to be one of the most reliable and precise medium for transferring time and frequency signals [102, 103].

At CSIR-NPL work is in progress about establishing an optical fiber link utilizing White Rabbit (WR) network [104, 105], which is capable of providing exceptional stability that time and frequency signals can be transferred over long distances within hundred picoseconds accuracy. White Rabbit technique [106] is significant for time

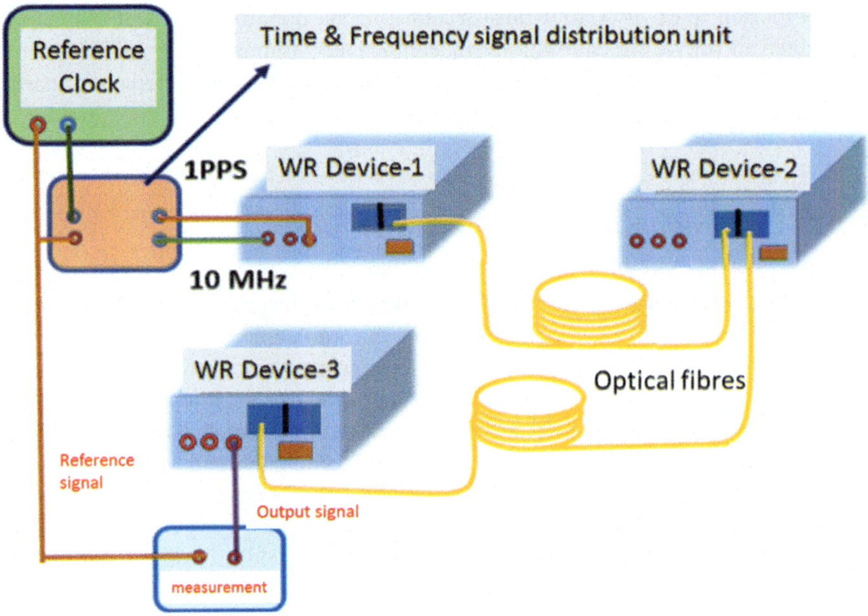

Fig. 4.19 Schematic representation of an optical fiber link through WR network

synchronization purpose because it is an ethernet based technique and follows protocols [107]. It solves the problem of limited bandwidth and the impossibility of dynamically evaluating of link delay. The leading technologies that are responsible for precise syntonization and synchronization in WR network are digital dual mixture time difference (DDMTD) [108] and wavelength division multiplexing (WDM). WR network consists of WR Nodes, single-mode optical fibers of different length, bidirectional small form factor Pluggable (SFP) transceivers, etc. A feedback loop of the fiber link has been utilized for monitoring the dynamic phase variation as well as time delay among the oscillators connected to the optical fiber link and enables to achieve extremely high stability of the optical fiber link. A schematic diagram of the optical fiber link has been shown in Fig. 4.19.

Time and frequency signals are being transferred through such optical fibers over a distance of several kilometers. WR network-based optical fiber link shows exceptional stability, which is quite suitable for comparing performance of remotely located ultra-precision atomic clocks.

4.4 Time and Frequency Metrology Services

Accuracy and precision are the two parameters that forms the benchmark for any measurements. Measuring and test equipment calibration for accuracy and precision is an important part of quality infrastructure in the country. The T&F metrology is engaged in providing calibration, traceability, and training programs to various sectors and services. The calibration and traceability services rendered are impartial and confirm to highest international standards.

4.4.1 Calibration, Traceability, and Training

For measurement uniformity across the country between various devices for a particular measurand (the physical quantity to be measured), the respective devices need to be periodically calibrated with respect to a source traceable to the national standard. The T&F metrology provides the national standard for time and related physical quantities. It provides calibration to accredited calibration laboratories which then provides calibration to various sectors or companies and so on. The T&F metrology also undertakes calibration services directly for different sectors requiring time synchronization.

The calibrated devices can be a standalone device providing calibration to industrial devices, till the expiry of the calibration validity, which then again needs to be brought to either T&F metrology or to accredited calibration laboratories. In some applications, however, there is a need for "24 $\times$ 7 real-time calibration" or so-called real-time traceability services. T&F metrology also provides real-time traceability services for different sectors based on their requirements. Such programs require elaborate procedures and infrastructure to provide continuous traceability services. In the "Product development" section, some of the indigenously developed products used in time dissemination architecture and associated components are described in detail. In the final section "National Synchronization programs", the nation-wide time distribution network which will provide such continuous traceability services have been discussed. Time being an important parameter in realizing units such as length, mass, candela, calibration services are provided to units of NPL which are maintaining such SI units.

4.4.1.1 Calibration

As mentioned in earlier sections, T&F metrology has currently 11 calibration and measurement capabilities (CMCs), traceable to international standards. The parameters are measurement time, gate time, and 1 Pulse per second (PPS) amplitude. The measurement standard uncertainties are reported usually with a confidence of 95%.

Armed with such measurement capabilities and some additional derived parameters developed in-house, the T&F provides calibration services to various laboratories, MSMEs, and large-scale industries. Some of the common devices are atomic clocks (Rubidium and Cesium frequency standard), Time interval/frequency counter, Hour meter, Tele-clock, GPS timing receivers, and Stopwatch. There is a dedicated calibration division at NPL, called Centre for calibration and testing (CFCT) [109],

Table 4.3 Calibration services provided by Time and Frequency metrology

S. No.	Device under calibration	Stakeholder/client	Calibration/test parameter	Accuracy of calibration
1	Stop watches	CSIR-central leather research Institute, Chennai (Department of Science and Technology)	Time interval, (s)	±140 ms
2	Cesium frequency standard	Air force station (Department of Defence), Space application center (Department of Space)	Frequency, (10 MHz)	10^{-15}/day
3	Rubidium frequency standard	BNN SPEAG Test Calibration Laboratory, Sahibabad	Frequency, (10 MHz)	10^{-15}/day
4	GPS receiver	Bharat Electronics Limited (Department of Defence)	EPOCH, (ns)	50 ps
5	Frequency counter	Standardization Testing and Quality Certification Laboratory (STQC), Ministry of Electronics and Information Technology	Frequency, (10 MHz)	10^{-15}/day

which acts as an interface between the customers and the T&F metrology. Table 4.3 list the details for some of such devices under calibration (DUC), sector, and the test parameter.

Time is an important parameter in realizing units such as length, calibration services are provided to units of CSIR-NPL which are maintaining such SI units.

4.4.1.2 Traceability Services

The T&F metrology actively provides traceability services to various sectors. The traceability is required for two cases:

i. Periodic calibration of the clocks or time/frequency devices and
ii. Real time synchronization of the clocks, time scale, network devices like computational terminals.

The periodic calibration of the clocks is provided through providing calibration certificates, which has details for the measurement (best estimate + standard uncertainty), measurement procedure, details about the uncertainty budget (random effects and systematic effects), and the calibration of the measuring device itself. The whole process is carried out at predefined intervals as per standard operating procedures and has traceability to international standard UTC.

For real-time synchronization of devices that requires 24 × 7 traceability services like in the case of navigation, telecom, finance, and energy grid sectors it is of utmost importance to provide traceability services through dissemination schemes which are reliable and robust.

There are broadly two ways of disseminating such services using:

(i) Satellite technologies
(ii) Terrestrial technologies

(i) Time synchronization through satellites is carried out through CV/AV GNSS technique and TWSTFT method as described in Sects. 4.3.2.4, 4.3.2.5.
Currently, such services are being provided to the Indian Regional Navigation Satellite System (IRNSS–ISRO) through both CVGNSS and TWSTFT methods [110]. The CVGNSS method of time traceability will also be used in the dissemination of time locally from the upcoming five timing centers at regional reference standards laboratory (RRSL) located across India [111].
(ii) Terrestrial time synchronization technologies include land-based methods like telephone lines, long wave radio (low frequency) signals, Internet protocol-based time transfer methods like Network Time Protocol (NTP), and Precision Time Protocol (PTP).
A telephone line-based time transfer technique having accuracy in the ms range has been developed in-house using specially designed transmitters and receivers (Fonoclock) to disseminate time. The signals generated from the transmitters are traceable to IST. Similarly, a project is underway to disseminate time through low-frequency signals transmitted from tall towers thorough time encoded signals. The transmitters themselves will be synchronized to IST either though internet-based NTP servers or through GNSS receivers linked with IST [112].
The third mode of terrestrial time transfer is through web of internet/ethernet network infrastructure. Data packets carrying timing information is sent through the network using standard internationally recognized protocols- NTP and PTP. The T&F metrology is providing time services from its NTP servers, which are synchronized to IST, for public use. Such services provide accuracy in the range of ms. Any computer terminal can synchronize its clock to the NTP servers at NPL by sourcing their time from:

time.nplindia.org ′or′ time.nplindia.in

Also, the services can be availed through the NTP web time display on the NPL webpage. This is a web-browser based tool to synchronize computer clocks to the NPL NTP server.

For customers requiring higher accuracy for their applications, data packets-based services using PTP protocol is in the works and is going to be launched soon. This service can provide accuracy up to few 10's of ns and requires dedicated ethernet connectivity and PTP compliant hardware like routers and switches.

4.4.1.3 Training Program

As part of CSIR-NPL's outreach program for MSMEs, large-scale industries, PSUs, and general public, regarding the importance of time synchronization in various sectors, the T&F metrology in recent years has started training programs for various sectors and society at large.

The training programs are broadly categorized into two categories:

i. Training for the education of school children and their teachers.
ii. Technical training for MSMEs, industries, and PSU's.

The T&F metrology actively participates in the JIGYASA programs held at CSIR-NPL. The program is designed to imbibe curiosity and develop scientific temper in school children and their teachers. CSIRhas collaborated with Kendriya Vidyalaya Sangathan (KVS) to implement this project in all CSIR laboratories. During JIGYASA program visits, the T&F metrologists interact with the students through atomic clock models and a tour of the laboratories. There are question and answer sessions arranged to arouse the inquisitiveness of the budding scientists. The scheme is highly successful and will be expanded in a phased manner.

The time metrologists also impart technical and classroom teaching for MSMEs and industries. Some of the sectors, which have undergone such training at CSIR-NPL, are the reference standards/accredited calibration laboratories, representatives from banking and telecom sectors. The topics of the training are customized according to the need of the specific sector or client. The T&F metrology collaborates with human resource development of the CSIR-NPL to frame the course curriculum designed specifically for training people of a particular sector [113]. They include the basics of time and frequency measurements, international standards, regulatory bodies, and impact of time synchronization in respective sectors. Such training is regularly arranged throughout the year and under special request, can be arranged in a short notice with number of trainees adjustable on a case-to-case basis.

Apart from regular training exercises, the T&F metrology also provides technical consultancy program for various sectors. It actively provides technical know-how and time-related design services to customers requiring such services. Recently, it has provided such technical consultancy to regional reference standard laboratories.

4.4.2 *Product Development*

In addition to providing services like calibration, traceability, and training, various products (instruments, technologies, etc.) for commercialization are developed through R&D in T&F metrology. Full development cycle of the product from R&D to hand-holding industries/MSMEs through various technology readiness levels are supported. The technologies or products that are ready for commercialization are advertised through the Industrial Liaison Group (ILG) on NPL website [114]. ILG acts as a bridge between the inventor and the industry interested in acquiring the know-how/technology/product from CSIR-NPL. Some of the products/technologies, currently in various phases of their development cycle, are being listed in this section.

4.4.2.1 Fonoclock

The Fonoclock is the name of the technology developed for the dissemination of IST through landline telephone lines [115]. In this technology, as discussed in Sect. 4.4.3, IST is disseminated over public switched telephone network (PSTN) through indigenously developed transmitters and receivers. The entire hardware and the software have been developed at NPL. This technology is a robust telephone time dissemination service with accuracy at par with international standards. It provides the time synchronization accuracy of ±10 ms (more than 100 times better than its predecessor Teleclock) by incorporating communication delay compensation mechanism. The final goal is to transfer the know-how to industry for commercializing compact and affordable time receivers for accessing IST using the easily accessible telephone lines.

Major consumers expected to avail such services defence, telecommunication, banking, and educational sectors. Due to the easy availability of dense landline telephone network, the service can be accessed by remote corners of the country. It has advantages over network time dissemination techniques as internet is still not readily available everywhere to make use of NTP kind of service. It is better than even GNSS (GPS etc.) based timing receivers, which are prone to jamming and spoofing and require clear access to receive the satellite signals (requires antennae to be mounted on the roof). All developed nations, e.g. the USA, UK, Germany, and Japan still offer telephone time dissemination services, and this will hold relevance for the coming decade.

The Fonoclock service has been rigorously tested at CSIR-NPL with co-located transmitters and receivers, as shown in Fig. 4.20.

Four transmitters have been indigenously developed and installed at NPL with IST reference signals and dedicated telephone lines. The receiver is also connected to a telephone line and is programmed to call the transmitter. On-call connection, IST is transmitted to the receiver over the telephone line and is displayed on the display attached to the receiver (shown in Fig. 4.20). Along with 19-inch rack-mounted receivers, compact Fonoclock receivers have also been developed. The

Fig. 4.20 Transmitter and receiver set-up for FonOclock

salient features of the technology are: complete transfer and display of Time-of-Year information, the transmission link delay is calibrated every time the receiver is synced, and the

receiver module can work as a source for further time dissemination. Once the receiver is commercialized and installed in customer locations, telephone time dissemination services can be started from CSIR-NPL.

The salient features of the technology are: complete transfer and display of Time-of-Year information, the transmission link delay is calibrated every time the receiver is synced, and the receiver module can work as a source for further time dissemination. Once the receiver is commercialized and installed in customer locations, telephone time dissemination services can be started from CSIR-NPL.

4.4.2.2 Pulse Distribution Amplifier

One pulse per second (1PPS) is an electrical signal that has a typical width of a few μs and has a sharply rising or abruptly falling edge which repeats once per second accurately. It is typically used for time synchronization applications. One PPS pulse generating devices usually are atomic clocks, frequency standards, other precision oscillators, GNSS receivers, etc. Pulse distribution amplifiers are easily available commercially. But, none of these are manufactured in India. With the aim of Aatamnirbhar Bharat, it is necessary to locally develop and manufacture as many devices/equipment than importing.

1 PPS Pulse Distribution Amplifier system designed at CSIR-NPL is able to distribute one 1 PPS timing signal to five isolated output channels [116]. The prototype of the five-channel 1 PPS pulse distribution amplifier, as shown in Fig. 4.21, has been developed and successfully tested at NPL. The device is ready for technology transfer to industry for commercialization. NPL can help the industry to further improve the device. Such devices will find users in various industries like Timing Laboratories, Telecom ground, and base stations, Satellite ground stations, Network timing infrastructure, Time and frequency equipment manufacturers.

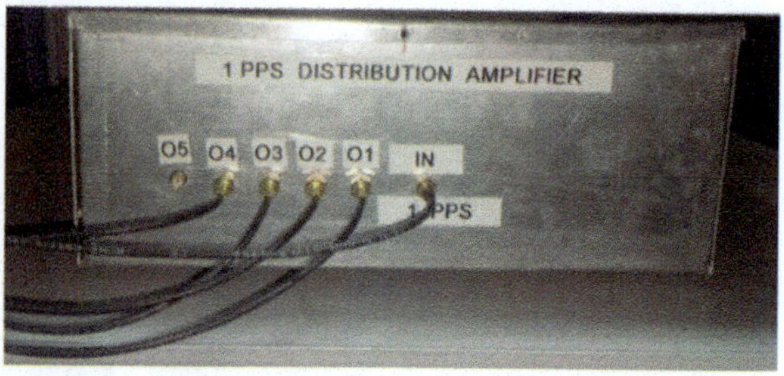

Fig. 4.21 Prototype of the five-channel pulse distribution amplifier

4.4.2.3 Mobile App for IST Dissemination

Another product under development is a mobile app for displaying IST as well as a delay in mobile time from IST. The IST is obtained by pinging on the NTP servers of CSIR-NPL. The android version of the App has been developed and is being tested at CSIR-NPL. Once tested thoroughly, the Mac version of the same will be developed and then the mobile app can be launched/dedicated in service of the nation. The App will serve the purpose of effective time dissemination with just a click on the app on the smartphone.

On similar lines, a general-purpose lock-in amplifier and PID controller has been developed and is awaiting commercialization [117]. Continuous efforts are put in order to provide all kinds of technical solutions in the T&F metrology to various sectors.

4.4.3 National Time Synchronization Programs

CSIR-NPL is working towards a nationwide time dissemination network through both satellite and terrestrial technologies. It will pave the way for reliable and assured time services across India. The network is based on timing requirements of end user and will accordingly use suitable dissemination technologies.

The T&F metrology is working towards the initial phase of implementation of such network. It has started the "National Time Synchronization Programs" in collaboration with major stakeholders.

4.4.3.1 Time Traceability to NavIC

The importance of time synchronization and traceability in Positioning, Navigation, and Timing (PNT) is explained in detail in Sect. 4.2.2. CSIR-NPL after an agreement with ISTRAC-ISRO is providing the UTC traceability to Indian Regional Navigation Satellite System (IRNSS) for PNT applications [110].

The IRNSS with the operational name as Navigation through Indian Constellation (NavIC) is a regional navigation system which has a satellite coverage over India and 1500 km across Indian boundaries. IRNSS consists of 3 Geostationary and 4 Geo Synchronous Satellites located at an orbital height of ~36,000 km. As per the first quarter of 2020 performance reports, has resulted in an average horizontal position accuracy of IRNSS is 2 m, whereas using GPS it is 8–15 m.

IRNSS broadcast the timing signals which are traceable to UTC through UTC(NPLI). The IRNSS system time is maintained by a stable time scale generated from weighted averaging of various atomic clocks and steered with respect to UTC(NPLI) located at ISRO Telemetry, Tracking and Command Network (ISTRAC) Bangalore and Lucknow. The two-timing centers are known as IRNSS Network Timing (IRNWT) centers which act as reference timing stations, for all the ground segment of the IRNSS which consists of INC, 17 IRIMS, IRCDR, IRSCF, etc., The two-timing centers IRNWT-I and IRNWT-II are synchronized to IST through common view GNSS (CVGNSS) and also through Two-Way Satellite Time and Frequency Transfer (TWSTFT) system as shown in Fig. 4.22. The IRNWT centers maintain the system time within 10 ± 1 ns at 95% confidence interval with respect to UTC(NPLI). Apart from the ground segment of IRNSS, a Navigation satellite normally consists of transponders for communication and highly stable space compatible atomic clocks to provide accurate timing signals.

The users of NavIC receivers will get the timing information in the navigation signal through a 27-bit binary number which contains Week Number (WN) (in 10 bits) and Time of Week Count (TOWC) (in 17 bits) with epoch from 00:00 UT 22nd August 1999. As the week number is a 10-bit binary number ($2^{10} = 1024$) there will be a WN roll over after every 1024 weeks, i.e., 7168 days or 7168/365.25 = 19.6249 years. The first NavIC WN rollover has occurred on 6th April 2019. The TOWC is a 12 s count number where it indicates the time elapsed from 00:00 UTC of Sunday for the next subframe to begin. To get the time of the week (TOW) in seconds the receiver multiplies the TOWC by 12 and to get the current time it will subtract 12 from TOW. The receiver gets the UTC time by converting the TOW from seconds to days, hours, minutes elapsed from 00:00 UTC Sunday. Adding 5 h. 30 min to this UTC time, the user gets IST [118]. The year, month, and date should be provided by the user during the installation of the receivers. The clock data of the IRNSS satellites is broadcasted every 48 s, the time offset IRNSS-UTC, IRNSS- GPS, and IRNSS- other GNSS is broadcasted every 20 min.

4.4.3.2 Secondary Timescales at Five RRSLs

Like many other countries Japan, China, Russia, etc., India also needs secondary time scales that provide time and frequency traceability services at regional level. CSIR-NPL reached an MoU with Division of Legal Metrology (DOLM), under Department of Consumer Affairs, which is the standard enforcing authority and also maintains the secondary standards in the country which are directly traceable to SI units at CSIR-NPL. DOLM has given a consultancy project to CSIR-NPL for setting

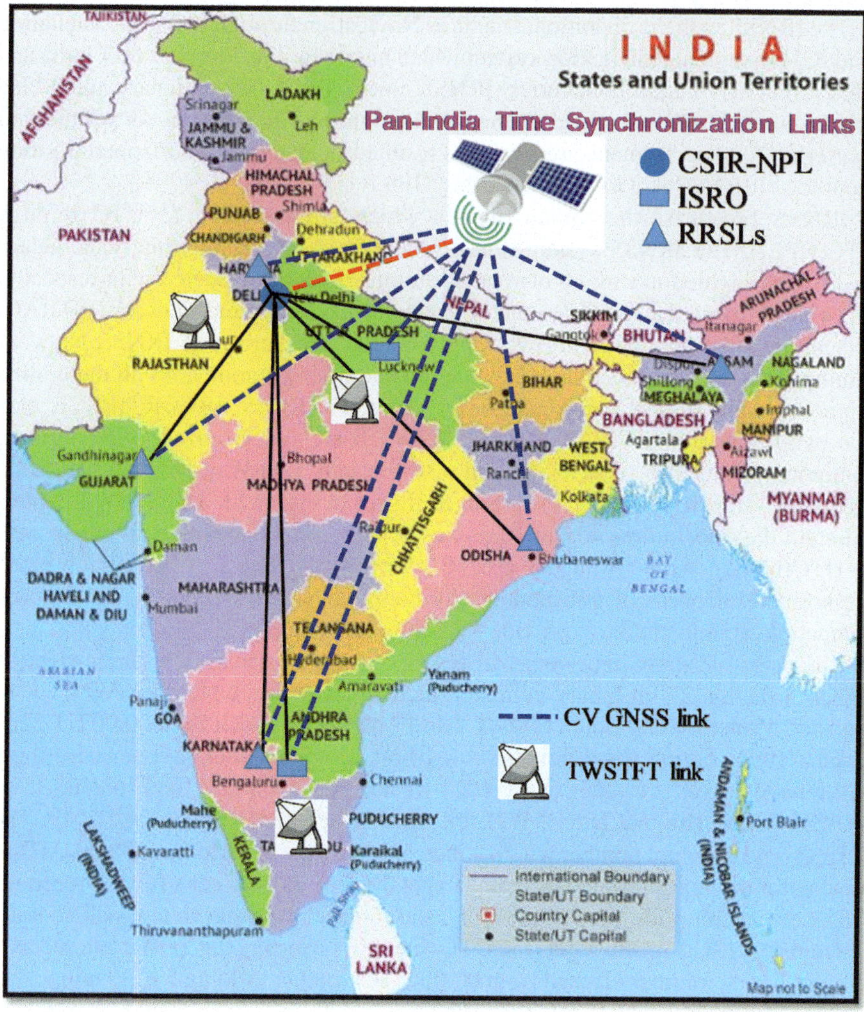

Fig. 4.22 Scheme for time traceability to ISRO and regional reference standards laboratory in India

up the secondary timescales at its five Regional Reference Standards Laboratories (RRSL) located in various cities across the country namely Ahmedabad, Bangalore, Bhubaneshwar, Faridabad, and Guwahati [111]. These secondary time scales will serve the regional end-users with all their requirements of time traceability and calibration.

Under this project, a separate lab facility for secondary time scales is being planned and developed at each RRSL, which need high precision temperature and humidity maintenance, an uninterrupted power supply system with the backup facility, entry of authorized personnel, with fire safety measures, etc. and highly secured internet connectivity for time dissemination to the users through NTP and PTP. The secondary time scales are planned to consist of a dual chain of measurements for redundancy with a common source of 4 high-performance atomic clocks. Hence, at each RRSL two timescales are generated for redundancy and reliability of time traceability services. These time scales are targeted to maintain within 20 ns accuracy with respect to IST at any point of time in a year via CVGNSS and NavIC Common View time transfer as shown in Fig. 4.22. This secondary timing centers will play a critical role in the country for their time traceability services to TSPs & ISPs (Telecom service providers and Internet service providers), calibration of various devices starting from atomic clocks, time interval, and frequency counters, stopwatches, etc., of various industries at regional level. For enforcing the legal time the Legal metrology officers can audit the time traceability through these secondary time scales. These secondary time scales act as a reference time source in providing the UTC traceable time for high precision time stamping applications like digital signatures, digital transactions, etc.

As a part of this project, a Disaster Recovery Centre (DRC) for the primary time scale of IST located in New Delhi is also planned and will be developed at RRSL Bangalore. Establishment of these five secondary time scales presently in the process is very important for the legalization of IST in India.

4.4.3.3 Using the Internet for Time Synchronization in Telecom Sector

Apart from satellite-based time dissemination service, work on terrestrial mode of time dissemination is under progress. One of the ways is based on data packets-based time transfer technologies using the internet or dedicated Ethernet networks, using NTP and PTP standards. Currently, T&F metrology is disseminating IST to civilians using its NTP servers and is building up a reliable dedicated IT infrastructure for customers requiring reliable and stable IST having accuracy in ms range through NTP.

Work is also under progress to develop PTP based time dissemination services from NPL for customers requiring accuracies in the tens of ns range. This service will also be started at regional reference standards labs, as PTP needs dedicated network connectivity, and hence the difficulty in long-distance transmission of PTP time packets. The customers for such services are envisaged to be telecom [119, 120], finance, and energy grid sector where presently they rely on GPS-based technologies for reference time information. The PTP based IST will be a reliable and robust time reference in case failure of satellite signals.

With increasing demand for mobile and internet connectivity, time synchronization in telecom sector plays a vital role. CSIR-NPL has made an agreement with Department of Telecommunications for IST dissemination to telecom and internet service providers (TSP/ISP). The requirement of telecom operators for time accuracy is in the sub-microsecond range, which can be fulfilled by PTP services. The implementation of the project will improve the telecom network efficiency by reducing packet loss due to reduced packet slips in a better synchronized digital network and creating a backup infrastructure in case of failure of satellite-based time services.

4.4.3.4 Traditional Channels of Synchronization

Apart from network data packet-based technologies, CSIR-NPL is also exploring Longwave radio (LWR) based time dissemination technology [112]. Due to the low frequency of the waves, they suffer from very less attenuation when propagating through the atmosphere. The project envisages installing two tall towers geographically distributed to cater to the whole of the country. This technology should help synchronize India within a second of IST. The plan is to erect two towers geographically located in such a way that coverage of the low-frequency signals (~100 kHz) can be accessed throughout India. The project will be implemented by M/s IFR information dissemination services.

Also, work is going on for an advanced version of telephone line-based time dissemination using dedicated transmitters and receivers, providing accuracy in the milliseconds range. The transmitters and receivers have been designed and developed in-house and are in the process of commercialization.

4.5 Summary

The chapter discussed the history of time measurement from mechanical clocks to the present-day high precision atomic clocks, international timekeeping and time synchronization with international reference time known as coordinated universal time (UTC). CSIR-NPL realizes and maintains the IST, which is kept traceable to UTC within 3 ns uncertainty. A discussion on various technologies available for time and frequency synchronization like telephone time dissemination, NTP, PTP, CV/AVGNSS & TWSTFT has been given to help the users to choose a proper mechanism for time transfer from CSIR-NPL to meet their requirements.

Along with time realization and dissemination services, R&D on primary standards is also carried out at CSIR-NPL. India's first Cesium fountain (NPLI-CsF1) was designed and developed at CSIR-NPL that was approved as a primary frequency standard by BIPM. With the ongoing efforts on the development of an optical frequency standard based on single trapped Ytterbium ion (one of the possible candidates for redefinition of SI second), CSIR-NPL will be ready for redefinition of SI second that may happen in 2026.

Presently, IST is being disseminated across Indian subcontinent through calibration, network time services, and satellite links. Especially, the time synchronization services by CSIR-NPL to provide IST traceability with ns accuracy to ISRO for the crucial NavIC program and the ongoing development of five IST traceable secondary timing centers at five RRSLs pan-India for Department of Legal Metrology are crucial programs towards pan-India time synchronization.

Acknowledgements The authors would like to acknowledge all the ex-colleagues, present colleagues Dr. Ashish Agarwal, Ms. Pranalee Thorat, Ms. Suchi Yadav, Mr. Mahavir Prasad, Ms.

Preeti Kandpal, Mr. Devender; and the students of CSIR-NPL for their contributions in establishing various activities of Time and Frequency Metrology. Authors would also like to thank the stakeholders such as Department of Legal Metrology, ISTRAC- ISRO, National Informatics Centre (NIC), DoT, and many others for their continuous support and collaboration in the national missions and endeavors for pan-India time synchronization.

References

1. The International System of Units (SI). BIPM. International Committee for Weights and Measures Ed. 8 (2006)
2. M.A. Lombardi, in *The Mechatroincs Handbook* ed. by R.H. Bishop (CRC Press, Boca Raton, 2002), Chap. 17
3. J. Jespersen, J.F. Randolph, *From Sundials To Atomic Clocks—Understanding Time and Frequency* (NIST, Washington DC, 1999)
4. F. Riehle, in *Frequency Standards—Basics and Applications* (Wiley-VCH, Weinheim, 2004), Chap. 1
5. F.G. Major, *The Quantum Beat: The Physical Principles of Atomic Clocks* (Springer, New York, 1998)
6. G. Kamas, M.A. Lombardi, *Time and Frequency Users Manual—NIST Special Publication 559* (NIST, Boulder, CO, 1990)
7. J. Vanier, C. Audoin, *The Quantum Physics of Atomic Frequency Standards* (Adam Hilger, Bristol, 1989)
8. J. Levine, Rev. Sci. Instrum. **70**(6), 2567 (1999)
9. E. Peik, The *Measuement of Time with Atomic Clocks* (2007)
10. M.A. Lombardi, T.P. Heavner, S.R. Jefferts, J. Meas. Sci. **2**(4), 74 (2007)
11. R. Wynands, in *Time in Quantum Mechanics*, vol. 2, ed. by J.G. Muga, A Ruschhaupt and A del Campo (Springer, Berlin Heidelberg, 2009), Chap. 13, p. 291
12. J.C. Maxwell, *Treatise on Electricity and Magnetism* (Clarendon Press, Oxford, 1873)
13. N.F. Ramsey, in *Biographical Memoirs*, vol. 62 (National Academy Press, Washington, D.C., 1993), Chap. II. Rabi, p. 315
14. N.F. Ramsey, H.B. Silsbee, Phys. Rev. **84**, 506 (1951)
15. C. Hackman, D.B. Sullivan, Am. J. Phys. **63**(4), 306 (1995)
16. L. Essen, J.V.L. Parry, Nature **176**, 280 (1955)
17. R.E. Beehler, Proc. IEEE **55**(6), 792 (1967)
18. A.O. McCoubrey, *IEEE International Frequency Control Symposium*, 1225 (1996)
19. N.F. Ramsey, Metrologia **1**(1), 7 (1965)
20. F.H. Reder, USASRDL Tech. Rep., 2230 (AD 265452) (1961)
21. A. Bagley, L. Cutler, Proceedings of the 18th Annual Frequency Control Symposium, 344 (1964)
22. J. Shirley, W. Lee, R. Drullinger, Metrologia **38**, 427 (2001)
23. S. Chu, Rev. Mod. Phys. **70**(3), 685 (1998)
24. C.N. Cohen-Tannoudji, Rev. Mod. Phys. **70**(3), 707 (1998)
25. W.D. Phillips, Rev. Mod. Phys. **70**(3), 721 (1998)
26. P. Forman, Proc. IEEE **73**, 1181 (1985)
27. H.J. Metcalf, P. van der Straten, *Laser-Cooloing and Trapping* (Springer, New York, 1999)
28. N.F. Ramsey, J. Res. NBS **88**, 301 (1983)
29. BIPM Annual Report on Time Activities, vol. 6, BIPM (2011)
30. F. Levi, D. Calonico, L. Lorini, A. Godone, Metrologia **43**(6), 545 (2006)
31. M. Kumagai, H. Ito, M. Kajita, M. Hosokawa, Metrologia **45**(2), 139 (2008)

32. T.P. Heavner, S.R. Jefferts, E.A. Donley, J.H. Shirley, T.E. Parker, Metrologia **42**(5), 411 (2005)
33. A. Takamizawa, S. Yanagimachi, Y. Shirakawa, K. Watabe, K. Hagimoto, T. Ikegami, *Proceedings of the 42nd Annual Precise Time and Time Interval Systems and Applications Meeting*, Reston, Virginia, 321 (2010)
34. K. Szymaniec, S.E. Park, G. Marra, W. Chalupczak, Metrologia **47**(4), 363 (2010)
35. S. Weyers, U. Hübner, R. Schröder, Chr. Tamm, A. Bauch, Metrologia **38**(4), 343 (2001)
36. V. Gerginov, N. Nemitz, S. Weyers, R. Schroder, D. Griebsch, R. Wynands, Metrologia **47**(1), 65 (2010)
37. J. Guena, M. Abgrall, D. Rovera, P. Laurent, B. Chupin, M. Lours, G. Santarelli, P. Rosenbusch, M. Tober, R. Li, K. Gibble, A. Clairon, S. Bize, IEEE Trans UFFC **59**(3), 391 (2012)
38. J. Guena, P. Rosenbusch, P. Laurent, M. Abgrall, D. Rovera, G. Santarelli, M. Tober, S. Bize, A. Clairon, IEEE Trans. UFFC **57**(3), 647 (2010)
39. P. Arora, S.B. Purnapatra, A. Acharya, A. Agarwal, S. Yadav, K. Pant, A. Sen Gupta, IEEE Trans. Instrum. Meas. **62**(7), 2036 (2013)
40. Y.S. Domnin, G.A. Eolkin, A.V. Novoselov, L.N. Kopylov, V.N. Baryshev, V.G. Pal'chikov, *Proceedings of IEEE IFCS and PDA Exhibition Jointly with the 17th EFTF*, 127 (2003)
41. J.M. López-Romero, S. López-López. E. de Carlos-López, N. Shtin, F. Jiménez-Tapia, Simposio de Metrología (2008)
42. F.U. Jun-Xian, C. Shuai, C. Xu-Zong, Y. Dong-Hai, W. Yi-Qiu, Chin. Phys. Lett. **18**(11), 1457 (2001)
43. T.Y. Kwon, H.S. Lee, S.H. Yang, S.E. Park, IEEE Trans. Instrum. Meas. **52**(2), 263 (2003)
44. P.T.H. Fisk, Rep. Prog. Phys. **60**(8), 761 (1997)
45. S.A. Diddams, Th Udem, J.C. Bergquist, E.A. Curtis, R.E. Drullinger, L. Hollberg, W.M. Itano, W.D. Lee, C.W. Oates, K.R. Vogel, D.J. Wineland, Science **293**, 825 (2001)
46. M. Takamoto, F.L. Hong, R. Higashi, H. Katori, Nature **435**, 321 (2005)
47. P. Gill, Metrologia **42**, 125 (2005)
48. S. Panja, S. De, S. Chakraborty, H. Kohli, N. Batra, D.K. daniel, A. Sinha, P. Arora, A. Agarwal, A. Sen Gupta, Proceedings of ADMET 2013, New Delhi (2013)
49. P. Arora, A. Awasthi, V. Bharath, A. Acharya, S. Yadav, A. Agarwal, A. Sengupta, Pramana -J. Phys. **82**(2), 173 (2014)
50. G. Panfilo, F. Arias, Metrologia **56**, 042001 (2019)
51. Le Systeme International d'Unites, 7th ed., Bureau International des Poids et Mesures, Sevres, France (1998)
52. E.F. Arias, D. Matsakis, T.J. Quinn, P. Tavella, IEEE Trans. Ultrason. Ferroelectr. Freq. Control **65**, 898 (2018)
53. J. Lodewyck, Metrologia **56**, 055009 (2019)
54. G. Petit, G. Panfilo, IEEE Trans. Instrum. Meas. **62**, 1550 (2013)
55. T.J. Quinn, Proc. IEEE **79**, 894 (1991)
56. Annual Report of the BIPM Time Section ~ Bureau International des Poidset Mesures, Sevres, France, 8, 17 (1995)
57. D.P. Sengupta, *Science and Culture*, January–February, 12 (2016)
58. M.P. Olaniya, S. Yadav, P. Kandpal, M. Dixit, A. Agarwal, IEEE Xplore, June 2019, https://doi.org/10.23919/ursiap-rasc.2019.8738334
59. A. Agarwal, M.P. Olaniya, S. Yadav, https://ieeexplore.ieee.org/author/37086861186
60. D.L. Mills, IEEE Trans Commun **39**, 1482 (1991)
61. IEEE 1588 home page. https://standards.ieee.org/standard/1588-2008.html. Accessed 10 July 2020
62. P. Arora, A. Acharya, S. Yadav, A. Sen Gupta, V.N. Ojha, *A New Time Dissemination Service with an Accuracy of* ± 10 *ms Over Public Switched Telephone Network.* 32nd URSI General Assembly 2017, Montreal, Canada
63. D.W. Allan, M.A. Weiss, *Proceedings of Frequency Control Symposium*, 334 (1980)
64. J. Azoubib, W. Lewandowsky, *7th CGGTTS Meeting*, 76 (1998)

65. D. Kirchner, H. Ressler, P. Grudler, F. Baumont, Ch. Veillet, W. Lewandowski, W. Hanson, W. Klepczynski, P. Uhrich, Metrologi **30**, 813 (1993)
66. https://www.bipm.org/kcdb/
67. P. Arora, S. B. Purnapatra, A. Acharya, A. Agarwal, S. Yadav, K. Pant, A. Sen Gupta
68. A. Acharya, V. Bharath, P. Arora, S. Yadav, A. Aagrwal, A. Sen Gupta, MAPAN
69. https://www.bipm.org/en/bipm-services/timescales/time-ftp/Circular-T.html#nohref,nos, 323, 335
70. A. Zhang, A. Sen Gupta, P. Arora, A. Acharya, A. Agarwal, in Proceedings of 28th European Time and Frequency Forum, Neuchatel, Switzerland (2014)
71. A. Sen Gupta, P. Arora, A. Agarwal, K. Pant, Current Sci. **100**, 1393 (2011).
72. P. Arora, S. Chowdhury, A. Agarwal, K. Pant, A. Sen Gupta, Ind. J. Pure App. Phys. **49**, 590 (2011)
73. K. Pant, P. Arora, S. Yadav, A. Sen Gupta, Mapan – J. Met. Soc. Ind. **26** 285 (2011)
74. P. Arora, A. Agarwal, A. Sen Gupta, Rev. Sci. Instrum. **82**, 125103 (2012)
75. P. Arora, A. Sen Gupta, Rev. Sci. Instrum. **83**, 046104 (2012)
76. P. Arora, S.B. Purnapatra, A. Acharya, R. Kumar, A. Sen Gupta, Mapan – J. Met. Soc. Ind. **27**, 31 (2012)
77. K. Pant, P. Arora, S. Yadav, A. Sen Gupta, Ind. J. Pure App. Phys. **50**, 593 (2012)
78. A. Bauch et al., Generation of UTC (PTB) as a fountain-clock based time scale. Metrologia **49**(3), 180 (2012)
79. G.D. Rovera et al., UTC(OP) based on LNE-SYRTE atomic fountain primary frequency standards. Metrologia **53**, S81 (2016)
80. S.G. Turyshev, *From Quantum to Cosmos: Fundamental Physics Research in Space* (World Scientific, Singapore, 2009)
81. T. Rosenband et al., Science **319**, 1808 (2008)
82. H. Katori, T. Ido, M. Kuwata-Gonokami, J. Phys. Soc. Jpn. **68**, 2479 (1999)
83. J. McKeever, J.R. Buck, A.D. Boozer, A. Kuzmich, H.C. Nägerl, D.M. Stamper-Kurn, H.J. Kimble, Phys. Rev. Lett. **90**, 133602 (2003)
84. C.W. Chou, D.B. Hume, J.C.J. Koelemeij, D.J. Wineland, T. Rosenband, Frequency comparison of two high-accuracy Al+ optical clocks. Phys. Rev. Lett. **104**, 070802 (2010)
85. M.S. Safronova, M.G. Kozlov, C.W. Clark, Precision calculation of blackbody radiation shifts for optical frequency metrology. Rev. Lett. **107**, 143006 (2011)
86. N. Huntemann, C. Sanner, B. Lipphardt, Chr. Tamm, E. Peik, Single-ion atomic clock with 3×10^{-18} systematic uncertainty. Phys. Rev. Lett. **116**, 063001 (2016) https://doi.org/10.1103/physrevlett.116.063001
87. Chr. Tamm, S. Weyers, B. Lipphardt, E. Peik, Stray-field-induced quadrupole shift and absolute frequency of the 688-THz 171Yb+ single-ion optical frequency standard. Phys. Rev. A, **80**, 043403 (2009). doi:https://doi.org/10.1103/PhysRevA.80.043403
88. G.P. Barwood, H.S. Margolis, G. Huang, P. Gill, H.A. Klein, Measurement of the electric quadrupole moment of the 4d 2D5/2 level in 88Sr+. Phys. Rev. Lett. **93**, 133001 (2004). https://doi.org/10.1103/physrevlett.93.133001
89. N. Huntemann, M. Okhapkin, B. Lipphardt, S. Weyers, Chr. Tamm, E. Peik, High-accuracy optical clock based on the octupole transition in 171Yb+. Phys. Rev. Lett. **108**, 090801 (2012). https://doi.org/10.1103/PhysRevLett.108.090801
90. L. Sharma, A. Roy, S. Panja, S. De, Atomic flux distribution from a low divergent dark wall oven. Rev. Sci. Instrum. **90**, 053202 (2019). https://doi.org/10.1063/1.5090199
91. L. Sharma, A. Roy, S. Panja, V.N. Ojha, S. De, Estimation of the ion-trap assisted electrical loads and resulting BBR shift. Sci. Rep. **8**, 16884 (2018). https://doi.org/10.1038/s41598-018-35234-5
92. N. Batra, S. Panja, S. De, A. Roy, S. Majhi, S. Yadav, A. Sen Gupta, Design and construction of a helical resonator for delivering radio frequency to an ion trap. MAPAN-J. Metrol. Soc. I. **32**, 193–198 (2017). https://doi.org/10.1007/s12647-017-0209-5
93. S. Panja, S. De, S. Yadav, Sen Gupta, Measuring capacitance and inductance of a helical resonator and improving its quality factor by mutual inductance alteration. Rev. Sci. Instrum. **86**, 056104 (2015). https://doi.org/10.1063/1.4919910

94. A. Rastogi, N. Batra, A. Roy, J. Thangjam, V.P. S. Kalsi, S. Panja S. De, Design of the ion trap and vacuum system for 171Yb-ion optical frequency standard. MAPAN-J. Metrol. Soc. I. **30**, 169–174 (2015). https://doi.org/10.1007/s12647-015-0140-6
95. N. De Batra, S. Chakraborty, S. Panja, A. Sen Gupta, Design of an ion trap for trapping single 171Yb+. Curr. Sci. **106**, 1348–1352 (2014)
96. N. Batra, A. Roy, C. Samal, S. Majhi, S. Panja, S. De, Automation of an optical frequency standard experiment. IEEE Conference, Proceedings of 2nd CIEC, pp 78–82 (2016). https://doi.org/10.1109/ciec.2016.7513823
97. N. Batra, A. Roy, S. Majhi, S. Panja, S. De, Singly charged ions for optical clocks. Asian J. Phys. **25**, 1069 (2016)
98. Chr. Balzer, A. Braun, T. Hannemann, Chr. Paape, M. Ettler, W. Neuhauser, Chr. Wunderlich, Electrodynamically trapped Yb+ ions for quantum information processing. Phys. Rev. A, **73**, 041407 (2006). doi: https://doi.org/10.1103/PhysRevA.73.041407
99. H. Dehmelt, Proposed 1014 $\nu/\Delta\nu$ laser fluorescence spectroscopy on T1+ monoion oscillator II. Bull. Am. Phys. Soc. **20**, 60 (1975)
100. W. Nagourney, N. Sandberg, H. Dehmelt, Shelved optical electron amplifier: observation of quantum jumps. Phys. Rev. Lett. **56**, 2797 (1986). https://doi.org/10.1103/PhysRevLett.56.2797
101. D.J. Wineland, C. Monroe, W.M. Itano, D. Leibfried, B.E. King, D.M. Meekhof, J. Res. Nat. Inst. Stand. Technol. **103**, 259 (1998)
102. V. Smotlacha, A. Kuna, W. Mache, *EFTF-2010 24th European Frequency and Time Forum*, Noordwijk, 1 (April 2010). https://doi.org/10.1109/eftf.2010.6533650
103. P. Krehlik, L. Sliwczynski, L. Buczek, J. Kolodziej, M. Lipinski, Metrologia **52**, 82 (2015). https://doi.org/10.1088/0026-1394/52/1/82
104. D, Grzegorz, W. Tomasz, *Proceedings of the 43rd Annual Precise Time and Time Interval Systems and Applications Meeting*, Long Beach, California, November 45 (2011)
105. N. Kaur, P. Tuckey, *European Frequency and Time Forum*, 1 (April 2016)
106. https://www.ohwr.org/projects/white-rabbit
107. T. Wlostowski, *Precise Time and Frequency Transfer in a White Rabbit network*, Master's thesis, Warsaw University of Technology, (May 2011)
108. P. Moreira, P. Alvarezl, J. Serrano, I. Darwezeh, *IEEE International Frequency Control Symposium*, 449 (June 2010). https://doi.org/10.1109/freq.2010.5556289
109. CSIR-NPL CFCT webpage (2020). http://www.nplindia.in/centre-calibration-testing-cfct. Accessed 5 July 2020
110. CSIR-NPL ISRO MoU webpage (2020). http://www.nplindia.in/csir-npl-and-isro-sign-mou-time-and-frequency-traceability-services. Accessed 5 July 2020
111. CSIR-NPL RRSL MoU webpage (2020). https://consumeraffairs.nic.in/sites/default/files/MonthlySummaryDecember2018.pdf. Accessed 5 July 2020
112. CSIR-NPL IFR MoU webpage (2020). http://www.nplindia.in/csir-npl-and-ifr-information-dissemination-services-pvt-ltd-sign-mou-time-and-frequency-traceability. Accessed 5 July 2020
113. CSIR-NPL HRD webpage (2020). http://www.nplindia.in/human-resource-development-group. Accessed 5 July 2020
114. CSIR-NPL ILG webpage (2020). http://www.nplindia.in/industrial-liaison-group. Accessed 5 July 2020
115. CSIR-NPL product development Fonoclock details (2020). http://www.nplindia.in/iglpdf/FonOclock.pdf. Accessed 5 July 2020
116. CSIR-NPL product development Pulse distribution amplifier details (2020). http://www.nplindia.in/iglpdf/Pulse-Distribution-Amplifier.pdf. Accessed 5 July 2020
117. CSIR-NPL product development Lock-in amplifier details (2020). http://www.nplindia.in/iglpdf/Lock-in-Amplifier-and-PID-Controller.pdf. Accessed 5 July 2020
118. ISRO-IRNSS-ICD-MSG-1.0, Signal in space interface control document for messaging service

119. CSIR-NPL and DoT MoU details (2020). http://www.nplindia.in/mou-was-signed-between-csir-national-physical-laboratory-and-department-telecommunication. Accessed 10 July 2020
120. Vigyan Prasar CSIR-NPL and DoT MoU details (2020). http://vigyanprasar.gov.in/isw/time_stamping_with_IST_story.html. Accessed 10 July 2020

Chapter 5
Time and Frequency Metrology

Part II: Application of Indian Standard Time for Safe Digital India

Bharath Vattikonda, Manoj Das, Trilok Bhardwaj, Subhasis Panja, Poonam Arora, Anurag Gupta, and D. K. Aswal

Abstract Time synchronization and frequency stability are crucial requirements in various cyber physical systems (CPS) representing e-governance, navigation, smart technologies, financial and strategic sectors that finally create a self-sufficient and cyber secure digital India. To maintain these two parameters within the tolerance limit, they require a continuous traceability with respect to Coordinated Universal Time (UTC), which is realized at CSIR-NPL as IST described in Chapter 4 (Part 1). At present, time is not a legal entity in India. IST is used as official time but has not been legalized yet. In the absence of legal time and its implementation, cyber safety with time audit mechanism is not available to all sectors and CPS applications based on precise time synchronization. A detailed road map for time accuracy and traceability requirements of various sectors has been discussed with an emphasis on cyber security and cyber physical systems in India. The steps needed have been listed to synchronize all clocks in the country to IST, starting from legalization and consequent implementation of IST in India. A new policy has been proposed for time accreditation, and audit mechanism by elaborating on usage of existing and new

B. Vattikonda · M. Das · T. Bhardwaj · S. Panja · P. Arora · A. Gupta (✉) · D. K. Aswal
CSIR-National Physical Laboratory, Dr. K. S. Krishnan Marg, New Delhi 110012, India
e-mail: anurag@nplindia.org

B. Vattikonda
e-mail: bharath.v@nplindia.org

M. Das
e-mail: manoj.das@nplindia.org

T. Bhardwaj
e-mail: trilok@nplindia.org

S. Panja
e-mail: panjas@nplindia.org

P. Arora
e-mail: arorap@nplindia.org

D. K. Aswal
e-mail: dkaswal.npl@nic.in

D. K. Aswal (ed.), *Metrology for Inclusive Growth of India*,
https://doi.org/10.1007/978-981-15-8872-3_5

infrastructure. As the future economy of the world would depend upon digitalization therefore "Time and Frequency Metrology" is going to be an essential component of quality infrastructure. A cyber secure digital India through IST synchronization will have a significant economic impact to the tune of 10% of GDP.

In a few years, all great physical constants will have been approximately estimated, and that the only occupation which will be left to men of science will be to carry these measurements to another place of decimals.

–James C. Maxwell (1831–1879)

5.1 Introduction

Digital India is a flagship program of Government of India which is based on a vision to transform India into a digitally empowered and knowledge economy [1]. The digital India program emphasizes on extensive use of information and communication technologies (ICT) to make most of the services electronically available to all the citizens. India's march towards 5 trillion-dollar economy [2] will be possible by new digital ecosystem—defined by various digital technologies along with robust, secure and synchronized devices and networks—supporting all the GOI services and strategic sectors. Any digital architecture depends on accurate time stamping for proper, smooth and safe functioning. Time synchronization to one common national reference time between all devices and networks is must for maintaining integrity, avoiding chaos and mishandling. Time synchronization with different time accuracies (ms to ns level) is needed for mission critical social and strategic applications in telecommunications, finance, power grid, broadcasting, navigation, defence, aviation, space, cyber security etc. Accurate time stamps are crucial in sorting events and digital transactions chronologically; maintaining authentic and confidential electronic records; proper diagnosis and management of system failures, cyber-attacks, audit trails and legal aspects [3, 4]. It is needless to say that accurate pan-India time synchronization to one common national reference time is crucial for safe Digital India.

At present, time is not a legal entity in India. Various sectors have their own time synchronization protocols with no law and implementation for traceability to a common national reference time. With increase in number of cyber-attacks, such situation is really alarming for the future based on digital foundations. For addressing the issue of building safe digital India, a national roadmap for pan-India time **synchronization** is required for identifying the focus areas such as institutional collaborations on effective time generation, transfer and disseminations, legalization and implementation of national reference time, etc. A policy for time accreditation and audit mechanism along with roles and responsibilities of various agencies have been proposed for the implementation of legalized IST. The infrastructure required for the proposed legalization of IST has also been identified along with the coordination required between NMI, various ministries and regulatory bodies of the mentioned sectors, and industries. This is exactly the focus of the present chapter.

Every aspect of time and frequency metrology has been presented with the ultimate goal to create understanding and identification of gaps in time synchronization for safe Digital India. A detailed roadmap, required for the legalization and implementation of IST in India, is presented in Sect. 5.2. All the efforts towards pan-India IST synchronization, legalization and implementation of IST will lead to significantly improve the efficiency and reliability of various sectors as well as a measurable economic impact. In Sect. 5.3, a first-hand estimate of the total impact on the economy is made, that turns out to be ~10% of GDP.

5.2 Road Map Towards Time Synchronization for Safe Digital India

For India to become a $5 trillion economy by 2024, digitalization is one of the key pillars. Digital economy involves the electronic governance, highly interconnected cities, town and villages empowering them to access the high-end technologies. The process of digitalization starts with digitization where the physical format/phenomena is converted into a digital format. This digital information is used for governance or process control or automation in various sectors. All these sectors must have a secure cyber-physical framework for its operation and reliability.

The road map towards time synchronization across pan India can be thought by considering various sectors as cyber physical systems which require time and frequency constraints to perform, their normal operation, maintenance and safety.

The time and frequency constraints are implemented in various sectors only when there is a legal time for all the users to be utilized. Hence, we propose an Indian Legal time act and a policy for time and frequency dissemination and audit. With legalization there also comes a responsibility of providing the time and frequency services to all the required sectors at various levels of accuracy. Hence, the infrastructure required for time and frequency traceability services is also discussed. After implementing the legal act with adequate infrastructure and time accreditation policy the impact of it on Indian economy at an appropriate approximation is presented.

5.2.1 Relevance of Cyber Physical Systems in India

Any government or private sector can be considered as a cyber-physical system where the governance or process control or automation or system monitoring can be controlled remotely. Cyber physical systems (CPS) are smart systems that connect the physical phenomena and computational capabilities through a network. There are other technologies which has some overlap with CPS like Industrial Internet, Machine-to-Machine (M2M), smart cities, Internet of Things (IoT) and others [5]. The CPS platform is a combination of hardware components like sensors, actuators, computing, communication channels and software components like device drivers and operating systems.

The modern cyber physical systems need interaction with multiple computational systems at higher rates which are synchronized very precisely. Time is the common parameter that computing and physical systems in CPS share. Time synchronization requirements in CPS depend on applications. Some applications may not require synchronization (scheduled events) between the nodes or sensors at all and some applications may require micro second to sub microsecond level as they have to perform time triggered actions.

Time and Frequency requirements [6] of a CPS are discussed below.

5.2.1.1 Sampling Frequency

Different applications in CPS require different sampling frequency (samples collected per second). Weather forecasting, electrical strike characterizing, human behaviour is difficult to predict as they occur at high frequency (short time interval). As per the Nyquist Theorem for proper reconstruction of the actual phenomena the sampling frequency should be at least twice the natural frequency of the phenomena. Hence the sampling frequency rate is one of the main criteria for sampling requirements in CPS. In case of multiplexing signals from various sensors and actuators the sampling frequency is again a critical parameter.

5.2.1.2 Chronological Sequencing

Information received from various sensors and actuators in a distributed CPS has their own precision and time scales. All inputs received from such systems must be merged such that all observers agree on sequence of events.

5.2.1.3 Simultaneous Events

Time triggered distributed CPS needs to act at the same time to all observers in the system. This will allow co-ordination and localization of the events. For example, in power grid for the fault location applications require measurements with microsecond precision to distinguish the events which appears to be occurred simultaneously.

5.2.1.4 Latency of Reporting

In time sensitive applications the information collected by the sensors may be valid only for a particular time interval. Latency requirements in a CPS depends on respond, react and execute time of the communication network. For example, in a 5G network or an industrial automation network the input and the response time latency is very low.

Secure Timing Source

Secure timing is one of the vital requirements of the CPS and hence it must be considered as a basic element in the design of CPS. As timing signals is one of the basic requirements of the CPS it must also be secured for safety, reliability, resilience. Most of the timing sources are from GNSS hence they must be protected from spoofing, and jamming.

The Time-Sensitive Networking (TSN) task group which is a part of IEEE 802.1 working on TSN develops standards that complement existing ones to enable real-time capabilities of Ethernet switches by introducing features such as system-wide clock synchronization, redundancy, and traffic preemption [7]. TSN targets to support real-time requirements with minimal latency. For such solutions, clock synchronization is a fundamental building block.

Various sectors like manufacturing Industry (4.0), Space Technology, Telecom Sector, Transportation, Smart Grid, Smart City, Smart Agriculture, Medical sciences etc., require complex and heterogeneous CPS that are time synchronized for delivering the desired outputs. The future Cyber Physical Systems (CPS) will require interoperability between elements and systems, coordinated by time framework and synchronization architectures.

A proper design methodology is needed in order to design robust, coordinated and time-sensitive CPS that meets the reliability and resiliency demands of the concerned sector. The timing requirements of various CPS vary widely from seconds to even sub-nanosecond level accuracy depending on application. It is required to understand and specify the timing requirements of various CPS for safe digital India.

5.2.2 Time Synchronization in Various CPS

Time interval/frequency, phase, epoch, Time-of-the-day (TOD) are very important requirements for various applications in our day to day life. Time synchronization is very important for Position, Navigation and Timing (PNT) applications like navigation in sea, air, space and on road (google maps), to the location based apps in our phone like Facebook check-in, ride sharing by Ola or Uber etc., Frequency synchronization (Syntonization) is very important in power grid for the synchronization of various generating stations frequency with the random variations of the load on the grid. Time/Phase synchronization and frequency stability with in tolerance limits are the most important requirement of telecommunications starting from 1 to 5G mobile technologies. Epoch time synchronization is important for global digital connectivity. An email received from a person in different time zone is time stamped according to the local time by the email server, as the epoch of the systems are synchronized.

The role of time and frequency synchronization in various sectors will be discussed in the following sub-sections.

5.2.2.1 Telecommunications and 5G

Mobile communication is a very basic day to day life requirement of a normal person in this present modern world. This is a fast-evolving sector especially important in this age of cloud storage and bulk data transmission. Many applications require telecommunication support starting from basic conversations to data-based voice and video call, digital transactions in banking and finance sector, online commerce, distance education and Tele-medicine. The technology for telecommunication has moved from circuit switched network to data packet-based network.

In our country, 4G is currently being widely used and trials are going on faster 5G technology. This requires high bandwidth and faster processing and a compatible telecom and internet infrastructure. Time synchronization plays an important role in the reliable operation of communication networks. It enables network nodes to maintain an accurate and precise time which results in accurate time stamping and ordering of the message traffic. In case of two different network operators having different reference time, and a customer in one network calling a user on the other network might experience voice lags and in extreme cases the call might get terminated, resulting in call drop. Similarly, a user calling a friend abroad and is on an international network and a different time zone might not be able to place the call itself, if the respective national references times (with which the local network operators are referenced to) are not traceable to global time standards, UTC. A disciplined time synchronized telecom sector helps in accurate forensics like call tracing and is essential for intelligence gathering.

The advanced 5G technology is an ultrafast wireless broadband technology: era of interconnected smart devices, unlock the transformational Impact of machine to machine (M2M) and Internet of Things (IoT) technologies. All the smart devices in the 5G network require a traceable reference time with greater accuracy for high speed data transmission.

Time synchronization plays a very important role in any generation of mobile communication i.e., Carrier synchronization in 1G, symbol synchronization in 2G, code synchronization in 3G and cell search in 4G mobile communications [8]. The 3

GPP2 C.S0010-B, 2010 standards specify that in 3G the Base Stations (BS) should recover the CDMA signal with an accuracy of ±3 μs with respect to UTC. It means the total time error budget i.e., Maximum time deviation (MTDEV) is ±3 μs with respect to UTC for a carrier signal generated from any base station. Whereas in case of 4G LTE-TDD this synchronization requirements are tightened to ±1.1 μs as shown in Fig. 5.1. Primary reference telecom clock (PRTC) acts as the source of the reference time for the telecom network at state level in the countries depending on the traffic and users they may be more in number. In 5G network the access nodes are distributed densely to provide significant improvement in number of connected devices, low end-to-end latency, and simultaneous operation of devices and energy efficiency which requires wider bandwidths and advanced radio access techniques. Timing synchronization requirements are very critical to achieve such improvements from 4 to 5G as shown in Fig. 5.1. The various synchronization requirements are shown in Table 5.1. The 5G wireless communication technology requires time synchronization ranging from 65 to 240 ns.

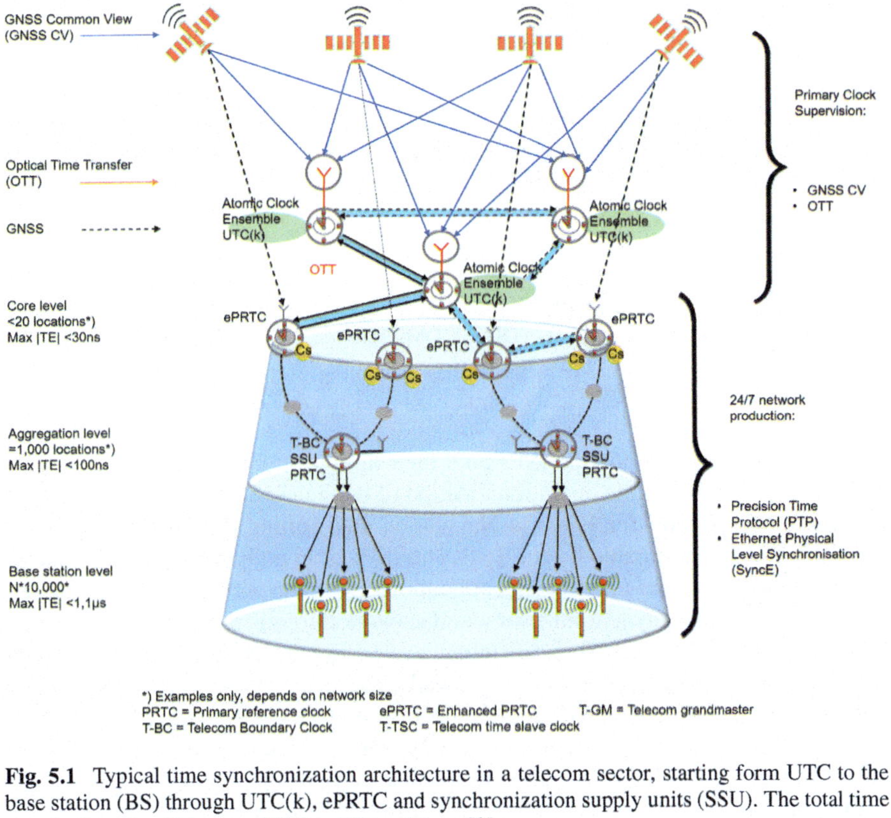

Fig. 5.1 Typical time synchronization architecture in a telecom sector, starting form UTC to the base station (BS) through UTC(k), ePRTC and synchronization supply units (SSU). The total time error budget should be from UTC to BS is <1.1 μs [9]

Table 5.1 Requirements of time synchronization in various telecom technologies [10]

Application	Precision needed
Wired (sources of timing)	
PRTC (primary reference time clock)	±100 ns with respect to coordinated universal time (UTC)
ePRTC (enhanced primary reference time clock)	±30 ns with respect to UTC
Wireless	
5G wireless communication	±65 to 240 ns
CDMA2000	±3 to 10 μs
TD-SCDMA	±3 μs
WCDMA-TDD (Node TDD mode)	±2.5 μs
W-CDMA MBSFN	±12.8 μs
LTE MBSFN	<±1 μs (spec. still under study)
W-CDMA (home node TDD mode)	Microsecond-level accuracy
WiMAX	±1–1.43 μs
LTE-TDD (wide area base station)	3 μs (small cell, < 3 km radius)
	10 μs (large cell, > 3 km radius)
LTE-TDD (home area base station)	3 μs (small cell, < 500 m radius)
LTE-TDD to CDMA handovers	±10 μs
IP network delay monitoring	±100 μs to ±1 ms

Currently, the TSPs and the ISPs pool their time from GNSS services like GPS, Glonass, Galileo. With the operationalization of indigenous NavIC satellite services by ISRO, which provides Position Navigation and Timing (PNT) services and is traceable to IST is available for the TSPs and ISPs as an indigenous reliable system. NavIC timing receivers for telecom applications are in the process of the developed. The telecom industry synchronizes it wired network through, precision time protocol (PTP) based on IEEE 1588. But at different levels down the hierarchy the use of stable clocks or dedicated channels for PTP to each base station is not economical hence, the low-cost technology of GPS disciplined oscillators is being used. However, to build a redundancy in the network in case of failure of satellite-based time services due to spoofing and jamming, terrestrial mode of time synchronization is equally important.

The international regulatory body ITU-T has provided various guidelines for time synchronization in telecom sectors, as shown in Table 5.2.

Table 5.2 Guidelines by ITU-T on various telecom time standards

ITU -T	Name of the standard
ITU-T G.811.1	Timing characteristics of enhanced primary reference clocks
ITU-T G.8261/Y.1361	Timing and synchronization aspects in packet networks
ITU-T G.8262/Y.1362	Timing characteristics of synchronous equipment slave clock
ITU-T G.8263/Y.1363	Timing characteristics of packet-based equipment clocks
ITU-T G.8264/Y.1364	Distribution of timing information through packet networks
ITU-T G.8265.1/Y.1365.1	PTP telecom profile for frequency synchronization
ITU-T G.8266/Y.1376	Timing characteristics of telecom grandmaster clocks for frequency synchronization
ITU-T G.8271/Y.1366	Time and phase synchronization aspects of telecommunication networks
ITU-T G.8271.1/Y.1366.1	Network limits for time synchronization in packet networks
ITU-T G.8272/Y.1367	Timing characteristics of primary reference time clocks
ITU-T G.8273/Y.1368	Frame work of Phase and time clocks
ITU-T G.8275.1/Y.1369.1	PTP telecom profile for phase/time synchronization with full timing support from the network
ITU-T G.8275.2/Y.1369.2	PTP telecom profile for phase/time synchronization with partial timing support from the network

5.2.2.2 Position, Navigation and Timing (PNT) Applications

Navigation has evolved from low accurate positioning mechanisms using sextant, compass, Orion, Polynesian stick chart etc., [11] to highly accurate radar and GNSS based systems. The world at present is very much dependent on GNSS (GPS, Glonass and Galileo) even for its day to day activity. In 2017 an advisory body issued a report estimating an economic impact of $1.25B to the United Kingdom for a five-day disruption to global navigation satellite systems (GNSS) [11]. Similarly, in 2019 the U.S Department of Commerce estimated an economic loss of at least $1 billion per day for loss of GPS for 30 days [12].

Location based apps like google maps, ride sharing in Uber and Ola, Facebook check-In, e-commerce apps etc., are gaining their importance in our day to day life. Apps in our smart phones initially were using triangulation method based on cellular towers for the location-based services which is less accurate and reliable. However, with the development of Qualcomm "Assisted GPS" (AGPS) technology which combines GNSS signals with cellular signals and Wi-Fi access points of the users can now provide accurate location of users to within a few feet [13].

Satellite based navigation needs accurate time for precise positioning and navigation. A GNSS receiver on the earth needs at least four satellite signals simultaneous to locate its position on the globe. The co-ordinates of a location (X, Y, Z) corresponds to latitude, longitude, and altitude of the GNSS receiver which can be calculated from the pseudo range measurements of the receiver from the three satellites (three equations and three variables). But, as the satellites and the receiver are not stationery with respect to each other (in a rotating reference frame of reference) there will be doppler shifts, tropospheric and ionospheric effects, multipath reflections of the received signals etc. All these realistic effects will dilute the precision of the location [14]. Hence, a fourth parameter time t, is required to estimate the time offset Δt_r between the receiver time (t_r) and, the GNSS time (t_g) at which the signal is broadcasted from the satellite. Thus, to solve these four parameters we need signals from at least four satellites simultaneously. The time offset Δt_r ($\Delta t_r = t_r - t_g$) calculated by the receiver at every epoch of the GNSS signal varies from tens of nano seconds to few microseconds depending up on the quality of the GNSS receiver. As these GNSS signals are electromagnetic in nature and travel at velocity of light, 10 ns time offset dilute the position accuracy at least by 3 m. A low-end GNSS receiver can update position and time at a rate of 20 Hz which corresponds to an interval of 50 ms [15].

Regarding the timing solutions from the satellite navigation systems each GNSS has three timescales (i) GNSS time (t_g) (ii) Satellite time (t_g) and (iii). UTC time (t_{UTC}). GNSS time is the time maintained by the control segment of the GNSS ex: USNO maintains the GPS time which is 18 s behind UTC. Satellite time is the time generated by the on-board atomic clocks of the satellites. The information about the UTC time in the form of Δt_{UTC} ($\Delta t_{\text{UTC}} = t_g - t_{\text{UTC}}$) is broadcasted as a part of the navigation message signal which also contains the leap second information. Hence the receiver has the receiver time offset at every epoch as.

$$\Delta t_r = t_r - t_g.$$

and it also has the information of the UTC as a component from the navigation message.

$$\Delta t_{\text{UTC}} = t_g - t_{\text{UTC}}.$$

.

The UTC time thus can be calculated from the above two equations as.

$$t_{\text{UTC}} = t_r - \Delta t_r - t_{\text{UTC}}.$$

.

Hence, the users can receive the UTC timing information through GNSS. Glonass time is aligned with the UTC and follow leap seconds, whereas the other constellations are continuous time scales without any leap seconds addition. Hence, GPS and Galileo lags TAI by 19 s, whereas BieDou lags behind 33 s [16]. With the help of the dual-frequency receivers the ionospheric dispersion on the signals can be estimated more accurately which reduces the time transfer uncertainty to 10 ns.

It takes 12.5 min to receive one complete navigation message of GPS which can be downloaded at a rate of 50 bit/s. The important parts of any GNSS signal are ephemeris, almanac data and satellite clock bias parameters. As per the Common

View Generic GNSS Time Transfer System (CGGTTS) standards for CV/AVGNSS systems the data format should be at 16 min data interval with 13 min averaging [17]. In CV/AV GNSS time transfer methods, the users exchange the files containing the data of Δt_r recorded at every epoch in CGGTTS and Receiver Independent Exchange (RINEX) formats for CV/AV GNSS time transfer and Precise Point Positioning (PPP) respectively.

At present there are two types of satellite navigation systems, e.g., global and regional. The Global Navigation Satellite Systems (GNSS) provide worldwide coverage. The others, known as Regional Navigation Satellite Systems (RNSS) provide coverage just to one region, which may be a single country or a sub-continent. GPS, GLONASS, GALILEO, COMPASS are the global navigational satellite constellation installed by US, Russia, European Union and China whereas QZSS, BDS and IRNSS are the regional navigational constellation installed by Japan, China and India respectively. Very recently India has introduced its reginal navigational satellites, and all of its onboard atomic clocks are traceable to international time i.e., UTC through UTC(NPLI).

5.2.2.3 Cyber Security

Any device or system which can communicate with other devices or systems through internet can be prone to a cyber-attack. For accessing the internet, which provides digital live communication platform throughout the world all the devices must be on the same time scale. We cannot access anything on internet if our system epoch time is not synchronized with the internet time. Basically, the epoch time represents the number of seconds elapsed since 1st January 1970 00:00:00 UTC. The number of seconds elapsed are represented by a 32-bit binary number in digital world. There will be a same problem as Y2K bug for all the 32-bit processor-based systems, when there is a role over after 2^{32} s which will be on 19th march 2038 [18].

There are normally two Internet protocols widely used for time Synchronization through Internet, Network Time Protocol (NTP) and Precision Time Protocol (PTP). The algorithm and architecture of these protocols are discussed in Sect. 4.3.2. Computers which work on windows operating system are default set to the Windows Time Service (w32time) provided by NTP servers of Microsoft, which serves as reference time to interact via internet.

NTP v4.0 has in built security feature of symmetric key authentication, and optional auto key feature which has public key authentication. The organisation, Network Time Foundation (ntp.org) very frequently updates the vulnerabilities on NTP based on research and experiences across the world. In case of PTP defined by IEEE 1588 some security features are defined in Annexure K, but they are not fully defined into a security protocol yet.

The threat analysis in a cyber-attack related to time protocols in the network are classified as follows [19].

a. **Packet manipulation**: the attacker tampers the time protocol packets. In this case the timing service is operational but provides and intentional inaccurate information.
b. **Spoofing**: the attacker convinces himself as a legitimate node by transmitting and generating protocol packets. This situation also leads to providing intentional inaccurate information as the time protocol is operational.
c. **Relay attack**: the attacker records the protocol and replays them after some time without any modification.
d. **Rouge master attack**: the attacker pretends to be as a legitimate grand master (Boundary clock) by manipulating the master election process like Best Master Clock Algorithm (BMCA).
e. **Packet intersection and removal**: The attacker intercepts the protocol packets and remove some of them or all of them from receiving the destination node.
f. **Delay attack**: the attacker delays the time protocol packets with fixed or variable delay.
g. **L2/L3 dos attacks**: when the Layer 2 and Layer 3 are attacked in the network the time protocol is affected accordingly.
h. **Cryptographic performance attacks**: when a timing service is provided by dedicated authentication and encryption protocols the attacker send fake protocol packets which requires very high utilization of the cryptographic engine.
i. **DoS attack against the time protocol**: the attacker sends a series of bulk time protocol packets to the clock which degrades the performance.
j. **Grand master time source attack**: when the grandmaster uses a GNSS based clock as reference the attacker can jam or spoof the GNSS signals.
k. **Network reconnaissance**: The attacker uses the time protocol packets for collecting the information of the IP addresses and locations of the nodes.

Out of all these the packet delay attack is more power full as it doesn't require any change in the message or creating a new one. The effect of such an attack on a network node causes complete network out of Synchronization [20]. With all the possible cyber secure vulnerabilities of time protocols many authentication and authorisation mechanisms are in place today. Authentication is the identity verification of the client or server and authorisation is to allow the it to perform the action it is intended.

5.2.2.4 Professional Broadcasting Industry

Live streaming of sports, news, interviews, gaming etc., on TV, Facebook and YouTube is an illusion of a series of images and audio signals transported with in a time frame which humans can perceive without any delay. For example, for a live broadcast of news requires audio, video recording then mixing of both in a production unit and broadcast by a nearby broadcasting station which requires microseconds of accuracy which humans cannot perceive. The access points of camera, mic and mixer are called media nodes. To maintain the illusion the maximum time difference between the two audio inputs (Ex: mic) is 10 μs, and between two video inputs

Table 5.3 Synchronization requirements of the professional broadcasting industry as per SMPTE—ST2059-2

Phase (time) accuracy with in facility	
For analog video	<100 ps
For SDI video	<1 μs
For IP video	<5 μs
Target	**<1 μs**
Phase (time) accuracy with respect to external world	±100 ms with respect to TAI
Quick lock time	<5 s

(Cameras) is 10 ms and Synchronization difference between the audio and video should be less than 10 ms and for colour proxy it is 10 ns [21]. In other words, the timing requirements for a professional live broadcast through a live audio–Video network are.

i. Maximum delay: The Maximum delay between a singer singing a song and listing the same song though a network is 10 ms. In this the processing delay in the digital signals processing (DSP), mixers, transit time of sound from speakers to the audience is 8 ms. So, the network gets 2 ms to broadcast the song to the live audience.
ii. Maximum Synchronization error: the maximum Synchronization error between an array of speakers is 10 μs and the network designers always target for 1 μs.

The thumb rule in professional media Synchronization is that all audio channels must be synchronized within 5–20 μs with each other and video can lead audio by 25 ms or lag audio by 15 μs. This is due to the human perception that late audio is taken as normal and early audio signal as un natural [22].

Society of Motion picture and Television Engineers (SMPTE) is the International standards Organization for broadcasting, film, audio recording, medical imaging etc., ST2059 is a standard published by SMPTE describes how to synchronize the video equipment over IP network. SMPTE profile for professional broadcast applications is based on IEEE-1588 PTP protocol which states that any two devices (media nodes) should be synchronized with in ±1 μs synchronization over 10 Boundary Clocks (BC) within 5 s of establishing connection [23].

Audio engineering society (AES) standards for audio applications of networks for high performance streaming audio (16-bit, 44.1 kHz) over IP interoperability standards define the Media profile based on IEEE 1588 [24]. The time synchronization standards of audio and video broadcasting are shown in Table 5.3.

5.2.2.5 Digital Signatures and Time Stamping

Digital signatures and time stamping are fundamental technique to prove authenticity of electronic documents, initiatives like Digital India applications. Applications include Towards Digital India; E-voting; Issuing forms and licenses; Filing tax returns online; Authentic archiving of judicial records; Submission of affidavits.

Digital signature is the cryptographic transformation of the data unit that proves the origin and integrity of the data protecting it from forgery by third parties. The electronics signatures basically give us "who" prepared "What" and a time stamp can present "when" and "What" was prepared. There are various mechanisms available for the cryptographic encryption and decipher [25].

i. Symmetric—Key Cryptography: The data is encrypted with a key to access the data. The sender and receiver use the same key which is shared only between them to encryption and decipher.
ii. Public—Key or asymmetric key Cryptography: The data is encrypted with two keys named public key and private key. The public key is distributed freely to encrypt the data but its pair private key is kept secret by the owner and is used for decrypting the data.
iii. Hash Function: No key is used in this method. A fixed value of alpha numerical number is generated using hash algorithms. There are various Hash functions available like the MD5, SHA-1, SHA-256 etc.

Presently the hash functions are used for cryptography and thus the digital data authenticity. All the digital signatures are required to be put along with a valid and traceable time stamp. Globally recognized standards are in place to provide such time stamps.

The ISO/IEC 18,014 standards specify the time stamping techniques. It mentions the requirements of the Time stamping Authority (TSA) that it should maintain it time source within 1 s with respect to UTC [26]. In India, Controller for certifying authority (CCA) under the Ministry of Electronics and information technology is the regulatory body for providing such digital signatures [27]. It sources the time part of the process from CSIR-NPL via NTP.

With online processing of government documents be it an insurance policy or a marriage registration certificate, digital signatures with IST traceable time stamp is used everywhere in the country. The Regulations of the CCA and ITU regarding time stamping are shown in Table 5.4.

Table 5.4 Standards and acts related to time stamping

Standards/Act's	Name of the standard or act
IT (CA)-2000-2001	Information technology act (certifying authority) rules and regulations
CCA	Time stamping guidelines for certifying authorities (CA) issued by CCA
ISO/IEC 18,014	Information technology—security techniques—time stamping services
ITU-R TF.1876	Trusted time source for time stamp authority

5.2.2.6 Financial Trading Platforms

A capital market is a place where people buy and sells securities. Securities in the sense means the rights of a company or institutions or authorities which will be sold to the public [28] like Equity Shares, Equity Derivatives, Currency derivatives, Interest Rate derivatives, commodities, bonds etc. All this kind of securities are further traded in a common platform called stock exchange which is also called secondary market. In India there are 7 equity derivative stock Exchanges, 5 Commodity Derivative stock exchanges and an International Stock Exchange INDIA-INX [29].

With the present available technologies, stock exchanges are providing ultra-low latency infrastructure which can execute lakhs of orders in a second which will increase the volume of trading per day. In a race to reach the stock exchange electronic order book, every day morning the brokers are also implementing the fastest trading software's which perform the trading automatically in fractions of seconds.

To have a fare and transparent trading system, all these transactions should be recorded with proper time stamp. For that all the trading systems should be synchronized to same reference time which should be highly accurate enough, otherwise the recorded data will doesn't make sense. When the different servers in the stock exchange are not synchronized with same reference time they work on their onboard oscillator clocks available. In such case an order may be recorded in wrong way i.e., order executed at 10:05:25 and order placed at 10:05:45 which can't be possible to execute the order 20 s before the order is placed. The same situation will arise even if all the systems are synchronized with respect to same reference time source which is not enough accurate than the trading latency of the stock exchange.

HFT plays a very crucial role in the stability of the securities market as the volume of the trade done by HFT is more than 60% whereas the HFT traders are less than 10% of the total traders [30]. The algorithms used by the HFT are very complex and cannot be controlled easily by humans if it went wrong causing a huge flash crash. These algorithms are also programmed to do trading based on the online media news, Facebook, twitter, news channels etc. For example in USA on 23rd April 2013 at around 13:07 to 13:12 the Associate Press twitter account was hacked and there was a false tweet that "***Breaking: Two Explosions in the white House and Barack Obama is injured***" the algorithm which is in built programmed to read all the online tweets has read this fake tweet and started selling all its shares causing to a flash crash of 200 Million dollars loss to investors [31].

Trading of shares can be done through various platforms provided by the stock exchange such as Colocation trading, algorithmic trading, trading through internet, mobile and from anywhere in the world. When all these devices and the stock exchange are not synchronized to UTC the time stamp for the share biding will not be in proper order. By the time we place an order to buy a share, there may be lot of shares purchased by colocation traders which increases the share price and we end up in buying a share at higher cost than we assumed to buy.

India has the world's fastest stock exchange "IndiaINX" which has a latency of ± 4 μs in executing and order [32] followed by Bombay Stock Exchange with ± 6 μs [33]. Which means the trading engines has a capability to execute more than one lakh

Table 5.5 Regulations of different countries for accurate and traceable time for business clocks

USA, FINRA regulatory notice 16–23, since Feb 20th 2017	Accuracy requirements	Reference time source
Computer clocks	50 ms	UTC(NIST)
Mechanical clocks	1 s	UTC(NIST)
European Union, MiFID II (RTS—25), since Jan 3rd 2018		
Manual orders	1 s	UTC
Automated orders (non HFT)	1 ms	UTC
HFT	0.1 ms	UTC
HFT time stamp resolution	1 μs	UTC
India, SEBI CIR/MRD/DP/09/2012, since March 30th 2012		
Stock exchange clock	1 ms	An atomic clock

transactions at least in a second. The Securities Exchanges Board of India (SEBI) which is the regulatory body issued "Broad guidelines on algorithmic trading" in 2012 [34] which states "*The stock exchange shall synchronize its system clock with the atomic clock before the start of market such that its clock has precision of at least* 1 *μs and accuracy of at least* ±1 *ms*". The same instructions are reiterated in 2017 in "*Fair and transparent access to data feed of the stock exchanges*" circular [34]. It doesn't specify the need for the traceability of the time to UTC through the traceability chain. The table mentioned below shows the regulations of the United states of America (USA) [35, 36] and European Union (EU) [37, 38] along with India for the time traceability requirements in financial sector. It can be seen that the in USA even the computer which is used for electronic trading should be synchronized with 50 ms with respect to their national standard and legal time maintained by National Institute of Science and Technology (NIST). The European Union has more stringent time requirements than USA up to 1 μs for High Frequency Trading (HFT). The regulations of various countries including India are shown in Table 5.5.

5.2.2.7 Smart Power Grid

Frequency Synchronization which is known as Syntonization is very crucial for operation of power grid. Mostly all the generating stations use synchronous generators known as Alternators for generation of power. The frequency of generation f depends on the speed of the turbine rotation N and number of poles P of the alternator as shown in the following equation.

The power generated by all the stations is distributed to various consumers located across the country through power grid in real time, as Alternating Current (AC) cannot be stored. When there is a sudden dip or demand of the load it will affect the rotating torque of the alternator causing making it non-synchronous and gets to halt. This situation is considered as Grid failure which happens due to non-isolation of the sudden jump in the behavior of the load on the grid. This changes in the load fluctuations should be measured with microsecond level of accuracy and need to be isolated in milliseconds. Thermal power plant needs 5–8 h to come back to the hot condition again provided the input from hydel power plants through grid, which can start by them self with in 10–30 min. Power grid failure causes lot of economic impact on industries.

Power grid failed two times in India once on 2nd January 2001 for 12 h and second time on 31st July 2012 for 12 h [39, 40]. Grid failure causes huge economic impact effecting the industries production, transportation, businesses and government operations. To retrieve the grid to the normal stage it will stage at least 1.5–2 days with some intermediate fluctuations this is what happened in 2012.

Time and frequency both parameters play a very critical role in power sector. Apart from time synchronization for digitalization of the measurements, power sector also requires frequency synchronization ("Syntonization") as a very critical parameter for power generation, transmission and distribution. According to the Indian electricity grid code the grid frequency should always remain within 49.90–50.05 Hz [41]. Indian power grid is around 1,42,989 ckt Km of Extra high voltage transmission lines spread across the length and breadth of India connecting 226 EHV AC and HVDC substations [42]. To synchronize such a wide-area network a common reference time is must and necessary for monitoring the faults in EHV transmission lines which are normally hundreds of kilometers lengthy carrying 100's of MW of power.

Phasor Measuring Unit (PMU) is a device that measures frequency, synchronized phasor and rate of change of frequency (ROCOF) from voltage, current and a time synchronization signals. The PMU's used in the power grid for wide area monitoring and fault detection in the transmission lines needs a reliable stable UTC traceable time source. Each measurement of PMU is time stamped with respect to UTC in three different forms namely: second-of-century (SOC) count, fraction-of-second count and a time quality parameter.

A phase error of 0.01 rad (0.57°) for a 50 Hz power frequency would cause $\pm 31\ \mu s$ error in time stamping [43]. Hence, the standards always prefer to have a time source of at least 10 times better i.e., time source must be accurate to at least $\pm 1\ \mu s$ so that the measurement can be accurately resolved with in specified 100 years. Time source of PMU should be accurate, traceable to UTC and must be aware of leap second corrections.

The common ethernet based time distribution mechanism used in power system protection, control, automation and data communications is IEEE 1588 Precision Time Protocol (PTP). The Intelligent Electronic Devices (IED) like digital relays, electronic multifunction meters etc., used in the digital substation must be synchronized to one reference time for Co-ordinated data collection and control. PTP attributes will consists of reference source UTC offset and leap second information in every PTP message.

The time distribution applications in a substation using PTP can be broadly classified into four categories:

Table 5.6 IEEE standards for time traceability requirements in power grid and substations

Standard	Name of the standard	Traceability requirement (μs)	Reference source of time
IEEE Std C37.118.1-2011	IEEE standard for synchro phasor measurements for power systems	±1	UTC
IEEE Std C37.238-2017	IEEE standard profile for use of IEEE 1588 PTP in power system applications	±1	UTC

i. General purpose time indicators (<±100 ms)
ii. Time stamping required for SCADA data and operational data logs (<±1 ms)
iii. IRIG-B applications (<±1 μs)
iv. Long distance applications with complete PTP support (<±1 μs).

The worst-case time error between the end device and the reference time source through a maximum of 16 network hops should be ±1 μs [44]. The time traceability requirements as per IEEE are shown in Table 5.6.

5.2.2.8 Railways

Many applications in Railways are time dependent, such as railway network timing, signalling systems, station clocks and passenger information and CCTV. Accurate, reliable and traceable time is essential to enable correct event time stamping, such as when a train passes a signal. This helps in proper diagnosis and provides a correct chronological audit trail of what happened, which is vital for safety information records and incident investigation. In order to monitor mobility of trains, track-to-train communications can also benefit from precise time; message transmission and reception must be precisely synchronized or the result will be of no use. Station management systems also need an accurate and reliable time source, such as CCTV, computer networks, public announcements and even auto-control of air conditioning systems [45]. Time synchronization requirements in Railways lie in ms to μs level.

Railways have a large number of PNT applications. They can be classified in two main categories: safety relevant and non-safety. The non-safety relevant category can be divided in two sub-categories: liability relevant and non-liability relevant applications. The non-safety and non-liability relevant applications are related to passenger information applications which are mostly addressed using global navigation satellite system (GNSS) such as GPS [46] The limitations for GNSS for Rail applications are signal obscuration (e.g. stations, tunnels), high safety requirements, position errors due to local effects (e.g. multipath, EMI, spoofing) and high accuracy time synchronization requirements for new and emerging techniques of rail management. Accurate synchronization will help to ensure optimal operation of converter stations.

Services based on location, autonomous driving all require precise time. Embedded sensors (on tracks, trains, stations etc.) are gaining the ability to communicate by means of IoT which is again based on precise time.

Owned and operated by the GoI through the Ministry of Railways, Indian Railways. As of today, no regulatory body has been formed for railways and it regulates itself. Need for a regulatory body for various functions and operations has been projected several times [47].

World's fourth largest railway network with 115,000 kms track length, 8,500 railway stations, 26 million passengers a day, more than one billion tonnes of annual freight, contributes to GDP (1–2%) through employment generation, freight collection, e-catering services etc. GPS based clocks are being used for stations, platforms and control rooms [48].

1474.1 IEEE Standard for Communications-Based Train Control (CBTC) Performance and Functional Requirements, IEEE Std., 2004. But it does not mention anything on precise time requirement [49]. Presently, GPS is most widely used but, it cannot be primary source as GPS does not have service level integrity required by railway safety and certification standards (ISO/TS 22,163:2017).

5.2.2.9 Civil Aviation

The major responsibility of the Air Traffic Controller (ATC) are monitoring the position of the aircraft and communicating the pilots. ATC implements maintenance of minimum separation among all the aircrafts at every point of time for avoiding any mishaps. ATC are responsible for providing services to all kind of aircrafts, including private jets, commercial aircraft as long as its operating within its define regions [50, 51]. Considering all of its parameter, ATC conveys its instructions or suggestion to the pilots to follow but the pilots have freedom to take ultimate decision.

In order to avoid any ambiguity about defining position or location of a flight, all ATC must have the same accurate time and that should be traceable to international time (UTC). CSIR-NPL maintains our National Time (IST) and its traceability with the International Time [52].

Aircrafts have several computer system and devices those are connected through a network and controlled by a master system. Each of the system maintain its own time through a local clock attached to it. All the devices follow the instruction comes from the master system along with its time and for that it is essential that all clock within the networks maintain same time. If the local times of the computers or devices are different, there will be problem during operation. The time differences among the local clocks may arises because of several reason like drift or jitter of the clock oscillators or its aging factor. In order to reduce this time difference between the systems connected in network, all the system should be synchronized regularly following some of the standard network-based protocol.

5.2.2.10 Disaster Management

Accurate time measurement and precise time synchronization play crucial roles in the field of disaster management [53]. Disaster management could only be as good as the information it has got for its decision by analyzing acquired data and all those data need to be on time, precise and accurate. Disaster management groups have data acquisition centers at different geographical location and all those centers need to be synchronized precisely (within 100 ms) to a common reference clock i.e., IST in order to extract reliable and usable information from those data. As per disaster data acquisition plan (DDPA) data are being integrated at real time for the use of rescue, police and relief organization but the most significant part is that accurate data can also be used for disaster prevention. As for example, Flood disaster or Tsunami disaster management requires very precise, accurate and actual data of the water surface. Airborne Lidar systems are used for scanning high water events and generate a digital elevation model. In the laser scanning technology, the distance between the laser emitter and the surface of area is measured by determining the travel time of the light pulse [54]. In order to generate a digital model of the topography, it is necessary to accurately record the position and latitude of the laser emitter, as well as the angel of the emitted beam. The differentiation of the first and last light echo following a single measuring pulse permits identifying and determining further properties of the terrain. Several emitters will be used for mapping the topology and for precise inter comparison of those data, a uniform and accurate time stamping is essential. Synchronizing all the emitters clock with respect to IST most useful data can be acquired for encountering disasters.

5.2.2.11 Defense

Timing and synchronization are very crucial in defence for many of their critical applications like surveillances as well as targeting or shooting any airborne and very fast flying objects.

Radars are being deployed at multiple locations along the border line of the country for the surveillance purposes [55]. Electromagnetic radio waves are shined from the radars aiming towards any object and it also monitors the reflected signal from the object. Measuring the travel time of the reflected signals, radars can estimate the distance of the object from their respective positions. In order to define the exact location and trajectory of that object in three dimensions, minimum three radars from different locations need to follow the same exercise. All those data have real time access from a central location and following a trilateration method precise position or trajectory of the object can be derived. In order to achieve positional accuracy of few ten metres, all those radars need to be synchronized within 100 ns and for that IST or UTC(NPLI) can be used as the reference time for each radar that an uniform time will be maintained and data from any radar can be correlated with others irrespective of their locations. Similarly, for shooting an object in the sky using ballistic missiles, the position and as well as its speed and trajectory need to be determined very correctly and all those can only be achieved with proper time maintenance and synchronization as already described in the case of surveillances [56].

Historically time division multiplexing (TDM) transport, dedicated wires and GPS like navigational satellite signals are used for achieving the synchronization function in defence systems. TDM transport enables frequency synchronization only, while other wired connection or satellite-based techniques provide both time and frequency synchronization. In modern time packet networks replace the TDM networks [57], thus eliminating the ability to relay on TDM transport for synchronization purposes. Satellite constellations like GPS, GLONASS or IRNSS can provide accurate frequency, phase and time but it needs infrastructural support and its performance can be degraded due to poor weather conditions, vulnerability to jamming and spoofing, poor signal reception in case of limited sky view, and requirement for high quality holdover oscillator. On top of everything other than IRNSS all satellite constellations, like GPS, GLONASS are controlled by the foreign countries like US or Russia, so for strategic reason those are not a preferable choice for defence applications. Reliable packet network-based synchronization methods like IEEE 1588 Precision Time Protocol (PTP or WRPTP) can be used for achieving precise time and frequency synchronization in defence applications. Utilizing those techniques times can be made traceable to IST and in addition as these methods are accurate, robust as well as secured. Navigational satellite, mainly from IRNSS can also be used as the backup and in case of temporary failure of the first methods.

5.2.2.12 Networks, Data Centers and Digital Transactions

A digital transaction can be defined as an online or automated transaction between people and organisation electronically without the usage of paper. A digital transaction is initiated through a website or through an application gets routed through a network to the data centres where various servers for authentication, data base etc., will process the transaction. A digital transaction can be an online purchase, bank transactions, e contract agreement etc., As the three entities involved in a transaction electronically, they all must be synchronized to a common reference time. For example in case of an online purchase or bank transaction through a mobile app, the mobile time is set by the telecom service provider, the data servers located in various parts of the country may have different time source and the network through which this transaction is routed may have different time source. The event logs of all these three entities may be in different formats ex: DD-MM-YY, UTC format, local time format etc. [58].

The most valuable information to investigate a security breach are time stamped event logs. Collecting the logs with accurate time stamps in a central location and reviewing them regularly for problems will lay the foundation for any security issues and quick forensic response. Many operating systems and network devices are capable of logging their own activities depending on their storage capacity [59]. These devices log basically includes.

i. System logs (e.g.: Windows Event logs, Unix syslogs)
ii. Application logs (ex: web server access logs, host-based antivirus logs)
iii. Authentication logs (ex: dial-up or VPN authentication logs)
iv. Network device logs (Ex: firewall logs, Netflow logs)

Storing the logs in various hosts like an application storing an authentication attempts in one log and network activity in another log will have high chances for time stamp miss match. Each host device or software generates the time stamped logs with reference to its internal clock. If the hosts clocks are inaccurate i.e., they are not able to distinguish to successive events or the time stamp or having an offset due to the drift in the clock the time stamps of the logs will be inaccurate. In a network due to shortage of the storage capacity most of the devices record the time stamps of the information which they consider important and send them to a centralised log server. For example, host 1 records the IP address but not the username, where as host 2 records the username but not the IP address. Hence, to co-relate the event logs only accurate time stamps are the basis. To store the event logs with common reference time of all the devices and their host servers of a complete network gives a wide scope for investigation and instant analysis of the target event.

In India there are various state level and national level networks for digital libraries, e-governance, banking etc., as categorised below [60].

General Network: INDONET (Data Network), NICNET, OPNET, VIKRAM, VIDYANET, I-NET, SAILENT, OILCOMNER, RAILNET, Indian Airlines Network etc.

Specialised Network: BTISNET (Biotechnology Information systems Network), CSIRNET (Council of Scientific and Industrial Research Network), DESINET (Defence Scientific Information Network), ERNET (Education and Research Network), INFLIBNET (Information and Library Network) etc., National Knowledge Network (NKN) is a state-of-the-art Pan-India network with multi-gigabit capacity connecting all universities, research institutions, libraries, laboratories, healthcare and agricultural institutions across the country.

Metropolitan Area Networks: Various other library networks include Ahmedabad Library Network (ADINET), Bangalore Library Network (BALNET), Bombay Library Network (BONET), Calcutta Library Network (CALINET), Knowledge Resource Centre (KRCNet) etc.

National Financial Switch (NFS): It is the leading multilayer ATM network in the country connecting 2.48 Lac ATMs (as of 30 April 2020) [61]. National Payments Corporation of India (NPCI) is an umbrella organisation for retail payments and settlements in India which does the online payments settlements through its various networks. Indian financial Network (INFINET) launched in 1999 is one of such networks which is the back bone of the Indian Baking network. It mainly handles interbank services, RTGS services, Government transactions, automatic clearing houses etc., As per the NPCI statistics in the month of June 2020 a total of 2486.74 Million transactions worth $11,938.62 Billion are carried out for retail payments [62]. NPCI facilitates the retails payments such as RuPay card, Immediate Payment

Services (IMPS), UPI, Bharat Interface for Money (BHIM), BHIM Aadhar, National Electronic Toll Collection (NETC Fastag) and Bharat BillPay.

In a data centre all the data servers must be very accurately synchronized with each other as they had to handle Gigabits of information in a second. If all the servers related to one application are not synchronized to one reference time source with sufficient precision the transaction cannot be complete. For example, during an online ticket booking if the application server and the payment authentication server are not synchronized to the same reference time, the ticket will not get confirmed even after the payment is completed.

In cloud storage of data which is used for cloud computing and analysis, the data is stored with proper timestamp of all events concerned to it like Last accessed, file created, last modified. If all the time stamps are traceable to UTC and in UTC format it will be a most valuable resource for network analysis and cyber forensics.

When the application servers and the payment authentication server are located in different countries in case of international transactions, i.e., different time zone then all the time stamps must be done with respect to UTC to have a synchronized network.

With the advent of online banking, banking sector heavily relies on very powerful IT systems and networks and these require a high level of availability, security and reliability. Dedicated data centers exist for most banks for performing critical operations. Telecom networks are a crucial component of the digital payment system. Beyond reliably carrying the transaction, the telecom networks can be an ally in the fight against fraud. For instance, many fraud checks are dependent on reliable identification of users, and devices through the telecom network. Additional factors, such as OTP, USSD messages are relied upon as well—all of which originate through the mobile network.

Payment card Industry Data Security standard (PCI DSS) is an information security standard. It must be followed by organisations that handle branded credit cards for the major card companies. PCI DSS requirement 10 mandates that all systems must maintain their event logs to centralized time servers. Only central time servers are allowed to receive time from external sources traceable to UTC. The 10th requirement states that the tracking and monitoring of all access to card holder data, network resources and includes specific requirement of NTP [63].

5.2.2.13 Internet of Things (IoT)

Internet of Things (IoT) is an ecosystem of connected physical devices that are web-enabled. IoT imagines a future wherein advanced physical elements can be connected, by suitable method so fin formation and communication technologies, to empower an entirely different class of utilizations and administrations. The machine-to-machine (M2M) communication emerged the concept of IoT i.e. the communication among machines without human intervention [64]. IoT is a network of sensors and millions of smart devices that connect systems, people and other applications to share and collect information. M2M offers connectivity that enables IoT. With just a limited degree of programmability and customizability, IoT systems link together highly specialized devices designed for specific purposes.

To connect the digital world from the physical world, IoT is equipped with a plethora of sensors and actuators. The layered architecture of IoT consists of three main layers such as the Sensing layer which consists of sensors for collecting information, Network layer where devices are connected through different technologies and application layer so forth [65]. The application layer of IoT provides services to users. The concept of IoT introduced different applications that make the physical world smarter by including IoT enabled Software Architectures and Middleware, Mobile Cloud Services, Data and Knowledge Management, Smart power grids, Smart homes, Industrial Internet, Smart agriculture, Smart Vehicles, Smart parking systems, Smart Shopping Malls, Smart Milk Industries, Connected Health (Digital Health/Telehealth/ Telemedicine), Smart Supply Chain etc.

Various technologies used for implementing IoT [66] are shown in Table 5.7 which depends on the application, like coverage area of the network required, data transmission capacity, network latency etc., IoT is leading to 4.0 Industrial revolution through Cyber Physical Systems.

Telecommunications Engineering Centre (TEC) of Department of Telecommunications (DoT) has issued 11 technical reports regarding the applications of IoT architecture and spectrum usage in smart cities, Automotive/Intelligent transport system, Safety surveillance, Utilities/energy [67], where various technologies need positioning and fast communication technologies. The role of time synchronization is very crucial as discussed in PNT section above. The various technologies used in IoT are need a latency of 1 s to few microseconds.

Table 5.7 Different wireless technologies used in IoT and their range of coverage

<table>
<tr><th>Technology/protocol</th><th>Proximity</th><th>Short range (10–100 m)</th><th>Short/medium range (100–1000 m)</th><th>Medium range (5–10 km)</th><th>Long range (up to 100 km)</th></tr>
<tr><td>Embedded chip based</td><td>NFC (EMV), RFID</td><td></td><td></td><td></td><td></td></tr>
<tr><td>Wireless personal area network (WPAN)</td><td colspan="2">Bluetooth LE, ZigBee, Thread (6LoWPAN), Z-wave, ANT+, Wireless HART, ISA 100.11a, EnOcean, Plus more</td><td></td><td></td><td></td></tr>
<tr><td>Wireless local area network (WLAN)</td><td colspan="3">802.11 a/b/g/n/ac, 802.11 af (white space), 802.11ah, 802.11p</td><td></td><td></td></tr>
<tr><td>Wireless Neighbourhood area network (WNAN)</td><td colspan="4">Wi-SUN (6LoWPAN), ZigBee -NAN (6LoWPAN)</td><td></td></tr>
<tr><td>Wireless wide area network (WWAN)</td><td colspan="5">Cellular 2G/3G/4G; LTE-MTC, 5G in the future, LPWAN, SIGFOX, LoRa, Telensa, PTC</td></tr>
</table>

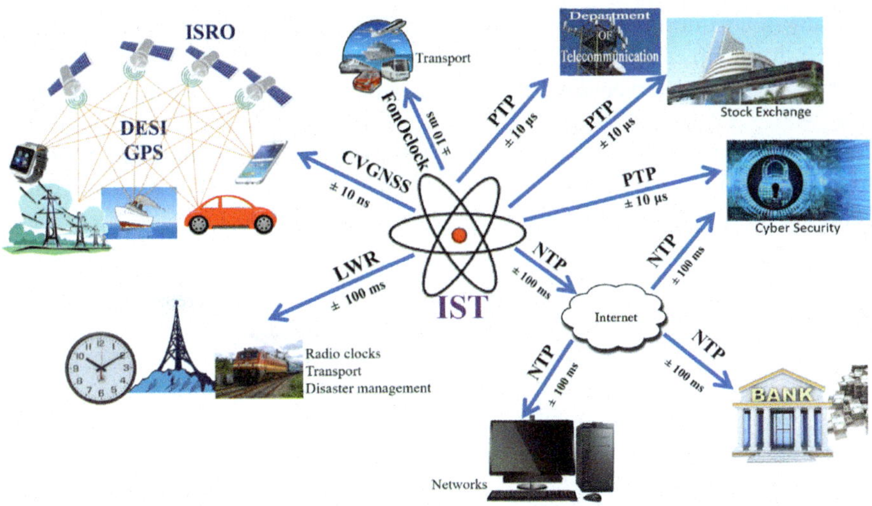

Fig. 5.2 IST traceability to various sectors through various technologies

As a conclusion from the above discussion regarding various sector Fig. 5.2 explains the various traceability methods of time and frequency transfer at different levels of accuracies to various sectors.

5.2.3 Legalization of IST

The IST synchronization of all the sectors discussed above cannot be done successfully without legalization and implementation of IST by GOI. The international scenario of the use of legalized time by few developed countries will be discussed first. This will be followed by a list of various relevant sectorial ministries and regulators that need to coordinate for implementation of IST synchronization pan-India. A new policy is proposed for time accreditation, and audit mechanism by elaborating on usage of existing and new infrastructure.

5.2.3.1 International Scenario of Legalized Time

A. **Uniform Time Act-United states of America (USA)**

In USA historically many laws were made relating to adoption and abolishment of day-light-saving time (DST) since 1883–2019. Presently the DST is adopted in some of the states of USA which is primarily intended to set the time one hour forward on the second Sunday of April at 2AM and backward by one hour on first Sunday of November at 2AM. This is done to adjust the length of the mean solar day as per the working hours during spring and fall seasons. The legal standard time is declared in the "Uniform Time Act-1966" and the U.S. Code, title 15. This act states to "promote the adoption and observance of uniform time with in specified standard time zone…" [68].

In USA there are three entities who are responsible for maintenance of primary frequency and time standards, dissemination and implementation as mentioned below.

a. National Institute of Science and Technology (NIST): NIST is the NMI of USA which maintains the national standard for frequency, time interval and time-of day officially in USA. It maintains the official US time in 12 time zones of USA [69] through various services like Internet time broadcast (NTP), low frequency (LF) and high frequency (HF) signals broadcasting, remote clock steering, calibrations etc.,
b. United states Naval Observatory (USNO): USNO maintains the national standard of Time (Epoch, Date), control of Naval T&F transmissions and eLoran signals. It is responsible for maintaining the standard time for Department of Defence (DOD) and other astronomical data which are essential for safe navigation at sea, in the air, and in space. GPS time is steered with respect to UTC(USNO) time on a daily basis.
c. Department of Transportation (DoT): Within the respective time zone the DoT is the implementation authority of legal time for all common carriers of interstate and foreign commerce. In all acts, statutes, orders, rules and regulations relating to the time of performance or non-performance of an act by one individual, it should be in so far, the standard time of that zone with in which the act is to be performed. DoT has the authority to shift an area from one time zone to another time zone [70]. DoT issues a public notice regularly for any such changes which should be approved by the congress.

B. Legal Time Act—Germany

The legal time act of Germany implemented from 25th July 1978 declared the Central European time (UTC + 1.0 h.) as the legal time of the country [71]. Physikalisch-Technische-Bundesanstalt (PTB) which is the NMI of Germany is charged with maintaining and disseminating the legal time. In July 2008 the new law "Acts on units of Measurement and Definition of Time" re states that CET is the legal time continuing the role of PTB to maintain and disseminating the legal time [72]. PTB maintains and disseminates the standard time and frequency signals by the low frequency ground waves at 77KHz named as DCF77, and by NTP services.

5.2.3.2 Ministries, Regulators and Policies

As discussed in Sect. 5.2.2, the time and frequency requirements of various sectors the details of the ministries they are associated with and their regulatory body [73] as shown in Table 5.8. The corresponding regulatory bodies should make the regulations for the IST traceability in their associated sectors.

Table 5.8 List of ministries, regulatory bodies of the sectors discussed

S. No.	Sector	Regulatory body	Ministry
1	Telecommunications and 5G	Department of Telecommunications	Ministry of Communications
2	Position, Navigation and Timing (PNT) applications	Department of Space	Department of Space
3	Cyber security	Cyber and Information security (C&IS) division	Ministry of Home Affairs
4	Broadcasting Industry	Telecom Regulatory Authority of India	Ministry of Information and Broadcasting
5	National data networks	National Informatics Centre	Ministry of Electronics and Information Technology
6	Banking network	Reserve Bank of India (RBI)	Ministry of communications
7	Digital transactions	Reserve Bank of India (RBI)	Ministry of Finance
8	Digital signatures and Time stamping	Controller of Certifying Authorities (CCA)	Ministry of Electronics and Information Technology
9	IoT and IIOT	TEC	Ministry of Communications
10	Trading platforms	Securities and Exchange Board of India (SEBI)	Ministry of Finance
11	Smart Power grid	Power grid corporation of India	Ministry of power
12	Railways	Railway regulatory authority of India	Ministry of railways
13	Civil Aviation	Director General of Civil Aviation	Ministry of civil aviation
14	Disaster management	National Disaster Management Authority (NDMA)	Ministry of Home Affairs
15	Defence		Ministry of Defence

5.2.3.3 Proposal for Indian Legal Time Act

IST is an official time as being generated from a Primary time scale, but need to be declared as legal time by the government of India through a Gazette. Department of Legal Metrology, under Ministry of consumer affairs is the implementation agency of standards in India. Indian Metrology Act, 2009 emphasized that the primary standards will be maintained and updated by NPL time to time, but doesn't mention anything about legal time [74]. CSIR-NPL alone cannot be able to synchronize all the sectors of India from one location. Hence, some infrastructure needs to be established and/or facilitate the existing infrastructure for outreaching the timing service at regional and state levels. There should also be an Act in place specifying the role and establishment of different agencies for time stamp certification, time audit, time assessment, time service provider etc., In this topic the intrigues of such a Legal Time Act are discussed.

We call this Act as "Legal time Act". It can be broadly classified on the basis of time traceability chain from UTC maintained at BIPM till the final end user as shown in Fig. 5.3. The traceability chain starts from BIPM providing traceability to National Timing Centre (NTC) which in turn provides traceability to three National time service providers (NTSP) to reach the final end user applications of various sectors.

CSIR-NPL maintaining the UTC(NPLI) i.e., IST, is the NTC which is directly traceable to UTC and provides further traceability to all sectors in India. CSIR-NPL will be providing time traceability directly to the NTSPs only as it is already said that it alone cannot cater the timing requirements of India directly, and such a greater accuracy is also not needed for all applications.

Depending up on the time transfer mechanism to NTU and end use applications the NTSPs are categorized into three major *entities.*

i. NavIC satellite constellation of ISRO can be used for satellite time transfer mechanism. Satellite based time synchronization is required for Wide Area Monitoring (WAM) applications like Navigation (on land and in sea), weather forecasting, protection of power transmission lines, Aviation etc.,
ii. The secondary time scales which are under establishment in five Regional Reference Standards Laboratories (RRSLs) in five different cities of India can serve the NTP and PTP requirements at regional level. These secondary laboratories maintain the secondary standards of time also providing the calibration services of various clocks and devices to the industries. All the TSPs, ISPs and commercial & public sectors like CCA, Banking requirements etc., can be catered from these five secondary laboratories at regional level.

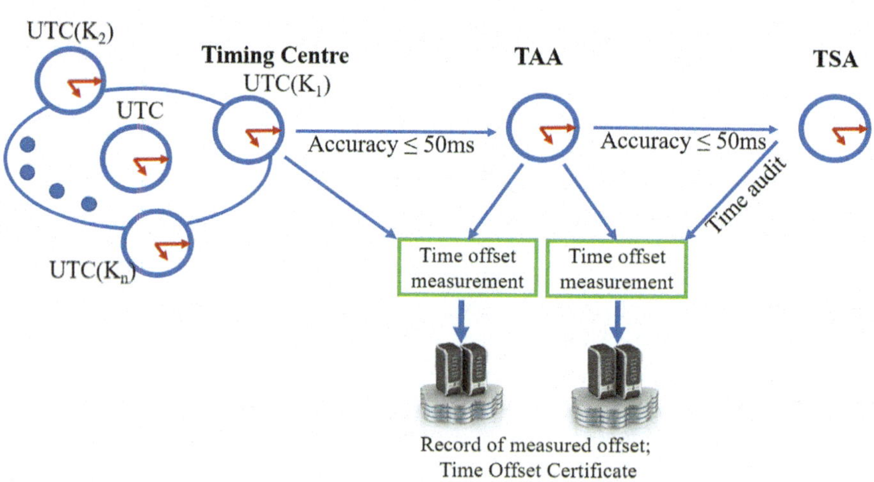

Fig. 5.3 Time traceability chain for time stamping

iii. Long wave Radio (LWR) signal is a mechanism by which, a low frequency ground signal is broad casted from either two or three locations to cover entire geographical area of India. These standard signals consist of timing information accurate to 100 ms. A small receiver chip in our wretch watch and wall clocks is sufficient to receive this signal and synchronize time with respect to IST. This mechanism is in process of establishment where CSIR-NPL has signed an MoU with M/s IFR for setting up the tower and developing the required receiver technology.

iv. India has various fiber networks laid across its length and breadth. Time can be transferred with greatest accuracy in the fiber optic medium using various technologies like active cancellation with electronic delays (ELSTAB) [86], and protocol-based encapsulations like white rabbit and PTP [87]. National fiber networks like National Optical Fiber Network (NOFN) connecting 2.5 Lakh gram panchayats in the country, fiber network of RailTel, Power grid, BSNL, Network for spectrum (NFS) of defence sector etc., can be used for time synchronization for strategic applications, National R&D Institutes, Universities and also for a terrestrial back up.

5.2.3.4 Time Accreditation and Certification Policy

Apart from providing the time synchronization signals across pan India there should be a mechanism in place to assess, monitor, audit and certify the legal time along the traceability chain till the end user. A brief introduction of such a policy is discussed here.

As the time synchronization is not only being done by internet based NTP timing service but also through satellite signals, low frequency ground wave signals and also at optical frequencies through fiber. Hence, apart from the Indian Information technology (Certifying authorities) rules 2000 and 2001 which mentions the time traceability requirement of the certifying authority for digital signatures [75], an elucidate and generalized guidelines for time stamping of all digital transactions is needed.

A time stamp basically registers the moment of time pertaining to "what and when" an event has occurred. The greater the accuracy i.e., seconds to microseconds with respect to the reference time, the better the sequence of events that can be reconstructed for any post analysis and also for real time tracing of the data packets.

Traceable time stamping and certification mechanism can be implemented through at least three agencies namely Timing center, Time Assessment Authority (TAA) and Time Stamping Authority (TSA). The ITU-R TF regulations define the role of these different agencies in detail [76]. The Timing center is the institute which maintains the UTC(k) time scale. The TSA is a Trusted Third Party (TTP) and timing center can function as a TAA or TAA could be a TTP. TAA has the role to audit and certify the traceable time maintained by TSA.

With the increase in the digital transactions around the world the number of TSAs is also gradually increasing. ITU/IEC 18,014—4 recommends that the timing accuracy should be at least 10 times better for each step moving towards UTC [26]. i.e., UTC(k) should maintain time 10 times better than TAA, and TAA should provide 10 times accurate time than TSA. As shown in Fig. 5.4. TAA should be accurate to 50 ms with respect to UTC(k), and TSA should be accurate to 50 ms with respect to TAA to issue a Time Stamp Token (TST).

As mentioned earlier that the time service provider is an entity which provides traceable time to the required agency through NTP/PTP/satellite signals/optical fiber from CSIR-NPL. But to certify the time offset a clear mechanism need to be defined in this policy. There should be a proper defined role, responsibilities, eligibility for TSAs and time service providers (NTSPs). Apart from the functionalities a proper established technical procedure for time traceability, maintenance of reference clocks, issuing time offset certificate and time traceability audit should be formulated in order to legalize time. The certification and accreditation policy should consider all this criterion and should clearly mention the standard procedure of operation and time offset measurements.

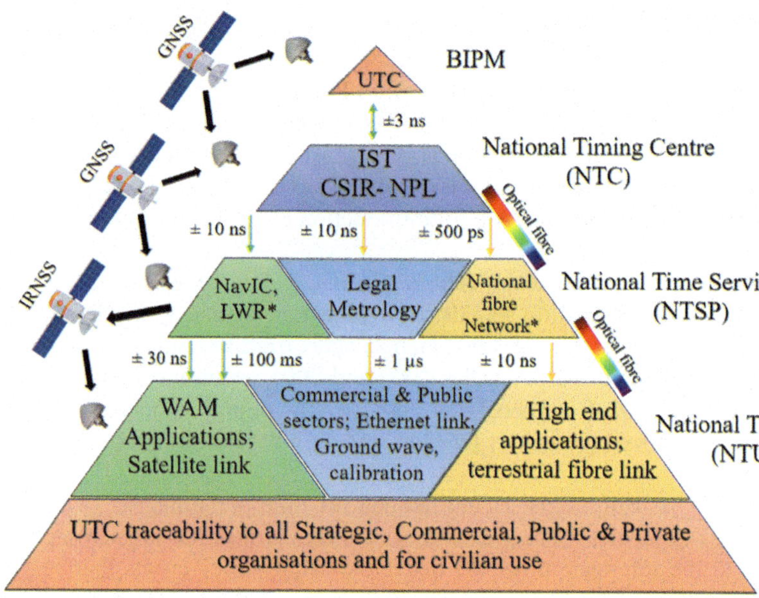

Fig. 5.4 Traceability chain from UTC to a common user through CSIR-NPL and other infrastructure of India (*not implemented yet)

5.2.4 *Infrastructure Required for Time Synchronized Digital India*

Firstly, if an Act is to be implemented, sufficient infrastructure is to be made available or established to serve the timing requirements of various sectors. As it is explained in the above section different sectors need different level of accuracy via traceability. As time is a dynamic parameter all the clocks are required to be synchronized as frequently as possible in near real time or through regular calibrations. So, here we need two different kinds of infrastructures one which provides near real time continuous service and other is which provides periodic calibration for clocks, timing counter, receivers etc., It is presented that the following infrastructure can be used to for time synchronization in India.

5.2.4.1 Secondary Timing Laboratories

To provide calibration services and real time traceability services at regional level in the country the secondary timing laboratories are required. Most of the developed countries has secondary timing laboratories traceable to their primary timing centre. The secondary timing laboratories for all the standards are maintained by Department of Legal metrology (DoLM) named as Regional Reference Standards Laboratories (RRSLs). They are located presently in five regions of the country Bangalore, Bhubaneshwar, Ahmadabad, Faridabad and Guwahati and two more are under construction at Nagpur and Varanasi [77]. CSIR-NPL has taken up the responsibility for setting up of secondary time scales in the five RRSLs centres, to enable the country with the capability of time and frequency metrology services to each end user (see Sect. 4.4.3 B).

These RRSLs play a very important role in providing the calibration and real time traceability services at regional level. Since, DoLM is the authority for the implementation of the standards in the country they are authorized to audit the time traceability at the end user application. They can provide real time traceability service to the TSPs and ISPs through NTP and PTP services. They can also provide calibration services of clocks, frequency counters etc., related to the time and frequency metrology.

5.2.4.2 Low Frequency Ground Wave Telecasting Stations

We can synchronize the computers through Internet via NTP and improve the navigation performance by precise atomic clocks on satellites but to synchronize each wall clock and wrist watch with the IST for day to day life standard time and frequency signals must be broadcasted through a ground wave. In various countries listed below in Table 5.9 they are in place from as early as 1965 [78]. This technology as explained in the early sections needs a chip receiver in the watches for automatically receiving the signals and synchronizing our clocks with in milliseconds of accuracy to IST.

Table 5.9 Various ground wave technologies used by countries for standard time and frequency signals broad cast

Call letters	Country	Frequency (kHz)	Radiated power (kW)	First time code
WWVB	United states	60	50	1965
MSF	United Kingdom	60	15	1974
DCF77	Germany	77.5	30	1973
JJY	Japan	40 and 60	12.5	1999 (60 kHz added in 2001)
BPC	China	68.5	NA	2002

CSIR-NPL has signed an MoU with M/s IFR for establishing required towers at three different locations in the country [79]. For implementing the legal time this technology will play a very critical role in the day to day life of every citizen of the country as the time is synchronized at home, at office, railway stations, bus stations, public places etc.,

5.2.4.3 Satellite Constellations

The first step towards the modern digital world for any country is having their own satellite-based navigation system. This will help the country in developing the PNT applications more accurately. Many wide area applications like the long power grid lines monitoring for fault detection and isolation, navigation in sea, in air etc., needs satellite based precise and accurate timing signals. Countries with global satellite navigation capabilities are GPS of USA, GLONASS of Russia, Galileo of European Union, Beidou of china and regional navigation capabilities are Qzss of Japan and IRNSS of India details as shown in Table 5.10 [80]. India is successful in establishing the IRNSS which can be now used for various time transfer technologies like CVGNSS, AVGNSS and one-way time transfer like the NavIC disciplined oscillators. CSIR-NPL is providing the traceability to IRNSS for the UTC traceability since 2018 by TWSTFT and also by CVGNSS. Hence, the NavIC time signals are traceable to IST which can be used for wide area applications in the country. The technical traceability audit mechanism should be developed.

5.2.4.4 National Fiber Networks.

Many applications need time and frequency traceability at a greater accuracy and stability. Optical fiber time transfer has the capability of providing such a precise traceability of time and frequency signals. With the help of the existing networks spread across India, time transfer by terrestrial means can be implemented which will be a backup for the satellite-based signals. Since satellite signals can be jammed

Table 5.10 List of satellite constellations used for time transfer applications and traceability

System	GPS	GLONASS	BeioDou	Galileo	NAVIC
Owner	United States	Russian Federation	China	European Union	India
Orbital altitude	20,200 km	19,130 km	21,150 km	23,222 km	36,000 km
Orbital period: (h/min/s)	11/58/2	11/14/30	12/52/4	14/4/41	23/55/59
Number of satellites	24–32 (at least 24 by design)	24–30 (at least 24 by design)	24–30 (and 5 GEO satellites)	24–30 (at least 24 by design)	7 (3 GEO and 4 GSO)
Signal frequency (MHz)	1575. 42 (L1), 1227.6 (L2 C), 1227.6 (L2 P), 1176.45 (L5)	1598.0625–1609.3125 (L1 C/A), 1242.9375–1251.6875 (L2 C), 1242.9375–251.6875 (L2 P), 1202.025 (L3 OC)	1561.098 (B1l) 1207.14 (B2l) 1268.52 (B3)	1575.42 (E1), 1176.45 (E5a), 1207.14 (E5b), 1278.75 (E6)	1176.45 (L5), 2492.028 (S)

or spoofed or may get effected by the solar storms in the space many countries have already implemented this terrestrial based optical fiber time transfer.

One of the major projects of Department of Telecommunications of India (DoT) is National Optical Fiber Network (NOFN) renamed as BharatNet which is implemented since 2011 to connect the 2.5 Lakh gram panchayats of the country with at least 100 Mbps speed to provide telecommunication services [81]. Another major project regarding the fiber network is the National Fiber Spectrum (NFS) of defence which is intended to provide an alternate "nationwide communication network" for defence services [82]. With the help of such networks time and frequency signals can be transferred to various critical applications like defence services, high pression science experiments, for department of space, 5G etc., Depending on the distance of transmission and accuracy requirements there are various technologies available for optical fiber time and frequency transfer. This may not require a complete dark fiber or not even a channel of the fiber, one DWDM (Dense wavelength Division Multiplexing) wavelength is also capable to transfer the time in nano seconds to as far as 3000 km with a small bandwidth of 10 Mbps.

The major fiber network services are provided by NOFN and NFS followed by Jio, Airtel, Vodafone networks as shown in the above Fig. 5.5 [83].

As a conclusion Fig. 5.6 shows the legal time act should include the three entities Infrastructure to be used of legal time dissemination, accreditation policy naming the authorities for time assessment, time audit, and time stamping and finally technical standards for traceable measurements as per international standards.

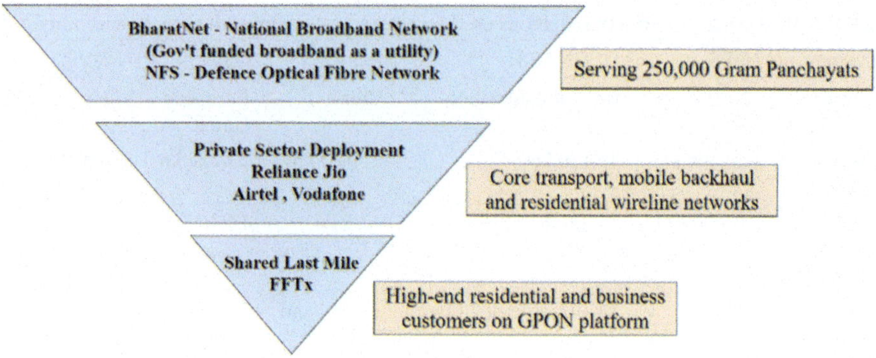

Fig. 5.5 Optical fiber networks provided by various agencies

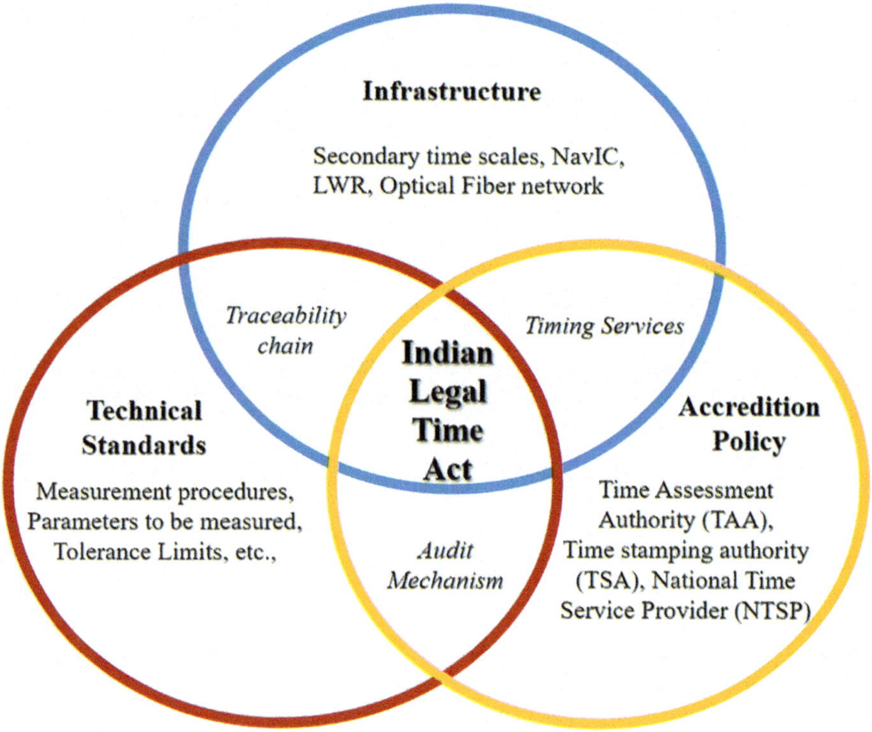

Fig. 5.6 The three entities: technical standards, accreditation policy, infrastructure required for the Indian Legal Time Act

5.3 Economic Impact of IST Synchronized India

Assuming that the proper legal time act is implemented in India and as per the required tolerance limits in all the sectors the timing sources are synchronized with IST, an order of magnitude estimation of its impact on Indian Economy was attempted. The component of economic losses in various sectors due to time and frequency synchronization failures was roughly calculated from the data available easily in public domain, which is summarized as economic impact in Table 5.11. The dominant reason for losses in various sectors that result due to absence/poor synchronization as reported in the sources used, are as follows:

i. Power grid: Grid failure in 2001 caused 107.1 Million business loss [39, 40] and in 2012, 22 out of 28 states are disconnected from power supply for 12 h.
ii. Stock Exchange: The NSE colocation broker scam of Rs. 50,000 Cr in during 2010–2014 by gaining and early access to the stock trading engine [84].
iii. Broadcasting: Component of loss due to the improper timing resulting poor quality broadcast.
iv. Cyber Security: 457% rise in Cybercrimes in India in last five years that can be resolved only by reliable time stamping.
v. Navigation through satellites: Component of loss due to inaccurate navigation due to inaccuracy of time signal.
vi. Disaster management: Component of loss due to time delayed predictions of Natural Disasters.
vii. Railways: Time synchronization of railways will reduce loss component resulting from accidents and low efficiency transportation.
viii. Digital Economy: Component of loss due to fraud in E-Governance and E-services in the absence of reliable time stamping.
ix. Defence: For instance, component of loss due to poor Missile communication and navigation because of inaccuracy in time signal.
x. CCA (Time stamping service): Component of Loss due to tampering Integrity of digital data in absence of reliable timestamps.
xi. Telecom and 5G: The telecom regulatory authority of India (TRAI) has imposed penalties for call drops and imposed a 2% cap for such loses in each telecom circle for a particular operator [85]. A recent article has projected an economic impact of Telecom and 5G at few % points of the GDP. McKinsey Global Institute suggests that the financial impact of IoT on the global economy may be as much as \$3.9 to \$11.1 trillion by 2025.

Needless to mention that for an accurate estimate of the economic impact of time synchronization in various sectors in India, an exhaustive survey and analysis of relevant data and international economic impact studies are required that is definitely warranted in future.

Table 5.11 Impact on Indian economy due to IST synchronization

S. No.	Sector	Reason for economic losses	Time synchronization requirement	Rough estimate of annual impact (INR)
1	Power Grid	Syntonization failure in power Grid	Power grid PMUs accuracy ~1 μs	~40,000 Cr
2	Finance-Stock Exchange	Trading scams by brokers	Stock exchange server clocks with an accuracy ~0.3 μs	~10,000 Cr
3	Professional Broadcasting Industry	Quality of broadcast	Wired: 6 s/wireless: 1 μs	Enormous
4	Cyber Security, Networks and Data Centres	457% (5 years) rise in cybercrimes, 76% (2018) business hit by cyber attacks	All network operations and network servers with an accuracy ~10 ms	~30,000 Cr and will increase
5	Navigation through Satellites	Inaccurate satellite navigation and communication	Master ground Stations with an accuracy ~20 ns	Limitless
6	Disaster Management	Delayed predictions of natural disasters	Remote sensing (e.g. LIDAR) with an accuracy ~1 μs	Huge (depends on circumstances)
7	Railways	Improper ticket booking and signaling, delays and accidents	Train control centres with an accuracy ~1 ms	Enormous
8	Digital economy	Fraud in E-Governance, E-services	Time stamping with an accuracy ~1 ms	Huge
9	Defence	Missile communication and navigation	Time stamping with an accuracy ~1 μs	Difficult to estimate but very high impact
10	Digital Signatures and Time Stamping	Tampering Integrity of digital data (e.g. e-tendering, financial certificate)	Time stamping with an accuracy ~1 s	Limitless
11	Telecommunications and 5G	Slow data rate, call drops and improper M2M communication etc	Primary reference time clocks ~30 ns (5G) and ~ 1 s (IOT)	~10% of GDP (future)

5.4 Summary

The chapter proposed and discussed the roadmap of accurate and traceable pan-India time synchronization for safe Digital India. All national sectorial applications demand a time-aware digital framework for its authentic, efficient and legally robust functioning. This can be ensured by each device and network synchronized to traceable common national reference IST in India. Presently, IST is being disseminated across Indian subcontinent through calibration, network time services and satellite links.

With digital transformations, most social and strategic sectors in India-like telecommunications, finance, power grid, broadcasting, navigation, defence, aviation, space, cyber security etc.—are examples of cyber physical systems. These CPSs have pertinent requirements of time accuracy (from ms to ns) and infrastructure for providing desired time synchronization for efficient and predictable functioning. However, IST is not yet the legal time for India and is not mandatorily used for time synchronization of these nationally important sectorial applications. A coordination of CSIR-NPL as NMI and relevant ministries and regulators, and identification of international standards required to implement legalized IST in India is proposed and discussed for various sectorial applications. Legalization and implementation of IST as common national reference time will help realize safe Digital India and result in a significant economic impact to the tune of 10% of GDP.

Acknowledgements The authors would like to acknowledge all the ex-colleagues; present colleagues Dr. Ashish Agarwal, Ms. Pranalee Thorat, Ms. Suchi Yadav, Mr. Mahavir Prasad, Ms. Preeti Kandpal, Mr. Devender; and the students of CSIR-NPL for their contributions in establishing various activities of Time and Frequency Metrology. Authors would also like to thank the stakeholders such as Department of Legal Metrology, ISTRAC-ISRO, National Informatics Centre (NIC), DoT and many others for their continuous support and collaboration in the national missions and endeavours for pan-India time synchronization.

References

1. https://digitalindia.gov.in/accessed. 28 June 2020
2. https://meity.gov.in/writereaddata/files/india_trillion-dollar_digital_opportunity.pdf. Accessed 03 July 2020
3. https://www.expresscomputer.in/news/why-a-national-accurate-traceable-legal-time-is-crucial-for-indias-digital-economy/22218/. Accessed 06 July 2020
4. https://cca.gov.in/sites/files/pdf/services/timestampingbrochure.pdf. Accessed 06 July 2020
5. NIST—Frame work for Cyber-Physical systems, vol 1, Version 1.0, NIST—special publication. https://doi.org/10.6028/NIST.SP.1500-201
6. Aviral Shrivastava et al., Time in Cyber-Physical Systems, 2016 International Conference on Hardware/Software Codesign and System Synthesis (CODES+ISSS), IEEE conference proceedings. http://dx.doi.org/https://doi.org/10.1145/2968456.2974012
7. IEEE Time—sensitive Networking (TSN) Task Group, https://1.ieee802.org/tsn/. Last accessed on 07.07.2020.

8. "Synchronisation requirements for 5G, An overview of standards specifications for cellular networks", Jia chin lin
9. Lukasz Sliwczynski et al., Fiber optic time transfer for UTC traceable synchronization for telecom networks. IEEE Commun. Stand. Mag. 1(1) 66–73 (2017)
10. Alan O'Connor et al., Economic benefits of the global positioning system (GPS) (Research Triangle, NC: RTI International, 2019), ES-4
11. National Geographic resource library, Encyclopedic entry, Navigation. https://www.nationalgeographic.org/encyclopedia/navigation/
12. G. Sadlier et al., *The Economic Impact on the UK of a Disruption to GNSS* (London Economics, London, UK, 2017), p. iii
13. Alan O' Connor et al., Economic benefits of the global positioning system (GPS) (Research Triangle, NC: RTI International, 2019), ES-4
14. B. Hampton, The history of GPS. Fleet Trax. https://fleettrax.net/the-history-of-gps/ (2016, August 1).
15. GPS positioning guide July 1995, Natural Resources Canada, Geodetic survey division. ISBN 0-660-15917-1
16. Khondokar et al., GNSS time synchronization in vehicular ad-hoc networks: benefits and feasibility. IEEE Trans. Intell. Transp. Syst. (99):1–10. https://doi.org/10.1109/TITS.2017.2789291
17. P. Defraigne, G. Petit, "CGGTTS—Version 2: an extended standard for GNSS time transfer, Metrologia 2015 52G1. Doi: http://doi.org/10.1088/0026-1394/52/6/G1
18. "Is the Year 2038 problem the new Y2K bug?". The Guardian. 17 Dec 2014
19. T. Mizrahi, Marvell, Security Requirements of time Protocol in Packet Switched Networks, RFC-7384, Internet engineering Task Force (IETF) (2014)
20. Elena Lisova et.al., Delay attack versus clock synchronization—a time chase, in IEEE International Conference on Industrial Technology (ICIT), 04 May 2017. https://doi.org/10.1109/ICIT.2017.7915522
21. Mike Ellis, why do broad casters care about timing, in International Timing and Sync Forum, Conference Proceedings ITSF (2017)
22. Johas Teener et al., Heterogeneous networks for audio and video; using IEEE 802.1 audio video bridging, in IEEE Proceeding, vol. 101, 11 Nov 2013. https://doi.org/10.1109/JPROC.2013.2275160
23. SMPTE standard, SMPTE profile for use of IEEE-1588 PTP in professional broadcast applications. SMPTE ST 2059–2:2015. https://doi.org/10.5594/SMPTE.ST2059-2.2015
24. AES-R16–2016: AES standards Report-PTP parameters for AES67 and SMPTE ST 2059-2 interoperability. www.aes.org
25. Understanding digital signatures, Security Tip (ST04–018), Cybersecurity and Infrastructure Security Agency (CISA), USA. https://us-cert.cisa.gov/ncas/tips/ST04-018
26. Information Technology—security techniques—time stamping services-part 4, traceability of time sources, ISO/IEC 18014-4 (2015)
27. Controller of certifying authorities, time stamping guidelines for CAs, V2.0 Sept 2016. https://cca.gov.in/guidelines.html
28. PhD thesis on "Role of Foreign Institutional Investors in Indian Stock Market"
29. Securities and Exchange Board of India, www.sebi.gov.in
30. Polansek, Tom, CFTC finalizes plan to boost oversight of fast traders: official. Reuters. Retrieved 8 July 2014
31. Dina ElBoghdady, 23 Apr 2013. https://www.washingtonpost.com/business/economy/
32. India International Exchange Limited (IndiaINX)
33. Bombay Stock Exchange Colocation trading
34. Broad Guide Lines on Algorithmic Trading, SEBI Circular: CIR/MRD/DP/09/2012. https://www.sebi.gov.in/sebi_data/attachdocs/1333109064175.pdf
35. Fair and Transparent Access to data feeds of the stock Exchanges, SEBI Circular: SEBI/HO/MRD/DP/CIR/P/2017/08. https://www.sebi.gov.in/sebi_data/attachdocs/1484915770520.pdf

36. Securities and Exchange Commission, Release no: 34-77196, File no. SR_FINRA-2016-005. https://www.sec.gov/rules/sro/finra/2016/34-77196.pdf
37. Federal Register/Vol 81, No. 72/Thursday, April 14, 2016/Notices. Release no. 34-77565; File No. SR-FINRA 2016-005. https://www.sec.gov/rules/sro/finra/2016/34-77565.pdf
38. European commission, Commission Delegated Regulation (EU), Brussels, 7.6.2016. C (2016) 3316 final. https://ec.europa.eu/transparency/regdoc/rep/3/2016/EN/3-2016-3316-EN-F1-1.pdf
39. https://www.wsws.org/en/articles/2001/01/ind-j09.html
40. https://en.wikipedia.org/wiki/2012_India_blackouts
41. System security aspects, Governor action (m), Indian Electricity Grid code (2010), Central Electricity Regulatory Commission
42. Government of India, Ministry of Power data base. https://powermin.nic.in/en/content/power-grid. Accessed on 14 June 2020
43. IEEE standard for Synchro phasor measurements for power systems, IEEE Std C37.118.1-2011
44. IEEE standard profile for use of IEEE 1588 Precision Time Protocol in power system applications. C37.238-2017
45. https://www.gsc-europa.eu/sites/default/files/sites/all/files/Report_on_User_Needs_and_Requirements_Rail.pdf
46. https://www.railway-technology.com/contractors/operation/time-frequency-solutions/
47. https://www.indianrailways.gov.in/railwayboard/uploads/directorate/infra/downloads/RDA-Regulator.pdf
48. https://rdso.indianrailways.gov.in/works/uploads/File/Handbook%20on%20Digital%20Clock%20with%20GPS%20Synchronization.pdf
49. https://standards.ieee.org/standard/1474_1-2004.html
50. P.N. Misra, Lincoln Laboratory J. **6**, 231–247 (1993)
51. A.K. Brown, Navigation. **35**, 23–40 (1988). https://doi.org/10.1002/j.2161-4296.1988.tb00938.x
52. D.C. Rickard, D.J. Sherry, S.J. Taylor, IEEE Xplore. 92 International Conference on Radar (1992)
53. Andre Zerger et al., Computers, environment and urban systems, **27**(2), 123–141
54. Zubaida Alazawi et al., 11th International Conference on ITS Telecommunications (2011) IEEE explore. https://doi.org/10.1109/ITST.2011.6060083
55. Q. He et al., IEEE J. Sel. Topics Sig. Process. **4**(1), 79–100 (2010)
56. Huang Yulin et al., International Conference on Communications, Circuits and Systems (2006). https://doi.org/10.1109/ICCCAS.2006.285118.
57. P. Yin et al., IEEE International Symposium on Phased Array Systems and Technology (2013). https://doi.org/10.1109/ARRAY.2013.6731859
58. Karen Kente et al., Guide to computer security log management, Recommendations of NIST, NIST special publications 800-92, September 2006
59. Hand book of Digital Forensics and Investigation, Chap. 4, Intrusion Investigation, Eoghan Casey et al., ISBN: 978-0-12-374267-4. https://doi.org/10.1016/C2009-0-01683-3
60. Shalini R. Lihitkar, Major Librabry and Information Networks in India, online document. https://eprints.rclis.org/14210/1/Network_Nagbhid_Shalini-1.pdf
61. National Financial Switch Product overview, National payments corporation of India. https://www.npci.org.in/product-overview/national-financial-switch-product-overview
62. Retail payment statics on NPCI platform, online data base. https://www.npci.org.in/statistics
63. Payment Card Industry Data Security Services (PCI-DSS), standards, guideline no. 10
64. ITU—T Y.2060 (6/2012), Overview of the Internet of things
65. Technical report on M2M gateway and Architecture, TEC—Department of Telecommunication (DoT), TEC-TR-S&D-M2M-001–01, Release 1.0, 2015
66. Technical report on M2M enablement in power sector, TEC—Department of Telecommunication (DoT), TEC-TR-S&D-M2M-002–01, Release 1.0, 2015
67. Technical report on M2M enablement in Intelligent Transport systems, TEC—Department of Telecommunication (DoT), TEC-TR-S&D-M2M-003–01, Release 1.0, 2015

68. The national measurement system for Time and Frequency, NBS special publication 445–1. https://tf.nist.gov/general/pdf/88.pdf
69. https://www.time.gov/#. Accessed on 27.06.2020
70. https://www.transportation.gov/regulations/time-act. Accessed on 27.06.2020
71. German Federal Law Journal (Bundesgesetzblatt, BGBI), 1978, Part I, pp. 1110–1111
72. https://www.gesetze-im-internet.de/me_einhg/index.html. Accessed on 29.06.2020
73. Government of India Directory. https://goidirectory.nic.in/ministries_departments_view.php
74. The Gazette of India, *Ministry of Consumer Affairs, Food and Public Distribution* (Department of Consumer affairs, National Standards Rules, 2011)
75. The Gazette of India, Information Technology (Certifying Authority) Rules—2000, Section 10.5 Audit Trails and Verification
76. ITU -R, TF.1876, Trusted time source for time stamp Authority
77. Department of consumer affairs, Legal Metrology, Overview. https://consumeraffairs.nic.in/organisation-and-units/division/legal-metrology/overview
78. A. Michael, *Lombardi* (Radio Controlled Clocks, NCSL International Workshop and Symposium, 2003)
79. CSIR-NPL IFR MoU webpage (2020). https://www.nplindia.in/csir-npl-and-ifr-information-dissemination-services-pvt-ltd-sign-mou-time-and-frequency-traceability. Accessed 5 July 2020
80. Arup Das Gupta, Article: Evolution of the global Navigation satellite systems (GNSS), Geo spatial world.net, 09.20.2016. https://www.geospatialworld.net/article/global-navigation-satellite-system-gnss/
81. Bharat Broadband Network Limited, Website: bbnl.nic.in.
82. Modernization of Communications, Indian Army Government of India, website: Indian-army.nic.in
83. fibre-Optic Co—deployment along the Assam High ways and Trans Asian Railways for E resilience: the cases of India and Bangladesh, United Nations, ESCAP. Oct 2018
84. F I R no: RC AC1 2018 A0011 Date: 28.05.2018, District: New Delhi, PS: CBI/AC-I
85. Call drop-TRAI sets new quality norms for 4G networks, The economic Times, Telecom news, 31 July 2018
86. Przemyslaw .K et.al. ELSTAB-Fiber -Optic time and Frequency distribution Technology: A General Characterisation and fundamental limits, IEEE Transactions on ultrasonic, ferroelectrics, and frequency control, 63(7), JULY 2016. https://doi.org/10.1109/TUFFC.2015.2502547
87. Namneet Kaur et.al. Time and Frequency transfer over a 500km cascaded white rabbit network, July 2017 Joint Conference of the European Frequency and Time Forum and IEEE International Frequency Control Symposium (EFTF/IFC), https://doi.org/10.1109/FCS.2017.8088808

Chapter 6
Physico-Mechanical Metrology

Part I: Impetus for Inclusive Industrial Growth

Sanjay Yadav, Nita Dilawar Sharma, S. S. K. Titus, S. K. Jaiswal, V. K. Jaiswal, Naveen Garg, Komal Bapna, and D. K. Aswal

Abstract Measurement standards and metrological activities provide a proper scientific and economically sustainable infrastructure to the country [1]. It play a pivotal role as an engine for industrial revolution, growth and support for the overall development of any nation. Metrology is essential to protect the planet, ensuring a dignified quality life for all people to achieve inclusive economic growth and prosperity. It is the main pillar for trade, scientific comparison, innovation of emerging technologies, technical cooperation and even for simple exchange of information. With the advent of better-quality products through metrological advancement, our industries can compete internationally by overcoming all the trade barriers/constraints to achieve the final exports targets. It ultimately translates into growth of industries through rapid industrialization, economic growth and societal upliftment. For industrial and scientific development of any country, the role of NMIs is significant and irreplaceable in the advancement of metrological fields and their related

S. Yadav (✉) · N. D. Sharma · S. S. K. Titus · S. K. Jaiswal · V. K. Jaiswal · N. Garg · K. Bapna · D. K. Aswal
CSIR-National Physical Laboratory, Dr. K.S. Krishnan Marg, New Delhi 110012, India
e-mail: syadav@nplindia.org

N. D. Sharma
e-mail: ndilawar@nplindia.org

S. S. K. Titus
e-mail: titus@nplindia.org

S. K. Jaiswal
e-mail: skjaiswal@nplindia.org

V. K. Jaiswal
e-mail: jaiswalvk@nplindia.org

N. Garg
e-mail: ngarg@nplindia.org

K. Bapna
e-mail: komal.bapna@gmail.com

D. K. Aswal
e-mail: dkaswal@nplindia.org

D. K. Aswal (ed.), *Metrology for Inclusive Growth of India*,
https://doi.org/10.1007/978-981-15-8872-3_6

technologies. Dissemination of measurement standards through apex level traceable calibration certificates, training and consultancy services to the industry, government, strategists and academia in different parameters are extremely important. The dissemination of measurement standards in SI units, now related with *fundamental constants of nature (FCN)*, is the prime responsibility of NMI of the country. As India wishes and approaching to become '*Atma Nirbhar or Self Reliant*', the NMI of India i.e. CSIR-NPL has to play a significant role, not only for providing traceability of measurement standards but also to work in close association with industries, especially with *Medium, Small and Micro Enterprises (MSMEs)* which contributes almost *38%* of the *Gross Domestic Product (GDP)*. It would directly correlate with India's economic growth and better quality of life for its citizens. The mandate of Physico-Mechanical Metrology (PMM) of CSIR-NPL is to establish, maintain, upgrade and disseminate the standards of mass, volume, density, viscosity; length, dimensions, angle, surface roughness and nanometrology; temperature and humidity; optical radiation (non-ionizing radiation-ultraviolet, visible and infrared region, LED lighting); force, torque and hardness; pressure, vacuum and ultrasonic; acoustic and vibration; and fluid flow. The PMM is the backbone and mainstay for Indian industries and other user organizations. As is evident, these measurements are ubiquitous and cross most strategic sectors and technologies. These parameters play a significant role in the industrial development and their traceable measurements to SI units provide basis for barrier free trade across the continent and the world. The services of these measurements underpin just about every area of human endeavour. The PMM of CSIR-NPL contributes enormously to the economic development of the country by providing traceability to almost all the strategic sectors, MSMEs, Public and Private Sector Undertakings, Government organizations, regulatory bodies, academic institutions as well SAARC nations and other countries through calibrations, testing, need based technical and consultancy services, organization of proficiency testing programmes, industrial trainings and development of academic expertise. The present Chapter of this book is mainly focused on highlighting the significance and importance of PMM, its role and impact in industrial and sectorial growths, applications and relevance with their global compatibility, and finally concludes with some futuristic perspectives.

An experiment is a question which science poses to Nature, and a measurement is the recording of Nature's answer.

– Max Planck (1858-1947)

6.1 Introduction

Physico-Mechanical Parameters (PMPs), play a crucial and pivotal role for the sustainable development as an important component of metrology which is essential to protect the environment, ensure a quality life for all citizens to achieve inclusive economic growth and prosperity. Though the PMPs are important in almost all walks of day–to-day life however, precision measurements of the same play a vital role in strategic sectors and industries. In order to attain self-sustenance, as per the Government plan, we need to provide the necessary metrological support to manufacturing industries for developing globally competitive products and services. Metrology would contribute significantly in such endeavours. Further, as India is aiming to become a 5.0 trillion economy, the in-house production of quality products must grow exponentially which is supported by the in-house manufacturing capability. The strong metrological and quality infrastructure is the one phrase answer for such sustainable growth. Over the years, though the PMM at CSIR-NPL is playing its part and offering reasonably good services, yet much more needs to be achieved.

The PMM, being backbone of the industrial growth has a very wide scope and its embodiment in a single chapter is a very difficult task. Though, one can easily write a separate chapter even on a single parameter, authors have tried to accommodate all the parameters, [mass, volume, density, viscosity; length, dimensions, angle, surface roughness and nanometrology; temperature and humidity; optical radiation (non-ionizing radiation-ultraviolet, visible and infrared region, LED lighting), force, torque and hardness; pressure, vacuum and ultrasonic; acoustic and vibration; and water and gas flow] into 4 Chapters as 4 Parts of the PMM.

There exists a close interaction between various PMPs, measurement techniques, and industrial and technological applications. Such interactions are extremely important when in present time worldwide international competition in industry and production engineering is huge. Any manufacturing process requires multiple measurements of multiple PMPs. For example, the manufacturers of force measuring instruments have started using more advanced and sophisticated electronics in their design for easy, effective, fast data acquisitions and collection. This not only requires the expertise and new capabilities of load cells to measure more than just force, clients demand load cells that can also measure barometric pressure, temperature, humidity, air density and location etc. Such devices of multiple functionalities help the users with additional benefits of simplified design and multiple data quality. Therefore, inclusion of the parameters and their understanding will hugely benefit the readers. While Part 1 (present Chap. 6) is of generic nature which describes the insinuations, impact and role of PMM in inclusive industrial growth, the next 3 Chaps. 7, 8 and 9, are focused on Mass and Length Metrology (Part II); Thermal, Optical and Acoustic Metrology (Part III) and Force, Pressure and Flow Metrology (Part IV), respectively. Figure 6.1 depicts the overview of all the chapters.

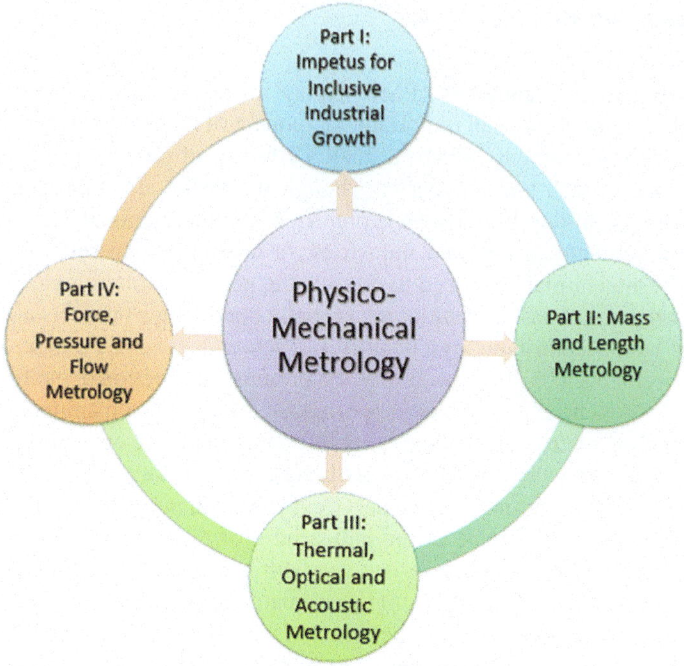

Fig. 6.1 Layout showing chapters of physico-mechanical metrology. Outer circle depicts the inter linkages of the all the parameters and chapters

6.2 Significance and Importance

With modern business becoming diversified, bigger and better, competition is higher than ever before. Therefore, it is incredibly important to ensure that a company/service provider, is offering top range of the products/services to consumers to keep up pace and competition with its rivals. The one particular way by which companies/service providers can ensure the excellence in deliverables, is by performing rigorous quality checks beforehand. This can only be achieved through a strong national quality infrastructure (NQI) as described in detail in Chap. 1, harmonization of international measurement system in Chap. 2 and harmonization of international and national dissemination mechanisms in Chap. 3 of this book.

The metrology has huge economic and social impact. It has strong role in NQI of the nation as one of the major pillars. The inaccurate measurement leads to cost of poor quality (COPQ) which may not be tangible but have enormous economic losses. The challenges of today's industries are driven by the demands of good quality products/services from the customers. The customers demand cost-effective wares of good quality that suffice their specific needs. For long term sustainability of any industry, the supply of its products must be stable, quality-competitive and cost-effective.

As discussed in previous Chapters, metrology is absolutely essential for make in India, trade, innovation and new technology developments which in the national context are the pillars on which the "*Atam Nirbhar Bharat*" would grow. The 5 pillars of the development as put forward by our honourable Prime Minister, viz. economy, infrastructure, system, demography and demand can be achieved only with precise measurements and their certification for product approval. Thus, the honourable Prime Minister's call for '*Vocal for Local*' leading to "*Atam Nirbhar Bharat*" is conceivable and for this purpose, metrological traceability and reliability in measurements would play a big role culminating in indigenous products with global acceptance and quality and it would finally result into quantum jump in economic growth.

The MSMEs are the major force for India to achieve the goals of 5.0 trillion economy and "*Atam Nirbhar Bharat*". These MSMEs heavily depend on locally available NQI services to support their participation in the regional and international markets. The stronger NQI would help them to have access to better services for consumer's benefit as well having improved product quality, health and consumer protection.

Customers want good quality products and services on very competitive rates that fulfil their specific needs. Therefore, the industrial outputs and products must be stable, competitive, innovative as well as indigenous. India is entering into the 4th industrial revolution which demands the digitization of the production industry. The bottlenecks to be overcome to adapt these new technologies for Indian industries, that have a clear archetype need for conformance in accordance to the standards, is to work towards consistent innovation. As the products/services are available a click away, the needs and satisfaction levels of the customers are also changing with the time. Therefore, all the stakeholders should be aware of the changing scenario and be always ready for adoption, innovation and competition to stay in business.

6.3 Role of PMM in Nation Building

The PMM systems, enabled with modern sensors, transducers, robust electronics and software, are the backbone of the Indian industries, especially in majority of the industrial sectors. As is evident, these precise measurements are ubiquitous and essential across most of the strategic and R&D sectors. Physico-mechanical parameters play a significant role in the industrial developments and its traceable measurement to SI units provide barrier free trade across the continents.

The precision measurement services of PMM at CSIR-NPL contribute hugely to the economic development of the Nation by way of providing traceability to the industrial sector, strategic sectors, MSMEs, Public Sector undertakings as well as government and regulatory bodies through apex level calibrations, technical and consultancy projects; and industrial training. On an average CSIR-NPL generates more than 2000 calibration/test reports and serves for more than 1500 customers annually only in PMPs.

The average direct cash flow from these services, range between 60–70 million Indian Rupees annually. However, the economic impact of these services, far outweight the expected returns on investment but include benefit to cost ratios and indirect contributions to the GDP of the Indian economy. In a recent client/service targeted survey, it was estimated that for the services to the industry, the social return ratio was as high as 180% for the direct beneficiaries and as high as 980% to the second level clients/beneficiaries [1].

Apart from providing these direct services, the CSIR-NPL has focused its objectives on the 5 pillars of the development as put forward by our honourable Prime Minister, as discussed above. With a vision to realize these goals, significant strides are being made. These new challenges in measurements are being met with the continuous up-gradation of measurement capabilities at CSIR-NPL which lead to competence at par with the other NMIs of the world and further contribute significantly to economic development. Figure 6.2 explicitly depicts the PMPs and the role

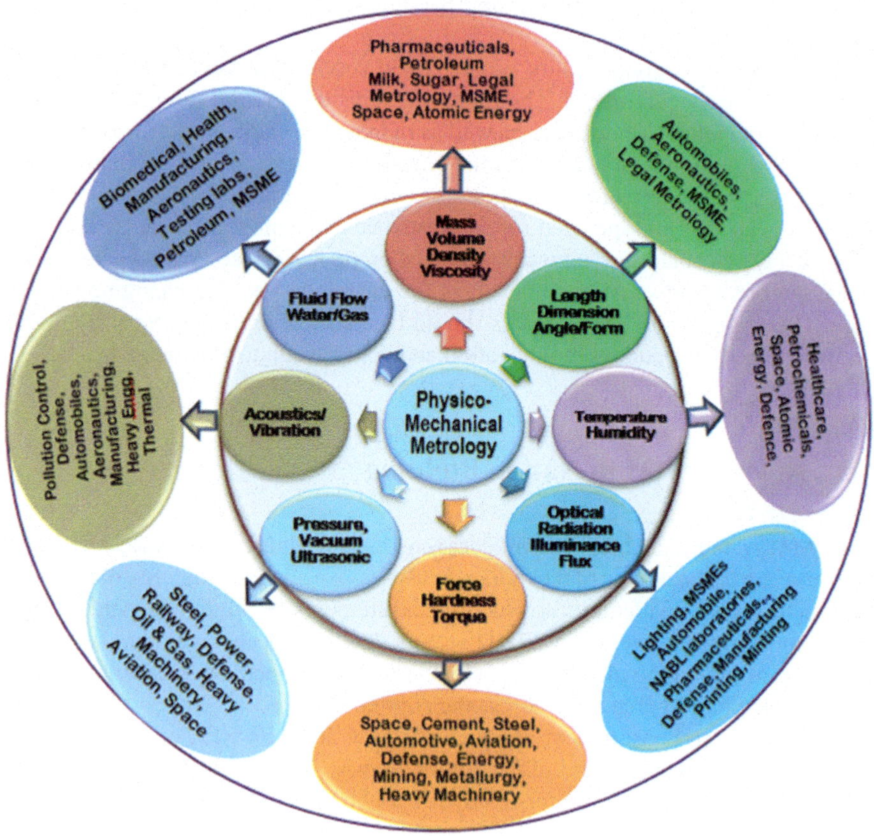

Fig. 6.2 The major physico-mechanical parameters and their role in different strategic and industrial sectors presented as a single window information

of its metrology in almost all major sectors of a Nation, which play significant role in its inclusive industrial growth.

As the SI units have been redefined in terms of fundamental constants, the leading NMIs are making efforts to realize the base units through quantum metrology. To keep pace with these NMIs, PMM is also endeavouring in the frontier areas of quantum metrology for realization of 4 base units through fundamental constants, such as realization of mass (kg) through Watt-balance, length (m) by frequency comb, thermodynamic temperature (K) by Boltzmann constant (k) and few photon metrology of light sources.

The country has a huge clientele base having more than 5000 calibration and testing laboratories in organized sector apart from more than 4 lakh laboratories in unorganized sectors. Sooner or later all these laboratories would require traceability from CSIR-NPL and major responsibility to serve these clients is of PMM and PMM is ready to accept the challenges and add new clientele base in coming days. The various socio-economic benefits of PMM are listed in Fig. 6.3 which clearly reflect the mirror image of the contribution of PMM in national growth.

The *Economic Infrastructure* of a country depends upon the industrial sectors which support the economy with livelihood, transport, water supply, health and sanitation and communication. These sectors are further reinforced by *Social Infrastructure* like academics, hospitals, legal system and enforcement. Being public goods, the economic and social infrastructures are regulated and governed by state agencies and

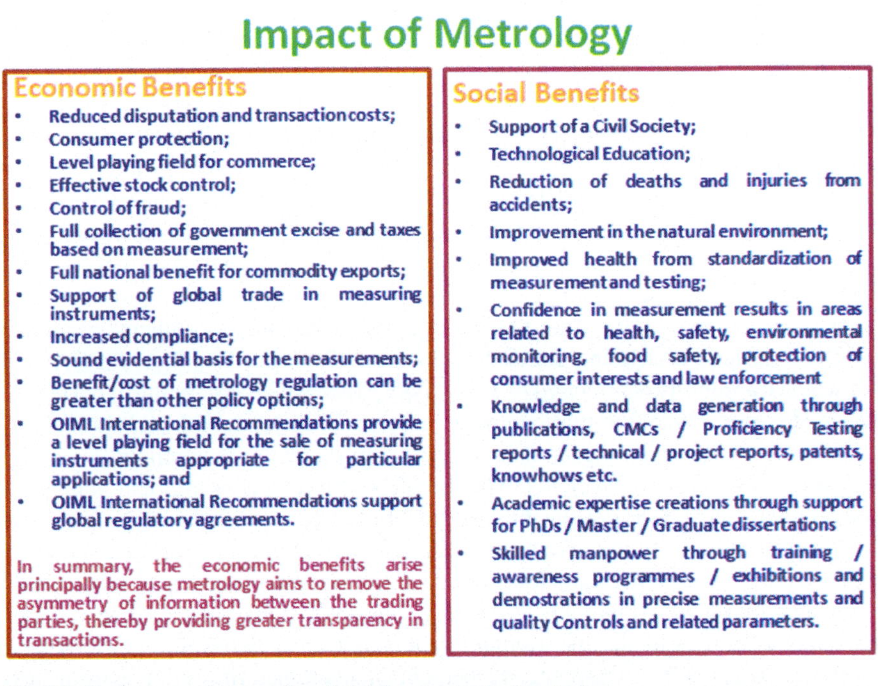

Fig. 6.3 Socio-economic benefits of physico-mechanical metrology

regulators. Apart from these socio-economic infrastructures, the *Technical/Scientific Infrastructure* also ropes the economy. Such infrastructure includes measuring instruments, standards, measurement techniques and methods, artefacts, standard reference materials which are used for efficient, authentic and valid scientific and engineering data, process for interfaces between manufacturing technologies such as factory controls, monitoring, recoding, automation and communication.

In this context, the role of PMM is substantial and paramount to support the consistency of measurements related to PMPs that are used as the basis for decision making both in industry, science and technology as described in next 3 Chapters. Also, such national metrological infrastructure includes the International System of units (SI) of physical quantities (Chap. 2), and harmonization of their dissemination mechanisms (Chap. 3) and technical and scientific details about the calibration and measurement facilities, maintenance of traceability chain, support services to different organizations, training imparted to measurement engineers and technicians (Chaps. 7, 8 and 9). The overall socio-economic benefits are already listed in Fig. 6.3.

6.4 Impact of PMM on Sectorial Growth

The main contribution of PMM is already highlighted in previous section. However, the major sectors and the stakeholders who are regularly taking traceability of PMPs are shown in Fig. 6.4.

One of such important sectors is *pharmaceutical industries which are making drugs and medicines*. According to some studies, during this pandemic situation, *Hydroxychloroquine (HCQ)* has shown strong antiviral effects on the corona virus infection and sudden demand for increased production of HCQ came up. *Zydus Group and IPCA Labs* have backward integrated HCQ making capability and the PMM at CSIR-NPL has provided traceability to their E2 class weights which are used in manufacturing of these medicines. Similarly, *paracetamol* is always in huge demand. Traceability has also been provided to *Ajanta Pharma.* As per the reports, Indian pharmaceutical companies would raise the production of HCQ to their peak capacity of 35 crore (350 million) tablets of 200 mg [2]. The government has already placed an order with leading domestic manufacturers Zydus Cadila and IPCA Labs for which traceability is provided by CSIR-NPL. The HCQ is also being exported and has made great impact internationally as India's visibility, strength and capabilities.

Dimensional measurements are extremely important in *automobile and automotive* sector apart from other fields. Automobile sector uses measurement intensively for product innovation, design models, materials and manufacturing processes at each and every stage of manufacturing to support trade with suppliers and end users. Calibration of automobile parts for measurements of surface roughness, straightness, roundness, angle measurements, coaxiality, centricity, flatness etc. are very important. In a liberal estimate, the *automobile industry* contributes almost 8% percent of India's GDP and a whopping 49 percent of manufacturing GDP with a large

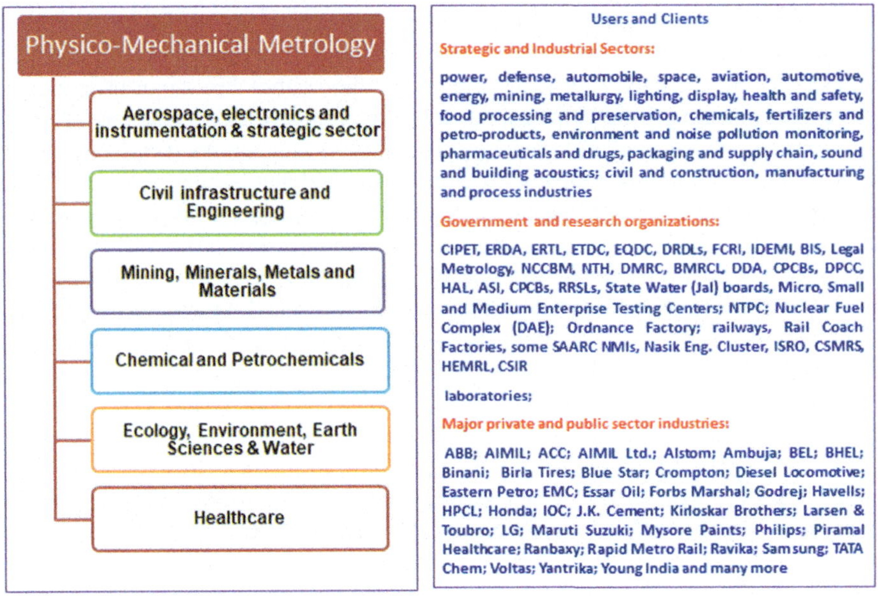

Fig. 6.4 Major sectors and stakeholders regularly getting traceability and other services for PM parameters

economic multiplier impact. CSIR-NPL provides traceability to major automobile giants like Maruti Suzuki, Tata Motors, Hero MotoCorp etc. [3].

Another important sector is *lighting industry*. LED based energy efficient illumination technology is paving way as a future lighting source phasing out the old inefficient lighting devices. CSIR-NPL has taken it up as a priority and endeavouring to establish testing and calibration facility for LED and LED based lighting as per national/international standards. Upon establishment of such services CSIR-NPL additionally, would be able to cater to the countries present need of evaluation of the photo-biological safety of lamps and lamp systems to safeguard users from potential hazards of radiation exposure to eyes and skin given by the increasing deployment of sources of invisible radiation in purification and sanitization industries. The LED market is accounting for about 60% of India's total lighting industry (approximately ₹ 376 billion) in 2020 [4].

Construction and civil infrastructure is the fundamental necessity which enable and serve a city or a country, to accommodate and provide all the operational functionalities to the people of the country. Cement, steel, aluminium, plastics and wood are the basic elements required by every construction sector for building and developing the desired Administrative, Business, Commercial, Domestic, Educational, Financial, Government offices, Healthcare and Industrial, infrastructure in the country. Therefore, the infrastructure sector provides direct and indirect employment to a large number of people promoting self-reliance, *contributing a major part to Gross Domestic Product (GDP)*. The Indian Cement Industry is one of the booming sectors of the Indian economy and therefore, cement industry is the most important basic primary industry for the economic development and next comes the iron and steel industry [5, 8]. CSIR-NPL provides the traceability, either directly or indirectly, to the testing machines used in the cement and iron industries across the country.

Accurate *pressure, vacuum, temperature, humidity and flow measurements* are not only important to improve the quality of products and increase the productivity but also crucial for the safety of plants, manpower and machinery. Such measurements are essentially performed during manufacturing/processes and contribute significantly to the advancement of technology. Being the most used process parameters, their measurements become a key to industry, research and society because of their extremely wide range of applications such *as optimization, process monitoring and controlling parameters in nuclear and thermal power plants, food sterilization, vessel manufacturing and production, machine components/parts designing and fabrication, autofrettage, chemical, petrochemical, drugs and pharmaceutical industries, water jet cleaning and cutting, hydro-static forming and isostatic pressing in automobile, diesel engineering, micro-and nanotechnologies and metal casting and forming*. In order to define efficiency and ensure safety of the processes, the pressure and vacuum are measured/monitored as process parameters. CSIR-NPL contributes significantly in these parameters to fulfil the customer's requirements of traceability of various equipment used in such applications and their training needs.

Oil and gas equipment-manufacturing sector is another noticeable field wherein record keeping, quality control and traceability are critical. As part of their company-wide quality protocol, various large oil and gas companies invest continually in a preventive maintenance programme that employs machine tool probes and calibration products to check and verify the accuracy of the CNC machine tools and measuring instruments.

The biomedical metrology has got boost after implementation of new *Medical Device Rules 2017* by Ministry of Health and Family Welfare, Government of India. As per the rules, now medical devices require testing and calibration for their optimum performance and for having confidence on measurement results of diagnostics. CSIR-NPL has established bio-medical laboratory for testing and calibration of various types of medical devices (Chap. 15). At present, PMM provides traceability to various biomedical laboratories in India by calibration of infusion pumps, infusion device analyzers, gas flow analyzers, ventilator testers and defibrillators and contributes effectively in the health mission of Govt. of India. Recently CSIR-NPL has also initiated for establishment of *Testing Facilities for gloves, gowns, masks, ventilators and thermal scanners as per National/International standards.* The project has been approved by CSIR and work is going on this direction. This would lead to self-reliance in biomedical field and saving of foreign currency spent on testing and calibration.

The applications of ultrasonic metrology in health, for both therapeutic and diagnostic, physiotherapy, cleaning, underwater navigation and communication, NDT, material quality and adulteration assessment, concrete testing, detection of liquid or gas leakages and material characterization etc., are well proved. Ultrasonic phased array, cleaning, acoustic emission, are the key technologies which are finding new applications in testing of materials for flaw and crack detection, cleaning purposes and for detection of leakage in the pipeline including liquid and gas.

Noise pollution is a growing issue in the country and therefore, *noise mapping* is significant in the present scenario, which is being adequately addressed by CSIR-NPL which has established several SODAR centres at different parts of the country. It is focused on developing noise mapping of various cities of India in collaboration with State Pollution Control Boards. Apart from focusing on such societal issues, CSIR-NPL has also planned to develop facilities for future metrological support in the country. Efforts are being made to develop the new facilities for calibration of sound intensity probes, primary free-field microphones, primary and secondary shock, audiometer and related testing.

The parameter specific contributions are described in the next 3 Chaps. 7, 8 and 9 of the book in detail.

6.5 Conclusion and Future Perspectives

From the metrological studies carried out over the years, it has been well understood that the precision and traceable measurements of physico-mechanical parameters such as mass, length, temperature, luminous intensity, force, hardness, pressure, vacuum, ultrasonic, acoustics, vibration, fluid flow, etc. are vitally important for sustainable infrastructure development and revolutionary industrial growth of the country leading to a better quality of life to its citizens in every sphere of human activity. It is therefore very much essential to realize all these physical parameters with a highest level of accuracy by way of establishing the national primary standards and systems. The measurement capabilities of such established primary standards are reaffirmed by periodically participating in the international inter-comparisons which enable us to obtain the global competence and recognition.

Every country needs such a robust metrology system in place to provide the necessary traceability in measurements to each and every sector of the economy. This process is inevitable for maintaining the uniformity in all the measurements for achieving sustainable growth in several fronts like science, engineering, technology, industry, commerce, business trade, etc. which ultimately enables us to attain self-reliance. Hence, it is also mandatory for the NMIs to continuously upgrade and maintain such systems at par with the other NMIs of world in accordance with the new technological advancements and challenges arising in the global scenario. The primary standards thus established and maintained provide traceable measurements of global equivalence which enables our county to overcome trade barriers, save a huge foreign exchange and the process time.

Over the years, CSIR-NPL has developed and achieved proven measurements capabilities in PMPs in its specific domain. The metrological activities have been established and maintained as per CIPM-MRA requirements, demonstrating international measurement equivalence and the calibration capabilities. It has also successfully participated in 81 international inter-comparisons and reaffirmed its CMCs which are registered in BIPM-KCDB database. So far it has registered 150 CMCs in BIPM-KCDB database which stands at 27th place in the world and 7th in APMP region. For maintaining these CMCs CSIR-NPL has been constantly maintaining and upgrading the standards at par with the other leading NMIs and is also in the process of expanding the scope of measurement activities by adding new parameters or measurement ranges depending upon the rising needs of the industries. These precision measurement activities of CSIR-NPL have also been extended for providing the required traceability to a chain of testing and calibration laboratories and industries across the county on a continuous basis and on the other hand, generating a substantial amount of funds for the organization.

The major beneficiaries of the calibration, testing, technical consultancy and training services in PMPs are several stakeholders in a variety of sectors such as manufacturing, lighting, pharmaceutical & healthcare, automobiles, aviation, etc. *This persistent apex level traceability support of CSIR-NPL for various PMPs has been the foundation for these Indian industries, and PMM is playing its significant role in providing traceability, technical and consultancy services, technical experts for several national and international committees related to metrology, accreditation, standardization which in turn results into over-all growth of the country.*

The apex level measurement and calibration capabilities thus developed at CSIR-NPL for 4 SI base units i.e. mass, length, temperature and luminous intensity; and further for their derived units namely, density, volume, viscosity, area, humidity, heat capacity, conductivity, illuminance, illuminance responsivity, luminous flux, luminance, colour temperature, colour coordinates, spectral irradiances, force, torque, hardness, pressure, vacuum, ultrasonic velocity, ultrasonic attenuation, ultrasonic power, sound pressure, vibration amplitude, sound transmission and absorption coefficients, totalized mass and volume, mass and volume flow rates in water and gas flow etc., have been playing their profound role in the growth of Indian economy through various industries as its stake holders, together with the flag-ship programmes of Govt. of India, namely the *Make in India,* and other initiatives.

The pharmaceutical industry in India is one of the largest in the world and ranks within five in the world, which caters to almost 70% of the domestic demand for bulk drug and around 40% of the total is exported to more than 200 countries in the world (including a large component for the US and UK). As per '*Pharma Vision 2020*', Department of Pharmaceuticals, Govt. of India, it is aimed to make India a major hub for end-to-end drug discovery and production which would help to further enhance the market share of India's domestic pharmaceutical industries, which was Rs. 1.4 lakh crore (US$ 20.03 billion) in 2019 in terms of turnover. As an example, affordable medicines under the *Pradhan Mantri Bhartiya Janaushadhi Pariyojana (PMBJP)* itself achieved a record sales turnover of Rs. 52 crores (US$ 7.38 million) in the month of April 2020. The CSIR-NPL's mass and its allied derived parameters are constantly contributing, providing traceability to pharmaceuticals, drugs, chemicals and fertilizers industries on a regular basis [2].

Lighting industry, with an estimated market share of about *Rs. 20,000 crores,* is also witnessing tremendous growth especially in LED segment alone with a growth of about 80%. Long lamp life, energy saving and environment friendly LED technology can surprisingly reduce the country's electricity demand by as much as 40 percent. The LED lighting market in India is projected to register a compound annual growth rate (CAGR) of 26.6% during 2017–23 and is expected to reach *$ 3758.74 million by 2022* with growing government initiatives encouraging the use of LED lights through, *Pradhan Mantri Sahaj Bijli Har Ghar Yojana—Saubhagya, Ujala scheme* etc. and development of smart cities under the *Smart City Mission.* It would also help in generating employment opportunities and would need to be supplemented with adequate number of professionals or skilled manpower in this vital field of lighting or LED or solar energy. The contribution of CSIR-NPL is extremely vital as traceability and other services are regularly being extended to lighting industries [4, 6].

State of the art test and calibration facilities to cater to the present and future calibration need of lighting industries as per the National/International standards for the next generation energy efficient LED based lighting products is also underway at CSIR-NPL. The project needs to be continued for achieving its objectives. It would help support the energy saving and green initiative of Govt. of India by not only eliminating sub-standard LED products from the market but would also help the nation in saving over Rs. 40,000 crores a year and reduction of approximately 2.7 crore tonnes of CO_2 every year.

Further, the automobile sector of India, which is one of the largest in the world and accounts for over *7.1%* of *India's GDP*, also contributes to nearly *22%* of the country's manufacturing GDP. The government of India aims to maintain the upward growth trend of the automobile industry and has launched the *Automotive Mission Plan 2016–26 (AMP 2016–2026)* endeavouring to make India among the top 3 automotive industries in the world and increase exports exponentially to reach *35–40%* of overall output. It also intends to increase its contribution to the GDP to over *12%*, by 2026. The traceability provided by CSIR-NPL in length, dimension, force, pressure and temperature is a noticeable contribution in the growth of automobile sector [3, 7].

Infrastructure and its development is the fundamental requirement that enables a country, to accommodate all the desired administrative, business, commercial, domestic, educational, financial, healthcare and industrial, infrastructure in the country. Therefore, the infrastructure sector provides direct and indirect employment to a large number of people promoting self-reliance, contributing a major part to GDP. As per an estimate, the *Indian Cement Industry* is one of the booming sectors of the Indian economy. The next comes the *iron and steel industry*. The market size of cement production is reached to *337.32 million tonnes in 2018–19* and stood at *278.79 million tonnes from April 2019-January 2020* [5]. Cement is the most commonly used building material due to its high compressive strength, which is tested using compression testing machines (CTM). These CTMs are duly calibrated either directly by CSIR-NPL or through NABL accredited laboratories whose traceability is mostly established from CSIR-NPL, providing the quality assurance related to the cement produced across the country.

Similarly, the manufacturing sector has seen its GDP contribution improved to 17.1% in the last 2 years, primarily due to the 'Make in India' drive and the inherent measurement traceability provided by CSIR-NPL towards the quality products [8].

In a nutshell, the traceability is disseminated through apex level calibration services to R&D laboratories, calibration & testing laboratories, strategic sectors, process and automobile industries, illumination and lighting, display, opto-electronics, defence, space, power, health and safety, pollution monitoring, manufacturing industries, government sectors, research organizations and some of the SAARC NMIs.

Every year more than 2000 calibration and testing reports are generated by CSIR-NPL in PMPs, for direct beneficiaries. These 2000 calibration reports are used by the stakeholders to further extend the services to huge number of their clients, ultimately resulting in percolation of NPL's traceable measurement scale to every consumer. For example, ERTL (North) gets its photometric measurement systems and dead weight testers calibrated from CSIR-NPL and automobile industry like Maruti Suzuki India Limited etc. gets its instruments calibrated from ERTL (North). These types of users are huge and contribute significantly to the economy of the country. Also, the product and services vendors called sub-contractors, supply different components and parts to these large companies (like Maruti Suzuki India Limited) and these big companies maintain the quality and compatibility of these parts through testing and calibration. Thus, the users of the calibration and testing facility for traceability, multiply enormously down the chain.

As per an estimate made in one of the studies carried out at CSIR-NPL [1] for one of parameters (pressure and vacuum), based on an industrial survey, an overall social rate of return SRR, an indicator of system efficiency, is found to be *180%* which is quite high as compared to the present rate of return, PRR (*8%*). In economic terms, the system is considered efficient if SRR is more than the PRR. This indicates that the activity is quite efficient, and future investment in the activity is safe and would provide excellent returns. Further, SRR increased from 180 to 980% in case of increase in number of beneficiaries at different levels as described in the above paragraph. This is an indicator of huge social impact of pressure and vacuum facilities of CSIR-NPL and services provided to industries. Admittedly, the study was carried out for only one parameter of the PMM, but in the same way it can be extended to other parameters and as a whole for the institution. The studies are very useful for broader understanding of the economic benefits and to improve understanding of the decision and policy makers about the relevance of metrology, and continuation of services and further investment in the activity.

The industries requiring all such traceability services from CSIR-NPL, are from all over India, and sometime far away from Delhi. They have to send their instruments to Delhi through air or road transportation, which at times, leads to delay and high cost, and occasional mishandling of the instruments during transportation. Sometimes these costs become higher and customer at his part avoids to send instruments for calibration and try to get calibration done locally by even compromising on the accuracy. In future, CSIR-NPL may adopt a policy to collaborate with some of the regional centres of NABL accredited laboratories or legal metrology laboratories, as reference laboratories for maintaining the sanctity of the unbroken chain of traceability and quality of services. Also, on-site calibration services may be augmented and encouraged to address the traceability needs for industries which are unable to bring their equipment to CSIR-NPL.

Keeping in view the significance, relevance and importance of metrology, the metrology programme of NPL requires sustained administrative and financial support on regular basis so that the existing CMCs can be maintained and further CMCs may be added at par with other advanced NMIs. Presently, some of our standards and measurement capabilities are now getting obsolete due to rapid technological advancements. Now, it is of utmost priority to upgrade, replace and establish new measurement standards including the upcoming quantum standards to maintain and enhance our CMCs to cater to modern days industrial demands. For enhancing and achieving measurement scopes, some of the PMM facilities need to be established or expanded to meet the new demands arising in the present industrial scenario. All such requirements are described in detail in respective Chaps. 7, 8 and 9.

The CSIR-NPL being a signatory of Metre Convention and key member of the APMP, it is essential that Scientists must attend the meetings of BIPM and APMP on regular basis, where the important metrology related polices are formulated and voted upon and international inter-comparisons are planned apart from discussions on other matters. Normally, most of the NMIs from APMP countries send large delegations (15–40 members) to attend these meetings. In the existing system, the CSIR-NPL finds it very difficult to send a single delegate to the respective TC (Technical Committee) meetings. The metrological activities of CSIR-NPL would be greatly benefitted by participation in such meetings. The acquired knowledge, sharing of ideas and collaborations with experts of different NMIs would help in future in enhancing and expanding the existing apex calibration capabilities, creating new calibration capabilities. *No country can be self-reliant if its metrology programme is not strong.*

This non-participation hampers the planned key and supplementary comparisons, decision/policy development and other related issues in various fields of metrology. Therefore, participation of CSIR-NPL in these techno-business meetings is highly important and essential. During the last several years, non-participation by CSIR-NPL in adequate numbers, in these meetings has not only resulted in a great loss in metrology per se but also loss in consistent credibility for India, being a giant nation among APMP region which possesses a great global standing in almost all disciplines of science & technology and requires to strengthen its emerging prospects of leadership among leading NMI members, like NIM, China, NMI, Japan, NMI, Australia, KRISS, South Korea and A*STAR, Singapore.

Apart from all the pro-active efforts towards establishment of new facilities and up-gradation of existing calibration facilities to the requirement of testing labs and industries, it is also important to have regular healthy interactions with the industries and the laboratories. This will enable us to work in harmony to address various metrological challenges faced by them, resulting in a better coordination between the NMI and its stakeholders working towards sustainable growth and self-reliance of the country.

Acknowledgements Authors are thankful to all the members and students of the PMM, past and present, who have directly or indirectly contributed in the development of the measurement standards and facilities. Thanks are due to all the funding agencies that have provided partial grants for the same time to time.

References

1. A. Zafer, S. Yadav, N.D. Sharma, A. Kumar, D.K. Aswal, Economic impact studies of pressure and vacuum metrology at CSIR-NPL, India. MAPAN J. Metrol. Soc. India **34**(4), 421–429 (2019). https://doi.org/10.1007/s12647-019-00356-2
2. https://www.ibef.org/industry/pharmaceutical-india.aspx
3. https://www.moneycontrol.com/news/business/markets/trouble-in-motown-auto-sector-which-is-49-of-manufacturing-gdp-looks-for-rescue-4264251.html
4. https://www.electronicsb2b.com/eb-specials/industry-report/indian-led-lighting-industry-poised-growth/
5. https://www.ibef.org/archives/industry/cement-reports/indian-cement-industry-analysis-january-2020
6. https://www.electricalindia.in/prospects-of-lighting-industry-in-india/
7. https://auto.economictimes.indiatimes.com/news/industry/automotive-mission-plan-2016-26-unveiled-here-are-the-key-highlights/48772090
8. https://www.thehindubusinessline.com/economy/manufacturing-sector-should-grow-contribute-25-to-gdp-expert/article9455444.ece

Chapter 7
Physico-Mechanical Metrology

Part II: Mass and Length Metrology

Sanjay Yadav, Goutam Mandal, Nidhi Singh, Santwana Pati, Rina Sharma, Girija Moona, Mukesh Jewariya, D. D. Shivagan, Komal Bapna, S. K. Jaiswal, V. K. Jaiswal, and D. K. Aswal

Abstract Measurement standards play a pivotal role in the scientific and industrial development of a country. The role of National Metrology Institute (NMI) is significant and crucial for realizing the national standards and disseminating the traceability

---

S. Yadav (✉) · G. Mandal · N. Singh · S. Pati · R. Sharma · G. Moona · M. Jewariya · D. D. Shivagan · K. Bapna · S. K. Jaiswal · V. K. Jaiswal · D. K. Aswal
CSIR-National Physical Laboratory, Dr. K. S. Krishnan Marg, New Delhi 110012, India
e-mail: syadav@nplindia.org

G. Mandal
e-mail: goutam@nplindia.org

N. Singh
e-mail: singhnidhi@nplindia.org

S. Pati
e-mail: santwana.pati@nplindia.org

R. Sharma
e-mail: rina@nplindia.org

G. Moona
e-mail: moonag@nplindia.org

M. Jewariya
e-mail: jewariya.mukesh@nplindia.org

D. D. Shivagan
e-mail: shivagand@nplindia.org

K. Bapna
e-mail: komal.bapna@nplindia.org

S. K. Jaiswal
e-mail: skjaiswal@nplindia.org

V. K. Jaiswal
e-mail: jaiswalvk@nplindia.org

D. K. Aswal
e-mail: dkaswal@nplindia.org

D. K. Aswal (ed.), *Metrology for Inclusive Growth of India*,
https://doi.org/10.1007/978-981-15-8872-3_7

to the entire country. The present chapter discusses the mandate of PMM of CSIR-NPL to establish, maintain, upgrade and disseminate the standards of mass, volume, density, viscosity; length, dimensions, angle, surface roughness and nanometrology. Mass and length are the fundamental physical quantities used since ancient times. This chapter discusses the past, present and future of mass, volume, density, viscosity; length, dimensional and nanometrology measurements at CSIR-NPL. The evolution of the SI units of mass and length in India from the historic times to present state of the art technology is discussed briefly. Participation in key comparison exercises by CSIR-NPL, potential services and future requirements are discussed in detail to serve the regional laboratories, industries and various stakeholders in the country. The mass and length parameters of PMM substantially contribute to the economic development of the country by providing traceability to strategic sectors, MSMEs, public and private sector undertakings, government organizations, regulatory bodies as well as the SAARC nations. The discussion is oriented towards describing the role of mass and length metrology for the inclusive industrial growth of the country as important PMM parameters.

We find no sense in talking about something unless we specify how we measure it; a definition by the method of measuring a quantity is the one sure way of avoiding talking nonsense.
– Sir Hermann Bondi (1919-2005)

7.1 Introduction

Mass and length are the fundamental parameters of PMM which play a very pivotal role in human endeavors. The establishment and maintenance standards of fundamental parameters are often carried out by NMIs. Thus, the role of NMI becomes very crucial not only to realize the national standards to the highest levels of accuracy and precision, but also to disseminate the traceability to the entire country. The mass and length measurements are required and performed in almost all the sectors and industries. Thus, the establishment of a strong national metrological and quality infrastructure is indispensable in economic growth and realizing the ambitions of becoming a 5 trillion economy and contributing towards the GDP of the country. This chapter describes the significance and relevance of mass and length metrology, established measurement standards and facilities, various services offered to different sectors, potential gaps and future perspectives in detail. The parameter-specific contributions and the various calibration and measurement capabilities of mass and length parameters, implications of redefinition of kilogram using Planck's constant, measurement techniques applied for dimensional measurements and facilities for nanometrology are described in the succeeding sections of this chapter.

7.2 Mass, Volume, Density and Viscosity Metrology

7.2.1 Introduction and Historical Aspects

7.2.1.1 Mass

In day-to-day life, "mass" and "weight" are often used for same purpose, but by definitions and applications, both are different. The mass of a body is defined as the quantity of matter contained in the body. Although there are two types of masses, namely gravitational mass (universal law of gravitation) and inertial mass (Newton's law of motion), for all practical purposes gravitational and inertial masses are same and mass is constant at all the places. On the other hand, weight of a body is force of attraction toward center of the earth which is proportional to the mass of that body. Conceptually, "mass" should be measured in kilogram (kg) and "weight" in Newton (N), but in metrological applications, "mass" is defined as "(true) mass" and weight as "conventional mass". Therefore, for both mass and weight, the SI unit is used as kilogram (kg).

A pictorial diagram is shown in Fig. 7.1 about realization of kilogram from French Revolution to modern era. Initially, it was defined by weight of 1 L of water at 0 °C in the year 1793, but in the year 1795, temperature was converted to 4 °C, since density of the water is maximum at 4 °C. In the same year, a brass prototype was made as provisional kilogram, since brass is a soft material. In year 1799, a prototype

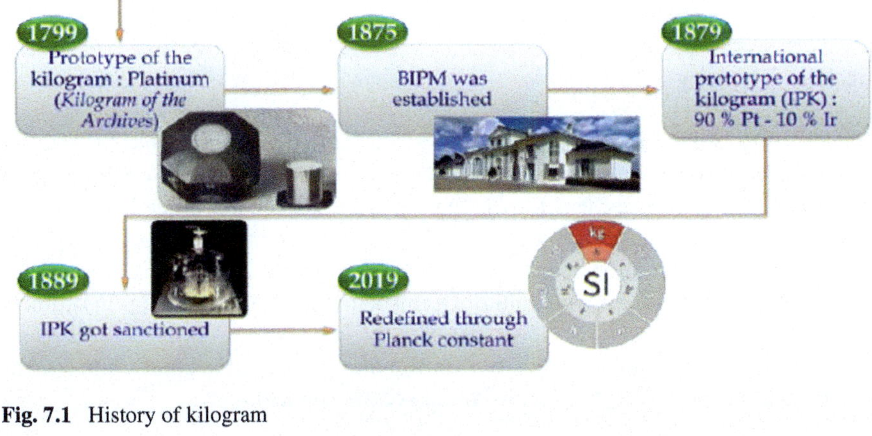

Fig. 7.1 History of kilogram

of the kilogram was made of platinum (kilogram of the archives), since platinum is noble metal. After signing metre convention in 1875, BIPM was formed. Although platinum is a noble metal, it is a soft metal too. In 1879, a number of prototypes of the kilogram were made of platinum-iridium alloy to increase hardness and long-term stability and best one was chosen as *International Prototype of the Kilogram (IPK)* and kept at BIPM. Till the year 2019, IPK served as primary standard of mass. During 2019, kilogram has been redefined in terms of physical constant of nature, i.e., Planck's constant, $6.62607015 \times 10^{-34}$ m^2 kg/s.

7.2.1.2 Volume

Volume is the quantity of three-dimensional space occupied by a substance (solid, liquid or gas). The volume of a solid (whether regularly or irregularly shaped) can be determined by fluid displacement. Displacement of liquid can also be used to determine the volume of a gas, where the volume of a liquid can be determined using volumetric measures (content type or delivery type). However, the SI unit of volume is cubic meter (m^3), but liter (symbol "L" or "l") is used as a special unit for liquid which is equal to 0.001 m^3.

7.2.1.3 Density

The density of a material is its mass per unit volume. An object made from a relatively dense material (e.g., iron, steel, etc.) will have less volume than an object of equal mass made from various less dense substance (e.g., water, petroleum ether, etc.). The density of a material varies with temperature and pressure. This variation is small for solids and liquids but much larger for gases. In some particular applications (for concentration of solutions), "specific gravity" is commonly used instead of density. Specific gravity is the ratio of the density of a material to the density of water. The SI unit of density is kilogram per cubic meter (kg/m^3).

7.2.1.4 Viscosity

The viscosity of a fluid is a measure of its resistance to flow. It describes the internal friction of a moving fluid. A fluid with large viscosity resists motion because its molecular structure gives it a lot of internal friction. On the other hand, a fluid with low viscosity flows easily because its molecular structure results in very little friction when it is in motion. For example, honey or syrup has a higher viscosity than water. Gases also have viscosity, although it is a little harder to notice it in ordinary circumstances.

The SI unit of kinematic viscosity is millimeter square per second (mm^2/s), and SI unit of dynamic viscosity is milli-Pascal second (mPa s).

7.2.2 *Significance, Applications and Relevance in Different Sectors*

7.2.2.1 Significance

Together with length and time, mass was probably one of the first physical quantities that human attempted to measure. Mass, a property of all matter, is unique in the sense that the concept is generally deduced from the motion of a body under the influence of an applied force. The mass of one object with respect to another is deduced from a comparison of the gravitational forces acting on the two objects. Mass has a longer history in trade and commerce than it has in research.

Volume is a quantity derived from the length unit; however, it is inseparably coupled with mass. The largest numbers of volume measurements are associated with determining the amount of liquid which is "contained" in, or can be "delivered" from, a variety of appropriate vessels, i.e., pipettes, burettes, flasks, metal containers or capacity measures, volumetric tanks, etc. When the container design is such that it can be weighed in the filled condition, the containment volume is determined by gravimetric methods. The container is first weighed in an empty condition and then filled to an appropriate level with a liquid of known density and reweighed. The volume of the liquid, determined from the definition of density, is assumed to be the volume of the container.

7.2.2.2 Applications and Relevance in Different Sectors

The nature of the services provided and the role of the mass, volume, density and viscosity parameters include support to the nation and a continuing effort to improve the efficiency of measurement. As per the requirement of quality system, it is assured that the reported results are adequate for the intended purpose. CSIR-NPL provides traceability and apex level test/calibration services for mass, volume, density and viscosity parameters to various sectors which includes Legal Metrology Department, ISRO, DAE, STQC, Heavy Alloy Penetrator Project (HAPP) under Ministry of Defence, DRDO, Indian Railway, Air India, Delhi Jal Board, Indian Oil Corporation, CPCB, BHEL, pharmaceutical industries, manufacturing industries, NABL accredited laboratories, etc.

The services provided by CSIR-NPL in these parameters have direct benefits to society, strategic sectors, etc. A few of them are highlighted as below:

i. Department of Consumer Affairs (Legal Metrology): As per Legal Metrology Act, 2009, and subsequent Legal Metrology (national standards) Rules, 2011, CSIR-NPL is custodian of "national standard" of mass, i.e., National Prototype of the Kilogram copy no. 57 (NPK-57). It disseminates value of NPK-57 to the Regional Reference Standard Laboratories (RRSLs) under the Department of Consumer Affairs (Legal Metrology) every three years to maintain the weights and measures system of the country. Only CSIR-NPL can provide this service as per rules.

ii. Pharmaceutical industries: Weights which are used as standards for making drugs and medicines are calibrated against CSIR-NPL mass standards.
iii. HAPP under Ministry of Defence: Density of heavy alloy rod is provided and this rod is used to verify the strength of weapons used in Defence.
iv. Delhi Jal Board: Dynamic viscosities of polyaluminium chloride (PAC) samples are provided and these samples are used for purification of water.
v. Milk industries: CSIR-NPL calibrated lactometers are used by the milk industries to check the purity of the milk by measuring its specific gravity.
vi. Sugar industries/petroleum industries: CSIR-NPL calibrated hydrometers are used to check the purity of petroleum product/sugar content.
vii. Paint industries: Ford cup calibrated by CSIR-NPL is used by paint industries for measuring the viscosity of paint which is used for surface protection of machinery parts.
viii. NABL accredited laboratories: Equipment used as standards by the NABL accredited laboratories in the areas of mass and related parameters are traceable to "national standard" of mass maintained at CSIR-NPL as NMI.

7.2.3 *Facilities and Global Status*

7.2.3.1 Measurement Standards, Techniques and Facilities

Measurement Standards [1]

At the apex level, copies of International Prototype of the Kilogram (IPK) were given to different National Metrology Institutes (NMIs) who are member of Metre Convention. These are known as *National Prototype of the Kilogram (NPK)*. India is having copy number 57 (NPK-57) which is kept at CSIR-National Physical Laboratory. NPK-57 serves as national standard of mass in India which is directly traceable to IPK with standard uncertainty of 11×10^{-9} kg at $k = 1$ (after redefinition of kilogram). It has undergone for periodic calibration in 1985, 1992, 2002 and 2012 so far at the BIPM. In CSIR-NPL, there are four 1 kg transfer standards which are periodically calibrated against the NPK-57 once in three years. Out of these four standards, two are made of stainless steel and other two are of nickel–chromium alloy.

The working standards are a set of stainless steel or nickel–chromium or brass weights in multiple or sub-multiple of 1 kg ranges from 20 kg to 1 mg. The working standards are calibrated once in two years or every year against transfer standards. These are used to calibrate the Reference Standard Weights of the Legal Metrology and for routine calibration.

The mass value of NPK-57 is disseminated to CSIR-NPL mass standards from 1 mg to 2000 kg, solid density standards and other derived parameters like pressure, vacuum, force, torque, hardness, ultrasonic, fluid flow, etc. Not only physical parameters, chemical parameters are also traceable to the mass standards.

Mass value of NPK-57 is also disseminated to the customers through calibration of weights of different accuracy classes as per OIML R111-1 standard received from them. This documentary standard contains technical (e.g., principal physical characteristics) and metrological requirements of weights used as standards for the calibration of weights and weighing instruments. As per this standard, weights with nominal values from 1 mg to 5,000 kg are classified in nine accuracy classes, i.e., E_1, E_2, F_1, F_2, M_1, $M_{1\text{-}2}$, M_2, $M_{2\text{-}3}$ and M_3. The reference weight should generally be of a higher accuracy class than that of test weight. However, for calibration of E_1 class weights, the reference weight should have similar or better metrological characteristics (magnetic properties, surface roughness, etc.) than the test weight. Under normal conditions of use, the surface conditions shall be such that any alteration of the mass of the weights is negligible with respect to the maximum permissible error.

Measurement Techniques

There are two methods each to establish the standards of mass, volume, density and viscosity. First method is primary method used to establish standards or to determine parameters at highest level of accuracy and precision and second method is used for dissemination of the standards. Brief descriptions of these methods are given below.

Mass

Subdivision (Multiplication) Method [2–5]

BIPM has provided only one mass standard of nominal value of 1 kg which is not enough to fulfill all the requirements of day-to-day lives. We need various mass standards with other nominal values from milligrams to kilograms, even in tons (1 t = 1000 kg). To establish values of nominal masses (unknown) other than 1 kg, substitution method is not suitable, since nominal values are different from each other. A special method needs to be followed which is called the subdivision method. In this method, a number of weights are combined together to make same nominal value and compare each combination. There are total 12 combinations of different nominal values. Applying principle of least squares, these 12 equations are converted into five linear equations to determine the mass value and associated uncertainty of each unknown weight.

For submultiples of 1 kg, typical weighing combinations for a set of masses of 500, 200, *200, 100, *100 g are followed against 1 kg standard for Decade I as shown in Fig. 7.2. In Decade II (combinations for a set of masses of 50, 20, *20, 10, *10 g), one of the 100 g masses is used as standard, and likewise, masses of all the weights are determined up to 1 mg. Same procedure is followed for multiples of 1 kg.

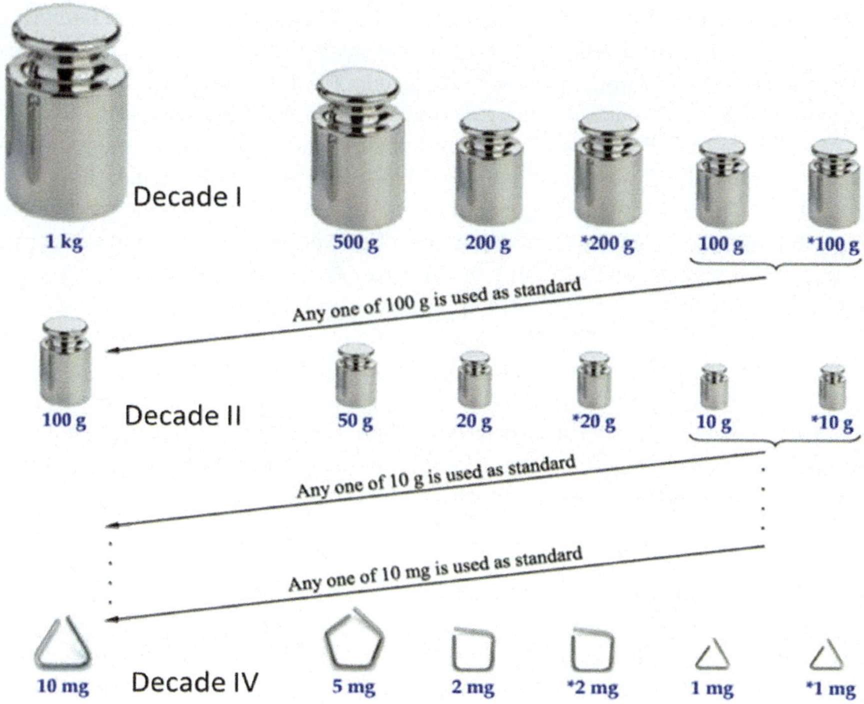

Fig. 7.2 Dissemination of mass standards using subdivision method

Substitution Method

In substitution method, calibration of a weight is carried out by comparing it against a reference weight (RW) of known mass whose nominal mass is equal to that of the test weight (TW). The comparison is done on a suitable weighing instrument. The indications of the weighing instrument are used only for calculating difference between the RW and TW. The comparison of two weights is always carried using ABBA or ABA or $AB_1 \ldots B_nA$ ($n \leq 5$) weighing cycle to eliminate linear drift.

Volume

Gravimetric Method

For precision determination of capacity of volumetric instruments, the gravimetric method is used. In this method, capacity is determined by weighing the quantity of distilled water that the measure contains or delivers at the temperature of the liquid and applying a correction to convert the weight of water to volume at the reference temperature (in India, reference temperature is 27 °C).

Volumetric Method

When a large number of vessels or volumetric instruments of very high capacities are required to be calibrated, a relative method is used. In this method, the capacity of the instrument under test is compared with that of the standard volumetric instrument of known capacity. Volumetric method is applicable where high degree of accuracy is not required.

Density

Hydrostatic Weighing Method (Cuckow's Method) [6, 7]

Hydrostatic weighing method (Cuckow's method) is the primary method for establishing traceability of reference hydrometers. In this method, apparent masses of a hydrometer are measured in air and in liquid, at desired stem scale point, using a balance by applying Archimedes' principle. In case of lower range of hydrometers, a sinker of known mass and volume is used. Silicon sphere is used as primary solid density standard which is traceable to mass standards and length standards (for volume) for determining density (mass/volume). The advantages of selection of silicon as primary standard are low density (very close to liquid density) and low thermal expansion coefficient; hence, air buoyancy correction is minimal. Liquid is selected as per desired stem scale point and density of that liquid is determined using silicon sphere. Once density of liquid is established, it is used to establish traceability of reference hydrometers.

Comparison Method

This method is easier and most commonly used for calibration of hydrometers received from customers. Calibration of a hydrometer is done using reference hydrometer whose scale errors are precisely known. The hydrometer to be calibrated is floated in liquid of appropriate density which is adjusted in such a way that the level of liquid surface intersects the graduation under examination.

Viscosity

Step-up Method

Distilled water is used as primary standard and kinematic viscosity of water is used to establish standard viscometers and standard viscosity oils through the step-up method. First, two master viscometers, having calibration constants in the range of (0.001–0.003) mm^2/s^2, are calibrated with water at 20 °C. The kinematic viscosities of two standard oils are measured at 40 °C in these two master viscometers. Corrections are made for buoyancy, temperature and surface tension, wherever applicable. Then two master viscometers, with calibration constants of 0.003 mm^2/s^2–0.009 mm^2/s^2, are calibrated at 40 °C with the two standard oils, and their calibration factors are calculated at standard conditions for water at 20 °C. Similarly, next series of master viscometers and standard viscosity oils are calibrated at 40 °C using the results of two master viscometers and two standard oils. Steps between successive calibration constants or viscosities increase by a factor of three or less until the desired viscosity range is covered. Oils are calibrated at other temperatures using the average result from at least two master viscometers.

Comparison Method

This method is used to calibrate viscometers received from customers using master viscometer of same/similar constant that has been calibrated by the step-up method. In this case, viscosity oils may be used as a medium only. However, test viscometers may also be calibrated using viscosity oil as a standard with known viscosity determined by the step-up method. Similar way, viscosity of oil may be determined using master viscometer of known constant.

Unique Facilities

CSIR-NPL has adequate facilities to establish and maintain "national standard" as well as to provide apex level services to the nation. A list of equipment is given in Table 7.1.

Few state-of-the-art facilities available at CSIR-NPL are at par to the facilities of leading NMIs.

Vacuum mass comparator (Fig. 7.3) is a unique and state-of-the-art facility that serves both the fundamental research and highly accurate measurement in mass. The computer-operated electronic vacuum mass comparator features a vacuum chamber and a load alternator with eight positions for holding masses as well as spheres (used as primary standards of density) and is capable of comparing mass against each other with highest accuracy. 1 kg vacuum mass comparator will be used to experimentally determine the air density by comparison weighing of buoyancy artifacts in air and vacuum.

A 1 kg mass comparator (Fig. 7.4) is also a unique and state-of-the-art facility in India, and it is used for highly accurate measurement in mass. The computer-operated electronic mass comparator features a load alternator with four positions for holding masses for comparing three masses against one. This comparator is suitable for comparing combination of masses up to 100 g using special designed plates as per weighing design of sub-division method. This is also used for calibration of weights of OIML E_1 class.

These mass comparators (vacuum and 1 kg) are designed to disseminate the mass value of national standard (NPK-57) and establish mass standards of sub-multiple of 1 kg. Special features of the comparator include electric weighing range to mitigate linearity error and self-centering to eliminate/mitigate eccentricity error.

A 50 kg mass comparator (Fig. 7.5) is an automatic mass comparator which is based on computer operation and has four positions for weight alternator. It has a modern and fast weighing system. Independent draft shields protect from air current effect. The self-centric mechanism reduces the eccentric load errors.

The special features of the mass comparators are electric weighing range to mitigate linearity error and self-centering to eliminate/mitigate eccentricity error.

Robotic mass comparator is the latest and most advanced facility addition in the CSIR-NPL. The comparator has recently been procured and is presently under installation and testing. As soon as the satisfactory results are obtained, the comparator will be dedicated to the national services.

Table 7.1 List of equipment at mass metrology

Name	Range
Mass standards	
National Prototype	1 kg
Transfer standard	1 kg (N)
Transfer standard	1 kg (Ń)
Transfer standard	1 kg (S)
Transfer standard	1 kg (Ś)
Working standard	10 kg to 1 g [224(2)]
Working standard	10 kg to 1 g [224(3)]
Working standard	1 kg to 1 mg [224(4)]
Working standard	20 kg - 9 Nos
Working standard	100 g to 1 mg [58211]
Working standard	100 g to 1 mg [58212]
Working standard	100 g to 1 mg [60027]
Working standard	100 g to 1 mg [60028]
Working standard	10 kg
Mass comparator/balance	
KE 2000	Capacity = 2000 kg Readability = 1 g
XP604KM	Capacity = 600 kg Readability = 0.1 g
PR10003	Capacity = 10 kg Readability = 1 mg
CC2001	Capacity = 2 kg Readability = 0.1 mg
Vacuum comparator	Capacity = 1 kg Readability = 0.1 μg
CC 1000 SL	Capacity = 1 kg Readability = 1 μg
CC 1200	Capacity = 1 kg Readability = 0.1 mg
AT201	Capacity = 205 g Readability = 10 μg
AT20	Capacity = 22 g Readability = 2 μg

(continued)

Table 7.1 (continued)

Name	Range
MT5	Capacity = 5 g Readability = 1 μg
Glass capillary viscometers	
Size no./identification	Nominal viscometer constant (mm^2/s^2)
OC_2	0.003
OC_3	0.003
I_1	0.01
I_2	0.01
I_{C1}	0.03
I_{C2}	0.03
II_1	0.1
II_2	0.1
II_{C1}	0.3
II_{C2}	0.3
III_1	1.0
III_2	1.0
III_{C1}	3.0
III_{C2}	3.0
IV_1	10
Standard oils	
Identification	Approximate kinematic viscosity @ 40 °C (mm^2/s)
VG 2	2
S 3	3
S 6	6
N 10	10
N 12	12
P 22	22
VG 25	25
N 32	32
N 35	35
J 73	73
K 75	75
N 100	100
N 220	220
VG 250	250
N 320	320
N 350	350

(continued)

Table 7.1 (continued)

Name	Range
K 427	427
N 460	460
VG 500	500
VG 600	600
S 600	600
J 910	900
S 2000	2000
N 4000	4000
N 15,000	15,000
Reference grade hydrometer	
Hydrometer	(0.6–2.0) kg/m^3

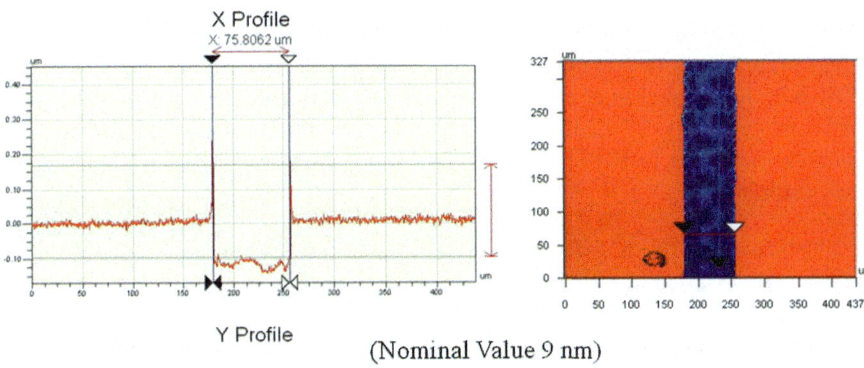

Fig. 7.3 Vacuum mass comparator

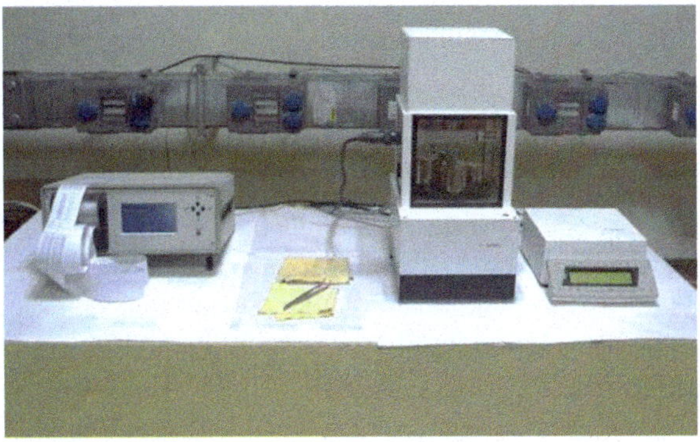

Fig. 7.4 A 1 kg mass comparator

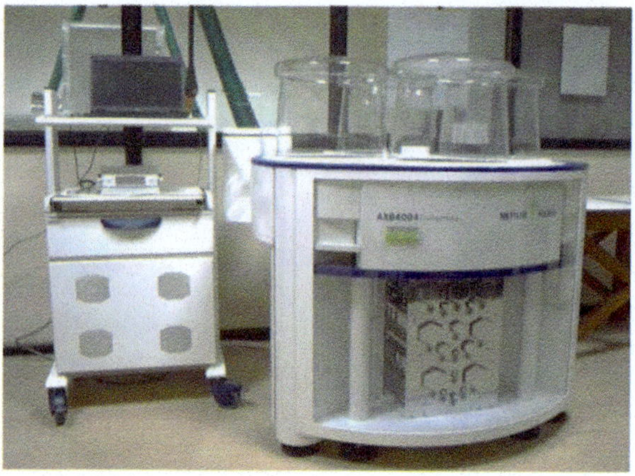

Fig. 7.5 A 50 kg mass comparator

7.2.3.2 Calibration and Measurement Capabilities

CSIR-NPL has successfully gone through the International Peer Review in 2004 and 2008, and presently, 27 CMCs related to mass, volume, density and viscosity are available in KCDB in BIPM Web site. All the CMCs are shown in Table 7.2 for ready reference.

7.2.3.3 Participation in International Key Comparisons

As per CIPM MRA, it is a mandatory requirement for any NMI to participate in the international intercomparisons at regular interval for proving technical competence. CSIR-NPL has participated in a number of intercomparisons on mass, volume, density and viscosity parameters. The list of these comparisons is shown in Table 7.3. The other comparisons are in pipelines.

7.2.3.4 Potential Services for Future Requirements

For last couple of decades, India has been moving toward economic reform through industrialization by transforming from primarily agricultural to manufacturing of goods, from individual manual labor to mass production by modern machine, and craftsmen are replaced by assembly lines. Conventional measurement instruments (glass hydrometers, glass viscometers, etc.) are gradually being replaced by digital equipment (digital density meter, digital rotational viscometer, rheometers, etc.). Although these equipments are convenient to use, yet they have limited usage owing to high price. In near future, demand of calibration/testing is to be increased, but calibration/testing of these equipments is not possible using old technologies and

Table 7.2 CMC of mass, volume, density and viscosity parameters of CSIR-NPL [8]

Calibration or measurement service			Measurand level or range			Measurement conditions/independent variable		Expanded uncertainty			
Class	Instrument or artifact: measurand	Instrument type or method	Minimum value	Maximum value	Units	Parameter	Specifications	Value	Units	Coverage factor	Level of confidence (%)
Mass	Mass standards	Comparison in air	1	1	kg			35	µg	2	95
Mass	Mass standards	Comparison in air	0.001	1	g			2.0	µg	2	95
Mass	Mass standards	Comparison in air	2	2	g			2.5	µg	2	95
Mass	Mass standards	Comparison in air	5	5	g			3.5	µg	2	95
Mass	Mass standards	Comparison in air	10	10	g			5	µg	2	95
Mass	Mass standards	Comparison in air	20	20	g			8	µg	2	95
Mass	Mass standards	Comparison in air	50	50	g			10	µg	2	95
Mass	Mass standards	Comparison in air	100	100	g			12	µg	2	95
Mass	Mass standards	Comparison in air	200	200	g			14	µg	2	95
Mass	Mass standards	Comparison in air	500	500	g			21	µg	2	95
Mass	Mass standards	Comparison in air	2	2	kg			0.58	mg	2	95

(continued)

Table 7.2 (continued)

Calibration or measurement service			Measurand level or range			Measurement conditions/independent variable		Expanded uncertainty			
Class	Instrument or artifact: measurand	Instrument type or method	Minimum value	Maximum value	Units	Parameter	Specifications	Value	Units	Coverage factor	Level of confidence (%)
Mass	Mass standards	Comparison in air	5	5	kg			1.4	mg	2	95
Mass	Mass standards	Comparison in air	10	10	kg			2.6	mg	2	95
Mass	Mass standards	Comparison in air	20	20	kg			6	mg	2	95
Mass	Mass standards	Comparison in air	50	50	kg			15	mg	2	95
Mass	Mass standards: *m*	Comparison in air	100	2000	kg	*m*	100, 200, 500, 1000 and 2000 kg	5 *m*, *m* in kg	mg	2	95
Density of solids	Solid density artifact: density *d*	Weighing in liquid	1000	22,000	kg m^{-3}	Temperature	15, 20 and 27 °C	0.03*d*, *d* in kg m^{-3}	g.m^{-3}	2	95
Density of liquid	Density measuring device	Weighing in liquid	600	2000	kg m^{-3}	Temperature	20 °C	0.4	kg.m^{-3}	2	95
Volume of liquid	Standard volume vessel: *V*	Weighing	0.01	50	l	Temperature	27 °C	0.1 *V*, *V* in l	ml	2	95

(continued)

Table 7.2 (continued)

Calibration or measurement service			Measurand level or range			Measurement conditions/independent variable		Expanded uncertainty			
Class	Instrument or artifact: measurand	Instrument type or method	Minimum value	Maximum value	Units	Parameter	Specifications	Value	Units	Coverage factor	Level of confidence (%)
Volume of liquids	Standard volume vessels	Weighing	100	500	1	Temperature	27 °C	10	ml	2	95
Volume of liquids	Standard volume vessels	Weighing	1000	2000	1	Temperature	27 °C	100	ml	2	95
Kinematic viscosity	Newtonian liquids	Capillary viscometers	0.001	0.01	mm^2/s^2	Temperature	20–100 °C	0.26	%	2	95
Kinematic viscosity	Newtonian liquids	Capillary viscometers	0.01	0.1	mm^2/s^2	Temperature	20–100 °C	0.30	%	2	95
Kinematic viscosity	Newtonian liquids	Capillary viscometers	0.1	10	mm^2/s^2	Temperature	20–100 °C	0.36	%	2	95
Kinematic viscosity	Newtonian liquids	Reference liquids	1	10	mm^2/s^2	Temperature	20–100 °C	0.28	%	2	95
Kinematic viscosity	Newtonian liquids	Reference liquids	10	100	mm^2/s^2	Temperature	20–100 °C	0.31	%	2	95
Kinematic viscosity	Newtonian liquids	Reference liquids	100	10 000	mm^2/s^2	Temperature	20–100 °C	0.35	%	2	95

Table 7.3 List of key comparisons participated by CSIR-NPL in mass, volume, density and viscosity parameters [7]

Name of the key comparison	Brief details	Year
APMP.M.M-K1	Comparison of mass standards	1999–2001
CCM.M-K5	Comparison of Mass Standards	2000–2003
CCM.V-K1.A	Viscosity measurements of standard liquids temperature: 20 °C	2002
CCM.V-K1.B1	Viscosity measurements of standard liquids temperature: 20 °C	2002
CCM.V-K1.B2	Viscosity measurements of standard liquids temperature: 40 °C	2002
CCM.V-K1.B3	Viscosity measurements of standard liquids temperature: 100 °C	2002
APMP.M.M-K2	Comparison of mass standards	2004–2007
APMP.M.M-K6	Comparison of mass standards	2005–2006
APMP.M.D-K4	Comparison of the calibration of density hydrometers	2007–2008
APMP.M.M-K2.1	Comparison of mass standards	2010
CCM.M-K4	Comparison of mass standards	2011–2012

following conventional method. Sophisticated technologies with advanced setup are required for calibration/testing of these equipments.

Another area where industries (petroleum, oil, etc.) are seeking support from NMI is volume measurement of overhead/underground storage tanks. It is not possible to measure them using conventional gravimetric method. To fulfill the requirements of the industries, special instrument like surveying total stations and their accessories for mapping the interior of a tank is required.

7.2.3.5 A Decade of Clients and Services

Calibration and Testing

CSIR-NPL is providing calibration services of mass, volume, density and viscosity parameters to various customers such as Legal Metrology Department, ISRO, DAE, STQC, Heavy Alloy Penetrator Project (HAPP) under Ministry of Defence, DRDO, Indian Railway, Air India, Delhi Jal Board, Indian Oil Corporation, CPCB, etc. During last 10 years, more than 1500 calibration reports were issued from these calibration services. The sector-wise customer distribution for these parameters is represented by pie chart shown in Fig. 7.6.

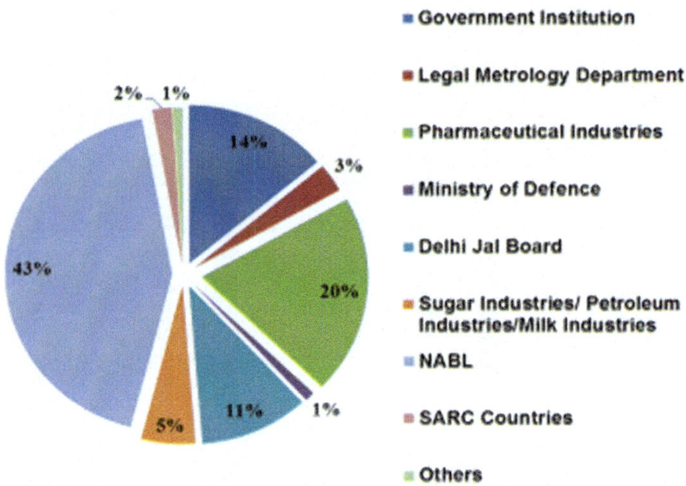

Fig. 7.6 Sector-wise customer distribution of mass, volume, density and viscosity metrology

Consultancy and Technical Services

It is possible to formulate both the problems associated with precise mass measurement and the suggestions for those with practical mass measurement problems in such a manner as to attract the interest of the scientific community and industry. There are few industries which are working with heavy capacity balances for specialized work. In such cases, consultancy service is provided to the customers who include calibration of heavy capacity balances along with technical suggestions related to precise measurements essential for maintaining uncertainty to its lowest.

Training and Academic Services

Several trainings are organized during the year for industry people to make them aware of metrology and quality system. Emphasis is given to them regarding mass and its related parameter measurements. Sometimes, special request is received from industry for specific kind of training which may not be in our plan. Such training is also conducted to benefit our customers. CSIR-NPL has also organized training programs for SAARC countries in association with PTB, Germany. In recent past, countries like Nepal, Bhutan, Sri Lanka, Bangladesh, Afghanistan, etc. have participated in such training programs. In the present scenario, country requires trained manpower which will improve the quality infrastructure of the nation. Continuous effort is being laid in this direction, and Precision Measurement and Quality Control course was launched where students are trained in almost all the parameters including quality system. This kind of program leads to continuous inflow of trained manpower in industries which will in turn improve overall countries measurement capabilities.

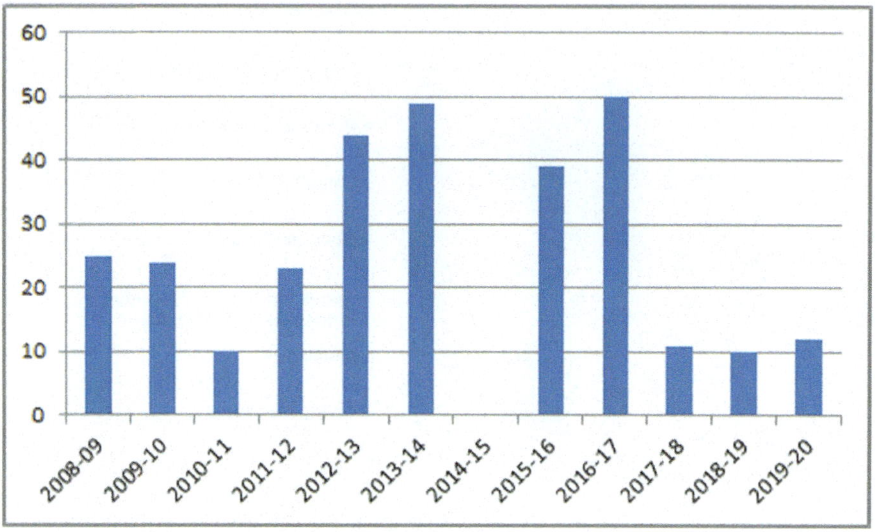

Fig. 7.7 Training provided to the participants from industry, legal metrology and abroad

During last 10 years, about 300 personnel were trained for mass, volume, density and viscosity parameters (Fig. 7.7).

Technology Development and Commercialization

Till 20th May, 2019, out of seven SI base units, "kilogram" was the only SI base unit that was based on an artifact (mass value of IPK). In 26th CGPM meeting in November 2018, all the member states unanimously agreed to redefine kilogram based on physical constant (Planck's constant) with replacement of IPK.

Few NMIs have already developed "Kibble balance" to redefine kilogram through "Planck's constant". In CSIR-NPL also, after successful demonstration of "Kibble balance" of 1 g, efforts are going on towards development of "Kibble balance" of 1 kg.

CSIR-NPL needs both financial and administrative support for developing "Kibble balance" of 1 kg to establish primary standard of mass. It is one of the prestigious projects of national pride in metrology.

7.2.3.6 Major Documentary Standards, Guides and SOPs

The major documentary standards, guides and SOPs used in mass, volume, density and viscosity metrology are given in Table 7.4.

Table 7.4 List of documentary standards and guides used in mass, volume, density and viscosity metrology

S. No.	Name of the documentary standard/guide/SOP	Brief description
1	OIML R111-1: 2004	Weights of classes E_1, E_2, F_1, F_2, M_1, M_{1-2}, M_2, M_{2-3} and M_3, Part 1: Metrological and technical requirements
2	OIML D 28: 2004	Conventional value of the result of weighing in air
3	OIML R 76-1: 2006	Non-automatic weighing instruments, Part 1: Metrological and technical requirements—Tests
4	IS/ISO 4787: 2010	Laboratory glassware—volumetric instruments—methods for testing of capacity and for use
5	ISO 8655–6: 2002	Piston-operated volumetric apparatus, Part 6: gravimetric methods for the determination of measurement error
6	ASTM D1193: 2006 (Reapproved 2018)	Standard specification for reagent water
7	ASTM D2162: 2017	Standard practice for basic calibration of master viscometers and viscosity oil standards
8	ASTM D445: 2017	Standard test method for kinematic viscosity of transparent and opaque liquids and calculation of dynamic viscosity
9	ASTM D446: 2012 (Reapproved 2017)	Standard specifications and operating instructions for glass capillary kinematic viscometers
10	ASTM D1200: 2010 (Reapproved 2018)	Standard test method for viscosity by ford viscosity cup
11	IS: 3104 (Part I): 1982 (Reaffirmed 2008)	Specification for density hydrometers, Part I: requirements
12	SIM MWG7/cg-03/v.00: 2016	Guidelines on the calibration of hydrometers—Cuckow method
13	EURAMET calibration Guide No. 18, version 4.0 (11/2015)	Guidelines on the calibration of non-automatic weighing instruments
14	EURAMET calibration Guide No. 19, version 3.0 (09/2018)	Guidelines on the determination of uncertainty in gravimetric volume calibration
15	EURAMET calibration Guide No. 21, version 2.0 (05/2020)	Guidelines on the calibration of standard capacity measures using the volumetric method

7.2.4 Possible Gaps and Suggestions for Their Redressal

CSIR-NPL is supporting industries on mass, volume, density and viscosity on an ongoing basis, but still there are a number of areas or range related to mass where there is a need for facility creation or improvement to fulfill requirements of industries. In some areas, enhancement/improvement in CMCs is required. For example, some of the accredited laboratories are taking traceability of E1 class weights from other NMIs because of demand of very fine uncertainty (in milligram range) from laboratories, and existing facilities of CSIR-NPL are not adequate to achieve the goal. CSIR-NPL has received robotic mass comparator which is presently under installation and establishing traceability. After achieving desired CMCs, services will be at par to the other NMIs in this range.

Similarly, capacity of the weighing range is also needed to be enhanced to fulfill the requirements of heavy industries.

7.3 Length, Dimension and Nanometrology

7.3.1 Introduction and Historical Aspects

Length measurement has been an indispensable part of human life since ancient to modern era. Length is the most revered measurement in all walks of life. Early length measuring methods were based on the use of human body parts like length and width of fingers, thumbs, hands, hand spans, cubits and body spans which seemed to have been popular choices [9–14]. CSIR-NPL, right from the beginning, had a mandate to maintain the standards of physical measurements and provide testing evaluation services. The standards maintained at CSIR-NPL are periodically compared with BIPM and other leading NMIs for international equivalence [15]. In this section of the chapter, an attempt has been made to discuss the brief history and advancements in length and dimensional measurement methods at CSIR-NPL.

It covers the evolution of SI unit of length through its definitions from the international prototype metre bar to the definition based on fundamental constant, i.e., speed of light in vacuum and the efforts being made at CSIR-NPL. The testing services at CSIR-NPL were started during 1953 under its weights and measure division. The *International Prototype Metre (IPM)* was set up about 1960s to realize SI meter and dissemination of traceability to this meter was initiated. A number of dimensional measuring standards, instruments like Hilger gauge block interferometer, universal measuring machine, transverse comparator, 4m longitudinal comparator, profile projector, Talysurf, sets of standard slip gauges and length bars, line standards of length, instruments for measurements of screw threads, roundness, angle, surface flatness, and angle standards, polygons, autocollimators, levels, etc. were received through *American aid program* and bringing India at par with other international NMIs [16]. With this, a very active work in dissemination of traceability started. As the definition of meter changed to the one based on Kr Lamp, CSIR-NPL also got its Kr Lamp to realize SI unit meter. At the same time, Dr. D. Sen at CSIR-NPL developed measurement techniques based on interferometers [17] which got acclaimed national and international appreciations.

CSIR-NPL also carried out developmental work for frequency stabilization of lasers and established iodine ($^{127}I_2$) frequency stabilized lasers indigenously by 1978 [18]. In year 1981, CSIR-NPL participated in the first international intercomparison of iodine frequency stabilized He–Ne lasers. It became one of the very few NMIs who were recognized to have capability to develop primary laser at that time. The definition of meter further changed in 1983 and BIPM recommended ways to realize this definition. CSIR-NPL was already prepared to realize it through its primary laser. During this period, worldwide efforts were going on to link optical frequency standard to the frequency standard used for time. For this, a number of laser workings at different wavelengths were being frequency stabilized.

As a continued and focused efforts, CSIR-NPL established cooperation with VNIIM Russia in the mid-1980s under *International Long-Term Program (ILTP)* of DST and CSIR-NPL scientists worked in VNIIM on methane stabilized laser which was later established at NPL India. At the same time, Indian industry had advanced quite a lot. There were plants of automobile manufacturers, heavy industries, laser machined parts, etc. that were grappling for traceability with better accuracy and for diversified parts/machine. Therefore, a major upgradation took place in 1989 under NPL-PTB cooperation, with aid from German Government. Many modern dimensional metrology equipments were donated to CSIR-NPL, and these included a robust iodine stabilized He–Ne laser, laser interferometer, coordinate measuring machine (CMM), electronic autocollimator, etc. [15].

With signing of CIPM MRA in 1999, the metrology activities of CSIR-NPL started efforts for implementing ISO/IEC 17025 standard. It developed the necessary documentation, including calibration strategy procedures, all required documentation for quality management system and record of all intercomparison for equivalence of measurement, and got a boost in terms of infrastructure, mainly for laboratories with better environment control and added some modern facilities. A major work was carrying out for the error analysis and calculating the uncertainty in measurements as per new ISO guidelines. With this, CSIR-NPL successfully conducted peer review and its CMCs were approved for publication in KCDB of BIPM in 2004 [8].

With emerging area of nano-science and nano-technology, CSIR-NPL undertook a project on generic development of nanometrology for establishment of facilities for nano-dimensional measurements. The peer review of laser frequency and dimension metrology CMCs took place again, and CMCs were approved in 2011. Today, CSIR-NPL holds two CMCs in laser frequency and 42 in dimension metrology.

An automatic flatness interferometer was added in 2008–09. The modernization of CMM and form measurement facilities were taken up during 2010–14. CSIR-NPL established a facility vision probe on CMM and has initiated calibration of portable CMM recently and is planning to set up a roughness and contour measurement facility and frequency comb in future.

Thus, it has been a journey of CSIR-NPL moving from material artifact to quantum standards. In addition to upgradation of standard and dissemination of traceability, CSIR-NPL has been helping Indian industries and accredited laboratories through training and consultancy. It has contributed to strengthen the NQI for length and dimension measurements by providing expertise, human resource development, writing technical guidance documents for standardization, accreditation and legal metrology.

7.3.2 Significance, Applications and Relevance in Different Sectors

Measurement is the most fundamental concept of science and technology that leads to innovations. The entire life cycle of a product is underpinned by accurate and precise dimensional measurements carried out for defining size, shape, form and to optimize the friction and wear. These measurements of lengths and angles as well as geometrical properties such as flatness, roundness and straightness are used by almost all sectors in one form or the other as these are vital in monitoring and controlling the variations inherent within any research, design, development and manufacturing process. This ranges from the tiniest tracks on computer chips present in any smartphone or television, up to the size and form of wings of airplanes and wind turbines. With rapid industrialization and liberalization, there is an obligation to the industries, to meet the global standards, and this requires control of measuring instruments.

Major user sectors of CSIR-NPL's traceability to laser, interferometers and other dimensional measurements are shown in Fig. 7.8. The list of customers includes Maruti Udyog, Tata Motors, BHEL, HAL, DEAL, DRDO, etc. who get their instruments calibrated at CSIR-NPL. So, in broad sense, it is said that the length and

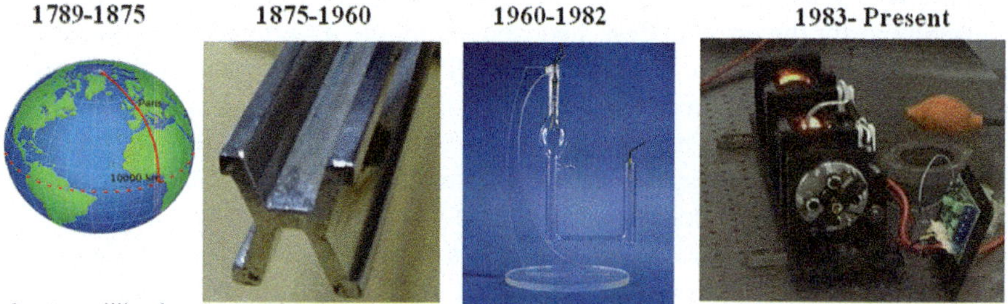

Fig. 7.8 History of meter [26–28]

dimension metrology services provided by CSIR-NPL have wider applications in industry, railways, avionics, defense, MSME, various other manufacturing processes and conformity assessment bodies.

Even a diminished negligence in dimensional measurements can cause catastrophic consequences in case of automobile and avionics industry. Currently, 75% of the components used by any manufacturer are produced by other industries; thus, it becomes essential to define and communicate the acceptance criteria in terms of dimensional characteristics to avoid rejections. A car or airplane undergoes hundreds of tests by calibrated instruments. Leading manufacturers deploy resources at design stage, and the analysis of measurement data is integrated with the production system, if needed real-time changes can be made whenever required. Design and manufacturing of wind turbines, aerodynamics of wings in a plane, parts of small wristwatch, dimensions of semiconductor chips, railway lines, smoothness of mobile phone screen, computers, TVs, robots, robotic assembly lines, etc. can be controlled to minimize losses due to rejection by accurate and precise dimensional measurements.

7.3.3 SI Unit "Metre" and Its Realization

In year 1793, metre had its origin as the unit of length to be 10^{-7} part of the earth's quadrant. Based on this several platinum-iridium (Pt-Ir), bars were embarked with two marks and the distance between these two marks was defined as the meter. This definition of meter was adopted till 1960, and later, it was defined in terms of wavelength of Kr radiation. With the discovery of lasers by Theodore Maiman in 1960, the development of the Michelson interferometer combined the ability to measure atomic wavelengths with higher accuracy and reproducibility took place. In year 1983, the SI unit of meter was redefined as "The meter is the length of path travelled by light in vacuum during the interval of 1/299 792 458 of a second" [19–28].

A brief progress in definition of meter is briefed in Fig. 7.8. At the time of adoption of modern definition of meter by the 17th CGPM, the CIPM drew up recommendations for practical realization of the definition. These have formerly been referred to as *mise en pratique (MeP)* of the definition of meter. In response to this invitation, CIPM adopted Recommendation 1 (CI-1983) (MeP) of the definition of meter (1997–2007) for realization of unit meter by means of length l of the path traveled by light in vacuum in a time t and using speed of light in vacuum $c_0 = 299\ 792\ 458\ \text{ms}^{-1}$; by means of the λ of light in vacuum whose frequency f is obtained from measured value or stated values in the list of recommended radiations given by MEP [26–28].

In this way, a laboratory standard of known frequency can be constructed using the specifications and operating conditions provided by the BIPM. These descriptions also indicate the error associated with this method of realization. CIPM gave a list of recommended radiations with assigned frequencies, wavelengths and uncertainties. CSIR-NPL has iodine stabilized He–Ne laser operating at 633 nm, which is used for the practical realization of SI unit "meter" and for calibration of commercial laser heads to be used with laser interferometers.

The iodine stabilized He–Ne laser at CSIR-NPL operates on 474.6 THz hyperfine components of iodine with a relative standard uncertainty u_r of $\pm 2.5 \times 10^{-11}$. In this primary laser, an iodine molecular cell is installed in the laser cavity to utilize the hyperfine lines of the R (127) 11–5 transition in $^{127}I_2$ for stabilizing the laser frequency [17–20]. This iodine stabilized He–Ne laser in use at CSIR-NPL is a commercially available one and was installed in October 2003. This laser ensures full compliance with CIPM 92 recommendations.

The definition of meter as revised in 2019 [29] is already included in Chapter 2 of this book. The *MeP-2019* [29] recommendation includes primary methods for the practical realization of the definition of the meter and it stipulates.

The fundamental equation underlying the definition of the meter is a direct relationship between a length, a time interval and the speed of light:

$$l = c \cdot \Delta t, \tag{7.1}$$

In which c is the fixed value for the speed of light in vacuum, $c = 299{,}792{,}458\ \mathrm{ms}^{-1}$, and Δt is the traveling time of the light along a geometrical path, of gap between length and l. Realization of the length unit, at a primary level, is thus linked to measurement of light traveling time; this may be achieved directly with high relative accuracy for long ranges (e.g., lunar ranging), but at typical macroscale ranges, indirect traveling time measurement offers better accuracy (due to the challenge of measuring a very short time interval). The MeP-2019 gives the primary method and also the secondary methods for realization.

7.3.3.1 Primary Methods

The definition of the meter can be realized in practice by one of the following primary methods.

Direct Measurement of Light Traveling Time (Time of Flight Measurement)

The direct measurement of the traveling time of light requires some form of modulation of the light in order to generate fiducial features used for the timing process. Any kind of modulation, even applied to monochromatic light, generates a superposition of light waves forming a wave packet. The path length of propagation of such a wave packet, e.g., of a laser pulse, can be determined as shown in Fig. 7.9. A light pulse is split into two parts so that two pulses are generated, one of which travels a short reference pathway and the other which travels the measurement pathway. The reflectors in both pathways are arranged such that the light is retro-reflected. After second passage through the beam splitter, the light pulse originating from the reference pathway first hits a light detector which sets a first trigger at a defined threshold,

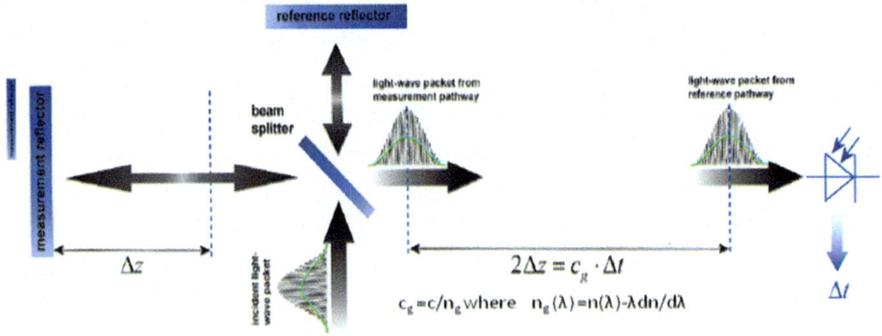

Fig. 7.9 Primary realization of the length by direct measurement of time delay between light-wave packets traveling pathways of different lengths before reaching a detector. The green curves inside the wave packets indicate the average light intensity that is sensed by the detector

defining a reference point in time. A second trigger is generated by the delayed light pulse originating from the measurement pathway. The two pulses must be clearly separated in time. Measurement of the time delay Δt between both the detector signals allows determination of the length difference, Δz, between measurement and reference pathways, which represents the length, l:

$$l = \Delta z = 12cg \cdot \Delta t, \tag{7.2}$$

In which cg is the group velocity of the wave packet. While under vacuum conditions cg is identical to c, under the influence of the atmosphere cg is obtained from $cg = c/ng$, in which ng is the group refractive index of air [29]. Such measurement is possible to apply for long distances such as the distance from the earth to the moon.

Indirect Measurement of Light Traveling Time (Optical Interferometry)

As per MeP 2019, for the realization of lengths below a few meters, but also for the most accurate realization of length in general, interferometric techniques are preferable. Optical interferometry is a measurement method based on the superposition (interference) of light as shown in Fig. 7.10.

The incident light is split and made to travel in two paths: The length of the reference pathway is assumed to be unchanged, while the length of the measurement pathway is assumed to be variable. The path length difference Δz is then strictly related to the phase difference $\Delta\phi$ [28].

$$l = \Delta z = \frac{\lambda}{2}\frac{\Delta\varphi}{2\pi} = \frac{1}{2}\frac{c}{n}\Delta t$$
$$\Delta z = \frac{1}{2\pi}\frac{\Delta\phi}{f} \tag{7.3}$$

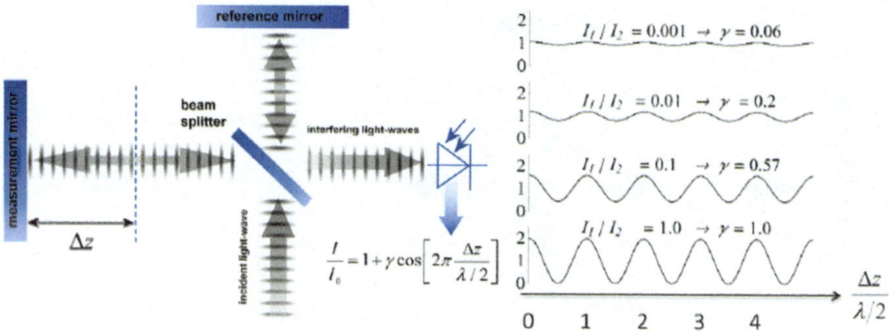

Fig. 7.10 Primary realization of the length unit by interferometry, i.e., by indirect measurement of the time delay between monochromatic light-wave traveling pathways of different lengths before reaching a detector. The interference contrast = 2 is related to the ratio of the individual intensities $a = I_1/I_2$

Knowledge of the frequency of the light, f, is an essential requirement for the realization of the unit of length. It provides the scaling factor between a measured phase difference and the length that is realized by interferometry. Often, the value of the "vacuum wavelength", which describes the distance between the wavefronts in vacuum under idealized conditions ($\lambda_0 = c/f$), is stated instead of the frequency [28]. For highest demands on the accuracy of the light frequency, a light source can be synchronized to the primary frequency standards by an appropriate technique. As an alternative to direct measurement of frequency or vacuum wavelength, CIPM list of recommended frequency standard values is used which also contains specifications relating to each frequency standard which are displayed after selecting a particular frequency [28]. CSIR-NPL uses iodine frequency stabilized laser as described in Sect. 7.3.4.1.

7.3.3.2 Secondary Methods of Realizing the Meter for Dimensional Nanometrology

The methods described above are convenient for implementation at larger scales. In case of dimensional nanometrology, the limitations arise due to fringe sub-division and periodic non-linearities in visible-wavelength interferometry. In view of the demand for nanometrology and future predictions for semiconductor industry, etc., nanometer or sub-nanometer scale measurements need to be traceable to SI. MeP 2019 makes a provision for secondary realizations, which was not there in earlier MeP. Agreed CODATA value of the Silicon Si {220} lattice spacing d_{220} = 192.015 571 4 × 10^{-12} m, with a standard uncertainty of ±0.000 003 2 × 10^{-12} m, (*i.e.*, $\Delta d/d$ = 1.67 × 10^{-8}) at a temperature of 22.5 °C in vacuum could be used. This value is for an ideal single crystal of natural-isotopically undoped silicon that is free of impurities and imperfections. The value will change with impurities and defects. An alternative approach for determining the lattice spacing is to use a lattice comparator, to compare

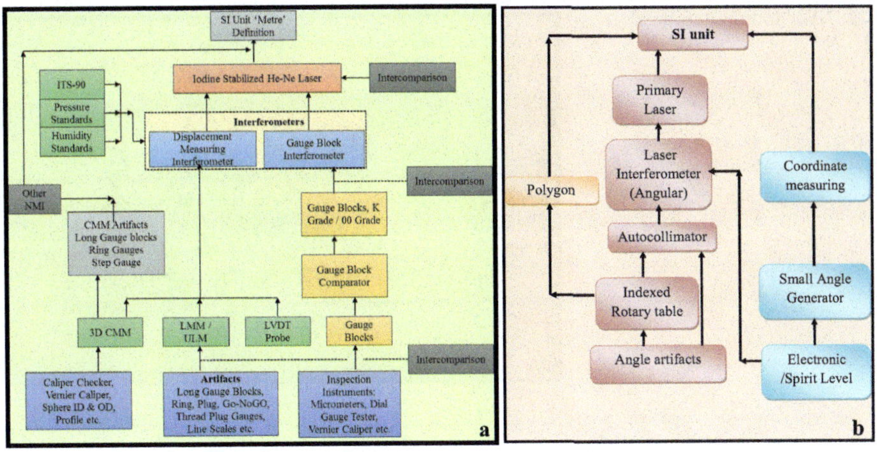

Fig. 7.11 Traceability chart for dissemination of SI unit meter and angle

silicon with unknown impurity concentration with a piece of silicon whose lattice spacing is known [29]. There are several examples of how a traceability pathway through the silicon lattice spacing is relevant for dimensional nanometrology. Three of these, namely X-ray interferometry for displacement metrology; calibration of TEM magnification; and step height standards based on the silicon lattice, are noted by MEP 2019.

CSIR-NPL practically realizes the SI unit "meter" and disseminates it across the country. The traceability using iodine stabilized laser maintained at CSIR-NPL is depicted in Fig. 7.11. For traceability to most of the items, an interferometer (indirect measurement of time technique) is the first link.

7.3.4 Facilities and Global Status

7.3.4.1 Measurement Standards and Facilities

CSIR-NPL realizes SI meter and is equipped with state-of-the-art instruments to disseminate to SI unit meter. These are described below.

An Iodine Stabilized He–Ne Laser

The primary optical frequency standard is a state-of-the-art facility for realization and dissemination of meter in CSIR-NPL, as shown in Fig. 7.12. Such lasers have been in use at CSIR-NPL since 1978. The present modern standard laser used for the calibration of laser frequencies was established in 2003. The intercomparisons

Fig. 7.12 Iodine stabilized He–Ne laser

were participated as early as in year 1981, and recent intercomparison participated was in year 2014. The laser meets the conditions prescribed by MeP-1992, and it can be locked to any hyperfine features of iodine frequencies. These frequencies were measured w.r.t frequency combs at BIPM in 2005 during intercomparison. CSIR-NPL has capability to calibrate other such primary lasers' frequency with CMC of relative expanded uncertainty 4.2×10^{-11} at $k = 2$. The traceability is disseminated to laser transducers and laser interferometers by calibration of laser frequency with CMC of relative expanded uncertainty 1×10^{-9} ($k = 2$).

Facilities for Dimension Measurements

The facilities for dimensional measurements are given in Table 7.5.

Facilities for Nanometrology

Nanometrology is the science of measurement at nano-scale level. Its importance has increased many folds as the existing methods of measurement do not hold well at nano-scale. The new technologies enabled by nanometrology go well beyond what is possible with conventional metrology. Also, the demand for highly accurate results with low uncertainty is important in these low level measurements. It is necessary to establish the highly sensitive tools with traceability for probing and characterizing nano-materials and nano-devices. Such tools are being established at various R&D organizations in India. However, there is a need to establish facilities for providing traceability to these tools. To achieve this, CSIR-NPL undertook project funded by Department of Information and Technology (DIT), now renamed as MeitY, and first step has been initiated by establishing metrology AFM and a 3D optical profiler for step height, line spacing and roughness measurements. Figure 7.13 shows a

Table 7.5 Some of the important facilities in dimension metrology

S. No.	Facility/instrument	Range	Major items for calibration
1	Twyman-green interferometer with phase-shifting capability and metrology software:	Up to 300 mm	Calibration of end standards by interferometers (e.g., 00, and K grade gauge blocks)
2	Fizeau interferometer with phase-shifting capability and metrology software	Up to 150 mm (optical flats)	Calibration of optically Flat surface (e.g., optical flat, super finished metals, ceramics and plastics)
3	Displacement measuring interferometers (heterodyne and homodyne)	Range up to 30 m	Calibrations of machine movements: length (displacement), way straightness, yaw and pitch Flatness of surface plates
4	Length measuring machines (LMM) (also used in conjunction with laser interferometer)	Up to 1 m and 4 m, respectively	Calibration of • End standards up to 1000 mm, e.g., Long gauge blocks, • Diameter standards—inner and outer diameter of ring gauges, plug gauges and many other artifacts, etc. • Dial gauge, etc.
5	3D coordinate measuring machine (CMM) with programmable tactile probing automatic thermal compensation	1000 mm × 900 mm × 600 mm	Versatile calibrations: of dimension, profiles and form based on the coordinates obtained by probing points on the surface Example—long gauge blocks, step gauge, check master, caliper checkers, piston, cylinder, contour, various complex profile
6	Form measurement	Up to 500 mm	Calibration of form standards for—roundness, cylindricity, squareness of cylinders/hemisphere/spheres
7	Autocollimator, index tables, small-angle generator and polygons	±1°	Calibration of angle artifacts and instruments

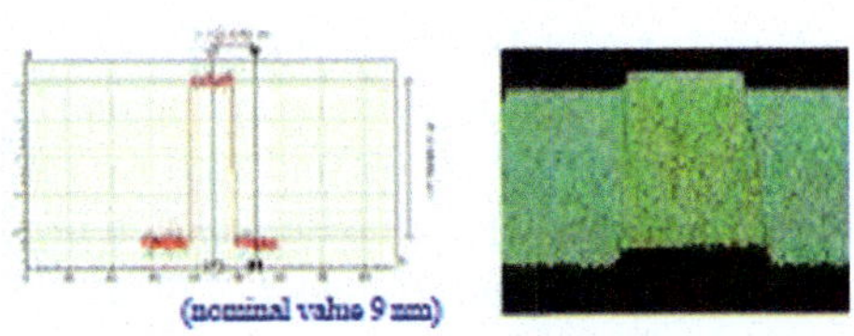

Fig. 7.13 Step height artifact

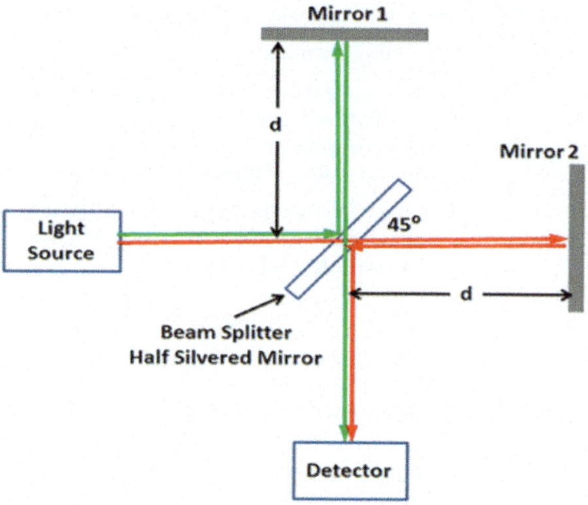

Fig. 7.14 Basic Michelson interferometer

measurement carried out at NPL for a 9 nm (nominal value) step height measurements. CSIR-NPL has piloted and participated intercomparison for step heights 9 nm to 10 μm successfully.

Techniques Applied for Dimensional Measurement:

Working principles of these advanced instruments are based on some of the most advanced techniques as described in upcoming sections.

Interferometry

As shown in traceability chart presented earlier and directed by MeP, laser interferometers are the first link between primary laser and the length and dimensional to realize length measurement in terms of indirect measurement of time and thus tracing it to SI unit "meter". In year 1881, by comparing an unknown distance in space against a known wavelength of light through Eq. 7.4 Albert Michelson introduced the first linear displacement measuring laser interferometer as shown in Fig. 7.14.

$$L = (N + f)\lambda \tag{7.4}$$

where L is the length being measured, N is an integer, f is a fraction, and λ is source wavelength. With many advances in interferometers, various setups have been reported worldwide and used for getting length and form information from the interferometer fringes. With addition of phase-shifting capabilities and automatics fringe analysis interferometer of today can reach nm and sub-nm resolution easily.

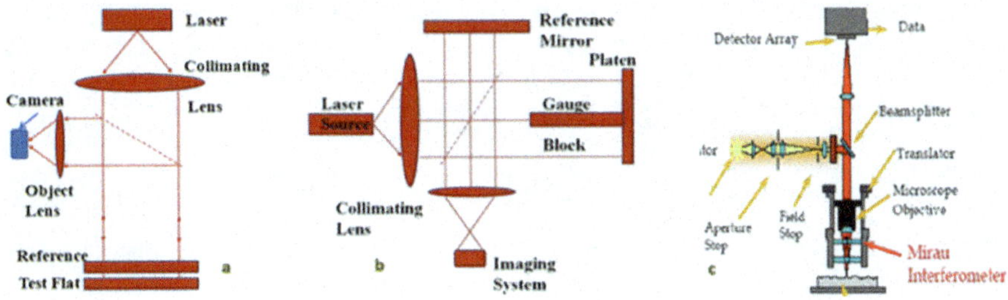

Fig. 7.15 A schematic diagram of **a** Fizau interferometer, **b** Twyman-Green **c** Mirau interferometer in optical profiler

As mentioned in Table 7.5, CSIR-NPL uses many types of interferometers in its quest to serve the nation with best possible methods. These include linear displacement measuring, Twyman-Green interferometer, Fizeau, Miraue and Linnik interferometers:

(a) A linear displacement measuring laser interferometer, which works on principal of Michelson interferometer and achieves high resolution and sense of direction by heterodyning methods or by using phase quadrature detection in case of homodyne interferometer. Velocity of light corrections for air and thermal expansion corrections is enabled by various sensors and software. Commercially available displacement interferometer with typical resolution of 0.1 and 0.01 μm is used at CSIR-NPL to provide traceability to heavy machines at the site and also at CSIR-NPL. These interferometers in a way are taking SI units to shop floor, thereby enabling manufacturers to achieve accuracy.

(b) A commercially available Fizeau interferometer Fig. 7.15 with phase shifting and automatic analysis of fringes is used at CSIR-NPL for determination of flatness errors, deviation from parallelism and small wedge angle. CSIR-NPL has participated in APMP intercomparison of flatness measurement with expanded uncertainty of 30 nm ($k = 2$).

(c) Twyman-Green interferometer is appropriate for surface shape and texture. The large field in such interferometer provides interference fringe pattern over whole surface of the test artifact, measuring surface form for up to 150 mm diameter flats at CSIR-NPL. The shape of fringes and spacing of fringes as shown in Fig. 7.16 is utilized to unfold undulations of flatness as shown in picture; here, the flatness deviation (PV) of this piece is 96.8 nm.

(d) A 3D optical profiler: The 3D optical profiler is equipped with a He–Ne (633 nm) laser interferometer for calibration of z-scan with traceability to SI unit meter. In VSI, white light LED along with interferometric objectives and phase shifting is used. The interferometric objective moves vertically to scan the surface at varying heights and scans through focus as the camera captures the frame of interference data at evenly spaced intervals. The system uses series of advanced computer algorithms to demodulate the interference data. Finally, the vertical position corresponding to the best focus, i.e., peak of the interference signal, is

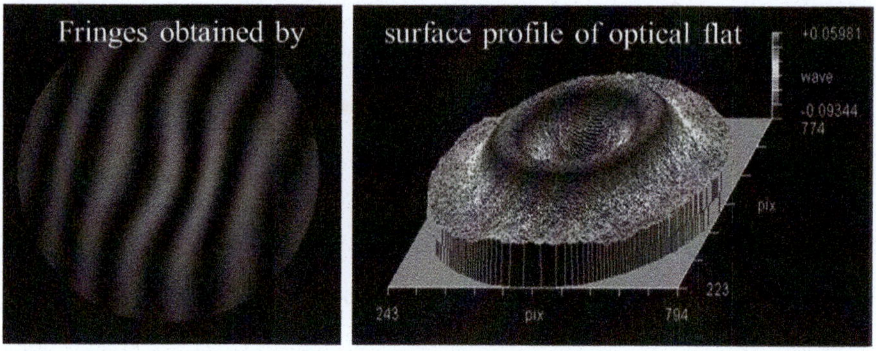

Fig. 7.16 Fringes and flatness profile as observed at CSIR-NPL

Fig. 7.17 Coordinate measuring machine

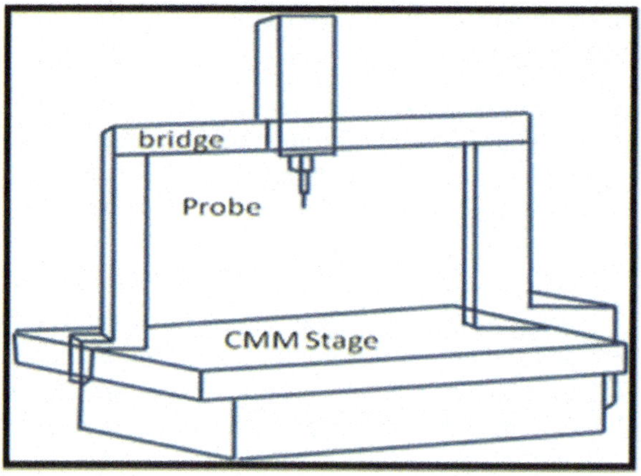

extracted for each point on the surface. This gives the information in the form of pixel by pixel data. This information is processed to obtain a three-dimensional picture of the test sample or further analysis. Software that accompanies the profiler calculates roughness parameters for the measured samples from this data. The vertical resolution of 3D optical profiler, available at CSIR-NPL, is 0.1 nm.

Coordinate Metrology

Coordinate metrology devices sense geometry of physical object by probing points on the surface, getting its coordinates and using these coordinates to get dimension information to get three-dimensional measurements. Coordinate measuring machines (CMMs) as shown in Fig. 7.17 are required to evolve with augmented accuracy and precision. A typical 3D "CMM" has provision for movement in axes, *X*, *Y* and *Z*, which are orthogonal to each other in a three-dimensional Cartesian coordinate system. Each axis has a sensor that monitors the position of the probe on that axis,

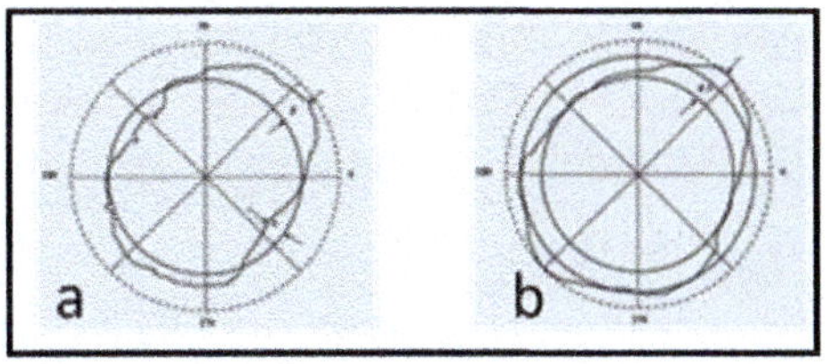

Fig. 7.18 Polar graph for roundness measurement **a** least square, **b** minimum zone circle method

when the probe contacts (or otherwise detects) a particular location on the object, thus measuring the location of point on the surface. CSIR-NPL has established a commercially available CMM with scanning probe head and bed to move part in X, Y, Z axis. The point cloud created by scan probes is analyzed by specialized software to give information on profile, dimension, distance, roundness, etc. Versatility of these CMM has been enhanced by CSIR-NPL by combining the use of vision probe and laser interferometer with the machine.

CNC Form Measurements

Form metrology is the process of estimating the overall shape of an object under observation. It is the most fundamental metrological aspect of industrial metrology and essential to ensure the correct performance of manufactured products and achieve customer satisfaction. Form metrology includes flatness, roundness, straightness, perpendicularity, cylindricity, roughness, etc. Some of the key aspects of form metrology are described below. Form testers for measurement of rotation of the component/indicator are necessarily coupled with the ability to measure change in radius. While rotating, the part is probed and the roundness data output obtained in the form on polar graph is analyzed for roundness profile of the workpiece surface due to stylus deflection. A form tester with fully automatic or a Digital Adjustment Table (DAT) function aided manual workpiece centering and leveling turns, powerful data analysis software for enhanced functionality has been installed at NPL. It has probes for roundness profile and roughness measurements Probe Z movement range is 500 mm, whereas measurement range is $\pm 400\ \mu m/\pm 40\ \mu m/\pm 4\ \mu m$. Low measuring force, 10–50 mN, ensures accuracy. To measure roundness, it is necessary to compare the part profile to an ideal circle or datum like least square reference circle (LSC), minimum zone reference circle (MZCI) as shown in Fig. 7.18, maximum inscribed circle (MICI) and minimum circumscribed circle (MCC). A series of roundness data taken while probing a part like cylinder can be analyzed to get cylindricity and perpendicularity (squareness) is shown in Fig. 7.19.

Surface Roughness

With higher precise demands of advanced engineering components, the control of surface texture parameters (height parameters, spatial parameters, hybrid parameters, functional parameters and feature parameters) along with dimensional accuracy has

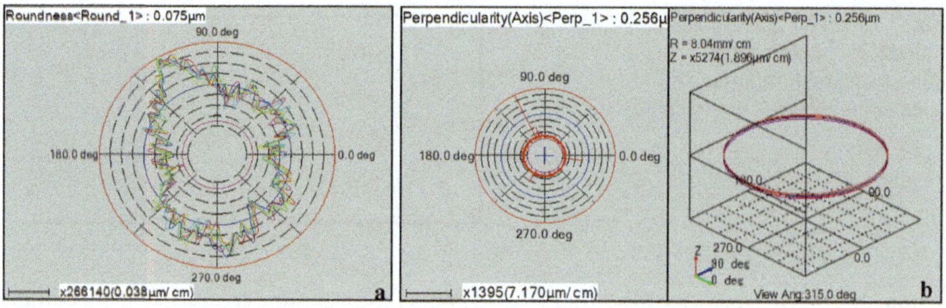

Fig. 7.19 Roundness profile for **a** circle, **b** cylinder

become extremely important. It has been realized that surface texture enormously influences the functioning of manufactured parts, by affecting the attributes such as appearance, corrosion resistance, wear resistance, fatigue resistance, lubrication, initial tolerance, ability to hold pressure, load carrying capacity, noise reduction, etc. Irregularities in a machined surface in form of continuous hills and valleys, varying in height and spacing are known as surface roughness/surface finish/surface texture/surface quality (represented as Ra, Rq, Rp, Rv, Rt, Rz, etc.).

Angle Metrology

Angle metrology is carried out for standard and instruments used for angle measurements. Full-circle calibrations of the angular divisions of polygon mirrors, indexing and rotary tables, and angle encoders can be accomplished without reference to separately calibrated reference artifacts using comparator techniques together with the principle of circle closure [https://www.ncbi.nlm.nih.gov/pmc/articles/PMC4890943/] based on the known fact that the sum of the angles around any point in a plane equals 2π rad (360°). Angle metrology is based on trigonometric principle. Assorted instruments such as angle gauges, protractors, sine bars, sine plates, sine blocks, sine tables, sine centers, true squares, spirit levels, clinometers, etc. are used to measure angles.

Autocollimator, polygons and laser interferometers are used at CSIR-NPL for angle metrology. Autocollimator is a telescope combined with a collimator, as shown in Fig. 7.20. It is used to measure small angle with high degree of resolution. An autocollimator might beam to a reflector, which is at an angle from the vertical plane. The reflected light beam is magnified and directed to a photodetector or an eyepiece. The angular tilt of reflector is represented by the deflection between incident and reflected beam. These advanced measurement techniques enable dimensional metrology instruments to cater expanding needs of industries and provide reliable measurement for fair global trade and customer satisfaction.

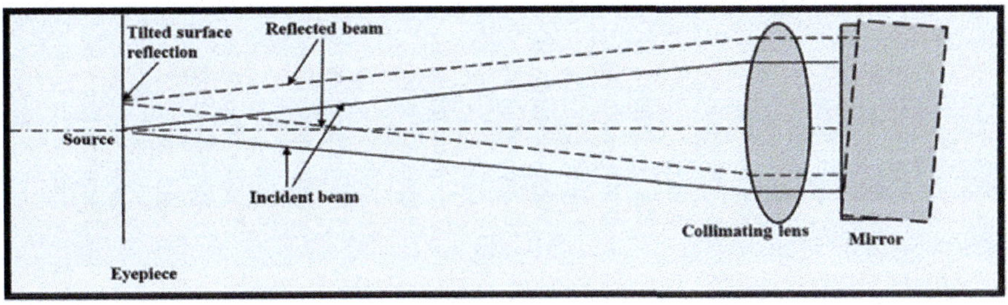

Fig. 7.20 Schematic diagram of autocollimator

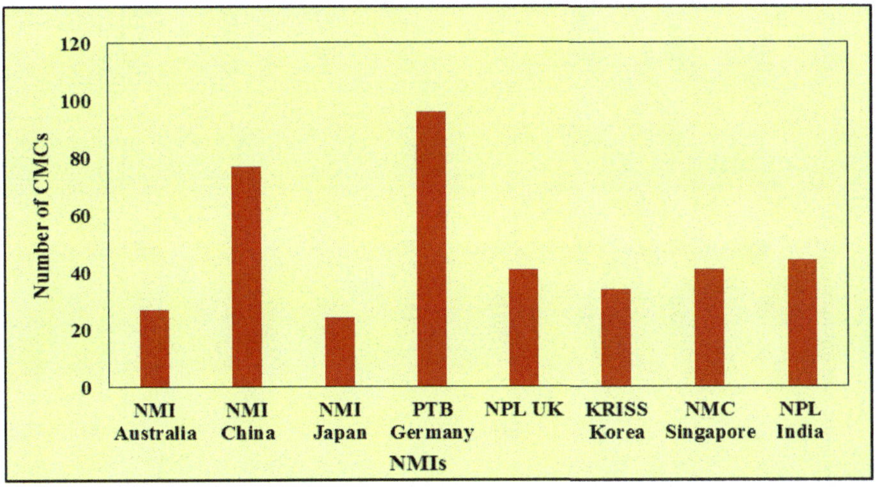

Fig. 7.21 CMCs of various NMI

7.3.4.2 Calibration and Measurement Capabilities

Length and dimension metrology of CSIR-NPL India is recognized globally. With hard work and competence of team, and successful peer reviews, the CMCs are recognized worldwide. CSIR-NPL's CMCs in area of length (laser frequency and dimension metrology) are compared with leading NMIs as shown in Fig. 7.21, and details of CMCs are shown in Table 7.6.

7.3.4.3 Participation in International Key Comparisons

To maintain the unbroken chain of traceability and prove international equivalence, CSIR-NPL continuously participates in international intercomparison, which are carried out through RMOs like APMP, EURAMET or SIM to achieve international equivalence of dimensional measurements. Length and dimensional metrology at

Table 7.6 Calibration and measurement capabilities (CMCs) in length and dimension metrology [29]

S. No.	Quantity	Instrument or artifact	Measurand value minimum maximum unit			Expanded uncertainty	
1	Length instruments	Length interferometer system	0.0	1.0	m	0.08	µm
2	Length instruments	Length measuring machines, error of indicated distance	0.0	30.0	m	Q[0.12, 0.5*L*]	*L* in m The uncertainty is expressed in µm
3	Length instruments	Length measuring machines, error of indicated distance	0.0	1.0	m	Q[0.04, 0.5 *L*]	*L* in m The uncertainty is expressed in µm
4	Length instruments	Gauge block comparator	0.0	150.0	mm	Q[0.032, 0.3*L*)	Q[0.032, 0.3*L*) The uncertainty is expressed in µm
5	Line standards	Precision line scale	0.0	1.0	m	Q[0.2, 0.3*L*]	*L* in m The uncertainty is expressed in µm
6	End standards	Gauge blocks, central length	0.5	305.0	mm	0.015 + 0.31*L*	*L* in m The uncertainty is expressed in µm
7	End standards	Gauge blocks, central length	0.5	100.0	mm	Q[0.04, 0.8*L*]	*L* in m The uncertainty is expressed in µm
8	End standards	Long gauge blocks	0.0	1000.0	mm	0.2 to 0.7	µm
9	End standards	Step gauge, face spacing	0.0	1220.0	mm	1 + 1.6*L*	*L* in m The uncertainty is expressed in µm

(continued)

Table 7.6 (continued)

S. No.	Quantity	Instrument or artifact	Measurand value minimum maximum unit			Expanded uncertainty	
10	End standards	Gap gauge, face spacing	0.0	100.0	mm	Q[0.5, 1.15*L*]	*L* in m The uncertainty is expressed in μm
11	End standards	Micrometer setting rod	0.0	1000.0	mm	0.2 to 0.7	μm
12	CMM artifacts	Ball (hole, bore) plate, ball or hole center coordinates	0.0	600.0	mm	0.5 + *L*/900	*L* in mm The uncertainty is expressed in μm
13	CMM artifacts	Ball bar	0.0	1000.0	mm	1 + 1.6*L*	*L* in m The uncertainty is expressed in μm
14	End standards	Feeler (thickness) gauge	0.0	1.0	mm	0.2	μm
15	Line standards	Engineer or machinist scale steel	0.0	1000.0	mm	Q[0.28, 0.3*L*]	*L* in m The uncertainty is expressed in mm
16	Line standards	Stage micrometers, line spacing	0.01	100.0	mm	Q[0.6, 2.5*L*]	*L* in m The uncertainty is expressed in μm

(continued)

Table 7.6 (continued)

S. No.	Quantity	Instrument or artifact	Measurand value minimum maximum unit			Expanded uncertainty	
17	Line standards	Measuring tapes, line spacing	0.0	30.0	m	Q[0.28, 0.12 *L*]	*L* in m The uncertainty is expressed in mm
18	Diameter standards	External cylinder (plug): diameter	0.1	150.0	mm	0.5 + *L*/900	*L* in mm The uncertainty is expressed in μm
19	Diameter standards	Internal, external cylindrical standards, diameter	3.0	400.0	mm	0.5–1.2	μm
20	Diameter standards	Internal cylindrical standards, diameter	3.0	300.0	mm	0.8–1.2	μm
21	Angle artifacts	Angle gauge block, included angle	0.0	180.0	°	0.5–1.0	″
22	Angle by circle dividers	Optical polygon	5.0	120.0	°	0.5–1.0	″
23	Angle by circle dividers	Index table	0.0	360.0	°	0.5–1.0	″

(continued)

Table 7.6 (continued)

S. No.	Quantity	Instrument or artifact	Measurand value minimum maximum unit			Expanded uncertainty	
24	Angle by circle dividers	Rotary table, rotary encoder	0.0	360.0	°	0.5–1.0	µm
25	Small-angle generators	Sine bar plate, sine table	0.0	1000.0	mm	$1 + 1.6L$	L in m The uncertainty is expressed in µm
26	Angle instruments	Autocollimator	−15.0	15.0	'	0.8–1.2	"
27	Angle instruments	Electronic level	−15.0	15.0	'	0.5–1.0	"
28	Angle instruments	Clinometer	0.0	45.0	°	0.5	"
29	Angle instruments	Spirit (bubble) level	−1.0	1.0	'	1	"
30	Angle instruments	Bevel protractor	−90.0	90.0	°	5	"
31	Angle artifacts	90 degree square, engineers square	90.0	90.0	°	$1 + 1.6L$	L in m The uncertainty is expressed in µm

(continued)

Table 7.6 (continued)

S. No.	Quantity	Instrument or artifact	Measurand value minimum maximum unit			Expanded uncertainty	
32	Roundness standards	Internal cylinder, roundness	0.01	100.0	µm	0.1–0.5	µm
33	Roundness standards	External cylinder, roundness	0.01	400.0	µm	0.1–0.5	µm
34	Roundness standards	Sphere (hemisphere)	0.01	100.0	µm	0.1–0.5	µm
35	Roundness standards	Magnification standard (flick, etc.)	0.01	100.0	µm	0.1–0.5	µm
36	CMM artifacts	Ball plates, ball center coordinates	1.0	550.0	mm	1.0–1.5	µm
37	Various dimensional	Micrometer head, error of indicated displacement	0.0	100.0	mm	1.0–3.0	µm
38	Complex geometry	Form measuring machine			(dimensionless)	0.07	µm
39	Complex geometry	Measuring projector, error of indicated distance	0.0	300.0	mm	Q[0.12, 0.5*L*]	*L* in m The uncertainty is expressed in µm

(continued)

Table 7.6 (continued)

S. No.	Quantity	Instrument or artifact	Measurand value minimum maximum unit			Expanded uncertainty	
40	Complex geometry	CMM, error of indicated distance	0.0	3000.0	mm	1 + 1.6*L*	*L* in m The uncertainty is expressed in μm
41	Various dimensional	Sieve, size of aperture	0.01	10.0	mm	2.0–4.0	μm
42	Form	Optical parallel, wedge	0.0	2.0	°	0.15	″
43	Laser radiations	Frequency stabilized laser	474.0	474.0	nm	4.2 E-11	Relative
44	Laser radiations	Frequency stabilized laser	474.0	474.0	nm	1.0 E-9	Relative

CSIR-NPL demonstrates the international equivalence of its measurement standards and calibration certificate through CIPM MRA. The CMCs for various dimensional parameters are peer-reviewed and approved after our participation in international intercomparisons. The international intercomparisons piloted/participated in last 10 years are shown in Fig. 7.22.

7.3.4.4 Potential Services for Future Requirements:

Length, dimension and nanometrology at CSIR-NPL has been involved in relevant research work for metrological growth in various fields such as image processing, moiré interferometry, characterization of dental implants, characterization of cutting tool inserts, Si wafer characterization for photovoltaic applications, hardness indenter characterization and application of Monte Carlo simulation for measurement uncertainty evaluation. Length dimension and nanometrology have achieved remarkable milestones using present measurement systems. Some of potential services for future requirements are discussed below:

- ***Calibration of lasers in visible and infrared regions using optical frequency comb***: Calibrations are required for many wavelength increasingly. A primary standard based on optical frequency comb offers versatility for calibrations of wavelengths over full octave in visible region for calibration of laser frequency.
- ***Preparing for services to Industry 4.0—calibration services with advanced interferometers, laser scanners and vision systems***
- ***Services for material metrology***: The techniques like SEM, TEM, HRXRD, SPM, SAXS, XRF, etc. that were traditionally used for material characterizations have now become tools for nano-dimension metrology. Specialized optical microscope-based dimensional measurements would be required to address the measurement needs of research and development organization, industries and academia. It is recommended that these two fields cannot work in isolation and pooling of interdisciplinary metrology is essential for growth of India.
- ***Line standard and mask inspection systems with nm range resolution and building competence to support other base and derived units of SI***
- These will be required for electronics industry, defense, vision metrology, Kibble balance, Si sphere measurements for mole and traceability lower uncertainty to pressure, force, torque, etc.
- ***Large-scale and complex coordinate Metrology***: World is moving ahead with portable CMMs with articulated arm, 5D CMM, etc. *Market surveys predicting market till 2038 have reported that CMM metrology will be one of the most demanding technology.*

A Decade of Clients and Services:

CSIR-NPL operates and maintains primary standard of length for realization of SI unit "meter", disseminates traceability for dimensional metrology at

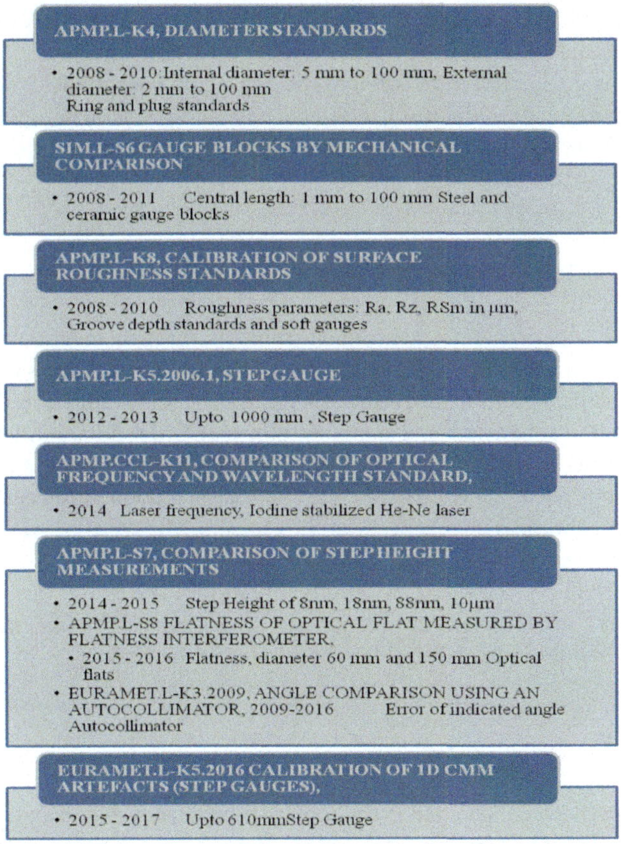

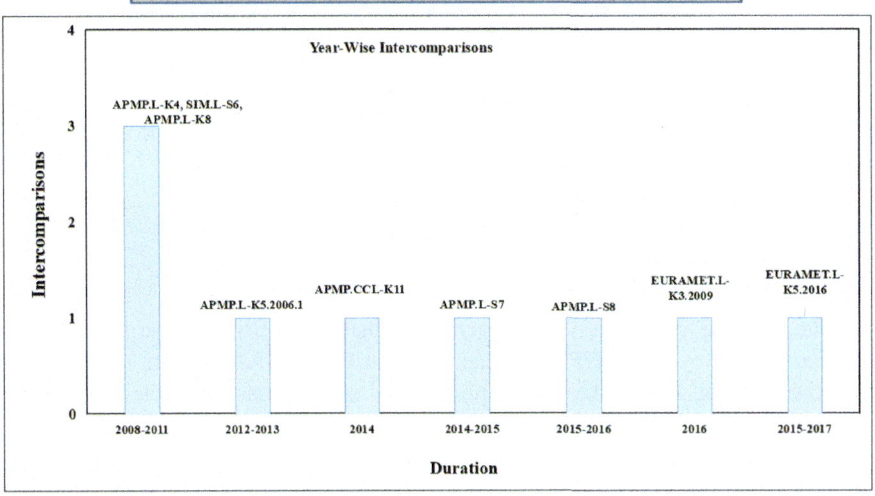

Fig. 7.22 Intercomparisons in last 10 Years

national/international (providing services to SAARC countries) level, caters the growing industrial needs and contributes to fair trade within the country and across the globe. Dimensional calibration services, trainings and consultancies foster innovation and support the mission of self-reliance, making local industry globally competent. Some of our appreciable contributions for socioeconomic upliftment of India in last decade are discussed.

Calibration and Testing

To ensure the accuracy of measuring instruments through reduction in measurement uncertainty, calibration of measurement instruments is essential to be carried out to quantify and control instrument errors and hence to improve the quality of product. Length, dimension and nanometrology at CSIR-NPL have continuously been contributing to nation's economic and social growth through apex level traceable calibrations services. More than 2300 calibration reports have been generated in last ten years for various clients of different s sectors. The types of services provided in last ten years are given in Fig. 7.23.

Figure 7.24 shows the distribution of calibration services for various sectors. It is evident that a major part of calibration services is provided to the manufacturing industries and calibration and testing laboratories in the country.

Consultancy and Technical Services

CSIR-NPL has been contributed in the form of consultancy and technical projects to the numerous industries/government organizations. These projects included implementation of ISO/IEC 17025 guidelines, measurement instructions, technical evaluations, measurement uncertainty evaluations, establishment of calibration setups and customized solutions for dimensional measurement related challenges. Some examples of dimension measurements helping economy and health are shown in Fig. 7.25.

Training and Academic Services

Various training programs on different dimensional metrology parameters have been conducted by CSIR-NPL, for many industries and SAARC nations for awareness creation and knowledge generation. Training programs conducted for various stakeholders in last ten years are presented below which include optical radiation metrology, lectures and lecture notes on basics of metrology, knowledge of few relevant standards, exposure of hands-on measurement on real experimental setups and evaluation of uncertainty involved in measurements. Since length and dimension metrology is a very specialized discipline of science, therefore, it finds place in

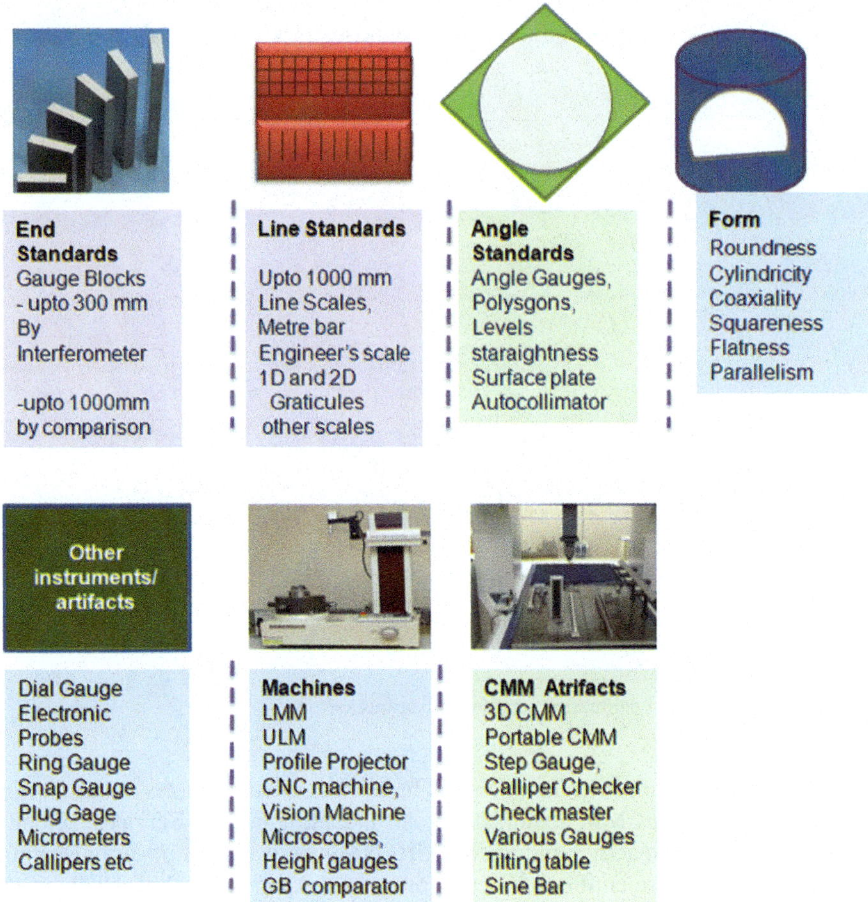

Fig. 7.23 Calibration services offered in dimension metrology

curricula of degree courses of countable number of universities/institution. It is therefore imperative for CSIR-NPL to conduct specialized regular courses on precision metrology to create a bank of trained human resource to cater the need of upcoming industry requirements in the field of dimension metrology. Currently, CSIR-NPL is running a one-year Post Graduate Diploma on Precision Measurement and Quality Control (PGD-PMQC) under the umbrella of Academy of Scientific and Innovative Research (AcSIR) in the line of skill development of Government of India's program. Length and dimension metrology is part of PMQC course wherein theory as well as practical is taught to the students. A course by name of advanced measurement techniques is run for PhD students wherein an introduction to definition of meter, its realization, interferometry and traceability is given. Courses are organized from time

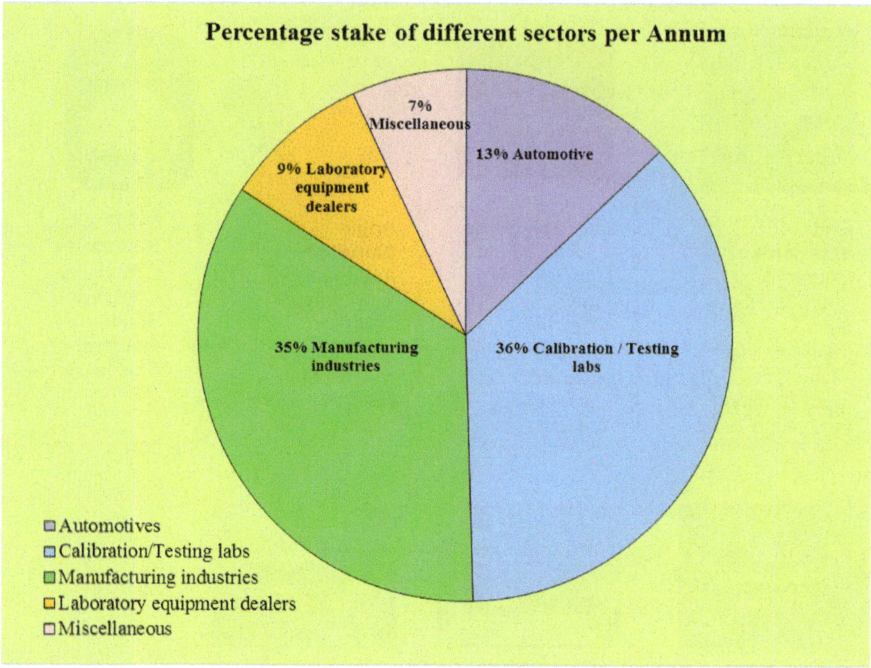

Fig. 7.24 Distribution of calibration services for various sectors

to time for various industries and demand-based sector-wise courses are also organized, e.g., course for CMM manufacturer, Air India Engineering Services, automobile manufacturer, capacity building, SAARC NMIs by way of short-term training, on-site training and scientific interaction programs. Apart from skill development program, students are also supervised for M.Tech and PhD programs.

7.3.4.5 Major Documentary Standards Guides and SOPs

The main standard documents used in length and dimension metrology are listed below, and the SOPs are based on these standards. Few guides for metrology have been published by EURAMET and available on their websites as listed in Table 7.7.

7.3.5 Possible Gaps and Suggestions for Their Redressal

- **Optical frequency standard**: A primary standard based on optical frequency comb offers versatility for calibrations of wavelengths over full octave in visible region for calibration of laser frequency. The existing iodine stabilized He–Ne

CSIR-NPL's Dimension Measurements helps Development of Dental Implant in India

Challenge: A leading institute in India was trying to dental implants indigenously. But to comply to standards requirements surface roughness measurements on the tapered part are needed.

Solution: CSIR-NPL carried out these measurements by using traceable 3D optical profiler for imaging and analysis with tilt compensation was done with Vision software.

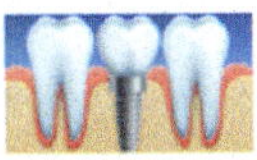

Impact: Successful development by them resulting in import substitution and thereby affordable health care in future

CSIR- NPL is providing Technical Services to MSME for setting up their Dimension lab by providing consultancy on implementation of ISO/IEC 17025

Dimension measurements help Innovation and Indigenous development

Challenge: RAMSME developed a instrument for measurement of Hardness. This requires dimensions of Depth to be measured for verification.

Solution: CSIR-NPL carried out these measurements developing strategy and designing jigs that could facilitate calibration with Interferometers.

Impact: Support to make in India and MSME

CSIR-NPL's Measures Dimensions of Artificial Rail Defect Standards

Challenge: Rail defect standards are being made in India. These standards need dimension of blind holes (sub mm and mm) to be measured.

Solution: CSIR-NPL carried out these measurements by using traceable methods and the rails could be verified for export

Impact: Support to make in India, export and safety

Fig. 7.25 Dimension measurements for economy and health

laser is limited to one wavelength only, not able to fulfill country's needs for other wavelengths and is not competitive to other NMIs. A frequency comb source is required because precise knowledge of frequencies or wavelengths requires complex "clockworks" to connect optical frequencies to those in microwave region.

- **Preparing for services to Industry 4.0—calibration services with better interferometers, laser scanners and vision systems**: The world of dimension metrology is changing very fast. With Industry 4.0 automations and artificial intelligence, entirely new set of skill in information technology, artificial intelligence will be required to cater increasing needs of industries. Laser scanners and interferometers are needed to lower the uncertainty. Envisaging this scenario, optical dimension metrology using multifaceted laser scanners; highly accurate laser interferometers, including three axis interferometers and double pass interferometers; vacuum comparators equipped with laser interferometers and precision software and vision systems is inevitable.
- **Services for material metrology**: The techniques like SEM, TEM, HRXRD, SPM, SAXS, XRF, etc. that were traditionally used for material characterizations have now become tools for nano-dimension metrology. Specialized optical

Table 7.7 Major documentary standards in length and dimension metrology

S. No.	Standard operating procedure/ISO standard
1	IS: 3651 (Part I, II and III)—specification for Vernier Calipers
2	IS: 2092—specification for plunger type dial gauges
3	IS: 11,498—specification for lever type dial gauges
4	IS: 2966—specification for internal micrometers (including Stick Micrometers)
5	IS: 2967—specification for external micrometer
6	IS:13,907—height setting micrometer and rising block
7	IS:1269 Part 2—metric steel tape (winding type)
8	IS:1269 Part I-material measure of length, Part 1: woven metallic and glass fiber tape measures
9	IS:15,372—measurement of roundness-terms, definitions and parameters of roundness
10	IS: 4213—engineering metrology—vernier depth gauges up to 300 mm: specifications
11	IS: 15,374—engineering metrology—methods for the assessment of departure from roundness measurement by two and three points
12	IS15635/ISO10360-2—performance assessment of coordinate measuring machines
13	EURAMET Cg-2—calibration of gauge block comparators
14	IS 10,936–1 optical and optical instruments—operation microscopes—part-1 requirements and test methods
15	IS: 2063/ISO 230 Part 1—test code for machine tools—geometric accuracy of machines Operating under no Load or Finishing Conditions
16	IS: 2285—engineering metrology—measuring equipment—cast iron surface plates—specification
17	IS: 7327—engineering metrology—measuring equipment—granite surface plates—specification
18	IS: 2554—specification for cast iron angle plates
19	IS: 2103—specification for engineer's squares
20	IS: 2220—engineering metrology—steel straight edges: specifications
21	IS: 4241—engineering metrology—engineer's parallels—specification
22	IS: 6952—engineering metrology—cylindrical squares
23	IS: 6973—engineering metrology—precision angle plates
24	IS: 4960—engineering metrology—universal and elongated type V-blocks
25	IS: 2984—specification for gauge blocks
26	ISO 3650—gauge blocks
27	IS: 2921—specification for Vernier height gauges
28	IS: 9483—measuring instruments-micrometer head
29	IS:7014—engineering metrology—length bars
30	IS:4440—engineering metrology—precision equipment—gauge block accessories
31	IS:4239—engineering metrology—mechanical level protractors
32	IS:5359—engineering metrology—sine bars

(continued)

Table 7.7 (continued)

S. No.	Standard operating procedure/ISO standard
33	IS: 5979—engineering metrology—sine centers
34	IS:5706—optical instruments—spirit levels for use in precision engineering
35	IS: 3073—assessment for surface roughness
36	IS:10,707—instrument for measurement of surface roughness by profile methods
37	IS:3510—engineering metrology—toolmaker's flats
38	IS: 3455—gauging practice for plain workpieces
39	IS:3485—engineering metrology—plain and master setting ring gauges (size range from 1 up to and including 315 mm)
40	IS: 9529—engineering metrology—taper plug gauges and taper ring gauges to check taper bore of gauge handles and taper shanks of gauging members
41	IS: 3179—engineering metrology—feeler gauges
42	IS 460 (Part I, II and III)—specification for test sieves

microscope-based dimensional measurements would be required to address the measurement needs of research and development organization, industries and academia. It is recommended that these two fields cannot work in isolation and pooling of interdisciplinary metrology is essential for growth of India.

- **Line standard and mask inspection systems with nm range resolution and building competence to support other base and derived units of SI:** A better detector is needed to reach at nanometer or sub-nanometer level to fulfill the demands of nanometrology, line-width measurements for semiconductor industry. Base units like Mol and kg depend on different high-end interferometry techniques. An example is interferometry for Kibble balance realization and for Mol, silicon sphere diameter measurements. Research is going on in leading NMIs like NMIJ, etc. utilizing double sided interferometers for Si sphere and NMI Canada is working on Kibble balance. Meticulous strategic planning would be required to get pool of talents and the funding. Setting up of these systems requires precision facilities for optical component manufacturing, experts of optical interferometer, image processing, electronics, mechanical fabrication and information technology.
- **Large-scale and complex coordinate metrology**: We want to develop methods and capabilities to support industries—including the aircraft, ship-building, construction, farm equipment and automotive. This needs to make measurements ranging from sub-meter to multiple-meter parts and structures with low, well-characterized measurement uncertainties for large-scale and complex mechanical parts. World is moving ahead with portable CMMs with articulated arm, 5D CMM, etc. **Market surveys predicting market till 2038 have reported that CMM metrology will be one of the most demanding technology**.
- **Non-destructive testing (NDT) metrology**: Non-destructive techniques have become important tools for the industries to initiate preventive maintenance programs. It has become possible to accurately predict manufactured product

life by knowing the true conditions of its internal parts or structure. X-ray tomography and for reverse engineering needed for boosting make in India using X-rays with coordinate measurements are the proposed solutions.

- **30 m long measuring facility machine**: Electronic distance meters (EDMs) used in metro railways, road, highways and meter tape widely used the structures of concrete around 30 m. At present, users need to get traceability from abroad, and thus, self-reliance is necessary to focus in this area.

7.4 Conclusions and Future Perspectives

The establishment of strong metrological and quality infrastructure highlighted in this chapter for mass and length metrology evidently plays a vital role in establishing a strong traceability chain and industrial growth of the country. The PMPs of mass, volume, density, viscosity, length, dimensions, angle, surface roughness and nanometrology are realized with the highest levels of accuracy and precision, which enables them to serve the strategic sector, MSMEs, public and private sector undertakings, government organizations, regulatory bodies as well as the SAARC nations. The measurement capabilities are reaffirmed by periodic participation in the international intercomparisons with other NMIs that enables us to acquire the global competence and recognition. The chapter briefly discussed the past, present and future of mass and length metrology measurements at CSIR-NPL. The evolution of the SI unit of mass, "kilogram" and length "meter" through its various definitions by BIPM is highlighted. CSIR-NPL's historical achievements to its present-day measurement capabilities like vacuum mass comparator, 1 kg mass comparator, 50 kg mass comparator, precision gauge blocks coordinate measuring machines, roundness, optical flat and parallel, line scale standards and angle metrology are discussed in detail. The work also describes a broad range of measurement technologies and calibration capabilities in CMC table for various parameters of mass and length metrology.

The implementation of new SI system based on redefined SI units on 20th May 2019, i.e., the World Metrology Day, has thus diverted the attention and efforts of every country toward the development of measurement methodology for realizing the SI unit of mass through the fundamental constant, invariant of time and space that has replaced the dependence on artifact. This has opened up the new era of quantum metrology in SI, linking all the seven base units to fundamental constants. CSIR-NPL is also continuously focused on realizing the kilogram using Planck's constant. A Kibble balance of 1 g for determining Planck's constant had been earlier successfully demonstrated. Till such time, the traceability chain of mass would be continued to be established through NPK-57. The length, dimension and nanometrology activity is also focused on establishing the primary standard of length using optical frequency comb for the frequencies in visible region and later extendable from ultraviolet to infrared region. Efforts are also focused on development of nanometrology and imaging standards so as to meet the challenging technological needs of the industries in India for measurement services.

Acknowledgements Authors are thankful to all the members and students of the PMM, past and present who have directly or indirectly contributed to the development of the measurement standards and facilities. Thanks are also due to all the funding agencies that have provided partial grants for the same time to time.

References

1. OIML R111–1: Weights of classes E_1, E_2, F_1, F_2, M_1, $M_{1\text{-}2}$, M_2, $M_{2\text{-}3}$ and M_3, Part 1: Metrological and technical requirements (2004)
2. R. Schwartz, Guide to mass determination with high accuracy. PTB-Bericht MA-40, Braunschweig (1995)
3. G.D. Chapman, Orthogonal designs for calibrating kilogram submultiples. NRCC25819. 27 April 1995, National Research Council Canada, Canada
4. E.C. Morris, Decade design for weighings of non-uniform variance. Metrologia **29**, 373 (1993)
5. J.M. Cameron, M.C. Croarkin, R.C. Raybold, Designs for the calibration of standards of mass, NBS TN 952 (1977)
6. J. Aguilera, J.D. Wright, V.E. Bean, Hydrometer calibration by hydrostatic weighing with automated liquid surface positioning. Meas Sci Technol. (2008). https://doi.org/10.1088/0957-0233/19/1/015104
7. S.V. Gupta, Practical density measurement and hydrometry. Meas Sci Technol. (2003). https://doi.org/10.1088/0957-0233/14/1/701
8. BIPM, France website htps://www.bipm.org
9. H.G. Jarrard, D.B. McNeill, A dictionary of scientific units. Chapman and Hall, p 85 (1964)
10. O. Arthur, *McCoubrey, NIST Special Publication 811—Guide for the Use of the International System of Units* (National Institute of Standards and Technology, Gaithersburg, MD, 1991)
11. N. Barry, *Taylor, NIST Special Publication 330—The International System of Units (SI)* (National Institute of Standards and Technology, Gaithersburg, MD, 1991)
12. D.A. Jennings, C.R. Pollock, F.R. Peterson, R.E. Drullinger, K.M. Evenson, J.S. Wells, J.L. Hall, and H.P. Layer, Direct frequency measurement of the I2-stabilized He–Ne 473-THz (633-nm) laser. Opt. Lett. 8(3) (1983)
13. T.J. Quinn, Mise enPratique, Metrologia **30**(5) (1994)
14. B. Chakrabarti, Fifty years of the metric system in India and its adoption in our daily life. Curr. Sci. **92**(3), 390–391 (2007)
15. CSIR-NPL, The Legacy continues, diamond jublee document. Thomson Press (I) ltd. (2007)
16. Dr. P C Jain, A Private Communication
17. P. Hariharan, D. Sen, "Radial Sheering Interferometrs. J. Sci. Instrum. 38–1–428 (1961)
18. V.D. Dandawate, P.N. Puntambekar, D. Sen: Frequency stability and reproducibility of iodine stabilised He-Ne laser at 633 nm. Pramana**22**, 573–578 (1984)
19. R. Davis, A short story on length. Nat. Phys. **14**, 868 (2018)
20. H.P. Layer, IEEE Trans. On Instrumentation and Measurements IM29 (4) (1980)
21. I. Shigeo, Weights and Measures in the Indus Valley. Encyclopedia of the History of Science, Technology, and Medicine in Non-Western Cultures (2nd edition) .edited by Helaine Selin, pp 2254–2255 (2008)
22. J.M. Kenoyer, Indus Valley Civilization. Ecyclopedia of India (vol. 2) edited by Stanley Wolpert, 258–266 (2006)
23. W. Ian, A Measure of all things: the story of man and measurement. Macmillan (2007)

24. R. Balasubramanium, New insights on metrology during Mauryan Period. Curr. Sci. **97**, 680–682 (2009)
25. Editor's Note Documents concerning the new definition of the metre. *Metrologia* **19**, 163 (1984). https://doi.org/10.1088/0026-1394/19/4/2004
26. Resolution 2 in ComptesRendus de la 15e CGPM (1975), 1976, p. 103, reported in "News from the Bureau International des Poids et Mesures", *Metrologia* **11**, 179–183 (1975). https://doi.org/10.1088/0026-1394/11/4/2006.
27. BIPM, The International System of Units (SI Brochure) [9th edition, 2019-Appendix 2], https://www.bipm.org/en/publications/si-brochure/
28. https://emi.epfl.ch.metrology/wyko_NT1100.php
29. BIPM. https://www.bipm.org/kcdb/cmc/search?domain=PHYSICS&areald=3&keywords=&specificPart.branch=12&specificPart.service=-1&specificPart.subService=-1&specificPart.individualService=-1&_countries=1&publicDateFrom=&publicDateTo=&unit=&minValue=&maxValue=&minUncertainty=&maxUncertainty

Chapter 8
Physico-Mechanical Metrology

Part III: Thermal, Optical Radiation and Acoustic Metrology

Sanjay Yadav, D. D. Shivagan, Komal Bapna, V. K. Jaiswal, Parag Sharma, Shibu Saha, Mahavir Singh, Naveen Garg, Kirti Soni, S. S. K. Titus, and D. K. Aswal

Abstract This chapter describes the three physico-mechanical metrology streams; temperature and humidity, optical radiation and acoustic and vibration metrology. Among the seven base SI units, temperature and optical radiation encompasses the two SI units while acoustic comprises only derived parameters. Temperature metrology includes realization and dissemination of temperature and humidity

S. Yadav (✉) · D. D. Shivagan · K. Bapna · V. K. Jaiswal · P. Sharma · S. Saha · M. Singh · N. Garg · K. Soni · S. S. K. Titus · D. K. Aswal
CSIR-National Physical Laboratory, Dr. K.S. Krishnan Marg, New Delhi 110012, India
e-mail: syadav@nplindia.org

D. D. Shivagan
e-mail: shivagand@nplindia.org

K. Bapna
e-mail: komal.bapna@nplindia.org

V. K. Jaiswal
e-mail: jaiswalvk@nplindia.org

P. Sharma
e-mail: sharmap2@nplindia.org

S. Saha
e-mail: saha.shibu@nplindia.org

M. Singh
e-mail: mahavir@nplindia.org

N. Garg
e-mail: ngarg@nplindia.org

K. Soni
e-mail: soniks@nplindia.org

S. S. K. Titus
e-mail: titus@nplindia.org

D. K. Aswal
e-mail: dkaswal@nplindia.org

D. K. Aswal (ed.), *Metrology for Inclusive Growth of India*,
https://doi.org/10.1007/978-981-15-8872-3_8

parameters, optical radiation metrology includes parameters like illuminance, illuminance responsivity, luminous flux, luminance, colour temperature, colour coordinates and spectral irradiances. Acoustic metrology includes sound pressure, vibration amplitude, sound transmission and absorption coefficients. All the parameters are inter-connected to fulfil the human needs and endeavours and have immense importance in the socio-economic growth of the country. The contribution, insinuation, impact, significance and relevance of these PMPs in different sectors like health, process and manufacturing, lighting industry, building acoustics, SODAR system and noise mapping are described along with established measurement standards and facilities, CMCs, participation in key comparisons, gap analysis and futuristic requirements and suggestions. All these endeavours are carried for the inclusive industrial growth of the nation.

Refining is inevitable in science when you have made measurements of phenomenon for a long period of time.

– Charles Richter (1900-1985)

8.1 Temperature and Humidity Metrology

This section describes the comprehensive information on the importance and status of temperature and humidity metrology, brief history about origin of temperature scale, its development, various techniques and principles associated with them, methods to realize these parameters at laboratory level, dissemination, applications in various sectors, potential services and future prospects, etc.

8.1.1 Introduction and Historical Aspects

Temperature and humidity are critical and essential parameters that need to be monitored and controlled in various fields such as in science, technology, medical, agriculture, pharmaceutical, industries and environmental applications. Industries and organizations need to have complete confidence in the performance of their measurements, over a wide range of temperatures and in extreme conditions. Therefore, it is important for the metrological communities to provide the accurate, precise and reproducible measurement of the temperature and humidity to face the globalization of industrial production and harmonization in doing trade.

Temperature, in general, is considered to be the degree of hotness or coldness. More specifically, it is the property required to describe the macroscopic state of a system which is in thermal equilibrium. Thermodynamic temperature is a fundamental physical quantity and is defined as one of the seven base quantities in the SI. To the most surprise, even being in SI unit, there was no proper definition of temperature until eighteenth century. Even today, only few people who are directly associated with thermometry, have an intuitive understanding of what is being measured. In 1869, Maxwell–Boltzmann gave a relation between the velocity or energy distribution of particles in a system at thermal equilibrium, given as $E = kT/2$. For a molecule of mass 'm' and velocity 'v', the average of the square of the velocity is then given by $v = \sqrt{(3k\text{T/m})}$. Boltzmann constant k thus provides the link between the mechanical quantity (quantum or classical) and thermal parameter.

Temperature is an intensive quantity which means it is non-additive. This tells that temperature scale is non-uniform and has to be measured in the whole temperature regime with maximum possible points. The measurement of temperature is termed as thermometry. The ability to precisely measure temperature strongly depends on the measuring sensor and method. The history of thermometry, since the first device which can measure the degree of hotness and coldness relative to ambient temperature was invented between 1592 and 1603 AD based on a glass bulb connected to a long tube immersed in the liquid. At the start of the seventeenth century, a need was realized to standardize these thermometers by temperature fixed points. In the 1724, D G Fahrenheit developed mercury-in-glass thermometers standardized using three fixed points: a mixture of ice, water and ammonium chloride, a mixture of ice and water and the human body and the corresponding temperatures were taken as 0 °F, 32 °F and 96 °F, respectively. Concurrently, the French scientist Amontons developed the gas thermometer by using air medium and found the ratio of greatest summer heat to greatest winter in Paris to be approximately five or six. Later, he concluded that the expansion coefficient of all gases is approximately the same and hence temperature is simply proportional to the pressure of a gas, and thus, only one fixed point is required to define the scale. Still Gas thermometry for a long time relied upon two fixed points using the equation:

$$p(T) = p(0) \cdot (1 + \alpha T), \tag{8.1}$$

where $p(0)$ is the reference fixed point.

In 1742, Anders Celsius developed mercury-in-glass thermometer and scaled by using freezing point (0 °C) and boiling point (100 °C) of water at standard atmospheric pressure (101.325 kPa). He divided this temperature scale into 100 equal parts, and the unit of the temperature was defined as degree Celsius (°C). At the start of eighteenth century, two approaches given by Fahrenheit and Amontons were followed for the further development of thermometry. In the beginning, the practical temperature scale was defined by ice point (0 °C) and steam point (100 °C). In 1854, William Thomson (Lord Kelvin) had proposed the new temperature scale by developing an idea of absolute zero temperature, where all the microscopic motions of constituent particles of matter completely cease, along with one additional fixed-point. Eventually, in 1948 at 9th CGPM, the triple point of water (TPW) was proposed as the primary reference point to define thermodynamic temperature. The TPW provides better accuracy than ice and boiling point of water, which were found to depend on atmospheric pressures. The thermodynamic temperature of the TPW is estimated by Gay-Lussac's gas law and calculated to be 273.16 K (0.01 °C). Thus the Eq. (8.1) remained valid until the unit of thermodynamic temperature was redefined by assigning a numerical value to just one fixed point, i.e. TPW. In 1954, 10th CGPM officially accepted the definition of SI unit of thermodynamic temperature, kelvin (K) based on TPW state, as the fraction 1/273.16 of the thermodynamic temperature of the TPW.

8.1.1.1 Primary Thermometry Methods

Thermodynamic temperature is the physical property which is intensive in nature and absolute measurement of temperature. The fundamental methods to determine the thermodynamic temperature is known as the primary thermometry methods. Primary thermometers are based on the well-understood physical system where the equation of state can be written explicitly without need of introducing unknown, temperature-dependent, constants [1, 2]. For instance, ideal gas law and Planck's law are the equations which satisfy this condition and are primary thermometry methods. When two objects having different temperatures come into contact, the only thing they share is the energy per degree of freedom. The thermodynamic temperature is a measure of energy per degree of freedom, but we can't express it in terms of Joule per degree of freedom because the temperature is expressed in kelvin. Moreover, Boltzmann constant is the fundamental constant, which appears as a constant in the thermal energy (kT). Therefore, the kelvin was proposed to be defined through thermal energy by fixing the value of Boltzmann constant. CGPM approved the new definition of thermodynamic temperature based on Boltzmann constant at its 26th meeting on 16th November 2018, and it was implemented from world metrology day, on 20th May 2019 [3]. Therefore, primary thermometric methods are playing a major role in the determination of Boltzmann constant with low uncertainty and dissemination of thermodynamic temperature (T). Also, the metrology community is making measurements of $T\text{-}T_{90}$, to come-up with the future practical thermometry scale (ITS-202X).

Primary thermometry methods for the determination of Boltzmann constant (k) and measurement of thermodynamic temperature are based on the fact that a macroscopic property is measured which can be related to the temperature at a known pressure e.g. density in Constant volume gas thermometry [1], dielectric constant in Dielectric constant gas thermometry [4], speed of sound in Acoustic gas thermometry [5, 6]. Johnson noise thermometry [7] and Doppler broadening thermometry [1, 8] are statistics and quantization methods. The Johnson noise thermometry is based on the relationship between thermodynamic temperature and generated fluctuating noise voltage or current in a resistor due to thermal motion of electron while the Doppler broadening thermometry is based upon the principle that Doppler broadening and the width of absorption line of the gas particles or atoms at thermodynamic equilibrium is proportional to the thermal motion of gas molecules. Total radiation thermometry is another primary method based on the Stefan-Boltzmann law of thermal radiation.

The fundamental equations for CVGT, AGT and DCGT are shown in Table 8.1.

Table 8.1 Fundamental equations for CVGT, AGT and DCGT

CVGT	AGT	DCGT
$pV = nRT$	$c = (\gamma RT/M)^{1/2}$	$p = kT(\varepsilon\text{-}\varepsilon_o)/\alpha_o$

Although these primary methods give very accurate way to measure the Thermodynamic temperature, but because of the complicated, time consuming and expensive measuring setup of this thermometry, it remained limited to the laboratory only and it demands a unifying method to realize temperature in a practical manner and to disseminate kelvin. Furthermore, these thermometers are not as reproducible as the platinum resistance thermometer in the most of temperature range.

CSIR-NPL has taken initiatives on development of Acoustic Gas Thermometry for the realization of Boltzmann constant based New Kelvin. This is one of the innovative projects, CSIR-NPL is carrying out towards the international efforts on re-definition of base units based upon fundamental constants. Under 12th Five-year plan project on Measurement Innovations in Science and Technology (MIST), the project was initiated. In the primary thermometry methods (equation of state-based methods) and in the laws of physics, the thermodynamic energy is proportional to the product kT, therefore if the value of 'k' is fixed, the thermodynamic temperature can be directly measured by measuring the energy (J/K). Among the primary methods discussed above, the best estimation of Boltzmann constant with the lowest uncertainty was attained by using the AGT. The AGT is one of the primary thermometry methods where velocity of sound is measured in the monoatomic gas in a temperature-stabilized acoustic resonator, where the resonating frequency is proportional to the dimension of resonator. In AGT, in terms of macroscopic measurable properties, the relation be (5, 6):

$$k = Mu_0^2/N_A\gamma_0 T_{\mathrm{TPW}} \tag{8.2}$$

where k is Boltzmann constant N_{A} is the Avogadro constant, γ_0 is the ratio of the principal heat capacities of the gas in the limit of low pressure, u_0 is speed of sound at low pressure limit M is the molar mass of the gas and T_{TPW} is the thermodynamic temperature at TPW. Nowadays, the speed of sound is measured in argon or helium gas medium by using quasispherical resonator.

The equation can be rewritten in terms of acoustic and microwave resonance frequencies as:

$$k = \left\langle \frac{3}{5}\frac{mc_0^2}{T_{TPW}} \left(\frac{Z_{nl}^{EM}}{Z_{nl}^{A}}\right)^2 \lim_{p\to 0}\left(\frac{\langle f_{nl}^{A} + \Delta f_{nl}^{A}\rangle}{\langle f_{nl}^{EM} + \Delta f_{nl}^{EM}\rangle}\right)^2 \right\rangle \tag{8.3}$$

AGT method was first developed by NIST USA and then successfully achieved below 1 ppm by NPL UK and LNE France, and useful for the realization of k and then dissemination of thermodynamic temperatures in wide temperature ranges.

We are developing AGT set up at our laboratory with in-house efforts. With design-development and fabrication of copper quasi-spherical cavity acoustic-microwave resonator in India and by in-house collaborations of temperature, acoustic, microwave, pressure, dimension and gas metrology groups, it is possible to develop the desired primary standard at CSIR-NPL. Part of the development of critical components such as copper cavity resonator, large volume high stability bath, capsule SPRTs and resistance bridge for simultaneous temperature measurement of resonator, pressure and flow accessories are developed under the 12th FYP project. The remaining acoustic, microwave and gas purification equipment and accessories will be developed to determine *k* and will be integrated and experimental parameters will be optimized to get the accurate value of Boltzmann constant and Thermodynamic temperatures. Figures 8.1 and 8.2 show schematic of AGT at CSIR-NPL and design & fabrication of Copper quasi-sphere, pressure vessel and vacuum vessel respectively.

The first measurement of *k* will be carried out at TPW temperature. Thermodynamic temperature at other temperatures can be determined by measuring the acoustic and microwave resonances and based on the ratio method.

$$\frac{T}{T_{tpw}} = \lim_{p \to 0} \left[\frac{f_{0n}(T)}{f_{0n}(T_{tpw})} \frac{\langle f_{\text{ln}}^{\sigma}(T_{tpw}) \rangle}{\langle f_{\text{ln}}^{\sigma}(T) \rangle} \right]^2 \tag{8.4}$$

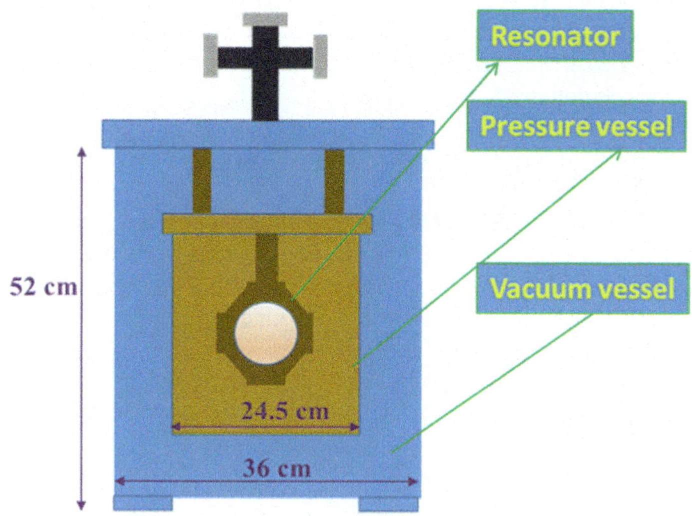

Fig. 8.1 Complete schematic of AGT at CSIR-NPL

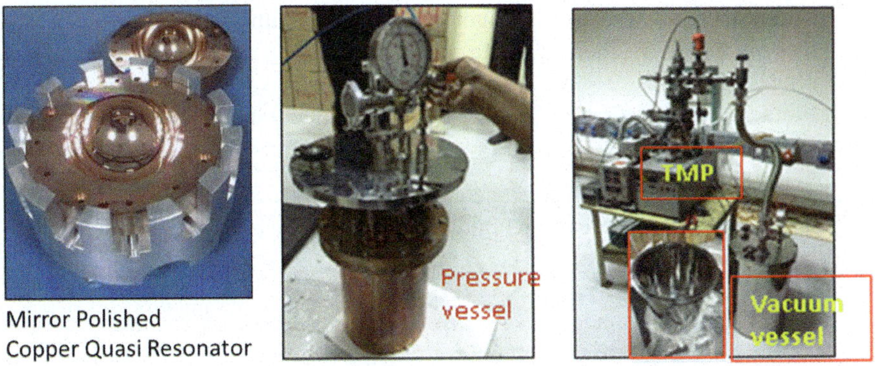

Fig. 8.2 Design and fabrication of Copper cavity resonator, pressure vessel and vacuum vessel

8.1.1.2 Practical Temperature Scales

As an alternative to the primary thermometry, international temperature scales (ITS) have been formulated for both realization and dissemination of temperature. Practical temperature scales such as Fahrenheit, Celsius and Reaumur are the foundations of the development of new unified practical temperature scale based on fixed points. At present, temperature is defined in terms of Boltzmann constant and major research has been going on to develop primary thermometry methods for direct realization of thermodynamic temperature. However, practical scale is required as well because primary thermometers are still difficult to operate for the day to day temperature calibration and traceability. The advantage of International Temperature Scales and the reasons for their development are [9–11].

- Easy for realization and maintenance.
- Give temperature close approximation to thermodynamic temperature
- Reproducibility is better than thermodynamic temperature scale realization.
- The definition range can be improved according to the technological need of the time.
- Unified and accepted worldwide, and hence eases the world trade.

International Temperature Scale (ITS)

To unify the temperature scale worldwide, International Committee on Weights and Measures (CIPM) adopted an international temperature scale (ITS) in 1927 on the basis of generalizations made by the consulting committee on thermometry (CCT). The scale was based on three different parameters:

(i) Reproducible and robust fixed points called the thermal equilibrium states (freezing points, boiling points and melting points of highly pure substances) to which numerical values were assigned.

(ii) Measuring instruments with highest degree of accuracy used as measurement standards (SPRT, Thermocouple, and pyrometer) in the different temperature ranges.
(iii) Mathematical relation is used to relate temperature with the measuring quantity of the instrument.

The platinum resistance thermometer, noble metal thermocouple (Platinum and 10% rhodium-platinum wires) and optical pyrometer were used for the low (−182.97 to 660 °C), medium (630 to 1064 °C) and high (1064 °C and above) temperature measurements, respectively.

Some changes were made in ITS-27 in 1948 in 9th CGPM and is termed as ITS-48. The ITS-48 was further amended as the IPTS-48 in 11th CGPM in 1960. The modification was formulated because in 1953 the TPW was designated as the unique fixed point which defines the unit of thermodynamic temperature. The ice point was replaced by TPW temperature (0.01 °C, 273.16 K) and FP of zinc (419.505 °C) was introduced as the fixed point in place of BP of Sulphur (444.6 °C). In 1968, 13th CGPM proposed the international practical temperature scale of 1968 (IPTS-68). Extensive changes in the IPTS-48 formulated the IPTS-68. It included new fixed points and changes in the assigned temperature of existing IPTS-48 defined fixed points to provide closer agreement with thermodynamic temperature. Further, the scale was amended in 1975 and 1976 provisional 0.5–30 K Temperature Scale (EPT-76).

The current scale of temperature is the International Temperature Scale of 1990 (ITS-90), which was adopted in 18th CGPM. The ITS-90 was made by choosing the set of highly reproducible states including melting, freezing, triple point and vapour pressure states of 17 very high pure substances from 0.65 to 1357 K, and is still open onwards. The Preston-Thomas et al. [12] describes the scale realization in ITS-90. The assigned temperatures to these defined fixed points are close to the thermodynamic temperatures.

ITS-90 provides one, unified and high precision temperature scale for the research and industrial users. The ITS-90 consists of many subranges, overlapping ranges and instruments which provides flexibility to user in scale realization to choose compatible instrument for accurate and reproducible measurement [12]. Fixed point defined in ITS-90 are listed in Table 8.2. The thermodynamic temperature to these phase transition fixed points were assigned by using the primary thermometric methods discussed. Apart from fixed points, the ITS-90 also defines instruments to be used for measurements between the defined fixed points as shown in Fig. 8.3.

NMI of each nation establishes the ITS-90 scale measurement facility. The defined protocols are applied to have the inter-comparisons amongst the NMIs to establish the equivalence of the measurements, and hence the worldwide physics research and industrial processes are made traceable to the unit definition of kelvin. In this way, the products and processes developed through these unbroken chains of calibrations are accepted worldwide.

Table 8.2 ITS-90 defined fixed points

S. No.	Fixed point	State of matter	T_{90}/K	T_{90}/°C
1	He	Vapour pressure point (VP)	3–5	−270 to −268.15
2	e–H_2	Triple point	13.8033	−259.3467
3	e–H_2 (or He)	Vapour pressure point	17	−256.15
4	e–H_2 (or He)	Vapour pressure point	20.3	−252.85
5	Ne	Triple point (TP)	24.5561	−248.5939
6	O_2	Triple point	54.3584	−218.7916
7	Ar	Triple point	83.8058	−189.3442
8	Hg	Triple point	234.3156	−38.8344
9	H_2O	Triple point	273.16	0.01
10	Ga	Melting point (MP)	302.9146	29.7646
11	In	Freezing point (FP)	429.7485	156.5985
12	Sn	Freezing point	505.078	231.928
13	Zn	Freezing point	692.677	419.527
14	Al	Freezing point	933.473	660.323
15	Ag	Freezing point	1234.93	961.78
16	Au	Freezing point	1337.33	1064.18
17	Cu	Freezing point	1357.77	1084.62

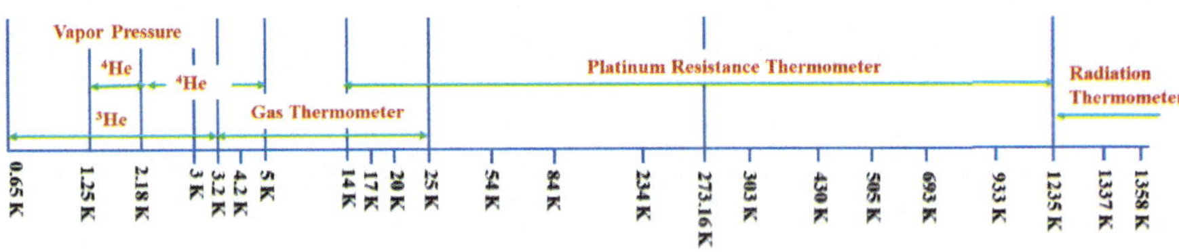

Fig. 8.3 ITS-90 defined ranges, subranges and interpolation instruments

ITS-90 Defined Measuring Instruments

Platinum Resistance Thermometry

In ITS-90, the Platinum resistance thermometer is defined primary interpolation device in the range from 13.8 to 1234.93 K. The resistance of platinum wire is used to define the temperature of a material. Qualities, such as, strong resistance to oxidation, chemically inertness in wide temperature range, high melting point, mouldable as fine wire in a high pure state make platinum appropriate for the resistance thermometry, which allows a unique resistance-temperature characteristics of platinum.

Temperature scales ITS-27, ITS-48 and ITS-68 were based on the relation between resistance and temperature of platinum resistance thermometer only. In ITS-90, the defining parameter is the normalized resistance obtained by taking ratios to the resistance at the TPW (273.16 K) and is given by:

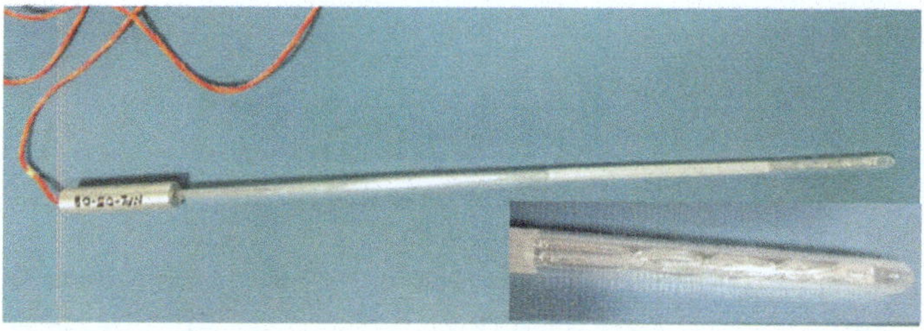

Fig. 8.4 Photograph of the long-stem SPRT fabricated at CSIR-NPL. Inset shows the expanded view of sensor part at the bottom

$$W(T_{90}) = \frac{R(T_{90})}{R(\text{TPW})} \tag{8.5}$$

where, $R(T_{90})$ and $R(\text{TPW})$ are resistances measured at T_{90} and TPW temperature, respectively. The measurement of resistance ratio $[W(T_{90})]$ eliminates the requirement for traceability to absolute resistance standards in the calibration of SPRT, and normalizes the drift effect induced due to thermal stress, mechanical shock, strain and contamination of platinum wire. The ITS-90 reference function and interpolation equation for platinum resistance thermometer is presented in the ITS-90 document. The criteria for SPRTs to be used as standard is given in ITS-90 and is based on the $W(T_{90})$ values at Ga (29.7645 °C) and Hg (−38.8344 °C) fixed Points [13, 14].

$$\text{W(Ga)} \geq 1.11807$$

$$\text{W(Hg)} \leq 0.844235$$

Three types of SPRTs are used for the temperature measurement based on the temperature range. Capsule SPRT (cSPRT) in temperature range from 13.8 K to 30 °C, which can be extended up to as high as 156 or 232 °C. Most commonly used long stem SPRTs are used between temperature range from T.P. of Argon (−189.3442 °C) to F.P. of Aluminum (660.323 °C) and the high-temperature SPRT (HTSPRT) designed to use temperature range from F.P. of Aluminum to F.P. of silver (961.78 °C). A typical schematic of SPRT is shown in Fig. 8.4.

Radiation Thermometry

Radiation thermometry provides a way to measure the temperature of a body in a non-contact manner. It is based on the Planck's law, according to which the spectral radiance of a blackbody cavity is the radiative power emitted per unit area per unit solid angle per unit wavelength, i.e.

$$L_{b,\lambda}(\lambda, T) = \frac{2hc^2}{n^2\lambda^5}\frac{1}{\exp\left(\frac{hc}{n\lambda kT}\right) - 1} \tag{8.6}$$

where the spectral radiation $L_{b,\lambda}(\lambda, T)$ is independent of the material and shape of the cavity wall, and function of the wavelength λ and temperature T. h stands for the Planck's constant, c is the speed of light, n for the refractive index of gas inside cavity and k is the Boltzmann constant. The unit of the spectral radiance is Wm^{-2} Sr^{-1} nm^{-1}.

In simplified form:

$$L_{b,\lambda}(\lambda, T) = \frac{c_1}{n^2\lambda^5} \frac{1}{\exp\left(\frac{c_2}{n\lambda T}\right) - 1} \tag{8.7}$$

where c_1 and c_2 are first and second radiation constants, respectively.

$$c_1 = 2hc^2 = 3.7417749 \text{ Wm}^2$$

$$c_2 = hc/k = 0.014388 \text{ Km}$$

The maximum spectral radiation wavelength shifted to shorter side as temperature increases [15].

Above silver fixed point (961.78 °C) the ITS-90 is defined by the radiation thermometry. In high temperature range, the T_{90} temperature is calculated by a spectral radiance ratio method relative to one of the ITS-90 defined primary blackbody fixed points (silver or gold or copper). Above copper fixed point (1084.62 °C) the uncertainty and temperature are calculated by using the extrapolation from these three fixed points. The unknown temperature ($T_{90,\,\mathrm{X}}$) can be determined by the spectral radiance ratio:

$$\frac{L_\lambda(\lambda, T_{90,\mathrm{ref}})}{L_\lambda(\lambda, T_{90,X})} = \frac{\exp\left(\frac{c_2}{\lambda T_{90,X}}\right) - 1}{\exp\left(\frac{c_2}{\lambda T_{90,\mathrm{ref}}}\right) - 1} \tag{8.8}$$

where, c_2 is the second radiation constant and defined as 0.014388 Km. $T_{90,\,\mathrm{ref}}$ stands for fixed-point temperature [Silver (961.78 °C), or Gold (1064.18 °C) or Copper (1084.62 °C)].

The spectral radiance ratio given in Eq. (8.8) by Planck's law is the idealized expression for the single wavelength. However, in practice, the radiation thermometer operates at monochromatic wavelength and has a finite bandwidth. Therefore, the expression for the practical implementation of ITS-90 is given by:

$$r = \frac{S(T_{90}(X))}{S(T_{90}(\mathrm{ref}))} = \frac{\int_0^\infty \varepsilon(\lambda, T_{90}(X)).L_\lambda(T_{90}(X)).R(\lambda)d\lambda}{\int_0^\infty \varepsilon(\lambda, T_{90}(ref)).L_\lambda(T_{90}(\mathrm{ref})).R(\lambda)d\lambda} \tag{8.9}$$

where r is the experimentally measured ratio of the detector signal, $S(T_{90}(X))$ and $S(T_{90}(\mathrm{ref}))$ are detector signals (voltage or current) at unknown blackbody radiator

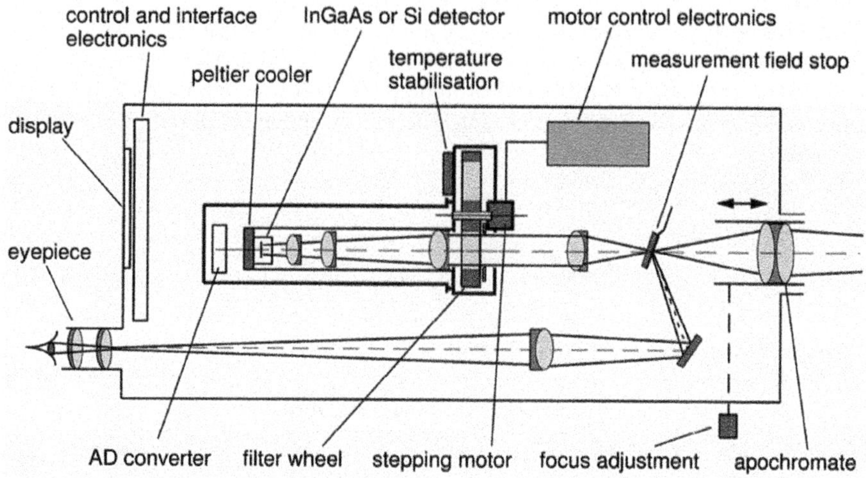

Fig. 8.5 Schematic of spectral linear pyrometer (LP4, KE Technologie GmbH) at CSIR-NPL

temperature and ITS-90 Fixed point (Ag, Au or Cu). $\varepsilon(\lambda, T_{90}(X))$ is the spectral emissivity of the unknown blackbody radiator at the temperature $T_{90}(X)$. $\varepsilon(\lambda, T_{90}(\text{Ref}))$ is the effective spectral emissivity of the ITS-90 fixed-point blackbody. $L_{\lambda}(T_{90}(X))$ is the spectral radiance concentration at a temperature $T_{90}(X)$, $R(\lambda)$ is the spectral responsivity of the radiation thermometer and λ is the wavelength of the radiation thermometer.

At CSIR-NPL, we are using spectral linear pyrometer (LP4) for the measurement and dissemination of T_{90} temperature and the schematic of LP4 is shown in Fig. 8.5. The measurement scheme to assign T_{90} to blackbody radiator using ITS-90 defined fixed point (Ag, Au or Cu) is shown in Fig. 8.6.

Thermocouple Thermometry: Secondary Interpolation Device

Thermocouples are based on the Seebeck effect, where two dissimilar metal or metal-alloy wires are joined at two junctions, one junction being at unknown temperature (T1) and another being at fixed reference such as ice point (T2). Due to the temperature gradient, there is flow of electrons from hotter junction to cooler junction. These thermally migrated electrons generates potential difference (emf or voltage) across the wires. A circuit for the thermocouple is shown in Fig. 8.7. The generated Seebeck voltage (dE_s) can be given by equation:

$$dE_s = s(T)dT \tag{8.10}$$

where, $s(T)$ is the Seebeck coefficient and dT is the temperature difference between two junctions.

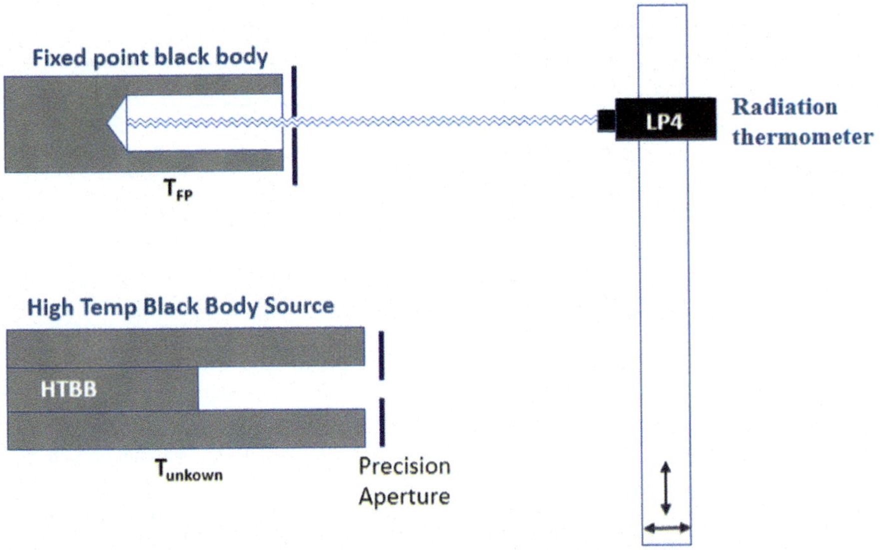

Fig. 8.6 Determination of unknown T_{90} temperature of blackbody source by a spectral radiance ratio measurement with respect to reference fixed point (Ag or Cu)

Fig. 8.7 Schematic of circuit for the thermocouple based on Seebeck effect

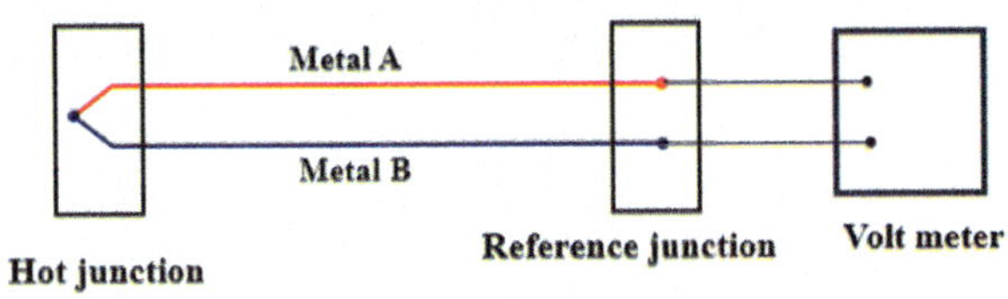

Thermocouples based on noble metal wires or its alloys such as Pt/Au, Pt/Pd, Pt/(Pt + 10%Rh)—Type-S, Pt/(Pt + 13%Rh)—Type-R, etc. are noble metal thermocouples and are used as reference thermocouples. Thermocouples made up of base metals are widely used in industries. These thermocouples are fabricated by many manufacturers with the standard reference tables [IEC (1977) (1982), CMEA (1978), GOST (1977), ASTM (1987b)]. The relative Seebeck effect was approved internationally for a number of different wire pairs. The type of standard thermocouple and their wire material are listed in Table 8.3.

Thermocouples are traceable to ITS-90 defined primary and secondary fixed points for different temperature ranges and subranges. Type-S and Type-R thermocouples are commonly used in laboratory to provide traceability to industries and other calibration labs. This provides a close approximation to T_{90} temperatures. Figure 8.8 shows the measurement scheme to realize fixed-point using thermocouple.

Thermocouples are calibrated by fixed-points primary and secondary fixed points as shown in Fig. 8.9. Ice point is used as the main reference point instead of TPW because thermocouples are not that accurate as platinum resistance thermometers and ice points provide accuracy about 1 mK which is sufficient for thermocouple

Table 8.3 Standard noble and base metal thermocouples

S. No.	Type	Materials	Temperature range (°C)	Operating atmosphere
1	Type-B	Platinum 30% rhodium/platinum 6% rhodium	300–1800 °C	Oxidizing, inert, vacuum for short periods
2	Type-E	Nickel–chromium alloy/copper-nickel alloy	−200 to 870 °C	Oxidizing, inert
3	Type-J	Iron/copper-nickel alloy	−40 to 760 °C	Oxidizing, inert, reducing in partial vacuum
4	Type-K	Nickel–chromium alloy/nickel-aluminum alloy	−200 to 1200 °C	Oxidizing, inert
5	Type-N	Nickel–chromium-silicon alloy/nickel-silicon alloy	− 200 to 1200 °C	Oxidizing, inert
6	Type-R	Platinum 13% rhodium/platinum	−50 to 1600 °C	Oxidizing, inert
7	Type-S	Platinum 10% rhodium/platinum	−50 to 1600 °C	Oxidizing, inert
8	Type-T	Copper/a copper nickel alloy	−200 to 350 °C	Oxidizing, inert, reducing in partial vacuum

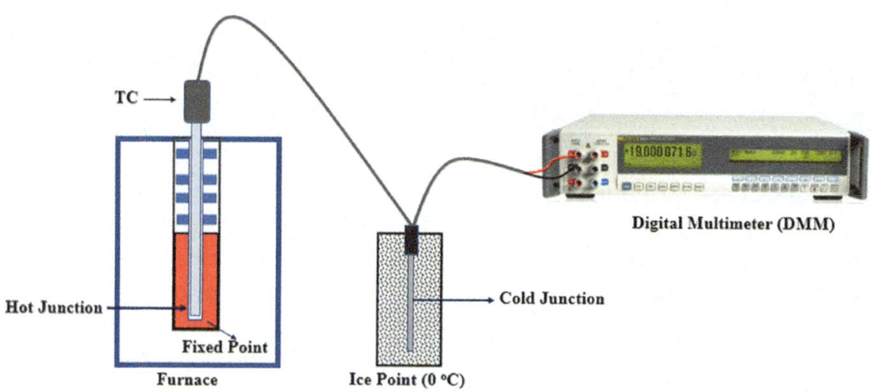

Fig. 8.8 Measurement scheme to calibrate thermocouple on ITS-90 defined fixed points

thermometer. Afterwards, thermocouples are calibrated on the ITS-90 defined fixed points up to Cu [1084.62 °C]. Above Cu point, Pd [1555 °C] and Pt [1768 °C] secondary fixed points are realized either by wire bridge method or miniature cells. We have developed Fe–C, Co–C and Ni–C eutectic fixed points for thermocouple calibrations above copper fixed point and improved our realization uncertainties.

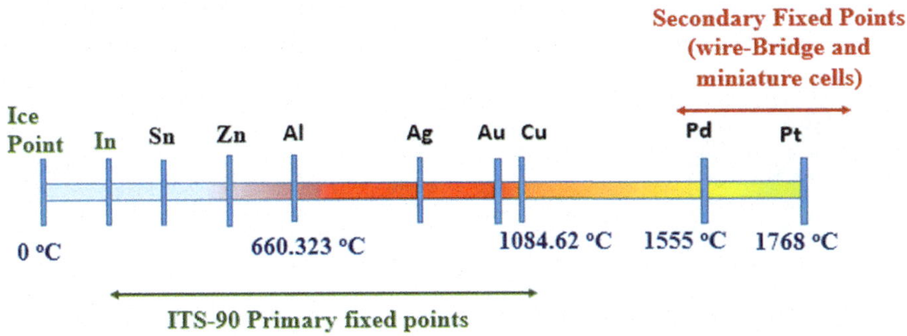

Fig. 8.9 Defined primary and secondary fixed points for thermocouple calibrations

Temperature Measurement by Liquid-In-Glass Thermometer (LIGT)

Liquid–in–glass thermometers (LIGT) are based on the linear expansion of material in a given temperature range. A typical glass thermometer is broadly composed of: a bulb (thin-walled cylinder that contains thermometric liquid like mercury or alcohol) and a stem (containing fine capillary where liquid expands or contracts). It is not possible to heat the liquid without heating the glass and the gas and these also respond to temperature changes. Mercury is the most widely used material in these devices over a wide range of temperature between its freezing point (−38.8 °C) and boiling point (357 °C). The range can be increased from −58 °C (by adding traces of thallium in the mercury) to as high as 500 °C by filling nitrogen gas in the upper portion of mercury column in the stem under pressure, thereby elevating the melting temperature. The main advantages of mercury-filled glass thermometer are cheap, handy to use and need no electric power to operate. However, because of its toxic nature WHO and Minamata convension on mercury has decided to phase-out the mercury use from the clinical applications by year 2020.

8.1.1.3 Development of Metal–Carbon Eutectic Fixed Points

CSIR-NPL has the facility for radiation thermometry up to temperature 3000 °C. The uncertainty of measurement at higher temperature is given by the extrapolation of Planck's Law from the Cu (1084 °C) to this temperature range. The extrapolation uncertainty largely contributes to the total uncertainty of pyrometers and thermal imager calibration, as high as ± 5 °C. To minimize these high uncertainty levels and to give direct traceability at high temperatures, International Thermometry community is researching on development of various metal–carbon eutectic fixed-point blackbodies from 1000 to 3000 °C. CSIR-NPL has indigenously developed the Fe–C [1153 °C] and Co–C [1324 °C] eutectic blackbody fixed points for the radiation thermometry applications, as well as for the thermocouple thermometry applications [16, 17]. The development process comprises design and fabrication of graphite crucible,

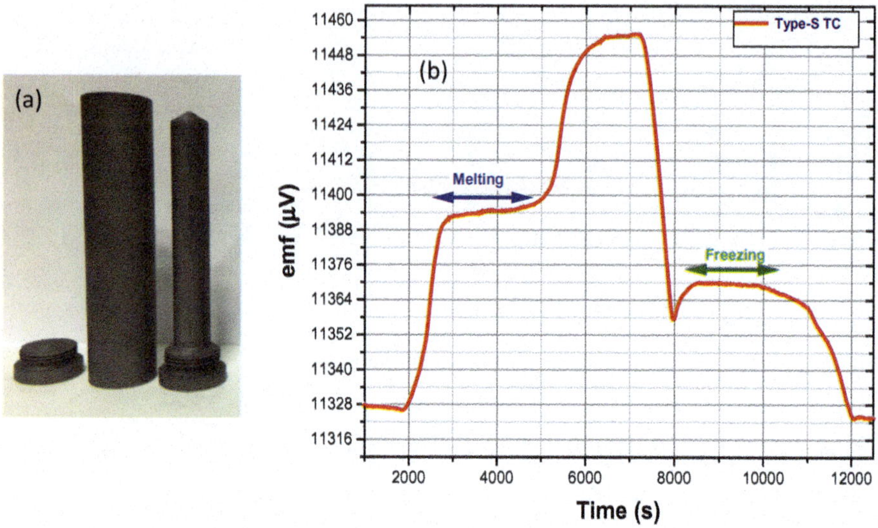

Fig. 8.10 **a** Blackbody designed at Temperature and Humidity Metrology, CSIR-NPL **b** typical melting and freezing cycle of Fe–C eutectic fixed-point measured by using Type S TC

optimization of eutectic composition of metal and carbon powder, assembly preparation, etc. The realization of blackbody fixed-point involves measurement of the melting plateau by using 650 nm spectral linear pyrometer (LP4). The LP4 gives the output in terms of photocurrent of the detector. The T_{90} temperature of the Fe–C cell was measured to be 1153.87 °C by LP4 as shown in Fig. 8.10. The repeatability and uncertainty parameters in realizing the cells are evaluated [16, 17]. CSIR-NPL is one of the few leading NMIs to achieve this metal–carbon eutectic technology.

8.1.1.4 Humidity Metrology

After having discussions about temperature metrology, we discuss about humidity metrology. Humidity plays an important role in our daily life. Relative humidity is defined as the amount of water vapour present in air expressed as a percentage of the amount needed for saturation at the same temperature. Humidity sensors are the devices that convert the amount of water vapour in some measurable parameters. Humidity sensors have wide range of applications in the industrial fields like pharmaceuticals, medical, chemical, agriculture, automobile, Grid transformers, etc. and their calibration is of utmost importance for the economy and the healthcare sector. There are various ways to generate humidity in a controlled manner. Some of these are:

- Saturated salt solution method
- Two-pressure, One-Temperature method

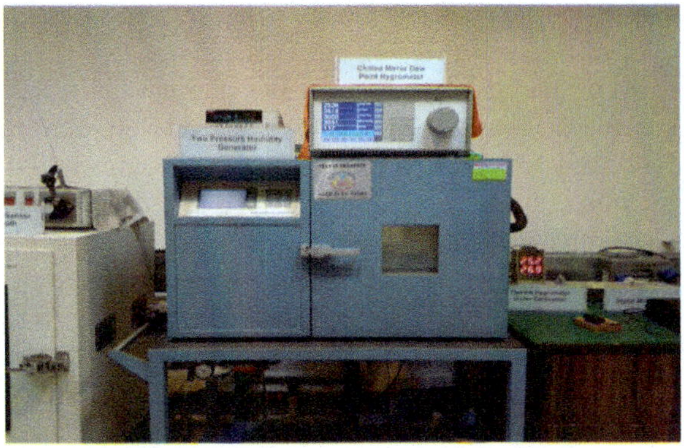

Fig. 8.11 Two pressure humidity generator (Thunder Scientific 2500) along with chilled mirror hygrometer (RH 370) established at CSIR-NPL

- 2 Pressure, 2 Temperature method
- Two-temperature method
- Mixed Flow method

CSIR-NPL has Thunder Scientific-2500 as Humidity generator based on "two-pressure, one temperature" principle. It is made traceable with CSIR NPL primary standard of temperature and pressure. It has a significant stability of better than 0.1% and capable of providing calibration in the range 10 to 95% RH with an uncertainty of 0.3% RH, and temperatures from 0 °C to 70 °C with the uncertainty of 0.02 °C. This system provides high precision humidity values for instrument calibration, and testing. Figure 8.11 shows the humidity generator facility at CSIR-NPL. The chilled mirror dew point hygrometer (RH 370) along with Thunder 2500 is used for the dew/frost measurements from −30 °C to 60 °C. This needs to be extended to −90 °C (Frost Point) to serve the low RH pharma and Transformer users in India.

8.1.1.5 Moisture Measurement

Moisture is a critical parameter for ascertaining the quality, storage and process conditions in agricultural, pharmaceuticals, polymers, food powders, and many others—during processing, storage and use. Moisture content above the permissible limit in the food grain and medicos result in rotting during the storage and has wider economic impacts. Moisture measurement in a bulk sample is a critical metrological task. The moisture content measurement in food grains is controlled under legal metrology in many countries. Moisture meters used under such transactions are under metrological scrutiny for traceable calibrations and verifications through a NMI.

At CSIR-NPL, we have developed two primary methods for the moisture measurements: Integrated Loss-on-Drying (LOD) method and Karl-Fischer Titrator method. CSIR-NPL has established and made extensive measurements of moisture content in food grains using the Loss-on-drying method traceable to primary standards of mass and temperature. The measurement procedure used to determine the grain moisture content refers to ISO 712 recommendations.

LOD method is based on the thermogravimetric principle. In this method, a substance is heated until it loses all its weight corresponding to surface and bulk H_2O, and becomes constant when it is completely dry. Weight of the substance is measured at the beginning and once it is dried. The final weight loss is calculated, and represents as moisture content of the sample. In this context, moisture refers to all the vaporizable matter within a sample, and hence include not only water but volatile solvents, and alcohols also.

The other method, Karl Fischer Titration is based on the Bunsen reaction responsible for water quantification in terms of oxidation of sulfur dioxide with iodine. The reaction could be modified to give water content in a non-aqueous system. Two types of titration methods are available: Coulometric and Volumetric. At Temperature and Humidity laboratory of CSIR-NPL, we have volumetric titration method. The main components of this method are cathode, anode and an analyte. In volumetric method, anode solution itself is used as the titrant solution. The titrant consists of an alcohol (ROH), base (B), SO_2 and a known concentration of I_2.

These moisture measurement facilities are being used for the calibration of various moisture meters made for different commodities such as cotton, rice, wheat, etc. using respective standard samples, and also for the testing of moisture content in several solid and liquid samples.

8.1.2 Significance, Applications and Relevance in Different Sectors

A major of the manufacturing industry depends upon the temperature and humidity measurement. Few important sectors are listed here:

- ***In Steel and metal, alloys Industries*** In steel mills there are many areas where temperature measurements have to be made. Monitoring temperatures during the steel-making process assures the desired mechanical properties as required. Hence accurate temperature measurement is a must for assuring robustness and the mechanical properties of the material. Mishra Dhatu Nigam Limited, few Steel plants, CSIR-CIMFR Dhanbad, etc. are our users in this area.
- ***Aircrafts*** use thermometers and hygrometers to determine if atmospheric icing conditions exist along with their flight path.
- They are also applied in ***meteorological and oceanographic applications*** to weather forecast method, temperature, relative humidity, dew/frost, etc. IMD, NIOT, CPCB, etc. are our users in this field.
- ***Healthcare sector*** Body temperature is a critical measurement parameter in the human health diagnostics. A number of diseases are characterized by a change in body temperature. Even an error of ± 1 °C can lead to wrong medication and treatment plan. Different types of clinical thermometers are being used to measure

febrile temperature. Mercury-in-glass, electrical and IR Forehead/Ear thermometers are widely used thermometers to measure febrile temperature. Each type of thermometer has to be calibrated to get an accurate and reliable temperature value. Similarly, in medical incubator to measure surface temperature, body temperature and relative humidity required to be adequately maintained as per the prevalent conditions of patients.

- ***Pharmaceutical sector*** Maintaining relative humidity level is an essential parameter for the quality and safety of medications from reagent, making to packaging stages. High humidity causes products to absorb the excess moisture in the air. Therefore the need for accurate measurement of RH, Dew/frost are very vital for this field.
- ***Agriculture sector*** Whether the humidity is too high or too low, the loss of quality reduces the selling price of crops and increases production costs, both of which reduce profits. Crop may get spoiled if is stored in an environment above the prescribed limits. Hence temperature and humidity are important parameters to be controlled in agriculture sector.

8.1.3 Facilities and Global Status

8.1.3.1 Measurement Standards and Facilities

The apex level calibration services of Temperature and Humidity Metrology are provided from −200 to 3000 °C for SPRTs, RTDs, Thermocouples and Radiation Pyrometers, Thermal Imagers and Blackbody sources and 10–95% RH, −30 to 50 °C dew/frost measurements, moisture measurements and mercury-free (electrical and IR based) clinical thermometers. We have established the state-of-the-art Primary Standards of Temperature and Humidity, using the ITS-90 defined fixed points, and giving the traceability services to Indian and SAARC users. The detailed description of the various facilities, measurement range and the associated uncertainties are listed in Table 8.4.

8.1.3.2 Calibration and Measurement Capabilities

The CMCs of temperature and humidity parameters published in KCDB are given in Table 8.5.

8.1.3.3 Participation in International Key Comparisons

CSIR-NPL has participated in several international comparisons of temperature. The list is given in Table 8.6.

Table 8.4 Calibration facilities in the present scope of Laboratory

S. No.	Calibration items	Temperature range	Uncertainty ($k = 2$, 95% CL)
1	SPRT/HTSPRT/PRT/RTD by Fixed-point methods (Using ITS-90 fixed-point cells and Resistance Thermometry Bridge)	(i) −196 to 0.01 °C (BP LN2, TP Hg, TPW) (ii) 0.01 to 660 °C (TPW, FP Sn, Zn, Al) (iii) 0 to 962 °C ITS-90 Fixed Points (TPW, FP Sn, Zn, Al, Ag)	±2 ±0.17 m °C ±0.17 ±4.0 m °C ±0.17 ±6.0 m °C
2	SPRT/SSPRT/PRT/RTD/Thermistors by comparison method (Using ITS-90 calibrated SPRT and Liquid Bath/Furnace/Calibrator)	−196 to 0 °C 0 to 660 °C	±40 to ±4 m °C ±4 to ±60 m °C
3	SPRT/SSPRT/PRT/RTD/Thermistors by comparison method (Using ITS-90 calibrated SPRT and Liquid Bath/Furnace/Calibrator) with indicator (Ohm, °C, K)	−196 to 0 °C 0 to 660 °C	±40 to ±4 m°C ±4 to ±60 m °C With additional uncertainty due to Indicator
4	Liquid-in-Glass Thermometers and Digital thermometers (With ITS-90 calibrated SPRT and Liquid Bath)	−90 to 0 °C and 0 to 300 °C	±0.02 to ±0.04 °C or depending upon Least Count
5	Noble metal thermocouples (Type-S, Type-R & Type-B, Pt/Au, Pt/Pd) by fixed-point method	0 to 1200 °C (FP of In to FP of Cu) 0 to 1600 °C ITS-90 Fixed Points FP of In to MP of Pd)	±0.31 to ±0.6 °C and ±0.31 to±0.9 °C
6	Noble metal thermocouple (Type-S, Type-R & Type-B), Base Metal and Tungsten Thermocouples by comparison method, **with and without Indicators** (With ITS-90 calibrated Thermocouples and Three-zone high stability furnaces)	0 to 1200 °C and 0 to 1600 °C	±0.4 to ±1.2 °C and ±0.4 to ±1.6 °C
7	Standard tungsten strip lamp (With standard lamps and LP4 pyrometer)	800 to 1700 °C and 1600 to 2200 °C	±0.9 to ±1.3 °C and ±1.0 to ±2.5 °C

(continued)

Table 8.4 (continued)

S. No.	Calibration items	Temperature range	Uncertainty ($k = 2$, 95% CL)
8	Pyrometer (LP4) and Thermal Imagers Comparison calibration (Using LP4 calibrated on ITS-90 fixed points Ag, Cu, Fe–C and Co–C (±0.45 to ±2.5 °C @3000 °C) and Variable temperature blackbodies of emissivity greater than 0.997)	50 to 1300 °C and 600 to 3000 °C	±0.6 to ±1.2 °C and ±1.5 to ±4.0 °C
9	Blackbody Sources	50 to 1600 °C (beyond on case-by-case)	±1 to ±2.5 °C
10	Thermal Sources ((Dry Block Calibrator/Furnace/ Oven/Heating Chamber/Climatic Chamber)	−196 to 1600 °C	Depending on the combination of sensors and measurement units used
11	Relative Humidity Hygrometer	10 to 95% at temperature from −10 to 60 °C (Humidity only)	±0.3% RH
12	Air Temperature Sensor/Hygrometer and Surface temperature sensor	−10 to 60 °C at humidity @ fixed RH (Temperature only)	±0.02 °C
13	Dew Point/Frost Point Temperature	−30 to 60 °C	±0.07 °C
14	Moisture Meters (Integrated loss-on-drying method and Karl Fischer Titration method)	As per the samples standard or sample basis (solids and liquids)	0.3% MC
15	Moisture sample testing	As per the customer requirement	0.3% MC
16	Clinical/Medical Thermometer (Contact Type) (Source uncertainty = 0.03 °C, stability 0.01 °C) Calibration and Testing for Model Approval	35– 43 °C, (95–109.4 °F)	Laboratory accuracy ± 0.1 °C/LIGT; ± 0.2 °C/Electrical or As per the Standard Used

(continued)

Table 8.4 (continued)

S. No.	Calibration items	Temperature range	Uncertainty ($k = 2$, 95% CL)
17	IR Ear/Fore Head (Non-contact type) (Source uncertainty = 0.06 °C, Stability 0.01 °C, source designed and developed as per International Standards) Calibration and Testing for Model Approval	35–43 °C, (95 to 109.4 °F)	Laboratory accuracy ±0.3 °C, as per the standard used
18	ITS-90 fixed-point cells- TPW, Other fixed-point cells for SPRT calibration-fixed-point cell with the user realization apparatus will be taken on case by case basis	−196 to 962 °C Simultaneous realization of NPL and DUC cell using NPL SPRT and Resistance Thermometry Bridge	±4 to ±0.3 m°C (TPW) to ± 80 m°C, depending upon the construction, impurity levels and plateau realizations

Table 8.5 CMCs of temperature and humidity parameters

Parameter	Instrument	Instrument type/Method	Range (°C)	Value
Temperature	Stem SPRT	Ga melting point cell	29.7646	1.36 mK
Temperature	LIGT, 0.05 °C	Comparison at water bath	5–95	0.02 °C
Temperature	LIGT, 0.05 °C	Comparison at oil bath	95–200	0.03 °C
Temperature	IPRTS	Comparison at oil bath & TPW cell	0.01–100	4–15 mK

8.1.3.4 Potential Services for Future Requirements

In ITS-90, the TPW cell is defined with the Vienna Standard Mean Ocean Water (VSMOW) with specific content of isotopes of Hydrogen and Oxygen. However, it is practically very difficult to maintain the purity level and isotope content in the water. Therefore, TPW becomes the material-dependent property or an artefact. The kelvin, earlier defined by the TPW, has been redefined in May 2019, by assigning an exact numerical value to the Boltzmann constant, as all the thermal energies are expressed as kT in the laws of physics. If k is fixed, the 'T-thermodynamic temperature" can be measured by computing the energy. This redefinition will ensure long-term stability and traceability of the unit for temperature by making it independent of any material substance.

CSIR-NPL is working in the direction of establishing facility to determine Boltzmann constant (k) and measurement of thermodynamic temperature by Acoustic Gas Thermometry. The combined acoustic-microwave copper spherical resonator has been fabricated to measure the velocity of sound to estimate the value of Boltzmann constant. The Boltzmann constant is the conversion factor between mechanical energy and temperature.

We have set up a calibration facility for IR clinical thermometers in India for the first time. This is a vital step for moving from the era of mercury thermometer to an era of mercury free thermometer to phase-out the mercury in healthcare applications. This standardization of mercury-free clinical thermometers is significant development in view of WHO and Minamata convention. Further, this is a major step in the present pandemic situation of COVID-19 where IR thermometers are playing a key role in screening the patients at the first level. These initiatives will help improve the quality of life and Society as a whole.

8.1.3.5 A Decade of Clients and Services

Calibration and Testing

CSIR-NPL provides the calibration service to several strategic sectors like DRDO, DAE-NPCIL (through other clients), ISRO (Bangalore, Ahmadabad, and Trivandrum), Air Force, Railway, BHEL, Oil–Gas and Petrochemicals,

Table 8.6 List of international comparisons participated by CSIR-NPL in temperature and humidity metrology

S. No.	Parameter	Range	Year	Name of Inter-comparison	Identification	Organized by
1	Temperature (Tungsten Strip Lamp No. 229A) Bilateral Comparison	1000–2000 °C	July–Aug. 1990	"Temperature unit standard inter-comparisons" of Bilateral Cooperation plan in the field of Temperature Metrology	Stage 3, Theme5.17	Bilateral Cooperation plan in Temperature Metrology between NPL, India and VNIIM Russia (USSR)
2	Temperature (Hg-in-Glass Thermometer) Among 7-Laboratories participated	0–50 °C	Dec. 2000	SADC-MET Regional Inter-laboratory Comparison. (SADC.T-P1)	SADC.T-P1	CSIR-NML, South Africa
3	Temperature (Type-R Thermocouple) 12-Laboratories participated	0–1100 °C	Sept–Dec. 2005	APMP Regional Comparison of Type-R Thermocouple	APMP-03	APMP Coordinated by NMIA, Australia
4	Pt/Pd Thermocouple on Co–C Eutectic fixed point	1324 °C	May 2016	APMP-T-S7 Supplementary Comparison, 7 NMIs	APMP-T-S7	APMP, Pilot-KRISS
5	Type-R Thermocouple on Cu, Co–C and Pd Fixed points	Up to 1554 °C	Nov 2017–2019 (Ongoing)	APMP-T-S16 Supplementary Comparison, 9 NMIs	APMP-T-S16	APMP, Pilot-NMIA
6	SPRT on Triple point of Mercury, Water, Gallium, Tin, Zinc (-38 oC to 420 oC)	−38.8344 to 419.527 °C	2000	Report available in App. C of KCDB of BIPM	APMP KC-3	Bilateral between VMI, Vietnam and NMI, South Africa

(continued)

Table 8.6 (continued)

S. No.	Parameter	Range	Year	Name of Inter-comparison	Identification	Organized by
7	SPRT on freezing point of Aluminum	660.323 °C	2004	Report available in App. C of KCDB of BIPM	APMP TK-4	APMP, Pilot-KRISS
8	SPRT on Freezing point of Silver	961.78 °C	2004	Report available in App. C of KCDB of BIPM	APMP TK-4	APMP, Pilot-KRISS
9	SPRT on Freezing Point of Zn to Triple Point of Ar	−189.344 °C −38.8344 °C 29.764 °C 156.598 °C 231.528 °C 419.527 °C	2018–2019	Ongoing	APMP TK-9	APMP, Pilot- NIM, China

STQC/ERTLs/ETDCs, IMD departments, Dairy, Ordinance Factory, private firms, intra and inter departments in CSIR-NPL and many more.

Consultancy and Technical Services

CSIR-NPL provides consultancy and technical service depending on the client's requirement. Several technical and consultancy service projects have been completed/undergoing. Out of them, one recent is sponsored by Legal Metrology Department to design, develop, fabricate and establish the testing and calibration facility for clinical thermometers (liquid-in-glass and electrical) with maximum device at 2 Regional Reference Standards Laboratories at Guwahati and Ranchi centers. This project will be completed by the mid 2021. Recently, a new technical service project is taken up by CSIR-NPL from DRDO-SASE for vetting of the items related to temperature, humidity and pyrometry services as part of this activity.

Training and Academic Services

Training program is conducted annually at CSIR-NPL to the officers of the different areas. We have given training to Legal Metrology officers, Air India officers, Indian Oil, ISRO, ARAI Pune at times which benefits them to understand various procedures adopted in the temperature and humidity metrology and help for accurate measurements. Various proficiency testing programs were organized by the Group in LIGTs and TCs in collaboration with NABL to disseminate the traceability of the laboratory to the NABL accredited laboratories and user industry as a part of metrology service to the nation.

Technology Development and Commercialization

Liquid temperature bath has been developed from −40 to 250 °C at CSIR-NPL and is transferred to M/s KV Scientific, Kirti Nagar, New Delhi.

8.1.3.6 Major Documentary Standards, Guides and SOPs

- Guide to the Realization of the ITS-90
- Fischer J, Battuello M, Sadli M, Ballico M, Park S N, Saunders P, Zundong Y, Johnson B C, van der Ham E, Li W, Sakuma F, Machin G, Fox N, Ugur S, Matveyev M (CCT-WG5 on radiation thermometry) (2003b) "Uncertainty budgets for realisation of scales by radiation thermometry" Working Document of BIPM Consultative Committee for Thermometry, 22nd Meeting, Document CCT/03–03.

8.1.4 Possible Gaps and Suggestions for Their Redressal

Realization of the Boltzmann constant based new kelvin at CSIR-NPL is essential to complete the traceability in temperature metrology at par [18]. We are also working to realize the new kelvin in radiation thermometry range by developing the new metal–carbon blackbodies, where ample opportunity is there to have in-house and improved traceability beneficial for thermometry and radiometry parameters. Besides this we lack behind the CMCs published in KCDB at present which has to be improved. We have to participate in international inter-comparisons and develop our facilities. There is yet no representation of India in TCT or APMP in Thermo-physical properties, such as thermal conductivity, thermal diffusivity, emissivity, etc. which need to be addressed, particularly for the new novel materials with the advancements in nano-sciences and nano-technology. This is an important area as most of the industries depend on this parameter for the quality check of their base materials. In radiation thermometry, emissivity measurement is a significant parameter and the accuracy of the temperature measurement hugely depend on this, however, we still need to develop facility for the emissivity measurement.

The group is also engaged in research and developmental activities for the creation of facility to realize Boltzmann constant based the new kelvin by developing the acoustic gas thermometry from about −80 to 100 °C, which can be enhanced later to about 1000 °C. Simultaneously, we are developing metal–carbon eutectic fixed points to realize the thermodynamic temperatures upto 3000 °C. With these two approaches, the CSIR-NPL possible traceability chart for the new kelvin is given in Fig. 8.12.

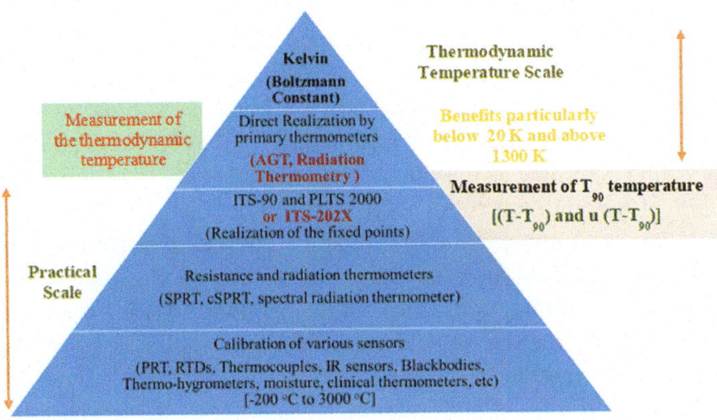

Fig. 8.12 Traceability chart

8.2 Optical Radiation Metrology

Radiation Metrology essentially covers metrology of all the electromagnetic waves or radiation fields which is studied under the branch of science called, 'Radiometry'. Therefore, radiometry is generally related to physical measurement of energy content and other properties of the electromagnetic radiation fields, including light or optical radiation, which is nothing but a very small visible spectral window of the electromagnetic spectrum. Radiations, on the other hand, obtained from radio nuclides, cosmogenesis and cosmic rays or artificially produced radiations, such as X-rays, gamma rays, etc. falls under 'Ionizing radiation', which have the tendency to ionize the medium present.

Photometry describes radiation effect of visible spectrum, perceived by the human eyes. It is science of measurement of physiological effects of electromagnetic radiation falling in visible spectral range, on the human eyes and can be defined in terms of photometric parameters, similar to those of radiometric parameters. The two fields of metrology namely, photometry and radiometry, and their parameters are closely linked and related through the current definition of the SI base unit, candela for luminous intensity [18]. The candela is the only SI base unit among the remaining six base units, related to a fundamental biological process 'human vision'.

The radiometric units (Q_e) can be converted to the photometric units (Q_v) using the following equation [19, 20]:

$$Q_v = K_m \int_{380nm}^{780nm} Q_e(\lambda).V(\lambda)d\lambda \tag{8.11}$$

where $V(\lambda)$ is termed as the spectral luminous efficiency function, which describes the relative spectral responsivity of the average human eye for photopic vision (light adopted) and scotopic vision (dark adopted). K_m is a constant giving the number of photometric units corresponding to 1 radiometric unit at 555 nm, peak wavelength of $V(\lambda)$ curve and is termed as maximum luminous efficacy function and its value is 683 lm/W [19, 20]. Radiometric and photometric parameters and their measurements are extremely vital for number of industries and numerous applications, ranging from lighting, automotive and colour industries, aviation, space, signage, displays and imaging, optical communication, semiconductor, photovoltaic to climate change. Other emerging fields include appearance, terahertz applications, photonics, and quantum-based information.

Traditionally, metrology in both radiometry and photometry is based on primary radiation sources (blackbody radiators, electron synchrotrons) and on primary radiation detectors (cryogenic radiometers). Recent advancements in the controlled manipulation of radiant flux at the level of individual photons show great promise of production of radiant flux of precisely counted number of photons in very near future. The emerging trend towards solid-state lightings also requires the development of new calibration methodologies adapted to the properties of these light sources.

In the present day scenario of global warming and alarming pollution levels, including light pollution, the focus has shifted towards energy-efficient lighting. As the lighting technology evolves, the standardization techniques are also upgraded for meeting the industry requirements. Current Covid-19 pandemic situation has developed researchers' interest towards the effective and contactless sanitization based on ultraviolet germicidal effect of radiation.

8.2.1 Introduction and Historical Aspects

As the mandate of various NMIs is to establish, maintain and upgrade national standards, the unit ***candela*** is also established, realized and is upgraded accordingly. There are two well-known ways to realize various radiometric and photometric units, (i) source-based radiometry and (ii) detector based radiometry [21]. In the source-based radiometry, sources radiating well-defined radiations are used to realize various photometric and radiometric quantities. Realization of the units requires a primary standard, a physical entity which obeys the definition and the value of the measurand is obtained with minimum uncertainty.

Historical developments of luminous intensity scale are depicted in Fig. 8.13. Initially, before 1948, the photometric standards were light sources such as specified candle, oil lamps and group of carbon filament lamps. From 1948 to 1979, the radiation from a blackbody at the temperature of freezing platinum was used to define candela [21]. Realization of photometric and radiometric units using radiation sources especially blackbody falls in the purview of source-based radiometry and photometry. Blackbody, a well-known source of radiation, which follows the Planck Radiation Law, is also called the primary standard of optical radiation. Source-based radiometry directly follows the Planck Radiation Law [21, 22]. However, various factors namely, its high cost, small lifetime limit its scope in optical metrology. In addition, inter-comparison of an artefact in the form of an incandescent lamp calibrated against blackbody gives higher uncertainty in the measurements and also

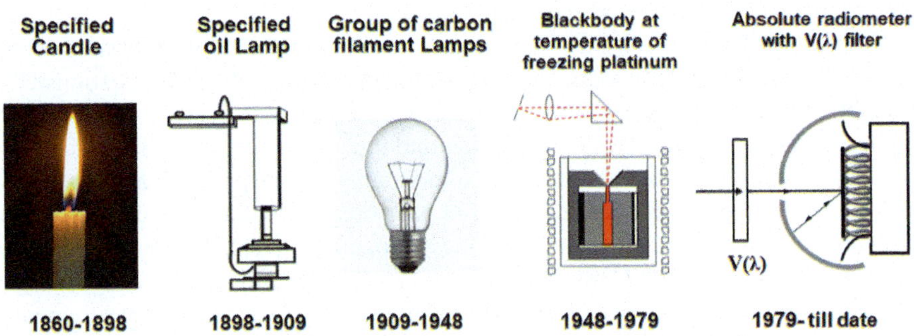

Fig. 8.13 Historical developments of luminous intensity standards

in the transfer of radiometric scales. In 1979, the 16th CGPM (1979) adopted a new definition of the candela following the difficulties in realizing a Planck radiator at high temperatures, and the new possibilities offered by radiometry, i.e. the measurement of optical radiation power, in which candela was expressed in terms of derived unit "Watt" [18–20].The present definition of the candela uses a fixed numerical value for the luminous efficacy of monochromatic radiation of frequency 540×10^{12} Hz, K_{cd}, adopted in Resolution 1 of the 26th CGPM (2018).

Candela (cd), the SI base unit of luminous intensity, in a given direction, of a source that emits monochromatic radiation of frequency 540×10^{12} Hz and that has a radiant intensity in that direction of 1/683 W per steradian. It follows that the spectral luminous efficacy for monochromatic radiation of frequency of 540×10^{12} Hz is exactly 683 lumens per watt, $K_m = 683$ lm/W $= 683$ cd.sr/W. This definition is given in terms of monochromatic radiation (540×10^{12} Hz) rather than the broadband radiation emitted by a blackbody. This new definition provided a basis for establishing, on firm forting, a detector based radiometry [21, 23–28]. A detector, which is a self-calibrating device, can be used as primary standard of optical radiation. Therefore, electrically calibrated radiometers, which are also referred to as absolute radiometers, make the basis for detector-based radiometry. The detector based radiometry overcomes the shortcomings that were apparent in source-based radiometry.

8.2.2 *Significance, Applications and Relevance in Different Sectors*

Artificial Lighting is used not only for the purpose of general illumination but for a plethora of applications spanning various sectors including medical, defence, space, aviation and industrial manufacturing. Being used in such areas of strategic applications the standardization and accurate measurement of lighting speaks its importance. Not being different from other sectors, medical sector also depends heavily on lighting applications not only for examination and facilitating complex medical procedures but also for therapeutic applications.

Specialized lighting is used as illuminant on a dentist's chair, examination room, operation theater, and procedure rooms. Apart from these light finds application in the form of laser employed in various surgeries, from phototherapy in infants and adults to antimicrobial treatments. Optical diagnosis and imaging procedures like endoscopy, ophthalmic imaging, optical coherence tomography, etc. depend on application of light. Recently, light has also been used in path-breaking fields of drug-delivery in the form of light-activated nano-medicine, optogenic therapies implantable opto-electronic devices, etc. [29, 30]. All these applications require proper dosage of light and hence a standard and scientifically proven protocol for accurate measurement.

Germicidal illumination has gained importance in the present era of Covid-19 pandemic where for the germicidal action, radiation in the UV region is used [31]. To eliminate inferior quality products from the market and provide standard products to the consumer, quality assessment of the UV illuminants is important. Recently, CSIR-NPL has augmented its spectral irradiance measuring capability for standardizing UV lights for catering the traceability services for germicidal applications.

In the area of defence, when technology plays a vital role in the modern warfare, the use of lighting has its own strategic importance. Right from using light for illumination along strategic areas and perimeter control, it is also used in weapon systems. Laser light is actively used in guided missile systems, satellite communication, range finders, jammers and communicators. Apart from these, Laser light itself is used as a potential weapon. Low light imaging and surveillance also remain an important defense application. Another important application is the night vision equipments involving light detection in the infrared region. In defense systems fiber-optic and laser gyros play an important role as inertial navigation sensors. Being of strategic importance in national defense, these systems require proper standardization to avoid on-field failures [29].

Lighting plays a very crucial role in the aviation sector as the modern-day aviation largely depends on it for critical activities. On an aircraft apart from the interior lighting, the exterior lighting plays crucial roles of position lighting, anti-collision lighting, navigation lights, landing lights and taxi lights. Another important aspect of lighting in the aviation sector is the runway lighting comprising runway edge lights, runway threshold lights, and runway end lights along with supplementary runway lightings. Testing and standardization of these lightings are of utmost importance for passenger and crew safety.

Space technology also depends on optical standardization. Vital components like satellite reflectors, laser gyros, and high-resolution satellite camera optics require apex level calibration. Upcoming space technologies like Laser satellite ranging and Light propelled spacecraft will be compelling for enhancement of the measurement capabilities.

Apart from the strategic applications discussed above, lighting plays an important role in the day-to-day activity of human society. Automobile lighting is one of them and requires proper testing and calibration of the illumination produced. Paint industry requires standardization in a respect other than illumination. Its requirement is limited for measurement of consistency with respect to colour. However, the general lighting industry remains concerned about both colour and different aspects of illumination.

In the past two decades the lighting, its various other applications and deployability with latest technological forms have evolved enormously which has propelled the requirement of upgradation of the photometric measurement protocols. With the growing importance of energy conservation leading to the increased manufacturing and usage of LED lightings, CSIR-NPL is in the process of establishing a facility, with funding from the Bureau of Energy Efficiency (BEE), Ministry of Power and Council of Scientific and Industrial Research (CSIR), to augment its state of the art photometry and radiometry facility for providing testing and calibration services to the entire nation and beyond the borders.

8.2.3 Facilities and Global Status

The quantification of photometric and radiometric parameters is of utmost importance in a wide range of industries including lighting, textiles, automotive and in health and medical industries. Few of the major existing facilities with Optical Radiation Metrology which are being used for catering services and to realize various parameters in the laboratory includes Blackbody, Cryogenic Radiometer, Goniophotometer, Integrating spheres, Fourier Transform IR Spectrometer, Double monochromator based spectral irradiance measurement setup, Spectrophotometers, Correlated photon metrology setup, Luminance measurement setup, Colour measurement setup, Near field Scanning Optical Microscope, various standard lamps, detectors, etc.

The CMCs on certain parameters and respective specific ranges are already published and available with their measurement uncertainties at BIPM-KCDB, also presented in detail in Sect. 8.2.3.2. To have an overall scenario and the global status of photometry and radiometry, selected facilities available across some NMIs are being presented based on their CMCs as published in the BIPM-KCDB.

National Institute of Standards and Technology (NIST) the NMI of USA [32], has a number of CMCs in the field of photometry and radiometry. To select a few, NIST has the capability of measuring Luminance (in the range 0.01–50,000 cd/m^2) using Integrating Sphere based source and has ability to provide Luminance responsivity of a Luminance Meter (in the range of 1–10,000 cd/m^2). Luminous intensity scale is provided for Tungsten lamps (in the range 0.01–10,000 cd) and for LED (in the range 0.01–1000 cd for white, red, green and blue LEDs) with uncertainties varying in different ranges. It has the capability of providing Illuminance Responsivity for Illuminance meters (over the range of 1 mlx–1 klx) using standard illuminant A. It can also provide scale for Illuminance over the range of 0.001–70,000 lx over a colour temperature of 2000–3200 K. NIST, USA has measurement capability of Luminous Flux for Tungsten Lamps (0.001–200,000 lm) and LED Lamps (0.001–10,000 lm for white LED) with uncertainties varying in different ranges. NIST also provides scale for colour coordinates (x, y and u, v) and Correlated Colour Temperature. Spectral Irradiance calibration is provided for Tungsten Lamps (250–2500 nm wavelength) and for Deuterium Lamp (200–400 nm) including UV, visible and IR ranges with uncertainties varying in different wavelength ranges. Spectral Radiance (225–2500 nm) and Spectral Radiant Flux (300–1100 nm) calibration services are also provided for Tungsten lamps. Calibration of detector responsivity is provided for various detectors.

The German NMI, Physikalisch-Technische Bundesanstalt (PTB) [33] also provides significant technical and scientific services in the field of photometry and radiometry. Calibration services for the parameter of Luminous Intensity is provided using a network of lamps and photometers over a range of 0.001–100,000 cd with one of the best uncertainty. It provides scale for Luminous Flux of Tungsten Lamps based on Goniophotometer measurements over a range of 0.001–100,000 lm with different range wise uncertainties. The capability of Illuminance measurement of Tungsten Lamp spans over a range of 0.001–100,000 lx. Services for calibration of averaged

Luminous Intensity for LED are also provided. It maintains the Spectral Irradiance Scale using the Deuterium Lamp and the Tungsten Lamp over respective wavelength ranges of 200–350 nm and 250–2500 nm with respective range wise measurement uncertainties. The spectral Radiance scale is maintained using the Deuterium Lamp (116–350 nm wavelength range) and the Tungsten Lamp (250–2500 nm wavelength range). However, the Spectral Radiant Intensity scale is maintained over the UV region of 116–350 nm through a Deuterium lamp. PTB, Germany also facilitates the calibration of detector responsivity for various detectors.

The National Physical Laboratory (NPL), UK [34] has Goniophotometer and Integrating Sphere based capability of measuring Luminous Flux of Tungsten Lamp over the range 1.0–20,000 lm. The Luminous Intensity measurement capability ranges over 1.0–10,000 cd for Tungsten Lamps having a CCT of 2000–3200 K. NPL, UK has the facility of measuring Illuminance over a range of 0.1–50,000 lx for Tungsten Source. The CMC for Luminance ranges over 1.0–20,000 cd/m^2. NPL, UK also has the measurement capability of Illuminance, Responsivity and Luminance Responsivity of respective meters. In the domain of colour measurements NPL, UK has the measuring capability of colour coordinates (*xy* and *u*'*v*') and Correlated Colour Temperature. The Spectral Irradiance, Spectral Radiance and Spectral Radiant Intensity scales are maintained using the Tungsten Lamp over 250–2500 nm wavelength range. It has the facility of measuring responsivity of various detectors and spectroradiometers.

8.2.3.1 Measurement Standards and Facilities

Primary standard is a physical entity which obey the definition of a measurand along with its unit. The uncertainty in the measurement of the unit is, therefore, minimum. One can make a perfect blackbody at a well-defined temperature and hence have a source of known radiance [22]. Or else one can develop an absolute radiometer which can measure irradiance directly without requiring a calibrating source [24, 32, 33]. Hence, a blackbody and an absolute radiometer are the primary standards for optical radiation measurements.

Blackbody: Source-Based Primary Standard for Optical Radiation

Blackbody is a source of well-defined radiation following the Planck Radiation Law. According to the Planck Radiation Law, the spectral radiance at a given wavelength λ and at a particular absolute temperature T, defined as [21, 30, 31]:

$$L_\lambda(\lambda, T) = \varepsilon_\lambda c_{1L}/(n_\lambda^2 \lambda^5(\exp[c_2/(n_\lambda \lambda T)] - 1))\mathrm{W/cm^3/sr} \qquad (8.12)$$

where ε_λ is the spectral emissivity (1 for ideal blackbody), c_{1L} is the first radiation constant and its value is equal to 1.191043×10^{-12} Wcm^2, c_2 is the second radiation

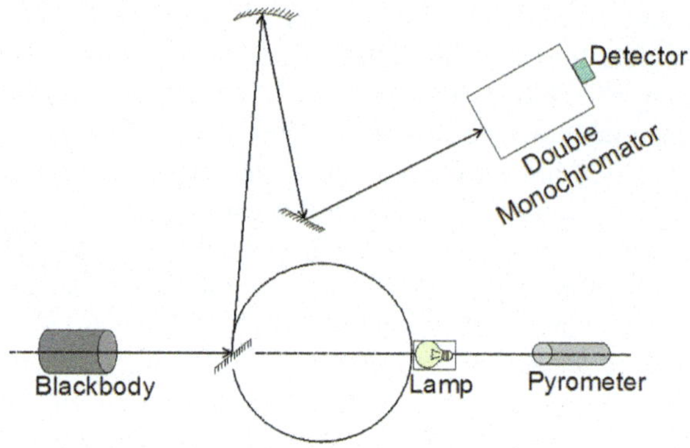

Fig. 8.14 Schematic diagram for spectral radiance measurement with blackbody

constant equal to 1.4388 cm K, n_λ is the refractive index of air at 15 °C and 1 atmospheric pressure.

Figure 8.14 depicts the measurement setup in schematic diagram for spectral radiance measurement setup. The temperature of the blackbody is measured by optical pyrometer to a minimum uncertainty and then it is used as a primary standard source whose spectral radiance is determined from Planck's blackbody equation. Further, the spectral radiance and irradiance scales are transferred to transfer standard sources, which are generally tungsten lamps [21, 22]. Primary standard of optical radiation in the form of a variable temperature (1800–3200 K) blackbody (Emissivity ~ 0.999) has been established at CSIR-NPL India. Blackbody is a thermal source emitting optical Planck's radiation in the ultraviolet, visible and near-infrared range, i.e. prominently in the wavelength range from 250 to 2500 nm. It features a radiator assembled of pyrographite rings with a cylindrical cavity heated by DC flowing directly through the radiator. The radiator with thermal insulation unit is mounted in a water-cooled chamber with an output window opened to the atmosphere and continually blown through with Ar gas. The chamber is also supplied with a system for preliminary air evacuation by means of vacuum pump. It also has a temperature stabilizing system based on optical feedback principle. In addition to 0.1% radiance uniformity, it also exhibits thermal stability in the range of $\pm$ 20 mK.

Initially, the blackbody is stabilized at certain temperature e.g. (2850 $\pm$ 0.02) K in the temperature stabilization mode with the help of optical feedback system and optical pyrometer. After stabilizing the blackbody at a particular temperature, the radiance values of the blackbody and the working standard lamp at various wavelengths in UV, visible and near-infrared region are measured using measurement system consisting of a double monochromator, photomultiplier tube (PMT), Si-photodiode and PbS-Te detectors. By comparing the spectral radiance of blackbody and working standard lamp, the spectral radiance scales are transferred for further calibrations.

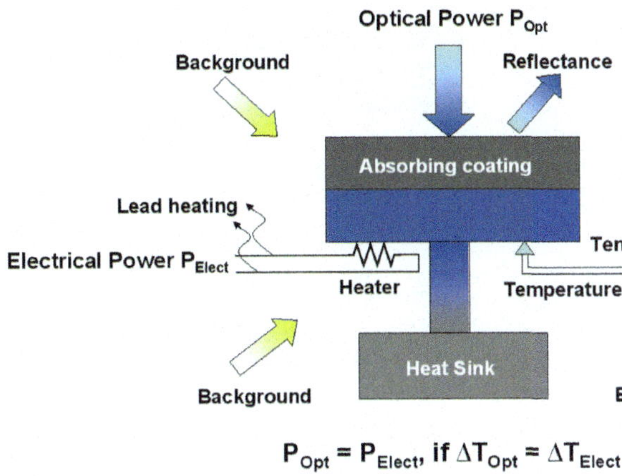

Fig.8.15 Schematics and principle of an ESR Radiometer

Cryogenic Radiometer: Detector Based Primary Standard for Optical Radiation

Absolute radiometers work on the principle of electrical substitution, in which optical power is linked with the electrical power [21, 25]. In the radiometers, absorbing cavity is used, which causes rise in temperature (ΔT) due to absorbed optical power, Fig. 8.15. Same rise in temperature is obtained by thermal heating with electrical power. When the two power inputs cause the same rise in temperature (ΔT) the optical (radiometric) and electrical powers are equivalent assuming negligible dissipation losses of energy. Hence, an electrically calibrated radiometer (ECR) uses the equivalence of radiant and electrical heating and therefore, the principle is also known as electrical substitution radiometry (ESR). An ECR has essentially 3 elements, (i) a radiation-absorbing element in which the absorbed power causes temperature rise; (ii) electrical heater with controllable and measurable electrical wattage; and (iii) a temperature sensor that is used to measure equality between radiant heating and electrical heating [25].

The measurement uncertainty in such ECRs can be reduced by keeping it at cryogenic temperature, which results in reduction in thermal noise upto a great extent as well as stability in measurements. Using cryogenic radiometers, the uncertainty can be reduced by factor > 50. Therefore, most of the NMIs prefers to realize SI base unit of luminous intensity, candela using cryogenic radiometer [27, 35]. CSIR-NPL India has also established the facility of cryogenic radiometer in 2014 for reducing the uncertainties in measurements, Fig. 8.16.

This cryogenic radiometer has a cavity absorbance 0.9998 at 632.8 nm wavelength and established to realize radiometric measurements in the visible region of spectrum. Cryogenic radiometer receiver cavity and heat sink replacement assembly has a receiver responsivity of (2 ± 0.25) K/mW, thermal response time < 5 s, and dynamic range 1.0 μW–1.0 mW. The noise, in this case, is as small as 2 nW and

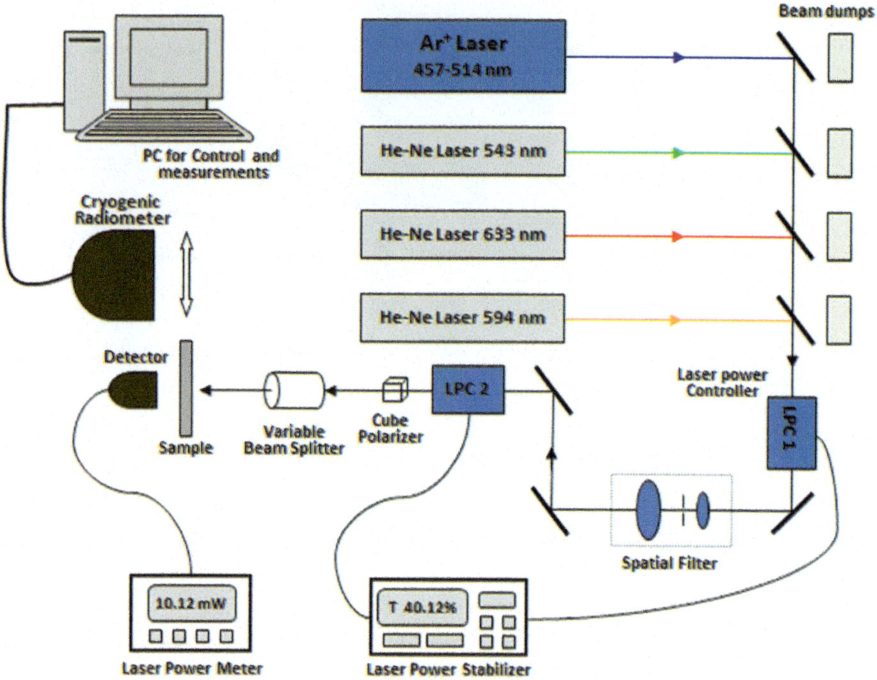

Fig. 8.16 Schematic diagram of the Cryogenic radiometer setup with laser alignments

absolute accuracy is of the order of 0.005%. The cavity is operated at the cryogenic temperature of 4.2 to 9 K, depending upon the optical power levels. The electronic control system exhibits two AC-Bridge thermometry circuits, two heater supply and measurement circuits for receiver and heat sink, multiple microprocessor controllers, temperature control circuit for the printed circuit boards, silicon diode thermometry readout for cryostat, calibration and system verification electronics. It also has a Brewster window assembly with transmission value of 99.96% at 632.8 nm, for laser measurements, kinematic window mount and liquid helium transfer line. Intensity stabilized Ar + ion along with other gas lasers, with frequency stability ± 1% or better, at various wavelengths is used for calibrating the radiometer at different wavelengths [27] in the visible spectral range. Turbomolecular pumping system provides oil-less operation for creation of ultra-high vacuum to reduce radiation and other losses inside the radiometer. High-quality trap detectors are used as transfer standards for transferring the scale of measurements [26, 28].

Radiometric and photometric measurements are of prime importance for a wide range of industries, including lighting, automotive, opto-electronics, telecommunications, space and health and safety. Precision measurements are required for quality control in industries and is key for new scientific findings. Optical radiation metrology at CSIR-NPL, encompasses traceable measurements in the spectral region of electromagnetic spectrum from 200 nm to 25 μm. The spectral region from

Table 8.7 Details of major calibration facilities with optical radiation metrology

Parameter	Range
Luminous intensity	1–10^3 cd
Luminous flux	1–2 × 10^4 lm
Illuminance	1–5 × 10^3 lx
Luminance	1–10^4 cd/m^2
Correlated colour temperature and chromaticity coordinates	2000–7000 K
Spectral irradiance (W/m^{-2} nm)	Wavelength range: 280–400 nm 400–800 nm 800–2500 nm
Spectroscopic measurements	
Spectral reflectance Diffuse Specular	380–780 nm
Spectral transmittance	380–780 nm

200 to 2500 nm is maximally used for measurement of photometric and radiometric parameters in many applications of interlinking disciplines. The core objective of the Optical Radiation Metrology at CSIR-NPL is to realize, establish, maintain, and upgrade the existing SI base unit of optical radiation, i.e. candela, and their derived units to provide apex level traceable calibration for various photometric parameters namely, luminous intensity, illuminance, luminous flux, luminance, detector responsivity, colour temperature and radiometric parameters namely, spectral transmittance, reflectance and spectral irradiance in the spectral range of 280–2500 nm. Table 8.7 shows the details of major calibration facilities with Optical Radiation Metrology.

8.2.3.2 Calibration and Measurement Capabilities

Currently, CSIR-NPL has 14 numbers of published CMCs for different photometric and radiometric parameters, namely luminous intensity, illuminance, illuminance responsivity, luminance, correlated colour temperature, luminous flux and spectral irradiance for their different respective ranges, which are available online at KCDB as listed in Table 8.8.

In addition, the measurement facilities for derived spectroscopic parameters namely spectral reflectance, spectral transmittance, absorbance, and polystyrene film calibration using FTIR technique are also available. Other calibration facilities of optical radiation metrology include calibration of thermo-vision camera, blackbody and NIR related measurements.

Table 8.8 List of calibration and measurement capabilities of CSIR-NPL India in the field of photometry and radiometry (PR)

Calibration or measurement service			Measurand level or range			Measurement conditions/independent variable		Expanded uncertainty				
Quantity	Instrument or artefact: measurand	Instrument type or method	Min. value	Max. value	Units	Parameter	Specifications	Value	Units	Coverage factor	Level of confidence (%)	Is the expanded uncertainty a relative one?
Luminous intensity	Tungsten lamp	Reference lamps and photometers	1	100	cd	Correlated colour temperature	2000–3000 K	1.6–1.4	%	2	95	Yes
Luminous intensity	Tungsten lamp	Reference lamps and photometers	100	1000	cd	Correlated colour temperature	2000–3000 K	1.4	%	2	95	Yes
Illuminance responsivity, tungsten lamp	Illuminance meter	Reference lamps and photometers			nA/lx	Correlated colour temperature	2000–3000 K	1.4	%	2	95	Yes
Luminous flux	Tungsten lamp	Integrating sphere, goniophotometer	1	100	Im	Correlated colour temperature	2000–3000 K	2.0 to 1.8	%	2	95	Yes
Luminous flux	Tungsten lamp	Integrating sphere, goniophotometer	100	20,000	Im	Correlated colour temperature	2000–3000 K	1.8–1.9	%	2	95	Yes
Illuminance	Tungsten lamp	Photometer	1	1000	Lx	Correlated colour temperature	2000–3000 K	1.6	%	2	95	Yes
Illuminance	Tungsten lamp	Photometer	1000	5000	Ix	Correlated colour temperature	2000–3000 K	1.6–2.0	%	2	95	Yes
Luminance	Tungsten lamp	Calibrated integrating sphere d iff user and telephotometer	1	10	cd/m^2	Correlated colour temperature	2000–3000 K	2.0–1.8	%	2	95	Yes

(continued)

Table 8.8 (continued)

Calibration or measurement service			Measurand level or range			Measurement conditions/independent variable		Expanded uncertainty				
Quantity	Instrument or artefact: measurand	Instrument type or method	Min. value	Max. value	Units	Parameter	Specifications	Value	Units	Coverage factor	Level of confidence (%)	Is the expanded uncertainty a relative one?
Calibration or measurement service			Measurand level or range			Measurement conditions/independent variable		Expanded uncertainty				
Quantity	Instrument or artefact: measurand	Instrument type or method	Min. value	Max. value	Units	Parameter	Specifications	Value	Units	Coverage factor	Level of confidence (%)	Is the expanded uncertainty a relative one?
Luminance	Tungsten lamp	Calibrated integrating sphere diffuser and telephotometer	10	1000	cd/m^2	Correlated colour temperature	2000–3000 K	1.8–1.6	%	2	95	Yes
Luminance	Tungsten lamp	Calibrated integrating sphere diffuser and telephotometer	1000	10,000	cd/m^2	Correlated colour temperature	2000–3000 K	1.6–2.0	%	2	95	Yes
Spectral irradiance	Tungsten lamp	Spectral irradiance comparator	0.001	0.2	$W/(m^2\ nm)$	Wavelength range	280–400 nm	6.0–1.9	%	2	95	Yes

(continued)

Table 8.8 (continued)

Calibration or measurement service			Measurand level or range			Measurement conditions/independent variable		Expanded uncertainty				
Quantity	Instrument or artefact: measurand	Instrument type or method	Min. value	Max. value	Units	Parameter	Specifications	Value	Units	Coverage factor	Level of confidence (%)	Is the expanded uncertainty a relative one?
						Bandwidth	10 nm					
Spectral irradiance	Tungsten lamp	Spectral irradiance comparator	0.001	0.2	W/(m^2 nm)	Wavelength range	400–800 nm	1.9	%	2	95	Yes
						Bandwidth	20 nm					
Spectral irradiance	Tungsten lamp	Spectral irradiance comparator	0.001	0.2	W/(m^2 nm)	Wavelength range	800–2500 nm	3.3–5.2	%	2	95	Yes
						Bandwidth	50 nm					
Correlated colour temperature	Tungsten lamp	Colorimeter	2000	3000	K			20	K	2	95	No

Table 8.9 List of various key comparisons performed by CSIR-NPL India in the area of Photometry and Radiometry (PR) and their current status

Comparison identifier	Metrology area	Sub-field	Description	Measurement period	Pilot institute	Status
APMP.PR-K3.a	Photometry and radiometry	Photometry	Luminous intensity	2012–2014	NMIJ AIST	Measurements in progress
APMP.PR-K4	Photometry and radiometry	Photometry	Luminous flux	2006– 2007	NIM	Approved for equivalence
APMP.PR-K3.b	Photometry and radiometry	Photometry	Luminous responsivity	1998–2000	NMIA	Approved for equivalence

8.2.3.3 Participation in International Key Comparisons

CSIR-NPL India, being a member of APMP, has been participating regularly in various key comparisons which are listed in Table 8.9.

Apart from this CSIR-NPL is also participating in the key comparison (APMP K1.a) for spectral irradiance within 250–2500 nm, to be piloted by the link laboratory NIM China, scheduled after the year 2020, as decided in the 33rd APMP General Assembly and Related Meetings, hosted by CSIR-NPL India during November 2017. Further, in order to establish new calibration and measurement capabilities (CMCs) in the area of LED metrology, CSIR-NPL India is also planning to participate in various other international inter-comparisons based on LED standards, post-installation of major measurement Systems under the LED testing and calibration facilities.

8.2.3.4 Potential Services for Future Requirements

As the challenges and threats towards safety of human life are ever-increasing, lighting technology is evolving and precision measurements and standardization techniques are being upgraded to meet such safety requirements from industries. In the current Covid-19 pandemic situation when the attention is again on radiation-based contactless ultraviolet germicidal lighting to contain the infection. Calibration and testing facility at CSIR-NPL India is being established for testing and standardization of such products to eliminate inferior quality products from the Indian market which may pose serious health hazard to the users. Because of energy efficiency, low power consumption, little heat, faster response, natural coupling for digital control and large lifetime (~50,000 h) Light emitting diode (LED) and LED-based lighting products are paving their ways in not only into lighting sector in a huge way but also into other spheres of general human life and other applications resulting in phasing out of the old energy-inefficient traditional lighting technology. Various countries, especially NMIs are putting their potential effort to explore its capability and bringing out more results relevant to exploit these energy-efficient sources in terms of customizing its design, engineering and luminous performance. A number of luminaries manufacturing giants in India are shifting their business focus and production towards the LED-based lighting, attributed to its high efficacy and longer life time.

Since the heavy shift of consumers' and production units' interest towards the next generation energy-efficient, long-lived, solid-state lighting (SSL) sources many substandard products are being launched into the Indian market at cheap prices which not only have substandard luminous performance but also may contain harmful blue radiations higher than the level fixed as per the international guidelines. Since these products are not quality tested they may pose high radiation risk to the society if not stopped with calibration/testing mechanism well in place.

Being an apex level measurement laboratory, understanding its responsibility CSIR-NPL is in the process of establishing state of the art, National test and calibration facilities as per the National/International standards for such next generation energy-efficient light products. This facility will not only support the energy saving initiative of Government of India via promoting next-generation energy-efficient lamps on 'green technology', but also eliminate sub-standard products from the market and to cater the need of lighting industries which are in the manufacturing of such lightings. On the other hand, successful implementation and the establishment of such facility would make CSIR-NPL India visible in its R & D efforts towards LED metrology at par with other NMI's.

The SI base unit ***candela*** the unit of luminous intensity of optical radiation, in a given direction is defined by taking the fixed numerical value of the luminous efficacy of monochromatic radiation of frequency 540×10^{12} Hz, K_{cd}, to be 683 lm/W, which, for a monochromatic source to be 1 cd at frequency of 540 THz, straightway translates into emission of 4.09×10^{15} photons per second, in a given direction per unit solid angle.

The recent development in photon counting technology has revolutionized the areas of electronics, biotechnology, quantum information processing, medical physics, space and military applications, and meteorology. Utilizing the advantages of quantum optical technologies, optical radiation metrology capabilities can be improved [36]. Hence, R&D towards metrology of photons is necessary to take step to realize standard few photon sources for futuristic quantum optical technologies. Since no single photon source has been realized so far that can fulfill the requirements of futuristic photonic technologies. Various leading NMIs including NIST USA, NPL UK, PTB Germany, INRIM Italy, etc., independently and jointly are involved in realization and standardization of photon sources and detectors [37]. CSIR-NPL India has also taken initiative in terms of creating few basic facilities to perform R&D in photon physics, and can take step forward in terms of basic research, which would help to meet the global objectives in few photon metrology.

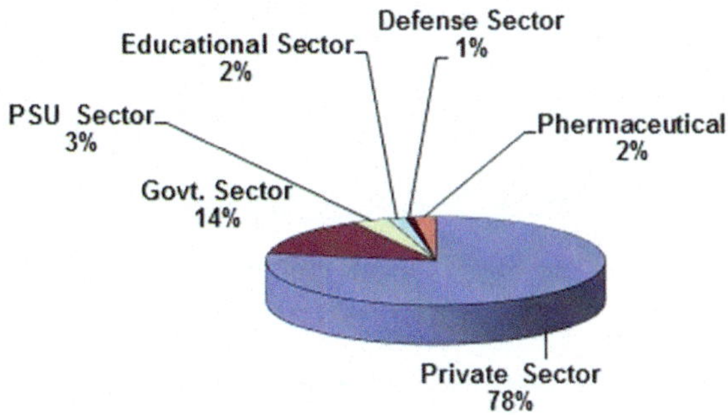

Fig. 8.17 Pie chart representation of sectorwise report generation over the spectrum of sectors in past one decade

A Decade of Clients and Services

As an apex measurement institute of the country, CSIR-NPL India since last two decades, is catering for traceability requirement related to optical radiation metrology, of a broad spectrum of organizations/entities including various Industries, MSMEs, Research and Development Institutes, Educational Institutions, Testing and calibration laboratories etc. Apart from realization, and being custodian of national standards of optical radiation, CSIR-NPL provides traceability in terms of calibration and measurement service, rendered to various major sectors.

The main parameters for photometry, radiometry and colorimetry, which need precision measurements and calibration for their accurate value requirement in various in-situ measurement applications or have effect in the product development or service quality, are luminous intensity, luminous flux, illuminance, colour temperature, colour coordinates, illuminance, luminance, illuminance responsivity, spectral irradiance, spectral transmission, spectral absorption, wavelength calibration, etc.

Following are some glimpses of data of calibration services rendered to the customers, Fig. 8.17, that includes spectrum of organizations and industries playing significant role in the overall growth of the country and raising a quality infrastructure to complement the society with their quality products or services. Optical Radiation Metrology has contributed significantly in catering to its services through dissemination of traceability to its customer and by generating more than 3700 calibrations and testing reports in past one decade.

The customer base of Optical Radiation Metrology encompasses large variety of stakeholders of Government Sector including few of Ministries also, namely, *Ministry of Electronics and Information Technology, Ministry of Micro, Small and Medium Enterprises, Ministry of Textiles, Ministry of Civil Aviation, Ministry of Heavy Industries and Public Enterprises, Ministry of Railways, State Electricity Boards and various Autonomous bodies working under ministries.*

Table 8.10 List of major calibration parameters with optical radiation metrology

Photometric, radiometric and colorimetric measurements		
Parameter	Range	Measurement uncertainty at coverage factor $k = 2$
Luminous intensity	1–10^3 cd	± 1.4–1.6%
Luminous flux	1–2 × 10^4 lm	± 1.8–2.0%
Illuminance	1 lx to 5 × 10^3 lx	± 1.6–2.0%
Luminance	1–10^4 cd/m^2	± 1.6–2.0%
Correlated colour temperature and chromaticity coordinates	2000–7000 K	± 20 K ± 0.002 unit
Spectral irradiance (W/m^{-2} nm)	Wavelength range: 280–400 nm 400–800 nm 800–2500 nm	± 1.9–6.0% ± 1.9% ± 3.3–5.2%
Spectroscopic measurements		
Spectral reflectance Diffuse Specular	380–780 nm	± 1.5%
Spectral transmittance	380–780 nm	± 1.5%

Calibration and Testing

Optical radiation metrology at CSIR-NPL, generally, covers traceable measurements in the spectral region of electromagnetic spectrum from 200 nm to 25 μm. The spectral region from 200 to 2500 nm is maximally used for measurement of photometric and radiometric parameters in many applications of interlinking disciplines. Such precision and traceable measurements are used in a wide range of industries including General lamp and lighting, automotive, aviation, space, telecommunication, defence, medical and health industries, etc. and are prime requirements for quality control in industries and in scientific research to arrive at novel results. Apart from research in frontier areas of radiation metrology, the mandate of optical radiation metrology at CSIR-NPL is to realize, establish, maintain, and upgrade the existing SI base unit of optical radiation, i.e. candela and their derived units to provide apex level traceable calibration for various photometric parameters namely, luminous intensity, illuminance, luminous flux, luminance, detector responsivity, colour temperature and radiometric parameters namely, spectral transmittance, reflectance and spectral irradiance in the spectral range of 200–2500 nm as discussed in Table 8.10.

Consultancy and Technical Services

Currently, the Optical Radiation Metrology (ORM) is actively involved in consultancy services to strengthen the development of quality product in compliance to standards guidelines, namely, the Section at present is involved in development of a UVC based sanitization application for MSME to help contain the Covid-19 pandemic infection, The ORM has also rendered its expert consultancy and technical services in past to various lighting industries and R&D organizations, through establishing critical measurement facilities at their sites.

Training and Academic Services

CSIR-NPL is also providing training and organize specialized short term courses on optical radiation metrology to personnel from industry or academic institutions. Such courses are organized, normally twice in each year, or as per need arises, covering hands-on training for optical radiation metrology, lectures and lecture notes on basics of metrology, knowledge of few relevant standards, exposure of hands-on measurement on real experimental set-ups and evaluation of uncertainty involved in measurements.

Since, metrology is a very specialized discipline of science which is why it finds place in curricula of degree courses of countable number of Universities/Institution. It is therefore imperative for CSIR-NPL to conduct specialized regular courses on precision metrology to create a bank of trained human resources to cater to the need of up-coming industry requirements in the field of metrology. Currently, CSIR-NPL is running a one-year Post Graduate Diploma on Precision Measurement and Quality Control (PMQC) under the umbrella of Academy of Scientific and Innovative Research (AcSIR) in the line of skill development of Government of India's programme. Optical Radiation Metrology is one of the key subject area of interest of students of this programme. More than two dozens of students have got their first breakthrough in Industries immediately after completing the course.

Apart from skill development program, guidance of students for their Post Graduate programme and PhD programme of AcSIR is also done.

Technology Development and Commercialization

Having fundamental knowledge and state-of-the-art facilities with Optical Radiation Metrology, CSIR-NPL is engaged in the upfront area of research and developing metrology of next-generation energy-efficient sources and other invisible sources. It is also active in solving societal problems and also working on new problems from industries that poses technological challenges during their commercialization or their reach to society.

8.2.3.5 Major Documentary Standards, Guides and SOPs

Optical Radiation Metrology parameter performs all of its calibration and testing activities according to the requirement of ISO/IEC 17025 and strictly adheres with quality management system. Few major National and International Standards, pertaining to carry out measurement and calibration activities of Optical Radiation Metrology are listed below in Table 8.11.

Table 8.11 Standard operating procedures/ISO standards used in optical radiation metrology

Sl. No	Standard Operating Procedure/ISO Standards
1	ISO 80000–7: 2019 (EN) Quantities and units—Part 7: Light and radiation
2	CIE S 017/E: 2011 ILV: International Lighting Vocabulary
3	ISO 23539:2005/CIE S 010:2004 Photometry—The CIE System of Physical Photometry
4	ISO/CIE 19476:2014(EN): Characterization of the performance of illuminance meters and luminance meters
5	EN 13032–1: 2004 Light and lighting—Measurement and presentation of photometric data of lamps and luminaires—Part 1: Measurement and file format
6	CIE 202: 2011 Spectral Responsivity Measurement of Detectors, Radiometers and Photometers
7	CIE 015: 2018 Colorimetry (4th edition) & CIE 015: 2004 Colorimetry (3rd edition)
8	ISO 11664–1:2007/CIE S 014–2: 2006 Colorimetry—Part 2: CIE Standard Illuminants
9	CIE 114/4–1994 CIE Collection in Photometry and Colorimetry—Distribution Temperature and Ratio Temperature
10	IES LM-79–2008 Approved Method: Electrical and Photometric Measurements of Solid-State Lighting Products
11	CIE S 025/E:2015 Test Methods for LED Lamps LED Luminaires and LED Modules
12	ANSI/IES LM-79–19 Approved Method: Electrical and Photometric Measurements of Solid-State Lighting Products
13	IS 10322 (Part-5/Sec-1): 2012 Luminaire, Particular requirements
14	IS 13383 (Part-1): 1992: Method of photometric measurement of luminaires for use in interior lighting
15	IS 13383 (Part-2): 1992: Method of photometric measurement of luminaires for road & street lighting
16	IS 16106: 2012 Method of electrical and Photometric Measurements of Solid State Lighting (LED) Products
17	IS 16107 (Part-5/Sec-1): 2012 Luminaires Performance
18	IS 16103 (Part-2): 2012 LED Modules for general lighting: Performance requirements
19	Bureau International Des Poids Et Mesures (BIPM), Principal Governing Photometry (2nd Edition), Rapport BIPM-2019/05
20	CIE 198: 2011 Determination of Measurement Uncertainties in Photometry

8.2.4 *Possible Gaps and Suggestions for Their Redressal*

Since the area of photometry and radiometry has enormous opportunity of playing vital role in various scientific as well as industrial applications, which directly affects the outcome/results or safety and security aspect of human life. Though CSIR-NPL India holds limited number of CMCs depending upon the traceability requirement of industries and testing and calibration laboratories, however, due to rapid industrialization, demand for quality and safety products and rise of high-quality R&D Institutions in the country need for having more and more CMCs in various other derived parameters are now unfolding in exponential way. To cater the present need, it requires CSIR-NPL to expand its measurement and calibration facilities setting up new metrological grade measurement facilities in tune with the rapid growth and demands, with not only as per the national and international standards but also at par with the world's leading NMIs, to suffice the present growing requirements. CSIR-NPL has already taken its step on the way of establishing new measurement facilities understanding the present requirements from industries, especially lighting industries, water purification and sanitization industries and the overall global metrology scenario and setup a quality infrastructure in the country. The upcoming LED-based testing and calibration laboratory at CSIR-NPL is a gamut of state-of-the-art facility which would fill a big gap rendering apex level measurement and traceability to country for not only energy-efficient solid-state lighting (SSL) based products but also for other conventional lighting technologies.

With the advancement of technology and exponential rise of industries, MSMEs to fulfil the consumers' demand, dearth of skilled manpower in the country is always felt. CSIR-NPL as per its capacity, and without compromising in the quality, is conducting specialized regular course on precision metrology and quality control (PMQC) to fill in the big void of manpower in metrology.

8.3 Acoustics and Vibration Metrology

8.3.1 *Introduction and Historical Aspects*

The Acoustics and Vibration parameter of CSIR-NPL, since its inception, has played a key role in the industrial growth. In accordance with the well-laid out objectives of CSIR-NPL to strengthen and to carry out advance physics-oriented research, the acoustic and vibration parameter has immensely contributed to reducing the air and noise pollution in the country. The major activities involved are calibration of acoustical instruments, evaluation of industrial products, acoustical materials, performance characteristics of audio equipment, auditorium acoustics and noise and vibration measurements and control. The calibration and other facilities available in CSIR-NPL in the area of Acoustics and Vibration are comparable with facilities in other countries. It is equipped with advanced instrumentation for measurement of sound and vibration and calibration of electro-acoustic equipment such as Sound Level Analyzer, Vibration Analyzer, Sound Intensity Probe and Impedance Tube system, specialized Reverberation and Anechoic Chambers for carrying out sound insulation and absorption studies of materials and diagnosing machinery noise. CSIR-NPL has been able to provide apex level calibration and testing services and technical advisory consultancy in architectural acoustics to the industries and institutions of the country.

This activity has not only provided calibration and testing services to the government bodies and industries but also provided consultancy and other technical services for reducing noise and vibration in machinery, transportation systems and ambient noise levels. The consultancy services provided to Delhi Metro, Bangalore Metro, Archaeological Survey of India, Delhi Development Authority, Central Pollution Control Board, etc. had been instrumental in tackling the noise and vibration issues associated with the transportation systems. The Acoustics and Vibration activity focuses on maintenance and up-gradation of two primary standards viz., the standard of sound pressure and standard of vibration amplitude. This activity is responsible to establish, maintain and upgrade continually the measurement standards of sound and vibration and for providing the apex level calibration services to the users across the country ensuring the measurement traceability in these parameters. The establishment of a self-reliant traceable sound and vibration systems and outstanding performance in recent key comparisons participated had been instrumental in strengthening the traceability chain. Till date, 34 Calibration and Measurement Capabilities (CMCs) in the field of sound and vibration has been included in the BIPM database. Efforts are targeted to increase the number of CMCs in the BIPM database to solve the problems and challenges faced by industries and other stakeholders.

To further strengthen research and developments in acoustics, in the early seventies, work in the area of atmospheric acoustics, the acoustic sounding of the lower atmosphere: SODAR (SOund Detection And Ranging) was initiated. It is one of the very promising and potential application-oriented activities of the laboratory.

8.3.2 Significance, Applications and Relevance in Different Sectors

The role of the Acoustics and Vibration parameter is multifaceted in various sectors such as automotive, avionics, heavy engineering industry, defence, space research and for environmental studies. The heavy engineering industry, avionics sector, automotive sector, and Public sector undertakings generally face the noise and vibration issues with machinery and products that require an engineering solution for meeting the international standards for ensuring the quality of products developed. The calibration, testing, and consultancy services provided in Acoustics and Vibration area had played a major role in helping all these sectors for solving their problems and challenges faced. The activity also contributed a lot in solving the environmental noise issues and the development of standards. The activity is working in collaboration with the Central Pollution Control Board (CPCB) and other State Pollution Control boards for noise mapping studies for controlling noise pollution of various cities in the country. The low-frequency primary vibration standard is capable of calibrating the seismic sensors (mass up to 200 g). The primary standard of sound pressure in the frequency range of 31.5 Hz to 25 kHz is used to calibrate the reference standard microphones (LS1P and LS2P). The activity also caters to the calibration, testing, and consultancy needs of customers in sound and vibration parameters. The important applications of acoustics and vibration parameters are as follows:

(a) Maintenance of primary standards of sound pressure and vibration amplitude,
(b) Calibration and testing of electro-acoustical equipment,
(c) Noise and vibration measurements and control,
(d) Acoustics materials characteristics measurement for building industries,
(e) Technical advisory consultancy in building acoustics,
(f) Noise barrier design and development for highway, railway, metro, etc.,
(g) Environmental noise measurement and control,
(h) Air Quality management using SODAR, and
(i) Formulation of Standards for noise and vibration control.

8.3.3 *Facilities and Global Status*

The Acoustics and Vibration activity is realizing the primary and secondary standards of sound pressure and vibration amplitude to provide the traceability of the regional laboratories, government bodies, and the industrial sector in the country. The description of these facilities, research, and development works towards reducing the uncertainty in the measurements and global status of measurement facilities is described in detail in the next section.

8.3.3.1 Measurement Standards and Facilities

The Acoustics and Vibration activity is focused on the maintenance, realization, and up-gradation of primary and secondary standards of sound pressure and vibration amplitude. The primary standard of sound pressure is realized using the Reciprocity method for the calibration of laboratory standard microphones (LS1P and LS2P) in the coupler cavity in frequency range 31.5 Hz to 25 kHz with an expanded uncertainty of ± 0.05 dB to 0.17 dB [38]. The Reciprocity calibration apparatus of CSIR-NPL, India is the B&K type 9699 system. The primary standard of vibration amplitude is maintained using laser interferometry and sine approximation method as per ISO 16,063–11 [39, 40] in the frequency range of 0.1 Hz to 20 kHz with an expanded uncertainty of ± 0.30 to 1.80%. Figure 8.18a and b shows the pictorial view of the primary standard of sound pressure and Primary vibration calibration standard using

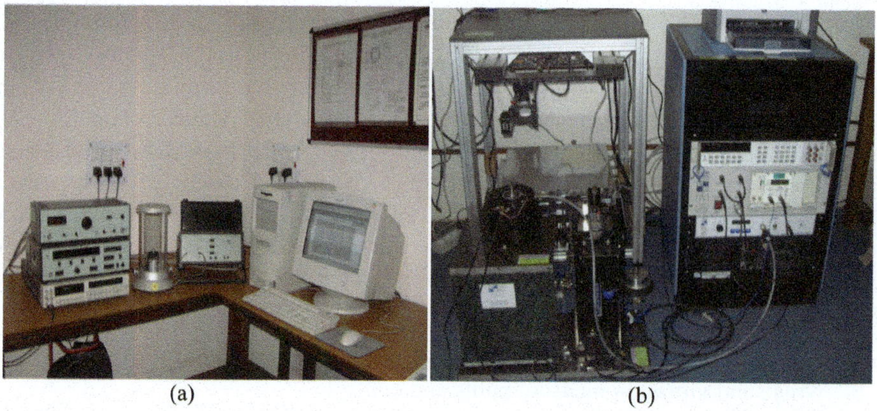

Fig. 8.18 **a** Primary microphone calibration standard using reciprocity method. **b** Primary vibration calibration method using a laser interferometer

a laser interferometer approach (TMS 9155D system) in the frequency range 0.1 Hz to 20 kHz.

The reciprocity method is usually employed by many NMIs in the world. The sensitivity of the microphone is obtained from measurements of the electrical transfer impedance and acoustical transfer impedance between the microphones in compliance with IEC 61094–2. The long coupler and short coupler are used to measure the sensitivity in the frequency range 31.5 Hz to 2 kHz and to record the sensitivity deviations. The measurement data is analyzed using computer software *MP.EXE* developed by Brüel & Kjær (B&K), Denmark as shown in Fig. 8.19 and sensitivity fitting exercise is performed for obtaining the final sensitivity [38].

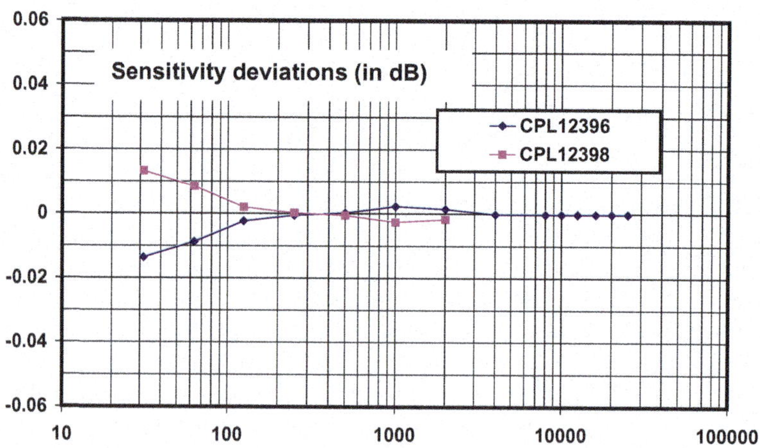

Fig. 8.19 Sensitivity deviations (in dB) for two couplers CPL 12,396 (length = 4.7 mm) & CPL 12,398 (length = 9.4 mm) of diameter 9.3 mm for B&K 4180 microphone in MP.EXE programme

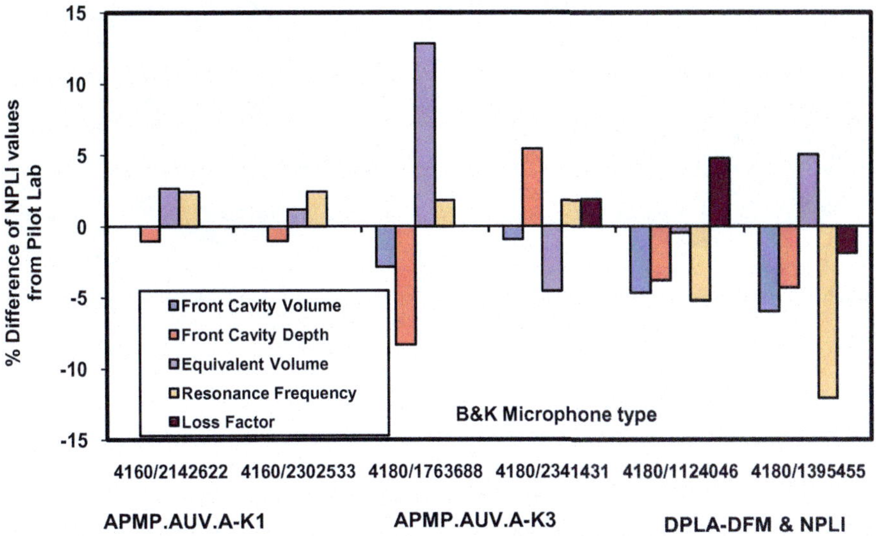

Fig. 8.20 Difference in measurement values (in %) of NPL, India with respect to the pilot laboratories (NMI, Japan, KRISS, Korea, and DPLA-DFM, Denmark) for each microphone parameters [42–44]

The validation of CSIR-NPL CMCs in the Reciprocity measurements has been confirmed from the degree of equivalence values in comparison to the other reputed NMIs for LS1P and LS2P microphones. The APMP.AUV.A-K1 comparison for LS1P microphones was piloted by NMI, Japan with nine participating laboratories [41]. NPL, India further participated in APMP.AUV.A-K3 comparison piloted by KRISS, Korea for LS2P microphones with 10 participating laboratories [42]. Figure 8.20 shows the percentage difference of NPL, India values with respect to the pilot laboratories for each microphone parameters [42–44]. The maximum deviation of 0.06 dB in sensitivity for LS1P microphone was observed at higher frequencies with respect to the pilot laboratory. In the case of the LS2P microphone, the maximum deviation of 0.11 dB in sensitivity determination with respect to the pilot laboratory was observed at 25 kHz [42]. The precise measurement of microphone parameters is very important for reducing the measurement uncertainty in the sensitivity determination of reference standard microphones, especially at higher frequencies. The conventional methodology followed was the use of nominal parameters specified by manufacturer and deviations up to 13% in equivalent volume and 6.4% in front cavity volume had been observed in Key comparison exercises. Figure 8.21 shows the parametric sensitivity analysis of microphone parameters affecting the sensitivity in dB re 1 V/Pa of reference standard microphones at 10 kHz using Analysis of Variance (*ANOVA*) method [44]. Thus, efforts are in progress to decrease the measurement uncertainty in sensitivity determination by conducting precise measurements for the front cavity volume of the reference standard microphone, which is one of the most critical parameters [44].

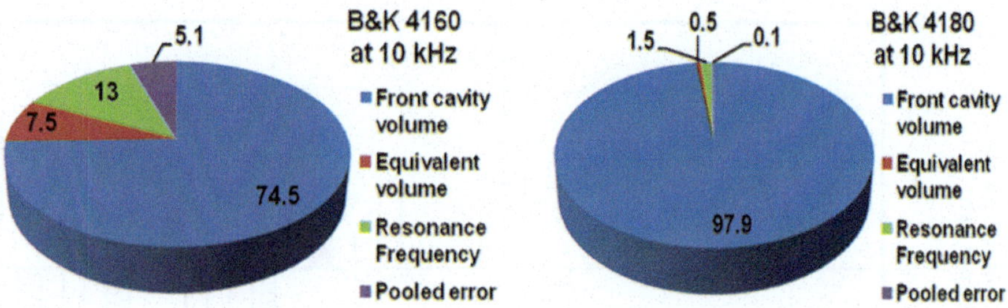

Fig. 8.21 Parametric sensitivity analysis of microphone parameters affecting the sensitivity in dB re 1 V/Pa of reference standard microphones at 10 kHz using the *ANOVA* method [44]

Also, with the precise dimensional characterization of front cavity depth and volume and the other microphone parameters, the measurement uncertainty in microphone parameters can be reduced especially at higher frequencies leading to reduced uncertainty in sensitivity determination of reference standard microphones. The precise dimensional characterization of microphone parameters in conjunction with use of latest signal processing techniques, inclusion of all corrections as recommended in IEC 61094–2 standard shall be crucial to reduce the random uncertainty and overall uncertainty in the measurements.

Thus, efforts are targeted to reduce the measurement uncertainty in primary sound standard and extension of the frequency range by focussing on the various aspects as follows:

- Dimensional characterization of the front cavity volume and precise determination of other microphone parameters of the reference standard microphones,
- Replacement of old hardware with a PULSE based system for reducing the signal-to-noise ratio, distortion, and random uncertainty,
- Inclusion of low-frequency option from 1 to 20 Hz for low-frequency calibrations of reference standard microphones and participation in a key comparison exercise with other NMIs.

These up-gradations shall help to realize the complete range of 1 Hz to 25 kHz with a measurement uncertainty ranging from ± 0.05 to 0.15 dB. The Primary vibration calibration standard is based on the principle of sine approximation method as per ISO 16063–11 and employs a laser interferometer for measurement of displacement of the transducer when excited at varied amplitudes and frequencies in the range 0.1 Hz to 20 kHz. CSIR-NPL, India realizes the complete frequency range from 5 Hz to 10 kHz using two systems: TMS 9155D system in the frequency range 5 Hz to 20 kHz and medium frequency B&K 3629 calibration system in the range 5 Hz to 5 kHz. The measurement uncertainty of the medium frequency system ranges from 0.7% at 160 Hz to 1.0% at 5 kHz [40]. The Calibration and Measurement Capabilities (CMCs) in range 40 Hz to 5 kHz had been earlier validated in an international key comparison exercise. The APMP.AUV.V-K1.2 with NIM China as a pilot laboratory and KIM-LIPI Indonesia as another participating laboratory was initiated in

2009. The degree of equivalence between NPL, India and NIM China at 160 Hz for single-ended accelerometer was observed to be 4.6×10^{-4} pC/ms^{-2} and that for back to back sensor was 4.2×10^{-4} pC/ms^{-2}. At 5 kHz, the degree of equivalence between NPL, India, and NIM China was observed to be 3.1×10^{-4} pC/ms^{-2} for single-ended accelerometer and 9.6×10^{-4} pC/ms^{-2} for back to back sensor [45]. With the installation of the TMS 9155D primary vibration calibration standard in 2011 as shown in Fig. 8.18b, the highest levels of accuracy and precision are realized in complete frequency range from 0.1 Hz to 20 kHz using an air bearing shaker. The hardware of the system comprises of an excitation and measurement subsystem. The excitation system contains a power amplifier, shaker (PCB Air bearing shaker 396C11), and function generator located on the PCI card for data acquisition. The long-stroke shaker APS 113AB is used for low-frequency calibration and provides a sufficient maximum stroke of 158 mm [46]. Continuous efforts are targeted to reduce the measurement uncertainty in the sensitivity and phase determination of the reference standard vibration transducer. The multi-point laser positioning approach is followed for the precise determination of the sensitivity of the transducer in the frequency range of 5 Hz to 20 kHz [39]. Figure 8.22a shows the multi-point laser positions on a PCB 396C11 shaker and Fig. 8.22b shows the random uncertainty observed in symmetrical laser positions used for the measurement of the sensitivity of the PCB 353B04 sensor. It can be observed that the maximum random uncertainty in the sensitivity calibrations using a multi-point laser positioning approach, was measured as 0.25% and that in phase calibrations was 0.13°. The basic motive behind the selection of symmetric laser positions is to restrain the effects of any non-rectilinearity of motion.

The secondary standards are used for providing calibration services for accelerometers, working standard microphones, sound calibrators, piston phones, multifunction acoustic calibrators, vibration meters, accelerometer calibrators, etc. The back to back comparison calibration as per ISO 16063–21 is done using a Spektra CS 18 MF system that utilizes an electrodynamic shaker, SE-10 for performing accelerometer calibrations in range 5 Hz to 10 kHz. The measurement uncertainty realized in the secondary accelerometer and vibration meter calibrations is 1.2–2.5% in the frequency range 5 Hz to 10 kHz. The transverse motion has been one of the significant sources of error in the vibration calibrations, that causes an error component

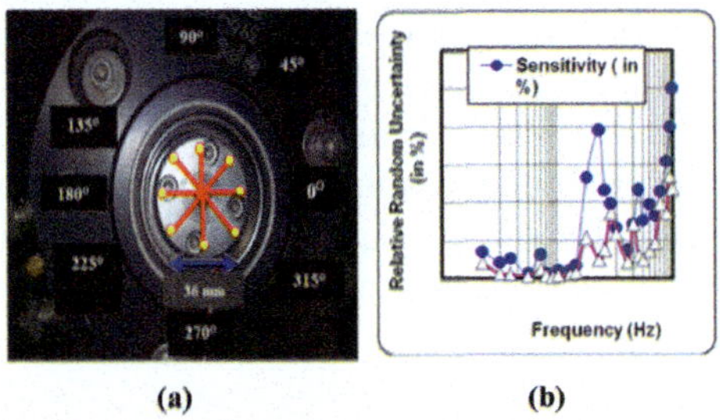

(a) (b)

Fig. 8.22 a, b Vibration sensitivity measurements using multi-point laser positions in the acceleration amplitude in conjunction with the transducer transverse sensitivity. Transverse motion limits are required by ISO 16063–11 to be less than 1% for frequencies below 10 Hz, less than 10% for frequencies below 1 kHz, and less than 20% for frequencies below 10 kHz. The secondary microphone calibration system of NPL, India is B&K 9721 that utilizes a traceable microphone for performing secondary microphone calibrations using an insert voltage method and frequency response measured using an electrostatic actuator. The measurement uncertainty achieved in the calibration of measurement microphones ranges from 0.1 to 0.35 dB in frequency range 31.5 Hz to 20 kHz. The uncertainty in measurement in the calibration of sound calibrators and multifunction calibrator range from $\pm$ 0.12 to $\pm$ 0.25 dB in frequency range 31.5 Hz to 16 kHz. The sound level meter calibration is carried out with an uncertainty of $\pm$ 0.30 dB to $\pm$ 1.0 dB in the frequency range 125 Hz to 10 kHz [47]. Participation in the APMP.AUV.A-S1 supplementary comparison for multifunction acoustic calibrator (B&K 4226) with NIMT, Thailand as the pilot laboratory has validated CMCs for sound pressure level determination. The maximum deviation of −0.13 dB for the sound pressure level of 94 dB and of −0.17 dB for the sound pressure level of 114 dB with respect to the Supplementary Comparison Reference values (SCRV) was observed [48].

Apart from the primary and secondary standards of sound and vibration parameters, the activity also undertakes various testing of noise and vibration and acoustical products with a theme of *Centre of Excellence for Noise and Vibration Control Facilities.* The Indian industry is in the process of developing indigenous technology wherein performance, evaluation of acoustical products, instruments of systems are very helpful. In this respect, acoustics and vibration activity undertakes the testing of a variety of acoustical products as per relevant Indian Standards. Major acoustical facilities available in the parameter are (i) Anechoic Chamber and (ii) Reverberation Chamber has been built to meet the ISO recommendations. Besides, there are various types of equipment such as the (iii) Building Acoustics measurement and Impedance Tube facility, (iv) Vibration measurement facility and (v) SODAR system for ABL (Atmospheric Boundary Layer) height measurement.

Anechoic Chamber

Precision acoustical measurements like sound power output and directional characteristics, calibration in electro-acoustic, psycho-acoustic experiments, etc. require a test space that is free from reflection from nearby surfaces and is sufficiently quiet. This is achieved artificially by treating the available surfaces of the room by special wedge absorbers, adopting double-wall construction, and isolation from other structures. The working space is 3.5 m $\times$ 3.5 m $\times$ 3.5 m and cut-off frequency is 70 Hz of the anechoic chamber available with CSIR-NPL shown in Fig. 8.23.

Fig. 8.23 Anechoic chamber facility at CSIR-NPL

Fig. 8.24 Reverberation chambers facility at CSIR-NPL

Reverberation Chamber

Absorption and transmission characteristics of acoustic materials and measurements like noise power levels of instruments are carried out in reverberation chambers. The essential requirements of a reverberation chamber are (i) absence of interfering noise, (ii) adequate sound diffusion and (iii) long reverberation time over the frequency range of measurement. These are achieved by non-parallel highly polished walls, duo-decahedral loudspeakers systems, and suspension of diffusing plates from the ceiling, oriented at random as shown in Fig. 8.24. The specifications of the reverberation chamber are listed in Table 8.12.

Table. 8.12 Specifications of reverberation chamber of CSIR-NPL

Volume of the chamber	257 m^3
Average RT	5.6 s
Cut off frequency	80 Hz
Type of construction	Double-wall, room inside a room
Ambient noise level	20 dB
No. of chambers	2 (Coupled to each other)

Fig. 8.25 Impedance tube facility at CSIR-NPL

Building Acoustic Measurement System and Impedance Tube Facility

The new facilities that had been established in CSIR-NPL, India include the Building Acoustic Measurement System and Impedance Tube facility used for the testing of a variety of acoustical products as per relevant Indian Standards. This includes Omni-directional sound source and pre-amplifier and dual-channel building acoustics analyzer. The parts of Building Acoustic measurement system conform to the following: Omni-directional sound source conforms to the ISO 140, ISO 3382 standards; Dual-channel real-time Analyzer for sound pressure level measurements conforms to the IEC 61672–1-2013 for sound insulation and reverberation time measurement and Airborne Sound Insulation measurement conforms to standards ISO-140 and RT measurement conforms to ISO-3382 standards. The Impedance tube system as shown in Fig. 8.25 is used for research and developing products by characterizing acoustic material properties and designing acoustic comfort in aircraft and vehicle interiors, by selecting optimal acoustic treatments and barriers. For testing specified material characteristics and verifying material compliance with regulations before materials are incorporated into manufactured assemblies such as components and for validating and calibrating computational methods such as acoustical modelling impedance tube facility is used. The uncertainty of measurement of different testing facilities is listed in Table 8.13. The subsystems of impedance tube facility are as follows:

Table 8.13 Uncertainty of measurement of different testing facilities

Parameters covered	Testing facilities available for	Range and uncertainty (Expanded) at $k = 2$ and 95% confidence level
Sound transmission class (STC)	Sound insulation materials	± 1.0 dB (100 Hz to 4 kHz)
Noise reduction coefficient (NRC)	Sound absorbing materials	± 5% (100 Hz to 4 kHz)

- AFD1000 Base Module Set (801022.5 Speaker Module, Amplifier, 971024.1, ¼ inch microphones)
- AFD1000 Absorption Tube 30 mm
- AFD1200 Transmission Tube 30 mm
- AFD1000 Absorption Tube 100 mm
- AFD1200 Transmission Tube 100 mm
- Measurement of sound absorption Coefficient AFD 1000—AcoustiTube®Analysis Software AFD 1001
- Measurement of Sound Transmission Loss AFD 1200—AcoustiTube® and Analysis Software AFD 1201

Measurements based on the two-microphone transfer-function method according to ISO 10534–2 and ASTM E1050–08 standards for absorption coefficient and ASTM E 2611–09 for transmission loss.

Noise and Vibration Measurement Facility

Excess noise and vibration are harmful to the human well being and stability of structures/machines. Efforts have thus to be concentrated to mitigate their harmful effects. CSIR-NPL was the first institute in the country to initiate the scientific evaluation of environmental noise pollution in the late 50 s with a systematic survey of prevailing noise levels in the cities of Delhi, Bombay, and Calcutta. The survey included noise due to vehicular traffic, trains, aircraft, factories, in and around shopping centres, residential areas, schools, hospitals, etc. both during day and night. Analysis has shown that the major contribution of city noise comes from vehicles like diesel buses, cars, and motorcycles. During the last five decades, the acoustics section of the laboratory has not only made the measurement of traffic noise but has also carried out measurements of noise and vibration in industry and has given suggestions/recommendations for remedial measures to avoid occupational hazard risks. This activity is also involved with several national committees in the country to formulate norms and standards to monitor and control noise in cities and industries.

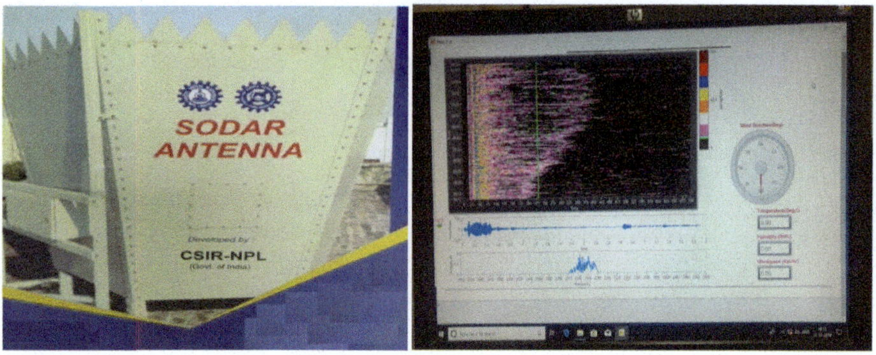

Fig. 8.26 Mono-static SODAR system

SODAR (SOund Detection and Ranging) System

Measurements related to many indiscernible physical parameters are also required in comparison to the physical quantity measurements like mass and length. Therefore, NPL's activities diversified into higher frequency ranges leading to studies on troposphere and lower atmospheric science by the mid-1970s. The design of SODAR was undertaken to study the atmospheric turbulence to map the thermal structures in the lower boundary layer. A SODAR is essentially an acoustic radar that provides a pictorial view of the thermal structure dynamics of the atmospheric boundary layer. SODAR was firstly designed, developed and operated in CSIR-NPL, India. In 1973, SODAR system using the reflector horn antenna as a receiver and a square array antenna as the transmitter and operating at 2 kHz became functional. A phased array wind profiling SODAR, capable of measuring boundary layer atmosphere winds, was also developed as shown in Fig. 8.26. Numerous developments had been carried out on SODAR which enabled the development of high-resolution SODAR with higher efficiency to carry out the research effectively. Table 8.14 shows the characteristics of CSIR-NPL Monostatic SODAR.

8.3.3.2 Calibration and Measurement Capabilities

The Acoustics and Vibration activity has 34 Calibration and Measurement Capabilities (CMCs) enlisted in the BIPM database. 23 CMCs are for the sound parameter, while 11 CMCs are for the vibration parameter. The pressure sensitivity determination of LS1P, LS2P and working standard microphones in the frequency range of 31.5 Hz to 25 kHz as per ISO/IEC 61094–2 constitutes 14 CMCs, sound pressure level measurements as per ISO/IEC 60942 constitutes 4 CMCs, charge sensitivity determination of vibration transducers as per ISO 16063–11 and ISO 16063–21 in the range 40 Hz to 5 kHz constitutes 6 CMCs and acceleration level (modulus) constitute 2 CMCs. Figure 8.27 shows the parametric description of Calibration and

Table 8.14 Characteristics of CSIR-NPL (New Delhi) Monostatic SODAR

Transmitted power (electrical)	90 W
Transmitted power (acoustical)	15 W
Pulse width	100 ms
Pulse repetition period	6 s
Operational range	1000 m
Receiver bandwidth	50 Hz
Frequency of operation	2250 Hz
Acoustic velocity	340 m/s (average)
Receiver gain	80 dB
Transmit –receive antenna	Parabolic reflector dish surrounded by a conical acoustic cuff
Receiver area	2.5 m^2
Preamplifier sensitivity	A fraction of a micro-Volt

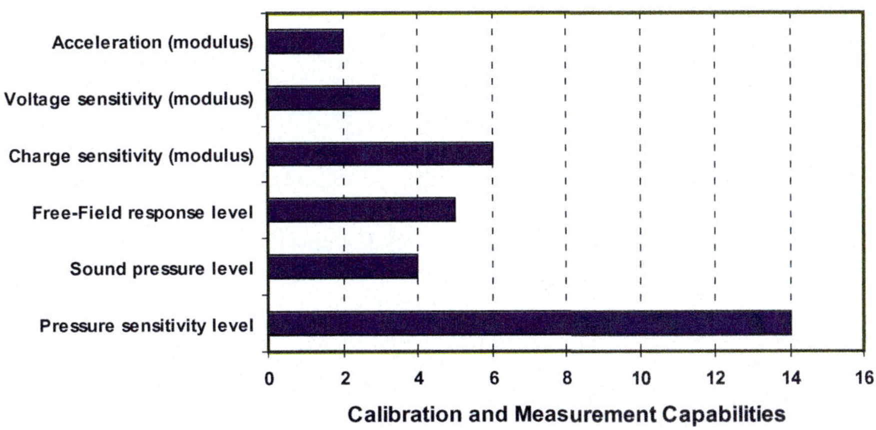

Fig. 8.27 Parametric description of calibration and measurement capabilities of acoustics and vibration area of CSIR-National physical laboratory, India in the BIPM database

Measurement Capabilities of Acoustics and Vibration Standards of CSIR-National Physical Laboratory in the BIPM database.

Table 8.15 List of key comparisons participated in sound and vibration parameter

S. No.	Key comparison details	Artefact	Result/comments
1	NPL-DPLA Bilateral Comparison	B&K 4160 and 4180 microphones	Successfully participated in the frequency range of 31.5 Hz to 25 kHz
2	APMP.AUV.A-K1	B&K 4160 microphones	Successfully participated in the frequency range of 63 Hz to 8 kHz
3	APMP.AUV.A-K3	B&K 4180 microphones	Successfully participated in the frequency range of 31.5 Hz to 25 kHz
4	APMP.AUV.A-S1	B&K 4226 Multifunction Acoustic Calibrator	Successfully participated in the frequency range of 31.5 Hz to 16 kHz
5	CCAUV.V-K1.1	B&K 8305 accelerometer	Successfully participated in a frequency range of 40 Hz to 2.5 kHz
6	APMP.AUV.V-K1.2	B&K 8305 accelerometer	Successfully participated in a frequency range of 40 Hz to 5 kHz
7	APMP.AUV.V-K3.1	Single-ended quartz-flexure accelerometer SA 704	Successfully participated in a frequency range of 0.1 Hz to 40 Hz

8.3.3.3 Participation in International Key Comparisons

Table 8.15 enlists the participation in the Key Comparison exercises in sound and vibration parameters. Participation in key comparison exercises for sound and vibration parameters has led to the inclusion of CMCs in the BIPM database. The recent participation in low-frequency vibration comparison, APMP.AUV.V-K3.1 in a low-frequency range of 0.1–40 Hz will be helpful for the inclusion of low-frequency CMCs also in the BIPM database [49].

8.3.3.4 Potential Services for Future Requirements

The continuous up-gradation of the primary and secondary standards with the latest technology is very essential for reducing the measurement uncertainties. Various services are planned to be developed in the future for serving the industry and regional laboratories in India as follows:

- Noise mapping studies for controlling the noise pollution of various cities for various pollution control boards,
- Development of sound power level standard (unit Watt in airborne sound),
- Facility for calibration of microphones in the low-frequency range from 1 to 20 Hz,

- Primary and Secondary Free-Field Calibration Standard,
- Facility for complete electrical and acoustical calibration of sound level meters as per ISO/IEC 61672.
- Development of primary vibration standard in ultra-low frequency range using a very long-stroke shaker and in the high-frequency range (5–50 kHz),
- Realization of Shock standards in range 10–1000 g using laser interferometer for providing traceability to defense, space research, and the automotive sector in India.

8.3.3.5 A Decade of Clients and Services

Calibration and Testing

The Acoustics and Vibration activity has provided calibration and testing services to various clients for the past few decades. Some of the major clients are:

- Shriram Institute for Industrial Research, Delhi,
- National Thermal Power Corporation Ltd., Korba,
- Maruti Suzuki India Ltd., Gurugram,
- Hindustan Aeronautics Limited, Lucknow,
- Fluid Control Research Institute, Palakkad, Kerala,
- Central Power Research Institute, Noida,
- Central Institute of Mining and Fuel Research, Dhanbad,
- Central Mechanical Engineering Research Institute (CSIR-CMERI), Durgapur,
- Electronics and Quality Development Centre (EQDC), Gandhinagar,
- Automotive Research Association of India, Pune,
- National Thermal Power Corporation Ltd., Auraiya, U.P,
- G E Power India Ltd., Noida,
- National Research and Technical Consortium, Parwanoo,
- Escorts Ltd., Faridabad,
- MKU, Kanpur,
- Central Pollution Control Board, New Delhi,
- Bharat Heavy Electricals Limited, Haridwar,
- Opto Electronics Factory, Raipur,
- Ramco Industries Ltd, Mylapore, Chennai,
- Rapid Metro Rail Gurgaon Limited, Gurgaon,
- Envirotech System Pvt. Ltd., Noida,
- Diamond International Index Private Limited, Gurgaon,
- Shivathene Linopack, Industrial Area, Parwanoo,
- Fluid Control Research Institute, Palakkad–(Kerala),
- USG Boral Building Products (India) Pvt. Ltd Gurgaon,
- Everest Industries Limited, New Delhi,
- ALP AEROFLEX India Pvt. Ltd., Road, Rudrapur, Udham Singh Nagar,
- Magneti Marelli UM Electronic Systems Pvt. Ltd. Gurgaon,
- RITES, Chennai,

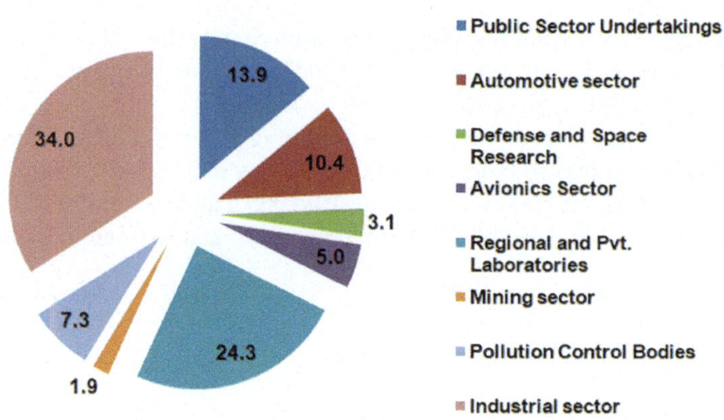

Fig. 8.28 Pie chart describing the percentage distribution of clients from various sectors

- Delhi Metro Rail Corporation Ltd. Pragati Vihar, New Delhi,
- Advance Paints Pvt. Ltd. Andheri (E), Mumbai,
- Pooja Elastomers, Mumbai,
- Shahsahib Woodwool Enterprises, New Delhi,
- BBM Acoustics India Pvt. Ltd. Pune,
- Haama Limited, India Office, Bangalore,
- Vibgyor Paints & Chemicals Pondicherry, Chennai,
- Sound Wud, Sattvaa Enterprise, Bangalore,
- Diamond International Inex Private Limited Gurgaon,
- Milton Industries Ltd., Ahmedabad,
- Envirotech Systems Pvt. Ltd., Noida,
- Ecotone Systems Pvt. Ltd. New Delhi,

The percentage distribution of clients from various sectors is shown in Fig. 8.28. It can be observed that regional and private laboratories and industrial sectors are the major clients.

Consultancy and Technical Services

The Acoustics and Vibration activity provides consultancy and technical services to various government and private organizations for solving noise and vibration issues. The major customers are Delhi Metro Rail Corporation, Bangalore Metro Rail Corporation Ltd., Archaeological Survey of India, Delhi Development Authority, Central Pollution Control Board, Dhyuthi Synthtex LLP, Soundworks, New Era Interior Collection Pvt. Ltd., Armacell India Pvt. Ltd., Annutone Acoustics Limited, Rashtrapati Bhawan, Power Grid Corporation Limited, UP Samaj Kalyan Nirman Nigam Ltd, Maple Consultants, Vijay Power Electricals, etc. The consultancy services provided in this area have been very helpful in solving the problems related to noise and vibration faced by these organizations.

Acoustics and Vibration Metrology activity offers institutional consultancy services in various fields like building acoustics, noise and vibration control, monitoring boundary layer in the lower atmosphere, installation of SODAR systems in industrial complexes to monitor air pollution & other boundary layer parameters, etc. Besides these, the activity also participates in the programmes of the Bureau of Indian Standards (BIS) and the Department of Environment in their work of laying down suitable standards for acoustical products and noise, etc. and in carrying out experimental investigations where necessary in the drafting of such standards.

Training and Academic Services

The activity is also responsible for providing training to the industry and government bodies on acoustic and vibration metrology, type approval testing of Diesel Gensets, fundamentals of noise and vibrations. This activity contributes to the Certification in Precision Measurement and Quality Control (PMQC) course every year by providing lectures and practical demonstrations to the young students for imparting knowledge about the measurement methodology and uncertainty evaluation as per the international guidelines. The activity also provides practical demonstrations and lectures to engineering students from the Indian Institute of Technology, National Institute of Technology, and Delhi Technological University. Also, the students in Physics and Engineering stream are guided for B. Tech, M. Tech, and Doctoral dissertations and summer internships. Trainings to officers of CPCB, New Delhi; staff of Boral Gypsum India and staffs of ERTL (North), and National Test House (NIT) had been provided. Also, workshops on noise and vibration control, celebrations of International Noise Awareness Day and World Hearing day, National and International conferences on Acoustics had been organized with an objective of *Make Listening Safe: A CSIR-NPL initiative to promote safe listening practices.*

Technology Development and Commercialization

Technology development and commercialization play a very significant role in the development of high quality and economical products that not only benefit the industries, but also the end users. The development of the portable secondary sound standard is one of technology that can be transferred to the relevant industries, laboratories, and manufacturers for developing low-cost systems with reduced uncertainty levels traceable to NPL. The Acoustic & Vibration Metrology has developed various technologies and commercialized them. Two such technologies are *Acoustical Light weight Interior Dry Wall Panel for High Sound Insulation* and *Noise Absorptive barrier for Metro/Railway/ Highway/Airport Noise Abatement.*

The main objective of this development is to develop a new partition panel with higher insulating value and higher STC rating value by involving new design and combinations and employing novel processing techniques. Figure 8.29 shows CSIR-NPL developed panel for this purpose.

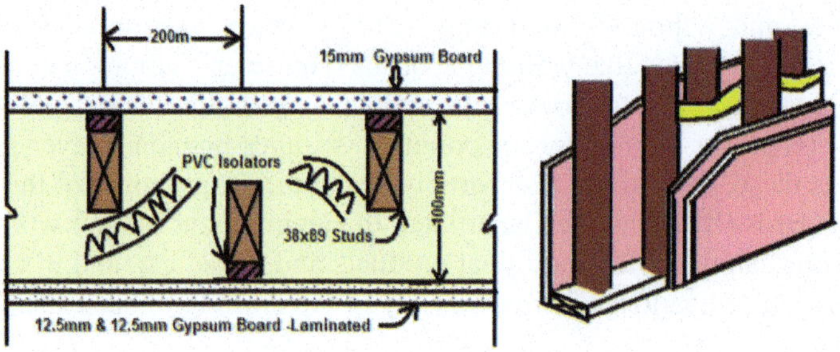

Fig. 8.29 Technology of acoustical light weight interior dry wall panel for high sound insulation to M/S. Sehgal doors, New Delhi

Fig. 8.30 Technology of noise absorptive barrier for metro/railway/highway/airport noise abatement to M/S Reliable Diesel Engineers (P) Ltd, Gurugram

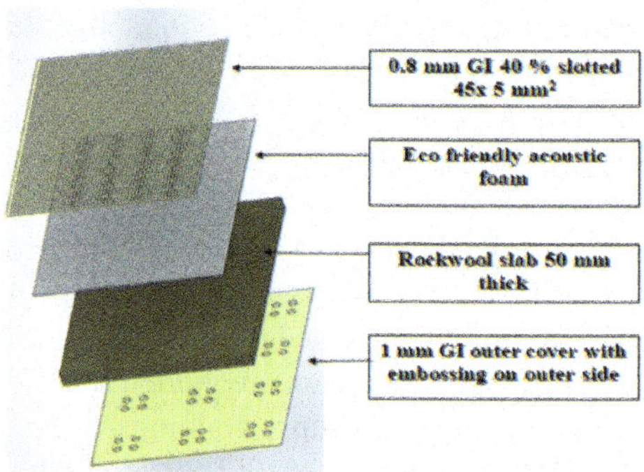

The problem of sound in buildings and residential areas through Metro, Railways, Highways, and Airports is growing nowadays. Noise barriers are the most effective method of mitigating noise from various sources such as road, railway, metro, airport and industry. Figure 8.30 shows CSIR-NPL developed Noise Absorptive barrier for this purpose.

8.3.4 Major Documentary Standards, Guides, and SOPs

The major documentary standards, guides, and Standard operating procedures are listed in Table 8.16 as follows:

8.3.5 Possible Gaps and Suggestions for Their Redressal

The major hurdle faced in the development of new standards and measurement methodologies is the economic constraints and participation of the private sector

Table 8.16 Major documentary standards, ISO Standards used in the realization of primary and secondary acoustics, and vibration standards

S. No	Documentary standard/SOP/Guide	Applications	Comments
1	IEC 61094–1, IEC standard: Measurement Microphones-Part 1-specifications for laboratory standard microphones, International Electrotechnical Commission, Switzerland, 2000	Microphones	International Standard
2	IEC 61094–2, Measurement Microphones-Part 2-Primary method for pressure calibration of laboratory standard microphones by reciprocity technique, International Electrotechnical Commission, Switzerland, 1992	Microphones	International Standard
3	IEC 61094–5, Measurement Microphones-Part 5: Method for pressure calibration of working standard microphones, by comparison, International Electrotechnical Commission, Switzerland, 2001	Microphones	International Standard
4	IEC 61094–6, Measurement microphones-Part 6: Electrostatic actuators for determination of frequency response, International Electrotechnical Commission, Switzerland, 2001	Microphones	International Standard
5	IEC 60942 Electroacoustics-Sound Calibrators, International Electrotechnical Commission, Switzerland, 1997	Sound Calibrators	International Standard, Corresponding Indian Standard IS 15059:2001

(continued)

Table 8.16 (continued)

S. No	Documentary standard/SOP/ Guide	Applications	Comments
6	IEC 61672–1: Electroacoustics—Sound level meters-Part 1: Specifications	Sound Level Meters	International Standard Corresponding Indian Standard IS 15,575 (Part 1):2005
7	ISO 16063–21:2003, Vibration calibration by comparison to a reference transducer, International Electrotechnical Commission, Switzerland	Vibration transducers	International Standard
8	ISO 16063–11:1999, Methods for the calibration of vibration and shock transducers-Part 11: Primary vibration calibration by laser interferometry	Vibration transducers	International Standard
9	IS-9901(Part-III)-1981, DIN: 52210, Part VI-1989, ISO:10140(Part II)-2010/ASTM-E-90	Sound Transmission Class (STC)	International Standard
10	IS:8225–1987 Equivalent to ISO: 354–1985 and ASTM 423–90	Noise Reduction Coefficient (NRC)	International Standard
11	ISO 3744–1994 €, ISO 8528–10:1998 €; GSR 97€29th Jan 2018 Ministry of Environment, Forest and Climate Change	DG Sets	International Standard
12	ISO 3744–1994 €, ISO 8528–10:1998 €	Mechanical Electrical Machine	International Standard
13	IS: 8225–1987 Equivalent to ISO: 354–1985 and ASTM 423–90	Multipurpose halls/Theatres	International Standard
14	ISO 20816-Part 1: 2016	Ground Vibration	International Standard
15	ISO 1996-2:2017	City Noise Mapping	International Standard

in R&D towards the development of new measurement facilities. However, efforts are focused on overcoming these hurdles by taking the following measures:

- Submission of Grant-in-aid and Consultancy projects for the involvement of private sector and overcoming funding constraints,
- Interacting with the industry for jointly working on the various issues for solving noise and vibration problems,

- Organizing Workshops/Conferences and training services every year for developing mutual collaborations and awareness.

8.4 Conclusions

It is now evident from the discussion in the chapter that the establishment of metrological standards and facilities for temperature; optical radiation; and acoustics and vibration metrology play a vivacious part in establishing traceability chain and disseminations for exclusive industrial growth of the country. The PMPs related to temperature, optical radiation and acoustics and vibration are realized with the highest levels of accuracy and precision, which enables them to serve the strategic sector, MSMEs, public and private sector undertakings, government organizations, regulatory bodies as well. The measurement capabilities are reaffirmed by periodic participation in the international inter-comparisons with other NMIs that enable to acquire the global competence and recognition.

The implementation of new SI system based on redefined SI units on 20th May, 2019, i.e. the World Metrology Day has thus diverted the attention and efforts of every country towards the development of new measurement methodology and strategies for realizing the SI units through the fundamental constant, invariant of time and space. Realization process of 2 SI base units, i.e. *kelvin* and *candela*, are described in details alike 2 SI base units, kilogram and metre in Chap. 7. CSIR-NPL is also continuously focused on realizing the temperature using the *Boltzmann constant* by *Acoustic Gas Thermometry*. The new calibration and testing facilities developed for IR thermometers at CSIR-NPL is a welcome and vital step for moving away from the era of mercury thermometer to an era of mercury-free thermometer in healthcare applications as mercury devices need to be phase-out as per WHO and *Minamata convention* so as to improve the quality of life.

It is now evident that the precision and traceable measurements of related photometric, radiometric, colorimetric and spectrometric parameters become utmost importance as they lie in the visible region of electromagnetic spectrum and thus play a vital role in human life. Efforts being made at CSIR-NPL in this direction are highlighted for the establishing state of the art, national test and calibration facilities as per the International standards for next-generation energy-efficient light products. These initiatives will not only support the energy saving by promoting the energy-efficient lamps, on 'green technology', but also eliminate sub-standard products from the market and cater to the needs of lighting industries. Also, the efforts made towards the development of photons metrology are emphasized. Photons metrology is crucial to realize standards for few photon sources for futuristic quantum optical technologies so as to meet the global objectives.

The acoustics and vibration activity is continuously focused on developing the new measurement standards and facilities for providing calibration, testing, training, academic, consultancy, and technology development services to the industry, laboratories, and other government organizations in the country. Efforts are in progress for the development of national standard facilities for calibration of sound intensity probes, noise monitoring terminals, free-field microphones, primary and secondary shock calibration and testing. The development of facility for Audiometer calibration for providing traceability to the medical institutions of the country is also at under planned stages. Also, in light of the recent orders of National Green Tribunal to CPCB for noise mapping of the cities, focussed efforts are being made on developing the noise maps of the cities in collaboration with CPCB and SPCB's and devising remedial action plans for reducing noise pollution in the country. Efforts are targeted to establish SODAR centers in different parts of the country for air pollution studies.

To date, there are total 52 CMCs registered in the KCDB of BIPM in all these PMPs, 4 in thermal, 14 in optical radiation and 34 CMCs in acoustic and vibration. Keeping in view of the vast potential of these parameters in various sectors, more efforts are continuously targeted to enhance the number of CMCs so that more industries can be benefitted from calibration, testing, and consultancy services. The overall exertions in conjunction with the strengthening of NQI, in no doubt, would play a vibrant role in economic growth of the country and realizing the ambitions of government of India for making *'Atma Nirbhar Bharat or Self Reliant India'*.

Acknowledgements Authors are thankful to all the members and students of the PMM, past and present who have directly or indirectly contributed in the development of the measurement standards and facilities. Thanks are also due to all the funding agencies that have provided partial grants for the same time to time.

References

1. T.J. Quinn, *Temperature* (Elsevier, 1990)
2. B. Fellmuth, C. Gaiser, J. Fischer, *Determination of the Boltzmann Constant—Status and Prospects* (Measurement Science and Technology, 2006)
3. J. Fischer, The Boltzmann constant for the definition and realization of the Kelvin. Ann. Phys. **531**, 1800304 (2019)
4. M.P. White, D. Gugan, Direct measurements of the dielectric virial coefficients of 4He between 3 K and 18 K. Metrologia **29**, 37–57 (1992)
5. L. Pitre, L. Risegari, F. Sparasci, M.D. Plimmer, M.E. Himbert, A.P.A. Giuliano, Determination of the Boltzmann Constant K From the Speed of Sound in Helium Gas at the Triple Point of Water (Metrologia, 2015)
6. L. Pitre, F. Sparasci, D. Truong, A. Guillou, L. Risegari, M.E. Himbert, Measurement of the Boltzmann Constant kB using a Quasi-spherical acoustic resonator Int. J Thermophys (2011)
7. J.B. Johnson, *Thermal Agitation of Electricity in Conductors* (Nature,1927)
8. C.J. Bordé, *Atomic Clocks and Inertial Sensors* (Metrologia, 2002)
9. J.A. Hall, *The Early History of the International Practical Scale of Temperature* (Metrologia, 1967)
10. J.A. Hall, C.R. Barber, *The Evolution of the International Practical Temperature Scale* (Metrologia, 1967)
11. B. Fellmuth, *Supplementary Information for the ITS-90* (BIPM, 2013)
12. H. Preston-Thomas, The International temperature scale of 1990 (ITS-90). Metrologia **27**, 3–10 (1990)
13. https://www.bipm.org/utils/common/pdf/ITS-90/Guide_ITS-90_1_Introduction_2018.pdf

14. https://www.bipm.org/utils/common/pdf/ITS-90/Guide-ITS-90-Platinum-Resistance-Thermometry.pdf
15. K. Anhalt, *Radiometric Measurement of Thermodynamic Temperatures During the Phase Transformation of Metal-Carbon Eutectic Alloys for a New High-Temperature Scale Above 1000 °C* (2008)
16. U. Pant, H. Meena, G. Gupta, D.D. Shivagan, Development and long-term stability assessment of Co–C eutectic fixed point for thermocouple thermometry, Int. J. Thermophy **40**, 80 (2019)
17. U. Pant, G. Gupta, H. Meena, K. Bapna, D.D. Shivagan, Development and realization of Fe C and Co–C eutectic fixed-point blackbodies for radiation thermometry at CSIR-NPL, Int. J. Thermophy **41**, 101 (2020)
18. Definition agreed by the 26th CGPM (November 2018), implemented 20 May 2019, BIPM, website https://www.bipm.org
19. CIE, *The Basis of Physical Photometry*. CIE Publication No. 18.2 (Central Bureau of the Commission Internationale de I'Eclariage, 1983)
20. G. Wyszecki, W.S. Stiles, *Color Science: Concepts and Methods Quantitative Data and Formulae*. (Wiley-Interscience Publications, New York, 1982)
21. F. Grum, R.J. Becherer, *Optical Radiation Measurements. Volume 1: Radiometry* (Academic Press, New York, 1979)
22. V.I. Sapritsky, Black-body radiometry. Metrologia **32**, 411–417 (1995)
23. J.E. Martin, N.P. Fox, P.J. Key, A Cryogenic radiometer for absolute radiometric measurements, metrologia. **21**, 147–155 (1985)
24. M. Krumrey, C. Herrmann, P. Muller, G. Ulm, Synchrotron-radiation-based cryogenic radiometry in the X-ray range, Metrologia. **37**, 361–364 (2000)
25. G. Zhizhong, W. Zhenchang, P. Dazhi, M. Shihua, Y. Chiuhong, Realization of the candela by electrically calibrated radiometers. Metrologia **19**, 85–92 (1983)
26. E. Ikonen, P. Karha, A. Lassila, F. Manoochehri, H. Fagerlund, L. Liedquist, Radiometric realization of the candela with a trap detector. Metrologia **32**, 689–692 (1995)
27. S.P. Morozova, V.A. Konovodchenko, V.I. Sapritsky, B.E. Lisiansky, P.A. Morozov, U.A. Melenevsky, A.G. Petic, An Absolute cryogenic radiometer for laser calibration and characterization of photodetectors. Metrologia **32**, 557–560 (1995)
28. S. Castelletto, M.L. Rastello, Luminous and photon standards by trap detectors. IEEE Trans. Instrument. Measur. **47**, 900–906 (1998)
29. Committee on Optical Science and Engineering, *National Research Council, Harnessing Light: Optical Science and Engineering for the 21st Century* (National Academy Press, Washington, USA, 1998)
30. S.H. Yun, S.J.J. Kwok, Light in diagnosis, therapy and surgery. Nat. Biomed. Eng. **1** (2017)
31. M.E.R. Darnell, K. Subbarao, S.M. Feinstone, D.R. Taylor, Inactivation of the coronavirus that induces severe acute respiratory syndrome, SARS-CoV. J. Virol. Methods **121**, 85 (2004)
32. NIST, USA Website: https://www.nist.gov/
33. PTB, Germany website: https://www.ptb.de/
34. NPL, UK website, https://www.npl.co.uk/
35. J.Y. Cheung, C.J. Chunnilall, E.R. Woolliams, N.P. Fox, J.R. Mountford, J. Wang, P.J. Thomas, The quantum candela: a re-definition of the standard units for optical radiation. J. Mod. Opt. **54**, 373–396 (2007)
36. https://www.bipm.org/utils/en/pdf/CCPR-strategy-document.pdf
37. https://www.ptb.de/emrp/siqute-home.html
38. N. Garg, M. Singh, O. Sharma, V. Mohanan, Current status of acoustic measurement standards at National Physical Laboratory of India (NPLI), New Delhi, Part 1: sound pressure. MAPAN-J. Metrol. Soc. India **22**(2), 77–90 (2007)
39. N. Garg, K. Soni, A. Kumar, T.K. Saxena, Applications of Laser Interferometry in providing traceable vibration measurements in India. MAPAN-J. Metrol. Soc. India **30**(2), 91–104 (2015)
40. N. Garg, M. Singh, O. Sharma, V. Mohanan, Current status of Acoustic measurement standards at National Physical Laboratory of India (NPLI), New Delhi, Part 2: acceleration amplitude. MAPAN-J. Metrol. Soc. India **22**(2), 91–101 (2007)

41. R. Horiuchi, H. Takahashi, T. Fujimori, S. Sato, Final report on key comparison, APMP.AUV.A-K1. Metrologia **44**, 09001 (2007)
42. H.S. Kwon, S.J. Suh, J.G. Suh, Final report on the key comparison, APMP.AUV.A-K3. Metrologia **49**, 09002 (2012)
43. S.B. Figueroa, L. Nielsen, K. Rasmussen, Report on the bilateral comparison between DPLA-DFM and NPLI. Metrologia **44**, 09005 (2007)
44. N. Garg, S. Maji, Analysis of the effect of microphone parameters in reciprocity calibrations using the taguchi method. MAPAN-J. Metrol. Soc. India **30**(3), 179–190 (2015)
45. Q. Sun, A. Liu, Report on the key comparison, APMP.AUV.V-K1.2. Metrologia **47**, 09004 (2010)
46. N. Garg, O. Sharma, A. Kumar, M.I. Schiefer, A novel approach for the realization of primary vibration calibration standard by homodyne laser interferometer in the frequency range of 0.1 Hz to 20 kHz. Measurement **45**, 1941–1950 (2012)
47. N. Garg, O. Sharma, Measurement accuracy of secondary standards of sound pressure in comparison to primary standards, MAPAN-J. Metrology Society of India **27**, 219–229 (2012)
48. V. Plangsangmas, S. Leeudomwong, A. Scott, Y. ZhongBo, Y. Huang, Final report on the regional supplementary comparison APMP.AUV.A-S1. Metrologia **51**, 09004 (2014)
49. T.U. Tsung-Hsein, W. Sheng-Han, Y. Lifeng, P. Rattanangkul, Y.B. Lee, I. Veldman, M. Singh, N. Garg, Final Report of APMP.AUV.V-K3.1: *Key Comparison in the Field of Acceleration on the Complex Voltage Sensitivity, APMP.AUV.V-K3.1* (2019)

Chapter 9
Physico-Mechanical Metrology

Part IV: Force, Pressure and Flow Metrology

Sanjay Yadav, S. S. K. Titus, Rajesh Kumar, Indu Elizabeth, Nita Dilawar Sharma, Ashok Kumar, P. K. Dubey, Afaqul Zafer, S. K. Jaiswal, Naveen Garg, Komal Bapna, and D. K. Aswal

Abstract Accurate and precise measurements of physico-mechanical quantities, such as Force, Torque, Hardness, Pressure, Vacuum and Fluid Flow are extremely important and crucial in the several manufacturing and process industries as well as

---

S. Yadav (✉) · S. S. K. Titus · R. Kumar · I. Elizabeth · N. D. Sharma · A. Kumar · P. K. Dubey · A. Zafer · S. K. Jaiswal · N. Garg · K. Bapna · D. K. Aswal
CSIR-National Physical Laboratory, Dr. K.S. Krishnan Marg, New Delhi 110012, India
e-mail: syadav@nplindia.org

S. S. K. Titus
e-mail: titus@nplindia.org

R. Kumar
e-mail: kumarr@nplindia.org

I. Elizabeth
e-mail: induelizabeth20@gmail.com

N. D. Sharma
e-mail: ndilawar@nplindia.org

A. Kumar
e-mail: ashok553@nplindia.org

P. K. Dubey
e-mail: dubeypk@nplindia.org

A. Zafer
e-mail: afaqul.zafer@nplindia.org

S. K. Jaiswal
e-mail: skjaiswal@nplindia.org

N. Garg
e-mail: ngarg@nplindia.org

K. Bapna
e-mail: komal.bapna@gmail.com

D. K. Aswal
e-mail: dkaswal@nplindia.org

D. K. Aswal (ed.), *Metrology for Inclusive Growth of India*,
https://doi.org/10.1007/978-981-15-8872-3_9

for advancements in R&D for technological developments. Realization and dissemination of measurements of these parameters are very important for the growth of the industries and the quality of the products/services rendered by these industries. CSIR-NPL, being the NMI of the country, is responsible and instrumental for realizing these physical quantities and disseminating the measurement scale to industries, test laboratories for maintaining the harmony in measurements and give a boost to the front-line R&D measurements. This chapter comprising of three broad sections and relevant subsections detailing about the standards, state of the art measurement facilities, realization of parameters, CMCs, key comparisons, consultancies, trainings, academic services and technologies developed.

In many spheres of human endeavour, from science to business to education to economic policy, good decisions depend on good measurement.

– Ben Bernanke (1953–)

9.1 Force, Torque and Hardness Metrology

9.1.1 Introduction and Historical Aspects

CSIR-NPL in terms of Force, Torque and Hardness Metrology deals with the realization of these physical parameters using the best possible fundamental principle following the internationally recognized and accepted procedures. After realizing these quantities with the lowest possible uncertainties, the measurement capabilities are reaffirmed through international intercomparisons. The standard values are disseminated across the country to several industries and stakeholders down to the level of working standards in order to establish uniformity in measurements.

Precise measurement of mechanical parameters such as Force, Torque and Hardness in various engineering applications such as testing of various engineering components made of different materials like metals, composites, plastics, polymers, etc., and testing of a concrete block of building or bridge are very much essential to ensure the quality of any designed engineering product or structure. This also ensures the fulfillment of quality control exercise demand by a respective ISO/IS standard. The performance evaluation of assembled machines is also guaranteed by the way of testing every component rigorously with the help of force testing machines commonly known as Universal Testing Machines (UTM) used by the quality control laboratories of large engineering production houses.

After rigorous mechanical testing of all individual components, they are assembled into complete machines as per the design of the system by applying the required torque through calibrated precision torque wrenches. For example, in the assembly line of automobiles, like cars, trucks, two-wheelers and household appliances like washing machines, mixers, and grinders, etc., torque wrenches are used for proper tightening of the parts to achieve optimum efficiency of the product. These applied torques should be within the prescribed limits, it should not be over-tight or loose, as over-tightening causes additional stresses leading to failure of nut and bolt assembly, whereas loosely tightened parts would lead to disengagement of components resulting in accidents. Therefore to get optimum performance, torque should be applied in a controlled manner.

The hardness of various components assembled together to make a machine is also crucial for achieving a guaranteed operational life of the machine claimed by the manufacturer. If the required hardness values are not maintained as per the design requirements, the parts will wear out and cause failure at an earlier stage. For example, the piston and cylinder assembly will fail if the hardness of the piston and cylinder bore is not meeting the desired hardness criteria. Therefore precision measurements of force, torque, and hardness values are very much important for all mechanical machines.

Keeping in view the importance and demands of force, torque and hardness metrology by the Indian industries, the first initiative was taken in early 1960, by establishing an applied mechanics laboratory within the workshop premises of CSIR-NPL in which a dial-type force testing machine installed initially for calibration of force dynamometer at CSIR-NPL for the customers. Further, in the year 1967, a three-ton cast iron deadweight force machine was indigenously fabricated and installed for the first time at the laboratory. Later, in the year 1975, a twenty-ton and a hundred-ton comparator type Reference Transducer based Force machine of Swiss make were installed and commissioned at the applied mechanics' laboratory of CSIR-NPL. Thus the initiative started to provide force measurements, calibration along mechanical testing to the Indian industries by CSIR-NPL.

After some time, a separate dedicated building was constructed during the year 1980 which was the "Force Standard Laboratory" in which the commissioning of 1 MN Dead Weight cum Lever Multiplication force machine and a huge physical balance of Schenck make was installed with the aid of PTB, Germany. The applied mechanics laboratory from the workshop premises was shifted to this new building named now as "Force and Hardness Standard". Further development and establishment of other facilities were made according to the growing demands of CSIR-NPL customers from time to time as detailed below.

In the year 1993, a 1 MN Hydraulic Multiplication machine was designed at CSIR-NPL and fabricated in collaboration with FIE, Ichalakaranji, Maharashtra and was established. In 1995, a Torque Standard machine was indigenously developed and installed. In the year 1996, another high capacity 3 MN Force machine was installed in compression mode with the help of PTB, Germany. In the year 2003, a 50 kN deadweight force machine and a deadweight-type Rockwell hardness machine were procured from Morehouse, USA and WAZAU, Germany respectively and were installed. A Torque Primary Standard Machine was procured from GTM, Germany and installed in the year 2005. During the year 2007, a dead weight type

Vicker Hardness Standardizing machine from LTF, Italy was installed and commissioned. During the year 2009–10, a 1MN dead weight cum lever multiplication force standard machine of GTM make was installed. In the year 2010, a Brinell Hardness Standardizing machine manufactured by Foundrax, UK was established. Later, during the 2014–15 Rockwell, Vickers and Brinell Secondary Hardness Standardizing Machines from Zwick Roel, UK, were installed. In the year 2019 a 5 kN Dead Weight Force Machine was installed and commissioned by an Indian company called Hemetek.

9.1.2 Significance, Applications and Relevance in Different Sectors

The role of CSIR-NPL in the field of Force, Torque and Hardness Metrology is very important to achieve the quality infrastructure in various scientific, engineering and technology sectors of the country such as Infrastructure, Space, Automotive, Aviation, Defense, Energy, Mining, Metallurgy, etc. Just to name some of the areas, where precision force measurements are indispensable, are testing of materials, weighing and balancing of heavy structures (aircrafts, ships, engineering structures, etc.) measurement of propulsion thrust of engines (jets, Automobiles, aircrafts, spacecrafts, etc.), monitoring forces on turning tools, hardness testing machines, tactile sensors, microelectronic systems, impact force, etc. In all of these applications, force transducers (artefact) are used to measure the static forces generated in the system. In the case of torque measurements, precision calibrated torque sensors are used for calibrating torque wrenches which are then widely used in automobile, aircraft and several other process industries for the appropriate tightening of the components in the assembly line for obtaining the optimum efficiency. Similarly, in the case of hardness measurements, standard hardness blocks calibrated in different hardness scales against the respective primary hardness machines are used for the verification of the hardness testing machines which are mainly used to test the quality of workpieces/components/parts manufactured in the industries for their suitability to match with the other moving parts in a system for minimizing the wear and tear between them in order to obtain the desired durability.

High precision transfer standards (artefacts) are calibrated against the reference standards/primary standards of higher accuracies maintained at the laboratory and disseminated to the user organizations across the country down to the working standards through an unbroken chain of calibrations to provide the traceability. This ensures uniformity in measurement with a defined uncertainty limits of all measurements made in the respective parameter. This is essential to produce reliable quality products at economical prices, to enhance their global competitiveness and to improve the quality of life of the citizens. The beneficiaries of such measurements are Government and Semi-government organizations, Regulatory bodies, Academia, Calibration

Laboratories, Industries and Commercial sectors, to name a few, Defence, Railways, ISRO, HAL, RRSL, RDSO, DGCA, Cement industries, Steel industries, etc.

9.1.2.1 Government Sector

For attaining the uniformity in the force and torque measurements throughout the country, CSIR-NPL made a plan and worked in consultation with the Department of Legal Metrology, some years ago, to design, develop and fabricate force standard machines to be used as force reference standards in all the five Regional Reference Standards Laboratories (RRSL) under this department. All of these standard machines were planned to be traceable to CSIR-NPL by their direct comparison using appropriate artefacts. These organizations can then cater to the demand for maintaining the national traceability in these parameters with the user industries in the respective regions. One of the RRSLs was designated as a reference laboratory (RRSL-Ref) having a force calibration machine based on first principles to provide the traceability to the other four laboratories having the force comparison machines which in turn can provide the traceability to the user industries. The complete traceability chain from CSIR-NPL to regional laboratories and to the user organizations and industries was thus established.

9.1.2.2 Infrastructure and Cement Industries Sector

Cement is the most commonly used building material due to its high compressive strength, which is tested after casting into concrete cubes of specific size and then they are subjected to compressive load using a compression testing machine (CTM) in civil engineering laboratories following a defined loading sequence. These CTM are first verified for their accuracies of generated forces against a set of reference force transducers which are duly calibrated either directly by CSIR-NPL or through some NABL accredited laboratories. This enables them to maintain an unbroken chain of measurement traceability to the primary standard established at CSIR-NPL and also provides the quality assurance related to the cement produced across the country.

9.1.2.3 Automotive and Aviation Sectors

In the automotive and aviation industries, every nut and bolt has to be tightened to predefined calculated torque values specifically designed for the assembly of different components of cars, aircraft, earthmovers, etc. Both under-tightening and over-tightening of parts can have a catastrophic failure, therefore it is important to tighten each component to the manufacturer's specified values. CSIR-NPL is maintaining the primary standard of torque from 1 to 2000 Nm range for calibrating torque meters and torque transducers to provide measurement traceability to handheld torque wrenches used at the automotive companies like Maruti, Eicher tractors, Hyundai, Honda cars and scooters, etc. and all sorts of mechanical components assembled are tightened to their optimum values.

9.1.2.4 Metals and Metallurgy Sector

Due to the great demand for steel and iron in the country of different varieties and compositions to fulfill the requirements of various engineering products manufacturers, many steel rolling plants, steel extrusion plants, steel forgings plants have come in the country to provide good quality of steel and iron. The desired mechanical properties like strength (tensile and compressive) and hardness parameters are controlled by heat treatments. The desired hardness value is important to resist the wear and tear of the machined components acting in an assembly during the operational conditions to achieve the optimal life as per the design criteria. CSIR-NPL is maintaining the hardness standards as per different IS/ISO standards in Rockwell, Vickers and Brinell hardness for the Indian industries and disseminating to Indian industries in the form reference hardness blocks to verify the hardness testers which are further used in real-time testing of different engineering products at quality control lab of production houses.

9.1.3 Facilities and Global Status

9.1.3.1 Measurement Standards and Facilities

As described above, precision measurements of force, torque and hardness play an important role in several mechanical-based industries, research and academic institutes which are actively engaged in stress measurement and characterization of materials. Precise measurement of these parameters helps us to optimize processes, enhance reliability and safety, improve the quality of products and efficiency in various sectors. CSIR-NPL has been continuously involved in establishing and developing primary standards in all these parameters in order to maintain the National standards and also to provide the necessary traceability for maintaining harmony in the measurement scale. This group is engaged in research and developmental activity for augmenting the standards at par with other leading NMIs to establish equivalence to the international system of measurements.

Force Metrology

With the advancement of science and technology, the accurate measurement of force has become essential in several industries through the required level of accuracy may vary from a few percent to a few parts per million. Keeping this in mind, the

Fig. 9.1 1MN force primary standard machine

international standard procedures for the verification of material testing machines and the calibration of force instruments along with their classification based on their measurement uncertainty are formulated. This has called upon the national NMIs to lower the uncertainty associated with force realization so as to be able to provide national traceability in the calibration of reference standards with lower uncertainties. In the last few years, a few NMIs have globally established either new facilities or have upgraded the existing facilities to lower the relative expanded measurement uncertainty values in the range of about 1×10^{-5}–5.0×10^{-5} in the realized forces using deadweight force standard machines (FSMs) in the lower range. For higher magnitudes of forces, deadweight cum lever or hydraulic amplification machines have been established with relative expanded uncertainty values in the range of 1×10^{-4} to 2×10^{-4}.

A 1 MN force machine established at CSIR-NPL serves as the primary standard for force which works on the dead weight cum lever multiplication principle as shown in Fig. 9.1. The machine now consists of a deadweight part with a 100 kN nominal force and a lever part with a 1 MN nominal force, amplified by means of a lever with a ratio of 1:10, supported by strain-controlled hinges [1]. The machine construction was carefully designed and engineered so that the machine frame received different supporting beams for a much higher stiffness. Dead weights are stacked in the required denominations to generate the desired calibration forces. These masses with their coupling system were designed suitably to reduce the risk of mass contacts and for the reason of improving the smooth asymptotic load change. The lever was also designed appropriately to allow higher stiffness and with regard to excellent stability of the lever ratio, further, the lever was covered by an enclosure to reduce ambient influences. The denominations of the deadweights are such that with the combination of the successive weights, it is possible to apply all the force steps at 10% intervals sequentially in ascending or descending order to calibrate force transducers of a capacity of 10, 20, 50 and 100 kN on the dead weight side, coupling the masses suitably to one another through self-aligned stubs. On the lever side, the force is multiplied by a factor of 10 by the lever resting on strain-controlled bearings for generating the higher forces to calibrate force transducers of a capacity of 100, 200, 500 and 1000 kN as per ISO 376-2011. The operation of the machine is controlled using a PC and calibrations can be performed according to international standards by applying the required force steps.

As there is a great demand for the calibration of force transducer of higher capacity in the market, CSIR-NPL has established a force calibration machine of 3000 kN capacity using a build-up system by comparison method in order to carter to the need of such industries and to provide the necessary traceability in measurement in such ranges. This force calibration facility is comprising of a hydraulic unit housed with a 3 MN build-up system in comparison mode. The 3MN system integrated with the machine is based on the latest design, consisting of loading pads and a balanced platen, developed by HBM, Germany and also calibrated by PTB, Germany. The performance evaluation of the established force calibration facility was carried out by a PTB, Germany calibrated 3000 kN KTN type precision force transducer procured from GTM, Germany. This 3000 kN KTN type transducer was calibrated as per ISO 376: 2011 procedure and the mean values obtained are compared with the average values reported in the PTB certificate. From these measurements and the ratios determined, it was observed out that the estimated measurement capability of the machine is well within 0.05%.

Torque Metrology

Precision torque measurements are very much in need among engineering sectors such as, in the automotive industry for testing of engines, gear-boxes, electric motors, etc., in the process industry for monitoring and control of mixing process, granule processing, rolling mills, manufacture of foils and laminates, etc., in material testing for shear and bending modulus/strength, tightening of bolts using torque wrenches, in prosthetics and robotic devices, etc. Demand for higher accuracy in torque measurement is increasing in order to improve upon the performance of the products and the processes to make them more economically competitive, efficient, safe and reliable. In practice, torque is most precisely realized by applying a dead weight force perpendicularly on a rigid lever arm. Two methods have generally been used to mount or suspend the lever arm on its axis, viz. the use of a supported beam or an unsupported beam.

CSIR-NPL has established Torque Standard machines of which one is a supported lever—dead weight torque machine and the other is an unsupported lever—dead weight torque machine. The supported lever torque machine uses very precise mechanisms to realize torque with high repeatability and hence is being used as a primary

Fig. 9.2 Torque primary standard machine

standard. The expanded uncertainty of this torque primary standard machine is evaluated to be ±0.01%, considering the uncertainty associated with the individual parameters contributing to the realized torque. It was also reaffirmed through the participation in the CCM-TK1.1 key comparison using two 1000 Nm artefacts at 500 Nm and 1000 Nm torque steps in both clockwise and counterclockwise directions. The unsupported lever torque machine is a simple design that has a long rigid beam suspended on the axis of the DUC. This machine has higher repeatability deviation and hence is being used as a secondary torque standard machine up to 2000 Nm with a claimed uncertainty of ±0.05%, evaluated against the primary standard machine using precision transfer standard torque transducers. Salient features of the primary supported lever deadweight machine are discussed below. The primary torque standard machine (2000 Nm) established is a lever-dead weight machine, comprising of the Main Frame, 2-m long lever, Strain Controlled Elastic Hinges, torque balance drive, flexible torque couplings, etc. The torque transducer under calibration can be mounted between the measuring side (lever-dead weight) and the torque balance drive-side using special hydraulic clamping bushes called ETP couplings. Strain signals produced in the hinges due to imbalance in the bending moments are brought to null to make the lever horizontal. The machine is capable of performing calibration of precision torque transducers according to the BS/DIN standards which is depicted in Fig. 9.2 [2].

Hardness Metrology

Hardness is a measure of resistance offered by a material when the material is subject to any scratch or indentation. The precision measurement of hardness plays a vital role whenever we deal with the wear and tear of moving parts, particularly with close tolerances, which are concerned in our daily life. For instance, in the case of automobiles, the tyres which constantly experience frictional forces with the surface of their contact get worn out with time, the operation of the clutch mechanism slowly loses its performance due to wear in the contact parts, the power and efficiency of engines get reduced due to the wearing out of the piston rings, etc. Development and characterization of novel materials having higher strength to weight ratio for potential applications in high-end technological industries like automobiles, aeronautics, railways, etc. are also supported by the hardness measurements. It is therefore inevitable to have a thorough knowledge of hardness values of the various materials or parts required to construct a system to have a longer, efficient operation life of the same which brings global competence and acceptance. The information from a hardness test can be used to provide insight into the durability, strength, flexibility, etc. of a variety of components, materials and products.

Various kinds of materials such as metals, alloys, ceramics, plastics, rubber, etc., are employed by industries for developing their products, depending upon the applications and accordingly the desired hardness values of these materials are identified for that purpose. The hardness values are generally expressed by hardness scales like Rockwell, Vickers, Brinell, Shore, etc. These hardness scales are established by religiously adhering to the universally defined, accepted and documented procedures to ensure uniformity of all the hardness measurements in the industry. CSIR-NPL has established hardness standardizing machines for realizing Brinell, Vickers and Rockwell hardness, scales in accordance with the respective internationally recognized ISO standards as well as the corresponding Indian standards for dissemination of the hardness scale to the user industries in order to provide the national traceability in hardness measurements.

Brinell Hardness Scale

The Brinell Hardness Testing is basically an indentation testing wherein an indenter of a carbide ball of diameter D (1, 2.5, 5 or 10 mm) is forced onto the material with an applied load F (varying from a minimum of 1 kgf up to a maximum of 3000 kgf) for a specified length of time and measuring the diameter (d) of the resulting indentation or impression. The Brinell hardness is proportional to the ratio between the test force and the surface area of indentation as expressed below.

$$\text{HB} = \text{Constant} \times (\text{Test force}/\text{surface area of indentation}),$$
$$\text{HBW} = \frac{1}{g} \times \frac{2F}{\pi D^2\left(1 - \sqrt{1 - d^2/D^2}\right)} \tag{9.1}$$

where the constant is $1/g$ ~0.102, F is the test force measured in newton, d is the mean diameter determined in mm and D is the indenter ball diameter used (expressed in mm) for making the indentation. The various Brinell hardness scales that are used with different applied forces with different diameter ball indenter are given Table 9.1.

The machine consists of two separate systems, A and B, for making the desired indentations on a given hardness block. The system A is used for the application of test force in the range of 1–62.5 kgf and the system B is used for the application test forces

Table 9.1 Test forces for Brinell hardness scales

Brinell hardness scale	Nominal value of test force *F* (*N*)	Brinell hardness scale	Nominal value of test force *F* (*N*)
HBW 10/3000	29,420	HBW 2.5/187.5	1839
HBW 10/1500	14,710	HBW 2.5/62.5	612.9
HBW 10/1000	9807	HBW 2.5/31.25	306.5
HBW 10/500	4903	HBW 2.5/15.625	153.2
HBW 10/250	2452	HBW 2.5/6.25	61.29
HBW 10/100	980.7	HBW 1/30	294.2
HBW 5/750	7355	HBW 1/10	98.07
HBW 5/250	2452	HBW 1/5	49.03
HBW 5/125	1226	HBW 1/2.5	24.52
HBW 5/62.5	612.9	HBW 1/1	9.807
HBW 5/25	245.2		

Fig. 9.3 Primary Brinell hardness standardizing machine

in the rage of 187.5–3000 kgf for making the required Brinell hardness indentations accordingly. In order to make the indentations on a standard hardness block as per the requirements of the standard, two separate control panels are used with a centralized computer system. These two machines can perform the indentation process through a controlled load cell feedback system following the predefined load-time cycle, indenter velocity, etc. as prescribed in the standard that are included with these systems which are centrally controlled using a personal computer. The entire Brinell hardness scale given in Table 9.1 is established using the Brinell Hardness Primary standard machine as shown schematically in Fig. 9.3 according to the requirement of the standard.

The frame of these two systems (*A* and *B*) include a proper loading hanger, consisting of an upper plate and a lower plate connected through three tie roads for generating the required test forces and a suitable platform is provided in between to support an anvil on which the standard hardness block under test is mounted. On the upper plate of the loading hanger, proper provision is made at the centre for fixing the selected indenter appropriately using suitable indenter holders. A mass stack is connected to the lower plate of the loading hanger for generating the required force step for specific hardness scale given in Table 9.1 according to IS 1500:2013. The Brinell hardness machines are also provided with a measuring system for determining the indentation diameters created by the indenters on the standard hardness blocks. The measuring system is comprising of a microscope with a CCD camera output that is connected with the personal computer through a compatible software for reading the input data, computing hardness value and evaluating the measurement uncertainty.

Vickers Hardness Scale

The Vickers Hardness testing is performed by forcing a diamond indenter having the shape of a right pyramid with a square base subtending an angle of 136° between opposite faces at the vertex. This would make a rhombus-shaped indentation on the surface of a test block and the diagonals of the indentation left on the surface are measured after removal of the test force, *F*. The Vickers hardness is proportional to the ratio between the test force and the surface area of indentation as expressed below

$$\text{HV} = \text{Constant} \times (\text{Test Force/surface area of indentation}),$$

$$HV = 0.102 \times \frac{2F \sin \theta/2}{d^2} = 0.1891\frac{F}{d^2} \tag{9.2}$$

The constant is $1/g = 0.102$. Test force is measured in Newton and the diagonal length is measured in mm. The various Vickers hardness scales that are used with different applied forces with the appropriate pyramid indenter are given Table 9.2.

Based on the requirements of the international and Indian standards, the Vickers Hardness scale mentioned in the above table is established which involves two systems, one for making the desired indentation on the surface of a test block and the other one for measuring the diagonals of the indentation made on the block and

Table 9.2 Test forces for Vickers hardness scales

HV scale	Nominal force *F* (*N*)	HV scale	Nominal force *F* (*N*)	HV scale	Nominal *F* (*N*)
HV 1	9.807	HV 20	196.1	HV0.05	0.4903
HV 2	19.61	HV 30	294.2	HV0.1	0.9807
HV 3	29.42	HV 50	490.3	HV0.2	1.961
HV 5	49.03	HV 100	980.7	HV0.3	2.942
HV 10	98.07			HV0.5	4.903

to determine the Vickers Hardness value of the test block. However, in the case of microvickers scales, the indenter and the objectives are fixed on a turret system and measurements are made after making the required indentation the block.

For the Vickers hardness scales of HV1 to HV100, the force application system consists of a loading basket having an upper frame connected to a lower frame through three tie rods. The required calibrated masses are selected on the lower frame of the basket to apply the appropriate test loads given in Table 9.2. A suitable indenter is mounted with a guide rail fixed in the upper frame of the basket then the basket is moved vertically up and down along with the guide rail mounted vertically in an air-bearing shaft. A rigid anvil is located at the centre of the base of the machine below the indenter frame on which the test blocks are placed for making the indentation. The required test load is applied on the hardness block by operating the loading basket as per the standard procedure through the computer-controlled software or using a joystick.

The test block is carefully placed on a measuring system, after making the indentations, and the diagonal lengths are measured to determine the Vickers hardness of the block. A Leica microscope consists of two eyepieces for viewing the object field is used for acquiring the image of the indentation. For displaying the acquired image on a LCD monitor, a high definition CCD is also provided in the system. A Heidenhain linear encoder with 25 mm stroke having the required resolution is mounted on the axis coinciding with the plane containing the indentation to be measured. A turret having five objective lenses of magnification 2.5x, 5x, 10x, 20× and 50× is mounted to ensure the correct resolution and accuracy of the measurement.

Rockwell Hardness Scale

An indenter is forced into a test block in two steps under specified conditions as depicted in Fig. 9.4. The permanent indentation depth created on the standard test block is measured and expressed in unit of 0.002 mm/0.001 mm which is defined as '*e*' from which a number known as Rockwell hardness is derived.

$$\text{Rockwell Hardness HR} = N - h/S \tag{9.3}$$

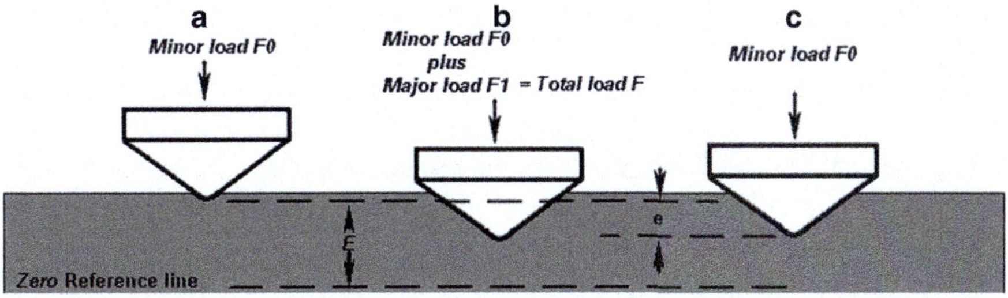

Fig. 9.4 Principle of Rockwell hardness measurement

Table 9.3 The required forces for various Rockwell hardness scales

Rockwell hardness scale	Preliminary test force F_0 (N)	Additional test force F_1 (N)	Total test force F (N)	Rockwell hardness
A	98.07	490.3	588.4	100-e
B	98.07	882.6	980.7	130-e
C	98.07	1373.0	1471.0	100-e
D	98.07	882.6	980.7	100-e
E	98.07	882.6	980.7	130-e
F	98.07	490.3	588.4	130-e
G	98.07	1373.0	1471.0	130-e
H	98.07	490.3	588.4	130-e
K	98.07	1373.0	1471.0	130-e
15N	29.42	117.7	147.1	100-e
30N	29.42	254.8	294.2	100-e
45N	29.42	411.9	441.3	100-e
15T	29.42	117.7	147.1	100-e
30T	29.42	254.8	294.2	100-e
45T	29.42	411.9	441.3	100-e

where N is number, specific to a scale (100 for HR A-C-D scales and 130 for B-E-F-G-H-K scales); h *is* permanent indentation depth and S *is* scale unit specific to the scale (0.002/0.001). The different forces required to be applied on the hardness block for establishing the entire scale is given in Table 9.3.

As illustrated above, the indenter is brought into contact with the hardness block test surface and the preliminary test force F_0 is applied without shock, vibration/oscillation of the test force. The duration of the preliminary test force is not less than 1 s nor more than 10 s. The measuring device is set to its datum position (reference level) and the initial reading is noted down with the measuring device. After recording the initial reading, the additional test force F_1 (as given in Table 9.3) is applied in not less than 2 s nor greater than 8 s without shock, vibration or oscillation. The duration of the total test force F is 2–6 s.

The additional test force F_1 is then removed after the specified duration of time and the final reading is taken immediately retaining the preliminary test force. The permanent indentation depth created by the additional test force F_1 is measured and is divided into units of 0.002/0.001 mm as applicable which is defined as '*e*'. Rockwell hardness value of the block is determined by subtracting the value of '*e*' from a constant as given in Table 9.3 for the respective scales.

9.1.3.2 Calibration and Measurement Capabilities

The CMCs established and available for providing the services to the user industries in the country related to Force, Hardness and Torque are given in the following Table 9.4.

Table 9.4 Calibration and measurement capabilities of force, hardness and torque metrology

Calibration and measurement service			Measurand range			Measurement uncertainty		
Class	Instrument or artefact: measurand	Instrument type or method	Min. Value	Max. Value	Unit	Value	Cov. factor	Level of conf. (%)
Force compression and tension	Force measuring device	Deadweight	10	1000	N	0.008%	2	95
Force compression and tension	Force measuring device	Deadweight	250	5000	N	0.008%	2	95
Force compression and tension	Force measuring device	Deadweight	500	50,000	N	0.007%	2	95
Force compression and tension	Force measuring device	Deadweight	1	100	kN	0.002%	2	95
Force compression and tension	Force measuring device	Lever amplification	10	1000	kN	0.009%	2	95
Force compression and tension	Force measuring device	Hydraulic amplification	100	1000	kN	0.020%	2	95
Torque	Torque measuring device	Lever dead weight	2	2000	Nm	0.01%	2	95
Hardness	Standard rockwell hardness block	Rock-well A ISO 6508-3	20	88	HRA	0.3 HRA	2	95
Hardness	Standard rockwell hardness block	Rock-well B ISO 6508-3	20	60	HRB	0.6 HRB	2	95
Hardness	Standard rockwell hardness block	Rock-well B ISO 6508-3	>60	100	HRB	0.5 HRB	2	95

9.1.3.3 Participation in International Key Comparisons

The different intercomparisons in the force, torque and hardness parameters are listed in Table 9.5.

The result of the APMP M.F.-K2 force comparison of the primary standard force machine at 50 kN and 100 kN force steps are depicted in Fig. 9.5 to give a feeling for the reader about the accuracy and the international standing reputation of these standards. It also reveals the importance of such inter-comparison which in fact

Table 9.5 List of Intercomparisons Participated by force, torque and hardness metrology

Parameter	Name	Participated NMIs	Range	Period	Date of report
Torque	CCM.T-K1.1	METAS, Switzerland; CEM, Spain; KRISS, Korea; NMIJ, Japan; INMETRO, Brazil; CENAM, Mexico; NPL, UK; PTB, Germany; NPL, India	0–500–1000 Nm	Jan–July 2007	April 2010
Force	PTB-NPL	PTB, Germany; NPLI, India.	(10–100) kN, (100–1000) kN	2010	Sep. 2011
Hardness	APMP.MH-S3	NIMT, Thailand; PTB, Germany; SCL, Hong Kong; KRISS, Korea; NPLI, India.	HRA HRB	October 2010	May 2012
Force	APMP. M.F.-K2	KRISS, Korea; NML-SIRIM, Malaysia; NIS Egypt; NPLI, India; A*Star, Singapore; SCL, Hong Kong; VMI, Vietnam; ITRI, Taipei; NIMT, Thailand; KIM-LIPI, Indonesia; KEBS, Kenya; NIM, China; NMIJ, Japan.	50kN 100kN	March 2011	August 2018

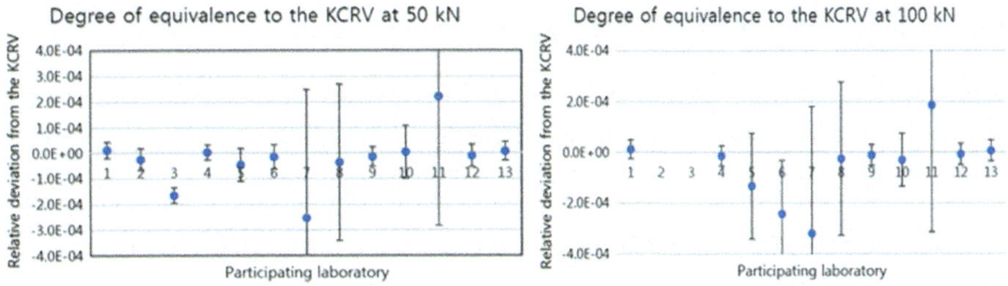

Fig. 9.5 Intercomparison data at 50 and 100kN. The list of NMIs participated in the above intercomparison is: 1. KRISS (Korea), 2. NML-SIRIM (Malaysia), 3. NIS (Egypt), 4. NPLI (India), 5. A* Star (Singapore), 6. SCL (Hong Kong), 7. VMI (Vietnam), 8. ITRI (Taipei), 9. NIMT (Thailand), 10. KIM-LIPI (Indonesia), 11. KEBS (Kenya), 12. NIM (China), 13. NMIJ (Japan)

requires continuous maintenance for achieving global competence. The results of the other intercomparison carried out in the other parameters mentioned above also look similar in terms of its closeness to the key reference value of the comparison through which the CMC is reaffirmed.

9.1.3.4 Potential Services for Future Requirements

In Force and Hardness Metrology, CSIR-NPL has established the primary standards for the realization and dissemination of force, torque and hardness scales. Presently, CSIR-NPL does not have any facility for performing the impact testing of materials and also the calibration of force proving instruments in the 5 MN force range. It is time for us to take the necessary initiative to establish these facilities in order to provide traceability to industries for the same. The demand for force measurement in the 5 MN range is increasing day by day due to its application and requirement in material testing, heavy engineering projects such as Metro rail, bridge construction, ISRO, etc. Therefore, it is important to establish a facility for calibration of forces in the higher range up to 5 MN to provide the necessary traceability for the customers which will be very much useful to them so that they need not seek any other NMIs for traceability which would also huge foreign exchange.

Impact testing of materials is very useful for predicting the likelihood of brittle fracture accuracy which plays a vital role in automobiles, hydropower sector, material testing industries, etc. It is to determine whether the material is brittle or ductile in nature and is often used in quality control applications where it is a fast and economical test. Impact test determines the amount of energy absorbed by a material during fracture. It is most commonly used on metals, however, it is also used on polymers, ceramics and composites. The purpose of impact testing is to measure an object's ability to resist high-rate loading. It is usually thought of in terms of two objects striking each other at high relative speeds. The absorbed energy is a measure of a given material's toughness and acts as a tool to study temperature-dependent brittle-ductile transition.

9.1.3.5 A Decade of Clients and Services

CSIR-NPL has been providing the apex level traceability in force, torque and hardness scale to several customers across the county such as Government organizations, Public and Private sectors, academia, Industries, etc. It gives the calibration services in force ranging from 1 N–3 MN and in torque 1–1000 Nm, having the best possible accuracy at par with other NMIs of the world. It has also been providing the calibration of standard hardness blocks in the Brinell, Vickers and Rockwell scales to various customers. CSIR-NPL issues more than 500 reports per annum in force, torque and hardness parameters.

Calibration and Testing

The Force, Torque, and Hardness Metrology have been providing the calibration services to various Government, public and private sector customers for the last four decades. Some of our major customers are ARAI and Automotive industries like Maruti Suzuki, Escorts Ltd., Mahindra, TATA, Bajaj, etc., Air Force, Defense organizations and Ordnance Factories, Bureau of Indian Standards, Cement Industries, Civil Aviation, FCRI, Palakkad, Heavy Industries like BHEL, NTPC, ISRO, Shri Harikota Space Centre, Nuclear Fuel Complex of India and Central power research Institute, Indian Railways, Steel Industries, Organizations like NCCBM, CPWD, PWD, National economics like Saudi Arabia, Nepal, Kuwait, Oman, Bangladesh, Sri Lanka, Bhutan, etc. are also served for measurement traceability by NPL. The number of reports generated through the calibration services is more than 5000 in the last decade.

Consultancy and Technical Services

The group provides consultancy and technical services for the set-up of force, torque, and hardness laboratory for the industry. Consultancy provided in the past, to establish force and torque standard machines at the various Regional reference standard laboratory, (Ministry of Consumer Affairs Food and Public Distribution) at their centers of Ahmedabad, Bangalore, Bhubaneswar, Faridabad, Guwahati with technical manuals, training and guidance to implement ISO-17025 to make the RRSLs competent in taking up force and torque calibration independently. CSIR-NPL also established for private-sector NABL accredited laboratory, i.e. DVG laboratory, Gurgaon, by supplying two deadweight force machines in the range of 5–50 kN, fabricated and tested at National Physical Laboratory, to enhance force and torque calibration capability outside CSIR-NPL.

Technical services provided to M/s Suzlon Energy Limited, Vadodara, involved in the manufacturing of windmills for the generation of power in dry and rocky areas. The windmill bolts, used to connect the big windmill blades, were tested for tensile stresses and tangential stresses. Technical services are provided to the Indian Air force by carrying out load testing of the cargo lifting cables and slings used with Air force helicopter operations for the lifting of heavy loads at remote hilly areas.

Training and Academic Services

CSIR-NPL participates regularly in the training and academic services conducted by NPL HRD group to train industry people as well as technical personnel of NABL accredited laboratories seeking technical expertise with hands-on experience related to calibration/verification of force testing machine, torque wrenches, hardness, and impact testing machines. Many such training programmes have been organized for Indian participants including people from neighboring countries like Sri Lanka, Bangladesh, Nepal, Bhutan, etc. CSIR-NPL is also contributing to academic services by offering a certification course in Precision Measurement and Quality Control (PMQC) course every year by providing theoretical lectures as well as practical demonstrations to the young engineering graduates/Diploma students to enhance their technical skills and understanding related to precision measurements. This certificate course enables them to find a place in metrology related industries and to contribute good laboratory practices to that organizations

Technology Development and Commercialization

CSIR-NPL has provided the know-how for the development and characterization of strain gauge based force transducers up to 500 kN to few private sector entrepreneurs. (Figure 9.6). The DST/industry sponsored projects are also undertaken for the development of force and torque sensors based on different materials like carbon

Fig. 9.6 Technology transfer of force transducer

and silicon-based MEMS transducers. The continuous efforts are going on for the various type of designs like the diaphragm, column, multiple bending, and shear beam force-based designs using different materials like aluminum and alloy steels. Studies/Investigation on the effect of temperature and strain gauge parameter over performance characteristics of Force Transducers are also carried out in the group with a view to optimize the performance of the transducers.

9.1.3.6 Major Documentary Standards, Guides and SOPs

The main documents required in the force, torque and hardness metrology are shown in Table 9.6.

9.1.4 Possible Gaps and Suggestions for Their Redressal

The possible scope for future expansion in the Force, Torque, and Hardness area are emerging due to the technological developments and demands in the low and the high ranges of measurements to meet with the new technical challenges in the global scenario which are as follows.

- Facility creation in low force measurement range, i.e. milli and micro newton range for experimental work at a subatomic level, Medical appliances, Robotic sensors, tactile sensors, microelectronic systems, textile, security tape, etc.
- Up-gradation of 200 kN force calibration facility to provide better accuracy in measurement to our customer in order to raise the international reputation in this range of force dissemination.
- High force range, i.e. up to 5 MN range for Weighing, balancing and Installation of bigger and heavier machines, heavy engineering structures (1 to ~30 MN)
- Torque ranges up to 5000 Nm for pneumatic pulsating torque systems, for power engineering, off-shore industry, aircraft and ship industries, etc.
- Primary Rockwell and Vickers hardness machine covering all hardness scales including superficial and micro Vickers scales.
- Impact Testing machine, a new facility to address the rising need among the industries

The economic constraint is the main hurdle along with the least participation of the private sector in the R&D towards the development of the above-required facilities. However, funding constraints can be solved by Grant-in-aid and involvement of the private sector through consultancy projects, jointly working with the industry, seeking collaborative work with the help of awareness programs, training, workshop, and seminars, etc.

Table 9.6 Major documentary standards required in the force, torque and hardness metrology

Sl. No.	Standard/guide	Brief description
1	IS 4169: 2014	Method For calibration of force-proving instruments used for the verification of uniaxial testing machines
2	ISO 376: 2011	Metallic—materials—calibration of force-proving instruments used for the verification of uniaxial testing machines
3	ASTM E-74 2013a	Standard practice of calibration of force-measuring instruments for verifying the force indication of testing machines
4	IS1828: 2015	Metallic materials—verification of static uniaxial testing machines
5	BS 7882: 2008	Method for calibration and classification of torque measuring devices
6	DIN 51309: 2005	Calibration of static torque measuring devices
7	ISO 6789: 2017	Assembly tools for screws and nuts—hand torque tools
8	ISO 6506: 2014	Metallic materials—Brinell hardness test—part 1: test method, part 2: verification and calibration of testing machines, part 3:c of reference blocks
9	ISO 6507: 2014	Metallic materials—Vickers hardness test—part 1: test method, part 2: verification and calibration of testing machines, Part 3: calibration of reference blocks
10	ISO 6508: 2015	Metallic materials—Rockwell hardness test—part 1: test method, part 2: verification and calibration of testing machines, part 3: calibration of reference blocks
11	IS 1500: 2013	Metallic materials—Brinell hardness test—part 1: test method, part 2: verification and calibration of testing machines, part 3: calibration of reference blocks
12	IS 1501: 2013	Metallic materials—Vickers hardness test—part 1: test method, part 2: verification and calibration of testing machines, part 3: calibration of reference blocks
13	IS 1586: 2012	Metallic materials—Rockwell hardness test—part 1: test method, part 2: verification and calibration of testing machines, part 3: calibration of reference blocks
14	ASTM E10–15a	Standard test method for Brinell hardness metallic materials
15	ASTM E384–11	Standard test method for Knoop and Vickers hardness materials
16	ASTM E18–16	Standard test method for Rockwell hardness metallic materials

9.2 Pressure, Vacuum, Ultrasonic and Raman Metrology

The Pressure, Vacuum, Ultrasonic and Raman metrology (PVUM) of CSIR-NPL has the responsibility for measurements in the physical quantities related to pressure, vacuum, ultrasonic at the highest possible and available accuracy. As mentioned, the definition of the SI units has been changed and now it is based on the fundamental physical constants. The unit of pressure, Pascal, is one of the derived units, now

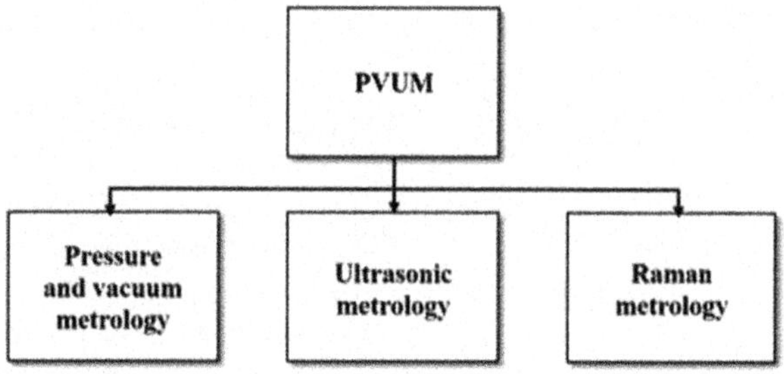

Fig. 9.7 The classification of PVUM at CSIR-NPL

realized by various NMIs using the quantum optical interferometer manometer or deadweight pressure testers.

CSIR-NPL has a very strong foot in the field of research and development which is beneficial for the nation's economy and constitutional obligation following CSIR-NPL charter of developing and maintaining primary and national standards of pressure, vacuum and ultrasonic measurements. The PVUM activity provides the calibration and measurement services and awareness related to the metrological aspects of science to grass root level industries, academicians and government laboratories and guides them about the various processes and procedures for realizing the CMCs. At present, the PVUM encompasses 3 sub-activities, as shown in Fig. 9.7.

9.2.1 *Pressure and Vacuum Metrology (PVM)*

9.2.1.1 Introduction and Historical Aspects of PVM

Pressure is defined in two ways, either as force per unit area or alternately as the height of a column of mercury in a barometer. Hence,

$$P = F/A \tag{9.4}$$

where F is the force defined as mass × gravity, and A is the area over which this force is applied. Pressure is also given as

$$P = \rho.g.h \tag{9.5}$$

where ρ is the density of the pressure transmitting fluid, g is the gravity and h is the height of the column, usually of mercury. Both the definitions of pressure are equivalent and give the unit of measurement as Pascal. The basic equipment used for both types of pressure measurements is shown in Fig. 9.8. The pressure ranges measurable by a barometer is limited by the height of the liquid column and hence for pressures usually higher than 360 kPa, piston gauges are normally used. In a piston gauge, the pressure applied at the bottom of a piston, rotating in a fitting cylinder is

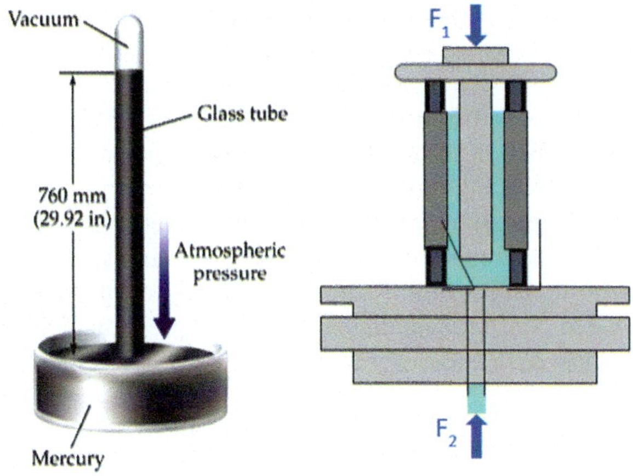

Fig. 9.8 A liquid column manometer and a piston gauge for measurement of pressure

balanced by applying weights on the top of the piston and the pressure is calculated accordingly.

Pressure can be measured either as absolute pressure or gauge pressure. Absolute pressure is measured with respect to the absolute zero pressure (or vacuum), while gauge pressure is measured with respect to the atmospheric pressure. For measurement purposes, when pressure is measured below the atmospheric pressure it is known as vacuum, while pressure above the atmospheric pressure can be either pneumatic which is driven by gas, or hydraulic which is driven by oil.

Historically, this activity started in the early seventies, named High Vacuum Group for the creation and measurement of high vacuum. Notable R&D works were carried out in this parameter and in the year 1980 the CSIR-NPL recognized the outstanding contribution by an award for the creation of facilities for Ultra High Vacuum. During 1981, a UNDP project was initiated which continued up to the year 1987. The activity was then renamed as Vacuum and Pressure Standards. Several primary standards of vacuum and pressure were established under this UNDP project. These primary and secondary standards include Ultrasonic Interferometer Manometer (UIM), Series Expansion System (SES), Orifice Expansion System (OES), Controlled Clearance Type Hydraulic and Pneumatic primary pressure Standards and also some pneumatic and hydraulic secondary standards. Most of these standards are still in service today with international compatibility.

Another important project received was the Indo US Project which continued from the year 1993 to 1996. The transfer leak standards were established in this project. Almost at the same time during the year 1994, another project was received to establish advanced surface analytical facilities (XPS, AES, SAM, etc.) which later merged with sophisticated measurement facilities of CSIR-NPL. This project continued for the next 3 years up to year 1996. Some of the facilities related to Raman metrology were also established during this period and upgraded in subsequent years through DST and CSIR sponsored projects. Ultrasonic metrology was merged with Pressure and Vacuum Metrology in the year 2017 and the activity was renamed as Pressure, Vacuum and Ultrasonic Metrology. The activity has also taken lead in adopting the quality system during 1990 and the international peer reviews of the facilities were first completed in year 2003 and again in year 2009. Over the years, modernization and up-gradation were relentlessly pursued along with building confidence in the measurements by participating in International key comparisons in various ranges of measurement.

The pressure and vacuum measurements play a vital role either directly or indirectly in numerous aspects of our lives. In our daily activities, accurate and precise measurements of pressure are essential for macroscopic as well as the microscopic world. Since the inception of modern pressure and vacuum technology, it is used in almost each and every walk of life. The classical examples are as follows:

(i) metrological/research/academic institutes where it is an essential part of laboratory to maintain and monitor environmental conditions which include temperature, barometric pressure, and humidity,
(ii) semiconductor industries which are the backbone of microelectronic industries, where fabrication of the films at the atomic and molecular level require very precise pressure and vacuum measurement depending on their fabrication conditions,
(iii) space and defense research where the precision vacuum and pressure/thrust is needed for successful completion of the projects, heavy industries, nuclear and power plants, etc.

To meet the calibration requirements of precision equipment worldwide, various primary pressure and vacuum standards have been developed by NMIs across the globe. At present, CSIR-NPL (NPLI) is among the various NMIs in the world and has the capability to calibrate pressure measuring equipment from micro Pascal to Giga Pascal pressure range using state of the art primary pressure and vacuum standards.

According to the protocols issued by the Bureau of weights and measures (BIPM), apex level instruments are categorized into various standards like primary, secondary, working, reference, transfer standards, etc. The most accurate standards which are directly traceable to the SI units are known as primary standards. CSIR-NPL has many primary pressure and vacuum standards which can meet the national or international requirements of calibration. The standards traceable to primary pressure standards are called secondary pressure standards and have higher uncertainty than that of primary pressure standards. The primary pressure standards which are at the top of the traceability chain, ideally have zero uncertainty. The uncertainty increases down in the chain and similarly, the instrument having higher uncertainty is traceable to the instrument having lower uncertainty.

The industries can only produce quality products if they have established measurement capabilities and their traceability from the NMIs for their sophisticated equipment having precise measurement and very low uncertainty. However, it is also worthwhile that low precision equipment can be traceable to any third party accreditation laboratories. This process leads to a state of the art industrial environment for a quality product which can compete with any product available worldwide in the same categories. For a trillion dollar growing economy, reliable, reproducible, and sustainable measurements are essential components, in which for measuring capabilities, NMIs are placed at the top. The roles of NMIs are not only indispensable for new precise measuring equipment development and traceability but also in providing the measurement to existing standards, equipment, devices and traceability to the national accreditation bodies in the country.

Many research/academic institutions, laboratories, private and government organizations come to CSIR-NPL for calibrating their pressure measuring instruments so that the product they are making for the market will have more reliability and quality. The accuracy in measurement of any instrument gains more confidence of the user/customer if the instruments are calibrated against the highest accuracy instruments such as primary and secondary standards but this process is also time-consuming and costly. The establishment, maintenance, and dissemination of primary standards are strategic activities that require rigorous calibration procedures and take time. Due to the calibration cost for the precision measurement, many laboratories/institutions do not come forward for the calibration at CSIR-NPL which leads to and results in a compromise in the quality measurement and hence lowers the confidence level. This chapter sensitizes the need to recognize the significance of the precise and accurate pressure measurements, important metrological applications, and disseminate the information about the CMCs of CSIR-NPL and present services being provided to Indian industries and other users.

9.2.1.2 Significance, Applications and Relevance in Different Sectors

The legal authorization of precise and accurate measurements in pressure is necessary in exporting any product for fulfilment of efficiency, quality and safety. The authorization of values is important to enhance the quality of product as well as productivity and safety of plants, manpower and machinery. The measurement of pressure performed during the manufacturing process contributes significantly to the advancement of technology. Since the activity has a wide range of pressure measurement from micro Pascal to Giga Pascal, it contributes significantly to the various technological and strategic applications [3–5] in the whole range of pressure. PVM is playing a pivotal role for industries and research institutions. The following are a few sectors relevant to the pressure and vacuum metrology:

(i) **Processes of goods and services for Industries:** For high efficiency, apex level safety, high quality, and branding of final products, pressure and vacuum are measured/monitored as process parameters. Some of the important areas are: High-pressure metal forming industries, automobile sectors, electric and diesel engines, petrochemical and pharmaceutical industries, heavy-ship vessel production, component designing and fabrication, micro- and nano-technologies, food storage and sterilization processing units, water jet cleaning and cutting machines.

(ii) **Production of renewable and non-renewable source of energy**: For silicon solar cells, nuclear power plants, thermal power plants and hydroelectric power plants, transportation, and storage of natural gases and petroleum products. High precision pressure sensors in gauge, absolute, and differential pressure modes are required in off-shore industries. Efficient energy production and utilization are directly linked with pressure metrology traceability.

(iii) **Non-invasive blood pressure (NIBP) and diagnostic and therapeutic parameters of the human body:** Type approval, clinical trials, validation, and testing of various invasive and non-invasive BP instruments need calibration using sophisticated blood pressure testing devices.

(iv) **Atmospheric pressure measurements for hazardous gases and clean environments:** A proper weather prediction along with gas leak measurements for the industrial wastage monitoring processes.

(v) **Transportation, aviation, and aerospace industries:** It involves pressure measurement in automobiles braking mechanisms and other parts; stagnation pressure measurement and atmospheric pressure measurement for speed and altitude control in the aviation sector; vehicle's tyre pressure maintenance for proper passenger safety, etc.

(vi) **Realization of quantum-based new definitions of SI units:** The determination of the Boltzmann constant and a new thermodynamic realization of temperature scale in the new and revised definition of SI unit Kelvin, necessitates accurate absolute pressure measurement with very low uncertainty in the range of 1 ppm.

(vii) High and ultra-high vacuum are an important aspect of surface science, thin-film research and related fields.

(viii) **Innovative technologies for societal and industrial applications:** Static, dynamic and differential pressure modes are widely used in strategic sectors such as automobile industries, microelectronics industries, Micro, Small, and Medium Enterprises (MSME) industries, flexible and power electronics based on pressure applications, agro-food processing and sterilization, surface coatings, high energy fission and fusion experiments, high energy particle physics (Large Hadron Collider), etc.

9.2.1.3 Facilities and Global Status

The PVUM at CSIR-NPL, has developed many primary standards having high accuracy and low uncertainty which have the capability to calibrate and authorize the standards kept at other industries, government as well as private institutions. There are apex level standards that are at par with the world-class standards and are used to provide calibration services in a pressure range of μPa to GPa for which CSIR-NPL has done extensive R&D and published several research papers in reputed journals [6–76]. CSIR-NPL has implemented the quality management system (QMS) which led to the fact that the standards are to be disseminated in the traceability chain through the calibration and testing process. The periodic calibration and testing of the instruments against the standards at CSIR-NPL give the quality check which is defined by the change in the uncertainty of measurements. Over the years, PVUM has earned the international status of maintaining the pressure standards and the calibration and testing facilities.

The PVUM possesses many primary standards as well as other standards, as shown in Table 9.7, with a large pressure range and its associated uncertainty. The most accurate pressure standard of this group is UIM which is placed at the top of the pressure standards chain in India. UIM was established in year 1982 in collaboration with NIST, USA and maintained with the uncertainty of 7.2 ppm at $k = 2$. However, the range of UIM is very limited, i.e. 1 Pa–130 kPa, therefore one more standard, i.e. air piston gauge (APG) was established in 2002, which has the pressure range of (6.5–360) kPa with an uncertainty of 12 ppm at $k = 2$. Since the handling of UIM is a tedious task, therefore an automatic forced balanced piston gauge (FPG) pressure standard in the range of 1 Pa–15 kPa was established in 2010 having the uncertainty of 25 ppm at $k = 2$. Two primary vacuum standards in the range of micro Pascal to several Pa, designated as OFS and SES, have also been established with technological upgradation in due course of time.

In higher pressure range, pressure balance/piston gauges/dead weight type primary pressure standards were established in 1999 (NPLH1), 2001(NPLH2), 2009 (NPLI-P1 and P10), 2013 (NPLH3) from 1 MPa to 1.4 GPa range, for both hydraulic and pneumatic pressure ranges. The piston gauges for pressure standards are usually of 3 types depending upon the pressure range. These are simple piston gauges, re-entrant type piston gauges and controlled clearance type piston gauges. For the pneumatic pressure ranges, the primary as well as secondary standards are usually of simple type while for higher pressure ranges in hydraulic piston gauges, primary systems are controlled clearance piston gauges (CCPG) while others are either re-entrant type or free deformation type simple designs.

Figure 9.9 shows the 3 types of piston gauges used as standards. For pneumatic pressures, a simple piston gauge is normally used. However, with increasing pressure ranges, the gap or the clearance between the piston and the cylinder becomes a limitation due to distortion in the piston and cylinder. To circumvent this issue, re-entrant and controlled clearance type piston gauges are used which give better control over the piston-cylinder clearance and hence can be used for higher, usually hydraulic pressure ranges. The above-mentioned primary pressure standards provide traceability to a number of other transfer/secondary standards. All these instruments are calibrated periodically with either primary or other secondary standards having lower uncertainty to uphold the quality of measurements.

Measurement Standards and Facilities

As mentioned, the available pressure and vacuum standards are in the range of μPa–GPa. The standards are divided into three categories (i) barometric and high vacuum, (ii) pneumatic pressure, and (iii) hydraulic pressure. A full list of the standards being established and used at CSIR-NPL is shown in Table 9.7. Some of these standards are primary, which are used as a reference standards to calibrate secondary and other standards with high accuracy and precision. From Table 9.7, it can be seen that

Table 9.7 List of measurement standards and facilities available

Pressure regions	Instruments	Range
Barometric and high vacuum [6–23]	Ultrasonic interferometer manometer (UIM)—primary standard	1 Pa–130 kPa
	Air piston gauge (APG)—primary standard	(6.5–360) kPa
	Forced balanced piston gauge (FPG)	1 Pa–15 kPa
	DPI—145—secondary standards	(100, 130, 400) kPa
	Capacitance diaphragm gauge—secondary standards	(133.32–0.013) kPa
	Static expansion—primary standard (SES)	1000–10^{-4} (Pa
	Orifice flow standard—primary standard (OFS)	(0.1–10^{-6}) Pa
	SRG–transfer standard	(1–10^{-4}) Pa
	Spinning rotor gauge—transfer standard	(10^{-3}–10^{-5}) Pa
	Ionization gauge—transfer standard	(0.1–10^{-6}) Pa
	Extractor gauge—transfer standard	(10^{-2}–10^{-6}) Pa
	Standards helium leaks	10^{-9} std.cc/s to 10^{-4} std.cc/s
Pneumatic [24–32]	Large diameter piston gauge—primary standards	
	NPLI-P1	(0.1–1) MPa
	NPLI-P10	(1.0–10) MPa
	Ruska (USA) model# 2465—secondary standards	
	NPLI-105	(1.4–105) kPa
	NPLI-4	10 kPa–4 MPa
	Ruska (USA) model# 2470—secondary standards	
	NPLI-12	10 kPa–12 MPa
	Desgranges Et Huot (France) model# 5502—secondary standards	
	NPLI-8	(0.4–8) MPa
	NPL-20	(0.8–20) MPa
	NPLI-40 Differential dual-piston gauge	(0.8–40) MPa
	Digital pressure transducers	(0.1–4) MPa (1.0–10) MPa

(continued)

Table 9.7 (continued)

Pressure regions	Instruments	Range
Hydraulic [33–76]	Controlled clearance piston gauges—primary standards	
	NPLH1	20 MPa–1 GPa
	NPLH2	(10–200) MPa
	NPLH3	(5–500) MPa
	Ruska (USA) model# 2450—secondary standards	
	NPL28MPA	(0.07–28) MPa
	NPL140MPA	(0.24–140) MPa
	NPL280 MPA	(0.48–280) MPa
	Desgranges Et Huot (France) model# 5306—secondary standards	
	NPL10MPA	0.1 MPa–10 MPa
	NPL20MPA	0.04 MPa–20 MPa
	NPL50MPA	0.1 MPa–50 MPa
	NPL100MPA	0.2 MPa–100 MPa
	NPL100MPN	0.2 MPa–100 MPa
	NPL200MPN	0.4 MPa–200 MPa
	NPL500MPA	1.0 MPa–500 MPa
	NPL500MPN	1.0 MPa–500 MPa
	Differential dual-piston gauge	0.1–50 MPa (For both as line and static pressures)
	Digital pressure standards	0 MPa–100 MPa 0 MPa–200 MPa

The numbers shown in the parenthesis are the cited references wherein the details of the respective measurement standards are published

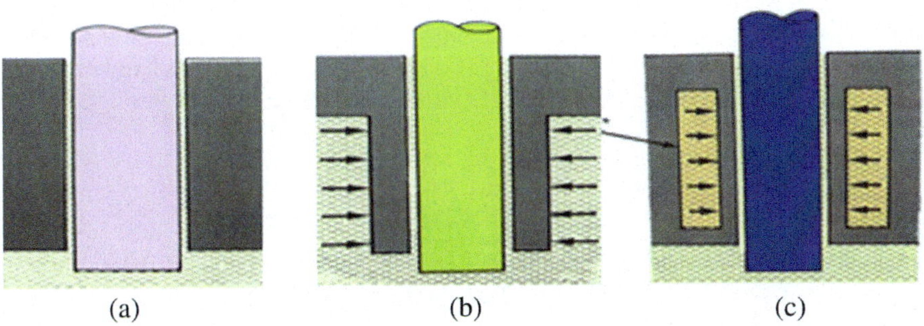

Fig. 9.9 Three types of piston gauges: (**a**) Simple, (**b**) re-entrant and (**c**) controlled clearance type

PVM has 10 primary standards in varying vacuum and pressure ranges and many more secondary/transfer standards which provide apex level calibration and measurement capabilities to various industries. Although, the details about the primary and high precision pressure and vacuum standards established at NPL have already been published in many highly reputed journals, however, for readers understanding, we have explained some of the primary standards, such as UIM, OFS, SES, and piston gauge.

The UIM [7, 13, 22] is based on the principle of measuring pressure obtained by the product of mercury column height, the density of mercury, and acceleration due to gravity. The column height is obtained using the time of flight technique at 44 different frequencies in the ultrasonic range. The phase measurement combined with the time of flight measurement using the exact fractional algorithm provides highly accurate column length and hence the respective pressure values. After considering all the factors for measuring pressure, the pressure indicated by UIM is given as,

$$P_{\mathrm{UIM}} = \frac{\rho_{20} g \{L_2 - (L_1 + L_3)/2\}}{1 + \alpha_d (T_{\mathrm{UIM}} - 20)} \tag{9.6}$$

where ρ_{20} is the density of mercury at 20 °C and is 13,545.855 kg/m^3 [55], α_d is the temperature coefficient of the density of mercury with a value of 181.15×10^{-6} $°C^{-1}$ [56], L_1, L_2, and L_3 are lengths of the mercury column in the side column 1, centre column and side column 2, respectively, T_{UIM} is the temperature inside UIM, and g is the local acceleration due to gravity at the location of UIM.

The OFS is the primary high vacuum standard and is used for the calibration of vacuum gauges in the molecular flow region, i.e. when the mean free path of the gas molecules is large compared to the characteristic dimension of the vessel. This condition sets the upper limit of pressures to about full scale, and the lower limit is set by the ultimate pressure achievable in the vacuum chamber and the limits of the flow measuring devices. The calibration is done by direct comparison of the gauge under test with a resident calibrated spinning rotor gauge.

For OFS, the gas used to calibrate the device under test is pumped through the flow meter, which is passed from the upper chamber of OFS to lower chamber via an orifice with known conductance (C) connected across two chambers under equilibrium conditions [9, 10, 14–16]. If Q is the flow rate of the gas throughout then

$$P = \frac{Q}{C}\left(1 + \frac{C}{S_P}\right) \tag{9.7}$$

where P is the pressure upstream of the orifice or the pressure in the calibration chamber, and S_p is the pumping speed available in the chamber below the orifice.

The third standard, SES is based on the principle of static expansion of the test gas. In this method, gas at a known high pressure is enclosed in a small volume. It is then allowed to expand, under isothermal conditions, into a much larger volume which has been previously evacuated and isolated from the pumps. Thereafter, making use of the gas equation, the pressure generated in the large volume after the expansion is calculated. If P_s is the initial pressure of the gas in the small volume, then the pressure P_L generated after the expansion is given by

$$P_L = \frac{P_s T_L}{RT_s(1 + B'' P_s)} \tag{9.8}$$

where $R = (V_L + V_s)/V_s$ is the ratio of the final volume to the small vessel volume, V_L is the volume of the large vessel, V_s is the volume of the small vessel, T_s is the absolute temperature of the small volume and T_L is the absolute temperature of large volume. The parameter $B'' = B/R_m T$ is the second virial coefficient to take into account the deviation of the gases from ideal gas behaviour, R_m is the molar gas constant.

Lastly, the piston gauges (PGs) are used as primary pressure standards for barometric, pneumatic, and hydraulic pressure ranges. The general working principle of all the piston gauges is to measure pressure by calculating force per unit effective area of the piston-cylinder assembly. The general equation of calculating pressure by piston gauges ($P_{\text{DWT/PG}}$) in controlled clearance mode, is given by the equation,

$$P_{\text{DWT/PG}} = \frac{\sum_i mi \cdot g \cdot (1 - \rho_{\text{air}}/\rho_{mi}) + \gamma \cdot C}{A_0 \cdot (1 + \lambda P_n + \lambda' P_n^2) \cdot [\{1 + (\alpha_p + \alpha_c)(T - T_r)\} \cdot \{(1 + d(p_j - P_z)\}]} \pm \Delta P \tag{9.9}$$

where $1 - \frac{\rho_{\text{air}}}{\rho_{m_i}}$ is the air buoyancy correction, ρ_{air} and ρ_{m_i} are densities of air and mass respectively, γC is the force exerted on the piston by surface tension of the transmitting fluid, $\{ 1 + (\alpha_p + \alpha_c)(T - T_r)\}$ is the thermal correction factor, $(1 + \lambda P_n + \lambda' P_n^2)$ shows the change of effective area with pressure, $\{1 + d(p_Z - p_j)\}$ is the jacketed pressure cylinder distortion correction term, g is the local acceleration due to gravity near the pressure standard, α_p and α_c are thermal expansion coefficients of piston and cylinder materials in piston gauge or deadweight tester, γ is the surface tension, C is the circumference of the piston where it emerges from the fluid, A_0 is the zero pressure effective area, m_i is the mass of ith mass combination applied on piston-cylinder (p-c) assembly, T is the temperature of the p-c assembly T_r is the temperature at which A_0 was calculated, λ and λ' are first and second-order pressure distortion coefficients, d is the cylinder distortion coefficient, p_z is jacket pressure coefficient, p_j is actual applied jacket pressure and ΔP is the head correction pressure. In case of simple and re-entrant modes, the term $\{1 + d(p_Z - p_j)\}$ becomes unity as there is no jacket pressure applied. In case of pneumatic deadweight testers, the term γC is not considered as surface tension of gas is considered almost negligible.

In the pneumatic pressure ranges, purified, moisture-free nitrogen gas is generally used as the pressure transmitting fluid while in case of higher pressures in hydraulic ranges, sebacate oil stored in reservoirs is used to drive the system with pressure.

The laboratory is thus fully equipped with expertise, measurement techniques, knowledge, and instruments to provide the following calibration services to the users;

(i) Various vacuum gauges, e.g. Pirani, Quartz Bourdon, thermocouple, Penning, ionization, capacitance diaphragm, inductive diaphragm, spinning rotor, blood pressure measuring machines are being calibrated.
(ii) Pressure Standards and gauges e.g. piston gauges, deadweight testers, pressure balances, transducers, pressure transmitters (in absolute, gauge, and differential pressure modes) in both hydraulic and pneumatic pressure regions, dial gauges, etc.

Calibration and Measurement Capabilities

CMCs in the pressure and vacuum ranges were approved after rigorous international peer reviews conducted by internationally reputed technical experts in the area and the approved CMCs are listed in the KCDB database of the BIPM website [77]. As of date, there are 17 CMCs in the various ranges of pressure and vacuum which would be further increased in the next planned peer review. At present, the uncertainties of our standards are at par with other NMIs across the globe and are shown in Table 9.8.

Participation in International Key Comparisons

The CIPM-MRA [78] provides the international equivalence of the calibration and measurement certificates issued by NMIs. It also provides the guidelines to NMIs for registering CMCs and also the opportunity to participate in the key comparisons exercises for establishing metrological equivalence. The CSIR-NPL has successfully participated in 19 international, bilateral key and inhouse comparisons [6, 7, 16, 18, 19, 25–29, 35, 40, 42, 46, 48, 49, 51–55, 64–66] in pressure and vacuum ranges for which details are shown in Table 9.9. Therefore, the services provided are internationally acclaimed and compatible [71, 72].

The global status of CMCs is shown in Fig. 9.10, which are modified from time to time, depending upon the improvement of pressure measuring capabilities and peer review process. The figure shows that India is among the top 6 nations with the highest number of CMCs. This number is further going to increase in the next planned International peer review. CSIR-NPL follows the ISO/IEC 17025:2005 and recently has initiated the process to switch over to ISO/IEC 17025:2017 for maintaining its quality system. At present, we are almost in the last phase of implementation of ISO/IEC 17025:2017. The metrological activities of CSIR-NPL are peer-reviewed through RMO, APMP, an Indian CAB. The PVM group has achieved the global status by the expertise and diligent pursuit of improvement over the years. To maintain the status of NMI and increasing requirement in the society, it is an urgent need and a constant endeavour, to regularly update the standards which this group is having like extend; standards such as extending the pressure range and lower the uncertainties.

The achievements of PVM at CSIR-NPL can be further seen by the comparison of results of the key comparisons in which this group has participated in several Key Comparisons like CCM.P.K7 in 2005 [52] RMO key comparisons, APMP.M.P.K7,

Table 9.8 The current CMCs of the pressure and vacuum standards are approved and listed in the KCDB database of the BIPM. The uncertainties are presented at $k = 2$

Quantity	Measurand minimum value	Measurand maximum value	Unit	Type of uncertainty	Uncertainty equation	Comment uncertainty equation
Absolute pressure: gas medium	0.001	130.0	kPa	Absolute	$[(9.2\text{E}{-}03)^2 + (7.2\text{E}{-}06p)^2]^{1/2}$	p pressure in Pa
Gauge pressure: gas medium	0.001	130.0	kPa	Absolute	$[(9.2\text{E}{-}03)^2 + (7.2\text{E}{-}06p)^2]^{1/2}$	p pressure in Pa
Absolute pressure: gas medium	6.5	360.0	kPa	Absolute	$(1.4\text{E}{-}01)^2 + (12\text{E}{-}06p)^2]^{1/2}$	p pressure in Pa
Gauge pressure: gas medium	0.036	4.0	MPa	Absolute	$22\text{E}{-}06p$	p pressure in MPa
Gauge pressure: gas medium	4.0	8.0	MPa	Absolute	$26\text{E}{-}06p$	p pressure in MPa
Gauge pressure: gas medium	8.0	12.0	MPa	Absolute	$32\text{E}{-}06p$	p pressure in MPa
Gauge pressure: gas medium	12.0	20.0	MPa	Absolute	$33\text{E}{-}06p$	p pressure in MPa
Differential pressure, gas medium	30.0	150.0	kPa	Absolute	$(2.5\text{E}{-}06p_l)^2 + (2.9\text{E}{-}05D_p)^2]^{1/2}$,	p_l line pressure in Pa and D_p differential pressure in Pa The uncertainty is expressed in kPa
Gauge pressure: liquid medium	200.0	500.0	MPa	Absolute	$135\text{E}{-}06p$	p pressure in MPa
Gauge pressure: liquid medium	100.0	200.0	MPa	Absolute	$50\text{E}{-}06p$	p pressure in MPa

(continued)

Table 9.8 (continued)

Quantity	Measurand minimum value	Measurand maximum value	Unit	Type of uncertainty	Uncertainty equation	Comment uncertainty equation
Gauge pressure: gas medium	20.0	360.0	kPa	Absolute	12E−06*p*	*p* pressure in Pa
Gauge pressure: gas medium	20.0	40.0	MPa	Absolute	36E−06*p*	*p* pressure in MPa
Gauge pressure: liquid medium	500.0	1000.0	MPa	Absolute	250E−06*p*	*p* pressure in MPa
Gauge pressure: liquid medium	50.0	100.0	MPa	Absolute	48E−06*p*	*p* pressure in MPa
Gauge pressure: liquid medium	0.1	50.0	MPa	Absolute	45E−06*p*	*p* pressure in MPa
Absolute pressure: gas medium, vacuum	0.05	10.0	Pa	Absolute	4E−03*p*	*p* pressure in Pa
Absolute pressure: gas medium, vacuum	3.0E−6	0.1	Pa	Absolute	2E−02*p*	*p* pressure in Pa

in 2005 [53], APMP-SIM.M.P.K7 in 2006 [54], CCM.P.K13 in 2012 [64], and APMP.M.P.K13 in 2015 [65]. The comparison of the values of all the participating institutes in the comparisons like APMP.M.P-K6, APMP.M.P-K1c, APMP.SIM-M.P-K1c, CCM.P-K13, APMP.M.P.K13 and EURAMET.M.P.K13, and CCM.P-K3 are shown in Fig. 9.11a–d [16, 27,28, 29, 64–66], respectively. In all, CSIR-NPL has participated in 19 comparisons in various ranges of vacuum and pressure and established its capability in each range. Figure 9.11 shows the degree of equivalence achieved by CSIR-NPL in various international comparisons vis a vis the other participating countries and it is apparent that excellent results were achieved and demonstrated. It is also clear from Fig. 9.11 that PVM activity CMCs are comparable to the topmost NMIs of the world.

Table 9.9 List of participated international intercomparisons by the PVM of CSIR-NPL

S. No.	Comparison Identifier	Description	Parameters	Measurement period	Pilot institute	Status
1	CCM.P-K9	Pressure measurements		1981–1987	KRISS Korea	Approved for provisional equivalence
2	CCM.P-K11	Pressure measurements (gauge mode)		1981–1989	LNE France	Approved for provisional equivalence
3	CCM.P-K10	Pressure measurements		1983–1995	NPL	Approved for provisional equivalence
4	CCM.P-K8	Pressure measurements in oil (gauge mode)		1996–1999	LNE France	Approved for provisional equivalence
5	APMP.M.P-K1.c	Pressure measurements in gas (gauge mode)	Pressure: (0.4–4.0) MPa	1998–2001	NPLI	Approved for equivalence
6	CCM.P-K4	Pressure measurements in gas (absolute mode)		1998–1999	NIST USA	Approved for equivalence
7	APMP.M.P-K6	Pressure measurements in gas (gauge mode)	Pressure: (20–105) kPa	1998–2001	NPLI	Approved for equivalence
8	CCM.P-K3	Pressure in gas, ultra-high vacuum (absolute mode)		1998–2002	NIST USA	Approved for equivalence
9	APMP.M.P-K7	Pressure measurements in oil (gauge mode)		2002–2004	NMIJ Japan	Approved for equivalence
10	CCM.P-K7	Pressure measurements in oil (gauge mode)	(10–100) MPa	2003–2004	PTB Germany	Approved for equivalence
11	APMP.SIM.M.P-K1.c	Pressure measurements in gas (gauge mode)	(0.4-4.0) MPa	2003–2004	NPLI	Approved for equivalence
12	APMP-SIM.M.P-K7	Pressure measurements in oil (gauge mode)	(40 200) MPa	2003–2004	NPLI	Approved for equivalence

(continued)

Table 9.9 (continued)

S. No.	Comparison Identifier	Description	Parameters	Measurement period	Pilot institute	Status
13	APMP.M.P-S2	Pressure measurements		2005	NPLI	Approved
14	SIM.M.P-S1	Pressure measurements (absolute mode)		2007–2008	NIST USA	Abandoned
15	CCM.P-K12	Leak rates	Temp.: 23 °C	2007–2009	PTB Germany	Approved for equivalence
16	CCM.P-K13	Hydraulic gauge pressure	(50–500) MPa	2008–2010	PTB Germany	Approved for equivalence
17	APMP.M.P-K9	Pressure measurements (absolute mode)		2009–2013	KRISS Korea	Approved for equivalence
18	APMP.M.P-K13	Hydraulic gauge pressure		2010–2013	NMIJ Japan	Approved for equivalence
19	APMP.M.P-K4	Pressure measurements in gas (absolute mode)		2015–2016	KRISS Korea	Measurements completed

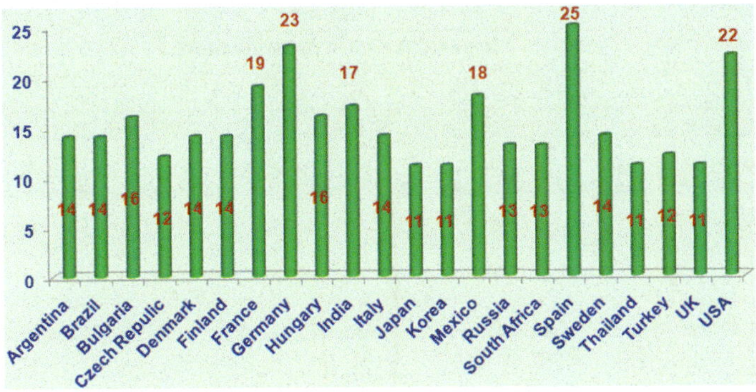

Fig. 9.10 A comparison of the CMCs of 23 NMIs across the globe who have published at least 10 or more CMCs in the KCDB database of the BIPM

Potential Services for Future Requirements

PVM capabilities are for measuring the pressure from micro Pascal to Giga Pascal with the lowest possible uncertainty related to the medical as well as industrial applications. One of the primary standards in this group, i.e. UIM, is having mercury, which is a toxic material. Most of the NMIs have banned mercury-based products. Therefore, the PVM group has established APG as a reference standard in the range of (6.5–360) kPa using rarefaction gas dynamics and dimensional metrology and for vacuum and ultra-high vacuum using static SES and OFS [79–81]. At the global level the other NMIs like NIST, USA, NMIJ, Japan, etc. have developed Fabry Perot cavity-based optical interferometer manometer (OIM) pressure standards [82, 83]. The CSIR-NPL has also conceptualized the project of OIM and is committed to establish the same in the coming years to compete with the international community and quantum realization of Pascal.

In the defense sector, there have been recent demands for extended calibration in pneumatic pressure ranges which is one of the major targets for the near future. Further significant improvement in our CMCs is expected both in numbers as well as improved uncertainties. Further, new vistas are being explored for areas such as dynamic pressure measurements, design and development of indigenous pressure transducers with lowest possible uncertainties, establishment of extended facilities, etc.

A Decade of Clients and Services

The PVM issues more than 250 calibration certificates annually at present and organizes many seminars and technical programs and also provides technical and consultancy services from time to time. There are some regular clients of PVM like BHEL, ERTL, ETDC, ISRO, FCRI, GAIL, IOCL, HPCL, SAIL, RRSL, FCRI, Indian Airlines, BIS, MSME, NTH, ORDINANCE Factories, Bharat Forge, NCCBM,

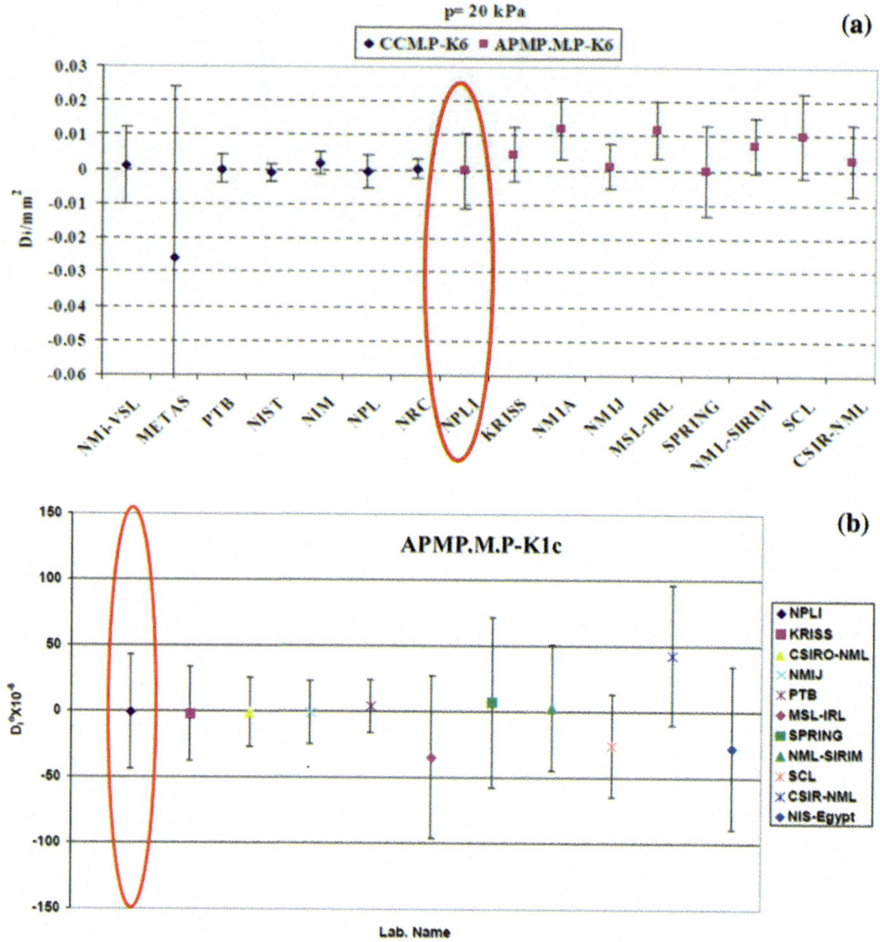

Fig. 9.11 Recent global degree of equivalence (Di) in key comparisons (**a**) CCM.P-K6 and APMP.M.P-K6 at 20 kPa, (**b**) APMP.M.P-K1c in the pressure range of 0.4–4.0 MPa and (**c**) CCM.P-K13, APMP.M.P.K13 and EURAMET.M.P.K13 at 100 MPa, (**d**) key comparison (CCM.P-K3) of mean gauge pressure reading (p_{jU}), with primary standards set at 9×10^{-4} Pa, for each NMI and expanded ($k = 2$) uncertainty in mean gauge reading shown as error bars

IDEMI, NTPC, BSF, DRDL, CIPET and many more private, NABL accredited laboratories, PSUs, MSMES, intra and inter departments in CSIR-NPL, etc.

Calibration and Testing

As explained earlier the PVUM does the calibration and testing for various government and other industries using the standards listed in Table 9.7. The instruments of the industries are calibrated in the pressure ranges with the respective uncertainties listed in Table 9.8. Every year demands from the industries are increasing. During

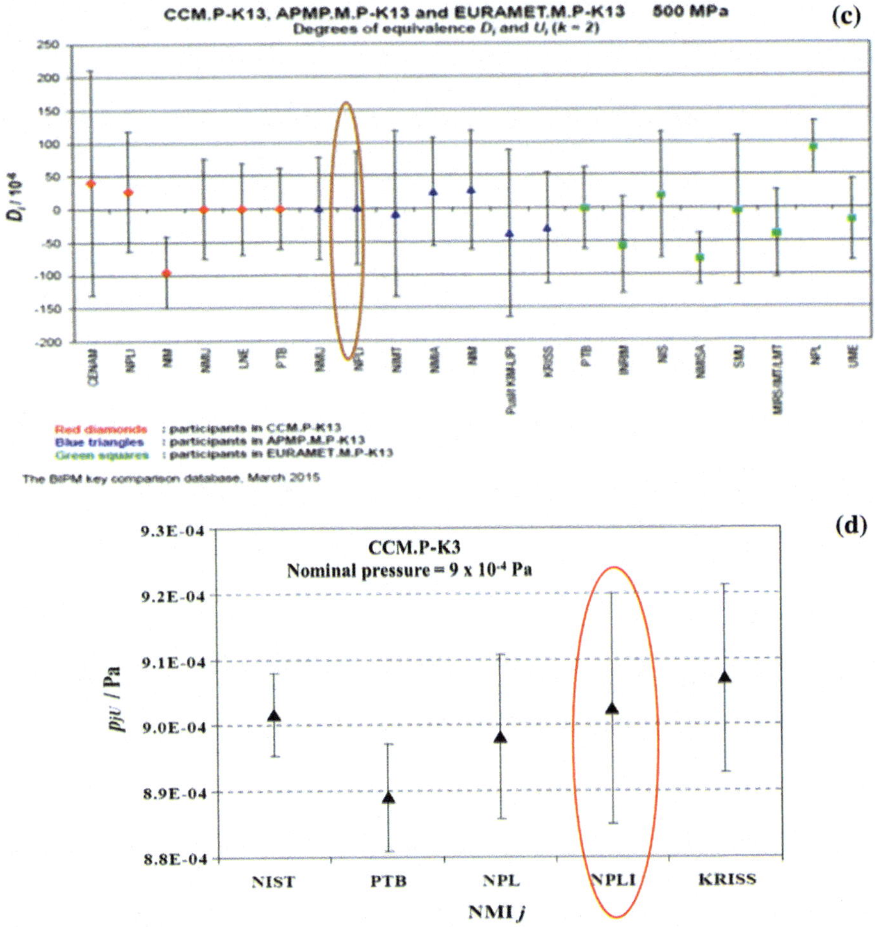

Fig. 9.11 (continued)

the last 10 years, more than 1600 calibration reports were generated by providing calibration services to various customers and stakeholders.

Consultancy and Technical Services

In addition to calibration and testing services, the PVM provides consultancy and technical services. These services include the development of skilled manpower via training as well as research and development (R&D) in pressure and vacuum metrology. The PVM is fully equipped with, expertise, various measurement techniques, and standards/instruments to provide the consultancy services to the users such as; (i) Metrological characterization of pressure balances in pneumatic as well as hydraulic, in the pressure ranges from high vacuum to Giga Pascal; (ii) To establish the laboratories which can provide calibration in pressure and vacuum measurements; and (iii) Reference standards for calibration of blood pressure are also available for measuring instruments in static and dynamic conditions.

While providing these services, the group has published several research papers and completed various projects to the respective clients. These projects were primarily for the characterization of piston gauges, which apart from regular calibration, include estimation of the effective area of the piston-cylinder assembly, estimation of distortion coefficient and the associated measurement uncertainty.

Training and Academic Services

The training is provided to the fresh personnel in the field of pressure and vacuum metrology and produce scientists, engineers and technical officers. CSIR-NPL also helps Ph.D. scholars, research associates, and B.Tech as well as M.Tech research interns to learn the mechanism and calibration of the pressure and vacuum standards. The research fellows are provided a deep understanding of calibration procedure with complete protocols, uncertainty evaluation, handling of the standards, etc and followed by practical hands-on training. During the last 5 years, more than 300 personnel from different organizations like BSB, Bhutan, NBSM, Nepal, BSTI, Bangladesh, MUSSD, Sri Lanka, Legal metrology officer, Air India, ISRO, DRDO, BHEL, Sri Lanka Standards Institution, Sri Lanka, calibration laboratories, Instrument manufacturers, etc. including students have been trained.

Technology Development and Commercialization

The PVM has also developed a cross floating valve (CFV) which is to be commercialized. The cross-floating valve is an important component of the hydraulic cross floating of piston gauges and plays a significant role in improving accuracy of measurements as well as saving the time required for calibration. The details of CFV developed in the pressure and vacuum lab are given in (Table 9.10).

Major Documentary Standards, Guides and SOPs

The major documentary standards, guides and SOPs used in pressure and vacuum metrology are given in Table 9.11.

9.2.1.4 Possible Gaps and Suggestions for Their Redressal

In the recent past, there has been an increase in the awareness level in the Indian industries about the importance of calibration to improve the quality of their products and also the required infrastructure to execute the calibration process. Still, we can see that the volume of industries/companies coming to CSIR-NPL is meager compared to

Table 9.10 Technology developed in pressure and vacuum metrology

Name of technology	Description
Cross floating valve (CFV) [75]	High pressure hydraulic cross floating valve(CFV) is the best-suggested way to minimize the time and help to improve the accuracy of pressure calibration of pressure balances. This manually operated indigenously developed valve at CSIR-NPL has various conventionally available pneumatically actuated constant volume valves. The CFV has novelties and simple design. The technology is now ready for commercialization for the interested industry
Environmental conditions monitoring module (EMMRM) [73]	Environmental conditions play a vital role in the metrological and calibration laboratories. These conditions have direct impact on the sensors, transducers and other metrological standards. Recently CSIR- National Physical Laboratory, India has indigenously developed, high precision environment monitoring device. The environment monitoring device has the capability to continuously monitor and record the ambient temperature, barometric pressure, relative humidity and air density

Table 9.11 List of documentary standards, guides and SOPs used in pressure and vacuum metrology

S. No.	Name	Description
1	JCGM 100:2008	Evaluation of measurement data—guide to the expression of uncertainty in measurement
2	NABL 141	Guidelines for estimation and expression of uncertainty in measurement
3	JCGM 101:2008	Evaluation of measurement data—supplement 1 to the "Guide to the expression of uncertainty in measurement"—propagation of distributions using a Monte Carlo method
4	IS/ISO/IEC 17025:2017	General requirements for the competence of testing and calibration laboratories
5	EURAMET cg-3 Version 1.0 (03/2011)	Calibration of pressure balances
6	EURAMET Calibration guide no. 17 Version 4.0 (04/2019)	Guidelines on the calibration of electromechanical and mechanical manometers
7	DKD-R 6-1	Calibration of pressure gauges

the enormous Indian market. However, the expected surge of the industries interested to have traceability in their measurements from an apex level body such as CSIR-NPL is not increasing at the desired pace. It may have contributions from a number of factors such as the complex administrative structure of calibration, time taken for calibration, and, most importantly, the total cost involved during the calibration process. A company having industries far away from Delhi needs to meet the cost of required transportation of the equipment and proper handling during transportation, as well as the calibration charges, etc. Sometimes these costs are higher than the real cost of the equipment; hence due to one or the other reason, they do not approach CSIR-NPL for calibration. In the future, CSIR-NPL may collaborate with some of the regional centers of NABL accredited laboratories or legal metrology laboratories for quality calibration. Also, on-site calibration services are expected to address the traceability needs for industries that are unable to bring their equipment to CSIR-NPL.

9.2.2 Ultrasonic Metrology

9.2.2.1 Introduction and Historical Aspects

Ultrasonics is termed as the field in which mechanical vibration waves are applied and are used above the human audible frequency range that is above 20 kHz. The ultrasonic waves called ultrasound are applied in various fields like industry, medical (both diagnostic and therapeutic), and domestic, civil and marine communication. In medical, the intensity of ultrasound being used for the diagnostic applications must be limited and known for the patient's safety [84–86]. In industries, non-destructive testing (NDT) is used to check the quality of material and the presence of defects in the structures. CSIR-NPL maintains national standards related to ultrasonics and non-destructive testing. It maintains the primary radiation force balance (RFB) based total ultrasonic power measurement facility for ultrasonic power transducer. Apart from the primary RFB, it also provides services that include calibration of ultrasonic standard reference blocks such as IIW-V1, IIW-V2 and step wedge. Calibration of ultrasonic thickness gauge and reference delay rod is also performed by CSIR-NPL. For ultrasonic reference blocks calibration; ultrasonic contact and immersion both techniques are used. All the dimensional and material related parameters are assessed according to IS: 4904: Clauses 3 and 4. In this section, various facilities of ultrasonic metrology of CSIR-NPL are described. Some of the future plans that need to be established particularly as per the recent demands of Indian society are also discussed.

9.2.2.2 Significance, Applications and Relevance in Different Sectors

In the past few decades, applications of ultrasound in various fields have drastically increased. The application of ultrasound in medical diagnostics, physiotherapy, cleaning, medicine, chemical reactions, underwater communications, non-destructive testing, adulteration detection, material quality assessment, proximity detection, concrete testing, liquid or gas leak detection in pipelines and material characterization, etc. are well known and established. The application of ultrasound-based on its power level is broadly categorized as; low power (few milliWatts) and high power (above 1 W). In low power ultrasound application, the low energy content is utilized to assess and characterize the medium. Whereas, in high power, it is intentionally applied to disturb the medium. The low power application is also in general termed as Non-Destructive Testing. It is characterized in terms of ultrasonic propagation velocity and ultrasonic attenuation in the medium.

9.2.2.3 Facilities and Global Status

The International electro-technical committee (IEC-61161) recommends radiation force balance (RFB) technique for the measurement of total ultrasonic power generated by the transducer [87]. CSIR-NPL has indigenously developed this primary system and successfully automated in order to reduce the chances of operator errors. The improved system has been validated by the participation in BIPM/CIPM international intercomparison (CCAUV.U-K3.1). In this method, the basic objective is to precisely measure the total amount of power output that is generated at a particular excitation voltage and frequency. The ultrasound generated by the transducer is directed towards a target that is either absorbing or reflecting and is attached to a microbalance (Fig. 9.12) [88, 89]. The radiation force exerted on the target gives a change in apparent mass by which ultrasonic power can be estimated [90, 91].

Measurement Standards and Facilities

At CSIR-NPL, continuous efforts had been made to improve, upgrade and fully automate the primary standard of total ultrasonic power measurement. The system has been improved in various aspects for full automation including the automatic measurement of RF voltage fed to the transducer [92]. Figure 9.12 shows the schematic of the fully automated radiation force balance (RFB) setup. The present system measures ultrasonic output in the range of 10 mW to 20 W in the frequency range of 1–20 MHz. The ultrasonic transducer whose power is to be measured is fixed at the bottom of a water container with its length, breadth, and height as 300, 300, and 170 mm, respectively. The container has a removable lid to cover water and minimize disturbances. The absorbing target is hanged on to the highly sensitive microbalance to record the variation in mass due to the radiation force of transducer.

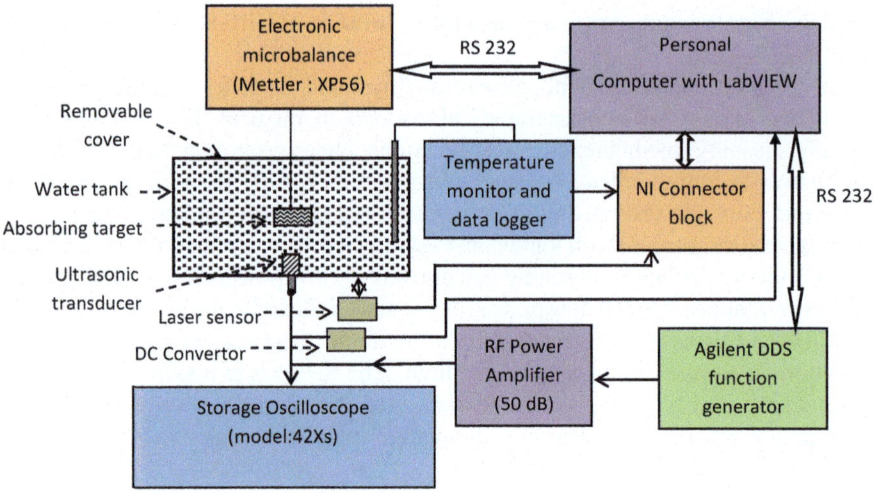

Fig. 9.12 Schematic diagram of the NPLI developed fully automatic radiation force balance setup showing the major contributing devices

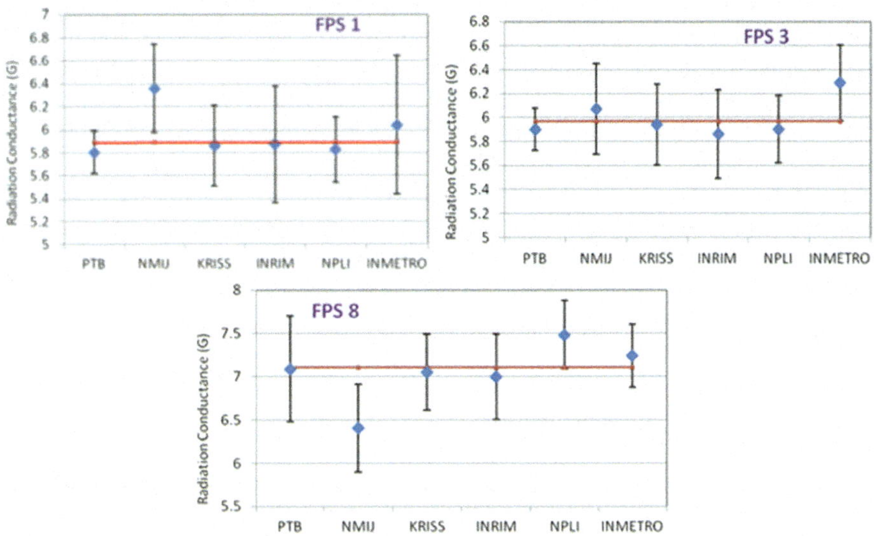

Fig. 9.13 Measurement results on reference transducer by participating counties at (FPS1) 2.0150 MHz and excitation amplitude 1.25 V, (FPS3) at 2.0150 MHz and excitation amplitude 50.0 V and (FPS8) at 15.8942 MHz and excitation amplitude 3.70 V

The radio frequency signal to be used for the transducer excitation is generated by a direct digital synthesis (DDS) based signal generator (Agilent model: 33250A). The signal generator can generate amplitude below 10 V and if needed a fixed RF power amplifier is used for voltage level upto 55 V. The voltage fed to the transducer is measured directly at the transducer node with the help of an indigenously CSIR-NPL developed DC convertor module. The module provides analog output which is then measured by high accuracy 6 ½ digital multimeter. The multimeter is connected to the computer to receive data. A highly sensitive microbalance is attached to the target to measure the apparent change in mass. The microbalance output is continuously monitored by acquiring the output into a computer via a serial (RS 232) port. The temperature of water is monitored by using LM35 based temperature monitor and data logger. The accuracy of temperature measurement is maintained at ±0.2 °C.

The ultrasonic propagation velocity in water is considered from standard literature which is universally accepted at the required temperature. The ultrasound generated by the transducer exerts a proportional force on the target which in turn is measured in terms of change in the mass (Δm). The total ultrasonic power is given by the following equation [93–97].

$$W = \Delta mgc(T) \tag{9.10}$$

where c is the ultrasonic propagation velocity in water at the measured temperature T and g is the acceleration due to gravity. The radiation conductance (G) of the ultrasonic transducer is the ratio of total ultrasonic power to the square of the applied voltage. So,

$$G = W / V_{\mathrm{rms}}^2 \tag{9.11}$$

Manufacturers of ultrasonic equipment can easily estimate the amount of ultrasonic power it will generate at the specified voltage using radiation conductance value. This needs to be understood that the value of G is expected to remain constant at a particular frequency. CSIR-NPL provides this primary facility to its customers whenever required.

Participation in International Key Comparisons

The primary standard of ultrasonic power measurement being maintained at the CSIR-NPL was greatly improved during 2009–2014. The improvement was in terms of automation of the system in order to reduce the operator errors, improvement in the RF voltage measurement and transducer excitation system. The capability of this improved system was verified by participating in the BIPM international key comparison (CCAUV.U-K3.1). The international key comparison was organized with an objective to judge the ultrasonic radiation conductance measurement capabilities of various NMIs. The time-averaged ultrasonic total power in the power range from 10 mW to 15 W was considered. The frequency range was from 2 to 16 MHz. The

Table 9.12 The frequency and power setting protocol used during the BIPM Key comparison (CCAUV.U-K3.1)

Measurement number (frequency power setting)	Operating frequency (MHz)	Transducer excitation level	RF Voltage level (Vrms)
1	2.0150	Very low	1.25
2		Medium	13.5
3		High	50.0
4	6.7513	Very low	1.20
5		low	4.0
6	11.3318	Very low	1.25
7		low	4.00
8	15.8942	Low	3.70

ultrasonic power was the basic intercomparison parameter, but the radiation conductance was preferred to be measured in order to be independent of the applied voltage [98–101]. The radiation conductance of an ultrasonic transducer is the characteristic property of the transducer and generally remains the same at a particular frequency. Table 9.12 shows various frequency power settings decided for the measurement. Figure 9.13 shows the measurement results of various participant countries including India. From the figure, it is clearly indicated that CSIR-NPL results are excellent at all the measurement points for low power, high power and high frequency and well within the uncertainty as compared to the weighed mean.

Potential Services for Future Requirements

Ultrasonic metrology activity of CSIR-NPL deals with the primary facility of total ultrasonic power measurement related to medical as well as industrial applications, whereas, other metrological services are almost related to industrial applications. Ultrasound is a vast subject area and is used in numerous domains wherein the metrology plays a key role [102, 103]. For instance, the new technology of phased array ultrasonic testing of materials for flaw and crack detection needs metrological support.

Ultrasonic cleaning is widely used in industries and also for domestic applications. Effective cleaning depends on the amount of ultrasonic energy released by the ultrasonic cleaning system [104, 105]. Therefore, the cleaning systems need to be calibrated for their effectiveness and amount of energy level. Further acoustic emission testing is the concept used for the detection of leakage in the pipeline including liquid and gas. CSIR-NPL has planned to develop all these facilities in the future to cater to the critical requirements of industry.

A Decade of Clients and Services

Ultrasonic metrology has been providing continuously its calibration services to the government sector, public sector undertaking units (PSUs) and private industries. The services include calibration of ultrasonic flaw and crack detection devices (UFD), ultrasonic transducers, thickness gauges, various standard blocks, etc. Some of the regular customers of these services are Indian Ordnance Factories, Ministry of Defence; Gun Carriage Factory, Hindustan Aeronautics Ltd, Bangalore; Indian Railways; Jindal Steel Plant; Bhilai Steel Plant; Various Private industries and NABL Accredited Laboratories.

Calibration and Testing

CSIR-NPL provides calibration services of the various instruments and parameters in ultrasonics as given in Table 9.13. These equipment include ultrasonic flaw detector, ultrasonic thickness gauge, ultrasonic transducer and ultrasonic standard blocks. Specific parameters include ultrasonic velocity, ultrasonic attenuation and ultrasonic power. During the last 10 years, more than 500 test and calibration certificates have been generated from these testing and calibration services.

The detection of critical inherent cracks in the railhead is a major challenge for the railway industry. The functionality of the rail testing system is verified by reference test rails which are ultimately calibrated and certified by CSIR-NPL for the presence of various kinds of artificial defects as per RDSO guidelines. Ultrasonic NDT plays

Table 9.13 list of major equipments and parameters that can be calibrated from ultrasonic metrology activity

S. No.	Item name/parameter	Standard procedure/documentation
1	Ultrasonic flaw detector	BS: 4331-II/ASTM E-317
2	Ultrasonic flaw detector	IS: 12666 Clauses 4 and 5
3	Calibration block A2 (IIW-VI)	IS: 4904: Clauses 3 and 4
4	Calibration block A4 (IIW-V2)	IS: 4904: Clauses 3 and 4
5	Step wedge	Comparison approach
6	Cylindrical FBH	ASTM E-127
7	Ultrasonic NDT transducer/probe	ASTM E-1065
8	Ultrasonic thickness gauge	Comparison approach
9	Test rail	Comparison NDT
10	Ultrasonic output power (Transducer)	IEC-61161
11	Ultrasonic velocity	Pulse echo approach
12	Ultrasonic attenuation	Pulse echo approach

Fig. 9.14 Ultrasonic testing of concrete block

a key role in upstream, and downstream oil and gas industries for ultimate safety and productivity. It is used for inspection of welds, pipes, cladding, and risers for cracks, corrosion, or defects. The ultrasonic flaw detectors (UFD) are widely used in the industries for such applications. For the correct and more precise estimation of such defects, the UFDs are calibrated and certified for its functionality at CSIR-NPL for its range (depth) and amplitude detection. Various ultrasonic standard reference blocks such as IIW-V1, IIW-V2 and Step are used to cross-check the functionality of ultrasonic devices on site. Such blocks are certified by CSIR-NPL. Ultrasonic pulse propagation velocity in the concrete structure gives useful information about the strength and the quality of the structure. Higher the ultrasonic propagation velocity through the concrete, the better will be the quality and strength. Generally, the ultrasonic velocity above 3500 m/s shows an excellent quality of structure (Fig. 9.14). All ultrasonic pulse velocity measuring instruments need to be traceable to the apex level national standard for its accuracy of measurement and the correct grading of concrete structures. CSIR- NPL also plays a vital role in providing the calibration service for the reference ultrasonic pulse velocity (UPV) blocks. The calibrated blocks for the specific time delay are used as a reference to calibrate the UPV measuring instruments before testing the actual concrete for its grading.

Consultancy and Technical Services

CSIR-NPL provides consultancy services in the domain of ultrasonic non-destructive testing (NDT) as well as development of ultrasonic and scientific equipment. It has provided a number of consultancy services to government departments and private industries in the various areas of ultrasonic.

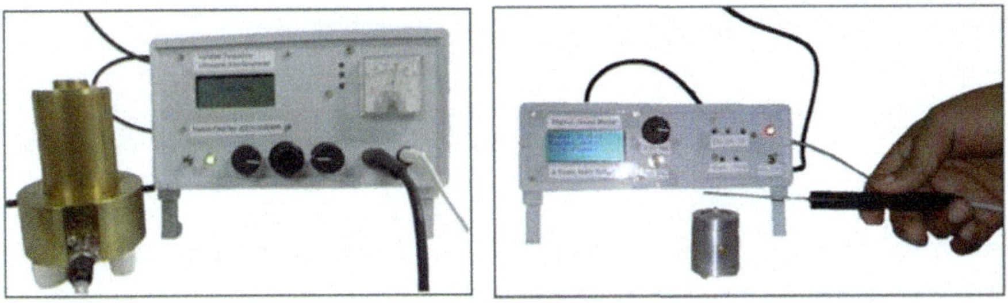

Fig. 9.15 Left—improved variable frequency ultrasonic interferometer and right—3 T auto-range digital Gaussmeter with probe and the permanent magnet being tested

Training and Academic Services

The ultrasonic metrology activity also provides training in the specialized area of ultrasonic NDT and metrology. The department provides academic training to carry out projects/dissertations both at B.Tech, M.Sc and M.Tech levels. The department also provides PhD programme in the area of ultrasonic metrology and electronic instrumentation.

Technology Development and Commercialization

The ultrasonic metrology activity has been involved in the development of ultrasonic related technologies and products as per the requirements of the society and industry. Some of the products that have been developed and commercialized are (i) Ultrasonic Normal Contact Probe, (ii) Ultrasonic Angle Probes (example 45°, 60°), (iii) Ultrasonic humidifier, (iv) Ultrasonic Nebulizer and (v) Ultrasonic interferometer for liquids. Some of the latest technologies that are available online for commercial transfer of technology on NPL website are "Improved variable frequency ultrasonic interferometer" and "3T Auto-range Digital Gauss Meter" (Fig. 9.15).

Major Documentary Standards, Guides and SOPs

Some of the major documentary standards used in ultrasonic metrology are given in Table 9.14.

Table 9.14 List of major documentary standards used in ultrasonic metrology

S. No.	Standards	Details
1	ASTM E-317	Standard practice for evaluating performance characteristics of ultrasonic pulse-echo testing instruments and systems without the use of electronic measurement instruments
2	IS: 12666	Performance assessment of ultrasonic flaw detection equipment
3	IS: 4904	Calibration blocks for use in ultrasonic non-destructive testing
4	ASTM E-127	Standard practice for fabrication and control of flat bottomed hole ultrasonic standard reference blocks
5	ASTM E-1065	Standard practice for evaluating characteristics of ultrasonic search units
6	IEC -61161	Ultrasonics-power measurement-radiation force balances and performance requirements

9.2.2.4 Possible Gaps and Suggestions for Their Redressal

Applications of ultrasound are widely spread in numerous domains wherein the metrology plays a key role. Ultrasonic phased array is the new technology of ultrasonic testing of materials for flaw and cracks detection which needs urgent metrological support in the country. Ultrasonic cleaning is yet another method widely used in industries and domestic purposes. The effective cleaning depends on ultrasonic energy level generated by the ultrasonic transducers in the cleaning system. It is of utmost importance to have the cleaning systems calibrated for its effectiveness and amount of energy release. Acoustic emission testing is the concept used for the detection of leakage in the pipeline including liquid and gas. CSIR-NPL has planned to develop all these facilities in the future to cater to the critical requirements of industries.

9.2.3 Raman Metrology

Ever since its discovery by C. V. Raman in year 1928, the Raman spectroscopy has come a long way and transformed from purely scientific discovery to a modern-day investigative technique which encompasses critical and strategic areas such a diagnostic applications in the bio-medical field, gem identification, purity and API concentration in pharmaceuticals, mineralogy as well as immense uses in device characterization. However, the scope of Spectroscopy study gets limited due to its qualitative nature as well as the fact that different Raman spectrometers, depending upon the experimental conditions employed, produce varying outputs in terms of intensity, peak positions, instrument resolution, standard reference materials, etc. Research on Raman spectroscopy in areas of quantification and comparability across labs and Raman instruments is the current focus area.

Consequently, the metrology and traceability of the measurements come into play as soon as we talk about the quantitative aspect. Raman Metrology is concerned with the optimization of various parameters. The main facet of the Raman Metrology is the need to have Relative intensity correction [106], testing of Resolution [107], calibration of Scanning as well as CCD detectors and shift standards for Spectrometer calibration [108]. Correction in intensities of the signals obtained from test samples is important because when Raman spectra are obtained using different instruments, they show notable differences in the peak intensities and the differences become large when different wavelengths are used in excitations. For interoperability of spectra obtained from different spectrometers, when spectra are to be transferred among different systems, and when spectrometer is operated at different excitation wavelengths, testing of spectrometer resolution becomes important factor to be checked.

To the best of our knowledge, work on Metrology of Raman measurements is not being pursued in India so far and NPL India has initiated studies in this direction with a focus to develop/identify reference materials for metrology of Raman spectrometers with regard to spectrometer calibration, intensity and resolution at present. Figure 9.16 shows the triple stage laser Raman spectrometer established at CSIR-NPL.

It has been known that the spectra of a given sample produced by different Raman spectrometers from different manufacturers as well as the laser wavelengths used, show measurable shifts in the peak position which makes the compatibility and comparison of the data a non-trivial issue [109]. Hence, there is a need for standard methods and reference materials to be used for the Raman spectrometer which will make the measurements traceable to the SI units. Without spectrometer calibration, the result of any measurement will vary from the standard result which will make the data questionable. The traditional way to calibrate the Raman instrument is by using a mercury arc-lamp which is very time-consuming. Recently, the ASTM guideline E1840-96 (reapproved 2014) has proposed seven chemicals for wavenumber calibration which are employed due to their stability over time and no polymorphism and known relative wavenumber shift [110].

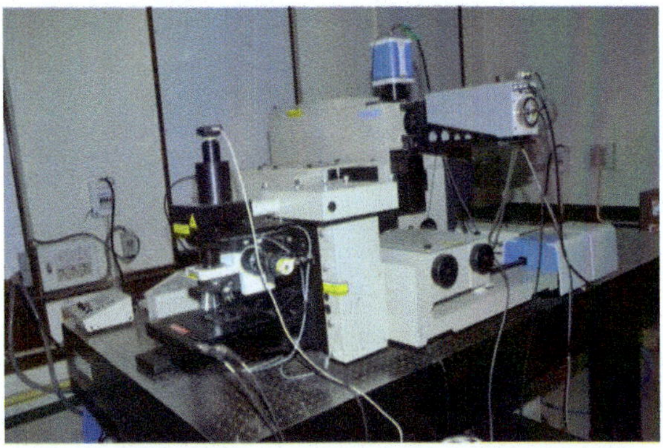

Fig. 9.16 Triple stage laser Raman spectrometer at CSIR-NPL

At CSIR-NPL, standard chemicals such as Naphthalene, 4-Acetaimidophenol, 1,4 bis (2-Methylstryl) Benzene, cyclohexane, and Benzonitrile, which are mentioned in the ASTM guidelines, were used to investigate the Raman spectrometer calibration. Along with these materials, we also used Diamond, standard silicon wafer, and standard GaN for the present study. Calibration was done using the most intense peak of each standard chemical following which the Raman spectra of all the other mentioned chemicals were recorded and compared. The whole spectral range was divided into three regions namely the lower spectral region, middle spectral region and higher spectral region. In each of the regions, most intense peaks are identified and used for the calibration (Fig. 9.17).

Initially, the most intense peaks of the various chemical in lower spectral region were used to calibrate the spectrometer and spectra of cyclohexane was recorded after each calibration. The recorded spectrum was compared with the standard spectrum and deviation in the peak position is calculated. The deviation is plotted with the peak position and the results are shown in the figure. From the results shown in the bar chart, it was inferred that for the lower wave number peaks the deviation was very small however, the deviation was seen to increase as we moved towards the middle and higher wave number peaks. Similarly, it has been found that for the Calibration using middle wave number peak, the peak position shift at lower and higher wave number values increased. And likewise, higher wave number calibration shifted the peak values at lower wave number and middle wave numbers. From these observations, it can be concluded that the spectral range of calibration in a Raman Spectrometer is an important factor for measurements that need to be taken into account.

The other facet of the Raman Metrology is the resolution of the spectrometer. Resolution of the spectrometer mainly depends on the four factors namely spectrometer focal length, diffraction grating used, laser wavelength and the detectors [110]. In the Triple Raman Spectrometer T64000, the diffraction grating remains

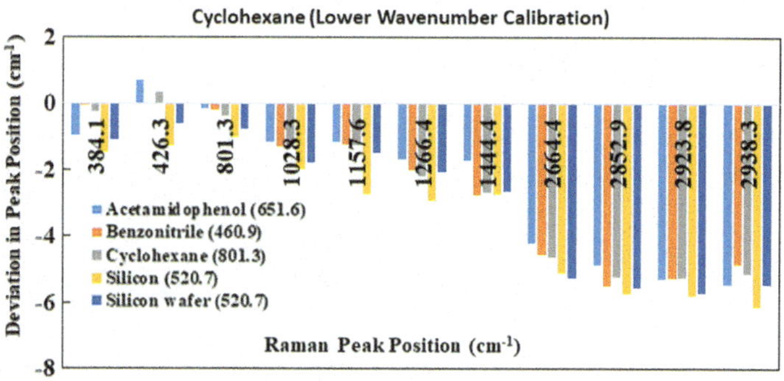

Fig. 9.17 Deviation in the peak position with respect to the peak position for lower wavenumber calibration for cyclohexane. The figures in parenthesis indicate the peak position used for the calibration of the spectrometer

fixed, i.e. at 1800 gr/mm. The Ar ion laser was used for the excitation and Charged coupled device was used as the detector.

As per ASTM guidelines, pure Calcite was used to test the Resolution of the Raman system because Calcite shows desirable optical properties for Resolution Standard. The Peak at 1085 cm^{-1} was considered for testing the resolution as it is centred in the Raman spectra for fixed grating systems. The Calibration Relation for nominal resolution of Spectrometer under test is $S_{resolution} = (Bw_{1085}–0.684)/1.0209$ [107]. The spectra of the calcite were recorded keeping the intermediate and exit slit fixed and varying the entrance slit size. This configuration was arrived at after investigations using various configurations such as keeping entrance and exit slit fixed and varying the intermediate slit or keeping the entrance and intermediate slits fixed and varying the exit slit size. Thus, the effects of various slits were optimized and a representative result is shown in the graph (Fig. 9.18). As seen from the graph, as the size of the entrance slit decreases the resolution of the spectrometer increases. The resolution of the spectrometer is found to be 2 cm^{-1} which was the best specification given by the manufacturer and adds to the confidence in measurements. These are the preliminary studies carried out on the resolution.

Additionally, work is being carried out to deduce the resolution using the Photomultiplier tube, keeping all other parameters constant. We are also working on how the various other factors affect the resolution of the spectrometer quantitatively so that we can find out the uncertainty in the resolution of the spectrometer. Table 9.15 shows major documentary standards used in Raman metrology.

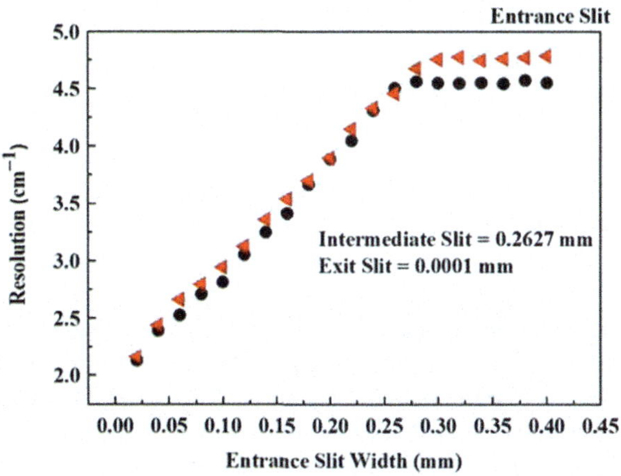

Fig. 9.18 Variation of the resolution with the entrance slit size

Table 9.15 List of major documentary standards used in Raman metrology

S. No.	Documentary Standard	Details
1	E1683–02	Standard practice for testing the performance of scanning Raman spectrometer
2	E1840–96	Standard guide for Raman shift standards for spectrometer calibration
3	E2529–06	Standard guide for testing the resolution of a Raman spectrometer
4	E2642–09	Standard terminology for scientific charge-coupled device (CCD) detectors
5	E2911–13	Standard guide for relative intensity correction of Raman spectrometers

9.3 Fluid Flow Metrology

9.3.1 Introduction and Historical Aspects

The fluid flow measurements, both in liquid and gas, are essentially required in all walks of human life. The various types of flowmeters such as electromagnetic, ultrasonic, vortex, differential pressure, positive displacement, turbine, Coriolis mass, etc. are used for flow measurements. These flowmeters require frequent calibration against appropriate measurement standards [111]. The history of flow measurement is very old. The hydraulic structures are understood to be existed before recorded history. The irrigation systems in Mesopotamia, check and diversion dams on the Arabian Peninsula were identified by Archaeologists about 5800 B.C. [112]. The water level on Nile river was recorded at several locations of Nile river between Nubia and the Nile delta about 3050 B.C. using gauges. Romans first introduced a method for water flow which was based on pipe areas. This method was used for tariff purposes for water supplied to citizen's residential areas. They were not fully aware of hydraulic discharge principle. Famous Greek Scientist, Hero was the first inventor to describe the correct relationship of flow with time, cross-sectional area, velocity and volume during the first century A.D. in his book *Dioptra.* This book, later on, became the basis for flow measurement. His findings went unnoticed for almost 1500 years until Leonardo da Vinci and Castelli used these findings to express continuity equation. Even their work remained unnoticed for almost 200 years until his manuscript become public in 1690. In year 1970, The German engineer, Reinhard Woltman, invented a spoke-vane current meter with a revolution counter for the measurement of water flow in the river.

During the eighteenth and nineteenth centuries, weirs and flumes were developed and installed which made flow measurements possible on irrigation canals. The gauging stations were constructed on many rivers to provide records of flows [112]. These techniques of flow measurement are in open channels or rivers. The weighing and volumetric techniques are used for the water flow measurements in closed conduits. Much attention and emphasis were placed on the closed conduit/channel measurements during the twentieth century. During this period, improved measurement techniques and standards were developed and error and uncertainty estimation for both open and closed channels were implemented. The closed conduit/channel measurements are generally performed in the laboratory.

9.3.2 Significance, Applications and Relevance in Different Sectors

The main applications of fluid flow include process control in industries; nuclear, thermal and hydro-electric power generation; manufacturing of drugs and pharmaceuticals, fertilizers, beverages, distilleries, petrochemicals; safety and health hazards and in aviation, agricultural, irrigation and dairy plants, etc. The ever-growing industrialization, urbanization, population and depleting groundwater levels are severely affecting the availability of water. Similar conditions are existing for the petroleum products, therefore, the role of metrologists becomes very important to develop measurement techniques for the accurate and precise measurement of quantity of fluid for its optimum use. It helps to quantify the exact quantity, calculate the correct tariff and to control its misuse.

For application in **environmental pollution monitoring sector**, various types of flowmeters such as mass flow controllers, mass flow meters, laminar flow meters, sonic nozzles, volume samplers, orifice flowmeters, piston flow calibrators, air flow calibrators, dry gas meters, root meters, etc. are used. One might have encountered the problem of different pollution measurement data even when two measurement stations are very closer. This is due to non-calibrated flowmeters/incorrect calibration of flowmeters. The range of pollution parameters is high in India whereas those in abroad, their range is small. The instruments calibrated at low operating range abroad do not suit Indian environmental conditions. They should be designed and calibrated as per Indian conditions. For application in **biomedical sector**, various types of flowmeters such as gas flow analyzers, ventilator testers, infusion pumps, infusion device analyzers, etc. are used. The ventilator tester and gas flow analyzers are further used to calibrate ventilators. Infusion pumps are used in precise and accurate drug dosing for cancer patients in chemotherapy and other important applications. If the infusion pump is not properly calibrated/un-calibrated, the wrong dose of medicines will lead to death of the patients. Similarly, ventilator is used as a life support system and providing treatment to the patients. If the ventilator is not properly calibrated, it will lead to the wrong amount of pressure and flow thus endangering human life. Most of the deaths in case of ventilator are its improper use. Thus, these instruments are directly related to human health and safety. For application in **aerospace sectors**, various types of flowmeters such as rotameters, mass flow meters, mass flow controllers, etc. are used. For example, Air India uses rotameters and mass flow meters in Jet Engine Overhaul Complex where repair and servicing of Aeroplanes are done. For the optimum performance of the jet engine and other components, it uses calibrated standards.

For application in **water and process control sectors**, various types of flowmeters such as domestic water meters, bulk type water meters, electromagnetic flowmeters, Coriolis mass flowmeters, ultrasonic flowmeters, vortex flowmeters turbine flowmeters, etc. are used. The various flowmeters manufacturers, Sugar Mills, NABL accredited testing and calibration laboratories (C&I Calibration Pvt. Ltd., Aquajet Flowmeters Lab., IDEMI, Mumbai, etc.) and State Water Boards (Delhi Jal Board, Tamilnadu Water Board, NDMC, Uttarakhand Jal Sansthan, UP Jal Vikas Nigam) uses our testing and calibration services for verification of their products and their calibration facility.

9.3.3 Facilities and Global Status

9.3.3.1 Measurement Standards and Facilities

As a brief description of development of Fluid Flow measurement facilities at CSIR-NPL, the Primary Water Flow Calibration Facility (comprising of DN50 and DN200 Test Rigs as per ISO 4185 standard) was established during 1992–1998, though envisioned and started during 1990, under NPL-PTB Technical Collaboration program, which was the state-of-the-art and fully automatic system [113–115]. The facility was fully operational during 1998–2001 with little intermittent maintenance problems, obvious and as usual in such systems. Further, there was not adequate and appropriately trained manpower to maintain such a huge facility. As the routine maintenance problems could not be adequately addressed, the facility finally became non-operational in year 2001. Since fluid flow parameters are backbone of several industries, the sustained demands from the users made the CSIR-NPL rethink and reconsider its decision. Accordingly, the efforts were made to revive the facility during 2008-09 using standard cards and their traceability was the main problem. However, its maintenance and operations were not financially and technologically viable. As a part of capability building, a prototype Water Flow Calibration Standard of DN100 size was designed and developed as per ISO 4185 standard indigenously during 2010–13 and enough knowledge and confidence were gained to design and develop such systems on our own. This system was developed using in-expensive instruments thus it was having coarser measurement uncertainty (0.1% at k=2) and it was not suitable to maintain NMI status in water flow area [116].

Further, it was planned to upgrade the facility by using state-of-the-art instrumentation and control systems to make it comparable to advanced NMIs and provide traceable measurements to Indian users. The fund for this upgradation was allocated in 12th Five Year Plan Project of CSIR. Thus, this facility was upgraded using state-of-the-art instrumentation and control systems during 2017–2019 and it became fully functional since 6th July 2019. Thus, NMI status in water flow measurement is maintained now. The facility is being used for providing calibration services for various types of flowmeters such as Coriolis mass, electromagnetic, ultrasonic, vortex, turbine, bulk types, etc.) in the flow range of 0.1 m^3/h to 650 m^3/h using static-gravimetric method as per ISO 4185. The uncertainty in flowmeter calibration for *totalized mass* and *totalized volume* parameters is in the range of 0.01–0.02% (at $k = 2$) whereas for *mass flow rate* and *volume flow rate* parameters is in the range of 0.03–0.05% (at $k = 2$) which are at par to the international level [117, 118].

Water Meter Testing Facility based on the standing start and standing finish method for testing of domestic water meters of sizes DN15 to DN50 was established during 2001-02. Here, water meters were being tested as per IS 779, IS 6784 and ISO 4064 standards. The testing facility was based on volumetric vessel. It was also upgraded, replacing volumetric tanks (20, 100 and 1000 L) by weighing scales along with SS tanks (Mettler weighing scales of 150 and 1500 kg capacity). The complete SS piping, new anti-rust platform, new temperature and pressure sensors and indicators, new pump of 1 HP were installed and integrated. Now, the facility is used to test and calibrate 3 water meters upto DN50 size in a single run as per IS 779, IS 6784 and ISO 4064 standards. Also, a dedicated line is provided for calibration of rotameters. The calibration services using this upgraded system were started in July 2017.

The planning for establishment of Gas Flow Calibration Facility was carried out during 2009–10 to cater the industrial as well as in-house requirements from several groups of CSIR-NPL, i.e. Chemical Metrology group, Radio and Atmospheric Science Division, Solar Energy group, etc. The external users include Pollution Control Boards, pharmaceutical industries, biomedical laboratories, petrochemical industries, environmental and pollution monitoring equipment manufacturers, R&D laboratories, NABL accredited laboratories, gas flow meters (mass flow controllers, mass flow meters, rotameters, digital flow calibrators) manufacturers, aerospace industries, etc. The fund was allocated for this purpose in 11th Five Year Plan Project of CSIR.

The Gas Flow Calibration Facility became fully operational since 31st May 2014 and it is now being used for providing calibration services of various types of flowmeters such as mass flow controllers (MFCs), mass flow meters (MFMs), laminar flow meters, sonic nozzles, rotameters, dry gas meters, air flow calibrators, orifice flowmeters, volume samplers, piston flow calibrators, ventilator testers, gas flow analyzers, digital flowmeters, turbine meters, rotary positive displacement (RPD) gas meters, etc.

Currently, Fluid Flow Metrology at CSIR-NPL has established 3 measurement facilities, which are being operated: (i) Primary Water Flow Calibration Facility; (ii) Primary Water Meter Testing Facility; (iii) Gas Flow Calibration Facility (Reference/Working/Transfer Standard) as briefly described as follows.

Primary Water Flow Calibration Facility

The Primary Flow Calibration Facility (WFCF)/Water Flow Measurement Standard (WFMS) works on the weighing method which is known as primary method. In this method, flow is realized in fundamental units of mass, length and time. In a weighing method, water is collected for a predetermined time and weighed [114, 120]. During the measurements, water flowing through the meter under test (MUT) is approximately under steady conditions of flow, pressure and temperature. The standard flow is determined through weighing along with the average of the flow indicated by the MUT during the collection interval and accordingly, the error of the MUT is determined. The flowmeter performance can be also determined in other parameters like meter K-factor apart from flowmeter error. An additional flowmeter, known as a reference flowmeter, is normally used to set the flow in the line and to monitor its stability. The weighing method is again categorized into static weighing and dynamic weighing techniques.

In a static weighing method, the mass of the collected water is measured after achieving the steady-state condition of the collected water. Most of the WFMSs worldwide (including CSIR-NPL WFCF) is based on static weighing due to their high accuracy.

In a dynamic weighing method, the mass flow rate is determined dynamically using two positions of lever equilibrium on classical one beam scale and timer actuation periods. In the modern technique, the water falling on the weighing tank is continuously weighed at high speed (at equal interval of approximately 50 ms or higher up to 200 ms) between the start and the end of the collection interval and large numbers of weighing readings are collected. The buoyancy correction is applied to the weighing readings. From the slope of mass-time curve, the mass flow rate is determined. One of the important conditions of measurement is stable flow and optimum speed of the mass indicating device for recording and plotting the data. The impact of falling water should not affect the measurements. If it is affecting the measurement, it should be properly modelled to take into account the effect [121]. Generally, the accuracy of dynamic weighing system is coarser than static weighing. However, due to the use of modern weighing system employing load cells and computer-based data acquisition, algorithm, modelling and simulation, the accuracy of dynamic weighing standard has been improved up to the same level of uncertainty as in the case of static weighing [121].

The four parameters measured with Primary WFCF are *total water mass, mass flow rate, total water volume* and *volume flow rate*. In the static-gravimetric systems, the mass flow rate, q_m, is determined as follows [120];

$$q_m = \frac{m}{t} \tag{9.12}$$

where m is the mass of water collected during time interval t. The volumetric flow, q_v, is derived from Eq. (9.12) of mass flow rate by using known water density, ρ, as follows;

$$q_v = \frac{q_m}{\rho} = \frac{m}{t \cdot \rho} \tag{9.13}$$

Thus, the sources of errors and uncertainties for mass flow rate measurement are mass and time. In the case of volume flow rate measurement, these sources are mass, time and density. The relative combined standard uncertainty associated with mass flow rate measured by the WFCF is calculated by root-sum-square (RSS) method using different uncertainty components as given below;

$$u_r(q_m) = \sqrt{u_r^2(m) + u_r^2(t)} \tag{9.14}$$

where $u_r(q_m)$ is the relative combined standard uncertainty of the mass flow rate; $u_r(m)$ is the relative standard uncertainty in collected water mass and $u_r(t)$ is the relative standard uncertainty in collection time. The relative combined standard uncertainty associated with volumetric flow rate measured by the WFCF is given below;

$$u_r(q_v) = \sqrt{u_r^2(m) + u_r^2(t) + u_r^2(\rho)} \tag{9.15}$$

where $u_r(q_v)$ is the relative combined standard uncertainty of the volumetric flow rate and $u_r(\rho)$ is the relative standard uncertainty in water density measurement. The $u_r(q_m)$ and $u_r(q_v)$ are the relative combined standard uncertainties assigned to WFCF. In the calibration of MUT, the volume flow rate indicated by MUT ($q_{v\mathrm{MUT}}$) is compared against calculated by the standard ($q_{v\mathrm{STD}}$) and error (e) of the MUT is determined from the following equation [117];

$$e = \frac{(q_{v\mathrm{MUT}} - q_{v\mathrm{STD}})}{q_{v\mathrm{STD}}} \times 100 \tag{9.16}$$

The flowmeter error is also determined in terms of standard volume as follows:

$$e = \frac{[(q_{v\mathrm{MUT}} \times t) - V_{\mathrm{STD}}]}{V_{\mathrm{STD}}} \times 100 \tag{9.17}$$

Flowmeters have frequency (pulse)/and current outputs. The flowmeter outputs are read by a calibrated frequency counter and digital multimeter during automation for averaging or totalizing the output as flowmeter readings are instantaneous and are continuously changing. Therefore, the uncertainties of data acquisition devices (i.e. digital multimeter and frequency counter) are also included in the uncertainty budget apart from repeatability (Type A uncertainty) of MUT. Some flowmeters are calibrated in terms of the meter K-factor, which is defined as follows;

$$\text{meter } K - \text{factor} = \frac{N}{V_{\mathrm{STD}}} = \frac{f}{q_{v\mathrm{STD}}} \tag{9.18}$$

where N is the no. of pulses output by MUT, f is the frequency output of the MUT, V_{STD} and $q_{v\mathrm{STD}}$ are the volume and volume flow rate measured by WFCF.

The upgraded Primary Water Flow Calibration Facility is fully functional for calibration of different types of flowmeters of sizes DN2–DN200 in the flow range up to 650 m^3/h using *weighing method* as per ISO 4185 Standard. The new system was completely designed by CSIR-NPL where five pipe lines (DN25, DN50, DN100, DN150 and DN200) have been installed and connected to 3 nos. of high accuracy weighing systems (300, 3000 and 6000 kg capacities) through fishtail, nozzle and diverter systems. A separate line of DN15 size has been also derived for calibration of DN2 to DN4 size mass flowmeters using 12 kg weighing scale employing standing start and standing finish method. These weighing systems are of *electromagnetic force compensation* type. A new direct pumping header of DN150 size was added using 3, 5 and 10 HP pumps for calibration of flowmeters upto DN100 size in the flow range upto 175 m^3/h. These pumps were integrated with variable frequency drives (VFDs) for flow control and energy saving. Out of 3 nos. of 50 HP pumps, 1 no. of 50 HP pump was integrated with VFD for flow control and energy saving. The *totalized mass, totalized volume, mass flow rate* and *volume flow rate* parameters of flowmeters are measured through this Primary Water Flow Calibration Facility. The uncertainty in flowmeter calibration for *totalized mass* and *totalized volume* parameters is in the range of 0.01–0.02% (at $k = 2$) whereas for *mass flow rate* and *volume flow rate* parameters is in the range of 0.03–0.05% (at $k = 2$) which are at par to international level. Thus, the NMI status in water flow area is maintained now by CSIR-NPL. Figure 9.19 shows the schematic diagram and photograph of new Water Flow Calibration Facility.

Global Status in Water Flow Measurement

The highly accurate WFCFs are globally based on static weighing with flying start and flying finish method as per ISO 4185. It is preferred over the dynamic weighing due to its comparatively improved accuracy. The WFCF of PTB, Germany has been designed for use in both static and dynamic weighing. In this system dual-type balances, i.e. strain gauge type for fast dynamic response and electromagnetic force compensation type for static conditions for long-term stability and reproducibility, have been used. The whole unit is embedded into a single functional unit.

The NIST, USA has implemented the dynamic weighing standard into the static weighing standard. The uncertainty achieved with this dynamic weighing method is found comparable to the static weighing method. The main advantage of this method is that the complex diverter system is not required. Most recently, a fully automatic dynamic weighing based water flow standard was developed by NIST in the flow range 0.22–15 kg/s having associated measurement uncertainty of ±0.021% at a coverage factor $k = 2$ [121, 122].

In a static weighing method, various sources of errors and uncertainties are collected mass, collection time and liquid density. Most of these errors and uncertainties contributions are improved by the use of advanced instrumentation but improvement in the diverter error (considered as a part of collection time) is a slightly complex process. It is not possible to improve diverter error only using better instrumentation but it requires special techniques for this purpose. Various NMIs, i.e. NIST,

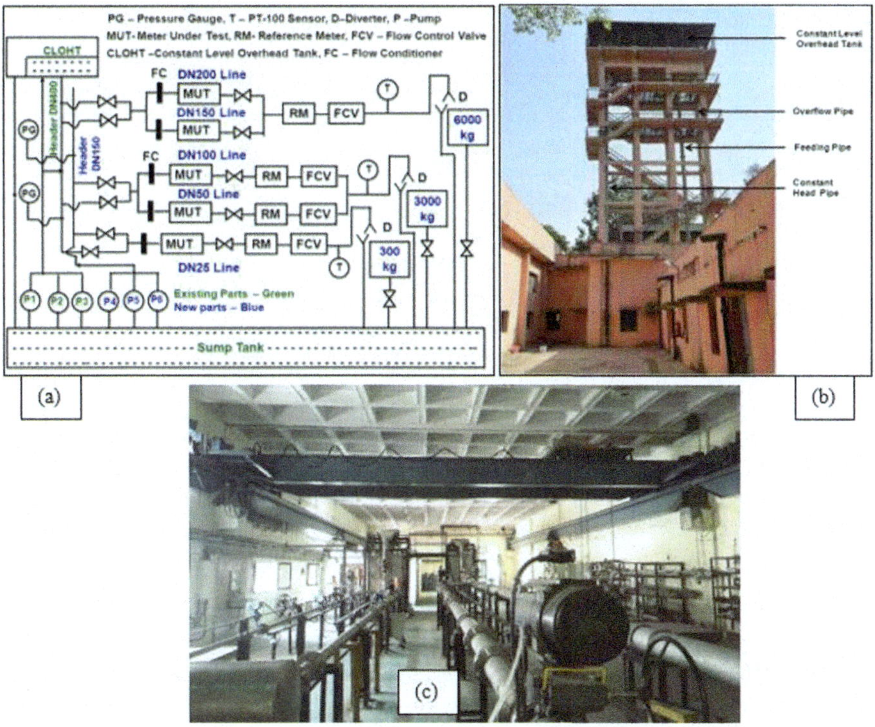

Fig. 9.19 New water flow calibration facility after upgradation, (**a**) Schematic diagram of New WFCF through upgradation, (**b**) Constant level overhead tank assembly and structure of primary WFCF (**c**) New 5 lines along with associated instrumentations

USA; PTB, Germany and NMIJ, Japan have carried out R&D on the improvement of diverter error.

During the year 2007, PTB, Germany established high accuracy WFMS with an expanded uncertainty of total flow as 0.02% which is a result of dedicated work of 9 years starting from 1998. In this system, a new diverter design was adopted in which an angular encoding transducer and electric diverter actuator with suitable electronics and software have been used to achieve targeted diverter timing error. This new flow facility of improved uncertainty has not been peer-reviewed so far and PTB CMC is still 0.04% for the whole flow range (upto pipe size DN400).

The NIST, USA has also worked during the same period on reducing the diverter error and developed a new WFMS based on self cancelling diverter (called uni-diverter) achieving diverter error less than 0.01%. The uni-diverter moves in the same direction against liquid jet both at the beginning and at the end of the water collection. Uni-directional travel of the divider reduces the error due to asymmetry in the divider actuated motion, the liquid jet velocity profile and the position of the diverter trigger. The uncertainty of flow measurement achieved by NIST is in the range of 0.03–0.05% at $k = 2$ for pipe size upto DN100.

Table 9.16 International status on water flow

S. No.	Name of the NMI/DI	Pipe/meter size (mm)	Uncertainty at k= 2 (%)
1	NIST, USA	Up to 100	0.03–0.05
2	CENAM, Mexico	Up to 300	0.04
3	PTB, Germany	Up to 400	0.04
4	NEL, UK	Up to 1000	0.10
5	NMIJ, Japan	Up to 400	0.04–0.06
6	CMS, Taiwan	Up to 250	0.03–0.06
7	KRISS, Korea	Up to 150	0.06–0.09
8	VSL, The Netherland	Up to150	0.02–0.05
9	SP, Sweden	Up to 300	0.06–0.10
10	CMI, Czech Republic	Up to 150	0.05–0.10
11	CSIR-NPL, India	Up to 200	0.03−0.05% (Upgraded system)

The NMIJ, Japan has developed WFMS in which diverter error of 0.003% has been achieved. For this purpose, double-wing method has been used in diverter design wherein both the wings of the diverter move in the same direction against the liquid jet at the beginning and end of the measurement. This design is being used in NMIJ WFMS. The uncertainty achieved in flow measurement is in the range of 0.04–0.06% for pipe size upto DN400. This design has limitations of adjusting the start position for the next diversion. To solve this problem, a novel rotating double wing (RDW) type diverter has been developed achieving diverter error <0.002% which is best in the world. This RDW type diverter is being used in hydrocarbon flow calibration facility at NMIJ, Japan in which uncertainty achieved in mass flow rate and volume flow rate is 0.02% and 0.03% respectively [121]. Table 9.16 shows the international status of water flow measurement.

Primary Water Meter Testing Facility

Primary Water Meter Testing Facility is used for testing domestic water meters of sizes DN15 to DN50 as per IS 779, IS 6784 and ISO 4064 standards and it is also used for calibration of rotameters. It is based on the standing start and standing finish method. It uses two nos. of Mettler weighing scales of 150 and 1500 kg capacity. It has 1, 3 and 7.5 HP pumps for supplying required flow in the range of 30–30,000 L/h for testing of water meters and calibration of rotameters. The flow in the line is monitored by electromagnetic flowmeters and rotameters. The uncertainty of facility in totalized volume and volume flow rate is 0.10 and 0.25% (at k= 2)

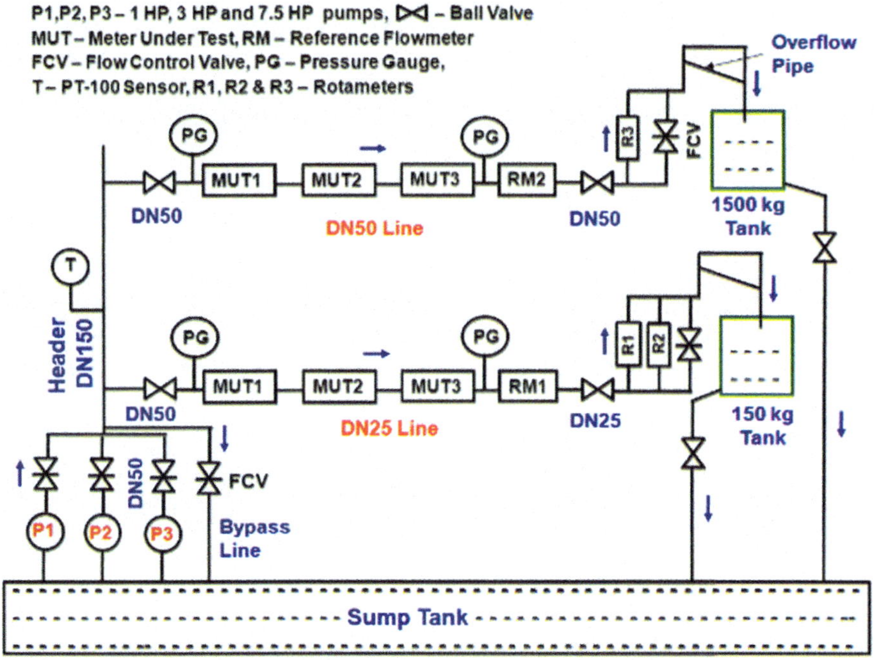

Fig. 9.20 Primary water meter testing facility

respectively. The rotameters are calibrated by weighing method at low flow rate (upto 1000 L/h) whereas at higher flow rate (above 1000 L/h), they are calibrated by comparison method employing electromagnetic flowmeters of DN25 and DN50 size as reference standard. The accuracy of water meters/flowmeters tested/calibrated in this facility are in the range of 2–5%. The water meters (called bulk water meters) above DN50 and upto DN200 size are tested in Primary Water Flow Calibration Facility. Figure 9.20 shows the schematic diagram of the facility.

Gas Flow Calibration Facility

The Gas Flow Calibration Facility is based on laminar flow elements and sonic nozzles.

The mass flow rate of a compressible fluid in laminar flow element is realized using Poiseulle's Law [118–124] as follows:

$$q_m = \frac{(P_1 - P_2) \times \rho_{(P,T)} \times \pi \times R \times h^3}{\eta_{(P,T)} \times 6 \times L} \tag{9.19}$$

where, q_{m} = mass flow rate (kg/s), P_1 = upstream absolute pressure (Pa), P_2 = downstream absolute pressure (Pa), $P = (P_1 + P_2)/2$ = average of upstream and downstream pressures (Pa), T = absolute temperature of gas (K), $\rho_{(P,T)}$ = gas density under P, T conditions (kg/m^3), R= flow passage radius (m), h = gap between piston and cylinder (m), η = dynamic gas viscosity under P, T conditions (Pa.s) and L = length of laminar flow (m). The gas density $\rho_{(P,T)}$ is calculated by the following equation:

$$\rho_{(P,T)} = \rho_N \times \frac{P \times T_N \times Z_N}{P_N \times T \times Z_{(P,T)}} \tag{9.20}$$

where, ρ_N, P_N, T_N, Z_N and $Z_{(P,T)}$ are the standard gas density, standard pressure (101.325 kPa), standard temperature (273.25 K), gas compressibility factor under standard conditions and gas compressibility factor under P, T conditions, respectively.

By defining:

$$C_D = \frac{\pi \times R \times h^3}{6 \times L} \tag{9.21}$$

and which becomes C_G (i.e. experimentally determined geometrical constant) by calibration, the Eq. (9.21) becomes

$$q_m = \frac{P \times (P_1 - P_2) \times \rho_{(N)} \times T_N \times Z_N}{T \times Z_{(P,T)} \times \eta_{(P,T)} \times P_N} \times C_G \tag{9.22}$$

Thus, for laminar flow element, the flow is proportional to the differential pressure ($P_1 - P_2$), created across the laminar flow gap.

The sonic nozzles also known as critical flow venturies (CFV), measure the flow using a geometrically specified flow path where the flow rate is the maximum based on the upstream conditions of the gas. As long as the ratio of downstream pressure to upstream pressure is low enough, the flow is "choked", i.e. the speed of the gas is limited to approximate local sonic velocity. The ratio of downstream pressure to upstream pressure is called backpressure ratio (BPR). The sonic nozzle is under "choked" condition if the value of BPR is less than 0.7. Once this condition is met, the mass flow rate of a compressible fluid in sonic nozzle is realized using the following Formula [119, 124]:

$$q_m = \frac{A^* \times C \times C_R \times P_0}{\sqrt{(R/M)T_0}} \tag{9.23}$$

where q_m, A^*, C, C_R, P_0, R, M, and T_0 are the mass flow rate (kg/s), throat area (m^2), discharge coefficient, critical flow coefficient, absolute stagnation pressure (Pa), universal gas constant (J/kg-mole-K), molecular mass (kg/kg-mole) and absolute stagnation temperature (K), respectively. In a sonic nozzle, the flow is increased or decreased by changing the upstream pressure to the nozzle using either a variable pressure regulator or a variable restriction.

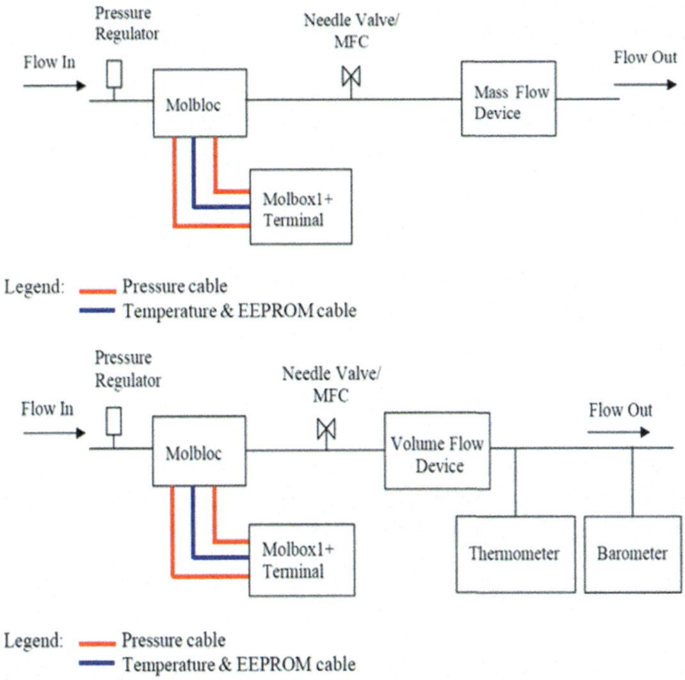

Fig. 9.21 Gas flow calibration facility (upper) calibration set-up of mass flow device (bottom) calibration set-up of volume flow device

The Gas Flow Calibration Facility consists of laminar flow elements, sonic nozzles, high precision flow display device named as mass flow terminal, molstic for mounting laminar, sonic and device under calibrations along with associated instrumentation such as barometric pressure monitor and digital thermometer (PT-100 sensor with indicator). There are four numbers of laminar flow elements (i.e. 100 sccm, 1 slm, 10 slm and 30 slm) which cover the flow range of 0.8 sccm to 55 slm (standard condition being 20 degrees centigrade and 101.325 kPa absolute) with operating pressure of 100 kPa absolute to 525 kPa absolute. There are three numbers of sonic nozzles (i.e. 2E2−S, 5E2−S and 2E3−S) which cover the flow range of 10 slm to 1200 slm with operating pressure of 50 kPa absolute to 600 kPa absolute which uses vacuum pump for realizing the flow below 30% of full scale (FS) rating of sonic nozzle. The laminar flow element and sonic nozzle are known as molblocs in Fluke terminology. Each molbloc has 2 nos. of built-in PT-100 sensors. The molbloc has one EEPROM chip which stores information like flow range, calibration coefficients, different types of calibration, etc. The output of laminar and sonic nozzle is read by mass flow terminal. Figure 9.21 shows Gas Flow Calibration Facility.

The mass flow terminal is a high precision display device which consists of two nos. of high precision absolute pressure sensors (reference pressure transducers) for reading the upstream and downstream pressure of laminar and sonic flow elements.

The terminal has two nos. of PT-100 sensor indicators for gas temperature measurement. It also shows average of these temperatures. The pressure, temperature and NIST gas reference property data are used to realize the flow. The mass flow terminal is essentially a flow computer which displays the flow in different units such as sccm, slm, vlm, vccm, kg/s, mg/s, etc. The uncertainty of the system in the flow range 10 sccm to 1200 slm is 0.20% at $k = 2$.

9.3.3.2 Calibration and Measurement Capabilities

The group is seriously working on getting CMCs of the fluid flow registered in KCDB. For this purpose, necessary arrangements and planning are going on. As a first and foremost requirement, the facilities have been covered under the quality management system of the laboratory as per ISO/IEC17025 standard. Further, step towards CMCs is participation in international intercomparison to prove technical competence. We have participated in one of the comparisons recently and have planned to participate in another soon. Table 9.17 shows the claimed CMCs in water flow whereas Table 9.18 shows the proposed CMCs in low-pressure gas flow. The claimed CMCs in gas flow would be further improved after the establishment of Primary Gas Flow Calibration Facility.

9.3.3.3 Participation in International Key Comparisons

CSIR-NPL has participated in *APMP Keycomparison (APMP.M.FF-K6.2018)* of low-pressure gas flow piloted by KRISS, Korea. The artefact for this comparison is a rotary gas meter. Total 9 NMIs are participating in this comparison. CSIR-NPL finished the measurement during January-March, 2020 and submitted the measurement results to pilot laboratory. For this key comparison, a fully automatic set-up was developed. Now we can calibrate any gas meter having frequency/pulse output. Also participating in *APMP Supplementary Comparison (APMP.M.FF-S3.2020)* of water flow using Coriolis Mass Flowmeter as artefact. In this international intercomparison, 7 NMIs are participating and NIMT, Thailand is pilot laboratory. The artefact is scheduled to arrive CSIR-NPL in February-March, 2021 for measurement. The group is also scheduled to participate in upcoming APMP Keycomparison *(APMP.M.FF-1KP)* in Water Flow range 30 to 200 m^3/h and APMP Keycomparison *(APMP.M.FF-K6P)* in low-pressure gas flow range 2 mL/min to 10 L/min which are in planning stage and waiting for completion of ongoing CCM key comparisons.

9.3.3.4 Potential Services for Future Requirements

There is a huge requirement for calibration of hydrocarbon flowmeters by petroleum industries (like HPCL, IOCL, BPCL, ONGC, etc.), Petrol Pumps, GAIL India Limited, Indian Air Force, etc. Mostly, mass flowmeters and turbine flowmeters are

Table 9.17 Proposed CMCs of CSIR-NPL in water flow

Sl. No.	Parameter/quantity measured	Standards used	Flow range	Expanded uncertainty at $k=2$ ± (%)	Remarks/ method used
1	Totalized liquid mass	12 kg Weighing system	0.03– 0.6 t/h	0.010–0.020	Standing start and finish method using digital indication of MUT
2		300 kg Weighing system	0.1–17 t/h	0.010–0.020	
3		3 Ton Weighing system	10–175 t/h	0.010–0.020	
4		6 Ton Weighing system	30–330 t/h	0.010–0.020	
5	Liquid mass flowrate	300 kg Weighing system	0.1–20 t/h	0.030–0.035	Flying start and finish method/same uncertainty for totalized liquid mass using pulse output of MUT
6		3 Ton Weighing system	10–200 t/h	0.035	
7		6 Ton Weighing system	100–650 t/h	0.04	
8	Totalized liquid volume	12 kg Weighing system	0.03– 0.6 m^3/h	0.010–0.020	Standing start and finish method using digital indication of MUT
9		300 kg Weighing system	0.1–17 m^3/h	0.010–0.020	
10		3 Ton Weighing system	10–175 m^3/h	0.010–0.020	
11		6 Ton Weighing system	30–330 m^3/h	0.010–0.020	
12	Liquid volume flowrate	300 kg Weighing system	0.1–20 m^3/h	0.035– 0.04	Flying Start and Finish method/same uncertainty for totalized liquid volume using pulse output of MUT
13		3 Ton Weighing system	10–200 m^3/h	0.04	
14		6 Ton Weighing system	100–650 m^3/h	0.04–0.05	

Table 9.18 Proposed CMC of CSIR-NPL in low-pressure gas flow (50 kPaa-600 kPaa)

Sl. No.	Parameter/quantity measured	Standards used	Range(s)	Expanded uncertainty at k=2 ± (%)	Remarks/method used
1	Totalized mass/mass flowrate	Laminar flowmeter	9.24×10^{-4} g/min to 1.16×10^{-2} g/min (0.8 sccm to 10 sccm)	2.50– 0.22	Comparison method (air and nitrogen medium)
2	Totalized mass/mass flowrate	Laminar flowmeter	1.16×10^{-2} g/min to 63.78 g/min (10 sccm to 55 slm)	0.22	Comparison method (air and nitrogen medium)
3	Totalized mass/mass flowrate	Sonic nozzle flowmeter	11.58 g/min to 13.92×10^{2} g/min (10 slm to 1200 slm)	0.22	Comparison method (air medium)
4	Totalized volume/volume flowrate	Laminar flowmeter	0.8 mL/min to 10 mL/min	2.50–0.25	Comparison method (air and nitrogen medium)
5	Totalized volume/volume flowrate	Laminar flowmeter	10 mL/min to 55 L/min	0.25	Comparison method (air and nitrogen medium)
6	Totalized volume/volume flowrate	Sonic nozzle flowmeter	10 L/min to 1200 L/min	0.25	Comparison method (air medium)

used in various refineries, oil pipelines, custody transfer terminals. The establishment of Hydrocarbon (Oil) Flow Calibration Facility in the flow range upto 300 m^3/h would help to serve this sector and contribute to the economy of the country. Efforts are being made to arrange funding for this facility from different sources. Efforts are also being made to have collaboration with industries and MSME sectors. As such an effort, discussions are on the cards to have collaboration with GAIL India Limited in establishment of their Primary Standard of Natural Gas (i.e. PVTt [125, 126] or Gas Oil Piston Prover (GOPP) [127] type) and provide traceability for natural gas calibration. At present, GAIL is getting the traceability of its Natural Gas Meter Prover Facility, Surat from the VSL, The Netherlands (i.e NMI of The Netherlands) and spends approx. Rs. 1.0 Crore each year for this purpose. If GAIL establishes such in-house Primary Standard in association with CSIR-NPL, it would reduce the frequency of calibration of the Meter Prover Facility and further reduce the uncertainty of the system apart from saving of foreign currency, thus meeting the goal of the Government of India of *Make in India*/*Aatamnirbhar Bharat*. Similarly, at present, our capability of Gas Flow Calibration Facility is upto 1200 L/min which we want to extend upto 2000 L/min for providing complete service to Environmental Pollution monitoring sector. Our present system is working on pressure mode. If we also develop facility for suction mode, it will serve to the environmental sector in a better way.

9.3.3.5 A Decade of Clients and Services

CSIR-NPL Primary WFCF was inactive since 2001. Also, the Water Meter Testing Facility based on volumetric method was in a very crude state. Thus, Fluid Flow Metrology contribution to National Mission was minimum. This was a big set back to the image of CSIR-NPL as NMI. The momentum for NPL contribution came only after the establishment of Gas Flow Calibration Facility in May 2014 and Prototype WFCF of DN100 size in 2015-16. As customers become aware of new Fluid Flow facilities of CSIR-NPL, we started providing them testing and calibration services. We upgraded the Primary Water Meter Testing Facility based on weighing method in July 2017 for better accuracy and reliability of services. After upgradation of Primary Water Flow Calibration Facility in July 2019, we are providing testing and calibration services of various types of gas flowmeters, water flowmeter and water meters to various customers in the country and we are now really fulfilling the mandate of being the NMI of the country.

Calibration and Testing

During the last 5 years (i.e. 2015–2019), 412 test and calibration reports were issued for gas flowmeters, water flowmeters and water meters by testing and calibration services. The sector-wise main customers of the testing and calibration services are as follows.

Environmental and Pollution Monitoring Sector

Central Pollution Control Board (CPCB), Delhi; Lata Envirotech Calibration Laboratory, Greater Noida; Ecotech Instruments, Greater Noida; Envirotech Instruments Pvt. Ltd., New Delhi; Perfact Researcher Pvt. Ltd., New Delhi; Shree Scientific Equipment and Calibration, Greater Noida; Greentech Testing and Calibration Laboratory, New Delhi; Spectro Analytical Labs, Greater Noida, Shriram Institute for Industrial Research, New Delhi; Belz Instruments Pvt. Ltd., Faridabad; Sigma Test and Research Centre, New Delhi; Central Power Research Institute (CPRI), Noida; SA Associates, Delhi; Gujarat Environment Management Institute (GEMI), Gandhi Nagar; Techno Con Industries, Thane; Rave Innovations, Noida; etc.

Biomedical Sector

Godrej and Byce Mfg. Co. Limited, Mumbai; Tektronix (India) Pvt. Ltd., Mumbai, Wipro GE Healthcare Pvt. Ltd., Bangalore, Philips India Limited Healthcare, Gurgaon; BEC Healthcare (I) Pvt. Ltd., Kochi; Inspior Biomed, Vapi (Gujarat);

Measure Techno Lab., Kolkata; Christian Medical College (CMC), Vellore; Desun Test and Calibration Laboratory, Kolkata; Sodexo Technical Services India Pvt.Ltd., Kolkata; Matrix Lab, Vapi (Gujarat), CSIR-Central Scientific Instruments Organization, Chandigarh; Apex Quality Certification Services Pvt.Ltd., Jaipur; Meditronic Engineering and Innovation Centre, Hyderabad; Life Force, Delhi; Reliable Techno Solutions, Pune; Suntech Medical Systems, Vijayawada; Global Medical Systems, Bangalore; Criticam Medical Systems, Mumbai; Jai Research Foundation, Vapi (Gujarat); Electronic Regional Test Laboratory (North), New Delhi

Strategic Sector

Hindustan Aeronautics Limited, Bangalore; DRDO-Defence Bio-engineering and Electro Medical Laboratory (DEBEL), Bangalore, Bhabha Atomic Research Centre, Mumbai; CSIR-Indian Institute of Petroleum, Dehradun; Air India, Delhi; Ordnance Factory, Kanpur;

Water, Energy, MSME and Process Industry Sector

Endress+Hauser (India) Pvt. Ltd., Mumbai; Micro Precision Products Pvt. Ltd., Dudhola, Palwal; Triveni Engineering and Industries Ltd. (Sugar Mill), Amroha (U.P.); BalrampurChini Mills Limited, Barabanki (U.P.); Bajaj Hindustan Sugar Mills, UP; Sherwani Sugar Mills, UP; Institute for Design of Electrical Measuring Instruments (IDEMI), Mumbai; UP Jal Nigam, Ayodhaya; Uttarakhand Jal Sansthan, Dehradun; Bharat Heavy Electricals Limited (BHEL), Haridwar; Techmark Engineers and Consultants, Delhi; Aquajet Electronics Instrument Pvt. Ltd, Delhi; Nagman Instruments and Electronics Pvt. Ltd., Chennai; NHPC Ltd., Kullu; SITI Energy Ltd., Moradabad (U.P.); C&I Calibrations Pvt. Ltd., Jaipur; Iota Flow Systems Pvt. Ltd., Delhi; Delhi Jal Board, Delhi; New Delhi Municipal Corporation (NDMC), New Delhi; Rama Industries Ltd., Mohali; Padmini VNA Mechatronics Pvt. Ltd., Gurgaon; Elementar India Pvt. Ltd., Gurgaon; Complete Dewatering System, Delhi; Capital Power Systems Ltd., Delhi; Goodyear India Ltd., Faridabad; Toshniwal Industries Limited, Udaipur; Central Institute of Plastics Engineering and Technology (CIPET), Haldia (West Bengal); Aarohi Embedded System, Rajkot; Al-Aziz Plastics Pvt. Ltd., Mumbai; Bureau Veritas CPS (I) Pvt. Ltd., Noida; CSIR-Central Institute of Mining and Fuel Research, Dhanbad; BM Meters Pvt. Ltd., Jalandhar (Punjab); Allport International Pvt. Ltd., Gurgaon; Indian Society for Technical Education (ISTE), Delhi; Akshat Test Lab and Calibration Services, Sahibabad; Metallizing Equipment Co. Pvt. Ltd., Jodhpur; Aforeserve Lab. Pvt. Ltd., Noida; URS Products and Testing Pvt. Ltd., Noida; Suez Projects Pvt. Ltd., Delhi; Air Force Bal Bharti Public School, Delhi, Renewable Energy Test Centre, Deenbandhu Chhotu Ram University of Science and Technology, Sonipat (Haryana) etc. Figure 9.22 shows the sector-wise distribution of customers for Fluid Flow Metrology.

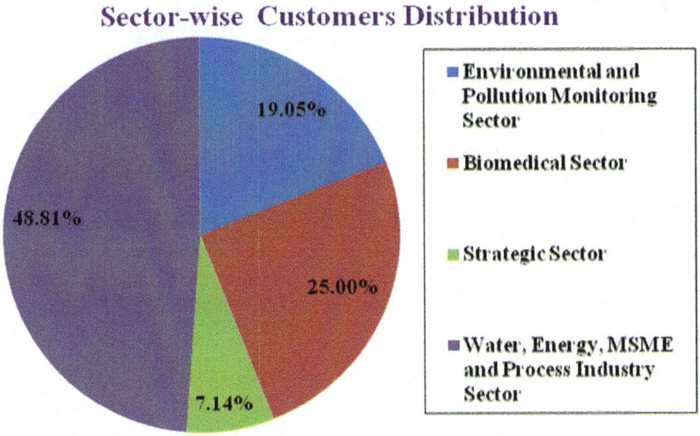

Fig. 9.22 Pie Chart showing sector-wise customers distribution for fluid flow metrology

Consultancy and Technical Services

The main focus of Fluid Flow Metrology at CSIR-NPL during the last one decade was to strengthen its metrological activities by R&D. Except Gas Flow Calibration Facility, development of the prototype Water Flow Calibration System, Upgradation of Primary Water Meter Testing Facility and Upgradation of Primary Water Flow Calibration Facility were completed through R&D efforts. Now, we are focusing on the consultancy services to various customers related to Fluid Flow parameters.

Training and Academic Services

The Fluid Flow Metrology has conducted 5 training courses since July 2017. In these training courses, 42 participants were trained. Also, Fluid Flow parameters are the important contents of one of the modules of the Certificate Course (now upgraded to Postgraduate Diploma Course) in Precision Measurement and Quality Control (PGD-PMQC) since its inception in 2017. During these periods, total 58 students are trained/being trained. The trained manpower has already got placement in Industries, Testing and Calibration Laboratories, etc.

Technology Development and Commercialization

The existing facilities are established for water flow calibration and water meter testing as per international standards. It is also planned to develop some technology for PVTt type Primary Gas Flow Calibration Standard and Primary Hydrocarbon Flow Calibration Standard. The development of these systems would not only enhance CSIR-NPL capabilities but would also have technological expertise for their commercialization.

9.3.3.6 Major Documentary Standards, Guides and SOPs

The documentary standards used in fluid flow metrology are given in Table 9.19.

Table 9.19 List of documentary standards used in fluid flow metrology

Sl. No.	Standard/Guide	Brief description
1	ISO 4185:1980	Measurement of liquid flow in closed conduits—weighing method
2	ISO 9368-1:1990	Measurement of liquid flow in closed conduits by the weighing method—procedures for checking installations—part 1: static weighing systems
3	ISO 5168:2005	Measurement of fluid flow—procedures for the evaluation of uncertainties
4	ISO 4006:1991	Measurement of fluid flow in closed conduits—vocabulary and symbols
5	IS 779:1994	Water meters (domestic type) – specification, 6th revision, 2004
6	IS 6784:1996	Methods for performance testing of water meters (domestic type), edition 3.1, 2nd revision, 2002
7	ISO 4064-1: 2014	Water meters for cold potable water and hot water—part 1: Metrological and technical requirements
8	ISO 4064-2: 2014	Water meters for cold potable water and hot water—part 2: test methods
9	ISO 4064-3: 2014	Water meters for cold potable water and hot water—part 3: test report format
10	ISO 4064-4: 2014	Water meters for cold potable water and hot water—part 4: Non-metrological requirements not covered in ISO 4064-1
11	ISO 4064-5: 2014	Water meters for cold potable water and hot water—part 5: installation requirements
12	OIML R 49-1: 2013	Water meters for cold potable water and hot water, part 1: metrological and technical requirements.
13	OIML R 49-2: 2013,	Water meters for cold potable water and hot water, part 2: test methods
14	OIML R 49-3: 2006	Water meters for cold potable water and hot water, part 3: test report format
15	ISO 9300:2005	Measurement of gas flow by means of critical flow Venturi nozzles

9.3.4 Possible Gaps and Suggestions for Their Redressal

Currently, Fluid Flow Metrology at CSIR-NPL is having a reference/working standard of Gas Flow based on laminar flow elements and sonic nozzles. The establishment of primary standard of gas flow is essentially required. At present, the existing facility has traceability of reference standards from abroad. There are four types of primary standards, i.e. piston prover, bell prover, PVTt standard and weighing/gravimetric standards. Except PVTt standard, all the other standards are commercially available. PVTt standard is available only with selected NMIs. It requires expertise to design and develop such a high accuracy primary standard. There is full potential and expertise available which could be best utilized to develop the PVTt standard indigenously, not only saving the huge foreign currency but also at almost one-third of the cost if imported from abroad. This would also be as per government policy of 'Make in India' and 'Aatmanirbhar Bharat'. The proposed primary standard would cover the flow range of 10 mL/min to 2000 L/min. To cover this huge flow range, we would require 2–3 Primary PVTt standards of different flow ranges.

Also the present uncertainty, i.e. 0.20% of the system will be improved and it would be about 0.10% to 0.12% and at par to leading NMIs of the world. As discussed in 9.3.3.1, the Gas Flow Calibration System is operated in pressure mode. The Gas Flow Calibration System based on vacuum mode for sonic nozzles would be more suitable for environmental applications because most of the MUTs/DUCs work in blower/vacuum mode. Also keeping in view of a huge requirement of calibration of hydrocarbon flowmeters by petroleum industries (like HPCL, IOCL, BPCL, ONGC, etc.), Petrol Pumps, GAIL India Limited, Indian Air Force, etc., there is urgent need to establish Hydrocarbon (Oil) Flow Calibration Facility in the flow range upto 300 m^3/h to serve this sector and contribute in the economy of the country.

In water meter testing, currently, we are able to perform tests as per the old edition of ISO 4064, i.e. 1993. There are very stringent requirements in the latest edition, i.e. 2014. Our facility is not suitable for performing all the tests as per ISO 4064:2014 edition. We can only perform accuracy and pressure loss tests. It would require funds to build additional set-ups and instrumentations to meet the testing requirements of water meters as per the new standard, i.e. ISO 4064:2014.

9.4 Conclusions

In this chapter, we have described the derived parameters belonging to physico-mechanical metrological activities such as Force, Torque, Hardness, Pressure, Vacuum, Ultrasonics, high-pressure Raman spectroscopy and Fluid Flow. For the sustainable infrastructure development and revolutionary industrial growth of the country, precision measurement of these derived parameters are also important for a better quality of human life in every activity. Therefore, all these physical parameters are realized at CSIR-NPL with the highest level of accuracy by establishing the national primary standards and systems.

These metrological activities have been established and maintained as per CIPM-MRA requirements, demonstrating international measurement equivalence and the calibration capabilities. CSIR-NPL has also successfully participated in several international inter-comparisons in these PMPs and reaffirmed its CMCs which are registered in BIPM-KCDB database. So far it has registered 26 CMCs in BIPM-KCDB database. For maintaining these CMCs, CSIR-NPL has been constantly maintaining and upgrading the standards at par with the other leading NMIs and also in the process of expanding the scope of measurement activities by adding new parameters or measurement ranges depending upon the rising needs of the industries.

Out of these metrological activities, more than 800 calibration and testing reports are issued to the stakeholders per annum to further disseminate the apex measurement scale to a huge number of manufacturing industries and testing and calibration laboratories, ultimately resulting in percolation of traceable measurements to every consumer. For example, ERTL (North) gets its deadweight tester calibrated from CSIR-NPL and automobile industry like Maruti Suzuki India Limited, etc. gets its instruments calibrated from ERTL (North). These types of industries and laboratories are huge in numbers which contribute significantly to the economy of the country. Further, the product and services providing ancillary units, called sub-contractors, manufacture and supply different components and parts to these big industries (like Maruti Suzuki India Limited) which finally results in quality products (e.g. a vehicle). Thus, the end-users of the calibration and testing facility for traceability multiply enormously.

Acknowledgements Authors are thankful to all the other members and students of the PMM, past and present who have directly or indirectly contributed to the development of the measurement standards and facilities. Thanks are due to all the funding agencies that have provided partial grants for the same time to time. Thanks are also due to our colleagues Dr. Sushil Kumar, Mr. Jasveer Singh and Mr. V.N. Thakur for their help from time to time.

References

1. S.K. Jain, H. Kumar, S.S.K. Titus, F. Tegtmeier, N. Prenzlow, D. Schwind, Metrological characterization of the 1 MN force standard machine of NPL India. Measurement **45**(3), 590–596 (2012)
2. S.S.K. Titus, S.K. Jain, Establishment and development of torque metrology in CSIR-NPL for providing the traceability in torque measurements to user industries. MAPAN J. Metrol Soc India **28**(1), 11–16 (2012)
3. F. Pavese, G. Molinar, in *Modern Gas-Based Temperature and Pressure Measurements* (Plenum Press, Boston, MA, 1992)
4. P.L.M. Heydemann, W.E. Welch, in *Piston gauges in experimental thermodynamics volume II*.(London-Butterworths, 1975)
5. R.S. Dadson, S.L. Lewis, G.N. Peggs, in *The Pressure Balance: Theory and Practice*. (London HMSO, 1982) p. 290

6. A.C. Gupta, D.R. Sharma, J.K.N. Sharma, Intercomparison of pressure standards in the barometric region. MAPAN J. Metrol. Soc India **11**(1), 3–6 (1996)
7. P. Mohan, D.R. Sharma, A.C. Gupta, Comparison of an ultrasonic interferometer manometer and a static expansion system using a capacitance diaphragm gauge. Metrologia (1996)
8. P. Mohan, A.C. Gupta, Use of a calibrated leak to calibrate vacuum gauges. Vacuum **48**, 515–519 (1997)
9. P. Mohan, Vacuum gauge calibration at the NPL (India) using orifice flow method. Vacuum **51**, 69–74 (1998)
10. P. Mohan, Vacuum measurements by some primary and transfer standards. MAPAN J. Metrol. Soc. India **17**, 119–132 (2002)
11. N. Dilawar, D. Varandani, A.K. Bandyopadhyay, A.C. Gupta, Characterization of a pneumatic differential pressure transfer standard. Metrologia **40**(2), 74–83 (2003)
12. D.R. Sharma, Pressure drop during measurements in a liquid displacement gas flowmeter of improved design. Vacuum **70**, 563–573 (2003)
13. D.R. Sharma, D.A. Vijayakumar, Re-evaluation of measurement uncertainty of ultrasonic interferometer manometer, a primary pressure standard established at NPL, India. MAPAN J. Metrol. Soc. India **21**, 37–46 (2006)
14. P. Mohan, H. Kumar, Static expansion primary vacuum standard—Part 1: determination of the volume ratio of the expansion stage. MAPAN J. Metrol. Soc. India **24**(2), 101–109 (2009)
15. P. Mohan, H. Kumar, Static expansion primary vacuum standard—Part 2: characterization of two spinning rotor gauges. MAPAN J. Metrol. Soc. India **24**(2), 111–118 (2009)
16. D.A. Olson, P.J. Abbott, K. Jousten, F.J. Redgrave, P.Mohan, S.S. Hong, Final report of key comparison CCM.P-K3: absolute pressure measurements in gas from 3×10^{-6} Pa to $9\times^{-4}$ Pa. Metrologia (2010)
17. D.A. Vijayakumar, O. Prakash, R.K. Sharma, Establishment of a force balanced piston gauge for very low gauge and absolute pressure measurements at NPL, India. J. Phys: Conf. Ser. **390**(1), 012013 (2012)
18. K. Jousten et al., Final report of key comparison CCM.P-K12 for very low helium flow rates (leak rates). Metrologia (2013)
19. N.D. Sharma, D.A. Vijayakumar, J. Singh, A. Kumar, D.R. Sharma, A.K. Bandyopadhyay, Establishing a continuous chain of traceability for pressure measurements up to 40 MPa. NCSLI Meas. **8**(1), 56–65 (2013)
20. V.N. Thakur, S. Yadav, A. Kumar, Evaluation of uncertainty in the effective area and distortion coefficients of air piston gauge using monte carlo method. MAPAN **34**(3), 371–377 (2019)
21. A. Kumar, V.N. Thakur, A. Zafer, N.D. Sharma, S. Yadav, D.K. Aswal, Contributions of national standards on the growth of barometric pressure and vacuum industries. MAPAN J. Metrol. Soc. India **34**(1), 13–17 (2019)
22. A. Kumar, V.N. Thakur, R. Sharma, H. Kumar, Omprakash, S. Yadav, Uncertainty evaluation and phase variation of ultrasonic interferometer manometer: a primary pressure and vacuum standard. Vacuum **165**, 232–238 (2019)
23. V.N. Thakur, R. Sharma, H. Kumar, Omprakash, D.A. Vijaykumar, S. Yadav, A. Kumar, On long-term stability of an air piston gauge maintained at national physical laboratory India. Vacuum **176** 109357 (2020)
24. J.K.N. Sharma, K.K. Jain, A.K. Bandyopadhyay, Characterization of a controlled clearance piston gauge using different working fluids up to 5 MPa. Jpn. J. Appl. Phys. **27**, 843–848 (1988)
25. J.K.N. Sharma, K.K. Jain, A.K. Bandyopadhyay, J. Jager, International intercomparison of pressure standards in the pneumatic pressure region 0.4-4.0 MPa between NPL (India) and PTB (FRG). J. Phys. E. (1988)
26. J.K.N. Sharma, K.K. Jain, A.K. Bandyopadhyay, Intercomparison of pneumatic differential pressure measurements at 30–150 kPa at high line pressure upto 7.5 MPa. High Temp. High Press. **21**, 627–636 (1989)
27. A.K. Bandyopadhyay et al., Results of the APMP Pressure key comparison APMP.M.P-K1c in gas media and gauge mode from 0.4 MPa to 4.0 MPa. Metrologia (2003)

28. R.G. Driver, D.A. Olson, NDilawar, A.K. Bandyopadhyay, Final report on key comparison APMP.SIM.M.P-K1c: bilateral comparison between NIST (USA) and NPLI (India) in the pneumatic pressure region 0.4 MPa to 4.0 MPa. Metrologia (2007)
29. A.K. Bandyopadhyay, S.Y. Woo, M. Fitzgerald, J. Man, A. Ooiwa, M. Jescheck, W. Jian, C.S. Fatt, T.K. Chan, K. Moore, Final report of APMP pressure key comparison (APMP.M.P-K6) in gas media and gauge mode from 20 kPa to 105 kPa. Metrologia **45**, 07001 (2008)
30. S. Dogra, J. Singh, A. Lodh, N.D. Sharma, A.K. Bandyopadhyay, A comparative approach for the characterization of a pneumatic piston gauge up to 8 MPa using finite element calculations. Meas. Sci. Technol. **22**(2), 025103 (2011)
31. J. Singh, N.D. Sharma, O. Prakash, Establishment of high pressure pneumatic standard up to 40 MPa at NPLI. MAPAN J. Metrol. Soc. India **27**(2), 113–121 (2012)
32. J.K.N. Sharma, K.K. Jain, V.E. Bean, B.E. Welch, R.J. Lazos, Effects of viscosity, temperature, and rate of rotation on pressure generated by a controlled-clearance piston gauge. Rev. Sci. Instrum. (1984)
33. A.K. Bandyopadhyay, P. Hilsch, J. Jaeger, Use of controlled clearance high pressure balances with highly viscous pressure transmitting media. PTB-Mitteilungen Forschen und Prufen (1987)
34. J.K.N. Sharma, K.K. Jain, G.F. Molinar, Use of strain gauge pressure transducer as working pressure standard up to 500 MPa. Sens. Actuators (1987)
35. K. Jain, C. Ehrlich, J. Houck, Intercomparison of hydraulic pressure measurements to 28 MPa using a single-piston gauge in the controlled-clearance, reentrant, and simple configurations. Rev. Sci. Instrum. **63**(5), 3127–3135 (1992)
36. J.K.N. Sharma, K.K. Jain, H.N.P. Poddar, High pressure measurement with simple piston gauge in static condition. Pramana **42**(3), 271–282 (1994)
37. S. Yadav, D.A. Vijayakumar, A.C. Gupta, Computer software for calibration of industrial and master simple/reentrant type piston gauges. MAPAN J. Metrol. Soc. India **12**, 101–104 (1997)
38. D.A. Vijaykumar, S. Yadav, R.K. Sharma, A.C. Gupta, Studies on pressure measuring equipment form 100 mbar to 5000 bar both absolute and gauge pressures. J. Instrum. Soc. India **27**(4), 269–277 (1998)
39. Y. Kumar, V. Kumar, K.K. Jain, S.C. Kashyap, A capacitive pressure gauge as a reliable transfer pressure standard. Sens. Actuators B Chem. **55**(2–3), 217–221 (1999)
40. J.K.N. Sharma, K.K. Jain, C.D. Ehrlich, J.C. Houck, D.B. Ward, An intercomparison between NPL (India) and NIST (USA) pressure standards in the hydraulic pressure region upto 26 MPa. J. Res. Nasl. Inst. Stand. Technol **99**, 725–729 (1999)
41. APLAC Pressure inter laboratory comparison, APM006-Jan, APLAC pressure program (1999)
42. A.K. Bandyopadhyay, A.C. Gupta, Realization of a national practical pressure scale for pressures up to 500 MPa. Metrologia (1999)
43. S. Yadav, D.A. Vijayakumar, A.K. Bandyopadhyay, A.C. Gupta, A proposal for the establishment of measurement uncertainty of pressure dial gauges and transducers Cal Lab. Int. J. Metrol. 32–43 (2000)
44. S. Yadav, J. Matsuda, L.P.L. Chitarage, Studies on uncertainty evaluation in straightness measurement. J. Robot. Mechatron. **13**, 643–650 (2001)
45. S. Yadav, A. Bandyopadhyay, V. Gupta, A.G. India, A reliable quartz digital transfer pressure standard for high pressure measurements up to 275 MPa. J. Instrum. Soc. **30**(1), 43–51 (2001)
46. S. Yadav, A.K. Bandyopadhyay, N. Dilawar, A.C. Gupta, Re-establishment of measurement uncertainty in pressure measurement through in-house laboratory intercomparison of national hydraulic pressure standards up to 500 MPa. MAPAN J. Metrol. Soc. India 170–177 (2001)
47. S. Yadav, A. Bandyopadhyay, V. Gupta, A.G. India, On the traceability and stability of a hydraulic transfer pressure standard up to 140 MPa, MAPAN J. Metrol. Soc. India **17**(1), 3–10 (2002)
48. S. Yadav, A.K. Bandyopadhyay, N. Dilawar, A.C. Gupta, Intercomparison of national hydraulic pressure standards up to 500 M Pa (Meas, Control, 2002)

49. S. Yadav, A.K. Bandyopadhyay, N. Dilawar, Intercomparison of national hydraulic pressure standards up to 500 MPa. Meas. + Control **35**, 47–51 (2002)
50. S. Yadav, R. Agarwal, B. Surekha, A.K. Bandyopadhyay, A.C. Gupta, Modern instrumentation techniques in pressure metrology under static conditions. MAPAN J. Metrol. Soc. India **18**, 57–82 (2003)
51. S. Yadav, A.K. Bandyopadhyay, A.C. Gupta, Characterisation of national hydraulic pressure standards in the pressure ranges up to 100 MPa, 200 MPa and 500 MPa. Callab- Int. J. Metrol. **1**, 28–35 (2003)
52. W. Sabuga et al., Final report on key comparison CCM.P-K7 in the range 10 MPa to 100 MPa of hydraulic gauge pressure. Metrologia (2005)
53. T. Kobata et al., Final report on key comparison APMP.M.P-K7 in hydraulic gauge pressure from 10 MPa to 100 MPa. Metrologia (2005)
54. R.G. Driver, D.A. Olso, S. Yadav, A.K. Bandyopadhyay, Final report on APMP.SIM.M.P-K7: bilateral comparison between NIST (USA) and NPLI (India) in the hydraulic pressure region 40 MPa to 200 MPa. Metrologia (2006)
55. A.K. Bandyopadhyay, S. Yadav, N. Dilawar, Current status of pressure standards at NPLI and our experiences with the key comparison data base (KCDB). MAPAN J. Metrol. Soc. India **21**, 127–145 (2006)
56. A.K. Bandyopadhyay, D.A. Olson, Characterization of a compact 200 MPa controlled clearance piston gauge as a primary pressure standard using the Heydemann and Welch method. Metrologia (2006)
57. S. Yadav, O. Prakash, V.K. Gupta, A.K. Bandyopadhyay, Studies on the stabilities of various types of industrial pressure measuring devices. J. Sci. Ind. Res. (India) **65**(9), 721–724 (2006)
58. S. Yadav, Characterization of dead weight testers and computation of associated uncertainties: a case study of contemporary techniques. Metrol. Meas. Syst. **14**, 453–469 (2007)
59. S. Yadav, O. Prakash, V.K. Gupta, A.K. Bandyopadhyay, The effect of pressure-transmitting fluids in the characterization of a controlled clearance piston gauge up to 1 GPa. Metrologia (2007)
60. S. Yadav, M. Singh, O. Prakash, A.K. Bandyopadhyay, V.R. Singh, Ultrasonic studies on high pressure transmitting fluids used for hydrostatic pressure measurements up to 1.0 GPa. Int. J. Acoust. Vib. **13**(3), 125–131 (2008)
61. S. Yadav, V.K. Gupta, A.K. Bandyopadhyay, Investigations on measurement uncertainty and stability of pressure dial gauges and transducers. Meas. Sci. Rev. (2010)
62. S. Dogra, S. Yadav, A.K. Bandyopadhyay, Computer simulation of a 1.0 GPa piston-cylinder assembly using finite element analysis (FEA). Meas. J. Int. Meas. Confed. (2010)
63. S. Sadana, S. Yadav, N. Jha, V.K. Gupta, R. Agarwal, A.K. Bandyopadhyay, T.K. Saxena, A computer controlled precision high pressure measuring system. Measurement Sci. Rev. **11**, 198–202 (2011)
64. W. Sabuga, D.A. Olson, J.C. Torres, S. Yadav, Y. Jin, T. Kobata, P. Otal, Final report on key comparison CCM.P-K13 in the range 50 MPa to 500 MPa of hydraulic gauge pressure. Metrologia Tech. Suppl. **49**, 07006 (2012)
65. I. Kocas et al., Final report on key comparison EURAMET.M.P-K13 in the range 50 MPa to 500 MPa of hydraulic gauge pressure. Metrologia (2015)
66. H. Kajikawa et al., Final report on key comparison APMP.M.P-K13 in hydraulic gauge pressure from 50 MPa to 500 MPa. Metrologia (2015)
67. V. Kumar, S. Yadav, Studies on use of lead iron tungstate-lead titanate relaxor binary system as a pressure sensing material. Sens. Actuators A Phys. **258**, 101–104 (2017)
68. A. Zafer, S. Yadav, A. Sanjid, L. Kumar, R.K. Sharma, Volume ratio and pressure drop on hydraulic dynamic pressure calibration system. J. Mech. Sci. Technol. (2017)
69. https://kcdb.bipm.org/appendixb/appbresults/ccm.p-k7/m.p-k7_may11.pdf. (Accessed 06 Feb 2018)
70. A. Zafer, S. Yadav, Design and development of strain gauge pressure transducer working in high pressure range of 500 MPa using autofrettage and finite element method. Int. J. Precis. Eng. Manuf. **19**(6), 793–800 (2018). https://doi.org/10.1007/s12541-018-0095-y

71. S. Yadav, A. Zafer, A. Kumar, N.D. Sharma, D.K. Aswal, Role of national pressure and vacuum metrology in indian industrial. MAPAN J. Metrol. Soc. India **33**(4), 347–359 (2018). https://doi.org/10.1007/s12647-018-0270-8
72. A. Zafer, S. Yadav, N.D. Sharma, A. Kumar, D.K. Aswal, Economic impact studies of pressure and vacuum metrology at CSIR-NPL India. MAPAN J. Metrol. Soc. India **34**(4), 421–429 (2019). https://doi.org/10.1007/s12647-019-00356-2
73. R. Kumar, P.K. Dubey, A. Zafer, A. Kumar, S. Yadav, Development of remote wireless environmental conditions measurement monitoring and recording device for metrological and other scientific applications. MAPAN J. Metrol. Soc. India **35**(1), 1–10 (2020). https://doi.org/10.1007/s12647-020-00368-3
74. V.N. Thakur, A. Zafer, S. Yadav, A. Kumar, Ferroelectric-dielectric composite pressure sensor. Sens. Actuators A: Phys **297**, 111536 (2019). https://doi.org/10.1016/j.sna.2019.111536
75. S. Rab, S. Yadav, R. Sharma, L. Kumar, V.K. Gupta, A. Zafer, A. Haleem, Development of hydraulic cross float valve. Rev. Sci. Instrum. **90**, 085102 (2019). https://doi.org/10.1063/1.5089953
76. V.N. Thakur, B.P. Singh, S. Yadav, A. Kumar, Giant pressure sensitivity in piezo/ferro-electric ceramics. RSC Adv. **10**, 9140–9145 (2020)
77. The BIPM key comparison database. [Online] Available https://www.bipm.org/exalead_kcdb/exa_kcdb.jsp?_p=AppC&_q=India&x=68&y=9
78. BIPM - CIPM MRA documents." [Online]. Available: https://www.bipm.org/en/cipm-mra/cipm-mra-documents/. Accessed 16 Jun 2020
79. A.H. Cook, Precise measurements of the density of Mercury at 20 degrees C II content method. Philos. Trans. R. Soc. A Math. Phys. Eng. Sci. **254**(1038), 125–154 (1961)
80. K.-D. Sommer, J. Poziemski, Density, thermal expansion and compressibility of Mercury. Metrologia **30**(6), 665–668 (1994)
81. Details of calibration rates. [Online] Available http://nplindia.org/calibtest/calibration-charges-d506-pressure-vaccumstandards
82. P.F. Egan, J.A. Stone, J.H. Hendricks, J.E. Ricker, G.E. Scace, G.F. Strouse, Performance of a dual fabry-Perot cavity refractometer. Opt. Lett. **40**(17), 3945 (2015)
83. Y. Takei, K. Arai, H. Yoshida, Y. Bitou, S. Telada, T. Kobata, Development of an optical pressure measurement system using an external cavity diode laser with a wide tunable frequency range. Measurement **151**, 107090 (2020)
84. K.M. Swamy, F.J. Keil, Ultrasonic power measurement in milliwatt region by the radiation force float method. Ultrason. Sonochem. **9**, 305–310 (2002)
85. S. Rajagopal, A. Shaw, The buoyancy method-a potential new primary ultrasound power standard. Merologia **49**, 327–339 (2012)
86. P.K Dubey, Y. Kumar, R. Gupta, A. Jain, C. Gohiya, Measurement of total ultrasonic power using thermal expansion and change in buoyancy of an absorbing target. Rev. Sci. Instrum. **85**(5), 054905 (2014)
87. IEC 61161: Ultrasonics – Power measurement – Radiation force balances and performance requirements, International Electrotechnical Commission (2013)
88. J.-F. Aubry, High intensity therapeutic ultrasound: metrological requirements versus clinical usage. Metrologia **49**, S259–S266 (2012)
89. J. Bercoff, M. Tanter, M. Fink, Supersonic shear imaging: a new technique for soft tissue elasticity mapping. IEEE Trans. Ultrason. Ferroelectr. Freq. Control **51**(4), 396–409 (2004)
90. IEC Standard 61161, Ultrasonics-power measurement-radiation force balances and performance requirements, International Electrotechnical Commission, 2nd edn (2006)
91. T. Kikuchi, T. Uchida, Calorimetric method for measureing high ultrasonic power using water as a heating medium. J. Phys. Conf. Ser. Adv. Metrol. Ultrasound Med. (AMUM 2010) **279** 012012 (2011)
92. P.K. Dubey, A. Kumar, Y. Kumar, R. Gupta, D. Joshi, Primary measurement of total ultrasonic power with improved accuracy in RF voltage measurement. Rev. Sci. Instrum. **81**(10), 104904 (2010)

93. B. Zeqiri, G. Zauhar, S. Rajagopal, A. Pounder, Systematic evaluation of a secondary method for measuring diagnostic-level medical ultrasound transducer output power based on a large-area pyroelectric sensor. Metrologia **49**, 368–381 (2012)
94. R.B.P. Costa-Felix, A.V. Alvarenga, R. Hekkengerg, A fully automated system for ultrasonic power measurement and simulation according to. J. Phys. Conf. Ser. Adv. Metrol. Ultrasound Med. **279**, 012004 (2011)
95. M.A. Perkins, A versatile force balance for ultrasound power measurement. Phys. Med. Biol. **34**(11), 1645–1651 (1989)
96. K. Beissner, Minimum target size in radiation force balance. J. Acoust. Soc. Am. **76**(5), 1505–1511 (1984)
97. S.E. Fick, Long-term stability of the NIST standard ultrasonic source. J. Res. Nat. Inst. Standards Technol. **113**, 281–286 (2008)
98. M.C. Ziskin, Specification of acoustic output level and measurement uncertainty in ultrasonic exposimetry. IEEE Trans. Ultrason. Ferroelectr. Freq. Control **50**(8), 1023–1034 (2003)
99. P.A. Lewin, N. Barrie-Smith, M. Ide, K. Hynynen, M. Macdonald, Interlaboratory acoustic power measurement. J. Ultrasound Med. **22**, 207–213 (2003)
100. M. Greenspan, F.R. Breckenridge, C.E. Tschiegg, Ultrasonic transducer power output by modulated radiation pressure. J. AcoustSoc. Am. **63**(4), 1031–1038 (1978)
101. S. Umchid, K. Prasanpanich, Ultrasound power meter with a three axis positioning system for therapeutic applications. Proc. World Congr. Eng. **2**, 3–5 (2013)
102. For HAM-A datasheet: Precision Acoustics, see www.acoustics.co.uk
103. National Physical Laboratory, Kaye and Laby Tables of Physical and Chemical Constants (2005)
104. K. Beissner, Primary measurement of ultrasonic power and dissemination of ultrasonic power reference values by means of standard transducers. Metrologia **36**, 313–320 (1999)
105. B. Zeqiri, in *Output Measurements for Medical Ultrasound*. ed. by R. C.Preston (Springer, 1991), Chap. 3, pp. 41–44
106. Standard Guide for Relative Intensity Correction of Raman Spectrometers (E2911–13)
107. Standard Guide for Testing the Resolution of a Raman Spectrometer (E2529– 06)
108. Standard Guide for Raman Shift Standards for Spectrometer Calibration E1840-96 (Reapproved 2014)
109. EURAMET, metrology for Raman spectroscopy report 2012
110. https://www.horiba.com/uk/scientific/products/raman-spectroscopy/raman-academy/raman-faqs/what-factors-affect-spectral-resolution-in-a-raman-spectrometer/
111. D.W. Spitzer, in *Flow measurement: Practical guides for measurement and control*, 2nd edn. (Research Triangle Park, NC, ISA, Practical guide series, 2001)
112. J.F. Ruff, Flow measurement: history. in *Encyclopedia of Water Science*, ed. by S.W. Trimble, 2nd edn. (CRC Press, Boca Raton, London, New York, 2008), pp. 388–393
113. J.N. Som, R. Singh, V. Babu, I.S. Taak, Environmental conditions required for fluid flow measurement laboratory with special reference to water flow measurement set-up at NPL. MAPAN J. Metrol. Soc. India **13**, 3–8 (1998)
114. ISO 4185, in *Measurement of Liquid Flow in Closed conduits—Weighing Methods*, 1st edn. (1980), pp.1–21
115. ISO 9368-1, in *Measurement of Liquid Flow in Closed Conduits by Weighing Methods—Procedure for Checking the Installations, Part 1: Static Weighing System*, 1st edn. (1990), pp. 1–22
116. ISO 5168, in *Measurement of Liquid Flow in Closed Conduits—Procedure for the Evaluation of Uncertainties*, 2nd edn. (2005), pp. 1–65
117. S.K. Jaiswal, S. Yadav, R. Agarwal, Design and development of a novel water flow measurement system. Measurement-A J. IMEKO **105**, 120–129 (2017)
118. C. Singh, S.K. Jaiswal, S. Yadav, Upgraded Primary water flow calibration facility at NPLI. in *6th National Conference on Advances in Metrology (AdMet-2020)*, Delhi, January 4–6 (2020)
119. S.K. Jaiswal, I.S. Taak, C. Singh, Establishment of the gas flow calibration facility at CSIR-NPL India. In *Proceedings NCSLI 2015 Workshop and Symposium*, Grapevine, USA, July 19–23 (2015)

120. NIST Special Publication SP 250-73, NIST Calibration Services for Water Flowmeters, Water Flow Calibration Facility, NIST, Gaithersburg (2006)
121. S.K. Jaiswal, S. Yadav, A.K. Bandyopadhyay, R. Agarwal, Global water flow measurement and calibration facilities: review of methods and instrumentations. MAPAN J. Metrol.Soc. India **27**(2), 63–76 (2012)
122. J.G. Pope, A.N. Johnson, J.B. Filla, J.T. Boyd, C.J. Crowley, V.E. Bean, NIST's fully dynamic gravimetric liquid flowmeter standard. in *Proceedings The 9th International Symposium on Fluid Flow Measurement,* Arlington, USA, April 14–17 (2015)
123. NEL, National Measurement System, Good Practice Guide–The Calibration of Flowmeters. Available at linkhttps://docplayer.net/8589009-Good-practice-guide-the-calibration-of-flowmeters.html [The link last Accessed on 29 Jun 2020]
124. M. Bair, "Technical Note 2011TN06A: Uncertainty analysis for flow measured by molboc-L and molbloc-S mass flow transfer standards" 2003 Feb 17, Revised May 18 (2004)
125. NIST Special PublicationSP 250-63, Gas Flowmeter Calibrations with the 34 L and 677 L PVTt Standards
126. NIST Special PublicationSP 250-1046, Gas Flowmeter Calibrations with the 26 m^3PVTt Standard
127. M.P. van der Beek, R. Van den Brink, Gas oil piston prover, primary reference values for gas-volume. Flow Measurement Instrum. **44**, 27–33 (2015)

Chapter 10
Electrical and Electronics Metrology: From Quantum Standard to Applications in Industry and Strategic Sectors

Atul S. Somkuwar, Satish, Praveen K. Siwach, Ajeet Kumar, Ajay K. Shukla, Sunil S. Kushvaha, Rajib K. Rakshit, Priyanka Jain, R. S. Meena, Sangeeta Sahoo, R. P. Aloysius, Anjana Dogra, Manju Singh, M. A. Ansari, Suraj P. Khanna, Jiji Pulikkotil, H. K. Singh, VPS Awana, and J. C. Biswas

Abstract CSIR-NPL is the custodian of National standards of electrical and electronic parameters. These include DC parameters such as voltage, current and resistance; low frequency and high frequency impedance related quantities such as capacitance, inductance and AC resistance; AC/DC high voltage and AC high current; AC power and energy; and quantum standard which includes quantum hall resistance

---

A. S. Somkuwar · Satish · P. K. Siwach · A. Kumar · A. K. Shukla · S. S. Kushvaha · R. K. Rakshit · P. Jain · R. S. Meena · S. Sahoo · R. P. Aloysius · A. Dogra · M. Singh · M. A. Ansari · S. P. Khanna · J. Pulikkotil · H. K. Singh · V. Awana · J. C. Biswas (✉)
CSIR-National Physical Laboratory, Dr. K. S. Krishnan Marg, New Delhi 110012, India
e-mail: jcbiswas@nplindia.org

A. S. Somkuwar
e-mail: assomkuwar@nplindia.org

Satish
e-mail: singhsp3@nplindia.org

P. K. Siwach
e-mail: siwachpk@nplindia.org

A. Kumar
e-mail: kumarajeet@nplindia.org

A. K. Shukla
e-mail: shuklaak@nplindia.org

S. S. Kushvaha
e-mail: kushvahas@nplindia.org

R. K. Rakshit
e-mail: rakshitrk@nplindia.org

P. Jain
e-mail: jainp@nplindia.org

R. S. Meena
e-mail: meenars@nplindia.org

D. K. Aswal (ed.), *Metrology for Inclusive Growth of India*,
https://doi.org/10.1007/978-981-15-8872-3_10

(QHR), quantum current (QC) and quantum nanophotonics (QN). The metrological traceability of the electrical and electronics parameters to SI units is derived from Josephson Voltage Standard (JVS), Quantum Hall Resistance (QHR) standard and frequency (time) standards; all of them are being maintained at CSIR-NPL with metrological precision at par with international standards. The traceability of the aforementioned parameters is disseminated through an unbroken chain of apex level calibrations and testing at par with international level to the industries and strategic sectors of the country to improve the quality of life, which in turn will lead to the inclusive growth of the country and economic development. R&D efforts on the development of quantum standards is a constant endeavour and continues to be at the forefront. Specifically, CSIR-NPL focuses on the development of quantum standards related to the unit of current (ampere), the unit of resistance (ohm) and quantum nanophotonics which aims for detection of few photons (or even single photons) using the novel concept of superconducting nanowire single photon detectors (SNSPD). Among these the research on QC needs special mention, as this will lead to realisation of the SI unit of electric current (ampere), the only unit out of the seven base units of SI system. The quantum current standard (QCS) realisation is bifurcated into two approaches, (i) based on the single electron tunnelling effect (SET) observable in semiconductor quantum dot (QD) structures and (ii) the quantum phase slip phenomenon (QPS) observable in superconducting nanowires of cross-sectional area of the order of coherence length of the system. The realisation of the resistance unit (ohm) is based on the quantum hall effect observable in semiconductor 2DEG structures such as GaAs/AlGaAs systems. Recently there has been tremendous evidence emerging for the use of monolayer graphene for the use of

S. Sahoo
e-mail: sahoos@nplindia.org

R. P. Aloysius
e-mail: alosrp@nplindia.org

A. Dogra
e-mail: anjanad@nplindia.org

M. Singh
e-mail: singhm@nplindia.org

M. A. Ansari
e-mail: ansari@nplindia.org

S. P. Khanna
e-mail: khanna@nplindia.org

J. Pulikkotil
e-mail: jiji@nplindia.org

H. K. Singh
e-mail: hks65@nplindia.org

V. Awana
e-mail: awana@nplindia.org

QHR metrology. CSIR-NPL also started the growth of "epitaxial graphene anchored on SiC" and have obtained encouraging results. The nano-photonics measurement research is also taken up actively for its applications to realize quantum standards for optical radiation and device fabrication.

The discovery of the Quantized Hall Effect (QHE) was the result of systematic measurements on silicon field effect transistors – the most important device in microelectronics.

– Klaus von Klitzing (1943-)

10.1 Introduction

Quantum metrology defines the process of measurement as a discrete phenomenon compared to the analogue process of classical metrology. As per the latest revision of SI units [1] all the base units are related to the respective fundamental constants of nature, thus defining the basis quantum metrology in all physical quantities and associated measurements. In the case of EEM, superconductivity plays a vital role in defining the constants of nature. For example, the unit of voltage is defined with respect to the well-known Josephson effect [2], wherein Cooper pairs tunnel through an insulating barrier sandwiched between two superconducting electrodes. Josephson also have predicted that if a superconductor-insulator-superconductor (SIS) junction is being irradiated with energy (usually in the microwave region) constant voltage plateaus appear in the current-voltage characteristics of the junction with the level of voltage plateaus directly proportional to the frequency of irradiated energy ($V = n.\ \Phi_0 \cdot f$). This becomes the basis for quantum voltage metrology, where in the voltage measurement is linked to fundamental constants of h and e ($\Phi_0 = h/2e$). Programmable Josephson voltage standards are now available with values of 1 and 10 V, with an uncertainty better than 10^{-9}. Similarly, the realisation of the unit of current (ampere) is based on the quantum mechanical phenomenon of coulomb blockade wherein tunnelling of discrete amount of charges takes place through quantum dot (QD) structure. The number of charges tunnelled across the dot depends on the cyclic potential difference applied on the gate terminals of the QD. The exploitation of quantum phase slip (QPS) phenomenon in a superconducting nanowire is also being pursued as an alternative for the SET effect. For the case of resistance metrology, the well-known quantum hall effect (QHE) observable in a 2D electron gas is being harnessed. Recently a renewed school of thought is emerging among metrologists to utilise the AC QHE effect to directly link the capacitance (impedance) standards, thus, reducing the uncertainty in the impedance metrology.

As the realisation of current, voltage and resistance are based on quantum effects, the rest of the EEM measurements and standards are also directly or indirectly is linked to quantum effects, thus quantum standards become the pillars of modern EEM. In a different approach superconducting nanowire single photon detectors (SNSPD) are used for counting of a few photons or even single photons. The SNSPD technology has the potential to revolutionise the way the single photons can be captured, manipulated or transported.

Though we have described very briefly about the quantum standards of EEM, it is worth mentioning about the practical realizations of the SI units related to electrical parameters such as ampere, volt and ohm as per the latest revision of SI units. These recommendations are adapted from the SI Brochure: The International System of Units (SI), Ninenth edition (2019): appendix 2 [1].

Practical realization of the ampere A: Realization of a unit generally means to establish the value and associated uncertainty of a quantity of the same kind as the unit that is consistent with the definition of the unit. In practice, the ampere A can be realized by,

1. Using Ohm's law, the unit relation $A = V/\Omega$, and using practical realizations of the SI derived units the volt V and the ohm Ω, based on the Josephson and quantum Hall effects, respectively; or
2. Using a SET or similar device, the unit relation $A = C/\mathrm{s}$, the value of e ($e = 1.602176634 \times 10^{-19}$ A s) given in the definition of the ampere and a practical realization of the SI base unit the second s; or
3. Using the relation, $I = C{\cdot}\mathrm{d}U/\mathrm{d}t$, the unit relation $A = F{\cdot}V/s$, and practical realizations of the SI derived units the volt V and the farad F and of the SI base unit second s.

SET implementations still have certain technical limitations and often produce larger relative uncertainties than some other competitive techniques. However, SET offer unique and elegant approaches to realize SI units. The uncertainties have improved in recent years, and continues to be better, in future. Alternatively, there is growing, in future. Alternatively, there is growing evidence of QPS which appears promising towards the realization of quantum current standards (QCS). QPS is observable in superconducting nanowires of cross-sectional area compared to the coherence length of the superconducting materials. As compared to Josephson effect, a superconductor nanowire exposed to microwave radiation is predicted to give constant current plateaus (dual to the voltage plateaus of the Shapiro steps in Josephson Junctions) as a function of applied frequency. The measurement of current becomes a measurement of frequency which can be realized with uncertainty to the level of better than 10^{-9}. CSIR–NPL is actively working towards the realization of QCS through QPS observable in superconducting nanowires. Efforts have also been started for the realization of SET devices and associated quantum phenomena of Coulomb blockade.

Practical realization of the volt, V: The volt V is the SI derived unit of electric potential difference (voltage) and electromotive force. The Josephson effect, with the Josephson constant K_J where $K_J = 483, 597.848\ 416\ 984$ GHz V^{-1} is used to realize the Volt, with $K_J = 2e/h$. The latter relation is strongly supported by several theoretical and experimental works. The values of both e and h and are given in the new adaption of SI units. Although "$2e/h$" is a universal constant and can obviously calculated to any significant digits, the recommended value is in error by 10^{-15}, thereby negligible to majority of applications. Having a particular value of K_J for practical use, ensures that all realizations of the volt based on the Josephson effect, would employ exactly the same value.

However, it may be noted that K_J is smaller than the value K_{J-90} = 483,597.9 GHz V^{-1}, the latter which was adopted by the CIPM in 1990 for the international realization of the volt using the Josephson effect. This amounts to a fraction of 106.665×10^{-9}, implying that the unit of voltage realized using K_{J-90} was larger than present SI unit.

Practical realization of the ohm, Ω: The SI derived unit of electric resistance and impedance is ohm (Ω) and is realized as follows:

1. By using the quantum Hall effect, consistent with the CCEM Guidelines [3] with von Klitzing constant R_K, (=25,812.8074593045 Ω) to 15 significant digits. A large body of experiments and theory shows that $R_K = h/e^2$, with values of e and h as per the new SI units based on quantum phenomena. Since h/e^2 represents ratio of two universal constants, the value is however, truncated to 1 part in 10^{15}, thereby providing negligible variation in majority of applications. The advantage of recommending a particular value of R_K for practical use is that it ensures, all measurements of the ohm are based on the quantum Hall effect.
2. By comparing an unknown resistance to the impedance of a known capacitance for example, a quadrature bridge, where, for example, the capacitance has been determined by means of a calculable capacitor and the value of the electric constant.

Note that the value of R_K is larger than the value $R_{K-90} = 25{,}812.807$ Ω, the latter which was adopted by the CIPM in 1990 for the international realization of the ohm using the Quantum Hall effect. The new adoption therefore differs by a fractional amount of 17.793×10^{-9}, implying that the unit of resistance determined by using R_{K-90} would be larger than the present h and e based SI unit. Thus, the numerical value of a resistance measured in terms of R_{K-90} with respect to R_K would be smaller by the same fractional amount.

This chapter discusses in detail about how accurate and precise measurements are realised for electrical and electronics parameters such as impedance, voltage, current, power and energy quantum hall resistance and quantum current. It includes impedance parameters up to radio frequencies, DC Voltage, Current, resistance, AC & DC high voltage and high current and power and energy.

The traceability of these parameters is being disseminated through apex level calibration and testing at par with international level to the power sector, public sector undertakings, MSMEs, electrical and electronic equipment manufacturers, strategic sector, accredited calibration and test laboratories and R&D organization and institutes.

In addition, fundamental and applied research in particular to the quantum metrology for JVS, QHR, realisation of quantum current and on-going research in field of nano-photonics metrology are further discussed. The R&D associated with QHR, QC and nano-photonics is the part of a continuous process to further enhance the measurement capabilities.

The impact of the electrical and electronics metrology on industrial growth, missions of the Government and quality infrastructure of the country is also discussed. Details of training and human resources development related to EEM are also briefly provided.

10.2 DC Metrology (Voltage, Resistance, and Current)

The traceability chart for the electrical quantities linking with the respective primary standards is shown in Fig. 10.1. The primary standard of voltage and resistance are JAVS/JVS and 1 kΩ reference standard respectively, from where the unit of volt and current are being disseminated. The unit of ampere is derived from voltage and current using the Ohm's law of electrical conductivity.

10.2.1 *Metrological Traceability*

The calibration and measurement capabilities (CMC) of the DC metrology are listed in Table 10.1. The laboratory have DC voltage measurement capability in the range from 10 μV to 1000 V with a maximum uncertainty of ~5 μV, DC current in the

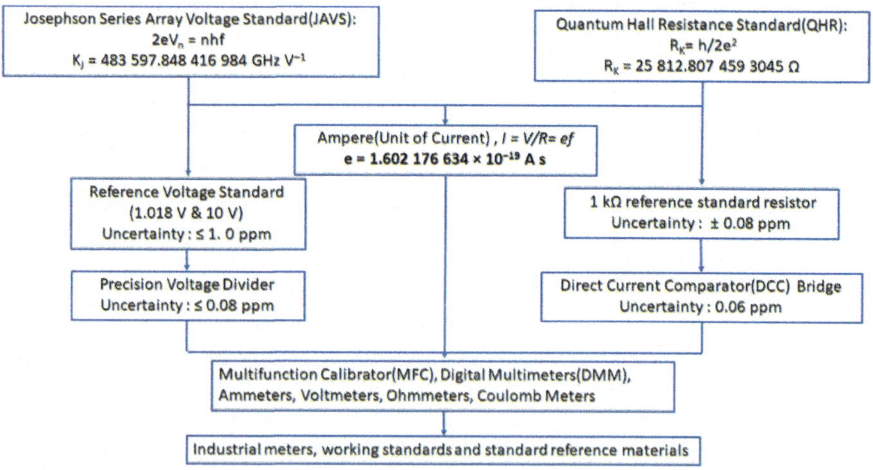

Fig. 10.1 Traceability chart for the electrical quantities linking with JVS and QHR

Table 10.1 Calibration and measurement capabilities of DC metrology (Voltage, resistance and current)

Parameter	Range	Uncertainty levels
DC voltage	10 μV–1000 V	(1.5–4.5) μV
DC current	1 μA–20 A	(5.5–100) ppm
DC resistance	0.1 μΩ–1 GΩ	(1.5–1000) ppm

range from 1 μA to 20 A with a maximum uncertainty of 100 ppm and DC resistance in the range from 0. 1 μΩ to 1 GΩ with an uncertainty of 1000 ppm. There are about 27 CMCs related to the DC metrology which are approved under CIPM-MRA and are listed at BIPM key comparison database (KCDB) [4].

The traceability of DC voltage is primarily dependent on the DC Voltage Reference standard with output voltages of 1.018 and 10 V. The stability of 10 V output is 0.3 ppm/month and 2.0 ppm/year. Similarly, the measurement traceability of DC resistance in the range from 0.1 μΩ to 1 TΩ and DC current is disseminated with standard reference resistors and high precision multifunction calibrators and digital multimeter respectively. Ratio measurements for voltage and resistance are being performed with the help of voltage dividers and direct current comparator (DCC) bridges.

10.2.2 R&D and Future Road Map

Recently there has been tremendous interest in realisation of very low values of voltage, current and resistance for upcoming applications such as IoT and radiation dosimetry. Looking into these aspects and after receiving many industry feedbacks the laboratory started extending and enhancing the measurement capabilities to the limit of low values. For DC current, the range has been extended up to 1 pA utilizing the method of capacitance charging with known value of voltage ramp. The current measurement is performed using an electrometer and pre-amplifier [5]. For a capacitor of known capacitance value C, and for applied voltage ramp of $\mathrm{d}V/\mathrm{d}t$, the current is generated in the capacitor as $I = C{\cdot}\mathrm{d}V/\mathrm{d}t$. Capacitance values of less than 1 nF are suitable for this technique, that too with gas or air dielectric as the medium. To implement the technique a suitable ramp generator (AD549) is fed by constant voltage pulses from the multifunction calibrator. The current developed across the capacitor is measured using the electrometer. The generation and measurements of currents in the range from 1 to 100 pA respectively has been demonstrated with and uncertainty of less than ppm.

10.3 Impedance Metrology

Electrical impedance, Z is a complex quantity and can be expressed in terms of resistance (R), reactance (X); where, the reactance can be capacitive or inductive or a combination of both.

$$Z = R + jX$$
$$X_C = (\omega C)^{-1} \text{ or } X_L = \omega L$$

The frequency behaviour of impedance is severely affected by the parasitic and stray components. The electrical impedance at a specified frequency is defined as the complex ratio of phasor voltage across the terminals and the phasor current flowing between the terminals of a component. This is a frequency-dependent complex quantity and dependent on the material and dimensions of component/device.

While AC resistance differs from DC resistance as it is affected by the skin effect, which increases the effective resistance of conducting wire. Capacitance is related to the electric field and it is affected by the fringing field that occurs at the boundaries of the electrodes separated by dielectric material. Inductance is related to the magnetic field and is therefore susceptible with environment conditions. A variety of bridge circuits are used to measure the impedance parameters and mostly are discussed further in detail.

10.3.1 Metrological Traceability

The traceability chart for the respective impedance parameters are shown in Fig. 10.2. The coaxial bridges discussed below are generally used for the measurement of impedance, which includes the ac resistance, capacitance and inductance parameters.

10.3.1.1 Primary Standard of Capacitance

Thompson-Lampard theorem (1956) allows the calculation of capacitance directly from dimension. Referring to Fig. 10.3, The relationship between capacitance C_1 and C_2 for infinite length system in vacuum [6, 7] can be written as

$$e^{-C_1 \frac{\pi}{\varepsilon_0}} + e^{-C_2 \frac{\pi}{\varepsilon_0}} = 1$$

For symmetrical cylindrical electrodes $C_1 = C_2$, then capacitance per unit length can be defined as:

$$C_0 = \varepsilon_0 \frac{\ln 2}{\pi}$$

$$\varepsilon_0 = \frac{1}{\mu_0 c^2}$$

with c defined as the velocity of electromagnetic wave in vacuum, which is a constant.

In practice, the cylindrical electrodes are not fully symmetrical but if the $C_1 - C_2/(C_1 + C_2/2)$ is less than 10^{-4} then uncertainty of 1 part per 10^8 can be achieved. The capacitance of the calculable cross capacitor was found to be frequency independent, but self-inductance and mutual inductance of the electrodes are frequency dependent and linearly proportional to the inverse square root of the frequency in the

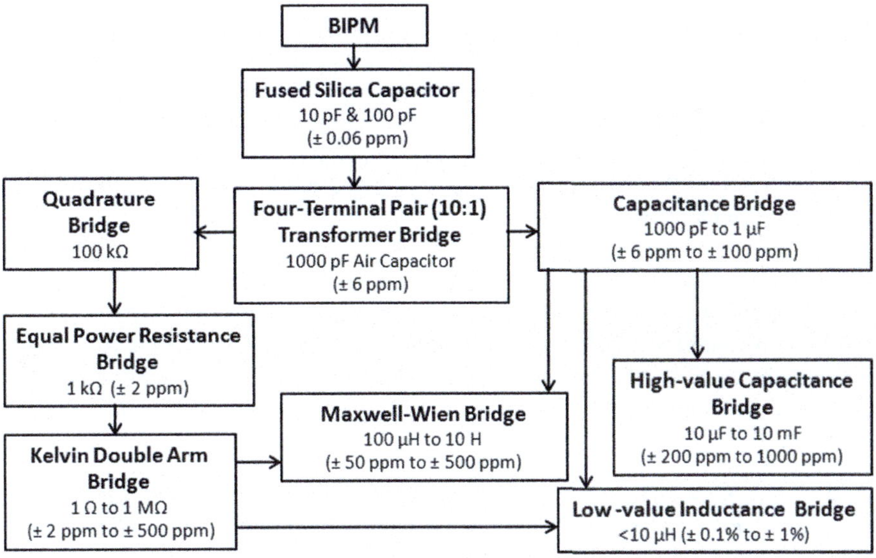

Fig. 10.2 Traceability chart for the impedance metrology laboratory at CSIR-NPL

Fig. 10.3 Cross-section of cylindrical electrodes

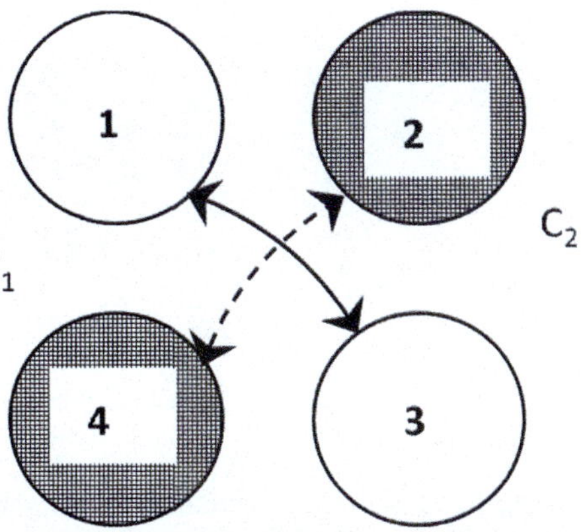

examined frequency range. Cross-calculable capacitor is a complex system and its capacitance is limited to the fractions of pF. It also requires the frequency correction term in the uncertainty calculation as it is defined for electrostatic case.

10.3.1.2 Fused-Silica Capacitor

The fused-silica capacitor standards of 10 and 100 pF drives the traceability of impedance standards and the scaling of capacitance is performed through the various capacitance bridges. The quadrature bridge is used to assign the resistance standard in terms of capacitance standards. Thereafter, the inductor standards are defined in terms of resistance and capacitance standards through Maxwell-Wein Bridge. The laboratory utilizes the high value capacitance bridge for 10 to 10 mF values while low value inductance-bridge is used to measure inductance values equal to below 10 μH.

10.3.1.3 Quadrature Bridge

The quadrature bridge is a product bridge (Fig. 10.4), which compares two resistors R_1 and R_2 with two capacitors C_1 and C_2 at a single angular frequency ω [8]. The equilibrium condition does not rely on voltage or current ratios given by electromagnetic dividers. Quadrature bridge is balanced using the balance condition provided by the equation below

Fig. 10.4 Transformer-based quadrature bridge assembly to define resistance standard in terms of capacitance at CSIR-NPL for audio frequency range. The assembly includes the stable oscillator as signal generator, inductive-voltage dividers, frequency counter, chokes, precise capacitance standards, oil bath, signal detector and unknown resistances

$$\omega^2 R_1 R_2 C_1 C_2 = 1$$

A quadrature bridge is based on the known fact that the phase angle of the impedance, ratio between a resistor and a capacitor is $\pi/2$. Implementation of two such ratio circuit give a total phase shift of π. Ratios with phase angle near to π, can be measured precisely with Inductive-voltage-divider.

10.3.1.4 Transformer Ratio Bridge

Transformer ratio bridges are versatile bridge circuit with wide measurement range and these bridges are used for precision measurement of impedance parameter especially in audio frequency ranges. Primary and secondary transformer ratio in the source and detector side may be tapped to provide the variable internal standards. This mainly removes the requirements of variable resistor and capacitor. These bridges can be used for precise measurements up to 1 MHz. It offers wide measurement range for resistance, capacitance, and inductance. The typical schematic diagram of transformer-ratio bridge and two terminal pair bridge are shown in Figs. 10.5 and 10.6, respectively. The balance condition can be written as (Fig. 10.5).

$$G_X = \left(\frac{N_X}{N_S}\frac{n_X}{n_S}\right)G_S$$

$$C_X = \left(\frac{N_X}{N_S}\frac{n_X}{n_S}\right)C_S$$

The accuracy of the transformer-based bridge is limited by the stray components, electrostatic and electromagnetic field nearby. Electrical and magnetic shielding is

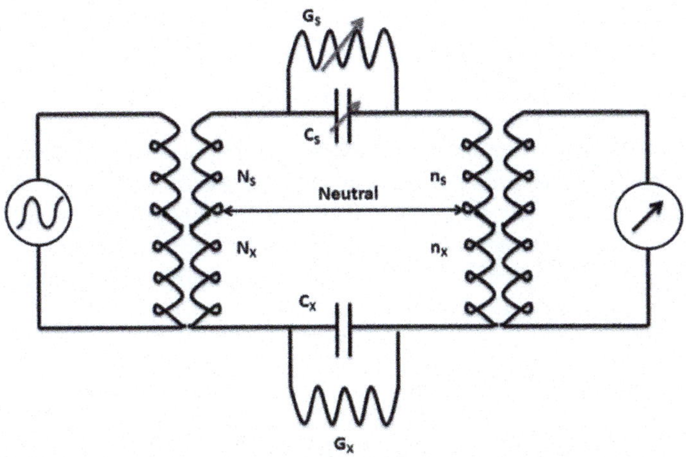

Fig. 10.5 Schematic diagram for transformer ratio bridge circuit

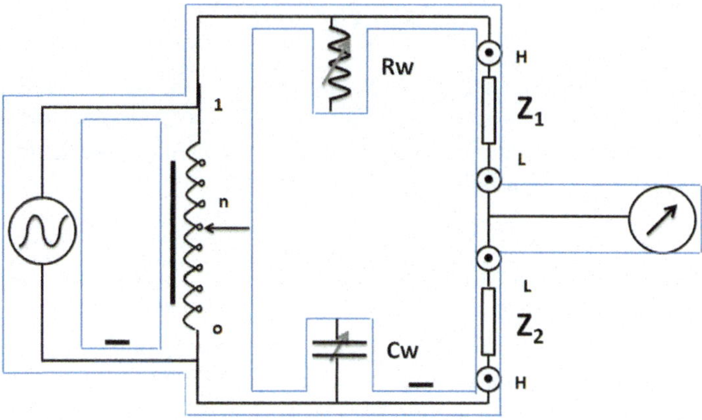

Fig. 10.6 Typical two terminal pair (2TP) IVD based bridge with Wagner balance and current equalizer

the common method that is used to minimize the impact of fields nearby the circuit and devices [8]. Shielding is essential in both DC and AC bridges. Guarding of bridge circuits is also important to reduce the capacitive coupling between the signal and low path. Coaxial circuit helps to confine the electrical and magnetic field well within the inner and outer conductor of coaxial system.

10.3.1.5 Coaxial Bridge

AC coaxial bridges are mainly derived from the conventional DC Wheatstone Bridge. It replaces battery from oscillator (AC source) while galvanometer from sensitive AC detector (tuned detector, vibration galvanometer, electro-mechanical meter, headphone, oscilloscope). Both magnitude and phase are required to be adjusted independently in order to balance the AC bridge. This balance condition is also the function of frequency. Thus, AC bridges are more complex than DC bridges. The bridge circuits can be classified as shown in Fig. 10.7.

The accuracy of the coaxial bridges is limited by the stray and residual components. The contribution from stray capacitance can be minimized by shielding the bridge arms that may reduce the effect of stray capacitance. Wagner balance is also used to eliminate some effects of stray capacitance in a bridge circuit [9]. The laboratory maintains the transformer-based coaxial analog bridges to measure impedance parameters at par with leading NMI's of the world.

10.3.1.6 Maxwell-Wien Bridge

This transformer-based bridge is used to define the inductance in terms of standard resistance and standard capacitance. This bridge can achieved uncertainty better than

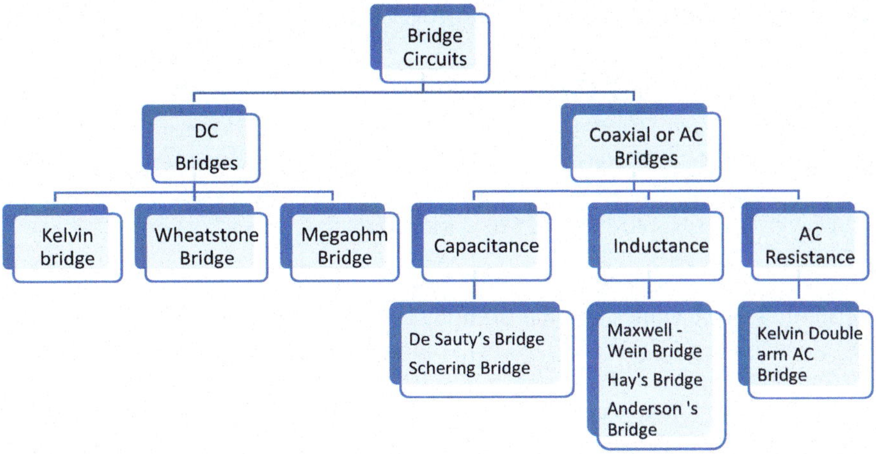

Fig. 10.7 Classification of DC and AC bridge circuits

10 ppm for the 10 and 100 mH inductor at 1 kHz. It requires additional shielding and Wagner balance to adjust the shield potential to detector [10, 11]. The requirement of internal shielding is stringent. Maxwell-Wien bridge is slow and difficult to operate. The connecting leads should be as short as possible, and precautions must be taken to minimize the connection error especially while connecting the unknown with bridge.

10.3.1.7 High Frequency Impedance

The ultra-high-precision air dielectric coaxial transmission lines utilized as national standard of capacitance up to radio frequency range. The set of seven coaxial reference air-lines (Type 900-LZ series) exhibiting capacitance values varying from 2 to 20 pF have been realized as a capacitance standard at high frequency (1–200 MHz). The dimensions of coaxial air-lines; the outer diameter of inner conductor, inner diameter of outer conductor and geometrical length of air-line have been measured precisely and are traceable to the primary standard of length being maintained at the laboratory.

10.3.2 Calibration Measurement Capability

The mandate of the laboratory is to maintain and disseminate the unit's related capacitance, inductance, AC resistance, ac voltage ratio and transformers ratio. All these quantities are linked to the respective primary/national standards through an unbroken chain of traceability. Traceability of impedance parameters are maintained through coaxial bridges, which are at par with international standards. The measurement

Table 10.2 Measurement capabilities of LF, HF impedance metrology laboratory at CSIR-NPL

Parameters	Range	Exp. uncertainty ($k = 2$)
AC resistance	1 Ω–1 MΩ @1 kHz	5–300 ppm
Capacitance	10 pF–1 μF	0.6–100 ppm
Inductance	1–100 mH	30–100 ppm
AC voltage ratio (IVD)	0–1 (1–50 V) @57 Hz, 1 and 10 kHz	0.05–1 ppm

capabilities of quantities are approved under CIPM-MRA and are listed at BIPM key comparison database (KCDB) [4]. The uncertainty associated with impedance parameters are given in Table 10.2.

The laboratory has participated in various international inter key comparison [12–14] with different NMIs for the degree of equivalence in impedance and associated parameters are listed below (Table 10.3).

Table 10.3 List of key comparison of Impedance parameters CSIR-NPL

Parameter	Artifact and range	Year	Country/program
Capacitance	10 pF Quartz capacitor (at 1592 Hz) 10 pF Quartz capacitor (at 1592 Hz) 10 pF Quartz capacitor (at 1592 Hz) 1 nF Air capacitor (at 1592 Hz) 10 pF Air capacitor (at 1592 Hz) 100 pF, Air capacitor (at 1592 Hz) 1000 pF Air capacitor (at 1592 Hz) 10 pF Quartz capacitor (at 1000, 1592 Hz) 100 pF Quartz capacitor (at 1000, 1592 Hz) 10 pF Quartz capacitor (at 1000 Hz) 10 pF Quartz capacitor (at 1592 Hz) 10 pF Quartz capacitor at 1000 and 1592 Hz 100 pF Quartz capacitor at 1000 and 1592 Hz High frequency capacitance standards up to 10 MHz from 1 to 1000 pF	1986 1990 1984 1984 1994 1994 1994 2004 2004 1997 1997 2011 2011 2018	VNIIM (USSR) VNIIM (USSR) APMP NML (South Africa) PTB (Germany) BIPM BIPM NIM China (*APMP EM-S15*)
Inductance	10 mH (at 1000 Hz)	1984 1988 1998	VNIIM (USSR) VNIIM (USSR) PTB (Germany)
AC resistance	1 kΩ Resistor (at 1592 Hz)	1986	VNIIM (USSR)
AC ratio	Inductive voltage dividers at 1 kHz and 55 Hz	2001	CCEM K7
DC resistance	1 Ω and 10 k Ω 1 Ω Standard 1 Ω and 10 kΩ	2013 1987 1986	BIPM (France) Indo-Soviet Prog. APMP
DC current	1 pA, 10 pA and 100 pA	2012	KRISS (Korea)
DC voltage	1000:10, 100:10 (Voltage ratios)	2011	IEN Italy

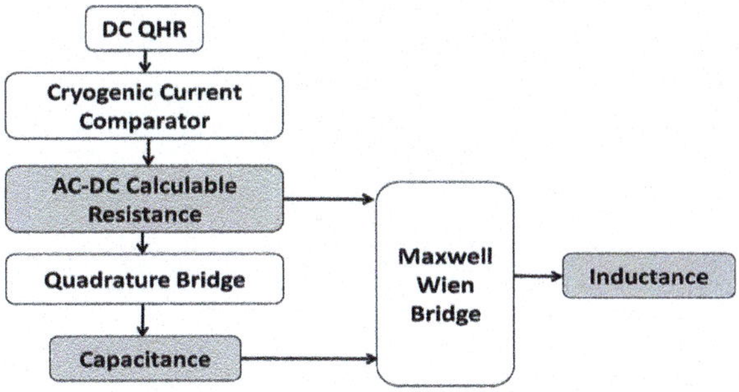

Fig. 10.8 Proposed traceability chart of AC impedance through calculable resistance standard

10.3.3 R&D and Future Roadmap

10.3.3.1 Development of AC–DC Calculable Resistance Standard

Cross calculable capacitance standard based on Thomson-Lampard theorem, was being used as a primary standard of impedance in leading NMI's. These capacitance standards were able to provide the uncertainty of the order of 10^{-7} but are bulky and complex in operation and it has other limitations as well. The laboratory is working on to establish traceability of impedance standards from AC–DC calculable resistance standard that will be traceable to QHR as shown in Fig. 10.8. It is compact, simple to design, and easy to operate and its uncertainty is better than the cross- calculable capacitance standards, considering these factors it has become the first choice for a primary standard.

10.3.3.2 Frequency Characterization of Four Terminal Pair Air Capacitance Standards

Frequency Characterization of four-terminal-pair (4TP) air capacitance standards have been explored using extrapolation technique from 1 to 1000 pF up to 30 MHz. The approach used for the evaluation is based on the determination of capacitive and inductive residual components of electrical equivalent circuit model of four-terminal-pair air capacitance standards [15–19]. S-parameter measurement was performed to determine the series and parallel resonances. A bilateral comparison for capacitance standards is in progress with National Metrology Institute of China under APMP. Figure 10.9 shows the procedure for the evaluation of 4TP capacitance standards and S-parameter measurement of 4TP capacitance standards using vector network analyzer (VNA).

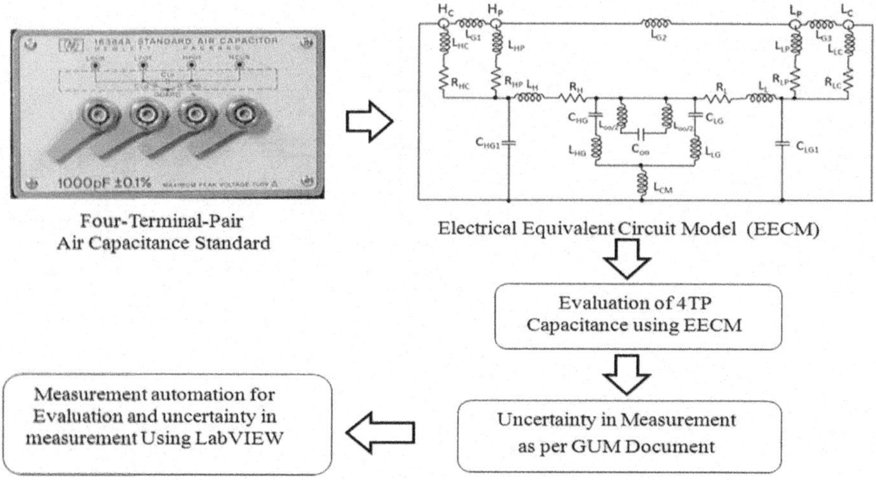

Fig. 10.9 Procedure for the frequency characterization of 4TP capacitance standard

10.4 High Voltage and High Current Metrology

Nowadays, share of renewable energy and its integration with power grids are rapidly increasing. It appears that role of high-voltage direct current (HVDC) is more promising for efficient and reliable bulk power transmission over long distance [20, 21]. However, in such an application, massive breakdown failure may occur due natural or artificial events such as ice storms, thunderstorm, hurricanes or cyber-attack which requires restoration of system within a time. In this regard, maintaining stability and ensure continuous supply of electrical power at various stages such as transmission and distribution network are required to avoid such a kind of major breakdowns. A periodic calibration of instrument transformer used in such a high voltage application such as voltage transformer (VT) and current transformer (CT) are undoubtedly required and are briefly discussed in detailed.

10.4.1 DC High Voltage

The resistive dividers are the primary standard of DC high voltage. They are generally used to measure DC ratio of 100,000/10 with measurement uncertainty of 10 ppm. The traceability of DC high voltage measurement from 1to 100 kV with an uncertainty of 20–100 μV is linked to JVS through resistive voltage dividers shown in Fig. 10.10.

The DC high voltage (100 kV) and high voltage resistance divider requires careful electrostatic shielding to minimize stray capacitance effect. The top of the divider is covered by high voltage grading ring so that the capacitance between individual

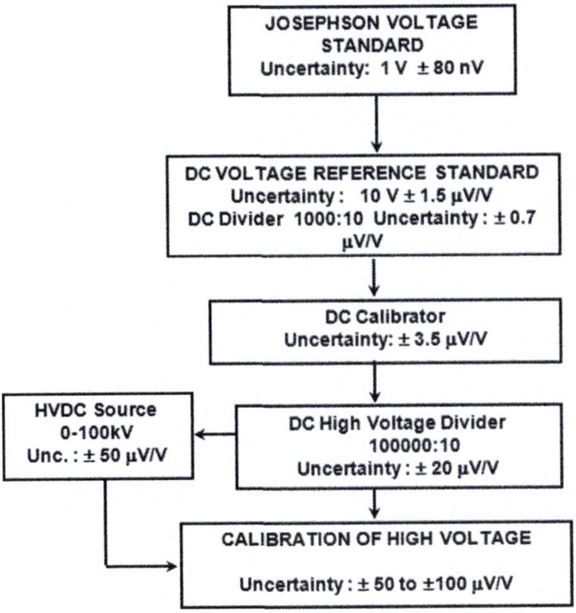

Fig. 10.10 Traceability chart of DC high voltage metrology

resistance units of the divider to earth is equal to the capacitance of the resistance units to the grading ring. The presence of the ring greatly reduces errors, which would be caused by capacitance leakage to earth. The guard circuit is used to reduce the leakage by corona along the whole length of the divider. The guard circuit takes the form of a resistance chain connected in parallel with the main divider having ten times its resistance. The divider is normally used in a high voltage area inside suitable screens and kept clean to keep the surface leakage to a minimum.

10.4.2 AC High Voltage and High Current

The calibration of instrument transformers is essential for precise measurement of bulk power. It is necessary to calibrate the CT and VT for the ratio error and phase displacement as per the IS/IEC standards [22–27]. The measurement of capacitance and tan δ (dissipation factor) are important for determining the deterioration and ageing effects of insulation of cables, bushings, transformers, etc. It may be noted that the increase tan δ values of the insulation of electrical equipment indicates its deterioration. This may be because of ageing, chemical deterioration due to temperature and time, contamination by moisture, carbon deposits, dirt, and leakage through cracks.

The laboratory is maintaining the national standards of AC high voltage (HV) ratio measurements up to 100 kV using HV ratio measuring system (HVRMS) and AC high current (HI) ratio measurements up to 5 kA at power frequency of 50 Hz

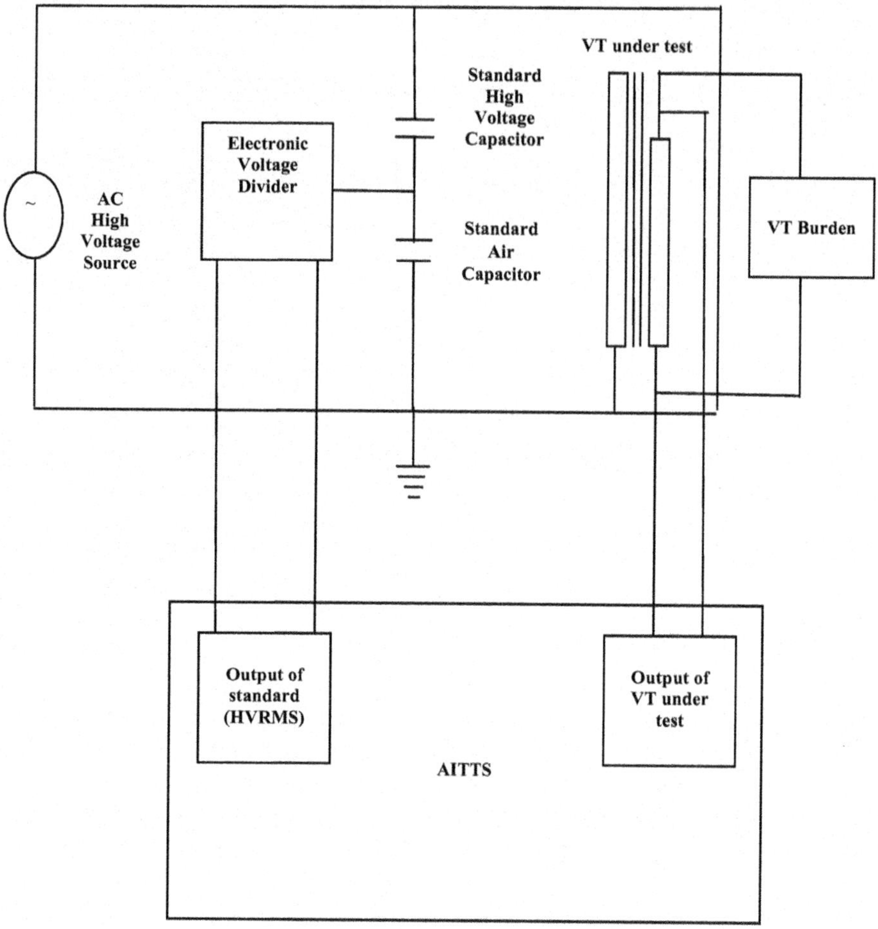

Fig. 10.11 Schematic diagram for voltage transformer ratio measurements using HVRMS

using standard current comparator (standard CT) [28, 29]. The automatic instrument transformer test set (AITTS) is used along with these standards in voltage and current mode, respectively. The traceability of HVRMS, standard CT and AITTS is ensured through periodical calibration from PTB, Germany. The laboratory also maintains the national standards of HV capacitance and tan δ up to 200 kV using high precision C and tan δ bridge, set of standard HV and air capacitors along with standard tan δ [30]. The C and tan δ bridge are periodically calibrated from NRC, Canada. Capacitors are traceable to the primary standard of calculable cross capacitor.

10.4.2.1 AC High Voltage Ratio Measurements

The AC high voltage ratios of the VTs are measured for ratio error and phase displacement as per specified burden. The ratio error of VT can be expressed in terms of percentage as,

$$\varepsilon = \frac{(K_t V_s - V_p)}{V_p} \times 100\%$$

where, k_t is the transformation ratio,
V_p is the actual primary voltage
V_s is the actual secondary voltage

The phase displacement of VT is the difference between the primary and secondary voltage phasors, the direction of the phasors being so chosen that the angle is zero for an ideal transformer.

The accurate and precise measurement of VT is carried out by comparison method as shown in Fig. 10.11. It consists of AC HV Source (300 kV/300 kVA), HVRMS and AITTS. The HVRMS comprises of standard HV capacitor (SF_6 gas filled) of 100 pF/100 kV, standard air capacitor of 10,000 pF/1 kV and standard Electronic Voltage Divider (EVD). The same nominal ratio is selected in HVRMS for the VT ratio to be measured. The required burden is connected in parallel on the secondary of the VT. The output of EVD and VT is connected to the AITTS which is used in voltage transformer test set (VTTS) mode. The VTTS compares the difference in voltages between the HVRMS and VT, and thus determines the ratio error and phase displacement of the VT. The measurements are carried out at 80, 100 and 120% of the rated primary voltage of the selected ratio as per IS 16227 (Part 3): 2015 [26].

10.4.2.2 AC High Current Ratio Measurements

The AC high current ratios of the CTs are measured for ratio error and phase displacement at specified burden. The ratio error of CT can be expressed in terms of percentage as,

$$\varepsilon = \frac{(K_t I_s - I_p)}{I_p} \times 100\%$$

where, k_t is the rated transformation ratio,
I_p is the actual primary current
I_s is the actual secondary current

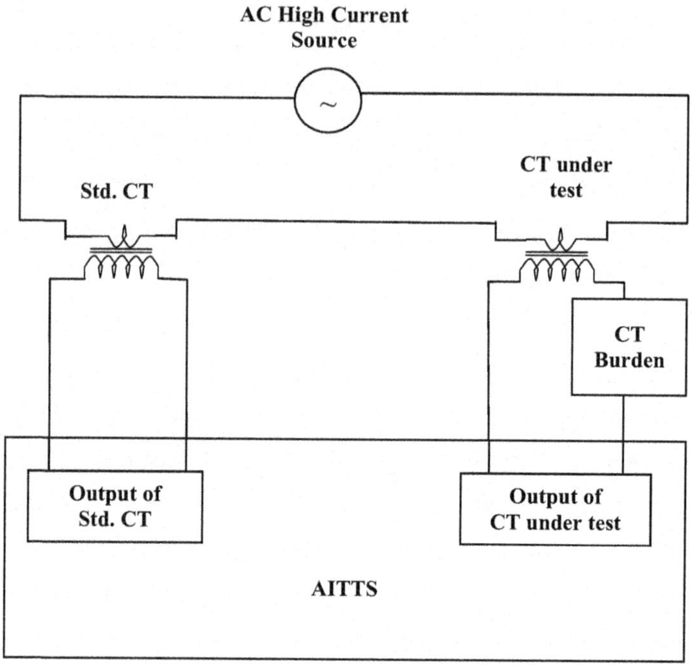

Fig. 10.12 Schematic diagram for current transformer ratio measurement using standard CT

The phase displacement of the CT is the difference in phase between the primary and the secondary current, the direction of the phasors being so chosen that the angle is zero for an ideal transformer.

The precise measurement of CT ratio is carried out by comparison method as shown in Fig. 10.12. It consists of AC HI source (5 kA/15 kVA), standard CT, AITTS, electronic programmable current burden (EPCB) and high current cables. The standard CT is having 310 standard ratios ranging from the lowest ratio of 5 A/1 A, 5 A to the highest ratio of 5 kA/1 A, 5 A. The EPCB provides the burden in the range from 1–75 VA at 0.5–1.0 Power Factor.

The standard CT and CT under test are connected in series using high current cables to a AC HI source. The AITTS is used in current transformer test set (CTTS) mode. The same nominal ratio is selected in both the CTs. The EPCB is connected in series with the secondary of the CT under test. The CTTS compares the output of the CTs and gives the ratio error and phase displacement. The measurements are carried out at 5, 20, 100 and 120% of the rated primary current of the selected ratio as per IS 16227 (Part 2): 2016 [27].

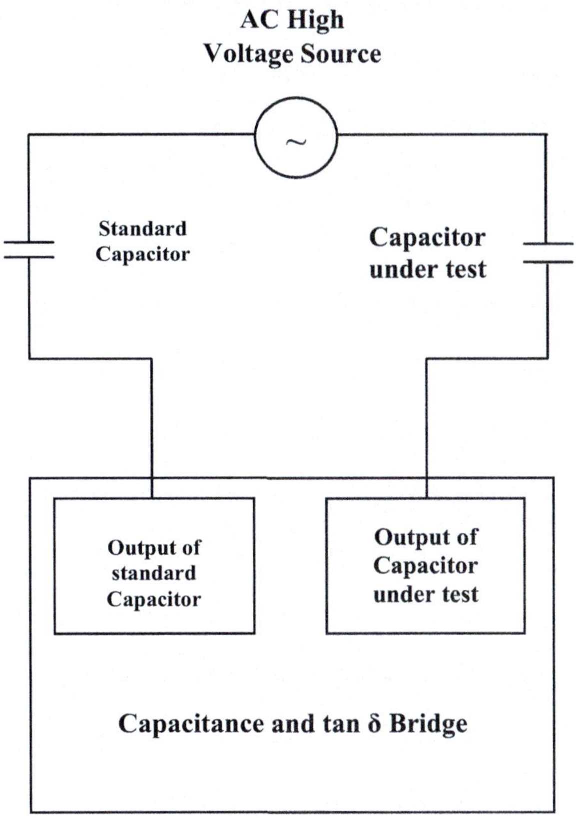

Fig. 10.13 Schematic diagram of high voltage capacitance measurement

10.4.2.3 AC High Voltage Capacitance and Tan δ (Dissipation Factor) Measurements

The AC high voltage, capacitance and tan δ measurements is carried out using high precision capacitance and tan δ bridge, a set of standard HV and air capacitors along with standard tan δ. This set of capacitors consists of HV capacitors of 100 pF/200 kV and 1000 pF/30 kV, standard air capacitors of 100 pF/2 kV and 1000 pF/2 kV along with tan δ standard ranging from 5×10^{-5} to 1×10^{-3}. The commercial C and tan δ bridge used is based on double vector–meter method, which relies upon measurement of current through standard capacitor and the current through capacitor under test as shown in Fig. 10.13. The measurement of capacitors up to 2 kV is carried out using with standard air capacitors. Similarly, measurement of HV capacitors or dividers from 2 to 200 kV is carried using SF_6 standard HV capacitors of 30 and 200 kV.

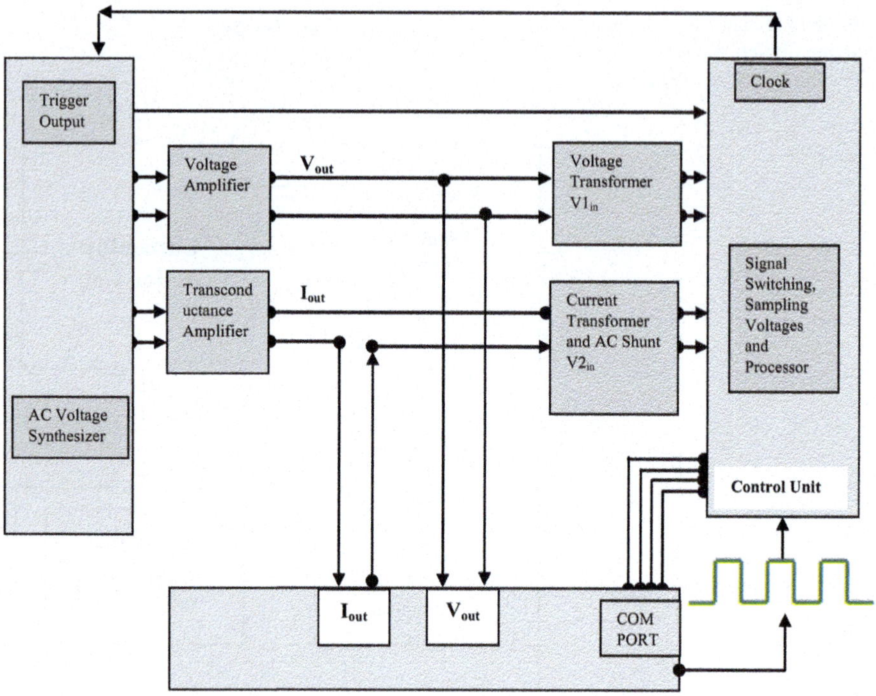

Fig. 10.14 Schematic diagram for primary standard of power and energy

10.4.3 Calibration and Measurement Capabilities (CMC) and Industrial Support

The measurement uncertainty of AC HV ratio error and phase displacement for voltage range up to 100 kV at 50 Hz are ± 60 μV/V and ± 70 μ rad respectively. Similarly, measurement uncertainty of AC HI ratio error and phase displacement for current range up to 5 kA at 50 Hz are ± 40 μA/A and ± 50 μ rad respectively. Whereas measurement uncertainty of high voltage (up to 100 kV) and current (up to 5 kA) are ±130 μV/V and ±120 μA/A respectively. The measurement capabilities of these quantities are approved under CIPM-MRA and are listed at BIPM key comparison database (KCDB) [4]. The measurement uncertainty and range associated with these parameters are given below in Table 10.4.

The traceability of HV and HI measurement parameter is disseminated through apex level calibration services to the power sector in general and power utilities, electrical equipment manufacturers, accredited testing and calibration laboratories in the country. In particular, it includes Current Transformers, Current Transformer Testing Sets, High Current Sources, Clamp Meters, Current Probes, CT Burdens, Voltage Transformers, Voltage Transformers Test Sets, HV Probes, Electrostatic Voltmeters (ESVM), HV Break Down Test Sets, Voltage Transformer Burdens, High

Table 10.4 CMC for AC High voltage and high current metrology

Quantity/parameter	Minimum value	Maximum value	Specifications	Expanded uncertainty at $k = 2$
AC High voltage: source	10 V	100 kV	50 Hz	±130 μV/V
AC High voltage: meter	10 V	100 kV	50 Hz	±130 μV/V
AC High voltage ratio: ratio error	−0.02	+0.02	10 V–100 kV (50 Hz)	±60 μV/V
AC High voltage ratio: phase displacement	−20 m rad	+20 m rad	10 V–100 kV (50 Hz)	±70 μ rad
AC High current: source	1 A	5 kA	50 Hz	±120 μA/A
AC High current: meter	1 A	5 kA	50 Hz	±120 μA/A
AC High current ratio: ratio error	−0.02	+0.02	1A–5 kA (50 Hz)	±40 μA/A
AC High current ratio: phase displacement	−20 m rad	+20 m rad	1A–5 kA (50 Hz)	±50 μ rad

Voltage Sources, HV Capacitors, HV Dividers, kV Meters, Capacitance and tan δ Bridges.

The traceable measurement helps the electrical manufacturing industries to develop accurate and reliable measuring equipment, which in turn translates in winning the consumer confidence in the equity of electricity trade. The traceability in HV Power measurements in support of bulk power measurements, provides a basis to the industries and utilities deeply engaged in the generation, transmission and distribution of electric power. It supports electrical power industries and utilities to be globally competitive.

10.4.4 R&D and Future Roadmap

10.4.4.1 Conventional Instrument Transformers

The voltage transformer (VT) is designed for a specified burden and its use at different burden which affects the performance. The amount and nature of the burden makes a significant influence on the ratio error and phase displacement of VTs. Burdens imposed by different instruments varies widely i.e. small for indicating instruments whereas large for recording and operating instruments. During the measurements on VT, burden equivalent to actual load in regular working conditions is connected in parallel on secondary of the VT.

To understand the effect of burden on accuracy aspect of VT, measurements have been carried out on VT of ratio 11 kV/110 V with leads burden (for laboratory grade

PT) and at different burdens ranging from 5 VA to 40 VA [31]. The AITTS has been used for determining the ratio error and phase displacement at 80, 100 and 120% of the rated voltage. The ratio error at 100% of rated voltage is increased to ~ 0.18% whereas, phase displacement increased to ~ 0.63 min for 40 VA burden. The results show that laboratory grade VTs should be used at rated burden. In case, such VTs are used for higher burden application, it will lead to higher errors in measurements.

Similarly, measurements were carried out on 500 A/1 A CT for ratio error for various burden values against the multi ratio standard CT having ratios up to 5 kA/1 A, 5 A ratio [32]. The burden of 2.5, 5 and 10 VA was connected in series with CT under measurement. The AITTS has been used for comparing the ratio error at 100% of the rated current. It was observed that there is substantial increase in the ratio error from leads burden to 10 VA.

10.4.4.2 Non-Conventional Instrument Transformers for High Current Measurement

In another context, non-conventional instrument transformers (NCIT), based on optical Faraday effect and Pockels Cell effect with digital readout are gaining lot of popularity in high current measurement for high voltage applications [33, 34]. NCIT's potential are well known over conventional one for operational safety, environmental benefit, compact in size and lightweight, measurement fidelity for voltage and current [33]. However, commercially available current sensor has limitation of measurement accuracy of 0.1% for 500 kA (DC) and 63 kA rms. The traceability for non-conventional sensor based current transformer are not yet commercially available; many NMI's are working together to achieve uncertainty below 100 μA/A for medium and high voltage application.

Future Scope: The electrical power transmission in the country is increasing towards extra HV and ultra HV. The laboratory is making continuous efforts to establish the AC HV ratio, HV capacitor and HV divider facility up to 300 kV in near future to meet the need of power industries and utilities. It also has a long-term plan to establish an ultra HV laboratory with state-of-the-art facilities for AC HV measurements up to 800 kV and AC HI ratio up to 20 kA.

10.5 AC Power and Energy Metrology

Accurate and reliable power and energy measurement has a huge economic impact, which plays crucial role in management, maintenance and distribution of electric infrastructure to protect earth's natural resources. Looking at current scenario, the entire ecosystem and economic growth relies heavily on the power sector. These includes renewable energy sources such hydroelectric power, biomass gasifiers, urban and industrial waste power, solar energy and wind energy. The government is taking various initiatives in power sectors by launching new policies and regulations such as program on National Smart Grid Mission (NSGM), Smart Meter National Programme (SMNP). To support government initiatives, the laboratory is trying to cater industrial needs at highest level.

10.5.1 Metrological Traceability

The traceability of Power and Energy is realised through Precision Power Calibration System (PPCS-Primary Standard) which is traceable to DC voltage (10 V), resistance (10 Ω) and frequency (10 MHz). The primary standard is based on digital sampling technique and is designed for high precision active, reactive and apparent power/energy measurements [35]. The basic idea is to utilize single clock frequency which is derived from high precision digital sampling voltmeter for sampling and synthesizing test signals. (shown in Fig. 10.14). The distortion free and highly stable test source (V_{out}, I_{out}) amplified from voltage amplifier and transconductance amplifier are fed to meter under calibration and device itself through voltage transformer and current transformer. The respective voltages $V1_{in}$ and $V2_{in}$ proportional to V_{out} and I_{out} are sampled alternately by Digital Sampling Voltmeter (DVM) using signal switch. Any cause of uncertainty arises is minimize by synchronizing sampling and switching signal. The complex amplitude ($V1_{in}$, $V2_{in}$) can be measured from sampled database using Discrete Fourier Transform (DFT).

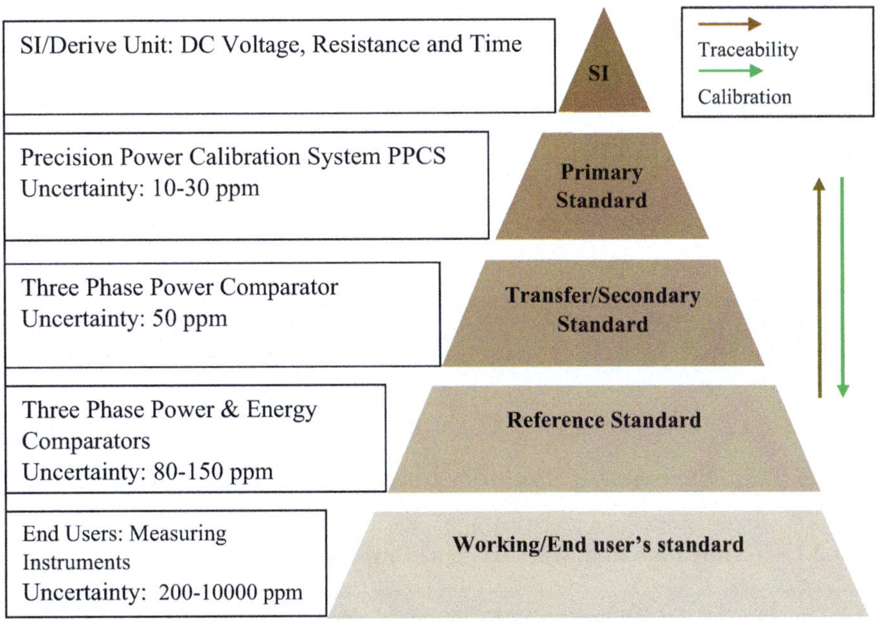

Fig. 10.15 Metrological traceability for power and energy

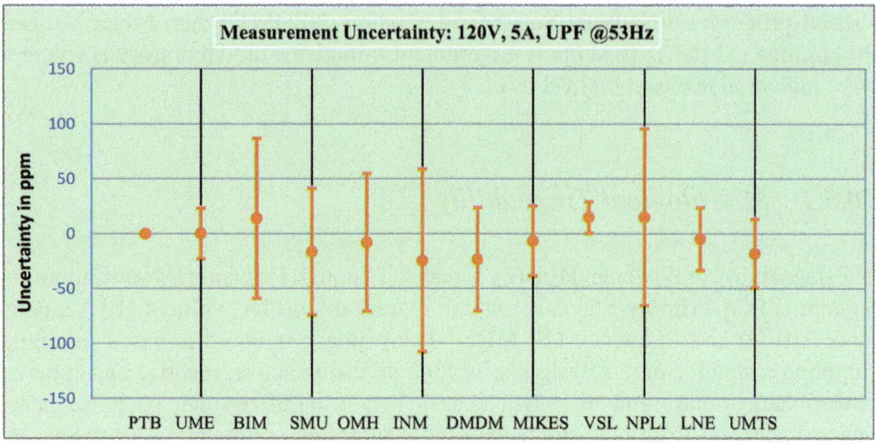

Fig. 10.16 Inter-comparison of NPLI with Other NMI (Normalized with PTB Germany) @120 V,5 A, UPF, 53 Hz

The apparent, active and reactive power can be calculated from the ratio between test and measured signal as (Fig. 10.14)

$$\text{Ratios, } R1 = \frac{V1_{\text{in}}}{V_{\text{out}}} \text{ and } R2 = \frac{V2_{\text{in}}}{I_{\text{out}}}$$

$$\text{Apparent Power, } S = V_{\text{out}} I^*_{\text{out}} = \frac{V1_{\text{in}}}{R_1}\frac{V2^*_{\text{in}}}{R^*_2} = \frac{V_1}{R_1}\frac{V_2}{R_2} e^{i(\varnothing_1 - \varnothing_2 - \arg(R_1) + \arg(R_2))}$$

$$S = \text{Mod}\{S\} = \frac{V_1}{R_1}\frac{V_2}{R_2}$$

$$\text{Active Power, } P = \text{Re}\{S\} = \frac{V_1}{R_1}\frac{V_2}{R_2}\cos(\varnothing_1 - \varnothing_2 - \arg(R_1) + \arg(R_2))$$

$$\text{Reactive Power, } Q = \text{Im}\{S\} = \frac{V_1}{R_1}\frac{V_2}{R_2}\sin(\varnothing_1 - \varnothing_2 - \arg(R_1) + \arg(R_2))$$

Traceability: The unbroken chain of power/energy measurements traceability maintained at the laboratory. The traceability of power/energy standard is ensured by calibrating it against reference power/energy standards as shown in Fig. 10.15.

10.5.2 Measurement Capability and Metrological Services

10.5.2.1 Calibration and Measurement Capabilities (CMC)

The mandate of the laboratory is to maintain and disseminate the unit's related to power and energy and associated measurements from frequency 40 to 400 Hz

Table 10.5 Calibration and measurement capabilities for AC power and energy

Parameter	Range	Expanded uncertainty ($k = 2$)	
		Single phase	Three phase
Active power	10–576 V 10 mA–120 A PF: 0.01 (lag/lead) to 1.0	50–70 μW/VA (40–70 Hz) 70–300 μW/VA (70–400 Hz)	65–150 μW/VA (40–70 Hz)
Apparent power	10–576 V 10 mA–120 A 40–70 Hz PF: 0.01 (lag/lead) to 1.0	80–150 μW/VA (40–70 Hz)	80–150 μVA/VA (40–70 Hz)
Reactive power	10–576 V 10 mA–120 A 40–70 Hz PF: 0.01 (lag/lead) to 1.0	90–150 μW/VA (40–70 Hz)	90–150 μvar/VA (40–70 Hz)
Active energy	30–576 V 10 mA –120 A PF: 0.25 (lag/lead) to 1.0	50–70 μW/VA (40–70 Hz) 70–300 μW/VA (70–400 Hz)	65–150 μWs/VA (40–70 Hz)
Apparent energy	30 –576 V 10 mA–120 A 40–70 Hz PF: 0.25 (lag/lead) to 1.0	80–150 μW/VA (40–70 Hz)	80–150 μVAs/VA (40–70 Hz)
Reactive energy	30–576 V 10 mA–120 A 40–70 Hz PF: 0.25 (lag/lead) to 1.0	90–150 μW/VA (40–70 Hz)	90–150 μvars/VA (40–70 Hz)

for Voltage range 10–576 V, Current range 10 mA–120 A and Power Factor 0.01 (lag/lead) to 1.0. The measurement capabilities of these quantities are approved under CIPM–MRA and listed at BIPM key comparison database (KCDB) [4]. The uncertainty associated with these parameters are given below in Table 10.5.

10.5.2.2 Metrological Services

Calibration Service: The laboratory provides apex level calibration services for power and energy measurements for those who needs the best accurate and reliable measurement. Generally accredited laboratory, power utility, energy meter manufactures require such a highest class of the reference power/energy traceability for stable, accurate and reliable measurements. Single (three) phase power (energy) comparators, power converters, power quality and energy analyzer and meter test equipment are calibrated at the laboratory. The methods and procedure used in the high precision measurements are internationally recognized at par with other leading NMIs. The laboratory also periodically validates its own capabilities by participating in various international key comparison.

Testing Services: The performance of energy meters are tested at par international/national standards such as IEC: 62053-21, IEC: 62053-22, IS: 13779, IS-14697, IS: 13010 and CBIP-88 [36–41]. Various tests are performed on energy meters before rolling out into actual field such as

- Test of accuracy requirements
- Test of electrical requirements
- Test for electromagnetic compatibility
- Test for climatic influence
- Test for mechanical requirements
- Test of influence quantities: Influence of voltage variation, frequency variation, phase sequence, Third harmonics and Fifth harmonics, voltage unbalance should not affect accuracy class of meter as per standard specifications.
- Test of influence of AC/DC magnetic inductions

i Test of influence of AC magnetic inductions
AC abnormal magnetic induction of 10 milli Tesla ±5%
AC abnormal magnetic induction of 0.2 T ±5%
ii Test of influence of dc magnetic inductions
DC stray magnetic induction of 67 milli Tesla ± 5%
DC abnormal magnetic induction of 0.27 T ±5%
DC abnormal magnetic induction of 0.5 T ±5%

- Tamper and fraud protection Tests: Single Phase Energy Meter

  i Interchange of phase and neutral wires
  ii Reversal of line and load terminals
  iii Reversal of line and load terminals with phase and neutral exchange
  iv Load Earthing: The meter should keep on registering energy even when the load is not terminated back to the meter and instead current is drawn partially or fully through earth.
  v Earth load (Forward): The supply to the meter is connected as normal but the load is returned to earth.
  vi Earth load (Forward) with phase neutral exchanged
  vii Earth load (Reverse): The line and load terminals are reversed, and load is returned to earth.
  viii Earth load (Reverse) with phase neutral exchanged
  ix Earth load with neutral disconnection
  x Any other tamper feature is also checked/tested as per customer's requirement.

- Tamper and fraud protection Tests: Single Phase Energy Meter: Three Phase Energy Meters

  i Interchange of Phase and neutral
  ii Missing Potentials
  iii CT Polarity Reversal

iv Phase sequence Reversal
v The working of the energy meter is checked in unbalanced condition
vi The detection of CT short/open condition
vii Tampering against magnetic field
viii The meter functioning for accuracy on all phases individually
ix All single-phase meter tampering conditions applied

However, tampering condition are unknown to all, nevertheless tamper events can be tested and verified as per customer's requirement.

Consultancy Services: The technology which is being adopted for energy meter design and specification are varying in many different ways, resulting difference in their functionality, operation and performance. From Metrological aspect, the performance of energy meter must withstand under influence of electrical and environmental condition; accordingly, it is necessary to optimize each component for accurate measurement and tariff billing. Furthermore, at distribution side, power loss arises due to Transmission and Distribution can be considerable but power theft, the non-technical losses, is a common concern and serious offence. Replacement of electronic meter over electro-mechanical one, technically reduces some sort of meter tampering condition, but in practice, there are numerous ways to tamper electronic meter. In turn, the distribution companies suffer enormous amount of economic problem at power management side. Utility, distribution boards and consumers require technical assurance for quality of product. Being the NMI of the country, the laboratory provide technically assistance to metering industries, power utilities, distribution boards and consumers for parametric optimization for advance design and measurements. The laboratory also supports Bureau of Standards (BIS) technically for policy implementation as when requires standardization in Metering Industry.

10.5.3 R&D and Future Roadmap

10.5.3.1 International Key Comparison

The laboratory also participates in various key comparison, one such a comparison organized by EURAMET under project no. 687 for measurement of power at low frequency is discussed here. [42] total of 12 NMIs, UME Turkey, PTB Germany, NCM Bulgaria, SMU Slovakia, OMH Hungary, INM Romania, DMDM Serbia, MIKES Finland, VSL Netherlands, CSIR-NPL India, LNE France and UMTS Ukraine were participated. HEG C1-2 Power Converter was used as a traveling standard for measurement of power proportional DC voltage (10 V at full range). The measurement taken at 120 V, 5 A, Power Factor 1/0.5i/c/0i/c (i: inductive, c: capacitive) and 53 Hz of frequency. Since the PTB Germany was pilot laboratory, the measurement results of PTB Germany considered as a reference to link each other participant. The measurement uncertainty DC output voltage proportion to 120 V, 5 A, UPF @ 53 Hz power at a coverage factor $k = 2$ shown in Fig. 10.16.

The laboratory is also developing new calibration and testing method for precise measurement power and energy. Laboratory has upgraded measurement capability of power/energy from voltage 0.1 to 640 V; from current 5 mA to 160 A; from power factor: 0.01(lag/lead) to 1.0; from frequency 40 to 70 and 400 Hz with measurement uncertainty ± 10 to ± 30 ppm. Furthermore, laboratory is involved in improving power/energy measurement capabilities beyond 400 Hz to cater industrial needs.

Besides, there are two critical issues which are of concern in metering industries that cannot be forsaken; first influence of electromagnetic field in energy data storage; second load analysis, load forecasting and load management.

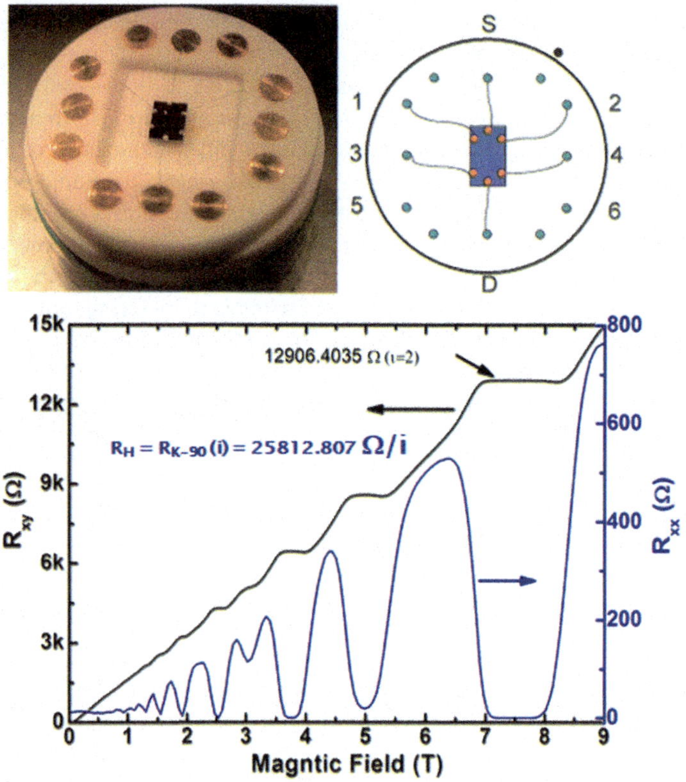

Fig. 10.17 The top panel shows the actual device on To8 holder with contacts and contact schematic (right). Typical specifications of the GaAs/AlGaAs based QHE devices used at CSIR-NPL are as follows: GaAs-$Al_xGa_{1-x}As$ ($x = 0.3$), GaAs (1.5 eV) is lightly p-doped, AlGaAs (2.2 eV) is *n*-doped, ~10–20 T^{-1}, n ~ 3–6 × 10^{15} m^{-2}, B ($i = 2$) ~7–9 T. The graph in the lower panel depicts longitudinal (R_{xx}) and transverse (R_{xy} or the Hall) resistance of a GaAs-$Al_xGa_{1-x}As$ ($x = 0.3$) device measured at 1.3 K

10.5.3.2 Influence of Electromagnetic Field on Energy Meters

The performance of energy meter under influence of conducted electromagnetic influence cannot be neglected, as reported in early publications [43, 44]. The feeders, active infeed converters (AIC) or active front ends (AFE), are widely used to transmit energy to the grid for ensuring better power quality. However, lack of the electromagnetic interference (EMI standards), between 2 and 150 kHz, possibly increase interference level between feeders and grid can cause error in energy measurement [43]. The performance of energy meter has to be tested for conducted emission, radiated emission, immunity to electromagnetic HF field, fast transient burst test, test of electrostatic discharge, surge immunity test, voltage dips and short interruptions and, flicker and harmonics measurements at par IEC 62052-11, CISPR 22, IS 16444 (Part 1) and IS 6842 for IS 16444 (Part 2) guidelines [45–47]. Considering effect of EMI, the laboratory is engaged in establishment of electromagnetic interference and compatibility measurements capability up to 40 GHz for compliance testing on energy meter.

10.5.3.3 Smart Energy Metering

A real time energy monitoring and digitalization of substation are really required to accomplish goals of smart grid infrastructure. During past decade, smart meters (SM), an energy monitoring unit, with bi-directional communication capability were introduced. The number of parameters associated with it such as load analysis, load forecasting and load management cannot be underestimate. Numerous extensive researches have been put forward for handling data set of energy consumption, but yet it is not up to the mark for big data analysis. In fact, real time analysis of big data is essential for load forecasting and load management. Also handling these data involves potential security risk from metering to head end system. There is substantial need of development in privacy challenges, privacy laws, policies, regulations, standards and privacy-enhancing technologies. Number of protocols have been used for liable data exchange with inter-operability such as ZigBee, Modbus, M-Bus, DLMS/IEC62056, IEC61107 and ANSI C.12.18. DLMS/IEC62056 [48–52] accepted in India due its efficiency; considering the above facts the laboratory is engaged in strengthening testing capability for smart metering to address above issues.

10.6 Quantum Metrology

10.6.1 Quantum Hall Resistance Metrology

The measurement of electrical resistance is very fundamental to electrical/electronic metrology. Quantum Hall Resistance Standard (QHRS) provides an invariant quantum standard of resistance in terms of the fundamental constants. The Quantum Hall Effect (QHE) is a characteristic of a perfectly quantized 2-dimensional electron gas (2DEG) system realized in Si-MOSFETs, GaAs/AlGaAs quantum wells, Graphene, etc. at low temperatures and intermediate to high magnetic field. The QHE was discovered by K von Klitzing, provides an invariant reference for resistance linked to fundamental constants "e" and "h" ($R_H = h/ie^2$) and is being used as the primary standard of resistance and its comparison since 1990. In order to facilitate international comparisons and conservation of the ohm, a conventional true value R_{K-90} of R_K was fixed by the CIPM on 1st January 1990. Numerically, R_{K-90} equals exactly to 25,812.807 Ω. The QHE (at 1.3 K) and a magnetic field of 5–14 T supplies a quantum Hall resistance standard (QHRS) of value R_K/i ($i = 2$ or 4), with a relative reproducibility better than 10^{-10}. The bridge used for comparison of resistance standards is based on a cryogenic current comparator (CCC) or room temperature direct current comparator (DCC); the former having much better accuracy (uncertainty ~10^{-9} in terms of R_{K-90}). The most popular QHE device currently used in metrology is the GaAs/AlGaAs. 2DEG quantum well where a 2DEG system is composed of modulation doping in a heterostructure, in which the barrier provides the carriers in the quantum well or hetero-interface of semiconductor multilayer. The quantized Hall resistance is given by $R_H = h/ie^2 = R_{K-90}/i$, where '$h$' is the Planck constant, 'e' is the electron charge, and 'i' is an integer. The voltage V_J obtained from the Josephson Effect and the resistances R_H obtained from the QHE are directly linked to two phenomenological constants, the Josephson constant K_J assumed to correspond to the ratio $2e/h$, and the von Klitzing constant R_{K-90} taken to be the quantum resistance h/e^2.

10.6.1.1 SI Traceability and Measurement Capabilities

The primary standard of resistance in terms of the QHE is maintained at laboratory and has been peer reviewed in 2005 and 2010 by international technical experts. The $i = 2$ plateau with 129,06.4035 Ω is taken as the primary standard and a 1000 Ω resistor maintain in constant temperature enclosure is compared with it. For a 1000 Ω standard measured against the QHRS using a direct current comparator (DCC) bridge, the combined expanded uncertainty at coverage factor $k = 2$ is U_c $(k = 2) =$ 80 ppb. DCC bridge is capable of measuring resistance ratios (1:1, 10:1 and 13:1) in the range 100 mΩ–100 k Ω. The CMC of 1000 Ω resistor at fixed temperature with a combined expanded uncertainty of $U_c = 80$ ppb is listed at the BIPM website [4] (Figs. 10.17, 10.18).

10.6.1.2 R&D in Quantum Hall Resistance Metrology

Typically, GaAs/AlGaAs devices based on QHE operated at 1.3 K, 8–10 T magnetic fields and breakdown current of ~ 10–40 μA [53–59]. Operation at lower magnetic

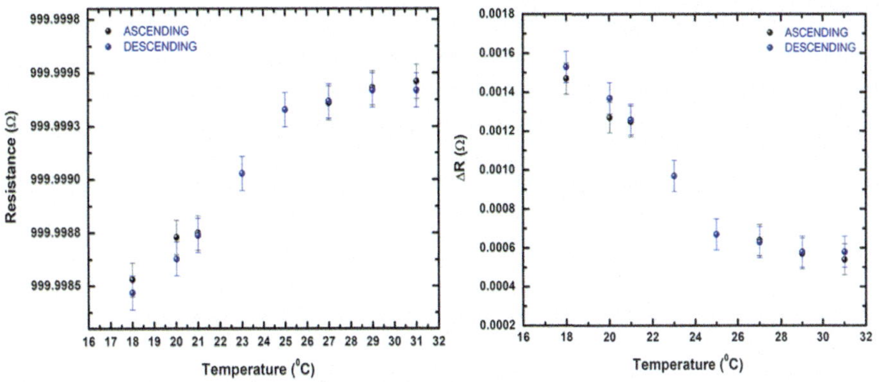

Fig. 10.18 Results of the comparison of a 1000 Ω standard resistor with R_{K-90} $(i = 2) =$ 129,06.4035 Ω using a high precision DCC bridge. Each measurement was done at fixed temperature in temperature ascending as well as descending cycle. The left panel of the figure shows the measured value while the right panel gives the deviation from the nominal value

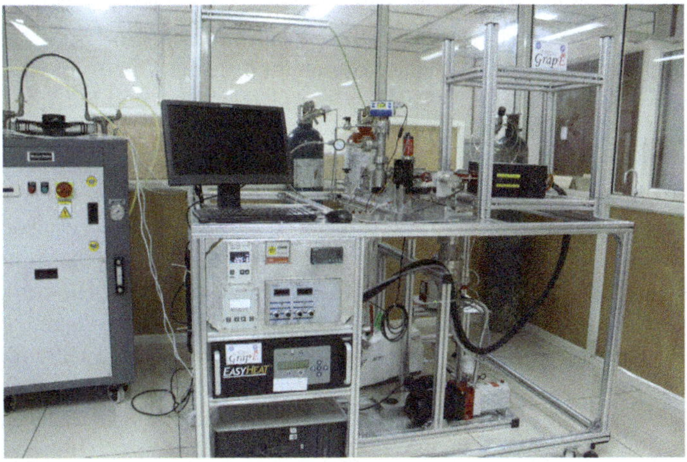

Fig. 10.19 Indigenously developed system for growth of epitaxial graphene on SiC

fields, higher temperatures with larger breakdown current may result in ease of realization and economically cheaper QHRS. It is also desirable to test the universality and material independence of QHRS. Considering these aspects, there has been ongoing efforts to realize QHRS in other system than GaAs/AlGaAs. Among all the new two dimensional electron gas (2DEG) systems, graphene has stolen all the lime light ever since its discovery in 2004 [60]. The origin of the exceptional physical attributes of graphene properties lies in the intriguing crystal structure, nature of chemical bonding and special electronic band structure of graphene [61]. Consequent to the discovery of integer QHE (IQHE) in graphene [62], global R&D efforts have been focused on the realization of the resistance standards under relaxed conditions. Realization of reliable QHRS based on graphene critically depends on the quality and behavior graphene device. There are three most popular synthesis methods to obtain graphene, namely, exfoliation of graphite, chemical vapor deposition (CVD) on metallic foils, and sublimation of SiC. Even though QHE has been observed at room temperature in the device made using exfoliated monolayer graphene, [62] it could not qualify for reliable QHRS due to the limitation of quantization accuracy arising due to small flake size and charge impurities related very low breakdown current [63]. Generally, graphene grown on Cu foil using CVD method is cheaper and easier. However, structural defects due to polycrystalline nature and transfer to Si/SiO2 substrate hinder the observation of dissipation less QHE, a prerequisite to QHRS [64]. Therefore, it is important to have single crystalline graphene with minimum structural defects and homogeneous monolayer thickness over large area, ideally in millimeter scale. Although, epitaxial graphene grown on SiC is much more costly and technically challenging; it can be grown in millimeter scale with highest structural quality [65–69]. Availability of large area single crystalline epitaxial graphene allows easier fabrication of QHE devices exhibiting much higher breakdown current and thus, better signal to noise ratio [57, 64, 69]. After the first demonstration of QHE in Graphene [62], NPL-UK took the lead and made the first demonstration of the quantum Hall effect in Graphene in quantum resistance metrology [53–56, 70]. The recent reports (NPL-UK, LNE-France, PTB-Germany and NIST-USA) in this direction [53, 58, 59, 68–70] clearly establish the fact that devices based on epitaxial Graphene grown on SiC are the most suitable for quantum resistance metrology. The agreement between quantized Hall resistance in Graphene and GaAs/AlGaAs with an ultimate relative uncertainty of 8.2×10^{-11} [57] is also very encouraging. It should also be noted that epitaxial graphene based QHRS can operate at much relaxed conditions compared to conventional GaAs/AlGaAs QHE devices i.e. at higher temperatures, lower magnetic field and with larger breakdown current, [57, 58] and it has resulted in the possibility of realizing QHRS in a table-top cryogen free system [70].

Since it is becoming clearer now that epitaxial graphene based QHRS may replace the conventional GaAs/AlGaAs based QHE devices due to its better performance, it becomes imperative on the part of the laboratory to have a focused program on the development of the quantum resistance metrology devices and measurement techniques based on epitaxial graphene. We envision developing indigenous technology/devices for the realization of graphene based QHRS through concerted, committed and continuous research focused on indigenous development of 2DEG systems to achieve leadership and global visibility. It will be part of national capacity building in the field of quantum metrology. Looking at ourselves as a self-reliant nation, it is essential to focus on advanced quantum material system like graphene so that desired goals are achieved in the area of quantum electrical metrology. Graphene growth is being performed in custom built growth system. This indigenous customized epitaxial graphene growth system Fig. 10.19 has been designed and fabricated and commissioned at the laboratory. Base pressure of growth system is 1

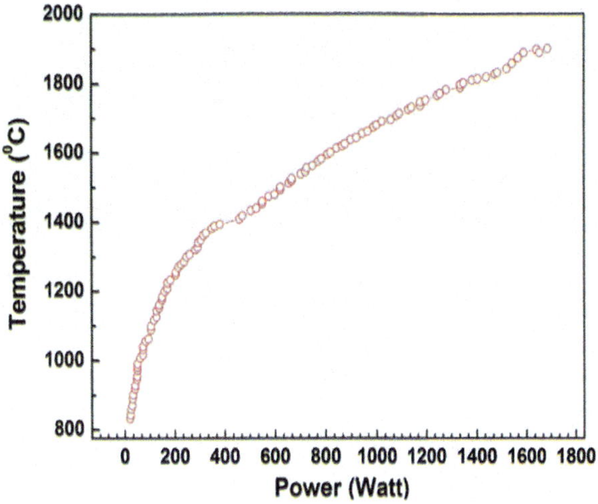

Fig. 10.20 Temperature of graphite enclosure as a function of RF-induction heating power

$\times\ 10^{-7}$ mbar and it can reach to ultra high vacuum (2×10^{-8} mbar) after overnight pumping and without baking.

Growth system is equipped with turbo molecular based pumping system and two gas lines equipped with mass flow controllers and various valves for attaining controlled gas atmosphere. Specially designed graphite enclosure containing 4H-SiC piece (2×4 to 7×7 mm) is heated using RF-induction heating to attain temperatures ~ 2273 K. In situ C-type thermocouple measures the temperature which are cross-checked by a suitable optical pyrometer. Near equilibrium growth is achieved by confining the sublimating Si vapor in a tight enclosure and let it to escape in controlled manner. Atomically smooth SiC is heated to 1673–2273 K in high vacuum as well as Ar atmosphere using RF induction heating to produce highly homogeneous, single crystalline (area near 1–4 mm^2) monolayer graphene. Exhaustive growth parameter (growth temperature, gas pressure, and heating rate and Si vapor escape rate from the close confined graphite enclosure) tuning is being done to achieve best quality monolayer graphene suitable for QHRS applications. Figure 10.20 shows the performance of the growth system in terms of achieved temperature of graphite enclosure and corresponding heating power.

AFM images of monolayer epitaxial graphene grown on 4H-SiC (1000) are shown in Fig. 10.21. It is evident from Fig. 10.21a that graphene covers underneath SiC surface like a carpet and height profile exhibited in Fig. 10.21c shows clearly that unit cell height (1 nm) of 4G-SiC is intact even after the growth. Phase contrast image shown in Fig. 10.21b demonstrates the uniformity of grown graphene.

Raman spectroscopy is a very reliable tool to ascertain the presence of graphene and to determine its layer thickness. Figure 10.22 exhibits the Raman spectra of graphene grown in Ar gas environment, vacuum conditions (~5×10^{-6} mbar) nearly 1773 K. There are two important features to look for graphene in Raman spectrum, *G* (1580–1590 cm^{-1}) peak associated with in-plane vibration mode of graphitic carbon and 2D (~2700–2770 cm^{-1}) peak associated with double phonon scattering. *G* peak

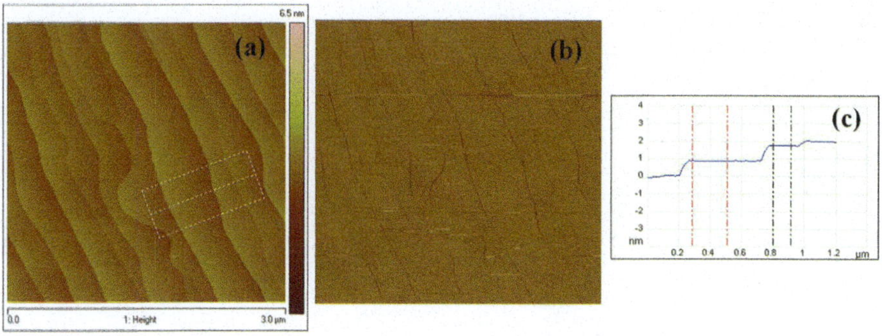

Fig. 10.21 AFM image (3 × 3 μm), **a** topography image and **b** phase contrast image of epitaxial graphene grown on 4H-SiC (1000) at 1773 K, 1030 mbar Ar gas pressure along with **c** height profile across dashed line shown in **a**

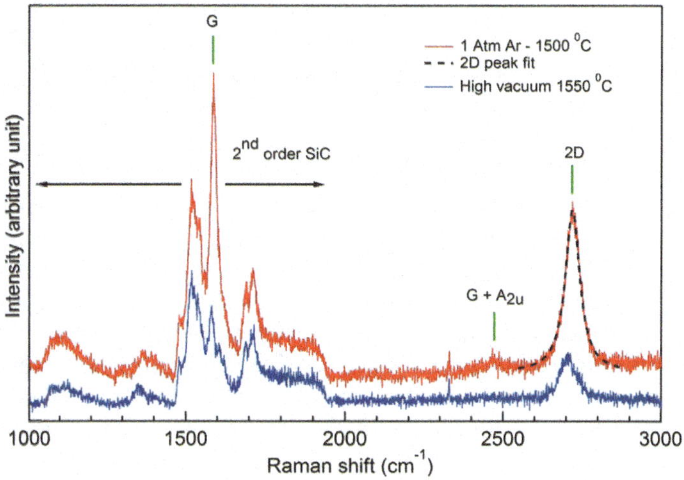

Fig. 10.22 Raman spectra of epitaxial graphene grown at Ar gas and high vacuum environment

is not very sensitive to the layer thickness of graphene. However, position and full width at half maximum (FWHM) of 2D peak are very sensitive to doping and layer thickness, respectively of graphene. In Fig. 10.22, we can observe both *G* and 2D peaks for both the samples along with multitude of features at lower wavelength range (1000–2000 cm^{-1}) originating from underlying SiC substrate. For graphene grown at vacuum environment, *G* and 2D peaks are diminished compared to growth at Ar gas environment. It shows growth at Ar gas environment is favourable to better quality graphene.

2D peak of graphene grown at Ar gas environment can be fitted with single Lorentzian component as shown in Fig. 10.22, whereas more than one component (not shown here) is required to fit 2D peak for graphene grown at vacuum condition,

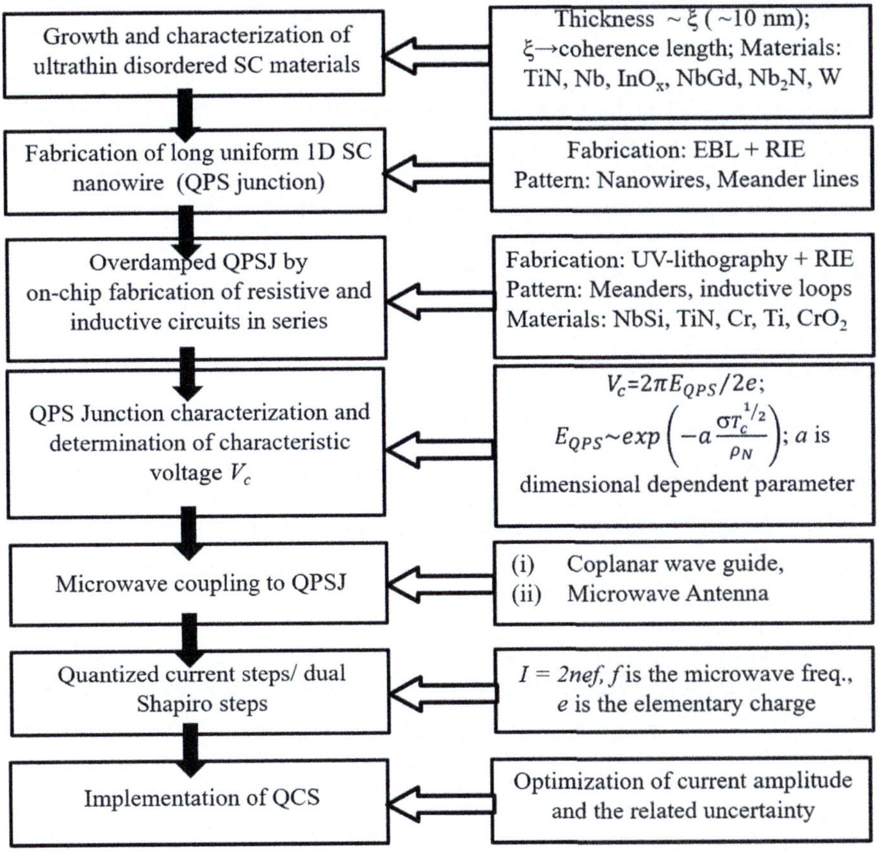

Fig. 10.23 Process flow chart for QPS based QCS

indicating larger layer thickness. Peak position and FWHM of 2D peak of graphene grown at Ar gas environment turns out to be $2715 \pm 6\ \text{cm}^{-1}$ and $50 \pm 0.45\ \text{cm}^{-1}$, respectively and these numbers support the monolayer thickness of grown graphene.

Further optimization of growth parameters and more characterization is going on to achieve reproducible, highest quality epitaxial graphene over a large area, preferably $4 \times 4\ \text{mm}^2$. Later on, metal contact will be optimized with aim to have sub-ohmic contact resistance and 6/8 contact Hall bar QHE devices will be fabricated to realize graphene based QHRS.

10.6.2 Quantum Current Metrology

In the absence of direct quantum current standard, the practical realization of the ampere using the established quantum standards of volt based on Josephson effect and ohm from quantum Hall effect via Ohm's law has become almost universally accepted [71]. However, being the only base unit in the electrical metrology, the ampere has to be established by direct realization of quantum current standard with sufficient accuracy so that the interrelation with other derived units via fundamental laws of physics can be verified. For example, one of the important and unresolved issues related to the quantum electrical metrology is the quantum metrological triangle (QMT) experiment which can be made with the quantum standards of voltage, resistance and current [72, 73]. The closure of the QMT is one of the important tasks in quantum metrology. From the QMT, not only one can inter-relate the three SI units (V, I and R) but also the validity of Ohm's law at the quantum limit can be verified [74]. Currently, the QMT needs to have a better accuracy for the current standard so that the other two established quantum standards, namely, the voltage standard based on Josephson Effects (JE) and the quantum resistance standard based on the Quantum Hall Effect (QHE), can be placed with the quantum current standard and their inter relation through Ohm's law can be verified without deviating the accuracy [73].

In the revised SI units effective from May 20, 2019, a new way of defining the base units has been adopted by using a set of seven defining constants that range from the fundamental constants of physics and other constants of nature [1]. In the new definition, the aforesaid defining constants are assigned to fixed numerical values leading to lower uncertainty with better accuracy. For example, by definition, electric current is the transportation of electrical charge per unit time and as per the new definition in revised SI, the unit of current 'ampere' is defined by fixing the value of the fundamental constant, the elementary charge '*e*'. The new definition follows as: "The ampere, symbol *A*, is the SI unit of electric current. It is defined by taking the fixed numerical value of the elementary charge '*e*' to be 1.602 176 634 × 10^{-19} *C* when expressed in the unit *C*, which is equal to As, where the second is defined in terms of $\Delta\upsilon_{Cs}$, the unperturbed ground-state hyperfine transition frequency of the caesium 133 atom, to be 9192 631 770 when expressed in the unit Hz, which is equal to s^{-1}" [1]. Therefore, from the new definition, one ampere is a measure of the electric current in which [1/(1.602 176 634 × 10^{-19})] (or 6.2415 × 10^{18}) elementary charges (e) move across a given point in one second. So, by controlling one electron at a time and by defining the charge on the electron as a fixed number, we can create a current which is precisely known and self-calibrated.

However, as the discoveries associated with quantum phenomena have brought revolutionary changes in the world of metrology, the quantum electrical metrology comes into play with enhanced measurement capability by exploiting quantum strategies for improved sensitivity and accuracy of measurements. There are two most popular quantum phenomena, viz. (i) single electron tunnelling (SET) effect and (ii) quantum phase slip (QPS) effect, by which the current can be defined in terms of the fundamental unit of electronic charge, '*e*'. Direct measurement of ampere is possible by counting the number of moving electronic charges over a period of time using SET devices. Here, one has to make sure that the counting is error free so that all the tunnelled electrons should be collected at the receiver end with no backward tunnelling. Further, in order to have a reasonable amount of current, the transit of electrons should be very fast, i.e. the frequency of pumping out of individual electrons from a device should be very high. Here come the experimental and measurement challenges that are being addressed by national metrological institutes and other high-end laboratories in the world. Meanwhile, researchers are looking into another exciting quantum phenomenon, quantum phase slip (QPS), which occurs in one-dimensional superconducting nanowires. The QPS is exact dual to the Josephson Effect (JE) which leads to the quantum standard of voltage. Quantized current steps are predicted to appear when a QPS junction is excited with microwave radiation and these steps, known as Shapiro steps, may lead to the realization of quantum current standard. For QPS based devices, the quantized steps do not deal with counting of individual electrons and the amplitude of current can be higher by tuning the frequency of the microwave excitation. However, there are enormous experimental challenges to execute and implement it successfully up to the level of quantum current standard.

10.6.2.1 Single Electron Tunneling (SET)

To realize this new definition of ampere, a standard transistor electron pump is being used to transport and electron across a device, which is based on the principle of SET phenomenon. This phenomenon states that "a quantum mechanical particle has a finite probability of crossing a barrier even if its energy is less than the barrier". It consists of three electrodes; source, drain and gate connected with an island. Source and drain electrodes are coupled resistively to the island and gate electrode are coupled capacitively. Island has finite number of electrons and charges on the island are quantized. Tunnelling of an electron with charge e, or rather a known number N of electrons Ne in each operation of a single-electron pump that is cyclically repeated at frequency f, generates an output DC current, ideally, equal to Ne $\times$ f. The first single-electron transistor was reported theoretically by Likharev and D.V. Averin [75] and further experimentally demonstrated by Fulton and Dolan [76]. After that Geerlings did the experiment on the linear array of small tunnel junctions with rf signal, which opened the way for higher frequency current standards [77]. Later, H. Pothier demonstrated the reversible turnstile operation under which the electron transfer path is determined by the phase of gate voltage [78].

SET as an Electron Pump: The needs of precision metrology generally state that this operation has to be performed at a relative error level not larger than 10^{-8} and at the same time the current level needs to be several hundreds of pico amperes. The precision of the single electron pump was first measured numerically and analytically by Dalsgaard et al. [77]. In which they demonstrated that cotunneling, thermal activation, and operating system at very high frequency limits the accuracy. Co-tunnelling can be suppressed by increasing both the resistance of the tunnelling island and the number of tunnel junctions. Keller et al. [79] reported single-electron pumping using seven tunnel junctions arranged in series in 1996. This experiment used a 5 MHz signal and yielded an error rate of 15 parts in 10^9. A four junction SET traps was made by Fulton and Gammel and Dunkleberger et al. had maximum trapping time of

1 s [80]. Dresselhaus et al. used seven junction trap of Al/AlO$_x$/Al tunnel junctions to store electron for 2 h. Nine junction traps were also made by Krupenin et al. [81] and they have trapped electron for more than 8 h. For having the control on the electrons Ono et al. fabricated a device which is the combination of MOSFET and SET, this structure shows good long term stability, but the frequency is limited by RC time constant [82] at that time it is difficult but now with the advancement of fabrication technology it's possible. Wright et al. [83] and Giblin et al. [84] found the effect of the magnetic field on the pumping accuracy, Giblin showed the uncertainty of 1.2 ppm for 150 pA at 14 T. Seo et al. showed that the shape of the potential barrier also helps to improve pumping, they found the uncertainty of 0.1 ppm for 80 pA [85]. At laboratory, we are currently fabricating silicon nanowire based SET devices for electron pumping and counting.

In the low current range uncertainty increases due to the lack of sensitivity of the measurement techniques. Brun-Picard et al. reported a programmable quantum current generator, which used a highly accurate cryogenic amplifier [71] which gives the uncertainty of 10^{-8} however its current range is limited to 10 mA to 1 μA. Recently at PTB Germany ultra stable low-noise current amplifier has been developed which gives ppm gain at 100 pA [86]. Material wise Graphene and carbon nanotubes could be good candidates for single-electron pumping, in terms of maximizing the frequency.

One more important thing to discuss is the noise, if in the environment there is thermal noise or any other kind of noise we would not able to see our device performance accurately even though it is doing good. So for this, it's important to have appropriate filtering, for covering wide frequency range we can combine the RC, LCR, LC filter with powder filters [87].

10.6.2.2 Quantum Phase Slip (QPS)

So far in the world scenario, the efforts have been focused in realization of quantum current standard (QCS) mainly by single electron tunnelling (SET) through Coulomb blockade effects in a single electron pump which relates the electrical current with the elementary charge and the pumping frequency. However, to perform the quantum metrological triangle (QMT) experiment the QCS should be able to deliver the current at least in the nano-ampere range and hence the pumping frequency of the SET device should be much faster than that achieved so far or there should be alternate strategy to combine many such SET pumps in parallel so that final output current is in the desired range. Obviously, the electronic circuitry and fabrication of parallel SET pumps become much more complex and the issues of reduced uncertainty and low reproducibility limit the realization of practically relevant QCS.

An alternative approach, known as quantum phase slip (QPS) phenomenon in one-dimensional superconducting nanowires, has been taken up as the challenge for the realization of QCS by many leading groups worldwide [88, 89]. In 1D limit, a small region of superconducting nanowires can block the flow of supercurrent momentarily and hence switches to the normal state due to constant quantum fluctuations of the order parameter and this allows the phase of the order parameter to slip by 2π. This phenomenon is known as quantum phase slip (QPS). With appropriate coupling to the microwave excitation, equally spaced quantized current steps, I = 2nef, are expected from a voltage biased QPS junction. Here, f is the microwave frequency n is an integer. Similar to the measurement of Shapiro steps observed for the Josephson junction (JJ) based voltage standard, under microwave irradiation the quantized current steps from a voltage biased QPS could in principle lead to the QCS with accuracy limited by the accuracy of the measurement of frequency [90].

Theoretically it has been predicted that quantum phase slip (QPS) process in superconducting nanowires is dual to the Josephson Effect (JE) leading to quantum standard of voltage which is now used and established in world metrology for measuring voltage [91]. Similarly, there are now unprecedented attentions around the globe towards exploration of QPS in superconducting nanowires for the establishment of quantum current standard. The simplicity in the concept with better performance for the QPS based current standard than the existing routes makes it a novel and attractive candidate for the realization of QCS. However, there are several challenges which should be addressed in order to have successful experimental realization of the QPS based QCS.

As mentioned, the various crucial steps in the journey from fabrication of superconducting nanowire based QPS junction to reach to the QCS are technically very much challenging and very rich in quantum physics. For example, the demonstration of dual nature of QPS to the JE itself is very much complicated. Further, the microwave coupling to a voltage-biased QPS junction needs high level of instrumentation and engineering. Fundamental physics behind QPS, its dependence on external parameters like temperature, magnetic field, electrical biasing are the backbones of the realization of QCS by using QPS. However, in the present scenario, establishing a quantum standard of current, capable of delivering at least nano-ampere range current with uncertainty better than one part in 100 million, by using QPS junction is the utmost requirement. In order to achieve that, first and foremost priority is to demonstrate QPS effect experimentally which and is a challenging task by itself. There are major challenges mainly related to

(i) Fabrication of long 1D superconducting nanowires with dimensions: $\sqrt{\sigma} \sim \xi$; σ = cross section and ξ = coherence length,
(ii) On-chip designing of inductive and resistive circuitry in series with the QPS junction (QPSJ) to provide the necessary damping and the high impedance environment favourable for the QPS events to occur,
(iii) Determination and tuning of the characteristic voltage (V_C) for the QPS junction in order to obtain the QPS energy E_{QPS},
(iv) Fabrication of on-chip low-loss microwave coupler circuit to the QPS junction for obtaining the quantized constant current steps desired for the establishment of QCS.
(v) Retrieving ultralow current signals from the proposed QPS device using sophisticated current meter or using ultra-low noise current amplifiers.

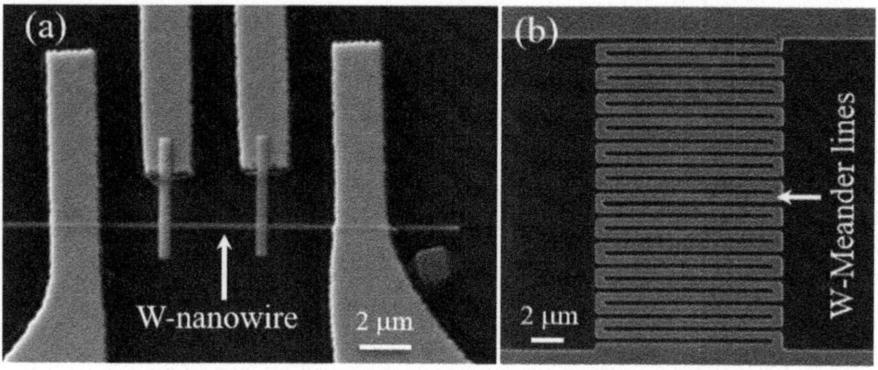

Fig. 10.24 Examples of representative nanostructures fabricated at CSIR-NPL using the existing FIB facility for the QPS study: a FIB-fabricated **a** W nanowire in 4-terminal geometry and **b** W-meander lines. Pt contact leads in **a** are fabricated by FIB too

A process flow chart for QPS based QCS is shown in Fig. 10.23. The first step is to grow ultrathin homogeneously disordered superconducting thin films with thickness of the order of the coherence length (ξ) or less. Some of the promising superconducting (SC) materials as reported in the literature are TiN, InO_x, Nb, NbGd, NbSi, Nb_2N, NbN, W etc. [92–101].

Here, selection of materials is very important as the phase slip rate depends strongly on the material properties like the normal state resistivity (ρ_N), [102] superconducting critical temperature (T_c) etc. The immediate next step after the thin film growth is to fabricate long uniform one-dimensional (1D) nanowire which acts as the QPS junction (QPSJ). This is one of the most important and crucial steps as it needs state of the art facilities and skilled expertise. Two main facilities, viz., electron-beam lithography (EBL) and reactive ion etching (RIE) systems are must to achieve 1D superconducting nanowire based QPSJs.

Further, recent studies demonstrated the use of focused ion beam (FIB) for the fabrication of nanowires [103]. FIB is also used to fabricate tungsten nanowires with a superconducting transition temperature ~ 5.2 K which is also a promising candidate for QPS study [101]. However, FIB may not offer the most desired dimensions and purity levels (Ga contamination) of the fabricated nanowires suitable for the QPS study.

In addition to the conventional 1D nanowires, superconducting meander lines are also used to study QPS [101, 104, 105] as meandering offers much longer nanowire in small space and eventually with increased normal state resistivity (ρ_N) [104]. The representative two types of structures, viz. nanowire and meander lines, from FIB-fabricated superconducting W are shown in Fig. 10.24.

These nanostructures are patterned by using the FIB facility. The length can contribute to the kinetic inductance which provides the inductive energy of the QPSJ. For QPS study the normal state resistance (ρ_N) should be higher or of the order of quantum resistance [89] and by using long nanowire with diameter of the order of the

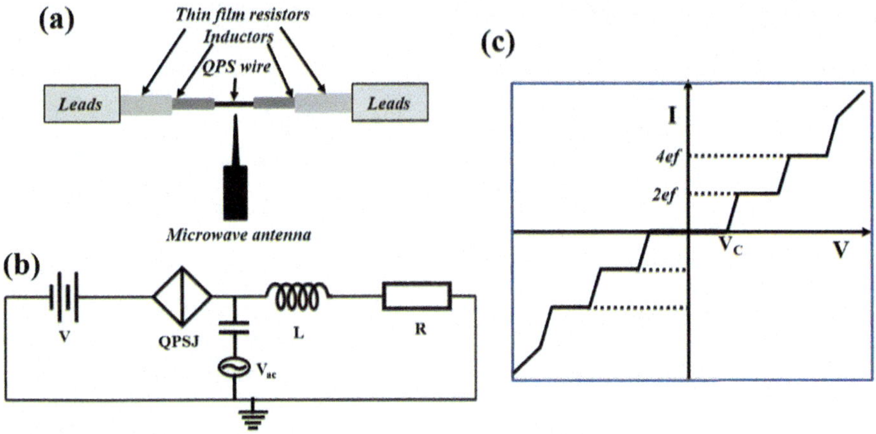

Fig. 10.25 Circuitry of QPS based quantum current. **a** Schematic presentation of QPS junction assembly with rf antenna, **b** the equivalent circuit diagram and **c** quantized current plateau under microwave excitation with frequency *f*

superconducting coherence length one can achieve higher (ρ_N). In addition, on-chip fabrication of resistive and inductive circuitry, in series with the superconducting nanowire based QPSJ, is needed to provide high impedance environment leading to an overdamping of circuit and hence suppression of charge fluctuation which is highly desirable for QCS [88, 97].

Figure 10.25 represents schematically the device geometry along with the microwave coupling scheme via an rf antenna, its equivalent circuit diagram and quantization of current under microwave irradiation with frequency f. As Cooper pair tunnelling in JJ is equivalent to phase fluctuation/slip in QPS junction, analogous to a resistively and capacitively shunted JJ, an equivalent QPS junction can also be described by a resistively and inductively series coupled junction (RLSJ) as shown in Fig. 10.25a and b. The essential components of QPSJ include a voltage source, a nonlinear serial shunt resistor that has different values of resistance in different phases of operation including normal to superconducting transition as a function of the voltage across the device, an inductor that represents the inductance of the nanowire and the ideal QPSJ device in series. The time dependent voltage across the QPSJ can be obtained through [88, 89, 91],

$$V(t) = V_C \sin\left(\frac{2\pi q}{2\mathrm{e}}\right) + L\frac{\mathrm{d}^2 q}{\mathrm{d}t^2} + R\frac{\mathrm{d}q}{\mathrm{d}t}$$

Here, V_C is the critical voltage of the QPSJ and it is related to QPS energy E_{QPS} as, $V_C = \left(2\pi E_{\mathrm{QPS}}\right)/(2e)$, q is the charge of the device at that instant, L is the inductance of the device, R represents the non-linear resistance of the device describing different resistance regions of the QPSJ before, during and after superconducting transitions which is a function of the voltage, *V(t)* is the voltage across the junction as a function

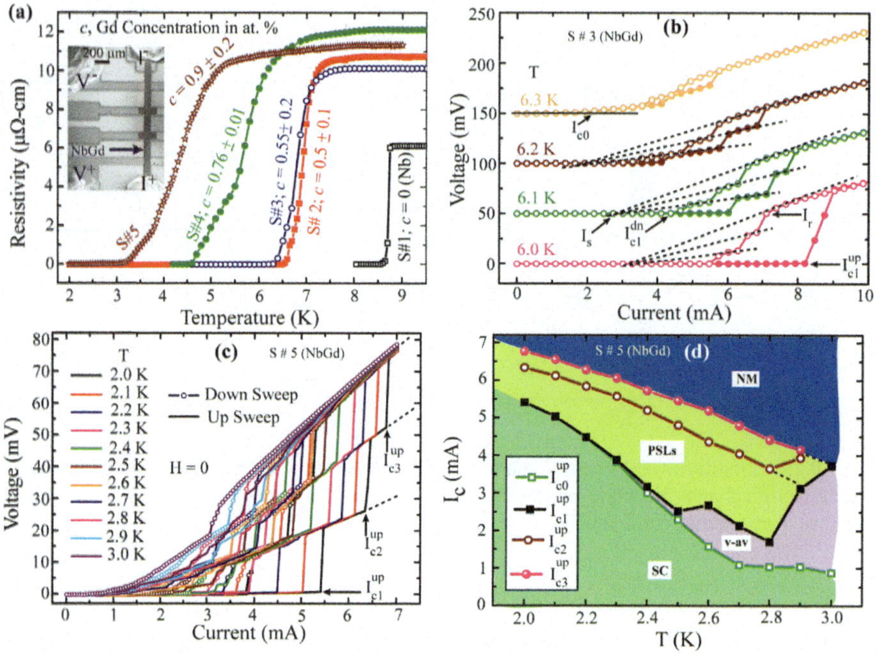

Fig. 10.26 Demonstration of phase slip events in NbGd thin film with Gd concentration varying in the range between 0 and 1 at.%

of time. In the above equation the Cooper pair current, $I = \mathrm{d}q/\mathrm{d}t$, is different from zero for only time dependent q. For a QPSJ, q, the charge, is a good quantum variable and further improvement is obtained by employing the overdamped junction so that the charge fluctuation is minimized.

Importantly, the equation for E_{QPS} given below is valid when it is larger than the inductance energy E_L, $E_L = \varphi_0^2/2L$, with ϕ_0 and L are the flux quantum and the inductance, respectively. Further [106],

$$E_{\mathrm{QPS}} \sim \left(-a\frac{\sigma T_c^{1/2}}{\rho_N}\right)$$

where, σ is the cross-sectional area of the QPS nanowire, 'a' is a dimension-dependent parameter, e is the electronic charge, ρ_N is the normal state resistivity and T_c is the superconducting critical temperature. A large value of V_C corresponds to large QPS energy (E_{QPS}) which is favourable for the QPS to be observed experimentally.

Under microwave excitation with suitable range of frequency, QPSJ is expected to exhibit quantized current steps as, $I = 2enf$, known as dual Shapiro steps. Here, f is the microwave frequency, $2e$ is the charge on Cooper pair and n is an integer representing the quantization number. Figure 10.25c represents the quantized current steps expected to appear in the current voltage characteristics (CVCs) of a voltage biased QPSJ coupled with microwave. With large E_{QPS}, Bloch-type oscillations are expected with a frequency $f_B = I/2e$ for QPS junctions and when these oscillations are synchronized with external microwave drive with frequency f, quantized current steps appear due to resonance at matching current. Therefore, Bloch-type oscillations and resonance at plasma frequency, which is related to the QPS energy, are the favourable conditions for the observation of quantized current steps or the Shapiro steps [91]. The range of excitation frequency can be determined from the junction characteristics of the QPSJ. For example, the characteristic voltage, V_C can be obtained from measured CVCs as shown in Fig. 10.25. Further, as explained previously, V_C relates to the QPS energy E_{QPS} which relates to the frequency as E_{QPS}/h with 'h' being the Plank's constant. Therefore, every circuit element in the QPSJ should be controlled very efficiently so that experimental observation of quantized current can be achieved.

10.6.2.3 Current Status and Developments

Being the national metrological institute (NMI), the has the mandate to maintain and establish the metrological standards. laboratory have started working on the realization of quantum current standard (QCS) by quantum phase slip (QPS) phenomenon. In the previous section, we have discussed about how to establish QCS by using QPS effect in superconducting nanowires. Here, we discuss about the on-going work towards the establishment of QCS including the achievements till date and the future plans.

In this direction, we have tried doping of a conventional superconductor like Nb with magnetic material in very dilute concentration limit of about 1 at.% or less. By magnetic doping with Gd of $\leq$ 1 at.% into superconducting Nb, phase slip events are shown to be triggered and the composite NbGd material can be a promising candidate for the QPS study. The experimental demonstration of phase slip process in NbGd thin film is summarized in Fig. 10.26 which represents the temperature dependent resistance *R(T)* measurements for few selective samples with variations in Gd concentration and current voltage characteristics (CVCs) measured at low temperature. Figure 10.26a highlights a strong dependence of metal-superconductor (NM–SC) phase transition on the Gd concentration. The (NM-SC) transition curve shifts towards lower temperature with wider transition region for increasing Gd concentration. Figure 10.26a and b present the CVCs for two different samples varying in Gd concentration. The phase slip lines are evident for both the samples by the appearance of the intermediate resistive states. Finally, in Fig. 10.26d, the phase diagram, based on the temperature dependent critical current for the sample shown in Fig. 10.26c, is displayed. Various dynamical regions are illustrated in the phase diagram including the specific phase slip regions as PSLs. The related research and the results are very promising towards the application of these composite materials for phase slip studies, Josephson effects, proximity effects etc. and are highly significant towards their application in the field of NanoSQUID, quantum computations, single photon detection, quantum current standards etc. [95, 96].

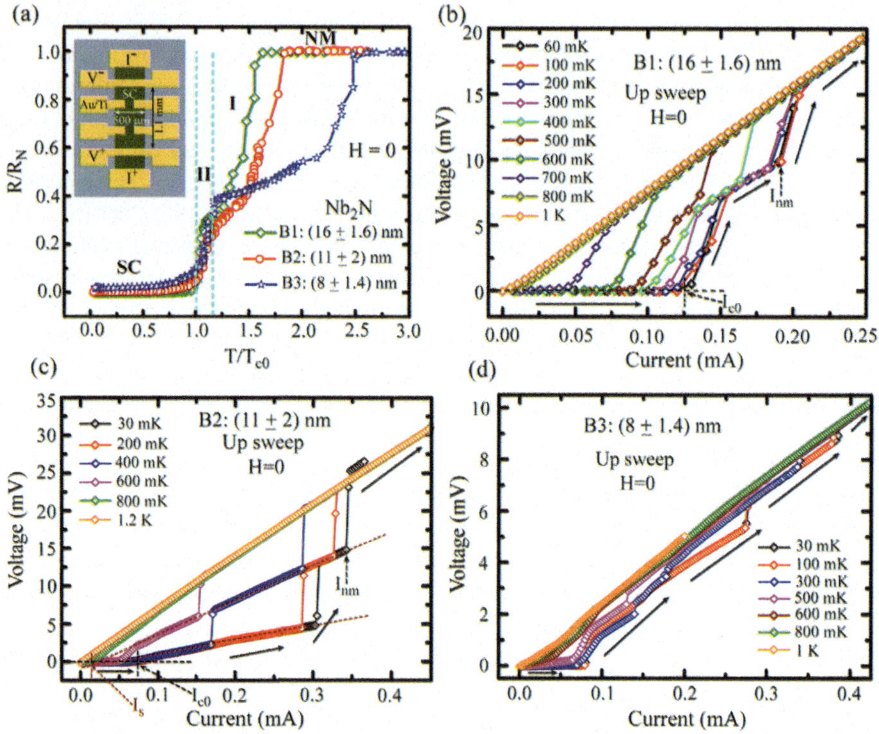

Fig. 10.27 Observation of phase slip events in Nb_2N thin films. **a** Normalized resistance (R/R_N) with respect to reduced temperature (T/T_{c0}) for three representative Nb_2N samples differing in thickness. Here T_{c0} is the superconducting critical temperature corresponding to zero-resistance. The isothermal CVCs for those three samples are displayed in **b**, **c** and **d**, respectively. Inset in **a** represents the device geometry

Recently, we have demonstrated a novel nitridation technique to transform transition metal Nb to one of its stable phases, Nb_2N, which has been further explored by studying its exciting superconducting properties. We simply deposit Nb thin films on Si_3N_4/Si substrate at high temperature and continue heating the substrate for a couple of hours and finally we end up having Nb_2N from the deposited Nb. Here, the Si_3N_4/Si substrate acts as the only source of nitrogen which is available when Si_3N_4 decomposes into Si and N atoms by high temperature annealing. Among the members of niobium-nitrogen family, cubic NbN has been explored extensively in the field of superconductivity, however, Nb_2N is mainly known for its hardness and other mechanical properties. Superconductivity is not been explored in that extent for this phase of niobium nitride. By performing low temperature transport measurements down to mK temperature range, we show here that indeed Nb_2N possesses interesting superconducting properties and more specifically it may be considered as a potential candidate to study phase slip related quantum phenomena and the related applications. Further, we show that this material can serve as a model macroscopic system to study the phase slip phenomenon beyond 1D limit.

In Fig. 10.27, we have presented few selected low temperature transport measurement data that actually exemplify the superconducting properties as well as the material's suitability for phase slip study. Three different samples, varying in thickness, have been presented here and the metal-superconductor transition has been observed for all of them in their *R(T)* measurements presented in Fig. 10.27a. The CVCs are enriched with phase slip lines and the evolutions of phase slip lines with temperature and with sample thickness have been illustrated in Fig. 10.27b, c and d, respectively, in a order of reduced thickness. With very broad transition width, transition at mK temperature as presented in *R*(*T*) measurements along with the phase slip lines in the CVCs, the material indeed can be a very promising candidate for future studies on QPS and other superconductivity related applications [98].

So far, we have been discussing mainly on superconducting thin films for the use in QPS study. As the next step obviously is to move towards 1D nanowires, here we discuss about our ongoing work on nanowires and meander type of structures specifically based on FIB-deposited W in dirty limit.

We have established a protocol for preparation of nanowires and meander structures of *W* deposited by focused ion beam (FIB) induced techniques. The T_c of the material is ~5 K, and the nanowires exhibited phase slip behaviour. The transport measurements data for three selective W-meanders are presented in Fig. 10.28.

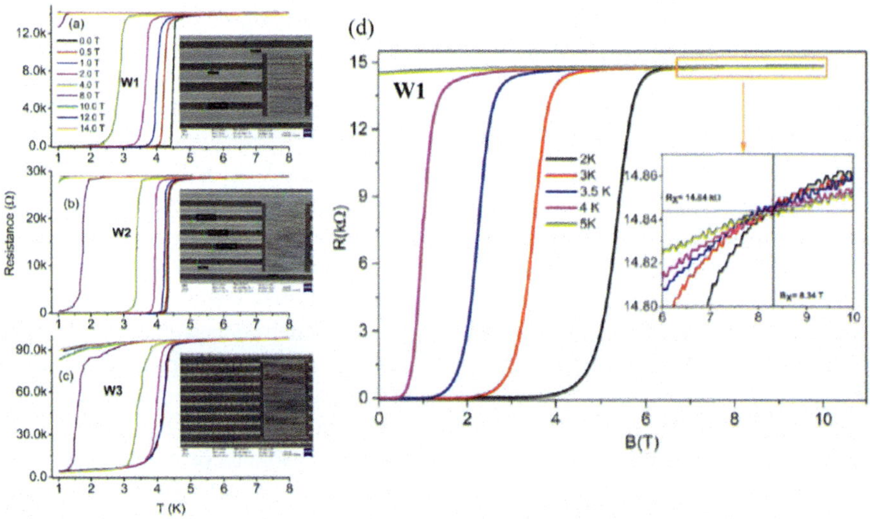

Fig. 10.28 Transport measurements on FIB-deposited W-meanders with varying dimensions. **a**–**c** Temperature dependent resistance [*R*(*T*)] data for three different samples. The dimensions are shown in the insets. **d** Magnetic field dependent resistance [*R*(*B*)/*MR*], measurements for sample W1. Inset: crossing of MR isotherms at single critical field B_x indicates the onset of field induced superconductor-insulator transition (SIT)

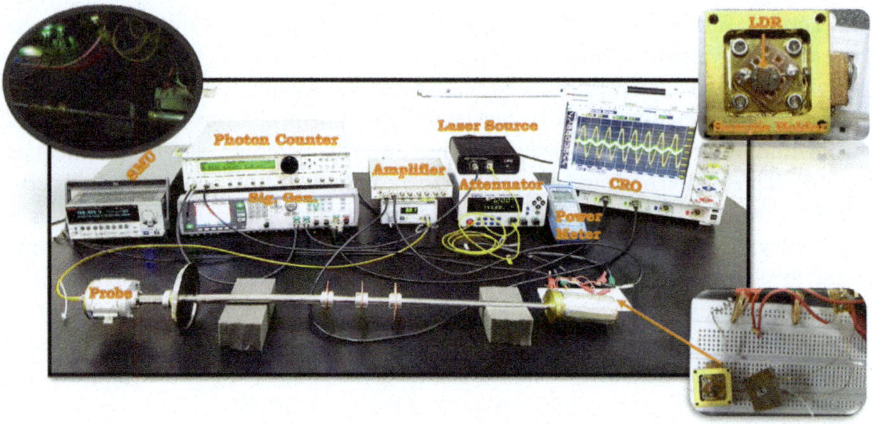

Fig. 10.29 Real time simulation setup operating at visible wavelength

The $R(T)$ curves, displayed in Fig. 10.28a–c, show the changes with the dimension. However, the T_c remains in the range of 4–4.5 K and with reduced width-the transition gets wider with residual resistance tailing at temperature below the transition temperature. This indicates the dominance of phase slip events for the reduced width. In Fig. 10.28d, we have shown magnetic field dependent resistance (MR) data for sample W1 and interestingly, the MR isotherms cross at a particular field, B_x, at which the MR becomes independent of temperature. The crossing of MR at a single point indicates the onset of superconductor-insulator transition (SIT) which is another criterion for the observation of QPS. Therefore, we believe that the W-based meanders and nanowires could be very important material for the future QPS related studies [101].

Finally, in accordance with revised SI, the primary standard for the only base unit in electrical metrology, the ampere, needs to be established by using the laws of physics dealing with elemental charge, e, as the ampere relates to the flow of elementary charges per unit time [71]. Conventionally, two different quantum phenomena have been adopted by the scientific community for the realization of quantum current standard. Single electron tunnelling (SET) is the one which deals with direct transport of electrons by using a pump probe technique. However, the amplitude of current and the uncertainty need to be improved and there have been continuous efforts worldwide to resolve the issues so that QMT experiment can also be performed. The other quantum phenomenon, which is totally different from its principle than that of SET, relates to continuous phase fluctuation in 1D superconducting nanowire and it's known as quantum phase slip (QPS) process. As phase and charge are canonically conjugate quantum variables, they are dual to each other. And correspondingly, a QPS junction is shown as the analogue or dual to a Josephson junction which leads to the quantum standard of voltage. Therefore, a QPS junction with the correct combination of inductive and resistive circuitry can be developed as the future quantum standard of current in a suitable electromagnetic environment. Further, QPS junctions are capable of transporting larger currents with higher accuracy than that with SET pumps [107]. However, high level of instrumentation and interfacing between many sophisticated low-noise electronic instruments are must to achieve the demonstration of QPS based quantization of current steps to be used for the quantum current metrology.

As the modern era heavily depends on electrical accessories and appliances requiring a wide range of driving current for their functioning, the accurate and precise measurement of electric current is not only a mandatory requirement but it has an enormous influence on the nation's economic growth and infrastructure. The realization and establishment of quantum current standard is thus a very important step towards the growth of the country and its self-reliance.

10.6.3 Quantum Nano Photonics Metrology

10.6.3.1 Superconducting Nanowire Single Photon Detector (SNSPD)

The fantasy of capturing and analysing any matter that exists in this world leads to an opening of a new era in the field of physical science and related technology. One such era has evolved from capturing a source of light in early days to now grabbing a few photons. Detecting a single photon from millions of photons that exists in a ray of light is now not only limited to a research topic but is slowly finding applications in almost every field. This includes medical care, space communication, cyber security, metrology, quantum cryptography, quantum computing, quantum key distribution etc. [108, 109]. Photo multiplier tube (PMT)-a photon detector and its better counterpart Avalanche photo diode (APD) possesses some limitations at telecom wavelength. Efficiency of InGaAs based photo detector is better at infrared wavelength range but it suffers from high dark count rate, large time jitter, after-pulsing effect etc. To overcome these limitations, new materials and technology are explored. Superconducting nanowire-based detectors is turning to be a good solution for such problem. It's been proved and demonstrated that the superconducting nanowire single photon detector (SNSPD) has low dark count rate, high spectral response over a large wavelength range, less timing jitter etc. [110]. SNSPD operation is based on the concept of hotspot formation [111] whereby, a meander shaped superconducting nanowire is biased below or close to the critical current. After photon absorption, superconductivity is lost and we get a voltage pulse that corresponds to a photon count. After a short interval superconductivity reappears and detector is ready for the detection of next photon. Besides single photon, there are many such exciting research areas involving physical quantities and are being developed or enhanced for futuristic applications. For most of the physical quantities, the laboratory have basic and derived SI units that help us to achieve standardization. The SI units had undergone a change and now are mostly being realized in terms of fundamental constants [112]. This had paved the way to realize quantum standards for each of the existing non-quantum standards particularly for SI units. Working on these aspects, the laboratory is developing photon-based quantum standards for optical radiation.

10.6.3.2 Fabrication

Single photon detection requires a nanoscale size chip and patterning. The laboratory have dedicated multi chamber sputtering facility and a PLD (Pulsed Laser Deposition) chamber for depositing superconducting thin films, ultra-thin films and devices. Superconducting thin films of different materials such as NbN, NbTiN, VN, YBCO etc. had been deposited using DC magnetron sputtering and PLD system under different growth conditions. They exhibit superconducting behaviour and are further being investigated for their application as single photon detector.

10.6.3.3 Measurements and System Integration

To characterize deposited thin films was done using a stable low temperature cryostat with the maximum drift of ~4 mK for electrical insert and ~7 mK for optical insert [113]. The individual characterization of the thin films gives us an idea about its initial performance in terms of diffusivity, coherence length, critical magnetic field etc. that are crucial for fabricating photon detectors. These parameters affect detection efficiency, timing jitter, dark counts etc. that marks the performance of a single photon detector. Once the fabrication concludes, the time is to test its real performance i.e. the photon count. The testing of single photon detector by shining the light requires a precise and accurate instrumentation setup. Some of the complex process in instrumentation includes light beam alignment, reading of the pulse, noise elimination etc. We are into the process of developing a setup that caters to all these issues. Figure 10.29 shows one such setup using a photo diode at visible and infrared wavelength.

10.6.3.4 Traceability Chain

Cryogenic radiometer will serve as the primary standard and will be at the top of the traceability chain for single photon detection measurements.

For visible wavelength, Si trap detector will serve as the transfer standard whereas for infrared wavelength, thermopile detector will be a transfer standard.

Finally, working standard will be optical power meter that will be used to calibrate single photon detectors efficiency.

With the advent of single photon detectors and related devices, the need arises of having a traceability chain for them, to ensure that the measurements made using detectors are accurate and are traceable to some SI standard. We propose a traceability setup based on single photon detector [114] as shown in Fig. 10.30.

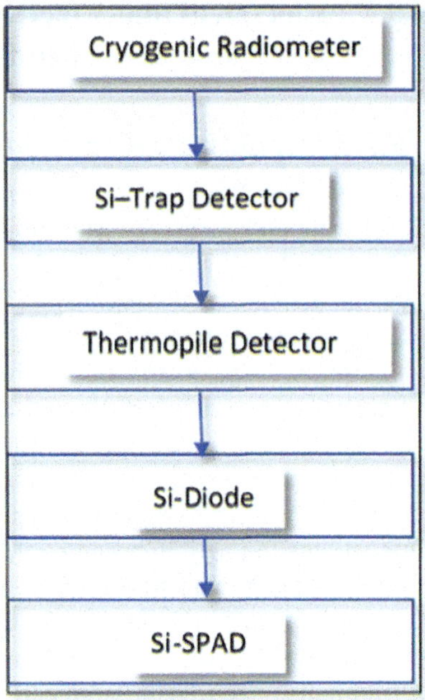

Fig. 10.30 Traceability chain for single photon detector [PTB Germany]

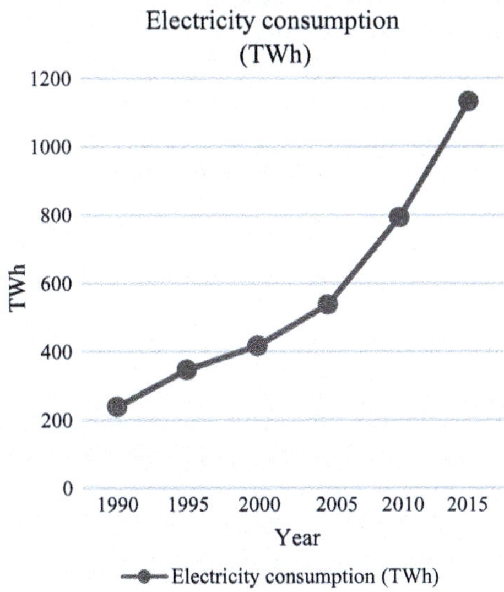

Fig. 10.31 Electricity consumption (TWh). *Source* [117]

10.7 Electrical and Electronics Metrology for Quality Infrastructure in India

10.7.1 Training and Human Resource Development

Electrical and Electronics Metrology is associated with the design and development of electrical networks, electronic circuits, communication equipments, monitoring devices, navigational equipments, data interface with computers. Human resources often are to be built in the fields of calibration and testing, product-quality evaluation and equipment maintenance and repair. With the advent of Industry 4.0, expertise in electrical and electronic automation processes are also on demand. A generalized system for automatic current and voltage measurements consisting of a set of programmable instruments interfaced with proper software has been on high-end metrology research. If such attempts are integrated with portable solutions, calibration and testing activities can be performed on on-site bulk equipments. Hence, the future for electrical and electronics metrology demand in for various specialization related to power generation, energy storage, control and distribution of electrical energy. It is also noted that materials science also would play a vital role. Materials for long distance transmission cables (superconducting wires), and oxide materials and graphene for realization of Quantum Hall Effects at or near room temperature are active fields of research.

Having understood that electrical and electronic metrology plays an important role in daily lives, CSIR-NPL has integrated in its AcSIR affiliated diploma programmes, the following modules—Introduction to AC power and energy measurements, testing and calibration of AC power and energy measuring instruments, transformers, bulk energy metering and allied equipments, kA meters, high voltage capacitors and others.

A survey over different job advertisement websites, shows more than 3000 job offers in the field of electrical and electronic metrology. These are associated with calibration technician and engineers. In common, the job responsibilities include calibration and testing of electrical equipments at customer site. The desired qualification includes experience of 2–3 years, with a prior diploma or degree from any university or All India Council for Technical Education (AICTE) approved institutes. Besides, familiarity with computerized systems, various plant and laboratory equipments, and experience in process control are also emphasized. Experience in National Accreditation Board for Testing and Calibration Laboratories (NABL) accredited calibration laboratories are at times considered as an additional advantage in the job selection process.

Given the job scenario, as above, India lack the appropriate skill sets due to the under-developed laboratory access to the trainees. Such human resource training centers fail to realize the practical side of the metrology aspects associated with the electrical and electronic measurement traceability. A possible solution to the gap is to have CSIR-NPL scientists and technicians as consultants and trainers to the industries, where practical knowledge can be imparted to the industry recruits based on their needs and requirement.

10.7.2 National Mission and Programs

10.7.2.1 Smart Grid and Smart Meter National Programme

For better energy efficiency and energy management, Advanced Metering Infrastructure (AMI) can be integrated with current grid infrastructure. The installation of smart energy meters, part of AMI, can convey complete information of energy usage almost in a real time. This information can be utilized to monitor, analyse, and control the substation for resolving energy peak demand problem. Thus, the installation of smart meters will resolve problem like billing efficiency and reduction in commercial and aggregate technical losses. The integration of smart energy meters not only help in boosting per capita income, but also will lead to accurate billing through real time monitoring usage. In effect, consumer get better knowledge on management of power usages because of reliable measurement and transparency in reading. It can provide complete solution to energy disputes problems, power wastage and indirect reduce in carbon emission. Following Energy Efficiency Services Limited (EESL), has taken up the charge to replace 250 million conventional meters with smart energy meters [115] in phases and by 2027 under SMNP, Integrated Power Development Scheme, Ujwal DISCOM Assurance Yojana (UDAY) scheme. Until now, more than 12 lakhs smart energy meters have been installed including 9.84 lakhs in Uttar Pradesh, 1.23 lakhs in Haryana, 0.57 lakhs in NDMC New Delhi and about 0.28 lakhs in Bihar.

Related but in another context, the transformation of electrical grid into digital one requires special attention on development of energy management using supervisory control and data acquisition systems (SCADA) with energy management systems (EMS) and distribution management systems (DMS), reduction in AT&C losses, automatic metering infrastructure, monitoring and automation of substation. In this regard, numerous programs were launch by Indian government such as national smart grid mission (NSGM, renewable generation program, national mission on electric mobility, restructured-accelerated power development and reforms program (R-APDRP). Number of regulatory, standardization, administrative, the private, and the wide range of bodies working together to achieve the objectives of the 24 $\times$ 7 Power for All.

The natural resources for power generation are limited and unevenly distributed in different regions, the optimum utilization of the power is necessary by transferring power from the resource centric to load centric regions. Moreover, with the introduction of renewable energy sources, the HV transmission grids are crucial for transportation of the electricity generated by large-scale renewable set-ups. Thus, synchronization of all regional grids is very important, which in turn facilitates the electricity trade across regions. The traceable measurements play a vital role in the synchronization of all regional grids in achieving One Nation-One Grid-One Frequency.

In support of national bodies, CSIR-NPL, keeps on targeting to improve and develop new measurement capability to ensure traceability chain for Electrical Power and Energy measurements.

10.7.3 Electric Vehicle (EV): The Next Generation Transport

EV is going to be next generation transport, which is needed to be adopted in mass scale in coming future. The benefit of EV is the reduction of fossil fuel dependency on large scale and thereby by reducing the CO_2 emission levels. Government of India (GoI) has adopted several schemes like FAME (Faster Adoption and Manufacturing of (Hybrid) and Electric Vehicles), which incentivises the manufacturing and implementation of the industries associated with the EV adoption. In general, EV runs on the chemical energy available from Li-ion battery. The battery needs to charge before they get completely drained off. So, the successful adoption of EVs also depends on the implementation of mass charging stations across the country. Under the Fame-II scheme, GoI aims for implementation of more than 3000 charging stations. To augment the developments of EV ecosystem, CSIR-NPL is planning to implement standardisation schemes for charging stations and EV battery testing facility: specifically, for Li-ion battery (the battery chemistry of Li-ion technology is completely different from other battery technologies). The developments in this direction are at a very infancy stage, needs an amalgamation of EV industries, MSMEs and GoI at large. CSIR-NPL is already providing support for adoption of EV standards in India through the ETD-51 technical committee on EV standards.

10.7.4 Quantum Technologies for Enhancing the Metrological Capabilities

The top priority for CSIR-NPL is to pursue fundamental R&D in quantum phenomena which will enable realisation of SI units linking to the constants of nature. This will also lead to development of quantum technologies which are beneficial for societal and economic developments and will lay the foundations for futuristic quantum technologies. The key benefit of incorporation of quantum technologies in metrology is the unprecedented accuracy and very low noise levels, that the classical counterpart cannot produce. Some of the quantum technologies relevant to metrology are realisation of the unit of resistance by the quantum hall effect, voltage through Josephson effect, current through the single electron tunnelling etc. The Josephson effect has also led to the development of superconducting quantum interference device (SQUID), through which the unit of magnetic flux quanta is realised. Now there is considerable evidence suggesting the use of proximity Josephson junctions and SQUIDs as quantum bits (qubits) which will lead to the realisation of quantum computation. Josephson effect is also linked to the development of quantum noise thermometry for measurement of the Planck's constant k and for defining the unit of temperature. R&D on SNSPD can change the way of counting/detection and even manipulation of photons. A large body of evidence is coming out for the use of SNSPD technique for experiments related to quantum teleportation and quantum computation. Finally, when voltage, current and ampere are realised using quantum technologies, the test of constancy of Ohm's law at the quantum limit is desirable, which gives the direct proof of consistency of the constants of nature. The future of quantum technologies demand much more than that of the realisation of fundamental units and determination of the constants of nature. For example, determination of physical quantities related to single atoms or molecules (spin, charge and flux etc.) demand cross disciplinary technologies. Thus, the use of quantum technology in metrology has led to the new definition of units such as volt, current and resistance. The unit of impedance also can be replaced through AC quantum Hall effect. Thus, the whole spectrum of electrical units in near future will be replaced by their quantum counter parts.

10.7.5 Economic and Social Impacts

10.7.5.1 Environmental Concern

During the past 20 years, the India economy has grown very rapidly [116] as a result of government's strong emphasis on industrial development. However, today's dependency on electrical energy consumption [117], carbon emission [118] of environmental concerns are posing serious economic and social impact which exert negative impression on environment (Figs. 10.31, 10.32, 10.33). In fact, the data shown in Fig. 10.34 are of great concern; which represents rising energy demand in near future. The data is estimated by Partial End User Method (PEUS) shows unprecedented rise in demand for energy. A detailed survey on long term analysis of electricity demand forecasting has already conducted by Central Electricity Authority (CEA), Ministry of Power and reported in a 19th Electric Power Survey (EPS) Report [119].

Improving energy efficiency and reducing energy demands are the two dominants factors considered as the most promising, fastest, cheapest and safest ways to reduce climate change [121]. Until now numerous efforts have been explored in understanding relation between electricity use and economic activity [122]. However, putting limits on reduction of energy demand is not a good sign considering economic consequences. The only remain challenge is to identify and improvise energy efficiency limitation [123]. Once it is known then remain is to develop new technology, implement policy and market framework.

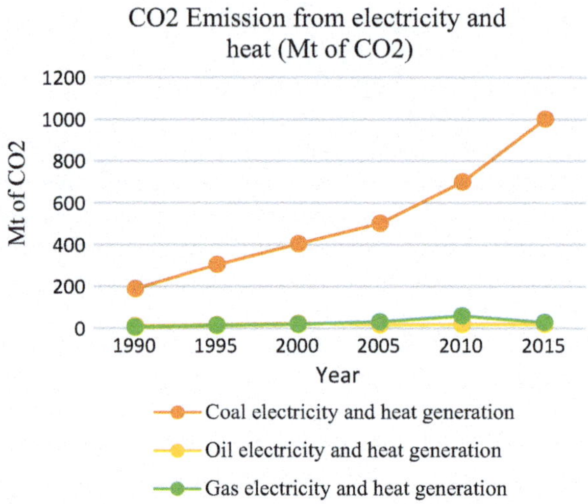

Fig. 10.32 CO_2 Emission from electricity and heat (Mt of CO_2). *Source* [118]

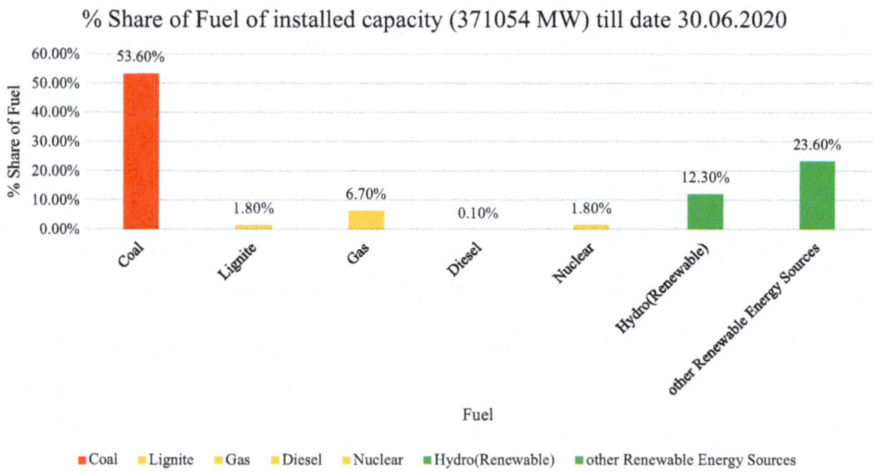

Fig. 10.33 % Share of fuel of installed capacity (371,054 MW) till date 30.06.2020. *Source* [120]

CSIR-NPL has a significant role to play in above said areas by bringing technologies and measurements traceable to SI units as well as by incorporating new measurements and methods for assessment of real-time consumption, dynamic load response and estimation of losses. There are several technologies with direct dependence on CSIR-NPL such as smart grid, smart metering for real-time energy monitoring, electric vehicle grid integration, technologies for distributed energy generation, electrification of the transport sector, implementation of zero energy infrastructures, high voltage DC transmission, transformer loss measurement. Metrological tools are

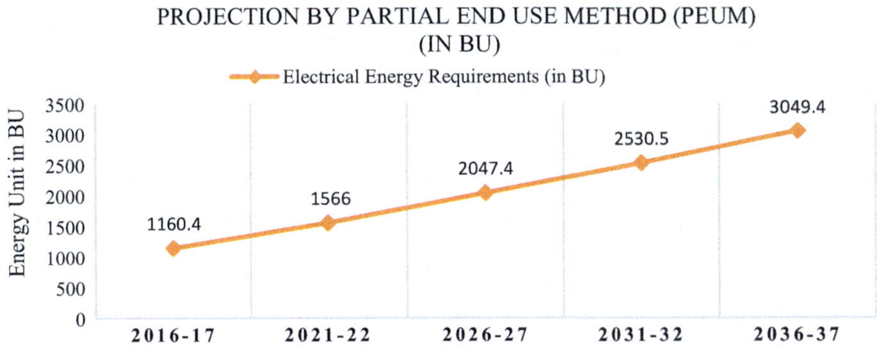

Fig. 10.34 Energy consumption projection by partial end use method (PEUM)(in BU). *Source* [119]

needed for efficient characterisation of grid transformers, motors and for the estimation of idle power at a given point of time. As transportation sector moves into DC mode, there would be a demand for measurement standards and techniques for DC energy consumption and metering. For the case of electrification of the transport sector, there several emerging battery technologies with different cell chemistry and charging characteristics. CSIR-NPL has a direct role play in the assessment of the performance characteristics of the battery technologies, by introducing measurement techniques and procedure which are traceable to the IS/IEC standards.

10.7.5.2 Health Care Concern

The health care industry at present and that which evolves for future applications undoubtedly depend on metrological parameters such as voltage and current. The role of CSIR-NPL to health care industry is not only to provide measurements traceable to SI units but also to provide better measurement techniques and diagnostic methods for betterment of human health and diagnosis in certain specific areas of diseases. Certain specific areas of direct involvement in health care industry are providing measurement traceability of voltage from 1 to 5 mV for Electrocardiogram (ECG) system, radiation dosimetry standards (low ionisation current as result of high energy ionising radiations etc.), electro-surgery analyzer, electrical safety analyzer. Also, there are a host of imaging and measurement techniques (MRI, CT Scan, tissue engineering, cancer diagnostic tools, targeted drug delivery to name a few) which directly and indirectly are dependent on EEM in many ways starting from traceable measurement standards, artifacts etc. and to develop and innovative measurement techniques [124–126]. One of aspects in the coming age (fighting against COVID-19 pandemic) is that the requirement of real time diagnosis and greater use of non-invasive measurement tools and diagnostic procedures that avoids/minimises physical contact with patient [127–129]. The IoT (Internet of Things) techniques and instrumentation can greatly help for such real time diagnosis and investigation tools. In addition, risk associated with safety of medical devices is a primary concern. Some of medical electrical hazards are power source, voltage, current leakage, insulation, earthing and grounding, separation of parts, and electromagnetic compatibility (EMC) [130]. The major concern is to provide measurement traceability for medical electrical equipment at various stages of safety requirements [131–134]. CSIR-NPL provides and interface between traceable measurements and for development of measurement procedures. Thus, the role of CSIR-NPL in health care sector is to provide quality assurance, traceable measurements, competitiveness with cutting edge technologies to support design and manufacturing at uniform across the globe.

10.8 Conclusion

The Electrical and Electronics Metrology (EEM) at CSIR-NPL maintains the national standards for electrical and electronics parameters traceable to SI units at par with leading NMIs. The EEM involves impedance, voltage and current, power and energy and quantum hall resistance metrology. The EEM continuously maintains and upgrades its facilities towards improved CMCs, new CMCs and range extension to meet the needs of the growing industries. The traceability of these parameters are disseminated through apex level calibration and testing services to power utilities, electrical and electronic equipment manufacturers, MSMEs, PSUs, strategic sector, STQC labs, NABL accredited calibration and testing laboratories, etc. The accurate and precise measurements support Indian Industries and business to innovate and improve their product to be globally competitive. This supports the government of India's missions such as Make in India, Smart Cities, Shashakt Bharat, Smart Grid, etc. as for "Aatmanirbhar Bharat".

The EEM is working towards establishment of compliant measurement facility for EMI/EMC up to 40 GHz, National facility creation for smart energy meters, national facility creation for 300 kV AC High voltage and 20 kA high current measurements, traceability of impedance standards using ac-dc calculable resistance standard traceable to QHR, automated IVD calibration procedure based on virtual instrumentation for enhancement of measurement capability promoting governmental missions and policies.

Further, as the NMI of the country, the EEM of CSIR-NPL is also involved in fundamental research at forefront concerning various electrical parameters such as JVS and QHR for quantum standards of voltage and resistance. Current emphasis is mainly on graphene for QHR and superconductor materials for QPS. The focus is on to realize quantum current standard (QCS) through SET, QPS. We also envisage single photon detection/counting through SNSPD based detectors.

As a leading NMI, the CSIR-NPL is focused on developing and enhancing its metrological capabilities for better service to the industries with the aim of achieving over all development of the country for becoming self reliant. This in turn helps the Indian industries to enhance their manufacturing/export and other services, thus contributing to the GDP of the Nation.

Acknowledgements The authors would like to acknowledge Mr. Anoop Singh Yadav, Mr. Prakash Singh, Ms. Usha Kiran, Ms. Poonam Sethi Bisht, Ms. Ashmeet Kaur Uppal, Mr. Sachin Kumar, Mr. Harish Kumar, Mr. Kul Bhushan Ravat, Mr. L. Sridhar, Mr. M. K. Tamrakar, Mr. Shrikrishan, Anish M Bhagav and Rajeev Sharma for their continuous efforts in maintaining traceability chain for electrical and electronics parameters. We deeply appreciate their continuous efforts in providing metrological services to the public and private users. We would like to thank other staff members and students of CSIR-NPL for their contribution to scientific society which is unforgettable. Finally, the authors also gratefully recognize the needful discussion with and insight from Dr. Sushil Kumar and Dr. D K Aswal for finalizing this chapter.

References

1. 9th Edition The International System of Units (2009)
2. B.D. Josephson, Possible new effects in superconductive tunnelling. Phys. Lett. **1**(7), 251–253 (1962). https://doi.org/10.1016/0031-9163(62)91369-0
3. F. Delahaye, B. Jeckelmann, Revised technical guidelines for reliable dc measurements of the quantized Hall resistance. Metrologia **40**(5), 217–223 (2003). https://doi.org/10.1088/0026-1394/40/5/302
4. BIPM: Key Comparison Database KCDB. https://www.bipm.org/kcdb/
5. B. Ehtesham, P.S. Bist, T. John, Erratum to: development of an automated precision direct current source for generation of pA currents based on capacitance charging method at CSIR-NPL. MAPAN **32**(1), 23–23 (2017). https://doi.org/10.1007/s12647-017-0204-x
6. A.M. Thompson, D.G. Lampard, A new theorem in electrostatics and its application to calculable standards of capacitance. Nature **177**(4515), 888–888 (1956). https://doi.org/10.1038/177888a0
7. W.K. Clothier, A calculable standard of capacitance. Metrologia **1**(2), 36–55 (1965). https://doi.org/10.1088/0026-1394/1/2/002
8. L. Callegaro, in *Electrical Impedance*. CRC Press (2012)
9. S.A. Bryan Kibble, J. Schurr, in *Coaxial Electrical Circuits for Interference-Free Measurements*. IEEE (2011)
10. R.N. Jones, in *The measurement of lumped parameter impedance: A Metrology Guide*. National Bureau of Standards (1974)
11. F.L. Hermach , R.F. Dziuba, in *Precision Measurement and Calibration: Selected NBS Papers on Electricity—Low Frequency*. National Bureau of Standards (U.S.) (1968)
12. L. Marais et al., Comparison of standards for the calibration of voltage, current and resistance meters. Metrologia **55**(1A), 01005–01005 (2018). https://doi.org/10.1088/0026-1394/55/1A/01005
13. N. Fletcher et al., Bilateral comparison of 1 Ω and 10 kΩ standards (ongoing BIPM key comparisons BIPM.EM-K13.a and 13.b) between the NPLI (India) and the BIPM. Metrologia **56**(1A), 01007–01007 (2019). https://doi.org/10.1088/0026-1394/56/1A/01007
14. M. Saleem, M.A. Ansari, A.K. Saxena, N. Fletcher, R. Goebel, M. Stock, Bilateral comparison of 10 pF capacitance standards (ongoing BIPM key comparison BIPM.EM-K14.a) between the NPLI and the BIPM, July 2010–May 2011. Metrologia **52**(1A), 01015–01015 (2015). https://doi.org/10.1088/0026-1394/52/1A/01015
15. S. Babita, B. Khurana, S. Kumar, A.K. Saxena, Evaluation of four-terminal-pair capacitance standards using electrical equivalent circuit model. Measurement **73**, 121–126 (2015). https://doi.org/10.1016/j.measurement.2015.05.003
16. S. Singh, T. John, Automated module for characterization of reference standards of capacitance by impedance-matrix method. Turkish J. Electr. Eng. Comput. Sci. **25**, 4142–4148 (2017). https://doi.org/10.3906/elk-1611-39

17. S. Babita, B. Khurana, T. John, Measurement automation to implement evaluation procedure of four-terminal-pair capacitance standards using S-parameters. MAPAN **32**(3), 175–181 (2017). https://doi.org/10.1007/s12647-017-0211-y
18. S. Singh, S. Kumar, Babita, T. John, Realization of four-terminal-pair capacitors as reference standards of impedance at high frequency using impedance-matrix method. IEEE Trans. Instrum. Meas. **66**(8), 2129–2135 (2017). https://doi.org/10.1109/TIM.2017.2673098
19. Satish, S. Kumar, Babita, T. John, A.K. Saxena, Evaluation of air dielectric four-terminal-pair capacitance standards using resonance frequency of impedance elements. Measurement **100**, 176–182 (2017). https://doi.org/10.1016/j.measurement.2016.12.052
20. J. B. Shaik, V. Ganesh, A power system restoration method using voltage source converter–high-voltage direct current technology, aided by time-series neural network with firefly algorithm. Soft Comput. (2019). https://doi.org/10.1007/s00500-019-04459-5
21. M.J. Pérez-Molina, D.M. Larruskain, P. Eguía López, G. Buigues, Challenges for protection of future HVDC grids. Front. Energy Res. **8**, (2020). https://doi.org/10.3389/fenrg.2020.00033
22. "IEC 61869-1 (2007) Instrument transformers part 1: general requirements (2007). https://webstore.iec.ch/home
23. "IEC 61869-2 (2012) Instrument transformers —part 2: additional requirements for current transformers (2012). https://webstore.iec.ch/home
24. "IEC 61869-3 (2011) Instrument transformers—part 3: additional requirements for inductive voltage transformers (2011). https://webstore.iec.ch/home
25. IS 16227 (2016): Instrument transformers: part 1 general requirements (2016). https://www.services.bis.gov.in:8071/php/BIS/PublishStandards/published/
26. IS 16227 (2015) Instrument transformers: part 3 additional requirements for inductive voltage transformers (2015)
27. "IS 16227 (2016) Instrument transformers: part 2 Additional requirements for current transformers (2016). https://www.services.bis.gov.in:8071/php/BIS/PublishStandards/published/
28. S.R. Gupta, Calibration and measurement facilities for AC high current and high voltage ratio standards at NPL. MAPAN **24**(1), 29–39 (2009). https://doi.org/10.1007/s12647-009-0006-x
29. M.A. Ansari, L. Sridhar, M.K. Tamrakar, Shrikrishan, V.N. Ojha, Present status and future plan of ac high voltage and high current standards at CSIR-NPL. in *The Abstracts of 9 th International Conference on Advances in Metrology* (AdMet–2016), p. 31 (2016)
30. M.K. Mittal, L. Sridhar, M.K. Tamrakar, Shrikishan, R.P. Agarwal, S.S. Rajput, Innovative system for precision measurement of high voltage capacitance and tan δ. Int. J. Phys. Appl. Sci. **2**(2), 41 (2015). [Online] Available http://ijmr.net.in/current-ijpas.php
31. M.K. Tamrakar, L. Sridhar, Shrikrishan, M.A. Ansari, Effect of burden on potential transformer. in *The Abstracts of 4th National Conference on Advances in Metrology* (AdMet-2015) (2015)
32. L. Sridhar, M.K. Tamrakar, Shrikrishan, M.A. Ansari, Effect of burden on current transformer, in *The Abstracts of 9th International Conference on Advances in Metrology* (AdMet–2016) p. 132 (2016)
33. R.M. Silva et al., Optical current sensors for high power systems: a review. Appl. Sci. **2**(3), 602–628 (2012). https://doi.org/10.3390/app2030602
34. R. Thomas et al., Non-conventional instrument transformers enabling digital substations for future grid. in *2016 IEEE/PES Transmission and Distribution Conference and Exposition (T&D)*, May 2016, pp. 1–5. https://doi.org/10.1109/tdc.2016.7519930
35. E. Mohns, G. Ramm, W.G. Kürten Ihlenfeld, L. Palafox, H. Moser, The PTB primary standard for electrical AC power. MAPAN **24**(1), 15–19 (2009). https://doi.org/10.1007/s12647-009-0004-z
36. IEC 62053-21 (2003) Electricity metering equipment (a.c.)—particular requirements—part 21: Static meters for active energy (classes 1 and (2003). https://webstore.iec.ch/home
37. "IEC 62053-22 (2003) Electricity metering equipment (a.c.)—particular requirements—part 22: static meters for active energy (classes 0,2 S and 0,5 S) (2003). https://webstore.iec.ch/home

38. "IS 13779 (1999) AC static watihour meters, class 1 and 2—specification (1999). https://www.services.bis.gov.in:8071/php/BIS/PublishStandards/published/
39. "IS 14697 (1999) AC static transformer operated, watthour and var—hour meters, class 0.2 S and 0.5 S—Specification" (1999). https://www.services.bis.gov.in:8071/php/BIS/PublishStandards/published/
40. "IS 13010 (2002) AC watthour meters, class 0.5, 1 and 2," Specification (2002). https://www.services.bis.gov.in:8071/php/BIS/PublishStandards/published/
41. "CBIP Technical report No.88 (with latest amendments) Specification for AC static Electrical Energy Meters." [Online] Available http://www.cbip.org/
42. H. Çayci, Final report on key comparison EURAMET.EM-K5.1 (EURAMET Project No. 687): comparison of 50/60 Hz power. Metrologia **48**(1A), 01009–01009 (2011). https://doi.org/10.1088/0026-1394/48/1A/01009
43. F. Leferink, C. Keyer, A. Melentjev, Static energy meter errors caused by conducted electromagnetic interference. IEEE Electromagn. Compat. Mag. **5**(4), 49–55 (2016). https://doi.org/10.1109/MEMC.2016.7866234
44. G. Leal Xavier, J.R. Macedo Junior, R. Ferreira Barata de Oliveira, L.T. Silva Oliveira, I. Nogueira Gondim, Performance analysis of active energy meters under non-sinusoidal conditions. IEEE Lat. Am. Trans. **18**(04), 808–815 (2020) https://doi.org/10.1109/tla.2020.9082225
45. "IEC 62052-11 (2003) Electricity metering equipment (AC)—general requirements, tests and test conditions—part 11: metering equipment (2003). https://webstore.iec.ch/home
46. CISPR/CIS/I (2008-09-24, 6.0) Electromagnetic compatibility of information technology equipment, multimedia equipment and receivers. https://www.iec.ch/dyn/www/f?p=103:7:0::::FSP_ORG_ID:1444
47. "IS 6842 (1997) Limits for electromagnetic interference (1997). https://www.services.bis.gov.in:8071/php/BIS/PublishStandards/published/
48. "IS 16444 (2017) AC static transformer operated watthour and var—hour smart meters, class 0.2 S, 0.5 S and 1.0 S: part 2 specification transformer operated smart meters." https://www.services.bis.gov.in:8071/php/BIS/PublishStandards/published/
49. "IS 16444 (2015) AC static direct connected watthour smart meter class 1 and 2—specification (2015)
50. "IS 15959 (Part 1), Data exchange for electricity meter—reading, tariff and load control—companion specification, vol. 4 (Electrotechnical Department, 17.220.20, 2011). https://www.services.bis.gov.in:8071/php/BIS/PublishStandards/published/
51. "IS 15959 : Part 2 (2016) Data exchange for electricity meter reading, tarif and load control—companion specification: part 2 smart meter (2016). https://www.services.bis.gov.in:8071/php/BIS/PublishStandards/published/
52. "IS 15959 : Part 3 (2017) Data exchange for electricity meter reading, tariff and load contr—-Companion specification: part 3 smart meter (Transformer Operated KWh And KVarh, Class 0.2 S, 0.5 S And 1.0 S) (2017). https://www.services.bis.gov.in:8071/php/BIS/PublishStandards/published/
53. A. Tzalenchuk et al., Towards a quantum resistance standard based on epitaxial graphene. Nat. Nanotechnol. **5**(3), 186–189 (2010). https://doi.org/10.1038/nnano.2009.474
54. T.J.B.M. Janssen et al., Graphene, universality of the quantum hall effect and redefinition of the SI system. New J. Phys. **13**(9), 093026 (2011). https://doi.org/10.1088/1367-2630/13/9/093026
55. T.J.B.M. Janssen et al., Precision comparison of the quantum hall effect in graphene and gallium arsenide. Metrologia **49**(3), 294–306 (2012). https://doi.org/10.1088/0026-1394/49/3/294
56. T.J.B.M. Janssen, A. Tzalenchuk, S. Lara-Avila, S. Kubatkin, V.I. Fal'ko, Quantum resistance metrology using graphene. Reports Prog. Phys, **76**(10), 104501 (2013). https://doi.org/10.1088/0034-4885/76/10/104501
57. R. Ribeiro-Palau et al., Quantum Hall resistance standard in graphene devices under relaxed experimental conditions. Nat. Nanotechnol. **10**(11), 965–971 (2015). https://doi.org/10.1038/nnano.2015.192

58. F. Lafont et al., Quantum hall resistance standards from graphene grown by chemical vapour deposition on silicon carbide. Nat. Commun. **6**(1), 6806 (2015). https://doi.org/10.1038/ncomms7806
59. M. Kruskopf, R.E. Elmquist, Epitaxial graphene for quantum resistance metrology. Metrologia **55**(4), R27–R36 (2018). https://doi.org/10.1088/1681-7575/aacd23
60. K.S. Novoselov, Electric field effect in atomically thin carbon films, Science (80-.) **306**(5696), 666–669 (2004). https://doi.org/10.1126/science.1102896
61. A.C. Ferrari et al., Science and technology roadmap for graphene, related two-dimensional crystals, and hybrid systems. Nanoscale **7**(11), 4598–4810 (2015). https://doi.org/10.1039/C4NR01600A
62. K.S. Novoselov et al., Room-temperature quantum hall effect in graphene. Science (80-.) **315**(5817), 1379–1379 (2007). https://doi.org/10.1126/science.1137201
63. J. Guignard, D. Leprat, D.C. Glattli, F. Schopfer, W. Poirier, Quantum hall effect in exfoliated graphene affected by charged impurities: metrological measurements. Phys. Rev. B **85**(16), 165420 (2012). https://doi.org/10.1103/PhysRevB.85.165420
64. F. Lafont et al., Anomalous dissipation mechanism and Hall quantization limit in polycrystalline graphene grown by chemical vapor deposition. Phys. Rev. B **90**(11), 115422 (2014). https://doi.org/10.1103/PhysRevB.90.115422
65. C. Virojanadara, M. Syväjarvi, R. Yakimova, L.I. Johansson, A.A. Zakharov, T. Balasubramanian, Homogeneous large-area graphene layer growth on 6 H-SiC. Phys. Rev. B **78**(24), 245403 (2008). https://doi.org/10.1103/PhysRevB.78.245403
66. K.V. Emtsev et al., Towards wafer-size graphene layers by atmospheric pressure graphitization of silicon carbide. Nat. Mater. **8**(3), 203–207 (2009). https://doi.org/10.1038/nmat2382
67. W.A. de Heer et al., Large area and structured epitaxial graphene produced by confinement controlled sublimation of silicon carbide. Proc. Natl. Acad. Sci. **108**(41), 16900–16905 (2011). https://doi.org/10.1073/pnas.1105113108
68. M. Kruskopf et al., Comeback of epitaxial graphene for electronics: large-area growth of bilayer-free graphene on SiC. 2D Mater. **3**(4), 041002 (2016). https://doi.org/10.1088/2053-1583/3/4/041002
69. Y. Yang et al., Epitaxial graphene homogeneity and quantum hall effect in millimeter-scale devices. Carbon N. Y. **115**, 229–236 (2017). https://doi.org/10.1016/j.carbon.2016.12.087
70. T.J.B.M. Janssen et al., Operation of graphene quantum hall resistance standard in a cryogen-free table-top system. 2D Mater. **2**(3), 035015 (2015). https://doi.org/10.1088/2053-1583/2/3/035015
71. J. Brun-Picard, S. Djordjevic, D. Leprat, F. Schopfer, W. Poirier, Practical quantum realization of the ampere from the elementary charge. Phys. Rev. X **6**(4), 041051 (2016). https://doi.org/10.1103/PhysRevX.6.041051
72. S. Sassine et al., The quantum metrology triangle experiment: quantization tests of an electron pump. Sba Control. Automação Soc. Bras. Autom. **21**(6), 609–615 (2010). https://doi.org/10.1590/S0103-17592010000600006
73. H. Scherer, B. Camarota, Quantum metrology triangle experiments: a status review. Meas. Sci. Technol. **23**(12), 124010 (2012). https://doi.org/10.1088/0957-0233/23/12/124010
74. N.-H. Kaneko, S. Nakamura, Y. Okazaki, A review of the quantum current standard. Meas. Sci. Technol. **27**(3), 032001 (2016). https://doi.org/10.1088/0957-0233/27/3/032001
75. D.V. Averin, K.K. Likharev, Coulomb blockade of single-electron tunneling, and coherent oscillations in small tunnel junctions. J. Low Temp. Phys. **62**(3–4), 345–373 (1986). https://doi.org/10.1007/BF00683469
76. T.A. Fulton, G.J. Dolan, Observation of single-electron charging effects in small tunnel junctions. Phys. Rev. Lett. **59**(1), 109–112 (1987). https://doi.org/10.1103/PhysRevLett.59.109
77. P. Delsing, K.K. Likharev, L.S. Kuzmin, T. Claeson, Time-correlated single-electron tunneling in one-dimensional arrays of ultrasmall tunnel junctions. Phys. Rev. Lett. **63**(17), 1861–1864 (1989). https://doi.org/10.1103/PhysRevLett.63.1861

78. H. Pothier, P. Lafarge, C. Urbina, D. Esteve, M.H. Devoret, Single-electron pump based on charging effects. Europhys. Lett. **17**(3), 249–254 (1992). https://doi.org/10.1209/0295-5075/17/3/011
79. M.W. Keller, J.M. Martinis, N.M. Zimmerman, A.H. Steinbach, Accuracy of electron counting using a 7-junction electron pump. Appl. Phys. Lett. **69**(12), 1804–1806 (1996). https://doi.org/10.1063/1.117492
80. T.A. Fulton, P.L. Gammel, L.N. Dunkleberger, Determination of Coulomb-blockade resistances and observation of the tunneling of single electrons in small-tunnel-junction circuits. Phys. Rev. Lett. **67**(22), 3148–3151 (1991). https://doi.org/10.1103/PhysRevLett.67.3148
81. V.A. Krupenin, S.V. Lotkhov, D.E. Presnov, Instability of single-electron memory at low temperatures in Al/AlOx/Al structures. J. Exp. Theor. Phys. **84**(1), 190–196 (1997). https://doi.org/10.1134/1.558147
82. Y. Ono, Y. Takahashi, Electron pump by a combined single-electron/field-effect- transistor structure. Appl. Phys. Lett. **82**(8), 1221–1223 (2003). https://doi.org/10.1063/1.1556558
83. S.J. Wright et al., Enhanced current quantization in high-frequency electron pumps in a perpendicular magnetic field. Phys. Rev. B **78**(23), 233311 (2008). https://doi.org/10.1103/PhysRevB.78.233311
84. S.P. Giblin et al., Towards a quantum representation of the ampere using single electron pumps. Nat. Commun. **3**(1), 930 (2012). https://doi.org/10.1038/ncomms1935
85. M. Seo et al., Improvement of electron pump accuracy by a potential-shape-tunable quantum dot pump. Phys. Rev. B **90**(8), 085307 (2014). https://doi.org/10.1103/PhysRevB.90.085307
86. D. Drung, C. Krause, U. Becker, H. Scherer, F.J. Ahlers, Ultrastable low-noise current amplifier: a novel device for measuring small electric currents with high accuracy. Rev. Sci. Instrum. **86**(2), 024703 (2015). https://doi.org/10.1063/1.4907358
87. K. Bladh et al., Comparison of cryogenic filters for use in single electronics experiments. Rev. Sci. Instrum. **74**(3), 1323–1327 (2003). https://doi.org/10.1063/1.1540721
88. J.E. Mooij et al., Superconductor–insulator transition in nanowires and nanowire arrays. New J. Phys. **17**(3), 033006 (2015). https://doi.org/10.1088/1367-2630/17/3/033006
89. Z.M. Wang, J.S. Lehtinen, K.Y. Arutyunov, Towards quantum phase slip based standard of electric current. Appl. Phys. Lett. **114**(24), 242601 (2019). https://doi.org/10.1063/1.5092271
90. A.J. Kerman, Flux–charge duality and topological quantum phase fluctuations in quasi-one-dimensional superconductors. New J. Phys. **15**(10), 105017 (2013). https://doi.org/10.1088/1367-2630/15/10/105017
91. J.E. Mooij, Y.V. Nazarov, Superconducting nanowires as quantum phase-slip junctions. Nat. Phys. **2**(3), 169–172 (2006). https://doi.org/10.1038/nphys234
92. I. Schneider, K. Kronfeldner, T.I. Baturina, C. Strunk, Quantum phase slips and number-phase duality in disordered TiN nanostrips. Phys. Rev. B **99**(9), 094522 (2019). https://doi.org/10.1103/PhysRevB.99.094522
93. O.V. Astafiev et al., Coherent quantum phase slip. Nature **484**(7394), 355–358 (2012). https://doi.org/10.1038/nature10930
94. W. Zhao, X. Liu, M.H.W. Chan, Quantum phase slips in 6 mm long niobium nanowire. Nano Lett. **16**(2), 1173–1178 (2016). https://doi.org/10.1021/acs.nanolett.5b04473
95. A. Bawa, R. Jha, S. Sahoo, Tailoring phase slip events through magnetic doping in superconductor-ferromagnet composite films. Sci. Rep. **5**(1), 13459 (2015). https://doi.org/10.1038/srep13459
96. A. Bawa, A. Gupta, S. Singh, V.P.S. Awana, S. Sahoo, Ultrasensitive interplay between ferromagnetism and superconductivity in NbGd composite thin films. Sci. Rep. **6**(1), 18689 (2016). https://doi.org/10.1038/srep18689
97. C.H. Webster, J.C. Fenton, T.T. Hongisto, S.P. Giblin, A.B. Zorin, P.A. Warburton, NbSi nanowire quantum phase-slip circuits: Dc supercurrent blockade, microwave measurements, and thermal analysis. Phys. Rev. B **87**(14), 144510 (2013). https://doi.org/10.1103/PhysRevB.87.144510
98. B. Gajar et al., Substrate mediated nitridation of niobium into superconducting Nb2N thin films for phase slip study. Sci. Rep. **9**(1), 8811 (2019). https://doi.org/10.1038/s41598-019-45338-1

99. J.C. Fenton, J. Burnett, Superconducting NbN nanowires and coherent quantum phase-slips in DC transport. IEEE Trans. Appl. Supercond. **26**(3), 1–5 (2016). https://doi.org/10.1109/TASC.2016.2531005
100. J.T. Peltonen et al., Coherent flux tunneling through NbN nanowires. Phys. Rev. B **88**(22), 220506 (2013). https://doi.org/10.1103/PhysRevB.88.220506
101. R.P. Aloysius et al., Superconducting properties of tungsten nanowires fabricated using focussed ion beam technique. Nanotechnology **30**(40), 405001 (2019). https://doi.org/10.1088/1361-6528/ab2d6d
102. A. Jafari-Salim, A. Eftekharian, A.H. Majedi, M.H. Ansari, Stimulated quantum phase slips from weak electromagnetic radiations in superconducting nanowires. AIP Adv. **6**(3), 035209 (2016). https://doi.org/10.1063/1.4944482
103. B. Bhattacharyya, V.P.S. Awana, T.D. Senguttuvan, V.N. Ojha, S. Husale, Proximity-induced supercurrent through topological insulator based nanowires for quantum computation studies. Sci. Rep. **8**(1), 17237 (2018). https://doi.org/10.1038/s41598-018-35424-1
104. H. Bartolf et al., Current-assisted thermally activated flux liberation in ultrathin nanopatterned NbN superconducting meander structures. Phys. Rev. B **81**(2), 024502 (2010). https://doi.org/10.1103/PhysRevB.81.024502
105. C. Delacour, B. Pannetier, J.-C. Villegier, V. Bouchiat, Quantum and thermal phase slips in superconducting niobium nitride (NbN) ultrathin crystalline nanowire: application to single photon detection. Nano Lett. **12**(7), 3501–3506 (2012). https://doi.org/10.1021/nl3010397
106. K.Y. Arutyunov, J.S. Lehtinen, T. Rantala, The quantum phase slip phenomenon in superconducting nanowires with high-impedance environment. J. Supercond. Nov. Magn. **29**(3), 569–572 (2016). https://doi.org/10.1007/s10948-015-3298-9
107. S.P. Giblin, M.-H. Bae, N. Kim, Y.-H. Ahn, M. Kataoka, Robust operation of a GaAs tunable barrier electron pump. Metrologia **54**(3), 299–306 (2017). https://doi.org/10.1088/1681-7575/aa634c
108. H. Takesue et al., Quantum key distribution over a 40-dB channel loss using superconducting single-photon detectors. Nat. Photonics **1**(6), 343–348 (2007). https://doi.org/10.1038/nphoton.2007.75
109. R.H. Hadfield, Single-photon detectors for optical quantum information applications. Nat. Photon **3**(12), 696–705 (2009). https://doi.org/10.1038/nphoton.2009.230
110. G. N. Gol'tsman et al., Picosecond superconducting single-photon optical detector. Appl. Phys. Lett. **79**(6), 705–707 (2001). https://doi.org/10.1063/1.1388868
111. C.M. Natarajan, M.G. Tanner, R.H. Hadfield, Superconducting nanowire single-photon detectors: physics and applications. Supercond. Sci. Technol. **25**(6), 063001 (2012). https://doi.org/10.1088/0953-2048/25/6/063001
112. S. Yadav, D.K. Aswal, Redefined SI units and their implications. MAPAN **35**(1), 1–9 (2020). https://doi.org/10.1007/s12647-020-00369-2
113. M. Singh, R. Chaujar, R.K. Rakshit, Cryogenic measurement set-up for characterization of superconducting nano structures for single-photon detection applications. Curr. Sci. **115**(6), 1085 (2018). https://doi.org/10.18520/cs/v115/i6/1085-1090
114. M. López, H. Hofer, S. Kück, Detection efficiency calibration of single-photon silicon avalanche photodiodes traceable using double attenuator technique. J. Mod. Opt. **62**(20), 1732–1738 (2015). https://doi.org/10.1080/09500340.2015.1021724
115. Smart Meter National Programme (SMNP). https://www.eeslindia.org/content/raj/eesl/en/home.html
116. World Economic Outlook. https://www.imf.org/external/index.htm
117. IEA World Energy Balances (2019). https://www.iea.org/
118. Country's Share of CO2 Emissions. https://www.ucsusa.org/resources/each-countrys-share-co2-emissions
119. 19th Electric Power Survey (EPS) Report. [Online] Available http://cea.nic.in/psandlf.html
120. Power Sector at a Glance ALL INDIA. https://powermin.nic.in/en/content/power-sector-glance-all-india

121. S. Sorrell, Reducing energy demand: a review of issues, challenges and approaches. Renew. Sustain. Energy Rev. **47**, 74–82 (2015). https://doi.org/10.1016/j.rser.2015.03.002
122. R.F. Hirsh, J.G. Koomey, Electricity consumption and economic growth: a new relationship with significant consequences? Electr. J. **28**(9), 72–84 (2015). https://doi.org/10.1016/j.tej.2015.10.002
123. D.M. Grueneich, The next level of energy efficiency: the five challenges ahead. Electr. J. **28**(7), 44–56 (2015). https://doi.org/10.1016/j.tej.2015.07.001
124. "New NIST Service: Extending Traceable Measurements Inside the Human Body." https://www.nist.gov/news-events/news/2018/03/new-nist-service-extending-traceable-measurements-inside-human-body
125. "Magnetic Resonance Measurements for MRI Biomarkers." https://www.nist.gov/pml/applied-physics-division/products-services/magnetic-resonance-measurements-mri-biomarkers
126. "Magnetic Resonance Imaging (MRI) Biomarker Measurement Service." https://www.nist.gov/programs-projects/magnetic-resonance-imaging-mri-biomarker-measurement-service
127. R.S.H. Istepanian, S. Hu, N.Y. Philip, A. Sungoor, The potential of internet of m-health things "m-IoT" for non-invasive glucose level sensing. in *2011 Annual International Conference of the IEEE Engineering in Medicine and Biology Society*, August 2011, pp. 5264–5266 (2011). https://doi.org/10.1109/iembs.2011.6091302
128. A. Alarcón-Paredes, V. Francisco-García, I.P. Guzmán-Guzmán, J. Cantillo-Negrete, R.E. Cuevas-Valencia, G.A. Alonso-Silverio, An IoT-based non-invasive glucose level monitoring system using raspberry pi. Appl. Sci. **9**(15), 3046 (2019). https://doi.org/10.3390/app9153046
129. R.P. Singh, M. Javaid, A. Haleem, R. Suman, Internet of things (IoT) applications to fight against COVID-19 pandemic. Diabetes Metab. Syndr. Clin. Res. Rev. **14**(4), 521–524 (2020). https://doi.org/10.1016/j.dsx.2020.04.041
130. S. Ramakrishna, L. Tian, C. Wang, S. Liao, W.E. Teo, Safety testing of a new medical device. in *Medical Devices* (Elsevier, 2015) pp. 137–153
131. "IEC 60601-1-11:2015 Medical electrical equipment—part 1-11: general requirements for basic safety and essential performance—collateral standard: requirements for medical electrical equipment and medical electrical systems used in the home healthcare e." https://www.iso.org/home.html
132. "IEC 61010-1:2010 Safety requirements for electrical equipment for measurement, control, and laboratory use—part 1: general requirements." https://webstore.iec.ch/publication/4279
133. "ISO 80601-2-12:2011 Medical electrical equipment—part 2-12: particular requirements for basic safety and essential performance of critical care ventilators." https://www.iso.org
134. "ISO 80601-2-12:2020 Medical electrical equipment—part 2-12: particular requirements for basic safety and essential performance of critical care ventilators." https://www.iso.org/

Chapter 11
Electromagnetic Metrology for Smart Technologies

Part I: Microwave Metrology

S. K. Dubey, Saood Ahmad, C. K. Suman, Sandhya M. Patel, Sudhir Husale, Anurag G. Reddy, Anurag Gupta, and D. K. Aswal

Abstract Electromagnetic (EM) Metrology at CSIR-NPL comprises of various measurement facilities in frequency range from 1 Hz to 110 GHz of electromagnetic spectrum for devices from household electronic appliances to advanced strategic communication and instrumentation. Part-I of the chapter focuses on the Microwave range of the EM spectrum. With the advent of smart technologies, Microwave has reached each house and every individual, which makes precise and traceable electromagnetic measurements and hazardless interconnectivity very important. CSIR-NPL (NMI of the country) provides measurements for microwave frequencies from 9 kHz to 110 GHz frequency range that are precise, accurate, reliable, internationally recognized and traceable to the SI units through Microwave Metrology. The Microwave based National Standards and measurement capabilities are realized, established,

S. K. Dubey · S. Ahmad · C. K. Suman · S. M. Patel · S. Husale · A. G. Reddy · A. Gupta (✉) · D. K. Aswal
CSIR-National Physical Laboratory, Dr. K. S. Krishnan Marg, New Delhi 110012, India
e-mail: anurag@nplindia.org

S. K. Dubey
e-mail: dubeysk@nplindia.org

S. Ahmad
e-mail: ahmads@nplindia.org

C. K. Suman
e-mail: sumanck@nplindia.org

S. M. Patel
e-mail: patelsm@nplindia.org

S. Husale
e-mail: husalesc@nplindia.org

A. G. Reddy
e-mail: g.reddy.anurag@nplindia.org

D. K. Aswal
e-mail: dkaswal.npl@nic.in

D. K. Aswal (ed.), *Metrology for Inclusive Growth of India*,
https://doi.org/10.1007/978-981-15-8872-3_11

maintained, and upgraded at CSIR-NPL. These standards are disseminated to reference laboratories across India to provide traceability to various sectors: strategic, defence, manufacturers, testing industries, government regulators and research institutions. Microwave metrology at CSIR-NPL has a unique combination of comprehensive capabilities of various parameters of Electromagnetics such as attenuation, microwave power, E-Field and specific absorption rate (SAR), along with various free space measurements parameters. Upcoming 5G technology is not just a routine technological change but a platform to enable several smart technologies such as smart banking, smart city, smart village, smart healthcare smart automobile and many more. This shows the overall impact of Microwave metrology on country's upcoming technological needs. Based on impact analysis of dissemination of microwave metrology by the NMI of India, its role on technological revolution will be discussed in detail. This chapter comprises of closed system microwave parameters such as attenuation, impedance, insertion loss followed by free space microwave measurements such as Electric Filed, SAR, Shielding effectiveness, dielectric measurements along with advanced Quantum E-Field measurements traceable to Plank Constant. Each parameter is discussed in terms of its associated primary standards, their calibration and measurement capabilities (CMCs) and their degree of equivalence with the leading NMIs of the world, and on-going research for advanced measurements including quantum standards. Next section of all microwave metrology parameters deals with their impact on quality infrastructure creation at strategic, defence and industrial fronts. How ministries, regulators, manufacturers, industries, academia and research institution directly/indirectly benefit from these parameters in terms of calibration, testing, training, human resource development and technical consultancy is discussed in detail. Questions related to the issues such as—the measurement challenges, effect on industrial certification and technical consultation, challenges for regulators to catch up with upcoming series of modified standards—associated with the advent of technology in electromagnetic domain are dealt with. The importance of a strong link amongst the ministries, regulators, NMI (CSIR-NPL) and industries evolving as a game changer to enhance export and improve the quality of import will be discussed. The impact of electromagnetic metrology on overall economic growth of the country and 'Aatmanirbhar Bharat' are outlined.

One accurate measurement is worth a thousand expert opinions.

– Grace Hopper (1906-1992)

11.1 Introduction

'Electromagnetic' spectrum is a term used for referring to electromagnetic radiation, for frequency range from 1 to 10^{20} Hz, which can further be classified into two parts, non-ionizing (1 Km to 100 nm) and ionizing (100 nm and up to 1 Å) [1], based on the energy EM wave carry at a particular frequency. Electromagnetic metrology at CSIR-NPL has calibration and measurement capabilities for 1 Hz–10^{11} Hz of electromagnetic spectrum. These calibration and measurement capabilities cater to almost all sectors of electrical and electronics industries as well as other interdisciplinary industries such as the automobile sector, biomedical devices, industrial RF heating instrumentation, and many more. Electromagnetic metrological services at apex level in the country cover parts of EM spectrum having day to day interaction in human life. The non-ionizing EM spectrum (1 Hz–110 GHz) covers area of services from household appliances to advance communication system, defence to strategic sector, electrical and electronics to automobile industry, aviation to naval dockyard, advance material characterization to biological liquids and magnetic fields to electric field generation and measurement as per their broad applications. The influence of EM metrology can be ascertained from the huge economic impact it can have on various industries it caters to.

Electrical and electronic industry has shown an optimistic growth since 2014, but the deficit between import and export has side lined the serious efforts made by regulators and manufacturers under 'Make in India' initiative [2]. As per available reports [3, 4, 5] in the public domain, the total electronic manufacturing in India has grown by a minimum of threefold since 2014. A figurative comparison of electronics export between India and China as shown in Fig. 11.1 explains the export growth scenario very clearly. Indian export reached up to 90 billion INR in 2019, almost double that of 2018 [6].

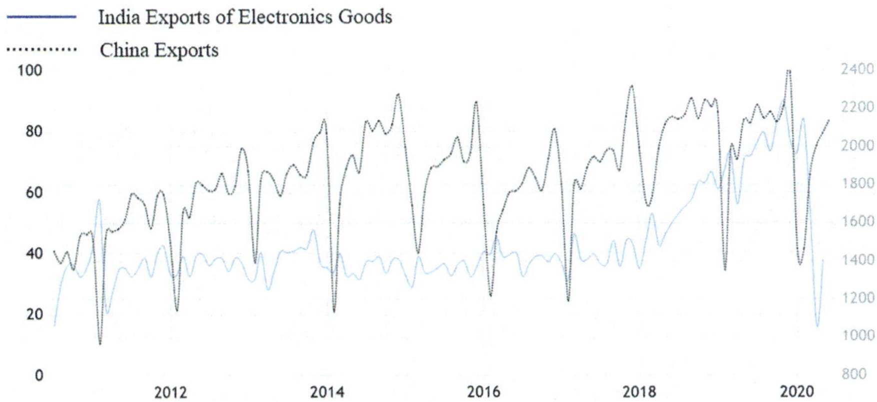

Fig. 11.1 Export comparison between India and China (*Y*-axis is in INR Billion) [6]

Similarly, as per Indian Electrical and Electronics Manufacturers' Association (IEEMA), the apex Indian industry association of manufacturers of electrical, industrial electronics and allied equipment concluded, based on their compiled data, that the electrical and electronics industry has witnessed a record double-digit growth of 12.8% in 2017–18 [7]. Potential growth associated with electric and electronics industry based on home appliances is still very bright. As per the latest reports [8], the five primary categories in the appliances and consumer electronics market—ACs, TVs, audio, refrigerators, and washing machines - is projected to have a compounded growth rate of 11.7% between 2018/19 and 2024/25. Similar situation will be present in almost every sector of electrical and electronics post Covid-19 new normal and 'Aatmanirbhar Bharat' [9] framework.

The third industrial revolution was about connecting each and every device with internet to control them via remote access and fourth industrial revolution is about converting a technology to smart technology; and it will be driven by IoT based interconnectivity from smart household electrical appliances to smart automobiles, smart offices to smart biomedical assistance and will be available on individual's fingertip via a smart phone. This fourth generation of industry will be derived from very sensitive smart sensors based on low voltage/current or magnetic field and their interconnectivity with IoT based smart phones, new upcoming devices via smart communication devices connected to smart base station, which will get power from a smart grid within assigned exposure EM limits. The interconnectivity of these devices with decision making capabilities to sustain the ecosystem with minimum interference to coexisted devices will be the key challenge in upcoming time and will be discussed in detail in following sections.

The true potential of the electrical and electronics sector can be reaped only via a sound and efficient quality infrastructure in India, which is reflected in the policies of the Indian government. The government of India launched National Policy on Electronics 2019 (NPE 2019 [10]) on 19 February 2019 with a target of $400 billion in domestic electronics manufacturing by 2025, of which $190 billion is expected to be contributed by exports with one crore job opportunities. One key target of this policy is to boost mobile manufacturing to one billion units, with exports targeted up to 0.6 billion units. Similarly, a committee by the honourable prime minister to make India an electronic manufacturing and export hub in the world, chaired by NITI Aayog CEO with several international members and world-leading manufacturers has been setup to come up with suggestions [11] to reduce India's 'disability' to export or high production costs. Key recommendations of this committee as well as NPE 2019 are very similar and explain the importance of quality infrastructure in the country. Following are the points of concern, as per committee's recommendations:

(a) Cost-effective solution
(b) Quality Assurance of the product
(c) Certification as per the world's advance standardization to enhance export in Europe and USA
(d) Robustness of the products
(e) Certification without any trade barrier.

Electromagnetic Metrology addresses the suggestions of the NPE 2019 to boost the export by providing framework to ensure quality infrastructure and support industries by enabling testing and certification domestically at par with international standards to match export level quality. EM Metrology at CSIR–NPL can be further categorized into major metrology parameters based on EM frequency and their applications: starting from microwave metrology followed by LF and HF voltage, current and PMU metrology, magnetic metrology, quantum voltage metrology and FIB application for metrology In the present Part-I of the chapter, we discuss the microwave metrology in detail. The next Part-II of the chapter will discuss remaining significant parts of Electromagnetic Metrology covering the details of LF and HF voltage, current and PMU metrology, magnetic metrology, quantum voltage metrology and FIB application for metrology.

Microwave Metrology at CSIR-NPL [12] exhibits an international degree of equivalence for various measurement parameters such as Microwave Power, Attenuation, Impedance, E-Field, Shielding Effectiveness, Radiated power density, Specific Absorption Rate in the frequency range from 9 kHz to 50 GHz with fifteen registered CMC's and seven international inter-comparisons. Indigenized instrumentation is the crucial aspect of this parameter, where primary standards for Attenuation, E-Field, SAR and shielding effectiveness have been developed in-house. With upcoming facilities such as EMI/EMC, IoT based devices testing, Vector SAR and Rydberg Atom based measurement and Quantum Communication, the metrology capabilities will extend to almost all advanced electronics manufacturing sectors supporting Defense, DGCA, MEITY, MoCA and DoT with TEC. The total impact of only the telecom industry is estimated to be a minimum 8.2% of India's GDP. A detailed discussion about the above points will be followed in the respective section.

From the discussion above, it is self-explanatory that the Microwave metrology can support various national missions such as Make in India, Electric Vehicle Policy 2019, NEP 2019, National Digital Communications Policy-2018 and National Mission on Quantum Technologies and Applications (NM-QTA) by establishing a quality infrastructure in the country that can provide all technological solution to the measurement requirement from these various sectors. The chapter further explains each parameter associated with Microwave Metrology in detail including the associated primary standards, their calibration and measurement capabilities with the international status. Also, the role of the Microwave metrology in quality infrastructure with identified sectors, regulators and ministries along with the economic impact analysis is also provided.

11.2 Microwave Metrology

Microwave Metrology at CSIR-NPL provides apex level calibration services in microwave attenuation, impedance and E-field strength parameters to the industry and user organizations of the country. In order to meet the growing demand for calibration of new age wireless devices in this era of continuously growing technology, the focused research and development activities at CSIR-NPL towards the microwave metrology also grows continuously and can be classified into three main domains of Microwave Engineering:

- Instrumentation and measurements (antenna measurement, attenuation and impedance measurement up to 40 GHz, E-field standards, Radiated power density standards, SAR measurement, dosimetric E-field and SAR probes)
- Analytical modelling and design (microstrip antenna, material characterization in microwave, computational modeling of SAR)
- Research towards Quantum Standards for direct SI traceability against universal constants.

CSIR-NPL (NMI of India) disseminates the traceability from various range of parameters ranging including microwave power, attenuation, impedance up to 40 GHz, along with measurements for E-Field strength and radiated power density up to 6 GHz and several other derived parameters and measurement capabilities which will be discussed in detail in following sections.

11.3 Measurement Capabilities and SI Traceability

This section outlines the measurement capabilities and their traceability to SI Units has been discussed in detail. Each parameter has been elaborated with their scientific basis and their application in measurements.

11.3.1 Primary Standards

Primary standards associated with microwave metrology, both of derived units as well as of independent parameters is discussed in following section.

11.3.1.1 Power Measurement

A 2.4 mm coaxial microcalorimeter system based on thermoelectric principle has been realized as the national standard of microwave power at CSIR-NPL [13]. The design is based on two symmetrical and thermally isolated transmission lines, one connected to power standard and the other connected to an identical power standard used as a thermal reference. This complete system is immersed in water for thermal stability as shown in water vat section in Fig. 11.2. The submerged system is illustrated mathematically in Fig. 11.3. The main function of the system is to determine the temperature variation between the two power standards, which is of the order of few milli-Kelvin, using a specially designed thermopile. The coaxial microcalorimeter along with the thermocouple power sensor will provide traceable measurements from 1 MHz to 50 GHz.

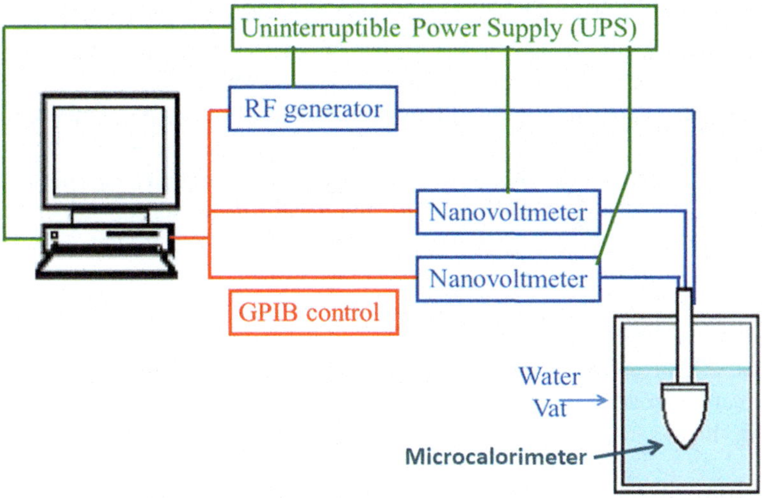

Fig. 11.2 Block diagram of the complete calibration system

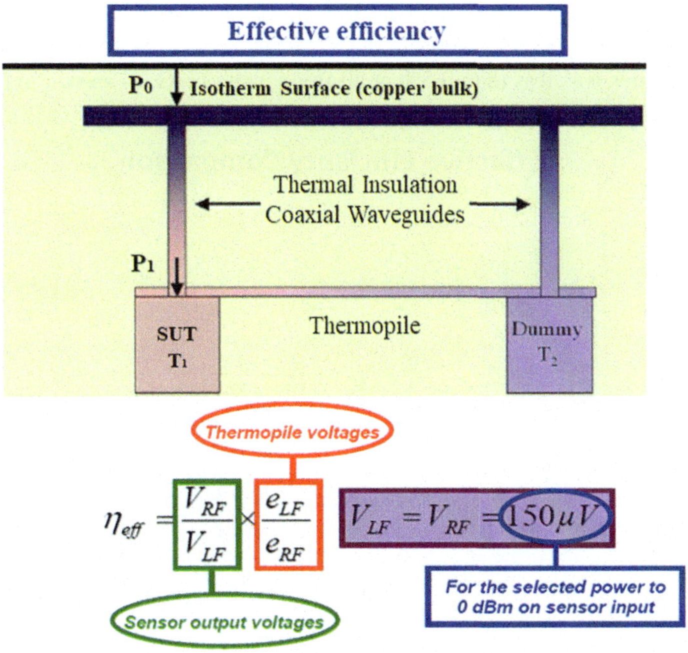

Fig. 11.3 Mathematical model for evaluation of effective efficiency

i. **Efficiency**

The efficiency is obtained from the first reference measurement in low frequency (200 Hz). Alternately, the LF signal and the RF signal, at the calibration frequency, is injected in the thermopile. The level of the RF signal is adjusted to obtain a sensor thermocouple output voltage similar to the one obtained with the LF signal. Where, e_{LF} and e_{RF} are the microcalorimeter thermopile output voltages and, VLF and VRF are the microcalorimeter sensor output voltages, when the LF and RF signals are applied respectively. The efficiency uncertainty ($k = 1$) is obtained from mentioned in Fig. 11.3.

ii. **Effective Efficiency**

The effective efficiency is obtained from both the efficiency and the thermal isolation coaxial line attenuation (A(dB)) using the expression shown in Fig. 11.3.

iii. **Measurement results**

An interlaboratory measurement comparison of microwave power for the validation of the 2.4 mm coaxial microcalorimeter system has been carried out between CSIR-NPL and Laboratoire National de Métrologie et d'Essais (LNE) France. The difference between the effective efficiency evaluated by the two laboratories was less than 0.5% at all frequency points, as shown in Fig. 11.4.

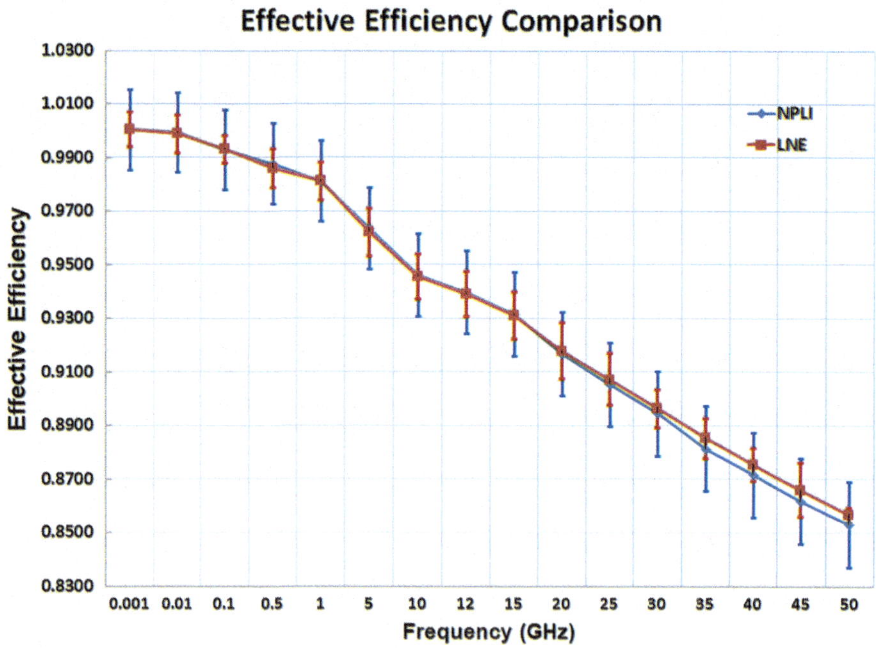

Fig. 11.4 Effective efficiency measurement comparison

The normalized error value of CSIR-NPL for effective efficiency varies between −0.23 and +0.09 with respect to LNE. The result shows good agreement in assigning the effective efficiency to power sensor among the two labs within their claimed expanded uncertainty. It proves the degree of equivalence in measurements between two national metrology institutes (NMIs).
The absolute value of the effective efficiency has been assigned to the microwave power standard from 1 MHz to 50 GHz using coaxial microcalorimeter and Vector network analyzer. The expanded uncertainty at 50 GHz is ±1.9%. The system has been validated with the measurement results of LNE France (Fig. 11.4). The maximum difference observed is 0.0042 between the values of LNE France and CSIR-NPL at 45 GHz. The effective efficiency assigned lies within the uncertainty of both the NMIs.

iv. **Traceability Chart of Microwave Power Measurement**
The traceability chart for microwave power measurement is a bit complex, as it relies on SI unit from dimensional metrology and impedance metrology driven from Josephson Voltage Junction and Quantum Hall Resistance again a SI traceable system (shown in Fig. 11.5). From the chart it is obvious to say that the sources of error will increase with each level down the traceability chart and at the end it reaches to microwave power sensor, the most common diode base device to measure microwave power since 1970.

11.3.1.2 Attenuation Measurement

Attenuation can be understood as the insertion loss generated when a device is inserted in between a perfectly matched the generator and load. It is measured in two steps. First, the power is measured when only impedance matched source is connected to the power sensor and this value is taken as reference. Second, a Device Under Test (DUT) is placed in between the source and power sensor and power received is recorded. The difference between the recorded value from the reference value is the attenuation of the DUT. The power from the sensor is measured by a power meter. Attenuation is thus given as:

$$A(dB) = 10\log_{10}\left(\frac{P_1}{P_2}\right) \tag{11.1}$$

where, P_1 is the power indication without the attenuator in line, and P_2 is the power indication with the attenuator in line.

A dimensional artefact based 30 MHz Waveguide below Cut-Off (WBCO) attenuator is the first primary standard of attenuation at CSIR-NPL, with uncertainty of $\pm[0.004 + 0.0001 \times \Delta\,A]$ for a range up to 60 dB [14]. The WBCO is a hollow tube of brass (perfectly conducting cylinder) which acts as a waveguide below cut-off transmission line at 30 MHz, which allows propagation of only resonant mode (TE11) with the help of mode filter. The basic illustration of the WBCO is given in Fig. 11.6,

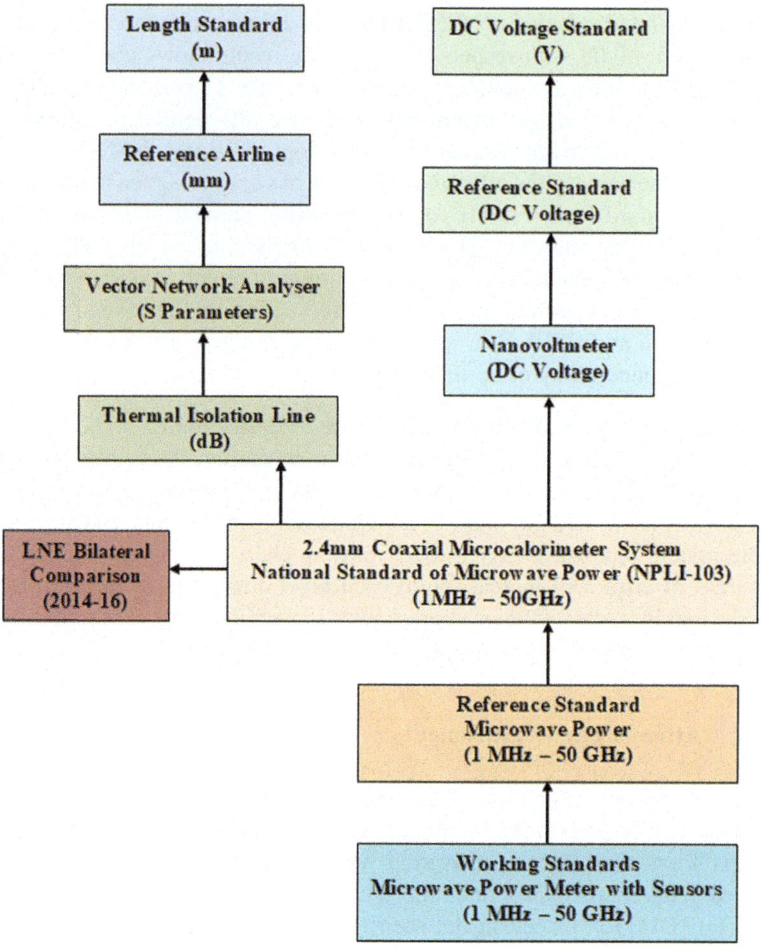

Fig. 11.5 Microwave power measurement traceability chart

Fig. 11.6 Inner Schematic of WBCO attenuator

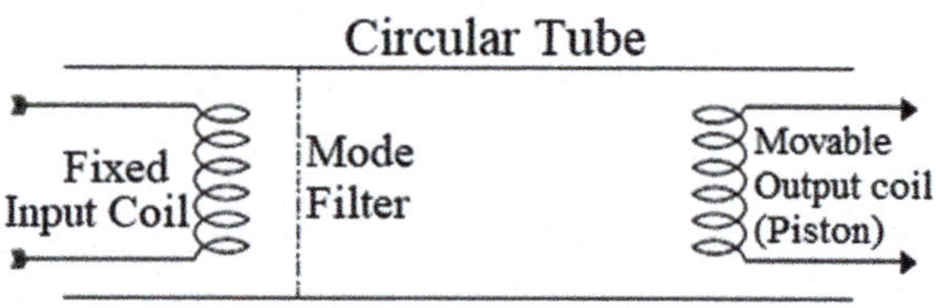

which explains the role of distance between two inductive coils to generate a fixed value of attenuation at a certain distance followed by the role of grid to propagate dominant mode only.

For perfectly conducting cylinder walls with a coil separation d_l to d_2, the attenuation in dB may be found from following expression in terms of dominant mode S_{mn} and wavelength λ,

$$\alpha p = 8.8682\pi(d_1 - d_2)\left[\left(\frac{S_{mn}}{2\pi r}\right)^2 - \left(\frac{1}{\lambda^2}\right)\right]^{0.5} \quad (11.2)$$

Before 2012, this facility was traceable to NMI Australia, now with the help of IF substitution technique this measurement is traceable to inductive voltage divider at 1 kHz IF frequency.

i. **IF Substitution Technique**
Intermediate Frequency (IF) substitution technique measures the attenuation at various frequencies with the help of mixers, as shown in Fig. 11.7. In the IF substitution technique, the system compares the attenuation through the device under calibration with an IF attenuation standard. A mixer along with an IF receiver is introduced into the measurement system to generate the desired IF frequency. To perform this substitution method, two different RF sources are used. A mixer with calculated non-linearity is deployed between the RF source and Local oscillator. The mixer is a three-port device, the two RF sources act as an input to the mixer, and the difference between their frequencies will generate an intermediate frequency on the output port. In the measurement, 30-MHz substitution technique is deployed because of the 30-MHz tuned receiver (VM7), shown in Fig. 11.7. Here, VM7 is a 30-MHz tuned receiver with the in-built local oscillator [15].
In the last 5 years, the microwave attenuation and impedance measurement systems are completely automated at microwave metrology at CSIR-NPL. An attenuation measurement system, which employs a traceable Inductive Voltage Divider (IVD) at 10 kHz as the reference standard and dual-channel Intermediate frequency (IF) substitution method, is developed as the attenuation standard to calibrate the 30 MHz WBCO in the frequency range 30 MHz–18 GHz with an uncertainty of ±0.005 dB/10 dB. This measurement method requires a precise

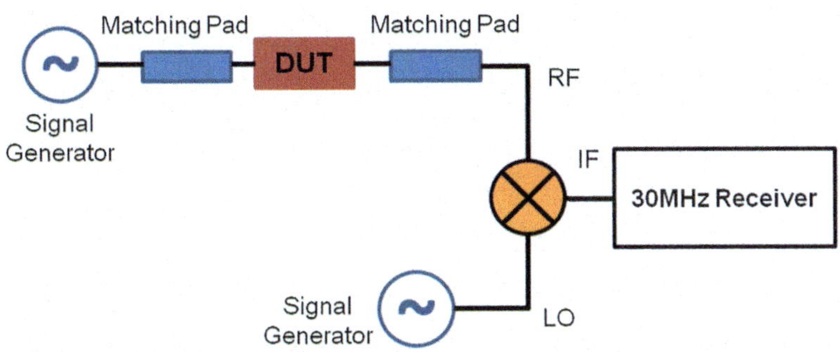

Fig. 11.7 Schematic diagram of 30 MHz IF substitution technique

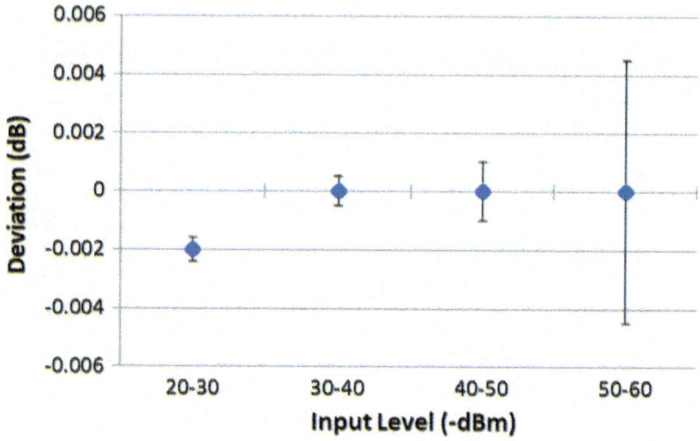

Fig. 11.8 Linearity of mixer designed at CSIR-NPL

mixer which can down convert the 30 MHz signal to 10 kHz signal for the IF substitution. A mixer was designed and indigenously made at CSIR-NPL, which can down convert 30 MHz signal to 10 kHz signal. The linearity of the designed mixer at different power levels is shown in Fig. 11.8, which is within the ±0.006 dB range.

The calibration of WBCO using the dual channel IF substitution technique was carried out at CSIR-NPL and the recorded values are compared with the values reported in earlier certificate of WBCO calibrated by NMI Australia. The measured values from both the measurement techniques are in degree of equivalence and are shown in Fig. 11.9.

In the past few years, fully automated in-house calibration of the primary standard is established based on inductive voltage divider (IVD) based technique. Microwave Metrology provides apex level calibration services for attenuation from 0.05 to 90 dB at 30 MHz and 1 to 50 GHz with an uncertainty of ± 0.005 dB/10 dB.

ii. **Traceability Chart for Attenuation Measurements**

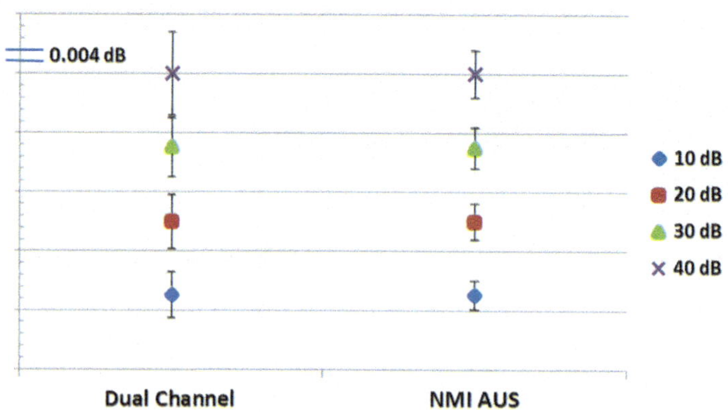

Fig. 11.9 Attenuation measurement comparison with NMI Australia

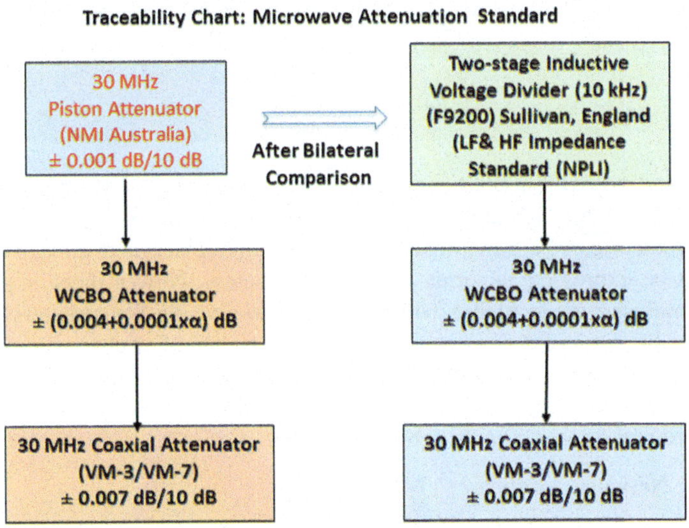

Fig. 11.10 Traceability status of attenuation measurement

As explained earlier, the attenuation parameters at microwave metrology was traceable to NMI Australia via calibration of WBCO against their primary standard. But, now the WBCO calibration facility was established in house against the registered CMC of 1 kHz IF substitution measurement with the help of IVD (Fig. 11.10). Further dissemination of attenuation parameters is being carried out with the help of VM7 receivers and commercially available attenuators.

11.3.1.3 Scattering (S) Parameters Based Impedance Measurements

Scattering parameters define the forward and reverse wave propagation through a network. Abundant literature [15] is available on the theoretical elucidation of *S*-parameters. A brief description here will help in understanding the basic concept of *S*-parameter and its measurement. Considering the example of a two-port network, shown in Fig. 11.11, *S*-parameters define the relation between the input and outputs from the network. In general, the *S*-parameters for a two-port network can be given as:

Fig. 11.11 Two-port device in terms of *S*-parameters

$$b_1 = S_{11}a_1 + S_{12}a_2, \quad (11.3)$$

$$b_2 = S_{21}a_1 + S_{22}a_2, \quad (11.4)$$

where, a_1 and a_2 are the signals entering the Port 1 and Port 2 of the two-port network and b_1 and b_2 are the signals leaving the respective ports.

S-parameters can be generalised for any n-port device and all network properties can be expressed in terms of its *S*-parameters. Now a days, sophisticated network analysers are available, which can directly give the *S*-parameters (and other network properties) of a device connected to its port(s). For characterizing a two-port network for impedance using its *S*-parameters, a VNA comprising of a RF source, bridges/couplers, mixers and tuned receiver is required in order to measure the forward and reverse signals at both the ports.

i. **Vector Network Analyzer (VNA):**

Vector Network Analyzer is a device which measures different network parameters of electrical networks based on transmission and reflection of EM waves in transmission line. It consists of a set of source and receivers; the source generates signals which pass through the device under test and routes it to receivers which analyse the received signal and decides the losses occurred due to the DUT. In general practice, a VNA is calibrated by the common techniques such as SOLT, TRL, multiple short and broadband, etc.[16].

CSIR-NPL being NMI of India cannot apply these common methods of VNA calibration, as the dimensions and other properties of these co-axial devices may change or degrade with time. Coaxial Air-lines with air in between their inner and outer conductors are used as primary standard for VNA calibration at CSIR-NPL [17]. The VNA calibration consists of a calibration of diameters of inner and outer connectors of all co-axial devices from dimension metrology and impedance calibration from DC resistance metrology, this process is represented in the traceability chart.

In the measurements through VNA there are multiple sources of errors and uncertainties, which need to be evaluated carefully. A plane or reference is to be identified in which the electrical length plays the major role to identified several uncertainty sources such as Directivity, Source match, Reflection tracking, Transmission tracking and load match, Isolation, Drift, Test port cables, Non-linearity, and Repeatability based on ten term error model [16] or ripple method [17]. Our clients such as ERTL, ISRO, IAF get their co-axial devices and calibration kits such as open, short, load, broadband etc. calibrated against our primary standard. The calibration of these devices is done using standard techniques mentioned above.

ii. **S-parameter Measurement**

The key comparison named *SIM.EM.RF-K5b.CL* [18] was held between NMIs of the world for S_{21} measurement of a 3 dB attenuator at 9 GHz. CSIR-NPL had participated

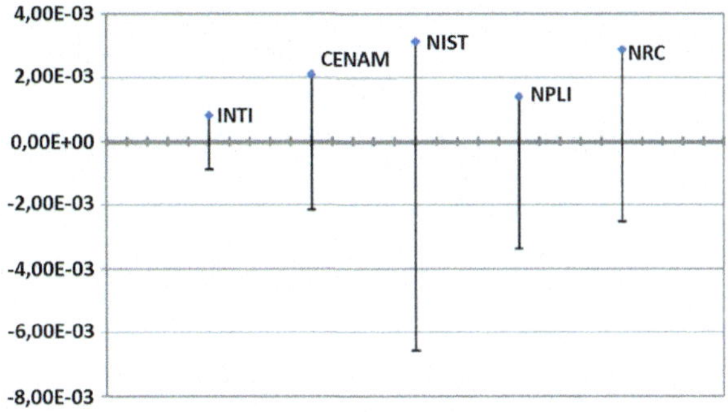

Fig. 11.12 S_{21} Measurement results of key comparison SIM.EM.RF-K5b.CL [18]

in this and the measurement data reported on BIPM website is shown below in Fig. 11.12. The measurements were in degree of equivalence with the world leading NMIs. Similarly, other key comparisons have also been participated in and their results are reported in CMC section below.

iii. **Traceability Chart for Impedance Measurement**

Traceability of impedance standard (Fig. 11.13) is derived from dimensional metrology (for meter (*m*)) and quantum Hall resistance (Ω). Electrical length of airline and resistive and reactive values from Hall resistance are used to derive the

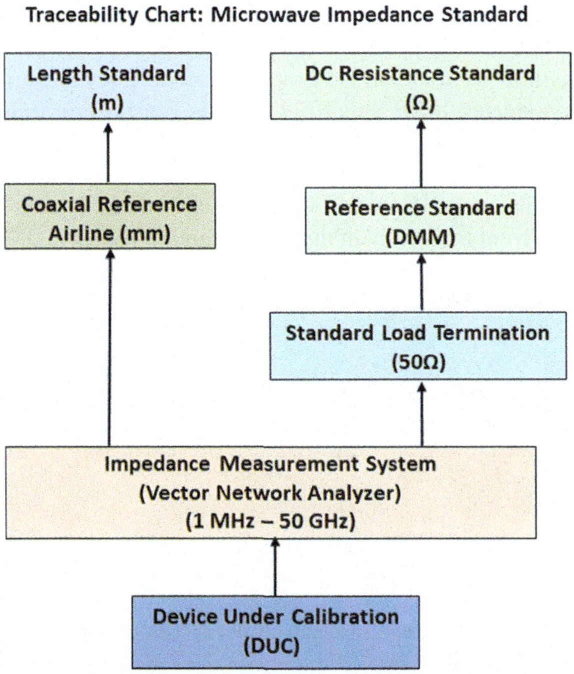

Fig. 11.13 Traceability chart for microwave impedance standard

polynomial of resistance, capacitance and inductance. These polynomials are then used to evaluate the cal-kit parameters calibrated at CSIR-NPL for various organizations. These cal-kits are then used to calibrate various instruments and thus provide traceability to the measurements by respective organizations. Thus, the chain of traceability is established.

11.3.1.4 Electric-Field Strength and Radiated Power Density Measurement

E-Field strength is the essential parameter for all free space measurements in RF frequency. It can be used to derive several parameters such as radiated power density, antenna factor, antenna gain, radiation pattern and intensity for antenna, shielding effectiveness, free space measurement and SAR. Microwave metrology at CSIR-NPL has established the in-house calibration facility for the E-field strength generation and measurement based on completely indigenous instrumentation. The E-field strength standard at CSIR-NPL relies on measurements performed by DUT probes inside the TEM cell, where uniform E-field is created by a RF source. One may refer to [19] for further insight into the measurement setup and procedure.

The Transverse/Gigahertz-Transverse Electromagnetic Cell (TEM/GTEM cell), as shown in Fig. 11.14, can be used to test the minuscule RF devices such as mobile phones, integrated circuits, PCBs, micro-electromechanical systems, and also the E-field probes. A thorough analysis of all the important parameters that affect the working of a TEM cell had been carried out while designing the cell at CSIR-NPL. The complete precise measurement methodology adopted for an isotropic probe [19] is now used as the calibration procedure for Electric Field strength measurement parameter. IEEE Std. 1309–2013 is consistently followed for calibration of E-field probes, which then can be used to measure the radiated power density from the different devices and cellular towers.

The uncertainty in E-field strength measurement is ±0.58 V/m per 10 V/m for different frequency in the range from 700 MHz to 3 GHz. Indigenous E-field probes are also designed and fabricated in-house, which are technically competent with the commercially available probes. E-field strength parameter in the microwave range is

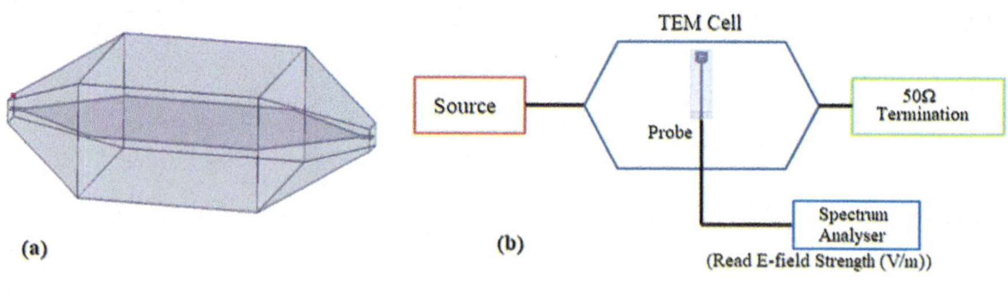

Fig. 11.14 Schematic diagram of **a** basic TEM cell **b** probe calibration inside TEM cell [20]

Fig. 11.15 Cross sectional view of TEM cell

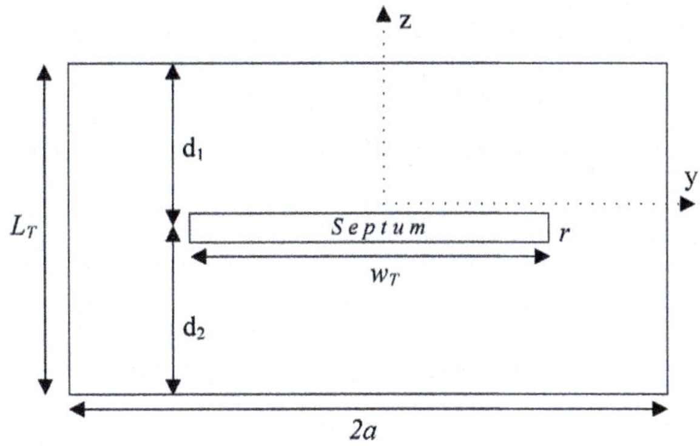

of great importance for the compliance testing of wireless communication devices and has great socio-economic impact.

The schematic diagram of TEM is shown in Fig. 11.14 with a cross-sectional view in Fig. 11.15, where the Electric Field inside TEM depends on the separation between the plates (conductors) and area of the plates i.e., the width and the length of the plates. The characteristic impedance inside the GTEM based on between plate capacitance and fringing capacitance can be defined as [19]:

$$Z_o(\Omega) \cong \frac{94.15}{(\varepsilon)^{1/2}\left[\frac{w_T}{L_T(1-r/L_T)} + \frac{C^f}{0.0885\varepsilon_r}\right]} \tag{11.5}$$

where, $C^{f'}$ is the fringing capacitance in pF/cm, ε_r is relative dielectric constant, r, L_T and W_T are shown in Fig. 11.15.

Other expressions that are useful for characterizing a TEM cell are Cell resonant frequency,

$$f_{res} = \sqrt{f_c^2 + \frac{(c)^2}{2l}} \tag{11.6}$$

where f_c is the cell cut off frequency and l is the resonant length of the cell.

and Cut off frequency,

$$f_c = \frac{75}{a}\sqrt{1 + \frac{2aL_T}{d_1 d_2 \ln\left(\frac{8a}{\pi g}\right)}} \tag{11.7}$$

where, g is the distance between the side walls of the cell and the septum.

Theoretically, the generated E-field strength measured by an ideal isotropic probe inside the TEM cell is calculable and given by Eq. 11.8:

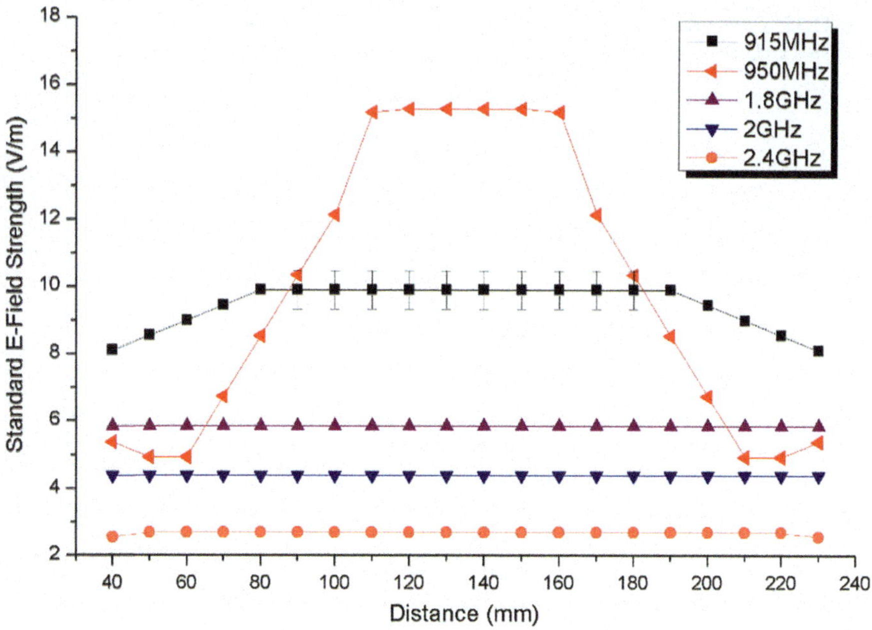

Fig. 11.16 E-Field inside fabricated GTEM cell [19]

$$E = \frac{\sqrt{PZ_r}}{a}, \tag{11.8}$$

where, P is the input power to the cell, Z_r is the real part of the characteristic impedance and a is the vertical distance between the septum and the inside wall of the cell. The performance of designed TEM is reported in Fig. 11.16.

i. **Design, fabrication and characterization of E-field sensors**

For characterizing the microstrip antenna as an E-field probe, its gain, antenna Factor (AF) and E-field strength are measured and compared with a calibrated isotropic E-field probe. A novel way of using AF to evaluate the accuracy of E-field measurement is eventually devised. It is realized that accuracy of such probes needs to be studied in terms of return loss, AF, probe linearity and mismatch. When an antenna is used as an E-field probe, it becomes advisable to evaluate the AF of the antenna, which can be used as a multiplying factor for evaluating the E-field strength from the measured voltage at the probe. The antenna factor is given by [20],

$$AF = 20\log_{10}\left(\frac{E_i}{V}\right) \tag{11.9}$$

where, E_i is the incident E-field and V is the received voltage.

For the wave propagating in free space, the E-field is related to radiated power density S (in W/m^2) and the intrinsic impedance η of free space (in ohms) as

$$|E_i| = \sqrt{\eta S}. \tag{11.10}$$

Also, the received power P_r is dependent on received voltage V and load impedance Z_{load} as

$$P_r = S \times A_e = \frac{V^2}{Z_{\text{load}}}, \tag{11.11}$$

where, A_e is the effective aperture dependent on the operating wavelength λ and gain G

$$A_e = \frac{\lambda^2 G}{4\pi}. \tag{11.12}$$

E-field strength 9.99 V/m at 915 MHz with an expanded uncertainty of ±0.58 V/m at fed power +18 dBm has been reported for the TEM cell fabricated at CSIR-NPL [19].

A comparative study has been performed for various E-filed measurement standards of several NMI's based on literature and presented in Table 11.1 below. The reported values of radiated power density, calibration method and the associated uncertainties are reported in the Table 11.1.

Data in Table 11.1 are sourced from inter-comparison EURAMET.EM.RF-S25 [21] for 10 V measurements and CCEM.RF-K7.a.F.2 [22] for minimum sensitivity level. CSIR-NPL did not participate in any of these intercomparison and the data are compiled to outlay the degree of equivalence of its measurement capabilities.

ii. **Traceability Chart for E-field Measurement**

Table 11.1 Comparison of E-Field and radiated power measurements among leading NMI's

Country	NMIs	Calibration method	Uncertainty in E-field strength* (V/m)	Radiated power density limit
UK	NPL UK	TEM cell and Tapered GTEM Cell	±0.35	100 μW/m^2
France	LNE, France	GTEM Cell	±0.60	1mW/m^2
Germany	PTB, Germany	Micro TEM Cell	±0.69	2mW/m^2
Korea	KRISS, Korea	Micro TEM Cell	±0.70	2mW/m^2
USA	NIST, USA	Micro TEM Cell	±0.45	500 μW/m^2
India	NPL, India	Micro TEM Cell	±0.60	1mW/m^2

CSIR-NPL is in the process to register its E-Field strength measurement capability as a CMC at BIPM and this measurement has yet not been registered. A detailed traceability chart is given in Fig. 11.17, explaining degree of equivalence in the E-Field measurements being ensured with world leading NMI through participation in various international inter-comparisons on other key parameters and other traceable standards.

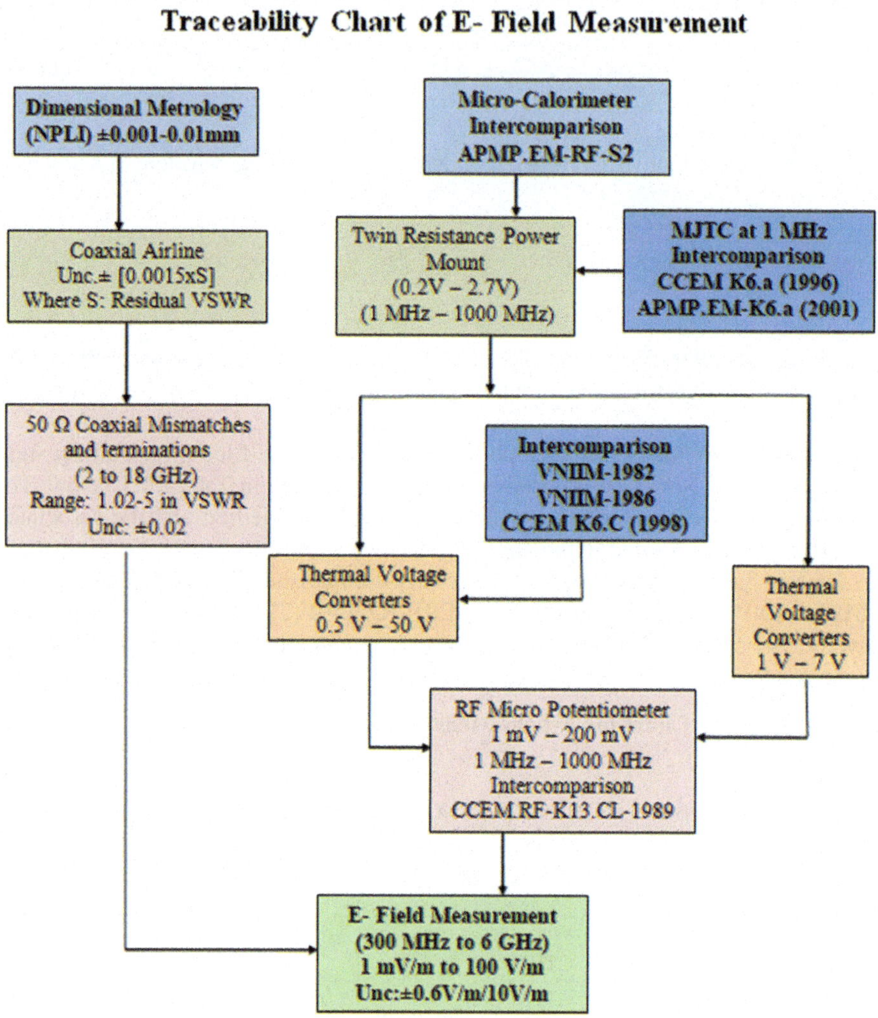

Fig. 11.17 Traceability chart for E-field measurement

11.3.1.5 Specific Absorption Rate

Specific absorption rate (SAR) is the measure of the heating value of radiated RF energy on a body or material such as human tissue. The absorption of electromagnetic (EM) wave is mainly due to the dissipation of its energy, i.e., conversion of its energy into another form, such as heat. The rate at which any material absorbs the incident EM wave is generally expressed in terms of SAR. It is one of the most fundamental dosimetric parameters used by various telecommunication regulatory bodies for specifying the maximum limits of exposures to minimize the adverse effects of the smart wireless devices on humans.

Specific absorption rate value, given in terms of Watt/kg can be defined as the rate of absorption of RF power in a material,

$$\mathrm{SAR} = \frac{\sigma(\omega)\left|E^2\right|}{\rho}, \tag{11.13}$$

where, σ is the conductivity of the material (S/m), E is the Electric field strength (V/m) and ρ is the mass density of the material (kg/m^3).

Another definition of SAR is given in terms of rate of increase of temperature in a material, given by,

$$\mathrm{SAR} = \left.\frac{c\Delta T}{\Delta t}\right|_{t=0}, \tag{11.14}$$

where, c is the specific heat capacity (J/kg °C), ΔT is the change in temperature (°C) and Δt is the duration of exposure (s). For this measurement, "ideal" non-thermodynamic circumstances are made i.e. no heat loss by thermal diffusion or heat radiation. The typical SAR measurement setup is schematically shown in Fig. 11.18a.

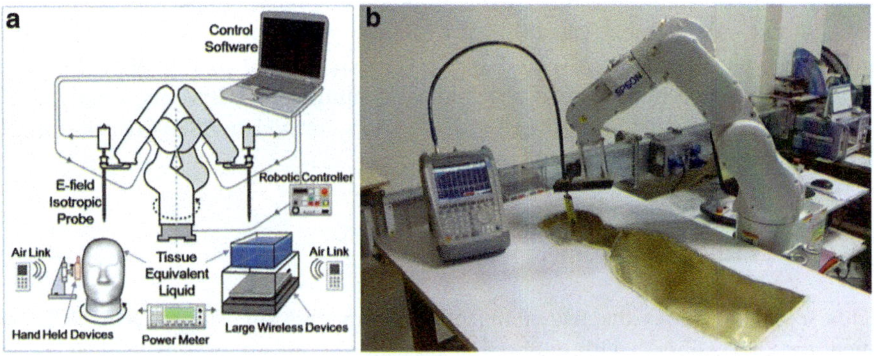

Fig. 11.18 **a** SAR Measurement setup schematic and **b** Actual photograph **of** SAR measurement setup at CSIR-national physical laboratory

A Specific Absorption Rate (SAR) evaluation system is indigenously developed by CSIR- NPL (Fig. 11.18b). In this setup E-Field Sensor probe, Tissue equivalent liquid, Robotic arm (EPSON C8-A701S) and a controlled GUI are being indigenously developed [23]. The system is capable to evaluate SAR up to 3 W/kg with an expanded uncertainty of ±0.25 W/kg per 1.6 W/kg.

As per Eq. 11.13, it is obvious that SAR measurement is derived from E-Field measurement and conductivity of tissue or tissue equivalent liquids. The conductivity tissues/TEL is defined at a particular frequency as per IEEE 1528–2013 standard and can be calculated from imaginary part of S-parameters. Hence, traceability of SAR measurement is combination of two traceability charts: one E-Field and second impedance. Same is reflected in Fig. 11.18. Till date, no inter-comparison for SAR measurement standard has been carried out or yet proposed, as per KCDB [24] database. But for establishing the degree of equivalence in measurement, CSIR-NPL has now become a part or EURAMET consortium on project SAR measurement using vector probes (Project Number: 16NRM07) so that traceability for any upcoming standards can be obtained in future for Mobile and wireless compliance for Indian Telecom industry. It will also provide the technical capability to address 4th industrial revolution as per smart and wearable wireless devices are concerned.

11.3.2 Calibration and Measurement Capabilities (CMC)

11.3.2.1 CMC at BIPM

To register CMC of any parameter at BIPM [24], an activity of NMI should have primary calibration system with minimum degree of freedom to have minimum uncertainty contributions. The associated sources of uncertainties of that parameter should be identified and properly evaluated against various measurements. A validation method is also essential to evaluate the claims of the primary standard. After having all these facilities, an NMI can participate in inter-comparison for the calibration and measurement capabilities. If the measurement results are well within degree of equivalence, then the activity is reviewed by a peer appointed by RMO and later gets accepted by chairman of technical committee at BIPM and finally CMC gets registered at BIPM website. The parameters for which CSIR-NPL has CMC registered at BIPM are given in Table 11.2.

11.3.2.2 List of Measurement Capabilities

This section includes the list of the other parameters for which CSIR-NPL can measure and calibrate with calibration procedure with uncertainty budget as per established methods but not yet peer reviewed and approved by BIPM.

As a result of continuous research in the field of RF and microwave metrology CSIR-NPL has upgraded all existing measurement and calibration facilities to the

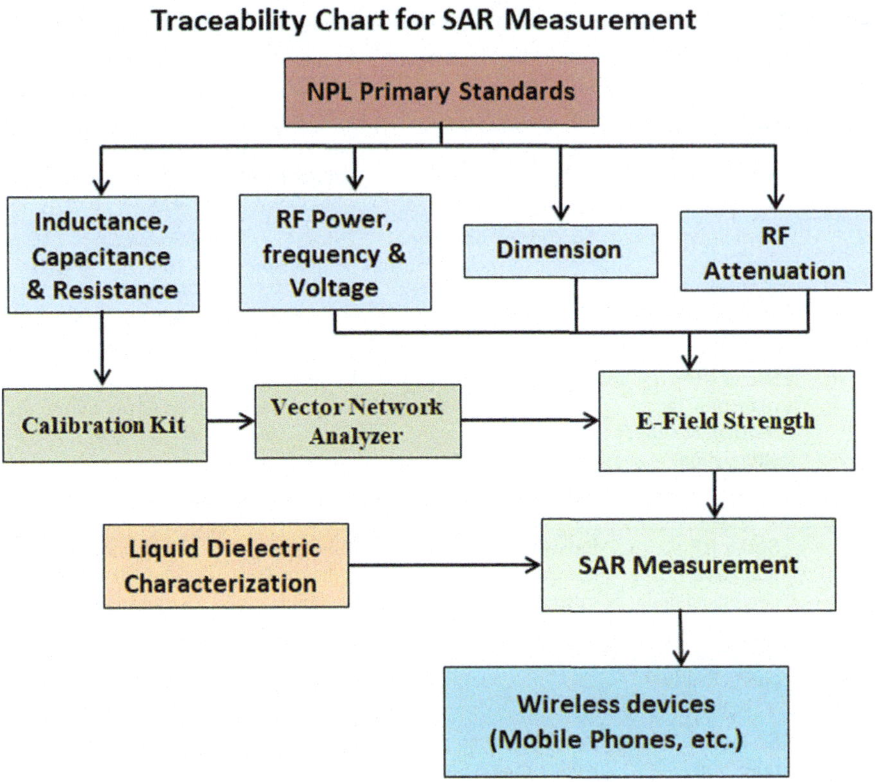

Fig. 11.19 Traceability chart SAR measurement

values reported in Table 11.3 for various parameters. Uncertainty budget for all the parameters have been made and are reported to APMP GA 2017 for CMC registration.

11.3.2.3 List of Inter-Comparison

Microwave metrology at CSIR-NPL keeps participating in inter-comparisons from time to time to upgrade our existing measurement capabilities add new measurement capabilities. Some of the inter-comparisons are listed in Table 11.4.

From above table of intercomparison it is obvious to say that microwave metrology is committed to have a degree of equivalence with leading NMI such as NIST- USA, PTB-Germany, NPL- UK, NRC-Canada, and AIST-Japan on almost all measurement capabilities so that Indian manufactured product could have a free trade as per all technical certification requirements. Some degree of equivalence in measurement with above listed NMI's is given below for reference from [18, 28].

Table 11.2 CMCs of microwave metrology for CSIR-NPL registered at BIPM [25]

TAG	Quantity	Instrument or artefact	Measurement method	Parameters
1	Radio frequency power: absolute power in coaxials	Power meter, power source	Direct measurement or comparison	Frequency: 1 MHz–18 GHz Connector: N, APC-7
2	Radio frequency power: calibration factor and efficiency in coaxials	Thermistor, barretter mount and power sensor	Direct comparison	Frequency: 10 MHz–18 GHz Connectors: N, APC-7
3	Scalar RF reflection coefficient and attenuation: VSWR in coaxials	Passive device: matched termination and standard mismatch	Coupled sliding load technique	Frequency: 2 GHz–18 GHz
4	Scalar RF reflection coefficient and attenuation: reflection coefficient in waveguides	Passive device: matched termination and standard mismatch	Tuned reflectometer technique	Frequency: 5.8 GHz–18 GHz (spot frequency)
5	Scalar RF reflection coefficient and attenuation: attenuation in coaxials	Passive device: fixed/variable attenuator/attenuator and signal calibrator	Substitution technique	Frequency: 30 MHz
6	Scalar RF reflection coefficient and attenuation: attenuation in coaxials	Passive device: fixed/variable attenuator	IF substitution technique	Frequency: 1 GHz–18 GHz
7	Radio frequency voltage and current: RF-DC difference	RF micro-potentiometers	Comparison	Frequency: 1 MHz–1 GHz
8	Radio frequency voltage and current: RF-DC difference	Thermal voltage converter	Comparison	Frequency: 1 MHz–1 GHz Impedance: 50 Ω

(continued)

Table 11.2 (continued)

TAG	Quantity	Instrument or artefact	Measurement method	Parameters
9	Radio frequency voltage and current: RF voltage sources	RF generator	Comparison	Frequency: 1 MHz–1 GHz Impedance: 50 Ω
10	Radio frequency voltage and current: RF voltage meters	RF voltage meter	Comparison	Frequency: 1 MHz–1 GHz
11	Scalar RF reflection coefficient and attenuation: attenuation in waveguides	Passive device: fixed/variable attenuator	Intermediate frequency substitution technique	Attenuation [0.1–60] dB Frequency: 1 MHz–40 GHz

Table 11.3 List of parameters CSIR-NPL is capable of measuring for the range reported against the respective parameter

Parameter	Range	System	Uncertainty
Attenuation	0.005–107 dB	30 MHz Coaxial system	±0.007 dB/10 dB
	0.005–90 dB	9 kHz–40 GHz Coaxial system 5.85–40 GHz Waveguide system	±0.02 dB/10 dB
Reflection coefficient/ VSWR	1.02–5	9 kHz–50 GHz Coaxial system	±0.02 in VSWR ±0.002– ±0.003 in $\|\Gamma\|$
	0.01–0.2	5.85–40 GHz Waveguide system	
Phase	−180°–180° Phase	9 kHz–50 GHz in coaxial system	±0.2° Phase
	0°–180° Phase	5.85–40 GHz Waveguide system	
E-field (V/m)	1 μV/cm–100 V/cm	From 1 Hz–6 GHz	± 0.6 V/M per 10 V/m
Shielding effectiveness	60–10 dB	From 30 MHz–40 GHz in reverberation chamber	±0.02 dB in 10 dB
Dielectric measurement	For Liquid and solid samples	From 8.2 GHz–40 GHz	As per scattering parameters

Table 11.4 Inter-comparisons listed in which CSIR-NPL have participated

S. No.	Pilot lab	Type/Identification/ pilot lab	Parameter/s	Participant countries	Status
1	CSIR-NPL	Supplementary/P1-APMP.EM.RF-S3/NPL India	Impedance (Complex reflection coefficient)	KRISS, CMS/ITRI, CSIR-NML	[26]
2	NPL UK	Key/CCEM.RF-K5.b.CL/NPL UK	Complex S-parameters	NPL, LNE, NIST, PTB	[27]
3	NIM China	Key/APMP.EM.RF-K19.CL/NIM China	Attenuation	NIMT, NMIA, KRISS, NMIJ	[28]
4	NMI Argentina	SIM.EM.RF-K5.b.CL/NMI Argentina	S-parameters by Broad-Band Methods	KRISS, NIST, CENAM	[18]
5	CSIR-NPL	Supplementary/P1-APMP.EM-S8	DC voltage, current, resistance, AC voltage and current	NIMT, NMIA, NIS, NML-SIRIM	in process, 2019
7	NMI Japan	APMP.EM.RF-K8.CL	APMP Comparison of power in 50 Ω coaxial lines (0.01 to 18 GHz)	NMIJ, KRISS, NIM, NMIA	in process, 2019
8	NMI Japan	CCEM.RF- K26	Attenuation measurements (18, 26.5, and 40 GHz) Max. att. 90 dB	KRISS, LNE, NIM, NPL, PTB	Draft A 2020
9	NMI Japan	APMP.EM.RF-S5.CL	Dimensionally derived characteristic impedance, nominal value: 50 ohms	KRISS, PTB, NIST, LNE, NPL	Data submitted

From Fig. 11.19, it is self-explanatory that microwave metrology represents CSIR-NPL with equivalent measurement capabilities against all leading NMI's. Both two-port coaxial measurement and the one port coaxial measurement at CSIR-NPL have same degree of equivalence with similar order of uncertainties (Fig. 11.20).

11.3.3 Derived Measurement Capabilities

11.3.3.1 Missile System Testing

The calibration and testing services for missile guidance system of IAF have been provided by CSIR-NPL, during the crucial time when it was not possible to send the calibration equipment to its manufacturer for testing. The measurements for shielding, transparency and signal strength for the calibration radome and system radome of the missile testing system were performed at CSIR-NPL on various frequency bands.

11.3.3.2 Radiated Power Density Probe Testing for Mobile Tower Radiation Exposure Assessment

RF exposure and safety concerns about usage of mobile and wireless devices have been getting increased attention in recent years. There is continuous media attention and general public concern towards harmful effects of radiations from mobile towers in residential areas. Government has taken this concern into account and specified the exposure limits in line with international regulations. As per revised national telecom policy 2012, the radiation exposure limits are expressed as Tables (11.5 and 11.6).

DoT has laid norms for timely assessment of radiation from mobile towers installed in residential and commercial areas. These assessments are done by various types of probe and meters, which are calibrated by microwave metrology of CSIR-NPL.

This measurement is done by measurement in a closed chamber TEM cell, which produces a uniform field in the range 700 MHz–6 GHz. Standard primary/secondary probe and the DUC probe are placed in the uniform field of particular frequency and measured as shown in Fig. 11.21.

The GTEM cell is fabricated in-house at CSIR-NPL and is calibrated against the primary commercial canonical GTEM cell.

11.3.3.3 Dielectric Measurements for Liquids and Solids:

We provide dielectric assessment for parameters dielectric constant and loss parameters for solid materials like ceramics, sheets, etc. for strategic sectors (CEL,

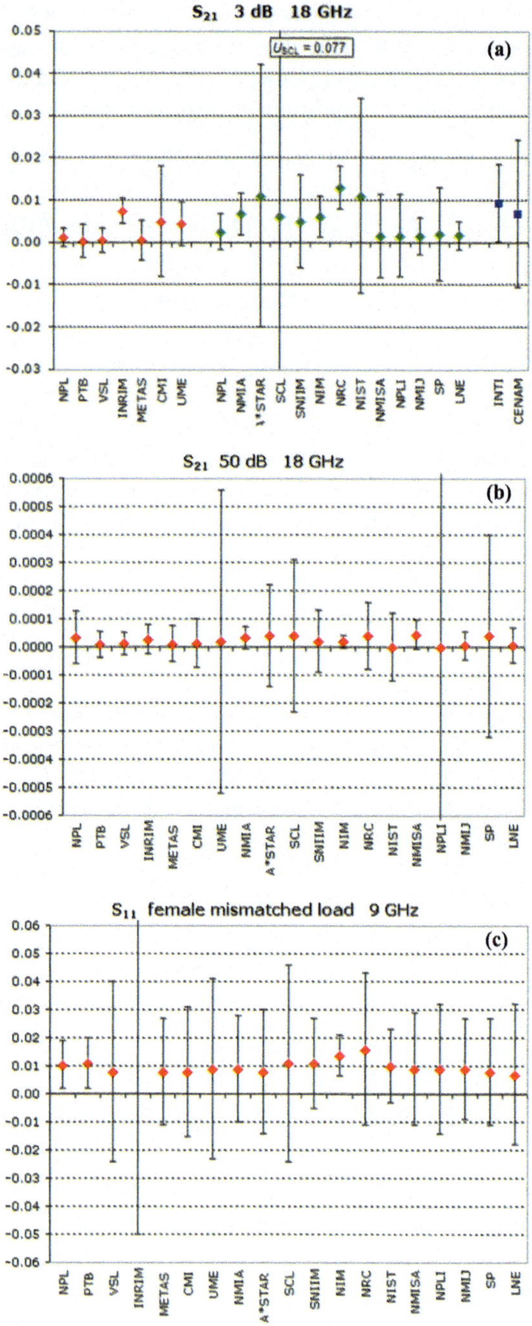

Fig. 11.20 **a** S_{21} Measurement of 3 dB attenuator at 18 GHz **b** S_{21} Measurement of 50 dB attenuator at 18 GHz **c** Reflection coefficient measurement of mismatch (0.01) at 9 GHz

Fig. 11.21 Radomes calibrated at CSIR-NPL for IAF

Table 11.5 EMF radiations limits at different frequencies EMF radiations

Sr. No.	Frequency (MHz)	ICNIRP Radiation limit (W/m^2)	Revised DoT norms for radiation limit w.e.f. 01.09.2012 (W/m^2)
1	900	4.5	0.45
2	1800	9.0	0.9
3	2100	10.5	1.05

Table 11.6 SAR limits at different frequencies EMF radiations

Sr. No.	Frequency	ICNIRP SAR limits (Watt/kg)	Revised DoT norms for SAR limit w.e.f 01.09.2012 (Watt/kg)
1	10 MHz–10 GHz	2.0	1.6

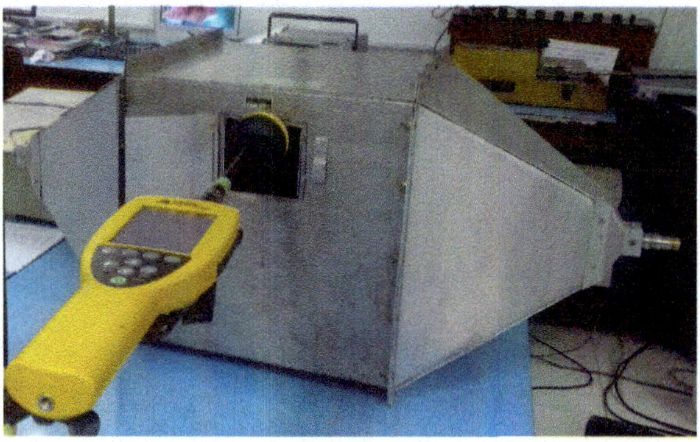

Fig. 11.22 Secondary probe measurement for radiated power density in closed GTEM cell

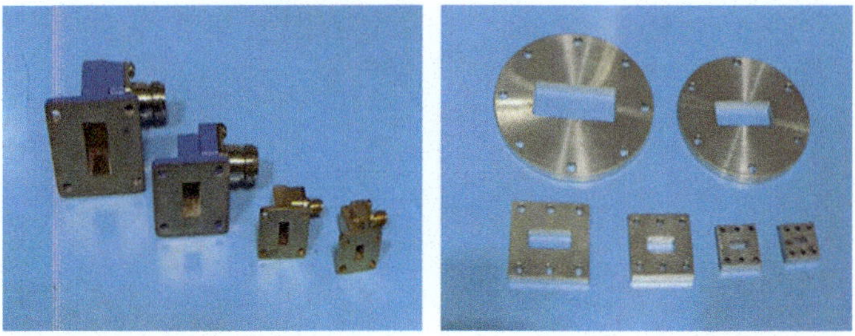

Fig. 11.23 Fabricated adapters and sample holder for solid dielectric material measurements

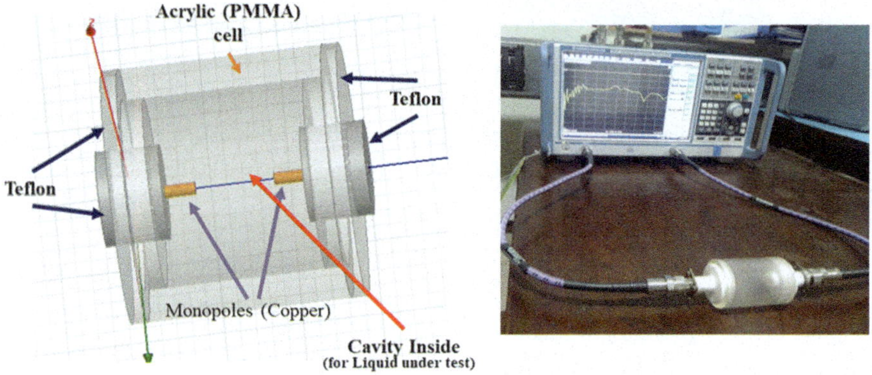

Fig. 11.24 Design for fabricated sample cell for liquid dielectric materials

Sahibabad) and manufacturers of various materials. Similarly, measurement for dielectric constant, loss and bulk conductivity of liquid materials such as biological liquids, lossy liquids, etc. are provided for industry and research institutes. Advancements in measurement schemes for these evaluations are also under process (Figs. 11.22 and 11.23).

11.3.3.4 RF Transparency and Shielding Effectiveness for "Covering Sheets" for Strategic Sector Industries

CSIR-NPL has provided Shielding effectiveness measurement for manufacturer of covering sheets for defence personal and vehicles. These are useful for anti-radar and stealth operation (Fig. 11.24). The measurements are done using two port free space method for effective shielding by the sheet for signal of particular frequency in microwave range in *X*—band and Ku band and *K* band using horn antennas at a RF noise-controlled room noise floor below −80 dB. Small samples can be tested in the waveguide systems for exact RF transparency with ±3% measurements same can be done in the anechoic chambers for these types of sheets. But to feel real situation

Fig. 11.25 RF transparency measurement using open air antenna method for X—band Frequencies (Front and back view)

an indoor environment with several reflectors along with working instrumentation, RF sources and multiple reflections is being used on even tilted sheets to evaluate its shielding performance of these sheets in best—and worst-case scenario.

11.3.4 R&D in Microwave Metrology

11.3.4.1 Research for Quantum E-Field Measurement

Recently, International Bureau of Weights and Measures (BIPM) has decided to redefine all the basic SI units in terms of the physical constants which describe the natural world. This decision, initiated research in all branches of science to make all the derived units traceable to either SI units or to physical constants. The metrological institutes across the world have proved that the atom-based metrological techniques provide more accurate, precise and reproducible results as compared to the conventional techniques. The atom-based standards have been accepted worldwide for measurement of time, frequency, gravity, magnetometry, and length. To date, the RF E-field sensing and calibration techniques does not have traceability to physical constants and have complex traceability path. The dependency on the materials under use, limitation of bandwidth of antennae under use, the surrounding parameters at the time of measurement, and the incorrect evaluation of uncertainties are some of the limitations of current techniques which makes measurable quantity vulnerable to the errors. These measurements will lead to establish the quantum communication and quantum sensing calibration facilities at CSIR–NPL for upcoming smart communication devices.

The technique discussed in this section has many promising benefits over the conventional techniques such as the dependency of dipole antennae (having limited bandwidth of operation) as a reference probe can be replaced by atomic cell which can sense RF E-fields from few MHz to 500 GHz and beyond. The atoms provide self-calibrating measurements due the invariance in atomic resonances, enhanced sensitivity can be achieved of the order of 1 μV m^{-1} Hz$^{-1/2}$. The field will not be perturbed by the metallic probes as the field is sensed by atomic transitions in this technique.

This technique is based on the interaction of the highly excited alkali atoms with RF energy and utilizes the concepts of quantum interference phenomena like Electromagnetically Induced Transparency (EIT), and Autler-Townes Splitting (ATS). The Rydberg atoms have a huge response to the external E-field which helps in converting the E-field amplitude measurement to the optical-frequency response. The discussed work may find its application in accurate measurement of E-field amplitude which plays a crucial role in the determination of many other parameters such as Specific Absorption Rate (SAR) in biomedical application, EMI/EMC testing of RF and electronic devices, non-invasive RF based detection and diagnostic medical devices, etc.

To measure microwave amplitude via Rydberg atoms a Rb vapor cell is placed as shown in the schematic of the experimental setup in Fig. 11.25a. The counter-propagating linearly co-polarized laser beams interact with the atoms in the presence of microwave (MW) illuminated from the transverse direction to the line of direction of propagation of laser fields by any microwave source with the help of antenna. The weak probe laser of ~780.24 nm tunes the ground state $5S_{1/2} \rightarrow 5P_{3/2}$. Another laser of ~480.044 nm couples the atomic states $5P_{3/2} \rightarrow nD_{5/2}$ (any desired Rydberg state with principal quantum number 'n') and will be referred to as control laser throughout this section. Both the lasers should be locked so that they should not drift with time. The transmission and absorption of the probe laser is continuously monitored with photodetector. Figure 11.25b shows the atomic ladder model depicting the transitions.

Experimental setup for atom-based E-field sensing is shown in Fig. 11.26. The cylindrical cell placed at top of white platforms act as an antenna. At specific conditions mentioned in [29] the absorption profile of probe beam splits in two peaks giving a transparent window at centre of absorption which is known as EIT. The

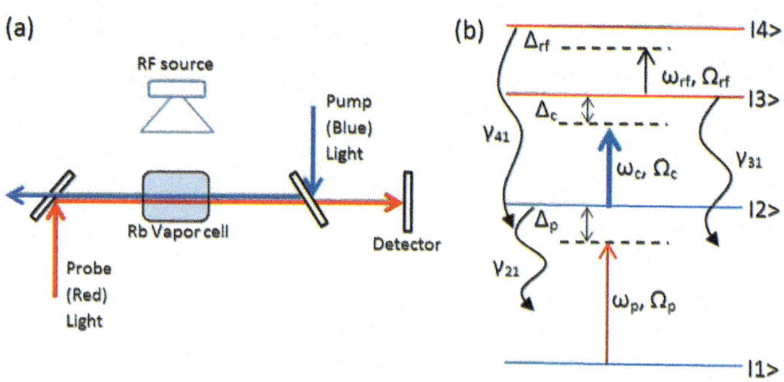

Fig. 11.26 **a** Representational setup for the atomic transitions-based E-field metrology; **b** Energy level structure of the atom

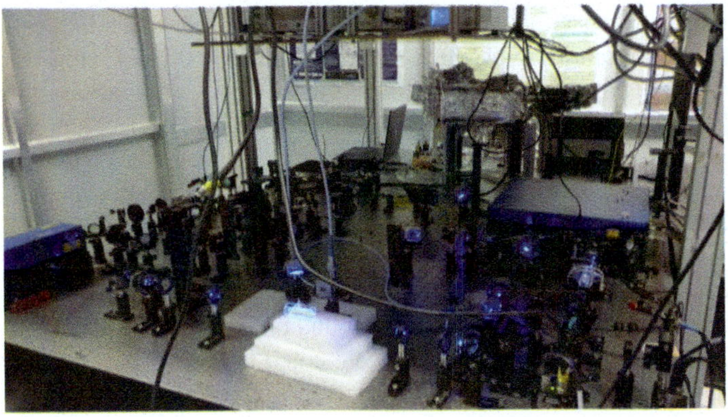

Fig. 11.27 Experimental setup for RF E-field strength measurement using Rydberg atom-based quantum sensor (this setup is at NISER, Bhubaneswar; CSIR-NPL and NISER have worked in collaboration for initial set of experimental data)

Radio-frequency creates additional interferences resulting in increase in difference between the two split peaks, known as ATS in EIT regime [30]. This behaviour of absorption profile is utilized in RF E-field strength measurement. As the power of incident microwave increases, the split between the two peaks increases which gives us the E-field amplitude by Eq. 11.15:

$$E_\mu = 2\pi \Delta f \frac{\hbar}{d_\mu} \tag{11.15}$$

where $\hbar$ is the Plank's constant, d_μ is the dipole moment of microwave transition, and Δf is the difference between the peaks.

In addition to E-field amplitude measurement, the vector RF E-field determination can also be performed using the same atomic sensor. Rawat et al. [31] have shown that the orientation of microwave E-field vector changes the nature of probe field absorption.

11.3.4.2 Vector SAR Assessment System and Probe Calibration

In the current age of fast evolving communication and being present online virtually always, has made us surrounded by communication equipment everywhere, be it mobile phones, tablet and other wireless devices and the upcoming IoT (Internet of Things) based wearable devices, smart homes, household equipment to smart sensors. We are dependent on these devices for all our day to day work and these are always radiating microwave of other RF radiation. International Standards require that each of these wireless devices to be tested individually, so as to ensure its radiation is within the specified and internationally accepted "safe limit" [32]. For this these devices need to be tested at each individual frequency on which it operates and for all possible configurations in which the devices may be used. However, this is a tedious and time consuming task to test all these devices, as per existing standard namely IEEE-1528 (2013) [33] and ICNRP guidelines [34]. Also, with continuously upgrading communication technology and increasing number of devices, the task becomes more complex.

Regularity bodies have been working to upgrade these standards and trying to come up with faster assessments methods to accommodate these multiple tests and such large number of devices. Suggested methodology is to make use of time domain probes and arrays which map the amplitude and phase simultaneously and then map the E-field generated by the device. These method uses time-domain Vector probes and MIMO antennas for faster measurements by employing simultaneous amplitude and phase measurements [35]. However, before making these measurements acceptable worldwide the standard for calibration of these probes needs to be established. CSIR-NPL has joined the EURAMET Program 2017 as member in collaboration with LNE, France and other participating NMIs. This program is launched by NPL, UK to address these issues and finally come up with guidelines, procedures and methodologies for SAR assessment using vector probes and their calibration schemes. CSIR-NPL, India has collaborated with coordinator (LNE) to validate their individual measurements and procedures. This collaboration will lead towards comprehensive methods, software tools and datasets required for traceable calibration and uncertainty analysis of vector probe array systems. It will also help incorporate new SAR assessment procedures in the upcoming standards (IEC 62,209–1528/D5 2019 [36] and the latest IEEE Std. C95.1 2019 [37]). These upgradations of existing SAR standards are essential to address the technological change upcoming with smart wireless technologies.

This project aims to address this issue and establish national facility for vector SAR measurement (First of its kind in country) with measurement equivalence to world's leading NMI for IEC 62,209–3 TC. This will serve as a precise calibration system for the Vector SAR assessment systems proposed. With India emerging as Second largest producer of mobile equipment in recent 2–3 years, there is huge scope to provide SAR assessment to the manufacturers in country itself and save lot of time and cost by having testing at par with new international standards. Microwave Metrology has taken this opportunity and responsibility, being the NMI of India.

11.4 Microwave Metrology for Quality Infrastructure in India

Metrology forms an integral part of Quality Infrastructure which in turns results in High Quality of Life as represented in Well-functioned quadruple helix (QH) [38]. To usher into the phase of 'Aatmanirbhar Bharat', i.e., Complete Self-Reliance, we must utilise of resources sustainably to their maximum potential. The gap in the potential growth and actual growth achieved is often due to weak Quality Infrastructure.

11.4.1 Metrological Services for Sectorial Growth

11.4.1.1 Calibration and Testing

Microwave metrology is a promising prospect for private players given the global wireless testing market is expected to grow at a CAGR of 6.7% to USD 14.4 billion by 2024 from USD 10.4 million in 2019. Recent advancements in fields of 4G, 5G, IoT and Smart sensors are pushing the expansion of testing market by adoption of manufacturing techniques that are virtually human intervention-less at the same time maintaining high quality of products [39].

Microwave Metrology offers calibration and testing services for numerous parameters for various types of items spanning across RF attenuation, Reflection coefficient, E-field, and Shielding Effectiveness etc[25]. Special price packages can also be negotiated based on the bulk orders of the customer or based on the custom request from the clients (Table 11.7).

11.4.1.2 Technological and Technical Consultancy

CSIR-NPL can use its expertise in the niche area of metrology in form of guidance and provide consultancy services to various players i.e., Government Bodies, Autonomous Bodies, Research Institutions, Private Industries etc.

i. ***GTEM with E-Field Sensor Under MPL (Major Lab Project) 160,432***

Radiations from mobile towers, Mobile Phone, Wi-Fi router and Microwave Oven can cause harmful effect on human being which need to be limit as per their radiation is concern. Government of India has made several guidelines regarding maximum radiations from mobile towers and electronic gadgets. To make these measurements traceable in country, CSIR NPL have developed GTEM along with E-Field sensor which is traceable to SI unit. This system will be utilized for the calibration of any E-Field probe or sensor used in wireless communication or Mobile Tower Radiation Measurements. It is a compact, lightweight and small volume system and word wide very few National Metrology Institute has developed this facility. CSIR-NPL India has developed a GTEM cell with E-field sensor based on IEEE Standards 1309–2013. The E-field strength standard is based on Gigahertz Transverse Electromagnetic (TEM) cell. The uncertainty in E-field strength is ±0.58 V/m per 10 V/m for different frequency in the range from 300 MHz to 8 GHz. Indigenous E-field probes are also fabricated for Ultrawide band(3–8 GHz), GSM frequencies (700–950 MHz, 1.8–2.45 GHz).

Table 11.7 Various calibration and testing parameters for various Item types at microwave metrology [25]

S. No	Parameter	Item name	Item type	Alias name	Range
1	Attenuation	Coaxial	Attenuator	Attenuation pad	0.1–60 dB
2	Attenuation	Waveguide fixed attenuator	Attenuator	Attenuation pad	0.1–60 dB at 5.8–18 GHz
3	Attenuation	Coaxial step attenuator	Attenuator	Variable attenuator	0.1–60 dB at 30 MHz and 1–40 GHz
4	Attenuation	Rotary vane attenuator	Attenuator	Variable attenuator	0.1–60 dB at 5.8–18 GHz
5	Attenuation	30 MHz WBCO attenuator	Attenuator	Standard piston attenuator	0.05–60 dB at30 MHz
6	Attenuation	Coaxial directional coupler	Coupler	–	0.1–60 dB at 30 MHz and 1–18 GHz
7	Attenuation	Waveguide directional coupler	Coupler	–	0.1–60 dB at 5.8–18 GHz
8	Attenuation	Dual directional coupler	Coupler	–	0.1–60 dB at30 MHz and 1–18 GHz
9	Attenuation	Attenuator and signal calibrator	30 MHz Receiver	Receiver with built-in 30 MHz attenuator	0.1–60 dB at30 MHz
10	Attenuation	Power splitter	Splitter	Power divider	0.1–60 dB at30 MHz and 1–18 GHz
11	Reflection coefficient/VSWR	Coaxial mismatch	Termination	Coaxial termination	1.02–5 at 2–18 GHz
12	Reflection coefficient/VSWR	RF Transfer standard	Sensor	Coaxial thermistor mount	1.02–5 at 2–18 GHz
13	Attenuation	VSWR Meter	1 kHz Receiver	RF Ratiometer	0.1–30 dB at 1 kHz

(continued)

Table 11.7 (continued)

S. No	Parameter	Item name	Item type	Alias name	Range
14	Reflection coefficient/VSWR	SWR bridge	Bridge	–	0.1–60 dB at 1–18 GHz, 1.02–5 at 2 -18 GHz
15	Reflection coefficient/VSWR	Waveguide standard mismatch	Termination	Waveguide match load	0.01–0.2 at 5.8–18 GHz
16	Reflection coefficient	Coaxial short/open circuit	Reflection standard	Offset short circuit	0.96–0.99 at 2–18 GHz
17	Reflection coefficient/VSWR	Coaxial airline	Impedance standard	Beadless Airline	0.01–0.05 at 2–18 GHz
18	Phase	Rotary vane phase shifter	P hase shifter	Variable phase shifter	0–180 degree (X-Band)
19	E-Field (V/m)/ radiated power density (mW/cm2)	Probe	Antenna	Microwave leakage detector	Probe as per TEC Norm TEC/TP/EMF/001/01.SEP-2009
20	Shielding effectiveness	MRI tester	Planar material	RF transparency sheet	-

ii. ***Rydberg Atom-Based Quantum-E-Field Standard Sponsored by R&S India Pvt. Ltd.***

CSIR-NPL is continuously working towards the realization of RF E-field strength measurement by utilizing the characteristics of Rydberg alkali atoms. This novel technique relies on the quantum phenomena like Electromagnetically Induced Transparency (EIT) and Autler-Townes splitting (ATS) which are observed in Rydberg alkali atoms under the continuous exposure of laser and RF fields. The idea is that the atoms inside an atomic vapor cell act as antennae (under controlled conditions) which can sense the fields ranging from few MHz to 500 GHz and above. The numerical and analytical simulations of the atomic model have been successfully realized at CSIR-NPL. The experimental results have also been successfully realized by working in collaboration with NISER, Bhubaneswar. The fully functioning facility once established at CSIR-NPL will provide calibration for the E-field probes with traceability to the physical constants such as Plank's constant and atomic constants. Rohde & Schwarz India Pvt. Ltd. (R&S) has participated with CSIR-NPL in Prime Minister Fellowship for Doctoral Scheme by funding a Ph.D. scholar in this project.

iii. ***Establishment of Conducted and Radiated Emission and Electrostatic Discharge (ESD) as a Part of National Facility Creation Project for EMI/EMC Standards***

Conducted and Radiated Emission standard for EMIEMC compliance as per CISPR 16–2, CISPR 14–1 and CISPR 14–2 are carried out on any household electronic device with a constant signal strength of 10 V/m at different modulation depth for a frequency range of 80 MHz–6 GHz at 10 V/m in CSIR-NPL. In addition, ESD Immunity test as per IEC 61,000–4–2 are also conducted at CSIR-NPL. To calibrate ESD against CSIR-NPL's primary standard an indigenous instrumentation setup is under development which will have traceability from AC-DC difference measurement of Electromagnetic Metrology.

11.4.1.3 Training and Human Resource

Microwave Metrology is also engaged in wide range of human resource related activities in various areas of core competence of the laboratory in terms of training for various stakeholders of the industry and academia from Technical Staff to research scholars/students not only from India but also from SARC nations, gulf countries and African countries as well.. The basic objective behind training and human resource activities is to generate a competitive, productive and useful pool of human resource who are better equipped scientifically, technically and administratively to render their services for the society and the country.

Numerous Industrial Training Programs (ITPs) are organized at CSIR-NPL for various fields of Microwave Metrology. The training program generally encompasses various physical parameters in the area of Metrology/Standards, as well as on other specialized topics. These programs are primarily meant for the personnel belonging to various industries, Testing and Calibration laboratories and other S&T organizations such as ERTL, ETDC, Defence Labs and Industries. The training programs consist of theory lectures on various scientific and technical aspects of the training course including different types of microwave measurement techniques, associated uncertainty followed by practical demonstration and hands-on training on the related instruments/apparatus / machines including uncertainty calculations and ISO17025 Quality system related to microwave metrology. The participants are generally B.E./B.Tech. or M.Sc. as well as Science Graduates or Engineering Diploma Holders. Microwave metrology also engages in summer training programs and internships to undergraduate pre-final year students, graduate training programs along with guiding PhD students.

11.4.2 Future Metrological Road Map

With Ministry of Civil Aviation planning modernisation of Air Traffic Control, Automation of Air Traffic Services and Air Navigation Systems through GPS Aided Geo Augmentation (GAGAN) Project aiming at efficiency and optimum utilisation of the air space, Microwave Metrology can provide a strong technical background for augmenting policies like 'AAI Start-up Policy' and Public Procurement policy linked with Local Content (PP-LC) [40 , 41].

Similarly, with respect to telecom sector, Telecommunication Engineering Centre (TEC), is in process to join hands with CSIR-NPL to develop indigenous testing instrumentation for mobile phones and wearable devices based on IEEE-1528 2013 and IEC-62209–1,2 standards. In area of 5G and advance communications, CSIR NPL is also a member of EURAMET consortium to develop vector SAR measurement system in the country for the conformity assessment of IEC 62,209–3, 2019 and upcoming modified IEEE 1528 standards which have to be implemented in 2022. Conformity assessment to each and every telecom equipment, either imported or manufactured in India, is essential to ensure quality of these products by TEC under DoT with the help of conformity assessment (CAB) labs. Microwave metrology will not only act as conformity lab for quality assurance, but it will also establish the national measurement traceability for testing equipment used by CABs. Apart from that, with upcoming vector SAR measurement facility, EMI/EMC National Facility and indigenous instrumentation for quantum sensing will address the interconnectivity among various devices within the safe RF exposure limits as per Fourth industrial revolution.

In area of connected devices, around 24 billion devices are projected to be present in the world by around 2020 end according to GSMA (GSM Association) and Machina Research, World's leading advisor on M2M(Machine to Machine), Internet of Things(IoT) and Big Data, with a combined business impact of around US $ 4.3 trillion. This projection was further increased later by CISCO/Ericsson / ITU to be around 50 billion [42]. With such huge projections, private players and industries are attracted towards them to explore and exploit wide range of opportunities that the M2M communication/IoT concept offers. This in turn enables novel business cases, efficient and improved quality of life. Such networked devices do find their application across various sectors such as Automotive, Power, Health care, Safety and Surveillance, Water management, Waste management, Smart homes, Environment monitoring and pollution control, Intelligent buildings, Smart Cities etc. In Indian context alone, there may be around 2.6 billion connected devices by 2020. Power sector (smart metering and smart grid) and Automotive Sector will have major deployment of M2M devices. CSIR-NPL is continuously striving to establish a Quality Infrastructure to cater to the metrological needs of IoT/M2M Industry by constantly striving to develop new and relevant standards.

With reference to EMI/EMC testing, a national facility having national primary standards with international traceability for immunity to electrostatic discharge, immunity to radiated RF, immunity to conducted disturbance induced by RF, immunity to fast transients (burst), immunity to surges are being established at CSIR-NPL. This established facility will not only save foreign exchequer but also cater to different standardization needs of the country by disseminating the traceability chain and fulfil broad industrial requirements.

To address the above-mentioned diverse sectors, these sectors can be classified as per their precise measurement requirements and based on those measurement requirements, sectors can be broadly divided into the following types of standards based on their applicability:

(a) CISPR standards
(b) IEC standards
(c) ISO standards
(d) SAE Electromagnetic Compatibility (EMC) Standards committee
(e) European standards concerning unwanted electrical emissions
(f) European standards concerning immunity to electrical emissions
(g) American standards (FCC, MIL-STD 461, MIL STD-464, MIL-STD 469).

In above two sections a detailed discussion is being carried out on diverse sectors and their classification for their standards based on their applicability. In Indian scenario, these sectors belong to ministries and which in turn have regulators to monitors the current situation and to implement policies. As a part of India's NMI, microwave metrology is capable to equip itself with any international standards as and when required by regulators and/or ministries. CSIR-NPL is constantly in the pursuit of developing similar and relevant standards suitable for Indian conditions and thus will be the torchbearer for a successful Make in India program.

In development of these standards suitable to Indian conditions, some of the stake holders, but not limited to, are as follows:

(a)	Telecommunication electronics centre (TEC)	(i)	Ministry of electronics and information technology (MeitY)

(continued)

(continued)

(b)	Telecom regulatory authority of India (TRAI)	(j)	Telecom equipment manufacturer association of India (TEMA)
(c)	Telecom enforcement resource and monitoring (TERM)	(k)	BEL (Bangalore, Chennai, Panchkula, Ghaziabad, Kotdwar)
(d)	Department of telecommunications (DoT)	(l)	Cellular Operators Association of India
(e)	Wireless planning and coordinate wing	(m) (n)	ETDC (Bangalore, Chennai) DEAL, Dehradun
(f)	Prasar Bharti	(o)	ERTL (North, South, East, West)
(g)	Airport authority of India	(p)	SAC, Ahmedabad
(h)	All India Radio	(q)	13 BRD, Palam

The future metrological roadmap of CSIR-NPL can be better visualised in terms of support to various sectors and various National Missions and programs CSIR-NPL is being a part of.

11.4.2.1 Sectorial Support

Microwave metrology provides a strong base for support to many national plans spread across various ministries and departments such as Directorate General of Civil Aviation, Department of Space, Department of Telecommunication, Ministry of Electronics and Information Technology, Defence Ministry, to name a few.

i. ***Civil Aviation***: Despite the incredible growth seen in last decade, India continues to be a small player in the international arena. The per capita trips still remain very low in India at 0.04 even by the standards of other emerging markets, such as China (0.15), Brazil (0.25) and Malaysia (0.54). China's domestic traffic is 5 times that of India's despite having a population a mere 15% larger [43]. Domestic Traffic can be further augmented with the transit traffic to increase trips per capita. This can be achieved by developing India as an aviation hub between Dubai and Singapore. According to SWOT analysis, by Ministry of Civil Aviation, one of the key weakness to achieve this potential include Lack of R&D; high dependence on foreign suppliers [43]. CSIR-NPL as an NMI can play a pivotal role in addressing this challenge by providing high quality conformity assessment.
ii. ***Electrical and Electronics Manufacturing Industry***: Post COVID-19 situation, when a major chunk of manufacturers is keen on relocating their manufacturing units away from China, many countries are eying for setting up their new base of operations in India. India can attract such investments in India by means of favourable tax structures, cheaper land lease, Ease of doing Business etc. NITI Aayog CEO Amitabh Kant's committee has been setup to come up with suggestions (on lines of Vietnam model) to reduce India's 'disability'—or higher production costs—in comparison with China and Vietnam [44]. However, an

important issue still remains to be addressed is that of Quality of the manufactured products. Many such shortcomings are reflected in data of electronics exports and imports. Electronics exports jumped over 38% year-on-year in FY19, but this growth has been majorly on a low base thus still affecting trade balance unfavorably. According to DGIS data, in FY19, India's electronics export was around $8.4 billion whereas electronics imports grew by approximately 8% to $55.5 billion. Additionally, since most of the industry works on the model of importing components completely or semi knocked-down conditions and shipping them out after assembly, local value addition is very low at about 15–18%, according to industry executives [44]. The lack in growth as compared to competitor nations such as Vietnam (Fig. 11.27) can be attributed to lack of Quality Control Environment. Similar trend has been observed in Export of Telecom Equipment as well (Fig. 11.28).

iii. ***Telecommunication***: A positive step has been taken in this direction by Department of Telecommunication. Prior mandatory testing and certification must be undergone for every telecom equipment, as provided by The Indian Telegraph (Amendment) Rules, 2017. The testing is to be carried out by Indian Accredited Labs, which are designated as CAB (Conformance Assessment Body) by TEC

Fig. 11.28 Electronics export of India vis a vis Vietnam [46]

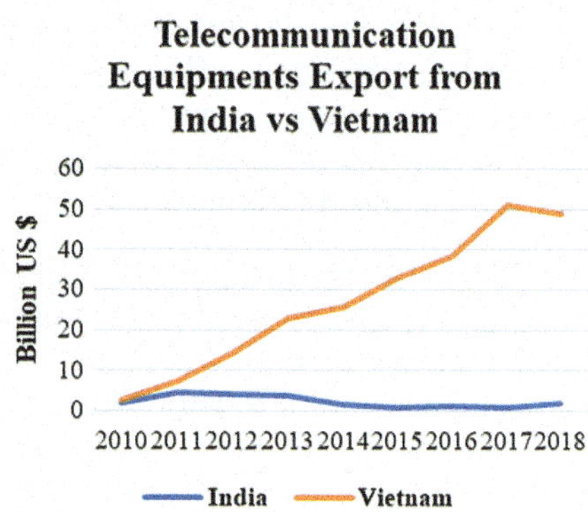

Fig. 11.29 Telecommunications equipment export India versus Vietnam [47, 48]

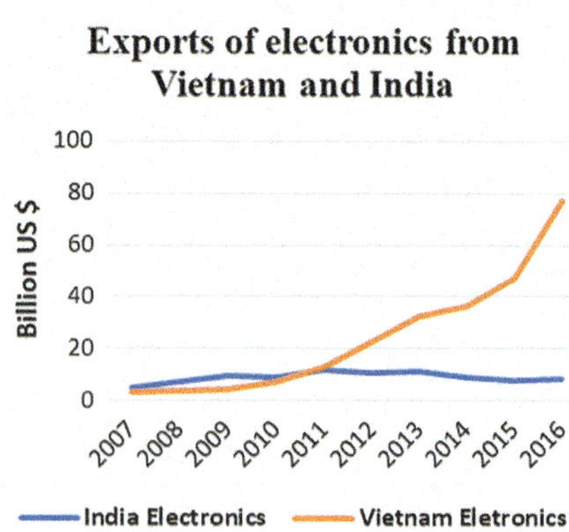

and TEC shall issue certificate based upon their test reports in conformance to essential requirements for the equipment. Department of Telecommunication's (DoT) Technical wing is ISO 9000:2008 certified, Telecommunications Engineering Centre (TEC) [45].

TEC has been identified as the nodal department for implementing the provisions related to procurement of goods, services or works related to the telecommunication sector by Department of Industrial Policy and Promotion (DIPP) [49]. TEC provides technical support to DoT and other government departments and formulates Generic Requirements (GR) for telecom equipment, networks, systems and services to be deployed in Indian Telecom Network, on the basis of International Standards [49]. TEC has published list of Essential requirements under Mandatory Testing and Certification of Telecom Equipment (**MTCTE**), test procedure for EMF measurement, Electromagnetic Interference and Electromagnetic Compatibility (EMI/EMC) compliance testing and so on. Addressing the safety concerns related to microwaves, and for the safety of the public, DOT has prescribed a certain EMF exposure from base station installations in accordance with International Commission on Non-Ionizing Radiation Protection (ICNIRP) guidelines. To give effect to this, TEC has published 'Test procedure for measurement of electromagnetic fields from base station antenna', which serves as detailed procedure for the certificate of compliance of EMF exposure norms by the Telecom Service Providers (TSPs) and audited by the Licensed Service Area (LSA) Units of the Department of Telecommunications (DOT) [50]. Under MTCTE (Mandatory Testing and Certification of Telecom Equipment) program from Department of Telecommunications (DoT) it is mandatory to test and certify all telecom equipment which are produced in India or imported from abroad. In this scenario CSIR-NPL will prove to be an important collaborator of TEC to establish a Quality Infrastructure in India in the field of telecom equipment testing and certification.

iv. ***Electronics***: National Policy on Electronics (NPE) envisaging a US$ 400 billion electronics manufacturing industry in the country by 2025 [51]. Through this Policy India is envisioned to be positioned as a global hub for Electronics System Design and Manufacturing—(ESDM) by encouraging and driving capabilities in the country for developing core components, including chipsets, and to create an enabling environment for the industry to compete globally. Simultaneously, NATRiP (National Automotive Testing and R&D Infrastructure Project) is setting up state-of-the-art Testing [52], Validation and R&D infrastructure in the country for Automotive Sector synergizing Make in India and FAME II (Faster Adoption and Manufacturing of Electric Vehicles) efforts.
v. ***Defense***: Role of Quality Infrastructure becomes more important keeping in mind the new Defence Procurement Procedure (DPP) and policy 2020, released by The Ministry of Defence (MoD) [53]. In line with government's "Make in India" initiative, in a bid to reduce reliance on imports and targeting to boost indigenous defence capability, DPP amends existing military procurement rules

accordingly. It has proposed enhancement of indigenous content requirements in foreign procurements too.

11.4.2.2 National Mission and Programs

Following national program can be supported by microwave metrology based on its technical capabilities:

i. ***Aviation***: Keeping in mind the vast market potential of civil aviation, Airport Authority of India (AAI) has recently set up a Civil Aviation Research Organization (CARO) at Hyderabad to significantly boost in-house research capabilities within AAI [40]. On similar lines, AAI is planning to establish Test Equipment Calibration lab at Hyderabad. CSIR-NPL, which is an apex body in this field of calibration, can play an active role by having a collaboration to help AAI establish and get NABL accreditation for the Test Equipment Calibration lab being established at Hyderabad to meet the calibration requirements of AAI Communication, Navigation and Surveillance (CNS) test equipment to comply with the DGCA's Civil Aviation Requirements (CAR). AAI calibration lab at Hyderabad to the extent that all equipment used by 29 ATC centers across India could have measurement traceability through NMI including ISO/IEC 17,025:2017 accreditations from NABL, SOPs for Calibrators and calibration process, Inter laboratory comparison (ILC)/ Proficiency Testing (PT) on established parameters to build confidence in measurement with manpower training to operate, control and calibrate instrumentation from the lab.
ii. ***Telecommunication***: MTCTE program of TEC is based on CABs which are private partners under TEC (PPP model). Success of MTCTE program will be highly dependent (Fig. 11.29) on TEC vigilant procedures. However, as NMI of India, microwave metrology can play an essential role to implement this program successfully with its measurement capabilities on those parameters. Microwave metrology will not only monitor the measurement capabilities of these designated CABs but also provide them with assistance in terms of measurement traceability, trained manpower, and surveillance, thus saving foreign exchequer. A surveillance mechanism forms an essential component to check the quality of products on random basis under MTCTE program and also under proposed model lab of TEC. A flow diagram for traceable surveillance is provided in Fig. 11.30.

CSIR-NPL can plan together with TEC regarding CAB Traceability with following technological output:

(a) Tissue Equivalent Liquid required for Specific Absorption Rate Testing and Calibration will be ready can be prepared by CSIR- NPL under BND Program
(b) Gigahertz Transverse Electromagnetic Cell and Chambers can be developed exclusively for testing of EMI/EMC Measurement and Certification for existing and upcoming smart telecom instrumentation

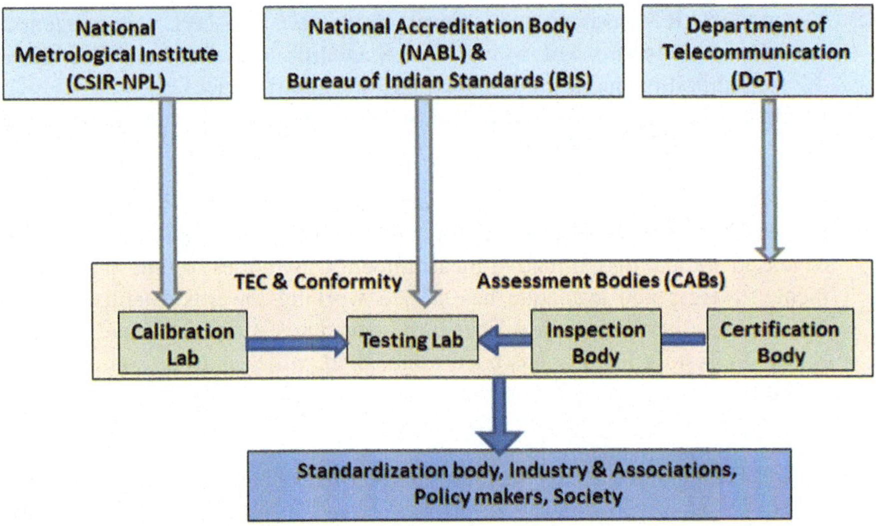

Fig. 11.30 Role of CSIR-NPL as NMI in Indian telecom quality infrastructure under MTCTE program

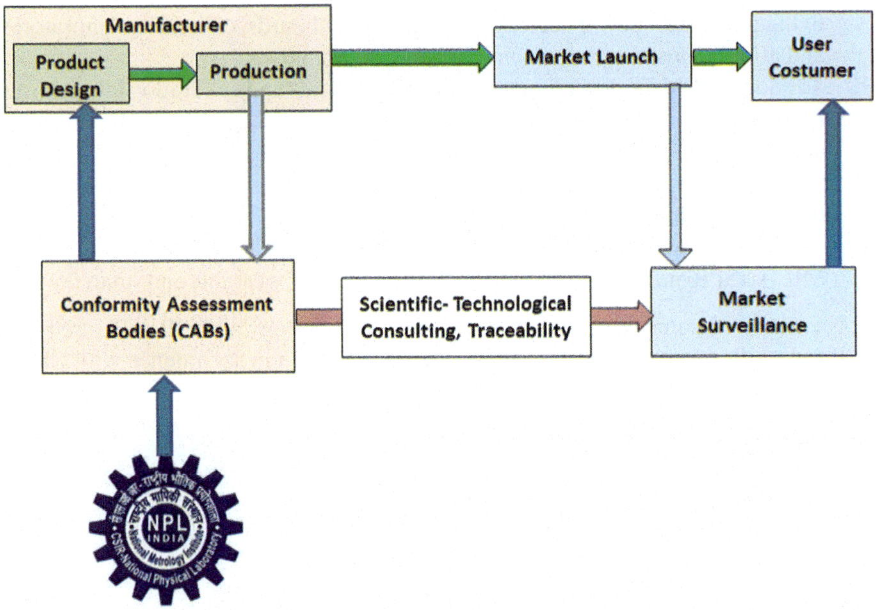

Fig. 11.31 Role of CSIR-NPL as NMI in the legal metrology and market surveillance

(c) Complete SAR Measurement System assembled at CSIR–NPL with indigenous SAR Probe, Software, TEL as per IEEE Standards 1528–2013 and IEC62209-1,2 for calibration and testing of wearable and smart devices.
(d) Complete E-Field and Radiated Power Density traceable measurement for accurate radiation exposure from Mobile Towers (Fig. 11.31).

iii. ***EMI/EMC National Facility for Primary Standards***: CSIR-NPL in its endeavor towards India's prosperity and quality of life through continuous research and development and disseminating measurement capabilities to industry, government, strategic and academia has started working towards establishment of National Facility for primary EMI/EMC Standards. Conducted and Radiated Emission Testing Facility has been established, which is to be soon followed by Electrostatic Discharge Testing facility and other parameters such as Surge, Burst, Harmonics, Flicker etc. within next 1 year.

National Facility for primary standards for EMI/EMC testing has the following objectives:

(a) To develop, maintain and upgrade national primary Electromagnetic Interference and Electromagnetic Compatibility (EMI/EMC) standards of international level and disseminate traceability of measurement through apex level calibration to the industries- power sectors, state electricity boards, calibration laboratories and allied industries in India and to SAARC countries.
(b) To add new CMCs as per requirements of industries and calibration laboratories.
(c) Participation in international intercomparisons with other NMIs to ensure compatibility with international standards.
(d) Add CMC for these parameters to BIPM so that international trade barrier can be removed to promote Make in India.
(e) All International Standards will be complying with these standards such IEC, MIL, CISPR and IEEE for calibration and testing of wearable and smart devices.

National Programs and policies of microwave metrology at CSIR-NPL are based on the classification of devices based on their application in accordance with CISPR (International Special Committee on Radio Interference):

(a) General purpose applications (Classes A or B, Group 1)
- Laboratory equipment
- Medical electrical equipment
- Scientific equipment
- Industrial process measurement and control equipment
- Semiconductor manufacturing equipment etc.

(b) ISM RF applications (Classes A or B, Group 2)
- Microwave-powered UV irradiating apparatus
- Microwave lighting apparatus
- Induction cookers
- Dielectric heating equipment
- Microwave ovens

Table 11.8 Standards as part of microwave metrology at CSIR-NPL

Test name	Standards
E-Field	IEEE 1309 -2013
Radiated power density	IEEE C95.1, 2019
Specific absorption rate (SAR)	IED 62,209–1,2,3
Conducted and radiated emission	CISPR 32 (2015) or CISPR 22 (2008) Class-B
Immunity to electrostatic discharge	IEC-61,000–4-2 (2008)
Immunity to radiated RF	IEC 61,000–4-3 (2010)
Immunity to fast transients (burst)	IEC 61,000–4-4 (2012)
Immunity to surges	IEC 61,000–4-5 (2014)
Immunity to conducted disturbance induced by radio frequency fields	IEC 61,000–4-6 (2013)
Immunity to voltage dips and short interruptions	IEC 61,000–4-1 1 (2004)

- Medical electrical equipment
- Electro-discharge machining (EDM) equipment etc.

Following Table 11.8 contains a non-exhaustive list of standards that are tested at CSIR-NPL or are at final stages:

iv. ***Defence***: CSIR-NPL aims to play the pivotal role through development of standards similar to MIL-STD of USA in field of EMI/EMC, SAR evolution, Antenna calibration, radome testing, anechoic chamber calibration, aerospace equipment testing E-Field probes and sensors calibration, and strategic material characterization so as to achieve the objective of making India self-reliant and a global manufacturing hub as envisioned by our honorable Raksha Mantri [54]. CSIR-NPL will continuously strive to support Indian government to develop the eco-system for indigenous defence production and thus empowering the Indian private industry including MSMEs. CSIR-NPL will be the catalyst for improving the international competitiveness for Indian Industries and thus give impetus to India's economic growth and realize our global ambitions.

11.5 Economic Impact

An International Civil Aviation Organization (ICAO) study attributes over 3.6% of global GDP to the air transport component of civil aviation [55]. In the Indian context, in last ten years, the civil aviation sector has grown at a tremendous pace and India has emerged as the 9th largest civil aviation market in the world [43]. According to another ICAO estimate, for every USD 1 of gross value added (GVA) directly created by the air transport sector, USD 3.8 of economic activity was supported elsewhere and approximately 6.4 jobs are dependent on every person directly employed in the aviation sector and in tourism made possible by aviation [55]. New industries/services for post covid-19 situation, wherein wireless tracking devices are going to play a major role for contact tracing especially for foreign nationals are the additional job creators linked to aviation sector.

The market for microwave/wireless devices encompasses IoT devices, Bluetooth devices, wireless telecommunication devices, Wi-Fi devices, RFID devices, smart sensors and so on, each witnessing an ever increasing CAGR over the last decade. With an expected growth of approximately 31 times from 0.06 billion IoT devices in India (base year of 2016) to 1.9 billion devices in 2020 end, it translates into a revenue opportunity of $9 billion by 2020 from the 2016 number of $1.3 billion, nearly 700% increment [56]. Government of India based on its draft IoT policy is targeting to create a $15 billion IoT industry in India by 2020 end [57]. To this end GoI has established Centre of Excellence for IoT in Bangalore in collaboration with software technology parks of India (STPI) with industry participation from companies such as Intel, NXP, Qualcomm, Texas Instruments, and others [58].

From the prospect of wireless devices providing internet, the number of internet subscribers in the country increased at a CAGR of 45.74% during FY06-FY19 to reach 636.73 million in 2018–19, second largest online market in the world, ranked only behind China. The internet subscribers reached 687.62 million till September 2019. Total wireless data usage in India grew 10.58% year-on-year to 19,838,886 terabytes between July–September 2019 [4]. However, the overall internet penetration is still at 31%, skewed in favour of urban India, which has ~70% internet penetration. Rural India is still underpenetrated with only 17% penetration. With a total rural population of ~906 million, approximately 750 million users still do not use internet [59]. Over the last few years, increase in internet penetration has been driven mostly by availability of cheaper smartphones. More than 92% of rural users and 77% urban users have been using mobile as primary device for accessing internet through cellular data connections [59].

National Primary Facility for EMI/EMC can help with required EMI/EMC certifications to broadly cater for Automobile Sector under FAME 2 (Faster Adoption and Manufacturing of Electric Vehicles in India Phase 2) [60], Telecom Equipment (under MTCTE) and electronic equipment especially IoT, robotics and drone-based devices. Currently, nationwide more than 300 labs are engaged in EMI/EMC work, with an expenditure from each lab of at least 30 Lakh INR annually on EMI/EMC Parameters Traceability ($300 \times 30 = 9000$ Lakhs). This is the result of majority of calibration laboratories across India using purely imported equipment and their traceability certificate is also imported due to lack of traceability on essential telecom parameters.

All these electronic industries need a dedicated, internationally traceable EMI/EMC standards and certification infrastructure to be export ready. This will give impetus to Make in India Initiative, FAME initiative by NITI Aayog, Export capability for Indian ESDM (Electronics System Design and Manufacturing) Sector by reducing technical barriers to trade, National Strategy for Artificial Intelligence and National Plan for Smart Electrical Metres, indirect employment generation and save at least 15 million USD foreign exchequer annually.

India has the 3rd largest armed forces and 10th largest Defence expenditure in the world. India spends around 1.5% of its GDP to Defence Budget (without the pensions) [61]. The Budget for Defence has increased by almost 5.8% to reach 3.37 lakh crore as the defence budget for 2020–21, the current financial year. India currently procures approximately 70% of its equipment needs from abroad. With a Compound annual growth rate (CAGR) of 18%, Indian Defence and Civil Aviation market would put the country, in very near future, among one of the top five Defence and Civil aviation markets. To strengthen the dedicated defence manufacturing corridors, internationally traceable EMI/EMC standards and certification infrastructure will not only make indigenously developed critical sub-components more reliable, but also export ready.

11.6 Conclusion

In the field of measurements in Microwave, CSIR-NPL deals with various parameters such as microwave power, attenuation, impedance, and shielding effectiveness for the frequencies ranging from 9 kHz to 110 GHz of electromagnetic spectrum. These parameters serve as a base for rapid progress in the field of electrical and electronics devices, and their interconnectivity that propel us towards Fourth industrial revolution. The concept of IoT driven wireless interconnectivity of smart homes, smart offices, smart cities and smart health care including upcoming quantum communication revolution will only be successful if CSIR-NPL will equip to provide traceability and certification on these smart technologies. Their applications are visible in almost every sector such as manufacturing industries, educational institutes, strategic sectors, biomedical devices, hospitals and households. Surprisingly, most of these advanced devices are imported into India and are not locally manufactured. In spite of government's all efforts to 'make India an electronics manufacturing hub which triggered several national programs and policies such as NEP (2019), MTCTE program of DoT, DPP policy (2020), Make in India policy and Aatmanirbhar Bharat, Indian manufacturers are still struggling to reduce import export ratio. The reason for such import imbalance lies with unavailability of high-end sophisticated instruments, which in turn is due to lack of quality infrastructure and cost ineffectiveness. Microwave metrology is capable to provide traceability, calibration and technical services to all manufacturing units in India associated with electrical, electronics and telecom via reference labs so that the quality of electronics systems could be enhanced and thus reduce the trade deficit.

Microwave metrology deals with defence, space, telecom and microwave driven industry related testing and calibration parameters as per IEEE, IEC and MIL standards to cater these sectors as per international norms. CSIR-NPL serves these sectors as per existing standards for microwave power, attenuation, impedance to E-Field, radiated power density and SAR; at the same time, it is upgrading itself in term of SI traceable Quantum E-field measurement, vector SAR and measurement standards for 5G so that future standardisation challenges could be addressed proactively. Microwave metrology is capable to provide technical/technological consultancy in the domain of E-field generators, E-Field sensors, SAR probes, tissue equivalent liquids and SAR measurement systems as per IEEE 1528 and IEEE 1309 standards. With regard to various national programs and policies, microwave metrology can play a vital role in MTCTE of TEC, NEP program of MeitY, ATC certification program of DGCA and defence calibration required for radiated EM parameters. Microwave metrology can provide technical and technological support in the form of indigenously designed and developed GTEM, tissue equivalent liquid and SAR measurement system. In area of advance telecom standards such as 5G and beyond, the microwave metrology at CSIR-NPL is part of European consortium and is in process to establish vector SAR measurement system indigenously so that Indian telecom industries could be served as per upcoming standards 62,209–3-2019 applicable from 2022.

Finally, the projected impact of Microwave metrology on India's GDP is around 12–18% and with its indirect contributions encompassing R&D support to the manufacturing sector, certification to the end products and saving foreign exchequer by developing various indigenous parameters, it could supplement the GDP by as much as 26%. Thus, Policy makers in active collaboration with CSIR-NPL for Microwave metrology at policy formulation stage can ensure a robust quality infrastructure for the nation within safe EM exposure limits.

Acknowledgements The authors would like to acknowledge all the ex-colleagues; present colleagues *Ms. Sunidhi Madan, Mr. Anurag Katiyar, Ms. Mandeep Kaur, Mr. Anish Bhargav, Ms. Archana Sahoo, Ms. Jyoti Chauhan, Ms. Swati Kiumari, Mr. Mange Ram and Mr. Amreek Singh; and students* of CSIR-NPL for their contributions in establishing Electromagnetic Metrology. We would also like to thank the stake holders Department of Legal Metrology, SAC-ISRO, PGCIL, STQC Labs, TEC-DoT, AAI, LM, M/s R&S India Pvt Ltd., M/s Valliant Comm. Pvt Ltd. and many others for their continuous support and collaboration in the National mission endeavours for Electromagnetic Metrology dissemination pan-India.

References

1. G. Elert, The Electromagnetic Spectrum, The Physics Hypertextbook (2016)
2. Business Standard, Focus on 'Make in India (2014). [Online] Available https://www.business-standard.com/article/government-press-release/focus-on-make-in-india-114092501206_1.html
3. Electronic Industries Association of India, Industry Overview: Growth of Electronics Sector (2019). [Online] Available https://www.elcina.com/overview.php
4. India Brand Equity Foundation, Indian telecom industry analysis. India Brand Equity Foundation, Mar (2020). [Online] Available https://www.ibef.org/industry/indian-telecommunications-industry-analysis-presentation
5. Economic Times, Appliances and consumer electronics industry expected to double to Rs 1.48 lakh crore by FY25. Nov (2019) [Online] Available https://economictimes.indiatimes.com/

industry/cons-products/durables/appliances-and-consumer-electronics-industry-expected-to-double-to-rs-1-48-lakh-crore-by-fy25/articleshow/72183954.cms?from=mdr#:~:text=Appliances%20and%20consumer%20electronics%20indust
6. Tradingeconomics.com, India exports of electronic goods (2020). [Online] Available https://tradingeconomics.com/india/exports-of-electronic-goods
7. Eletimes.com, Indian Electrical equipment industry records highest growth of 12.8% in seven years," 18 Jun (2018). [Online] Available https://www.eletimes.com/indian-electrical-equipment-industry-records-highest-growth-of-12-8-in-seven-years
8. G. Das, Make in India: consumer electronics manufacturing to grow at 18%. 7 May (2020). [Online] Available https://www.businesstoday.in/current/economy-politics/make-in-india-consumer-electronics-manufacturing-to-grow-at-18/story/390876.html
9. U. Mishra, PM Modi's Aatmanirbhar Bharat Abhiyan economic package: here is the fine print. 14 May (2020). [Online] Available https://indianexpress.com/article/explained/narendra-modi-coronavirus-economic-package-india-self-reliance-6406939/
10. Deity, Cabinet approves the proposal of national policy on electronics 2019, Feb (2019). [Online] Available https://pib.gov.in/PressReleaseIframePage.aspx?PRID=1565285
11. E. Bureau, "India unveils Rs 50,000 crore schemes to attract electronics makers," 3 Jun (2020). [Online] Available https://economictimes.indiatimes.com/tech/hardware/govt-starts-inviting-application-for-rs-50000-cr-electronic-incentive-schemes/articleshow/76153644.cms?from=mdr
12. CSIR-NPL, https://www.nplindia.in/data/npldjvsec.pdf. [Online]
13. S. Ahmad, M. Charles, D. Allal, P.S. Negi, V.N. Ojha, Realization of 2.4 mm coaxial microcalorimeter system as national standard of microwave power from 1 MHz to 50 GHz. Measurement **116**, 106–113 (2018)
14. R. Swarup, J.R. Anand, P.S. Negi, On 30 MHz TE11 mode piston attenuator. Rev. Sci. Instrum. **72**, 1858 (2001)
15. S. Dubey, N. Narang, P. Negi , V. Ojha, Microwave measurement systems. In: LabVIEW Based Automation Guide for Microwave Measurements, Springer Briefs in Electrical and Computer Engineering (Singapore, Springer, 2018)
16. EURAMET, Guidelines on the Evaluation of Vector Network Analysers (VNA) EURAMET Calibration Guide No. 12 Version 3.0 (Euramet, Germany, 2018)
17. R. J. Collier and A. Skinner, Microwave measurements vol 12. (IET , United Kingdom, 2007)
18. G. Monasterios, H. Silva, Key comparison SIM.EM.RF-K5b.CL: scattering coefficients by broad-band methods, 2 GHz–18 GHz — type N connector," Metrologia (2016)
19. N. Narang, S. Dubey, P.S. Negi, V. Ojha, Accurate and precise E-field measurement for 2G and 3G networks based on IEEE Std. 1309–2013. Microwave Opt. Technol. Lett. **57**(7), 1645–1649 (2015)
20. N. Narang, S.K. Dubey, P.S. Negi, V.N. Ojha, Design and characterization of microstrip based E-field sensor for GSM and UMTS frequency bands. Rev. Sci. Instrum. **87**(12), 124703 (2016)
21. Final report on supplementary comparison EUROMET.EM.RF-S25 (EURAMET project 819): Comparison of electrical field strength measurements above 1 GHz," Metrologia (2011)
22. M. Kanda et al., International comparison GT/RF 86–1 electric field strengths: 27 MHz to 10 GHz. IEEE Trans. EMC **42**, 190–205 (2000)
23. N. Narang, Computational Modeling and Measurement of RF Specific Absorption Rate—PhD thesis," (AcSIR, CSIR National Physical Laboratory, New Delhi, 2017)
24. BIPM, "KCDB BIPM," [Online] Available https://www.bipm.org/kcdb/
25. CSIR-NPL, New Delhi, "Calibration Charges: D6.02c, LF&HF voltage, current and microwave metrology," [Online] Available https://www.nplindia.in/calibration-charges-d602c-lf-hf-voltage-current-and-microwave-metrology-wef01042019
26. K. Patel, P. S. Negi and R. Swarup, "Final report on APMP supplementary comparison P1-APMP. EM. RF-S3: APMP comparison of 50 ohm coaxial mismatches," Metrologia **49**(1A) (2012)
27. C. Eio, Final report on key comparison CCEM. RF-K5b. CL (GT-RF/92–3): scattering coefficients by broad-band methods, 2–18 GHz, Type N connector. Metrologia (2000)

28. G. Qiulai et al., Final report on APMP attenuation key comparison APMP. EM. RF-K19. CL: attenuation at 60 MHz and 5 GHz using a type N step attenuator. Metrologia (2010)
29. H.S. Rawat, S.K. Dubey, V.N. Ojha, Distinction between double electromagnetically induced transparency and double Autler–Townes splitting in RF-driven four-level ladder 87Rb atomic vapor. J. Phys. B: At. Mol. Opt. Phys. **51**(15) (2018)
30. C.L. Holloway et al., Gord broadband rydberg atom-based electric-field probe for SI-traceable, self-calibrated measurements. IEEE Trans. Antennas Propag. **62**(12), 6169–6182 (2014)
31. H.S. Rawat, S.K. Dubey, V.N. Ojha, Polarization dependence of interferences inside rubidium atomic vapor governing microwave vector E-field metrology. JOSA B **36**(12), 3547–3554 (2019)
32. I. Standards, IEEE Std C95.6™ —IEEE standard for safety levels with respect to human exposure to electromagnetic fields, 0–300 GHz.," IEEE Standards (2019)
33. I. Standards, IEEE 1528-IEEE recommended practice for determining the peak spatial-average specific absorption rate (SAR) in the human head from wireless communications devices: measurement techniques. IEEE Standards (2013)
34. I. Standards, ICNIRP, "Guidelines for limiting exposure to time-varying electric, magnetic and electromagnetic fields (up to 300 GHz)," Health Phys. (1998)
35. A. Djamel, "SAR measurement Using Vector Probes Short Name: Vector SAR, Project Number: 16NRM07," (2017). [Online] Available https://www.euramet.org/research-innovation/search-research-projects/details/?tx_eurametctcp_project[project]=1478&tx_eurametctcp_project[controller]=Project&tx_eurametctcp_project[action]=show
36. I. Standards, IEC/IEEE draft international standard—recommended practice for measurement procedures for the assessment of specific absorption rate (SAR) of human exposure to radio frequency fields from hand-held and body-worn wireless communication devices (2019). [Online] Available https://standards.ieee.org/project/62209-1528.html
37. W.H. Bailey et al., Synopsis of IEEE Std C95.1TM-2019 IEEE standard for safety levels with respect to human exposure to electric, magnetic, and electromagnetic fields, 0 Hz to 300 GHz. IEEE Access **7**, 171346–171356 (2019)
38. D.K. Aswal, Quality infrastructure of India and Its importance for inclusive. MAPAN **35**(2), 139–150 (2020)
39. Wireless testing market by offering (Equipment, Services), technology (Bluetooth, 2G/3G, 4G/5G, Wi-Fi), application (Consumer Electronics, Automotive, IT and Telecommunication, Medical Devices, Aerospace and Defense) and region—global forecast to 2024 (2018). [Online] Available https://www.marketsandmarkets.com/Market-Reports/wireless-testing-market-128820153.html;www.marketsandmarkets.com.
40. Airport Authority of India, AAI Startup Initiativ–Innovate For Airports (2018). [Online] Available https://www.aai.aero/en/important-links/aai-startup-initiative-innovate-airports
41. Airport Authority of India, AAI draft policy for public procurement linked with local content (2017). [Online] Available https://www.aai.aero/en/corporate/resources
42. CISCO, The internet of things: how the next evolution of the internet is changing everything. [Online]. Available https://www.cisco.com/c/dam/en_us/about/ac79/docs/innov/IoT_IBSG_0411FINAL.pdf
43. Ministry of Civil Aviation, India, Strategic Plan. [Online] Available https://www.civilaviation.gov.in/strategic-plan
44. K.B. Pattanayak, R. Ranjan, Financial Express, 19 Aug (2019). [Online] Available https://www.financialexpress.com/economy/kant-panel-aims-for-vietnam-like-model-for-mobile-phone-exports/1679103/
45. Telecommunications Engineering Center, "Mandatory Testing and Certification of Telecom Equipments (MTCTE)." [Online] Available https://www.tec.gov.in/mandatory-testing-and-certification-of-telecom-equipments-mtcte/
46. S. Kantha, "Make-in-India lessons: why are vietnam's electronics exports ten times that of India's?." 22 May (2017). [Online] Available https://thewire.in/economy/india-vietnam-trade-exports

47. United Nation Conference on Trade and Development, Vietnam Exports. [Online] Available https://www.ceicdata.com/en/indicator/vietnam/exports-ict-goods;www.ceicdata.com
48. United Nation Conference on Trade and Development, India Exports. [Online] Available https://www.ceicdata.com/en/indicator/india/exports-telecommunication-equipment;www.ceicdata.com
49. Telecommunications Engineering Centre, "Public Procurement (Preference to Make in India) Order (2017)." [Online] Available https://www.tec.gov.in/public-procurement-preference-to-make-in-india-order-2017/
50. Telecommunications Engineering Centre, "Test procedure For Measurement of Electromagnetic Fields," Jun (2018). [Online] Available https://www.tec.gov.in/test-procedure-for-measurement-of-electromagnetic-fields/
51. Business Standard, "Cabinet approves proposal of National Policy on Electronics 2019," Business Standard, Feb (2019). [Online] Available https://www.business-standard.com/article/news-cm/cabinet-approves-proposal-of-national-policy-on-electronics-2019-119022000282_1.html
52. National Automotive Testing and R&D Infrastructure Project (NATRiP), Vision and Mission (2011). [Online] Available https://www.natrip.in/
53. P. DAS, "India's defence procurement policy 2020," Observer Research Foundation, Apr (2020). [Online] Available https://www.orfonline.org/expert-speak/indias-defence-procurement-policy-2020-old-wine-in-a-new-bottle-64673/#:~:text=The%20Ministry%20of%20Defence%20(MoD,%E2%80%9CMake%20in%20India%E2%80%9D%20initiative
54. Ministry of Defence, "Defence Procurement Procedure 2020," Press Information Bureau of India, Mar (2020). [Online] Available https://pib.gov.in/PressReleaseIframePage.aspx?PRID=1607400
55. International Civil Aviation Organization (ICAO), "Aviation Benefits Report 2019, International Civil Aviation Organization (ICAO)," Sept (2019). [Online] Available https://www.icao.int/sustainability/Pages/IHLG.aspx
56. Deloitte, Technology, Media and Telecommunications Predictions 2018—India edition (2018). [Online] Available https://www2.deloitte.com/content/dam/Deloitte/in/Documents/technology-media-telecommunications/in-tmt-predictions-2018-noexp.pdf
57. Department of Electronics and Information Technology, Draft Policy on Internet of Things (2016). [Online] Available https://www.mygov.in/sites/default/files/master_image/Revised-Draft-IoT-Policy-2.pdf
58. Economic Times, "STPI Announces IoT Centre of Excellence in Bengaluru," Dec (2019). [Online] Available https://economictimes.indiatimes.com/small-biz/startups/newsbuzz/stpi-announces-iot-centre-of-excellence-in-bengaluru/articleshow/72383263.cms?from=mdr
59. Internet and Mobile Association of India , Internet in India—An Iamai and Kantar IMRB Report (2016). [Online] Available https://cms.iamai.in/Content/ResearchPapers/a7d38c59-689b-4764-9c34-f90ec17f5a2c.pdf
60. Department of Heavy Industries, Ministry of Heavy Industries and Public Enterprises, FAME India Phase 2 (2019). [Online] Available https://fame2.heavyindustry.gov.in/content/english/11_1_PolicyDocument.aspx
61. Economic Times, "Marginal increase in defence budget could mar new acquisitions," Economic Times, Feb (2020). [Online] Available https://economictimes.indiatimes.com/news/defence/marginal-increase-in-defence-budget-could-mar-new-acquisitions/articleshow/73839394.cms#:~:text=Without%20the%20pension%2C%20the%20defence,crore%20is%20the%20revenue%20head

Chapter 12
Electromagnetic Metrology for Smart Technologies

Part II: Current, Voltage and Magnetic Metrology

S. K. Dubey, Saood Ahmad, C. K. Suman, Sandhya M. Patel, Sudhir Husale, Anurag G. Reddy, Anurag Gupta, and D. K. Aswal

Abstract In continuation to the Part I in the previous chapter, the Part II comprises of electromagnetic metrological parameters apart from microwaves, namely LF & HF voltage, LF current, magnetic metrology and Josephson voltage standard along with the application of Focused Ion Beam (FIB) in Metrology. Each parameter is first discussed in terms of its associated primary standards, their calibration and measurement capabilities (CMCs) and their degree of equivalence with the leading NMIs of the world, and on-going research for advanced measurements including quantum standards. Next, the impact of all the parameters on quality infrastructure creation at strategic, defense and industrial fronts is outlined. How ministries, regulators, manufacturers, industries, academia and research institution directly/indirectly benefit from these parameters in terms of calibration, testing, training, human resource

S. K. Dubey · S. Ahmad · C. K. Suman · S. M. Patel · S. Husale · A. G. Reddy · A. Gupta (✉) · D. K. Aswal
CSIR-National Physical Laboratory, Dr. K. S. Krishnan Marg, New Delhi 110012, India
e-mail: anurag@nplindia.org

S. K. Dubey
e-mail: dubeysk@nplindia.org

S. Ahmad
e-mail: ahmads@nplindia.org

C. K. Suman
e-mail: sumanck@nplindia.org

S. M. Patel
e-mail: patelsm@nplindia.org

S. Husale
e-mail: husalesc@nplindia.org

A. G. Reddy
e-mail: g.reddy.anurag@nplindia.org

D. K. Aswal
e-mail: dkaswal.npl@nic.in

D. K. Aswal (ed.), *Metrology for Inclusive Growth of India*,
https://doi.org/10.1007/978-981-15-8872-3_12

development and technical consultancy is discussed. The chapter addresses questions related to the issues such as the measurement challenges, effect on industrial certification and technical consultation, challenges for regulators to catch up with upcoming series of modified standards-associated with the advancement of technology in electromagnetic domain. The importance of evolution of a strong link among the ministries, regulators, NMI (CSIR-NPL) and industries vis-a-vis these EM parameters to enhance export, improve the import quality and on overall economic growth of the country and 'Aatmanirbhar Bharat' are brought out.

"What science has failed to notice is that the measurement has become more real than the thing being measured."

– R.A. Delmonico

12.1 Introduction

The general definition of full spectrum of Electromagnetic Metrology, and its economic impact through Electrical and Electronic industries was already introduced in Part I of the chapter. In continuation in the present Part II of the chapter, the following activities will be discussed in detail:

(a) LF & HF Voltage and LF Current and Phasor Measurement Unit Metrology
(b) Magnetic Metrology
(c) Programmable Josephson Voltage Standard
(d) Focused Ion Beam Application in Metrology

LF & HF Voltage and LF Current and Phasor Measurement Unit metrology deals with establishing, maintaining, and updating the national standards of these parameters. To maintain the international traceability, a number of international & bilateral intercomparisons under the umbrella of BIPM and APMP have been carried out. Apex level calibration services are provided as per ISO/IEC 17,025 standards and the measurement traceability is disseminated to the industries, STQC labs, ISRO, DRDO, BEL, 13 BRD Air Force, etc. At CSIR-NPL, PMU-CAL system traceable to its national standards carries out a comprehensive calibration of a PMU as per IEEE Standard C37.118.1 [1]. It plays a vital role in power distribution sector through the PMU calibration and measurement system. The total impact of power sector alone on the GDP is expected to be around 4.1%.

Magnetic metrology deals with maintaining and upgrading National Standard related to magnetic parameter like Magnetic flux Density, Magnetic flux, turn area of search coil, power loss measurement of electrical steel, through continuous development. Calibration and testing facility are provided for these parameters to industries and institutions as per ISO/IEC: 17,025 guidelines which are traceable internationally. Traceability is maintained and disseminated by means of Calibration of permanent Magnet, Calibration of AC/DC Gaussmeter, Calibration of Magnetic field indicator, Calibration of low-field magnetometer, Calibration of Helmholtz Coil, Calibration of Flux meter, Calibration of Turn area of search coil, Testing of AC/DC electromagnet, Power loss measurement of Electrical Steel, Magnetic Radiation Measurement, etc. It may be roughly estimated that the contribution of Magnetic metrology on the GDP is ~10% directly or indirectly.

Electromagnetic metrology also includes several metrology activities directly depending on microwaves for traceability, such as Josephson Voltage Standard, which is one of the Quantum standards established at CSIR-NPL. 'Programmable Josephson Voltage Standard' (PJVS) system, serves as the primary standard of voltage. This system also plays an important role in electrical metrology as it is used to disseminate unit 'Volt' throughout the nation to maintain traceability. The quantum accuracy of voltage levels in this system is derived from the 'Josephson Effect', due to which the superconducting junctions in the PJVS chip produces a voltage precisely proportional to the frequency of the applied microwave bias signal. Presently this system is used for dissemination of unit 'volt' in India through calibration of national standards (bank of Zener-diode based dc reference standard) at 1.018 and 10 V as per the ISO/IEC 17,025:2005. These Zener standards further provide traceability to all dc instruments to this primary standard.

The fabrication of various novel structures/patterns and prototype devices for electromagnetic metrology is also carried out at CSIR-NPL through the FIB (Focused Ion Beam) Laboratory. The FIB instrument at CSIR-NPL consists of dual-beam system, i.e., focused ion beam and scanning electron microscope which enable the precise micro-nano imaging and machining of samples with nanometer-scale control. Dimensional measurements (lateral X and Y) performed using this instrument are accurate and traceable to SI unit (meter). FIB laboratory also produces superconducting junctions, weak links, nanostructures and nanodevices that are used in the electromagnetic metrology's program for developing new standards and technologies.

The chapter further explains the metrological parameters mentioned above in detail including the associated primary standards, their calibration and measurement capabilities with the international status. Also, the role of these metrological parameters in quality infrastructure with identified sectors, regulators and ministries along with the economic impact analysis is provided.

12.2 LF and HF Voltage, LF Current and PMU Metrology

Precise and traceable measurements are important across all areas of life to ensure the performance and reliability of the measurement results. CSIR-NPL measures electrical and electromagnetic quantities that are accurate, reliable and traceable to the SI units. The National standards are realized, established, maintained, upgraded and disseminated to support industry, govt. sector and research. We run one of the comprehensive capabilities of LF & HF Voltage, LF Current and Phasor Measurement Unit (PMU) metrology parameters. To remain competitive globally, international comparisons are essential for the validation of CMCs and measurement techniques. This sub section contents are based on above-mentioned parameters, their significance, CMCs, traceability, measurement technique, GOI ministry involved, role of regulation bodies and their economic impact.

12.2.1 Measurement Capabilities and SI Traceability

CSIR-NPL also provides **apex level calibration services** in LF & HF Voltage, LF Current and PMU parameters to the industry, strategic sectors, STQC labs & private organizations of the country by maintaining their respective primary standards. We are disseminating measurement traceability in the above-mentioned parameters along with the other derived parameters and their measurement capabilities, which are discussed in the coming respective sections. We have 22 registered CMC's and participated in more than 13 BIPM/APMP key/Supplementary comparisons to establish mutual degree of equivalence with the corresponding standards of the leading NMIs having internationally recognized traceability.

12.2.1.1 Primary Standards

LF Voltage and Current Standard

Alternating voltage is one of the most important parameters, which is essential for the development, production, and reliable operation of most of the precision electrical and electronic instruments used in communication, professional electronics, strategic areas etc. The SI traceability for all of the electrical parameters is through Ampere, which is realized with the help of Quantum based DC Josephson Standard and Quantum Hall Resistance, popularly known as metrology triangle [2]. Several methods are available for the accurate measurement of alternating voltage; these methods involve the use of the voltage of a dc standard as the reference voltage. The transfer instruments and techniques which enable the measurement of alternating voltage directly in terms of known direct voltages without necessitating correction for frequency are standards of the rated accuracy in the specified frequency range. Direct substitution of alternating voltage has low residual harmonic content and results in high accuracy. There is no absolute standard for AC Voltage/Current worldwide till date. It is evaluated by computing the AC-DC transfer difference which is basically an error ratio between AC and DC Voltage applied to a thermal device. Banks of Multi-Junction thermal converters (MJTC) are used for assigning AC-DC transfer

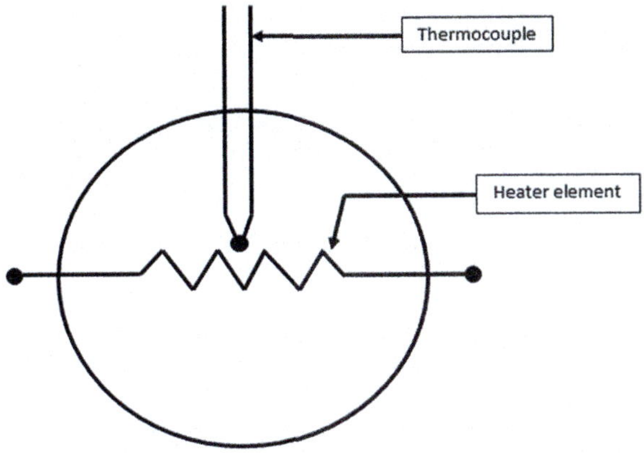

Fig. 12.1 Single junction thermal

difference to the primary standard (MJTC) at 2 V and 5 mA with the best uncertainty of ±5 ppm [3].

The technique of LF voltage standardization makes the use of transfer principle involving the substitution of unknown LF voltage with the variable DC voltage to obtain the same thermal response in terms of the thermo emf produced. The thermo emf in the two cases is made exactly equal, i.e., RMS value of the applied LF voltage is taken to be equal to the corresponding dc voltage applied; assuming that the harmonic content in the LF signal is not appreciable. From considerations of accuracy, these transfer techniques should essentially yield highly repeatable response to both alternating and direct voltages with maximum sensitivity and precision should have ideally zero ac/dc transfer difference and a flat response over a wide frequency band. The instrument/devices associated with the above said transfer technique is known as Electrothermic instruments. Electrothermic devices used as Primary/Reference Standard worldwide for evaluating the AC-DC transfer difference are given below:

I. **Single Junction Thermal Converter**

It is a simple device consisting of a short heater mounted between two thick support wires; the rise in temperature of the heater is sensed by a thermocouple placed at the midpoint of the heater which is also the maximum temperature point. The whole assembly is placed in an evacuated glass bulb [4]. The schematic diagram is shown in the Fig. 12.1. When direct current flows through the junction formed by the heater wire and its supports, the Peltier effect causes one of the junctions to heat and the other to cool. Also, because there is temperature gradient along both halves of the heater, Thomson effect will cause heat to be absorbed in one half and liberated in the other half. Both Peltier and Thomson effect influence the temperature distribution along the length of the heater causing it to be asymmetric about the center. When the thermocouple is placed exactly at the midpoint, the Peltier and first order Thomson effect cause negligible asymmetry. The heat thus generated due to the temperature difference, is sensed by the thermocouple, which can be measured with help of a sensitive meter.

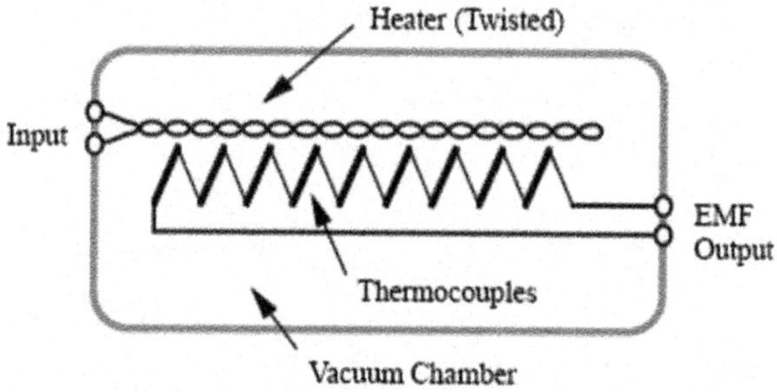

Fig. 12.2 Multi-junction thermal converter

Fig. 12.3 Bank of multi-junction thermal converter

II. Multi–Junction Thermal Converters (Primary Standard of LF Voltage & Current)

The single junction thermal converter is dependent on the sensing of the temperature at one point on the heater. As both Peltier and Thomson effects act in opposite senses in two halves of the heater, it would seem that a better indication of Joules heating would be if we can integrate the temperature rise all along the length of the heater as shown in Fig. 12.2. In Multi-junction thermal converters (MJTC) precisely the same process mentioned above, i.e., integrate the temperature rise all along the length of the heater is done. MJTCs are used in very high-accuracy ac-dc transfer metrology because they have very small ac-dc difference, follow the RMS law of excitation, and produce high output EMFS. MJTC are used as primary standards for calibrating and measuring Low frequency (LF) Voltage and Current in frequency range from 10 Hz up to 1 MHz [5]. The measurement capabilities of Primary and Reference Standards are given below. MJTCs are the most accurate standard providing link between LF voltages with DC Voltage. At CSIR-NPL, we have a bank of four MJTCs as shown in Fig. 12.3 with identical characteristics. The measurement capabilities of primary standard and reference standards are shown in Tables 12.1 and 12.2 respectively.

MJTCs are the most accurate standard providing link between LF voltages with DC Voltage. At CSIR-NPL, we have a bank of four MJTCs with identical characteristics so that absolute AC-DC transfer difference [6] can be assigned to each of

Table 12.1 MJTC primary standard CMC

MJTC (primary standard)	Frequency	Expanded uncertainty
2 V	10 Hz to 1 MHz	±5 to 12 ppm
5 mA	40 Hz to 10 kHz	±5 to 25 ppm

Table 12.2 TTS reference standard CMC

TTS (reference standard)		
Range	Frequency	Expanded uncertainty
1 mV to 1000 V	10 Hz to 1 MHz	±7 to 1000 ppm
1 mA to 20 A	40 Hz to 10 kHz	±10 to 92 ppm

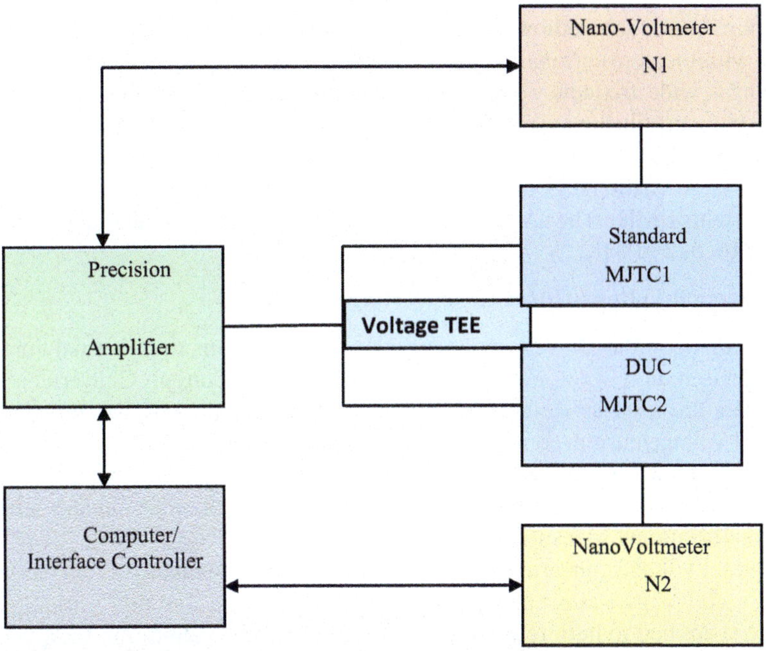

Fig. 12.4 Measurement setup to calibrate MJTC

them by comparing each one of them against the other in a round robin manner. The measurement setup used to assign AC-DC transfer standard to MJTC is shown in Fig. 12.4. The thermal transfer standard is a precision AC/DC transfer standard consisting of a fluke RMS sensor. It is capable of measuring any Voltages between 2 mV to 1000 V RMS in frequency range from 10 Hz to 1 MHz [7]. The primary standard MJTC cannot be used for day-to-day measurement and calibration work because it is the national standard which is to be used for the purpose of providing traceability to all others as voltage measuring. Its maximum voltage limit is 2 V. Therefore, the thermal transfer standard or thermal converters are designed to cover voltage range of 1 mV volts to 1000 V in the frequency range 10 Hz to 1 MHz.

III. **Low Voltage Thermal Converter (LVTVC)**

Accurate and precise calibration of low AC voltages below 1 V presents several challenges that are basically not encountered at voltages above 1 V. The accurate measurement of low alternating voltages is essential as a large number of electrical and electronic equipment such as AC milli voltmeter, precision ac calibrators very often operate in micro volt and milli volt ranges. The normal calibration procedures employing signal generators, calibrators, milli voltmeters and other associated equipment generally result in erroneous measurements and uncertainties as large as 100%. The low signal-to-noise ratios at such low levels is the major contributor to this large uncertainty and the corrective attempts to achieve better accuracies render these procedures even more tedious and expensive. Because of the above deficiencies associated with the conventional methods, a voltage source of precisely known voltage covering a wide frequency range and having minimum components is needed. A simple but versatile device meeting all the above requirements of such a source were devised in as LVTVC, Low Voltage Thermal voltage converter is designed by Holt has met these challenges covering voltages from 1 to 200 millivolts at frequencies from 6 Hz to 1 MHz The LVTVC provides seven outputs Voltage ranges: 200, 100, 50, 20, 10, 5, 2 mV [8].

IV. **Thermal Voltage/Current Converters**

Depending upon the above terminology the first single junction thermal converters were devised and were named as Thermal Voltage/Current Converters. These converters are used as Reference Standards for evaluating AC-DC transfer error. Thermal Voltage/current converters are the building blocks for LF voltage and current measurements across the world. It laid the foundation for the techniques of ac-dc transfer and provided the first theoretical basis for the thermal transfer structures used in all National Measurement Institutes [9].

Thermal voltage converter (TVC)/ Thermal Transfer Standards (TTS) to be calibrated is compared against a reference standard TVC/TTS. Initially a nominal rated voltage is applied to both reference and the device under calibration for a period of 15–30 min depending upon the voltage rating. The applied voltage should not exceed the lower one of rated voltages of two TVCs connected in calibration setup [10]. This is done to stabilize both the devices thermally. To eliminate the influence of drift in the output of TVCs, the following sequence of measurements is used: AC, DC^+, DC^- and again AC. The desired LF voltage signal is applied to both the TVCs and response in terms of emf output measured by the nano-voltmeters is recorded. Then DC^+, DC^- & AC signals are applied, and their responses are recorded respectively. The time delay between two measurements is kept more than the time constant of TVC used. After the above measurements are taken 10-times for a particular frequency, mean corrected AC-DC difference and its standard deviation are calculated. After taking these measurements the AC source is then set for the next frequency and AC-DC difference measurements are repeated as mentioned above [11].

The AC or DC signal is applied to both the thermal converters simultaneously and corresponding emfs are recorded. Then AC-DC transfer difference δ_x assigned to TVC is calculated using:

$$\delta_x = \frac{S_{\text{ac}} - S_{\text{dc}}}{n_s S_{\text{dc}}} - \frac{X_{\text{ac}} - X_{\text{dc}}}{n_x X_{\text{dc}}} + \delta_s \tag{12.1}$$

Where, S_{ac} and X_{ac} are the mean emf outputs of the standard and device under calibration (DUC) when AC is applied. S_{dc} and X_{dc} are the mean emf outputs of the standard and device under calibration (DUC) when DC is applied.

Differences are usually expressed in parts per million (ppm) and a positive sign signifies that more ac than dc was required for the same output response.

V. **Traceability chain for LF Voltage and Current**

As aforesaid in sections above, that for calibration of LF Voltage and Current, CSIR-NPL has a well-established and traceable measurement system. To be at par with other NMIs in order to maintain degree of equivalence CSIR-NPL regularly participates in International Inter-comparisons. The traceability chain for LF voltage and current starting from the primary standard to the reference standard is shown in Fig. 12.5a and b. This traceability chain also shows the international traceability with the help of international intercomparisons. CSIR-National Physical Laboratory (NPL India) is having a responsibility of disseminating measurement traceability as per the needs of country. CSIR-NPL maintains the Indian National Standards. Accurate and precise measurements are essential to drive the growth engines of Indian Science & Industry as it removes chaos and prompts innovations, which in turn, would save precious lives, resources and time [12]. The mission of the CSIR-NPL are developing India's measurement standards that are internationally accepted and disseminating the measurement capabilities to industry, government, strategic and academia that underpin the India's prosperity and quality of life so that India remains on par with international measurement laboratories. The visions of the CSIR-NPL is to be "Giant in measurements" to support needs of the country by enhancing and expanding the existing apex calibration capabilities through traceable measurements [13].

HF Voltage Standard

High frequency (HF) voltage is one of the important parameters in electrical metrology. CSIR-NPL HF voltage standard is realized by a twin resistance coaxial thermistor mount fitted with a tiny resistor which provides the proper termination for 50 Ω at the end of a coaxial line in the frequency range of 1–1000 MHz. The calibration technique and the traceability in assigning the RF DC transfer difference to HF voltage primary standard has been established for accurate measurements. The

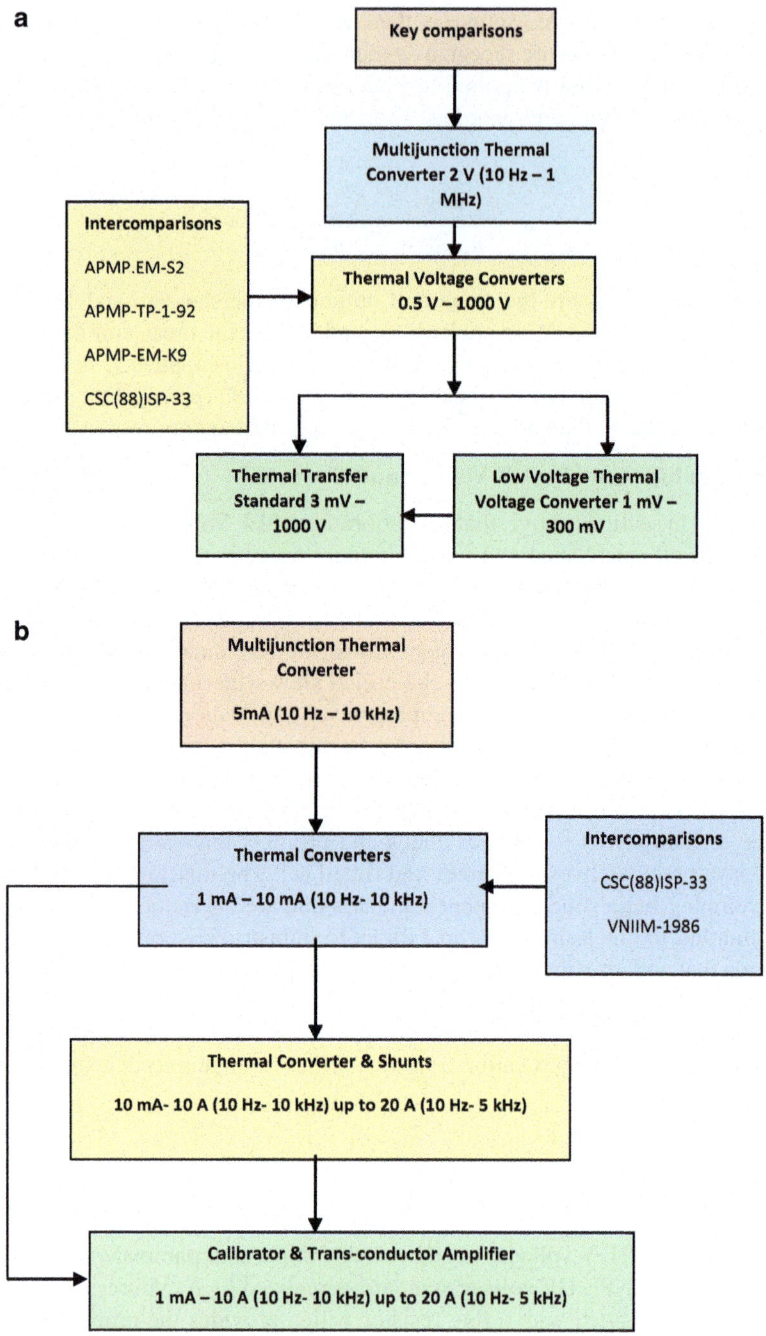

Fig. 12.5 **a** Traceability chain for LF voltage, **b** traceability chain for LF current

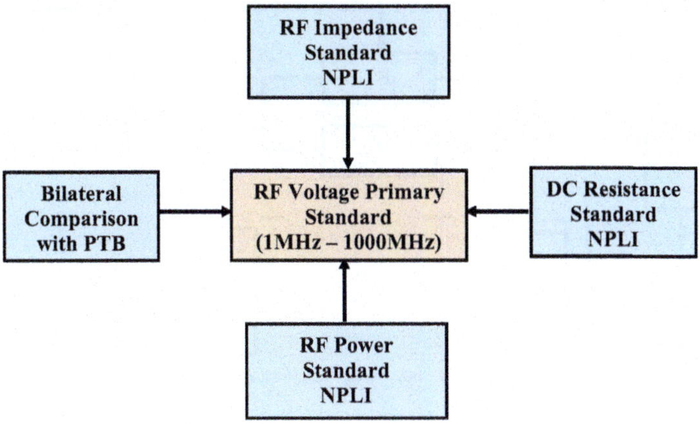

Fig. 12.6 Measurement traceability of HF voltage primary standard

calibration technique and the traceability in assigning the RF DC transfer difference to RF voltage primary standard and its traceability is obtained through the determination of its effective efficiency, RF impedance and the DC resistance against each of the parameter's reference standard as a function of frequency as shown in Fig. 12.6.

All these measurements are done at the input reference plane of the primary standard. Therefore, measurement uncertainty in assigning the RF DC transfer difference to the primary standard is the cumulative effect of the uncertainty in determination of effective efficiency, RF impedance and the DC resistance. The overall uncertainty in assigning RF DC transfer difference has been estimated to be within ±0.53% in the frequency range up to 1 GHz. With the establishment of the RF voltage primary standard, one can calibrate and assign the RF DC transfer difference to the transfer standard thermal voltage converters [14]. HF Voltage standard is used to calibrate the RF Voltage meters, high frequency thermal converters using the well-established techniques [15].

It consists of two resistive loads R_1 and R_2 which are nominal 50 W and are of nearly equal value. These resistances are thin film resistances with negligible reactive component up to 1 GHz. Both these resistances have nearly zero temperature coefficients. In thermal contact with R_1 and R_2 are two resistances R_3 and R_4 respectively which are highly temperature sensitive (thermistor). R_3, R_4 along with R_5, R_6 and R_7 form a Wheatstone bridge as shown in Fig. 12.7. R_7 is a variable resistance and is used for initial balancing of the bridge. When high frequency power is applied to the R_1, it is heated up and in turn the temperature of R_3 also rises thereby changing its resistance [16].

The bridge gets unbalanced and to restore the balance, DC power is applied to R_2 and is adjusted till the balance is re-established. The RF power applied to R_1 is then replaced by accurately known dc power and balanced condition is restored by adjusting this dc power. Thus, RF power is measured in terms of the dc substituted power fed to R_1. In addition to the RF power being absorbed by the standard, some

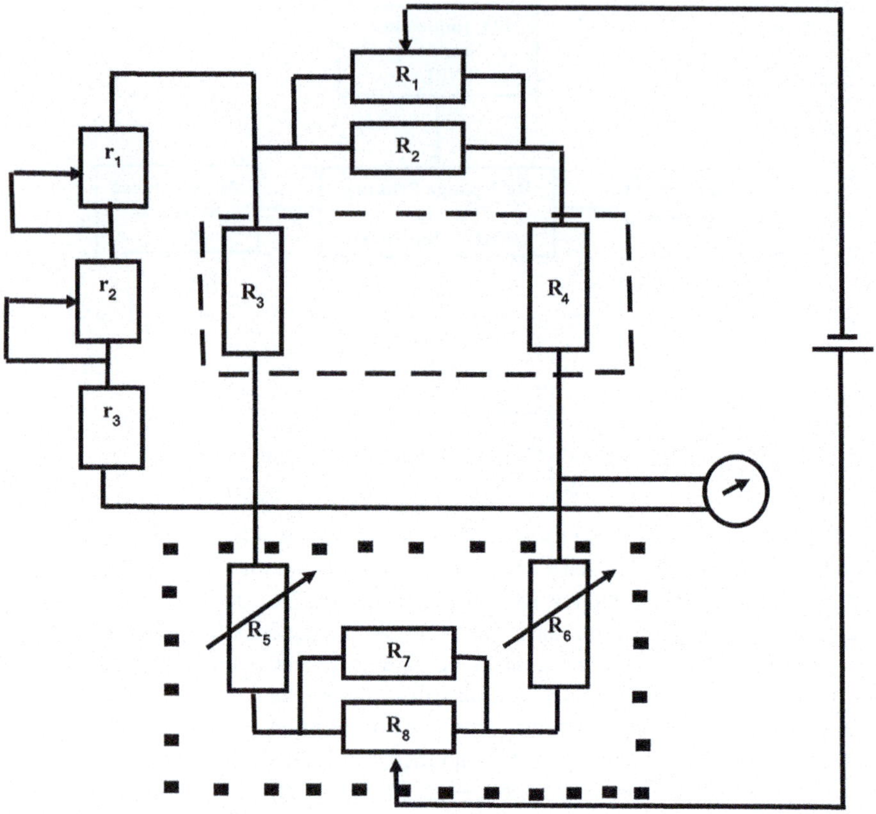

Fig. 12.7 Schematic diagram of HF voltage primary standard

power is lost in the wall of the standard and also in the transmission line connecting the terminating resistance R_1 and the input terminal, and the error in substituting the RF power by dc power is called the RF DC transfer difference, which has not been completely transferred in terms of DC power. Resistors R_5, R_6 and R_7, are external to the power mount. All these measurements are done using indigenously developed automation software [17].

Phasor Measurement Unit (PMU) Standard

Accurate and precise measurements are essential to drive the growth engines of Indian Science & Industry as it removes chaos and prompts innovations, which in turn, would save precious lives, resources and time [18]. Today's smart grid relies on phasor measurement units (PMUs) to deliver real-time critical data on the voltage, current, frequency and phase mainly within the power grid. It is mandatory that PMUs

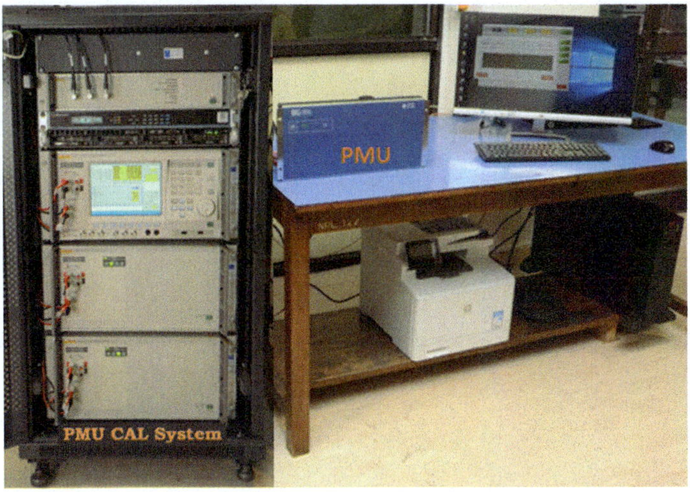

Fig. 12.8 CSIR-NPL phasor measurement unit (PMU-CAL) system

must be calibrated from a traceable calibrator system to ensure its consistency, accuracy and reliability. The CSIR-NPL Phasor Measurement Unit Calibration System (PMU-CAL) as shown in Fig. 12.8 has been installed at CSIR-NPL, New Delhi, which is capable of providing perfect solutions for PMU users, manufacturers, electrical power utilities and organizations associated with electrical power transmission [19]. The integrated and automated PMU-CAL system fully complies with the IEEE C37.118.1a-2014 [20], IEEE Synchro phasor Measurement Test Suite Specification-Version 2–2015 and C37.242–2013 standards for PMU operation and verification [21]. The compliance testing of Phasor Measurement Unit (PMU) from 50 to 150 V and 1 to 5 A at 50/60 Hz with a frame rate of 10, 25 and 50 frame/seconds has become possible with the establishment of PMU-CAL system serving the nation through power sector. The CSIR-NPL PMU calibration system has an uncertainty of ±0.005 to ±0.010%. Besides its full compliance with IEEE standard, system is fully automated, and the measurement period is shortened to a few hours as compared to the conventional methods which took weeks to complete the tests. Voltage and current are generated with up to six digits precision. We use PMU-CAL system to generate a wide variety of complex signals, including flicker, harmonics. This feature gives us enormous flexibility to calibrate a wide variety of PMUs by different manufacturers.

Using the PMU calibration system, PMU manufacturers can now perform the verification of each new PMU; users can validate their PMU before deployment and perform periodic calibration whenever required. Apart from this, to perform IEEE compliance measurements, the PMU calibrator shall be traceable to national standards of the country and it is generally recommended that uncertainty of calibrator should be less than 10% of the allowed IEEE limits. It is a challenge for a manual PMU calibrator to meet these high-accuracy requirements, as the compliance test

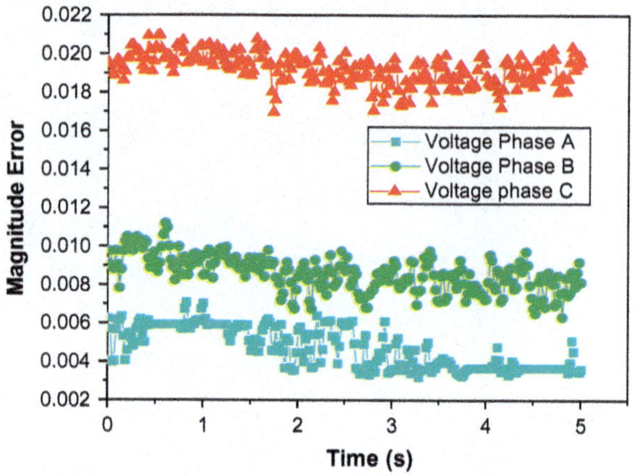

Fig. 12.9 Magnitude error of PMU at three different phases of voltages

procedure itself is a very time-consuming and complicated task. All these limitations of conventional PMU calibrators have been removed by an automated high precision PMU calibration system (PMU-CAL) [22]. The CSIR-NPL PMU-CAL system calibrates PMU by simulating the actual voltage and current conditions that occur in a power grid as per the IEEE standards. The system is fully automated, controls each and individual equipment once the initial configurations are done and acquired the measured data from PMU. The PMU automation software evaluates the measured data to calculate total vector error (TVE), frequency error (FE), and rate of change of frequency error (RFE) and then compares the results with the original signals.

PMU calibration system consists of **three-phase Electrical Power Standard, System Timing Unit**, **GPS receiver and Controller as** shown in Fig. 12.8, with an integrated test connection. **Electrical Power Standard i**ncludes one master unit and two auxiliary Units, which provides three-phase voltage and current signal to the PMU under calibration. **System Timing Unit c**ontrols timing and synchronizes the tests throughout the calibration. **GPS receiver** gives the PMU-CAL system and the PMU under calibration with a source of Universal Coordinated Time (UTC), time stamping. System c**ontroller works as** an application controller, receive the commands from the computer to control the calibration system [23]. This system handles all the steady-state, dynamic-state tests for both M class and P class configured PMU at all desired frame rates. The PMU calibration results for amplitude variation at three different phases of voltage in graphical form are shown in Fig. 12.9 and PMU results (TVE in %) as per IEEE limits is shown in Fig. 12.10. PMU can only be calibrated by CSIR-NPL in the country as it requires a calibrated PMU-CAL system, which is traceable to CSIR-NPL primary standards. This system works as per IEEE standard having several static and dynamic individual tests. Besides its full compliance with IEEE standard, PMU calibrator is fully automated, and the measurement period is shortened to a few hours with much higher performance.

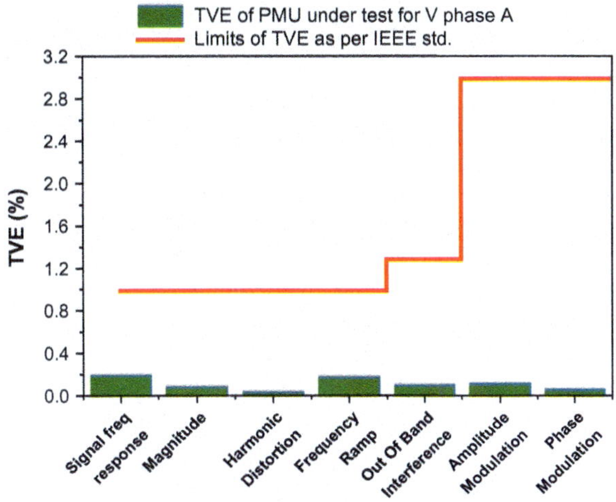

Fig. 12.10 PMU calibration results and its IEEE limits

For an NMI, it is their prime responsibility to maintain the traceability for each and every measured parameter, for which calibration is provided by them. Figure 12.11 shows the traceability chart for the PMU calibrator.

As we can see in chart, PMU calibrator is traceable to following primary standards:

(a) Primary standard of DC voltage, i.e., Programmable Josephson voltage Standard at CSIR-NPL
(b) Primary standard of DC resistance, i.e., Quantum Hall Resistance Standard at CSIR-NPL
(c) Primary standard of AC voltage, i.e., Thermal converters at CSIR-NPL
(d) Primary standard of Frequency, i.e., Cesium atomic clocks at CSIR-NPL
(e) Primary standard of AC-DC current reference standard.

All the above-mentioned primary standards are maintained at CSIR-NPL and are used to calibrate secondary/reference standards which are further used to calibrate the instruments from consumers. Thus, we achieve the traceability of PMU calibrator installed at CSIR-NPL directly from our primary standards that are at par to international level.

12.2.1.2 Calibration and Measurement Capabilities (CMCs)

CMCs at BIPM

CSIR-NPL, being the National Metrology Institute by an act of parliament, is the path for measurement traceability to the SI units in India and it is in the mandate of the laboratory. It is the custodian of “National Standards” with a responsibility of disseminating measurement traceability as per the needs of country. Establishing new standards, developing, maintenance and up-gradation of existing standards for

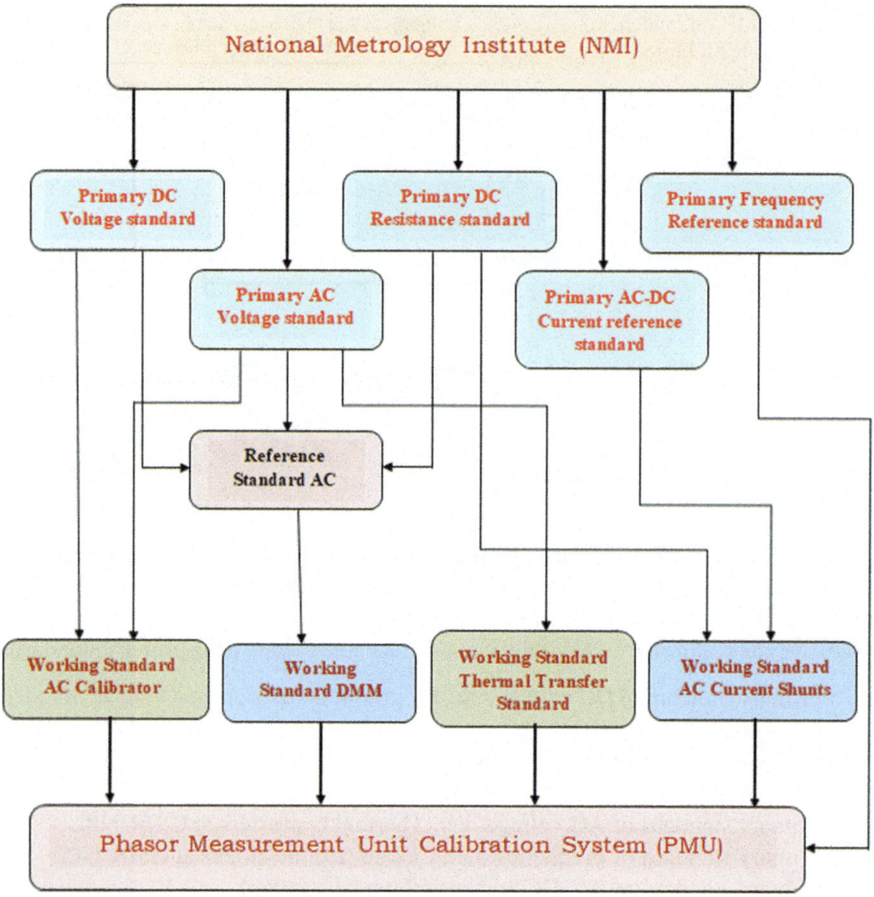

Fig. 12.11 Traceability chart for CSIR-NPL PMU-CAL system

improvement of ranges and uncertainties are essential to meet the requirements and demands of various industries, all regional calibration and test laboratories and others to sustain their calibration and measurement capabilities. Since a great majority of measurements in any field now-a-days rely on electrical measurements in one form or the other, calibration service we are providing is essential in supporting the development of research, commerce and industries in India. As per our mandate we cater to emerging needs of calibration in these areas because of the availability of relevant expertise and infrastructure. CSIR-NPL maintains the Indian Standards with accurate and precise measurements, essential for the growth of the country. Internationally accepted India's national standards have been developed by CSIR-NPL and maintained at par through international Intercomparison. Being an NMI, it is the prime responsibility of CSIR-NPL to disseminate the measurement traceability

Table 12.3 CMCs at CSIR-NPL registered at BIPM (BIPM) [18]

Quantity	Instrument or artifact	Measurand range	Measurement condition	Expanded uncertainty
AC/DC voltage transfer: AC/DC transfer difference at low voltages (typically ≤0.5 V)	Micro-potentiometers	1–500 mV	Frequency, 10 Hz to 1 MHz	±10 to ±1000 μV/V
AC/DC voltage transfer: AC/DC transfer difference at medium voltages (typically 0.5–5 V)	Thermal voltage converter	0.5– 5 V	Frequency, 10 Hz to 1 MHz	±10 to ±35 μV/V
AC/DC voltage transfer: AC/DC transfer difference at higher voltages (typically >5 V)	Thermal voltage converter	>5–1000 V	Frequency, 10 Hz to 1 MHz	±10 to ±100 μV/V
AC voltage up to 1000 V: sources	Multifunction calibrator	0.003–1000 V	Frequency, 10 Hz to 1 MHz	±10 to ±1000 μV/V
AC voltage up to 1000 V: meters	Transfer standard and digital multimeters	0.003–1000 V	Frequency, 10 Hz to 1 MHz	±20 to ±1000 μV/V
AC/DC current transfer: AC/DC transfer difference	Thermal converter and shunt	0.005–20 A	Frequency, 40 Hz to 10 kHz	±20 to ±100 μA/A
AC current up to 100 A: sources	Transconductance amplifier	1–0 A	Frequency, 40 Hz to 10 kHz	±100 to ±200 μA/A
AC current up to 100 A: meters	Digital multimeters	0.001–1 A	Frequency, 40 Hz–10 kHz	±100 μA/A
HF Voltage up to 50 V	HF Voltage Measuring Devices/HF TVC	1 mV to 50 V	Frequency, 1 MHz to 1 GHz	±17 mV/V to ±35 mV/V

to private industry, government organizations, STQC labs and academic institutions to improve the quality of life.

The Low Frequency Voltage and Current, HF Voltage metrology is well-established at CSIR-NPL (National Physical Laboratory-India) with a total of 22 CMCs registered with BIPM (International Bureau of Weights and Measures), France. The CMCs are as shown in tabular form given below. The parameters for which CSIR-NPL has CMC registered at BIPM are given in Table 12.3.

Table 12.4 Parameters CSIR-NPL is capable of measuring for the range reported against the respective parameters [18]

S.No	Parameter	Item name	Item type	Alias name	Range
1	AC Voltage and Current/*AC-DC Voltage Transfer Difference	Multimeter	Digital Multimeter		1 mV to 1000 V 1 mA to 20 A
2	AC Voltage and Current/ *AC-DC Voltage Transfer Difference	Calibrator	Voltage & Current Calibrator	Multifunction/Multi Product Calibrator	1 mV to 1000 V 1 mA to 10 A
3	AC Current	Transconductance Amplifier	Transconductance Amplifier	AC Current Source	Up to 20 A
4	AC-DC Current Transfer Difference	Shunt	Current Shunt		1 mA to 20 A
5	AC Voltage	AC Measurement Std	AC Measurement Std	AC Measurement Std	1 mV to 1000 V
6	PMU Calibration	M Class	Phasor Measurement Unit	PMU	50 V to 150 V 1A to 5A 50/60 Hz 25/50 fps
7	PMU Calibration	P Class	Phasor Measurement Unit	PMU	50 V to 150 V 1A to 5A 50/60 Hz 25/50 fps

List of Measurement Capabilities

This section includes the list of the parameters for which CSIR-NPL can calibrate with calibration procedure as per established methods but not yet peer reviewed and not approved by BIPM are shown in Table 12.4.

Intercomparison

LF Voltage and Current metrology at CSIR-NPL keeps participating in inter-comparisons from time to time to maintain our existing measurement capabilities having a degree of equivalence with the leading NMIs. It has piloted one of the latest intercomparison in LF Voltage and Current having an artifact as 8.5 DMM. This intercomparison is registered at APMP as S8. Some of the inter-comparisons are listed in Table 12.5.

Table 12.5 List of Inter-comparisons recently participated by CSIR-NPL

S. No.	Pilot lab	Type/identification/pilot lab	Parameter/s	Participant countries	Status
1	SP (NMI Sweden)	EURAMET.EM-K11	DC voltage, current, resistance, LF voltage and current	INRIM, METAS, CEM, SP, LNE, PTB, VNIIM, CSIR-NPL, UME	Report published on BIPM website, Jan 2011
2	CSIR-NPL	Supplementary P1-APMP.EM-S8	DC voltage, current, resistance, LF voltage and current	NIMT, NMIA, NIS, NML-SIRIM	Draft B report submitted, 2019
3	NMI Australia	Supplementary P1-APMP.EM-S12	DC voltage, current, resistance, LF voltage and current	KRISS, NMIA, NIMT, NPL India	Report Published in Metrologia, 2019
4	NMI Japan	APMP.EM.RF-K8.CL	APMP Comparison of Microwave Power in 50 Ω coaxial lines (0.01 to 18 GHz)	NMIJ, KRISS, NIM, NMIA	Draft A, 2019

(a) Supplementary Comparison on Digital Multimeter **(P1-APMP.EM-S8)** initiated in 2011. The intercomparison was **piloted by CSIR-NPL**, India.
(b) Comparison of Standards for the Calibration of Voltage, Current and Resistance Meters **(P1-APMP.EM-S12)** initiated in 2015. The intercomparison is piloted by NMI, Australia and the results were announced in 2018.
(c) The international status of LF Voltage and LF Current parameters can be demonstrated in plots of intercomparisons, one is from **EURAMET** (EURAMET.EM-K11) comparison piloted by NMI Sweden and other one is from **APMP** intercomparison piloted by CSIR-NPL India as shown in Fig. 12.12a and b.
(d) Being an NMI of India, to support the quality system and CMCs we also participate in international peer reviews. LF & HF Voltage, LF Current and PMU Metrology has been peer reviewed twice by NMI Australia and NIST USA.

The intercomparisons results shown in Fig. 12.12a and b, demonstrate the capability of CSIR-NPL in LF voltage and Current and show the degree of equivalence in measurement uncertainty with the other leading NMIs of the world.

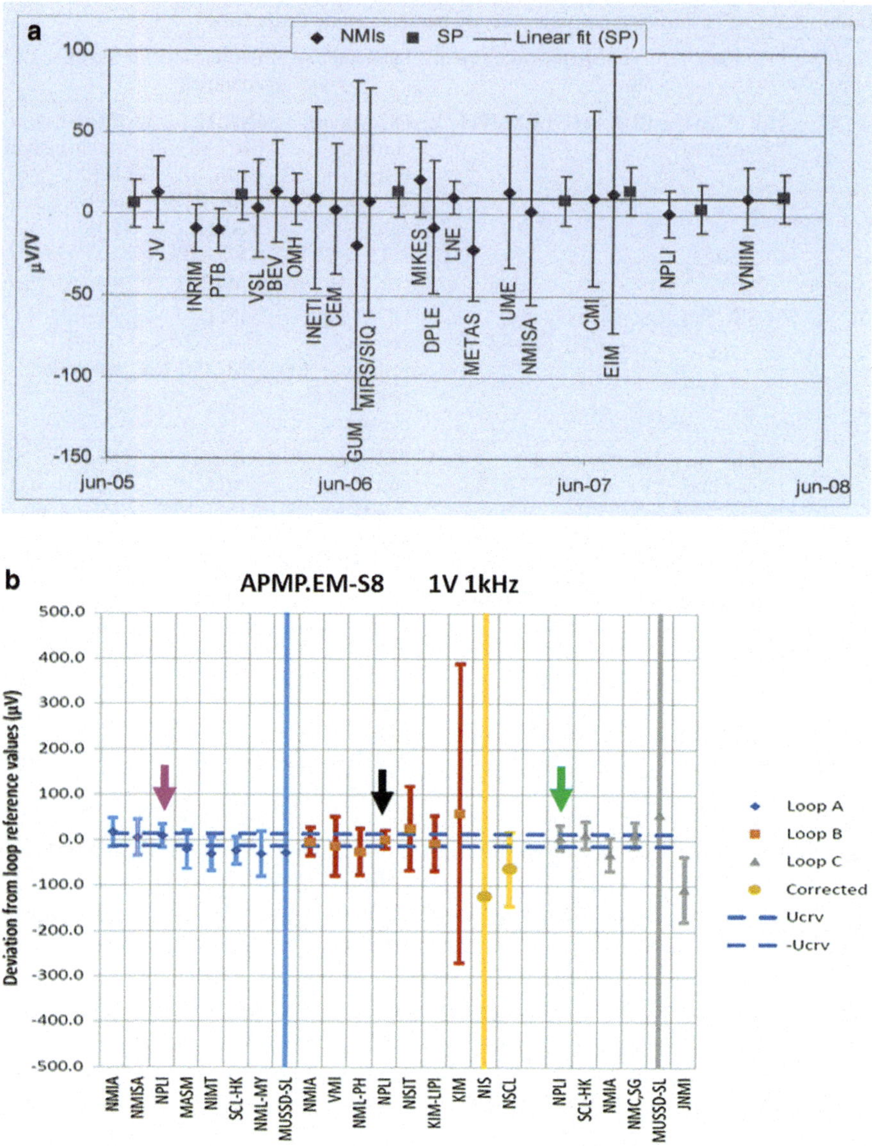

Fig. 12.12 **a** EURAMET Inter-comparisons results for LF Voltage, **b** APMP Inter-comparisons results for LF Voltage

Table 12.6 List of parameters CSIR-NPL is capable of measuring for CRO calibration

Quantity	Instrument or artifact	Measurand range	Measurement condition	Expanded uncertainty (%)
AC voltage	Oscilloscope	0.001–1000 V	10 Hz–1 MHz	±1
Frequency	Oscilloscope	10 Hz–2 GHz	10 Hz–2 GHz	±1
Bandwidth	Oscilloscope	10 Hz–2 GHz	10 Hz–2 GHz	±1
Rise time	Oscilloscope	10 Hz–2 GHz	10 Hz–2 GHz	±1
DC Gain accuracy	Oscilloscope	1–1000 V	–	±1

12.2.1.3 Derived Measurement Capabilities

CRO Calibration

An oscilloscope is a type of electronic test instrument that graphically displays varying signal voltages, usually as a two-dimensional plot of one or more signals as a function of time. Oscilloscopes display the change of an electrical signal over time, with voltage and time as the *Y*- and *X*-axes, respectively, on a calibrated scale. The waveform can then be analyzed for properties such as amplitude, frequency, rise time, time interval, distortion, and others. Application of oscilloscopes is very wide in sciences, medicine, engineering, automotive and the telecommunications industry. The oscilloscope observes the changes in the electrical signals over time. The calibration setup for oscilloscope calibration is different as per its traceability with the different parameters has been explained below. For CRO calibration, an oscilloscope needs to be calibrated in both vertical and horizontal deflection by applying following calibrated signals of DC & AC Voltages. The measurement capability of CRO calibration with their uncertainty parameters are given in Table 12.6. The measurement traceability for CRO calibration is from DC standard, LF Voltage and Current Standard, Microwave Standard and Time and Frequency standard of CSIR-NPL as shown in Fig. 12.13.

The traceability of CRO by DC in vertical deflection is performed by applying a traceable DC signal to the oscilloscope. For performing measurements either a calibrated DC calibrator or a calibrated Digital Voltmeter is used. An oscilloscope can be calibrated either directly through a traceable DC Calibrator or indirectly through a traceable reference voltmeter via an oscilloscope calibrator. The traceability by AC in vertical deflection is performed by applying a traceable AC signal to the oscilloscope. For performing measurements either a calibrated AC calibrator or a calibrated Digital Voltmeter is used. An oscilloscope can be calibrated either directly through a traceable AC Calibrator or indirectly through a traceable reference voltmeter via an oscilloscope calibrator. Applying a traceable RF signal to the oscilloscope performs the RF traceability in vertical deflection. The procedure for amplitude measurement in RF is done. The time-base calibration is performed using a stable and accurate periodic signal. The measurements can be done either using a Signal generator as

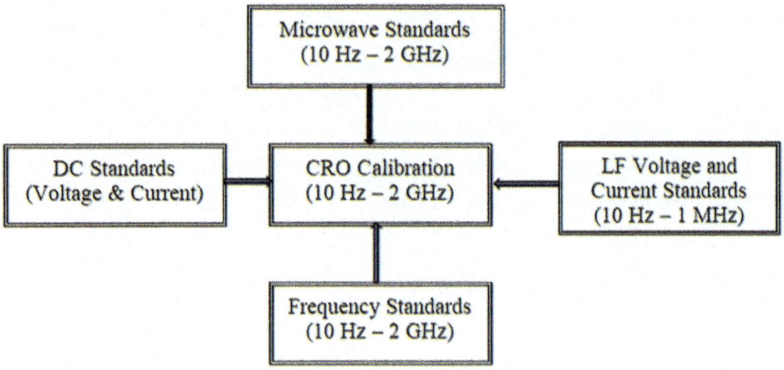

Fig. 12.13 Measurement traceability of CRO calibration

Reference Standard directly or indirectly by a Universal counter as reference standard using 10 MHz reference signal.

12.2.1.4 R&D in LF Current Metrology

Research for LF High Current Measurement

CSIR-NPL is supporting strategic sectors of the country by providing traceability and involved in research and development work in order to upgrade existing calibration facility. Thermal current converters or thermal transfer standard with AC–DC current shunts act as the reference standards for accurate and precise measurements of low frequency current in the frequency range from 40 Hz to 10 kHz [24]. At present CSIR-NPL, India has the AC–DC current transfer difference calibration facility up to 20 A. In supporting the strategic sectors like IDEMI and BHEL we are trying to extend our AC current calibration range from 20 to 100 A. In this regard, a current Tee for AC high current using LC connectors has been indigenously designed and developed at CSIR-NPL workshop as shown below in Fig. 12.14. This current Tee is used for high current measurements up to 100 A for assigning AC–DC current transfer difference to current shunts. The current Tee for AC high current has solved the connectivity issue of standard and device under calibration, i.e., current shunts.

Using the high current Tee, AC-DC current transfer difference has been assigned at 30 A with TTS against a 20 A current shunt with TTS. The uncertainty has been evaluated for the measurement setup at 40 Hz for 30 A. Research and development work is still going on to performs the high current measurements above 30 A up to 100 A. The uncertainty estimated due to current Tee for AC high current is very low because within the desired frequency range it has a good match and low losses. The estimated uncertainty of current Tee is ±8 ppm.

Fig. 12.14 A pictorial view of high current tee

12.2.2 Voltage, Current and PMU Metrology for Quality Infrastructure in India

Metrology forms an important part of Quality Infrastructure which in turns results in good Quality of Life. There should be a well dedicated dissemination of measurement traceability across all the calibration labs of the country for the implementation of quality system and support to the industries for their economic growth. Metrological activities are working in the line with various national missions such as Make in India, Swasth Bharat, Power for all, one nation one grid etc. to establish a quality infrastructure that can provide all technological solution to the requirement from various sectors.

12.2.2.1 Metrological Services for Industrial Growth

Calibration and Testing

LF & HF Voltage, LF Current and PMU Metrology offers calibration and testing services for numerous parameters for various types of items like AC calibrators, Digital multimeters, thermal voltage converters and AC measurement standards, RF Voltage meters, and PMUs. The calibration facility for AC Voltage is from 1 mV to 1000 V in frequency range 10 Hz to 1 MHz and for AC Current is from 1 mA to 20 A in frequency range 40 Hz to 10 kHz. The RF voltage calibration facility is from 1 to 50 V in frequency range of 1 to 1000 MHz. The compliance testing of Phasor Measurement Unit (PMU) is available from 50 to 150 V and 1 to 5 A at 50/60 Hz with a frame rate of 10, 25 and 50 frame/seconds as per IEEE synchrophasor standards. The details of the parameters are given in Table 12.7.

Table 12.7 Parameters CSIR-NPL is capable of doing testing and calibrating [18]

S. No.	Parameter	Item name	Item type/group	Alias name	Range
1	AC Voltage & Current / *AC-DC Voltage Transfer Difference	Multimeter	Digital Multimeter		1 mV to 1000 V 1 mA to 20A
2	AC Voltage & Current/ *AC-DC Voltage Transfer Difference	Calibrator	Voltage & Current Calibrator	Multifunction/Multi Product Calibrator	1 mV to 1000 V 1 mA to 10A
3	AC Voltage	Reference Source	AC Reference Source		
4	AC Current	Transconductance Amplifier	Transconductance Amplifier	AC Current Source	Up to 20A
5	AC-DC Voltage Transfer Difference	Potentiometer	Micropotentiometer	Low Voltage TVC	1 mV to 200 mV
6	AC-DC Voltage Transfer Difference	Thermal Converter/AC Measurement Standard/	Thermal Voltage Converter		1 mV to 1000 V
7	AC-DC Voltage Transfer Difference	Thermal Transfer Standard	Thermal Transfer Standard		250 mV to 1000 V
8	AC-DC Voltage Transfer Difference	Thermal Transfer Standard	Thermal Transfer Standard		2 mV to 1000 V
9	AC-DC Current Transfer Difference	Shunt	Current Shunt		1 mA to 20A
10	AC-DC Voltage & Force	Amplifier with Load cell	Dynamic Strain Meter		0–5 V and up to 100 kN
11	AC Voltage	AC Measurement Std	AC Measurement Std	AC Measurement Std	1 mV to 1000 V
12	AC-DC Current Transfer Difference	Current Shunt	AC Measurement Std/ + Current Shunt	AC Measurement Std/ + Current Shunt	1 mA to 20A

(continued)

Table 12.7 (continued)

S. No.	Parameter	Item name	Item type/group	Alias name	Range
13	PMU Calibration	M Class	Phasor Measurement Unit	PMU	50 V to 150 V 1A to 5A 50/60 Hz 25/50 fps
14	PMU Calibration	P Class	Phasor Measurement Unit	PMU	50 V to 150 V 1A to 5A 50/60 Hz 25/50 fps

Technical Consultancy

Technical services and consultancy for Phasor Measurement Unit (PMU) performance evaluation and testing provided to M/s Valiant Communications Limited, New Delhi. Valiant Communications Limited is one of the first companies in India, and only one of the few companies in the world designing Phasor Measurement Unit (PMU). M/s Valiant Communications required technical consultancy in performance evaluation of PMU under development PMU as per IEEE synchrophasor standards using CSIR-NPL PMU System.

The technical service rendered towards the performance evaluation of under developing PMU from a traceable PMU-CAL system had enabled client with the IEEE compliance certified equipment with national traceability for both P and M class. Traceable measurements and our expertise have helped M/s Valiant Communications in performance evaluation of their PMU under development. The IEEE compliant calibration certificate will help the client in their business of exporting the items. Technical services offered had helped, Valiant Communications Ltd. in not only manufacturing world-class PMUs in India, but also to become one of leading companies of the world.

Training and Human Resource

CSIR-NPL is also providing training to industry and academia from Technical Staff to research scholars/students not only from India but also from SARC nations and APMP. The aim is to transfer scientific and technical knowledge in the field of metrology to create manpower well educated scientifically and technically for the country. The training programs consist of theory lectures on various scientific & technical concepts, followed by practical demonstration and hands-on training on the equipment.

CSIR-NPL always encourages professional and student trainings. Here is the list of Professional or student trainees that are imparted trainings:

These include professional trainings to the metrologists of BSTI Bangladesh, PNG Papua New Guinea, KIMLIPI Indonesia and NIS Egypt on LF Voltage and LF current calibration techniques.

Apart from this we are also providing academic training to the engineering students of B. Tech, M. Tech and Diploma PMQC course. To support the Level II labs, various kind of Metrology workshops are also organized at different times.

12.2.2.2 Future Metrological Road Map

The Primary Standards and measurement capability realized, established, maintained, upgraded and disseminated at CSIR-NPL provides a strong base for support to many national plans spread across various ministries and departments such as Airport Authority of India, ISRO, Ministry of Power, Ministry of Electronics and Information Technology, DRDO, MSME etc.

Electronics is a key area of defense technologies and become a vital component of nearly all the weapon systems, platforms and equipment designed and developed for defense purpose. As per ELCINA [25], the production of Strategic Electronics has grown a long way, since 2017 there has been focus on increasing defense production, R&D, indigenization as well as support to MSMEs and start-ups. However, govt has set a target of Rs. 170,000 cr by 2025 and exports of Rs. 35,000 cr. This requires growth rate of 17–18% annually over the next 7 years. The total global electronics hardware industry is about US$2 Trillion, out of which, India's Production was about US$ 47 billion during the year 2016–17 [26]. The domestic consumption in India was about $ 86.4 billion during the year 2016–17, while exports were about US$ 6 billion. As a result of various initiatives taken by the Government and efforts of Industry, production of electronics in India has shown significant growth of 22% during the last three years. The government has placed electronics manufacturing on high priority [27] with a major focus on initiatives such as Digital India, Make in India and supportive FDI policies to bolster electronic manufacturing. As a result, domestic production is expected to grow at a CAGR of 27% during 2016–2020 to reach $ 104 billion in 2020, as compared to the CAGR of 9.6% during 2010–2016 [28].

National Policy says that electronics industry is the world's largest and fastest growing industry and is increasingly finding applications in all sectors of the economy. India is a signatory to the Information Technology Agreement (ITA-1) of WTO and Free Trade Agreements (FTAs) with various countries/trading blocs such as ASEAN, Korea and Japan [29]. However, the electronics hardware manufacturing sector faces lack of level playing field vis-à-vis competing nations on account of several disabilities which render domestic electronics hardware manufacturing uncompetitive. This includes lack of adequate quality infrastructure and inadequacies in skill development. Manufacturing base of electronics products in the country is grossly inadequate in comparison to demand of such goods. Even in cases where products are manufactured in India, the extent of domestic value addition is low. One of the critical constraints in the way of attracting investments for manufacturing is lack of availability of reliable and quality infrastructure for growth of industries and its dispersal across the regions. With a view to promote industrialization and growth of electronics manufacturing in the country; the Government notified Electronic Manufacturing Clusters (EMC) Scheme to provide support for creation of world-class infrastructure for attracting investments in Electronics Systems Design and Manufacturing (ESDM) sector. CSIR-NPL as an NMI can play an important role along with STQC labs under MeitY and therefore proposes to introduce a quality infrastructure for electronic equipment having traceability in LF & HF Voltage and LF Current and quality assurance to compliment the efforts put-in by the Government to make India an Electronics Manufacturing Hub as envisaged under "Digital India" and "Make in India" initiatives [30].

"24 × 7 Power for All" is a Joint initiative of Government of India (GoI) and State Governments with the objective to ensure availability of 24 × 7 power supplies to all households, industries, commercial businesses, public needs & any other electricity consuming entities and adequate power to agriculture consumers as per the policy. However, with the help of quality infrastructure, traceable measurements, affordable, reliable and quality power supply can be achieved with the availability of 24 × 7 power supplies. Hence, Indian govt. is implementing their dream projects of smart grids and smart cities, which fully depend upon PMUs reliable functioning. PMUs are being installed at electric power grids to monitor their static and dynamic behavior. The measurement data they record is important to detect problems on these systems and to prevent blackouts.

With the traceable measurements, it is expected that the PMUs deployed for monitoring purpose in the national grid system will provide accurate and uniform results that will lead to good efficiency, reliability and quality. This will also ease the implementation of smart grid system in India thereby attaining more economical and more reliable delivery of power throughout the nation. If all the PMU testing laboratories are traceable to national standards along with the trained operators, the measurement results will be comparable and within their uncertainties. Uncalibrated PMU measurement results can lead to the misleading reports and analysis based on such reports will result in severe financial losses and safety of the power grid. Around 3000 PMUs have already been deployed in the national power grid and their growth is expected to increase significantly in the near future. Though, calibrated PMUs are costly but their potential undoubtedly, will play a major role in providing the required real-time critical data in realizing futuristic GOI funded esteemed projects of smart grids & smart cities [31].

Sectorial Support

Metrological activities are working in the line with various national missions such as Make in India, Swasth Bharat, Power for all, one nation one grid etc. to establish a quality infrastructure that can provide all technological solution to the requirement from various sectors. With the rapid advancement of industrialization in the country, the area of electrical and electronic instrumentation has made tremendous progress. The growth of instrument industry is also keeping pace with the recent development. Therefore, industry required highest level of standards and calibration facilities from

CSIR-NPL. The standards of electrical and electronic parameters such as Voltage, Current and PMU have been established, maintained and periodically updated to meet all of the industries requirements and support.

In the field of electronic technology and metrology, the LF Voltage and Current measuring/generating instruments are of extreme importance to almost all advanced electronics manufacturing sector. They are used to calibrate high precision multimeters, Voltmeters, Ammeters, Measurement standard etc. The major components of industrial sectors such as Information Technology sector, Telecommunication sector, Automobiles sector, Electronic appliances sector and Medical equipment sector are electronics and hardware [32]. The electronic sector is normally bifurcated into consumer electronics, electric utilities, and general electronics. But consumer electronics drive most of the growth in this sector. Consumer electronics has evolved over a period with multiple industries and sectors, because of the development of new software, apps, robotics, artificial intelligence and personalized health care. The future of consumer electronics growth is driven by rising demand for smartphones, artificial intelligence, and voice recognition technology. Within the consumer electronics sector, companies that focus on emerging technology are driving significant growth and include manufacturers of smart watches, smart home products, and smart speakers [33]. All these electronic industries need a dedicated internationally approved measurement traceability from an NMI through apex level calibration services and good quality infrastructure to be export friendly. CSIR-CSIR-NPL is rigorously working for welfare of its customers by providing apex level metrological services. Being a National Metrology Institute (NMI) of India and a part of Asia Pacific metrology Program (APMP), it provides calibration to regional laboratories like STQC labs, DRDO, ISRO, BEL, IDEMI, 13 BRD Air force, Private Industries etc.

State-owned power transmission company Power Grid Corporation of India Ltd will spend Rs. 1 trillion over four years to expand its network and connect areas affected by left-wing extremism, facilitating Prime Minister Narendra Modi's goals on village electrification, uninterrupted power for all. PGCIL is also planning to expand overseas by bidding for global transmission projects jointly with private players. The idea is to scale up the global presence from consultancy services at present to engineering, design, procurement and construction. Eventually, it will explore the possibility of ownership of transmission assets abroad. Phasor Measurement Units are being installed at electric power grids to monitor their static and dynamic behavior. The measurement data they record is important to detect problems on these systems and to prevent blackouts. The PMUs major advantage is time synchronization of the data taken across the grids. National Primary Facility of CSIR-NPL PMU-CAL System can help with the required IEEE synchrophasor standard certifications of PMUs to cater the needs of Indian power sector.

National Mission and Programs

The technological boom and new start-ups in India have led to a tremendous increase in the power demand. The grids are being made smart, more efficient and reliable to meet not only the growing demands but also to prevent blackouts. The synchrophasor technology is evolving, explored and used by the power sectors for real-time measurements with Phasor Measurement Unit (PMU) as the main elemental device [19]. But there is a big challenge for them to ensure that the measured PMU data is precise, reliable and have IEEE C37.118.1a-2014 compliance, as the corrective action depends on these measured data. All the PMUs mounted at power grid must be calibrated to confirm that they are reliable and accurate and have same response. In order to meet the IEEE compliance, PMUs must be calibrated from a traceable PMU-CAL system. Indian power system is facing high Losses, poor distribution network, wide demand—supply gap of energy, poor asset management etc. Smart grid technology which further depends upon calibrated PMUs will bring solutions to all of the mentioned problems and sustainability by way of demand side management, demand response, outage management and reduction in losses and improved customer satisfaction. Large investment is expected for Smart Grid Applications in distribution in future Plan, which will provide huge business, prospects in coming years.

India's electricity sector faces many issues; one of the important is "Lack of timely information on load and demand". Intraday graphs at 15-min or more frequent intervals are required to understand the short comings of the power grid with respect to grid frequency, including comprehensive data collected from PMUs for all grid-connected generating stations and load data from all substations [34]. If the grid's supply and demand is not perfectly matched, frequency imbalances can cause stress on the grid, which is a potential cause for power outages. CSIR-NPL being an NMI (National Measurement Institute) of the country by an act of parliament which has the responsibility to establish national standards and provide apex level calibration services to the nation can play an important role in providing the measurement traceability to PGCIL.

However, PGCIL is playing an important role in implementing the govt. policies of smart grid and smart cities through maintain and operating PMUs but they lack in quality infrastructure and measurement traceability's to the NMI of the country to have best possible measurement uncertainties. Hence, we can work in co-ordination with PGCIL in providing the IEEE compliant certification and measurement traceability to their installed PMUs regularly, verification of the imported PMUs and maintaining the quality infrastructure. We can provide them IEEE compliant NMI calibration certificate which will boost their confidence in world trade and help them implementing the PMUs in any new govt. project across the country. In this way, we are trying to implement quality infrastructure in power grid for their enormous growth. As any event occurring analysis depends upon the PMU measured data, its precision and accuracy is utmost important. This can be realized once we get our PMUs calibrated regularly. The main aim of our future program is to provide apex level calibration and dissemination of standards for maintaining traceability of measurement to strengthen the overall development of metrological services in the country. CSIR-NPL as a govt. entity will always support Indian government programs to develop the country through industrial production which needs mission like power for all and continuous power and thus empowering the Indian economy and realize our dreams.

12.2.3 Economic Impact

In the next 5–7 years India's technology sector is expected to produce \$1 trillion opportunity that include \$400 billion of hardware, \$350 billion of software and \$250 billion of telecom and Internet of Things. As per the report published by Department of Electronics and IT (DeitY), the 25% of India's GDP will potentially come from the technology sector. The electronics and hardware industry in India's GDP play an important role in the growth of the other industries which will impact on Indian economy. The electronics and hardware industry are growing very rapidly due to the manufacturing of the semiconductors. The encouraging conditions, favorable policies with terms and conditions have made India one of the leaders in electronics and hardware. However, "Global Electronic Components Market 2020 Research Report," says that the international market for electronic components is projected to grow at a compound annual growth rate (CAGR) of about 4.8% from 2020 to 2025 [32].

The most important and crucial component of country's infrastructure for economic growth and welfare of nation is power. The infrastructure existence and its continuous development are needed for the growth of the Indian economy. 24 × 7 power supply plays an important role in providing quality life to the nation. India is the world's third largest producer and third largest consumer of electricity. As of today, the national electric grid has an installed capacity of 370 GW as per the data provided by PGCIL. Total FDI inflow in the power sector reached US\$ 14.65 billion recently accounting three per cent of the total FDI inflow in India. GDP is the measure of the economic development and the quality of life in a country that mostly depends upon the power and energy consumption. The Industry sector which includes electricity and power consumption contribute around 29.6% of the Indian GDP [31].

12.3 Magnetic Metrology

Magnetic metrology has its technological implication to every aspect of life such as transportation (Automotive, Railway Marine) or Consumer Electronics, wearable devices, gaming consoles, drones, also in healthcare, aerospace and defense, ATM and many more. The key reason to use magnetic sensing on above-mentioned sectors is the capabilities of these magnetic sensors in various areas such as speed sensing, position and navigation sensing, metal detection and flow rate sensing [35]. These different types of sensing are possible via magnetic materials based on different physics phenomena on which these sensors can be classified as either Hall Effect Sensors or Magnetoresistive Sensors or SQUID Sensors or Fluxgate Sensors. These sensors can be further classified based on their range i.e., <1 Microgauss (Low-Field Sensors) or Microgauss–10 Gauss (Earth Field Sensors) or > 10 Gauss (BIAS Magnetic Field Sensors). From above classifications, it is clear that the variety of magnetic sensors and their application in day-to-day life make magnetic metrology an essential subfield in electromagnetic metrology. Apart from the above potential application, the magnetic metrology plays a key role in calibration of Magnetic resonance imaging (MRI) instruments which is sensitive non-invasive medical imaging tool.

CSIR-NPL is committed to continuously develops and commission the capabilities to the range of measurement techniques and support the stake holder by providing calibration and testing service facilities to our Industries in accordance with ISO 17025. The important element in most of the industries as per their demand "National standard for magnetic measurements" was decided to establish in 1994 and by the end of 1998 the magnetic standard laboratory started working with the facilities of apex level calibration of magnetic field, H-sensors, AC and DC measurements on soft magnetic materials and magnetic measurements on permanent magnetic materials [36].

The magnetic properties measurements facilities of electrical steel Epstein samples was established as per the IEC 60404–2 standard for core loss, peak value of magnetic induction, specific apparent power as a Primary Standard for Power Loss in Soft Magnetic Materials [37]. To carry out the Industrial measurements of intrinsic properties of magnetic materials, the Vibrating Sample Magnetometer Facilities was established for extremely sensitive open circuit testing of solid, thin film, powder and liquid.

12.3.1 Measurement Capabilities and SI Traceability

The prime role of the Magnetic Metrology is maintaining and upgrading National Standard related to magnetic parameter like Magnetic flux Density, Magnetic flux, turn area of search coil, power loss measurement of electrical steel, through continuous development and providing calibration/test facilities of such parameters to industries and institutions as per ISO/IEC: 17025 guidelines, which are traceable internationally/with national standards [38].

12.3.1.1 Primary Standards

The unit of magnetic flux density, Tesla (T) is realized using NMR magnetometer (Primary standard of CSIR-NPL), a large pole diameter (25 cm) electromagnet with programmable current source for field generation. The traceability chart of magnetic flux density to the SI units are shown in Fig. 12.15. The NMR magnetometer intrinsically measures the precession frequency of the proton (used as NMR probe), more precisely the Larmor precession frequency, which is directly proportional to the exposed magnetic flux density times the gyromagnetic ratio of the species involved [39]. In the atomic or nuclear resonance type magnetometers, the value of magnetic flux density is measured as the resonance frequency of transition (Larmor precession frequency) of a species whose energy level separation has been obtained by an applied magnetic field. An externally applied magnetic field creates a stable population difference between the ground state and the excited state in the species. This population difference can be tapped by the application of an external energy in the form of radio frequency (RF) wave that has the same energy value as that of the energy splitting between the ground state and the excited state. A suitable RF detector (frequency counter) are used to detect this absorption frequency and are used as the direct measurement of magnetic flux density unit. Thus, the measurement of the magnetic flux density is reduced to the measurement of frequency (or time-seconds), which is a base SI unit. The NMR magnetometer covers a magnetic flux density ranging from 0.043 to 1.5 T with the help of different probes of proton and deuterium.

The NMR Teslameter with different probes is calibrated from NPL, India in the field range of 0.05 to 1.5 T and associated uncertainty ± 5 ppm ($k = 2$)). The NMR is traceable to the Time & frequency division of NPL (India) [40]. The calibrated NMR Teslameter serves as the primary standard for the transfer standard Gaussmeter, reference magnets, DC electromagnets The calibration of primary standard (here a NMR Teslameter) is by direct comparison with the values measured by the external frequency counter [41]. The block diagram of the calibration setup is shown in Figs. 12.16 and 12.17 shows the actual measurements equipment setup of CSIR-NPL, New Delhi. For magnetic fields of the order of 1 T, NMR frequency lie in

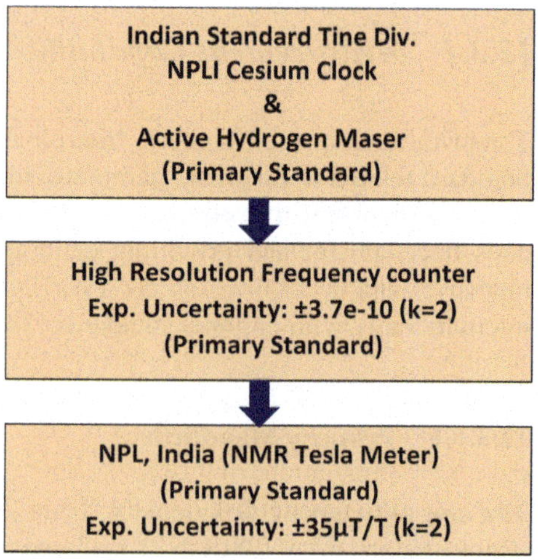

Fig. 12.15 The traceability chart for magnetic flux density

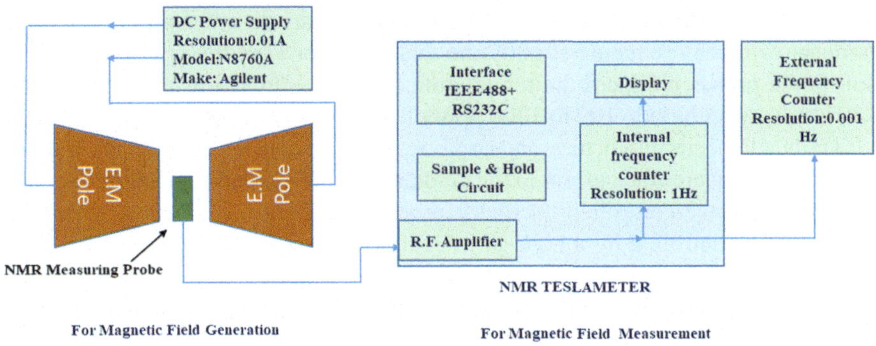

Fig. 12.16 Schematic of Setup for the calibration of the primary standard (NMR teslameter) against high-resolution frequency counter

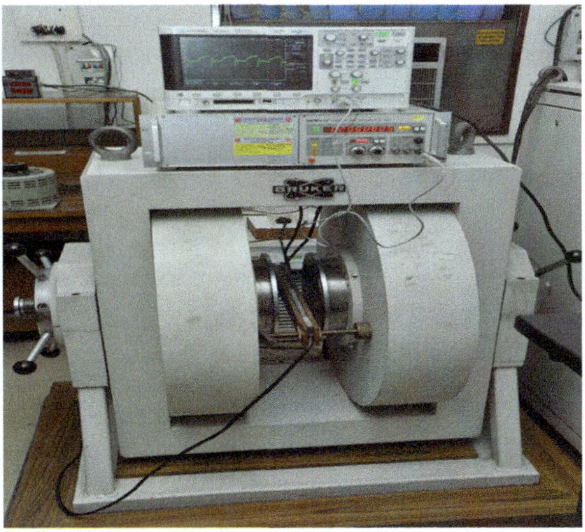

Fig. 12.17 Photograph of primary NMR measurement setup equipment at CSIR-NPL

the radio frequency region. The NMR frequency for protons and deuterons are γ_P, ^{1}H = 42.57608 (12) MHz/Tesla (for protons) and γ_d, ^{2}H = 6.53569 (2) MHz/Tesla (for deuterons), respectively. For detecting the proton magnetic resonance, a small water filled coil is placed in a static field *B*, with its axis perpendicular to *B*. the magnetic moments of the protons in the water molecules point preferentially in the direction of *B*, i.e., the lower energy states are more populated than the higher ones. Therefore, if the transitions are induced with an alternating magnetic field, those from lower to higher energy states are more frequent than the contrary. The protons absorb more energy from the alternating field than they supply to it and the difference between the populations of the two energy states is reduced. The thermal equilibrium populations are re-established due to spin–lattice interactions, at a rate described by the spin lattice relaxation time T_1. This is the reason why protons continuously absorb energy from the alternating field if the coil is driven at the proton resonance frequency, thereby reducing slightly the quality factor Q of the coil. Briefly, a flat coil in the probes produces the modulating field B_{mod}. Its frequency is from 30 to 70 Hz and its amplitude 100 to 1000 ppm of B_0.

The NMR electronics detects and amplifies the nuclear resonance signals of the LC circuit and measures the current in the modulating coil at the instant when resonance occurs. A voltage-controlled oscillator produces the radio frequency voltage. Its frequency is controlled by a high-gain feedback loop to ensure that the resonance occurs at the instant when B_{mod} crosses zero. Therefore, this frequency f, equals the proton resonance frequency of B_0 and automatically follows any changes in B_0. The LC circuit is automatically tuned to the applied by means of varicap diode. An internal frequency counter measures the NMR frequency, which is sent to the probe, the result is displayed in Tesla with a resolution of 0.1 μT (1 mGauss) or on frequency with a resolution of 1 Hz.

12.3.1.2 Calibration and Measurement Capabilities

Magnetic Metrology is committed to the establishment, realization, development and maintenance of standards of magnetic quantities and associated calibration procedures traceable to National Standards (Reference/Transfer). There are ten calibration and measurements capabilities (CMCs) applied in BIPM by the magnetic metrology. Moreover, the decision for the CMCs is awaited. The Table 12.8 represents important measurements parameters with units and method with expanded uncertainty claimed as CMCs in BIPM.

12.3.1.3 Derived Measurement Capabilities

Magnetic Metrology provides [42] solution to industrial problems and industrial consultancy. Some of the important calibration & test facilities available are Calibration of permanent Magnet, Calibration of AC/DC Gaussmeter, Calibration of Magnetic field indicator, Calibration of low-field magnetometer, Calibration of Helmholtz Coil, Calibration of Fluxmeter, Calibration of Turn area of search coil, Testing of AC/DC electromagnet, Power loss measurement of Electrical Steel and Magnetic Radiation Measurement. Apart from the above-mentioned calibration services magnetic metrology also provide services for the industries required measurements like magnetic field mapping of yoke, the testing of magnetic materials and other magnetic related measurements [43]. The technical background and discussion for calibration of some of these equipment's are as follows:

Table 12.8 The list of measurements capabilities applied as CMCs in BIPM

S. No	Parameter	Range	Unit	Expanded uncertainty (%)	Method
1	Magnetic Flux Density	0.05–1.5	T	1	Direct Comparison with NMR Magnetometer
2	Magnetic Flux Density	1–2000	μT	1.6–0.5	Helmholtz coil and Fluxgate magnetometer
3	Turn area of search coils	0.0002–0.2	m^2	0.2	NMR Magnetometer, Electromagnet
4	Magnetic flux (Fluxmeter)	0.0001–10	V.s	0.005–0.15	Direct Volt-second pulses from the Volt-second Generator
5	Specific total Power loss	0.1–10	Wkg^{-1}	0.2	BH loop tracer 25 cm Epstein frame strips (IEC 60404–2)
6	Peak Value of AC Magnetic polarization	1–1.7	T	0.2	BH loop tracer 25 cm Epstein frame strips (IEC 60404–2)
7	Peak Value of Magnetic field Strength	5–10,000	Am^{-1}	0.2	BH loop tracer 25 cm Epstein frame strips (IEC 60404–2)
8	Specific Apparent Power	0.5–250	$Vakg^{-1}$	0.2	BH loop tracer 25 cm Epstein frame strips (IEC 60404–2)

i. **Calibration of Hall Effect Gaussmeter (Transfer Standard) against the NMR Magnetometer (Primary Standard) for Range: 0.05 to 1.5 T and Stated uncertainty: ±1% ($k = 2$)**

The traceability of the transfer standard magnetometer (Hall Effect Gaussmeter) with respect to the primary standard of magnetic flux density measurement, i.e., an NMR magnetometer (PT2025) is established for Range: 0.05 to 1.5 T with Stated uncertainty: ±% ($k = 2$) to develop a methodology for the calibration of the Gaussmeter against the NMR magnetometer (National Standard). The calibrated Gaussmeter serves as the Transfer standard for the secondary standard Gaussmeter (Working Standard), reference magnets, DC/AC electromagnets etc. The calibration of Transfer standard magnetometer (here a Hall Effect Gaussmeter) is by direct comparison with the values measured by the NMR magnetometer. The hall coefficient for semiconductors magnetic field sensing probe are presented by Eqs. (12.2) and (12.3), where the carriers are generally both the electrons and holes which may be present in different concentrations and have different mobilities [44, 45]. For moderate magnetic fields the Hall coefficient is

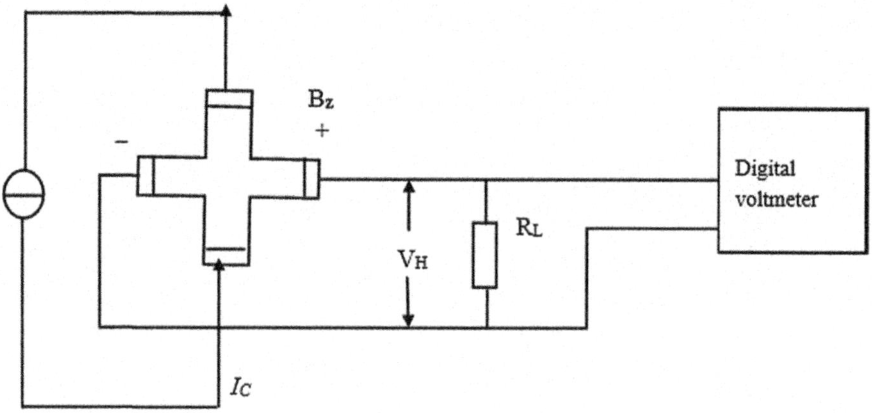

Fig. 12.18 Schematic view of a hall magnetometer circuit with DC control current I_c

$$R_H = N_d = \frac{1}{1 + l_m/l_g} \tag{12.2}$$

where 'n' is the electron concentration, p the hole concentration, μ_e the electron mobility, μ_h the hole mobility and 'e' the absolute value of electron charge. For large applied fields the simpler expression analogous to that for a single carrier type holds.

$$R_H = VH \tag{12.3}$$

With, $b = V_H = -\frac{IB}{\text{ned}}$, A stabilized current source provides the constant control current i_c and the Hall voltage $V_{H,}$ detected on the transversely placed contacts, is fed into a precession digital voltmeter, possibly through a pre-amplification stage. The role of the load resistance R_L is one of the counter acting, when required, the possible non-linear behavior of V_H, in balance with variation of internal resistance of the Hall element as measured between the Hall contacts. Note that the Hall semiconducting plate electrically connects the current source circuit and the output circuit. Consequently, no other direct connections can be tolerated without disrupting the potential at the four contacts on the Hall plate. No common ground can exist between the input and the output circuits. Deviations from perfect material homogeneity, slight asymmetry of the plate and misalignment of the contacts, thermal electromotive forces and the signal conditioning circuitry itself can give rise to offset voltages. These are normally compensated for, preliminary to any measurement, either by acting on a potentiometer of a zero-balance circuit or, frequently, in recent commercial products, by exploiting automatic self-zeroing of the setup itself. Because of the presence of earth's magnetic field, it is expedient to make zeroing with the hall probe inserted in a zero-gauss chamber (null chamber).Besides offset voltage and non-linearity sources

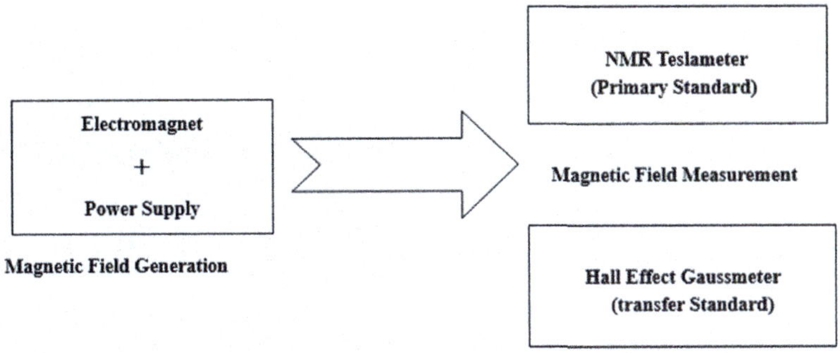

Fig. 12.19 Schematic block diagram setup for the calibration of the transfer standard Gaussmeter (Hall Effect) against the Primary Standard (NMR Teslameter)

of uncertainty in the field measurement by a Hall gaussmeter come from the temperature dependence of R_H and the uncertainty in the hall plate orientation. Figure 12.18 shows the schematic Hall magnetometer circuit with DC control current.

The thermal variations of the hall voltage are kept around ±0.05%/K in quality high commercial products. The Hall plate is placed during the measurement, normal to the field direction, a position associated with the maximum instrumental reading. With an angular uncertainty of ±1°, an uncertainty of ±0.01% is achievable in the magnetic field reading. When the field has a strong planar component, the so-called planar Hall effect introduces a small contribution to V_H. It is actually an effect deriving from the anisotropy of magnetoresistance and is proportional to the difference $\rho_{\|}$-$\rho_{\perp}$. It can be disregarded under normal measuring conditions. The block diagram for calibration setup are shown in Fig. 12.19.

The main components of uncertainty associated with these measurements are spatial inhomogeneity field, Probe angle Error, Gaussmeter Uncertainty, Probe Uncertainty (Linearity), Magnetic field instability due to current, Uncertainty of NMR, deviation from NMR value, Uncertainty-Repeatability-NMR and Uncertainty—Repeatability of Gaussmeter. The combined uncertainty is represented as the square root of sum of all the associated parameters.

ii. **Calibration Procedure for Reference Magnets as magnetic flux density source**

The Calibration procedure is applicable for the calibration of reference magnets and for the determination of the magnetic field strength at specific points or on the pole faces of permanent magnets in the Range: 0.01 to 1.5 T and Stated uncertainty: ±1% ($k = 2$). Reference magnets are magnetic flux density sources that can be used for the calibration of industrial Gaussmeter. It is a direct calibration process in which the magnetic flux density values in the air gap of the reference magnet is measured using a calibrated Hall probe. The ability of a permanent magnet to provide a defined and stable magnetic field in a certain region of space depends besides, the material properties, on the specific features of the magnetic circuit employed. This means that the shape of the magnet must be optimized for required application, often on

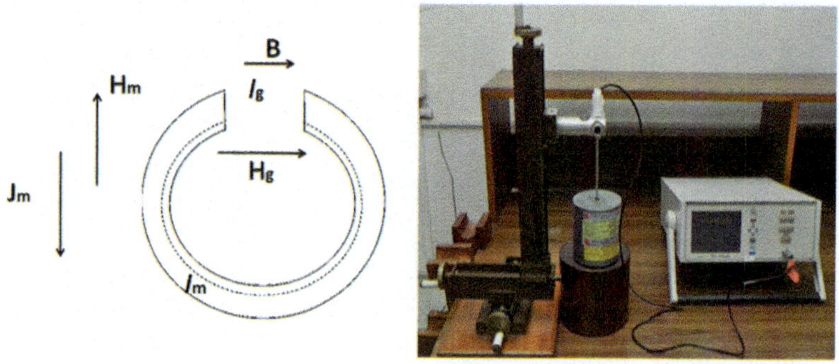

Fig. 12.20 Representation of a permanent magnet as an open ring with homogeneous field in the gap and the measurement setup of reference magnet with Gaussmeter at CSIR-NPL

the combination with flux guiding soft magnetic parts [46]. A classical permanent magnet source can be schematically represented by gaped ring as shown in figure below (Fig. 12.20). It is assumed that the magnetic flux is uniform along the whole circuit. The continuity equation for the flux Φ imposes the same value of magnetic induction in the material and in the gap, i.e., $B = B_m = B_g$. On the other hand, the line integral of the field along the average circumferential length $l = l_m + l_g$ is, according to Ampere's law, equal to zero, which means $H_g\ l_g = -H_m\ l_m$. Where H_m andH_g are the fields in the material and the gap respectively. H_m is the demagnetizing field directed against the polarization J_m in the material. A classical permanent magnet source cab be schematically represented by gaped ring as shown in Fig. 12.20.

The flux generated by cylindrical parallelepiped permanent magnet is conveyed in the gap by a pair of soft pole pieces. Normally these poles are made of very soft Fe having very high saturation polarization and very high permeability [47]. The measurements setups of reference permanent magnet with Gaussmeter at CSIR-NPL, New Delhi are shown in Fig. 12.20. The calibration process of the reference magnet is to measure the magnetic flux density in the air gap of the permanent magnet using a calibrated Gaussmeter. The main components of uncertainty associated with these measurements are Gaussmeter Error, Spatial Inhomogeneity.

iii. **Calibration Procedure for Turn area of a search coil**

Calibration of search coils for the determination of its turn area using the NMR magnetometer and an integrating fluxmeter in the range of 2×10^{-4} m^2–0.2 m^2 (2 cm^2–2000 cm^2) and with expanded Uncertainty: 0.2% ($k = 2$). The method of calibration of standard search coils was established for the determination of its turn area (area turns) and to link the traceability of the measurement to the primary standards of magnetic flux density. The magnetic flux density is measured using an NMR magnetometer. The block diagram of the calibration setup for the search coil is shown in Fig. 12.21.

The NMR Magnetometer with probe 1–5 has been calibrated from NPL, India in the field range from 0.05 to 1.0 T. So, the traceability link of the measurement

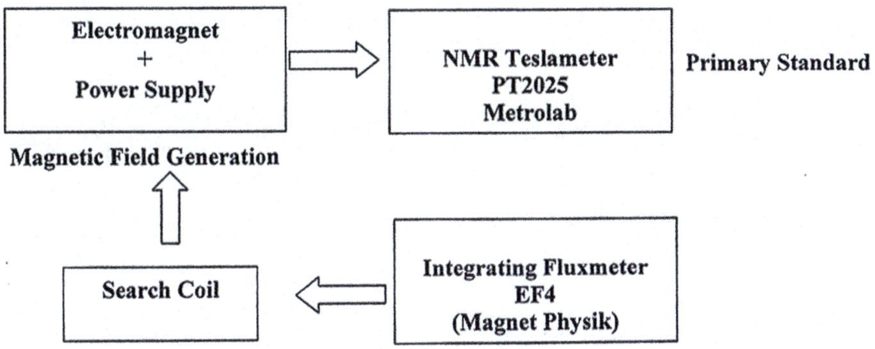

Fig. 12.21 Calibration setup for the calibration of the search coil area turns

of magnetic flux density in the range from 0.05 to 1.5 T is established. The main components of uncertainty associated with these measurements are Repeatability, Uncertainty in the NMR measurement, Uncertainty of the fluxmeter, Uncertainty originating from the angle of insertion, Uncertainty in the magnetic field non uniformity, Uncertainty in the Turn area (Physical). The combined uncertainty is the sum of all components square root as already shown in other process.

iv. **Calibration of low-field magnetometers**

Calibration of magnetometers such as fluxgate, magnetic field indicators (pilot compass) digital compasses, and hall effect gaussmeter in the low field range up to 2 mT are discussed here in the range of 1–2000 μT with Expanded Uncertainty: ±1.6 to 0.5% ($k = 2$). This method of calibration for magnetometers was developed using a Helmholtz coil of known coil constant. The coil constant is the ratio of magnetic flux density to the current which produces the flux density ($k = B/I$). The traceability link of the magnetic flux density standards using physical artifacts such as Helmholtz coils is equally important for precise measurements of low magnetic field [100].

As the range intended for the low-field magnetometer calibration is from 1 μT to 2 mT, one need to cancel out the earth's magnetic field component and other magnetic noises in the region of field generation. This is usually achieved by placing the standard Helmholtz coil inside the larger tri-axial Helmholtz coil pairs, which are driven independently by high precision current sources. Required amount of current is passed through the tri-axial Helmholtz coil pairs so that cancelation of the background field can be achieved to the minimum level as possible. The schematic diagram of the calibration setup and the experimental setup of low-field calibration at CSIR-NPL are shown in Fig. 12.22a and b.

v. **Calibration of Fluxmeter**

The process was developed and established for the measurement of magnetic flux using Flux meters in the range of range: 0.0001–10 V-seconds (10–1000 kilo Maxwell turns) with Stated Uncertainty: ±0.005–0.15% ($k = 2$). The SI unit of magnetic flux is

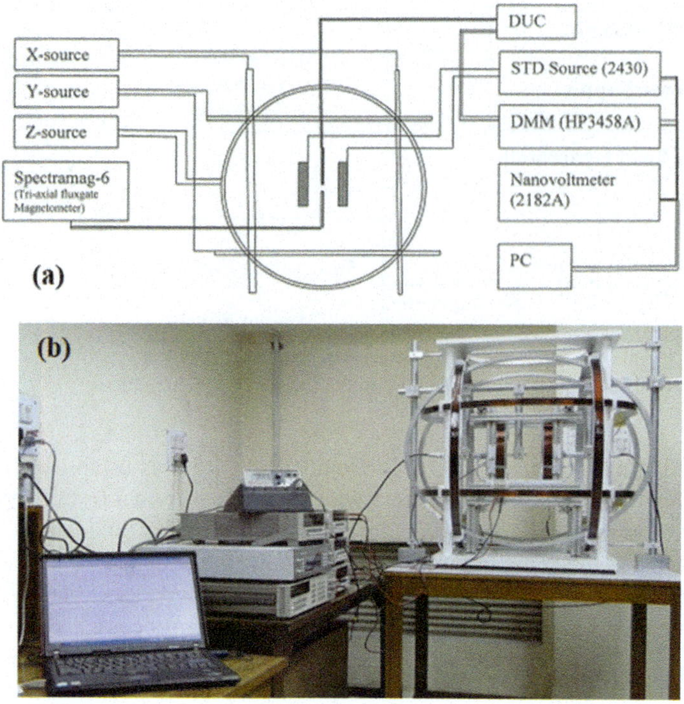

Fig. 12.22 **a** schematic diagram of Calibration setup for low-field magnetometers calibration, **b** Photograph of the low magnetic field calibration setup at CSIR-NPL

Weber (volt-second). Most fluxmeters measures the total change in flux of a changing magnetic field by integrating the voltage induced in a search coil. Fluxmeters are calibrated in Maxwell turns (kilo Maxwell turns) or in milli Weber. The volt-second generator or the fluxmeter calibrator produces output pulses in terms of volt-second. In other words, the integral of the pulse is simply height × width, or volt-seconds. The relationship 1 Wb = 1 V-second = 10^8 Maxwell turns, provide a method of calibration of fluxmeters, by directly applying volt-second pulses on the fluxmeter input. The traceability of the unit of Magnetic flux to volt-second pulses are shown in Fig. 12.23.

vi. **Calibration Procedure for Magnetic Properties of Electrical Steel Strips**

The measurements of ac magnetic properties of grain oriented and non-oriented electrical steel strips, up to 400 Hz was established as per IEC60 404–2 Standard for the range of Peak Value of AC magnetic Polarization 1–1.7 T, Peak value of Magnetic Field strength: 5–10,000 A/m and Specific apparent power 0.5–250 VA/kg. The a.c. magnetic properties are determined for sinusoidal induced voltages, for specified peak values of magnetic polarization and for specified frequency. The

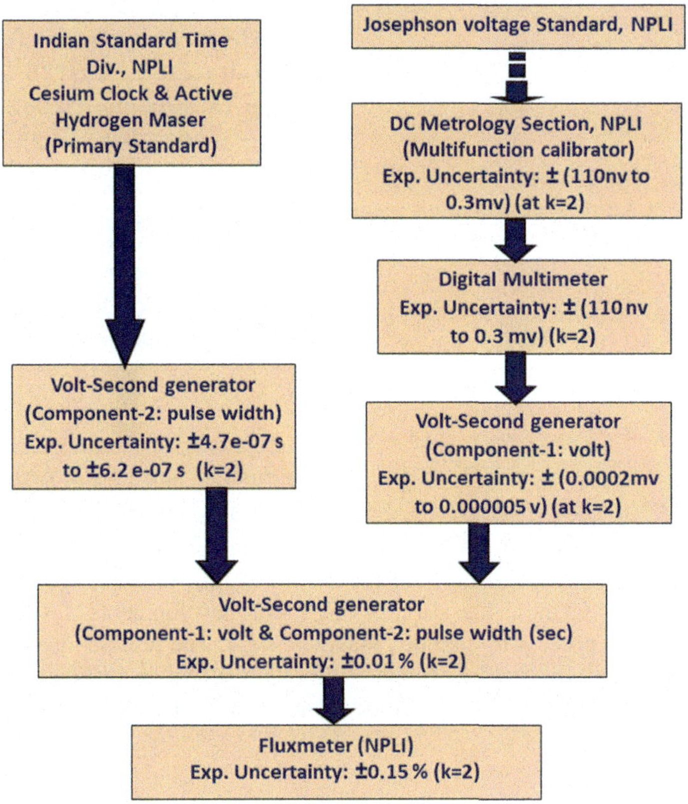

Fig. 12.23 The Traceability chart for Magnetic flux

test method fulfills the requirement of the user industry and other national institutes that require quality control for electrical steel as per the standard test procedure, IEC 60404–2 [37]. Presently the system is directly traceable to PTB Germany through the set calibrated Epstein strip samples. However, to establish the link of traceability independently International Inter-comparisons are proposed at regular intervals (Fig. 12.24).

12.3.1.4 R&D in Magnetic Metrology

The focus of research and developments is to develop the process to improve the uncertainty of all the existing measurements parameters. Especially the low magnetic field need to compensate the noise of background magnetic field. The magnetic field in the range of micro or nano tesla it is crucial to cancel out the background noise. The effort is going on to develop EMI shielding materials or the structure to perform such sensitive measurements. CSIR-NPL is also involved for the developments of

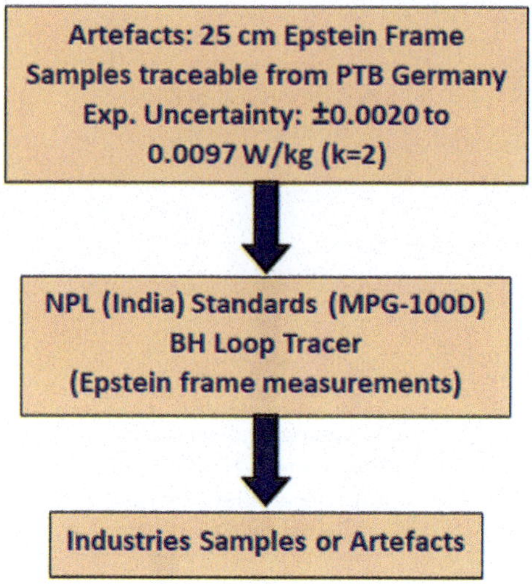

Fig. 12.24 The traceability chart for magnetic properties of electrical steel strips

frequency dependent magnetic field measurements for broad range of frequency facilities. Apart from these, it is also working for the developments of Nano magnetic mapping and the process to map this magnetic field. The precise measurements of Spintronic based device, which includes different resistive magnetic sensors and the upcoming technology, are part of our current research and development activities [48, 49].

12.3.2 Magnetic Metrology for Quality Infrastructure in India

The magnetic metrology is dedicated for the quality infrastructure of India as per ISO/IES 17025 guidelines. The Telecom Regulatory Authority of India, Central Electricity Regulatory Commission make and implement the regulation of magnetic based device and technologies. As an NMI, the role of CSIR-NPL is unavoidable bigger to be in the picture of quality control of these regulatory bodies. At the same time, the auto sector and its environment friendly engine and emission regulation are the important untouched problem where the quality infrastructure of India need to be NMI pro. Magnetic metrology is looking forward to being the strong pillar for the traceable measurements of magnetic based technology or product for quality infrastructure.

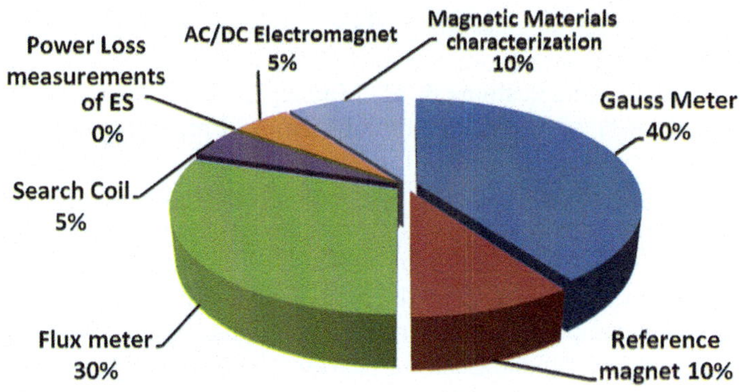

Fig. 12.25 The measurements parameter/equipment wise demand of industries

12.3.2.1 Metrological Services for Sectorial Growth

Calibration and Testing

The traceability to the costumer is provided through the unbroken chain of Calibration as the Primary/Reference standards maintained in Metrological Activities are inter-compared from time to time (once in two years) with Primary/Reference Standards of the other National Metrology Institutes in the world and degree of equivalence is ensured. The Reference/Transfer Standards (National Standards) are maintained and Calibrated against Primary/Reference Standards once in two years. The Secondary Standards are maintained and Calibrated against national standard at least once a two year. These are used for the Calibration/Testing of the instruments received from the various customers. The periodic calibrations are used in assigning current values and uncertainties associated with secondary standard. All other instruments used for calibration are traceable to National Standards. In special cases, with the prior written permission of the competent authority, directly calibrate precision instruments cab against National Standard/Primary/Reference standard as the case may be. The records of all the calibrations, inter-comparisons etc., are maintained and updated from time to time. The calibration of these parameters Magnetic Flux Density, Resonance Frequency, Magnetic Field, Magnetic Flux, Volt.Sec, Magnetic Power Loss, Permeability & Coercivity and Turn Area with different range are available with the magnetic metrology for the industries. The parameter/equipment wise application and demand from the industries are shown in Fig. 12.25.

The testing parameter available with the magnetic metrology, are B-H Curve/M-H Curve & M-T Test/Curie Temperature for magnetic materials, Magnetic Field of AC/DC electromagnet and Helmholtz coil.

12.3.2.2 Future Metrological Road Map

The Ministry of Communications & IT, Ministry of Electronics and Information Technology, Ministry of Micro, Small and Medium Enterprises are related to the magnetic based components quality for telecommunication and the regulation of magnetic field exposure limit. The transformer industries or the power transmission sector is directly linked to the Ministry of Power and Ministry of New and Renewable Energy. The Ministry of Civil Aviation, Ministry of Defence, Ministry of Railways, Department of Space with regulatory body about different magnetic based sensors or magnetic based components. Some of the important regulatory body are Telecom Regulatory Authority of India, Central Electricity Regulatory Commission.

The linking gap among the Ministries-regulatory bodies of government-NMI-Industries need to be strengthen in view of the dream project or program of government "Sashakt Bharat & Samarth Bharat". It is expected to touch the business of Rs. 16.16–18.18 trillion by Indian automotive industry including component manufacturing till 2026 [50]. The approval for the faster adoption and manufacturing vehicle (FAME-II) scheme with a fund requirement of Rs 10,000 crore has been granted by the Government of India for FY20-22 [51]. The government aims to make India global manufacturing hub and CSIR-NPL is committed to support the industries and contribute in economy. The avenue and the sectorial requirements are discussed here.

Sectorial Support

The cohesive relation between the NMI and the Industries is needed to implement a good quality structure. The regulatory body is expected to make the bridge and implements the quality structure for industrial growth and quality products. The sector wise relevance and their significance are discussed in sectorial support. A cohesive relation between the NMI India and their industries has always been the benchmark of CSIR-NPL. The calibration and testing services provided by NMI India is already listed above apart from that there are many sector and industries are still untouched. The power loss measurements of magnetic materials traceable to Indian national standard is paving the way for export of transformer and standing firmly with Indian industry. The magnetic metrology of CSIR-NPL is providing the facilities for power loss measurements traceable to SI units. The Indian power and distribution transformer market is expected to reach approximately Rs. 220 billion by 2022. The power as well as distribution network is getting expanded and strengthened across the country and the industry is facing its challenge to use secondary Cold rolled grain oriented (CRGO) material by manufacturers. The main raw material of transformers is CRGO steel that covers around 20–30% of transformer manufacturing cost. There is almost no manufacturing of CRGO in India, so the transformer industries completely depend on its imports. Being the largest consumers of CRGO, the efforts must be needed to facilitate the measurements and quality structure indigenously to support the industries. The quality control of the transformer has been ensured by standard IS 1180 to improve energy efficiency, confidence and reliability of distribution transformers. Central Electricity Authority (CEA) is taking care the standardization of power transformers. The power loss measurements of magnetic materials traceable to Indian national standard is paving the way for export of transformer and standing firmly with Indian industry. The magnetic metrology of CSIR-NPL is providing the facilities for power loss measurements traceable to SI units.

In the automotive and Aviation industry, the properties of magnetic material need to be accurately measured with traceability. A normal car or vehicle may contain hundred and more magnets and many of them may be in use for critical safety applications. The magnets and soft magnetic materials are being used in hybrid and electric propulsion motors and hence the traceable measure of these important component and its materials may certainly improve its efficiency and may limit the carbon efficiency. These magnetic based components have huge impact and contribution on our GDP or our economy. Similarly, in the aerospace industry, magnetic based components are under use of harsh environmental condition. It is needed to characterize their properties and have measurements as per established standards. Many of the strategy sector industries are based on magnetic components and the confidence in the product can come only by quality control of the components. CSIR-NPL is continuously supporting directly and indirectly to those industries by providing traceable measurements. Apart from the medical equipment's and its components calibration, the exposure of magnetic field levels produced by domestic and industrial appliances must be monitored by calibrated equipment for good health and safety norms. Simply, the manufacturers to maintain the EMC exposure limits to magnetic fields under exposure regulations must carefully monitor the MRI and any wireless power transfer.

The development of MRI calibration process is one of the future targets of the CSIR-NPL; the protocol and its research has already been initiated by developing the process for the generation of strong magnetic field and gradient magnetic field. These two field components have the main use in an MRI for taking picture of the body organs or anatomy. The optimization parameters of an MRI are free induction decay (FID) area, echo symmetry, signal amplitude, spectrum bandwidth and image distortion. These parameters are directly related to the strong static magnetic field and gradient magnetic field that must be as homogeneous as possible within the selected field of view (FOV) for reliable performance. The field homogeneity is influenced by the construction of the superconducting magnet, the susceptibility of receiving coil and the susceptibility of the sample itself. Adjustment of the homogeneity is still a challenge not only for the scanner manufacturers, but also for the operators of MRI instruments. The calibration of these mention parameters makes the MRI system more reliable. Two other medical applications of superconducting magnets are compact cyclotrons for charged-particle radiotherapy and in nuclear medicine.

National Mission and Programs

As per existing calibration and testing parameters, magnetic metrology is capable to cater traceable precise measurements for healthcare of all imported medical devices including MRI, high quality transformer productions, high-end household home appliances as well as advance public distribution system, digital payment security, automotive testing infrastructure to electric cars, wireless charging of various wireless devices including phones and cars, and various industrial applications.

Transformer manufacturing requires about 70–80% towards material cost. So, the MSME industry may enter into the materials production sector of these industries. The government's schemes such as the Deendayal Upadhyay Gram Jyoti Yojana (DDUGJY), the Integrated Power Development Scheme (IPDS) and the recently launched Sahaj Bijli Har Ghar Yojana (Saubhagya) have generated a spurt in demand for transformers [52]. The Ministry of New and Renewable Energy under Green Energy Corridor (GEC) scheme has planned to invest Rs 430 billion for enhancing transmission network which also support the transformer manufacturing [53]. The Ministry of Defence, Government of India, has approved the "Strategic Partnership" model for manufacturing submarines, fighter jets, helicopters and armored vehicles. All these strategic equipment or items need magnetic based sensors or components hence magnetic metrology may/must participate in these national mission programs for ensuring the quality through traceable measurements.

12.3.3 Economic Impact

The magnetic sensor market itself is very huge and it is expected to be valued at ~ Rs. 40,000 crores by 2024, growing at a CAGR of 7.5% between 2019 and 2024 [35]. The manufacturing and industry sector of India is the volumes of approximately Rs. 2,853,986 crores and Rs. 5,042,587 crores as per gross value-added share of 16 and 29% respectively for financial year 2018–2019 [54]. Similarly, the other sector where the magnetic based sensors, equipment or materials are involved may contribute for the economic impact of magnetic metrology. The magnetic metrology somehow contributes in these sectors by providing the calibration and testing facilities traceable to national standard and continuously helping to support in international trade.

12.4 Quantum Voltage Metrology

12.4.1 Measurement Capabilities and SI Traceability

Microwave applications of superconductor technology play a fundamental role in quantum metrology. One such microwave driven quantum standard Programmable

Josephson voltage standard (PJVS) system is established in CSIR-NPL. The Quantum Voltage Metrology maintains this system for the realization of SI electrical units 'Volt. It involves the precision measurements of microwave generated voltage levels that are intrinsically accurate and repeatable for the realization of voltage and assign the measurement values and uncertainties based on fundamental constants.

A detailed explanation of the operation of the Programmable Josephson voltage standard and the microwave optimization of the PJVS chip to provide traceability for voltage metrological applications along with the description of a few important measurement results are presented.

12.4.1.1 Primary Standards

Josephson Array Voltage Standard (JAVS) serves as the primary standard of voltage for metrological applications. Such standard is based on a quantum effects predicted by Josephson [55], which directly relates frequency of microwave irradiation of the superconducting tunnel junctions (known as Josephson Junctions) and the generated quantized voltage steps through fundamental constants by relation,

$$V_n = \frac{nf}{K_{J-90}} \tag{12.4}$$

here, n is an integer, and $K_{J\text{-}90}$ is the Josephson constant. This Josephson constant, depends on the fundamental constants and is given by the relation,

$$K_{J-90} = \frac{2e}{h} \tag{12.5}$$

where e is the elementary charge and h the Planck constant. The quantized voltage steps were later experimentally confirmed by Shapiro and were called 'Shapiro Steps' [56] that form the basis of the operation of this quantum voltage standard. The conventional value of this constant was accepted to a value of 483 597.9 GHz/V worldwide in 1990 [57]. Later, in the CODATA 2006 adjustment, a new value of this constant was evaluated to be 483,597.891 GHz/V [58].

The JAVSs generally require large series arrays of Josephson junctions (JJs) to generate a useable output voltage levels depending on the applied microwave frequency. The hysteretic JJs based Conventional Josephson Voltage Standards (CJVS) have many operational difficulties due to the step instability, difficulty in step selection and the step transition noise. These factors limit the fast and easy measurements for the quick automated analysis of signals in CJVSs. But with the advances in the material developments and the fabrication techniques of JJs, the arrays are devised in a way to produce damped and non-hysteretic current–voltage curve (IVC) as shown in Fig. 12.26. Such stable junctions are based on SNS and SINIS type junctions' This led to the development of second generation Josephson

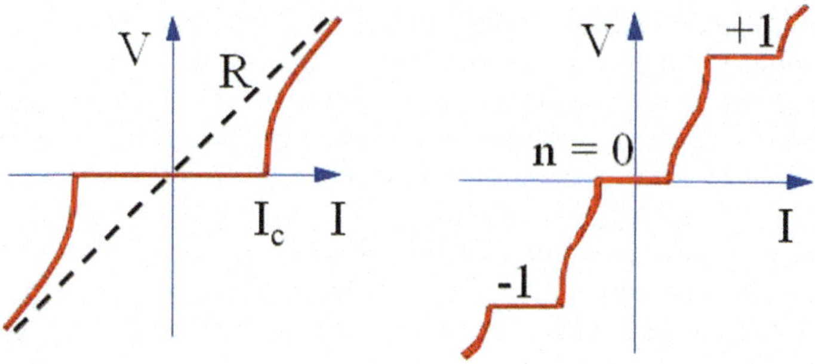

Fig. 12.26 The current–voltage (I-V) curve of a non-hysteretic junction (left without microwave and right with microwave applied)

standards called PJVS [59, 60, 61, 62]. The PJVS system being programmable, the step numbers (whether $n = +1$, 0 or -1) for all the junctions in the series can be digitally programmed. This makes possible complete system automation and easier control of the electrical response, bias current range and output voltage of all series arrays in the system.

Nowadays, PJVS systems are the primary standard of voltage in most of the NMIs. And one such PJVS system which has been established in CSIR-NPL to provide traceability for dc metrology application is shown in Fig. 12.27. The NIST designed '10 V PJVS chip' is the heart of this system, which is able to generate voltage levels up to 10 V. This chip contains a total of 256,116 Josephson junctions which are further divided into smaller subarrays. These JJs are made of Nb electrodes and NbxSi1-x barriers to produce non hysteric characteristics. All these junctions are connected in series and integrated in a microwave transmission line and further distributed into 32 microwave coplanar waveguide lines. The operating frequency of these junctions is from 17 to 21 GHz. The CPW transmission lines are designed in such a way to eliminate the need for through-substrate vias and reduce package dispersion unlike in Conventional JVS.

Fig. 12.27 Photograph of programmable Josephson voltage standard system at CSIR-NPL

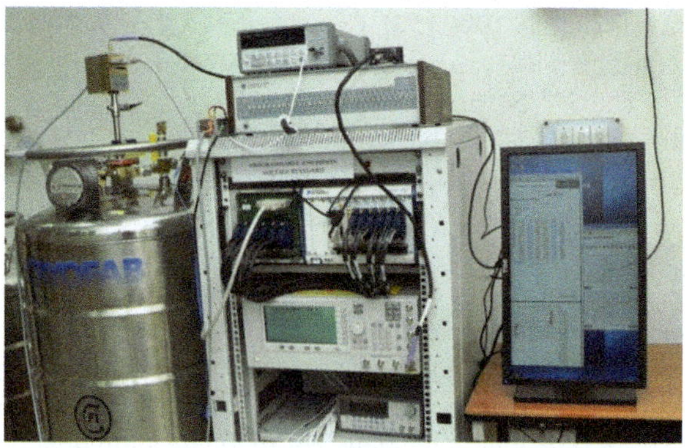

The proper operation of a Josephson array voltage standard has been exclusively mentioned in [63] which emphasizes the selection of a frequency at which the array and microwave source work well together. The microwave power and frequency must be adjusted to obtain an *I-V* curve shown in Fig. 12.28. The microwave coupling to the chip is done precisely to provide the uniform distribution of microwaves to a large number of Josephson junctions. Too much or too little microwave power to the junctions changes the current range of a junction's constant voltage step which results in smaller total current margin for an entire array of junctions [64]. The current margin is the parameter which defines the robustness of the system to an external current which is typically considered to have a value of 1 mA or more for a particular subarray bias configuration. To test the performance of bias parameters.

for a wider margin, flat spot measurements are carried out. In this measurement, individual subarrays are polarized for any of the N O or P state and the smallest step width of the active (*n* or *p*) subarray is test for all the subarrays to check the step flatness. In order to get a consistent operation, the flat spot should be centered around zero, with at least 0.3 mA symmetrical width on each side. One such optimization is shown in Fig. 12.28 a when flatspot were measured at 18.645 GHz with 0 dBm power level with combination of "*n*/*p*" state selection of subarrays junctions.

12.4.1.2 Calibration and Measurement Capabilities

CSIR-NPL provides calibration services to disseminate the primary voltage standard. The 10 V PJVS system is capable of providing traceability at 1.018 and 10 V levels. This system is primarily used for the calibration of dc reference standards which serves as secondary standard of voltage for further dissemination to industries. Two number of CMCs (details given in Table 12.9) mentioned with Institution Service code as NPLI/1–2 registered in BIPM [65] are maintained.

12.4.1.3 Derived Measurement Capabilities

DC Voltage Disseminating from PJVS to a Zener Reference Standard

Zener based DC reference standards are monitored and periodically calibrated through PJVS system. The comparison method has been used for providing the traceability. For calibration of unknown Zener reference standard (DUT), it is connected in series opposition to the Josephson array and difference between the array and the DUT voltage is measured by a precision digital voltmeter (DVM). The unknown voltage of the DUT V_{DUT} can be derived from the array voltage V_a and the DVM reading VDVM from Eq. (12.6) given below

$$V_{DUT} = V_a - V_{DVM} \tag{12.6}$$

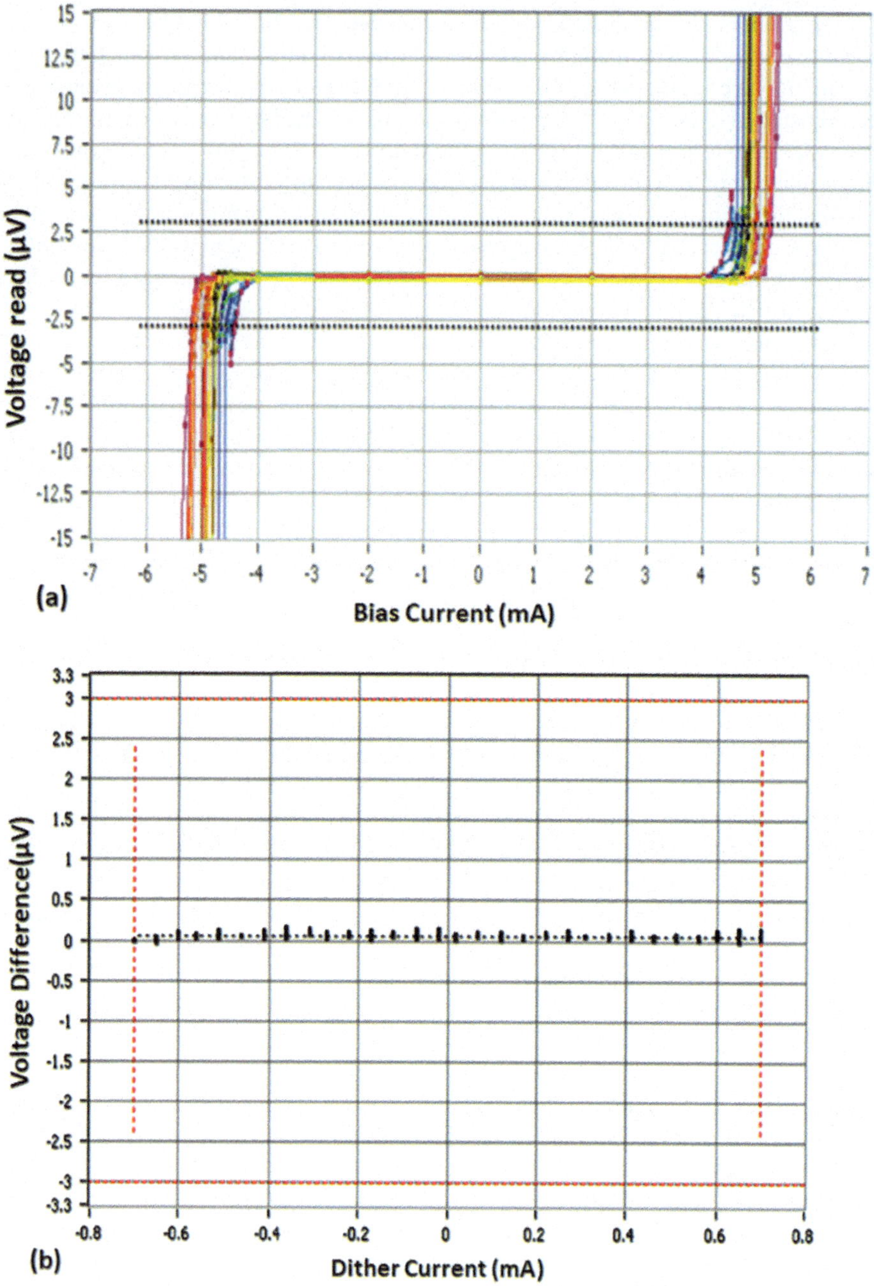

Fig. 12.28 Optimized Margin and DC flat spots at 18.645 GHz and 0.0 dBm

Table 12.9 CMC

Quantity	Instrument or artifact	Measurement method	Parameters	Expanded uncertainty
DC voltage sources: single values	Solid state voltage standard	Comparison	[1–10] V	[8.0E1 to 3.5E2] nV

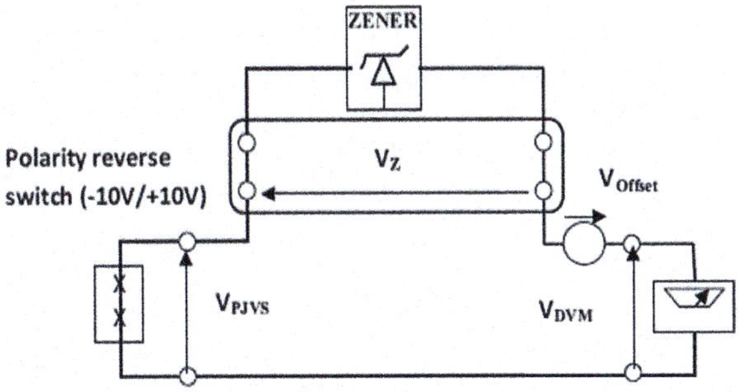

Fig. 12.29 Basic measurement circuit for calibration of Zener reference standard

Table 12.10 Uncertainty table for the calibration at 1.018 V level

PJVS uncertainty parameters			
Frequency reference	0.30 nV/V =	0.3	nV
Leakage	1.10 nV/V =	1.1	nV
Zero Offset	50.0		nV
Repeatability	0.000155897		nV
Expanded Uncertainty at 1.018 V		102.5670512	nV

Figure 12.29 shows basic principle of calibration of Zener reference standards against the Josephson Voltage Standard.

For final reporting of the results for a Zener reference standard, a number of samples per measurements (SPMs) are executed. From these SPMs, repeatability in the measurement (Type A uncertainty) is calculated. The other uncertainty parameters like zero-offset uncertainty, reference frequency uncertainty and leakage uncertainty are also assigned. The Table 12.10 below shows the uncertainty table for the calibration of an arbitrary Zener at 1.018 V level.

Another important application of the primary Josephson voltage standard is the calibration and linearity measurement of high precision digital voltmeters. Since PJVS operation involves the in-situ DVM calibration for gain and linearity parameters. Recently, the gain and the linearity parameter of the DVM have been calibrated using this primary PJVS system. These calibrated parameters have been used to provide traceability from primary standard for the microwave power metrology in coaxial micro calorimeter system.

12.4.1.4 R&D in Programmable Josephson Voltage Standard Metrology

One of key subjects in electrical metrology is to synthesize waveform of calculable RMS values so that the devices can be used for calibration of thermal voltage converters (TVC), transfer difference of TVCs and development of power standards. PJVS system, in addition to applications requiring accurate and stable dc voltages, is been applied to ac metrology by exploiting the ability to generate waveforms approximated by a staircase series of quantized voltages. Since ACJVS is used for voltages limited to few hundred mV rms thus to get higher amplitudes, Stepwise approximation for the generation of AC waveforms using PJVS system is used and this whole unit is ACPJVS system. Differential method with integrating sampler to determine the amplitude and phase of sine wave voltages. This unit forms the basis for the development of Quantum AC power standard.

12.4.2 Metrological Services for Sectorial Growth

12.4.2.1 Traceability and Calibration

The PJVS system is used for the dissemination of unit 'Volt' in India through the apex level calibration of national standards. The national standards are maintained in the form of bank of Zener-diode based dc reference standards at 1.018 and 10 V. These national standards are further used to calibrate the secondary standards of different metrology groups and the working standards of different user organizations to maintain the traceability chain to the primary Josephson standard. CSIR-NPL also transfers traceability to the private and public sector and also to the government and industrial users and to the developers and suppliers of high performing voltage products. MSME namely IDEMI Mumbai, private firms namely FLUKE and Agilent are getting benefitted from CSIR-NPL. The PJVS system is used to provide the traceable measurements for the gain and linearity of high precision digital voltmeter and is provided to the microwave power metrology at CSIR-NPL for their primary calibration.

12.4.2.2 Training and Human Resource

CSIR-NPL is involved in imparting good practice guide towards the operation of quantum voltage standards through training programs for generation of highly skilled human resource. Several laboratory visits are conducted under various programs to exchange the experiences and expertise with the experts from various nations for the maintenance of this quantum standard. Different workshops and training programs are organized to create awareness among the user organizations towards the importance of quantum traceability for industrial growth. Many educational organizations are also benefited too from Quantum Voltage Metrology. Academic and vocational trainings are also imparted to the under-graduate and post-graduate students and research scholars timely.

12.4.3 Economic Impact

Quantum Voltage Metrology provided traceability of SI Unit Volt through calibration of national and secondary standards maintained as of Zener-diode based dc reference standard and also to customer organizations with most demanding requirements for accuracy. These Zeners further link to the measurement of electrical dc voltage to various industries and user organizations, and in this manner the traceability for all these electrical measurements is being conclusively provided from the PJVS system. The PJVS acts as a backbone for the electronic and electrical industries and various other sectors. These sectors, which are the vital contributors towards the GDP and economic growth of the nation, are finally reliant on the quantum traceability of this system.

12.5 Focused Ion Beam Application for Metrology

Microwave measurements have two key parameters Impedance and E-field, which were discussed above. Both parameters have traceability from dimensional metrology against coaxial-air-lines and GTEM, which are the primary standards for both parameters, respectively. With the increasing frequency the dimensional metrological requirements for these standards is below 10 microns with accuracy in nm range. For that purpose, FIB is an excellent tool which can easily produce such dimensional artifacts (DA). In addition, FIB can be used to develop precise electromagnetic devices, like superconducting junctions, to study quantum metrology.

The focused ion beam (FIB) uses finely focused beam of gallium (Ga+) ions to scan the sample, as the gallium (Ga+) ion beam hits the sample surface, depending on the energy of ions, a small amount of material gets sputtered or dislodged from the surface. The dual-beam feature (ion and electron beam) offers an effective combination of a high-resolution scanning electron and ion beam microscope which enables the precise controlling of the milling process either by removing thin slices of material or cutting the material and perform nano scale imaging simultaneously. A combination of additional attachments such as e-beam writer (for e-beam lithography), energy dispersive X-ray spectrometers (EDS), electron backscattered diffraction (EBSD) systems and micromanipulators increases the flexibility of the instrument and transform it into a powerful tool.

In dual-beam FIB system, the electron beam column is mounted in a vertical position on the vacuum chamber and the ion column is at an angle of around 54° to the electron column. Here both the electron and ion beams are coincident on the same region of the specimen, provided the area of interest lies in the eucentric plane [66]. Preferably a gallium ion source is used in the ion column and Ga^+ ions are generated from a liquid metal ion source (LMIS) consisting of a tungsten needle that is mounted below a reservoir of liquid gallium. The LMIS provides a very high brightness beam. The tip of the needle, which is coated with a thin film of gallium, is situated just above the extractor. The extractor is held at a higher voltage (typically~6 keV) relative to the source, and this produces an intense electric field at the source tip. The field ionizes the gallium and draws the liquid metal into a fine tip. Ion emission occurs due to field evaporation at this tip and the gallium ions are emitted from the tip and accelerated down the column. The usual acceleration voltage used in current generation FIBs is ~30 keV. These Ga ions are pulled out and then focused beam is scanned over the sample surface by lenses.

12.5.1 Fabrication, Measurement Capabilities and SI Traceability

The FIB instrument mainly works in three modes: imaging, milling or pattern writing and deposition. The method used in FIB imaging is similar to that in the scanning electron microscope, a Ga + beam is focused onto the sample and rastered across the surface. FIB imaging is normally performed using a smaller aperture where the beam current is small (1–10 pA) and no or minimal sputtering of the specimen occurs. FIB images are particularly useful due their ability to display detailed surface topography.

The milling is the most powerful and versatile capability of the FIB system. When a beam of energetic particles like Ga ions with energy between 5 and 30 keV, strikes a solid surface, a fraction of the particles are backscattered from the surface layers, while the others are slowed down within the solid [67]. Usually large ions find a difficult time penetrating the sample by passing between individual atoms. Their size increases the probability of interaction with atoms, causing a rapid loss of energy. When an ion impinges on a solid, it loses kinetic energy through interactions with the sample atoms. As a result, ion-atom interaction mainly involves the outer shell, resulting in atomic ionization and breaking of the chemical bonds of the surface atoms. In the FIB, the Ga ion scan easily disrupt the placement and alignment of the atoms causing lattice damage and phonons which are dissipated as heat and create a sputtering effect, this leads to localized sputtering and the formation of trenches in the specimen surface. Upon impact, the ion-atom collision is either elastic or inelastic in nature. Elastic collisions result in the excavation of surface atoms (i.e., sputtering) and are the primary cause of the actual modification of the material surface. Inelastic collisions occur when the ions transfer some of their energy to either the surface atoms or electrons. This process produces secondary electrons (electrons which become excited and are able to escape from their shell), along with X rays (the energy released when an electron drops down into a lower shell). Secondary ions are also produced through the collision, seemingly after secondary electrons have been emitted. By controlling the sputtering such as rate, location, and depth, the FIB system demonstrates remarkable milling abilities. The milling through digital pattern generation allows user to modify the material surface, create cross sections, and carve materials into any shape we desire.

The dimensional measurements (*X–Y*) are mostly carried out using FIB based systems where the back scattered secondary electrons in combination with the detectors are used to capture the image of the sample. This process is very fast and topographical image of the sample can be acquired in less than a minute. This digital image often consists of 728 × 512 pixels along lateral (*X* × *Y*) axes. The standard reference materials (SRM) are used to calibrate these axes and magnification of the microscope. Usually these SRM are nothing but the lithographically fabricated nanodimensional artifacts (periodic patterns of squares, lines nested loops etc.) measured using the metrological AFM at a national measurement institute which is directly traceable to SI unit of meter. To calibrate the FIB instrument, SRM loaded in the chamber of the instrument and detectors captures the image of pattern at certain magnification. The capture image shows the patterns of squares or lines which already have assigned periodicity (dimensional *X–Y*) values. These traceable values are further used to calibrate the instruments scale bar. The topographical images obtained using ion or electron beam are often used for the dimensional measurements (lateral *X* and *Y*) which are accurate and traceable to SI unit (meter). The best lateral (assuming the optimum conditions) resolution can be achieved using electron beam is 1.0 nm @ 15 kV and 1.9 nm @ 1 kV whereas the ion beam imaging gives resolution of about 3 nm @ 30 kV. The magnification, probe current, and acceleration voltage range for SEM and FIB are 12x-1000kx, 4pA-20 nA, 0.1–30 kV and 300x-500kx, 1pA-50 nA, 1–30 kV respectively. Practically the SEM resolution of about 5 nm @ 5 kV, writing single pixel line or dot milling down to ~ 20–0 nm (substrate dependent) can be easily achieved as shown in the Fig. 12.30a and b.

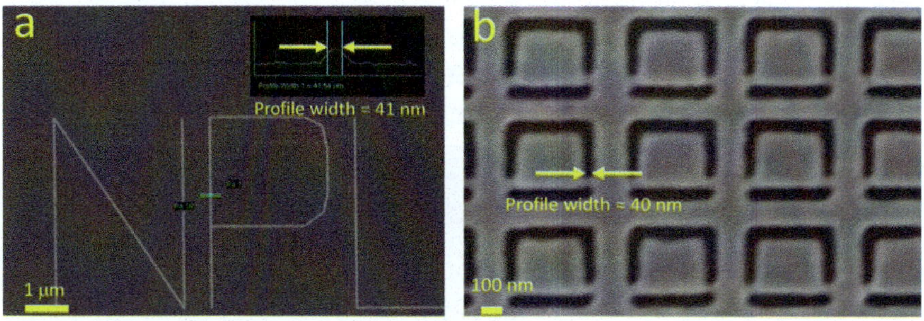

Fig. 12.30 **a** and**b** The FIB milled pattern and SEM scan profile measures width down to 40 nm

12.5.2 Dimensional Artifacts

The nanotechnology-based products and devices are now very common and attractive in modern society and have applications that includes almost all the sectors whether it is space research or nanomedicine. Due to continuous developments new nano-based technologies or products are coming into the market with high performance parameters and miniaturization of device dimensions. Further miniaturization either enhance or blemish the performance of the products or devices where Nano dimensions are very critical and slight variations in their parameters are not acceptable.

The instruments like SEM and AFM are widely used for the dimensional characterization of micro-nano size structures in various sectors which includes R&D as well as industries. These instruments should be calibrated regularly using a reference material traceable to SI unit. It has been observed that the magnification error of an SEM can vary from 5–10% and nanoscale measurements demand much more accurate results where strict traceability and very accurate calibration is a big concern [68].

To fulfill the requirement of reference artifacts for the calibration of instruments (SEM and AFM), recently we have fabricated a prototype reference artifact using the FIB nanofabrication facility (as shown in the schematics: Fig. 12.31a. The actual FESEM image of the artifact made on Si/SiO_2 substrate is shown in the Fig. 12.31b where the periodic repetitive structure of either squares or lines of equal width appears at equal intervals. The dimensional calibration can be done by using pitch of a periodic structure which is defined as the distance measured from start of a feature (fabricated metallic element, e.g., square or line arrays, Fig. 12.31b) till the next

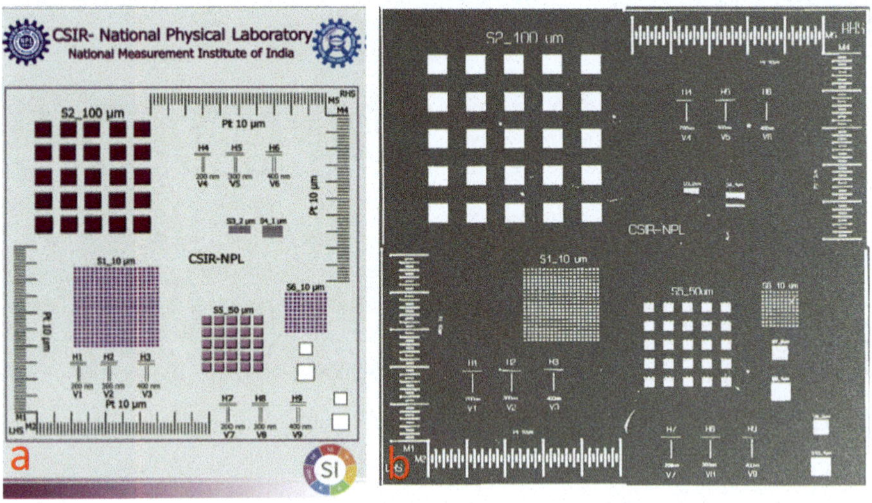

Fig. 12.31 **a** The schematics of the nanodimensional artifact and **b** the actual FESEM image of nanodimensional artifact fabricated using FIB nanofabrication facility.

periodic feature appears. Thus *x*- and *y*-axis calibrations are performed for SEM and *X*, *Y* and *Z* (height) can be performed for AFM. FIB laboratory has demonstrated the fabrication of 10 μm size arrays with an accuracy 10.02 ± 0.03 μm (*x* axis) and 9.87 ± 0.03 μm (*y* axis). The minimum single pixel line or dot milling accuracy is mentioned above in the instrument's capability section. Thus, this artifact will work with high and low accelerating voltage and at high and low magnifications, with AFMs, SEMs and optical microscopes as well. Using FIB facility, CSIR_NPL has also fabricated new SEM resolution test samples on a user friendly SiO_2/Si and carbon nanotube substrates which will help the present and future SEM users.

Microwave spectroscopy is very well known for probing the rotation of the molecules, detecting dynamic phenomena originating from the charges or spins at GHz frequencies. In the microwave spectroscopy studies, FIB setup can be used for micromachining of the NSMM (nearfield scanning microwave microscopy) tip into desired shape and size. The best high spatial resolution can be obtained with the help of a sharp tip, i.e., radius of curvature of the sharp end is about 20 nm. The reference sample for the microwave imaging, e.g., metal squares ($2 \times 2\ \mu m^2$) and interdigitated capacitor patterned (width & gap of 1 μm each) and on a Si substrate can be also fabricated using FIB machine. The contrast observed in the images comes from the changes in the capacitance values sensed by the tip [69].

In industry (e.g., semiconductor), FIB used for defect analysis, circuit modification, mask repair, sample preparation of site-specific locations on integrated circuits, the trimming of thin film magneto-resistive heads, device modification, process control, and failure analysis etc. Some advance research applications include surface structure modifications and TEM sample preparation for imaging and analysis. Further this technology also used for the localized deposition of conductors and insulators with very high precision as per requirements in the electronic circuit.

12.5.3 Training and Human Resource

Nanotechnology has become a very popular research topic or subject for study at many research institutes. It is now being taught at either graduate or undergraduate level where students are more curious to learn about sophisticated instruments. FIB instrument is a very handy tool for training the new users through programs like workshops and hands-on experience. Since FIB is a high-end sophisticated instrument and not many such instruments are available in India. Thus, many users are always finding interest in joining programs where they can learn and use highly sophisticated nanofabrication instruments to execute their real-world application. Another program that FIB laboratory can conduct is the high-resolution imaging using ion beam and electron beam to learn the morphological details of user's sample. These are unique one-week programs where participant can learn the nanofabrication by using FIB microscope and make the use of instrument to carry out own interest sample. Through CSIR_NPL's outreach program, user can also come with their own sample and get the work done with the help of instrument experts. So far about 20 summer research fellows/scholars have become familiar with the nanofabrication and have learnt basics of FIB instrument.

12.5.4 Future Metrological Road Map and Economic Impact

The FIB fabricated micro-nano dimensional reference materials/artifacts play an essential role during the calibration of SEM, AFM, optical microscope and similar instruments. There are more than 2500 SEM and AFMs are installed in India and are routinely performing the nanodimensional measurements which are traceable to imported traceable reference material (TRM) costing Rs. 100,000 based on NI certification. So estimated industrial demand is more than 2500 units/2 years. In present scenario all TRM are imported which can be saved with help of indigenous nano dimension artifact will save Rs. 12.5 Cr per year as import substitution of TRM. Hence there is utmost requirement for the fabrication of high magnification reference standards in India so that it can be easily available with traceability certificate from an NMI.

12.6 Conclusion

In the field of measurements in electromagnetics, besides Microwaves, CSIR-NPL deals with various other parameters such as AC/DC difference, voltage, current, PMU, magnetic flux, magnetic field density, inductive field, for the frequencies ranging from 1 Hz to 100 MHz of electromagnetic spectrum.

LF & HF Voltage, Current metrology have the facility which can calibrate almost all the parameters—LF voltage and current are valid for frequency range 1 Hz to 1 MHz along with PMU (IEEE C37.118.1a 2014) at 50/60 Hz—required for electrical household appliances. AC-DC calibrators are the key instruments to provide traceability to electrical manufacturers for their voltage, current and impedance parameters directly from NPL or from a reference lab. PMUs are smart devices, already installed in national power grid, that need traceability against primary standards to perform better in case of sudden power drop or enhanced demand of power. Technical solutions to Indian PMU manufacturers are also provided by this CSIR-NPL to enhance MSME technical capabilities in the area of PMU manufacturing under Make in India initiative.

An individual's day-to-day life is driven by smart sensors based on various technologies out of which magnetic field-based sensors are very important due to their application in household appliances to biomedical application, scanning system to advance electronic devices. Magnetic metrology is capable to measure magnetic flux density, magnetic flux, specific total power loss, specific apparent power in a range of 1 micro Tesla to 1.5 T. It is also capable to contribute in characterization of magnetic materials, calibration of magnetic probes, flux meters and magnetic coils. With these measurement capabilities the magnetic metrology can serve various national missions on smart technologies such as wireless power transfer, electrical vehicles and FAME-II schemes.

Microwave driven quantum standard of voltage known as programmable Josephson voltage standard at CSIR-NPL is one of SI traceable system. This system is the backbone for providing apex traceability, where voltage is the parameter, to all kind of electrical and electronic instruments manufactured in India. Additionally, the advanced nano-fabrication tool FIB in the electromagnetic metrology is capable to design various dimensional artifact that can serve as traceable reference material, junction devices and primary standards for microwave-based metrology.

The projected impact of LF & HF voltage, LF current, magnetic and Josephson voltage standard metrology on India's GDP is roughly 10%. Considering the indirect contributions by support to the manufacturing sector, certification to the end products, the GDP impact could be as high as 20%. Thus, involving CSIR-NPL for LF & HF voltage, LF current, magnetic and Josephson voltage standard metrology at national policy formulation stage can be of an advantage to 'Aatmanirbhar Bharat' and 'Make in India' initiative.

Acknowledgements The authors would like to acknowledge all the ex-colleagues; present colleagues *Ms. Sunidhi Madan, Mr. Anurag Katiyar, Ms. Mandeep Kaur, Mr. Anish Bhargav, Ms. Archana Sahoo, Ms. Jyoti Chauhan, Ms. Swati Kiumari, Mr. Mange Ram and Mr. Amreek Singh; and students* of CSIR-NPL for their contributions in establishing Electromagnetic Metrology. We would also like to thank the stake holders Department of Legal Metrology, SAC-ISRO, PGCIL, STQC Labs, TEC-DoT, AAI, LM, M/s R&S India Pvt Ltd., M/s Valliant Comm. Pvt Ltd. and many others for their continuous support and collaboration in the National mission endeavors for Electromagnetic Metrology dissemination pan-India.

References

1. E. Kenneth, Martin, Synchrophasor measurements under the IEEE Standard C37.118.1–2011 with amendment C37.118.1a. IEEE Trans. Power Deliv. **30**(3) 1514–1522 (2015)
2. NIST, [Online]. Available: https://www.nist.gov/si-redefinition/ampere/ampere-quantum-metrology-triangle
3. BIPM, [Online]. Available: https://www.bipm.org/kcdb/cmc/quick-search?includedFilters=cmcBranches.AC+voltage%2C+current%2C+and+power&excludedFilters=&page=0&keywords=INDIA
4. B.D. Inglis, Standards for AC–DC transfer. Metrologia **29**, 191–199 (1992)
5. I. Budovsky, et al., APMP international comparison of AC-DC transfer standards at the lowest attainable level of uncertainty. IEEE Trans. Instrument. Measure. **54**(2) (2005)
6. T. Funck et al., Determination of the AC-DC voltage transfer difference of high-voltage transfer standards at low frequencies. IEEE Tran. **54**(2) (2005)
7. B. Pal, Automation and evaluation of two different techniques to calibrate. MAPAN **28**(1), 31–36 (2013)
8. I. Budovsky et al., Reference micropotentiometer resistors for Ac-dc transfer and phase error measurements. in *CPEM* (2016)
9. K. Takahshi et al., AC DC voltage transfer difference due to seeback effect in thermal converters. IEEE Trans. Instrument. Measure. **48**(2) (1999)

10. S.A. Rasha, Comparison between two different designs in the AC voltage measurement. Measurement **44**, 1539–1542 (2011)
11. I. Budovsky, Algorithm and uncertainty of AC–DC transfer measurements. in *Metrology Society of Australia*. Gold Coast (2001)
12. M. Halawa et al., A step-down technique to calibrate AC current down to 10 lA using a precision 10 mA current shunt. MAPAN **27**(3), 139–142 (2012)
13. .M. Halawa et al., Integrated calibration system for accurate AC current measurements up to 100 kHz. MAPAN **27**(3) (2012)
14. .S. Ahmad et al., Assigning RF DC transfer difference to RF voltage primary standard from 1 to 1000 MHz. in *AdMet*. Bangalore (2011)
15. .J.R. Kinard et al., Intercomparison of thermal converters at NIM, NIST, PTB, SIRI and VSL from 10–100 MHz. IEEE Trans. Instrum. Measure. **1M**(42), 618–621 (1993)
16. S. Ahmad et al., Realization of low frequency power standard at NPLI. MAPAN **29**(4), 285–288 (2014)
17. H. Gierke et al., Automatic RF voltage calibration with a primary voltage standard up to 1 GHz. IEEE Trans. Instrument. Measure. **1M**(42), 519–523 (1993)
18. CSIR-NPL, New Delhi, "Calibration Charges: D6.02c, LF & HF Voltage, Current and Microwave Metrology," [Online]. Available: https://www.nplindia.in/calibration-charges-d602c-lf-hf-voltage-current-and-microwave-metrology-wef01042019
19. A. Bhargav, S. Ahmad et al., Technical evaluation and optimization of phasor measurement unit using CSIR-NPL PMU calibrator system to ensure reliability. MAPAN (2019) https://doi.org/10.1007/s12647-019-00346-4
20. I. Explore. [Online]. Available: https://ieeexplore.ieee.org/document/1611105
21. M. Adamiak, Synchro phasors: definition, measurement, and application. in *59 Annual Georgia Tech Protective Relaying*. Atlanta.
22. P. Komarnicki et al., Practical experience with PMU system testing and calibration requirements. in *IEEE Power and Energy Society General Meeting*. USA (2008)
23. D. Novosel, Y. Hu, Progress in PMU testing and calibration. in *Electric Utility Deregulation and Restructuring and Power Technologies*. China (2008)
24. S. Luthra, Development and implementation of current tee for AC high current calibration. MAPAN **33**(1), 29–32 (2018)
25. Electronics Industries Association of India, Strategic Electronics, [Online]. Available: www.elcina.com
26. www.elcina.com, "https://www.elcina.com/current_update.php. [Online]
27. Invest India, Electronics Manufacturing, [Online]. Available: www.investindia.in
28. www.investindia.gov.in, https://www.investindia.gov.in/sector/electronic-systems. [Online]
29. Ministry of Electronics and Information Technology, Ministry of Electronics and Information Technology, [Online]. Available: meity.gov.in
30. meity.gov.in, https://meity.gov.in/content/policies-0. [Online]
31. www.ibef.org, [Online]. Available: https://www.ibef.org/download/Power-February-20181.pdf
32. business.mapsofindia.com, https://business.mapsofindia.com/india-gdp/industries/electronics-hardware.html. [Online]
33. Investopedia. [Online]. Available: https://www.investopedia.com/ask/answers/052515/what-growth-rate-electronics-sector.asp
34. Wikipedia, Indian Electricity Sector, [Online]. Available: www.wikipedia.org
35. www.MarketandMarkets.com, 2020. [Online]. Available: https://www.marketsandmarkets.com/Market-Reports/magnetic-field-sensors-market-521.html
36. https://www.nplindia.in/data/npldjvsec.pdf, [Online]
37. IEC, https://webstore.iec.ch/publication/2063, 28 3 1996. [Online]
38. CSIR-NPL, https://www.nplindia.in/data/npldjvsec.pdf, [Online]
39. K. Borer, The nuclear magnetic resonance system for the CERN muon storage ring. Nuclear Instrum. Methods **2**, 143 (1977)

40. F. Piquemal, B. Jeckelmann, L. Callegaro, H.J., "Metrology in electricity and magnetism: EURAMET activities today and tomorrow. Matrologia **R1**, 54 (2017)
41. [Online]. Available: https://ee44ll.blogspot.com/2015/01/hall-effect.html
42. CSIR- National Physical Laboratory, Magnetic Metrology. [Online]. Available: www.nplind ia.org
43. J. Sievert, H. Ahlers, The magnetic metrology of materials–a review. Bull. Mater. Sci **17**, 1393–1399 (1994)
44. D.L. Rode, C.M. Wolfe, G.E. Stillman, Magnetic-field dependence of the hall factor for isotropic media. J. Appl. Phys **54**, 10–13 (1983)
45. E.H. Hall, On a new action of the magnet on electrical current. Amer. J. Math. **2**, 287–292 (1879)
46. G. Ashcroft, M. Hall, Traceable measurement of magnetic moment and DC susceptibility. In *Conference on Precision Electromagnetic Measurements* (2004)
47. F. Fiorillo, Measurements of magnetic materials. Metrologia. **47**(2) (2010)
48. S. Tamanski, Modern magnetic field sensors-a review. Przeglad Elektrotechniczny **89**, 12 (2013)
49. M. Irena , W. Fabienne, L. Yoann, Magnetic metrology for iron oxide nanoparticle scaled-up synthesis. RSC Advances **4** (2014)
50. Media Reports, Press Releases, Department for Promotion of Industry and Internal Trade (DPIIT), Automotive Component Manufacturers Association of India (ACMA), Society of Indian Automobile Manufacturers (SIAM), Union Budget 2015–16, Union Budget 2019–20
51. Policy documents from Department of heavy Industries. Government of India website (2020)
52. Power Ministry, https://powermin.nic.in/, Government of India, Ministry of Power, 2020. [Online]
53. Ministry of new and renewable energy, https://mnre.gov.in/, Government of India, Ministry of new and renewable energy, 2020. [Online]
54. The National Statistical Office (NSO), First Advance Estimates of National Income, 2019–20, Ministry of Statistics and Programme Implementation, Government of India. The National Statistical Office (NSO), Ministry of Statistics and Programme Implementation, 2020
55. B.D. Josephson, Possible new effects in superconductive tunnelling. Phys. Lett. **1**, 251 (1962)
56. S. Shapiro, Josephson currents in superconducting tunneling: the effect of microwaves and other observations. Phys. Rev. Lett. **11**(80) (1963)
57. B.N. Witt, T. J. Taylor, New international electrical reference standards based on the Josephson and quantum hall effects. Metrologia **26**(1) (1989)
58. B.P. Kibble, I. A. Robinson, J. H. Belliss, A Realization of the SI Watt by the NPL Moving-coil Balance. Metrologia **27**(4) (1990)
59. S.P. Benz, Superconductor-normal-superconductor junctions for programmable voltage standards. Appl. Phys. Lett **67**, 2714 (1995)
60. H. Daniel , F.R. Dolata, R. Popel, Development of Sub-micron SNS Ramp-type Josephson Junctions. IEEE Trans Appl Supercond. **11**(1) (2001)
61. H. Schulze, R. Behr, F. Müller, J. Niemeyer, Nb/Al/AlOx/AlOx/Al/Nb Josephson junctions for programmable voltage standards. Appl. Phys. Lett. **73**(996) (1998)
62. H. Yamamori, M. Itoh, H. Sasaki, A. Shoji, S. P. Benz, All-NbN digital-to-analog converters for a programmable voltage standard. Supercond. Sci. Technol. **14**(12) (2001)
63. Operation of NIST Josephson array voltage standards. J. Res. Natl. Inst. Stand. Technol. **95**(3) (1990)
64. P. Dresselhaus, M. Elsbury, C. Burroughs, Microwave optimization of 10 V PJVS circuits. IEEE (2010)
65. BIPM, https://www.bipm.org/kcdb/cmc, [Online]
66. P.R. Monroe, The application of focussed ion beam microscopyin material sciences. Materi. Character (2009)
67. Focussed Ion Beam Milling, [Online]. Available: mx.nthu.edu.tw

68. M. Postek, A. Vladár, Does your SEM really tell the truth?—how would you know? J. Scan. Microscopi. 355–361 (2013)
69. S. Geaney, D. Cox, Hönigl-Decrinis, T. Shaikhai, Near-field Scanning Microwave Microscopy in the Single photon Regime. Sci. Rep. **9**(1), 1–7 (2019)

Chapter 13
Metrology for Atmospheric Environment

Part I: Atmospheric Constituents

Chhemendra Sharma, Tuhin Kumar Mandal, Sachchidanand Singh, Govind Gupta, Monika J. Kulshrestha, Prabha Johri, Ashish Ranjan, Arun Kumar Upadhayaya, Rupesh M. Das, Daya Soni, Sumit Kumar Mishra, Senthil Kumar Muthusamy, Sudhir Kumar Sharma, Preetam Singh, Shankar Gopala Aggarwal, Soman Radha Radhakrishnan, and Manoj Kumar

Abstract Metrology for atmospheric environment has assumed great importance for India as country aspires to undertake rapid sustainable economic growth for amelioration of living conditions of its citizens. The enhancement of economic activities impacts the status of environment which needs to be precisely & accurately monitored for formulation and undertaking of required policy measures. CSIR-NPL

C. Sharma (✉) · T. K. Mandal · S. Singh · G. Gupta · M. J. Kulshrestha · P. Johri · A. Ranjan · A. K. Upadhayaya · R. M. Das · D. Soni · S. K. Mishra · S. K. Muthusamy · S. K. Sharma · P. Singh · S. G. Aggarwal · S. R. Radhakrishnan · M. Kumar
CSIR-National Physical Laboratory, Dr. K.S. Krishnan Marg, New Delhi 110012, India
e-mail: csharma@nplindia.org

T. K. Mandal
e-mail: tuhin@nplindia.org

S. Singh
e-mail: ssingh@nplindia.org

G. Gupta
e-mail: govind@nplindia.org

M. J. Kulshrestha
e-mail: monikak@nplindia.org

P. Johri
e-mail: pjohri@nplindia.org

A. Ranjan
e-mail: aranjan@nplindia.org

A. K. Upadhayaya
e-mail: upadhayayaak@nplindia.org

R. M. Das
e-mail: rupesh@nplindia.org

D. Soni
e-mail: dsoni@nplindia.org

D. K. Aswal (ed.), *Metrology for Inclusive Growth of India*,
https://doi.org/10.1007/978-981-15-8872-3_13

being the national metrology institute of India has started to develop capabilities for providing apex level traceability for atmospheric measurements. In this chapter, a detailed description is provided about the atmospheric species and properties for which testing and calibration facilities are required to be built in addition to the available test and calibration facilities for some of the atmospheric parameters. CSIR-NPL has undertaken development of various gas standards for providing SI traceability to pollution and greenhouse gas measurements in the country. Efforts have also been undertaken to cater to the national needs of testing and calibration of sensor-based monitoring systems for atmospheric/industrial pollution.

Our experience shows that not everything that is observable and measurable is predictable, no matter how complete our past observations may have been.

– Sir William McCrea (1904–1999)

13.1 Introduction

Earth has a unique atmosphere with a precise mixture of gases, moisture and temperature conditions necessary to sustain life and our existence. The Sun which is the ultimate source of energy transfers its radiation through the Earth's atmosphere and interacts with the air, land and ocean and their various constituents to give rise to environment, weather and climate. The environment consists of all the things like air, water, land, animals and plants that make up our surroundings and affects our

S. K. Mishra
e-mail: mishrask@nplindia.org

S. K. Muthusamy
e-mail: senthilmk@nplindia.org

S. K. Sharma
e-mail: sudhircsir@gmail.com

P. Singh
e-mail: singhp@nplindia.org

S. G. Aggarwal
e-mail: aggarwalsg@nplindia.org

S. R. Radhakrishnan
e-mail: rkrishnan@nplindia.org

M. Kumar
e-mail: manojkumar@nplindia.org

lives on the Earth. Weather is the short-term or day-to-day behaviour of environment and atmosphere on a scale of minutes to weeks. Climate, on the other hand, is the prevailing weather or meteorological conditions, including temperature, humidity, precipitation, cloudiness and wind that characteristically prevail in a region over a long period of time (~30 years or more). 'Climate System' constitutes of a complex interaction between Sun, Atmosphere, Hydrosphere, Solid Earth, Cryosphere and Biosphere and exchanges energy and moisture amongst these six spheres.

Climate Change is referred to as a change in the state of the climate, which can be identified through measurements or statistics by changes in the mean properties that persist for a prolonged period of time, typically on a scale of decades or longer. The change in climate may occur on different temporal and spatial scales and may be caused by natural changes or as a result of human interference. A change in climate or its variable over time or spatial scale beyond the weather event is called a Climatic Variability. It is measured by deviations, called anomalies and can typically be derived through statistics on a time scale of month, season or year. If climate variability is caused due to natural internal processes within the climate system then it is called internal variability. On the other hand, if it is due to the variations in external factors, natural or anthropogenic then it is called external variability. The difference between climate variability and climate change lies in the persistence of anomalous conditions. In the former case, the events or an anomaly that used to be rare starts occurring frequently in the later. Because the climate variability goes back centuries before the instrumental records started (only a couple of centuries ago), scientists decipher and reconstruct the past climates by using indirect evidence. It comes from natural records of seafloor sediments, glacial ice, fossil pollen, tree-growth rings, historical documents, etc.

There are natural as well as human-induced causes of climate change. The natural mechanisms of climate change are unrelated to human activities and are caused due to (i) Plate tectonics (rearranging Earth's continents, moving them closer to or farther away from the equator and the poles), (ii) Volcanic activity (changing the reflectivity and composition of the atmosphere), (iii) Variations in Earth's orbit (the natural, cyclic change in our planet's orbit, axial tilt, and wobble), (iv) Solar variability (the Sun varies in its radiation output and sunspots number affect these output). The human impact on climate change started even before modern industrialization started (around 1750). For thousands of years, humans have modified the environment through large scale fire and extensive overgrazing, which led to changes in surface albedo, wind patterns and rate of evaporation. Two major Anthropogenic Impacts on climate change can be categorized as (i) Increase in carbon dioxide and other greenhouse gases (methane, nitrous oxide, chlorofluorocarbons, etc.) to the atmosphere and (ii) Addition of human-generated aerosols (carbonaceous, non-carbonaceous and particulate matters) to the atmosphere. Unprecedented use of natural resources, increased population and rapid industrialization has led to significant change in climate and environmental constituents.

The environmental constituents in the lower atmosphere and their impacts are intrinsically coupled with the upper atmosphere and vice-versa. Coupling of atmospheres predominantly affects the dynamics of different layers of the atmosphere. It has been found through observations and models that cooling of the upper atmosphere due to Greenhouse Gases (GHGs) is also a consequence of an increase in anthropogenic sources in the lower atmosphere. So the estimation of climate change crucially depends on the identification and efficacy of these responsible sources, the sturdiness of climate data, and the uncertainties in measurements therein. Metrology is now playing an important role in climate study by providing improved measurement capabilities and traceability of data.

Atmospheric composition influences climate, human and animal health, terrestrial and aquatic ecosystems, water balance, agricultural productivity, weather forecasting, renewable energy production and so on. The variations in atmospheric composition change the atmospheric radiation budget and drive climate change. To understand the radiative forcing, it is important that we have the global observations of climate forcing agents and an understanding of the processes related to their interaction with radiation. The monitoring of reactive trace gases, greenhouse gases, ozone, water vapour and aerosols are essential to characterize the global impact of these constituents on the global radiation budget. Therefore, it is extremely important to have high quality, traceable measurements with specified uncertainties for all the measurements concerning environment, atmosphere and climate.

To understand the complexity of the physico-chemical processes that control the atmospheric composition on different scales, the quality controlled and traceable measurements /estimations of atmospheric constituents are essential. National Measurement Institutes (NMIs) play a vital role in estimation of atmospheric constituents. To address the needs of essentiality of calibration and traceability in the area related to atmospheric measurements, CSIR-NPL focused on the various parameters like GHGs, O_3, aerosols, reactive gases, atmospheric deposition, UV radiation, etc. Quality Assurance and Quality Control (QA/QC) of these atmospheric constituents are disseminated through a hierarchical process of primary standards to working standards. To make all the measurements traceable to SI, a hierarchical chain of calibration and standards with stated uncertainties are used (Fig. 13.1).

In everyday life, a reliable and traceable measurement with international acceptability is the key to sustainable development, not only in the area of trade and commerce but also for the energy, health and environment. The precise and accurate measurements of physical environment that refer to the conditions of temperature, humidity, rainfall, pressure, sunlight, gases, aerosols, etc. is extremely important on day to day, seasonal and climatic (long-term) basis. Apart from this, monitoring of environment and different pollutants are crucial to the organism health, impacts on environment and the sustainable development.

In this chapter, the issues related to atmospheric constituents and their metrology are discussed along with the sensor, gas & particle metrologies (Fig. 13.2) on which CSIR-NPL is currently focusing its efforts on serving the national needs.

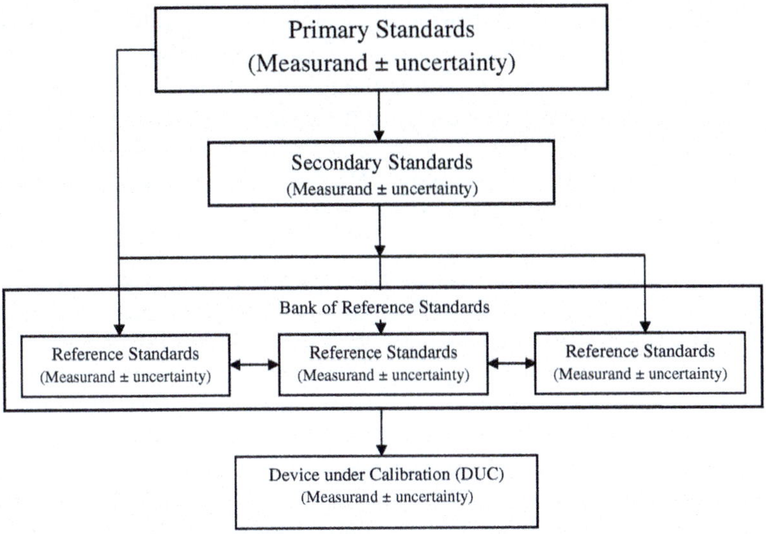

Fig. 13.1 The hierarchical traceability chart of atmospheric parameters

Fig. 13.2 Focus areas of environmental metrology of CSIR-NPL

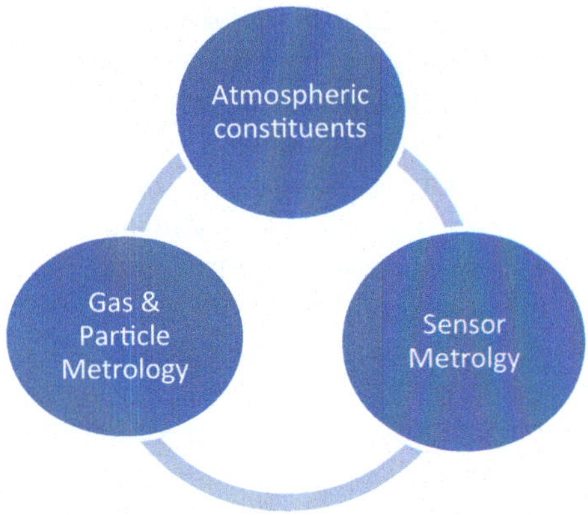

13.2 Metrology for Atmospheric Constituents

India is an agrarian economy mainly dependent on monsoon particularly the summer monsoon which brings 80% of precipitation. Due to seasonal shifting of thermally produced planetary belts of pressure and winds under continental influences summer monsoon occurs. During winter, high pressure over central Asia and northern south Asia and low pressure over Indian Ocean facilitates low-level air flow from the north-east towards India to Indian Ocean. Normally northeast monsoon brings dry conditions, but southern part of India receives additional rainfall during this period. India has continuous and homogenous meteorological record of rainfall since 1871 onward with at least one rain gauge per district, although fewer stations have continuous and reliable records till 1913. India can assess the variability of all Indian rainfall and monsoon using instrumentally measured rainfall data. Monsoon circulation pattern can be assessed by another meteorological parameters, i.e. surface air temperature. Like rainfall, seasonal and annual surface air temperature over Indian region show a significant warming of 0.4 °C per century [1]. However, the historical instrumental record of surface air temperature starts from 1901 only.

Nowadays, satellite data are augmented in the global circulation models (GCM) to better understand the climate changes, however, these data are still validated using well-calibrated measurements at surface stations. GCM can predict accurately the monsoon pattern and rainfall depending on how accurate the measurements of meteorological parameters are. Since any change in monsoon pattern affects food security, human health and finally economy of a country, an accurate prediction of monsoon and rainfall is the key to Indian economy. It is therefore important that all the weather and climatic parameters (rainfall, surface pressure, surface air temperature, etc.) are measured accurately and traceability of instruments is established at least through an NMI. Similarly, the columnar O_3 is naturally low in equatorial regions and allows substantial amount of UV in this region. Changes in UV-B dosage can affect crop yield and human health particularly skin and eyes in tropical region, on the other hand, it also helps in the formation of vitamin D in humans [2]. Changes in stratospheric ozone are not trivial whereas, increase in tropospheric ozone due to increase in precursor gases may have large effect on crop yield as well as acts a potential greenhouse gas. Low-level haze due to presence of aerosol particles reduces the photosynthesis and crop yield [3] and significantly affects the radiation balance through radiative forcing [4]. Aerosol concentration has also reached above alarming levels even in smaller cities. Climate change due to anthropogenic activities projected as worrisome trend in the food basket in the Indo-Gangetic plain (IGP) of India.

Automatic weather system (AWS) measures above-mentioned parameters including pressure, temperature, humidity, radiation, rainfall (volume, time) and wind speed (length, time) and wind direction. CSIR-NPL has primary standard for calibrating the above-mentioned parameters. India Meteorological Department (IMD) apart from several private agencies operates almost 600 surface automatic weather stations and more than 50 upper air observations over India. At each upper air observatories IMD launches radio sounding at least twice daily. These soundings are all imported. Calibration facility of all these parameters will allow India of economic resilience leading to "atmanirbhar". Reliable surface observation of meteorological parameters by using AWS is part of environmental monitoring. Meteorological data traceability and comprehensive performance of AWSs (a calibration facility equipped with temperature and pressure reference sensors) directly traceable to SI will allow well-documented calibration uncertainty.

13.2.1 Driving Forces to Climate Changes: Atmospheric Constituents

India is a densely populated country with 130 billion people where the urban population is increasing rapidly. Due to this, megacities are expanding the peripheral area for accommodating increasing urban population. The increasing population and growth in economy has led to a growth trajectory in industries associated with higher energy consumption and emissions of gaseous and particulate pollutants. Atmospheric composition like long-lived greenhouse gases (CO_2, CH_4, N_2O, CFCs, etc.) and short-lived gases like CO, NO_x and SO_2 as well as particulate matter are increasing continuously. If we can measure these gaseous and particulate pollutants and their sources accurately, it would be easy to mitigate them. GCM simulates monsoon circulation in response to increased GHGs concentration and particulate matter. To assess climate changes and to understand their driving forces one needs to measure climate parameters, their concentration and sources of gaseous and particulate pollutants accurately.

Clean air is a basic requirement for wellbeing of all inhabitants on the globe. However, it has undergone a persistent degradation in the past few decades due to increased population density, industrial expansion, construction work and vehicular traffic, etc. Several atmospheric constituents are contaminated due to anthropogenic activities and continue to pose a significant threat to health worldwide [5]. The anthropogenic release of trace chemicals into the atmosphere has multiplied significantly over the past many decades. Their abundance in the atmosphere is determined both by the growth in their sources and removal rates [6]. Species such as CO_2, CH_4, CFCs, N_2O, etc. are long-lived and well-mixed with a significant lifetime and influence the climate change. In contrast, short-lived trace species with lifetimes ranging between few minutes to days, i.e. CO, NO_x, VOCs, O_3 etc, have highly variable distributions on both spatial and temporal scales. These species influence the atmospheric chemistry [7], radiation energy balance of the earth-atmosphere system [8], ecology as well as human health [9].

Increasing concentrations of radiatively active trace gases affect the thermal flux to space resulting in changes in surface temperatures. Many trace gases are also chemically reactive in nature and alter atmospheric chemistry by reacting in the presence of sunlight, to form secondary air pollutants such as tropospheric ozone [10]. The alterations of atmospheric constituents by the anthropogenic intervention are the major threat to Earth and environment. A number of contemporary issues arise due to implications of atmospheric trace constituents including regional climate [11] (aerosol and ozone exhibit a strong regional effect in climate forcing); Global climate change (for example, CO_2, N_2O, CH_4 and other greenhouse gases lead to Global warming) [12]; Ecosystem Imbalance (impacting crop yield, loss of biodiversity) [13]; Human Health Effects (leading to respiratory problems, cancer, allergies, etc.) [14]; National heritage deterioration (degradation of buildings, monuments, etc.) [15] etc. Though atmospheric constituents are a measure of the anthropogenic perturbation of the natural atmospheric state, it has to be considered in a wider context of atmospheric composition and climate. These constituents are available in trace level in the atmosphere; hence their accurate and SI traceable measurements are essential for further climate and other related studies. The accurate measurement of these constituents depends on the testing and calibration of measuring equipments by national and international standards through unbroken chain of traceability with stated uncertainty.

13.2.2 Other Driving Forces: Atmospheric Coupling

The coupling of atmospheres and the lower atmosphere impacting the state of the upper atmosphere and its dynamics have been profoundly researched in the last two decades. Recent studies have conclusively shown that there could be a strong correlation between the meteorological phenomena concerning the variation in the winter-time polar stratosphere temperature termed as the sudden stratospheric warming (SSW) which brings large-scale variations throughout the atmosphere. The polar vortex shifts off the pole during SSWs leading to weakening of Zonal winds and an abrupt increase in stratospheric temperature (which could be as large as 70 K). This type of warming is designated as minor stratospheric warming whereas; the event is called a major SSW event when the vortex breaks up resulting in a change in the direction of the zonal wind. Studies have also shown that temperature, wind, chemistry, and wave activity in the middle atmosphere could be affected by SSWs events. Apart from this, SSW can also impact vertical thermo-dynamical coupling in a wide range of altitudes and latitudes which in turn can lead to changes in the ionosphere and exosphere.

In a number of studies it has been reported that during SSW event ionospheric F2 region vary in terms of anomalous semidiurnal signature in vertical plasma drift, total electron content (TEC), perturbations in electric fields, electron and neutral densities, electron and ion temperature, foF2, NmF2, and hmF2 besides its occurrence with equatorial counter electrojet [16–18]. This variability in ionospheric F2 region has been attributed to perturbations brought into by the lower atmosphere [16, 19] and its coupling with ionosphere via nonlinear interaction between quasi-stationary planetary waves and migrating tides [20–22] modulating E region dynamo and transferring energy and momentum to the ionosphere.

It is more or less recognized through the recent studies that the seismic activity causes disturbances in the atmosphere and to the extent in height up to the ionosphere [23–25]. Furthermore, perceptible ionospheric perturbations (foF2) are observed before the onset of an Earthquake [22, 26] and thus reflect the possibility of seismo-ionospheric coupling through atmospheres. Subsequently, various lithosphere-atmosphere–ionosphere coupling mechanisms have been formulated which incorporate anomalous vertical electric field generation which perturbs the ionosphere [27], radon radioactivity and leakage of carrier gases such as CO_2, CH_4, H_2, and He [28] which ionize the near-ground atmosphere. Alternate mechanisms proposed were based on the generation of acoustic pressure waves, the release of positive holes, and underground emission of aerosols emitting electromagnetic radiation, which can change the ionospheric electron density distribution [29].

13.2.3 Measurement Principles and Techniques

The surface level of various trace gases (CH_4, N_2O, O_3, CO, NO_x, SO_x, NH_3, NMHCs, and VOCs, etc.) are measured using online analyzers of the respective gases with high accuracy and repeatability [30]. The equipments used for the measurement of these trace gases are summarized in Table 13.1. The CH_4 and N_2O are potential greenhouse gases (GHGs), generally measured using GC-FID. These gases are also measured with various commercial analyzers working on FID as well as electro-chemical sensors. Surface O_3 are measured with a UV-photometric analyzer operating on the principle that ozone molecules absorb UV radiation at a wavelength of 254 nm. Carbon monoxide (CO) is measured using the trace level gas filter correlation (GFC) analyzer, based on the principle that CO absorbs infrared (IR) radiation at a wavelength of 4.6 μm. Atmospheric NO and NO_2 ($NO_x = NO + NO_2$) are measured using NOx-analyzer (with or without a photo-catalytic converter) upto sub

Table 13.1 Trace gas measurement techniques and methods

S. No.	Parameters	Instrument used	Methods
1	Methane (CH_4)	Gas chromatograph, Gas Analyzer	GC-FID Electro-chemical sensor
2	Nitrous oxide (N_2O)	Gas chromatograph Gas Analyzer	GC-FID Electro-chemical sensor
3	Chlorofluorocarbons (CFCs)	Electron capture gas chromatograph (EC-GC)	GC-ECD
4	Ozone (O_3)	O_3-Analyzer	UV absorption photometry
5	Carbon monoxide (CO)	CO-Analyzer	NDIR gas filter correlation
6	Nitrogen oxides (NO_x)	NO_x-Analyzer	Chemiluminscence
7	Sulphur oxides (SO_x)	SO_2-Analyzer	UV Flourescence
8	Ammonia (NH_3)	NH_3-Analyzer	Chemiluminscence
9	Total non-methane hydrocarbon (NMHCs)	Hydrocarbon Analyzer	GC-FID
10	Volatile organic compounds (VOCs)	Thermal desorption gas chromatograph	GC-FID / Mass spectrometer
11	Particulate Matter ($PM_{2.5}$, PM_{10}, TSPM)	Gravimetric, Beta Attenuation Method	Particle Sampler

ppb (parts per million) level operating on chemiluminscence method. NO_x is also estimated by chemical trap method which is further analyzed using UV–visible spectrometer (absorption method). Similar method is also employed for the estimation of SO_x ($SO_2 + SO_3$). Sulphur dioxide (SO_2) on the other hand, is measured using SO_2 analyzer operating on the ultraviolet fluorescence principle. Ammonia (NH_3) is analyzed by chemical trap method as well as by online NH_3-Analyzer, which is working on Chemiluminscence principle [31].

Non-methane hydrocarbons (NMHCs) and CH_4 are measured by using NMHC-Analyzer, based on GC-FID. VOCs/NMHC species (isoprene, benzene, toluene, ethyl benzene + p-xylene, 1,2,4-trimethylbenzene, methyl vinyl ketone, acetic acid, dimethyl sulfide, propanamide/ butylamine, hydroxyacetone, $C_4H_6O_2$, phenol, $C_5H_8O_2$, styrene, methanol, acetonitrile, acetaldehyde, acetone, monoterpenes, formaldehyde, propyne, propene, isocyanic acid, formamide/ dimethylamine, formic acid, acrolein/methylketene, and acetamide/ trimethylamine, etc.) are estimated using gas chromatograph including thermal desorption system [32].

Based on the manner in which sampling is done, atmospheric chemical composition measurement techniques can be broadly categorized as active sampling, passive sampling, and remote sensing techniques. Active sampling employs moving parts such as pumps to force the air sample/sampling medium through detector or a sampling device and measurements can either be made continuously or at least quasi-continuously with short integration times. Whereas, in passive sampling, which has the same underlying principle (flow of molecules/ions from the air/water on to collecting medium) as that of active sampling, diffusion of air to the sampling devices is used for sample collection. It should be emphasized that they provide "time-averaged concentrations" of chemical constituents with the averaging time being determined by the period (which can be daily, weekly, monthly, etc.) they are exposed to ambient air. These passive samplers are of course not very suitable for determining individual peak values or recording the temporal variations over short time intervals when real-time measurements are desired. Passive samplers can be used for the measurement of a variety of species such as ambient SO_2, NO_2, NH_3, and NMHCs concentrations. In addition to this, we have remote sensing techniques in which both active sampling (e.g. LIDAR) as well as passive sampling (Spectrometers) methods can be employed. Active sensors in LIDAR uses internal stimuli to gather data by measuring the time difference between the projection and reverse traversing of the Laser beam after being reflected from the target. In spectrometers, there are passive sensors that respond to external stimuli and the most common source of radiation detected by theses sensors is the reflected/emitted sunlight from the target. In remote-sensing techniques, the detector is placed at a location different than that of the location from where the air volume is being analyzed.

For the examination and quantification of trace constituents in so collected atmospheric samples, spectroscopic and chromatography based methods are used. There can be no single best technique which can be utilized in all the possible situations. Based on the state of samples, type of constituents, probable concentrations, and the size range of trace elements different techniques are employed. Analytical techniques which are commonly used can be summarized as follows:

Spectroscopic methods: refer to the measurement of changes in the intensity of radiation after interaction with a molecule or aerosol particle as a function of wavelength. The origin of spectral lines in molecular spectroscopy is the absorption, emission, and scattering of a photon when the energy of a molecule changes. In both absorption and emission, the energy of radiation absorbed or emitted is equal to the difference between the excited and ground states, and therefore the measurement of these absorbed/emitted radiations enables us to determine the characteristics energy levels and thereby identification of constituent molecules/ions. In the microwave region, the transition between rotational energy levels of molecules takes place therefore rotational spectroscopy employs microwave radiation. Vibration levels have energy differences corresponding to infrared frequency, therefore vibrational spectroscopy is performed using infrared radiation. Electronic transitions require further higher energy content which matches the frequency of UV–Vis part of the EM spectrum.

***Gas chromatography* (GC)**: GC is widely used for the identification and quantification of almost every type of organic compound. It is a dynamic process of separation in which the components of a sample partition between two phases namely stationary (liquid or solid) and a mobile phase (Gas). The mobile phase flows through the stationary phase which is held immobilized inside the chromatographic column. The partition coefficient/distribution constant (Kd) controls the flows of elements across the GC column. The partition coefficient is the ratio (Kd = As/Am) of concentration of a compound in the stationary phase (As) and a mobile phase (Am) and therefore differences in the partition coefficients allow for the chromatographic separation of compounds. Major components of the GC instrument are Gas Carrier, Column, Injector, and detector. The inert Gas-filled in the GC column is known as Carrier Gas which allows the transport of the sample without interacting with it. The injector is the place where the sample is volatilized and quantitatively introduced into the carrier gas stream. The GC column, which is the heart of the system, contains a liquid stationary phase which is adsorbed onto the surface of an inert solid. The separation of sample components takes place in the GC column. A chromatography detector then senses a physicochemical property of the analyte and enables to visualize components of the mixture being eluted off the chromatography column. Successively, a response from the detector is converted into an electronic signal to produce a chromatogram after proper signal amplification. Detectors used in chromatography can be broadly categorized as general purpose and high sensitivity detectors. Flame Ionization Detector (FID), Thermal Conductivity Detector (TCD), are general-purpose destructive detectors that can analyze a wide range of compounds by performing continuous transformation of the column effluents. In contrast, high-sensitivity detectors such as Electron Capture Detector (ECD) and Flame Thermionic Detector (FTD) are capable of detecting only specific types of compounds but with very selectivity and sensitivity.

Mass spectrometry (MS): Mass spectrometry is measuring the mass-to-charge ratio (*m/z*) of one or more molecules present in a sample. A mass spectrometer first converts the compound under study into gaseous ions (by electron ionization). Once ionized, the ions are sorted and separated according to mass-to-charge (*m/z*) ratios using Mass Analyzer and then sent to an Ion Detection System where the relative abundance of different ions is recorded. A mass spectrum of the molecule is thus produced by plotting of *m/z* ratios of the ions present in sample vs. their intensities. The MS can be used to determine the elemental composition of a sample, the masses of particles and molecules as well as to elucidate the chemical structures of compounds.

Aerosol measurements/characterization: Aerosols are tiny liquid or solid particles that are present in the atmosphere having diameters in the range from 1 nm to over 10 μm. Most of the aerosol mass lies in the range from about 0.1 to 10 μm. Their production mechanism can be (1) Direct injection of primary aerosols, and (2) Indirect, by transformation of inorganic and organic gaseous precursors into secondary aerosols (fine <1 μm). Aerosols play an important role in atmospheric chemistry, have effects on human, animal and plant health and welfare, and also influence the climate (directly and indirectly). The direct effect includes the reflection and absorption of the solar radiation and also the absorption and emission of some terrestrial infrared radiation by the aerosol particles. Indirectly, they influence the properties and processes of clouds, and, possibly, also change the heterogeneous chemistry of reactive greenhouse gases. A variety of measurement and analytical techniques can be employed to determine the physical, optical and chemical composition of the size-resolved aerosol particles. Some of these are tabulated in Table 13.2.

OP-FTIR: Open Path Fourier Transform Infrared Spectroscopy (OP-FTIR) is a non-destructive analytical technique (Fig. 13.3). It can be used for qualitative and quantitative analysis of organic and inorganic functional groups. The instrument with near-infrared (NIR) source with liquid nitrogen cooled Mercury-Cadmium-Telluride (MCT) detector can be used for aerosol analysis. It scans the sample from 500–4700 cm^{-1} and gives well-resolved peaks of functional groups present in the sample. Before analysis, the instrument is purged with nitrogen gas to remove CO_2 and water vapours. In order to remove the background signature of PTFE filters from each sample, background PTFE spectra of unused filters are subtracted from IR spectra of each sample. No sample preparation is required for this instrument, the filters can be analyzes as such. This instrument is very fast, easy to use, low cost, and non-destructive. The analyzed filters can be used for any other analysis also.

Simulation of optical properties of aerosols: The optical response of the aerosol particles is governed by their shape, size, chemical composition, and mixing configuration. The physico-chemical properties of the aerosol particles are observed using aforementioned techniques and the optical properties (SSA, g, and Q_{ext}) are calculated with the help of various optical models. The calculated optical properties are an essential input to the radiative model which gives radiative forcing (Fig. 13.4). Amongst the optical models, Mie code is used for the optical properties simulation of the spherical particles while the core–shell model is an advanced version of Mie code where optical properties of the core–shell structured spherical particles are calculated. T-matrix, FDTD, SIEM and DDA models give the opportunity to simulate optical properties for some non-spherical particles.

Air Quality Monitoring Station (AQMS) at Hilly region, Palampur: An air quality monitoring station has been established at CSIR-IHBT, Palampur, Himachal Pradesh in 2014 under the CSIR 12th plan to study the ambient air quality over the

Table 13.2 Characterization of aerosols

S. No.	Parameters	Instrument used	Methods
1	BC	Aethalometer	Optical
2	OC and EC	OC/EC analyzer	Thermo-optical
3	WSOC and WIOC	TOC analyzer	NDRI
4	Trace elements	X-ray fluorescence/AAS/ICP-MS/ICP-OES	Fluorescence/spectroscopy
5	Water-soluble inorganic ions	Ion chromatograph	Chromatography
6	Water-soluble organics	GC–MS	GC–MS
7	Shape and Size/Morphology	SEM/TEM/EPXMA/SEM–EDS/ FIB-SEM–EDS	Optical
8	Surface Characterization	X-ray Photoelectron Spectroscopy (XPS)/OP-FTIR	
9	Stable C and N isotopes	IRMS	Isotopic ratio
10	Particle numbers	Optical Particle Counter/Differential mobility analyzers (DMAs)/(DMPS)/(SMPS)	Optical
11	AOD	Radiometers/sunphotometers	Optical
12	Aerosol Scattering/extinction coefficient	Nephelometer/Photoacoustic spectroscopy (PAS)	Optical
13	Single scattering albedo	PSAP (particle soot absorption photometer)/multi-angle absorption photometer (MAAP)	Optical
14	Cloud condensation nuclei	CCN counter	Optical
15	Remote sensing of Aerosols	LIDARs	Optical

north-west Himalayan region of India (Fig. 13.5). The concentrations of ambient NO, NO_2, CO, O_3, SO_2, NH_3, HCs, PM, BC as well as the meteorological parameters are being measured since its inception for time-series observations. The GHGs like CH_4 and CO_2 are also being measured using the cavity ring-down spectroscopic (CRDS) technique for its climatic impacts. A Raman lidar system is also installed at this station to study the vertical distribution of aerosols and clouds. It consists of two Mie channels at 355 nm and a Raman nitrogen channel at 387 nm to study the various physical and optical properties of the aerosols including aerosol/cloud depolarization and lidar ratio, in the troposphere and lower stratosphere.

Measurements at Antarctica and Arctic under Indian Polar Research Program: CSIR-NPL is one of the important scientific partners of National Centre for Antarctic and Ocean Research Laboratory (NCAOR), Ministry of Earth Sciences and participated right from the first Indian Scientific Expedition to Antarctica in the

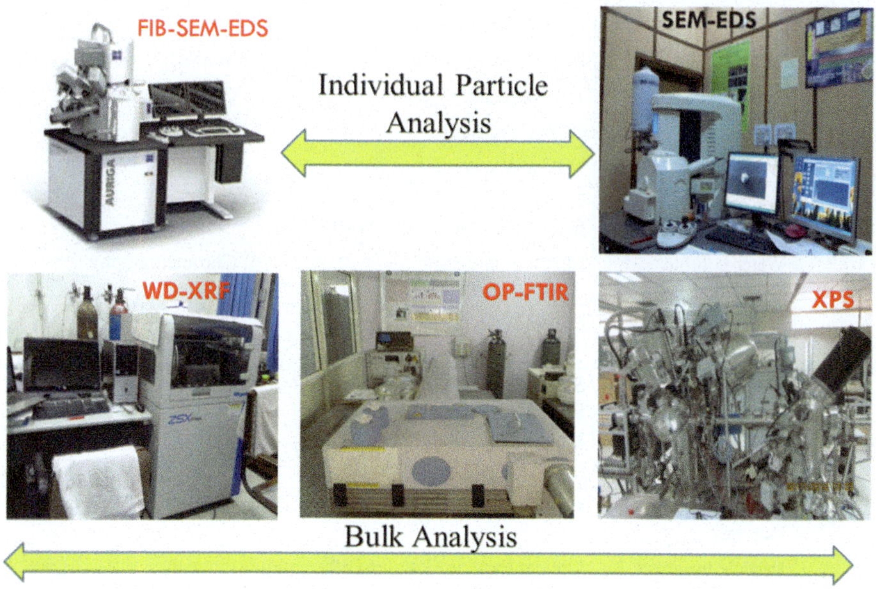

Fig. 13.3 Physico-chemical characterization techniques: FIB-SEM–EDS, SEM–EDS, WD-XRF, OP-FTIR, and XPS

year 1981 and till now performed various scientific investigations related to climate change and space environment system. In view of this, CSIR-NPL deployed various states of art facilities like SODAR, Gas Chromatogram, LASER Heterodyne System and various analyzers for monitoring of atmospheric trace gases. Also, CSIR-NPL deployed various experimental facilities to monitor the space environment system like Diameter, Very Low-Frequency Receivers Advanced Digital Ionosonde and Global Ionospheric Scintillation and Total Electron Content Monitoring System. CSIR-NPL also provided national leadership to the four Indian Scientific Expeditions to Antarctica (ISEA), i.e. 8th, 10th, 14th and 31st ISEA. The 8th and 31st Indian Scientific Expeditions were marked as important milestone for Indian Antarctic mission when the two permanent scientific bases, i.e. Maitri (70° 46′ S, 11° 43′ E) and Bharati (69° 24′ S, 76° 11′) were established by India and joined the elite club to have more than one scientific base in Antarctica (Figs. 13.6, 13.7 and 13.8). During 28th ISEA (2008–2009), CSIR-NPL established a state of art Indian Polar Space Physics Laboratory (IPSPL) at Indian Permanent Research Base "Maitri", Antarctica (Fig. 13.6) on the occasion of International Polar Year (IPY) for continuous and real-time monitoring of high latitude ionosphere to address the scientific interest of high latitudinal ionospheric consequences caused by the modulation of near-earth space environmental conditions.

The Ionospheric Laboratory consists of Canadian Made Advanced Digital Ionosonde (CADI) system and Global Ionospheric Scintillation and TEC Monitoring

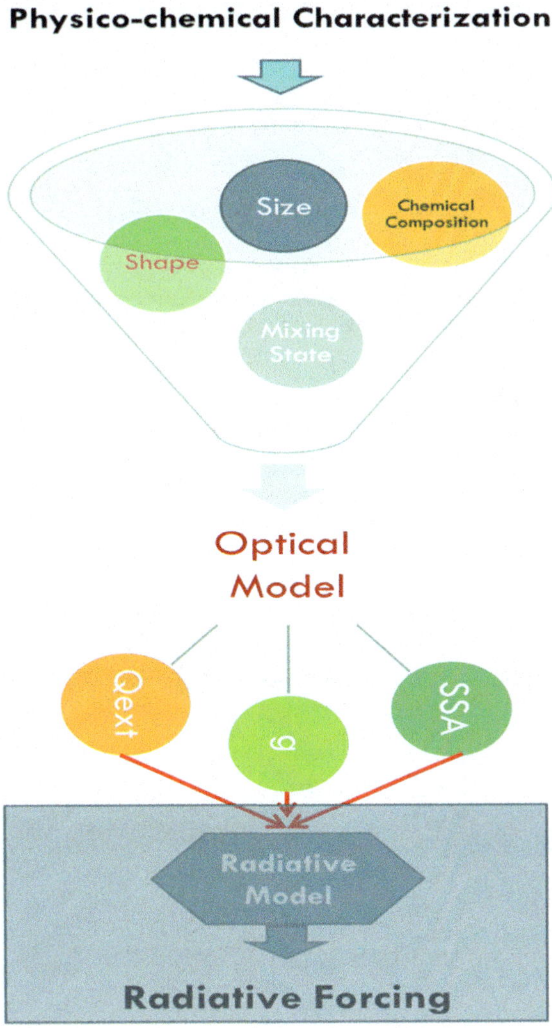

Fig. 13.4 Inputs to optical and radiative models

(GISTM) System. These systems are not only useful for monitoring the ionized atmosphere but also able to provide first observations on sun-earth coupling processes which is helpful for forecasting of upper atmospheric irregularity generated due to sun driven sources.

CSIR-NPL is regularly participating in various scientific programme coordinated by International Scientific Committee on Antarctic Research (SCAR) under the code of SCAR-72 "*How does space weather influence the polar ionosphere and what are the wider implications for the global atmosphere?*" & "*How will the polar space-atmosphere-interaction region (SAIR) change due to global climate change and what is the impact of those changes on the lower atmosphere?* The objective of the

Fig. 13.5 Atmospheric Monitoring Station of CSIR-NPL at Palampur, Himanchal Pradesh

Fig. 13.6 Photograph of Maitri Station

programme is to exchange of data, knowledge and technical competence between the countries involved in polar research activities for societal benefits. Under this programme, CSIR-NPL utilizing its facilities to monitor the polar upper atmosphere on round the clock and real-time basis since 2015 and generated a time series data. The investigations reveal various aspects of adverse space weather conditions and its impact on sub-auroral ionospheric region, i.e. over Maitri like the ionization over sub-auroral region depends upon the combination of different electro-dynamic processes, e.g. solar zenith angle, particle precipitation, auroralelectrojet and direction of the meridional wind. The observations also evident close association of negative and

Fig. 13.7 Photograph of Bharati Station

Fig. 13.8 Photograph of CSIR-NPL Ionospheric Observatory, Maitri

positive ionospheric response to the south-ward and north-ward turning of interplanetary magnetic field component Bz, respectively. The study and monitoring of these disturbances are important for communication and navigation applications. The CSIR-NPL disseminates such information's through its "Regional Warning Centre" (RWC) which is a part of the international network and member of "International Space Environment Services" (ISES).

CSIR-NPL is also actively participating in the Indian Scientific Expedition to Arctic and performed scientific activities related to the polar climate change program. CSIR-NPL mainly focused to study the snow pack production of carbon monoxide and its diurnal variability at Indian Arctic Station Himadri, Ny-Ålesund, Svalbard (78° 54′ N, 11° 53′ E) (Fig. 13.9). The results reveals positive co-relation between ambient CO concentration and UV radiation when atmospheric conditions were clear

Fig. 13.9 Photograph of Himadri station

sunny with calm winds. This co-relation indicates the direct photoreactions of organic compounds and indirect photoreaction involving other oxidants which may probably responsible for CO production and diurnal variability in Polar Regions. The study provides and important input to the scientific community involved in climate change related study. The CSIR-NPL continuously working with NCAOR, Ministry of Earth Sciences, Govt. of India to elevate the national polar scientific mission and improves its international visibility.

13.2.4 Measurement Standards, Quality Assurance and Traceability

Data quality of the measurements of the trace gases (CH_4, N_2O, O_3, CO, NO_x, SO_x, NH_3, NMHCs, and VOCs, etc.) as well as aerosols are achieved by testing and calibration of the measuring instruments through the unbroken chain of traceability [33]. To assure the traceability of trace gases, we performed the calibration of trace gas analyzers/equipments with respective trace gas standards. Before performing the span calibrations of any trace gas analyzers/equipments, the zero base line of the each range of the analyzers are calibrated using pure air (Pure Air Generator: accuracy $\leq \pm$ 1 ppb or better). Then span calibration of each range has been performed using the standard gas mixture (assigned value $\pm$ uncertainty) with the help of multigas gas calibrator and pure air generator as mentioned in the flow chart (Fig. 13.10). Examples, for calibration of NOx-analyzer, standard gas mixture of NO_x (assigned value $\pm$ uncertainty) are used for calibration. Similarly, the standard gas mixtures of CH_4, N_2O, CO, NO_x, SO_x, NH_3, NMHCs, and VOCs are used to calibrate the respective analyzers / equipments [34].

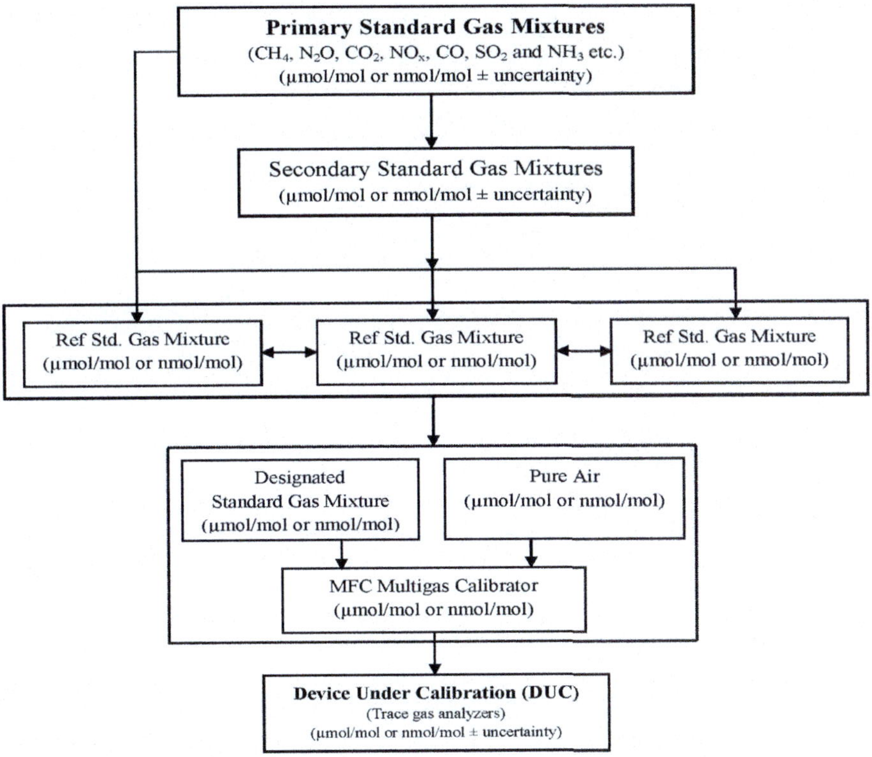

Fig. 13.10 The hierarchical traceability chart of trace gases calibration

To make all the measurements traceable to SI a hierarchical chain of calibration and standards with stated uncertainties are used. Such chain is shown in Fig. 13.10 with examples of ambient trace gases. The traceability chain has essentially two parts, i.e. vertical and horizontal traceability. The vertical traceability is achieved through calibration whereas horizontal traceability through inter-comparison at the same level (equivalence). Calibration and inter-comparison are ways to establish traceability and to ensure reliable measurement of components, parts, etc., to make them compatible for their utility worldwide. In the global market place, the demand is for international acceptance of product calibration/testing and conformance and this requires international traceability. NMIs play an important role in maintaining the quality assurance, measurement traceability and a role of interface between national and international measurement systems (SI units). The quality assurance of measured datasets of trace gases (CH_4, N_2O, O_3, CO, NO_x, SO_x, NH_3, NMHCs, and VOCs, etc.) may only be achieved through testing and calibration of the measuring equipments with stated uncertainty [33].

The Primary Ozone Standard (Standard Reference Photometer, SRP 43) has been setup at CSIR-NPL to provide the facility for APEX Level calibration and traceability of surface O_3 (Fig. 13.11). The comparison between the CSIR-NPL SRP43 and

Fig. 13.11 Primary Ozone Standard (SRP 43) at CSIR-NPL, India

the reference standard BIPM-SRP27 was maintained at the BIPM (via the transfer standard SRP0 maintained by the NIST). The common reference standard SRP27 and the transfer standard SRP0 were first compared at the BIPM in January 2007. Then SRP0 was compared with the SRP43 at the CSIR-NPL in February 2009. After successful completion of the international key comparison and peer review the Calibration and Measurement Capability (CMC) for this facility was obtained in 2012 and re-reviewed by CCQM in 2015 and approved in 2016 January [35].

The CSIR-NPL is involved in measuring various atmospheric trace gases including greenhouse gases and particulate matter to study their chemical and physical properties and identifying their roles in influencing the atmosphere and climate through state of art instruments and models. Best practices for accurate measurement of atmospheric trace species are being developed for adoption by different agencies and institutions in India engaged in atmospheric monitoring. The improvement in measurement quality will also have societal benefits as that will foster better policy formulations for amelioration of air quality and mitigation of climate change.

13.3 Gas Metrology

Gas metrology activity of the NMI has a direct role to play on the economic growth of that country. Therefore, this activity is an inherent and major part of any NMI in the world, and at BIPM, CCQM has a separate consultative committee on the Gas Analysis Working Group (CCQM-GAWG). The details of internationally accepted mission of this activity are available on BIPM website (https://www.bipm.org/en/committees/cc/wg/gawg.html). Gas metrology directly relates to the quality assurance of air and emission, health, energy, food, etc. and hence indirectly influence the life style of common people, trade, economy, and finally to the economic growth. CSIR-NPL, the NMI of India has a major activity of *Gas Metrology* which includes realization of mole in gas measurements through gravimetric preparation of gas standards to provide the traceability for emission, energy, GHGs and air quality measurements as per National Ambient Air Quality Standards. The mandate of the activity is to prepare the Primary Reference Gas Mixtures (PRGMs) for greenhouse gases, pollution gases, energy and emission gases, particulate matter standard; participation in inter-comparison study and organise proficiency testing; calibration of particulate matter samplers, related testing and project jobs; air quality/emission related instrument development/ transfer standards; research and development for reference methods, standards and instruments; workshop, training and skill development programs, etc.

13.3.1 Industrial Gas and Economy

Industrial gases are atmospheric gases and the gases which are produced in the industry for various industrial applications. These gases are nitrogen (N_2), oxygen (O_2), carbon dioxide (CO_2), nitrous oxide (N_2O), nitrogen oxides (NO), helium (He), argon (Ar), acetyline (C_2H_2), hydrogen (H_2), sulphur dioxide (SO_2), hydrogen sulphide (H_2S) etc. There are a series of gases, which are used as fuel gases. Due to air pollution, now many of pollution gases are also under this category as those are required in many industrial processes. These gases are produced in industrial-scale having their unique technology and have impact on the economy. For example, atmospheric gases are produced by air separation technique in large-scale having high quantity of impurities in it. But there is also demand for high purity gases for various industrial and scientific applications. This has led to the development of high purity industrial gas required in many sectors. Oxygen is used in steel/ copper industry, engineering fabrication and gasification. It has medical applications and is used for cleaning water and hazardous waste. There are many more examples of industrial as well as ultra-high purity gases used for the fabrication of computer, TV to digital circuits and several gadgets of modern life; health care; manufacturing sector; material processing; food preservation and processing; beverages; chemicals and fertilizers, etc. Gas industry is however requires a very high capital investment to start the industry and meet the day to day running costs. The safety concerns and stakes are also very high in this industry. But this sector generates a lot of employment not only in the gas plant but also in the distribution chain having middle-level investment requirements.

The global gas industry is estimated to be of the order of 83–91 billion USD in the year 2020. It is projected to grow at the compounded annual growth rate (CAGR) in the range 4.5–6.74% for the duration 2020–2027 [36, 13, 37]. At present India's population is around 17% of the world's population. In the coming times if the Indian industrial activity invigorates in the county this will increase the demand and consumption of industrial gases here. This industry in India may achieve its rightful place, of at least 15% of the total industrial gas market, or an annual turnover of around USD 12 billion. The global calibration gas mixture market is valued as 590 million USD and is expected to grow at a CAGR of 5.3% during 2020–2026 [38]. There will be consistent increase in the market demand of speciality gases, i.e. Standard reference gas mixtures or commonly called calibration gases to the tune of 10 million USD in India also.

Various other gases also find applications in large number of industries/processes and scientific explorations, which has led to the challenges for quantifying gases in various matrixes. Apart from industrial applications of gas measurement, other important areas, which have ample demand to quantify gases accurately, are health care; emission measurements; environmental pollution and green house gas determination for climate change studies. Such challenges created high demand for gas metrology which is defined as the science of measurements of gases or its suspended constituents, embracing both experimental and theoretical determinations at any level of specified uncertainty in this field. In order to do the reliable measurement of gases, two important factors play key role: the measurement has to be traceable to SI units and the process must follow quality management system for measurement. It requires that the amount of substance is measured in the units of moles which is achieved through Gas Metrology route.

13.3.2 SI Unit of the Amount of Substance (The Mole)

The mole, (symbol—mol), is the SI unit of amount of substance. It is defined in Chap. 2 of the SI Brochure (8th edition), Sect. 2.1.1.6 (BIPM) [39]. The mole may be realized by a variety of primary methods. The attributes of primary method of measurement were defined in the CCQM 1998 [40, 41]. Primary method of measurement produces results always traceable to SI units and is of smallest achievable uncertainty. It has the highest metrological qualities. Its operations can be described completely in the procedure in sequential steps and it can be operated and executed without any doubt and ambiguity. Each step has been completely understood and a complete uncertainty equation can be written down and estimated in SI unit. The results of primary methods are accepted without reference to a standard of the quantity being measured. Primary methods are normally implemented only by the NMIs. It is normally not implemented by the routine testing or field laboratories. The benefits of the primary method are that the measurements are stable, repeatable, independent of locations. The results from different measurement methods are comparable. This is achieved as all the measurements are traceable to SI units. The mole realization through PRGM is applicable in Gas Metrology is described below.

13.3.3 Realizing the Mole, SI Unit of Amount of Substance, through Primary Reference Gas Mixture (PRGM) Standard Preparation

The gravimetric preparation of primary reference gas mixture (PRGM) provides SI traceability to gas measurements. This method involves mixing of pure gases or pure liquid components by mass in the gaseous phases one after another or simultaneously. Relevant documented standards for the gravimetric preparation of PRGM are as following

ISO 6142–1: Gas analysis—Preparation of Calibration gas mixtures—Part 1: Gravimetric Method for Class I mixture

ISO 6143: Gas analysis—Comparison method for determination and checking the composition of calibration gas mixtures

ISO 6141: Gas analysis—Requirement for Certificates of Calibration gases and gas mixtures

Principle of PRGM preparation

Primary Reference gas mixtures are prepared by transferring pure gases, pure liquids or gravimetrically prepared mixtures (called premixture) of known composition quantitatively into a cylinder in which the calibration gas will be contained.

As described in ISO 6142, the traceability to SI, of amount fraction of these mixtures arises from the correct execution of the following steps:

(a) The determination of the masses added.
(b) The conversion of added masses to amount of substance, by knowledge of their chemical purity and appropriate relative atomic and or molecular masses.
(c) The verification of the final mixture against independent reference gas mixtures.

The correctness of the prepared mixture depends upon how accurately the smallest mass is transferred in the container cylinder and how precisely it is weighted on a very sensitive balance.

Theoretical calculations for masses of the component gases to be taken in the gravimetric preparation is done using ideal gas equation for two-component mixture

$$PV = nRT \tag{13.1}$$

where
P is pressure
V is the volume
n is number of moles
R is the gas constant (8.3145 J/mol k)
T is the temperature in K (298 K)
for more than two component mixtures

$$X_i = \frac{n_i}{\sum_1^i n_i} \tag{13.2}$$

where
X_i = mole fraction of ith component
n_i = number of mole of ith component

The PRGMs are generally prepared in 10 or 6 L aluminium cylinder at the pressure 120/100 bar by the NMIs. Here for the calculation volume of the cylinder is taken 10 L. The pressure up to which the diluent gas is filled 120 bar at 298 K (i.e. 25 °C) temperature. Whatever is the value of volume and the pressure of the proposed gas mixture are to be considered in mass calculations using Eq. (13.1) as given above.

Then the amount of pure gas (component 1) required to achieve the desired concentration (in diluent gas) is calculated by using following formula

Required mass of component 1 (W_1)

$$W_1 = \frac{PVM_1}{RT} \tag{13.3}$$

where
P = Pressure in Pascal = 12,000,000
V = Volume of the cylinder in m^3 = 0.01
R = Gas constant (8.3145 J/mol k)
T = Temperature in K (298 K)
M_I = Molecular weight of the component 1

Before preparation of gas mixture in the aluminum cylinders, its valve and gas transfer system material integrity for high-pressure handling have to be ensured. It must not react with the gas components. Purity analysis of the component gases is a critical step, so it must be completed before gravimetric mixing of the components as after that it cannot be ensured or analyzed.

13.3.4 Purity Analysis and Weighing using Balance for Component Gas

Purity analysis of the component gases is a critical step in the preparation of gravimetric PRGM standards. Purity analysis is carried out for all the starting gases and liquids to be used in the PRGM preparation in accordance to ISO 19229:2015. This must be completed before gravimetric mixing of the components as after that it cannot be ensured or analyzed. The starting materials should have lower amount fraction of minor impurities present in it. The materials used for the PRGM should be of ultra-pure grade. Impurity analysis report should mention amount fractions of each impurity present along with associated uncertainties. Primary method cavity ring down spectroscopic (CRDS) analyzers for the minor impurity analysis facilities are available at the CSIR-NPL.

Gravimetric method of preparation of PRGM is weighing of components in cylinders very accurately and precisely. The challenge here is that the mass of the gas or liquid component is attempted to be kept small to the tune of 20 g or less. The mass of the aluminum cylinder container is 10 to 12 kg and the cylinder hanger mass 2.5 to 3 kg. Hence mass of the component gas is weighed in a container with cylinder hanger having mass 12 to 15 kg. The uncertainty of the weighing process of the component is required to be in the range of 20 mg. In some of the cases steel cylinders are used for PRGM preparation, which are much heavier than the aluminum cylinder for example of the 25–28 kg having water capacity 10 L. The mass and the volume of the cylinder is very large in comparison to the actual component mass. There will be large factor of buoyancy in normal weighing process. In the gravimetric preparation processes weighing of components in cylinders are done in accordance to high accuracy mass metrology principles. The weighing of components is carried out using reference cylinder which is similar to the sample cylinder. It is taken from same lot of cylinder manufacturing. In order to cancel the buoyancy due to the volume of the cylinder preparation of PRGM is carried out using equal arm double pan balances. In the weighing process on one side an empty reference cylinder is kept and the other side the sample cylinder. Before weighing empty pan reading is always noted and accounted in the weighing. The weighing cycle involves as empty pan-cylinder-empty pan as one set of weighing and this process is repeated at least thrice. Each weighing step whether empty pan or with cylinder is carried out seven times. This process of weighing is very tedious and time-consuming. The cylinders are loaded to the balance with minimal jerks and vibrations. Recording of result is done after sufficient time was given to damp down all the vibrations and swings of the balance beam. One can get very accurate and precise mass of the component with very low uncertainty in the range of 10 mg or so. But the weighing process will take long time to the tune of 7–9 h. But if one wish to have uncertainty in the range 20–30 mg weighing process may take 2–3 h. PRGM preparation weighing balance at CSIR-NPL India is shown in Fig. 13.12. The maximum weighing capacity of the balance is 25 kg with sensitivity of ± 1 mg. All the gas components are weighed using this balance. Liquid component is weighed using high sensitive balance separately. The liquid weighing is done by differential weighing process, i.e. the required amount of component is taken in liquid-tight syringe and is transferred into the cylinder through gas line and weighed. Mass of component liquid transferred is calculated as difference in the mass before and after transfer.

The smaller mass component challenge is taken care by doing successive gravimetric dilutions from a pre-mixture to achieve smaller mass of targeted component. One can use different balances for weighing minor targeted component on a highly sensitive balance in a smaller cylinder having lower uncertainty. This may reduce the uncertainty of the prepared PRGM. Successive dilution scheme is shown in Fig. 13.13.

The quest to achieve lower uncertainty PRGMs and the tedious time-consuming weighing process has led to the development of automation and technological improvement by several NMIs of the world. Federal Institute for material Testing (BAM) Germany tried to automate its double pan balance weighing process. Van Swinden Laboratory (VSL), The Netherlands developed automatic cylinder exchange mechanism in its single pan balance for gas weighing. National Institute for Metrology for Japan (NMIJ) developed rotary exchange mechanism in its

Fig. 13.12 PRGM preparation weighing cycle at NPL-I

top pan balance. Korean Research Institute for standards and Science (KRISS) developed linear exchange mechanism for the top pan balance. These improvements really reduced uncertainty to 10 mg or less. It takes less time in weighing process and is person independent.

Expression for Mixture composition: In the gravimetric method of PRGM preparation, the value of masses of the components are measured in the weighing process. These masses are supposed to be converted to number of mol of that component. To do that one needs to know the relative molar masses of the species or the component. It can be calculated using equation given below.

$$X_i = \frac{{}^{m_i}/_{M_i}}{\sum_1^i {}^{m_i}/_{M_i}} \tag{13.4}$$

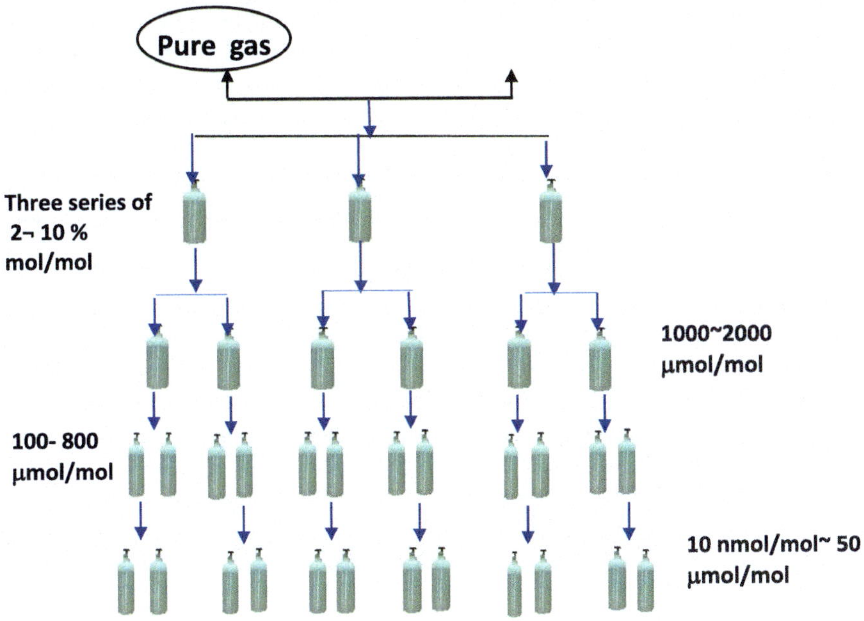

Fig. 13.13 Successive dilution scheme for gravimetric preparation of PRGMs

X_i = mole fraction of *i*th component,
m_i/M_i = number of mole of *i*th component (n_i)

Estimation of Gravimetric Uncertainty: The Uncertainty of the amount of fraction of prepared PRGM is calculated taking account of the Eq. (13.4) given above. The factors like uncertainty due to balance and weighing process, mass pieces uncertainty and buoyancy correction, repeatability, reproducibility, calibration of balance, environmental condition of the room such as temperature, humidity, barometric pressure, purity of the component, cylinder expansion due to gas filling at high pressure, cylinder handling losses, residual gas in the cylinder after evacuation, CO_2 concentration in the balance room for buoyancy calculation is also taken into account. In case if the PRGM preparation involved smaller transfer cylinder for minor component or in the case of volatile liquid used high sensitive balance uncertainty and the weighing process uncertainty has to be further accounted for in the gravimetric uncertainty calculations.

Verification and Stability study: The gravimetrically prepared PRGMs are put to homogenization. Gravimetrically prepared mixtures are analyzed using various techniques for the verification of its molar fraction. The experimental verification of the PRGM composition is done to show that it is in agreement with the gravimetric preparation value. This confirms that there are no errors in the gravimetric preparation process in the individual cylinder PRGM and in the entire successive dilution scheme. There has been complete transfer of components and no reaction in between the components or with the cylinder material, etc. Only after the verification of PRGMs the composition of the cylinder is considered to be SI traceable. Usually at the National Metrology Institutes during the new development of standards no appropriate references are available. Thus, the prepared PRGMs are verified for its composition among the similar prepared PRGMs in its successive dilution preparation processes in consistence to ISO 6142 methods.

At the other reference material producer laboratories verification is carried out using appropriate reference gas mixture having better uncertainty than the prepared one. Results of PRGM analysis is reviewed according to ISO 6143 methods. The stability of the prepared PRGMs are determined time to time to review any rate of drift from the amount of fraction of the gravimetric value. Stability study is planned taking account of the amount fraction and reactivity of the components in the mixture along with the minor impurities present. For example, a PRGM of stable gas two-component mixture is not required to be tested in short interval for the stability check. In this case it may be checked after one year or so.

13.3.4.1 Metrological Traceability of PRGM

In the field of measurement or in metrology in chemistry challenge is how to do a comparison of results of over time or by different methods or by different procedures in same laboratory or at different geographical locations. This is achieved by making all the measurements traceable to SI units. The benefits of SI traceable measurements are that it is stable and comparable to each other at different time zones and scales. For measurement with more than one input quantity in the measurement process, each input quantity should itself be metrologically traceable and the calibration hierarchy involved may form a branched structure or a network. The effort involved in establishing metrological traceability for each input quantity value should be commensurate with its relative contribution to the measurement. The metrological traceability of PRGM is schematically represented in Fig. 13.14

13.3.4.2 International Inter-Comparison for Comparability

In gas metrology, primary standard gas mixtures (PSMs) are used to maintain the measurement hierarchy in traceability chain. The collaborative work on gas analysis started in 1993 with the formation of Cooperation on International Traceability in Analytical Chemistry (CITAC). The programme for the collaborative work among the national metrology laboratories was designed in the first meeting of the Consultative Committee for Amount of Substance (CCQM) held in 1995 [42]. The first international key comparison of PSMs was registered at Bureau International des Poids et Mesures (BIPM) as CCQM-K1.a-g in which except three natural gas compositions, the other selected compositions were simple with one minor component in nitrogen. CCQM-K1 key comparison showed that the coordinated gas activities in the field of

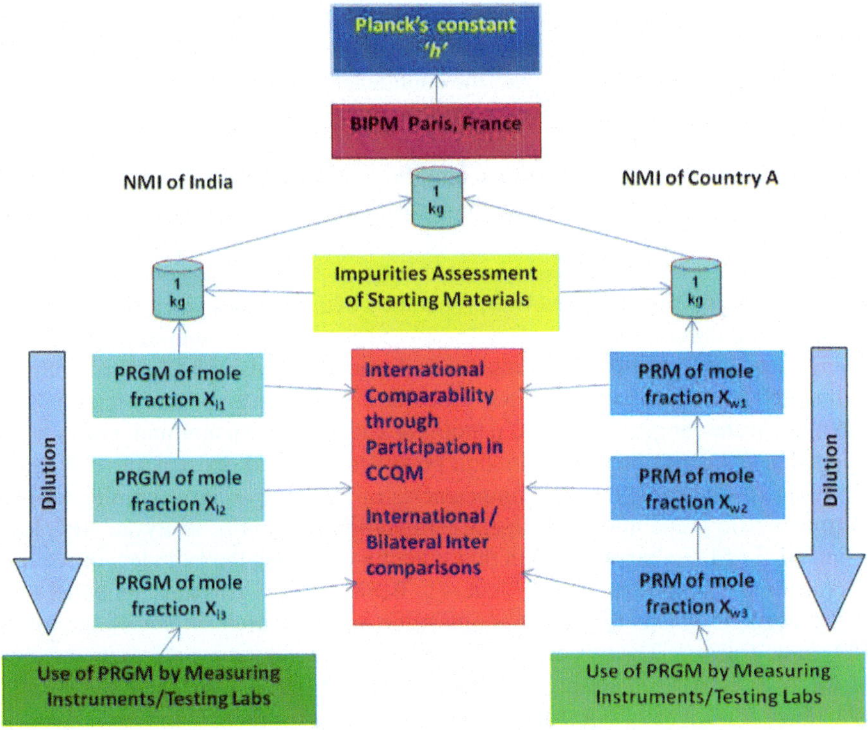

Fig. 13.14 Metrological traceability of primary reference gas mixture

gas mixtures generally lead to international comparability within 1% of the reference values [43–45]. Since then, CCQM Working Group on Gas Analysis (GAWG) is continuously working on improving the harmonization of measurement methods and improvement in measurement uncertainty. At present BIPM and all the regional metrology organizations (RMOs) are organizing the key comparisons and regional international inter-comparisons/ pilot studies to provide the metrological traceability linkage and CMCs which are published at BIPM official website [40].

NPLI has created basic requirement facilities to develop the Gas metrology area for providing traceability working for the dissemination of traceability in Gas area. Recently, NPLI has successfully participated in CCQM-K120 key comparison coordinated by BIPM and other inter-comparisons coordinated by the national metrology institutes (NMIs) in the APMP region (APMP-QM.S9.2017, APMP-QM.S7.1, APMP-QM.S15) to establish the international measurement comparability and traceability linkage for different minor component in matrix gas (nitrogen/ air). Dissemination of traceability is one of the mandates of NPLI. The key comparisons are usually selected keeping in mind all the areas of application and most commonly used measurement methods. To participate in any inter-comparison, scientists in the gas metrology activity at NPLI registered in the inter-comparison proposed by APMP or BIPM and then prepare the gas mixtures in the required mole fraction and perform the analysis of the cylinder sent by the coordinator NMIs against their own gravimetrically prepared PRGMs. This gives the confidence in the gravimetric preparation of gas mixtures as well as the measurement method analysis.

CCQM-K120.a and b Carbon dioxide in Air: CCQM-K120.a Carbon dioxide in Air at background level (380–480) μmol/mol and CCQM-K120.b Carbon dioxide in Air at urban level 480–800 μmol/mol [46] key comparison was coordinated by BIPM France and NIST USA jointly and fourteen laboratories participated in this study. This key comparison was scheduled in April 2016 and the final report came in 2018. CCQM K120.a and b involves preparation of carbon dioxide in air so that these standards are fit for purpose for the atmospheric monitoring community and climate change studies. These comparisons are designed to evaluate the level of compatibility of NMI preparative capabilities for carbon dioxide in air primary reference gas mixtures in the range of 380–480 μmol/mol and 480–800 μmol/mol, respectively. The balance gas for the standards shall be either scrubbed dry real air or synthetic air. It also tests-core skills and competencies of a participating laboratory required in gravimetric preparation, analytical certification and purity analysis. NPL India being NMI of India participated in this study by preparing the cylinders of required composition and sending three cylinders to BIPM France. This key comparison has stringent requirements for matrix composition and measurement uncertainty of the CO_2 mole fraction. The preparative compositional requirements as per the protocol of the key comparison are given in Table 13.3 and preparation scheme of CO_2 in synthetic air is given in Fig. 13.15.

Measurement Results of NPLI Standards at BIPM: CSIR-NPL prepared three standards as per the composition given in protocol and sent to the BIPM for measurement. BIPM did the measurements using Fourier Transform Infrared Spectroscopy (FTIR) and Isotope Ratio Infrared Spectrometer (IRIS) systems and verified with Gas Chromatography-Flame Ionization Detection (GC-FID). Figs. 13.16, 13.17 and 13.18 represent the measurement comparison results of CSIR-NPL prepared in gravimetric standards with other NMIs at BIPM. The figures are taken from the CCQM-K120 final report published [46] at BIPM site. **CCQM-K120a:** inter-comparison

Table 13.3 CCQM-K120.a and b Matrix gas composition limits

Species	Ambient-level mole fraction	CCQM-K120.a Matrix gas composition limits (380 μmol/mol and 480 μmol/mol CO_2 in air)		CCQM-K120.b Matrix gas composition limits (480 μmol/mol and 800 μmol/mol CO_2 in air)	
		Min	Max	Min	Max
N_2 (mol/mol)	0.780876	0.7804	0.7814	0.7789	0.7829
O_2 (mol/mol)	0.2093335	0.2088	0.2098	0.2073	0.2113
Ar (mol/mol)	0.0093332	0.0089	0.0097	0.0078	0.0108
CH_4 (nmol/mol)	1900	0	1900	0	1900
N_2O (nmol/mol)	330	0	330	0	330

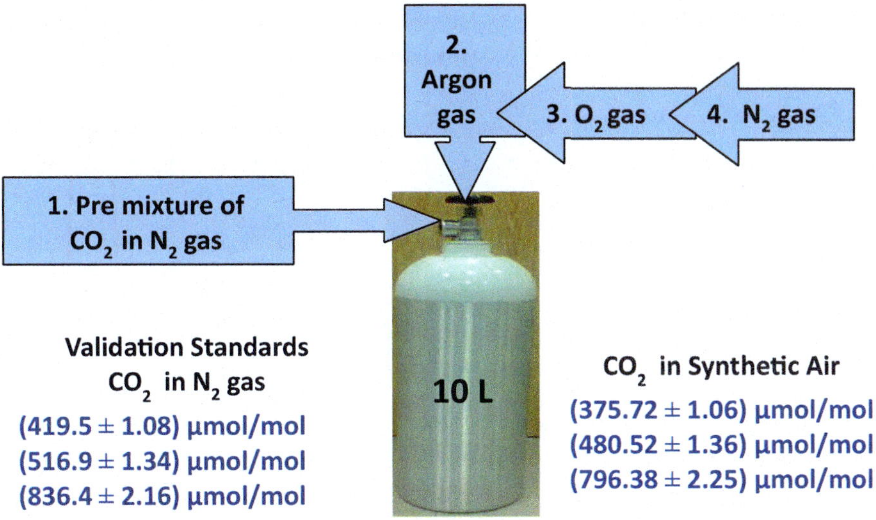

Fig. 13.15 Preparation Scheme of CO_2 in synthetic Air (300–800 µmol/mol)

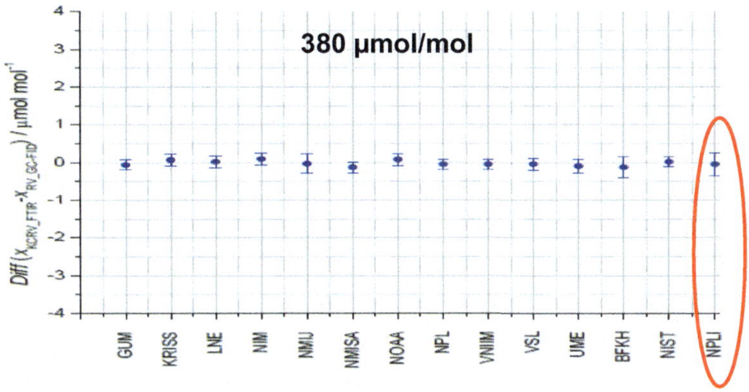

Fig. 13.16 Difference between KCRV values obtained using FTIR and GC-FID reference values at 380 µmol/mol. The error bar represents the expanded uncertainty at a 95% level of confidence (© Bureau International des Poids et Mesures. Reproduced by permission of IOP Publishing. All rights reserved)

has been performed for CO_2 in the air in mole fraction range of 250 µmol/mol to 520 µmol/mol. Whereas, CCQM-**K120.b** inter-comparison has been performed for CO_2 in air and nitrogen in a mole fraction range 500 mmol/mol.

APMP-QM.S9.2017:*Carbon monoxide in Nitrogen (100 µmol/mol)*: This inter-comparison was coordinated by KRISS to establish the CMC claims of NMIs on CO/N_2 for industrial application. The first key comparison on carbon monoxide in nitrogen in 1993 (CCQM-K1a) [46] was related to nominal value of 100 µmol/mol

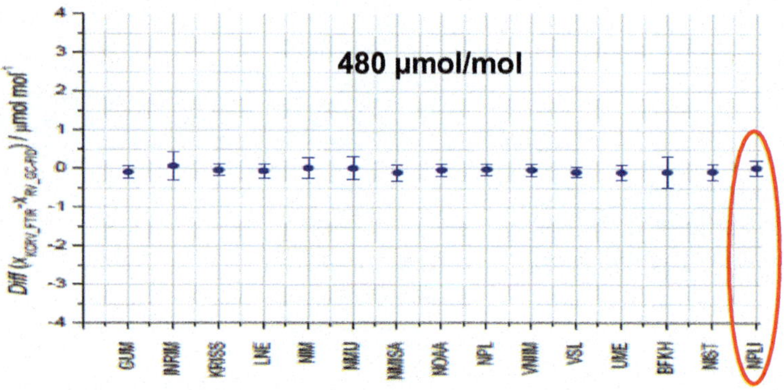

Fig. 13.17 Difference between KCRV values obtained using FTIR and GC-FID reference values at 480 μmol/mol. The error bar represents the expanded uncertainty at a 95% level of confidence. (© Bureau International des Poids et Mesures. Reproduced by permission of IOP Publishing. All rights reserved)

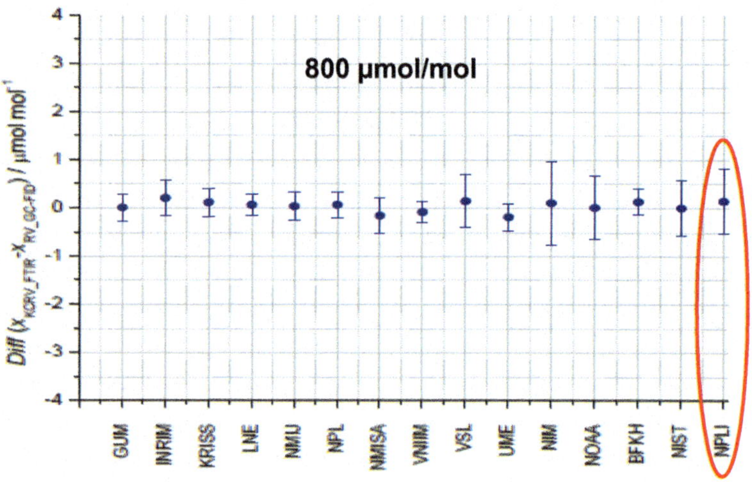

Fig. 13.18 Difference between KCRV values obtained using FTIR and GC-FID reference values at 800 μmol/mol. (© Bureau International des Poids et Mesures. Reproduced by permission of IOP Publishing. All rights reserved)

CO in nitrogen. Later GAWG had carried out two more CO comparisons and RMO including COOMET and APMP had performed their regional comparisons [under review for draft B report of the APMP.QM-S9]. The others' intercomparisons were on ambient level (5 μmol/mol) of CO in nitrogen for the CCQM K51 or trace level (350 nmol/mol) of CO in air for the CCOM K84. This intercomparison result was presented in Draft A report [47]. The NMIs participated in this intercomparison were

CSIR-NPL, Kazakhstan Institute of Metrology (KAZ Kazakhstan), National Institute of Metrology (NIMT Thailand), National Metrology Centre (NMC Singapore), Pusat Penelitian Metrologi (LIPI Indonesia), Korea Research Institute of Standards and Science (KRISS Korea). Using the reported values of NMIs in the Draft Report, the degree of equivalence (DoE), d_i, of an individual NMI/DI result, x_i, and the reference value, x_{ref}, and its standard uncertainty $u(d_i)$ are calculated using Eqs. (13.5), (13.6) and (13.7), in accordance with the CCQM guidance note 2013. The results of the intercomparison are shown in Fig. 13.18.

$$d_i = x_i - x_{ref} \tag{13.5}$$

$$u(d_i) = \sqrt{u^2(x_i) + u^2(x_{ref})} \tag{13.6}$$

Expanded uncertainty, $U(d_i)$ is calculated at a coverage factor of $k = 2$ as follows:

$$U(d_i) = k \cdot u(d_i) \tag{13.7}$$

Results taken from report are plotted in Fig. 13.19. This supplementary comparison will support the measurement capability, which can be used to support CMC claims, for carbon monoxide in nitrogen from 50 μmol/mol to 200 mmol/mol.

APMP.QM-S7.1*:*CH*$_4$ *in nitrogen (nominal concentration 2000 μmol/mol): This intercomparison was coordinated by KRISS Korea for the NMIs of APMP region. It was scheduled in 2017 and completed in 2019. Methane is one of the major greenhouse gases that affect climate. The measurement of methane emission from fossil fuel combustion is important to mitigate anthropogenic CH_4 emissions effectively. Therefore, it is important for NMIs to have the capability of an accurate CH_4 emissions measurement. All sample cylinders were prepared gravimetrically and verified by KRISS. The gravimetric value was used as the comparison reference value. Each cylinder with its own reference value and associated expanded uncertainty was sent to the participating laboratory. Besides KRISS Korea the three other participant laboratories were National Metrology Institute of South Africa (NMISA), National

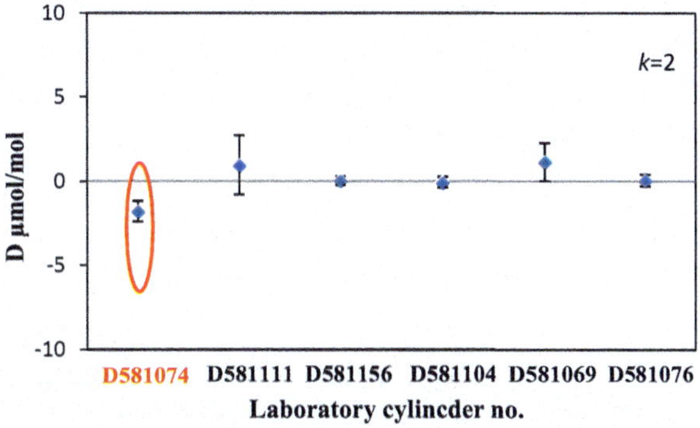

Fig. 13.19 Degree of equivalence of all participants NMIs of APMP.QM-S9.2017 with their associated uncertainty U(d_i). The error bars represent the expanded uncertainty at 95% level of confidence. The circled values represent CSIR-NPL

Metrology Institute of Malaysia (NMIM) and National Physical Laboratory India (CSIR-NPL). On the basis of results reported by all participant NMIs and the reference cylinder value given by the coordinating NMI in APMP.QM-S7.1 Draft B Report [48] the DoE and the associated uncertainty is calculated according to Eqs. (13.5), (13.6) and (13.7). The result is plotted in Fig. 13.20 in which the NPLI result is shown in red circle. Results of this comparison can be used to support CMC claims for CH_4 in air and/or nitrogen over amount-of-substance fractions from 0.05 cmol/mol to 0.5 cmol/mol.

APMP-QM-S15; Carbon dioxide in Nitrogen (1000 μmol/mol): The first key comparison of carbon dioxide in nitrogen was CCQM-K1b published in 1999 [49]. Since then GAWG and three RMOs had carried out CO_2 comparisons at emission level (CCQM-K3 APMP.QM-K3, EUROMET.QM-K3, COOMET.QM-K3) [50–53]. The amount of CO_2 for those comparisons were around from several to a few tens mmol/mol in nitrogen. Almost after one decade, APMP-QM S15 was proposed in 2017 to support CO_2/N_2 measurement capability for controlling CO_2 emissions from industry as well as automobiles. Recently, NMIs in the APMP region have focused on developing emission standards for regulating CO_2 released by various powered vehicles such as automobiles (motor cars). This inter-comparison was coordinated by KRISS and provides a CMC for APMP regional NMIs to develop CO_2/N_2 CMC claim. Total six laboratories participated in this supplementary intercomparison; CSIR-NPL, KAZ Kazakhstan, NIMT Thailand, NMC Singapore, LIPI Indonesia and KRISS Korea.

A set of mixtures of carbon dioxide in nitrogen at approx. 1000 μmol/mol prepared by KRISS Korea and gravimetrically prepared mixtures after verification and certification sent to all the participant laboratories included NPL India. Their gravimetric value was adopted as comparison reference value (RV) through the comparison. Stability of CO_2/N_2 cylinders were known as invariable. The summary of the comparison of APMP.QM-S15 and degree of equivalence in Fig. 13.20. On the basis of results reported by all participant NMIs and the reference cylinder value given by the coordinating NMI (KRISS, Korea) in APMP.QM-S15 Draft A Report [54], the

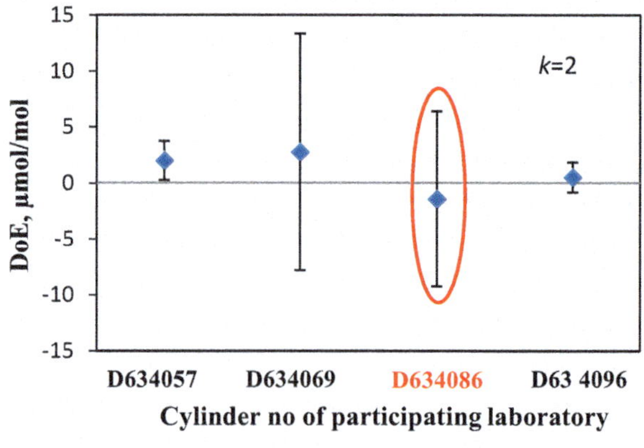

Fig. 13.20 Degree of equivalence of all participants NMIs of APMP.QM-S7.1with their associated uncertainty $U(d_i)$. The error bar represents the expanded uncertainty at 95% level of confidence. The circled values represent CSIR-NPL

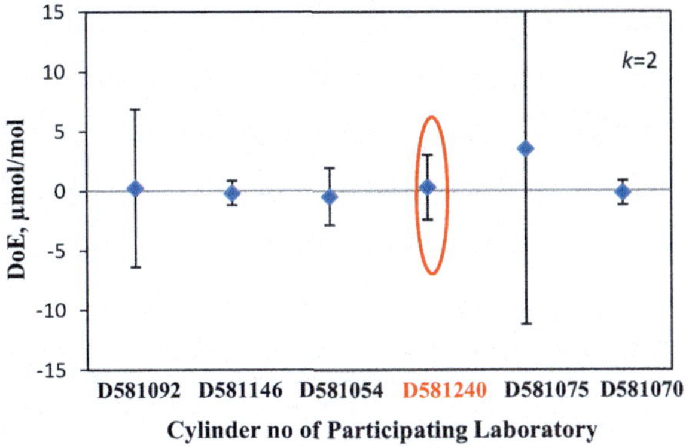

Fig. 13.21 Degree of equivalence of all participants NMIs of APMP.QM-S15 with their associated uncertainty U(d_i). The error bar represents the expanded uncertainty at 95% level of confidence. The circled values represent CSIR-NPL

DoE and the associated uncertainty is calculated according to Eq. (13.5), (13.6) and (13.7). The result is plotted in Fig. 13.20 in which the CSIR-NPL result is shown in red circle. As shown in figure, all results were consistent within uncertainties Fig. (13.21).

Results shown in table are presented as the degrees of equivalence (DoE) for the APMP.QM-S15 in figure. As shown in figure all results were consistent within uncertainties. This supplementary comparison will support the measurement capability, which can be used to support CMC claims, for carbon dioxide in nitrogen from 500 µmol/mol to 200 mmol/mol.

13.3.5 Aerosol/Particulate Metrology: Primary National Facility for Calibration of PM Samplers

Aerosols are the solid or liquid (or heterogeneous particles) particles suspended in air. In general, we deal with two types of aerosols in our day-to-day life, medical purposes (e.g. inhaler for drug delivery) or industrial interest aerosols (e.g. fire distinguisher) and ambient aerosols. Once aerosols are collected or deposited on a surface, we called them particulate. CSIR-NPL initiated aerosol metrology activity in the year 2010 [55]. At present, we provide services for particulate matter sampler cut-off size calibration to several manufacturers (https://www.nplindia.in/calibration-charges-d302-gas-metrology-wef01042019-0).

For this service, a particulate matter (PM) wind-tunnel has been established at CSIR-NPL. This is a first and unique facility in the country to test and calibrate specifically PM_1, $PM_{2.5}$ and PM_{10} samplers. In addition to this Andersen type samplers consisting of several stages for size segregated sampling can also be calibrated for their cutoff sizes. Using this facility, now all the PM samplers manufactured in the country can be calibrated. Also, all the imported PM samplers can be tested for their suitability under Indian conditions. The CSIR-NPL certified samplers can be exported without any technical barrier to trade as CSIR-NPL is a signatory of

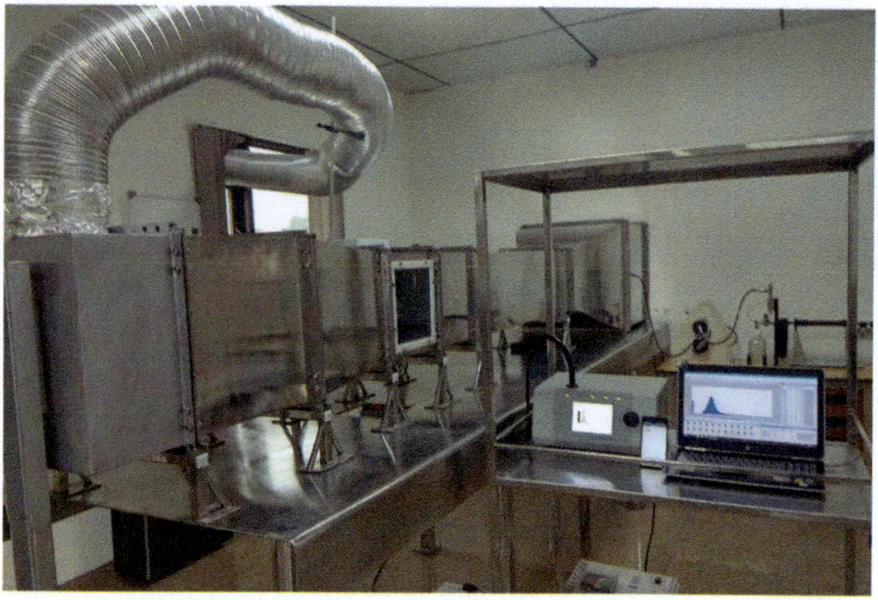

Fig. 13.22 Photograph of wind-tunnel at gas metrology, CSIR-NPL

CIPM-MRA. Figure 13.22 shows the photograph of the wind tunnel. This is an open circuit tunnel, and can generate monodisperse homogeneous cloud of particles for PM sampler testing and calibration, it has novel specifications, e.g. size: length 4.1 m and height 0.68 m, cross-section: 0.34 m × 0.34 m, air-velocity: 2–24 km/h. The SI traceability in size measurement is obtained through particle sizer spectrometer (Aerosol Particle Sizer, APS), which is calibrated with particle standards.

Following parameters of the sampler can be tested and calibrated using this facility:

(i) Intake efficiency: This wind-tunnel test provides information (for an example) in the capability of a sampler to capture all $PM_{2.5}$ particles present in known volume of air at given flow rate (16.7 lpm) in the tunnel.

(ii) Cutoff size: The particle segregation device (impactor/cyclone) should be capable enough to segregate coarse sized particles (>2.5 μm) and allows particles with $\leq$ 2.5 μm size ($PM_{2.5}$) to be collected on filter which is placed after this device. Cutoff size of a segregation device is the aerodynamic particle size at 50% of penetration efficiency.

(iii) Sharpness of cutoff: The slope of this efficiency curve, i.e. efficiency versus particle size should ideally be between 1 – 1.3.

CSIR-NPL also proposed to coordinate a pilot study on the $PM_{2.5}$ cutoff size and mass concentration at APMP in 2016, first time in the world. This pilot has been accepted by TCQM and now its protocol is being prepared after setting up this wind tunnel facility.

At CCQM, GAWG has organised first "Particulate Workshop" in April 2015 where CSIR-NPL has made a presentation on particulate metrology and proposed the possibility of this pilot study (https://www.bipm.org/en/committees/cc/ccqm/workshops.html). In the same year, the CSIR-NPL paper which raises the issue of traceability of $PM_{2.5}$ and PM_{10} was got selected for CITAC award 2013, https://www.citac.cc/ [56]. Recently in 2017, at CCQM Gas Analysis Working Group, NPL UK has coordinated a key comparison, CCQM-K150 on "Particle number and concentration comparisons" and the comparison is at the stage of Draft B report.

Other two activities on which work is ongoing at Gas Metrology, CSIR-NPL include the performance evaluation of Beta Attenuation Monitor (BAM) under Indian conditions, and the development of certified reference material (CRM) for particulate matter as well as the calibration solutions for elements mentioned in NAAQS, i.e. As, Pb and Ni which are discussed in the following sections.

13.3.5.1 Performance of Beta Attenuation Monitor (BAM) and Tapered Element Oscillating Microbalance (TEOM)

According to National Ambient Air Quality Standards (NAAQS) notified by nodal agency CPCB, three different methods are to be used in order to monitor the PM concentrations across the country, namely Gravimetric Method (GMM), Beta Attenuation Monitor (BAM) and Tapered Element Oscillating Microbalance (TEOM). BAM and TEOM are the real-time measurement methods whereas GMM is the primary method based on the gravimetric measurement of filter used before and after sampling. Real-time methods can provide measurement data on a time scale of minutes whereas the gravimetric method works on a time scale of days. Therefore, the continuous measurements can provide an instant input for policy makers and central agencies to take appropriate mitigating steps in case of severe pollution events.

BAM is the most commonly used real-time PM measurement monitor in India by the different monitoring agencies in Continuous Ambient Monitoring System (CAMS). However, some studies have raised the concern of level of reliability of the data produced by it as the technique has certain limitations when used in different weather conditions. The factors such as relative humidity, high mass loading and chemical composition of particulate matter have been reported to affect its performance [57, 58]. BAM either over-estimate or under-estimate the PM mass concentration levels since the humidity (30 to > 95%) and temperature (minus degree to > 45 °C) varies widely across the country depending on season and region. This leads to significant variation in measurement data when compared with the reference method (GMM). The major factors that are reported to cause inconsistency in PM concentrations include (1) the cutoff aerodynamic diameter and the penetration curve of the inlet [59, 60] (2) heavy particle deposition on the impaction surface or inner surface of instrument which may reduce particle penetration [61] (3) the deliquescence of aerosols [62, 63] and (4) the evaporation loss of volatile species of aerosols [57]. Variations in the aerosol chemical composition affect the water content of particles and thus induce differences between BAM and GMM measurements. Water content of aerosols depends on the ability of the particles to hold water under given RH conditions, i.e. their hygroscopicity. Also, the water content has been reported to be governed by the relative abundance of inorganic salts in aerosols as inorganic particles are more hygroscopic in nature than the organic particles [64]. This leads to overestimation in mass measurements. This further indicates that performance of BAM is affected differently by both low and high levels of RH. In operation, the inlet line of BAM is heated up to a temperature of 30 °C before the sample reaches the filter tape to reduce the relative humidity to below 60%. This feature minimizes the water content of particles to lower the bias from high RH, but it may also cause a loss of volatile and semi-volatile components (such as NH_4NO_3) of particulate matter present in the ambient air sample, thus underestimating the mass concentration. But during high ambient RH conditions (60–90%), the heating of inlet at this temperature is not enough to evaporate water in aerosols leading to high mass measurements [65]. These differences can significantly affect quantitative agreement between the BAM and the GMM measurements. Therefore, there is a need to explore these areas and find a solution to overcome bias in BAM measurements. At CSIR-NPL, we are performing a comparative study for measuring $PM_{2.5}$ mass concentrations by BAM and gravimetric method under different conditions as stated above. Some preliminary data have already been reported [66].

13.3.5.2 Development of Certified Reference Material for Particulate Matter

The degrading air quality across the globe has instilled a vast research scope in this field. Since air pollution is a very sensitive issue owing to its harmful health effects on humans and environment, accurate monitoring and measurements become an important aspect of control strategies. Maintaining quality and reliability in the data produced from a laboratory requires certain tools such as use of calibration standards and certified reference materials. CRMs are required to ensure the quality, reliability and inter-comparability of the analytical data generated from any analytical process or instrument operation [56, 67]. Also, the use of CRMs ensures the data generated to be at par and acceptable at national and international levels. The instruments used in quantifying the characteristics of airborne particulate matter can be calibrated or quality assured using CRMs of other materials but the use of a representative matrix CRM of the measured characteristic is much more helpful in ensuring data quality and comparability. Accuracy of air pollution studies holds great importance since the control and abatement strategies depend largely on them. Also, such studies provide inputs for governmental action in severe pollution events to take appropriate steps.

Few of the NMIs of the world have been engaged in developing particulate matter CRM since 1970s. As a result, several of them are available with different certified properties of particulate matter. NIST-USA; National Institute of Environmental Studies (NIES), Japan; Institute for Reference Materials and Measurements (IRMM), Belgium are three such NMIs that have been able to develop and provide such a facility to maintain data quality and meet international standards. CRMs are one such tool to ensure the measurement data quality and laboratory performance. Currently, no suitable CRM is available representing an Indian ambient condition that certifies the contents of particulate matter. Therefore, at CSIR-NPL, we are working actively to fill this gap. Although, development of CRM is a tedious process, especially CRM for particulate matter as it takes several years to collect a sufficient amount. A collection facility has been set up for collection of particulate matter to be developed as particulate CRM.

On the other hand, CSIR-NPL has successfully organized a pilot study at APMP for calibration solutions of the elements listed in NAAQS for their determination in particulate matter, i.e. APMP.QM-P30 "Calibration Solutions of Pb, As, Ni and Fe".

13.4 Sensor Metrology

Sensor technology is an advanced tool for the detection, measurement and monitoring of environmental pollutants. The key advantages of sensor technology lie in its high-sensitivity, small-size, power-efficiency, long-life and low-cost. Currently, for real-time monitoring of air pollution due to hazardous gases, a large number of continuous emission monitoring systems (CEMS) has been installed at various locations across India. The CEMS are composed of various technologies like chemiluminescence, UV-florescence, NDIR-spectroscopy, etc. to provide permanent record of emissions or process control parameters continuously for real-time air pollutant measurements/monitoring. Each of these technologies has its own shortcomings and a single technology alone cannot be used to detect all the atmospheric pollutant gases at a time. Hence, a set of technologies are used collectively for the detection of various atmospheric pollutant gases, which leads to several constraints with CEMS implementation like technology selection, calibration, field performance testing, data acquisition, bulky system, huge manpower, high cost, etc. making it difficult for public use.

In this scenario, sensor technology has become attractive for air pollution measurements accessible to individuals. The development of multi-sensor elements, device miniaturization and their potential integration with smart electronics and wireless networks offer versatile applications in various fields including air quality monitoring, medicine, automobiles, food industry and military through data capturing, interpretation and communication. Hundreds of millions of gas sensors are being built into intelligent detection systems for several real-world applications to provide improved health, safety and security for people and it is fuelling the rapid growth of global sensor market. A recent study projects that the gas sensor market will cross over $ 2B margin by 2026 opening up new opportunities for the expansion of sensor manufacturing industry [68]. Especially, in developing countries like India, there is an enormous scope for the production of indigenous gas sensors through encouragement of start-ups and micro, small and medium enterprises (MSMEs) as gas sensor fabrication is an affordable technology. In practice, gas sensors require regular maintenance to ensure valid measurements during continuous operation over a prolonged period since sensor characteristics vary due to physical changes of sensing element caused by their routine functioning and environmental operating conditions. Hence, a periodic calibration is a must to ensure the right functioning of gas sensors. Here, sensor metrology plays a critical role in achieving accurate and precise data of pollutant measurements with uncertainty and traceability. Currently, CSIR-NPL is taking initiatives to establish sensor metrology for testing and calibration of gas sensors.

13.4.1 Gas Sensor Technology

Sensor technology comprises of three major components such as sensor structure, manufacturing technology and signal processing as represented in Fig. 13.23. Gas sensors are generally categorized as chemical detectors and convert a desired chemical reaction into a measurable electronic signal such as change in the resistance, frequency, current, or voltage. The performance of gas sensor is evaluated by several parameters including sensitivity, selectivity, limit of detection (LOD), resolution, accuracy, precision, reversibility, recovery time and response time. Advanced manufacturing technology and embedded signal processing unit have made gas sensor devices more versatile enabling their implementation for various applications.

Gas sensors can be divided based on their working principles namely as electrical, electro-chemical, optical, thermal, electro-mechanical, acoustic and chromatography gas sensors. Among these gas sensor technologies, semiconducting metal oxide and electrochemical based electrical sensors are currently enjoying a global market share of more than 82% due to their commercial feasibility. Global market share of various gas sensor technologies is given in Fig. 13.24 [68].

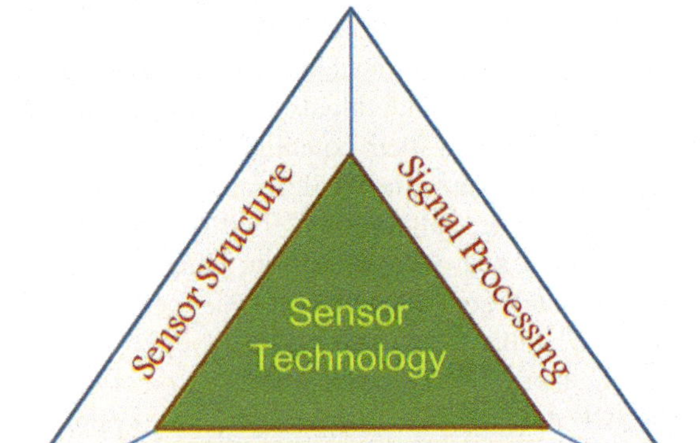

Fig. 13.23 Schematic depiction of the foundation of sensor technology

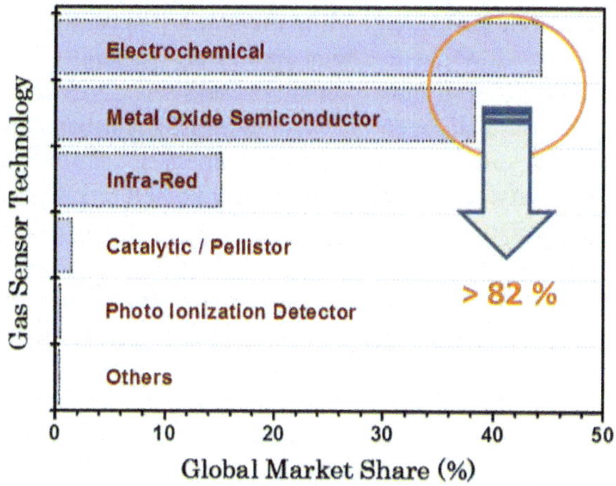

Fig. 13.24 Global market share of various gas sensor technologies

Metal-Oxide Semiconductor (MOS) Gas Sensors: The basic characteristic of MOS gas sensors is the reversible reaction of the gas with the surface of a solid-state material, which causes a change in the electrical conductivity of MOS upon gas–solid interaction. These reactions are influenced by various material properties such as electronic structure, crystal orientation, chemical composition, microstructure, etc. The great interest of industrial and scientific world on MOS based gas sensors comes from their numerous advantages, namely, small sizes, high sensitivities in detecting very low concentrations (at levels of ppm or even ppb), possibility of on-line operation and bench production at low cost. A typical fabrication process of MOS gas sensor is schematically presented in Fig. 13.25. MOS sensors are capable of detecting various gases like O_x, CO_x, NO_x, SO_x, H_2S, NH_3, VOCs, etc. In recent years, there has been a significant improvement in sensor design along with the advances in micro- and nano-fabrication technology and the use of newly designed materials, leading to the development of high-performance gas sensors. Moreover, in order to improve the detection limit and response time, Chemical field effect transistor (ChemFET) and high electron mobility transistor (HEMT) devices with MOS functionalized gate electrodes are being explored. MOS gas sensors employ both n-type and p-type oxide materials. Most used n-type metal oxides for gas sensor fabrication are SnO_2,

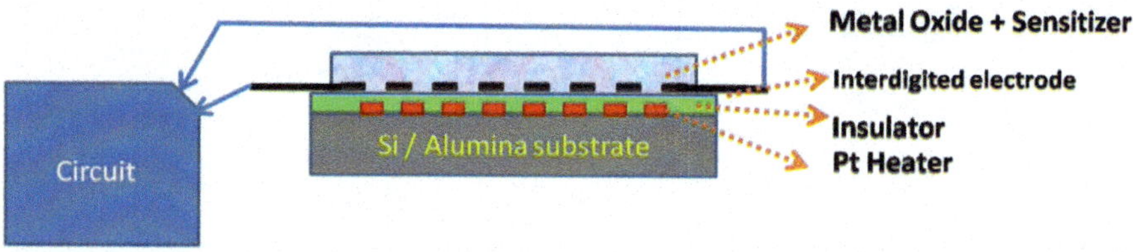

Fig. 13.25 Schematic diagram of typical structure of MOS gas sensor devices

ZnO, TiO_2, WO_3, In_2O_3, Fe_2O_3, etc. while the major p-type metal oxides comprise of CuO, NiO, Co_3O_4, Cr_2O_3, etc. Metal oxides are generally doped/mixed with noble metals/metal oxides to improve sensitivity and selectivity towards a particular analyte. Recently, metal chalcogenide, organic and carbon-based low-dimensional materials are also being employed for the development of chemiresistive gas sensors with improved sensitivity and selectivity for a designated target gas.

In most MOS gas sensors, the sensing occurs at high temperatures of 100–400 °C because the increase in thermal energy will lead to the generation of free electrons that flow across the grain boundaries of metal oxide. So, oxygen can easily adsorb the free electrons from the metal oxide surface because oxygen has a high electron affinity. This will lead to the formation of potential barrier at the grain boundaries of metal oxide as represented in Fig. 13.26a and defines the initial resistance of MOS layer. When the surface is exposed to a reactive gas, the adsorbed oxygen ion reacts with the gas molecules and gets desorbed from the MOS surface. In this process,

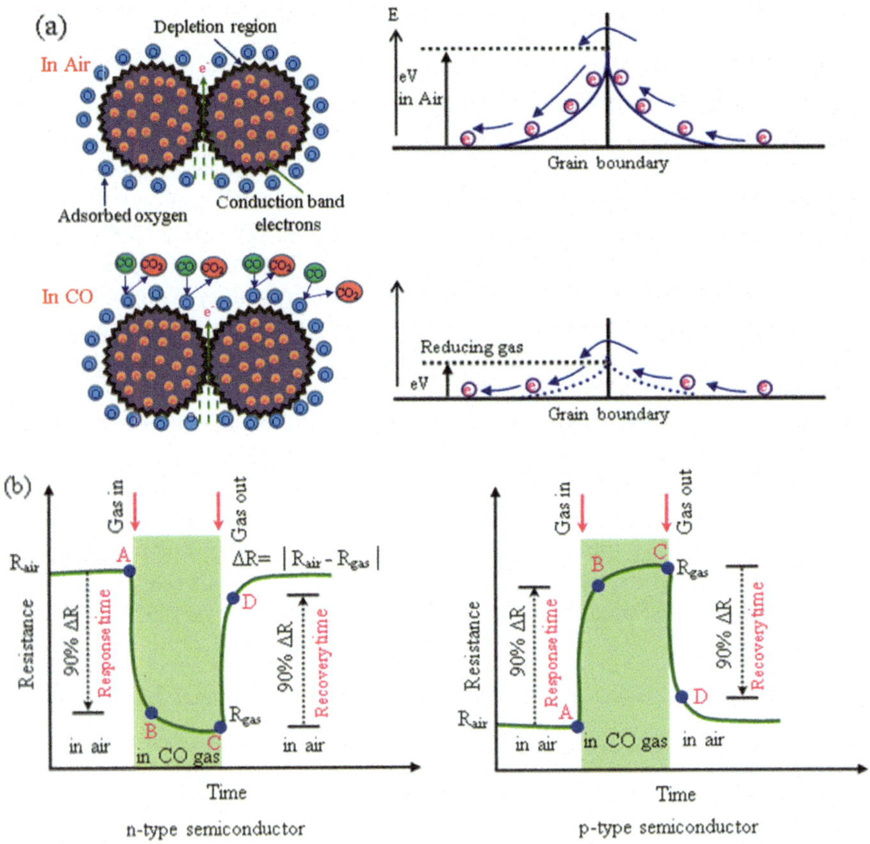

Fig. 13.26 **a** Schematic diagram of gas sensing mechanism of metal oxide semiconducting gas sensors. **b** Sensing response of *n*-type and *p*-type semiconductor in reducing gas (CO) environment.

the electron will be reverted back to the materials and the increased electron density reduces the potential barrier thereby reducing the electrical resistance to the original value [69].

The sensing response characteristics of *n*-type and *p*-type MOS films are reversed depending upon the nature of gas, i.e. reducing or oxidizing gas. For example, a *p*-type semiconductor gas sensor in the presence of reducing gas environment like CO will show an increase in its electrical resistance from the original baseline to a certain value, which is considered as the response of the semiconductor. In the presence of oxygen environment (upon removal of reducing gas), the film will retain its original electrical resistance from the value obtained during its response to the target gas. On the other hand, an n-type semiconductor will show a decrease in resistance upon exposure to reducing gas. This phenomenon is schematically depicted in Fig. 13.26b. When the environment has a presence of oxidizing gas like NO_2, the sensing response will be reversed for both n-type and p-type semiconductors.

Electrochemical Gas Sensors: Electrochemical gas sensors are gas detectors that are used to measure the concentration of a target gas by reducing or oxidizing it at an electrode and measure the resultant current. These gas sensors use an electrochemical cell consisting of a casing that contains a group of chemical reactants in contact with the surroundings through the terminals, namely anode and cathode, of similar composition. The schematic of electrochemical gas sensor is shown in Fig. 13.27. The electrochemical gas sensors are highly reliable and easy to use and are primarily used for detecting toxic gases and oxygen level in the atmosphere.

Electrochemical sensors produce electrical signals upon reaction with the gas of interest and the signal strength is proportional to the gas concentration. A typical electrochemical gas sensor is comprised of sensing electrode also referred as working electrode and a counter electrode both separated by an electrolyte. In the sensor device, the gas enters through a small capillary-type opening at the top and reaches the electrode surface by diffusing though a hydrophobic barrier. In this way, the proper amount of gas is controlled for the reaction at the sensing electrode to

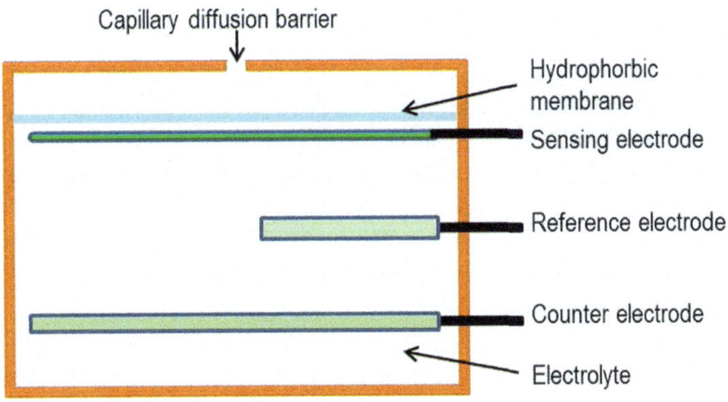

Fig. 13.27 Schematic diagram of electrochemical gas sensor

generate a sufficient electrical signal and prevents the leakage of electrolyte out of the sensor. The diffused gas then reacts at the surface of the sensing electrode through either an oxidation or reduction mechanism. The electrode material is selectively developed for the catalytic reaction with a specified gas. The current signal proportional to the gas concentration flows between the anode and the cathode through a resistor connected across them. By quantitatively measuring this current, the gas concentration is determined.

Performance Parameters of Gas Sensors

The performance of gas sensors is mainly characterized by the following parameters.

(i) *Sensitivity:* It is defined as the change in measurement signal per analyte concentration unit of the target gas. It is used to evaluate the variation of target gas concentration values in the surrounding environment.
(ii) *Selectivity:* It is defined as a parameter to determine the property of a gas sensor to respond selectively to a target gas.
(iii) *Response Time:* It is defined as the time required for the gas sensor to respond to a target gas, that is, 90% of the total response of the signal.
(iv) *Recovery Time:* It is defined as the time required for the gas sensor to recover in oxygen environment, that is, 90% of the original baseline signal upon removal of the target gas.
(v) *Stability:* It is defined as the ability of a gas sensor to provide reproducible results when it is continuously used for a prolonged period of time.

Recent Advances in Gas Sensor Technology

The current advances in sensor technology are powered by the development of new functional materials, improved sensor designs, innovative manufacturing processes, high-speed and low-cost electronic circuits and novel signal processing methods. The synergetic interaction of these developments in gas sensor technology provides promising technical solutions by improving the device quality, measurement reliability, large-scale manufacturing and economic productivity of gas sensors.

Device Miniaturization: Futuristic technologies constantly demand a device miniaturization in order to increase its functionality for various high-end applications like air quality monitoring, breath analysis, food monitoring, smart phone usages, car ventilation control, gasoline vapour detection, security devices, etc. Here, the advanced micro-and nano-fabrication technologies like micro electro-mechanical systems (MEMS) and nano electro-mechanical systems (NEMS) benefit the miniaturization of MOS gas sensor devices. These capabilities facilitate the development of miniaturized integrated gas sensor arrays with improved sensitivity, fast response time and low power consumption. For the miniaturization of electrochemical sensors, screen printing technology is successfully employed.

Electronic Nose: Electronic nose is basically an attempt to mimic human olfaction to sense the smell with better accuracy and timing. It is a collection of several individual sensors connected to electronics and an effective pattern recognition system and is used to detect, classify and, where necessary, quantify a variety of chemical analytes or odours of concern in the area of interest. For instance, electronic nose is widely used in medical industry for the diagnosis of diseases via breath analysis by detection and pattern-recognition of VOCs with its multiple sensor arrays.

Wearable Sensors: Wearable sensors are of enormous interest in these days for their unique applications in environmental and health monitoring. Especially, people involved working in hazardous ambient such as medicine, mining, etc. face a greater risk of getting exposed to toxic gases and require continuous monitoring of environmental air compositions, where wearable sensors can potentially help.

Internet of Things (IoTs): The gas sensor technology fuelled by IoT through efficient integration of signal conditioning circuits and wireless networks has brought new dimensions for sensor applications in diverse fields ensuring easy collection and sharing of data, better experiences to the users and lower expenses. The sensors installed in fields can collect and transfer the environmental data allowing remote monitoring of area of interest and also can take automatic and intelligent decision on certain events without needing any user interference.

13.4.2 Calibration and Testing of Gas Sensors

The performance of gas sensors is subjected to degradation over a period of their operation for various reasons that cause variation in the characteristic parameters of the sensors. This phenomenon is known as 'drift' and leads to uncertain results and false alarms. Therefore, these sensors need frequent calibration in order to ensure accurate and precise measurements or sensor replacement if required. The chemical stability of gas sensors is affected by their routine operation towards certain analytes under identical conditions causing temporal variation of sensor response, which is generally attributed to 'sensor aging' or 'thermo-mechanical degradation'. The environmental operating conditions also influence the chemical stability of sensing elements in a long run and pose challenge to maintain the required performance and life span. In addition, the sensor fabrication technology can also be responsible for the physical changes of gas sensors which alter the sensing and transduction mechanisms. Hence, the periodic testing and validation of gas sensors are of great requirement. The stability of sensing layer can be improved by proper optimization of sensor manufacturing procedures and tighter sensor quality control. Also, following stringent measuring procedures including sensor cleaning and conditioning of sensing surface regeneration would assist to achieve good repeatability of gas sensor characteristics. However, these measures may only help to prolong the validity of sensor performance but the recalibration is necessary at designated intervals as per the nature of individual sensors. In MOS sensors, the drift is mainly attributed to different structural and morphological variations, chemical diffusion of oxygen vacancies and degradation of the electric contacts. In case of electrochemical sensors, the nature of this technology itself causes fast aging and drift during their life in addition to the harsh operating environments resulting in frequent calibration. All sensors ultimately reach their end of life, when the performance no longer meets the requirements and the sensor needs to be replaced.

Currently, a large number of gas sensors are being imported in India for air pollutant measurements in public places, industries, houses, restaurants, etc. These sensors are generally calibrated at manufacturing industries following the standards of respective countries. Since the performance characteristics of these sensors are sensitive to the environmental operating conditions such as humidity, temperature and air composition, they need to be recalibrated as per the practical atmospheric conditions of specific locations. India has a broad range of climatic conditions owing to its large spread across low and high altitude regions. Therefore, imported gas sensors demand testing and recalibration before being employed for the domestic use. In addition, there are many Indian manufacturers of electrical gas sensors who require sensor metrology to test and validate their products for quality assurance. A certified product with traceability will not have any trade barriers and can easily be marketed globally. Also, in India, several academic and research institutes are actively involved in research and development of solid-state gas sensors. Here, sensor metrology will play a predominant role in authenticating the obtained results and in bringing out reliable technology for indigenous sensor production.

Gas sensors need to be calibrated and tested both at laboratory and field levels to ascertain their performance parameters and to validate them for specific applications. For the calibration and testing of gas sensors, quality management system has to be adopted following the ISO/IEC 17025:2017 guidelines with measurement traceability to SI units. CSIR-NPL being the National Measurement Institute (NMI) of India has apex test facility for all the parameters used in the measurement of sensing characteristics of solid-state gas sensors, i.e. (i) temperature, (ii) humidity, (iii) sensor resistance and (iv) target gas.

(i) ***Temperature***: Temperature is an important factor in the operation of solid-state gas sensors. Gas sensors exhibit response to target gas at some temperature. The responses increase and reach their maximums at a certain temperature, called operating temperature and then decreased rapidly with increasing the temperature, called recovery temperature. This tendency is commonly observed in metal oxide gas sensors. Thus, the precise and accurate measurement of temperature is essentially required. CSIR-NPL has the temperature standard with measurement capacity $\pm 0.17 \times 10^{-3}$ °C at triple point of water (TPW) TPW (0.01 °C).
(ii) ***Humidity***: Environmental humidity is also an important factor influencing the performance of solid-state gas sensors, which is related to the mechanism of sensing water vapor and pollution gases. The conduction mechanism is influenced by H^+ or H_3O^+ generated by the dissociation of adsorbed water molecules, which hops between adjacent hydroxyl groups. The reaction between the surface oxygen and the water molecules leads to a decrease in initial resistance of the gas sensor and results in a decrease of the sensitivity. Also, the adsorption of water molecules leads to less chemisorption of oxygen species on the metal oxide surface due to the decrease of the surface area that

is responsible for the sensor response. Further, water molecules also act as a barrier against C_2H_2 adsorption. The superficial migration of the C_2H_2 on the metal oxide surface becomes difficult, thus the sensitivity decreases and the response and recovery times increase. The prolonged exposure to humid environments leads to the gradual formation of stable chemisorbed OH^- on the surface causing a progressive deterioration of the sensitivity of gas sensors [70]. CSIR-NPL has the humidity standard with capacity 10–95% RH (±0.3%) range.

(iii) ***Sensor Resistance*****:** The metal oxide gas-sensing mechanism is based on the change in the resistance of the device-sensing component in the presence of reducing or oxidizing gases. In these sensors, the sensing layer is in direct exposure to the target gases. The change in the sensor resistance is directly proportional to the concentration of the exposed gas. The precise and accurate measurement is very important to produce the reliable data. CSIR-NPL has apex calibration facility for Voltage and Current with minimum capability of 1 μV (±0.1 μV) and 1 pA (±0.01 pA), respectively.

(iv) ***Target Gas*****:** Gas sensors require regular check involving testing and re-calibration of the sensors, for which traceable target gases are needed. CSIR-NPL has the capability of preparation of primary gas standards for air quality measurements.

The government of India has developed guidelines for air quality and ordinances to restrict emissions in an effort to control air pollution. Regulatory bodies also have to define the protocol for the industries/manufacturers to get NMI certification for the gas sensors. As of now, there is no established sensor metrology facility globally for testing and calibration of solid-state gas sensors based on semiconductor and electrochemical technologies by any NMIs. NPL, UK is currently involved in developing sensor metrology for portable gas sensors. Sensor metrology is an inevitable requirement for our country to realize valid air pollution measurements and quality assurance of indigenous gas sensor production. CSIR-NPL is in the process of establishing a "National Facility for Testing & Calibration of Solid State Gas Sensors" which will provide a complete and cost-effective solution for test, validation and certification to the Indian as well as foreign manufacturers of solid state gas sensors. This will largely enable India to effectively implement several national programs like Digital India, Make in India, Smart City, Start-up India, etc. so as to harvest vast benefits for societal and economic development.

Acknowledgements Authors are thankful to Ms. Beena Gupta, Mrs. Saroj Gandhi, Mr. Alok Mukherjee, Dr. Khem Singh, Ms. Smriti Tiwari Singh, Mr. Devesh Kumar Shukla, Mr. Lalit Goswami, Mr. Jitender Kumar, Ms. Abha Shukla, Mr. M.L. Arora, Mr. Suresh Chandra Yadav, Mr. Saket Vihari and Ms. Kamla for their support.

Referencess

1. O.C. *Change, Intergovernmental Panel on Climate Change* (World Meteorological Organization, 2007)
2. S. Singh, N.K. Lodhi, A.K. Mishra, S. Jose, S.N. Kumar, R.K. Kotnala, Assessment of satellite-retrieved surface UVA and UVB radiation by comparison with ground-measurements and trends over Mega-city Delhi. Atmos. Environ. **188**, 60–70 (2018). https://doi.org/10.1016/j.atmosenv.2018.06.027
3. W.L. Chameides, H. Yu, S.C. Liu, M. Bergin, X. Zhou, L. Mearns, G. Wang et al., Case study of the effects of atmospheric aerosols and regional haze on agriculture: an opportunity to enhance crop yields in China through emission controls? PNAS **96**, 13626–13633 (1999). https://doi.org/10.1073/pnas.96.24.13626
4. S. Singh, S. Nath, R. Kohli, R. Singh, Aerosols over Delhi during pre-monsoon months: Characteristics and effects on surface radiation forcing. Geophys. Res. Lett. **32**, L13808 (2005). https://doi.org/10.1029/2005GL023062
5. World Health Organization, Annual Report, 2005. No. WHO/MPS/07.01. (World Health Organization, 2006)
6. J.H. Seinfeld, S.N. Pandis, *Atmospheric Chemistry and Physics: From Air Pollution to Climate Change* (Wiley, 2016)
7. M.O. Andreae, P.J. Crutzen, Atmospheric aerosols: biogeochemical sources and role in atmospheric chemistry. Science **276**, 1052–1058 (1997). https://doi.org/10.1126/science.276.5315.1052
8. V. Ramanathan, R.E. Dickinson, The role of stratospheric ozone in the zonal and seasonal radiative energy balance of the earth-troposphere system. J. Atmos. Sci. **36**, 1084–1104 (1979). https://doi.org/10.1175/1520-0469(1979)036%3c1084:TROSOI%3e2.0.CO;2
9. M. Kampa, E. Castanas, Human health effects of air pollution. Environ. Pollute. **151**, 362–367 (2008). https://doi.org/10.1016/j.envpol.2007.06.012
10. A.J. Haagen-Smit, C.E. Bradley, M.M. Fox, Ozone formation in photochemical oxidation of organic substances. J. Ind. Eng. Chem. **45**, 2086–2089 (1953). https://doi.org/10.1021/ie50525a044
11. J.H. Seinfeld, G.R. Carmichael, R. Arimoto, W.C. Conant, F.J. Brechtel, T.S. Bates, T.A. Cahill et al., ACE-ASIA: regional climatic and atmospheric chemical effects of Asian dust and pollution. Bull. Amer. Meteor. Soc. **85**, 367–380 (2004). https://doi.org/10.1175/BAMS-85-3-367
12. R.E. Dickinson, R.J. Cicerone, Future global warming from atmospheric trace gases. Nature **319**, 109–115 (1986). https://doi.org/10.1038/319109a0
13. https://www.businesswire.com/news/home/20180426006864/en/India-Industrial-Gases-Market-2013-2018-2023-Product
14. P. Matson, K.A. Lohse, S.J. Hall, The globalization of nitrogen deposition: consequences for terrestrial ecosystems. Ambio **31**, 113–119 (2002). https://doi.org/10.1579/0044-7447-31.2.113
15. J.J. West, A. Cohen, F. Dentener, B. Brunekreef, T. Zhu, B. Armstrong, M.L. Bell et al., What we breathe impacts our health: improving understanding of the link between air pollution and health. Environ. Sci. Technol. **50**, 4895–4904 (2016). https://doi.org/10.1021/acs.est.5b03827
16. R.J. Stening, Electron density profile changes associated with the equatorial electrojet. J. Atmos. Sol. Terr. Phys. **39**, 157–164 (1977). https://doi.org/10.1016/0021-9169(77)90109-X
17. S. Sridharan, S. Sathishkumar, S. Gurubaran, Variabilities of mesospheric tides and equatorial electrojet strength during major stratospheric warming events. Ann. Geophys. **27**, 4125–4130 (2009)
18. C. Vineeth, T.K. Pant, R. Sridharan, Equatorial counter electrojets and polar stratospheric sudden warmings-A classical example of high latitude-low latitude coupling. Ann. Geophys. **27**, 3147–3153 (2009). https://doi.org/10.5194/angeo-27-3147-2009

19. H.L. Liu, A.D. Richmond, Attribution of ionospheric vertical plasma drift perturbations to large-scale waves and the dependence on solar activity. J. Geophys. Res. Space Phys. **118**, 2452–2465 (2013). https://doi.org/10.1002/jgra.50265
20. T. Matsuno, A dynamical model of the stratospheric sudden warming. J. Atmos. Sci. **28**, 1479–1494 (1971). https://doi.org/10.1175/1520-0469(1971)028%3c1479:ADMOTS%3e2.0.CO;2
21. A.K. Upadhayaya, K.K. Mahajan, Ionospheric F2 region: variability and sudden stratospheric warmings. J. Geophys. Res. Space Phys. **118**, 6736–6750 (2013). https://doi.org/10.1002/jgra.50570
22. S. Gupta, A.K. Upadhayaya, Morphology of ionospheric F2 region variability associated with sudden stratospheric warmings. J. Geophys. Res. Space Phys. **122**, 7798–7826 (2017). https://doi.org/10.1002/2017JA024059
23. M. Dunajecka, S. Pulinets, Atmospheric and thermal anomalies observed around the time of strong earthquakes in Mexico. Atmosfera. **18**, 235–247 (2005)
24. M.R.M. Daneshvar, M. Khosravi, T. Tavousi, Seismic triggering of atmospheric variables prior to the major earthquakes in the Middle East within a 12-year time-period of 2002–2013. Nat. Hazards. **74**, 1539–1553 (2014). https://doi.org/10.1007/s11069-014-1266-5
25. I. Mahmood, M.F. Iqbal, M.I. Shahzad, S. Qaiser, Investigation of atmospheric anomalies associated with Kashmir and Awaran Earthquakes. J. Atmos. Sol. Terr. Phys. **154**, 75–85 (2017). https://doi.org/10.1016/j.jastp.2016.12.018
26. H. Le, J.Y. Liu, L. Liu, A statistical analysis of ionospheric anomalies before 736 M6.0+ earthquakes during 2002–2010. J. Geophys. Res. **116**, A02303 (2011). https://doi.org/10.1029/2010JA015781
27. S. Pulinets, M. Dunajecka, Specific variations of air temperature and relative humidity around the time of Michoacan earthquake M8. 1 Sept. 19, 1985 as a possible indicator of interaction between tectonic plates. Tectonophysics **431**, 221–230 (2007). https://doi.org/10.1016/j.tecto.2006.05.044
28. S. Pulinets, D. Ouzounov, Lithosphere–atmosphere–ionosphere coupling (LAIC) model—an unified concept for earthquake precursors validation. J. Asian Earth Sci. **41**, 371–382 (2011). https://doi.org/10.1016/j.jseaes.2010.03.005
29. F. Freund, Earthquake forewarning-A multidisciplinary challenge from the ground up to space. Acta Geophys. **61**, 775–807 (2013). https://doi.org/10.2478/s11600-013-0130-4
30. S.K. Sharma, T.K. Mandal, M. Saxena, Inter-annual variation of ambient ammonia and related trace gases in Delhi India B. Environ. Contam. Tox. **99**, 281–285 (2017). https://doi.org/10.1007/s00128-017-2058-x
31. S.K. Sharma, A. Datta, T. Saud, T.K. Mandal, Y.N. Ahammed, B.C. Arya, M.K. Tiwari, Study on concentration of ambient NH_3 and interactions with some other ambient trace gases. Environ. Monit. Assess. **162**, 225–235 (2010). https://doi.org/10.1007/s10661-009-0791-2
32. R. Taipale, T.M. Ruuskanen, J. Rinne, M.K. Kajos, H. Hakola, T. Pohja, M. Kulmala, Quantitative long-term measurements of VOC concentrations by PTR-MS? measurement, calibration, and volume mixing ratio calculation methods. Atmos. Chem. Phys. **8**, 6681–6698 (2008). https://doi.org/10.5194/acp-8-6681-2008
33. S.K. Sharma, T.K. Mandal, C. Sharma, J.C. Kuniyal, R. Joshi, P.P. Dhyani, A. Sen et al., Measurements of particulate (PM 2.5), BC and trace gases over the northwestern Himalayan region of India. Mapan **29**, 243–253 (2014). https://doi.org/10.1007/s12647-014-0104-2
34. A. Sharma, T.K. Mandal, S.K. Sharma, D.K. Shukla, S. Singh, Relationships of surface ozone with its precursors, particulate matter and meteorology over Delhi. J. Atmos. Chem. **74**, 451–474 (2017). https://doi.org/10.1007/s10874-016-9351-7
35. J. Viallon, P. Moussay, R. Wielgosz, B.C. Arya, S.K. Mishra, A. Kumar, D.K. Shukla, J.E. Norris, F.R. Guenther. Final Report on the On-going Key Comparison BIPM.QM-K1: Ozone at Ambient Level, Comparison with NPLI, 2009. Metrologia **47**, 08015 (2010). https://doi.org/10.1007/s10874-016-9351-7
36. https://www.airproducts.in/~/media/Files/PDF/homepage-EPC-World-Magazine-July-pg40-in.pdf

37. https://www.researchandmarkets.com/reports/4520300/india-industrial-gases-market-by-product-by
38. https://www.marketwatch.com/press-release/calibration-gas-mixture-market-size-growth-analysis-2020-2026-covering-recent-trend-and-future-growth-feasibility-regional-outlook-and-future-forecast-2020-06-03
39. https://www.bipm.org/en/publications/si-brochure/mole.html
40. http://www.bipm.org/
41. Report on 6th Meeting of CCQM, (2000), https://www.bipm.org/utils/en/pdf/CCQM6-EN.pdf
42. https://www.bipm.org/utils/common/pdf/CC/CCQM/CCQM1.pdf, CCQM Report on 1st meeting, BIPM, (1995)
43. https://iopscience.iop.org/article/10.1088/0026-1394/37/1/5 Alink, BIPM Com. Cons. Quan. Mat., progress report on comparison 96–12, 97–12, 99–12.
44. https://iopscience.iop.org/article/10.1088/0026-1394/37/1/5/pdf, A link, CCQM-key comparison on primary standard gas mixtures, BIPM com. Cons. Quan. Mat. Report 99 (revised)
45. A. Alink, The first key comparison of primary standard gas mixtures. Metrologia **37**, 35–49 (2000). https://doi.org/10.1088/0026-1394/37/1/5
46. Final Report, CCQM-K120 a & b; Carbon dioxide in Air at background level (380–480) μmol/mol and at urban level (480–800) μmol/mol https://www.bipm.org/utils/common/pdf/final_reports/QM/K120/CCQM-K120.pdf
47. APMP.QM-S9.2017 Draft A report Comparison of measurement capability with 100 μmol/mol of Carbon monoxide in nitrogen. https://www.bipm.org/utils/common/pdf/final_reports/QM/S9/APMP.QM-S9.pdf
48. Draft B Report International Comparison APMP.QM-S7.1 Methane in nitrogen at 2000 μmol/mol
49. A. Alink et al., Final report for the Key Comparison CCQM-K1.b, (1999). https://kcdb.bipm.org/appendixb/appbresults/ccqm-k1.d/ccqm-k1_final_report.pdf.
50. A. van der Veen et al., CCQM key comparison CCQM-K3 of measurements of CO, CO_2, and C_3H_8 in N_2. Metrologia **39**, 121 (2002). https://doi.org/10.1088/0026-1394/39/1/18
51. L.A. Konopelko et al., COOMET.QM-K3: automotive emission gas measurements. Metrologia **44**, 08005 (2007). https://doi.org/10.1088/0026-1394/44/1A/08005
52. A.M.H. Van der Veen, J.I.T. Van Wijk, R.P. Van Otterloo, R.M. et al, EUROMET. QM-K3: automotive emission gas measurements. Metrologia**39**, 08005 (2002). https://doi.org/10.1088/0026-1394/39/1A/23
53. J.S. Kim, D.M. Moon, K. Kato, M. Maruyama, M.J. Kao, A. Botha, M. Dimashki, APMP. QM-K3: automotive emission gas measurements. Metrologia **40**, 08009 (2003). https://doi.org/10.1088/0026-1394/40/1A/08009
54. APMP.QM-S15, Draft A report, Comparison of measurement capability with 1000 μmol/mol of Carbon dioxide in nitrogen.
55. S.G. Aggarwal, Recent developments in aerosol measurement techniques and the metrological issues. Mapan **25**, 165–189 (2010)
56. S.G. Aggarwal, S. Kumar, P. Mandal, B. Sarangi, K. Singh, J.Pokhariyal, S.K. Mishra, S. Agarwal, D. Sinha, S. Singh, C. Sharma, P.K. Gupta, Traceability issue in PM 2.5 and PM 10 measurements. Mapan **28**, 153–166 (2013). https://doi.org/10.1007/s12647-013-0073-x
57. Y.H. Cheng, C.J. Tsai, Evaporation loss of ammonium nitrate particles during filter sampling. J. Aerosol Sci. **28**, 1553–1567 (1997). https://doi.org/10.1016/S0021-8502(97)00033-5
58. S.E. Shin, C.H. Jung, Y.P. Kim, Analysis of the measurement difference for the PM10 concentrations between beta-ray absorption and gravimetric methods at Gosan. Aerosol Air Qual. Res. **11**, 846–853 (2011). https://doi.org/10.4209/aaqr.2011.04.0041
59. M.B. Ranade, M.C. Woods, F.L. Chen, L.J. Purdue, K.A. Rehme, Wind tunnel evaluation of PM10 samplers. Aerosol Sci. Technol. **13**, 54–71 (1990). https://doi.org/10.1080/02786829008959424
60. C.J. Tsai, A field study of three collocated ambient PM10 samplers. Part. Part. Syst. Charact. **12**, 10–15 (1995). https://doi.org/10.1002/ppsc.19950120103

61. W. John, W. Winklmayr, H.C. Wang, Particle deagglomeration and reentrainment in a PM_{10} sampler. Aerosol Sci. Tech. **14**, 165–176 (1991). https://doi.org/10.1080/02786829108959480
62. Z. Meng, J.H. Seinfeld, P. Saxena, Y.P. Kim, Contribution of water to particulate mass in the south coast air basin. Aerosol Sci. Technol. **22**, 111–123 (1995). https://doi.org/10.1080/02786829408959731
63. C. Pilinis, J.H. Seinfeld, D. Grosjean, Water content of atmospheric aerosols. Atmos. Environ. **23**, 1601–1606 (1989). https://doi.org/10.1016/0004-6981(89)90419-8
64. S.G. Aggarwal, M. Mochida, Y. Kitamori, K. Kawamura, Chemical closure study on hygroscopic properties of urban aerosol particles in Sapporo. Japan. Environ. Sci. Technol. **41**, 6920–6925 (2007). https://doi.org/10.1021/es063092m
65. S.E. Shin, C.H. Jung, Y.P. Kim, Estimation of the optimal heated inlet air temperature for the beta-ray absorption method: analysis of the PM10 concentration difference by different methods in coastal areas. Adv. Environ. Res. **1**, 69–82 (2012)
66. K. Shukla, R. Agarwal, P. Patel, K. Singh, D. Soni, P. Johri, S.G. Aggarwal, V.K. Jain, *Some Preliminary Results of Particulate Matter Metrology*. (IJAPC, vol XX, No. 1–2, March and September 2020.)
67. I. Mori, Z. Sun, M. Ukachi, K. Nagano, C.W. McLeod, A.G. Cox, M. Nishikawa, Development and certification of the new NIES CRM 28: urban aerosols for the determination of multielements. Anal. Bioanal. Chem. **391**, 1997–2003 (2008). https://doi.org/10.1007/s00216-008-2076-y
68. https://www.gminsights.com/industry-analysis/gas-sensors-market-size
69. Nicola Donato, Mariangela Latino and Giovanni Neri, in Carbon nanotubes, ed. By Stefano Bianco (IntechOpen Limited, London, 2011) p. 229
70. C. Wang, L. Yin, L. Zhang, D. Xiang, R. Gao, Metal oxide gas sensors: sensitivity and influencing factors. Sensors **10**, 2088 (2010). https://doi.org/10.3390/s100302088

Chapter 14
Metrology for Atmospheric Environment

Part II: Environmental Governance

Chhemendra Sharma, Tuhin Kumar Mandal, Sachchidanand Singh, Govind Gupta, Monika J. Kulshrestha, Prabha Johri, Ashish Ranjan, Arun Kumar Upadhayaya, Rupesh M. Das, Daya Soni, Sumit Kumar Mishra, Senthil Kumar Muthusamy, Sudhir Kumar Sharma, Preetam Singh, Shankar Gopala Aggarwal, Soman Radha Radhakrishnan, and Manoj Kumar

Abstract The chapter provides an assessment of environmental governance in India which includes the need of environmental metrology, ideal quality infrastructure, interface between stakeholders and the role of National Metrology Institute (NMI). The chapter also discusses the training needs for developing human resources to implement best measurement practices for improving the quality of measurement

---

C. Sharma (✉) · T. K. Mandal · S. Singh · G. Gupta · M. J. Kulshrestha · P. Johri · A. Ranjan · A. K. Upadhayaya · R. M. Das · D. Soni · S. K. Mishra · S. K. Muthusamy · S. K. Sharma · P. Singh · S. G. Aggarwal · S. R. Radhakrishnan · M. Kumar
CSIR-National Physical Laboratory, Dr. K.S. Krishnan Marg, New Delhi 110012, India
e-mail: csharma@nplindia.org

T. K. Mandal
e-mail: tuhin@nplindia.org

S. Singh
e-mail: ssingh@nplindia.org

G. Gupta
e-mail: govind@nplindia.org

M. J. Kulshrestha
e-mail: monikak@nplindia.org

P. Johri
e-mail: pjohri@nplindia.org

A. Ranjan
e-mail: aranjan@nplindia.org

A. K. Upadhayaya
e-mail: upadhayayaak@nplindia.org

R. M. Das
e-mail: rupesh@nplindia.org

D. Soni
e-mail: dsoni@nplindia.org

D. K. Aswal (ed.), *Metrology for Inclusive Growth of India*,
https://doi.org/10.1007/978-981-15-8872-3_14

data. An attempt has also been made to estimate the potential economic impacts of improving the existing quality framework in India. Apex level test and calibration facilities for all of the environmental and emission monitoring systems are currently not available in the country. The lack of these facilities in India is not only a major constraint for industries dealing with manufacturing/production of environmental monitoring equipment but also to user organizations/agencies in respect of ascertaining the data quality. Realizing this need, the Government of India has designated the CSIR-NPL as the national agency for certifying instruments and equipments for monitoring emissions and ambient air. To cater to this need a 'CSIR-NPL India Certificate Scheme (NPLI CS)' has been developed to provide certification of performance evaluation of air pollution monitoring equipment. In addition, technologies developed by CSIR-NPL in the field atmospheric measurements are also discussed along with the capacity building efforts undertaken in this field by CSIR-NPL.

The wonder is, not that the field of the stars is so vast, but that man has measured it.

– Anatole France (1844-1926)

14.1 Introduction

The inclusive growth of a region and a country is driven by a combination of economic growth, social development and environmental protection. For sustainable growth, it

S. K. Mishra
e-mail: mishrask@nplindia.org

S. K. Muthusamy
e-mail: senthilmk@nplindia.org

S. K. Sharma
e-mail: sudhircsir@gmail.com

P. Singh
e-mail: singhp@nplindia.org

S. G. Aggarwal
e-mail: aggarwalsg@nplindia.org

S. R. Radhakrishnan
e-mail: rkrishnan@nplindia.org

M. Kumar
e-mail: manojkumar@nplindia.org

is important to protect the natural balance of our environment. The global environment is known to be largely impacted by the manmade (i.e. anthropogenic) activities leading to environmental degradation and climate change. The United Nations and its member states are already aware of the seriousness of this global problem and try to take remedial actions through various international programmes. For example, the United Nations Environment Programme (UNEP) founded in 1972 after the UN conference on human environment organizes and implements various measures aimed at improving and protecting the environment for the benefit of the society. The Intergovernmental Panel on Climate Change (IPCC) and the Earth System Governance Project (ESGP) are the international networks to assess the scientific information in the area of the climate change and the global environmental change, respectively. The Global Atmosphere Watch (GAW) programme of World Meteorological Organization (WMO) provides reliable scientific data and information on aerosols, greenhouse and reactive gases, ozone, ultraviolet radiation and atmospheric deposition etc. The International Bureau of Weights and Measures (BIPM), Regional Metrology Organization (RMO) and National Metrology Institutes (NMIs) are involved in providing all metrological services, including in field of climate change and environmental monitoring. The International Organization of Legal Metrology (OIML), on the other hand, coordinates the international network of NMIs and national legal metrology authorities to ensure that quality environmental and atmospheric measurements are available and help make the correct decisions to take care for the natural environment. The overall aim of these institutions is to provide SI traceable reliable, stable and comparable measurement standards and measurements related to the monitoring of environment and atmosphere on a short and long-term basis. The monitoring and measurements of atmospheric and environmental parameters require comparable measurement standards to provide reliable, stable and long-term measurements of various constituents. Different countries with the help of their NMIs and other agencies address the quality of measurements through necessary legislations. These measurements would be meaningful only when they are internationally recognized and are performed as per globally accepted measurement standards.

In developed economies, NMIs viz. NIST-USA, NPL-UK, PTB-Germany, NMI-Australia, NMIJ-Japan and KRISS-Korea etc. are continuously engaged in the developments of state-of-the-art standards and their dissemination for measurements of various atmospheric constituents. For example, NIST-USA is involved in measurements of GHGs, distributed emissions, stationary emissions, satellite calibration standards, carbonaceous aerosols, gas metrology, and sensor development for metrology and climate science, standard reference materials and nano-metrology. NPL-UK is working on gas and particle metrology, emission and atmospheric metrology, environmental monitoring, Earth observation, climate change and optical phenomenon. PTB-Germany has leading contributions in spectrometric gas analysis, aerosol and particle measurements, transport emissions and flue gas analysis etc. NMIJ-Japan contributes significantly in the area of standards for inorganic, environmental, gas and humidity, organic (primary and analytical), biomedical and particulate measurements. KRISS-Korea is involved in CRM for pollutants, measurement technologies and related properties, development of gas analysis for industry, real time hazardous chemicals in ambient and work place and smart sensors. Developed countries are quite capable to monitor parameters related to climate change due to maintenance of traceability chain, but same is not true for many developing countries. Over the time the developing economies' NMIs have also made significant progress in the area of environmental measurements and their standards.

CSIR-NPL in India, since its empowerment as the custodian of national standards in 1956 by the act of parliament is actively involved in establishment and maintenance of basic SI units and their derivatives. It has been contributing in the field of environmental and atmospheric metrology from the early days of its inception. Initially, it started to explore the characteristics of the atmosphere mainly in the upper atmosphere to sort out the issues related to radio communication and propagation. Focus had shifted to middle atmosphere during seventies and eighties to gauge reasons behind ozone depletion by the natural (volcanic eruption) and anthropogenic interventions (viz. emission from supersonic planes, usage of refrigerants etc.). After economic liberation of India in 1991, several multinational companies started to invest and set up industries in India. This led to increase emission from different sectors like industry, power, etc. Controversy over large methane emission from paddy fields in India(which was earlier estimated as 37.4 Tg/year) forced Indian scientists to prove the actual methane emissions from Indian paddy fields to be only of the order of 3.46 Tg/year [1, 2], that eventually laid the foundation of preparation of methane standard in CSIR-NPL.

In 1999, Indian Ocean Experiment (INDOEX) conducted in collaboration with different countries including, USA, UK, Germany, Netherlands and India found large cloud of brown haze over Indian Ocean [3, 4] called Atmospheric Brown Clouds (ABC). This was linked to pollution emissions from several types of sources in Indian subcontinent. Due to absence of regional inventory of emissions of pollutants, debate on the nature of pollutants and their potential sources facilitated the UNEP to establish several observational sites over South Asia. In addition, a couple of coordinated efforts were made later on as a part of land campaigns and cruises under the umbrella of ISRO/DOS [5–7]. However, the data qualities mostly dependent upon imported instruments, which were not suitable for the Indian meteorological conditions, become a major issue. During 11th Five Years plan (2007–2012), Govt of India (GOI) supported the *Chemical Metrology* programme in CSIR-NPL for making reference standards for Greenhouse Gases (GHGs) and PTB, Germany provided the required training. During 12th Five Years Plan (2012–2017), GOI further strengthened the activity. Such efforts are resulting in robust CSIR-NPL programme for development of Bhartiya Nirdeshak Dravya (BND®) as detailed in Chaps. 18 and 19.

It has also been realized that to further improve the quality infrastructure in the country in the field of atmospheric measurements to ensure the SI traceability of the ambient air quality and emission measurements, the country needs to have a robust testing, calibration, and certification facility for air pollution monitoring equipment. Keeping this in view, the Ministry of Environment, Forest and Climate Change

(MoEF&CC) has designated CSIR-NPL as national verification agency for Certifying instruments and equipment for monitoring emissions and ambient air vide its Gazette Notification No. 2758 dated August 22, 2019 and bestowed the responsibility of testing, calibration, and certifying environmental monitoring instrument to be used in atmospheric monitoring and industrial emissions to the CSIR-NPL. Key features of this programme would be (a) providing reliable and traceable measurement across India, (b) calibration, testing and certification of monitoring instruments of Indian and foreign manufacturers, (c) strengthen the gas and sensor metrology programme (d) providing reference standards required by the industries and academic institutes for reliable and accurate measurement. Based on these reliable and accurate data, mitigation policy will enable India to fulfil the commitment made to international community and certifying the home developed monitoring system particularly by MSME sector which would be engine for higher economic growth in India. CSIR-NPL is determined to bring together world-class measurement expertise at the forefront of innovation and contribute to our economic prosperity as envisaged in the 'Aswal Model of Inclusive Growth' and detailed in Chap. 1. Over the years, CSIR-NPL has emerged as a guardian to the nation in the field of quality data for environmental and atmospheric measurements.

14.1.1 Institutional Framework for Environmental Governance

The term Environmental governance signifies a combined effort in which concepts like sustainable development, environmental management and environmental sustainability are merged together in a holistic manner [8]. It refers to a complete set of regulatory processes, their mechanisms and specified responsible organizations through which the environmental actions and outcomes are influenced and regulated. The first major effort towards a system of global environmental governance was taken up in 1972 by the United Nations Conference on the Human Environment, also known as the Stockholm Conference. Afterward, India made the 42nd amendment to the Indian constitution and added Article 48-A and 51-A(g). According to the Article 48-A *the State shall endeavour to protect and improve the environment and to safeguard the forests and wildlife of the country* [9]. Further Article 51-A(g) stated "*it shall be the duty of every citizen of India to protect and improve the natural environment including forests, lakes, rivers and wildlife and to have compassion for living creatures*".

The most important environmental legislations created by the parliament under Art. 249 of the Constitution are: (i) the Water (Prevention and Control of Pollution) Act, 1974; (ii) the Air (Prevention and Control of Pollution) Act, 1981; and (iii) the Environment (Protection) Act, 1986. During the course of time the legislator of India made more the than 200 act/laws for protecting the environment with adequate provisions in the constitution of India. The details of some of the important acts/laws are given in Annexure I.

14.1.1.1 Water (Prevention and Control of Pollution) Act, 1974

The aim of this Act was to provide the prevention and control of water pollution and the maintaining or restoring of wholesomeness of water. Under this Act central government constituted a statutory body known as Central Pollution Control Board (CPCB). This Act also empowers the state governments to establish their own State Pollution Control Board (SPCB) for control of water pollution at state level. Both, CPCB and SPCB frame and enforce the standards for water quality and effluents discharge in water bodies.

14.1.1.2 Air (Prevention and Control of Pollution) Act, 1981

This Act was made in 1981 to provide for the prevention, control and abatement of air pollution over the country. The act also provide similar type framework as of the Water Act 1974 for providing authority to CPCB and SPCBs for maintaining and ascertaining the compliance of designated air pollution standards.

14.1.1.3 Environment Protection Act, 1986

After the Bhopal Gas tragedy, Environment Protection Act (EPA) 1986 came into the force to fill the many gaps in the existing laws. The Act provide framework for the protection and improvement of environment. Further, the Act framed for air, water and land pollution and also empowers the government for setting national ambient and emission standards, establishing procedures and regulations for inspection and checking pollution sources. The main objective of this act is to implement the decisions of United Nations Conference on the Human Environment (UNCHE) which was held at Stockholm in June, 1972.

14.1.1.4 Major Institutions in India for the Environmental Governance

In India, a strong linkage between the Central and State Governments has been made for smooth functioning of the Governance and implementation of various laws and acts. With regard to the environmental pollution control, the linkage between Central and State Governments are complementary. Government of India framed ministry, pollution control boards at central and state level to execute/implement the act/law/regulations related to various environmental issues (Fig. 14.1). The major institutions in India and their inter-linkages are depicted in Fig. 14.1.

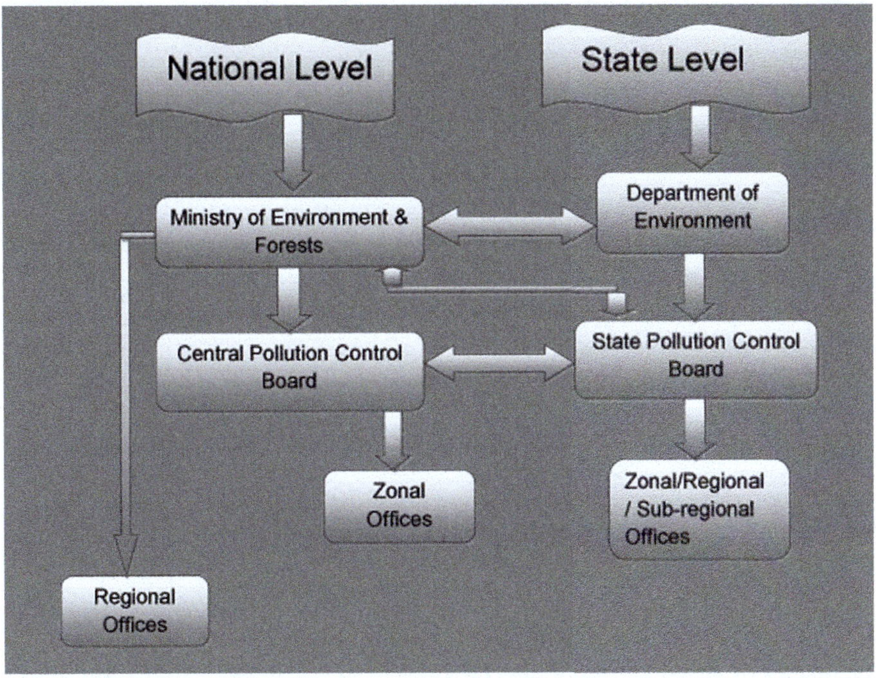

Fig. 14.1 Administrative structure for environment governance in India (*Source* Adopted from Gargava et al. 2000)

Ministry of Environment, Forests and Climate Change (MoEF&CC)

Ministry of Environment, Forests and Climate Change (MoEF&CC) is a nodal agency of the Government of India for planning, promotion, co-ordination and overseeing the implementation of India's environmental and forestry policies and programmes. The main objectives of the MoEF&CC are to implement the policies and programmers related to the conservation of the country's natural resources (including lakes and rivers, biodiversity, forests and wildlife, ensuring the welfare of animals) and the prevention and abatement of pollution. The MoEF&CC is also the nodal agency for the United Nations Environment Programme (UNEP), South Asia Co-operative Environment Programme (SACEP), and International Centre for Integrated Mountain Development (ICIMOD) and the United Nations Conference on Environment and Development (UNCED) (Fig. 14.2).

Central Pollution Control Board (CPCB)

The CPCB is constituted under section 3 of the Water Act, 1974 as a statutory body. Further, the powers and function of the CPCB was entrusted by Air (Prevention

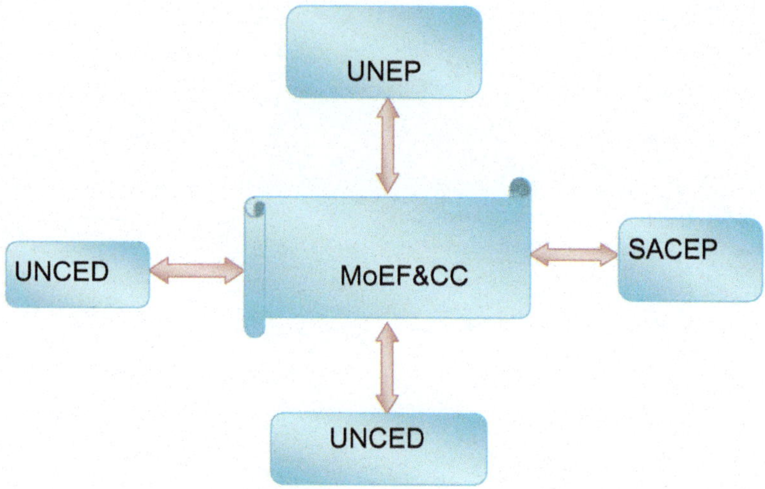

Fig. 14.2 MoEF&CC and its linkages with international organizations

and Control of Pollution) Act, 1981. It advises to central government on any matter concerning prevention and control of water and air pollution and improvement of the quality of air. Moreover, it also provides technical supports to the MoEF&CC on the provisions of the Environment (Protection) Act, 1986. The principal function of CPCB is to reduce pollution of water and air under the Water and Air Acts as well as National Air Monitoring Programme (NAMP). It also provides technical assistance, guidance and coordinates the activities of the SPCBs. The detail structure and function of CPCB are available at its official website (www.cpcb.nic.in).

State Pollution Control Boards (SBCBs)

The State Pollution Control Boards (SPCBs) were created at the State level with primary responsibility of issuing of consent or permit for operating an industry subject to certain evaluation and emission limits. Besides this, SPCBs also perform functions at state level similar to what CPCB does at National level. Similar to CPCP, it is an advisory body of the concerned state governments.

National Green Tribunal (NGT)

National Green Tribunal (NGT) was established on October 18th, 2010 under the National Green Tribunal Act 2010 and became fully operational on 5 May 2011 [10]. According to the Act, NGT was established "*for effective and expeditious disposal of cases relating to environmental protection and conservation of forests and other natural resources including enforcement of any legal right relating to environment and giving relief and compensation for damages to persons and property and for matters connected therewith or incidental there to*". Initially, the Supreme Court of India in three important judgments observed that a separate "Environment Court" should be established for fast disposal of environmental cases in context to three cases [11].

The Principal Bench of the NGT is located in New Delhi, with regional benches in Pune, Bhopal, Chennai and Kolkata. Each bench has a specified geographical jurisdiction covering several states in a region. The tribunal has jurisdiction to hear all the civil cases related to environmental issues and questions that are linked to the implementation of laws listed in Schedule I of the NGT Act [12]. Since its establishment, NGT has emerged as a strong force for the enforcement of Environmental Legislations in India. Moreover, the tribunal has given some notable decisions for the environmental protection, some of which are as follow:

- Ban on diesel vehicles more than 10 years In Delhi-NCR
- Protection of Indian rivers (Yamuna river)
- Cancellation of clearance of coal block of Chhattisgarh forest
- Imposing fine on burning waste in open.

Strategy for Pollution Control in India

For pollution control in India, the various tools, techniques and strategies are being used. These are as follow:

(i) Technological advancement
(ii) Judicial intervention
(iii) Environmental Impact Assessment
(iv) Environmental auditing.

(i) **Technological advancement**—A number of technological options have been implemented in India to improve the air quality. Many of them are aimed to mitigate vehicular emissions. Some of these are described below:

Improvement in fuel quality: Due to harmful effect of air pollution on human health, the fuel up graduation programme in India was started with a notification of vehicular emission norms for new vehicles in 1991 and further revised in 1996. The term Bharat Stage (BS) followed regulatory pathway for fuel quality and vehicle emissions standards. The fuel quality standards of India have been made successively strict since mid-1990s. The low quantity of lead was introduced in 1994 in four metro cities i.e. Delhi, Mumbai, Kolkata and Chennai. Moreover, in 2000, unleaded gasoline was mandated for all over India. In the same year, the benzene limits were also reduced progressively from 5% over the country.

Reduction of sulphur content: In 2010, the sulphur content in gasoline was reduced to 150 ppm nationwide and 50 ppm in BS-IV compliant cities. Further, octane number of fuel had been increased to 88 and 93 for regular and premium gasoline fuel respectively, under the Bharat Stage (BS)—II. Further, it increased to 91 and 95 for regular and premium fuels respectively, under BS-III. Moreover, the sulphur content has been further reduced to 10 ppm in gasoline for BS VI vehicles which were introduced from April 2020 in India.

Upgradation in motor vehicle engine: In 1996, the emission norm along with fuel quality specification required with some modification in the engine design particularly in regard to crankcase emissions and evaporative emissions control. Recently, the automobile industry of India had made rapid technological change in terms of both product characteristics and manufacturing processes. After the intervention of Supreme Court of India, the major steps have been taken in mid-1999 for emission norms for passenger cars and commercial vehicle for National Capital Region (NCR) called India-2000 (Euro-I equivalent) norms. Further, these norms were extended to the rest of the country in January of 2000. Subsequently, Bharat Stage-II (BS-II, which is Euro-II equivalent) norms introduced in NCR form 2000 for passenger cars and multi-utility vehicles. In 2001, it has been extended to other three major metros viz. Mumbai, Kolkata, and Chennai.

After the Auto Fuel Policy (AFP) of the government requisite the adoption of BS-III (Euro III equivalent) norms in eleven most polluted cities (Delhi, Mumbai, Chennai, Kolkata, Bangalore, Hyderabad, Ahmedabad, Pune, Surat, Kanpur and Agra) by 2005. Moreover, according to requirement of vehicle pollution abatement, AFP requires BS-IV (Euro-IV equivalent) adoption in the major metros by 2010 and BS-II norms extended the rest of the country.

Use of catalytic converters in vehicle: Catalytic converter is a device which converts harmful exhaust gases (CO, unburned HC, NO_x) into benign gases by using different types of catalysts [13]. Initially, India had adopted European emission standards for four wheeler. After implementation of National Auto Fuel Policy 2003, India introduced BS-II to VI standards which are similar to Euro 2 and 6 emission standards. The Indian Standards of emission summarized chronologically in Table 14.1.

(ii) **Judicial intervention**

The legal duty of state is to protect natural resources which are provided by nature to full fill the need of public. First time in India, the court applied this doctrine in case of M.C. Mehta (Supreme Court Lawyer) verses Kamal Nath (Minister, MoEF&CC) related to an environmental problem. In the past several years, Supreme Court took suo moto action for the protection of environment and its resource. Some of the landmark cases are narrated below:

Taj trapezium zone (TTZ): Taj Mahal attract tourist of all over word therefore it was declared as a UNESCO word heritage in 1983. Over the time of period, due to increased pollution level, the monument has started to develop a yellowish tinge and in some areas brown and black spots. Consequently, Mr. M. C. Mehta, an attorney in the Supreme Court of India, filed a Writ Petition in 1984 mentioning the adverse effects on the Taj Mahal due to industrial as well as vehicular emissions. The supreme court of India delivered a ruling on December 30, 1996, in response to public interest

Table 14.1 History of emission standards implemented in India for 4-wheeled vehicles (*Source* modified from [14])

Standard	Reference	Year	Region
India 2000	Euro 1	2000	Nationwide
Bharat stage-II	Euro 2	2001	NCR*, Mumbai, Kolkata, Chennai
		2003	NCR*, 13 Cities#
		2005	Nationwide
Bharat stag-III	Euro 3	2005	NCR*, 13 Cities#
		2010	Nationwide
Bharat stage IV	Euro 4	2010	NCR*, 13 Cities#
		2017	Nationwide
Bharat stage V	Euro 5	(Skipped)	
Bharat stage VI	Euro 6	2018	Delhi
		2019	NCR
		2020	Nationwide

*National Capital Region (Delhi)
#Mumbai, Kolkata, Chennai, Bengaluru, Hyderabad, Ahmedabad, Pune, Surat, Kanpur, Lucknow, Sholapur, Jamshedpur and Agra

litigation, seeking to the protection of Taj Mahal from the pollutants emitted from the industries and vehicles. Further, the court banned the use of coal/coke in industries under TTZ and ordered for switching over from coal/coke to natural gas as well as relocating them outside the TTZ or shutting down.

CNG implication in vehicle: The emission of air pollutants from vehicles were posing serious health risks to residents of Delhi-NCR, therefore a PIL was filed by environmentalist and lawyer Mr. M. C. Mehta in 1996. Thereafter, in 1998, the Supreme Court has issued a directive for the conversion of all public transport (buses, taxis and three-wheelers etc.) to CNG in Delhi-NCR. Further, it has been in implemented in Delhi since April 2001 [15]. CNG is helpful to reduce emissions of air pollutant. It has been estimated that it produce approximately eight times lower particulate matters in comparison to diesel combustion [16].

Delhi bus rapid transit system: The honourable Supreme Court passed an order to convert all diesel buses to CNG to reduce air pollution; however, the total suspended particle could not be reduced up to the prescribed limit of WHO. Therefore, there was need to build a new road or widen the existing roads. Consequently, the Delhi Bus Rapid Transit System (BRTS) has been introduced in Delhi to cater the city's transportation needs.

Graded response action plan: In December 2016, apex court ordered to MoEF&CC for preparing a Graded Response Action Plan (GRAP) in Delhi-NCR to control air pollution. Moreover, to control $PM_{2.5}$ concentrations government of India formulated and implements a GRAP [17]. The honourable Supreme Court has given so many

Table 14.2 Some important landmark cases filed in Supreme Court of India by Mr. M. C. Mehta related to environmental issues

S. No.	Name of cases	M. C. Mehta versus	Year
1	Delhi monuments case	archaeological survey of India and others	1996
2	Ganga pollution case	Union of India and others	1985
3	Protecting the Beas river	Kamal Nath and other	1996
4	Oleum gas leak case	Union of India and others	1985
5	Delhi hazardous industries	Union of India and others	1985
6	Introduction of CNG in India	Union of India and others	1985
7	Delhi ridge case	Union of India and others	1985
8	Gamma chamber case	Union of India	1987
9	Environmental education	Union of India	1991
10	Dust pollution case	Union of India	1992
11	Graded response action plan	Union of India	2016

historical decisions to protect our environment and ecology in response to various PIL and suo moto action. The list of some landmark PIL filed by M. C. Mehta related environmental issues are given in Table 14.2.

(iii) **Environmental Impact Assessment (EIA)**

EIA is a process of evaluating the environmental impacts of a proposed project or development, taking into consideration of inter-related socio-economic, cultural and human health impacts, both beneficial as well as adverse. Its purpose is to inform the decision-makers and to the public regarding the environmental consequences of implementing a proposed project. Thus, after identification of potential impacts of developmental project could reduce the adverse impacts on the environment and encourage for sustainable development. The role of EIA was formally recognized internationally in Earth Summit at Rio de Janeiro in 1992. In India, EIA was started first time in 1977–78 for evaluation of a river valley project [18]. Further, it has extended to mining, Industries, thermal power, port and harbours and other areas. Presently, EIA is compulsory for all types of projects. In our country, MoEF&CC is responsible for environmental clearance, therefore, through a notification, made it mandatory to prepare and submit an EIA, an Environment Management Plan (EMP), and a project report to MoEF&CC which had the option to consult a multidisciplinary committee of experts for appropriate decision.

(iv) **Environmental Auditing**

The Government of India, by the way of notification of March 13, 1992, has made environmental auditing mandatory and the industries are required to submit environmental statements every year starting from 1992–93 [19]. The aim of this auditing was to minimize consumption of natural resources and promoting use of clean technologies to abatement of pollution [19]. It is also used as to assess the quality of existing environmental management system for improving the environmental performance of a facility [20]. Further, it varies for different organizations, the audit would be planned as per the objectives of auditing. Furthermore, it should be incorporated with some important environmental tool like EIA and Environmental Management System for appropriate result. Consequently, effective implementation of environmental audit helps to reduce adverse effect on environmental risk at low cost.

Role of civil society: At the international level, environmental governance primarily deals with the international organization, the framework of international law, and a global financing mechanism [21]. In India, civil society strengthens environmental governance by encouraging human rights and contributing to sensitizing people, disseminating about the environment, and maintaining the ecosystem [22]. Further, the Indian civil society helped in environmental governance policy framework along with major legal development through Public Interest Litigation (PIL). Indian civil societies have made certain valuable contributions to the development on environmental governance. For example, the original of "Green Tribunal Act, 2010" was the result of a writ petition filed by a civil society organization called Indian Council for Enviro-Legal Action in the Supreme Court in 1996.

Guidelines for recognition of environmental laboratory by CPCB: Under the Water (Prevention and Control of Pollution), Act, 1974 and Air (Prevention and Control of Pollution) Act, 1981, for the pollution control, the central/state government has to establish central, state and private laboratories for analysis of samples. The detail guidelines for recognition of environmental laboratories under the Environmental (Protection) Act 1986, is given by MoEF&CC [23]. In this context, MoEF&CC has been implementing a programme for the recognition of environmental laboratory with the aim of increasing as well as upgrading the facilities for the analysis of environmental samples. The guidelines (2008) for the recognition of environmental laboratories guideline related to infrastructure requirement, sample handing, quality assurance and quality control etc. are described within six different chapters in Laboratory Analytical Techniques Series (LATS/9/2008–2009) of MoEF&CC. The schematic diagram of the procedure for evaluation of environmental laboratories is shown in Fig. 14.3.

14.1.1.5 Relevance for Stakeholders

The environmental monitoring involves a large number of stakeholders including common man, industries, research and academic institutions, equipment manufacturers, test laboratories, regulators and policy makers. It is important to realize the importance of metrology which would ensure the confidence in the measured values. The accurate and reliable measurements are the keys for understanding and answering the science and policy questions. The regulatory mechanisms also need robust measurements for ascertaining the compliances and developing appropriate policies for the amelioration of environmental conditions. There is a great need in

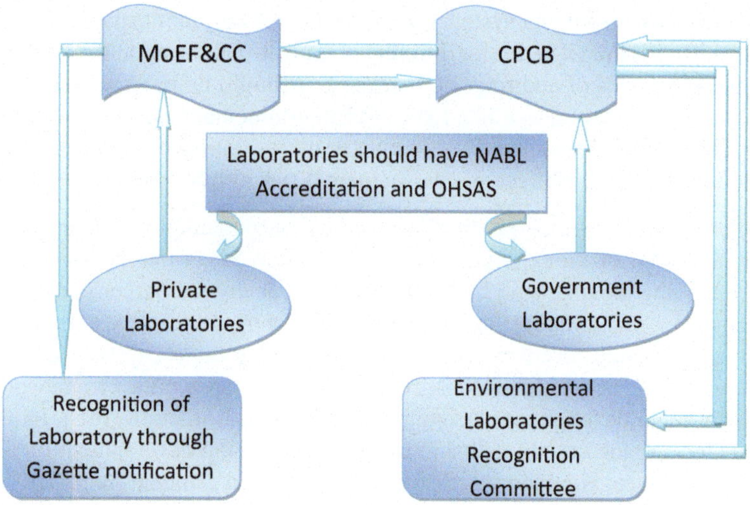

Fig. 14.3 Schematic flow diagram for recognition process of environmental laboratories

India to strengthen the test, calibration (for details refer to Chap. 13) and certification system of environmental monitoring equipment to support the manufacturers and entrepreneurs as the newer opportunities are becoming available due to vast market potential. This would also help a large number of other stakeholders who are using such equipment for research and monitoring purposes.

14.2 Metrology for Industrial Emissions

Industrial growth always accompanied with challenges to mitigates the environmental pollution e.g. air, water, soil pollution etc. These challenges may be overcome through implementation of national environmental governance system which includes policy formulations, enforcement of regulatory guidelines, legal binding of stakeholders etc. as elaborated in section "Central Pollution Control Board (CPCB)". In India, the regulatory agency (i.e. CPCB) enforces the industries to install Online CEMS (OCEMS) to monitor the concentrations of gases and particles which are emitted from the stacks on real time basis. Accordingly, seventeen category of industries in India such as Pulp and Paper, Distillery, Sugar, Tanneries, Power Plants, Iron and Steel, Cement, Oil Refineries, Fertilizer, Chloral Alkali Plants, Dye and Dye Intermediate Units, Pesticides, Zinc, Copper, Aluminium, Petro-chemicals and Pharma Sector are required to have OCEMS for monitoring of the their stack emissions. The OCEMS systems are composed of the equipment which are capable to measure the required range of gaseous emission concentrations, particulate matter emission rate etc. Since, these emissions have impacts on the ambient air quality, therefore ambient air quality data are also acquired through Continuous Ambient Air Quality Monitoring Systems (CAAQMS) for assessment of ambient air quality.

In both CEMS and CAAQMS systems the assurance of data quality is one of the major challenges. The roles of proper instrument and their calibration is also an important concern. In India, most of the instruments used are usually imported from abroad which comes with certifications from agencies like US-EPA, TUV, MCERT etc. These certificates are issued based on the environmental conditions of the certificate issuing country which are very different from the environmental conditions prevalent in India. Hence a reestablishment of traceability for these instruments are important to ensure the data quality. However, as of now, no certification system is available in India for environmental monitoring equipments. The traceability of measurement is also an integral part of generation of reliable data. Given increasing stress on the quality & transparency, the CAMS manufacturers in India need support particularly, regarding traceability in measurements and calibration of equipment. However, apex level calibration facilities in this area is currently not available in the country. The lack of calibration facilities in India is not only a major constraint for the manufacturing/production of environmental monitoring equipment but also to user organizations/agencies for ascertaining the data quality. CSIR-NPL being the NMI is uniquely positioned and working on to create a facility for calibration, testing and certification of OCEMS and CAAQMS.

The above said calibration and testing facility will be based on the international quality documents such as ISO 17025:2017 and ISO 17065: 2005 for the testing and calibration laboratories. The main focus of the test and calibration facility for OCEMS and CAAQMS can be divided in to two broad categories. First one is the product testing and certification (for manufacturers) facility which falls under Quality Assurance Level-1 (QAL-1). Second one is the regular calibration and validation of OCEMS and CAAQMS including post installation testing and certification facility known to be as Quality Assurance Level-2 (QAL-2).

Performance Standards and Test Procedures for Continuous Emission Monitoring Systems will be based on the EN 15267-1 and 2 regulations. The Performance Standards for Continuous Ambient Air Quality Monitoring Systems will be based on EN 14662-3 or equivalent. The design concept of this facility is based upon the testing of various species under the ranges of measurements specified by regulatory agency. The gases to be tested and its ranges are listed in Table 14.3.

Regarding CAAQMS, Currently, NPL provides apex level calibration facilities for O_3, CO, SO_2 and NO_2 which can be used for providing calibration of CAMS equipments. However, these facilities need to be upgraded to meet the magnitude of required calibration. Also, a facility for PM sampler calibration is operational. The facilities for remaining target species of CAMS viz. NH_3, C_6H_6 and Benzo (a) Pyrene (BaP) are required to developed. These trace gases are summarized in Table 14.4.

The type approval test (QAL-1)/product certification service to the CEMS consist of laboratory test and field test. The laboratory test will be conducted to ascertain the performance of the CEMS in various temperature, relative humidity and pressure levels. The performance of the system in various input voltage level, vibration

Table 14.3 Testable and their respective concentration ranges

Testable	Concentration range
SO_2	0–10,000 mg/m^3
CO	0–1,000 mg/m^3
CO_2	0–20% volume
NO	0–3,000 mg/m^3
NO_2	0–3,000 mg/m^3
HCI	0–2,000 mg/m^3
TOC	0–70 mg/m^3
O_2	0–25% volume
H_2O	0–45% volume

Table 14.4 Various gases to be tested in CAAQMS

Target parameters	Availability
O_3	Available at NPL
CO	Available at NPL
SO_2	Available at NPL
NO_2	Available at NPL
PM	Available at NPL
NH_3	To be developed
C_6H_6	To be developed
Benzo (a) Pyrene (BaP)	To be developed

level etc. also will be conducted. The type approval of the system is based on the calculation of the combined expanded uncertainty in the measurement result derived from the numerical values of the tested performance characteristics and compared to a prescribed maximum uncertainty.

The laboratory test will be conducted in controlled environment (six months). The field test will be conducted at a high altitude station and in a coastal environment (three months each). Proposed performance criteria for laboratory testing and filed testing are given in Tables 14.5 and 14.6, respectively as per MCERTS 2016 which would be finalized after due deliberations based on regulatory requirements [24].

The Performance Testing (QAL 1) for the CAAQMS System

The CAAQMS shall be able to measure instantaneous values in a range that is at least twice the upper limit of the certification range in order to measure the half-hour values. If it is necessary to use more than one range setting of the CAAQM to achieve this requirement, supplementary ranges require additional testing. Proposed performance criteria for the laboratory and field testing are given in the Table 14.7 as adopted from MCERTS 2016 [25].

Table 14.5 Proposed performance criteria for laboratory testing

Performance characteristics	Performance criteria
Response time	<400 s
Repeatability standard deviation at zero point	<2.0%
Repeatability standard deviation at span point	<2.0%
Lack of fit (linearity)	<2.0%
Influence of ambient temperature change from	20 °C within specified range at zero point <5.0
Influence of ambient temperature change	From 20 °C within specified range at span point <5.0%
For a pressure change	<2.0%
Influence of sample gas flow on extractive CEM for a given specification by the manufacturer	<2.0%
Influence of voltage, at –15% and at +10% from nominal supply	Voltage <2.0%
Influence of vibration	<2.0%
Cross-sensitivity	<4.0%
Excursion of the measurement beam of cross-stack in situ CEM	<2.0%

The percentage values are the percentage of the upper limit

Table 14.6 Proposed performance criteria for field testing

Performance characteristics	Performance criteria
Determination coefficient of calibration function, R^2	≥0.90
Availability	≥95.0%
Lack of fit	≤2.0%
Zero drift, within maintenance interval	≤3.0%
Span drift, within maintenance interval	≤3.0%
Reproducibility, R_{field}	≤3.3%
Minimum maintenance interval	8 days
Response time	<400 s

The percentage values are the percentage of the upper limit

14.2.1 CSIR-NPL INDIA Certification Scheme (NPLI CS)

The conformity assessment and product certification of air pollution monitoring equipment is an integral part of the environmental governance. Such product certification ascertains the highest level of data quality under the prevailing operational environment and further helps in formulation of better national level policy for problem mitigations. Keeping in view of this, many countries like United Kingdom

Table 14.7 Proposed performance criteria for the laboratory and field testing

Performance characteristics	Performance criteria for laboratory test	Performance criteria for field test
Repeatability standard deviation at zero point	≤1 nmol/mol	–
Repeatability standard deviation at hourly limit value	≤3.0 nmol/mol	–
Residual Lack of fit at zero	≤5.0 nmol/mol	–
Lack of fit	≤4% of the measured value	–
Sensitivity coefficient to sample gas temperature	≤1 nmol/mol/k	–
Sensitivity coefficient to sample gas pressure	≤2 nmol/mol/kPa	–
Sensitivity coefficient of surrounding air temperature	≤1 nmol/mol/k	–
Sensitivity coefficient of electrical supply voltage	≤0.3 nmol/mol/V	–
Converter efficiency	–	–
Interference by H_2O	≤10 nmol/mol	–
Interference by H_2S	≤5 nmol/mol	–
Interference by NH_3	≤5 nmol/mol at conc of 500 nmol/mol	–
Interference by NO	≤5 nmol/mol at conc of 500 nmol/mol	–
Interference by NO2	≤5 nmol/mol at conc of 500 nmol/mol	–
Interference by m-xylene	≤10 nmol/mol at conc of 200 nmol/mol	–
Interference by toluene	–	–
Interference by CO_2	–	–
Interference by N_2O	–	–
Averaging effect	≤7% of measured value	–
Reproducibility under field test	≤5% of the average over three month period	≤7% of measured value
Long-term drift at zero	≤4 nmol/mol	≤5% of the average over three month period
Long-term drift at span level	≤5% of the maximum certification range	≤4 nmol/mol
Short term drift at zero	≤2 nmol/mol over 12 h	–
Short term drift at span level	≤6 nmol/mol over 12 h	–
Response time (rise)	≤180 s	–
Response time (fall)	≤180 s	–

(continued)

Table 14.7 (continued)

Performance characteristics	Performance criteria for laboratory test	Performance criteria for field test
Difference between rise and fall time	$\leq$1%	–
Residence time in the analyzer	–	–
Period of unattended operation	3 months	3 months
Availability of analyzer	>90%	>90%
Total expanded uncertainty	$\leq$15%	$\leq$15%

(MCERTS), United States of America (US-EPA), France (ACIME), Germany (UBA Type approval system) etc. implemented product certification schemes not only to strengthen the environmental governance of their own country but also to generate a high-quality air pollution data which helps them in better policy making and implementation of required mitigation technique. The certification schemes like MCERTS, US-EPA & UBA are globally preferred schemes and captures almost 80% market share. The US-EPA verification system started in 1995 under the auspices of US-EPA with the objective of providing credible performance data for commercial-ready environmental technologies. Similarly, the UK Environment Agency launched its own Monitoring and Certification Scheme called MCERTS, in 1998. The Sira Certification Service (SCS) acted as the Certification Body to operate the MCERTS on behalf of the UK Environment Agency. Under the Sira Certification Service, CSA International is the MCERTS certification body and provides certification of equipment, personnel and inspection services. Similarly, in Germany, the UBA approval system for CEMs was started in 1975, which is responsible for the verification and performance of the product. It also checks the data provided by the monitoring system and certifies that it meets the minimum requirements of the German regulations. Over the time the UK Environment Agency and UBA agreed in the year 2000, through a Memorandum of Understanding (MoU) on the product certification/type approval of CEMs to mutually recognize the MCERTS and the UBA scheme.

Moreover, in India, the air pollution measurements are carried out with the help of the imported equipment which comes with certifications from agencies like US-EPA, TUV, MCERTS etc. However, till August 21, 2019, no certification system was is available in India for air pollution monitoring equipment. But on August 22, 2019, the MoEF&CC, Government of India exercise of the powers conferred by Sect. 14.3 of the Environment (Protection) Act, 1986 (29 of 1986) and designates the CSIR-NPL as a national verification agency for certifying instruments and equipments for monitoring emissions and ambient air. In view of this CSIR-NPL developed CSIR-NPL India Certification Scheme (NPLI CS) to cater the national need and bridge the identified gap area in national environmental governance structure.

14.2.1.1 Scope

The scope of the CSIR-NPL India certification scheme (NPLI CS) is to provide certification services conforming to international standards like ISO 17065, ISO 17067 for emission as well as ambient air pollution measurement equipment. This certification scheme will provide a complete and cost effective solution for certification to the manufacturers of these equipment/systems in India or abroad. Being a signatory to the CIPM-MRA, the certificates issued by CSIR-NPL will be acceptable worldwide. This will also help manufacturers to trade the certified product in international market as well besides helping in 'Make in India' programme.

The NPLI CS framework is the integral part of the Indian Environmental Governance System (IEGS). The NPLI CS framework is developed by CSIR-NPL in line of the Gazette notification issued by MoEF&CC, Govt. of India dated August 22nd 2019. The MoEF&CC, Govt. of India designated CSIR-NPL as national verification agency for certifying instruments and equipments for monitoring emissions and ambient air. The NPLI CS framework comply the requirements of ISO 17065:2012 and has provision of impartial attestation of the conformity assessment to fulfilment of specified requirement of air pollution monitoring equipment. The NPLI CS also includes the surveillance part; basically the scheme is combination of scheme type 1a and 4 as defined in clause 5 of ISO 17067. The NPLI CS product certification provides confidence to regulators, policy makers, consumers and other stakeholders related to specified requirements which include product performance, safety, sustainability and data quality. This also facilitates trade, market access, fair competition and consumer's acceptability of the air pollution monitoring equipments. The scheme is implemented in such way that the functional approach is in line with the ISO/IEC 17000:2004 which includes selection of product, determination through conformity assessment, review to fulfilment of requirements, decision on certification, attestation to fulfilment of specified requirements and surveillance. The scheme allows certification body to select two or more samples of the product from a batch for determination activity. These samples are representative of subsequent production items which must be identical in accordance with the certified type. The procedure of certification illustrated in the flow diagram shown in Fig. 7.30. The conformity assessment has been performed based on the determination of characteristics as per the guidelines of clause 5.2.2 of ISO 17067 which includes testing, inspection, design appraisal, assessment of processes and verification. In the developed NPLI CS the testing of product has been carried out as per the test programme generated by technical committee know as certification committee (comply the requirement of ISO 17065). The generation of the test programme is based upon review of the information & specifications provided by the client. The testing process divided into two parts i.e. the testing of product under control environment (Laboratory testing) and testing of product under claimed environmental condition (Field testing) for product verification. For laboratory testing, CSIR-NPL already initiated to establish a testing and calibration facility for emission and ambient air monitoring system in collaboration with National Physical Laboratory, United Kingdom (NPL-UK). The field testing will be carried out by outsourced agencies which compile the requirements of ISO 17025. The scheme also has provision for client's site inspection through its audit committee to assess the process and other requirements. The audit committee is the integral part of the scheme but independent from other committees to maintain the impartiality clause of ISO 17065. After performing the tasks like testing, inspection, design appraisal, assessment of processes and verification, the reports will be submitted to the certification committee for review (Fig. 14.4).

In NPLI CS framework, the certification committee is empowered to examine the reports generated during the determination stage to establish whether the product met the specified requirements or not. Further, the committee will submit its recommendation to certification body i.e. NPL India Certification Body (NICB) for attestation/issuance of certification. The NICB is the apex level committee of NPLI CS and empowered for issuance, maintaining, extension, reduction, suspension or withdrawal of certification. In the case of NPLI CS, NICB also authorized to grant the right of use of certificate/logo and issuing a certificate of conformity for a batch of

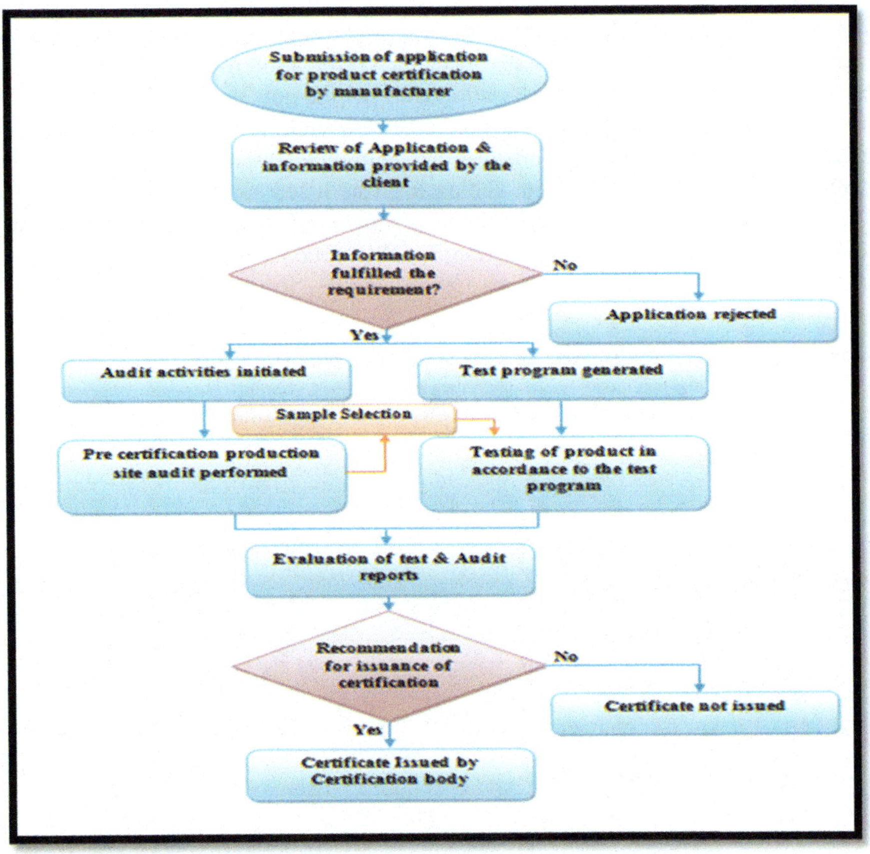

Fig.14.4 Flow diagram of certification procedure

product. In this certification scheme, the surveillance will be carried out by assessment of the production along with operation of the product. To deal with impartiality clause of ISO 17065, the NPLI CS framework has inbuilt but independent complaints redressal mechanism to address the appeals against the decision or complaints about the certification body.

This NPLI CS framework is developed to provide a cost effective and complete solution for Indian as well as International manufacturers those who are dealing with the production of Air pollution monitoring equipment. CSIR-NPL is a signatory of CIPM-MRA, hence the certificates issued by CSIR-NPL within the approved CMCs ranges will be acceptable world-wide. This will help manufacturers of these equipments to trade in international market as well besides helping in 'Make in India' programme.

14.2.1.2 Quality System

The NPLI CS framework is designed in way to have an established quality system to compete with the other internally available certification framework. The overall certification scheme NPLI CS is developed as per the guidelines of ISO 17067 and having a well-defined structure in accordance with ISO 17065. The scheme has five major components i.e. NICB, Certification Committee, Test and Calibration facility, Audit Committee and the Client. The each of the listed components complies the requirement of respective ISO guidelines as shown in Fig. 14.5. The NICB follows the requirements as mentioned in section-7 of ISO-17065 to monitor the various steps

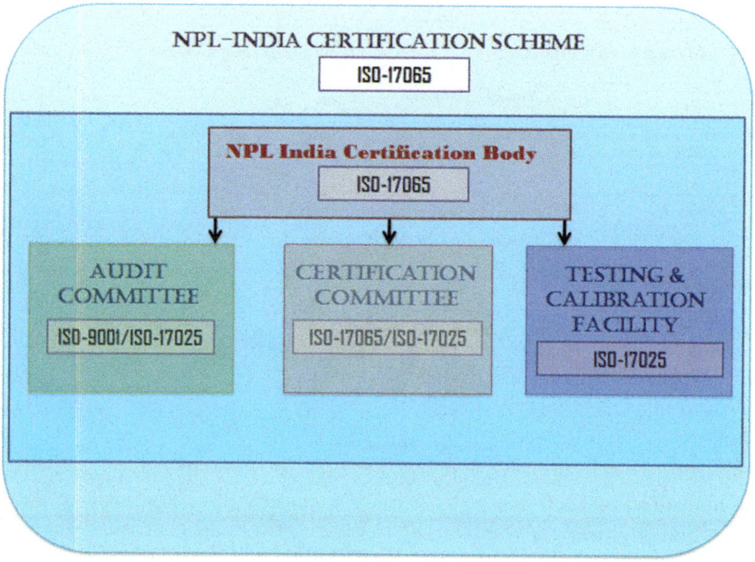

Fig. 14.5 Quality system NPLI CS framework

of the certification process. This includes receiving and reviewing of applications, ensuring the product, process, or appropriately evaluation of service, and persons independent of the evaluation review to make a final decision upon whether or not all certification requirements have been met. Similarly, the role and responsibilities of the certification committee fully comply with the requirements of ISO-17025 and other applicable international norms/standards. The main role of the certification committee will be to evaluate client application, design a suitable test programme and review the reports. However, the test and calibration facility will be the integral part of the CSIR-NPL and mandated to comply with the requirements of laboratory quality system i.e. ISO 17025. The facility will also comply the standard testing methods for different air pollution measurement techniques. On the other hand, the scope of audit committee will defined as per the requirements of ISO 17025 and ISO 9001. CSIR-NPL is the NMI of the country and comply with the international standard and obtain accreditation from bodies which have entered into MRA with equivalent bodies in other countries using the international standards. It has a long history of testing and calibration certification and already has a proper quality management system in place. However, in order to provide testing, calibration and certification to air pollution monitoring equipment, CSIR-NPL is continuously working to strengthen and enhance the level of quality management system in the laboratory. The normative references for NPL India Certification Scheme are listed in Annexure-II.

The global technological growth increases stress on the certification bodies to upgrade or identify the laboratories to have capabilities to verify the latest equipment with latest technology. Also, the laboratories must have to qualify or comply the requirements of quality guidelines with adequate trained manpower to handle the required testing work. At present, in India several unproven OCEMS and CAAQMS have been installed by various Indian agencies and industries. Such unproven monitoring systems are not only a major constraint for industries dealing with manufacturing/production of environmental monitoring equipment but also to user organizations/agencies in respect of ascertaining the data quality.

14.3 Potential Impact on National Economy

The global market of environmental monitoring is estimated to be about US$ 3997.8 million. Many market surveys anticipated its Compound Annual Growth Rate (CAGR) during the year 2017–2025 as 7.5% which means, that it is expected to become worth of US$ 71,187 million by the end of 2025. The air pollution monitoring system market in Asia Pacific region, Middle East and Africa is expected to have a huge opportunity due to fueling industrialization as well as establishment of strong environmental regulations. Traditionally the health of the national economy has been measured by GDP (Gross Domestic Product). However, GDP fails to fully capture the contributions of nature to economic activity and human well-being. To address this critical omission, the Gross Ecosystem Product (GEP) measurement has been developed that summarizes the value of ecosystem services in a single monetary metric [26]. In India also, HESCO (Himalayan Environmental Studies and Conservation Organization, based at Dehradun, developed the measure of Gross Environmental Product (GEP), which quantifies the health of the environment. GEP is based on various parameters such as groundwater table, soil erosion, air quality, dissolved oxygen in river water etc. The accurate and traceable measurements of these parameters only can provide accurate GEP values, as quantified products, and hence promote the environmental assets growth as economic growth.

The environmental monitoring in India has assumed a greater importance due to various regulatory requirements put in place for improving the environmental parameters like air quality, control of stack emissions etc. In addition, significant role is being played in India by the judiciary and civil society due to health related concerns. These requirements have resulted in use of sophisticated online equipments for monitoring various parameters entailing a huge investment. As per one of the estimate, amount to the tune of Rs. 5,000 crores (~USD 670 millions) have already been invested in India on installation and maintenance of various equipment like OCEMS, CAAQMS, effluent quality monitors (CEQMS), portable sensors etc. It is also envisaged that in near future another investment to the tune of Rs. 2000 to 3000 crores (~USD 26–40 millions) would be made in India to replace existing systems which will complete their life span and also to augment the monitoring requirements. In addition a large sum is also invested by various research organizations in India for procurement of various monitoring equipment for R&D purposes. It is extremely important that these instruments provide credible data which require a strong metrology support.

Political economy is the study of production and trade and their relations with law, custom and government; and with the distribution of national income and wealth [27]. Stuligross [28] relates Aspects of international trade and diplomatic interaction, Courts, NGOs (Non-governmental organizations) as few of the factors, which put politico-economic constraints on government and business control of environmental regulation. If global regulations were uniformly enforced in India without intelligent optimization, then it would put grave strain on small scale industries [28]. Only High-quality traceable data can bring neutral and balanced optimization in the environmental regulations, so that there is a synergy between the interests of each of the above factors for the larger common good.

14.4 Indigenous Technology Development

In India most of the equipment used for measurement and monitoring of environmental parameters are imported. However, for facilitation of environmental governance, a country like India need to have indigenous equipment which could provide a more affordable option and contribute in required quality data generation for robust decision making. This section describes the technologies developed at the CSIR-NPL with this aim. The potential lies largely in the ability of these technologies to address key issues in pollution measurement and mitigation. In general, the pollution measurements are being carried out with high-end expensive analyzers which are majorly imported from the developed countries. In India, very few companies are indigenously developing high-end pollution monitoring instruments. On the other hand, the sensor-based networks, with their relatively inexpensive monitoring nodes allow us the monitoring with finer spatiotemporal resolution. This resolution can be extremely helpful for more conclusive studies of air pollution's effect on socio-ecological justice and human quality of life. Recently, the air pollution monitoring paradigm is rapidly changing due to technological advances in (1) the development of portable, low-cost air pollution sensors which are capable of reporting data in near-real time with high-time resolution, (2) increased computational and visualization capabilities, and (3) wireless communication/infrastructure.

In the following sub-sections, the technologies developed at CSIR-NPL (for aerosol and gases monitoring) have been discussed.

14.4.1 PM Monitoring Technologies

Aerosol measurement field started after the second half of nineteenth century. Airborne dust particle measurements were started in the 1900s. Later the aerosol sampling methods like Konimeters, Cascade impactors, Impingers, precipitators etc. came in to existence. The above techniques can be called as classical techniques. Most preferred technique for the aerosol measurement among classical techniques was the particle counting and sizing. The technological developments and improvement in the laser, electronics and computer technologies led the development of new improved technologies for the sample and measurement of aerosols.

Measurement of aerosol mass and number concentration is very important for many reasons. The mass concentration measurements are important for the regulatory bodies as a deciding factor in the national ambient air quality standard. The aerosol particle size measurements are very important due to its impact on human health and climate. As the atmospheric aerosols contain several compounds its chemical composition measurement is complicated than size measurement.

14.4.2 3D Scanning Aerosol Lidar for Understanding the Structure of the Atmosphere

Light detection and ranging (LIDAR) is an unique technique in providing the vertical profile of physical and optical properties of aerosols and clouds with good spatial and temporal resolutions. A Mie 3D scanning portable lidar system is under development to study the structure of the atmosphere, by means of elastic scattering of laser light from atmospheric particulates. This LIDAR system will provide us a visualization of several parameters of the planetary boundary layer (PBL) and aerosol distribution. To achieve this, we have to obtain the vertical profiles of the aerosol extinction and backscattering coefficient by means of the LIDAR inversion technique. Apart from this we can obtain other aerosol parameters too i.e. size distribution, shape information etc.

This lidar system is utilizing the pulsed Nd:YAG laser as the transmitter. Nd: YAG laser is emitting 1 mJ pulse at 532 nm with 100 kHz pulse repetition rate. The laser beam is transmitted into the atmosphere after reducing the beam divergence (beam divergence <1 mrad) by a beam expander. The received light is collected by a 20 cm cassegranian telescope in coaxial configuration. Apart from this we are using an optical assembly in this lidar system to discriminate the unwanted light components. The optical assembly includes an orifice, collimator lens, interference filter and a polarization beam cube splitter. This whole assembly of transmitter and receiver is placed inside a scanning box and rotated with the help of a stepper motor having 0.1° resolution. Usually, the lidar receiver consists of an optoelectronic receiver and a data acquisition card. We are using two photomultiplier tubes as the optoelectronic receiver to receive both co and cross polarized light and a FAST COMTEC P7882 multi input scalar as the data acquisition card with 32 bit counting capacity for each bin. This data acquisition system can count up to 350 MHz with 100 ns/200 ns time resolution in single and dual mode respectively.

14.4.3 Twilight Photometer to Study the Upper Troposphere and Stratospheric Aerosols

Measurement of aerosol characteristics can be carried out by ground-based techniques such as sun photometer, lidar, particle impactors, air- borne techniques such as balloons, either with optical particle counters or impactors, and satellites. The continuous measurement on the stratospheric aerosols is sparse over the tropics. The vertical aerosol profile of the lower stratosphere and upper troposphere can be obtained from twilight measurements. Twilight is the scattering of sunlight that happens when sun goes beyond the horizon. The sky brightness from the layer above the shadow of the earth helps in the altitudinal sounding of twilight phenomena. The less expensive Twilight photometer provides the vertical profile of the aerosol in the upper troposphere and stratosphere with good spatial and temporal resolutions. Hence, it is decided to develop a Twilight Photometer to study the aerosol particles in the upper troposphere and stratosphere over Delhi.

14.4.4 Light Weight Scattering Coefficient Counter (LWSCC)

The aerosol scattering coefficients are one of the major inputs to the radiative models for assessing their radiative impacts. Traditionally, the integrating nephelometers are used for measuring the aerosol spectral scattering coefficients. Generally, for measuring the vertical profile of aerosol scattering, it is difficult to use this type of instruments on board airborne platforms (like UAV and small balloons) where payload mass is a big constraint. This is the reason that now the research focus is shifting toward the development of light weight instruments which could be suitable for airborne observations. In view of this growing demand of light weight instruments for airborne observations, the CSIR-NPL designed and developed a LWSCC [weight * 500 g, 7" × 3"] for the measurement of aerosol scattering coefficient at 633 nm wavelength [29]. The performance of LWSCC has been evaluated at the ground level. The performance of LWSCC has been compared with that of high-end AURORA 3000 Nephelometer for two different seasons (monsoon and winter seasons). A strong correlation has been observed for this comparison. In near future, LWSCC could be a potential instrument for generating the vertical profile of aerosol scattering coefficient using airborne platforms which have payload mass constraints. In present condition of LWSCC, high humidity issues have been taken care by using silica gel beads in the instrument but in near future we can use better drying solutions. LWSCC can be modified in future to measure spectral scattering coefficient. At present, it is restricted to single wavelength observation (i.e. 633 nm wavelengths).

14.4.5 Sensor Technology

Current gas sensor market combining various manufacturing technologies holds a billion dollar business globally and is anticipated to grow multi-fold in this decade thanks to the greatly increasing demand of smart and wireless sensors for various applications. Among available gas sensor technologies, MOS based gas sensors are most favoured due to its high sensitivity, simplicity and low-cost production. For developing countries like India, indigenous gas sensor production will be attractive for MSMEs and start-up companies only when the associated technology is commercially viable. Gas sensor technology can be very affordable if the device fabrication is made simpler without involving sophisticated, expensive instruments and complicated, time-consuming processes. In this context, CSIR-NPL is engaged in developing lithography-free metal oxide based chemiresistive gas sensors. The gas sensor device is constructed on thin alumina platform with an attached heater for sensor operation. The schematic of gas sensor structure and fabrication flow is given in Fig. 14.6. The sensor device development includes three major components namely (i) sensing element, (ii) hot plate and (iii) sensor electrode.

Sensing Element: It is the heart of a gas sensor device where the electrical gas detection signal is generated upon reaction with target gases due to change in total

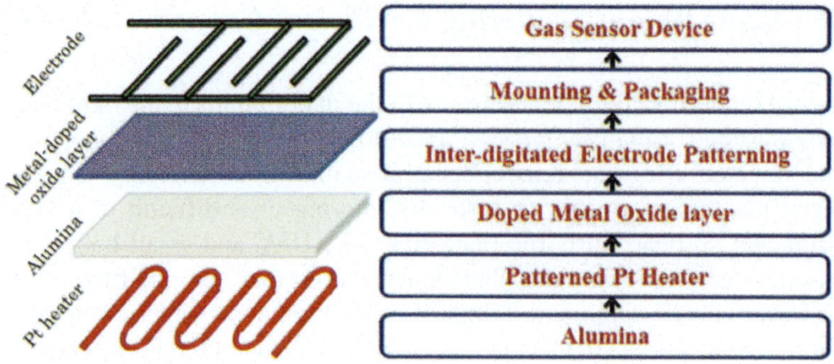

Fig. 14.6 Device structure and fabrication flow of MOS gas sensor developed at CSIR-NPL

charge carrier of the material. Semiconducting metal oxides are potential materials for gas sensing applications as they are very reactive to both reducing and oxidizing gases through their surface adsorbed oxygen ions and offer a high sensitivity. However, MOS gas sensors suffer a poor selectivity of targeted gas as they are reactive to various gases. In order to improve both sensitivity and selectivity, MOS gas sensors are normally doped with noble metals and/or metal oxides as it is believed that the doping enhances the catalytic reduction/oxidation process at microscopic sensor surface. At CSIR-NPL, metal oxide based thin films are prepared using physical vapour deposition processes, which can provide stable sensing materials due to clean deposition conditions. Stable MOS sensing elements are necessary to achieve a good repeatability for reliable measurements and a long life-time. Several metal oxides such as SnO_2, ZnO, CuO, PdO and WO_3thin films and nanostructures are developed for the detection of various target gases. For example, PdO surface-modified SnO_2 thin films are prepared by simple thermal evaporation or sputtering process on alumina substrate aiming at the detection of CO gas molecules. In addition, metal chalcogenide based two-dimensional (2D) materials such as MoS_2 are also prepared for the development of chalcogenide/MOS hybrid sensors.

Sensor Hot Plate: MOS thin films require a high operating temperature up to 500 °C for the efficient detection of gas molecules. High temperature increases the background electron concentration which can assist more oxygen adsorption on MOS surface through the supply of electron. MOS gas sensors are generally fabricated with a hot plate to provide high temperature. In lithography-less process, single-meander Pt based heater is fabricated on the backside of alumina plate using a shadow mask. The Pt deposition is usually performed with help of e-beam evaporation or sputtering technique. The generation of temperature of more than 500 °C is achieved on alumina plate as measured by a thermocouple or an optical pyrometer.

Sensor Electrode: In order to electrically communicate with sensor devices, interdigited finger electrodes (IDE) are formed on MOS thin films. IDE is known for efficient charge collection across the film surface for higher sensitivity and fast response.

The electrode structure is deposited using shadow masks by the evaporation of Au or Pt by physical vapour deposition process.

Gas Sensing: The total size of MOS sensors is determined to be about 4 mm × 4 mm, which is convenient enough to fabricate the device without employing lithography processes. The MOS based gas sensor devices are tested for their sensitivity towards a calibrated CO gas concentration of 915 ppm balanced in nitrogen gas. The sensitivity of the samples is calculated using the following formula [30].

$$\text{Sensitivity (\%)} = \frac{R_a - R_g}{R_a} \times 100 \tag{14.1}$$

where, R_a and R_g are the resistance of sensing layer under dry air and CO gas, respectively. The response/recovery time is determined by the time taken for reaching 90% of resistance change in the response/recovery signal during the presence/absence of CO gas. The PdO modified SnO_2 sensors exhibit a sensitivity of 52% towards CO gas of 915 ppm concentration with a fast response and recovery time of 34 and 46 s, respectively. The typical CO sensing characteristics is shown in Fig. 14.7. As of now, the performance of MOS gas sensors has been tested at laboratory level. Further research and development work to improve the device structure and sensing characteristics are underway.

Sensor Packaging: The final MOS sensor structure with attached heater and electrode are fixed on standard sensor mount and packaged with wire-mesh top cover that allows an easy flow of ambient air for analysis. The MOS sensing element with heater and electrode structures and the packaged discrete sensor device are shown in Fig. 14.8.

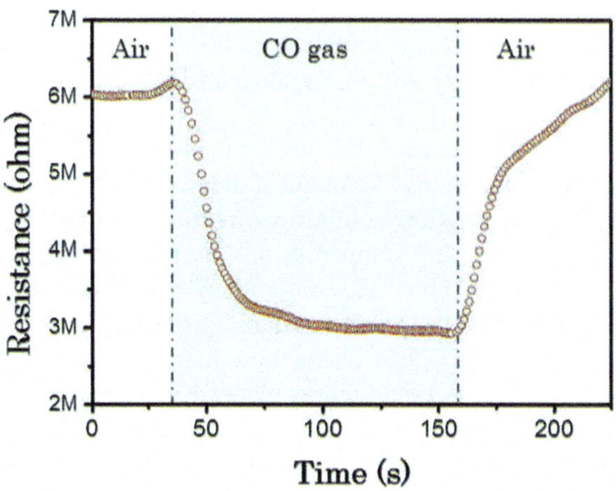

Fig. 14.7 Gas sensing characteristics of PdO/SnO_2 thin films for CO gas of 915 ppm concentration

Fig. 14.8 MOS gas sensor developed at CSIR-NPL. **a** Sensor heater, **b** Sensor element + electrodes and **c** Fabricated sensor device

Testing and Calibration: The major challenge lies in successfully converting the lab-scale sensor process into mass production for commercialization. Mostly, the laboratory-developed technology fails in repeatability and reproducibility of device performance in large-scale production due to process scaling-up. Hence, the stability of sensor elements and device structure needs to be ensured for long-term operation without any degradation of sensing characteristics. MOS sensor films developed at CSIR-NPL are rigorously tested for their temporal stability of gas response over a period of over a period of months. In addition, sensor metrology is to be adopted to realize accurate and precise measurements through sensor testing and calibration under required environmental conditions. Thus, the lab-level sensor technology backed with sensor metrology will likely have a higher success rate for industrial production.

The growth of gas sensor market is driven by the increased adoption of gas sensor products for various applications like medical diagnosis, emission level monitoring in industries, building automation, combustion monitoring, food quality control, etc. At present, in India, only a few companies are involved in manufacturing electrochemical based gas sensors, however, no company is commercially manufacturing MOS based gas sensors due to lack of availability of industry friendly technology. But, considering the larger population of the country, there is a massive requirement for gas sensor products for usage in places like industries, weather stations, offices, restaurants, homes, etc. and therefore, gas sensors has a potential market locally. Currently, a huge volume of solid-state gas sensors are being imported for the fabrication of various gas leak detectors and related products in India. Hence, the development of indigenous technology for gas sensor fabrication will enable the Indian industries to produce gas sensors domestically and thus can save a lot of foreign exchange.

14.4.6 Light Weight Gas Sensor System (LWGSS)

Traditionally, there are various analyzers available in the markets for measurements of gas of interest but if we are interested for the vertical profile of different gases, it becomes very difficult to accommodate these high-end analyzers in the payload due to payload mass constraints especially for UAV and light weight tethered balloons.

In view of this, NPL research group has designed and developed a LWGSS [350 g, 7" × 3"] equipped with low price electrolytic gas sensors [31]. LWGSS is capable of measuring major gaseous pollutants (such as CO, SO_2, NO_2, O_3 and H_2S) present in the atmosphere using Alphasense electrochemical gas sensors. LWGSS has been designed in such a way that it could be suitable for vertical observations using balloons/UAVs. Before using LWGSS in field observations, it becomes imperative to compare this with that of collocated high-end gas analysers. The performance of this laboratory developed instrument has been compared with that of high-end analyzers measuring different gases. For ozone, it has been compared with that of analyzer at CSIR-NPL while for other gases it has been compared with analyzers available at IMD, New Delhi. A good correlation was observed for the ozone sensor deployed in LWGSS against the high-end ozone gas analyser at CSIR-NPL. LWGSS can be modified in near future to accommodate more sensors for measurement of different gases of interest.

14.4.7 High-volume $PM_{2.5}$ Sampler Technology

Particulate matter (PM) is one of the most important parameter of ambient air quality standards. Depending on particle aerodynamic size, it is monitor as $PM_{2.5}$ and PM_{10}. However, in view of the impact to the health, $PM_{2.5}$ is of more concern. Further, according to air quality standards, low-volume sampling (with a flow rate 16.67 lpm, on filter size 47 mm diameter) is sufficient to perform $PM_{2.5}$ mass concentration measurement. Also this low-volume sampler is easy to be used even at remote sites. On the other hand, if particulate matter is also needed to be analysed for chemical specie or toxicology studies, or the sampling is to be performed where the particulate concentration is very low, such as pristine regions or polar regions, a high-volume sampling with a flow rate 1130 lpm and filter size 203 mm × 254 mm is essentially required which is the specification of high-volume sampler approved by US-EPA.

This technology of **High-Volume $PM_{2.5}$ Impactor Sampler** involves a novel design of $PM_{2.5}$ impactor sampler, which segregates PM >2.5 μm size and facilitate to collect ≤2.5 μm size particles suspended in ambient air on filter size 8" × 10" with a high-flow rate (1.13 m^3/min). The developed impactor design is simpler to

Fig. 14.9 High-volume $PM_{2.5}$ sampler launched by honourable minister Dr. Harsh Vardhan

be fabricated and such as it can be retrofitted to existing PM_{10} or TSP standard sampler mentioned in air quality standards of many countries including US-EPA. This technology has already been launched (Fig. 14.9) and commercialized. Overall this sampler technology has unique specifications and advantages for the regulatory measurement of $PM_{2.5}$ mass concentration:

1. Large filter size (8" × 10 ") and high flow rate >1 m^3/min to reduce sampling time, minimizing sampling errors caused by D50 cutoff shift and collection of enough material on filter, hence relatively low sensitive weighing balance can be used to correctly measure $PM_{2.5}$ mass concentration.
2. Collection of a huge amount of particulate material which is also needed to extract enough organics for identification, quantification and bioassay testing.
3. Designing of impactor should be cost effective and retrofitted to existing PM_{10} HVS.
4. Separation characteristics (D50 cutoff and GSD) should be equivalent to low-volume separators, such as WINS impactor and VSCC cyclone.
5. Separation characteristics should not change even at high particle mass loading condition.
6. Easy to maintain with no movable part.

14.5 Training and Academic Services

Skill and knowledge are the driving forces for the economic growth and social development of any country. The National Skill Development Mission has been developed by Ministry of Skill Development and Entrepreneurship (MSDE) to create convergence across sectors and States in terms of skill training activities. CSIR-NPL is disseminating knowledge and increasing global measurement skills and competence in the field of environmental metrology by providing training and academic courses. A wide range of technical training courses emphasizing on skill development are provided by the learned faculty of environmental metrology with practical and hands-on learning.

14.5.1 Training Courses

14.5.1.1 Training Programme on Air Quality Measurements

The Gas metrology faculty provided a skill development training on Air Quality Measurements which emphasizes on the national ambient air quality standards (NAAQS) parameters and working gas standard mixture preparation, calibration of analysers and estimation of uncertainty in the measurements. The course is developed keeping in mind about the employers who work in state and central pollution control laboratories, gas industries and air monitoring stations. The training course is taught in clear and well thought out manner with practical and hands-on training using air monitoring analysers. It includes the hands-on training on gas analysers, Beta Attenuation Monitoring (BAM) and particulate measurements. BAM is used for the detection of PM_{10} and $PM_{2.5}$ which are monitored by most of the air pollution regulatory agencies including centre and state pollution control laboratories and are the critical parameters of NAAQS. Technical aspects for quality assurance are also taught in the training.

14.5.1.2 Training Programmes in Gas Sensors Fabrication Technology

Sensor devices and metrology team can provide a skill development training course on the complete gas sensor device fabrication for practical application viz. integrating the sensor layer, sensitizing layer, inter-digitized electrode and micro-heater. This includes hands-on training on the following;

(i) Growth of thin films and nanostructures of different kinds of materials on various substrates by Physical Vapour Deposition and Chemical Vapour Deposition techniques for the development of sensing layers and sensitizer layers. Thin films and nanostructures materials were developed in order to improve/enhance the sensitivity, selectivity, response/recovery time and stability of gas sensor.

(ii) Fabrication of inter-digitized electrodes and the fabrication of Pt based micro-heater by Physical Vapour Deposition techniques for the measurement of sensing performance.

14.5.1.3 Training and Service for Characterizations of Materials Properties

The performance of these devices is dependent on the materials properties along with their fabrication methods and environment conditions. The interaction between the sensing layer and gas pollutants results in changes in the chemical properties of the sensing material at elevated temperatures, leading to a change in sensor device resistance. Therefore, the characterizations of material properties/sensor devices are essential for measuring and improving the sensing performance. Hands-on training can be provided on the characterization techniques used for measuring the materials properties/device performance (i) Sensing measurement set-up, (ii) X-ray photo-electron spectroscopy (iii) Photoluminescence (iv) Spectroscopic Ellipsometer, (v) Stylus Profiler (vii) UV-VIS-NIR Spectrometer, etc.

14.5.1.4 Training on Sensor Metrology

Gas sensor devices require regular maintenance to check and secure proper functioning since gas sensors are prone to degradation depending upon the sensing materials and environmental conditions. As a result, the performance parameters of gas sensor may not be reliable over a period of time. Validation and periodic testing of gas sensors are essential to ensure precise and accurate measurement for obtaining reliable data. This demands the requirement of sensor metrology. Training and services in sensor metrology to academic/research institutes will enhance the quality of related R&D activities, which in turn, will help to bring out reliable technologies for gas sensor manufacturing. Also, the related training and services to industries will benefit "Start-up India", "Made in India" and SME/MSME (Small and medium-sized enterprises/Ministry of Micro, Small and Medium Enterprises) programmes of Govt. of India.

14.6 Conclusions and Future Perspectives

The environment is an important issue for ecosystem as it has direct bearing on its existence. In present days, anthropogenic activities like industrialization, agriculture and urbanization are significantly impacting the environment resulting in deteriorating air quality, climate change and their adverse impacts on various components of ecosystem. These also have serious economic implications for individuals and countries. The key to develop proper understanding for all these causes and their influences are the accurate and reliable measurements of various physical and chemical properties that play major roles in the complex interplay of various mechanisms. World over, a number of efforts are being made to monitor and measure various parameters in this context. The metrology is an integral part of all these efforts as it will ensure reliability and accuracy of generated data.

In India, a large number of stakeholders including policy makers, regulators, judiciary, industries, academia, research institutions, civil society play their respective roles in environmental governance. For environmental monitoring, which is essential for both policy formulation as well as for policy compliance, a huge investment in terms of instrument and man power has already been made in India which is expected to grow many folds in future as the pace of economic development increases necessitating further augmentation for enhanced monitoring and compliance ascertainment. CSIR-NPL has been involved in measurement of a numbers of physical and chemical parameters of atmosphere over last few decades and participated in many national and international research programmes in this field. It is also providing assistance to stakeholders in the reliable and accurate measurements of certain atmospheric parameters through the primary reference gas standards BND® for various GHGs and pollution gases and Standard Reference Photometer (SRP) for ozone (Please refer to Chap. 13). However, in recent years, the CSIR-NPL is developing technical capabilities to meet the new challenges like testing, calibration and certification of air pollution monitoring equipment; testing and calibration of gas sensors etc. which would help in obtaining reliable and accurate data from various measurements besides helping industries in their efforts to develop indigenous equipment and getting benefitted from the vast market potential that exist for such instrument in India and abroad. CSIR-NPL is also putting its efforts to develop human resources to help in getting reliable and accurate environmental measurements by imparting skill training.

14.7 Annexure I

List of important Environment Legislation, Acts, Rules, Notifications and Amendments.

S. No.	Environmental amendment act/law	Year
1	Water (prevention and control of pollution) act	1974
2	The water (prevention and control of pollution) cess act	1977
3	Forest (conservation) act	1980
4	Air (prevention and control of pollution) act	1981
5	Environment protection act	1986
6	Air (prevention and control of pollution) amendment act	1987
7	Motor vehicles act	1988
8	Hazardous waste handling and management act	1989

(continued)

(continued)

S. No.	Environmental amendment act/law	Year
9	The public liability insurance act and rules and amendment	1992
10	The national environment tribunal act	1995
11	The national environment appellate authority act	1997
12	The biomedical waste (management and handling) rules	1998
13	The environment (sitting for industrial projects) rules	1999
14	The ozone depleting substances (regulation and control) rules	2000
15	Noise pollution rule	2000
16	The batteries (management and handling) rules	2001
17	The biological diversity act	2002
18	The noise pollution (regulation and control) amendment rules	2010
19	Plastic waste (management and handling) amendment rules	2011

14.8 Annexure-II

Normative references for NPL India Certification Scheme.

S. No.	Reference
1	ISO-17025: general requirements for the competence of testing and calibration laboratories
2	ISO-17021: conformity assessment—requirements for bodies providing audit and certification of management systems
3	ISO-17065: general requirements for the competence of certification body
4	ISO-9001: quality management systems—requirements
5	EN 14,211 (or Indian equivalent): ambient air quality—standard method for the measurement of the concentration of nitrogen dioxide and nitrogen monoxide by chemiluminescence
6	EN 14,212 (or Indian equivalent): ambient air quality—standard method for the measurement of the concentration of sulphur dioxide by ultraviolet fluorescence
7	EN 14,625 (or Indian equivalent): ambient air quality—standard method for the measurement of the concentration of ozone by ultraviolet photometry
8	EN 14,626 (or Indian equivalent): ambient air quality—standard method for the measurement of the concentration of carbon monoxide by non dispersive infrared spectroscopy
9	EN 14,662-3 (or Indian equivalent): ambient air quality—standard method for measurement of benzene concentrations—part 3: automated pumped sampling with in situ gas chromatography

(continued)

(continued)

S. No.	Reference
10	EN 15,267-2 (or Indian equivalent): air quality—certification of automated measuring systems—part 2: initial assessment of the AMS manufacturer's quality management system and post certification surveillance for the manufacturing process
11	EN 15,267-3 (or Indian equivalent): air quality—certification of automated measuring systems—part 3: performance criteria and test procedures for automated measuring systems for monitoring emissions from stationary sources
12	EN 15,267-1 (or Indian equivalent): general requirements for bodies operating product certification systems

Acknowledgements Authors are thankful to Ms. Beena Gupta, Mrs. Saroj Gandhi, Mr. Alok Mukherjee, Dr. Khem Singh, Ms. Smriti Tiwari Singh, Mr. Devesh Kumar Shukla, Mr. Lalit Goswami, Mr. Jitender Kumar, Ms. Abha Shukla, Mr. M.L. Arora, Mr. Suresh Chandra Yadav, Mr. Saket Vihari and Ms. Kamla for their support.

References

1. D.C. Parashar, A.P. Mitra, P.K. Gupta, J. Rai, R.C. Sharma, N. Singh, S. Koul et al., Methane budget from paddy fields in India. Chemosphere **33**, 737–757 (1996). https://doi.org/10.1016/0045-6535(96)00223-8
2. A.P. Mitra, Global change greenhouse gas emissions in India—methane budget estimates from rice fields based on data available up to 1995. Sci. Rep. UK **10** (1996)
3. M.G. Lawrence, J. Lelieveld, Atmospheric pollutant outflow from southern Asia: a review. Atmos. Chem. Phys. **10**, 11017 (2010). https://doi.org/10.5194/acp-10-11017-2010
4. V. Ramanathan, C. Chung, D. Kim, T. Bettge, J.T. Kiehl, W.M. Washington, Q. Fu, D.R. Sikka, M. Wild, Atmospheric brown clouds: Impacts on South Asian climate and hydrological cycle. PNAS USA **102**, 5326–5333 (2005). https://doi.org/10.1073/pnas.0500656102
5. S. Singh, B. Singh, B.S. Gera, M.K. Srivastava, H.N. Dutta, S.C. Garg, R. Singh, A study of aerosol optical depth in the central Indian region (17.3–8.6N) during ISRO-GBP field campaign. Atmos. Environ. **40**, 6494–6503 (2006). https://doi.org/10.1016/j.atmosenv.2006.01.033
6. T. Vinod, S.N. Tripathi, N. Chinnam, A.K. Srivastava, S. Dey, M. Manar, V.P. Kanawade et al., Measurements of atmospheric parameters during Indian space research organization geosphere biosphere program land campaign II at a typical location in the Ganga Basin: 2. chemical properties. J. Geophys. Res. Atmos. **111**, D23 (2006). https://doi.org/10.1029/2006JD007279
7. P.K. Gupta, K. Singh, C.K. Dixit, N. Singh, C. Sharma, S. Sahai, A.K. Jha, D.P. Singh, M.K. Tiwari, S.C. Garg, Spatial distribution in aerosol mass and size characteristics between Delhi and Hyderabad during land campaign in February 2004 (2007)
8. M.C. Lemos, A. Agrawal, Environmental governance. Annu. Rev. Environ. Resour. **31**, 297–325 (2006). https://doi.org/10.1146/annurev.energy.31.042605.135621
9. O.P. Dwivedi, B. Kishore, Protecting the environment from pollution: A review of India's legal and institutional mechanisms. Asian Surv. **22**, 894–911 (1982). https://doi.org/10.2307/2643803

10. S. Shrotria, Environmental justice: is the national green tribunal of India effective? Environ. Law Rev. **17**, 169–188 (2015). https://doi.org/10.1177/1461452915595548
11. G.N. Gill, Environmental justice in India: the national green tribunal and expert members. Transnatl. Environ. La. **5**, 175–205 (2016). https://doi.org/10.1017/S2047102515000278
12. J. Khandare, Role of national green tribunal in protection environment. Indian J. Appl. Res. **4**(12) (2015). https://doi.org/10.2139/ssrn.2713728
13. P. Bera, M.S. Hegde, Recent advances in auto exhaust catalysis. J. Indian I. Sci. **90**, 299–325 (2012)
14. G. Elavarasan, M. Kannan, D. Karthikeyan, History of emission standards in India—a critical review. IJRAR **6**, 28–35 (2019)
15. P. Goyal, Sidhartha, Present scenario of air quality in Delhi: a case study of CNG implementation. Atmos. Environ. **37**, 5423–5431 (2003). https://doi.org/10.1016/j.atmosenv.2003.09.005
16. N.O. Nylund, A. Lawson, Exhaust emissions from natural gas vehicles. IANGV Emission report, 1–19 (2000). https://citeseerx.ist.psu.edu/viewdoc/download?doi=10.1.1.201.5675&rep=rep1&type=pdf
17. S.H. Kota, S. Sharma, M. Khare, H. Zhang, Estimating the influence of feasible interventions on potential decrease in PM 2.5 concentrations in Delhi-NCR using a one-way nested regional and urban air quality modelling framework. AGUFM, A21G-2655 (2019)
18. V.S. Yadav, Environmental Impact assessment: a critique on Indian law and practices. Int. J. Multidiscip. Res. Dev. **5**, 1–5 (2018)
19. R.S. Mahwar, N.K. Verma, S.P. Chakrabarti, D.K. Biswas, Environmental auditing programme in India. Sci. Total Environ. **204**, 11–26 (1997). https://doi.org/10.1016/s0048-9697(97)00161-7
20. D.P. Tripathy, Environmental auditing for sustainable development of Indian industries. Asian J. Water Environ. Pollut. **8**, 9–20 (2011)
21. A. Najam, M. Papa, N Taiyab, Global environmental governance: a reform agenda (2006). https://agris.fao.org
22. N. Ahmad, The role of civil society institutions in environmental governance in India: post-colonial context and human rights challenges in the environmental justice. Int. J. Leg. Stud. Res. 16–39 (2018)
23. CPCB, Guidelines for Recognition of Environmental Laboratories under the Environmental (Protection) Act, 1986. Laboratory Analytical Techniques Serieslats/9/2008–2009 (2008). https://parivesh.nic.in
24. MCERTS performance standards for ContinuousEmission Monitoring Systems, Version 3.5 (2016), https://www.gov.uk/government/publications/mcerts-performance-standards-and-testprocedures-for-continuous-emission-monitoring-systems
25. MCERTS Performance Standards for Continuous Ambient Air Quality Monitoring Systems, Version 10 (2016). https://assets.publishing.service.gov.uk/government/uploads/system/uploads/attachment_data/file/532491/LIT_7050.pdf
26. Z. Ouyang, C. Song, H. Zheng, S. Polasky, Y. Xiao, I.J. Bateman, J. Liu, M. Ruckelshaus, F. Shi, Y. Xiao, W. Xu, Using gross ecosystem product (GEP) to value nature in decision making. PNAS **117**, 14593–14601 (2020).https://doi.org/10.1073/pnas.1911439117/-/DCSupplemental
27. V. Bladen, An introduction to political economy (1956). ISBN 978-1-4426-3210-3. OCLC 1013947543
28. Stuligross, The Political Economy of Environmental Regulation in India. Pac. Aff. **72**, 392–406 (1999). https://doi.org/10.2307/2672228
29. A. Ahlawat, S.K. Mishra, S. Gumber, V. Goel, C. Sharma, A. Wiedensohler, Performance evaluation of light weight gas sensor system suitable for airborne applications against co-location gas analyzers over Delhi. Sci. Total Environ. **697**, 134016 (2019). https://doi.org/10.1016/j.scitotenv.2019.134016

30. A.K. Mauraya, D. Mahana, P. Pal, S. Muthiah, P. Singh, S.K. Muthusamy, Effect of bulk and surface modification of SnO_2 thin films with PdO catalyst on CO gas sensing characteristics prepared by vacuum evaporation process. J. Alloys Compd. **843**, 155979 (2020). https://doi.org/10.1016/j.jallcom.2020.155979
31. A. Ahlawat, S.K. Mishra, S. Gumber, V. Goel, V.K. Soni, C. Sharma, Performance analysis of light weight scattering coefficient counter with AURORA 3000 nephelometer over Delhi. Mapan **35**, 213–219 (2020). https://doi.org/10.1007/s12647-019-00361-5

Chapter 15
Biomedical Metrology: Role in Nation's Healthcare Sector

Rajesh, Sumana Gajjala, Ved Varun Agrawal, and D. K. Aswal

Abstract Metrological assurance is decisive in medical equipment calibration and is indispensable in quality control and assurance, risk assessment, monitoring of the patients, diagnosis and treatment of patients. As a part of health care regulations, the biomedical equipment must demonstrate compliance with metrological traceability to SI unit chain through a chain of calibration, verification and certification on the performance of the equipment. In this context, CSIR-NPL is serving the country by providing calibration facilities traceable to SI units to support the quality regulation of Medical equipment. The Biomedical Metrology program at CSIR-NPL contributes enormously to the economic development and healthcare sector of the country by way of providing calibration and consultancy services to healthcare sector that includes accredited laboratories, regulatory bodies, hospitals and testing laboratories, etc. In this chapter, we have elaborated the role of regulatory bodies of India and National Metrology Institute, CSIR-NPL in context of the "New Medical Device Rule". We also critically analyse the importance and implementation of Medical Device policy and Metrology, its significance on healthcare sector and the quality infrastructure of Medical devices. The contribution of CSIR, NPL in strengthening and supporting the medical device policy, as well as the establishment of medical device calibration facilities and the development of diagnostic devices through R & D are highlighted and discussed briefly.

It amount to a say that progress in the practical arts of medicine in any of its branches whether preventive or curative, only comes from the growth of accurate knowledge as it accumulates in the laboratories and studies of various sciences.

– Walter Fletcher (1892-1956)

Rajesh (✉) · S. Gajjala · V. V. Agrawal · D. K. Aswal
CSIR-National Physical Laboratory, Dr. K.S. Krishnan Marg, New Delhi 110012, India
e-mail: rajesh@nplindia.org

D. K. Aswal (ed.), *Metrology for Inclusive Growth of India*,
https://doi.org/10.1007/978-981-15-8872-3_15

15.1 Introduction to Biomedical Metrology

It amounts to a say that progress in the practical arts of medicine in any of its branches whether preventive or curative, only comes from the growth of accurate knowledge as it accumulates in the laboratories and studies of various sciences—Walter Fletcher (1892–1956).

Implementation of Quality infrastructure boosts the healthy competition in the industrial sector and ensures the safety of the products, which facilitates the ease in international trade and the protection of consumers. Accuracy and precision in the functioning of medical devices used for diagnostic and treatment in the medical area are of paramount importance as the quality and maintenance of the biomedical devices are directly linked to the health care system of a Nation. The calibration for the medical devices used in the healthcare sector, where the measurement results are traceable to SI units are vital for the maintenance of proper healthcare system, prevention and control of health ailments and diseases. Calibration and metrology in this area result in reliable measurements and thus have innumerable benefits in the health sector. Therefore, the efficacy of quality control steps depends directly on the accuracy and precision of the testing and measurement procedure, the quality control and periodic calibration and testing of the measuring equipment. To prevent Medical device errors, it is essential to assess the performance of the equipment and get reliable measurements. In present times, a large portion of the global economy and health care sector depends on accurate and precise measurements and tests with stated uncertainty, which is trusted and accepted internationally [1]. The business and the national economy depend on the manufacturing and production of the quality-controlled products in the health sector such as medical electronics through sustained innovation, calibration, testing and accreditation traceable to international technical standards. Metrological assurance is very crucial in medical equipment calibration and is indispensable in quality control and assurance, risk assessment, monitoring of the patients, diagnosis and treatment of patients. As a part of health care regulations, the biomedical equipment must demonstrate compliance with metrological traceability chain through the calibration and certification of the performance of the equipment. These certificates will be valid if issued by a laboratory accredited for the activities concerned traceable to the SI units. In this context, the development of biomedical equipment standards supports the National economy by strengthening the quality infrastructure and thereby minimizing the dependence on other International NMIs for Quality assurance.

Indian Medical Device Industry sector is growing fast and its current market size is at least $11 billion. The sector growing rapidly at an unprecedented scale and the statistical estimates indicate that the market size is expected to reach $280 billion by 2025 mainly due to the building up of quality infrastructure in the nation [2]. In 2015, India permitted 100 percent FDI through automatic route in Medical device sector. This step resulted in the steep increase in the FDI to 430 million dollars in 2016. A total of USD 2.1 billion foreign direct investment (FDI) has come to India in the medical device sector during the last ten years [3]. The significant growth in FDI inflow in the recent past is certainly helping and encouraging new R&D activities, start-ups and manufacturing innovations, in the area of medical electronic products. The development of new advancements in the medical device technologies and their product quality implementation has led to the enhancement in the Indian medical tourism. During the last few years, there is significant acceleration in International patients travelling to India to avail the cost-effective and quality health care facilities. The accreditations to hospitals and clinics are provided by Joint Commission International (JCI) and National Accreditation Board of Hospitals (NABH) to maintain the quality healthcare structure and to achieve a safe environment for patients, staff and visitors.

Accreditation for medical electronics and clinical diagnostics is a boon and highly beneficial for hospitals in achieving their desired quality standards. Indian Health ministry has published New Medical Device Rule 2017 with effective medical equipment regulations on 31.01.2017 under Drugs and Cosmetics Act, which are being implemented with effect from January 1, 2018.

These regulations emphasize the critical requirement of quality control and periodic calibration checks of medical equipment [4]. In this context, CSIR-NPL, National Metrology Institute is serving the country by providing calibration facilities traceable to National standards to support the quality regulation of Medical equipment.

15.2 Medical Device Policy and Metrology

In India, various acts have been formulated time to time to strengthen the National quality infrastructure in healthcare sector and a few among them are Legal metrology act 2009, New Drugs and Clinical Trials Rules, 2019, Food Safety and Standards Rules, 2011, Food Safety and Standards Act, 2006 (Act 34 of 2006), etc. All these have served the very purpose for which they were formulated. However, in recent years, world has seen changes to meet their national requirements. India imports almost 80% of its medical instruments every year, and one of the reasons is the lack of quality infrastructure in medical sector.

In spite of above-said acts, the quality of imported equipment was not regulated and this resulted in a much-needed act on the regulation of medical devices. Development of medical device policy has been the most critical and challenging task, not just for our country but across the globe. As per the survey of WHO in the year 2010, among the low-income countries, 33% had a national policy for health technology, leaving out 67% of countries having no defined policy. Whereas, around 55% of the countries have a notified health regulation authority for the enforcement and implementation of medical device regulations and 85% of the countries have a designated unit within the health ministry which is responsible for managing the technical and quality issues of the medical devices [5]. One of the major challenges in formulating policy was that the medical device ranges from very sophisticated equipment to ordinary medimeter/ruler as measuring tool. Some are life-saving, while some have an indispensable role in operations. Till date, variety of different international classification systems is being followed for medical devices in the world. This irregular classification of devices is one of the prime reasons for not having a uniform defined policy globally. Indian medical device policy classifies them in four major categories, on the basis of their risk assessment and a detailed classification is discussed in section 'Role of Legal Metrology in Medical Device Sector'. However, this qualification is as per their intended use and not the technology involved in manufacturing, operation or calibration. Every country has its own norms for classification and challenges and required various degrees of skills for operation and even maintenance. However, considering its relevance in patient safety it is vital to achieve harmonization in the nomenclature of medical devices.

Proper functioning and calibration of the medical devices over time is another major challenge, where the metrological assurance plays its critical role. The functioning of healthcare quality infrastructure is represented in Fig. 15.1 showing the government policies such as the medical device rule 2017, which defines the role of primary service providers of quality infrastructures whose services are utilized by the industries and private labs to support quality health structure of the Nation. It is the unbroken chain of quality assurance that ultimately governs the quality of life. Thus, metrological traceability of medical device is an essential component that ensures the functioning of the device, and therefore has been given utmost importance in Medical device policy 2017.

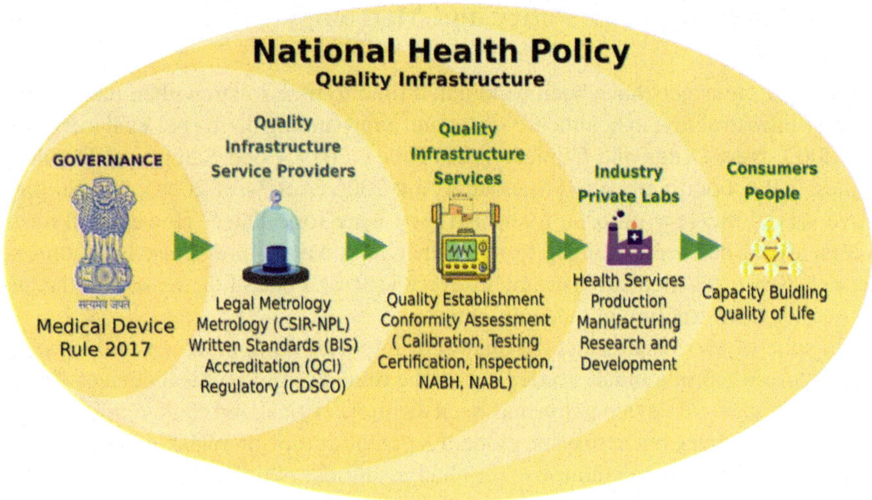

Fig. 15.1 Schematic of India's quality infrastructure in health sector demonstrating the impact of policies such as medical device rule 2017 on quality of life

15.2.1 Medical Device Quality Infrastructure

Keeping in view of rapid growth in Indian economy, its size and significance in global trade, it is highly desirable to establish a robust 'Quality Infrastructure' in India. This would stimulate the economic growth and boost the 'Made in India' program. The "standards" are considered as catalysts for technological development, industrial growth and well-being of the society. The role of standards and regulations have been well recognized in global trade through international agreements like World Trade Organization (WTO). In India, strategies have been formulated to coordinate the standardization work with technological, societal, and economic development at the national level and efforts are being made for global standardization through various regulatory bodies as discussed below.

15.2.1.1 Regulatory Bodies

The Indian medical equipment and medical electronics sector is regulated by four regulatory bodies (i) Ministry of Health and Family Welfare, (ii) Drug Controller General India (DCGI), (iii) Bureau of Indian Standards (BIS) and (iv)Atomic Energy Regulatory Board (AERB). However, In India, "Medical Devices" have no separate regulatory status and are under the Drugs and Cosmetics Act, 1940, and Central Drugs Standard Control Organization (CDSCO) is responsible for its regulation in coordination with Department of Consumer Affairs for labeling and declaration requirements for notified medical devices under the Legal Metrology Act, 2009 [6] and the Legal Metrology Rule (Packaged Commodities) Rules, 2011 and consequent legal metrology rule 2019 [7].

Role of CDSCO and Medical Device Rule 2017

Indian Medical Device sector is regulated by The Drugs and Cosmetics Act, 1940 ("DCA/Act") and its enforcement is the responsibility of the Central and the state Governments. Drugs Controller General of India ("DCGI") of "CDSCO" ascertains the formulation of policies and implementation of the act throughout the country by coordinating with State Drugs Licensing Authorities.

Prior to year 2006, the manufacturers from India and abroad could sell the medical devices in Indian market without any restrictions, due to the absence of necessary regulation. However, since 2006, as per the regulations of CDSCO, before marketing the notified medical equipment must comply with the Indian Medical Device Regulation described by CDSCO. This organization handles regulation and approval of new drugs, clinical trials, quality control of the drugs, coordinating with state drug control organizations by providing expert opinion about the uniformity and enforcement of Drugs and cosmetic act.

As per the CDSCO guidelines, the conformity assessment should be done and the guidelines are different for different classes of devices as per the classification. For Class A (low risk) and B devices (low moderate risk), the medical equipment manufacturers are permitted to perform their own conformity assessment procedures, whereas for other classes (Class C, moderate high risk and D devices, high risk), central licensing approval authority in consultation with the BIS notified the approved list of authorized laboratories or bodies to perform conformity assessment. There is a need to submit an application for the assessment of class C and D devices by the medical device manufactures to any of these authorized notified laboratories [8].

Since Medical Devices are high on innovation, India is becoming the world's preferred medical tourism place, therefore, defining an appropriate framework for standards and calibration of medical instruments becomes imperative. This has to lead the center to unveiled a Medical Device Rule 2017, which came in to force with effect from January 2018, for regulating manufacturing/Import/sale/Clinical investigation (*Rule-19 in chapter-III, of Medical Device Rule, 2017, prescribes that no Medical device Testing laboratory, shall be so designated unless it has been duly accredited by NABL or any other similar body*) and other related issues concerning Medical devices. CDSCO has introduced risk based classification of various medical devices as Class A, Class B, Class C and Class D and defined the product standards. Besides this, the organization has introduced single-window clearance and defined the timelines for issuing licenses to the Medical device companies and industries to facilitate the high-quality end products and accelerate "Make in India" Program. In addition, it also consolidated the requirement of obtaining import license and registration certificate into a single license for foreign manufacturers [4]. CDSCO has amended Medical Device rules to simplify the rules for easy regulation of Medical devices to obtain licensing and registration thereby accelerating the manufacture and export of indigenously made medical devices from India. On February 11, 2020, the Ministry of Health and Family Welfare (MoH and FW) issued two notifications in the Indian Gazette [9, 10] on registration rules of medical devices and new description of medical devices. The Medical Devices (Amendment) Rules, 2020, the latter amends the Medical Devices Rules, 2017, and has been effective from April 1, 2020. The Ministry after consultation with Drugs Technical Advisory Board (DTAB) has intended for a cumulative effect of both these notifications to ensure that all medical devices must meet certain standards, safety measures and quality and for mandatory registration of all medical devices as per the CDSCO enacted act 1940. Until February 11, 2020, only 37 medical devices have been notified as drugs by CDSCO [4], which are shown listed in Table 15.1. CSIR-NPL provides measurement traceability for physical and electrical parameters to some of the medical devices listed in the table are traceable to primary standards maintained at CSIR, NPL like disposable hypodermic syringes, blood monitoring devices, digital thermometers, defibrillators, MRI equipment and ultrasound equipment, whereas the measurement traceability to X-Ray machine are being provided by Atomic Energy Regulatory Board (AERB).

As per the new Medical device rule, Medical devices which come under the following classification (Fig. 15.2) will be regulated as 'drug' under the DCA and MDR.

Table 15.1 A list of 37 notified medical devices before 11th February, 2020

S. No.	Medical devices
1	Disposable hypodermic syringes
2	Disposable hypodermic needles
3	Disposable perfusion sets
4	Substances used for in vitro diagnosis including blood grouping sera
5	Cardiac stents
6	Drug-eluting stents
7	Catheters
8	Intra ocular lenses
9	I.V. Cannulae
10	Bone cements
11	Heart valves
12	Scalp vein set
13	Orthopedic implants
14	Internal prosthetic replacements
15	Ablation devices
16	Ligatures, sutures and staplers
17	Intra uterine devices (Cu-T)
18	Tubal rings
19	Condoms
20	Surgical dressings
21	Umbilical tapes
22	Blood/blood component bags
23	Organ preservative solution
24	Nebulizer
25	Blood pressure monitoring device
26	Glucometer
27	Digital thermometer
28	All implantable medical devices equipment
29	CT scan equipment
30	MRI equipment
31	Defibrillators
32	PET equipment
33	X-ray machine
34	Dialysis machine
35	Bone marrow cell separator
36	Disinfectants and insecticide
37	Ultrasound equipment

"All devices including an instrument, apparatus, appliance, implant, material or other article, whether used alone or in combination, including a software or an accessory, intended by its manufacturer to be used specially for human beings or animals which does not achieve the primary intended action in or on human body or animals by any pharmacological or immunological or metabolic means, but which may assist in its intended function by such means for one or more of the specific purposes of – (i) diagnosis, prevention, monitoring, treatment or alleviation of any disease or disorder; (ii) diagnosis, monitoring, treatment, alleviation or assistance for, any injury or disability; (iii) investigation replacement or modification or support of the anatomy or of a physiological process; (iv) supporting or sustaining life; (v) disinfection of medical devices; and (vi) control of conception."

Fig. 15.2 Definition of medical device as per amendment in the act

All the medical devices other than 37 will now be referred as "Newly Notified Medical Devices" under the new definition of medical devices. "The new medical device Rules 2020 has been introduced by CDSCO to implement quality healthcare structure in India to ensure that every medical device either manufactured in India or imported, have quality assurance before marketing. According to newly notified medical device rule, all the medical device manufacturers, whether indigenous or imported must mandatorily register for the certificate of compliance as per the ISO-13,485 guidelines. The Medical devices and Quality Management Systems and their Requirements for Regulatory Purposes are to be compliant with ISO 13,485 guidelines. The registered company or the manufacturers mandatorily require to maintain and implement the quality management system by proper documentation according to ISO 13,485 guidelines and by conducting periodic auditing.

For the licensing and registration, medical devices are categorized on the basis of the prescribed risk. The suitable timeline for the registration as per the risk-based category shall serve as a breather for pharmaceutical companies. It opens a wider scope to test and modify such devices, which fall under the ambit of the new definition, post-introduction upon the relevant market. Under section 3 of the BIS Act, 1985 (63 of 1985) [11], the quality guidelines indicate that all the medical devices shall be in conformity with BIS standards. However, in case of non-availability of any such standards, the medical devices shall conform to the International standards laid down by ISO and IEC [12] or any other pharmacopeia standards. The Act further mentioned that in absentia of written standards by international or national organizations, it must conform to the validated manufacturer's standards.

The overall objective of Drugs and Cosmetics Act, 1940 (DCA), and Medical Device Rules, 2017 (MDR) is to:

1. Regulate the manufacture, distribution and sale including import of notified medical devices.
2. Ascertain the availability of notified medical devices with standard quality to the end user.

The Medical Device Rules, 2017 under the Drug Control Authority, regulate the following categories of Medical devices:

(i) Notified Medical Devices are defined as those which are intended to use for the prevention of mitigation, diagnosis and treatment of diseases or disorders in human beings or animals. Under the DCA, these medical devices are notified by the government from time to time.
(ii) Devices intended to functioning or structure of the human body such as Mechanical contraceptives insecticides and disinfectants, etc.
(iii) Surgical Devices used for surgical applications such as staples, sutures, dressings, bandages, ligatures, blood and blood component collection bags, etc.
(iv) Medical Devices used for in vitro diagnosis.

Role of Legal Metrology in Medical Device Sector

The Legal Metrology (Packaged Commodity) Rules, 2011, notified under the Legal Metrology Act, 2009, regulates the packaging and labelling of pre-packed commodities. The notified medical devices are required to declare the particulars like the name of the country of origin or manufacture or assembly on the package with effect from 1st January 2018. National Pharmaceutical Pricing Authority (NPPA), a drug price regulator, notified 22 medical devices as 'drugs' that must carry MRP on packs. These devices included surgical dressings, heart valves, stents, disposable hypodermic syringes, condoms, orthopaedic implants, etc.

15.2.1.2 Role of Bureau of Indian Standard in Medical Metrology

BIS is the Indian counterpart of ISO and works in the documentation of quality standards as per the country requirements. BIS has dedicated division dealing with medical devices under the name "Medical Equipment and Hospital Planning (MHD)". At present, 20 committees as shown in Table 15.2 are operating under MHD and have published 1731 standards in this area till July 18, 2020 [13].

MHD mainly focuses on making written standards in diversified areas such as equipment relating to rehabilitation, diagnosis, dental implants, surgery, etc. Besides this, the organization also contributes to written standards in Bio-medical Waste Management and Infection Control in Hospital management, Medical Biotechnology and Nanotechnology, etc. Scientists working on biomedical metrology at CSIR-NPL are members of BIS committees of MHD and are contributing their expertise in formulating the written standards for various medical devices. To make the conformity assessment in actual practice as per the BIS standards, it has introduced a "Laboratory Recognition Scheme" with laboratories in India and abroad to cater to the needs of medical device manufacturers. The product certification scheme is also executed by BIS for ensuring conformity with Indian Standard. The manufactured quality of the Indian Products is certified with ISI mark by BIS to comply with its

Table 15.2 A list of total number of committees under MHD and standard published [13]

S. No.	Committee number	Committee name	Published standards
1	MHD 1	Surgical instruments	115
2	MHD 2	Orthopaedic Instruments, Implants and accessories	130
3	MHD 3	Obstetric and gynaecological instruments and appliances	64
4	MHD 4	Ear, nose and throat surgery instruments	82
5	MHD 5	Ophthalmic instruments and appliances	91
6	MHD 6	Thoracic and cardiovascular surgery instruments	53
7	MHD 7	Neurosurgery instruments Implants and accessories	46
8	MHD 8	Dentistry	174
9	MHD 9	Artificial limbs, Rehabilitation appliances and equipment for the disabled	116
10	MHD 10	Medical laboratory instruments	32
11	MHD 11	Anaesthetic, resuscitation and allied equipment	58
12	MHD 12	Hospital equipment and surgical disposal	137
13	MHD 13	Veterinary hospital planning and surgical instruments	17
14	MHD 14	Hospital planning	42
15	MHD 15	Eletromedical diagnostic imaging and radiotherapy equipment	92
16	MHD 17	Health informatics	88
17	MHD 19	Immuno-biological diagnostic kits	34
18	MHD 20	Medical biotechnology and nanotechnology	2
19	MHD 21	Hospital biomedical waste management and infection control	0
20	MHD 23	Anatomy and forensic sciences equipment	0
		Total no of published standards as on 18th July 2020	1373

standard. The compliance of several products with Indian standards is mandatory as per the Indian Government guidelines under various statues in the public interest.

Medical device rules, under chapter III, section 11, clearly state that national accreditation body shall not act as a notified body. The notified body herein refers to a body corporate or other legal entity, registered under rule 13, as a body competent to carry out audit of the manufacturing site, assessment, and verifications of specified category of medical devices for establishing conformity with standards. At present, various quality implementing and regulatory bodies have overlapping roles in the quality infrastructure pertaining to medical devices and therefore, they need to interact periodically to avoid the overlapping mandates by actively involving CSIR-NPL (the NMI of India) for the improvement and advancements in the quality infrastructure.

15.2.1.3 Role of Quality Council of India in Health Sector

QCI is a registered non-profit society and is responsible for the establishment, operation and promotion of national accreditation structure under the ministry of commerce and Industry. QCI was established in 1977 jointly by the Government of India, Confederation of Indian Industry (CII), Federation of Indian Chambers of Commerce and Industry (FICCI) and Associated Chambers of Commerce and Industry of India (ASSOCHAM). QCI is administrated by a Council of 38 members with equal representations of industry, government, and consumers [14]. The QCI accreditation structure is depicted in Fig. 15.3, wherein the organization functions through the executive boards in the specific areas, i.e., accreditation for:

(i) Healthcare Establishments
(ii) Conformity Assessment Bodies
(iii) Education and Vocational Training Providers.

Role of NABL in Medical Metrology

In India, NABL adopts the responsibility of assessment and accreditation of Testing and Calibration Laboratories for biomedical equipment as per the international standard ISO/IEC 17,025 and ISO 15,189 guidelines. Till March 2020, NABL has provided accreditation to 1206 biomedical equipment calibration laboratories based on the quality assessment as per the ISO guidelines of medical equipment [15].

Fig. 15.3 Accreditation boards under quality control of India

NABL lays down specific criterion for the laboratories to operate and demonstrate its competency to carry out calibration of medical devices in accordance with ISO/IEC 17,025:2005, ISO 13,485, IEC 60,601, IEC 62,353 [16]. The purpose of these general requirements are (i) to have a uniformity between the laboratories, assessors and assessment process in terms of maximum permissible error, calibration and measurement capability, measurement uncertainty, etc. in accordance with national/international standards; (ii) To achieve uniformity in selection of equipment, calibration methods, maintaining required environmental conditions and personnel with relevant qualification and experience.

Role of NABH

NABH provides accreditation to hospitals and sectors specific to healthcare on similar lines as NABL provides accreditation at laboratory level and relies on the competency of calibration and testing laboratories. NABH is also a constituent board of QCI. It was set up in the year 2006 to establish and operate accreditation program for healthcare organizations on the lines of International Accreditation Standards like JCI of USA and ACHS of Australia. These accreditations bodies are members of International Society for Quality in Health Care (ISQua) that grants approval to Accreditation Bodies in the area of healthcare as the mark of equivalence of accreditation program of member countries. NABH is a founder member of the Asian Society for Quality in Healthcare (ASQua), which is a sub-continent body of ISQua. It can be seen as a chain of accreditation bodies, ensuring uniform protocols across the globe. To cater to the needs of the consumers, the above bodies have been structured to set criterion for the growth of health care industry [17]. NABH in India is known for maintaining the highest benchmark standard for hospital quality infrastructure in the country. The various activities of NABH are to provide accreditation and quality promotion in healthcare facilities through offering healthcare quality courses, workshops and trainings in patient safety, as represented schematically in Fig. 15.4.

NABH is a designed accreditation board in the medical field and it highly focuses on quality medical care and safety, and quality in diagnosis. NABH accreditation ensures regular training, qualification and expertise of the hospital staff for patient safety and quality. Additionally, NABH insures improved quality of additional services provided by hospitals such as empanelment of Insurance and third-party administrator. It also stimulates continuous improvement and commitment to quality care. There are different levels of accreditation for hospitals like pre-entry level, surveillance or re-assessment, 3rd edition standard level and 4th edition standards. The accreditation for pre-entry level is valid for three years, but the surveillance should be done after two years or 18 months of pre-entry level accreditation. It is different from NABL as a formal recognition of the competency of a lab-based on Third-party assessment following international guidelines. At present NABH claims to have 102 standards with 636 objective elements grouped in ten main chapters. For any hospital to obtain NABH accreditation, it must ensure the implementation of

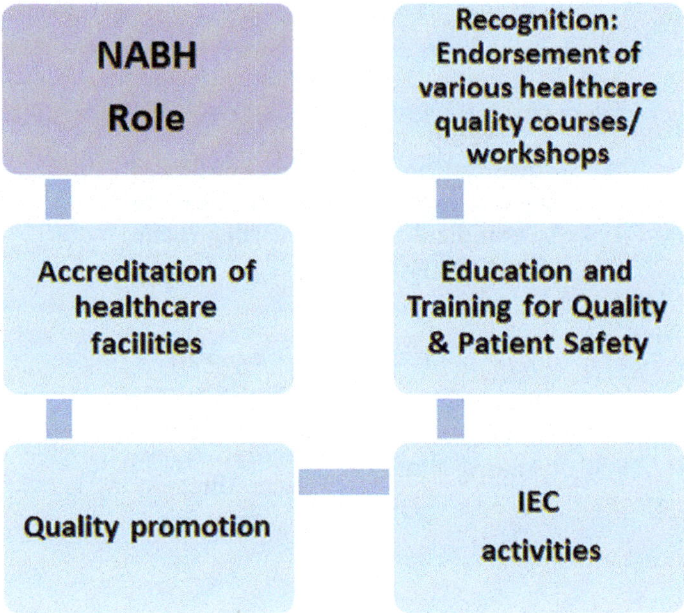

Fig. 15.4 Various activities of NABH in promoting hospital quality infrastructure in the country

above-said standards in their setup. As per NABH records currently over 350 hospitals in India are accredited and according to a report of "centre for disease dynamics, economics and policy" published on 20th April 2020 [18], around 70,000 hospitals in the country are NABH accredited, which is still an insignificant number considering the size of the Nation and population.

15.2.2 Role of National Metrology Institute (NMI)

NMIs occupy the top level of the calibration hierarchy and are officially entrusted with calibrations, and are responsible for the implementation and maintenance of the National metrology infrastructure. CSIR-NPL is designated National Metrological Institute of India, which has taken initiatives in establishing a biomedical metrology facility to cater to the need of medical device sector. To achieve its designated mandate in medical metrology, the laboratory has actively participated in Asia Pacific Metrology Program (APMP) by becoming a member of a dedicated medical metrology focus group with the primary scope as defined and represented schematically in Fig. 15.5.

APMP medical focus group formed with collaboration among NMI's of Asia pacific region has the following objectives:

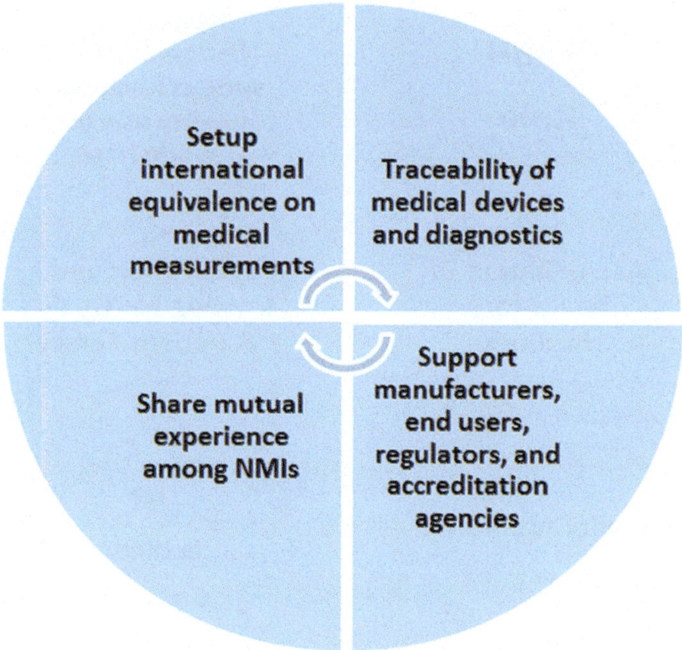

Fig. 15.5 Significant activities of APMP medical focus group

(i) Establishment of collaborations on progress of medical metrology techniques, instrumentation and facilities, etc.
(ii) Development and validation of calibration procedures for medical devices.

15.3 Biomedical Metrology and Instrumentation at CSIR-NPL

The economic development has created vast commercial opportunities in India in the area of medical sciences for various industries and other stakeholders. The medical devices in the country are getting measurement traceability from other NMIs abroad and therefore are shipped abroad every year for calibration. National Measurement Institutes (NMIs) from other countries are capturing Indian market and a huge amount of money is going abroad for the same. As NMI's of other countries have well established biomedical metrology group, the same is required in our country too to support the medical device sector, especially in the context of newly enacted 'Medical Device Rule 2017' which oversee the quality and condition of the medical devices used in the healthcare sector. It highlights the requirement of establishing Calibration and Testing centres for medical devices, thereby providing quality control regulation in health sectors. CSIR-NPL is uniquely positioned to cater to emerging needs of calibration in medical devices because of the availability of relevant expertise and mandate of the laboratory.

CSIR-NPL is promoting the quality measurements in the domains of biomedical metrology under its mission project through working with different stakeholders, for fulfilling the societal needs of the nation in healthcare sector with the following objectives:

(i) To establish and realize National Standards of measurements compatible to International standards in biomedical instrumentation through continuous research and development.
(ii) To provide Apex level calibration and dissemination of standards in the area of biomedical instruments as per ISO17025:2017 at one place.
(iii) Establishment of medical equipment standards in India.
(iv) Calibration and testing.
(v) Research and development.
(vi) To provide the above facility to various hospitals, clinical labs, industries, institutes and other stake holders.
(vii) Training on biomedical calibration and testing.
(viii) Preparation of standard documents to be used by regulatory bodies such as care NABH, NABL, CDSCO and BIS at National level.
(ix) Promote importance of standards in health care.

This activity could be further elaborated with a flow diagram as shown in Fig. 15.6. The activity is mainly focused on the establishment of a calibration facility for medical devices with an unbroken chain of measurement tractability to national/international standards, which in turn supports the accreditation bodies like NABL, NABH, NACB and BIS in implementing the medical device regulations. The metrological program is also providing training to healthcare personnel and supporting the other calibration laboratories, manufacturers and end users by providing calibration services to promote the quality infrastructure in Indian medical device sector.

The outcome of such an establishment would support the country healthcare sector in the following way:

(i) Excellence, accountability and uniformity in medical facilities throughout the country.
(ii) Trust building on medical health-care industry, resulting in India as major hub in international market for medical tourism.
(iii) Restriction on non-quality controlled equipments from grey market.
(iv) Such establishment would improve the health of public, support Indian industries to export their equipment with globally accepted norms, creation of jobs and start-ups.
(v) Providing high-quality control regulation through standardization
(vi) Biomedical equipments in health sectors, which in turn would contribute to healthy, risk-free and dispute free treatments to society.

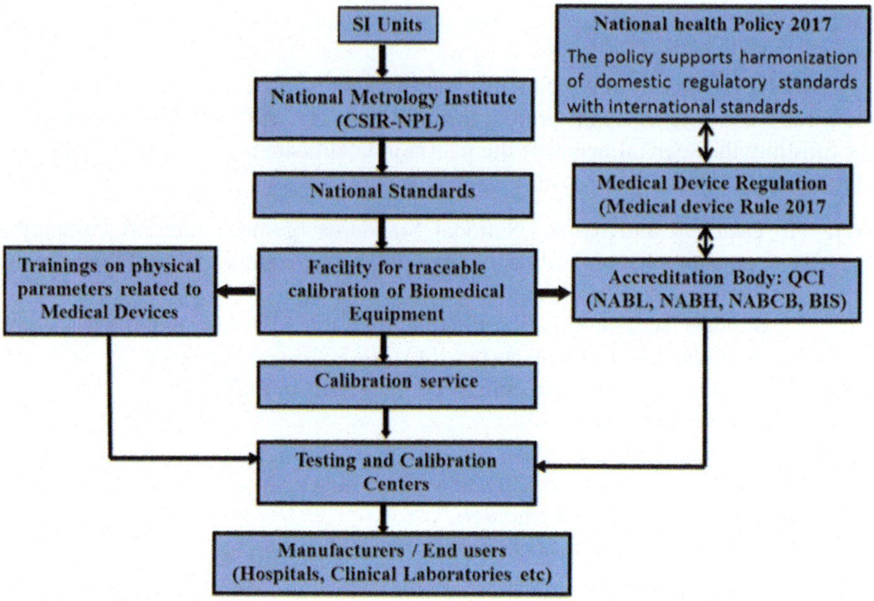

Fig. 15.6 A flow chart of biomedical metrological activities at CSIR-NPL

Taking concern of public health in our country, the scientists associated with biomedical metrology activity are also engaged in the R&D of fabricating indigenous point-of-care devices using novel biocompatible sensing materials for the diagnosis of chronic disease viz. renal dysfunction and heart disease by quantifying their markers and constantly working towards improved diagnostic systems.

15.3.1 Calibration services

Biomedical metrology program at CSIR-NPL is committed towards establishment and realization of National Standards for biomedical metrology. The Institute is developing measurements compatible to International standards in biomedical instrumentation through continuous research and development. So far the institute has developed national standard for medical devices such as defibrillator, blood pressure, clinical thermometer, magnetic resonance imaging devices and disseminating its services to nation through calibration.

In order to protect the traceable calibration certificate issued by the CSIR-NPL from being misused or copied by unethical means for unwanted reasons and to protect its genuineness and originality, a trademark as shown in Fig. 15.7, is secured for the above said calibration services.

Fig. 15.7 A composite trademark for the scientific and technological services provided for the calibration of biomedical equipment

15.3.1.1 Calibration Setup of Defibrillator Analyzer/Defibrillator

With the onset of new medical device rule in 2017, CSIR-NPL had initiated establishments of much-required standards as per the policy. In this regard, we have established an apex level calibration facility for defibrillator analyzer by installing a set-up of a primary standard of defibrillator with its physical parameters traceable to national standards [19]. This facility is the "first-of-its kind" in India to cater the needs of quality assurance in the area of healthcare. Defibrillator analyzer is used for the calibration of the defibrillators. Defibrillator is the most critical life-saving biomedical equipment used to restore the normal heart rhythm of a patient by applying an energy pulse during sudden cardiac arrest and for this reason it is mandatory to have defibrillator installed at public places, offices, airports, railway stations, ambulances, hospitals, etc. Defibrillator works by supply monobasic/biphasic pulse to patient with energy levels in range of 50–360 J. It is critical to supply defined energy to patient as per physiological conditions; else the treatment will not be effective and may cause death. Thus, it is extremely essential to have defibrillator analyzer calibrated and in turn the defibrillator with high precision, in stipulated time periodically, as per the accuracy requirement of the energy function, specified in the international standard (IEC 60,601-2-4) for medical electrical equipment. Traceability of the measurement result is linked with calibrated standards (voltage, resistance and time) used for the calibration of Defibrillator Analyzer (DA). A schematic setup of the calibration facility is shown in Fig. 15.8, wherein a defibrillator is used as an energy source which discharges a pulse of required energy to the defibrillator analyzer (unit under test). A specially designed high voltage divider is used to steps down the discharged

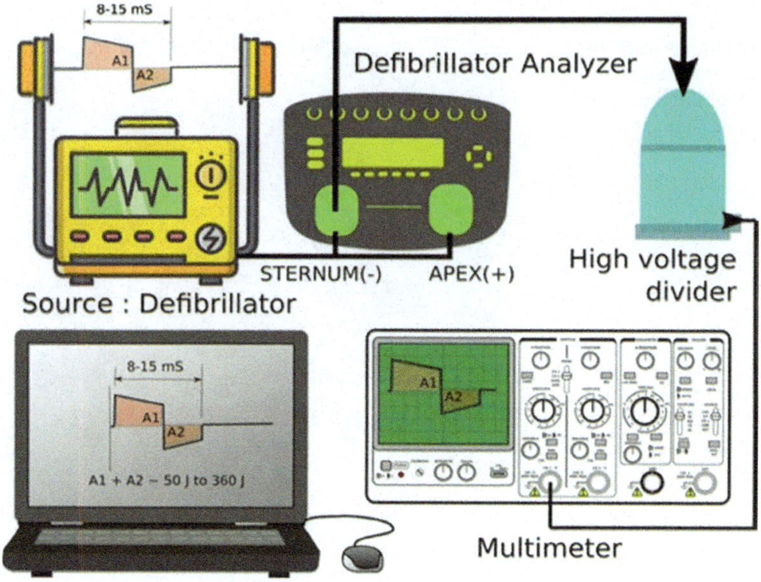

Fig. 15.8 A setup of calibration facility for Defibrillator analyzer and defibrillator

voltage in 1000:1 (V/V) ratio for calculating the energy through a high-speed digital storage oscilloscope (DSO).

The energy contained in a pulse of arbitrary wave shape, such as the biphasic waveform, is area drawn by biphasic curve, thus total energy is a sum of these two arbitrary shaped areas (A1 and A2) as shown in Fig. 15.8, and can be calculated [20] using Eq. 15.1.

$$E = \int_{0}^{T} v(t) \cdot i(t) \mathrm{d}t \tag{15.1}$$

where

E is the energy of the discharge pulse (J).
$v(t)$ is the voltage as a function of time.
$i(t)$ is the current as a function of time.
T is the time duration of the pulse.

The current in Eq. 15.1 is replaced with voltage according to ohm's law, wherein the energy dissipated through pulse across a physiological resistance (~50 Ω) is determined by using Eq. 15.2.

$$E = \int_0^T [v(t)]^2 \cdot \mathrm{d}t / R_L \tag{15.2}$$

The reference energy (source) is calculated by Eq. 15.3

$$E_{\text{calculated}} = \sum [(\text{data}(i) + V_c) \times V_r]^2 / R_L \times t_{\text{sampling}} \tag{15.3}$$

The corrections in the DA reading is estimated with respect to reference (source) energy by using Eq. 15.4

$$E_c = E_{\text{calculated}} - E_{\text{reading}} \tag{15.4}$$

where

E_c is the correction in UUT (DA) energy;
$E_{\text{calculated}}$ is the reference (source) calculated energy;
E_{reading} is the mean reading of UUT (DA);
Data (i) is the ith reading of the DSO (V);
V_{c} is the correction in DSO (V);
V_{r} is the ratio of the high voltage divider;
R is the internal resistance of UUT (DA) (Ω);
t_{sampling} is the sampling time of the DSO (s);

The actual set up of configuration for the above-said facility is shown in Fig. 15.9 for the accurate measurement of energy with a standard uncertainty $u(x_i)$ 0.64 J.

The combined uncertainty is calculated by taking into account the individual measurement uncertainty associated with all the possible sources (voltage, resistance and time) as envisaged by a fish bone diagram given in Fig. 15.10.

A flow chart to demonstrate the measurement traceability is shown in Fig. 15.11, wherein the physical parameters involved in the measurement are traceable to national

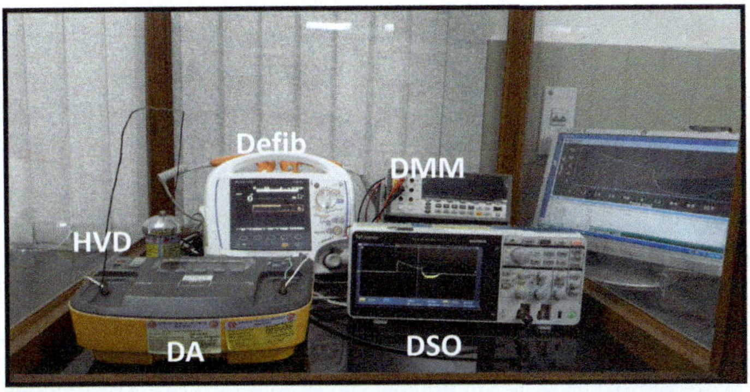

Fig.15.9 Actual setup of calibration facility for defibrillator analyzer [19]

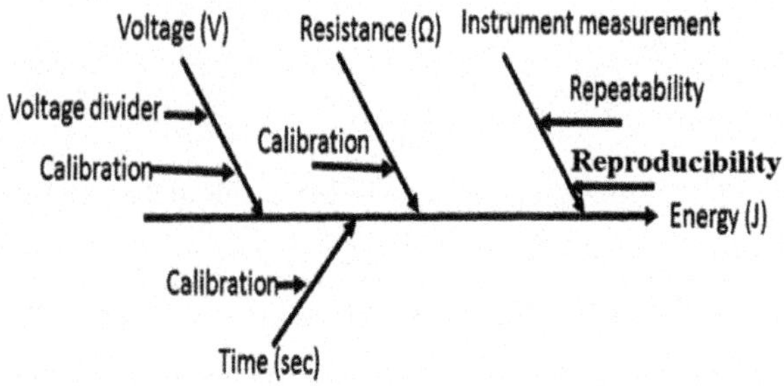

Fig. 15.10 Cause and effect diagram (fish bone diagram) for depicting uncertainty components involved in the energy measurement of defibrillator analyzer [19]

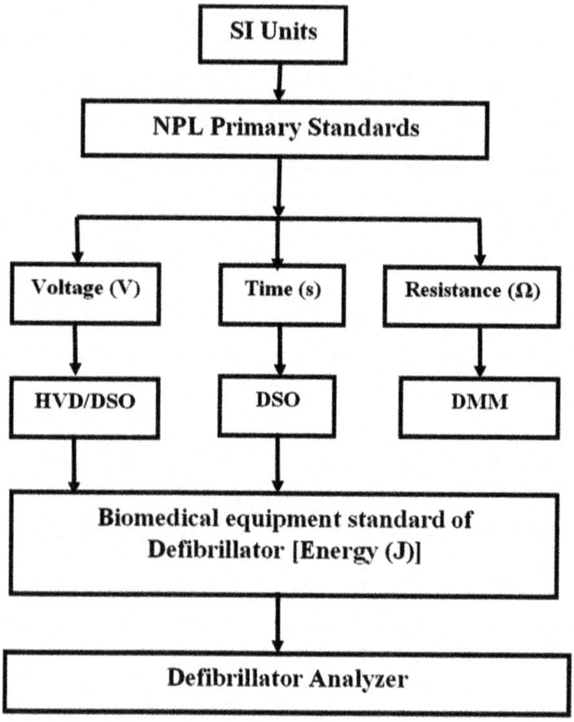

Fig. 15.11 Flow chart of national metrological traceability chain of measurement defibrillator analyzer is designed to measure the energy of the pulse discharged by defibrillator [19]

standards viz. voltage, resistance and time, maintained at CSIR-NPL, as described in Chaps. 3 and 5, through which a derived unit of energy (J) is assigned to the primary source (defibrillator).

This calibration facility is providing services to various stakeholders with effect from 1st September 2018. The recent beneficiaries of this service are the testing and calibration laboratories viz. Lawkim Motors Group (Godrej), Mumbai; Tektronix, Mumbai, TransCal Pvt. Ltd., Bangalore; APEX Laboratories, Jaipur, Rajasthan, Life-Force Pvt. Ltd. Faridabad, Haryana, Kirloskar Group, Bangalore, Inspire Biomed, Calcutta, West Bengal, KIMS Healthcare Management Ltd., Thiruvananthapuram, Kerala, etc.

15.3.1.2 Calibration of Syringe Pump and Infusion Pump Analyzer

A syringe pump is extensively used biomedical equipment. The prime application of the syringe pump is the accurate and precise delivery of drugs, blood and even nutrition, as volume (mass) is the most critical parameter in medical treatment. It also has wide applications in microfluidic-based biomedical devices and thus its traceable calibration is crucial for accurate measurements. Gravimetric method is the most preferred primary calibration method for infusion pump and syringe pump. A calibration setup of syringe pump based on Gravimetric method is shown in Fig. 15.12. Gravimetric principle relates the flow of a liquid to its volume delivered in the specified time. Further, the volume and density of the known liquid, at a given temperature, is directly associated with the weight of the liquid, which is measured using high precision balance, and thus obtained its traceability from mass. Though gravitation method, being a primary method, is mostly used by NMI's for the calibration of syringe and infusion pumps, the hospitals and secondary laboratories instead use infusion pump analyzer to disseminate the traceability of flow in infusion pumps with incremental uncertainty.

The schematic of the Infusion pump analyzer (IPA) calibration set up under development at CSIR-NPL with traceability chain is depicted in Figs. 15.13 and 15.14, respectively.

The calibration setup is designed with the following accessories: (i) a storage tank for water, (ii) infusion pump as a source of pressurized flow, (iii) standard pressure gauge, (iv) infusion pump analyzer, as a device under test and (v) standard high precision balance. The above individual parts are connected with 1/4" diameter tube for maintaining a consistent flow throughout the measurement.

In a typical procedure [21], the volumetric flow rate is calculated by directly converting the mass into volume (V_0) with respect to a reference temperature normally at 20 °C using Eq. 15.5, as per ISO/TR 20,461 ("Determination of uncertainty for volume measurements by gravimetric method"). However, IS 13,450 (Part 2/Sec 24): 2009 equivalent to IEC 60,601-2-24 describes a calibration procedure particularly for the medical electrical infusion pump device.

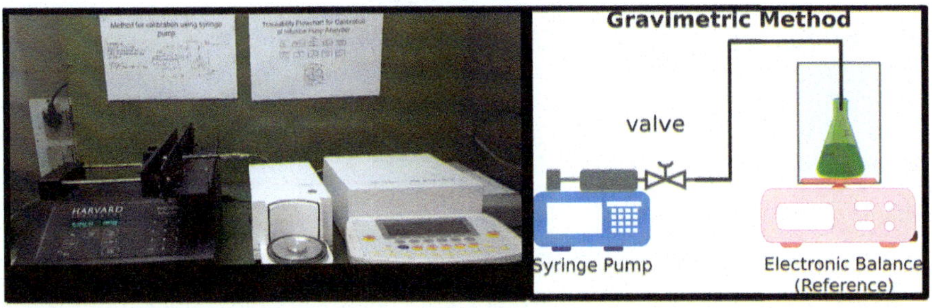

Fig. 15.12 **a** Calibration set up of syringe pump at CSIR-NPL; **b** schematic of primary gravimetric method of syringe pump

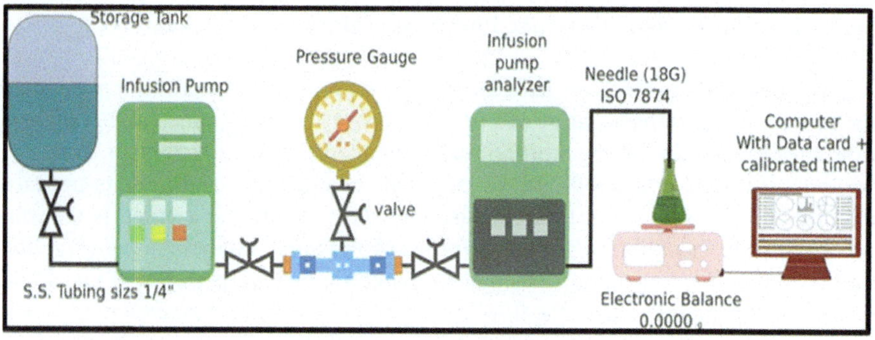

Fig. 15.13 Schematic of Infusion pump analyzer calibration set up at CSIR-NPL

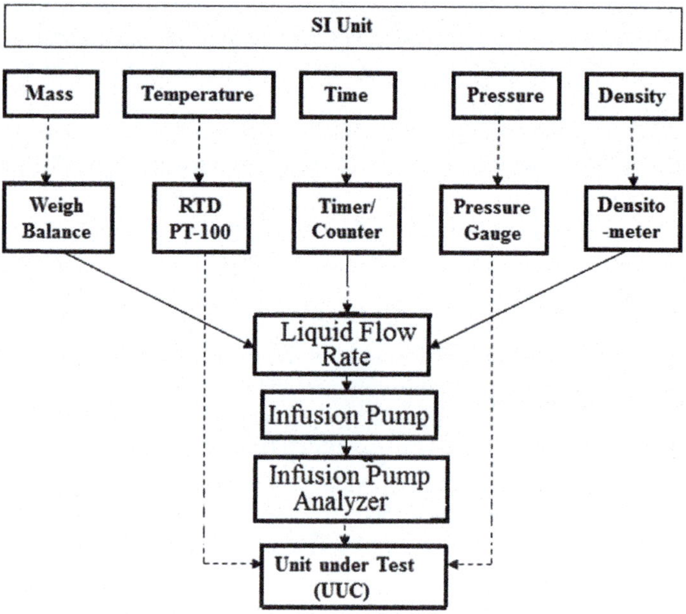

Fig. 15.14 National metrological traceability chain of equipment (IPA)

$$V_0 = m \times Z \times Y \tag{15.5}$$

where,

m is the balance reading of delivered water;
Z is the combined factor for buoyancy correction and conversion from mass to volume.
Y is the thermal expansion correction factor of the delivering device.

By using Eqs. 15.6 and 15.7, the buoyancy corrections and thermal expansion in volume is applied to obtain the measured volume (V_o), at a given temperature, as given in Eq. 15.8.

$$Z = \left[\frac{1}{\rho_W - \rho_a}\right] \times \left(1 - \frac{\rho_a}{\rho_b}\right) \tag{15.6}$$

$$Y = 1 - \alpha_c(t_d - t_0) \tag{15.7}$$

$$V_0 = (I_L - I_E) \times \left[\frac{1}{\rho_W - \rho_\alpha} \times \left(1 - \frac{\rho_a}{\rho_b}\right)\right] \times [1 - \alpha_c(t_d - t_0)] \tag{15.8}$$

wherein,

I_L is the balance reading of vessel with water, in grams;
I_E is the balance reading of empty vessel, in grams;
ρ_w is the density of water, at t_d °C, in grams per mL;
ρ_a is the density of air, in grams per mL;
ρ_b is the density of the standard weight used to calibrate the balance, in grams per mL;
α_c is the coefficient of cubical thermal expansion of the test material in reciprocal degrees Celsius;
t_d is the temperature of water used in testing, in degrees Celsius;
t_0 is the reference temperature (normally at 20 °C).

Volumetric flow (Q) of the test device can be obtained from Eq. 15.9 by taking difference between the volume at two instances ($t_{\text{initial}}, t_{\text{final}}$) using Eq. 15.8.

$$Q = \frac{1}{(t_f - t_i)}[I_f \times \left(1 - \frac{\rho_a}{\rho_b}\right)\frac{\lfloor 1 - \alpha_c(t_d - t_0)\rfloor}{(\rho_w - \rho_b)} - I_i \times \left(1 - \frac{\rho_a}{\rho_b}\right)\frac{\lfloor 1 - \alpha_c(t_d - t_0)\rfloor}{(\rho_w - \rho_a)}\delta V_{\text{evap}} \tag{15.9}$$

wherein, an additional compensatory term δV_{evap} has been added to overcome the uncertainties arises due to evaporation.

Figure 15.14 represented a traceability chart of a typical flow of uncertainties associated with the potential sources viz. mass, temperature, time and density and pressure, which are traceable to national standards maintained at CSIR-NPL, as described in Chaps. 3 and 4. However, the present setup is being tested and any further improvement in the setup may lead to modification in the traceability chart.

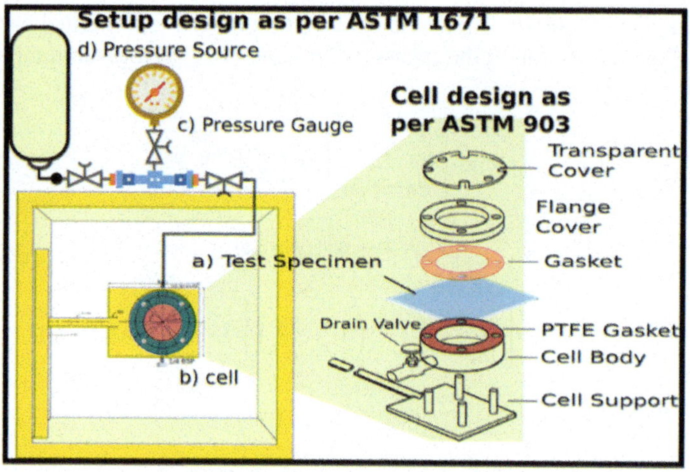

Fig. 15.15 Blood penetration test setup to test PPE at CSIR-NPL

15.3.1.3 Calibration Set Up for Blood Penetration Test on Personal Protection Equipment

CSIR-NPL has setup testing of personal protective equipment in the time of pandemic due to COVID 19 virus, which included blood penetration setup as per international protocols and the schematic of the fabricated set up is represented in Fig. 15.16. As per Indian standard (BIS 16,546:2016) a material used in PPE kits is placed in contact with pressurized synthetic blood, and the resistance offered by test specimen to synthetic blood at varying pressure is tested. A dedicated test setup as per ASTM 903 standard has been fabricated at NPL workshop. In the typical procedure, a test specimen (Fig. 15.15a) is placed in a leakage-proof cell assembly (Fig. 15.15b). Further, artificial blood is subjected to the cell backed with regulated air pressure (Fig. 15.15d) measured using a calibrated gauge (Fig. 15.15c). The pressure gauge gets its traceability from national standard, as described in Chap. 4.

15.3.1.4 Bharatiya Nirdeshak Dravya; BND (Indian Reference Material) of Artificial Blood

The resistance offered by PPE to artificial blood is considered as an acid test for their utility in pandemic situations like the worldwide spread of COVID19. The prime requirement of blood penetration tests is an artificial blood with its physical properties similar to human blood. CSIR-NPL has taken a lead in developing a standard for an artificial blood, particularly for testing PPE kits. The artificial blood must possess the following four properties: (a) Surface Tension (0.042 ± 0.002) N/m, (b) pH (7.3 ± 0.1), (c) Viscosity (2.7 ± 0.3) mPa.s and (d) Conductivity: (12.0 ± 1.2) mS/cm, as per BIS standard 16,546:2016. To obtain the aforementioned properties,

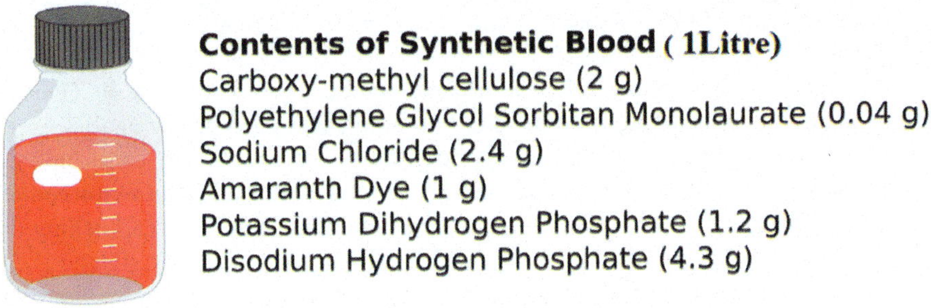

Fig. 15.16 The composition of ingredients used in the synthesis of artificial blood

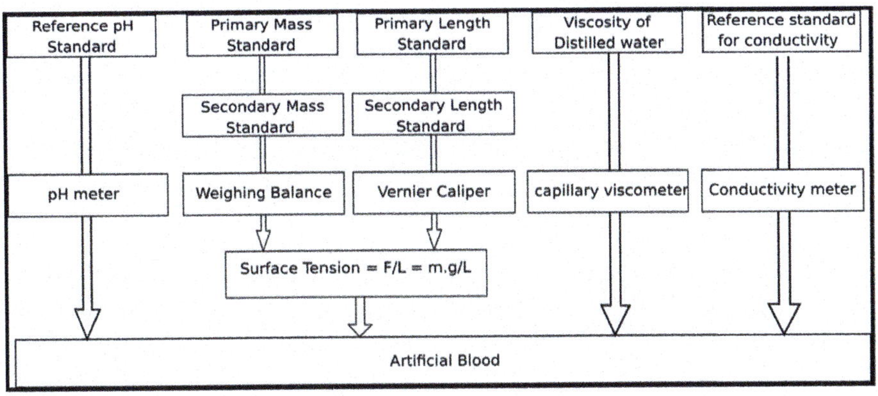

Fig. 15.17 National metrological traceability chain for physical parameters of artificial blood

the said standard also provides the composition of an artificial blood, as shown in Fig. 15.16.

To support the nation's need for seamless testing of PPE kits, CSIR-NPL has developed BND of artificial blood with its measurement parameters of physical and chemical properties traceable to national standards, as evident from the chain of traceability shown in Fig. 15.17.

15.3.1.5 Calibration Setup of Other Medical Devices

Apart from creating new calibration facilities for medical devices, as described above in the earlier sections, CSIR-NPL, has well-established calibration facilities for medical devices viz., clinical thermometer, blood pressure monitoring and magnetic resonance imaging devices and providing calibration services to the stakeholders of healthcare sector. The calibration measurement of physical and electrical parameters of clinical thermometer and blood pressure monitoring devices are traceable to primary standards of temperature and pressure, described in Chap. 3, whereas the primary standard of magnetic field for magnetic resonance imaging devices is described in Chap. 6 in detail. Furthermore, CSIR-NPL is also in a process of extending its calibration capability to other biomedical equipment like ECG simulator, incubator analyzer, multiparameter simulator, electrical safety analyzer, etc. These facilities will support the accreditation body such as NABL to set up new centers or help them in providing accreditation to more testing and calibration centers of medical devices based on national standards and contribute to strengthening the medical device regulation. In this way, the development of primary standards for biomedical equipment and its dissemination to industries, secondary laboratories and hospitals will have a huge impact on National healthcare sector in reducing the health complications due to medical device errors.

15.3.2 Research and Development on Biosensors

A biosensor is an analytical device that uses specific biochemical reactions mediated by isolated enzymes, antibodies, nucleic acids, organelles or cell receptors to detect chemical compounds usually by electrical, thermal or optical signal.

CSIR- NPL has been actively engaged in the development of biosensors of clinical relevance for the past several years. In this context, extensive work has been done on the development of glucose, urea, cholesterol, uric acid and DNA biosensors. Glucose biosensor, the ‘NPL glucosense’ was developed at CSIR-NPL, (Fig. 15.18) and the technology was transferred to three companies viz Gama Instrumentation Pvt. Ltd, Faridabad, Haryana; Pulsatum Health Care Pvt. Ltd., Bangalore, Karnataka and Transgeniks, Ahmedabad, Gujarat.

The laboratory is currently engaged in the development of point-of-care biosensors for the diagnosis of chronic and infectious diseases and a few of them are highlighted and discussed briefly in the subsequent sections given below.

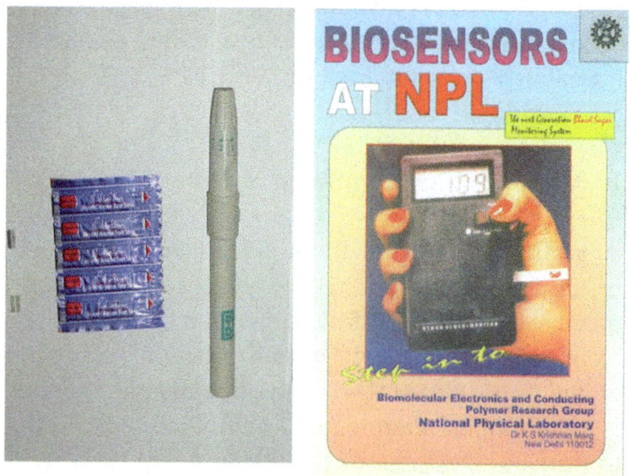

Fig. 15.18 Electrochemical glucometer developed at CSIR-NPL

15.3.2.1 Technical Development of Biosensors for Chronic Diseases

Scientists at CSIR-NPL have been engaged in the synthesis of new nano-objects and exploitation of their extraordinary properties in the field of sensors for biomedical applications. Nanostructured materials of Pt-dendrimer, carbon nanotubes and single-layered graphene were used as a suitable matrix for biomolecular immobilization for the development of "Lab-on-a chip" for biomedical applications, especially in the diagnosis of cardiovascular and infectious diseases. The sensing performance of the biomaterials was investigated by electrochemical/Field effect transistors (FET) characteristic studies towards the quantitative detection of protein biomarkers of chronic diseases.

A screen-printed carbon electrode was modified with Pt nanoparticles encapsulated polyamidoamine (PAMAM) dendrimer to develop an impedimetric biosensor, for the quantitative detection of human cardiac biomarker troponin-I (cTnI). PAMAM-Pt was electrochemically deposited over SPCE and its 128 terminal carboxyl groups were used as anchors for the site-specific biomolecular immobilization of protein antibody, anti-cTnI, as shown in Fig. 15.19.

A single-frequency impedance analysis technique was utilized for biomolecular sensing of cardiac biomarker, cTnI, by monitoring the changes in phase angle with immunoreaction (antigen–antibody interactions) occurring at the electrode surface. A maximum change in the phase angle was obtained at a frequency of 100 Hz upon immunoreaction with a given concentration of analyte at the electrode surface. The fabricated immunosensor exhibited an electrochemical response in the range of 1 pg mL^{-1} to 100 ng mL^{-1} of clinical relevance [22].

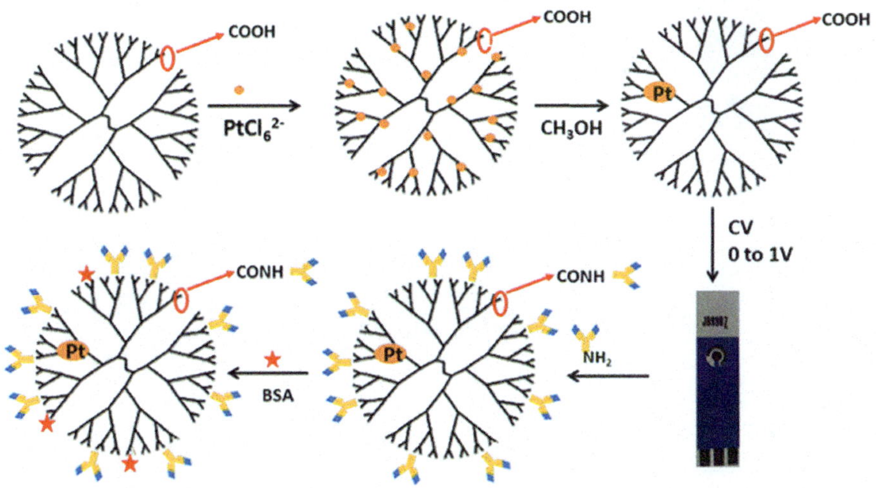

Fig. 15.19 Schematic representation of stepwise preparation of PAMAM-Pt and anti-cTnI-PAMAM-Pt/SPCE (Bio-SPCE) [22]

A biofunctionalized reduced graphene oxide (rGO)-modified screen-printed carbon electrode (SPCE) was prepared, as shown schematically in Fig. 15.20, as an immunosensor for C-reactive protein (CRP) detection, a biomarker released in early stage acute myocardial infarction. In the electrochemical impedance measurement, an optimized frequency of 10 Hz was obtained, where a maximum change in a phase angle was observed with antibody-antigen reaction at the electrode surface.

At this optimized frequency of 10 Hz, the concentration-dependent response of immunosensor to CRP with change in phase angle was observed in the range of 10 ng mL^{-1} to 10 μg mL^{-1} in PBS. Sensing of analyte at such a low ac frequency with high selectivity and sensitivity makes this system a potential sensor for clinical applications [23].

Paper-based electronics have potential for the fabrication of simple, sensitive and disposable new generic clinical diagnostics for efficient employment in the low-resource settings and underdeveloped countries. Considering the precincts in the present methods, efforts were made towards the development of a low-cost and rapid method for the modification of paper using a very simple electrochemical deposition approach. The patterned screen printed paper electrodes (SPPE) were designed and PANI was deposited onto the paper electrochemically, as shown in Fig. 15.21.

These matrices were utilized for the detection of troponin, a cardiovascular biomarker and this approach has potential applications towards the development of cost-effective diagnostics. Screen printed Whatman filter paper has been chosen

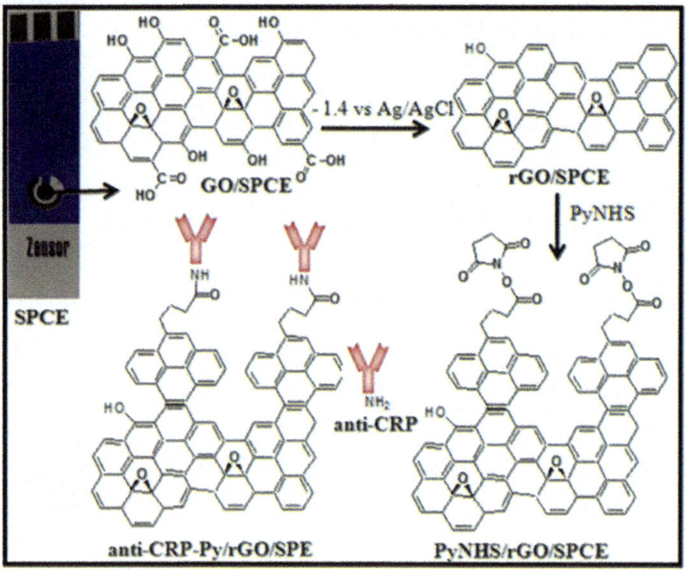

Fig. 15.20 Schematic representation of the stepwise fabrication of the bioelectrode [23]

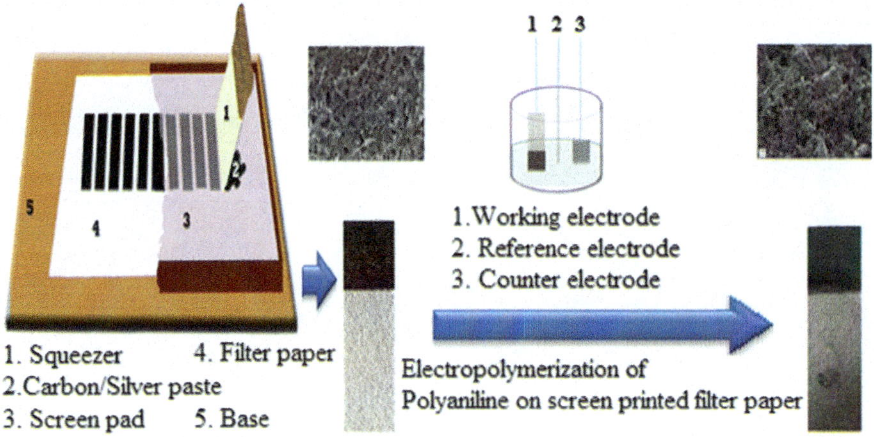

Fig. 15.21 Schematic representation of screen printed paper electrode for the detection of cardiovascular marker [24]

as the matrix for the biosensing applications. Polyaniline has been electrochemically polymerization using cyclic voltammetric technique. The cardiovascular antibody marker has been immobilized on to the polyaniline coated paper substrates using covalent immobilization technique using EDC-NHS chemistry. The substrates have been found to detect troponin in concentration range of 1–100 ng mL^{-1} with sensitivity 5.5 μA/ng mL^{-1} cm^{-2} and response time of 150 s [24].

An array of 52 graphene field-effect transistors (GFETs) was developed using high-quality chemical vapor deposited graphene decorated with antibody-functionalized platinum nanoparticles (PtNPs) over microfabricated gold electrodes on a SiO_2/Si wafer, for the quantitative detection of breast cancer biomarker HER3. HER3-specific, genetically engineered thiol-containing single-chain variable fragment antibodies (scFv) was immobilized with PtNPs/GFET hybrid nanostructure to realize a biosensor for HER3, as shown in Fig. 15.22.

The Dirac point voltage of pristine GFET is in the range −4.0 to +2.0 V, with an average carrier mobility (μ) of 1197 ± 56 cm^2 V^{-1} s^{-1} (Fig. 15.23a), showed a clean transfer of graphene over microfabriacted gold electrodes on SiO_2/Si wafer. An increase in the carrier mobility ($\mu = 1330 \pm 14$ cm^2 V^{-1} s^{-1}) with a positive shift in the Dirac voltage ($V_D = 26.4 \pm 1.0$ V) upon biomolecular immobilization of scFv (Fig. 15.23b), suggesting a decrease in charge scattering conforming an efficient attachment of biomolecule for sensing a target molecule.

The biosensor exhibited a concentration-dependent response to HER3 antigen as a function of change in Dirac voltage in the range 300 fg mL^{-1} to 300 ng mL^{-1} with a limit of detection of 300 fg mL^{-1}, showing its potential to be used as a label-free biological sensor for an early detection of breast cancer [25].

A Single-walled carbon nanotube (SWNT) based field-effect transistor (FET) was also developed using Polyamidoamine (PAMAM) dendrimer as anchor for site

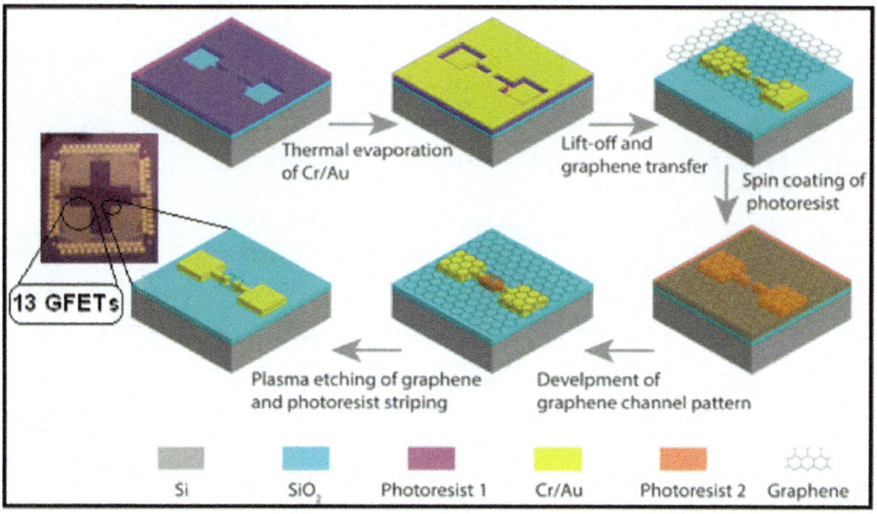

Fig. 15.22 Schematic representation of fabrication process for an array of 52 GFETs [25]

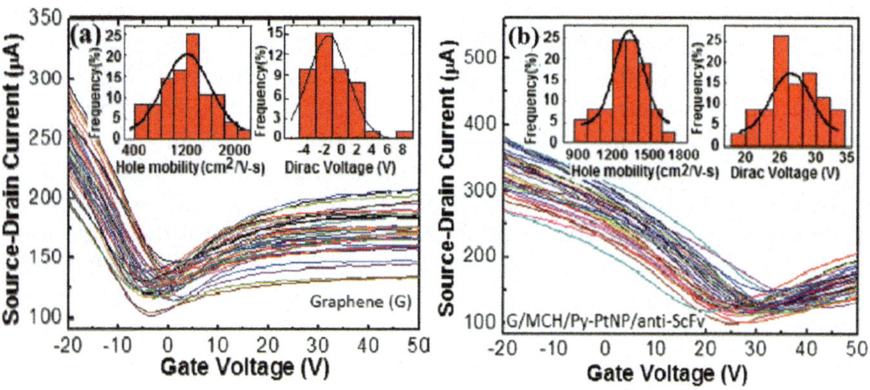

Fig. 15.23 Electrical characteristics of GFET devices with a representative set of 52 *I*-*V*g curves of **a** As-fabricated GFET; **b** scFv-PtNP/GFET nanohybrid device. Insets in each case are histograms of carrier mobility and Dirac voltage along with Gaussian fits [25]

specific biomolecular immobilization of protein antibody for the detection of cardiac biomarker, C-reactive protein (CRP), as shown in Fig. 15.24.

The source-drain current of the device decreases on immunoreaction with the target analyte of CRP in the regime of clinical significance, with a detection limit of ~85 pM [26].

The applications of microfluidic-based devices have grown exponentially due to excellent merits over traditional devices in terms of portability, bioanalysis at microscale, automation in drug discovery, and consumption of samples and reagents.

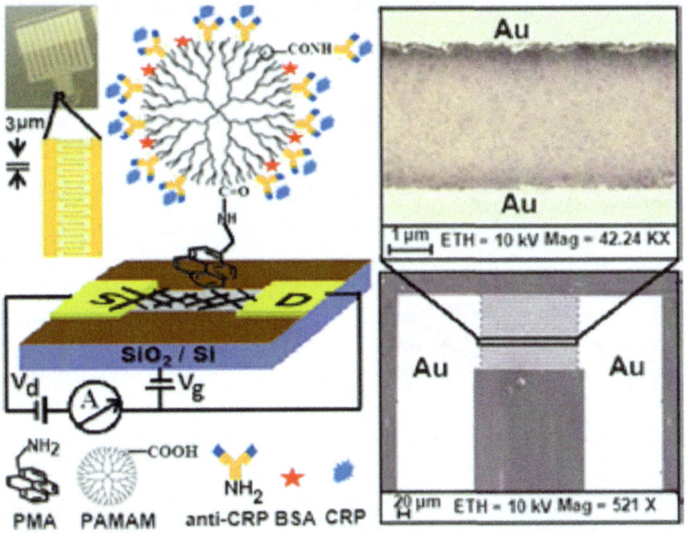

Fig. 15.24 Schematic illustration of the PAMAM modified SWNT-FET biosensor device with the SEM image of SWNT network between 3 µm apart gold source-drain electrodes [26]

Some examples of integration of nanomaterials with microfluidic devices (lab-on-a-chip) for the fabrication of platforms for biosensing applications are presented. In one such study nanocomposite of nickel oxide nanoparticles (nNiO) along with multiwalled carbon nanotubes (MWCNTs), was integrated with photolithographic constructed microchannels of polydimethylsiloxane (PDMS). This nanocomposite chip of nNiO-MWCNT Chip was functionalized with a cholesterol oxidase (ChOx) and cholesterol esterase (ChEt), via an amide bond. Schematic of same is shown in Fig. 15.25. The response studies on this coenzyme microfluidic platform have shown excellent reproducibility and selectivity, and a sensitivity of 2.2 mA mM^{-1} cm^{-2}. This integrated microfluidics biochip provides a promising low-cost platform for the rapid detection of biomolecules using minute samples [27].

Further, a label-free impedimetric lab on a chip (iLOC) is fabricated using protein (bovine serum albumin), and antiapolipoprotein B functionalized on similar nanocomposite system for detection of low-density lipoprotein (LDL). A cytotoxicity study on the synthesized CNTs, NiO nanoparticles, and their nanocomposite (CNT-NiO) in the presence of lung epithelial cancer A549 cell line using MTT assay has been carried out. The cell viability studies show promising response for CNT-NiO nanocomposite at a concentration of 6.5 µg/mL compared to individual CNTs. A chronocoulometry and impedance spectroscopic techniques were employed to study the binding kinetic and electrochemical activities of CNT-NiO-based iLOC. The iLOC shows excellent sensitivity of 5.37 kΩ $(mg/dL)^{-1}$ in a wide concentration range (5−120 mg/dL) of LDL. The binding kinetics of antigen–antibody interaction of LDL molecules show a high association rate constant of 8.13 M^{-1} s^{-1} [28].

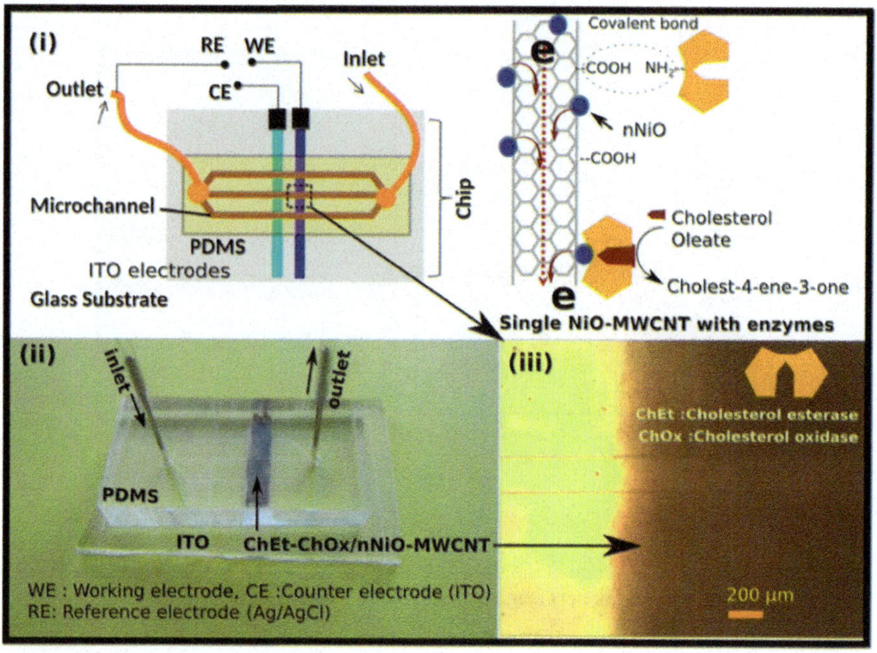

Fig. 15.25 **a** The schematic of the microfluidic biochip is used for total cholesterol detection. **b** The photograph of a microfluidic biochip for cholesterol detection and **c** the magnified view of the optical microscopic image of the biochip [27]

15.3.2.2 Technical Development of Biosensors for Infectious, Food Toxin and Water Borne Diseases

Gonorrhea is presently recognized as the second most prevalent sexually transmitted bacterial infection (STI) and is a co-factor for HIV infection. Worldwide, there is an estimated annual incidence of 62 million new cases of gonococcal infections. Untreated gonorrhea has major risk factors for acquiring pelvic inflammatory disease (PID) that can lead to complications of infertility and tubal pregnancies and can result in conjunctivitis and pneumonia in newborn infants exposed during passage through an infected birth canal. Traditional laboratory diagnosis of this infection is carried out by culture, microscopy and PCR techniques are time- consuming (14–72 h) and are not reliable. Therefore, efforts are being made towards the development of a rapid STD sensing device in collaboration with AIIMS, Delhi and Safdarjung Hospital, Delhi. The response studies of probe DNA immobilized onto various platforms such as conducting polymers, nanomaterials, gold matrix and gold screen printed electrodes have been studied for electrochemical hybridization using standard complementary, non-complementary sequences and clinical samples. The studies carried out on clinical samples on highly reproducible gold substrates based DNA biosensor indicate that the sensor can detect presence of *N. gonorrhoeae* from culture (0.5 McFarland) 10^8 cells/ml, spiked sample and *N. gonorrhoeae* culture positive swab sample of patients. The specificity of the gold substrate based DNA hybridization biosensor has been studied using the control strains of *N. gonorrhoeae*, other pathogenic Neiseria species and other GNBs [29]. The nucleic acid electrode has also been tested using patient sample under similar conditions. These results reveal that the electrode is highly specific and can distinguish presence of *N. gonorrhoeae* from other species like *N. meningitidis* and other GNBs, i.e., *Escherichia coli*. Under the optimum conditions, this nucleic acid sensor can be used to detect complementary target DNA concentration in the range [1×10^{-6}–0.5×10^{-18} M] with a detection limit of 1.0×10^{-18} M within 60 s of hybridization time at 25 °C. The sensor has been found to be very stable and the data is reproducible several times with improved detection limit. The biosensor has been validated using *N. gonorrhoeae* culture samples, spiked samples and positive swab sample of patients. Identification of the specific DNA sequence and the fabrication of biosensor with its performance has been patented in India and PCT.

The rapid growth in the development of biosensor has given a boost to the advanced evolution using biomaterials and improvements of sensing techniques for food-borne toxin Aflatoxin B_1 detection. Although traditional techniques are available for quantification of this mycotoxin, but these techniques are quite expensive and time-consuming. Therefore, efforts have been made at CSIR-NPL for the fabrication of biosensors for food toxin detection. Nanocomposites of Graphene Quantum dots and gold nanoparticles (GQDs-AuNPs), were utilized as biomolecular immobilizing matrix for the construction of a biosensor for AFB1 detection in food content. GQDs-AuNPs composites were transferred on ITO coated glass substrate by electrophoretic deposition technique, followed by antibody immobilization for the fabrication of biosensor, as shown schematically in Fig. 15.26.

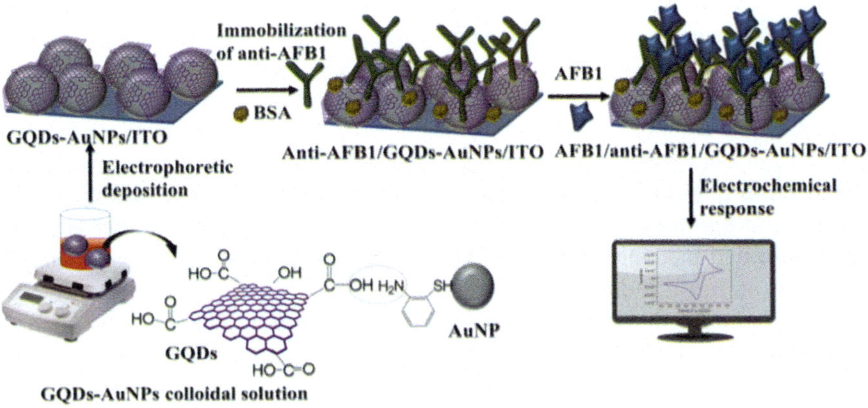

Fig. 15.26 Schematic representation of the synthesis of GQDs-AuNPs composites and its electrophoretic deposition onto ITO-glass electrode leading towards the fabrication of biosensor for food toxin, Aflatoxin B1 (AFB1) detection [30]

The biosensor has been tested with the real (spiked) maize sample using electrochemical impedance signal based transduction and was found to detect AFB1 concentration in the range of 0.1–2.5 ng mL^{-1} [30].

Bisphenol A (BPA) is a synthetic compound, extensively used in food packaging industries and has emerged as a threat to the human health and environment, thus to detect bisphenol in an analytical range of 0.01–100 μM, an amperometric sensor has been developed using enzyme tyrosinase immobilized on to rGO and Mn_3O_4 NPS for a synergistic effect, resulting in improved sensitivity of 93.2 μA nM^{-1} cm^{-2} with an analytical range from 0.01 to 100 μM, and detection limit of a 10 nM [31].

15.3.3 Training and Academic Program and Consultancy Services

The biomedical laboratory is actively participating in academic program with Academy of Scientific and Innovative Research (AcSIR) through research and development in the core area of nanomaterials for biomedical applications together with providing biomedical metrology training and consultancy services to the accredited laboratories in establishing the secondary level calibration facilities for medical devices in the country.

15.3.3.1 Skill Development Program

Hands-on training services are being provided to the various stakeholders about the National measurement system, its handling and calibration of the sophisticated biomedical equipment, and its importance in supporting the medical device rule. The training programs are designed to familiarize the participants with the concepts on the importance of biomedical metrology, the importance of accurate and precise measurements, medical device regulations in India, calibration procedures for biomedical equipment, maintenance and operation of the equipment and practical examples on the evaluations of associated uncertainties with measurements, etc. The beneficiaries of these training programs include doctors, industries, accredited laboratories and biomedical engineers, etc. We have also conducted industry-institute interaction and meetings in the past to sensitize people about the significance of metrology in medical devices for risk-free healthcare services to the Nation. Such kind of training-cum-awareness programs support startups and creation of jobs in the healthcare sector.

15.3.3.2 Academic Program

Research and developmental activities pertaining to biomedical instrumentation have been carried out at CSIR-NPL for the last few years, both in medical metrology and synthesis of novel biomaterials as well for the development point-of-care diagnostic system for infectious and chronic diseases. This activity has resulted into knowledge generation and supported the theses work of doctoral and post-graduate students from various universities and Indian Institute of Technologies and which in turn are providing their services in nation-building through serving in various academic and/or industrial set up within the country.

15.4 Conclusions and Future Prospects

Implementation of Quality infrastructure in Medical device sector in India is crucial for the National healthcare system. In this regard, CSIR-NPL is playing a pivotal role in providing primary calibration facilities for biomedical equipment. It is constitutional that the guidelines of Medical device rule 2017 are implemented and regulated with an effective outcome. This chapter highlights current regulations pertaining to medical device rules and the contributions of CSIR-NPL and Role of other regulatory bodies of India. CSIR-NPL is rapidly expanding its calibration facilities and engaged in fundamental research work to support the same, through national and international collaborations. CSIR-NPL is providing primary calibration services to medical equipment such as Defibrillator analyzer, Magnetic Resonance Imaging, Clinical thermometers and Blood pressure measuring equipment etc. besides providing training to manufactures, testing and calibration laboratories, legal metrology officers, other stakeholders in the area of biomedical metrology and biomedical sciences. The laboratory is further making efforts to enhance the calibration facilities for other life saving biomedical equipment and their calibrators such as neonatal incubators, Infusion pump analyzer and electrical safety analyzer. Apart from providing primary medical equipment calibration services, CSIR-NPL is also working towards the futuristic goals of creating secondary level calibration laboratories at different locations within India through providing traceability, which will ultimately help in the successful implementation of "Swasth Bharat Mission" of GOI. Efforts will also be made in future to establish a certification scheme for medical equipment after due consultation with CDSCO to support the indigenous Medical device manufacturers under 'Make in India' program. It has also been emphasized that the medical device regulatory bodies need to minimize their overlapping roles and work synergistically in consultation with CSIR-NPL for the improvement and advancements in the quality infrastructure.

Acknowledgements The authors gratefully acknowledge Mr. V. K. Tanwar, Mr. M. K. Pandey, Ms. Vishesh Garg and Dr. Vikash Sharma for their technical support.

References

1. S. Yadav, A. Zafer, A. Kumar, N.D. Sharma, D.K. Aswal Mapan, **33**, 347 (2018)
2. https://www.makeinindia.com/article/-/v/sector-survey-medical-devices. Accessed on 21 July 2020
3. Fact Sheet on Foreign direct investment from April 2000 to June 2019, https://dipp.gov.in/sites/default/files/FDI_Factsheet_4September2019.pdf
4. https://cdsco.gov.in/opencms/opencms/en/Medical-Device-Diagnostics. Accessed on 12 June 2020
5. First WHO global forum on medical devices: context, outcomes, and future actions, WHO publications, Version April 2011
6. https://www.indiacode.nic.in/handle/123456789/2102. Accessed on 16 June 2020
7. https://consumeraffairs.nic.in/acts-and-rules/legal-metrology/the-legalmetrology-act-2009. Accessed on 12 June 2020
8. N. Radhadevi, V. Balamuralidhara, T.M. Pramod Kumar, V. Ravi, Regulatory guidelines for medical devices in India: an overview. Asian J. Pharms. **6**, 10 (2012)
9. Gazette notification (CG-DL-E-11022020-216075) dated 11 February 2020
10. Gazette notification (CG-DL-E-11022020-216073) dated 11 February 2020
11. https://www.iso.org/standard/65529.html. Accessed on 15 June 2020
12. E. Costa Monteiro, L.F. Leon, J. Phys. Conf. Ser. **588**, 012032 (2015)
13. https://www.services.bis.gov.in:8071/php/BIS/PublishStandards/published/subcommtt?depid=NjQ%3D. Accessed on 18 July 2020
14. S. Schnorr, A.K. Srivastava, Overview of India's quality infrastructure: a guide to standardisation, conformity assessment, accreditation, market surveillance, and metrology Indo-German working group on quality infrastructure, knowledge series 1, Published by Federal Ministry for Economic affairs and Energy (2018)
15. D.K. Aswal, MAPAN J. Metrol. Soc. India **35**(2), 139–150 (2020)
16. Specific Criteria for Calibration of Medical Devices, NABL 126, Issued on 2 Jan 2018
17. https://www.nabh.co/introduction.aspx. Accessed on 16 June 2020
18. https://cddep.org/publications/covid-19-india-potential-impact-of-the-lockdown-and-other-longer-term-policies. Accessed on 21 July 2020
19. Rajesh, V.K. Tanwar, G. Sumana, V.V. Agarwal, V.N. Ojha, D.K. Aswal, Current Sci. **117**, 179 (2019)
20. S. Yang, Aaron, Y.K. Yan, J. Meas. Sci. **10**, 55 (2015)
21. E. Batistal, N. Almeida, E. Filipe, A. Costa, 16th International Congress of Metrology, 02007 (2013)
22. S. Singal, A.K. Srivastava, R.K. Kotnala, Rajesh, J. Solid State Electrochem. **22**, 2649–2657 (2018)
23. Rajesh, S. Singal, R.K. Kotnala, Appl. Biochem. Biotech. **183**(2), 672 (2017)
24. K.K. Jagadeesan, S. Kumar, G. Sumana, Electrochem. Commun. **20**, 71 (2012)
25. Rajesh, Z. Gao, R. Vishnubhotla, P. Ducos, M.D. Serrano, M.K. Robinson, A.T.C. Johnson, Adv. Mater. Interf. **3**, 1600124 (2016)
26. Rajesh, V. Sharma, N.K. Puri, A. Mulchandani, R.K. Kotnala, Appl. Phys. Lett. **109**, 243504 (2016)
27. M.A. Ali, S. Srivastava, P.R. Solanki, V. Reddy, V.V. Agrawal, C.G. Kim, R. John, B.D. Malhotra, Sci. Rep. **3**, 2661 (2013)
28. M.A. Ali, P.R. Solanki, S. Srivastava, S. Singh, V.V. Agrawal, R. John, B.D. Malhotra, ACS Appl. Mater. Interf. **7**, 5837 (2015)
29. R. Singh, G. Sumana, R. Verma, S. Sood, M.K. Pandey, R.K. Gupta, B.D. Malhotra, J. Biotechnol. **150**, 357 (2010)
30. H. Bhardwaj, M.K. Pandey, Rajesh, G. Sumana, Microchim Acta **186**, 592 (2019)
31. K.K. Reza, M.A. Ali, M.K. Singh, V.V. Agrawal, A.M. Biradar, Microchimica Acta **184,** 1809 (2017)

Chapter 16
Materials Metrology and Nanomaterials

Sanjay R. Dhakate, Bhanu P. Singh, Bipin Kumar Gupta, Kiran M. Subhedar, Vijaykumar Toutam, Sanjay K. Srivastava, Mahesh Kumar, Parveen Saini, Sushil Kumar, P. Prathap, Ritu Srivastava, Vandana, and Shailesh Narayan Sharma

Abstract Commercial impacts of technology on any nation's economy depends upon reliable measurements and tests that are accepted internationally, which helps in its continuous growth of gross domestic product (GDP). Materials play crucial

S. R. Dhakate (✉) · B. P. Singh · B. K. Gupta · Kiran M. Subhedar · Vijaykumar Toutam · S. K. Srivastava · M. Kumar · P. Saini · S. Kumar · P. Prathap · R. Srivastava · Vandana · S. N. Sharma
CSIR-National Physical Laboratory, Dr. K.S. Krishnan Marg, New Delhi 110012, India
e-mail: dhakate@nplindia.org

B. P. Singh
e-mail: bps@nplindia.org

B. K. Gupta
e-mail: guptabgupta@nplindia.org

Kiran M. Subhedar
e-mail: kms@nplindia.org

Vijaykumar Toutam
e-mail: toutamvk@nplindia.org

S. K. Srivastava
e-mail: srivassk@nplindia.org

M. Kumar
e-mail: mkumar@nplindia.org

P. Saini
e-mail: pksaini@nplindia.org

S. Kumar
e-mail: skumar@nplindia.org

P. Prathap
e-mail: prathap@nplindia.org

R. Srivastava
e-mail: ritu@nplindia.org

S. N. Sharma
e-mail: shailesh@nplindia.org

D. K. Aswal (ed.), *Metrology for Inclusive Growth of India*,
https://doi.org/10.1007/978-981-15-8872-3_16

role as building blocks in designing any new technology. Materials are the stepping stone for any technology but accurate and precise measurements with traceable calibration of equipment/devices provides the quality products for its commercial applications. For the wider acceptability of measurements of any materials and their properties demands an international collaboration associated with products which directly impact trade, innovation and hence competitiveness. On this aspect, Versailles Project on Advanced Materials and Standard (VAMAS) established in 1982. Now VAMAS has seventeen member countries particularly NMI of countries and India joined this group in 2008. The VAMAS activities conducted through Technical Work Areas (TWAs) covers most of the materials metrology. In this direction, CSIR-NPL has taken a major initiative on various types of materials development and their metrology for suitable applications which covers the aspects of synergetic materials designing as building blocks for technology development and to provide a quality infrastructure to support the major programmes of VAMAS. Moreover, CSIR-NPL is developing the national facility of quality infrastructure for materials and device testing for primary standard for solar cell including validation of solar cell performance, testing facility for body armour raw material testing. Apart from this, CSIR-NPL further focused on the development of newer materials including the emerging nanomaterials such as carbon nanotubes, electrospun nanofibres, graphene, graphene-based aerogel, organic metal complexes, quantum dots, plasmonics and metamaterials which are described in this chapter.

Numbers are the product of counting. Quantities are the product of measurement. This means that numbers can conceivably be accurate because there is a discontinuity between each integer and the next.

– Gregory Bateson (1904-1980)

16.1 Introduction

In the last few decades, rapid advances have been taken place in the area of materials technology and it is going to continue further all over the world with significant investment in new material development, processing and applications. The use of advanced materials infuses all manufacturing industries and has a strong persuasiveness on its competitiveness. However, to convert advanced materials into competitive and acceptable products, the manufacturer should have access to appropriate design methodologies which specify material property requirements. The quantitative assessment of materials behaviour and performance is important to achieve quality, reliability and safety of products. Therefore, materials metrology is a convenient way of bringing together an increasing awareness of the need for a rigorous approach to the measurements of material properties. The measurements most importantly required accurate degrees of precision and traceability that must be documented.

For the wider acceptability of the measurement of any materials and their properties demands an international collaboration associated with products which directly impact trade, innovation and hence competitiveness. On this aspect, Versailles Project on Advanced Materials and Standards (VAMAS) established in 1982 by G7 group of nations and representatives from the commission of the European communities. Currently VAMAS has seventeen member countries particularly NMI of countries and India joined this group in 2008. The VAMAS activities conducted through Technical Work Areas (TWAs) covers most of the materials metrology. In this direction, CSIR-NPL has taken a major initiative on various types of materials development and their metrology for suitable applications which covers the aspects of synergetic materials designing for making a building block to provide a quality infrastructure and to support the major programmes of VAMAS. Furthermore, CSIR-NPL is also establishing various quality infrastructures facilities i.e. Primary standard for solar cell performance, PV modules testing, body armour raw material testing, to validate the materials and products.

Recently emerging nanomaterials play lot of new breakthroughs in the direction of nanotechnology. Basically, the recognition of nanomaterials is established from exotic form of the carbon as carbon nanotube (CNT). With time newer allotropic forms of carbon have been discovered in twentieth century and more recently in the beginning of the twenty-first century, most popular allotropes in the form of CNTs and graphene. Since the discovery of CNTs in 1991, CSIR-NPL has also engaged in the development of nanomaterials and nanomaterials-based products and devices. CSIR-NPL has developed the process technology of synthesis of long length CNTs which are suitable for making CNTs buckypaper for variety of applications. The CNTs are in small object and is not in the continuous form. To overcome the drawback of CNTs, continues polymer or carbon nanofibres can be drawn by simple electrospinning technique and these nanofibres can be used in the various field.

Graphene is another form of nanomaterial with outstanding properties having numerous applications in various fields of science and technology. The most efficient synthesis method of high-quality single layer is chemical vapour deposition (CVD) technique by controlling the processing parameter particularly the deposition temperature, copper substrate quality and hydrogen flow rate. It is also possible to produce the bi-layer graphene by varying carbon vapour flux during growth. The graphene also produced from the chemical methods in bulk as compared to CVD method and graphene-based hydrogel and aerogels are possible to develop for different application.

Recently, the quantum dots, emerging organic metal complexes and plasmonic materials are found highly useful for future devices and the metrological applications in nanometrology. The certain artificially fabricated materials and crystals which can achieve optical properties that are not known to existing naturally occurring materials. Some initiative has been taken to use above said materials for futuristic optoelectronic applications.

16.2 Materials Metrology

Materials occupy a major share of the trade for any country which is a parallel outcome of gross domestic product (GDP). The success of any technology and hence the trade is ultimately linked to the quality of materials used for its development. The inconsistent variation in the materials properties and their quality can be a significant barrier for the trade between the countries. Hence, to have the barrierless trade in the materials it was essential to establish Materials Metrology which will have onus of standardization of the processing and testing of the materials. Keeping these aspects in view, a Versailles Project on Advanced Materials and Standards (VAMAS) was established following an economic summit in 1982 held at Versailles by the heads of state of the G7 group of nations and representatives from the commission of the European communities. The main objective of VAMAS is "*to promote world trade by innovation and adoption of advanced materials through international collaborations that provide the technical basis for harmonization of measurement methods, leading to best practices and standards*". VAMAS activities are managed through a Steering Committee (SC). The pre-standards research activities of VAMAS are organized into technical committees called Technical Working Areas (TWA), which are overseen by the SC and led by International Chairs. Since the inception of VAMAS, 45 TWAs have been set up, covering a wide range of pre-standards research needs for the applications of advanced materials. Till date over 500 individuals from many research groups including researchers from outside the original founding countries have participated in over 80 individual projects. "Usually, international collaboration takes place to agree on the technical basis for test method development and, using round robin intercomparisons, the test methods are checked to ensure that consistent results are obtainable at different laboratories, and provide precision (repeatability and reproducibility) data" [1].

Further to strengthen materials metrology, VAMAS has set up linkage with several international bodies. A memorandum of understanding (MOU) is signed with both International Organization for Standardization (ISO) and International Electrotechnical Commission (IEC) under which they publish Technology Trends Assessments (TTAs) based on the work of VAMAS to accelerate the development of standards in advanced materials. In July 2008, VAMAS and the World Material Research Institute Forum (WMRIF) signed a MOU for the information exchange and possible joint work items between the two organizations. A MOU on cooperation has been signed with the BIPM to facilitate the collaboration and to identify key metrological traceability issues affecting the comparability and accuracy of the measurement of materials properties. VAMAS with this networking and linkages with important metrological bodies and institutes facilitates the metrology related to materials especially with its association with ISO and IEC contributes to development of national and international standard basically through its primary activities like coordinating the pre-standards work in rapidly developing technical areas, establishing the basis of new standard technical committees, transfer of results to standards bodies leading directly to international standards, contributing to the development of reference materials, development of test methods and procedures, facilitating the increased proficiency of laboratories, including industrial laboratories, arranging for agreement of nomenclature, high-quality data generation via inter-laboratory comparison, to exercises Precision data statements and providing the reliable material properties.

To encourage materials metrology VAMAS has a well-established simple system where anyone with expertise and capability in the area of study can participate voluntarily and it is not obligatory that the participant is from the VAMAS member region. Participants have to fund their own involvement in the study. However, the test materials for the interlaboratory exercises are supplied free of charge. VAMAS website lists studies which are open for participation. Once an area of interest is identified one can contact the project leaders, TWA chair or the region Steering Committee Representative for more information for their participation. Once a new project or idea is formulated as per expertise of the participant one needs to check for the relevant TWA. If it is not available then one has to contact the secretariat for the establishment of a new TWA and if it already exists then TWA chair can be contacted for discussion of the proposal. The TWA chair and SC coordinator decide if the idea falls within scope of VAMAS and TWA. In that case the TWA chair submits a new project proposal to the VAMAS secretariat. The Secretariat then circulates to the SC for comments and then finally if a proposal satisfies then it gets approval. There are important points needs to be considered for writing new project like importance of the technical issue in terms of potential for advancement of technology, trade, sustainable growth, employment and safety, industry and user participation, involvement of advanced materials technology, significance for standards, inability to proceed with standards development on the basis of existing technology or current programme to develop new technology, Non-duplication of efforts already underway in another organization, probability of implementation of the results. Proposal prepared considering these points are further reviewed on the basis of technical focus/advanced materials technology, international participation (minimum 2 required), standardization needs/committees, non-duplication of efforts being undertaken elsewhere, timescales/implementation of results, any non-member country participation for approval.

16.2.1 VAMAS Activities at CSIR-NPL

CSIR-NPL is India's "National Measurement Institute" (NMI) by act of Parliament and is the custodian of "National Standards" with a responsibility of the dissemination of measurements to the needs of the country. In addition to maintaining standards, CSIR-NPL also focused on various materials metrology related activities. The materials metrology deals with measurement applied to the characterisation, understanding and development of materials. The characteristics of materials as a constituent of an engineered component must perform a given task and must do so in an economical and society acceptable manner. The characteristics of materials are categorized as intrinsic and extrinsic. Intrinsic characteristics are the materials composition and microstructure; it is an outcome of the processing and synthesis of

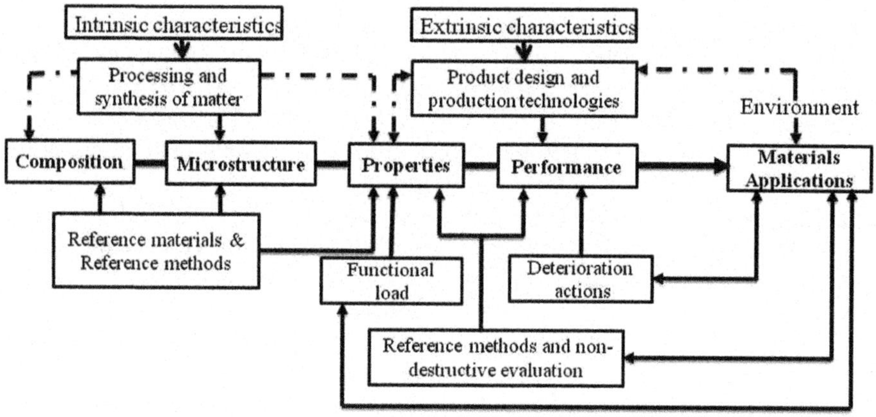

Fig. 16.1 Characteristics of material and metrology prospective

matter. Metrology and testing to determine these characteristics have to be supported by reference materials and methods. Extrinsic characteristics are the materials properties and performance; they are procedural characteristics and describe the response of materials and its products. Metrology and testing to determine performance characteristics have to be supported by reference methods and non-destructive evaluation. (Fig. 16.1).

CSIR-NPL Materials metrology programme was initiated after the visit of Prof. J. Cajoux, University of Rhiehms, France in 1986 and when he informed about the formation of VAMAS. In year 1990, Dr. Cedric Powell, Surface Science Division, NIST, USA visited CSIR-NPL and discussed about the VAMAS activities in line with available sophisticated instrument facilities and R&D activities, and proposed CSIR-NPL to become observing country in the area of surface chemical analysis (AES & XPS) of VAMAS TWA. In the year 1992 surface analytical facility project under United Nations Development Programme (UNDP) was sanctioned, under which secondary ion mass spectroscopy (SIMS) facility was established and round robin studies on depth profiling of phosphorus, arsenic, boron and nitrogen doped silicon were conducted using SIMS in collaboration with North Carolina State University in 1994 under the same project. The report of round robin test on depth profiling using SIMS is submitted to Dr. C. Powell NIST, USA and Chairman of VAMAS in 1995. In 2008 apart from G-8 countries, another six countries including India and China became members and CSIR-NPL has signed an MOU with VAMAS. The succeeding year in 2009 VAMAS signed MOU with BIPM and CSIR-NPL represented India during the 35th steering committee meeting that was held at NPL UK. Subsequently, the 36th steering committee meeting was held at CSIR-NPL in Feb 2010 for the first time in Asia on the theme "Measurements needs for emerging materials and technologies". The SC meeting was attended by 14 countries. To make the presence of CSIR-NPL in global scenario of VAMAS activities, CSIR-NPL attended 37th and 43rd SC meetings held at Rio, Brazil and BAM, Berlin.

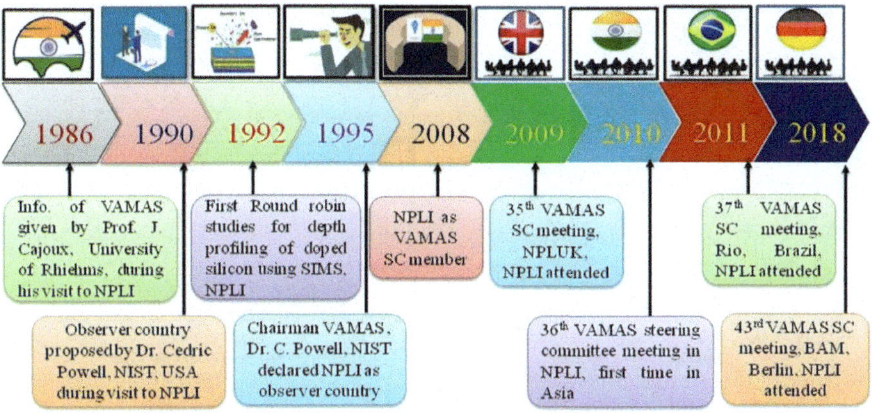

Fig. 16.2 Timeline of VAMAS activities at CSIR-NPL

Since the inception as member country, CSIR-NPL participated in several VAMAS activities and contributed to several projects related to the area of surface chemical analysis (TWA-2), nanoparticle population (TWA-34), polymer composites (TWA-33), and graphene and related 2D materials (TWA-41). The timelines of VAMAS activities of CSIR-NPL are shown in Fig. 16.2.

16.2.2 *Contribution to Technical Work Areas (TWAs) of VAMAS*

Under the umbrella of the VAMAS programme, CSIR-NPL contributed significantly during past decades and also contributed at present for different important TWAs through various activities like interlaboratory comparison, arranging meetings and workshops related to materials metrology.

16.2.2.1 TWA 2: Surface Chemical Analysis

Among these the TWA 2 [2] deals with surface chemical analysis and conducts pre-normative studies helping to ensure that international standards are robust, have inter-operability and are enduring. The importance of surfaces and interfaces for several applications across a wide range of industrial sectors from drug delivery, imaging of tissue to photovoltaics, nanoparticle catalysts, security applications and aerospace is ever increasing and a need for quantitative measurement of properties and functionalities is on high demand. Authentication of any such measurements needs a standard protocol along with a Bharatiya Nirdeshak Dravya (BND) [3] which is an Indian Certified Reference Material (CRM). Under TWA-2 namely surface

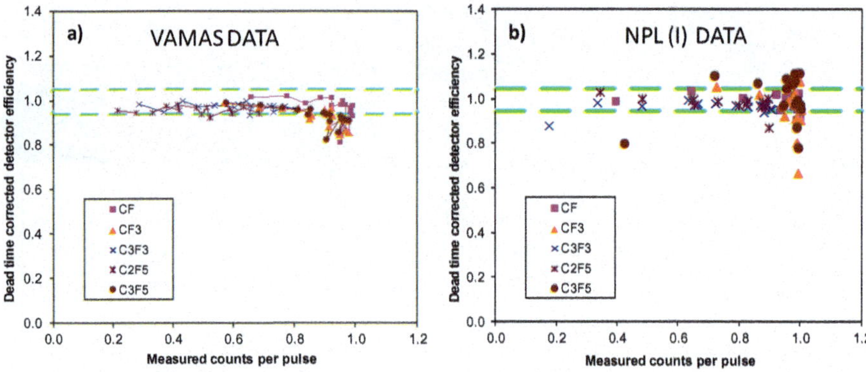

Fig. 16.3 Detector efficiency for CF^+, CF_3^+, $C_3F_3^+$, $C_2F_5^+$ and $C_3F_5^+$, calculated from the normalized intensity ratio with the corresponding isotope peak, versus counts per pulse for data with instrument software dead-time correction and upper and lower limits for linearity of the intensity scale. **a** VAMAS data, **b** CSIR-NPL data [5]

chemical analysis several projects for inter comparison were being undertaken by CSIR-NPL. An MOU signed with VAMAS in 2008, CSIR-NPL procured SIMS facility under UNDP project and continuously engaged in the VAMAS activity by collaborating with NIST, USA and NPL, UK. The main objective of this TWA was to establish standards in modern technologies involving surface treatment or depositing films, e.g. microelectronics, ion implantation, coating and plasma processing.

The first project under VAMAS was taken up by CSIR-NPL under the TWA 2, project no A 3(c)—Linearity of the intensity scale for static SIMS [4] led by NPL-UK. The standard sample of Poly-tetra-fluoro-ethylene (PTFE) supplied by NPL, UK was used for the linearity of the intensity scale for static SIMS measurements as per the protocol given in the NPL, UK report AS31 as shown in Fig. 16.3. The programme was successfully completed and results were then sent for its inclusion in ISO report.

In another project A3 (d) under the same TWA 2, organic depth profiling by cluster ion beams for static SIMS [5] was undertaken by CSIR-NPL and the standard sample for depth profiling (Irganox) was procured from NPL, UK. The depth profiling was carried out using Cs^+ and O_2^+ ion beams and the measurement was completed as per the protocol No. AS32 of VAMAS and the data was sent to NPL, UK. The main matrix of the organic sample is Irganox 1010 and the delta layer or the marker layer is Irganox 3114 having a thickness of 3 nm over the substrate material SiO_2/Si. Four such marker layers are there in the main matrix of Irganox 1010 with thickness 50, 50, 100 and 100 nm as shown in Fig. 16.4. As per VAMAS report most of the depth profiles have been done mostly by C_{60}^+ cluster ions and rarely by Cs^+ ion beam, but for the first time O_2^+ beam was used by CSIR-NPL and could detect all the four delta layers of Irganox 3114.

Another project, graphene layer number characterization by Kelvin Probe Force Microscopy (KPFM), Project no. A12 under TWA 2 was also taken up by CSIR-NPL

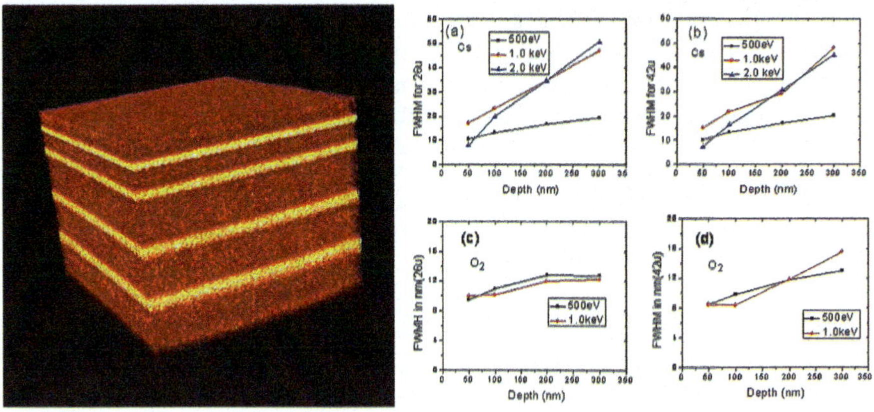

Fig. 16.4 SIMS data. **a** Total ion image showing all the four delta layers of Irganox 3114 inside 1010. **b** Variation of FWHM (depth resolution) w.r.to depth for Cs and O_2 at different energies [6]

during the year 2011. Samples of three different graphene layers on Au/SiO_2/Si and SiO_2/Si which can be clearly identified under optical microscope were supplied by National Institute for Material Science (NIMS), JAPAN. In this programme contact potential difference (CPD) measurements on graphene deposited on both gold coated and bare SiO_2/Si (001) substrates using Scanning Kelvin Probe Force Microscopy (SKPFM) were done. This could be an alternative technique to the existing techniques of layer number characterization and was carried out among 13 laboratories across the world to establish a standard protocol through international inter comparison and proficiency testing. This includes NIST USA, NIMS Japan, AIST Japan and NPL-UK along with CSIR-NPL among other laboratories. The objective of this round robin test is to establish the standard protocol for quantitative CPD measurements and to study the effectiveness of CRM generated through potential calibration.

Scanning Kelvin Probe Force Microscopy (SKPFM) is a noncontact variant of atomic force microscopy (AFM) for the direct measurement of contact potential difference (CPD) between an AFM tip and a sample surface with nanometre resolution. SKPFM has found diverse applications in material research fields as the surface potential relates to various surface phenomena.

In SKPFM, the amount of voltage to nullify the electrostatic force generated between AFM tip and sample surface due to their contact potential difference (CPD) which is the difference in work function of two surfaces in contact is estimated based on Fermi level equilibrium model. Measured CPD will also be affected by defects and residual charges present on sample surface [7, 8].

The electrostatic force generated between tip and sample due to applied bias is

$$F = \frac{q_s q_t}{4\pi \varepsilon_0 z^2} \frac{1}{2} \frac{\partial C}{\partial z} (U^2)$$

Where U is the applied potential given by $U = V_{dc} + V_{cpd} + V_{ac}\mathrm{Sin}\omega t$.

In SKPFM several scans with ac bias of different modulation voltage to the tip is done to verify miscellaneous effects due to tip contamination, substrate capacitance and nature of electrical contacts and for consistent measurement. The force/potential required to nullify the contact potential difference is monitored by ac signal when its amplitude reaches zero through counter bias. For semiconductors the CPD can be expressed as

$$V_{CPD} = \frac{1}{e}(\phi_m - \chi_s - \Delta E_{fn} - \Delta\phi)$$

whereas for metals the CPD [9] is simply the difference in work functions of tip and sample $V_{CPD} = \frac{1}{e}(\phi_m - \phi_s)$, where ϕ_m is the work function of the tip, ϕ_s is the work function of the metal sample, χ_s is the electron affinity of a semiconductor, ΔE_{fn} is the energy gap of the fermi level to the bottom of the conduction band and $\Delta\phi$ is the band bending due to surface states. Hence the comparison of SKPFM data with

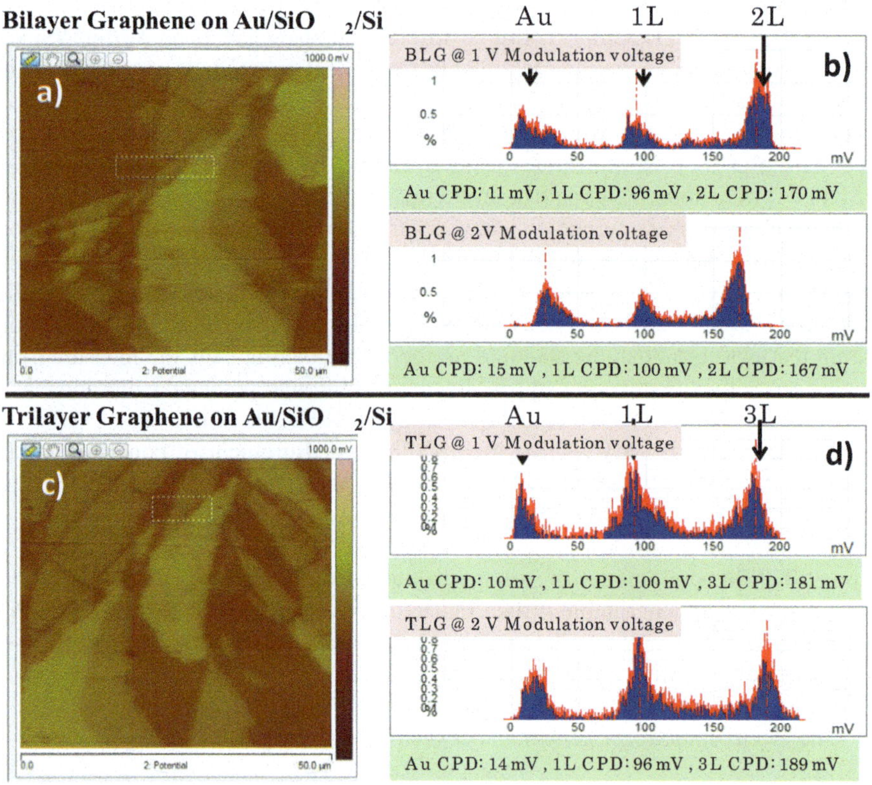

Fig. 16.5 KPFM data of Bilayer and tri layer Graphene on Au/SiO_2/Si. **a, c** Potential scan of BLG and TLG at modulation voltage ~1 V respectively and **b, d** Histogram plots of CPD for Au, BLG and TLG at 1 V, 2 V modulation voltage

two different substrates will establish the concept and is must for making reference material for graphene layer number characterization using present technique. All the results were sent to the project coordinator for intercomparison and the results were compiled as a report as shown in Fig. 16.5 which was then put in the annual report 2014 of TWA 2, VAMAS.

16.2.2.2 TWA-34: Techniques for Characterizing Morphology of Airborne Nanoparticles

Under TWA-34, Project 3: 'Techniques for Characterizing Morphology of Airborne Nanoparticles,' [6] samples Klebosol 30R25 and Klebosol 30R50 were received from KRISS Korea initiated by the project leader Dr. Charles Motzkus of LNE France. Samples both in liquid form and deposited on formvar/carbon coated TEM grids were received in the consecutive year 2011 and 2012. For liquid samples, aerosolization was done and particles were collected on the Cu grid mounted on Silicon substrate. These samples were analysed using both TEM (Fig. 16.6) and SEM for size and morphology characterization. The results were sent to the project leader of the project which got published under the corresponding TWA and the project is completed and the report is available on VAMAS webpage.

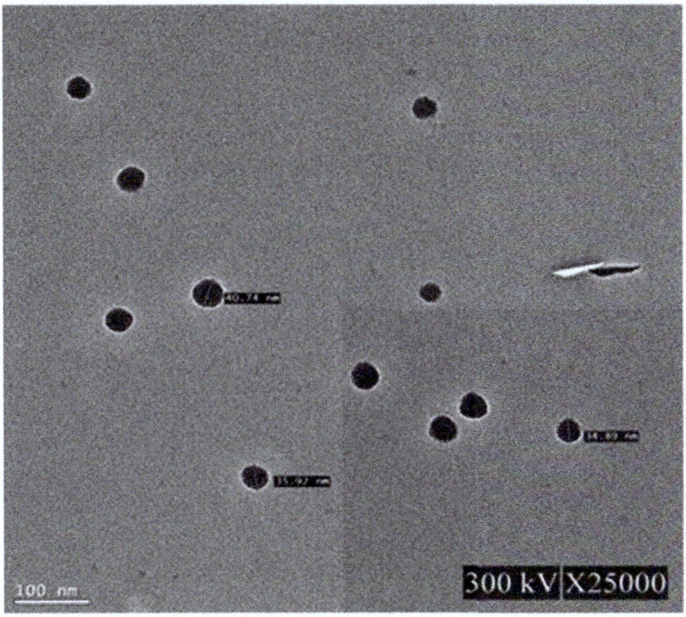

Fig. 16.6 TEM micrograph of Klebosol 30R25 sample recorded at 25KX [10]

16.2.2.3 TWA-33: Polymer Nanocomposites

The use of polymer nanocomposite (PNC) is growing rapidly in diverse industries e.g. packaging, transportation, electronics etc. But there are no internationally recognized standards for PNC testing. For the production of PNC various kinds of nanoparticle fillers are used which are very much different from commercially used fillers for the conventional composite materials. PNC exhibits remarkably improved properties when compared with the pure polymers or conventional composites because of dispersion of nanometre size filler in the polymer.

In TWA 33—Polymer Nanocomposites, two projects are continuing at present which are as follows:

1. Project 1: Determination of the shape, size and size distribution of nanofiller particles
2. Project 2: Chemical composition of nanofillers

A new project has been initiated on the mechanical properties measurement of polymer nanocomposites in which CSIR-NPL has also joined for the participation. In this TWA, graphene oxide based polyamide nanocomposites will be prepared by different processing technique and their mechanical properties will be investigated at different metrology institutes to derive a conclusion on the dispersion of graphene oxide in polymer matrix and their correlation with the mechanical properties. At CSIR-NPL, graphene oxide received from Materials Metrology Division, National Institute of Metrology, Quality and Technology—Inmetro, Rio de Janeiro, Brazil are dispersed in polyamide by using twin screw extruder and the test specimens are prepared with the help of micro injection moulding machine as per the ASTM D638-10 type V. The prepared samples have been tested for their tensile strength and Young's modulus and also circulated to other laboratories for intercomparison.

16.2.2.4 TWA-41: Graphene and Related 2D Materials

The graphene possesses excellent electronic and mechanical properties because of its unique 2D nature and its special band structure and hence finds application in a wide range of fields such as solar cell, biosensors, displays, composites, flexible electronics and energy storage etc. As a consequence of this the industry uptake on this material is continuously increasing. In order to enable the commercialization and promote the involved trade, the international standardization is critical. Moreover, for this highly explored material, different, equally important and feasible routes of synthesis have emerged. Hence reliable, accurate, and reproducible measurements are very much necessary in order to maintain quality in manufacture. So to cater the needs of the standardization specially for ISO TC229 'Nanotechnologies', jointly with IEC, including ISO/PWI 21356 'Structural characterisation of graphene', which focuses on determining the physical properties of graphene flakes and CVD-grown graphene, this TWA 41 has very much important role to play.

CSIR-NPL participated in Project 1: Structural characterisation of CVD-grown graphene: Coverage on substrate, number of layers and level of disorder. In fact, the quality of the device or product strongly depends on the graphene used for its processing. Raman spectroscopy is quite a reliable and sensitive tool to characterize the intrinsic properties both qualitatively and quantitatively. However, the technique being sensitive it needs well defined protocol with associated uncertainties to authenticate the information obtained and subsequent use of the technique.

The main objectives of this project are to validate methodology for characterisation of CVD-grown graphene supported by a substrate, using Raman spectroscopy, determine the uncertainties associated with the measurement and data analysis and input into ISO TC229 JWG2, for ISO/PWI 21356 'Structural Characterisation of Graphene. In round robins inter-laboratory comparison (ILC), CSIR-NPL participated to contribute for standardization and helps to develop best practice and understand the associated uncertainties in measurement. The CVD-grown graphene transferred onto SiO_2 is used for Raman spectroscopy measurements. The CVD-grown graphene which is used for the measurements in ILC is a single source.

Under TWA-41, CSIR-NPL is participating in another project no. 8, which is on characterization of bulk graphene materials and its derivatives which deals with studies of quantities of components, relative thermal stability and temperature for maximum weight loss rate by thermal gravimetric analyser.

16.3 Quality Infrastructure for Materials and Device Testing

CSIR-NPL is setting-up calibration and testing facilities for solar cells, PV modules, pyranometers, encapsulants materials, etc., to improve PV quality infrastructure in the country. The solar PV module testing facilities meeting the country-specific conditions would ensure its reliable performance as per the projected lifetime. The synergy of these activities would enhance the green energy yield with longer performance of PV products and also support the MSMEs in improving the product quality suitable for the prevailing weather conditions. The quality infrastructure is also going to be established for ballistic raw material testing.

16.3.1 Primary Standard for Solar Cell

Photovoltaic is one of the core research area of CSIR-NPL and large infrastructure for cell fabrication and measurements have been created. Recently, new initiative has been taken to create facilities for PV metrology and related activities. In recent past, government has taken huge initiative under "National Solar Mission". Under this initiative, across the country, major thrust on solar cell efficiency improvement along with its stability, new materials and process development is underway. Improved cell efficiencies realized in the labs with advanced cell concepts/materials need quality measurement to assess the cell parameters accurately. Large number of solar installations are being done to generate substantial amount of energy from renewable sources to address continuous energy needs. However, country is still heavily dependent on the import of solar cells and modules to achieve mission targets. These activities involve huge investments and country need PV quality infrastructure to ascertain the quality of cells/modules produced/imported for long-term performance in the field. An uncertainty in device efficiency measurement may lead to ambiguity in the product value assessment both in technical and financial terms. However, module failures in fields, low performance, poor reliability has witnessed poor national quality infrastructure (NQI) for photovoltaics. Cell measurements reliability is ascertained with the reference devices, which needs to be recalibrated at regular intervals, once procured by academic institutions, industries and research organisations in our country.

CSIR-NPL is working towards the improvement of NQI by the establishment of primary standards for reference solar cell calibration (funded by MNRE, Govt of India) which will be unique in terms of lowest measurement uncertainty in the world (0.35% at $k = 2$), large area industrial solar cell calibration and secondary reference solar cell measurement. The key quantity in the solar cell calibration is the short-circuit current generated by a reference solar radiation with 1 kW m^{-2} total irradiance and with reference solar spectral irradiance distribution [11]. The World Photovoltaic Scale (WPVS) provides a scale for PV performance measurements which was established through round robin calibration of a group of reference cells and is traceable to SI units [12]. There are only few WPVS labs worldwide. The facility which is being established is a laser based differential spectral response (L-DSR) system with feature of angular dependence and temperature effects study as per required climatic condition in the country and will be established in collaboration with PTB Germany (NMI and WPVS lab) [13]. Till now, no such facility exists in the country for traceable measurement of PV devices. CSIR-NPL is also working for the creation of solar cell efficiency test facility by utilising existing infrastructure and calibrating its subcomponents with CSIR-NPL's apex calibration facilities using the best practices. Once these facilities are established, CSIR-NPL will provide traceability to other laboratories, academic institutes, PV industries/test centres in the country for reliable PV device measurements. This will lead to self-sustainable quality measurements in the country which is essential for PV growth in the country, which will ultimately help in arresting adverse effect on the environment.

16.3.2 Validation of Solar Cell Performance

In India, the research on solar cells is being pursued in the scientific laboratories and industries; however, there is no centre for the validation of solar cell efficiency. Globally there are three recognized centres available for validation of solar cell efficiency namely NREL (USA), PTB (Germany) and AIST (Japan). In order to validate the efficiency of the fabricated solar cells, these have to be sent to one of the mentioned centres and this process is not only expensive but also time consuming. For the validation of efficiency of such devices it requires calibration of various individual parameters involved in measurement of efficiency such as light source, current/voltage meter, temperature sensors and active area of device. Being NMI of India, CSIR-NPL has standards and traceability to all SI units used in solar cell efficiency measurement. Thus, a 'National facility' for validation of solar cell efficiency was set up at CSIR-NPL (Fig. 16.7). This set-up is capable of measuring all the parameters of solar cells with the maximum possible accuracy. The facility is a potential service to nation, including academic institutions and research laboratories and industries across the country.

This established efficiency validation facility is available for all kinds of solar cells including wafer based silicon and amorphous/micro-crystalline thin films, CIGS/CZTS, dye-synthesized, organic, perovskite etc. based thin film. Standard procedure for efficiency validation of highly unstable organic and perovskite solar cells is also being developed at CSIR-NPL.

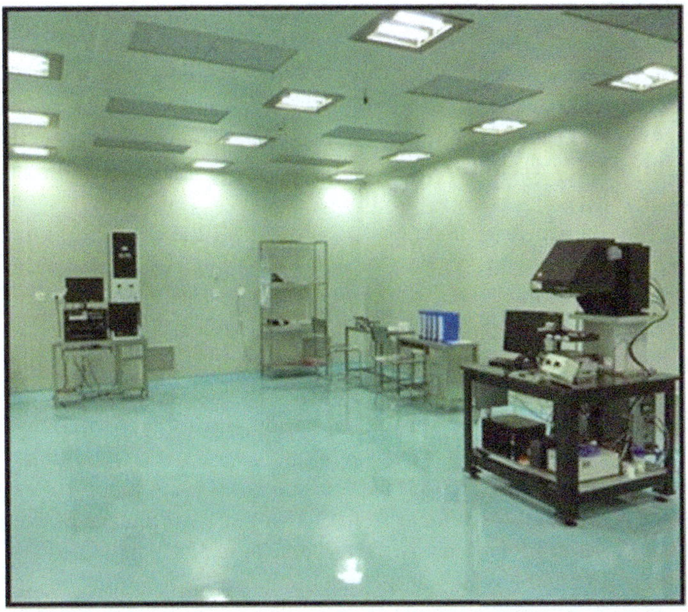

Fig. 16.7 Solar Simulators housed in clean room [ISO Class 7 (Class 10,000)] used for validation of efficiency of solar cells. Calibration of various parts of solar simulator carried out using NPL apex calibration facilities

16.3.3 PV Modules Testing

India has pledged to generate 175 GW of electricity from renewable energy sources by 2022 and reduction in carbon emission intensity by 33–35% by 2030. By 2022, Government of India plans to deploy 100 GW of solar PV installations to meet its enhanced energy demands. Almost 35 GW solar panels have already been installed. To achieve the target, deployment of PV modules is increasing at a rapid pace, although most of this is being met through imports. However, the market is hampered by underperforming, unreliable and failing products that create barriers to the development and enhancement of this renewable technology and its adoption. Few of the major factors for underperforming imported PV modules can be understood as below:

Presently, there are only few accredited labs in the country available for PV module design and performance evaluation test of solar panels as per international standards (IEC 61215/IEC61646 and IEC 61730–1 & 2) or equivalent Indian standards (IS 14286/16077/ 61730), These are: National Institute of Solar energy (NISE), Gurugram; TUV Rheinland (India) Pvt. Ltd, Bengaluru; UL, Bengaluru; Electrical Research and Development Association (ERDA), Gujarat; HiPhysix, Pune; Electronics Regional Test Laboratory (ERTL), Kolkata; Electronics Test and Development Centre (ETDC), Bengaluru. As per the new guidelines, testing labs must be accredited from NABL as per BIS standard, which is fulfilled by only few. TUV and UL test labs have their origin abroad in Germany and USA respectively. Their testing facilities are traceable to their NMI's. NISE, India is also getting traceability from other countries.

Also, in present scenario, testing capacities of all the centres are not sufficient to fulfil Indian market needs in view of expected spurt growth of PV installations. Keeping in view that there are around 700 module producers (from few MW to 1GW) in India, the number of production batches which can be tested is almost negligible. So, accredited testing and calibration facilities capacity in the country is limited and many more are needed to ascertain the quality and performance of the solar modules. At the same time, during the testing, the traceability chain involved in measurements is either missing or not mentioned. Moreover, the country is lagging in testing facilities that can provide authentic data in time, precise measurements, uncertainty and traceability in measurements as per requirement of the prevailing climatic conditions in India. This is most worrisome as PV modules are tested for design qualification and has lots of adverse financial implications on the end users due to uncertainty of measurements in the solar cell efficiency determination. To achieve 1GW power, ~ 4 Million number of PV modules with 15% efficiency is required and these panels have ~240 Million number of solar cells. To install such a high volume of module/cells, precise measurement to evaluate the cell and module along with quality assurance becomes very important parameter. The installation involves a huge amount of money and even a 1% lower efficiency (uncertainty or ambiguity in performance measurements) may lead to a revenue loss of INR ~20000 Million for 100 GW (@INR 20/Wp module cost). Moreover, there is an area loss of approximately 8% which means more space is required to install the modules for a certain lower level of power. Similarly, the stability of PV modules is guaranteed for 25 years (with 80% of peak power) which demands the degradation rate of <0.8%/year. In US and Japan, the stability is approaching 50 years with 90–95% of peak power demanding degradation rate of <0.4%/year. In India, however, a recent study of NISE, MNRE on PV module degradation, is frightening as the degradation rate is 2–5%, indicating the life of panels reducing to <10 years. This not only make achieving 100 GW net solar PV capacity a dream, but also create a huge waste in the immediate future that will be difficult to dispose-off and have detrimental impact on the environment. Some of the basic reasons for fast degradation of PV modules, are mainly the poor and inadequate quality infrastructures for the testing and calibration of PV cells and modules, non-traceability of associated testing and calibration parameters/equipments to the national standards in the country, and lack of skilled trained manpower (Source: IRENA, 2017). It is also due to the fact that PV modules are tested as per the conditions suiting the western part of the globe but not tested as per the Indian climatic conditions. In this context, CSIR-NPL is a potential partner in contributing the auspicious national programme to ascertain the module performance and reliability with the unique feature of traceability of parameters to national standards (SI units). CSIR-NPL being NMI of the country maintains all primary standards for SI units and is also establishing the primary standard for solar cell calibration as mentioned above. Govt. of India is promoting "Make in India", "Made in India" and "Skill India" programmes for country's growth and this is our collective responsibility to be self-reliant in the area which is of national importance to ensure that national mission objectives are accomplished. Therefore, our vision is to have test centres in India to be traceable to country's NMI (CSIR-NPL) in order to make country self-sustainable in long run in this sector. In this direction, first initiative is to establish test centre at CSIR-NPL where all the in-house standards facilities and quality infrastructure will be utilized for this mission. In view of the above, CSIR-NPL is in the process of establishing a "National testing centre for the traceable, precise and accurate measurements on SPV modules", which not only help Indian manufactures to test PV modules reliably in the country but also save lots of foreign exchange. In addition, this facility would be open for the International Solar Alliance (ISA) countries. Therefore, the test centres providing solar panel/module testing facilities would be traceable to SI units by utilizing apex facilities established and maintained at CSIR-NPL itself.

16.3.3.1 Potential-Induced Degradation: Implications and Strategy

The issues of reliable performance of PV systems are being crop-up due to global competition and a fast growing production capacity. In this scenario, PV metrology solutions address the issues in increasing the yield and the quality of the produced solar cells/modules. For instance, PV module materials, which play a crucial role in the long-term and reliable performance of the modules, should be assessed right at the beginning of the production chain to avoid aftermath financial implications, and the assessment should account the country-specific conditions.

Strengthening the PV quality infrastructure (QI) in the country is an antidote for all issues identified. In spite of the PV QI outweigh costs of PV systems [14], it helps to achieve a positive economic balance by increasing revenues for various stakeholders across the value chain, improving consumer protection, reducing the carbon footprint, etc. Therefore, implementation of a QI is paramount, since QI is part of a quality management system, which provides the confidence that quality requirements are fulfilled.

CSIR-NPL being the NMI of India and one of the pillars of QI, it is responsible for the national measurement infrastructure and its dissemination in the country. Since it has in-house accessibility for traceability for all required parameters, CSIR-NPL provides the best and reliable measurements. It is very important time to provide PV metrological services to the PV products being installed in the country for its reliable performance. CSIR-NPL is targeting to address the premature failures of solar PV modules by developing indigenous technologies, testing tools and related protocols. Therefore, CSIR-NPL is expanding its services in PV sector.

16.3.3.2 Potential-Induced Degradation (PID)

Failure of solar modules under the influence of high-voltage (DC) is named as potential-induced degradation (PID) and it is one of the most catastrophic risks for PV manufacturers and plant developers. A high-negative electrical potential between solar cells and the grounded module surface is a triggering factor, and the potentials are specific to installation conditions as per the electrical regulation. PID causes electric power loss, and also leads to the total failure of PV power plants [15]. The PID affected solar cells show a drastic drop of its shunt resistance. Though there are several degradation mechanisms responsible for the underperformance of PV modules, PID is known to be third most expensive technical risk with an average economic impact of €0.12 per Wp over the entire solar portfolio, following improper installation procedures and glass breakage [16]. PID effect accounts the PV module power loss > 40% within 5 years of the operational lifetime. For instance, a premature deterioration in performance, >30% in a few months/years as a result of PID has been noticed in the power plants of Delhi metro rail corporation, India [17].

The PID phenomena are not yet fully understood and it remains a big issue in the PV industry and deterrence to the investors and other stakeholders. It is a very severe "national level issue to be resolved", which results in huge financial implications that hamper the deployment of PV technology.

16.3.3.3 Strategy of CSIR-NPL

Addressing the PID effects for the development of PID resistant modules is a fourfold strategy as shown in Fig. 16.8, including test facility for module encapsulants,

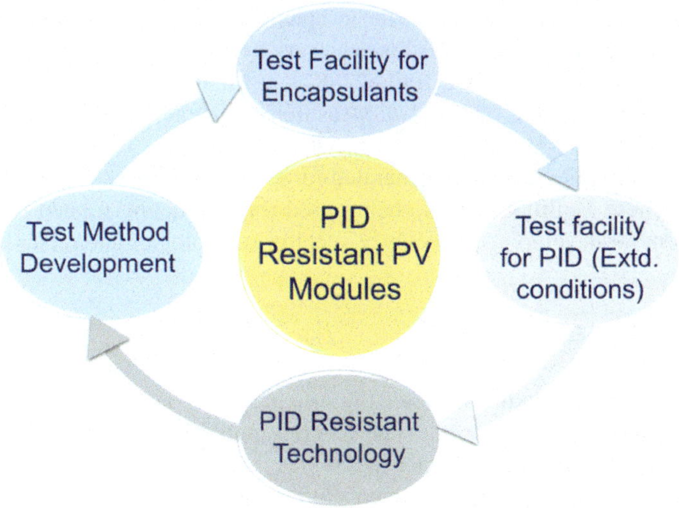

Fig. 16.8 Components involved in the development of PID resistant solar PV modules that sustain the country-specific conditions and to ensure its long-term performance as per the predicted lifetime

test facility for PID including country-specific extended test conditions, development of appropriate PID resistant technology for mitigating the PID effects followed by development of country-specific test method.

16.3.3.4 Testing Facilities for Module Materials

A large inconsistency in the design, construction and the "bill of materials" of PV modules is noticed due to rapid deployment of solar modules and competitive challenges in the cost of the PV modules. On the other hand, it has driven the manufactures to include the changes in the technology and compromising on the quality to meet short-term goals, which has a strong impact on the performance of the modules. Existing IEC standards for PV modules are only qualification standards, which can identify manufacturing-related defects only. The standards are not designed to test the lifetime performance of the modules in particular to extreme climatic conditions. This allows the use of low-cost and poor quality materials casing the immature failure risks and hence it adversaly impacts investment returns. It is therefore specification of materials and properties is critical to ensure the use of the right materials in the modules to minimize performance risks and meet specified lifetime. Several failure issues have been noticed in India as a result of encapsulant materials and appropriate testing methods are to be developed according to the requirement that ensures the reliable performance of the module.

CSIR-NPL committed to develop the test facilities as per the draft standards which have already been approved by IEC as new work items such as IEC 62775: Measurement of EVA crosslink density, IEC 62788–1-2: Measurement of encapsulant and back-sheet resistivity, IEC 62788–1-4: Measurement of encapsulant optical transmission, IEC 62788–1-5: Measurement of encapsulant shrinkage during processing, etc. The required facilities are being generated for establishing test facilities for module materials. These facilities are also being used for conducting related research and development for developing the country-specific technologies and related standards.

16.3.3.5 PID Testing Facility

To date, PID tests on modules (for testing solar modules on their PID susceptibility) are performed in accordance with the recently published IEC 62804–1 technical specification. The electrical contact to the glass surface of the modules is realized either by the humidity or by laying an aluminium foil on the module glass. The increase of temperature (and possibly of the humidity) is responsible for accelerating the degradation which is realized using large climate chambers. PID tests of solar modules on the laboratory scale cause correspondingly high investment and operating costs.

Small-scale PID tests on cell level were developed to test individual solar cells in the laboratory for PID. Non-encapsulated solar cells are used for this testing and hence both expenses on module production as well as tests in large climate chambers could be avoided. As a result of its economic advantage, it is gaining a large demand for standardization of PID tests on cell level. Therefore, test facility is being set-up as shown in Fig. 16.9 according to the SEMI Test Method **(SEMI Draft Document 5889)** on cell level for PID susceptibility of solar cells and module encapsulation materials.

The efficacy of the test set-up is demonstrated using the test structures of solar cells. PID test was carried out at a voltage stress of 1000 V for 0–6 h at the maintained temperature 80 $\pm$ 1 °C of the solar cell. The illuminated I-V characteristics of silicon solar cell before and after PID test are recorded and the results are shown in Fig. 16.10a. From the I-V curves, the solar cell parameters like open circuit voltage (V_{oc}), current density (J_{sc}), maximum power voltage (V_m), maximum power (P_m), shunt resistance (R_{sh}), series resistance (R_s), fill factor (FF), efficiency, power loss were evaluated for the cells before and after PID stress. Severe degradation in all the cells parameters was noticed as a result of voltage stress on the cells.

16.3.3.6 PID Resistant Technology

The module components such as glass, EVA, pottants, etc., are to be engineered and tested to improve the efficiency of these components against the PID effects. Several batches of solar cells/EVA/glass with impurity barrier are subjected to the PID test to verify its PID resistance using in-house developed PID test set-up with extended test

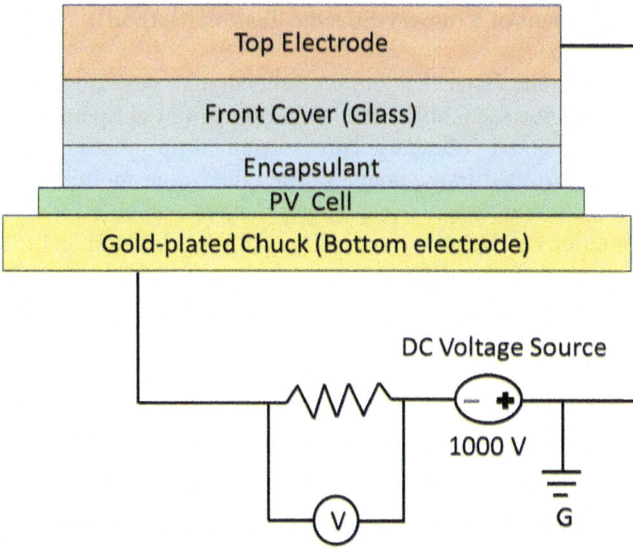

Fig. 16.9 Schematic of PID test setup

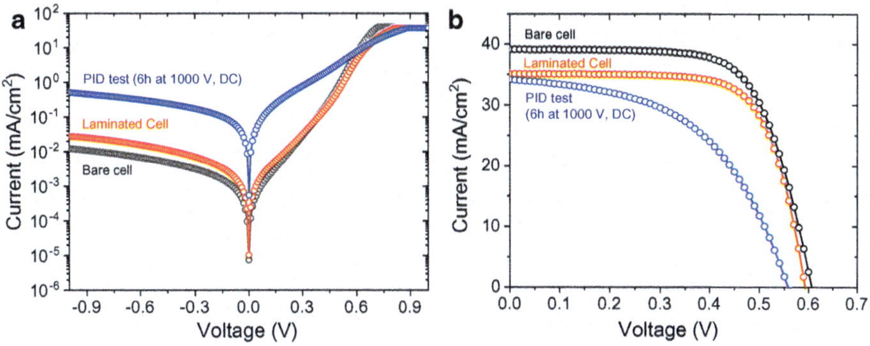

Fig. 16.10 I-V characteristics of the cells before and after PID test at ~80 ± 1 °C **a** dark and **b** illuminated measurements

conditions (85 °C, 85% RH and 1500 V). In the following, impurities diffused from different sources such as glass, EVA, etc., are being examined. As of now, there is a general consensus that Na^+ impurities diffused from glass cover sheet is responsible for the power loss of the PV module [18]. However, there are other elements that may diffuse from different sources of encapsulating environment. Detailed analysis is being made to improve the quality of the module materials.

16.3.3.7 Development of Country-Specific Testing Method

The PV modules degrade faster than the set limits though they are tested as per the standards, which do not account the real-time operating field parameters in addition to other environmental pollutants. This is essential for reliable performance of PV modules being installed in the country. Therefore, mini-module test structures are being tested employing appropriate testing methods suitable for the prevailing atmospheric conditions and high system voltages (85 °C, 85% RH and 1500 V). The studies would also supplement the findings of the IEC technical committee (TC-82) working on development of testing standard for potential-induced degradation (WG-2).

16.3.4 Body Armour Raw Material Testing

Body armour or personal armour is a protective gear specially designed to protect the wearer's vital organs from injury caused by bullets and stab attacks by absorbing their impacts while ensuring that the movement of the wearer is not unduly restricted. It was previously used only to protect military personnel, however in the current scenario of increased threat perception; it is also used to protect police, private citizens, private security or bodyguards.

Personal armour is made of either soft armour panels (SAP) or combination of soft and Hard Armour Panel (HAP). Soft Armour Panels are made up of stacked layers of woven polymeric fibres with high stiffness and flexibility such as para-aramid synthetic fibres (e.g. Kevlar, Twaran) and Ultra high molecular weight polyethylene (UHMWPE) fibres (Fig. 16.11). Upon impact of the striking bullet, the fabric absorbs the energy by stretching of the fibres while its stiffness ensures that the load is dispersed over a large area. This process slows down the bullet and ultimately stops it from penetrating the body. To provide greater protection against blunt trauma and higher velocity ammunition that cannot be provided by a SAP alone, HAP is introduced in the armour. It consists of a rigid facing comprising of ceramic inserts, steel or titanium panels with a ballistic fabric backing. In HAP with ceramic inserts, the kinetic energy of the projectile is absorbed and dissipated by the localized shattering of this ceramic tile and blunting of the bullet during its impact with the hard ceramic.

Thus, body armours are complex multi component system with unique structure of each component, and different types of bonding between different phases of the components. Hence for addressing the quality and reliability of such personal security equipments have to be achieved through dedicated raw material testing facility which is under systematic quality checks as per ISO 17025. These aspects are very important especially in the context of programme like "Make in India" wherein the quality of raw materials of the body armour, which constitute nearly 85% of the cost of the product, would primarily govern the final outlay of the body armour industry. Currently in India, a few industries are manufacturing body armours. However, their

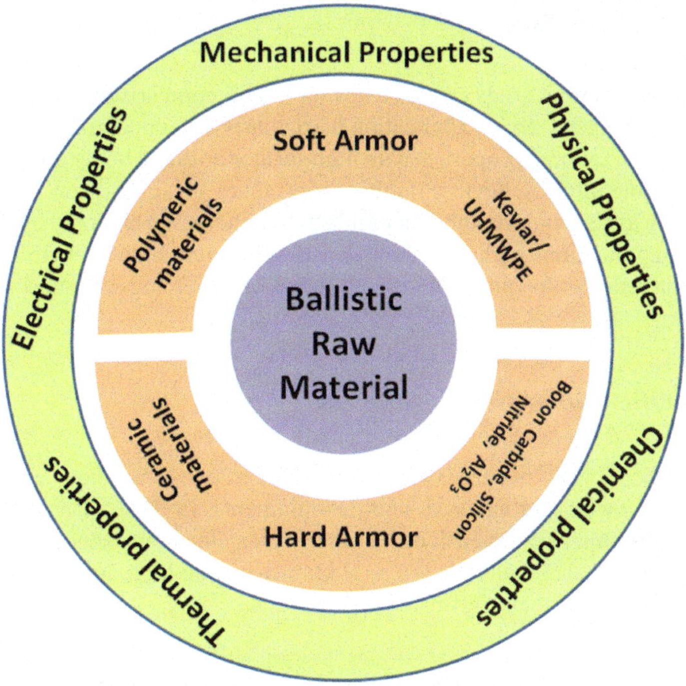

Fig. 16.11 Body armour materials and material properties required to test to insure quality

raw materials are mostly imported which leads to outflow of foreign exchange apart from their questionable quality in terms of their desired properties. Further, the ease of availability of the raw material is a critical issue, therefore, the government is considering by inviting international raw material manufacturers to establish manufacturing plant within the country under Make in India initiative.

The quality of body armour primarily depends on the properties of raw materials used for making the body armour. Hence, precise, accurate and reliable measurement of various properties of these raw materials is *Intrinsic and Intransigent*. Presently, in India there is no dedicated state of the art laboratory for ballistic raw materials testing. Therefore, CSIR-NPL is in process to establishing testing facility at par with international standards for testing of ballistic raw materials which will be developed within the country and also imported one. The precise and accurate measurement with respect to physical, structural, mechanical, thermal, electrical, optical and environment properties of ballistic raw materials as well as testing of all types of technical textiles and ceramics will be undertaken.

16.4 Nanomaterials

The impact of nanomaterials creates a new option in comparison with existing bulk material for advanced device application particularly; nanomaterials are highly efficient in terms of tailored properties with high surface area and better compactness for desired portable ultrafast devices. In this series, there are several new materials (1D and 2D) such as Carbon nanotubes, nanofibres, graphene, are explored by CSIR-NPL for various applications such as in the field of light weight high strength composites, high capacity electrode materials for batteries, EMI shielding, gas sensors etc.

16.4.1 One-Dimensional Material

One-dimensional (1D) materials are in nanoscale in two dimensions e.g. nanotube, nanofibre, nanowire and nanorod, and nanofilaments. In these materials, electrons are confined within two dimensions, in other words electrons cannot move freely in all directions. The 1D materials are well known for the high aspect ratio which is really very useful for many advanced applications.

16.4.1.1 Carbon Nanotube (CNT)

Carbon nanotube (CNT) is a seamless cylinder of few nanometre diameters and with a length ranging from a micrometre to several millimetre made of honeycomb hexagonal networks of carbon atoms. CNTs are of two major types considering the number of walls: Single walled carbon nanotubes (SWCNT) and multiwalled carbon nanotubes (MWCNTs). It is also classified on the basis of the way of folding of graphene layer into seamless cylinder, i.e. Armchair, zigzag and chiral nanotubes. In the initial studies, CNTs were synthesized by dc-arc discharge technique in which arcing between two electrodes takes place in controlled environment either in the presence of catalyst or in absence of the catalyst. In the presence of catalysts mostly SWCNTs form while in absence of catalyst MWCNTs form. The most important parameter for the growth of CNT is type and size of catalyst which support in the synthesis of desired diameter and length of synthesized CNT. The biggest challenge is production of particular type of CNT which can support for its applications in electronic industry. For semiconductor industry, armchair type of CNTs is required because of its semiconducting nature. If there is mixture of armchair and zigzag (or semiconducting and metallic type of CNTs), their utilization in semiconducting industry will not be possible. Generally at a time of synthesis a mixture of CNTs are formed but the success has been achieved by various researchers up to certain level of production of one type of CNT. Therefore, purification of one type of CNT is required from a mixture which is a big challenge for the present research community. At CSIR-NPL, CNTs have been synthesized by dc-arc discharge technique. In one of the study at CSIR-NPL, a process was developed to synthesize the SWCNT and MWCNT selectively at two different places of arc discharge reactor i.e. SWCNT in chamber and MWCNT at cathode [19, 20].

During arcing process, web-like deposits were formed on the inner walls of the arc chamber as shown in SEM micrograph of deposit (Fig. 16.12a). It shows the presence of SWCNT bundles along with the amorphous carbon and catalyst particles and called as chamber deposits. Rod-like microstructures were also found at the tip of the cathode and called as cathode deposits depicted in Fig. 16.12b. The inset in Fig. 16.12b shows the presence of sharp needle-like nanostructures which exhibited the MWCNT structure with an outer diameter of ~20–25 nm upon detailed electron microscopic examination (Fig. 16.12c). Raman spectra for both deposits are shown in Fig. 16.12d. Still further quantification of the type of CNT on the basis of chirality is required.

The biggest problem with arc discharge is the presence of large amount of catalyst in the final product sometime close to 40 wt. %. Therefore, for further application these need to be purified. During purification the structure of CNTs gets change which register into change in different properties. Therefore in recent time several other techniques have been explored for the synthesis of good quality CNT with less

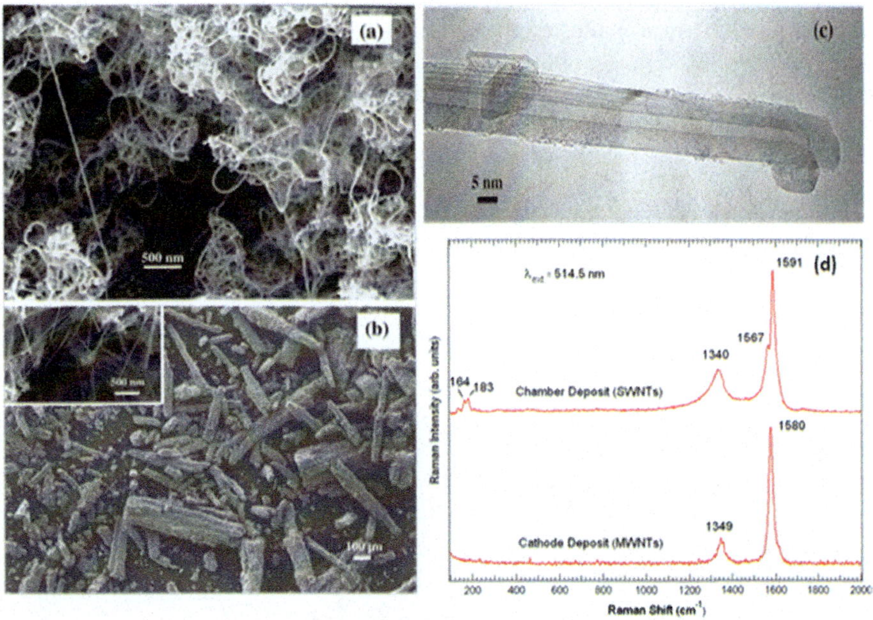

Fig. 16.12 **a** SEM micrograph of the chamber deposit showing the presence of long and flexible carbon nanotubes. **b** SEM micrograph of the cathode deposit showing the presence of rod-like microstructures. The inset figure shows the presence of needle-like nanostructures present within each microstructure **c** TEM micrograph of a single needle-like nanostructure, **d** room temperature Raman spectra of the chamber and cathode deposit. Reprinted from Carbon, 45, 132, copyright (2007) with permission from Elsevier [19]

amount of catalyst. Other than dc-arc discharge technique, several other techniques are used now a day for the synthesis of CNTs i.e. Chemical vapour deposition (CVD), Laser ablation, Fluidized bed chemical vapour deposition and High pressure carbon mono oxide decomposition (HiPCO) technique. At present for bulk production of CNTs, CVD technique is more suitable. At CSIR-NPL, an in-house CVD set-up is established for synthesis of CNTs up to 60gm/batch as shown in Fig. 16.13 using toluene as a carbon precursor and ferrocene as an organometallic catalyst [21].

Further, these as produced CNTs were characterized by SEM, HRTEM, XRD, and Raman spectroscopy and a low magnification SEM image of pristine MWCNT bundle is shown in Fig. 16.14a. The MWCNTs bundle length is approximately 450 μm revealed that as produced CNTs by CVD is long length CNTs. The long length CNTs are very important for various applications as high electrical conductivity in the polymer at low loading because of high aspect ratio, good reinforcement efficiency, applicable for making free-standing papers for energy storage devices. Figure 16.14b represents HRTEM image of MWCNT showing its diameter of 21 nm with inter-wall spacing of 0.342 nm (inset of Fig. 16.14b). XRD of MWCNTs showed a graphitic peak at the 2θ angle of 26.06 corresponds to (002) crystallographic plane (Fig. 16.14c). This shows the interlayer spacing of 3.41 Å which is close to the value for graphite. Raman spectrum shows peaks corresponding to D-band, G-band and 2D-band at 1356, 1581 and 2715 cm^{-1}, respectively (Fig. 16.14d). The ratio of G-band and D-band intensity gives the graphitic characteristic. In this study the I_G/I_D is 2.6, which represents the highly graphitic structure of MWCNTs. Now days, fluidized bed CVD technique is used in various industries for the synthesis of CNTs at large scale. Further, there is an inconsistency in terms of characteristics of CNTs in the form of their diameter, length, number of walls in case of MWCNTs, type of chirality in case of SWCNT, purity from industry to industry as well as from laboratory to laboratory. Therefore, there is a stringent requirement to develop standard operating procedure or protocols for the measurement of the characteristics of CNTs

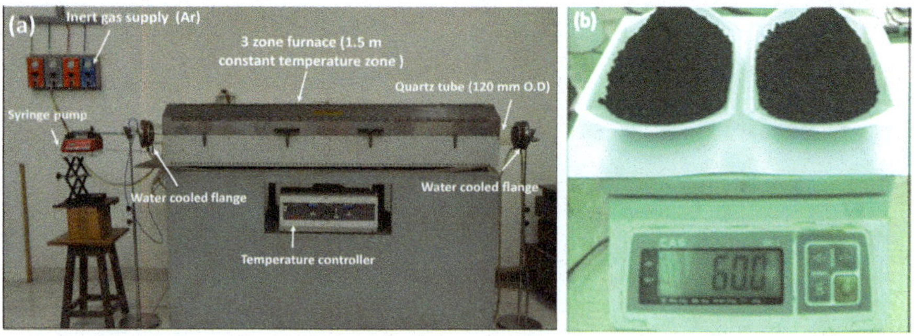

Fig. 16.13 **a** CVD set-up, and **b** Long length MWCNTs produced in single batch in CSIR-NPL. Reprinted from Carbon, 137, 104, Copyright (2018) with permission from Elsevier [21]

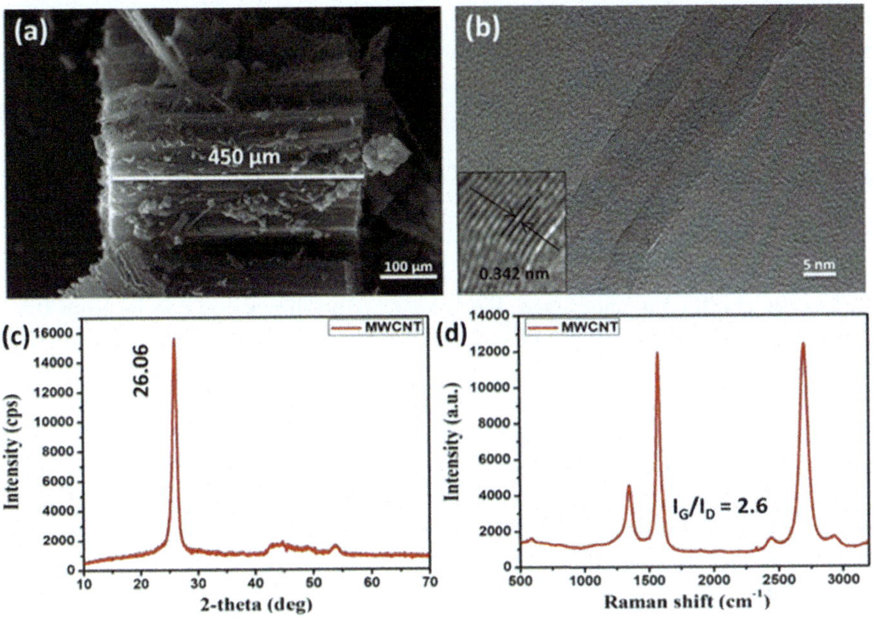

Fig. 16.14 **a** SEM of MWCNT bundle, **b** HRTEM of a MWCNT with inset showing its inter-wall spacing, **c** XRD pattern in scanning range of 2θ value 10–70° and **d** Raman spectrum of MWCNTs. Reprinted from Carbon, 137, 104, Copyright (2018) with permission from Elsevier [21]

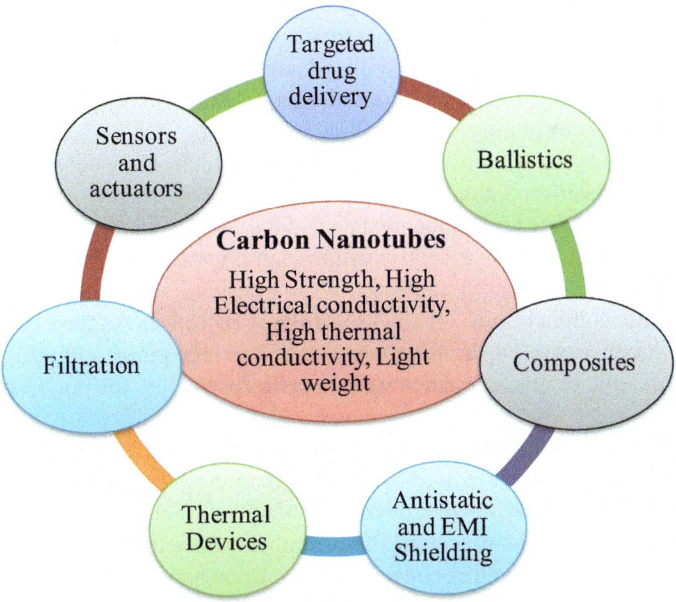

Fig. 16.15 Some of the important applications of CNTs in different field

either at the laboratory or in industries. This will help to reach into a final conclusion for searching of its end application. Fig. 16.15 show some of the important applications of CNTs.

16.4.1.2 Electrospun Nanofibres

The extraordinary properties of nanoscale materials have created tremendous motivation among the scientific community since the discovery of CNTs by Iijima [22]. In case of continuous nanofibres, the limit of nano dimension can be from 0.5 to 1 μm in contrast to > 100 nm of CNTs [23]. There are varieties of techniques for the synthesis of nanofibres and nanomaterial, these techniques produce relatively short nanofibres which are difficult to align, assemble and process into applications. Among these methods, electrospinning is a simple and versatile process to fabricate unidirectional (UD) nanoscale materials such as nanowires, nanofibres and nanotubes [24–26]. The traditional melt spinning process, spun the fibres of diameter between 5 and 200 μm, whereas electrospun fibres can have diameters in the nanometre range. In conventional fibre spinning method, mechanical force is applied at the end of jet whereas in electrospinning process, electric force acts on the elements of charged fluid. The polymeric chains are polarized and consequently attracted by the electric field creating the Taylor cone. At this point the pendant droplet of polymer solution at the needle tip is deformed into conical shape, typically referred as Taylor cone. This causes the ejection of a charged jet from a capillary tube which is elongated and accelerated by an external electric field. The polymer solution forms a continuous nanofibre when the electric forces overcome the surface tension of the fluid. If the voltage surpasses a critical value, the electrostatic force overcomes the surface tension and fine charged jet is ejected in the form of nanofibres [27–29].

The nanofibre morphology is controlled by two type of parameters, i.e. intrinsic parameters (molecular weight, concentration, surface tension, viscosity and conductivity) and processing parameters (applied voltage, flow rate of polymer solution, tip to collector distance and collector speed). So for different type of polymer-based nanofibre have been demonstrated for different applications such anode material for Lithium ion batteries and super capacitor, Water purification, drug delivery, nanocomposites, medical applications, etc.[30–36]. Figure 16.16a shows the electrospinning apparatus, it consists of DC high-voltage source, syringe pump, cylindrical metallic collector, metallic or plastic syringe with steel tip for loading the polymeric solution etc.

The different type of polymer nanofibre has been produced by electrospinning technique and this polymer nanofibre can be converted in to carbon nanofibres but it depends upon the ultimate content of carbon in the polymer. Generally the polyacrylonitrile (PAN) is used as precursor of carbon nanofibre due to high carbon yield. The PAN Electrospun nanofibre converted in to carbon nanofibre via stabilization and carbonization. Basically Pan is linear polymer which does not have stable structure during processing in to carbon nanofibres. Therefore, it has to convert in to cyclic structure by introducing the oxygen content in its structure during stabilization leads

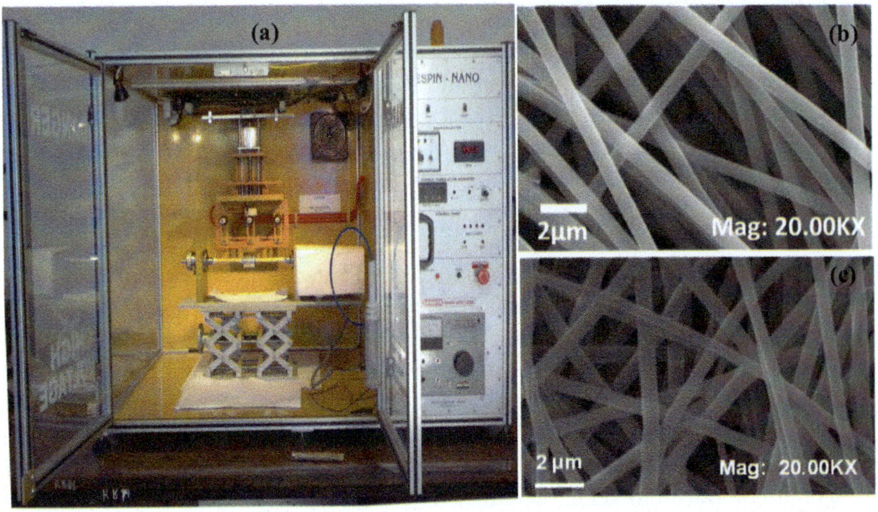

Fig. 16.16 **a** Electrospinning apparatus, **b** PAN nanofibre and **c** carbon nanofibre

to stable PAN fibre and further heat treatment in inert atmosphere above 1000 °C resulted in carbon nanofibres. Figure 16.16b and c, shows the PAN nanofibres, and carbon nanofibre heat treated at 1000 °C in inert atmosphere.

The electrospun nanofibres are now a day's used for different applications, some of the applications are depicted in Fig. 16.17. In the energy storage application, Polyacrylonitrile (PAN) based electrospun carbon nanofibres heat treated at temperature 1000, 1800 and 2200 °C were used as free standing anode material for Lithium ion battery [30]. The electrochemical performance of heat treated carbon nanofibres are evaluated by assembling CR2032 type coin cells and testing galvanostatically using

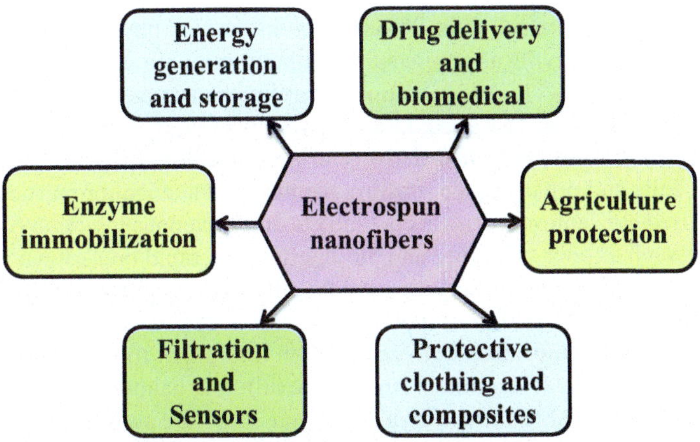

Fig. 16.17 Some of the applications of electrospun nanofibres

battery testing system. The carbon nanofibres surface area and pore size are changes on heat treatment and shows that it influences electrochemical performance. The carbon nanofibres heat treated at 1000 and 1800 °C has the better specific capacity 254 and 238 mAh/g as compared to 2200 °C heat treated carbon nanofibres (105 mAh/g) over 150 cycles at high discharge rate of 1C. The high capacity of nanofibres as compared to conventional graphite anode is due to preferable pore size, electrolyte wetting, and stable morphology of nanofibres. In most part of India, Arsenic (AsIII) is one of the contaminant in drinking water which led to massive epidemics of arsenic poisoning. To remove the arsenic from contaminated water efficiently, chitosan-polyvinyl alcohol composite nanofibres loaded with cerium (Ce-CHT/PVA) was reported [32]. The kinetic study data of adsorption reveals that more than 80% of As (III) can be removed within first 10 min time and adsorbs As (III) efficiently without presence of oxidizing agent. The Ce-CHT/PVA composite nanofibres membrane purifies water below prescribed limit of WHO/EPA up to 1500 μg/L. Besides, Ce-CHT/PVA composite nanofibres are nontoxic and can be used for water purification as such or after embedding them on non-woven fabrics, porous membranes or candles. Because of rapid and high adsorption of As (III), such materials can be useful in areas having no access to electricity.

One of the applications of electrospun nanofibres is to use as substrate for immobilization of catalyst or enzyme due to its high volume to surface ratio. Compared to free catalyst or enzymes in solution, immobilized catalyst or enzymes are more robust and more resistant to environmental changes. Specifically, the heterogeneity of the immobilized catalyst or enzyme systems allows an easy recovery of both catalyst or enzymes and products, multiple re-use, continuous operation, rapid termination of reactions and greater variety of bioreactor designs. The different types of enzymes are immobilized by different protocols on the polyacrylonitrile nanofibres membrane. *Thermomyces lanuginosus lipase* (TLL) is physically and covalently immobilized on the electrospun PAN nanofibres membrane. The immobilized TLL are used as biocatalyst for geranyl acetate synthesis with geraniol and vinyl acetate as substrates and their performance is compared with free enzyme [33]. It is found that the percentage conversion of ester is more than 90 and 66% in case of physically adsorbed and covalently bonded enzyme as compared to without immobilization. Also covalently immobilized TLL shows higher operational stability than native without immobilized and physically adsorbed TLL [33]. Drug delivery is broadly investigated because of strong demand for the controlled delivery of pharmacologically active compounds to tissues, cells and organs by controlling the physicochemical properties. As most regularly used conventional dosages are tablets or intravenous injection of drugs has drawbacks such as poor solubility and bio-distribution, rapid breakdown of the drug in vivo, and tissue damage on extravasations, the drugs rapidly lose their activity, show cytotoxicity, and lead to tissue break in addition to unwanted side effects. To overcome the problem of drug delivery through traditional routes, a lot of effort is made by using nanoscale materials. Among the different nanomaterials, electrospun nanofibres possess a large surface area and porous structure and it possible to spin in to membrane form. A biodegradable poly (vinyl alcohol) (PVA) and sodium alginate

(NaAlg) electrospun composite nanofibre-based transmucosal patch loaded with anti-diabetic drug insulin by active loading. It is found that oral formulation for an anti-diabetic drug could not only deliver it in the active form but also provide a sustained and controlled release profile via the sublingual route [34]. The same electrospun PVA–NaAlg composite nanofibres are loaded with antibiotic drug ciprofloxacin to use as transdermal patch for local delivery of antibiotic drug. The patch is used to wound healing and it is found that it control infections without any side effect and wound healing rate is higher as compared to controlled one [35].

Electropsun carbon nanofibres can also use as an interfacial layer in carbon fibre polymer composites to control the delamination between carbon fibre layers to improve the mechanical properties. The carbon fibre fabric hybrid polymer composites developed by integrating continuous semi-aligned electrospun carbon nanofibres layers in the interlaminar region between carbon fibre fabric plies [36]. It is found that interface layers with 1.1 wt% content of semi-aligned nanofibres significantly enhance the bending strength by 175% and the modulus by 200%. The greater interfacial adhesion of the high surface area of the nanofibres significantly increases the interlaminar shear strength by 190%. The improvement in the mechanical properties of the polymer composites is due to the modification of the matrix properties, crack deflection and debonding [36]. The high surface area and nano voids that form around the carbon nanofibres can absorb higher fracture energy. This is one of the promising approaches to utilized carbon nanofibres for improving the properties of carbon fibre polymer composites for different applications.

16.4.2 Two-Dimensional Material

Among recent trends in research, 2D materials are the prominent area for exploration due to its unique chemical, physical, optical, and electronic properties. The promise of the 2D material provides a new class of material of 2 dimensions which can be a better alternative as compared to bulk structure to design portable ultrathin device from metal, semiconductor (Molybdenum disulphide, MoS_2), insulator (Boron nitride, BN) to superconductor (Bismuth strontium calcium copper oxide, BSCCO).

16.4.3 Graphene

Graphene is the name for a honeycomb sheet of carbon atoms. It's striking physical, electronic and mechanical properties originate from the two-dimensional electron confinements within a one atom-thick-layer. Graphene is unique combination of extraordinary properties which are depicted in Table 16.1. This offers a fascinating material platform for the development of next-generation technologies in many areas some of them depicted in Fig. 16.18.

Table 16.1 Properties of graphene

S. No	Graphene properties	Feature
1	Atomic thickness	One atomic thick of about 0.335 nm
2	Electron mobility	With theoretical limit of 200,000 cm/(V.s)
3	Strength	~130 GPa
4	Toughness and stretchability	Relatively brittle can be stretched up to 25%
5	Stiffness	Young's modulus −1 TPa
6	High surface area	~2630 m^2/g
7	Impermeability	Impermeable even to Helium
8	Electrical resistivity	~1 × 10^{-4} Ω.m
9	Thermal conductivity	1500–2500 W/mk at RT
10	Transparency	Absorbs only 2.3% of the reflecting light

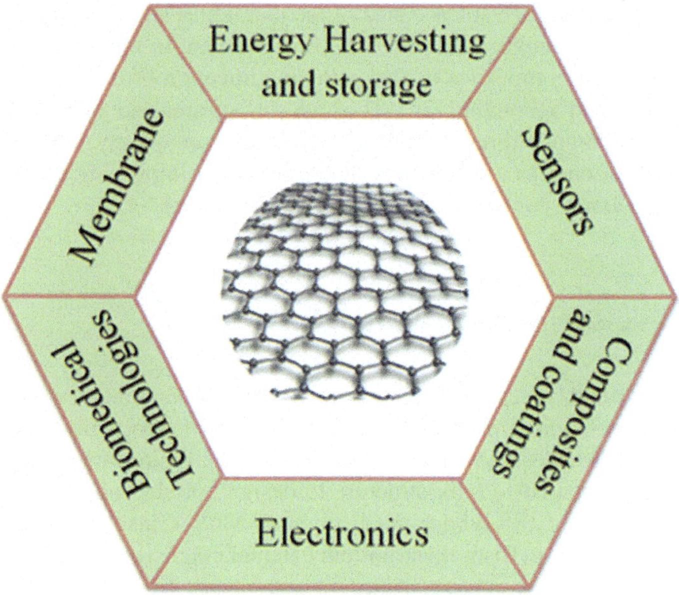

Fig. 16.18 Application of graphene in some of the prominent areas

Graphene is currently considered as the most promising potential successor to silicon for use in next-generation electronic devices. Graphene as well as transition metal dichalcogenides (TMDs) were prepared using mechanical exfoliation and CVD methods for comparative study at CSIR-NPL. Qualitative and quantitative studies have been performed using various characterization techniques. Among them, optical microscopy and Raman spectroscopy are primary techniques for quality check. The discovery of single-layer graphene with a simple but efficient micromechanical exfoliation method opened the era of a new class of material in the form of 2D materials [37].

Since then variety of techniques emerged for synthesis of single-layer graphene and among the CVD technique produces graphene with reasonably good quality and the growth process is quite versatile and scalable compared to the conventional and routinely used trial and hit based technique of micromechanical exfoliation which produces pristine graphene. In spite of it the quality of the CVD-grown graphene has not matchable to superior quality of the graphene produced by micromechanical exfoliation. Some of the critical issues associated with CVD growth of graphene which were till unexplored, have been taken up at CSIR-NPL and the growth process for high-quality single-layer graphene has been established [38–40] which was further employed for specific applications such as liquid crystal devices, MBE growth of GaN etc. A fit to purpose custom made low pressure chemical vapour deposition system (LPCVD) designed and fabricated at CSIR-NPL. To overcome the shortcomings of the conventional CVD the geometry of the reaction site is altered to tune the involved flow dynamics of the growth process for high-quality continuous graphene.

The catalyst surface plays an important role during CVD growth of graphene. The commercially available polycrystalline copper in the form of cold worked foil usually has induced stress, defects and other crystallographic imperfections and consequently leads to inferior quality of graphene with multilayers and huge grain boundaries. A novel method has been attempted to counter the said problem where the commercially available cheap polycrystalline foil was pre-treated by melting, resolidification and recrystallization which ultimately produces preferred (111) oriented copper with strain relaxed and smooth surface and leads to the high quality of CVD-grown graphene [38] as shown in Fig. 16.19.

Apart from single-layer graphene (SLG), the bilayer graphene (BLG) also offers interesting properties such as AB stacked BLG shows opening of energy gap under transverse applied electric field while the twisted stacked BLG offers intrinsic properties of SLG. Moreover, at specific angle about 1.1° so called a magic angle in BLG it shows interesting electron correlation effect and anomalous features such as unconventional superconductivity similar to cuprate superconductors [41–44].

The process has been established at CSIR-NPL to control the layer stacking in BLG by CVD growth and was studied with Raman spectroscopy. The two types of BLGs, one usual AB stacked BLG and other twisted or misoriented BLGs can be grown at specific CVD condition and a plausible mechanism for both of these stacking has been proposed [39]. Layer stacking in BLG can be controlled by varying carbon vapour flux during growth. Carbon vapour flux is controlled by using two different geometries of reaction sites. Geometry facilitated by a surface controlled process gives bernal stacked whereas diffusion dominated process gives twisted stacked graphene. Figure 16.20 shows Raman spectra and curve fitted for 2D peak of (a) twisted bilayer graphene (b) single-layer graphene (c) AB stacked graphene grown with different geometries of reaction site.

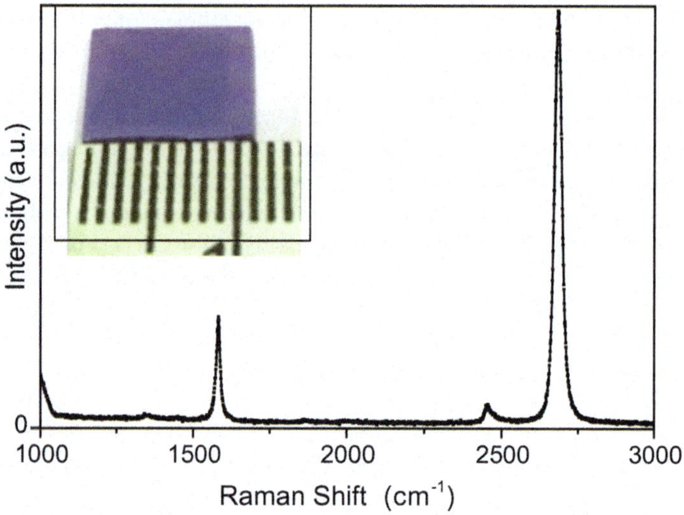

Fig. 16.19 Raman spectra of CVD-grown single-layer graphene on the Si/SiO_2 substrate with a prior process of resolidification and recrystallization (the inset is the corresponding photograph). Reprinted from Ref. [38] with permission from Royal Society of Chemistry [38]

Fig. 16.20 Raman spectra and curve fitted for 2D peak of **a** Twisted bilayer grapheme **b** single-layer grapheme **c** AB stacked graphene grown with different geometries of reaction site. Reproduced from Ref. [39] with permission from PCCP Owner Societies [39]

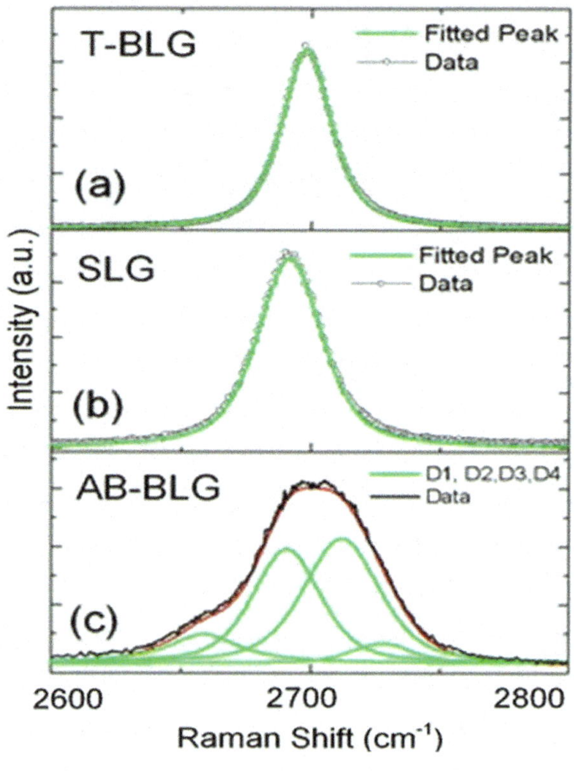

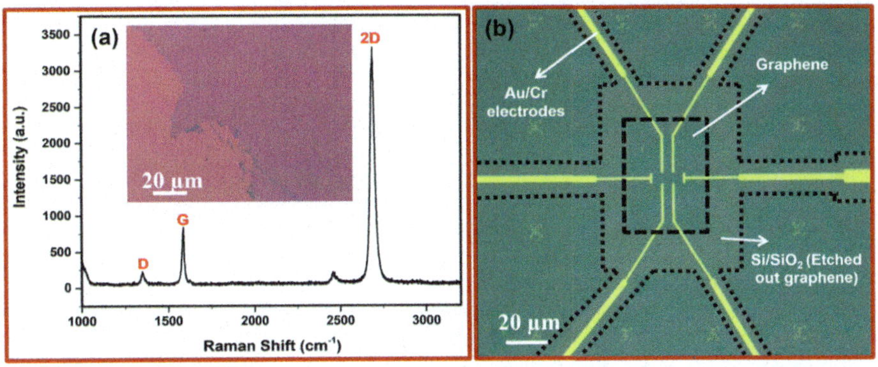

Fig. 16.21 **a** Raman spectra of CVD graphene with an optical image (inset) of graphene on Si/SiO_2 **b** quantum hall measurement device fabrication by sputtering and e-beam lithography on graphene

Further there is inconsistency of the properties across process to process as well as laboratory to laboratory with the same CVD-grown process of this extraordinary and unconventional immerging material. Hence it highly needs standard operating procedures (SOPs) and protocols for graphene synthesis and measurement of its properties so as to have the unanimous conclusion arrived from results contributing from different places on this with vast and highly explored material.

In another study graphene was grown by the CVD method with very good I_{2D}/I_G ratio of 5 using methane gas as a source gas, and hydrogen gas acts as co-catalyst at a high temperature of 1000 °C. Hence, low-cost, high quality, large size, and highly reproducible graphene can be obtained by using indigenously developed CVD at CSIR-NPL, New Delhi, and can be used for graphene-based quantum devices and metrology (Fig. 16.21).

The large specific surface area (planar nanostructure), tunable electrical properties, high mechanical strength, the possibility of chemical functionalization, and stability has made it most popular and suitable two-dimensional (2D) material to use it as sensing layer material for gas sensors at room temperature. These properties empower graphene-based gas sensors to perform well in terms of accuracy, fast response, high sensitivity, long-term stability, selectivity, low detection limits. Further alterations in graphene lattice through doping/functionalization additionally enhances the sensing performances and capabilities. Graphene-based gas sensors are also advantageous in terms of easy fabrication and low power consumption as compared to other traditional gas sensors. This provides basics of gas sensors, discussion on various sensor characteristic parameters (sensitivity, selectivity, stability, response/recovery time), details on graphene-based gas sensors, followed by sensing performance of graphene& its derivatives for two most studied gases NO_2 and NH_3 at room temperature as shown in Fig. 16.22. Finally, the current challenges and prospects of graphene for high sensing performances are presented.

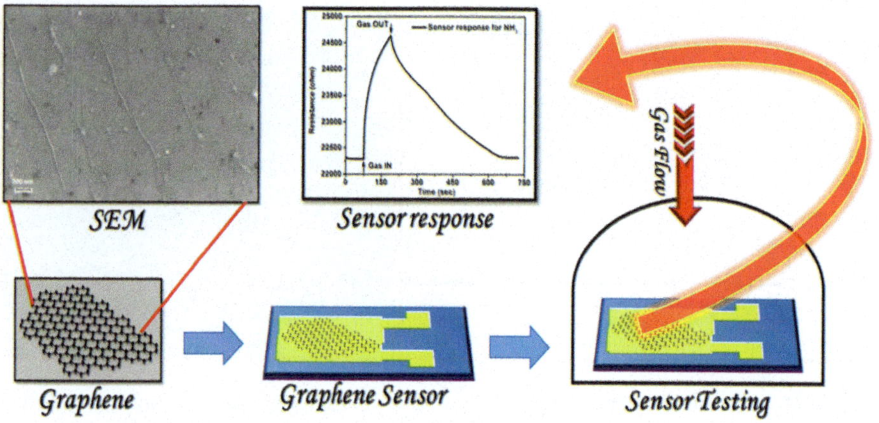

Fig.16.22 Schematic illustration of ammonia gas sensor with CVD-grown Graphene gas sensor

16.4.3.1 Organic Metal Complex

The organic compounds that emit light on excitation using UV light or electric field are known as the organic phosphors. Different types of organic phosphors are: (a) Metal complexes (b) Organic dyes and (c) Polymers Metal complexes synthesized as light-emitting materials are highly purified organic materials complexed with metal ions. The metal ions usually are rare-earth ions, transition metals, and metals from I, II, and III groups of the periodic table. Some complexes have more than one organic ligand and in such cases, the neutral ligand is used as second organic ligand.

Luminescent metal complexes have been attracting much concentration because of their possible application in flat panel displays based on organic light-emitting diodes (OLEDs). These materials are useful in OLEDs because of their relatively high stability and efficiency. In metal complexes, ligand centred π-π^* electronic transitions are dominant and responsible for colour emission. The metal ions are used to stabilize the structure of the luminescent ligand. The choice of metal ions for luminescent metal chelates is limited to those metals which do not exhibit d-d transitions that may interfere with the luminescence of the ligand. Therefore, Lithium (I), Aluminium (III), Boron (III), Beryllium (II), zinc (II), etc. are efficient for luminescent applications since these materials (except Zinc) have no d electrons. For zinc, d shell is filled and no d-d transition is possible in zinc complexes. Metal complexes are placed in the categories of organic materials (also known as organic semiconductors). The use of the name semiconductor is based on the extrinsic semiconducting properties of organic systems, i.e. the capacity to transport charge generated by light, injected by electrodes, or provided by chemical dopants. The term 'organic' includes small-molecule organic materials as well as polymeric materials that may be used to fabricate organic electronic devices. Metal complexes having organic ligand part also fall in the category of small-molecule organic materials. Small molecules may be quite large and may include some repeated units.

The core moiety of a dendrimer may be a fluorescent or phosphorescent small molecular emitter. The use of organic materials in electronic devices makes such devices relatively inexpensive, easy processability, and cost-effective over inorganic materials. Also, the inherent property of organic materials, such as their flexibility, may make them well suited for particular applications such as fabrication on a flexible substrate.

Metal complexes emit light across the whole visible spectrum, organic materials are readily incorporated in electrical devices that are inexpensive to fabricate on a large scale. Organic light-emitting devices based on metal complexes may be easily grown on glass or flexible transparent plastic substrates. They operate at low voltage and are Lambertian sources, meaning that the colour of organic displays is independent of viewing angle. Small molecular OLEDs have been found to allow better tailoring of the properties of the material such as band gaps and electron affinities, as well as purer materials without a contamination concern. CSIR-NPL has developed a novel method to designed and synthesized lithium metal complexes, zinc metal complexes, and iridium (III) complexes. The structural and optoelectronic properties of these synthesized complexes are well studied. Besides emissive, these materials show good electron-transporting properties. Among these materials, zinc complexes have been especially important because of the simplicity in synthesis procedures and wide spectral response.

White OLEDs have been fabricated by choosing two zinc complexes (Zn $(HPB)_2$ and ZnHPB(Me_2q)). Synthesis of Zn $(HPB)_2$ has been done according to the known literature method. The emissive layer of different thickness was placed to control the emission colour of OLED. ZnHPB (Me_2q) emit light in the oranges-yellow region and Zn $(HPB)_2$ emit light in the blue region of the visible spectrum. White light has been achieved by the combination of two complementary colours. These fabricated devices showed very good colour stability with applied voltages. The Commission International d'Eclairage (CIE) coordinates lied well within the white region [45]. Different WOLEDs developed at CSIR-NPL using fluorescent and phosphorescent metal complexes are shown in Fig. 16.23.

16.4.3.2 Luminescent Quantum Dots

The size dependent emission property for colloidal semiconductors like Cadmium Selenide (CdSe) and Indium Phosphide (InP) quantum dots renders it indispensable in the field of nanocrystal-based emitters such as light-emitting diodes, photovoltaic devices, lasers, biomedical tags, etc. For this reason, the control of the PL properties by controlling the size/surface modification of the semiconductor nanocrystals has been a major goal in the synthesis of colloidal semiconductor nanocrystals.

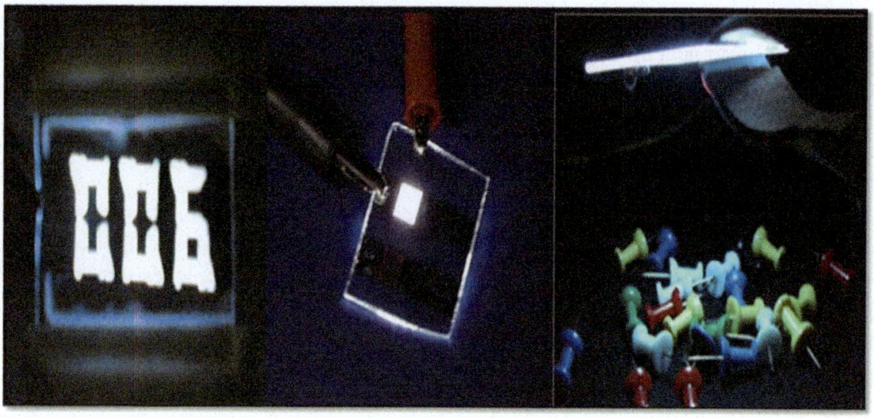

Fig. 16.23 White organic light emitting diode (WOLED) developed at CSIR-NPL

16.4.3.3 CdSe Quantum Dots

CSIR-NPL demonstrated a single-pot chemical route to synthesize CdSe quantum dots. The quantum dot sizes have been varied from 5–14 nm by changing the Cd: Se initial precursor ratio (0.5:1.0–3.0:1.0). On the basis of XPS and other complimentary techniques, it has been found that CdSe quantum dots produced by the chemical route upon variation of the input precursor ratio exhibits three compositional regimes, viz., Se rich, stoichiometric and Cd rich. For the stoichiometric regime (2:1) smallest sized (Eg ~ 2.33 eV), monodispersed and highly luminescent CdSe QDs (λ max ~510 nm) could be obtained where as for non-stoichiometric regimes (Cd/Se- 0.5:1.0, 1.0:1.0:1.0, 3.0:1.0), signature of polydispersity were visible and also Cd or Se related oxides were observed. From the XPS and PL measurements it was inferred that stoichiometric CdSe QDs were stable against oxidation (oxygen & light) with no evidence of surface states unlike the Se and Cd rich QDs. The surface of the CdSe quantum dots have been modified by changing the cap layer of the nanoparticles, either by organic (amines p-phenylenediamine (PPD)) or inorganic materials (ZnSe). 2:1 CdSe system is chosen as the parent material. The changes in the photophysical properties of CdSe nanocrystals arising from the interaction with PPD are compared with an aliphatic amine, n-butylamine (n-BA). The transient absorption measurements confirm the formation of PPD cation radical and subsequent formation of coupling product. Formation of such charged species at the surface extends the bleaching recovery over several microseconds which will aid in understanding the blinking behaviour of CdSe nanocrystals. An effective, simple and reproducible single-pot (SP) synthetic method has been developed for the formation of highly luminescent, stable and composition-tunable core–shell CdSe-ZnSe and ternary ZnxCd1-xSe semiconductor nanocrystallites. The emission wavelength across the visible spectrum can be tuned by merely varying the Zn content (0–40 wt. %) in core–shell and ternary quantum dots without much variation in the crystallite sizes (5.0–6.5 nm). The CdSe-ZnSe core–shell QD corresponding to Zn ~10 wt. % showed extremely high quantum yield ~11.0 times higher than that of standard Rhodamine 6G dye.

16.4.3.4 Highly Luminescent Blue-Emitting InP Quantum Dots

CSIR-NPL demonstrated a strategy of extremely small blue-emitting colloidal InP-based quantum dots (size ~2–4 nm) (In:P ~ 1:1–3:1). The method reported here is unconventional, quite rapid (~90 min), more viable, less expensive and relatively greener as compared to other conventional methods that employ tris(trimethylsilyl) phosphine (P(SiMe3)3) which is rare, expensive, flammable, highly toxic and even banned in a few countries. Highly luminescent InP QDs having bluish-green emission (λ ~ 490 nm) can be synthesized using this method without resorting to any post-synthesis etching to tune the emission to the blue region. Besides being the source of phosphorus and the particle size regulating agent, the efficacy of TOP is further realized during synthesis via its reduction of indium salt, which aids in the formation of indium metal and then subsequently in the development of InP QDs. The PL intensity of as synthesized InP QDs is further enhanced by growing a shell of wide band gap material, i.e. ZnS resulting in a concurrent increment in quantum yield from ~25% to ~38% respectively.

16.4.4 Plasmonics and Metamaterials

The creation of plasmonic materials is one of the latest technology-oriented research works which is highly useful for future devices and the metrological applications in nanometrology. This involves the generation of surface wave at the interface between a noble metal and a dielectric material because of strong interaction between incident electromagnetic wave and free surface electrons of metal. This surface wave resulting from collective electron oscillations under the influence of electromagnetic radiations is composed of quasiparticles known as surface plasmon polaritons (SPPs) and decay exponentially while propagating along the interface between metal and dielectric.

The coupling of the light by noble metal can be enhanced by confining the SPP into 2D (metallic nanowire) or 3D (metallic nanoparticle) through nanostructuring and thus, there is a strong electromagnetic field confinement for size $\ll$ wavelength resulting into the enhancement in optical field intensity. These localized SPPs show sensitivity towards the shape, size, and material properties of the metallic nanostructure. Thus, the surface plasmonic coupling can lead to the development of super-resolution imaging, efficient energy harvesting devices, faster and miniaturized forms of integrated optical devices with high performance etc.

There are certain artificially fabricated materials and crystals which can achieve optical properties that are not known to exist naturally. These materials are called metamaterials and are made up of the periodic or random arrangement of subwavelength structures called the unit cell. Most of the naturally occurring crystal or material have their defined electrical permittivity (ε) and magnetic permeability (μ) > 0. However, when the materials are fabricated with a periodicity in lateral dimensions it can yield the electrical permittivity (ε) and magnetic permeability (μ) to be < 0. This behaviour can be extremely exploited for the use of the creation of ultrafast and miniaturized devices.

Plasmonics made nano-optics feasible by beating the diffraction limit of light while metamaterials helped in achieving the properties like negative refraction, focusing through flat slab, chirality which are not found in nature. Plasmonics and metamaterials were two separate fields till the optical regime. In the optical regime the metamaterials stopped following the scaling low because of two reasons (a) metals started showing plasmonic affect and (b) inertia of electron came into picture. These complications led to saturation in resonance frequency above ~100THz. Later researcher found that an optimum change in dimension and shape of metamaterials in agreement with plasmonic property of metal can give new dimensions to nanoscience and nanotechnology.

Of late the plasmonic metamaterials have been developed in the form of various forms of resonating structures operating in the different frequency regimes especially in the Terahertz frequencies. Thus, they are highly useful for the detection, generation and utilization of Terahertz frequencies for various applications. Another useful application of plasmonic metamaterials is superlensing where these plasmonic metamaterials can enable sub-wavelength imaging.

The ease of fabrication of metamaterials is one of the key components in its success. These plasmonic structures can be formed either by exposing to acidic treatment leading to the rough surfaces composed of noble metal clusters and exposed part of the dielectrics or by using FIB, EBL or femto second(fs) -laser writing method. These structures will have a changed refractive index and more absorption in the VIS–NIR regions along with higher lifetimes. These parameters can be used to fabricate the photoconductive materials to be used as sensors along with energy harvesting and can be used as one of the steps for efficiency improvements in PV. The creation of sharp interfaces will be possible using ALD and PVD techniques.

Plasmonic metamaterials can also act as plasmon rulers where the change in the lateral distance of nanoparticles by just 1 Å can change the SPR by 5 nm. This property can be utilized to create plasmonic rulers in nanometrology. While the development of this technology is full of challenges but the availability of the nanofabrication technique like FIB and fs laser writing method can make it feasible.

The studies are performed on ultrafast transient spectroscopic measurements of the plasmonic structures and utilising the plasmon resonance energy transfer, Fortser resonance energy transfer and created the periodic structures using fs laser and etching using acid treatment [46, 47]. These structures have shown promising charge transfer and energy transfer processes. Also the metamaterials have been simulated theoretically which are also being fabricated by EBL and photolithography [46–48] (Fig. 16.24). The interaction of electromagnetic waves with various materials over a span of visible to NIR radiation is highly important. The need to explore the

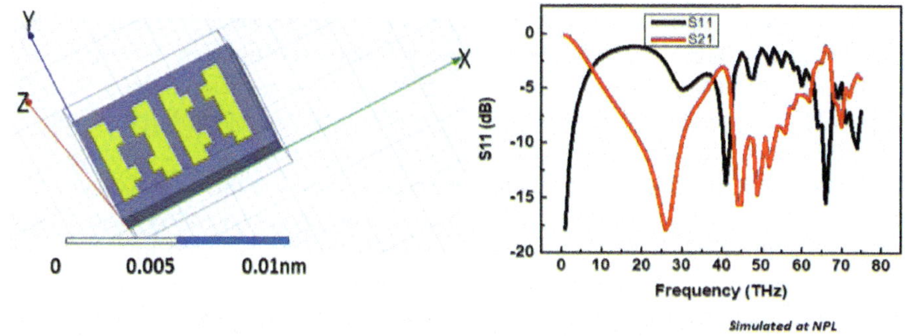

Fig. 16.24 Simulated structures for THz frequency using gold structures

science behind the interaction of these EM waves with materials and their interfaces will lead to the development of new EM devices; thus, it affects the metrological parameters as well for nanoscale dimensions to bulk materials. These include the GHz and THz frequency response of the new and advanced materials to develop into a new materials metrology area, which will include the parameters to optimize like temperature, dimensions, frequency etc. THz technology has direct impact on our quality of life in terms of health care (detection of dangerous chemicals, identifying cancers), public security (dangerous weapons and explosives identification) and quality control (medicine and food inspection etc.).

16.5 Conclusions and Future Perspective

CSIR-NPL is a member of VAMAS since 2008 and contributing in various VAMAS activities by participating in different TWA projects. In the line of materials metrology programme, CSIR-NPL is establishing national facility under quality infrastructure for materials and device testing such as primary standard for solar cell including validation of solar cell performance and body armour raw materials testing facility creation to insure the quality and reliability of ultimate personal protective gear used by lakhs of our security persons.

The CSIR-NPL is also working on various nanomaterials such as CNTs, graphene, electrospun polymer nanofibres, metal organic complex, quantum dots, plasmonic materials. CSIR-NPL is contributed in the synthesis of indigenous high-quality CNTs which are used for making self supporting buckypaper, nanocomposite, energy generation and storage devices etc. The high-quality single-layer graphene is synthesized in the laboratory for applications in electronic devices and metrology. The electrospun polymer nanofibre has been produced by controlling various processing parameters for different applications. Further, PAN based nanofibres are converted in to carbon nanofibre for energy storage devices applications.

The materials metrology programme is a stepping stone towards making the self-reliant India by addressing the problem of various fields through development of nanomaterials and related technology. Thus, the indigenous development of materials with their reliable properties is not only ensuring the quality but also enhance competiveness which can directly impact growth and hence GDP of country.

To make the global impact in the competitive world, industries are needed to demonstrate traceability of materials data used in design, traceability of measuring and test equipment, comply with regulations and standard. Government bodies and regulators are constantly required to adapt international good practices to insure materials that are internationally competitive to make country self-reliant. In continuation of importance of the materials metrology for the sustainable growth of materials, CSIR-NPL contribution in development of advanced materials for strategic and societal applications is emphasized in the next chapter.

Acknowledgements The authors are grateful to all technical staffs of Advanced Materials & Device Metrology for their support.

References

1. https://www.vamas.org/formation.html. Accessed on 30th July
2. www.vamas.org/twa2/index.html. Accessed on 30th July
3. www.nplindia.in/bhartiya-nirdeshak-dravya-bnd-indian-reference-materials. Accessed on 30th July
4. J.L.S. Lee, I.S. Gilmore, M.P. Seah, Surf. Interface Anal. **44**, 1 (2012)
5. B.R. Chakraborty, P. Mohan, S.M. Shivaprasad, D.R. Sharma, C. Anandan, A.C. Gupta, A.K. Raychaudhuri, Curr. Sci. **78**, 1523 (2000)
6. B.R. Chakraborty, A.G. Shard, M.K. Dalai, G. Sehgal, Surf. Interface Anal. **46**, 36 (2014)
7. I.D. Baikie, P.J. Estrup, Rev. Sci. Instrum. **69**, 3902 (1998)
8. B.R. Chakraborty, S.K. Halder, K.K. Maurya, A.K. Srivastava, V.K. Toutam, M.K. Dalai, G. Sehgal, S. Singh, Thin Solid Films **520**, 6409 (2012)
9. P.M. Bridger, Z.Z. Bandić, E.C. Piquette, T.C. McGill, Appl. Phys. Lett. **74**, 3522 (1999)
10. www.vamas.org/twa34/documents/2010_vamas_twa34_p3_airborne_nanoparticles.pdf. Accessed on 30th July
11. J. Metzdorf, S. Winter, T. Wittchen, Metrologia **37**, 573 (2000)
12. C.R. Osterwald et al., Prog. Photovoltaics Res. Appl. **7**, 287 (1999)
13. S. Winter, T. Fey, I. Kröger, D. Friedrich, K. Ladner, B. Ortel, S. Pendsa, F. Witt, Measurement **51**, 457 (2014)
14. www.irena.org. Accessed on 30th July
15. I. Lillo-Bravo, P. González-Martínez, M. Larrañeta, J. Guasumba-Codena, Energies **11**, 363 (2018)
16. D. Moser, M. Del Buono, U. Jahn, M. Herz, M. Richter, K. De Brabandere, Prog. Photovoltaics Res. Appl. **25**, 592 (2017)
17. www.delhimetrorail.com. Accessed on 30th July
18. W. Luo et al., Energy Environ. Sci. **10**, 43 (2017)
19. R.B. Mathur, S. Seth, C. Lal, R. Rao, B.P. Singh, T.L. Dhami, A.M. Rao, Carbon **45**, 132 (2007)
20. R.B. Mathur, C.L. Nil, T.L. Dhami, B.P. Singh, A.K. Gupta, J. Ghawana, US Patent 7,955,663, 2011

21. S. Sharma, A.K. Pathak, V.N. Singh, S. Teotia, S.R. Dhakate, B.P. Singh, Carbon **137**, 104 (2018)
22. S. Iijima, Nature **354**, 56 (1991)
23. J.F. Hagewood, A. Wilkie, Nonwoven World **12**, 69 (2003)
24. S. Ramakrishna, K. Fujihara, W.-E. Teo, T. Yong, Z. Ma, R. Ramaseshan, Mater. Today **9**, 40 (2006)
25. Z.-M. Huang, Y.Z. Zhang, M. Kotaki, S. Ramakrishna, Compos. Sci. Technol. **63**, 2223 (2003)
26. D.H. Reneker, I. Chun, Nanotechnology **7**, 216 (1996)
27. S.R. Dhakate, B. Singla, M. Uppal, R.B. Mathur, Adv. Mat. Lett. **1**, 200 (2010)
28. S.R. Dhakate, A. Gupta, A. Chaudhari, J. Tawale, R.B. Mathur, Synth. Met. **161**, 411 (2011)
29. S. Sinha, S.R. Dhakate, P. Kumar, R.B. Mathur, P. Tripathi, S. Chand, Biores. Technol. **115**, 152 (2012)
30. A. Gupta, S.R. Dhakate, P. Gurunathan, K. Ramesha, Appl. Nanosci. **7**, 449 (2017)
31. A. Gupta, P. Gurunathan, K. Ramesha, M. Singh, S.R. Dhakate, Energy Storage **1**, e89 (2019)
32. R. Sharma, N. Singh, A. Gupta, S. Tiwari, S.K. Tiwari, S.R. Dhakate, J. Mater. Chem. A **2**, 16669 (2014)
33. A. Gupta, S.R. Dhakate, M. Pahwa, S. Sinha, S. Chand, R.B. Mathur, Process Biochem. **48**, 124 (2013)
34. A. Sharma, A. Gupta, G. Rath, A. Goyal, R.B. Mathur, S.R. Dhakate, J. Mater. Chem. B **1**, 3410 (2013)
35. K. Kataria, A. Gupta, G. Rath, R.B. Mathur, S.R. Dhakate, Int. J. Pharm. **469**, 102 (2014)
36. S.R. Dhakate, A. Chaudhary, A. Gupta, A.K. Pathak, B.P. Singh, K.M. Subhedar, T. Yokozeki, RSC Advances **6**, 36715 (2016)
37. K.S. Novoselov, A.K. Geim, S.V. Morozov, D. Jiang, Y. Zhang, S.V. Dubonos, I.V. Grigorieva, A.A. Firsov, Science **306**, 666 (2004)
38. I. Sharma, S.R. Dhakate, K.M. Subhedar, Mater. Chem. Front. **2**, 1137 (2018)
39. K.M. Subhedar, I. Sharma, S.R. Dhakate, Phys. Chem. Chem. Phys. **17**, 22304 (2015)
40. M. Borah, D.K. Singh, K.M. Subhedar, S.R. Dhakate, RSC Advances **5**, 69110 (2015)
41. Y. Cao, V. Fatemi, S. Fang, K. Watanabe, T. Taniguchi, E. Kaxiras, P. Jarillo-Herrero, Nature **556**, 43 (2018)
42. Y. Cao, D. Chowdhury, D. Rodan-Legrain, O. Rubies-Bigorda, K. Watanabe, T. Taniguchi, T. Senthil, P. Jarillo-Herrero, Phys. Rev. Lett. **124**, 076801 (2020)
43. M. Huang et al., Nat Nanotechnol **15**, 289 (2020)
44. M. Yankowitz, S. Chen, H. Polshyn, Y. Zhang, K. Watanabe, T. Taniguchi, D. Graf, A.F. Young, C.R. Dean, Science **363**, 1059 (2019)
45. A. Kumar, I. Singh, R. Srivastava, *Synthesis and Opto-Electronic Characterization of Some Metal Complexes: Metal Complexes for Organic Light Emitting Devices, Applications* (Lambert Academic Publishing, LAP, 2016)
46. G. Kedawat, I. Sharma, K. Nagpal, M. Kumar, G. Gupta, B.K. Gupta, ACS Omega **4**, 12626 (2019)
47. A. Sharma, C. Sharma, B. Bhattacharyya, K. Gambhir, M. Kumar, S. Chand, R. Mehrotra, S. Husale, J. Mater. Chem. C **5**, 618 (2017)
48. A. Rana, N. Gupta, A. Lochan, G.D. Sharma, S. Chand, M. Kumar, R.K. Singh, J. Appl. Phys. **120**, 063102 (2016)

Chapter 17
Advanced Materials for Strategic and Societal Applications

Sanjay R. Dhakate, Bhanu P. Singh, Bipin Kumar Gupta, Kiran M. Subhedar, Sanjay Kumar Srivastava, M. Saravanan, Parveen Saini, Sushil Kumar, P. Prathap, Saroj Kumari, Rajeev Kumar Singh, Asit Patra, Vandana, Pankaj Kumar, Rachana Kumar, Priyanka H. Maheshwari, Bhasker Gahtori, Jai Prakash Tiwari, Kriti Tyagi, and Jeyakumar Ramanujam

Abstract Recently, the demand of the material manipulation with specific design is highly required for several strategic and societal applications such as in defence, nuclear, security, space, waste management, renewable energy generation and

---

S. R. Dhakate (✉) · B. P. Singh · B. K. Gupta · K. M. Subhedar · S. K. Srivastava · M. Saravanan · P. Saini · S. Kumar · P. Prathap · S. Kumari · R. K. Singh · A. Patra · Vandana · P. Kumar · R. Kumar · P. H. Maheshwari · B. Gahtori · J. P. Tiwari · K. Tyagi · J. Ramanujam
CSIR-National Physical Laboratory, Dr. K. S. Krishnan Marg, New Delhi 110012, India
e-mail: dhakate@nplindia.org

B. P. Singh
e-mail: bps@nplindia.org

B. K. Gupta
e-mail: guptabgupta@nplindia.org

K. M. Subhedar
e-mail: kms@nplindia.org

S. K. Srivastava
e-mail: srivassk@nplindia.org

M. Saravanan
e-mail: saranm@nplindia.org

P. Saini
e-mail: pksaini@nplindia.org

S. Kumar
e-mail: skumar@nplindia.org

P. Prathap
e-mail: prathap@nplindia.org

S. Kumari
e-mail: kumaris@nplindia.org

R. K. Singh
e-mail: rajivsingh@nplindia.org

D. K. Aswal (ed.), *Metrology for Inclusive Growth of India*,
https://doi.org/10.1007/978-981-15-8872-3_17

storage. To address these challenges, it is important to design the materials synergetically fit according to their demand in different applications. The impact of such materials is directly related to our daily life from health, wealth, energy and environment. In this direction, CSIR-NPL is designing new materials to indigenize the several new technologies such as luminescent security ink to curb against counterfeiting of passport, Indian currency note and protect secrete documents, carbon materials for fuel cell and battery applications, carbon composite fuel tubes for nuclear reactor applications, carbon fibre polymer composite to replace steel Ilizarov fixator for biomedical applications, lightweight carbon nanotubes polymer composites as EMI shielding materials etc. Moreover, in spite of this, CSIR-NPL is also developing the different technologies to manage the reuse of plastic, electronic, PV module and agriculture waste to ensure safeguard of environment and fully recycling of material for the societal applications. Hence, the central aim of CSIR-NPL is to design and develop the materials with metrological grade to convert into all laboratory research invention to innovation, which will directly help all industrial sectors to promote their quality products with world class infrastructure.

I care about how many people's lives I've positively impacted. That's the unit of measurement I measure myself by.

–Brock Pierce (1980-)

A. Patra
e-mail: apatra@nplindia.org

Vandana
e-mail: vandana1@nplindia.org

P. Kumar
e-mail: pankaj@nplindia.org

R. Kumar
e-mail: rachanak@nplindia.org

P. H. Maheshwari
e-mail: hedap@nplindia.org

B. Gahtori
e-mail: bhasker@nplindia.org

J. P. Tiwari
e-mail: tiwarijp@nplindia.org

K. Tyagi
e-mail: kriti.tyagi@nplindia.org

J. Ramanujam
e-mail: ramanuj@nplindia.org

17.1 Introduction

Human endeavour has created materials that nature did not make. It is now possible to design and develop new materials according to desired properties and specifications which are globally competitive, economic and environmental friendly. Materials research has a special role to play in technology prudence due to a new technology depending on the availability of raw materials. Initially more emphasis has been vested on the indigenous development of raw materials to cater the need of industry, which has grown stronger with time and decisive on advanced materials and devices development with focus on the range of materials for strategic, industrial and societal applications.

In the last few decades, rapid advances have taken place in the area of materials technology and it is going to continue further all over the world with significant investment in new material development, processing and applications.

In this chapter various efforts undertaken in the recent development of advanced material and devices are presented. These advanced materials include a range of fields from structural, electrical, optical and thermal applications. Carbon is an important class of material due to their unique physico-mechanical and chemical properties which is used in almost all branches of science and technological applications. Natural carbon occurs in two allotropic forms such as soft graphite and hard diamond. With time newer allotropic forms of carbon have been discovered in twentieth century and more recently in the beginning of the twenty-first century, most popular allotropes in the form of carbon nanotubes (CNTs) and graphene are discovered. There are different sources of solid carbon; one of them is coal tar. The coal tar is a by-product obtained during the destructive distillation of metallurgical coal processed by the steel industry for the production of coke. Coal tar is a complex mixture of phenols, polycyclic aromatic hydrocarbon and heterocyclic compounds. The coal tar is converted in to coal tar pitch by distillation process which has around 50% solid carbon. The coal tar pitch is a starting material for development of different value-added carbon products. The materials which are strategically important for the country's progress, but are not available at any cost due to the embargo of developed nations which is required indigenous development. Carbon fibre-carbon matrix composite i.e. Carbon–carbon composite is a lightweight high-temperature-resistant material for rocket nozzle and space vehicles applications.

Since the discovery of CNTs in 1991, CSIR-NPL has also engaged in the development of nanomaterials and nanomaterials-based products and devices. CSIR-NPL has developed the process technology of synthesis of CNTs buckypaper as self-supporting anode material in Li-ion batteries for energy storage devices. Due to superior thermo-mechanical properties of the CNTs, these are used as reinforcing material for the development of nanocomposite as well as hybrid composites. The CNT-based composite possesses superior mechanical and electromagnetic interference (EMI) shielding properties. On the other hand, another technologically important material is luminescent material that is able to emit light without heating in response to stimulus. There are two types of photoluminescent materials; fluorescence and phosphorescence. These materials has a wide range of applications in such as advanced optical displays, photovoltaic, biomedical, anti-counterfeiting, solid-state lighting etc,. CSIR-NPL is working on different type of luminescent material and device developments such as Lanthanide-based luminescent materials, radioactive-free long afterglow phosphor, bi-luminescent and magnetic ink, red fluorescence indelible ink and nanotaggants composed of nanophosphors incorporated in the polymer matrix to form a nanofibre-based mat for anti-counterfeiting applications. Additionally, the other materials such as zeolite as molecular sieves for gas separation of O_2/N_2 from air, CO_2/CH_4 from natural gas, CO_2/N_2 from flue gas etc.

The de-fossilization of the energy sector is a goal as per the Paris agreements. To ensure this, fossil sources used to produce energy must be reduced and with time these sources are depleting, therefore renewable energy sources to be adapted to save the mother earth. In the renewable energy sector particularly silicon photovoltaic (PV) solar cells, CSIR-NPL has been working in crystalline and multicrystalline silicon solar cells development since the late 1970s. With time due to advancement in the technologies, microcrystalline and amorphous thin-film silicon solar cells technologies have been developed. Recently black Si (both mono- and multicrystalline) solar cells with efficiency of >17% of solar cell size 156 × 156 mm^2 have been demonstrated. CSIR-NPL is also working on the organic PV solar cell and it is a rapidly emerging technology with improving cell efficiency and it has potential to manufacture on a roll-to-roll basis. Furthermore, a new initiative on perovskite material-based solar cells has been taken and work is under process. It is a low-cost and industry-scalable technology, in which perovskite materials offer excellent light absorption, charge-carrier mobility and lifetimes. The hydrogen-based energy technologies also become increasingly imperative, such as production of hydrogen and hydrogen storage as an energy source. Owing to wide range of structural characteristics of carbon, it has various applications such as in high-temperature nuclear reactors in the form of carbon–carbon composite and high density isotropic graphite. In this line high density isotropic graphite is being developed for nuclear reactor programme of atomic energy. The hydrogen-based fuel cell is expected to play a key role in future energy systems for various applications. The polymer electrolyte membrane (PEM) fuel cell is a clean source of energy, even though different types of carbon materials are used, such as carbon nanomaterials as catalyst support, porous conducting carbon paper as gas diffusion layer and graphite composite bipolar plate as hydrogen distributor over the gas diffusion layer and mechanically support membrane-electrode assembly. Indigenous porous conducting carbon paper and graphite composite bipolar plate of properties equivalent to imported one has been employed in the PEM fuel cell.

With increasing the utilization of various products and devices for the well-being of human life and on the other hand it generates a lot of waste, such as plastic, electronic, batteries, solar modulus waste as well as agriculture waste. CSIR-NPL has taken initiative for sustainable development through waste management by recycling and converting the waste into value-added products. India being agriculture dominated country, more than 600 million tons of biomass waste in the form of crop stubble is generated every year. To use waste biomass as suitable renewable energy source, CSIR-NPL has taken initiative to convert it into biocoal by torrefaction process of calorific value equivalent to coal which is used in the thermal power plant. India also generates around 10-15 million tonnes of plastic waste annually, of which 40% remains uncollected. The uncollected waste plastic is generally thrown away in open space that creates an environment as well as health problems. In this direction, a process technology for recycling of plastic waste into paver tiles, interlock tiles, roof tiles which can be used in footpaths, pathway in garden, fall ceiling etc. These initiatives solve societal as well as environmental problems arising due to plastic waste. As the solar PV installed capacity increases with time, in the same severity the solar PV module scrap will be generate after 20–25 years and disposal of it will become a paramount problem in coming decades. Thus, there is a need for environment friendly recovery of valuable materials from waste and used photovoltaic modules. In this direction, CSIR-NPL is working on establishing of recycling and reuse technologies to address key issue of our social responsibility. This will ensure to safeguard of environment and ensures the recycling of the material without any waste. CSIR-NPL has also initiated the work on recycling of silicon solar cells/modules.

17.2 Advanced Materials for Strategic and Energy Applications

CSIR-NPL also aims towards synergizing the research and development to develop state-of-the-art bulk and nano-scale materials, process technologies and devices for Strategic, Energy and Societal applications. The thrust is on the development of indigenous, economic and efficient carbon-based advanced materials and composites, thermoelectric materials and devices, luminescent materials, organic and inorganic photovoltaic materials and devices, and thin-film-based devices.

17.2.1 Materials for Strategic Sector

Design and development of strategic materials are on high demand in various industries and commercial sectors to fulfil the demand of strategic materials. CSIR-NPL has taken the challenge to address such issues in a vast material-based programme to build the nation ahead in this direction. Therefore, CSIR-NPL developed several strategic materials such as carbon composites for nose tips of missiles, fuel tubes for nuclear reactors, CNT-based ballistic materials, CNT and graphene-based polymer composites for microwave shielding, luminescent materials for security ink, luminescent quantum dots and organic metal complex.

17.2.1.1 Carbon-Based Materials

The carbon atom is a building block in nature and it is at the origin of life on our planet especially because of the complexity of its chemical bonds. Its special ability with various types of bonding and molecular hybridization results in numerous compounds. Some of its basic allotropes like diamond and graphite have been known from a long time. In past 70 years researchers have discovered several new forms of carbon such as carbon fibre, hard carbon, carbon nanotubes, diamond-like carbon, nano diamonds, graphene and its derivatives, carbyne, activated carbon, carbon coil, carbon/carbon composites etc.

Carbon–Carbon Composite

Carbon/carbon (C/C) composites are composite materials consisting of carbon fibres as reinforcement and carbon matrix derived from coal tar pitch or thermosetting phenolic resin [1, 2]. The C/C composite possesses low density, high mechanical strength, high thermal stability, high electrical conductivity, high thermal conductivity, low coefficient of thermal expansion and excellent fracture toughness. The C/C composites are the thermally stable material up to 3000 °C in inert or vacuum environment without any degradation in mechanical properties and as a result C/C composite is used as structural material in spacecraft and advanced aircraft applications [3]. The processing of the C/C composite consists of preparation of carbon fibre preform that may 2D, 3D, 4D and 5D. The carbon fibre preform impregnated by binder coal tar pitch followed by carbonization and graphitization. This process gives the porous C/C composites due to evolution of volatile products during the carbonization and graphitization process. Therefore, to achieve the desired bulk density it is required to impregnate by impregnating coal tar pitch followed by carbonization and graphitization under high pressure. After obtaining the desired density of C/C composite, it is machined into required shape and size for particular application. In 1983, Dr. A.P.J. Abdul Kalam (Ex-President of India), then Director, Defence Research and Development Laboratory, (DRDL), Hyderabad approaches CSIR-NPL for developing specific grade of C/C composites for making various components, particularly the nose tip of Agni missile. Figure 17.1a shows the scaled down nose tip was made in CSIR-NPL, which passed all the DRDL tests. This technology of C/C composite has been taken up by DRDL for scaling up and for developing associated components.

In nuclear energy sector, carbon is used as moderator and reflector due to its neutron interaction characteristic, stability under irradiation, mechanical strength, low atomic weight with high scattering probability and low capture probability. In India, Bhabha Atomic Research Centre, Department of Atomic Energy, perceiving the various indigenous nuclear reactors programme. The high-temperature reactor development programme was initiated because of a need to develop alternative energy carriers to substitute fossil fuel-based transport fuel [4]. Hydrogen production processes by thermo-chemical splitting of water need either electricity or process

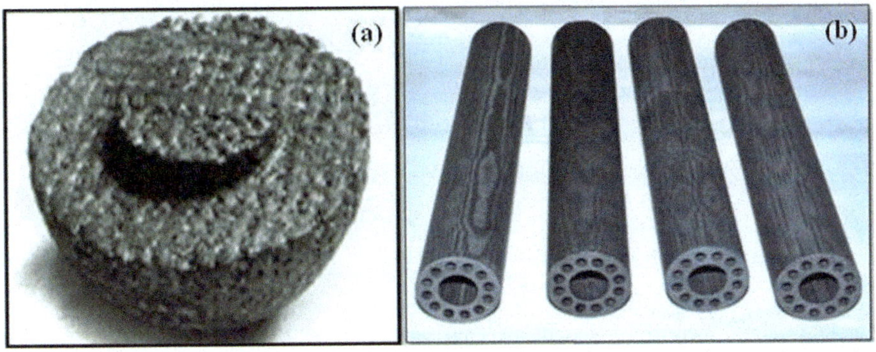

Fig. 17.1 **a** Scale down carbon–carbon composite nose, **b** Carbon–carbon composite fuel tubes

heat at high temperature, or both depending upon the process of hydrogen production selected. High-temperature nuclear reactors have a large potential for sustainably supplying energy for hydrogen production. The BARC has undertaken the development of a compact high-temperature reactor (CHTR), which has a prismatic core configuration consisting of moderator blocks [5]. These blocks contain centrally located fuel tubes. Fuel tubes would be made of high density carbon–carbon composite or high density isotropic graphite. Each fuel tube carries fuel compacts inside longitudinal bores made in its wall. Due to strategic nature, this material is not excess to country from any other sources. Therefore, indigenous development of these materials is absolutely necessary. CSIR-NPL in collaboration with BARC has developed the process of C–C composites tubes. These C/C composite fuel tubes were of size 500 mm length, 75 mm diameter and 35 mm inner diameter with the tubes having 12 number of 10 mm diameter holes up to 450 mm depth (Fig. 17.1b).

High Density Isotropic Graphite

High density isotropic graphite is a special type of carbon product derived from the processing of carbon raw material such as pitches. There are few methods available for the development of high density graphite including conventional filler and binder method, using self-sintering coke, and hot moulding. Out of these, filler and binder is the conventional method which uses coke as filler and pitch as binder, mixture of these two is isostatically pressed followed by carbonization. After carbonization, pores are created due to removal of volatiles present in the binder (pitch) so to fill the pores, product is impregnated by repeated cycles of impregnation and re-carbonization before graphitization up to 2700 °C. Other method involves self-sintering carbonaceous materials such as mesocarbon microbeads and semi-coke both are prepared from coal tar pitch followed by their compaction into product form without using any extra binder, carbonization and then graphitization. Use of mesocarbon microbeads is not economical as yield of mesocarbon microbeads which is synthesized from

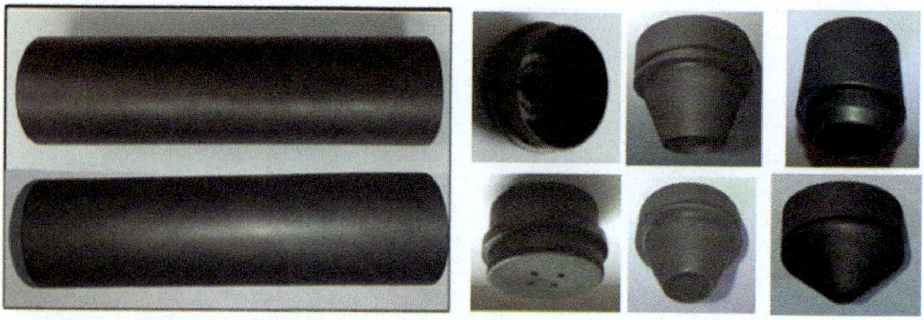

Fig. 17.2 High density isotropic graphite for nuclear reactor applications and high density graphite-based collector components of travelling wave tube for amplify the RF signal

coal tar pitch is very low ~20 to 30% and process of making mesocarbon microbeads itself is not very simple. Another carbonaceous material is semi-coke prepared from heat treatment of coal tar pitch with yield of 60–70%. This material is very good for making high density isotropic graphite products of superior properties without any extra binder. Semi-coke powder is compacted into iso moulds using isostatic press and carbonized at 1000 °C and then finally graphitized at 2700 °C to get graphitic structure of desired properties.

High density isotropic graphite is one of extremely useful carbon products as crucibles and moulds for molten metals, hot pressing dies and in blast furnace as linings and fixtures due to its stability at higher temperatures. Due to its low secondary electron emission coefficient, it is used as collector of travelling wave tube which is a microwave device used for amplify the RF signal as shown in Fig. 17.2.

These are being used in nuclear reactors as moderator, reflector and fuel elements because of its low absorption of neutrons with high strength at higher temperature and high thermal conductivity. Because of its high electrical conductivity, isotropic graphite is also used for electrodes and brushes in electric motors and electric discharge machining. It is also used for military applications for example nose tip of missiles, thrust tabs and heat shields. Further, it is used as anode of lithium ion batteries and lubricant to reduce friction and corrosion and has biomedical and solar energy applications

In another development programme, CSIR-NPL has taken initiative to develop the high density isotropic graphite for the application in high-temperature nuclear reactor in collaboration with BARC. The high density isotropic graphite is a special type of graphite which is made of coal tar pitch-based raw materials. This type of graphite is very expensive and not available to country for strategic applications. CSIR-NPL has developed the technology for production of high density isotropic graphite by cold isostatic pressing of carbon-based semi-coke of special characteristic route followed by carbonization and graphitization. Figure 17.2 shows the high density isotropic graphite cylindrical rod of diameter 75 mm and length 200 mm. The high density isotropic graphite of bulk density ~1.80g/cc is required around 325 tonnes for CHTR and IHTR programme of country. Based on the knowledge and capability developed

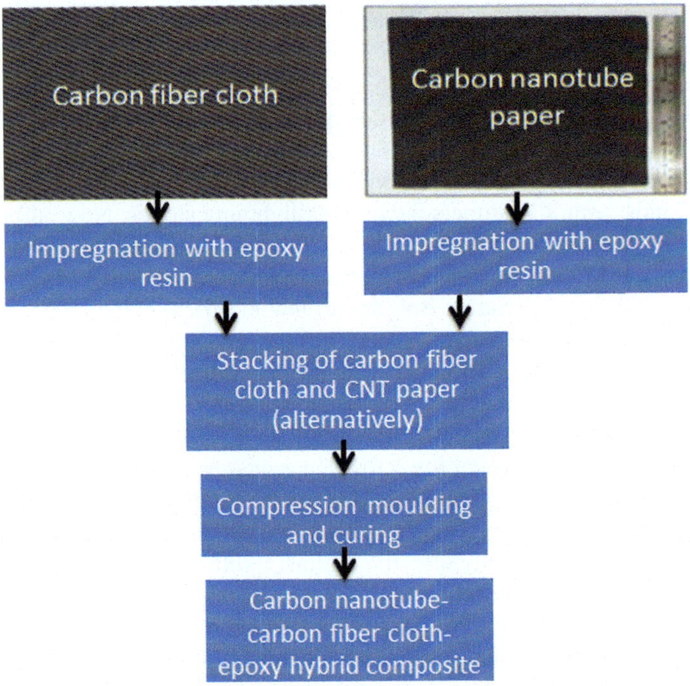

Fig. 17.3 Process for making CNT-carbon fibre polymer hybrid composites by sandwiching the CNT paper between carbon fibre cloths

at CSIR-NPL, it is possible to develop the material with help of industrial partner to cater the need of country nuclear programme.

Fibre Reinforced Hybrid Polymer Composite

Fibre reinforced composites are very important class of lightweight and high strength material used as structural, ballistic and lightning strike applications. CSIR-NPL has significantly contributed in development of carbon fibre-CNT-epoxy hybrid composites for structural and lightening strike applications and Kevlar-CNT-epoxy composites for ballistic applications. Carbon fibre reinforced polymer (CFRP) is class of polymer composite material with the advantages of lightweight, high strength, corrosion resistant which make them suitable for wide range of applications specifically in strategic sector where weight reduction is the prime requisite. CFRP layered composites are prepared by using several layers of polymer impregnated carbon fibre cloth but there is limitation of poor interlaminar shear strength which directly affect the mechanical properties of the final products. Therefore, at CSIR-NPL, a novel method is developed to fabricate the hybrid carbon fibre-epoxy composites reinforced by sandwiching with free standing CNT papers as shown in Fig. 17.3.

The flexural strength of the hybrid composites (45 ± 2 vol.% CF + CNT) was achieved more than 600 MPa which is more than 35% over pure carbon fibre/epoxy composites (50 ± 2 vol.% CF). These high-strength hybrid composites can be used in structural applications in automobile, defence and space, airframe, wind mill blades, sports industries etc. [6].

Lightning strike is a very common phenomenon in travelling aircraft through charged clouds, naturally attracts the thunder lightning strikes. Now, aircraft industries are changing from conventional structures based on metals to new advanced lightweight composites, which offers high specific strength compared to conventionally use metal-based structures. But, due to low electrical conductivity of CFRPs; these are not safe against the lightning strike. Therefore, CFRPs need an additional protection to make them safe against lightning strike. [7] In this direction, CNT paper interleaved CFRP laminates have been prepared. This interleaved CNT paper effectively enhanced the in-plane and through-thickness electrical conductivity of the CFRP laminate as well as it improves resistance against lightning strikes. The lightning test at 40 kA based on simulation results were conducted on pristine CFRP and CNT paper interleaved CFRP. Improved electrical conductivity in CNT paper-based composite proved to be highly effective against lightning strike. There is a sudden rise in the temperature around the area of the lightning strike in case of without CNT paper-based CFRP, which leads to epoxy evaporation and release of hot gases. The trapped gases can also cause significant damage. Therefore, an adequate safety from the lightning strike is required with high electrical conductivity to reduce the Joule's heat and should have good thermal stability to withstand temperature rise. No significant fire on the surface of CNT paper interleaved CFRP is observed; this can be because of the fire resistant property of the CNT paper [7] due to its high electrical and mechanical properties.

Kevlar fibre is also used as structural material due to its high tenacity and excellent specific mechanical properties, is a widely used material for reinforcement in polymer composite for automobile, aerospace, marine, safety helmet and ballistic applications. But, due to its high crystallinity and smooth surface, its adhesion with the matrix system is very poor. For the improvement of Kevlar fibre tapes, long lengths CNTs are used as secondary reinforcement to improve the interfacial interaction between Kevlar fibre and epoxy through bridging action. The maximum tensile strength, Young's modulus of 0.3 wt% of MWCNTs in epoxy resin composite showed an overall improvement of ~81%, ~56%, respectively over control composite tape [8, 9]. Kevlar-CNT paper epoxy hybrid composites have been developed with 8 layers of Kevlar and 7 layers CNT paper sandwiched alternatively. These composites showed 55% improvement in tensile strength, 64% in flexural strength and 31% in impact energy over the Kevlar epoxy composites. These hybrid composites showed their potential for lightweight high impact resistance helmets. Further NIJ type IIIA ballistic test has been performed which showed 30% reduction in back face signature (BFS) over without CNT paper interleaved composites (Fig. 17.4).

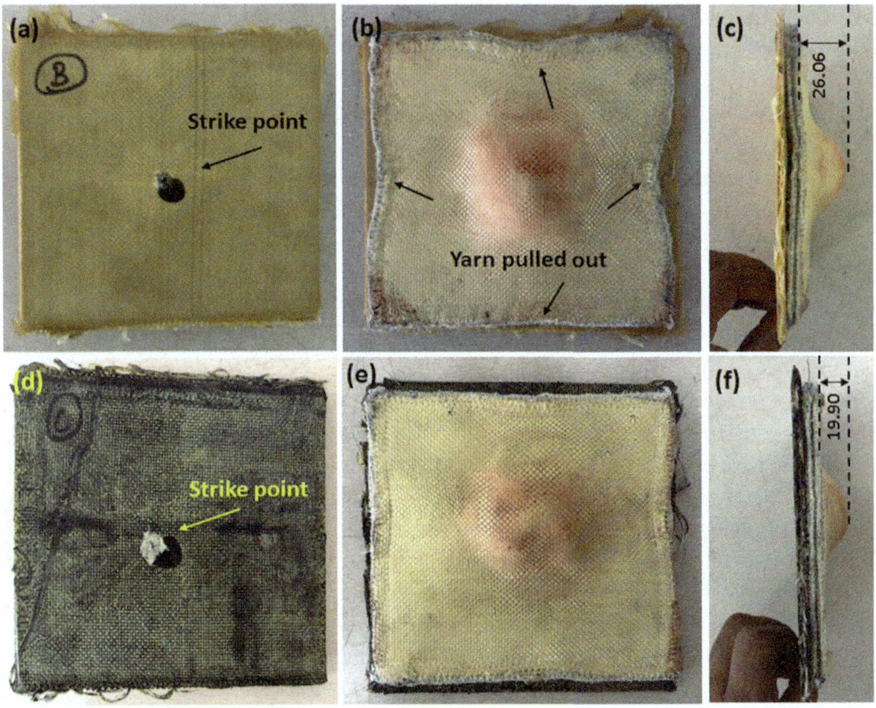

Fig. 17.4 Post failure Digital micrographs of **a** front, **b** rear and **c** lateral surfaces of Kevlar epoxy block and **d** front, **e** rear and **f** lateral surfaces of Kevlar epoxy—CNT Paper block following NIJ 0108.01 Level III-A

CNTs and Graphene-Based Polymer Composite

Electromagnetic interference (EMI) shielding materials are currently in huge demand for both commercial and defence purposes. Therefore, there is a stringent requirement for development of lightweight and efficient EMI shielding materials. The high electrical and thermal conductivity, high aspect ratio and excellent mechanical properties of carbon nanotubes (CNTs) and graphene make them a better choice to be used as conductive polymer composites for high performance EMI shielding in microwave region. The carbon nanomaterials (CNTs and Graphene) have been incorporated in different thermoplastic and thermosetting matrices for development of efficient microwave shielding composite materials.

In the thermosetting matrices, two different kind of CNT-epoxy composites are prepared [10, 11], firstly, the effect of length of MWCNT on the mechanical, electrical, and EMI shielding properties of MWCNT-epoxy composites up to 0.5 wt% loading is studied. It was shown that the long length MWCNT-based epoxy composites have higher electrical conductivity compared to short length MWCNT-based composites which also registered to high EMI shielding properties for long length

MWCNT-based epoxy composites (−16 dB) as compared to short length MWCNT-based epoxy composites (only −11.5 dB) at 0.5 wt% MWCNTs loading. For further improvement of EMI shielding value, long length MWCNT-based epoxy composites are fabricated using a novel technique which enabled dispersion of high loadings of CNTs (up to 20.4 wt%) uniformly. Compression moulded CNTs-epoxy prepreg papers are prepared and a value of 9 S cm^{-1} of electrical conductivity is achieved which helped in attaining high EMI shielding properties of composites i.e. ≥50 dB in wide frequency range (8.2–18 GHz).

CNT-Phenolic [12] composite thin films of thickness 140 μm are prepared and it demonstrated an EMI shielding effectiveness of −32.4 dB. Further high-strength composites are also required for various applications where merely CNT polymer composites cannot fulfil the requirement. Therefore, hybrid composites with CNT-CF cloth-epoxy [13], hybrid CNT-CF felt-epoxy [14], based are prepared. The EMI shielding effectiveness enhances from −29.4 (carbon fibre-epoxy) to −51.1 dB (MWCNTs-carbon fibre/epoxy) with high mechanical properties (flexural strength >500 MPa). The incorporation of MWCNTs on the carbon fibre felt produced a significant change in the EMI shielding properties which enhances from −21 to −27 dB for CF felt-MWCNT-epoxy hybrid composites.

The composites materials fulfil the requirements of high strength with a comparatively low weight, but their electrical conductivity is inhibited because most of the thermosetting polymer matrices used in structural components is based on epoxy matrix, which is insulating and cannot fulfil the criterion for EMI shielding. Polyaniline (PANI) doped with dodecylbenzenesulphonic acid (DBSA) linked with divinylbenzene (DVB) [PANI-DBSA-DVB]-MWCNT paper composites [15] have been prepared and their microwave shielding properties reached to −37.5 dBm. In these studies the thermosetting polymer composites, microwave shielding value varies from −16 to −60 dB which are suitable for various microwave shielding applications with better mechanical properties.

Thermoplastic polymers are important category of polymer which are used in number of applications and widely used in many industries because of ease of processing, large scale production and recyclability. In the thermoplastic matrices, CNT-acrylonitrile butadiene styrene [16], were prepared by twin screw extrusion followed by injection moulding technique and it shows EMI shielding effectiveness up to −39 dB for 10 wt% loaded MWCNTs. Further, CNT-polycarbonate [17], composites are prepared by solvent casting followed by compression moulding and demonstrated its EMI shielding effectiveness of −43 dB of a five-layered system (2 mm thickness) of as-synthesized-MWCNT/PC composite films at 20 wt% loading. CNT-low density polyethylene (LDPE) composites are also prepared by solvent casting followed by compression moulding and showed EMI shielding of −22.4 dB at 10 wt% MWCNT loading [18]. Polyurethane (PU) is one of the most versatile polymeric materials and it is widely used in coatings, adhesives, and thermoplastic elastomers and composites materials. CNT-polyurethane composite with functionalized and as produced CNT-based composites are prepared by solvent casting technique and its EMI shielding value of −29 dB (acid functionalized CNT) and −1.6 dB (as produced CNT) in X-band is reported [19, 20]. Further two different types of CNT-polymethyl methacrylate (PMMA) composites are prepared, thin film and layered structures [21, 22] and their EMI shielding value varies from −18 dB (single layer) to −40 dB (seven layers). These studies supports towards the development of microwave shielding materials with additional advantages of easy processability.

The EMI shielding effectiveness of a composite material is controlled by aspect ratio, intrinsic conductivity, complex permittivity and permeability of the filler. The primary mechanism of shielding is the refection loss resulting of interaction between the conducting particles in the conducting material and the EM shield while absorption loss depends on the value of permeability and permittivity. Therefore, different dielectrics and magnetic materials and their combination with CNTs and graphene are explored to achieve better absorption dominant microwave shielding properties i.e. MnO_2 decorated graphene nanoribbons [23]; Ni/Co decorated single walled carbon nanotubes (SWCNTs) [24], PANI-CNT-Graphene [25], Mesocarbon microbeads (MCMB)-CNT paper [26], carbon foam-CNT [27], γ-Fe_2O_3 decorated reduced graphene oxide in polyaniline core–shell tubes [28]. These materials can be incorporated in various matrices for the development of absorption dominant microwave shields.

17.2.1.2 Luminescent Materials and Devices

Fluorescent materials have found a large number of applications in all our daily life, including advanced optical displays, photovoltaic, biomedical, anti-counterfeiting, solid-state lighting, X-ray intensification, scintillation and many others. Therefore, enormous effort has been dedicated to exploring novel luminescent materials till date, which is the most important part for many applications in everyday life. These materials are explain as the emission of light in the visible range but can exist in other spectral regions such as ultraviolet or infrared with energy beyond ground state energy. The increasing demand for newer luminescent materials has encouraged scientific and technological efforts to improve the existing features of luminescent materials and to provide the design new effective luminescent materials having the required shape, size, morphology, and optical properties, etc. based on rigorous metrology. Luminescence is the emission of light from any substance when it is exposed to various excitation sources such as photons, electrons, and electric fields, etc. In another word, it can be explained by emission of light from a substance without heating, basically it is non thermal heating process. In general, a luminescent material absorbs energy in the form of photons and emits the visible light from electronically excited states of substance. The electrons of an atom are raised from the ground state (lower energy state) to its excited state (higher energy state) when it is exposed to some energy source such as photons and electric field etc. This excited state is not stable; therefore, electrons revert to its ground state (lower energy state) with the emission of energy in the form of light.

CSIR-NPL is working on the use of various organic and inorganic luminescent materials to be used in energy harvesting through detection and generation of light. The materials in the form of nanostructures to bulk materials are being maneuverer for the detection, guidance, manipulation, amplification, and management of light to be used in various applications like metrology, optoelectronic devices (OLEDs) and detectors, etc.

Rare Earth Doped Luminescent Materials

Recently, the lanthanide-based luminescent materials have been considered all over the world for their better chemical and optical properties originating from their unique electronic structures as well as due to their vast applications in photovoltaic, biomedical, anti-counterfeiting, solid-state lighting and display technologies, etc. Lanthanide-based luminescent materials are also known as phosphors. Generally, phosphors have one or more trivalent lanthanide ion-doped in the host lattice. In periodic table, there are 15 rare earth elements including scandium and yttrium which can be used as trivalent ions in lanthanide-based luminescent materials. The lanthanide-based luminescent materials associated with sharp emission due to their intra −4f or 4f–5d electronic transitions. It has many advantages over other luminescent materials (organic dye, quantum dots, plasmonic materials, etc.) like sharp emission, broad excitation spectrum, good thermal stability, low toxicity, a longer lifetime, better chemical stability and high photochemical stability, etc. These materials can show emission from the range of ultraviolet (UV) to near-infrared light region depending upon excitation wavelength.

Counterfeiting has curved out to be a major area of apprehension these days and is becoming a serious treat among the worldwide. Modern technological advances have made counterfeiting of refined products easy. Therefore, it is essential to develop new security ink features to combat counterfeiting applications [29–33]. CSIR-NPL has made efforts in last few years to design and developed advanced featured pigments to formulate the unclonable security ink which can be easy to detect but difficult to counterfeiting. CSIR-NPL developed several new features-based security ink such as mono-luminescent, bi-luminescent, single excitable dual emissive, colour shift ink, luminomagnetic ink, dual-mode polymeric nanofibre-based flexible mat to design nanotaggants for security paper authentication as well as transparent red fluorescence indelible ink to deter the double voting, which is described briefly in below sections.

(a) **Mono-luminescent Ink, Bi-luminescent Ink and Bi-luminescent Golden Ink**

Most of the security features are derived on luminescent materials derived pigments which are currently imported from abroad to design the luminescent security ink. To address this, import substitution and combat the anti-counterfeiting problem of Indian currency, CSIR-NPL has developed mono-luminescent materials to design the security ink for currency applications under MOU (2017) and granted a project on "Make in India-Colour shift ink" with Security Printing and Minting Corporation of India Limited (SPMCIL), Government of India enterprise. The mono-luminescent pigments can emit orange and red emission under 365 nm ultraviolet light. The specific size and shape are achieved by CSIR-NPL after rigorous metrological efforts

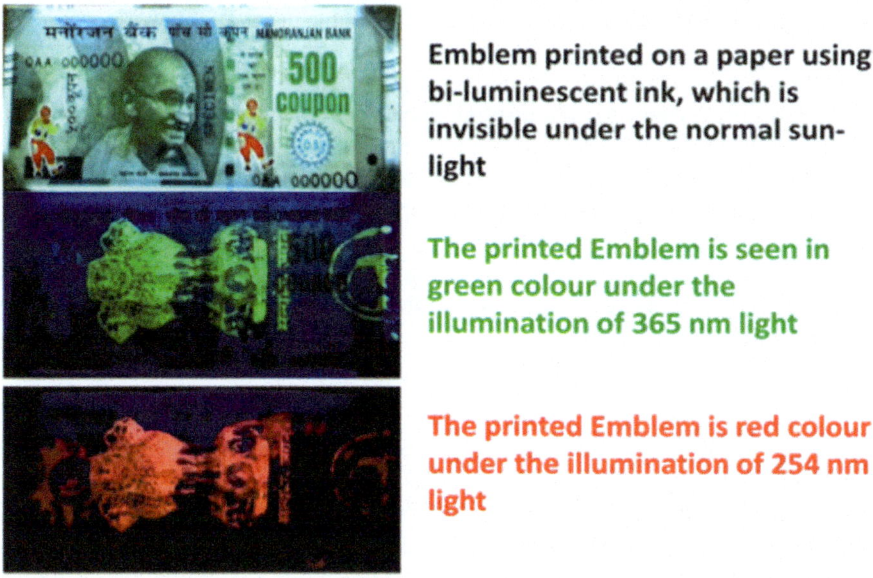

Fig. 17.5 Demonstration of indigenously development of bi-fluorescent pigments formulated security ink at CSIR-NPL for anti-counterfeiting applications [(IN201711037004) and (US2019011724A1)]

for materials to design a special type of morphology that can stabilize in a specific medium of ink without photobleaching effect. The advantage of this technology is that it is cost-effective with all its constituent chemicals easily available within country.

One of the major problems with the Indian paper currency is that they are prone to easy counterfeiting. Despite the introduction of new currency notes of the higher denomination with additional security features, this problem of counterfeiting is still prevalent. Therefore, it is imperative to introduce novel fool-proof security features in our currency notes, especially for higher denominations, to ensure that the counterfeiting becomes nearly impossible. With this premise, the first time in the world, CSIR-NPL has designed and developed a unique bi-luminescent material in which the higher quantum yield of two different transitions occur with exceptional brightness in a single compound without any quenching effect. The patent for this work has been published in India (IN201711037004) and the US (US20190111724A1). Images printed with ink formulated from this material, glow in prominent red and green colours on illumination with two different excitation sources as shown in Fig. 17.5. This newly formulated ink has found the real application none other than Indian Currency Note and it has been tested successfully for its prospective use for printing security features on Indian currency by Bank Note Press, Dewas, Security Printing & Minting Corporation of India Ltd (SPMCIL), an enterprise of Govt. of

India. This will be a unique feature and is not available in any of the world's currency notes.

CSIR-NPL has developed bi-luminescent golden security ink for the passport. Emblem printed using this security ink appears in golden colour when viewed in sunlight but it glows in two prominent colours, red and green when illuminated by two different excitation UV sources as compared to the existing single-colour glow under UV light. The advantage of this technology is that it not only combat the counterfeiting by this advanced technology but also a huge amount of money can be saved by this technology after using bi-luminescent golden ink in comparison to presently using expensive golden foil to emboss emblem on the passport with using separate single excitable security ink.

(b) **Colour Shift Intaglio Ink**

Colour shift ink (CSI) is another feature employed globally to protect the counterfeiting of the currency. The intaglio printed security features using this ink displays two distinct colours depending on the viewing angle of the currency. Although this feature is currently available on the Indian currency notes, the security ink at present is resourced largely through imports thereby leading to a large outflow of foreign exchange. The technology for this product is mainly monopolized by a few select European countries, which have protected this invention by filing several hundred patents and they charge an exorbitant amount for this CSI. Thus, the premise of this project is to indigenize the process technology for the production of CSI under the "Make in India" mission of India, which would not only lead to the huge savings but also limit a large outflow of foreign exchange for its import. This would also address the challenge that our Hon'ble Prime Minister of India had posed to the scientists and industrialists (in his address on the 80th foundation day of RBI on 1st April, 2015) for indigenizing the production of ink and paper used for the currency notes. CSIR-NPL has developed CSI ink that changes colours depending upon the viewing angle and this technology has been transferred to BNP, Dewas for scaled-up ink as shown in Fig. 17.6, which will lead to a huge saving of foreign exchange for Govt. of India. MOU was signed with Security Printing & Minting Corporation of India Ltd. (SPMCIL), an enterprise of Govt. of India, and received a project on "Make in India-Colour Shift Intaglio Ink".

(c) **Luminomagnetic for Security Ink**

A luminomagnetic material has also been developed by CSIR-NPL for the printing of MICR codes on bank cheques to generate unclonable security features. These MICR codes can emit two different colours at two different excitation wavelengths as well as exhibit magnetic property. Presently most of the bank's cheques have both features printed at two different locations on the cheque, which makes the security checking process cumbersome and time-consuming. The uniqueness in terms of bi-luminescent and magnetic features helps to overcome this drawback and provides additional security features to combat the act of counterfeiting.

Fig. 17.6 CSIR-NPL developed colour shift intaglio ink that changes colours depending upon the viewing angle for Indian Banknotes applications

(d) **Dual-Mode Polymeric Nanofibre-Based Flexible Mat for White Security Paper and Encrypted Nanotaggants**

Counterfeiting of important data, currency, stamp papers, branded products, etc., has become a serious security threat that could lead to big damage to the economy of any country. The consequences of such damage are persuasive for scientific community to innovate new unclonable security features to address such problems [30–32]. CSIR-NPL has developed a dual-mode flexible highly luminescent white security paper and nanotaggants composed of nanophosphors incorporated in the polymer matrix to form a nanofibre-based mat for protection against counterfeiting problems. This mat appears white in daylight while emitting strong red ($NaYF_4$: Eu^{3+}) and green ($NaYF_4$: Yb^{3+}, Er^{3+}) colours at excitation wavelengths of 254 nm and 980 nm, respectively as shown in Fig. 17.7. These luminescent nanofibres can also act as encryption for security paper as a novel nanotaggants.

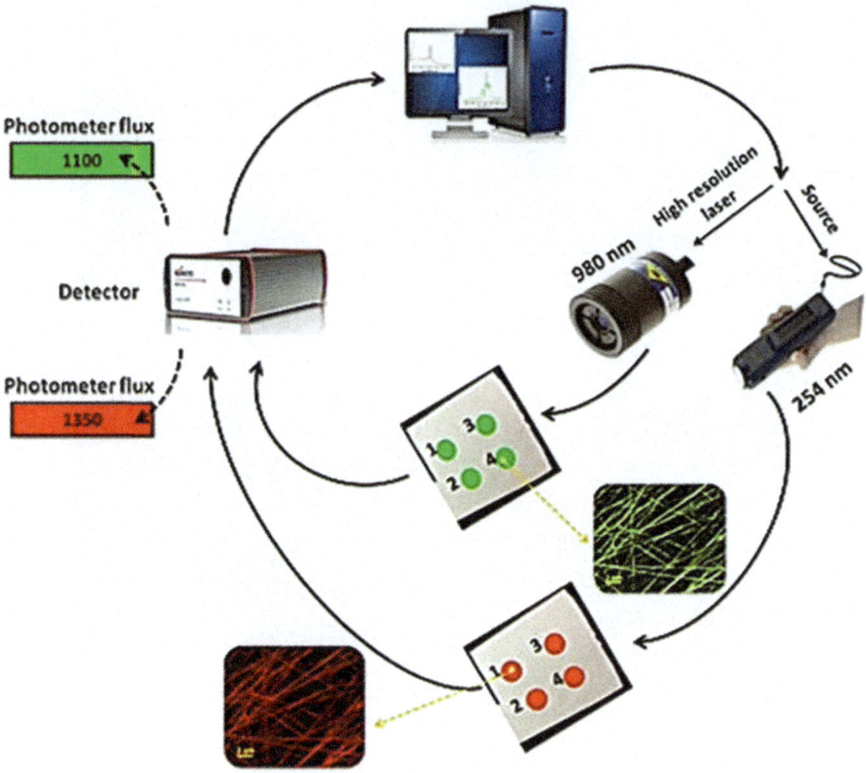

Fig. 17.7 Development of highly luminescent dual-mode polymeric nanofibre derived flexible mat for white security paper and encrypted nanotaggant applications [30]

(e) **Transparent Red Fluorescence Indelible Ink**

CSIR-NPL has recently developed the transparent luminescent ink, which emits red luminescence under 365 nm ultraviolet LEDs. India is a major exporter of indelible election ink to many countries, where they desire the election mark on the finger to be invisible. The synthesized luminescent pigment for this purpose easily makes the transparent colloidal with a protein-based linker solution, which is highly biocompatible and chemically stable against various bleaching agents up to 48 h (Fig. 17.8). Further research is under progress and developed ink has demonstrated to Mysore paint and varnish industry. CSIR-NPL has also designed and developed a lightweight portable LED-based excitation source for exciting this ink.

(f) **Bi-functional Luminomagnetic Materials for Cellular and MRI Imaging**

A nanoparticles having both magnetic and luminescent properties in a single compound is desired for many biological applications. A single compound exhibiting this combination of properties is meagrely exists. CSIR-NPL has developed a strategy

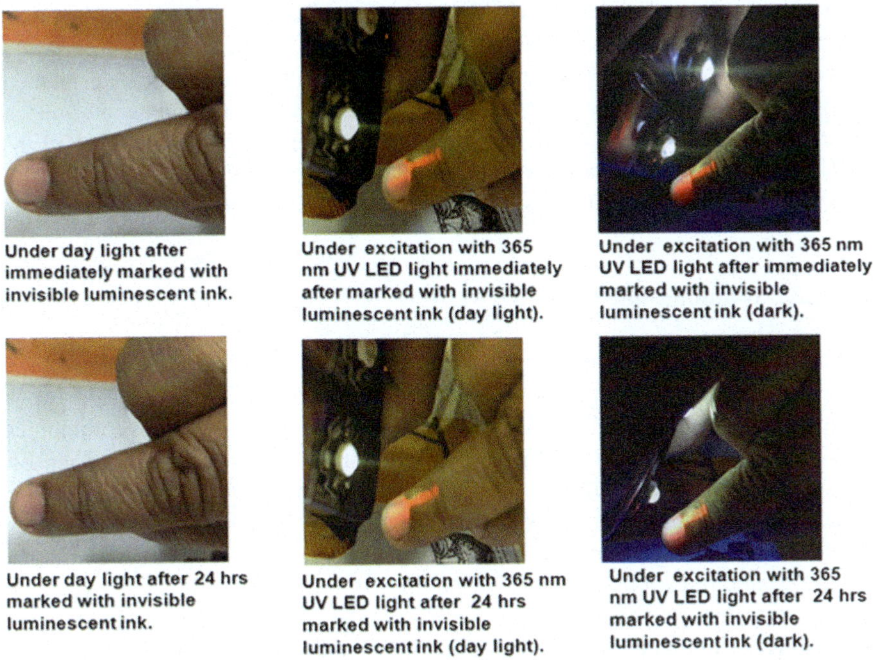

Fig. 17.8 Development of invisible ink (red luminescent visible under 365 nm ultraviolet LEDs) to be used as deterrent against voting twice

to synthesize of lumino magnetic Eu-doped Gd_2O_3 nanorods with millisecond photoluminescence lifetime by a base-catalysed hydrothermal method. This nanophosphor exhibits hypersensitive intense red emission at 610 nm with good brightness, which is required for high-contrast bioimaging applications, the efficiency of luminomagnetic nanorods as a biological nanoprobe synthesized by CSIR-NPL exhibits high-contrast cell imaging and highly biocompatible with two human breast cancer cell lines T47D and MDA-MB-231. CSIR-NPL has also successfully performed an in vivo experiment to assess the tissue distribution and fluorescence ability. Thus, this novel approach establishes a newly designed material which is highly biocompatible nanoprobe for high-contrast cellular imaging with magnetic tracking capability as well as for MRI applications [34].

Long Afterglow Phosphor

Long afterglow phosphor has been of great interest globally for many strategic applications like emergency and safety marking in dark conditions. Phosphors based on zinc sulphide with radioisotopes were so far in use for such applications. However, with a ban imposed on the use of radioisotopes, the development of a radioactive-free long afterglow phosphor has become very important. A big effort is being made on the international level to develop an efficient with sufficiently longer afterglow times to replace existing markers like beta lights etc. This is of immense commercial value. However, these materials have importance in our daily life and can be used as an alternative source of light in extremely dark conditions and power breakouts and lead to the improvement in the quality of life by conserving the energy. CSIR-NPL has developed a technology for the development of sunlight sensitized long afterglow phosphor powder and paint and successfully transferred to a private industry.

17.2.1.3 Diamond-like Carbon

Extensive research have been taken place globally towards development of diamond-like carbon coatings, because of the possibility of applications in a wide variety of areas which include Tribology (computer hard discs, optics, communication, machine tools), Microelectronics, Biocompatible coatings, Advanced sensors, Automobile industries, MEMS, Flat panel display technology and as an electronic/optoelectronic material for novel devices. CSIR-NPL has also worked extensively on this wonderful material for about two decades for realization of number of industrial applications; like DLC as antireflection (AR) coating for germanium optics, on watch glasses as hard and protective coatings, on shaving blade edges and stainless steel balls as hard and lubricant coatings, on inner surface of bottles as barrier coating and in realizing field emission devices. One of the most important achievements by CSIR-NPL was the large area Diamond-like Carbon (DLC) coatings as antireflection (AR) coating for germanium optics (in range of 8–12 μ) for DRDO's missile project NAG. Design and fabrication of plasma enhanced chemical vapour deposition (PECVD) process equipment was also done at CSIR-NPL and the same along with process know-how for DLC coating on large area germanium optical components was transferred to the Instruments Research and Development Establishment (IRDE, DRDO), Dehradun. Various PECVD systems, with different configurations, such as operating in RF (13.56 MHz), VHF (100 MHz), microwave (2.46 GHz) as well as in pulsed RF and VHF mode and also novel filtered vacuum arc system, were developed and a number of such equipment transferred to different laboratories as well as for in-house use.

CSIR-NPL worked with a very important relevant problem of residual stresses in hard hydrogenated amorphous carbon (a-C:H, also called Diamond-like Carbon, DLC) films, which is generally found in the range of 4–7 GPa. Such high stresses do not normally allow one to realize many exciting applications of these materials. Identification of deposition conditions to obtain stress relieved DLC films (otherwise which is very high) with reasonable hardness constitutes an important area of study in order to realize the various industrial applications. CSIR-NPL made an important contribution, both in identifying the cause and suggesting ways to significantly reduce the same. Specifically, CSIR-NPL handled the problem of high residual stresses by applying innovative approaches; like use of a very high frequency (100 MHz) discharge, pulse plasma technique, dual frequency (microwave/rf) plasma etc. [35–38]. After establishing successful plasma-based growth techniques to obtain stress relieved hard DLC films (20–30 GPa) sustained effort were made to utilize these films for various industrial applications as mentioned earlier in this section.

A vacuum plasma arc system to grow highly tetrahedral DLC films and tetrahedral amorphous carbon (ta-C) films was also built in CSIR-NPL. The development of carbon nanotubes through PECVD route and establishing an RF ion gun, saddle filed fast atom beam source for DLC deposition were also successfully realized. Cathodic arc as a process also offers many unique features such as non-hydrogenate DLC i.e. ta-C (tetrahedral amorphous carbon) films leading to higher sp^3 bonding with more diamond-like characteristics. Another technique called saddle field fast atom beam (FAB) source which was found to be an attractive technique due to its both dc and rf attributes and made to cover large areas for identified industrial applications. Using this source it became possible to deposit DLC films, simultaneously by neutral and ionic radicals as well as by their mixture. DLC films showed significantly low stress and reasonably high hardness as produced by this technique [39].

Further other novel methodology was developed to promote high adhesion of DLC films to non carbide forming substrates; otherwise this is always a serious problem. Basically, this involved designing of three layer structure namely, a-Si:H as adhesive layer, silicon incorporated a-C:H as bulk layer and pure a-C:H as surface layer. Other approaches for such problem involved investigations metal incorporated diamond-like carbon (DLC) and their various multilayer structures [40, 41], such as; Copper incorporated diamond-like carbon (Cu-DLC) films, Copper/hydrogenated amorphous carbon (Cu/a-C:H) multilayer structure, Titanium/hydrogenated amorphous carbon (Ti/a-C:H) multilayer structure, Nickel-chromium/diamond-like carbon (Ni-Cr/DLC) bilayer structure, Aluminium/diamond-like carbon (Al/DLC) bilayer structure, Copper/diamond-like carbon (Cu/DLC) bilayer structure, Titanium/diamond-like carbon (Ti/DLC) bilayer structure etc. Particularly, objective of this research is to obtain diamond-like properties and its adhesion for tribological applications for which CSIR-NPL is looking forward to transfer the technology to industry. Specifically for deposition of these multilayer DLC coatings a hybrid system with configuration of PECVD and sputtering processes was developed at CSIR-NPL. Understanding of correlation between plasma parameters with properties of DLC coatings was established using VI probe-based hydrocarbon plasma diagnostic tool [42]. Modified DLC coating was applied on inner surface of shaped objects. Figure 17.9 shows the schematic of experimental setup for DLC coating on inner surface of a container and coated bottle. These coatings on inner surfaces act as barrier coatings which will improve the shelf life of drink, brewages etc. [43]. CSIR-NPL has also made a serious study of disorder behaviour in DLC films. During these investigations, a correlation was observed to exist between stress and the inverse slope of the valence band edge i.e. Urbach energy in these films. From the observed correlation it was inferred that the Urbach energy in DLC (a-C:H) films below 140 meV may not be realizable [44]. Those conditions were explored under which DLC films show onset of photoconduction and it was one of the very early reports on photoconduction in these films [45].

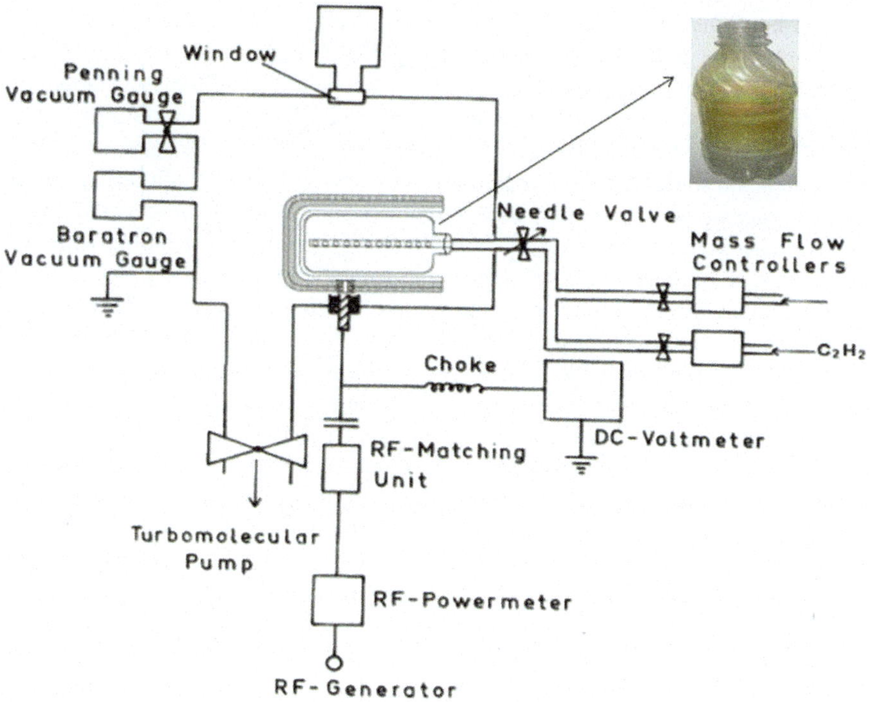

Fig. 17.9 Schematic of setup for DLC coating on inner surface of bottle and coated bottle

17.2.1.4 Conducting Polymer-Based Materials

Conducting polymers (CP) are special class of polymers, with extended conjugation leading to onset of electrical conductivity, which is otherwise considered mutually exclusive in context of polymers. The enormous significance of this path-breaking and revolutionary discovery is acknowledged by awarding the Nobel Prize in Chemistry for year 2000 to the discoverers; Prof. Hideki Shirakawa, Late Prof. Alan G. MacDiarmid and Prof. Alan J. Heeger [46].

It is important to distinguish between highly doped metallic CPs (better known as ICPs) and their relatively less conducting (insulating or semiconducting) counterparts i.e. undoped (or base forms), which in general simply termed as CPs. The subsequent advances and expansion of the field of CPs/ICPs, have led to a variety of materials and technologies (Fig. 17.10) with great potential for various techno-commercial applications [46–48]. CSIR-NPL has proactively contributed towards designing and development of CPs for various applications. Particularly, a pilot scale facility, having double/triple wall reactors, chillers and other processing facilities for kilogram scale synthesis of Polyaniline (PANI) analogues and their blends/composites has been established [49–55]. The formation of polyaniline-based systems by both direct

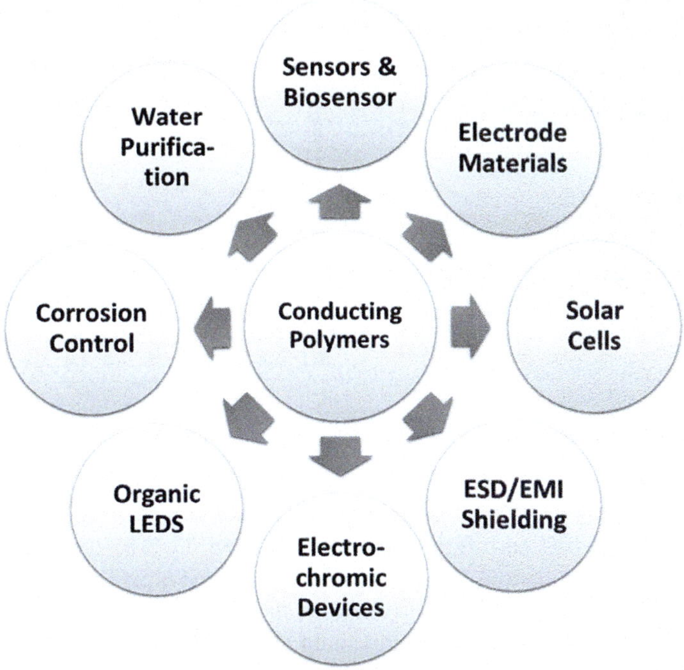

Fig. 17.10 Some potential applications of doped/undoped organic conducting polymers

and indirect doping routes; undoping process, as well as power of emulsion polymerization route, has been highlighted using various dopant systems, to demonstrate the regulation of polymer properties and their utility for specific applications [46, 49, 56–58]. In order to expand the already established wide applicability of CPs/ICPs systems, attempts have been made to form their blends, copolymers and composites/hybrids [46–50, 54, 59–64].

CP/ICP-based blends refer to special class of materials involving combination of undoped/doped CPs with another CPs, conventional insulating polymers (e.g. thermoplastic, rubber, elastomer etc.) or suitably functionalized derivative of carbonaceous materials, to generate a material with combination of properties of constituent phases. The use of CPs/ICPs as filler has been demonstrated to form the electroactive and/or electro-conducting blends, with applications in the areas like water purification, antistatics/electrostatic dissipation, EMI shielding, corrosion protection etc. [46–48, 52, 54, 65].

For example, PANI/polysulphone (PSF)-based porous films display efficient desalination performance maintaining good salt rejection even at high permeate flux, which scales with PANI concentration [65]. Similarly, PANI/LDPE melt blended system [61, 66] containing PANI loaded low density polyethylene (LDPE) display sufficient conductivity, so that PANI/LDPE film (Fig. 17.11a) can quickly dissipate

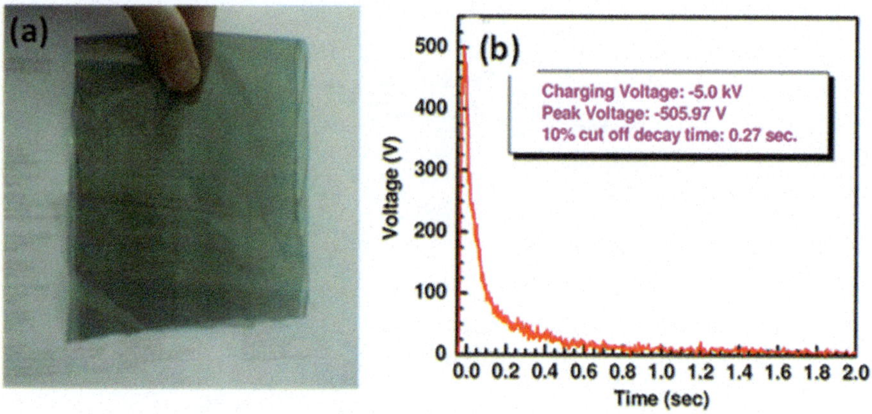

Fig. 17.11 **a** Optical image of PANI/LDPE melt blend-based blown-film and its **b** static charge decay profile

static charges with 10% cut-off time of only 0.27 s (Fig. 17.11b), thus useful for antistatic packaging applications. Likewise, PANI coated MWCNTs filler designed by our process display compatibility driven good dispersion within LDPE matrix, ultimately resulting in low percolating threshold and efficient EMI shielding response [51, 62, 63]

The CPs/ICP-based copolymers represent system wherein two different conjugated monomer units are distributed along the main or side chain, giving it set of properties which are governed by proportion of constituents and their arrangement. Such systems are formed by in situ polymerization route involving co- or sequential-addition of monomers or polymerization of monomer in presence of other conjugated polymeric phase. CSIR-NPL has demonstrated the formation and utility of polyaniline-based copolymers for applications such as antistatics, EMI shielding and corrosion control. It has been highlighted that the processability of ICPs can be improved by copolymerization approach to obtain trade-off between solubility, electrical properties and EM response [58, 59, 61].

CPs/ICP-based composites/hybrids belong to multiphase materials where suitable CPs/ICPs (or their copolymers) are combined with another organic/inorganic phase (e.g. conventional insulating polymer; fillers like graphene/CNT analogues; metallic, semiconducting, dielectric, ferroelectric, magnetic nanoparticles (NPs) and occasionally another CPs/ICPs) to form advanced systems with fascinating properties and wealth of applications. Such systems are formed by in situ polymerization or blending techniques, such as solution or melt processing routes. Number of ICP-based composites/hybrids have been developed at CSIR-NPL and demonstrated their applications in diverse areas such as corrosion control, water purification, EMI shielding, antistatics etc. The strength of carbonaceous fillers (e.g. graphite powder, colloidal graphite, CNTs etc.) and/or dielectric/magnetic particles (e.g. cubic/tetragonal $BaTiO_3$, γ-Fe_2O_3, Fe_3O_4 nanoparticles) containing PANI matrix-based composites/hybrid systems, for realization of combination of properties

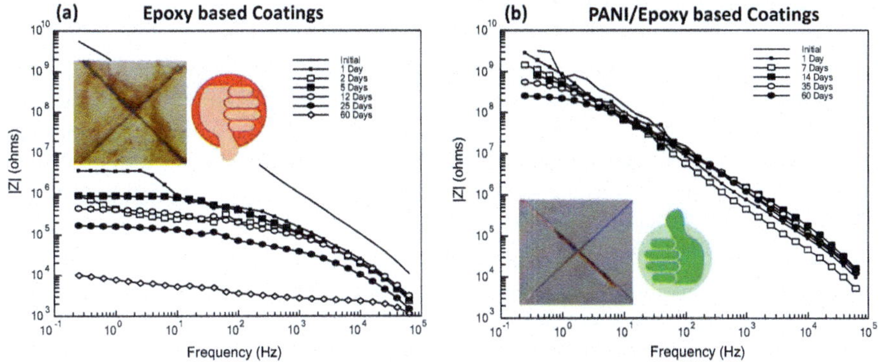

Fig. 17.12 Corrosion protection response of mild steel panels coated with paint containing **a** no PANI **b** PANI as filler, after 60 days exposure to 3% NaCl mist under salt spray test. Inset shows the actual panels after 60 days exposure

and increased effectiveness of system for targeted application has been demonstrated [46–49, 59–64, 66–69].

For example, PANI/$BaTiO_3$ or PANI/Fe_3O_4 NCs coated fabric-based hybrid systems are synthesized via in situ polymerization route, leading to uniform dispersion of $BaTiO_3$ or Fe_3O_4 NPs within coated conducting PANI (ICP) matrix, ultimately resulting in good static charge dissipation efficiency due to PANI phase and efficient EMI shielding response contributed by combination of conductivity (due to PANI), dielectric losses (due to PANI & $BaTiO_3$ phases) and magnetic losses (due to iron oxide phase). Similarly, PANI & particulates filled acrylic coating (Fig. 17.12) have been demonstrated for development of smart and self-healing corrosion protection coatings [58] that can withstand 60 days exposure to 3% NaCl mist under salt spray test. The technology was patented and subsequently transferred to M/s Krishna Conchem Products Pvt. Limited, Mumbai, India.

17.2.2 Materials for Energy Sector

Now a day's energy crisis is a big problem all over the world and energy demands are always more than energy generation. Therefore, there is need of an hour to produce new materials for energy generation and storage which can supplement for energy sectors such as PEM fuel cell, batteries, supercapacitors, thermoelectric devices, photovoltaic. CSIR-NPL has taken initiative to develop variety of materials for such applications to derive a new alternative in this direction.

17.2.2.1 Materials for PEM Fuel Cell

In view of the expected declining resources of fossil fuels in the coming decades and increasing environmental issues, it is essential to develop alternative sources of energy generation. At present hydrogen-based fuel cells is one of the most promising sources of alternative energy. A fuel cell is a electrochemical device that converts chemical energy into electrical energy. There are different types of fuel cell such as Polymer electrolyte membrane fuel cell, phosphoric acid fuel cell, direct methanol fuel cell, Solid oxide fuel cell. Among these, Polymer electrolyte membrane fuel cells (PEMFCs) are the most promising power sources for various portable electronic devices, stationary and mobile applications that operate at low temperatures between 70 and 80 °C. In PEMFC, hydrogen and oxygen used as fuel, electricity as main product along with water and heat are generated as other reaction byproducts. In this, proton from dissociated hydrogen atoms, permeate through a nonconductive membrane while the corresponding electron generates current by flowing through external circuit. The main components of the PEMFCs are electrolyte membrane assembly, gas diffusion layer, platinum catalyst, and bipolar plate. The carbon-based components in PEM fuel cell are bipolar plate, gas diffusion layer and catalyst support.

The bipolar plate and porous conducting carbon paper are the most important components of PEMFC, which costs for as much as 60–80% of the fuel cell stack weight and 40–50% of the total cost of the fuel cell stack.

Porous conducting carbon paper acts as the backing layer of the electrode in a PEMFC, thus providing mechanical support to the MEA. It not only assists in the conduction of electrons, but also provides reactant gas access from the flow field channels of bipolar plate to the catalyst layer and passage for removal of product water in the other direction. The above requirements impose stringent conditions on its properties, including high conductivity, permeability, strength and porosity with a uniform pore size distribution. The CSIR-NPL has developed porous conducting carbon paper [70–73] shown in Fig. 17.13 and properties of carbon paper are given in Table 17.1.

Graphite composites bipolar plate is a multifunctional component in the PEM fuel cell stack. It connects and separates the individual fuel cells in series to form a fuel cell stack with required voltage, its main function is to uniform distribution of fuel gas and oxygen over the whole active surface area of the membrane-electrode assemblies (MEA), conducts electrical current from the anode of one cell to the cathode of the next, facilitates water management within the cell, supports thin membrane and electrodes and clamping forces for the stack assembly etc,.

Essentially the composites bipolar plates are lightweight and electrically conducting plates which join together the anode of one cell to the cathode of another. CSIR-NPL has developed the graphite-based composites bipolar plates (Fig. 17.14) [73–75] of properties as mentioned in Table 17.2.

The densities of multicomponent graphite plate are in the range of 1.9–2.0 g/cc, further to reduce the density of bipolar plates which can ultimately able to reduce stack weight, the expanded graphite composite bipolar plates have been developed

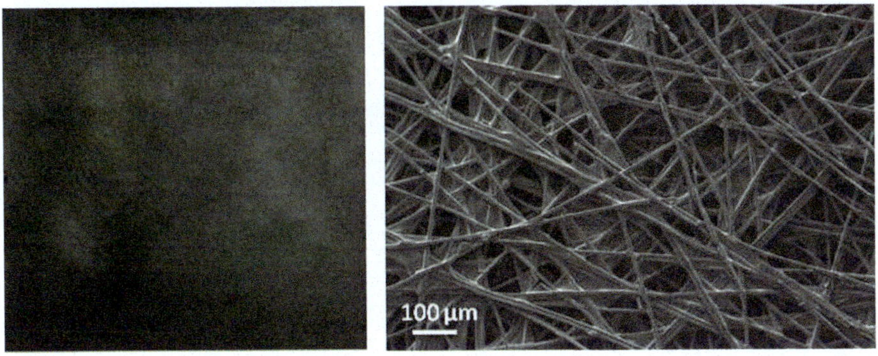

Fig. 17.13 Digital image of porous conducting carbon paper and SEM image of porous conducting carbon paper

Table 17.1 Properties of porous conducting carbon paper

S. No	Properties (unit)	Values
1.	Thickness (mm)	0.28 ± 0.02 mm
2.	Density (g/cc)	0.5 ± 0.02 g/cc
3.	Electrical conductivity (S/cm)	>200 S/cm
4.	Flexural strength (MPa)	>40 MPa
5.	Flexural modulus (GPa)	10 ± 2 GPa
6.	Porosity (%)	>75%
7.	Pore tortuosity	<1.4
8.	Gas permeability (ml cm^{-2} min^{-1} mm Aq^{-1})	25

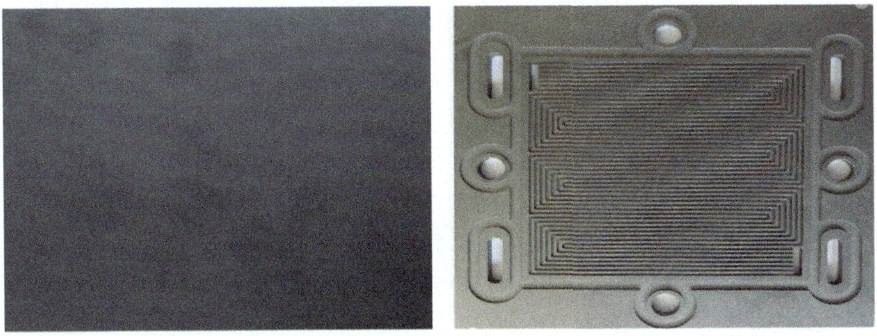

Fig. 17.14 Digital image of multi component graphite composite bipolar plate with flow field channel

Table 17.2 Properties of multi component graphite composite bipolar plate

Properties	Graphite composite bipolar plate
Bulk Density (g/cc)	>1.90
Bending strength (MPa)	>50
Modulus (GPa)	>15
Compressive strength (MPa)	>40
Shore Hardness	>55
Electrical conductivity in plane (S/cm)	>150
Electrical conductivity through plane (S/cm)	>40

of density 1.60–1.65 g/cc [76–78] with equivalent performance with that of graphite composite bipolar plate with the advantage of it possible to make in situ flow filed channel during processing of bipolar plates. Catalyst is used to catalyse the otherwise sluggish oxygen reduction reaction at the fuel cell cathode. Commercially Pt/C is used as a FC catalyst, wherein Pt being a precious metal adds to the cost of the cell. Hence reduction in Pt has been addressed in a series of experimentations [79–81], which includes comparing functionalized, as such and purified CNTs as catalyst support; and carrying out the reaction under different pH conditions and gaseous environments. pH/CNTs with CNTs purified by heat treatment and the reaction carried out at neutral pH under nitrogen atmosphere showed equivalent performance (with half the Pt loading) as compared to the commercial catalyst (supplied by Jhonson Matthey, UK) [81]. The Pt/CNTs electrode shows similar performance to a commercial Pt/C electrode with the same loading even in PEM FC. The lower Pt %wt content and the possible larger "void fraction" caused by the CNTs support generate thicker electrode than obtained with a standard Pt/C 20 %wt powder. The increased electrode thickness probably plays a positive effect in preventing acid loss from the membrane electrolyte.

17.2.2.2 Materials for Batteries

Battery is an important device for energy storage applications where the electrode materials play an important role for developments of high capacity battery. Nanotechnology has opened up new frontiers to meet the challenge for efficient energy conversion and storage by creating new materials, particularly carbon nanomaterials due to its physical and electrical properties. CSIR-NPL has developed a novel anode material by utilizing a novel route to attain high anode charge capacity by incorporating nano tin oxide in MWCNT and formation of free standing paper (Fig. 17.15a). The CNTs used in the process are synthesized in-house on the CVD setup. The specific capacity of NPL-based CNT paper is found to be ~500 mAh/g at 200 mAg^{-1} current density as shown in Fig. 17.15b which is much higher as compared to commercial

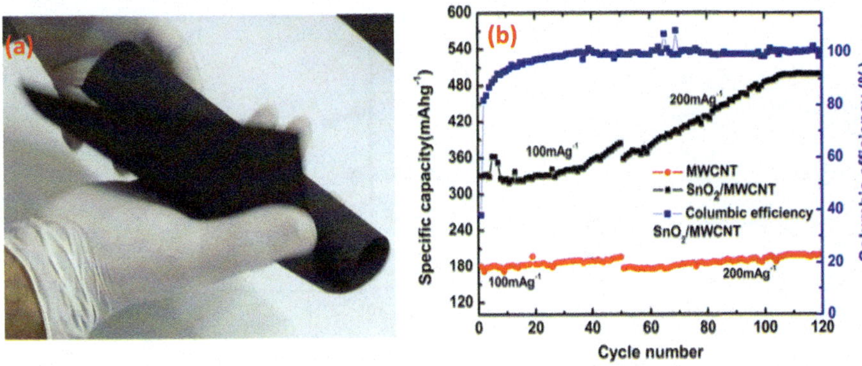

Fig. 17.15 **a** SnO_2-CNT-based free standing and flexible anode, **b** discharge capacity at 100 mAg^{-1}, 200 mAg^{-1} for pristine CNT paper and SnO_2-CNT paper. Reprinted from Electrochimica Acta 231, 255, copyright (2017) with permission from Elsevier [82]

batteries anodes graphite with additional benefit of removal of copper as a current collector [82].

17.2.2.3 Materials for Supercapacitors

Supercapacitors (SCs) are the modern energy storage systems that attracted enormous global attention due to properties such as pulse power deliverability, compact configuration and long cycle life, with wide applicability in the consumer electronics, power supplies, escalators, hybrid electrical vehicles and industrial systems [83]. Electrode materials plays a vital role in all the different kinds of SCs [e.g. electrical double-layer capacitors (EDLCs); pseudo-capacitors (PCs) and hybrid-supercapacitors (HSC) [84] and greatly affect their functioning mode, efficiency, safety, life, cost and field of applicability. Generally, pseudo-capacitive materials such as metal oxides, and conducting polymers, provide high power but have low cyclic stability; whereas EDLCs like carbon materials (e.g. activated carbon, mesoporous carbon, carbon onions, CNTs & derivatives, graphene analogues etc.), possess moderate power density with good life cycle. HSCs combine the relative advantages of EDLCs and PCs to realize better performance characteristics so as to achieve energy and power densities greater than EDLCs without losing the cycling stability and cost, that largely limited the success of PCs. CSIR-NPL has capability to synthesize conducting polymers, graphene analogues and their composites/hybrids with each other or nanomaterials like CNTs or metal oxides, to realize more efficient and effective electrodes and based SC devices [84].

CSIR-NPL has demonstrated better SC performance of hybrid electrodes based on polyaniline/CNTs composite/hybrid system formed by wet phase mixing, when tested through Potentiostat-Galvanostat facility, under three electrode configuration. It is found that maximum performance is realized only at optimum ratio of the phases.

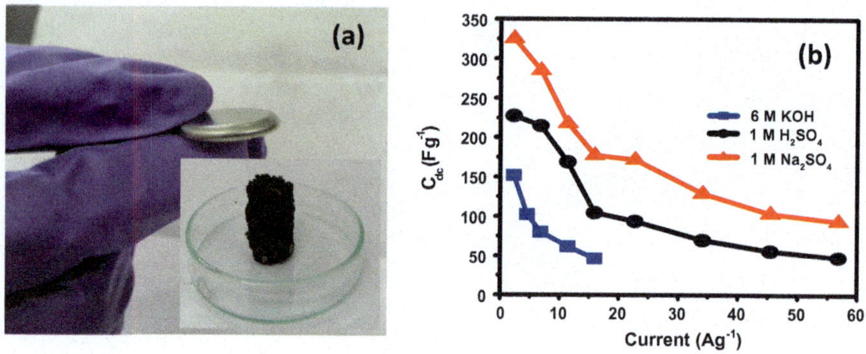

Fig. 17.16 **a** Chitosan reduced graphene oxide aerogel having iron oxide incorporation (inset) and based packed coin cell SC device; **b** specific capacitance versus current density plots for SC devices under different electrolytes

Graphene, with large specific surface area of 2630 m^2/g, charge storage capacity of 21 mF/cm^2 and high electrical conductivity, is considered as most promising carbon analogue candidate as EDLC electrode material. Again, hybrid systems are considered superior; therefore, a high rate-performance HSC containing symmetric electrodes based on in situ chitosan reduced graphene oxide with incorporated iron oxide nanoparticles has been demonstrated. The resulted aerogel-based SC cell (Fig. 17.16a) under symmetric device configuration, has been tested in presence of various electrolytes (Fig. 17.16b), and it has been highlighted that neutral electrolyte offers best storage performance with specific capacitance value of 326 Fg^{-1} [85]. Due to highly conducting nature, the free standing CNT paper has been explored in supercapacitors. The high specific capacitance of 270 F/g (2 mV/s) with superb stable behaviour (88% at 10,000 cycles) of CNT paper has promised as decent electrode in energy storage application [86].

Materials for Thermoelectric Devices

Thermoelectric power generators are stable devices with no moving parts; they are noiseless, durable, and scalable, realizing them perfect for waste heat recovery applications. The proven service life of more than 20 years without device failure of these thermoelectric power generators has shown its reliability and durability in hostile environmental conditions. The technology for the development of thermoelectric power generators in our country is currently in an exceedingly emerging stage, and also this technology would lead to the Low carbon technology deployment in the rapidly emerging area of waste heat recovery applications. CSIR-NPL has developed different process technologies for the development of n- and p-type thermoelectric materials with record figure-of-merit, employing several novel material processing techniques, including rapid solidification employing melt-spinning, one-step solid-state synthesis employing spark plasma sintering, liquid metallurgy using vacuum

Table 17.3 The thermoelectric materials developed at CSIR-NPL for the comprehensive spectrum of applications

S. No.	Applications	Materials Developed at CSIR-NPL	Figure-of-merit (ZT)	References
1	Low-temperature applications (<250 °C)	Bi_2Te_3	~1	–
		Bi_2Se_3	~0.7	[87]
2	Mid-temperature applications (250–600 °C)	Cu_2Se	~2	[88]
		$CoSb_3$	~0.7	[89]
		Mg_2Si	~0.7	[90]
		$MnSi_{1.73}$	~0.8	[91]
		$CrSi_2$	~0.32	[92]
3	High-temperature applications (>600 °C)	n-type Half-Heusler	~1.4	[93]
		p-type Half-Heusler	~0.8	[94]
		n-type SiGe	~1.5	[95]
		p-type SiGe	~1.2	[96]

melting, induction melting & arc melting and powder metallurgy using mechanical alloying employing high energy ball milling [87–96]. Using this strategy, the record value of thermoelectric performance in several materials have been demonstrated. Furthermore, the focus of research is to develop compatible n- and p-type thermoelectric elements for integration as a thermoelectric device. Table 17.3 shows the comprehensive spectrum of thermoelectric materials developed, for low-mid-high-temperature applications at our thermoelectric group with their figure-of-merit values.

The low-temperature applications, Bismuth Telluride (Bi_2Te_3) and Bismuth Selenide (Bi_2Se_3) material are known to be the best thermoelectric materials, and they are used in both energy generation and cooling applications. High Figure-of-merit Bi_2Te_3 and Bi_2S_3 are synthesized at CSIR-NPL and reported the thermoelectric compounds, employing spark plasma sintering (SPS) process. Furthermore, successfully synthesized a good quality and high thermoelectric performance p-type Bi_2Te_3 and n-type Bi_2Se_3 single crystals [87] with a specially designed ampoule employing a vertical Bridgman technique. The mid-temperature applications, silicides have proven to be one of the best candidates for an excellent thermoelectric generation, owing to its enhanced thermoelectric figure-of-merit, thermal stability, and abundance in earth crust associated with non-toxic nature [90–92]. The successful synthesis of n-type magnesium silicide and p-type manganese silicide materials not only strengthens capability. The non-toxic material synthesis for the device/module provides the green solution for the industries in terms of environmental point of view. The single-step reactive spark plasma sintering is employed to process single element and multi-element doped Mg_2Si. The significant thermoelectric figure-of-merit of ~0.7 at 873 K is gained in doped Mg_2Si [90]. Applying novel methodology, efficient and scalable, the highest thermoelectric Figure-of-merit (ZT ~ 0.8) thermoelectric material is synthesized at optimized process parameters [91]. The high temperature applications, both p- and n-type SiGe alloys and half-Heusler alloys by optimizing the process parameters of arc melting, high energy ball milling, and followed by spark plasma sintering are synthesized [93–96] and successfully integrated and fabricated in 2 × 2 paired p-n legs device at Bhabha Atomic Research Centre, Mumbai. The pellets are mounted in a zircar housing and interconnected using a silver stripe. The bonding temperature is ~750 °C and load applied is ~250 kg. The resistance of p- and n-type pellets were respectively ~3 mΩ, 8 mΩ and 1 mΩ, 1 mΩ and the fabricated module has a resistance of ~25 mΩ, indicating that contact resistance is ~12 mΩ. The fabricated module has a resistance of ~25 mΩ, indicating that contact resistance is ~12 mΩ. At a matched load of 25 mΩ, the fabricated module exhibited a power output of ~361 mW for a temperature difference of 553 K.

17.2.2.4 Photovoltaic Materials and Devices

Solar energy is one of most viable options for green energy as it is most secure and abundantly available. India is blessed with about 4–7 kWh per m^2 per day solar energy incident over its most of the land area and so has vast potential to harness it. Solar PV is the most favoured technology route for the conversion of solar radiation into electricity. The main advantages of SPV are: long life, low maintenance, high reliability, modularity in use and almost no emission of pollutants i.e. purely eco-friendly.

The crystalline silicon-based PV devices are considered to be first generation PV technology, second generation technology offers thin-film technology based on amorphous silicon(a-Si–H), CdTe or CIGS thin films and other semiconductor materials. The first generation technology holds a large market share (>95% production and deployment both grid-connected and off-grid-connected mode), while the second generation materials cover very small segment of the market though it is expected to boost its share in near future. The third generation technology covers photo-electrochemical, polymer/organic and nano-crystal material-based solar cells. Development of these materials is still in the laboratory research level. The fourth generation technology offers composites, multilayer/multiband gap-based solar cells. This segment is also in the stage of development. In silicon-based cells, some new approaches are also being tried in extensive manners. The concept of black silicon solar cells and advanced passivation schemes for low-cost and high-efficient PV devices are few of the recent ones and are very promising.

CSIR-NPL is involved in crystalline and multicrystalline silicon solar cells development since late 1970s and is the first laboratory to demonstrate the complete process from metallurgical grade silicon to solar grade polysilicon, fabrication of solar cells and demonstration of a solar PV system generating electricity to run a water pump (Fig. 17.17). Increasing efficiency and reducing cost of the solar cells have been the most important research consideration for the success of any solar cell technology for an efficient and economical harnessing of solar energy. The organization has mission to develop industrial compatible processes for low-cost production of solar

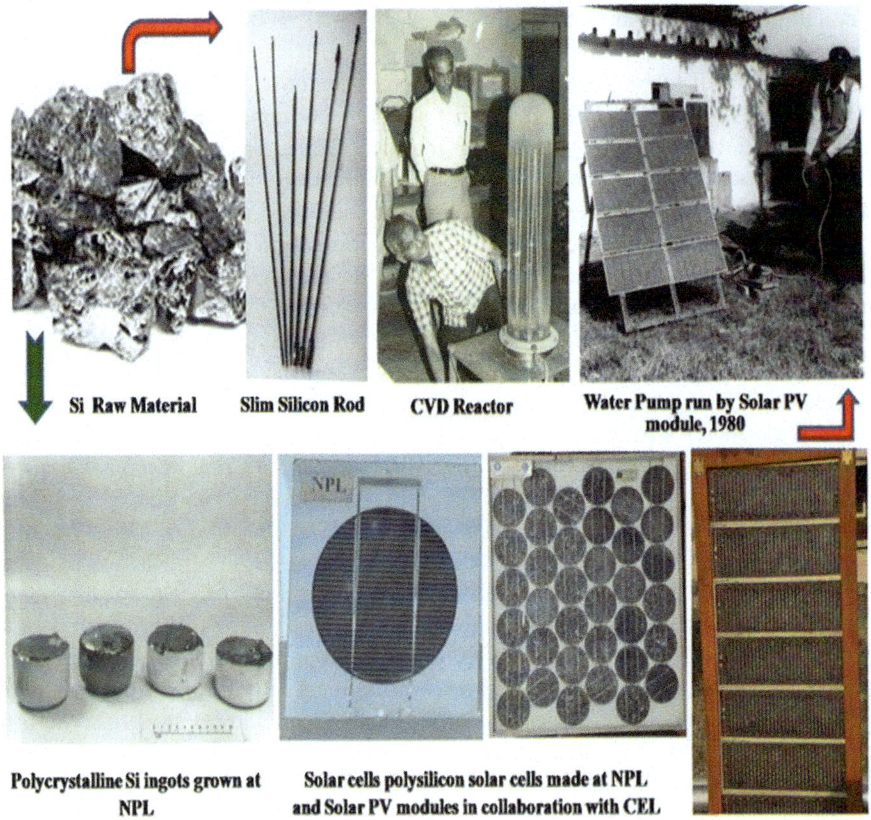

Fig. 17.17 A journey from raw materials to electricity generation via a solar PV system at CSIR-NPL

cell, materials, components, systems and to help PV industries in troubleshooting industrial processes. CSIR-NPL is involved in basic and applied research spanning from wafer-based silicon photovoltaic technology, thin films to latest concepts such as organic and perovskite-based photovoltaic materials and devices development, their testing and measurements following the metrological standards in respective area and have developed adequate expertise and basic infrastructures in the field. It is committed to the translation of our research outputs into beneficial applications and services. The institute works closely with both industry and government agencies on major initiatives to reduce the cost of PV devices for wide generation of solar electricity in India. Some of the recent developments are highlighted in the following section.

Crystalline Silicon Solar Cells

Silicon solar cell technology is the back bone of photovoltaic industry. Presently, it contributes more than 95% share in total cell production in global PV market and it is believed that trend remains in the near future. To further reduce the cost of PV electricity, continuous improvements with respect to the efficiency, manufacturing cost, and reliability of silicon wafer-based solar cells are required. CSIR-NPL's programme on silicon solar cell research covers multi- and mono-crystalline wafers, covering development of materials, processes, device architectures with industrial relevance and diagnostic tools.

In the past, the institute has developed process for industrial solar cells based on screen printing (1992–1996) wherein partially covered back contacts for economical Si solar cells was demonstrated. High efficiency c-Si and mc-Si wafer-based solar cells of efficiencies 16% and 13.5% respectively using screen printing approach have been successfully developed [97]. CSIR-NPL demonstrated porous silicon as antireflection coating (ARC) [98]. CSIR-NPL also designed and developed diagnosis tools such as Mechanical Load Tester (for PV modules of size $150 \times 70\ cm^2$): Simulated wind, snow, ice loads for solar energy centre (SEC, MNRE, India). It has recorded success in materials development also. For example, process was developed for polycrystalline silicon growth using tri-chlorosilane route (TCS) in early 1980s. mc-Si ingot growth up to 20 kg (1987–1990) using directional solidification approach through indigenously designed crucible lowering method (CLM) has been successfully demonstrated and a process for re-usable graphite mould (1988–1992) was also developed [99, 100]. In addition to this, mc-SiGe ingots growth (2000–2004) was demonstrated using the same method. Besides, a number of indigenous methods and diagnostic tools for materials and device characterization for fundamental studies of silicon solar cells have also been developed over the years. [101–108] The institute has significantly contributed in impedance spectroscopy as a tool to understand the physics of solar cells [109–112].

In the recent years, the institute has got expertise in development of value-added unit processes (reflectance and recombination losses) to fill the technological gaps for high efficiency solar cells development, standard BSF cell and passivated emitter and rear contact (PERC) cell, PV materials characterization, PV device measurements and calibration. Integration of value-added unit processes to develop process technology for high-efficiency solar cell compatible with industrial fab-line($156 \times 156\ mm^2$) is of prime objective. In addition, it is aimed to explore feasibility of thinner wafer (<150 μm) cells for cost-effective silicon PV technology–which is a future silicon PV technology. The most significant contributions besides developing new cell structures is the development of (1) rear surface passivation to enhance device efficiency, (2) nano-scale texturization for efficient light trapping process. The proof of concepts for both the technologies have been demonstrated in industrial environment.

(a) **Surface Passivation Schemes**

Electronic recombination of photo-generated charge carriers is a significant loss mechanism in silicon solar cells and needs to be suppressed for realization of higher conversion efficiency. Recombination occurs largely via defects in the bulk and at the surfaces, which create localized energy levels within the band gap [113]. The material quality of a semiconductor depends upon the type and concentration of these defect levels. The effective minority carrier lifetime (τ_{eff}) is a parameter which is used to monitor the material quality and its degradation during subsequent processing. The measured τ_{eff} has contributions from the bulk (τ_{b}) and the two surfaces (τ_{s}). In the last decade, thin Si wafer-based solar cells (<180 μm) are becoming popular for PV cost reduction, however the dominant recombination at wafer surfaces limit/lower the achievable efficiency. Surface recombination losses are associated with surface defects and could be reduced by a mechanism named as surface passivation. It could be realized in two ways, i.e. Chemical passivation and Field-effect passivation. Chemical passivation involves attachment of suitable groups to unsaturated dangling bonds whereas field-effect passivation reduces surface recombination by reducing concentration of one type of charge carriers near the surface by an electrostatic field. A number of lifetime measurement techniques are available and are used for mapping or to determine the effective lifetime [114–116].

Chemical passivation has been reported using iodine-ethanol (I-E), bromine-ethanol (B-E), quinhydrone solutions, sol-gel-based Al doped zinc oxide films etc. to study the silicon material quality [117–120]. We realized good passivation for a ranges of concentration, which are 0.07–0.12 M and 0.05–0.07 M for I–E and B–E solution, respectively. Better passivation is realized for I–E solution whereas improved stability is obtained for B-E solution [121]. Pre-conditioning of the silicon surface is an important procedure to realize good quality passivation and it is observed that presence of native oxide layer affects the Si-I and Si-Br bonding and hence affects the measured lifetime values [122]. CSIR-NPL has also reported that lifetime values from the literature could be compared with any study, only if their injection levels are matched [121].

For thinner silicon wafers, the increased surface recombination makes surface passivation essential for high efficiency solar cell structures. Thermally grown silicon oxide, PECVD deposited silicon nitride and amorphous silicon were commonly used materials for surface passivation of c-Si solar cells for a long time [123]. In the last decade, excellent level of passivation has been realized with atomic-layer-deposited (ALD) aluminium oxide films, mainly due to a high density of fixed charges (Q_F) present in the films providing field-effect passivation [124–126]. These studies led to a research path for new materials (aluminium oxide, hafnium oxide, AlO_x/Si_xO_y stacks, etc.) and new industrial solar cell structure (PERC, PERT, PERL etc.) other than the traditional Al-BSF (back surface field)-based cells. ALD is a conformal coating technique, low-temperature process and has excellent control over the film thickness and uniformity on larger areas.

An ALD system with thermal, plasma and ozone process (M/s Picosun) was installed at CSIR-NPL (Fig. 17.18a). A number of processes were optimized to

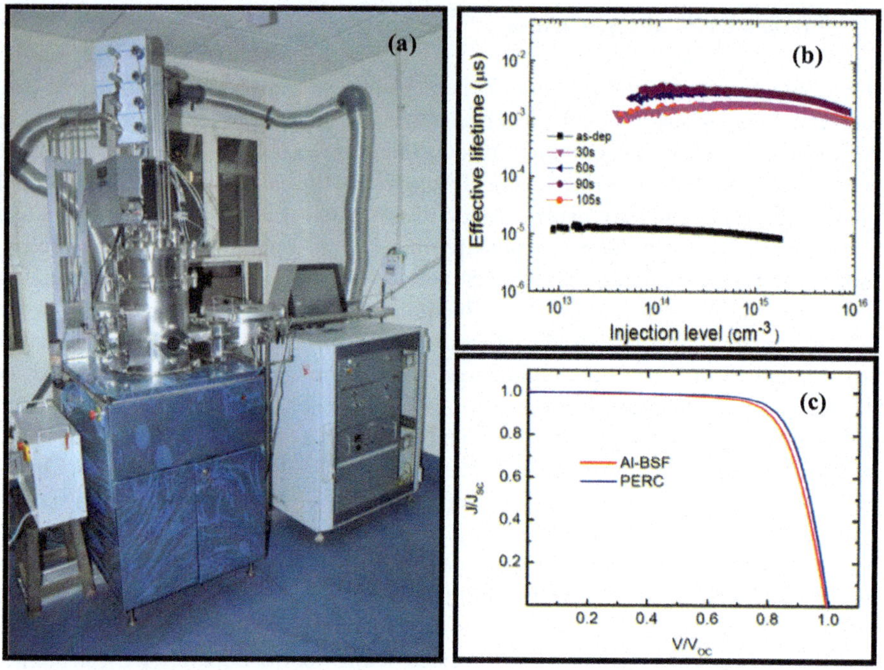

Fig. 17.18 **a** ALD system installed at CSIR-NPL **b** Measured injection level dependent τ_{eff} for plasma ALD AlO_x films, as deposited and annealed in Nitrogen ambient (400 °C for 30–120 s). The AlO_x films are grown for 50 ALD Cycles, 200 °C substrate temperature. An average growth rate of 1.2 Å/cycle at 200 °C is realized, **c** Normalized I-V curves of the champion Al-BSF and PERC cells prepared in collaboration with CEL, Sahibabad

realize good quality surface passivation. Thermal and plasma ALD-based aluminium oxide films have shown state-of-the-art surface recombination velocities (SRV < 10 cm/s) after short-duration annealing in nitrogen ambient using RTP (Fig. 17.18b). The rapid annealing makes it a low thermal budget process [127]. The annealing step sequence is crucial to extract the interface and film related true information for the aluminium oxide film-based MIS structure [128]. As-deposited films show passivation in thermal ALD, whereas no effective passivation is observed for plasma ALD films. Annealing improves passivation for thermal ALD process whereas it is essential to activate the surface passivation induced by plasma ALD aluminium oxide films. The interface defect density (D_{it}) is high ~10^{13} eV^{-1} cm^{-2} for as-deposited samples and reduces to ~10^{11} eV^{-1} cm^{-2} after activation highlighting the importance of chemical passivation in plasma ALD films. Hafnium oxide is not widely studied as surface passivation material and it has been explored by using thermal and plasma ALD process. In thermal ALD process, as-deposited HfO_x film (~10 nm) exhibits SRV ~100 cm/s which reduces to <20 cm/s upon annealing at 400 °C (for 10 min)

[129]. In plasma-ALD HfO_x films, SRV < 40 cm/s are realized after annealing (10–12 min) [130].The realized surface passivation is a trade-off between the field-effect (high Q_F) and chemical passivation (low D_{it}).

In addition to single layer dielectrics, stack of dielectrics are also explored for silicon surface passivation. SRV ~ 10 cm/s could be realized with a stack of HfO_x and AlO_x. films made by thermal ALD, where the individual layer thickness is varied from 3 to 7 nm. This SRV is lower by a factor of ~2.5 with respect to the single AlO_x layer (~10 nm). The improved field-effect passivation is due to the introduction of extra trap centres and interfacial dipoles at bilayer system interface, where the sequence of dielectric layer is also important [131]. A good quality surface passivation (τ_{eff} > 1.5 ms) is realized for thermal ALD deposited Al doped zinc oxide (AZO) thin films at intermediate bulk injection level under optimized annealing conditions of temperature and time (450 °C, 15–30 min, H_2 ambient). These films are prepared by introducing Al precursor between ZnO layers, achieved by a supercycle deposition approach. Annealing temperature activates the hydrogenation of interface defects which involves the transport of hydrogen atoms towards the interface and their interaction with dangling bonds at the Si surface and provides chemical passivation [132]. The importance of boosting mechanism in the delivery of low vapour pressure precursor to the reaction chamber is investigated working on the utilization of CV and CVS measurements to understand the interface has also been studied [133]. Thermal ALD of Al_2O_3 process is integrated with industrial silicon solar cell fabrication line; where a proof of concept of PERC cell is given in industrial environment at CEL Sahibabad (~18% efficient PERC solar cell). At CSIR-NPL a new method of printing/dispensing a pattern of an etching paste on the wafer was developed for locally removing the dielectric layer which is different than conventional layer-based pattering process. After locally removing the passivation layer, the rear & front side metallization is done in conventional industrial process. The normalized I-V curves of champion Al-BSF and PERC cells are shown in the Figure below (Fig. 17.18c). The curves clearly show an enhancement in the fill factor of the PERC cell.

(b) **Black Silicon Solar Cell Concept**

The institute has developed a simple processes for large area fabrication of nano-scale textured black silicon with very low surface reflectivity (R_λ < 2%) in broad spectral range which is equally effective for both crystalline and multicrystalline Si [134, 135]. The two-step wet chemical process is fast (<30 s) and could produce surface with ultra low reflectivity (<2%) which is only achievable in a double-layer antireflection coatings (ARCs) [136–138]. The Proof of Concept of "black" silicon solar cells (both on mono-crystalline and multicrystalline Si) have been demonstrated in the laboratory scale with improved performances. The laboratory-level concept has been demonstrated in industrial line on 156 × 156 mm^2 multicrystalline silicon wafers with champion cell efficiency >18.0% (Average PCE of ~17.75%).The performances of the black cells are better. The contributions have significant impact both in fundamental research as well as in applied fields [135–137]. The black silicon concept developed is a promising approach to eliminate front surface reflection (<2% in broad spectral range at normal and off-normal angles) in PV devices without the need for a conventional ARC (~80 to 90 nm by plasma enhanced chemical vapour deposition (PECVD), etc.) [137].

This can lead to enhancement of the efficiency as well as reducing the manufacturing costs of solar cells. The concept of black silicon fabrication by the developed wet chemical etching process is cheaper, more environmental friendly since it does not involve any toxic or hazardous gases, does not produce any green house gases, is free from uses of expensive tools such as PECVD (used for conventional AR coatings) and therefore requires less electricity and relatively safe (as compared to silane and ammonia gases used for silicon nitride) and has potential for higher efficiency than conventional AR techniques. An optimal trade-off between light absorption gain and electrical losses (due to recombination) is required to be maintained [135–137]. In addition, the institute has developed silicon nanowires and nanostructure-based effective light trapping schemes. The silicon nanowire-based black Si solar cell on bulk substrates and thin solar cells on cheap substrates like glass, initiated at CSIR-NPL [137, 139–145]. The team has established a simple and inexpensive process for controlled fabrication of vertical silicon nanowire arrays on large area using on Ag- induced selective etching of silicon wafers in aqueous HF solution at room temperature [137, 141, 143]. In addition to PV, several other technological applications of black Si, like in, thermoelectric systems, micro-electro-mechanical systems (MEMS), lithium batteries, chemical sensors and biosensors, field-effect transistors, photodetectors and light emitting diodes, and other optoelectronic and photonic devices, H_2 production by photoelectrochemical splitting of water, and as a "self-cleaning" surface and many more are being explored extensively [137]. The black silicon process, the representative results on enhanced light absorption and the black silicon solar cells developed are illustrated in Fig. 17.19.

Amorphous/Microcrystalline Thin-Film Silicon

Efficient hydrogenated amorphous silicon (a-Si:H)-based solar cells and fully integrated panels are realized at CSIR-NPL in year 1985. Participating in an all-India mission mode programme coordinated by the MNES (then Ministry of non-conventional energy sources, which is now renamed as Ministry of New and Renewable Energy, MNRE) the CSIR-NPL developed and demonstrated highly efficient single junction a-Si:H solar cells and fully integrated solar panels. Subsequently, a double tandem solar cells based on a-SiGe:H and a-Si:H of about 10% efficiency were also fabricated by laboratory. The Laboratory worked closely with the BHEL on the pilot line of their Amorphous Silicon Solar Cell Project (ASSCP) and rendered important technical services. Low noise p-i-n detectors for X-rays and Optically Addressed Spatial Light Modulators (OASLMs) have also been made using highly photoconducting a-Si:H layers [146]. The laboratory has also designed and developed plasma systems of different levels of sophistication and supplied to other laboratories and the industry. A multichamber large area PECVD system, with a facility for

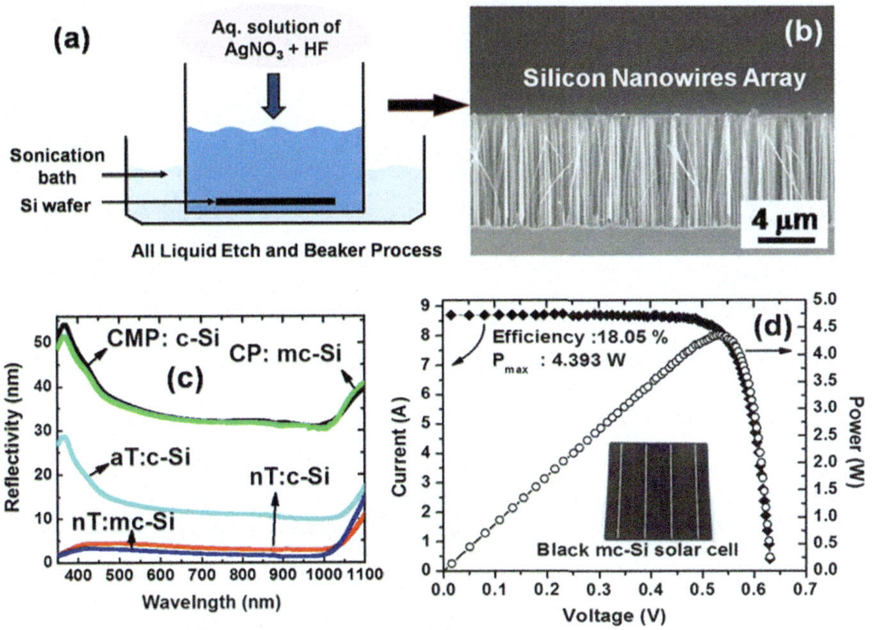

Fig. 17.19 **a** Schematic of Ag- assisted wet chemical etching setup for fabrication of nano-structured black Si surfaces [137], **b** a representative SEM image illustrating silicon nanowires array prepared by Ag-assisted etching process, **c** Reflectivity *vs* wavelength spectra of polished (c- and mc-) Si, KOH textured-Si, nano-textured c-Si and mc-Si surfaces (abbreviations; CMP: chemically mechanically polished; CP: chemically polished; aT-alkaline textured; nT: nano-textured) [134]. **d** PV characteristics (I-V and power) of a typical black mc-Si solar cell on 156 × 156 mm^2, processed in industrial fab-line in collaboration with M/s Tata Power Solar Systems, (**a** digital image of such black solar cell in the inset of (**d**))

cassette-to-cassette transfer of substrates, was designed and fabricated in association with ASSCP-BHEL and Hind High Vacuum Ltd., Bangalore.

Under a fresh initiative, CSIR-NPL worked on micro-/nano-crystalline silicon (μc/nc-Si:H)-based photovoltaic materials & devices which were developed by novel very high frequency (VHF, involving various frequencies 27.12, 60 and 100 MHz), pulsing of these frequencies, and microwave (2.45 GHz) PECVD processes [147–154]. Under this initiative related PECVD systems were developed for fast deposition (>10 A^0/sec) of these μc/nc-Si:H films (Fig. 17.20).

Under a MNRE sponsored project, efficient photoconducting sub-nano-crystalline silicon thin film grown at high rate (10 A^0/sec) by very high frequency (VHF) plasma enhanced chemical vapour deposition (PECVD) technique over 10 × 10 cm^2 substrate and made cells and mini modules as shown in Fig. 17.21. High growth rate of amorphous, micro-/nano-crystalline & ultra-nano silicon thin films using 27.12 and 60 MHz frequencies assisted PECVD processes needed for high throughput of solar cells production. Process to deposit photoluminescent nano-structured silicon thin films using plasma enhanced chemical vapour technique was also developed

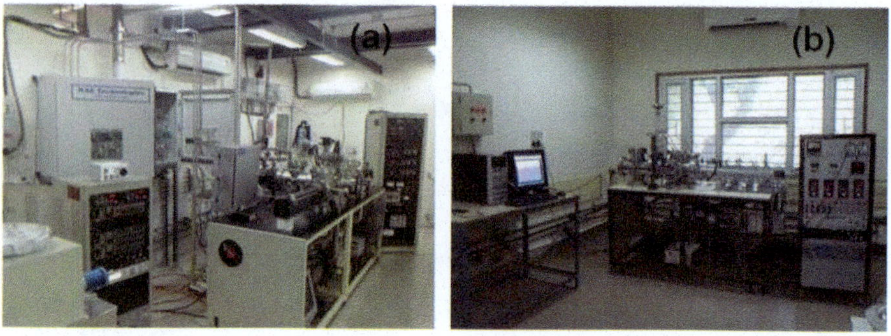

Fig. 17.20 PECVD systems for high rate deposition of micro-/nano-crystalline silicon (μc/nc-Si:H) developed at CSIR-NPL **a** Multichamber refurbished PECVD **b** Dual frequency PECVD

Fig. 17.21 Solar module (10 cm × 10 cm) involving high rate deposited μc/nc-Si:H layers using VHF-PECVD process

[155]. These films can find also use in electroluminescence devices like light emitting diodes (LEDs), LASER etc. Photoluminescence in nano-structured silicon films is very encouraging finding as there is hardly any photonic devices available made of silicon. Processes have been developed for growth of ultra-nano-crystalline silicon and photoluminescent nanostructure silicon thin films. A novel approach for investigating phase transition from a-Si:H to nc-Si:H using was also carried out using plasma impedance analysis [156, 157]. In recent past, simulation studies of efficient HIT solar cells, microcrystalline p-i-n silicon solar cells and DLC as window layer for efficient amorphous silicon p-i-n solar cells has also been performed at CSIR-NPL [158–161].

Organic Solar Cells

CSIR-NPL has been involved in activities of materials synthesis and device fabrication of excitonic solar cells (organic and perovskite solar cells) to meet the scientific and technological challenges to develop end-to-end solutions for efficient solar cells. CSIR-NPL has also established state-of-art facilities for organic and perovskite cells fabrication, and characterization techniques and achieved power conversion efficiency (PCE) ~ 9.0% in bulk hetero junction (BHJ) organic solar cells (~10 mm^2) [162]. Further, theoretical modelling in view of understanding charge transport mechanism of these devices has been done. A concept of extended Ca/Al combinational layer used as cathode in organic photovoltaics (OPVs) which improves the life time of device from ~70 h to ~400 h which further increase up to >1000 h on proper encapsulation has been established. [163] Organic materials such as donor, acceptor and hole transport materials (HTM) are the key components in organic solar cells. In this context thiophene-based semiconductor thiophene-based donor materials have been synthesis and indigenously developed was used for fabrication of BHJ solar cells. Electropolymerized poly(3,4-ethylene dioxy thiophene) (PEDOT) was developed for HTM in organic solar cells and PCE of ~2.67% and improved stability of ~2000 h has been achieved in device configuration of ITO/e-PEDOT/P3HT:PC_{60}BM/Ca/Al. PEDOT was also obtained by solid-state polymerization for HTM in organic solar cells device. Solution processable of CuBr, CuI, and CuSCN etc., as new family of interfacial layer for efficient and solution-processable OPV solar cells with PCE ~5.94% has been developed [164]. A cost-effective synthesis route for most popular acceptor material, i.e. Phenl C_{61} Butyric Acid Methyl Ester (PCBM) [165] and several other novel methanofullerene derivatives have been synthesized following this novel method which is given the name "amine assisted cycloaddition reaction" (AACA process) [166, 167]. The process has been patented in India as well as in USA. [168] Several novel non-fullerene n-type semiconductor materials have also been synthesized showing excellent electron conductivity as well as electron transport properties. Rylene materials which include perylenediimides (PDI) and naphthalenediimide (NDI) have been found to be exciting materials to be used as n-type semiconductor materials in excitonic solar cells.

Perovskite Solar Cells

CSIR-NPL has undertaken the development of printed perovskite solar cells (PSCs) and is working on their various aspects including high PCE, long-term stability and large area printing on flexible substrates like plastic sheets. Though the power PCE of PSCs has shown marvellous improvement in last few years but stability remains a serious issue. Unfortunately, these devices have shown to exhibit rapid degradation in their performance under actual operating conditions and as consequence these devices lack from commercialization. The stability of these devices is affected by several parameters like temperature, humidity, light intensity, electrical stress etc. To understand the degradation mechanisms and effect of different parameters on the stability of these devices, CSIR-NPL is carrying out detailed investigation on their degradation as per the International Summit on Organic Photovoltaic Stability (ISOS) protocols in different environmental conditions [169, 170]. It is observed that under continuous illumination of the white light source the top Ag electrode decomposed and reacted with the active layer and this decomposition and reaction of Ag electrode with active layer got accelerated under direct sunlight in outdoor storage. Even is dark condition active perovskite layer exhibited photo-oxidation and perovskite decomposition. The studies suggested that Ag is a good electrode material for efficient PSCs, but it is not good for their long-term stability, therefore use of some other suitable materials is required for the top electrode. The search for suitable electrode materials for PSCs, pointed out towards Cu to be one of the other suitable electrode materials as it has high work function and high corrosion resistance for non-oxidizing acids like HCI, HI, HBr and halogens. CSIR-NPL incorporated Cu as top electrode in PSCs and investigated their degradation mechanisms and photostability [171]. The solar cells exhibited PCE of 8.1% at SCT. As expected, Cu resulted in better device stability compared to Ag but these devices too exhibited degradation under illumination. Cu was observed to diffuse into other layers underlying and reacted with them. It was fund that Cu is also not a very stable electrode material for PSCs when it comes to operation under direct sunlight. CSIR-NPL is working out to find out some other suitable electrode materials, which could impart longer stability to PSCs.

CSIR-NPL is developing flexible PSCs on polyethylene terephthalate (PET) substrates to evolve the process of in-house fabrication of highly efficient and stable roll-to-roll printed PSCs. To start with, the solar cells were prepared on small areas ITO coated PET substrates (4 cm^2) in the structure PET/ITO/SnO_2/$CH_3NH_3PbI_{3-x}Cl_x$/Spiro-MeOTAD/Ag. These solar cells exhibited very high external quantum efficiency of ~80% and PCE of 12.7 $\pm$ 0.6% with champion cell efficiency of 13.3% (active area 0.096 cm^2) [172]. These solar cells were observed to be very robust and exhibited no deterioration in the performance over multiple bending cycles. Thought the flexible PSCs degraded much faster than those prepared on glass substrates, but to some extent their degradation mechanisms were similar. While stored in dark in ambient air, the flexible PSCs exhibited a half life of about 340 h, but when stored in direct sunlight they degraded very rapidly. Encapsulation is expected to enhance their stability; therefore a suitable material is being searched for their encapsulation without compromising their efficiency and flexibility.

CSIR-NPL prepared flexible solar modules in similar structure as mentioned above, in the area of 5 $\times$ 5 cm^2 (active area 15.84 cm^2) by integrating eight solar cells of 1.98 cm^2 in series together. The solar modules exhibited PCE of 5.6 $\pm$ 0.2% at STC with a maximum PCE of 5.8%. Figure 17.22a shows the photographs of one of the solar modules and Fig. 17.22b shows its current-voltage (I-V) characteristics. The inset of Fig. 17.22b shows the photovoltaic parameters of the module at STC. The modules exhibited considerable loss in PCE compared to small area devices and has been ascribed to higher series resistance of longer ITO electrodes used in the modules. CSIR-NPL is moving forward to develop the processes to scale

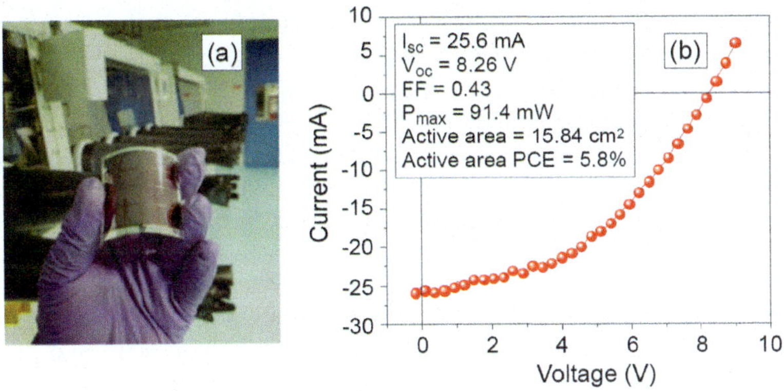

Fig. 17.22 **a** Photograph of the flexible PSC module (5 × 5 cm^2) and **b** I-V characteristics of the module at STC. Inset shows the photovoltaic parameters of the module [172]

up the fabrication of roll-to-roll printed PSCs via conventional printing techniques like doctor blade, slot die and screen printing. Investigations are being carried out to understand degradation mechanisms in printed PSCs and a suitable barrier layer is also being developed for their encapsulation to ensure their long-term stability in actual operating conditions.

In the research area of photovoltaics there have been standards for inorganic solar cells like silicon and others for characterization and measurements for reliable data. However, there has been no calibration standard for organic/perovskite solar cells (excitonic solar cells). This creates discrepancy in results from laboratory to laboratory or place to place for the same device. To avoid such non reliable data there is need to develop calibration standards in the area of excitonic solar cells. CSIR-NPL is dedicated towards fabrication of stable excitonic solar cell devices with the aim to develop standards for organic solar cells. At CSIR-NPL world class facility for fabrication and testing of perovskite solar cells is available. Also have capability of testing organic and perovskite solar cells under standard testing conditions as per IEC 60904/IS12762 protocol. I-V characteristics are measured using AAA solar simulator which is calibrated to standard testing conditions (STC) state by reference solar cell. CSIR-NPL is focusing towards preparation of stable organic and perovskite solar cells with lifetime >5000 h (t80 or t50) to be used as standard for calibration of solar simulator for excitonic solar cells with standard protocol for efficiency measurement will be delivered for nation.

Hybrid Solar Cells

Inorganic-organic hybrid nanocomposites integrate the useful inorganic and organic characteristics within a single composite. These nanocomposites are expected to exhibit unique electronic and optical properties. To understand the basic science of the synthesis, characterization and sensitization of various morphologies of CdSe quantum dots within poly [2-methoxy-5-(2'-ethylhexyloxy)-p-phenylenevinylene] (MEHPPV) and Poly(3-hexylthiophene (P3HT) polymer matrices respectively for its application in hybrid solar cells is very important.

In CSIR-NPL, Hybrid Organic (P3HT, MEHPPV)-Inorganic (CdSe) Nanocomposites are developed to use in optoelectronic devices. The effect of different surface passivation of CdSe QDs synthesized by the oleic acid (OA) and TOPO routes, on the structural and optoelectronic properties of their respective polymer nanocomposites have been demonstrated by dispersing synthesized OA-passivated CdSe QDs and TOPO-passivated CdSe QDs in MEHPPV and P3HT polymer matrices, respectively. It has been established that OA-passivated CdSe QDs (~7 nm) provide relatively better steric stability against coagulation, homogeneity and photostability of their respective polymer:CdSe QDs composites than TOPO-passivated CdSe QDs (~5 nm). It is observed that there exists a dominance of Forster energy transfer between host polymer (donor) and guest CdSe nanocrystals (acceptors) in polymer:CdSe(OA) composites and efficient charge transfer between polymer: CdSe(TOPO) composite constituents for its application in efficient electroluminescent and photovoltaic devices respectively. Upon ligand-exchange with pyridine (p), both CdSe QD's capped with OA and TOPO show evidence of different rates of PL quenching in their respective polymer:CdSe composites because of partial ligand-exchange and complete ligand-exchange in case of CdSe (OA/TOPO) QD's respectively. Transient absorption experiments were carried out using nanosecond laser flash photolysis. In case of pure MEHPPV, the depletion of MEHPPV absorption as indicated by the bleaching of the ground state of MEHPPV confirms its participation in the photoinduced chemical process. However, for pure CdSe QD's after pyridine treatment, the CdSe nanocrystals (OA/TOPO) remain stable during laser pulse duration as indicated by the absence of transient absorption. Only in the presence of polymer (MEHPPV), when the sample was excited, the formation of transient between 670-730 nm corresponding to CdSe-* radical was seen. The ultrafast electron transfer to CdSe (p-TOPO) as monitored from the quick bleaching recovery of MEHPPV and the formation of CdSe-* radical is completed with an average rate constant of $1.8 \times 10^{10} S^{-1}$. On the other hand, the slow recovery of the transient bleaching of MEHPPV in the presence of CdSe QD's (p-OA) indicates energy transfer for MEHPPV:CdSe (p-OA) nanocomposites. A model is proposed in a bid to understand the ligand-dependent properties of CdSe QD's, the mechanism of exciton dissociation and charge/energy transfer across MEHPPV:CdSe composites. The importance of Forster Resonance Energy Transfer (FRET) and charge transfer processes as observed for MEHPPV: CdSe and MEHPPV: CdSe nanocomposites capped with different ligands oleic acid and TOPO respectively makes them ideal candidates for efficient hybrid organic photovoltaic devices.

17.3 Materials for Societal Applications

Engineering for the development of the new materials fulfil the various aspects of the society in daily life. Thus the search for new materials is the real backbone of the society for improvement of quality of life. Therefore, CSIR-NPL is working on various aspects of new advanced materials which can be used for societal applications.

17.3.1 Carbon-Based Materials

Carbon is an important class of material due to its unique physico-mechanical and chemical properties which is used in almost all branches of science and technological applications. There are different sources of solid carbon, one of them is coal tar. The coal tar is a by-product obtained during the destructive distillation of metallurgical coal carried out by the steel industry for the production of coke. Coal tar is complex and variable mixture of phenols, polycyclic aromatic hydrocarbon and heterocyclic compounds and by distillation process, coal tar is converted into coal tar pitch which having around 50% solid carbon. The coal tar pitch is a starting material for development of different value-added carbon products such as impregnating and binder coal tar pitch, mesophase pitch, needle coke, mesocarbon microbeads, precursor for pitch-based carbon fibre etc. (Figure 17.23).

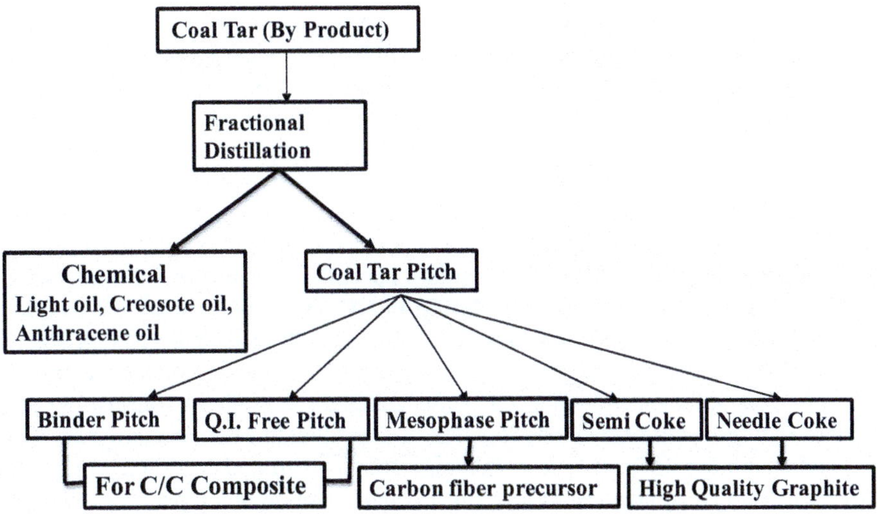

Fig. 17.23 Different value-added carbon products from coal tar

17.3.1.1 Coal Tar Pitches

Coal tar pitch (CTP) is a solid, shiny and black substance. It is an extremely important carbon precursor used in production of variety of conventional and newer carbon products. Conventional carbon products are graphite electrodes used for the production of steel in electric arc furnaces in which two types of coal tar pitch are used, binder grade and impregnating grade coal tar pitch, and prebaked anodes and Soderberg paste for the production of aluminium. In newer carbon products, coal tar pitch is the starting precursor for light weight and porous carbon foam, carbon fibres and spherical carbon known as mesocarbon microbeads (MCMB). Properties of all carbon products depend on quality and type of CTP. Quality of coal tar pitch is defined in terms of certain properties such as solubility in different solvents like quinoline, toluene, pyridine and benzene; softening point, carbon content, rheological properties, etc. and these properties depend on the initial source of coal and treatment processes. There are mainly four types of CTP, (1) Binder grade CTP, (2) Impregnating grade CTP, (3) Isotropic CTP and (4) Mesophase pitch. Binder grade CTP has high quinoline/toluene insolubles, carbon content and softening point as compared to impregnating and isotropic CTP. Brooks and Taylor defined the optically anisotropic intermediate phase of CTP as the "mesophase". The pitch like substance having mesophase characteristics is defined as mesophase pitch. Coal tar pitch is obtained from distillation of coal tar and coal tar is a brown/black, thick viscous liquid which is obtained as a by-product in the destructive distillation of metallurgical coal carried out by the steel industry. The distillation of coal tar results in the several oil fractions and coal tar pitch (CTP) as residue which is the main product. CTP is a complex mixture of chemically complex product containing thousands of organic compounds (poly aromatic hydrocarbons), made up of several fused benzene rings, their alkyl and hydroxyl derivatives, and the corresponding heterocyclic compounds containing oxygen, nitrogen or sulphur. The important characteristic of CTP is its high carbon content and it has ability to graphitize with heat treatment temperature which increases the thermal and electrical conductivities of the products. The electrical and thermal conductivities are directly related to the orientation and structure of carbon materials. The CTP can be converted into variety of carbon products (MCMB, mesophase, coke) due to chemical reactions such as polymerization and condensation during heat treatment and depends on heating conditions. When CTP goes under heat treatment around 400–450 °C, due to polymerization reaction, polyaromatic hydrocarbon molecules start joining each other and grow in size or form bigger molecules of higher molecular weight and release volatiles, results in the homogeneous nucleation of a new phase, called the mesophase. These molecules arrange themselves with the planes parallel to each other like nematic liquid crystals. This reaction keep on increasing with temperature and around 1400 °C, basic structure called graphene layers start forming. Further increase in temperature enhances the stacking of graphene layers with minimum defects to form graphitic structure of carbon and around 3000 °C, perfect graphitic structure can be obtained. Optical images of coal tar pitch with heat treatment temperature are shown in Fig. 17.24.

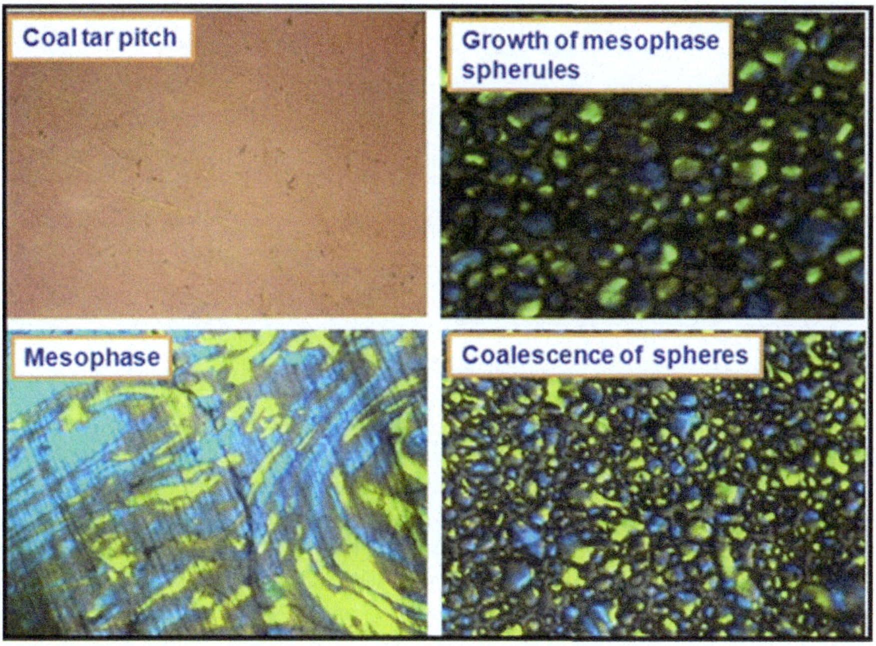

Fig. 17.24 Optical images of coal tar pitch with heat treatment temperature

17.3.1.2 Carbon Foam

Carbon foams (CFoams) are sponge-like high performance lightweight engineering material in which carbon ligaments are interconnected to each other, it possesses low density, large surface area with open cell wall structure, good thermal/electrical transport properties and mechanical stability. The carbon foam is generally developed from thermosetting and thermoplastic polymers by various techniques. The thermosetting polymer, i.e. phenolic resin derived carbon foam, is less electrically and thermally conductive, while coal tar pitch and mesophase pitch derived carbon foam are thermally and electrically conductive. There are different methods to fabricate the carbon foam. Most of them are based on foaming of pitches followed by oxidation stabilization, carbonization and graphitization. Foaming is achieved by either using blowing agent or pressure release employing the expensive high-temperature-high pressure autoclave systems. But technologists are always looking for relatively simple and inexpensive methods to produce the carbon foam. In this direction, CSIR-NPL has developed the carbon foam derived from coal tar pitch and well phenolic resin based by sacrificial template technique. In this technique, polyurethane foam is used as a template and it is impregnated by the coal tar pitch and in another case by phenolic resin [173, 174]. The coal tar pitch and phenolic resin impregnated polyurethane foams are converted into carbon foam by several processing steps and heat treatments up to 2500 °C. To improve the properties of

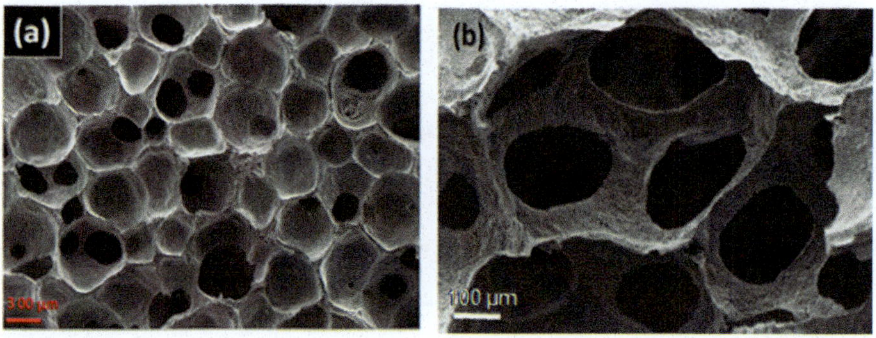

Fig. 17.25 Images of carbon foam derived from phenolic resin and coal tar pitch precursor

carbon foam developed from coal tar pitch and phenolic resin, different approaches are adapted such as incorporating the nanosize organometallic compound, multi-walled carbon nanotubes (MWCNTs), expanded graphite, magnetic and dielectric nanoparticles in the starting carbon precursor and decoration of multiwalled carbon nanotubes over the carbon foam surface by chemical vapour deposition technique [175, 176]. Figure 17.25 shows the images carbon foam derived from phenolic resin and coal tar pitch precursor. Due to light weight in nature, highly porous, high specific electrical and thermal conductivity and good mechanical properties, carbon foam can be in number of demanding application such as radar absorbing materials, power electronic cooling, heat sink, radiator, filters, anode material for Lithium ion batteries, electrode for lead acid batteries etc.

17.3.1.3 Carbon Fibre Polymer Composite Half Ring as Ilizarov Fixator

Ilizarov fixator is a type of external fixture used in orthopaedic surgery to lengthen or reshape limb bones and to correct angular deformity in the leg. It is a limb sparing technique to treat complex or open bone fractures and also in cases of infected non-union of bones that are not acquiescent with other techniques. The Ilizarov is named after the orthopaedic surgeon Gavriil Abramovich Ilizarov, who pioneered this technique. It is a specialized form of external fixator, a circular fixator modular in construction. It consists of stainless steel rings are fixed to the bone via stainless heavy-gauge wire. The rings are connected to each other with threaded rods attached through adjustable nuts. The circular construction and tensioned wires of the Ilizarov apparatus provide far more structural support than the traditional mono lateral fixator system. This allows early weight bearing. But the stainless steel ring possess the density ~8.0 g/cm^3, due its bulky nature, it is some time difficult for movement to patient. On the hand carbon fibre is a biocompatible material fibre have multiple potential advantages in developing high-strength biomaterials with a density close to bone for better stress transfer. To replace the steel ring external fixator, the carbon

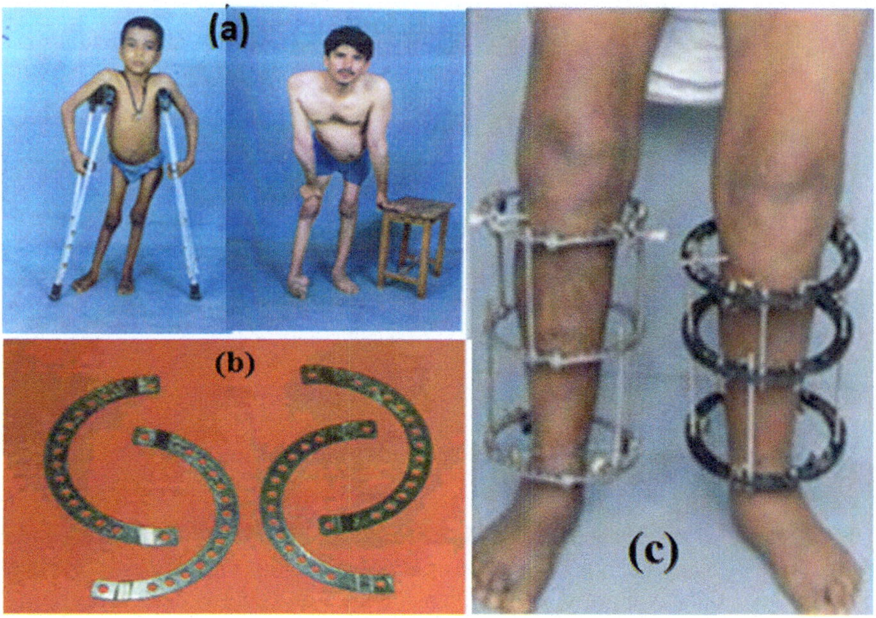

Fig. 17.26 **a** Deformation in legs, **b** carbon fibre polymer composite ring and **c** application of fixator to patient

fibre polymer composite-based half ring was developed by filament winding and compression moulding technique using thermosetting resin as polymer matrix of density 1.50 g/cm^3. These rings are light weight (density 1.50 g/cm^3), transparent to X-rays and have better elastic deformation behaviour compared to stainless steel rings. Figure 17.26 shows carbon fibre ring fixator.

The principle based on the theory of tensions, in which controlled and mechanically applied tension stress that can help to regenerate in a reliable and reproducible manner bone and soft tissue. As shown in Fig. 17.26, top ring allow force to be transferred through the external frame, bypassing the fracture site. Force is then transferred back to the healthy bone through the bottom ring and the tensioned wires. This allows the Ilizarov apparatus to act as a sort of bridge, both immobilizing the fracture site and relieving stress, while allowing for the movement of the entire limb and partial weight-bearing. Middle rings and tensioned wires act to hold the bone fragments in place and to give greater structural support to the apparatus and limb. However, the critical load bearing rings are the top and bottom rings which transfer the force from the healthy bone down to the healthy bone, bypassing the fracture site. The technology of these rings has been transferred to M/s. Agrawal Orthopaedic Hospital, Gorakhpur.

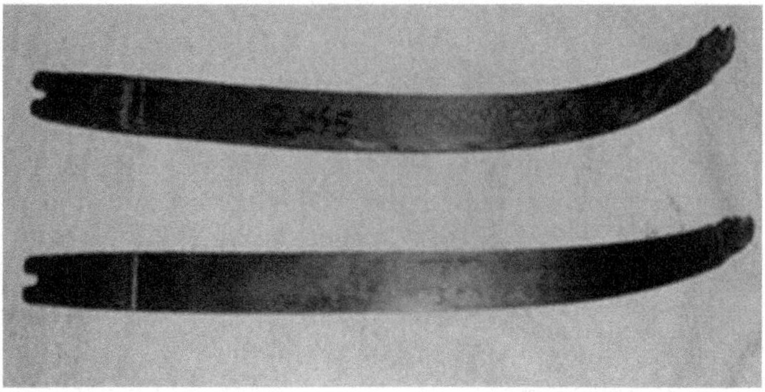

Fig. 17.27 Small scale down size carbon fibre composite archery bow limb

17.3.1.4 Carbon Fibre Polymer Composite Archery Bow

The archery bows have been used long back from prehistoric times for hunting and in warfare. New generation limbs are made of multilayer composite materials involving different layers like compression layer, tension layer and intermediate layer of high modulus material. The intermediate layer is made of polymer foam for providing high modulus. The indigenous production of the professional composite recurve bow will make the archery sport quite affordable. Considering the fact that limbs are main part in functioning on a recurve bow. Therefore, at CSIR-NPL efforts have been made to develop CFRP-based recurve bow limb of 20 inch as shown in Fig. 17.27.

Further upscaling of the size to actual size is under process. The indigenous production of the professional composite recurve bow and alloy or metal-based riser will make the archery sport quite affordable in the country. Considering the fact that riser and limbs are main part in functioning on a recurve bow. Therefore, in this line CSIR-NPL with MSME, PPDC sport complex Meerut is involved for development of cost-effective archery recurve bow.

17.3.1.5 Graphene-Based Materials

Since its discovery in 2004 and subsequent Nobel Prize for the same in 2010, Graphene has drawn enormous scientific attention due to fascinating properties and wealth of prevalent applications [177]. Over the years, a variety of top-down and bottom-up strategies viz. micro-mechanical exfoliation, chemical vapour deposition (CVD), electrochemical synthesis, and chemical oxidation route, have been exploited for synthesis of graphene and its analogues [177–179]. Among them, chemical method involving oxidation of suitable graphitic precursor, followed by exfoliation is most popular approach for scalable synthesis of its oxidized derivative i.e. graphene oxide (GO), which could be further subjected reduction step (by thermal, chemical, electrochemical or biological means) to realize reduced GO (RGO). At CSIR-NPL, a gram scale (up to 250 g batch size, scalable up to kg level) facility for safe synthesis of GO using modified Hummers' method has been established, wherein upon chemical oxidation of graphite, and adoption of suitable isolation strategy (e.g. centrifugation, vacuum filtration, vacuum drying systems, freeze drying etc.), GO can be obtained in different forms such as pulp, paper like mass or highly exfoliated fluffy powder. These GO-based solid forms could be readily re-dispersed inside water using ultrasonication route, to form aqueous colloidal solutions with excellent stability i.e., no setting, even after 8–10 months [178, 179]. In order to exploit the properties of graphene layers, extensive efforts have been made by various researchers to realize their macroscopic assemblies, viz. paper like materials (pseudo 2D system) and hydrogels/aerogels (3D systems), and their diverse applications is demonstrated. Again, GO is widely considered as cornerstone precursor for designing of such systems owing to its excellent re-dispersibility in water and ability to react/interact with variety of chemicals through its rich oxygen functionalities, often via sol-gel chemistry.

Particularly, graphene-based papers are considered as pseudo 2D macroscopic assemblies of graphene analogues layers arranged in the form of stack, to assume a membrane like structure. These papers are typically obtained either by filtration or by slow evaporation of solvent from graphene analogue's dispersion/solution, leaving behind the stacked layers. Such layers are bound together by physical interlocking, van der Waals forces or ionic interactions [180] which along with thickness determine the ultimate properties. CSIR-NPL has developed expertise in making GO and RGO-based papers by exploiting GO colloidal solution as precursor [180].

Various kinds of graphene-based papers such as based on GO, RGO and their hybrids/composites with other phases can be fabricated using solution technique from highly concentrated GO pulpy mass, wherein by controlling the volume of given concentration suspensions, papers of variable thickness can be achieved. The GO-based papers display characteristic brownish shade with stiff and strong nature, due to inherent stiffness of constituent GO layers and strong interactions between them. Such papers with hydrophilicity, semiconducting nature and rich oxygen functionality, are suitable for applications in water purification, static charge dissipation, selective transport of gaseous phases, electrochemical catalysis etc. The chemical reduction (e.g. using ammonia, ethylene diamine, hydrazine and HI/acetic acid vapours etc. as reducing agents) of GO paper to RGO paper visible change in colour to black, increase in buoyancy & hydrophobicity, good flexibility and improvement in electrical conductivity. Particularly, the low density, flexibility, hydrophobicity and high electrical conductivity of RGO papers demonstrate their promising potential to be used in shielding of EMI, dissipation of static charges, foldable electronics, energy storage devices, water purification etc. The physicochemical properties of such papers can be easily tuned as per application requirement by controlling the nature of graphene analogue, concentration and volume of suspensions, solvent type, presence of specific linkers and drying protocols. For example, a reduced GO-based paper prepared from GO paper by vapour phase reduction has shown good electrical conductivity with excellent flexibility, with shielding effectiveness (SE) value of 35.49 dB and specific shielding effectiveness (SSE) value of 221.8 $dBcm^3g^{-1}$, which far exceeds the value for binder-less expanded graphite-based sheet with good response at other frequency bands too [180].

Hydrogels and aerogels are a kind of special materials due to their unique structural properties including very high porosity, large specific surface area, low apparent density and dependent interesting set of mechanical, thermal and chemical properties. Particularly, graphene-based hydrogels and aerogels have shown tremendous potential from the technological perspective due to their specific properties derived from constituent graphene derivative layers (which are chemically stable, biocompatible, and inherently strong and electrically conducting) [181]. CSIR-NPL has developed expertise and facility for design and development of graphene-based hydrogels and aerogels via sol-gel chemistry and freeze drying-based protocols to prepare graphene-based hydrogels and aerogels. In a typical hydrothermal treatment process, graphene analogues-based dispersions are thermally treated in closed container (autoclave) in presence of suitable reducing agent, to form hydrogels, followed by purification strategy involving water soaking or solvent exchange operations.

The hydrogels are known to contain very large amount of trapped solvent and represent highly expanded structure with good mechanical strength, large interfacial area & specific surface functionalities. These traits are responsible for their application in diverse areas including but not limited to energy and environmental science [177]. For example, removal of lead from water where an ethylene glycol mediated graphene hydrogel (EGGH)-based fixed bed column exhibited 80% removal efficiency for 100 mL of an aqueous lead solution (100 ppm) has been demonstrated.

The controlled drying of hydrogels (e.g. freeze drying) transform it into corresponding aerogels, wherein solvent filled cells are drained out and replaced with air. Alternatively, aerogels can also be realized via template routes, involving coating of suitable graphene precursor (gaseous or chemical dispersion) over porous template (e.g. foams) via specific operation (chemical vapour deposition, dipping etc.) followed by sacrificial remove of the template (e.g. via chemical or thermal treatment), leaving behind graphene analogue-based framework [177]. Depending upon the graphene precursors, presence of secondary phase (e.g. specific incorporation) and adopted synthesis protocol, the as prepared aerogel are known to display properties such as extremely low density, high specific surface area, controlled porosity, desired resilience, tunable surface functionality, desired electrical properties etc. For example, magnetic aerogel powder (MAP) with loosely integrated aerogel flakes has been prepared in laboratory that displays instantaneous methylene blue (MB) dye adsorption response (complete discoloration in 30 s).Moreover, upon MB adsorption, dye laden MAP can be rapidly recovered from water system exploiting external magnetic field. Most importantly, MAP has very facile regeneration process and good recyclability with more than 75% capacity retention even after five regeneration-reuse cycles [181]. Macroscopically well integrated graphene-based aerogel (EGGA) blocks through hydrothermal treatment and subsequent freeze drying route has also been prepared; using GO as precursor, water as dispersion medium and organic medium as self-assembler-cum-reducing agent. The formed EGGA aerogel block display very low weight extremely low density (~3.5 $mgcm^{-3}$), very high surface area (~1100 m^2g^{-1}), residual oxygen functionalities and repeated compressibility, making it suitable candidate for several application especially in wastewater treatment. More specifically, EGGA-based fixed bed columns have shown excellent dye removal capacity (as evident from complete discoloration of MB solution in 100 s with facile regeneration using acidic ethanol solvent and excellent recyclability. Besides, applicability of other specifically designed aerogel systems, oil spill clean-up, EMI shielding and energy storage devices has also been demonstrated [177–182] e.g. high rate-performance supercapacitor (SC), using a chitosan-graphene-based hybrid aerogel [182].

17.3.2 Molecular Sieves for Gas Separation

Energy efficient separation and purification technology is the requirement of today for separation of gas and chemical mixtures. Molecular sieves, which comprises of zeolites, metal organic frameworks (MOF), carbon molecular sieves etc., have been proved to be energy efficient mode of molecular scale separation and purification of components like O_2/N_2 from air, CO_2/CH_4 from natural gas, CO_2/N_2 from flue gas etc. [183, 184]. Molecular sieve-based separation is rapidly becoming a main stream separation technology for air separation and production of nitrogen or oxygen enriched air. [185] Zeolites, the alumino-silicate minerals, having three dimensional framework of $[SiO_4]^{-4}$ and $[AlO_4]^{-5}$ (primary building units, PBU) tetrahedra with defined pore size. PBUs are connected to each other by shared oxygen atoms and form geometrical forms known as secondary building units (SBUs). Composite building units (CBUs) are also component of zeolite and can be in ring, cages of cavities forms. Most common CBUs are ring type and named on the number of tetrahedrons involved. The valency difference located on oxygen atom connected to aluminium cation is balanced by a cation usually an alkali ion. The cation can be exchanged for another ion by ion exchange technique. The O_2 and N_2 differ by ~0.3 Å in their molecule (3.75 Å vs. 4.07 Å). Considering the entropic selectivity advantage of molecular sieving materials by limiting the degrees of freedom of penetrants for the transport, components with such small dimensional difference can also be effectively separated. We have prepared the most applicable zeolite type used for O_2/N_2 separation from air is Linde Type -A (LTA) with Si/Al ratio ~1. The zeolites with pore size in the range of 4 Å can successfully restrict rotation of N_2 molecules making entropically unfavourable for N_2 diffusion and shorter O_2 passes through. Zeolites also get advantage from special ability to adsorb nitrogen more strongly than O_2 due to interaction between the quadruple moment of N_2 and cation in zeolite framework.

Zeolite A is synthesized using alumina and silica precursors in different composition ratio. The general formula for zeolite is represented by $M_{\frac{2}{n}}O : Al_2O_3 : XSiO_2 : YH_2O$ where, M represents a metal and n is its valence. For real application crystalline zeolite is converted to beads or pallets using a binder material, i.e. clay/alumina or polymers with diameter of 0.5–6.0 mm in common with meso or macroporous network facilitating transport of gas (Fig. 17.28).

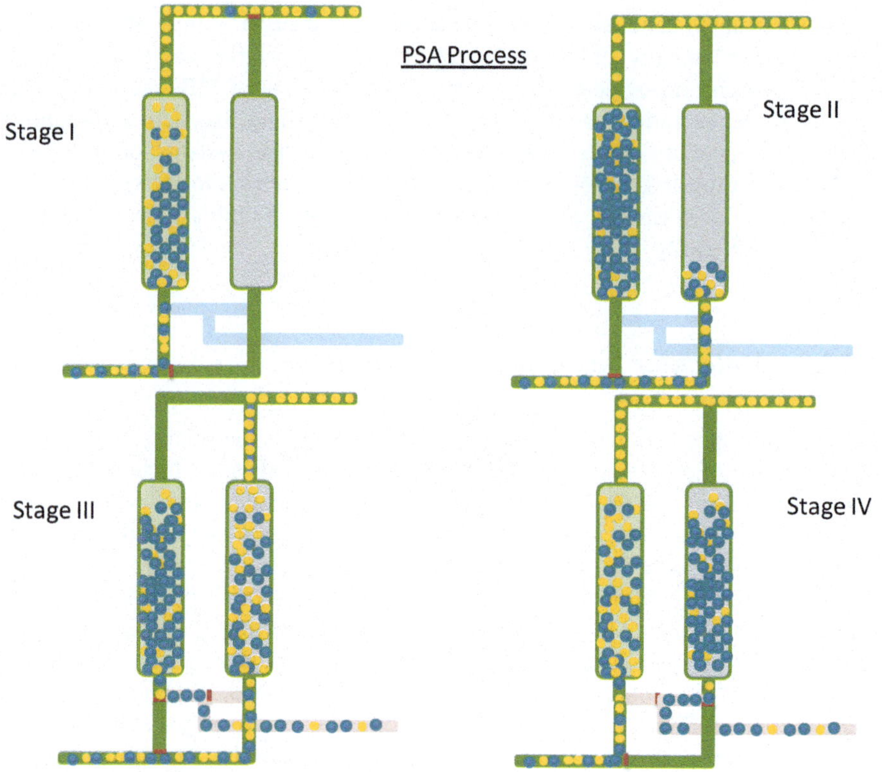

Fig. 17.28 Stages of PSA for oxygen concentrator, where blue and yellow balls represent Nitrogen and Oxygen molecules

In pressure swing adsorption (PSA) process-based oxygen concentrator design, two columns operate semi-continuously providing desired purity and flow rate in air separation. To the first bed with zeolite beads, compressed air is passed where nitrogen molecules are trapped while oxygen is allowed to flow. When the first bed is saturated with nitrogen molecules, feed is directed towards the second column of zeolite bed. While the nitrogen is adsorbed in the second column, first bed is depressurized and nitrogen is released in atmosphere. Now again the air is fed into the bed of first column and second bed is a depressurized releasing nitrogen molecule. This cycle keeps repeating to produce continuous flow of oxygen enriched air. Using LTA zeolites oxygen production with ~ 95% purity is obtained.

17.4 Waste Material Managements

With increasing the utilization of various product and devices for the well-being of human life and on the other hand it generates lot of waste, such as plastic, electronic, batteries, solar modulus waste as well as agriculture waste. CSIR-NPL has taken some initiative for sustainable development through waste management by recycling and converting the waste into value-added products.

17.4.1 Recycling of Plastic Wastes

Plastics are inexpensive, lightweight and durable materials, which can readily be moulded into a variety of products that find use in a wide range of applications. As a consequence, the production of plastics has increased markedly over the last 60 years. However, current levels of their usage and disposal generate several environmental problems. Around 4% of world oil and gas production, a non-renewable resource, is used as feedstock for plastics and a further 3–4% is expended to provide energy for their manufacture. A major portion of plastic produced each year is used to make disposable items of packaging or other short-lived products that are discarded within a year of manufacture. These two observations alone indicate that our current use of plastics is not sustainable. In addition, because of the durability of the polymers involved, substantial quantities of discarded end-of-life plastics are accumulating as debris in landfills and in natural habitats worldwide. Recycling is one of the most important actions currently available to reduce these impacts and represents one of the most dynamic areas in the plastics industry today. Recycling provides opportunities to reduce oil usage, carbon dioxide emissions and the quantities of waste requiring disposal. Disposal of plastic waste is a major problem. It is non-biodegradable & it mainly consists of low density polyethylene plastic bags, bottles etc. Burning of these waste plastic bags causes environmental pollution. According to disposal policy of government of India plastic waste has to go to landfill site. In landfill, plastic material may take 1000 years to completely degrade. Plastics in different states of degradation release toxic material that leaches into ground and pollutes ground water. Plastic material should not be burnt. A multitude of toxic gases are released when plastic is burnt. These include carbon monoxide, phosgene, nitrogen oxide, dioxin etc. In addition to this, burning one kilogram of plastic releases 3 kilograms of carbon dioxide, a gas that contributes to global warming.

Plastic industry is making significant contribution to the economic development and growth of various key sectors in the country such as: Automotive, Construction, Electronics, Healthcare, Textiles, and FMCG etc. Its demand has been growing rapidly at ~10% CAGR to reach 10 MnTPA by FY13. India is currently net importer of PolyEthylene (PE), however with the commissioning of IOCL plant at Panipat, the imports are expected to go down significantly. India observes significant regional diversity in consumption of plastics with Western India accounting for 47%, Northern

India for 23% and Southern India for 21%. Bulk of the consumption in Northern India is from the end use industries of Auto, packaging (including bulk packaging), plastic culture applications, electronic appliances etc. which are concentrated mostly in UP and Delhi- NCR (>50%). However, plastic processing in other parts like Rajasthan, Punjab, Haryana, Uttarakhand, J&K and Himachal Pradesh are expected to grow based on increased availability of feedstock and higher focus on manufacturing sectors. Polyethylene (PE) is the most largely used plastic raw material by Indian industry. Its demand has grown at 8% CAGR in last 5 years to reach ~3.6 MnTPA in FY13. Polypropylene(PP) is the second largest with consumption growing from 1.8 MnTPA in FY08 to 2.1 MnTPA in FY13 at a growth rate of 2% p.a. According to CPCB data Mumbai and Pune are among top 10 cities in terms of plastic waste generated. Four major cities of the Maharashtra state generated around 576 tonnes of plastic waste per day. Plastic waste constitutes nearly 8% of Nagpur's total municipal solid waste, higher than Mumbai in terms of percentage. States and Union territories directed to step up actions to meet 2022 goal of eliminating single-use plastic.

CSIR-NPL has developed a novel technology to utilize waste plastic scrap for designing materials for utilization of tiles in the building of structures and rooms for the general public for societal benefits. The main feature of the CSIR-NPL technology is to utilize waste plastic scrap for designing tiles in the building of structures and rooms for the general public for societal benefits. The various concerns with tiles such as mechanical strength, flame retardancy, water permeability, and UV- protection from sunlight, acid and alkali resistance and antistatic response are successfully addressed by uniqueness in the technology. Recycling of waste plastics are from polyethene bags, bottles, and other containers in the form of a tile, so that these tiles can be inserted in the panels (Fig. 17.29). The various challenges like mechanical strength, flame retardancy, water permeability and UV- protection from sunlight and

Fig. 17.29 The waste plastic tiles made from plastic waste and its utilization for different application

antistatic response is the novelty of the concept. The technology has intellectual property, Indian patent 201611025127 to safeguard the interest of the technology partners. This technology has been transferred to several startups and private industries across India. This technology has also been selected among top Smart Fifty innovations by DST and IIMC.

17.4.2 Recycling of Solar Modules

Solar power is now widely recognized as a cost competitive and reliable source of energy. Being one of the cleanest forms of energy, there has been an unprecedented growth in the PV Solar installations during last two decades worldwide. The World's Cumulative PV capacity reached 633 GW in 2019 and will reach 770 GW by end of 2020. In India's perspective, more than 33 GW PV capacity already achieved and 100 GW is targeted by 2022. World's Cumulative PV capacity is expected to reach 4500 GW by 2050 as per a report given by IRENA. This large amount of installations will leave behind huge volumes of solar panel waste. The expected volume of solar waste by 2050 is 78 million tonnes and will only going to increase with time. So there is an urgent need, to look out for the piling up of solar waste and to develop simple, efficient, eco-friendly process to safely recycle of the used solar panels. After the end of life of solar panels can be recycled and used as materials for new panels. Various layers of silicon solar modules and its life cycle are shown in Fig. 17.30.

Hardly any organization is working on recycling of solar cells/modules in India. Recently CSIR-NPL has initiated the work on recycling of silicon solar cells and modules. Various materials from silicon solar modules were extracted such as silicon wafer, Ethylene-Vinyl Acetate (EVA), metals, glass etc. These materials were extracted using combination of mechanical, chemical and thermal treatments. The MNRE (Ministry of New & Renewable Energy) is the nodal ministry of the government of India for all matters related to solar modules & its installation. Recently, the ministry has issued a blueprint for disposal of solar PV modules and particularly for glass containing Antimony. However, there is no clear guideline for recycling of solar modules in India so far. It is important to note that in some of other countries in Europe, there is well placed guidelines like the European Union included PV component as discarded electronic devices in the waste electrical and electronic equipment (WEEE) directive in July 2012.

17.4.3 Electronic and Battery Waste Management

Rapid development in the areas like semiconductor science, power electronics, electrical engineering and automation, etc., has led to global explosive growth in the production and consumption of electrical and electronic equipments (EEE), with proportionate increase in corresponding EEE waste (WEEE or simply E-waste), via

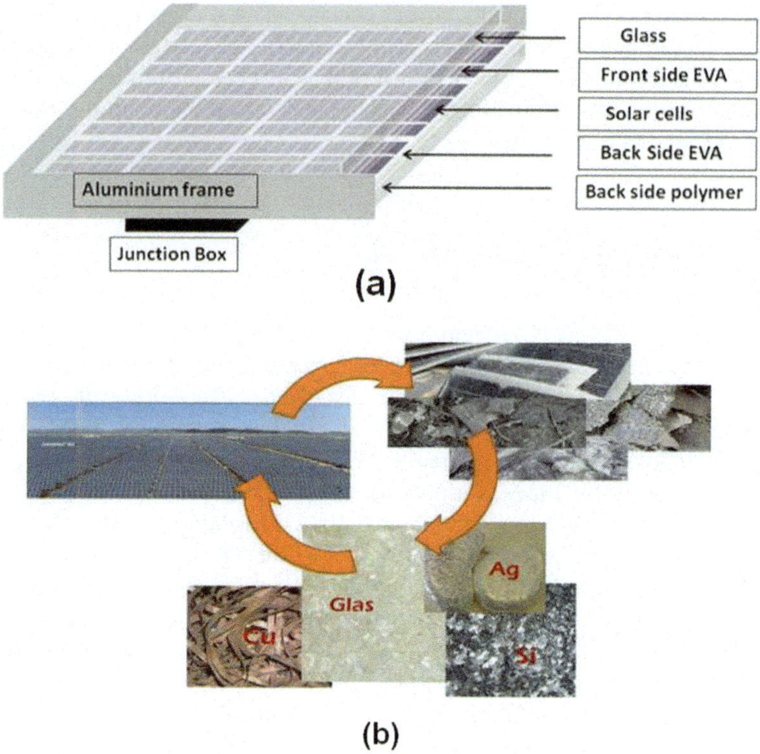

Fig. 17.30 **a** Various layers of silicon solar modules and **b** its life cycle

both regular as well as early loss scenario. The continuous developments in the areas also contribute towards end of life (EOL) EEE via obsolescence and upgradation due to short-lived technologies [186–189]. Indeed, E-waste is considered as fastest growing waste stream, with global volumes touching figures of 50 metric million tonnes per annum (MTA), with India, touch figure of 5 MTA (3rd highest generator) and growth rate of 10% per annum. The E-waste can be broadly categorized into three subclasses with overlap between them i.e. general waste electrical and electronic equipments (WEEE); solar systems waste (including solar panel, as discussed in previous section); and EOL energy storage systems (like batteries or supercapacitors) as shown in Fig. 17.31.

It is worth mentioning here that due to presence of heavy metals and toxic chemicals, E-waste has been recognized as hazardous waste, which if dumped alongside municipal solid waste, may leads to severe soil contamination and air/water pollution. The E-waste management problem is further complicated by inadequate recycling capability and unchecked E-waste import from other nations. Therefore, like other countries, Govt. of India (GoI) has also taken concrete steps to tackle emerging waste

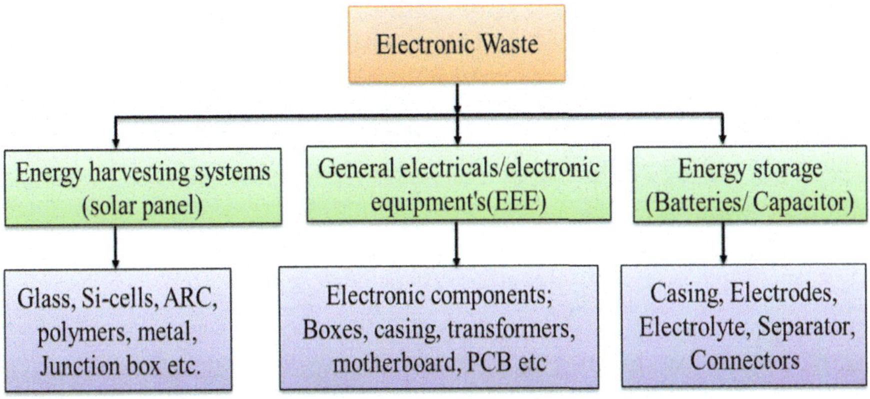

Fig. 17.31 Categorization and subcategorization of E-waste into specific and generalized streams

categories, growing waste volumes and their management/recycling, with notification of waste management rules E-waste (Management and Handling) Rules, 2016 (Amended in 2018) and Battery waste Management Draft Rules, 2020, with emphasis on waste segregation, proper collection; efficient recycling; resource recovery, material circular economics; producer responsibility; and monitoring authority. In this context, though number of initiatives like Swachh Bharat, Swasth Bharat, Make in India, Digital India, Skill India have been already initiated by GoI; however, still collection, transportation, disposal and recycling of E-waste is still predominantly done by informal sectors in India. Due to prevalence of waste handling in crude a non-scientific manner by informal sector, it lead to inefficient material recover, adverse effect health of workers, and cause soil, water and air pollution. Therefore, it is important to develop inexpensive, efficient and sustainable E-waste recycling technologies. In this context, initiatives have been taken for development of indigenous technologies for management of E-waste and demonstrated chemical route for recovery of metals (e.g. Silver and aluminium) form Si solar cells as also discussed in previous section and other facilities for bulk processing via size reduction; physical separation; chemical & thermal treatment are in establishment stage. As PCB is considered as heart of electronic systems and related WEEEs; strategies have also been established for the recycling of PCB. In one such route, thermal and mechanical treatment of PCB to depopulate various mounted components and solder was done. The bare PCB is then subjected to chemical processing, size reduction and screening operations, to separate epoxy powder, which is subsequently subjected to pyrolysis, and transformed into activated carbon. The high porosity, presence of graphitic domains and good high electrical conductivity of recovered porous carbon, make it useful for energy and environmental applications. Similar strategies have been established to transform other non-metallic materials, of electronic and battery waste into materials for environmental, societal and strategic applications. It has been realized that, globally and in light of India's ambitious Solar mission targeting 175 GW renewables (with ~20 GW off-grid/rooftop systems having dedicated battery storage) and also

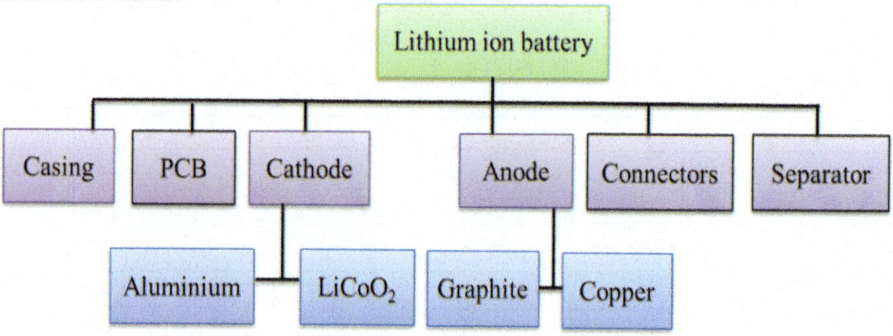

Fig. 17.32 Primary and secondary treatment steps in the recycling of EOL Li-ion batteries, for recovery of various components and compounds

India's strong commitment of total electrical vehicle-based mobility by 2030, *energy storage devices and allied systems will be playing major role.* Therefore, related waste generation through manufacturing process or through EOL of such articles (whose volumes are already touching lakh tonnes mark) will be major challenge, especially if dedicated efforts for their recycling have not been taken right at this stage itself. In this context, CSIR-NPL has also made dedicated efforts for recycling of energy storage devices and demonstrated *EOL* Li-ion batteries (LIBs) recovered from used mobile phones (Fig. 17.32). During the primary treatment, the neutralized battery is subjected to dismantling for recovery of various components including casing, connectors/circuit, PCB, separator, cathode and anode. The recovered cathode and anode are then subjected to secondary treatment to recover substrate (e.g. copper foil), active materials (e.g. graphite, $LiCoO_2$) and allied chemicals (e.g. Al and its compounds). By developing E-waste recycling technologies and creating awareness, CSIR-NPL is building support system for MSMEs and specifically informal sector, which indeed account for majority of recycling framework.

17.4.4 Agriculture Waste Management

India is agriculturally dominant nation and agriculture is regarded as the backbone of India's economy. The large population of the country is employed in the agricultural sector. Agriculture contributes ~15% of the gross domestic product (GDP). Over the past few decades, the contribution of the agricultural sector has decreased continually. This is due to the agriculture is not profitable business over the time and urbanization. Agriculture in India is described by a net sown area of 141 million hectares, witnessing diversification over the decades. India's agriculture is composed of many crops, rice and wheat, pulses, potatoes, sugar cane, oilseeds, and non-food items such as cotton, tea, coffee, rubber, and jute etc. Over the year from various crops around 620 Million tonnes of crop stubble or biomass generated in India. To

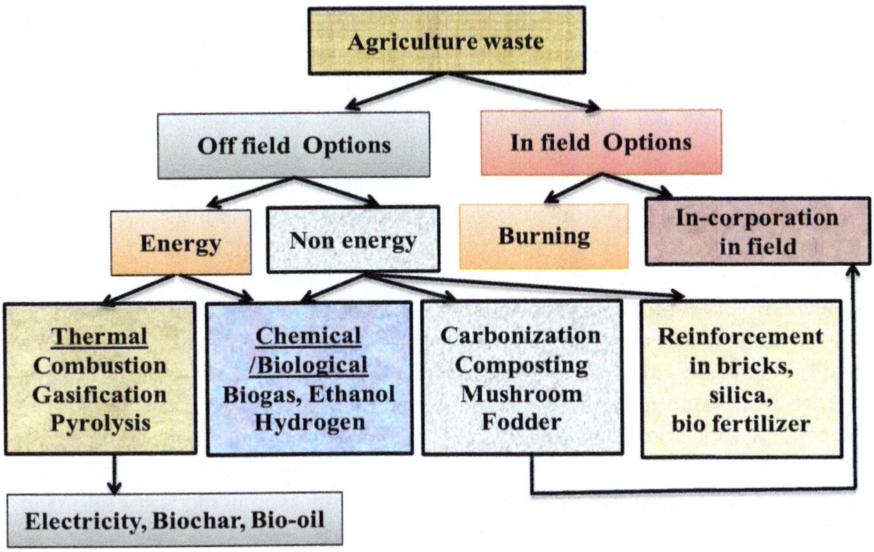

Fig. 17.33 Different process to manage agriculture waste

manage the agriculture bio waste, there are different processes which are depicted in Fig. 17.33 However, In India some quantity of agro-residues use as roofing material, animal feed, fuel and packing material, while most of it disposed by burning in the field to make field ready for next crop. Burning agro-residues in the field is considered as a cheap and easy means of disposal of excess residues.

Burning of crop stubble causes severe pollution of land and water on local as well as regional and global scales. The burning of crop residue results loss of nutrients from the farming soil such as organic carbon, nitrogen, phosphorus and potassium etc. This will adversely affects the nutrient composition of the soil. When crop residue is burnt existing minerals present in the soil are destroyed, which adversely hampers the cultivation of the next crop. The straw carbon, nitrogen and sulphur are completely burnt and lost in the atmosphere in the process of burning. The one tonne of rice straw burning is emitting 3.7 kg particulate matter, 1.0 kg volatile organic compounds, 3.1 kg NO_x, 0.7 kg SO_x and 34.7 kg of CO [190]. This result in the emission of smoke which when added to the gases present in the air like methane, nitrogen oxide and ammonia can cause severe pollution. These gaseous emissions can pose health risks, aggravating asthma, chronic bronchitis and decreased lung function. Burning of crop residue also contributes towards the depletion of ozone layer. The crop stubble is lignocelluloses biomass which is mainly composed of cellulose, hemicelluloses and lignin. It is an organic matter in which solar energy is stored in the chemical bonds of the cellulose, hemicelluloses and lignin molecule. When these bonds between adjacent carbon, hydrogen and oxygen atoms are broken this stored chemical energy is released. Biomass is renewable source and it can be utilized to produce heat and energy. Biomass absorbs CO_2 during growth and emits it during combustion; thus, biomass has a zero net greenhouse effect [191]. However, there are inherent problems with raw rice straw biomass materials compared to fossil fuel resources (low bulk density, high moisture content, hydrophilic nature, and low calorific value) render it difficult to use on a large scale [192]. These limitations greatly impact logistics and final energy efficiency. Due to its low energy density compared to fossil fuels, high volumes of biomass are needed, which multifaceted problems associated with storage, transportation, and feed handling at cogeneration, thermo-chemical, and biochemical conversion plants. High moisture in raw rice straw biomass is one of the primary challenges, as it reduces the efficiency of the process and increases fuel production costs [193]. To overcome these challenges and make biomass suitable for energy applications requires some form of pre-treatment. Among the different pre-treatment technique, torrefaction, which is a thermal pre-treatment process, is a viable technology that significantly alters the physical and chemical composition of the biomass. Torrefaction is defined as slow heating of biomass up to particular temperature in protective atmosphere. This process improves the physical, chemical, and biochemical composition of the biomass, making it useful material for its directly use in boiler or co-firing with the coal which is useful energy source in thermal power plant (Fig. 17.34) [194]. The use of torrefied product with sub bituminous coal up to 20 % does not influence the performance of the coal to higher extent.

Fig. 17.34 Process flow from agriculture waste from field to thermal power plant

The torrefaction process reduces the volume, increases the bulk density of biomass, reduce the moisture content, and increase the calorific value. This can reduce problems associated with rice straw burning in the agriculture field, environmental pollution and greenhouse emission, the dependency on fossil fuel. At present farmers are not paid for the crop stubble waste biomass and as a result they are simply burning it. With these advantages, farmers will be discouraged to burn the biomass in the field and they will store the waste biomass along side of field, once farmers get income from the waste, it will stop burning in the field.

17.5 Conclusions and Future Perspective

CSIR-NPL is working towards the development of new materials with synergetic properties which are highly required in various industries for their product development for strategic, industrial and societal applications.

This chapter mainly concludes the CSIR-NPL efforts on the development of newer materials for societal, strategic and energy sector applications such as, high-temperature-resistant carbon–carbon composite for countries defence sector as well nuclear energy area, high density isotropic graphite, lightweight carbon fibre reinforcement hybrid polymer composites, graphene-based polymer composites, luminescent materials as security ink, diamond-like carbon, conducting polymer, carbon foam, coal tar pitches, carbon fibre composite half ring for orthopaedic applications. Moreover, CSIR-NPL also initiated several new programs related to recycling of plastic waste, recycling of solar modules, recycling of electronic battery and recycling of agriculture waste which directly affect to our environment to make clean and protective from hazard wastes. This can also generate raw materials for the development of products and devices, replacement of renewable fuel source instead of fossil fuel etc.

Hence, CSIR-NPL has created a significant impact for strategic and societal applications which is not only impact the daily life but also creates an alternative for import substitution for country to make self-reliant. Further, design and development of materials which can address the burning challenges of health, energy, strategic and environment sector is needed. CSIR-NPL has to take up further more specific programmes in big way to make a paradigm shift of industries towards NMI laboratory as well as address the industrial challenges in various fields where materials play a key role to address such issues since industries are the backbone of countries economic growth as well as quality of life.

Acknowledgements The authors are grateful to all technical staffs of Advanced Materials & Device Metrology Division, CSIR-NPL for their continuous support.

References

1. S.R. Dhakate, R.B. Mathur, O.P. Bahl, Carbon **35**, 1753 (1997)
2. S.R. Dhakate, O.P. Bahl, Carbon **41**, 1193 (2003)
3. S.R. Dhakate, P. Bahl, P.D. Sahare, Journal of Materials Science Letters **19**, 1959 (2000)
4. P.P. Kelkar, I.V. Dulera, ed. by B. Newsletter (2010)
5. I.V. Dulera, R.K. Sinha, J. Nucl. Mater. **383**, 183 (2008)
6. B.P. Singh, S. Teotia, S.R. Dhakate, US Patent, (2019)
7. V. Kumar, S. Sharma, A. Pathak, B.P. Singh, S.R. Dhakate, T. Yokozeki, T. Okada, T. Ogasawara, Compos. Struct. **210**, 581 (2019)
8. S. Sharma, A.K. Pathak, V.N. Singh, S. Teotia, S.R. Dhakate, B.P. Singh, Carbon **137**, 104 (2018)
9. S. Sharma, S.R. Dhakate, A. Majumdar, B.P. Singh, Carbon **152**, 631 (2019)
10. B. Singh, V. Choudhary, P. Saini, S. Pande, V. Singh, R. Mathur, J. Nanopart. Res. **15**, 1 (2013)
11. B. Singh, K. Saini, V. Choudhary, S. Teotia, S. Pande, P. Saini, R. Mathur, J. Nanopart. Res. **16**, 1 (2014)
12. S. Teotia et al., RSC Adv. **4**, 33168 (2014)
13. B. Singh, V. Choudhary, P. Saini, R. Mathur, AIP Adv. **2**, 022151 (2012)
14. B. Singh, P. Bharadwaj, V. Choudhary, R. Mathur, Appl. Nanosci. **4**, 421 (2014)
15. S. Sharma, V. Kumar, A.K. Pathak, T. Yokozeki, S.K. Yadav, V.N. Singh, S. Dhakate, B.P. Singh, J. Mater. Chem. C **6**, 12396 (2018)
16. J. Jyoti, S. Basu, B.P. Singh, S.R. Dhakate, Compos. B Eng. **83**, 58 (2015)
17. S. Pande, A. Chaudhary, D. Patel, B.P. Singh, R.B. Mathur, RSC Adv. **4**, 13839 (2014)
18. B. Singh et al., J. Nanopart. Res. **13**, 7065 (2011)
19. T. Gupta, B. Singh, S. Teotia, V. Katyal, S. Dhakate, R. Mathur, J. Polym. Res. **20**, 1 (2013)
20. T.K. Gupta, B.P. Singh, S.R. Dhakate, V.N. Singh, R.B. Mathur, J. Mater. Chem. A **1**, 9138 (2013)
21. R. Mathur, S. Pande, B. Singh, T. Dhami, Polym. Compos. **29**, 717 (2008)
22. S. Pande, B. Singh, R. Mathur, T. Dhami, P. Saini, S. Dhawan, Nanoscale Res. Lett. **4**, 327 (2009)
23. T.K. Gupta, B.P. Singh, V.N. Singh, S. Teotia, A.P. Singh, I. Elizabeth, S.R. Dhakate, S.K. Dhawan, R.B. Mathur, J. Mater. Chem. A **2**, 4256 (2014)
24. B.P. Singh et al., J. Mater. Chem. A (2015)
25. T.K. Gupta, B.P. Singh, R.B. Mathur, S.R. Dhakate, Nanoscale **6**, 842 (2014)
26. A. Chaudhary, S. Kumari, R. Kumar, S. Teotia, B.P. Singh, A.P. Singh, S. Dhawan, S.R. Dhakate, ACS Appl. Mater. Interfaces. **8**, 10600 (2016)
27. R. Kumar, S.R. Dhakate, T. Gupta, P. Saini, B.P. Singh, R.B. Mathur, J. Mater. Chem. A **1**, 5727 (2013)
28. A.P. Singh, M. Mishra, P. Sambyal, B.K. Gupta, B.P. Singh, A. Chandra, S. Dhawan, J. Mater. Chem. A **2**, 3581 (2014)
29. P. Kumar, S. Singh, B.K. Gupta, Nanoscale **8**, 14297 (2016)
30. A.K. Gangwar, A. Gupta, G. Kedawat, P. Kumar, B.P. Singh, N. Singh, A.K. Srivastava, S.R. Dhakate, B.K. Gupta, Chem.—A Euro. J. **24**, 9477 (2018)
31. A.K. Gangwar, Kanika, G. Kedawat, G.S. Papanai, B.K. Gupta, J. Mater. Chem. C **7**, 13867 (2019)
32. A.K. Singh, S. Singh, B.K. Gupta, ACS Appl. Mater. Interfaces. **10**, 44570 (2018)
33. P. Kumar, K. Nagpal, B.K. Gupta, ACS Appl. Mater. Interfaces. **9**, 14301 (2017)
34. B.K. Gupta et al., Sci. Rep. **6**, 32401 (2016)
35. S. Kumar, P.N. Dixit, D. Sarangi, R. Bhattacharyya, Appl. Phys. Lett. **69**, 49 (1996)
36. S. Kumar, P.N. Dixit, D. Sarangi, R. Bhattacharyya, J. Appl. Phys. **93**, 6361 (2003)
37. S. Kumar, P.N. Dixit, D. Sarangi, R. Bhattacharyya, J. Appl. Phys. **85**, 3866 (1999)
38. S. Kumar, P.N. Dixit, C.M.S. Rauthan, R. Sharma, M.Y. Khan, R. Bhattacharyya, J. Mater. Sci. Lett. **19**, 2055 (2000)

39. S. Kumar, D. Sarangi, P.N. Dixit, O.S. Panwar, R. Bhattacharyya, Thin Solid Films **346**, 130 (1999)
40. N. Dwivedi, S. Kumar, H.K. Malik, J. Appl. Phys. **112**, 023518 (2012)
41. N. Dwivedi, S. Kumar, H.K. Malik, Appl. Phys. Lett. **102**, 011917 (2013)
42. Ishpal, S. Kumar, N. Dwivedi, C.M.S. Rauthan, Phys. Plasmas **19**, 033515 (2012)
43. S. Kumar, P.N. Dixit, C.M.S. Rauthan, US Patent No. #US 9260781 B2 (2016)
44. S. Kumar, P.N. Dixit, O.S. Panwar, R. Bhattacharyya, Diam. Relat. Mater. **12**, 1576 (2003)
45. P.N. Dixit, S. Kumar, D. Sarangi, R. Bhattacharyya, Solid State Commun. **90**, 421 (1994)
46. P. Saini, in *Fundamentals of Conjugated Polymer Blends, Copolymers and Composites*, ed. by P. Saini (Wiley, 2015), p. 1
47. P. Saini, in *Thermoset Nanocomposites*, ed. by D. V. Mittal, 2013), p. 211
48. P. Saini, M. Arora, in *In New Polymers for Special Applications*, ed. by A. De Souza Gomes (InTech: Croatia, 2012), p. 71
49. P. Saini, Indian Institute of Technology Delhi (2012)
50. P. Saini, V. Choudhary, S.K. Dhawan, Polym. Adv. Technol. **23**, 343 (2012)
51. P. Saini, V. Choudhary, J. Nanopart. Res. **15**, 1415 (2013)
52. S.K. Dhawan, S. Sathiyanarayanan, S. Azim, P. Saini, S. Radhakrishnan, India Patent 537/DEL/2010 (2010)
53. Parveen Saini, M. Arora, R.K. Kotnala, S.K. Barala, R.P. Pant, C. Puri, US 2891DEL2013 (2013)
54. P. Saini, C. Puri, M. Arora, US and India Patent 0012NF2014 (2014)
55. J. Kumar, M. Shahabuddin, A. Singh, S.P. Singh, P. Saini, S.K. Dhawan, V. Gupta, Sci. Adv. Mater. **7**, 518 (2015)
56. P. Saini, R. Jalan, S.K. Dhawan, J. Appl. Polym. Sci. **108**, 1437 (2008)
57. P. Saini, M. Arora, J. Mater. Chem. A **1**, 8926 (2013)
58. S.K. Dhawan, P. Saini, R. Jalan, India Patent 271553 (2016)
59. S.K. Dhawan et al., US 2009/0302263 (2012)
60. P. Saini, M. Arora, G. Gupta, B.K. Gupta, V.N. Singh, V. Choudhary, Nanoscale **5**, 4330 (2013)
61. P. Saini, V. Choudhary, J. Mater. Sci. **48**, 797 (2013)
62. P. Saini, V. Choudhary, B.P. Singh, R.B. Mathur, S.K. Dhawan, Mater. Chem. Phys. **113**, 919 (2009)
63. P. Saini, V. Choudhary, B.P. Singh, R.B. Mathur, S.K. Dhawan, Synth. Met. **161**, 1522 (2011)
64. P. Saini, V. Choudhary, N. Vijayan, R.K. Kotnala, J. Phys. Chem. C **116**, 13403 (2012)
65. R. Mukherjee, R. Sharma, P. Saini, S. De, Environ. Sci.: Water Res. Technol. **1**, 893 (2015)
66. P. Saini, V. Choudhary, IJPAP **51**(2), 112 (2013)
67. P. Saini, H. Bhandari, V. Choudhary, S.K. Dhawan, IJEMS **15**, 497 (2008)
68. P. Saini, V. Choudhary, S.K. Dhawan, IJEMS **14**, 436 (2007)
69. P. Saini, V. Choudhary, K.N. Sood, S.K. Dhawan, J. Appl. Polym. Sci. **113**, 3146 (2009)
70. P.H. Maheshwari, R.B. Mathur, T.L. Dhami, Electrochim. Acta **54**, 655 (2008)
71. R.B. Mathur, P.H. Maheshwari, T.L. Dhami, R.K. Sharma, C.P. Sharma, J. Power Sources **161**, 790 (2006)
72. S. Waseem, P.H. Maheshwari, S. Abinaya, A.K. Sahu, A. Saini, S.R. Dhakate, Int. J. Energy Res. **43**, 2897 (2019)
73. R.B. Mathur, S.R. Dhakate, D.K. Gupta, T.L. Dhami, R.K. Aggarwal, J. Mat. Process. Technol. **203**, 184 (2008)
74. S.R. Dhakate, R.B. Mathur, B.K. Kakati, T.L. Dhami, Int. J. Hydrogen Energy **32**, 4537 (2007)
75. S.R. Dhakate, S. Sharma, N. Chauhan, R.K. Seth, R.B. Mathur, Int. J. Hydrogen Energy **35**, 4195 (2010)
76. S.R. Dhakate, R.B. Mathur, S. Sharma, M. Borah, T.L. Dhami, Energy Fuels **23**, 934 (2009)
77. S.R. Dhakate, S. Sharma, M. Borah, R.B. Mathur, T.L. Dhami, Int. J. Hydrogen Energy **33**, 7146 (2008)
78. S.R. Dhakate, S. Sharma, M. Borah, R.B. Mathur, T.L. Dhami, Energy Fuels **22**, 3329 (2008)

79. C. Gupta, P.H. Maheshwari, S. Sasikala, R.B. Mathur, Mater. Renew. Sustain. Energy **3**, 36 (2014)
80. C. Gupta, P.H. Maheshwari, S.R. Dhakate, Mater. Renew. Sustain. Energy **5**, 2 (2016)
81. C. Gupta, P.H. Maheshwari, D. Sachdev, A.K. Sahu, S.R. Dhakate, RSC Adv. **6**, 32258 (2016)
82. I. Elizabeth, B.P. Singh, T.K. Bijoy, V.R. Reddy, G. Karthikeyan, V.N. Singh, S.R. Dhakate, P. Murugan, S. Gopukumar, Electrochim. Acta **231**, 255 (2017)
83. A. González, E. Goikolea, J.A. Barrena, R. Mysyk, Renew. Sustain. Energy Rev. **58**, 1189 (2016)
84. P. Saini, *Fundamentals of Conjugated Polymer Blends, Copolymers and Composites: Synthesis, Properties, and Applications* (Scrivener Publishing LLC, 2015)
85. N. Chadha, M.Y. Bhat, S.A. Hashmi, P. Saini, 13/NCSSI/2019, P-11, IIT Roorkee, India (2019), p. 39
86. B. Pandit, S.R. Dhakate, B.P. Singh, B.R. Sankapal, Electrochim. Acta **249**, 395 (2017)
87. S. Gupta, N. Vijayan, A. Krishna, K. Thukral, K.K. Maurya, S. Muthiah, A. Dhar, B. Singh, G. Bhagavannarayana, J. Appl. Crystallogr. **48**, 533 (2015)
88. B. Gahtori, S. Bathula, K. Tyagi, M. Jayasimhadri, A.K. Srivastava, S. Singh, R.C. Budhani, A. Dhar, Nano Energy **13**, 36 (2015)
89. R. Bhardwaj, K.K. Johari, B. Gahtori, N.S. Chauhan, S. Bathula, S.R. Dhakate, S. Auluck, A. Dhar, Intermetallics **123**, 106796 (2020)
90. S. Muthiah, B. Sivaiah, B. Gahtori, K. Tyagi, A.K. Srivastava, B.D. Pathak, A. Dhar, R.C. Budhani, J. Electron. Mater. **43**, 2035 (2014)
91. S. Muthiah, R.C. Singh, B.D. Pathak, P.K. Avasthi, R. Kumar, A. Kumar, A.K. Srivastava, A. Dhar, Nanoscale **10**, 1970 (2018)
92. N.K. Upadhyay, L.A. Kumaraswamidhas, B. Gahtori, S. Bathula, S. Muthiah, R. Shyam, N.S. Chauhan, R. Bhardwaj, A. Dhar, J. Alloy. Compd. **765**, 412 (2018)
93. N.S. Chauhan et al., J. Phys. Chem. C **124**, 8584 (2020)
94. N.S. Chauhan, S. Bathula, B. Gahtori, Y.V. Kolen'ko, A. Dhar, J. Electron. Mater. **48**, 6700 (2019)
95. S. Bathula, M. Jayasimhadri, N. Singh, A.K. Srivastava, J. Pulikkotil, A. Dhar, R.C. Budhani, Appl. Phys. Lett. **101**, 213902 (2012)
96. S. Bathula, M. Jayasimhadri, B. Gahtori, N.K. Singh, K. Tyagi, A.K. Srivastava, A. Dhar, Nanoscale **7**, 12474 (2015)
97. S.N. Singh, N.K. Arora, A. Prasad, G.C. Jain, V.K. Sharda, Indian Patent No. 156462
98. A. Prasad, S.K. Balakrishnan, S.K. Jain, S.N. Singh, N.K. Arora, G.C. Jain, Indian Patent No. 156459
99. P.K. Singh, P. Prakash, R. Kishore, S.N. Singh, B.K. Das, US Patent No. 5431869 (1995)
100. P.K. Singh, P. Prakash, R. Kishore, S.N. Singh, B.K. Das, Indian Patent No. 182633 (1999)
101. G.C. Jain, S.N. Singh, R.K. Kotnala, Solar Cells **8**, 239 (1983)
102. B.C. Chakravarty, S.N. Singh, B.K. Das, IEEE Electron Device Lett. **4**, 138 (1983)
103. S.N. Singh, R. Kishore, N.K. Arora, Solar Cells **14**, 13 (1985)
104. S.N. Singh, R. Kishore, P.K. Singh, J. Appl. Phys. **57**, 2793 (1985)
105. S.K. Sharma, S.N. Singh, B.C. Chakravarty, B.K. Das, J. Appl. Phys. **60**, 3550 (1986)
106. P.K. Singh, S.N. Singh, R. Kishore, Appl. Phys. Lett. **48**, 127 (1986)
107. S.N. Singh, P.K. Singh, IEEE Trans. Electron Devices **38**, 337 (1991)
108. S.N. Singh, D. Kumar, J. Appl. Phys. **103**, 023713 (2008)
109. S. Kumar, P.K. Singh, G.S. Chilana, S.R. Dhariwal, Semicond. Sci. Technol. **24**, 095001 (2009)
110. S. Kumar, P.K. Singh, S.R. Dhariwal, Appl. Phys. Lett. **96**, 162109 (2010)
111. S. Kumar, Vandana, C.M.S. Rauthan, V.K. Kaul, S.N. Singh, P.K. Singh, IEEE J. Photovoltaics **4**, 380 (2014)
112. J. Panigrahi et al., Sol. Energy **136**, 412 (2016)
113. A.G. Aberle, Prog. Photovoltaics Res. Appl. **8**, 473 (2000)
114. T.S. Horányi, T. Pavelka, P. Tüttö, Appl. Surf. Sci. **63**, 306 (1993)
115. M. Ichimura, A. Tada, E. Arai, H. Takamatsu, S. Sumie, Appl. Phys. Lett. **80**, 4390 (2002)

116. R.A. Sinton, A. Cuevas, M. Stuckings, in *25th IEEE Photovoltaic Specialist Conference* (1996), p. 457
117. A.W. Stephens, M.A. Green, Sol. Energy Mater. Sol. Cells **45**, 255 (1997)
118. H. M'Saad, J. Michel, J.J. Lappe, L.C. Kimerling, J. Electron. Mater. **23**, 487 (1994)
119. B. Chhabra, S. Bowden, R.L. Opila, C.B. Honsberg, Appl. Phys. Lett. **96**, 063502 (2010)
120. F. Khan, Vandana, S.N. Singh, M. Husain, P.K. Singh, Solar Energy Mater. Solar Cells **100**, 57 (2012)
121. N. Batra, Vandana, S. Kumar, M. Sharma, S.K. Srivastava, P. Sharma, P.K. Singh, Solar Energy Mater. Solar Cells **100**, 43 (2012)
122. N. Batra, Vandana, P. Kumar, S.K. Srivastava, P.K. Singh, J. Renew. Sustain. Energy **6**, 013121 (2014)
123. G. Dingemans, W.M.M. Kessels, J. Vac. Sci. Technol., A **30**, 040802 (2012)
124. B. Hoex, S.B.S. Heil, E. Langereis, M.C.M. v. d. Sanden, W.M.M. Kessels, Appl. Phys. Lett. **89**, 042112 (2006)
125. B. Hoex, J. Schmidt, P. Pohl, M.C.M. v. d. Sanden, W.M.M. Kessels, J. Appl. Phys. **104**, 044903 (2008)
126. J. Schmidt, A. Merkle, R. Brendel, B. Hoex, M.C.M. v. de Sanden, W.M.M. Kessels, Progress Photovoltaics: Res. Appl. **16**, 461 (2008)
127. N. Batra, J. Gope, Vandana, J. Panigrahi, R. Singh, P.K. Singh, AIP Adv. **5**, 067113 (2015)
128. Vandana et al., Phys. Chem. Chem. Phys. **16**, 21804 (2014)
129. J. Gope, Vandana, N. Batra, J. Panigrahi, R. Singh, K.K. Maurya, R. Srivastava, P.K. Singh, Appl. Surface Sci. **357**, 635 (2015)
130. R. Singh, Vandana, J. Panigrahi, P.K. Singh, RSC Adv. **6**, 97720 (2016)
131. J. Panigrahi, Vandana, R. Singh, P.K. Singh, Solar Energy Mater. Solar Cells **188**, 219 (2018)
132. J. Panigrahi, Vandana, R. Singh, C.M.S. Rauthan, P.K. Singh, AIP Adv. **7**, 035219 (2017)
133. S. Tomer, Vandana, J. Panigrahi, R. Srivastava, C.M.S. Rauthan, Thin Solid Films **692**, 137629 (2019)
134. S.K. Srivastava, D. Kumar, P. Sharma, P.K. Singh, in *26th EUPVSEC*, Hamburg, Germany, (2011), p. 1699
135. S.K. Srivastava, D. Kumar, Vandana, M. Sharma, R. Kumar, P.K. Singh, Solar Energy Mater. Solar Cells **100**, 33 (2012)
136. S.K. Srivastava, P. Singh, M. Yameen, P. Prathap, C.M.S. Rauthan, Vandana, P.K. Singh, Solar Energy **115**, 656 (2015)
137. S.K. Srivastava, C.M.S. Rauthan, V. Kumar, P.K. Singh, in *Advances in Nanomaterials*, ed. by M. Husain, Z.H. Khan (Springer India, New Delhi, 2016), p. 329
138. P. Singh et al., J. Mater. Sci. **50**, 6631 (2015)
139. S.K. Srivastava, D. Kumar, P.K. Singh, V. Kumar, in *2009 34th IEEE Photovoltaic Specialists Conference (PVSC)*, (2009), p. 001851
140. S.K. Srivastava, D. Kumar, P.K. Singh, M. Kar, V. Kumar, M. Husain, Sol. Energy Mater. Sol. Cells **94**, 1506 (2010)
141. D. Kumar, S.K. Srivastava, P.K. Singh, K.N. Sood, V.N. Singh, N. Dilawar, M. Husain, J. Nanopart. Res. **12**, 2267 (2010)
142. D. Kumar, S.K. Srivastava, P.K. Singh, M. Husain, V. Kumar, Sol. Energy Mater. Sol. Cells **95**, 215 (2011)
143. S.K. Srivastava, D. Kumar, S.W. Schmitt, K.N. Sood, S.H. Christiansen, P.K. Singh, Nanotechnology **25**, 175601 (2014)
144. S.W. Schmitt et al., Nano Lett. **12**, 4050 (2012)
145. S.W. Schmitt, G. Brönstrup, G. Shalev, S.K. Srivastava, M.Y. Bashouti, G.H. Döhler, S.H. Christiansen, Nanoscale **6**, 7897 (2014)
146. T. Seth, C. Mukherjee, C. Anandan, P.N. Dixit, O.S. Panwar, S.S. Bawa, R. Bhattacharyya, IETE J. Res. **43**, 131 (1997)
147. A. Parashar, S. Kumar, P.N. Dixit, J. Gope, C.M.S. Rauthan, S.A. Hashmi, Sol. Energy Mater. Sol. Cells **92**, 1199 (2008)

148. S. Kumar, P.N. Dixit, C.M.S. Rauthan, A. Parashar, J. Gope, J. Phys.: Condens. Matter **20**, 335215 (2008)
149. J. Gope, S. Kumar, A. Parashar, P.N. Dixit, C.M.S. Rauthan, O.S. Panwar, D.N. Patel, S.C. Agarwal, J. Non-Cryst. Solids **355**, 2228 (2009)
150. J. Gope, S. Kumar, S. Singh, C.M.S. Rauthan, P.C. Srivastava, Silicon **4**, 127 (2012)
151. J. Gope, S. Kumar, S. Sudhakar, C.M.S. Rauthan, P.C. Srivastava, Mater. Chem. Phys. **141**, 89 (2013)
152. S. Juneja, S. Sudhakar, J. Gope, K. Lodhi, M. Sharma, S. Kumar, J. Alloys Compds. **643**, 94 (2015)
153. S. Juneja, S. Sudhakar, J. Gope, S. Kumar, Mater. Sci. Semicond. Process. **40**, 11 (2015)
154. S. Juneja, S. Sudhakar, A.K. Srivastava, S. Kumar, Thin Solid Films **619**, 273 (2016)
155. S. Kumar, P.N. Dixit, C.M.S. Rauthan, US Patent No. #US 8586151 B2 (2013)
156. D. Chaudhary, M. Sharma, S. Sudhakar, S. Kumar, Phys. Plasmas **23**, 123704 (2016)
157. D. Chaudhary, M. Sharma, S. Sudhakar, S. Kumar, Plasma Chem. Plasma Process. **37**, 189 (2017)
158. M. Sharma, S. Kumar, N. Dwivedi, S. Juneja, A.K. Gupta, S. Sudhakar, K. Patel, Sol. Energy **97**, 176 (2013)
159. A. Rawat, M. Sharma, D. Chaudhary, S. Sudhakar, S. Kumar, Sol. Energy **110**, 691 (2014)
160. N. Dwivedi, S. Kumar, A. Bisht, K. Patel, S. Sudhakar, Sol. Energy **88**, 31 (2013)
161. N. Dwivedi, S. Kumar, H.K. Malik, J. Appl. Phys. **111**, 014908 (2012)
162. V. Gupta, V. Bharti, M. Kumar, S. Chand, A.J. Heeger, Adv. Mater. **27**, 4398 (2015)
163. F. Ali, A. Sharma, J.P. Tiwari, S. Chand, Indian Patent 943/DEL/2014 (2014)
164. R. Bhargav, N. Chaudhary, S. Rathi, Shahjad, D. Bhardwaj, S. Gupta, A. Patra, ACS Omega **4**, 6028 (2019)
165. R. Kumar, S. Naqvi, N. Gupta, S. Chand, RSC Adv. **4**, 15675 (2014)
166. S. Naqvi, N. Vasistha, M. Kumar, R. Kumar, New J. Chem. **43**, 15626 (2019)
167. N. Gupta, C. Sharma, M. Kumar, R. Kumar, New J. Chem. **41**, 13276 (2017)
168. R. Kumar, S. Naqvi, N. Gupta, S. Chand, US Patent US20160237018 A1 (2016)
169. A.K. Chauhan, P. Kumar, S.R. Pal, S.K. Srivastava, S. Muthiah, J. Mater. Sci. **52**, 10886 (2017)
170. A.K. Chauhan, P. Kumar, J. Phys. D Appl. Phys. **50**, 325105 (2017)
171. A.K. Chauhan, P. Kumar, J. Mater. Sci.: Mater. Electron. **30**, 9582 (2019)
172. P. Kumar, A.K. Chauhan, J. Phys. D Appl. Phys. **53**, 035101 (2019)
173. R. Kumar, S.R. Dhakate, P. Saini, R.B. Mathur, RSC Adv. **3**, 4145 (2013)
174. P.R. Agrawal, R. Kumar, H. Uppal, N. Singh, S. Kumari, S.R. Dhakate, RSC Adv. **6**, 29899 (2016)
175. P.R. Agrawal, R. Kumar, S. Teotia, S. Kumari, D.P. Mondal, S.R. Dhakate, Compos. B Eng. **160**, 131 (2019)
176. P. Rani Agrawal, N. Singh, S. Kumari, S.R. Dhakate, J. Hazardous Mater. **377**, 274 (2019)
177. R. Sharma, P. Saini, in *Diamond and Carbon Composites and Nanocomposites*, ed. by M. Aliofkhazraei (In Tech Open access, 2016)
178. P.K. Saini, R. Sharma, S. Akodia, World J. Textile Eng. Technol. **2**, 1 (2016)
179. R. Sharma, N. Chahda, P. Saini, Indian J. Pure Appl. Phys. **55**, 625 (2017)
180. P. Saini, S. Kaushik, R. Sharma, D. Chakravarty, R. Raj, J. Sharma, Euro. Phys. J. B **89**, 137 (2016)
181. R. Sharma, R.K. Kotnala, P. Saini, J. Environ. Biotechnol. Res. **6**, 146 (2017)
182. N. Chadha, P. Saini, IISF-2019, Kolkata, India, YSC-11849, p. 38
183. R. Kumar, C. Zhang, A.K. Itta, W.J. Koros, J. Membr. Sci. **583**, 9 (2019)
184. R. Kumar, W.J. Koros, Ind. Eng. Chem. Res. **58**, 6740 (2019)
185. R. Kumar, W.J. Koros, US Patent App. 13/906,143 (2013)
186. P. Saini, R. Sharma, N. Chadha, C. Kant, S. Kumar, IJWBME-2018, NPL, Delhi (2018), p. 27
187. N. Chadha, P. Saini, IJWBME-2018, NPL, Delhi, December (2018), p. 18
188. A. Thukral, R. Sharma, N. Chadha, P. Saini, ICPREM-2020, Hyderabad, March (2020), A11, p. 61.

189. M.S. Gautam, N. Chadha, S. Kumar, V.N. Ojha. P. Saini, ICPREM-2020, Hyderabad, March (2020), A16, p. 66
190. K.L. Kadam, L.H. Forrest, W.A. Jacobson, Biomass Bioenerg. **18**, 369 (2000)
191. X.H. Liang, J.A. Kozinski, Fuel **79**, 1477 (2000)
192. B. Arias, C. Pevida, J. Fermoso, M.G. Plaza, F. Rubiera, J.J. Pis, Fuel Process. Technol. **89**, 169 (2008)
193. Biomass torrefaction as a preprocessing step for thermal conversion: Reducing costs in the biomass supply chain, in *Evergreen Renewables, LLC.* (2009)
194. S.R. Dhakate, A.K. Pathak, P. Jain, M. Singh, B.P. Singh, K.M. Subhedar, S.S. Sharda, R.K. Seth, Curr. Sci. **116**, 831 (2019)

Chapter 18
Bharatiya Nirdeshak Dravyas (BND®): Indian Certified Reference Materials.

Part I: Terminology, Definition and International Equivalences

R. P. Pant, S. Swarupa Tripathy, D. K. Misra, Vidya Nand Singh, Arvind Gautam, N. Vijayan, G. A. Basheed, K. K. Maurya, S. P. Singh, and Nahar Singh

Abstract Bharatiya Nirdeshak Dravya (BND®) is the trademark of Certified Reference Material (CRM) of CSIR-NPL. BNDs are the primary standards, which ensure reliability and comparability of the measurements as a benchmark for the quality assurance achieved through international networking. Networking between the NMIs through bilateral cooperation, participation with international

---

R. P. Pant (✉) · S. S. Tripathy · D. K. Misra · V. N. Singh · A. Gautam · N. Vijayan · G. A. Basheed · K. K. Maurya · S. P. Singh · N. Singh
CSIR-National Physical Laboratory, Dr. K.S. Krishnan Marg, New Delhi 110012, India
e-mail: rppant@nplindia.org

S. S. Tripathy
e-mail: tripathyss@nplindia.org

D. K. Misra
e-mail: misradk@nplindia.org

V. N. Singh
e-mail: singhvn@nplindia.org

A. Gautam
e-mail: arvind.gautam@nplindia.org

N. Vijayan
e-mail: nvijayan@nplindia.org

G. A. Basheed
e-mail: basheedga@nplindia.org

K. K. Maurya
e-mail: kkmaurya@nplindia.org

S. P. Singh
e-mail: singhsp@nplindia.org

N. Singh
e-mail: naharsingh@nplindia.org

D. K. Aswal (ed.), *Metrology for Inclusive Growth of India*,
https://doi.org/10.1007/978-981-15-8872-3_18

bodies/organization establishes wide-ranging recognition of BND in India. BND is the heart of India's scientific and industrial development by establishing quality control in measurements. A detailed discussion to overcome barriers in international trade and standardization of measurement with high metrological aspect has been emphasized. BNDs are produced by CSIR-NPL: (i) in-house and (ii) in association with Reference Material Producers (RMPs) across the country as per ISO 17034 standard. A detailed emphasis on production of BNDs are given in Chap. 19. The BNDs produced by CSIR-NPL in different sectors are uploaded in the international COMAR database, i.e. COde d'Indexation des Matériaux de Reference (COMAR) for its global visibility. To enhance the impact of reference materials in the global economy, Bureau of Indian Standard (BIS) has recently established national mirror committee on reference materials which is a part of ISO-REMCO (Committee on reference materials). The main aim of this chapter is to make the readers acquainted with process and guidelines who are working in reference standard and looking for certification of their products. This chapter describes the terminology/definition and international networking. The metrological traceability of BNDs established are linked to SI units which have been elaborated in detail. Further, the metrological perspective, such as testing, calibration, method of validation, repeatability, reproducibility studies, control chart preparation, inter-laboratory comparison, quality control and quality assurance are also discussed for a broader readership. A detailed emphasis has been given on international key comparison, proficiency/round-robin testing, pilot study in various sectors like physico-mechanical, physico-chemical, foods/feedstuffs, biomedical, environmental, health-care, agricultural etc. in this chapter.

That which cannot be measured cannot be proven.
–Anthony W. Richardson (1899–1964)

18.1 Introduction

Quality infrastructure (QI) fulfills essential tasks in a country's economy. It enhances the competitiveness of industry and facilitates international trade, contributes to ensuring the safety of products and the protection of consumer's satisfaction. The quality products and services, including compatibility and performance are on high demand for the growth of economy. Understanding the criterion for harmonized measurement system among the NMIs through international networking is essential to improve the measurement capability. Therefore, a uniform measurement system is achieved by a reference standard which is acceptable anywhere in the world. These reference standards are maintained in any NMI of the country. It is necessary to ensure that these reference standards should reach to the common man. There are metrologically valid procedures and internationally accepted protocols to develop such standards to calibrate the analytical equipment's to the countryside. Therefore, in this era of modern technological word, reference standard has its own importance to improve the quality in measurements for the advance equipment/devices for the quality research which uplifts the technological advancements and safety of life. In this context, BNDs being a primary standard provide accurate and precise measurements to fulfill demand of the society for the quality assurance.

18.1.1 Importance of Reference Material

The modern technological world requires globally comparable measurements and test results through traceability to the SI units for ease of doing a business-driven economy. It ensures quality products for the end-users. Measurements are essential in our daily life and also for advanced science and technology for the current modern world. Without precise measurements [1], it is challenging to perform trade with other countries. The comparable measurements, i.e. 'once measured or tested, globally accepted' are bound to uplift the quality life of consumers and enhances the socio-economic fabric of the society through unhindered trade. Consumer satisfaction and trader's confidence in measurement results are of utmost importance in many aspects of modern society. Testing & calibration, analytical methods, and equipment need to provide the evidence on the quality of their processes and methods through traceable measurements. All around the world, there are millions of research going on for life-saving and other fields, and almost all of them are entirely dependent on the analytical process and measurements. In case the recorded measurement data is inaccurate, it leads to fatal consequences in the field of food, drug, toxicology and forensic areas, agriculture, etc. Hence, it is highly essential to develop a reference standard for adequate quality in research and development activities. To achieve reference standard, there are set metrological ISO guidelines for reliable measurements, comparable (traceable) results, accompanied by stated measurement uncertainty with a stipulated time. As per the International organization of standards (ISO), "*totality of characteristics of an entity that bears on its ability to satisfy stated and implied needs*" defines the quality. Therefore, each laboratory must adhere to test procedure of reference standard.

The laboratories demonstrate their quality by adhering to appropriate ISO quality standards, e.g., ISO9001 (Quality Management system), ISO/IEC 17025 [2] (requirements for the competence of testing and calibration laboratories), ISO 17034 (Reference materials production) which are accredited/certified by authoritative accreditation bodies. These ISO standards are required whenever quality assurance is to be demonstrated by the respective domains [3].

Quality and quality control (QC) together form the basis for quality assurance (QA), which encompasses operational techniques and activities that are required to fulfil requirements for quality. Quality assurance and quality control are two aspects of quality management. While some quality assurance and quality control activities are interrelated, the two are defined differently. Typically, QA activities and responsibilities cover virtually all of the quality system in one fashion or another, while QC is a subset of the QA activities.

In general, quality assurance comprises of all those planned and systematic actions that are undertaken by the organization, which are necessary to provide adequate confidence that a product or service will satisfy the given requirements for quality or not. The detailed information about various CRMs produced by various NMIs can be found in the COMAR database (see Sect. 18.1.2.4). As atoms are the building blocks for materials, similarly, measurements are the building blocks of products/devices. For material to be perfect, material metrology is required to establish standards to calibrate the instruments. The materials produced through the valid metrological procedures are the standards, which are comparable internationally, which is an essential requirement for establishing global trade to strengthen the economy. For doing the international trade, we need international networking among the NMIs/regularity bodies [4, 5]. It is worth mentioning that achieving standard reference materials has been found somehow difficult. However, by using a higher level of standard protocol, one can approach closer to the true reference standard. There are many types of standards available which help in calibrating the instrument and evaluating the measurand value.

The confidence in the accuracy and reliability of the measurements arises when the result is linked with a SI unit. The reference standards developed at CSIR-NPL are traceable to SI units. Verification of specified properties of the developed reference standards is being done by using analytical equipment with a properly validated method and primary reference materials (traceable to SI unit). To establish the uniform measurement system, India being the member of *Bureau International des Poids et Mesures* which is commonly known as International Bureau of Weights and Measures (BIPM) & the APMP is maintaining standards at par with international standards. In addition to this, CSIR-NPL is producing/developing several Indian reference materials by following the metrologically valid procedures for use in various sectors. It is continuously improving its measurement capabilities through international key comparison, pilot studies, peer reviews to support the various industries in India. With these capabilities of measurement, also provides its services to industries, strategic and academia that support the India's prosperity and quality of life.

18.1.2 International Network

This section provides an overview of the BND activity linking to the international platform which underpins the metrological capability and the status of the BND on the international level. In order to build international comparability among the NMIs, it is needed to create wider networking in the international forum which supports to enhance the role of metrology under the domain. The networking with the international organization accelerates global harmonization in testing/calibration and measurement standards. Networking among the NMIs through bilateral cooperation, participation with international bodies/organization establishes wide-ranging recognition. To overcome barrier in international trade, standardization of measurement with high metrological aspect is required. Due to increasing competitiveness in the global market there is an urgent need to produce metrologically developed reference standards which ultimately promote international trade.

The development of BND follows with the major organizations such as BIPM, ISO-REMCO committees, APMP committees/focus group, COMAR database to strengthen its quality-infrastructure in the field of testing and calibration [6].

Traditionally, trade was regulated through bilateral treaties between two nations. International trade is the main source to enhance the global economy where supply and demand will decide the prices, which affect the business globally. Currently, 164 countries are dealing with international trades through the World Trade Organization (WTO), an intergovernmental organization that is concerned with the regulation of international trade among nations. WTO devotes significant attention to good standardization and regulatory practices so that standards, regulatory and conformity assessment procedure requirements don't become technical barriers to trade.

In this context, the main objective of CSIR-NPL is to develop contemporary standards/BNDs by international networking forum, i.e. mutual recognition agreement (MRA). The importance of the networking with the international organization for dissemination of BNDs and international trade scenarios are illustrated in Figs. 18.1 and 18.2, respectively. Figure 18.2 provides a relation in trade between international accreditation laboratory committee (ILAC), standardization (ISO) and regulation (WTO). These processes are essential to establish trade internationally.

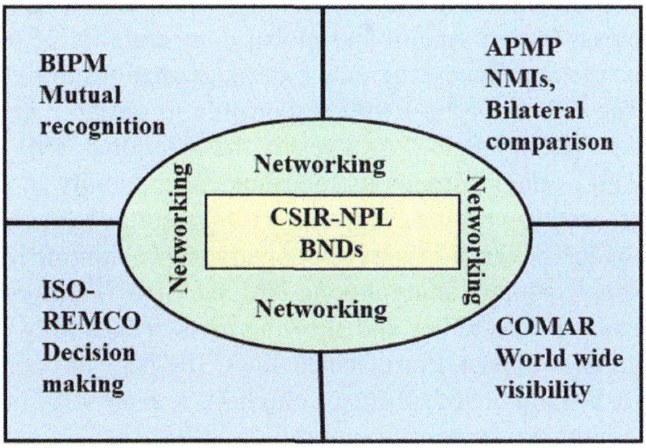

Fig. 18.1 Networking with international organization

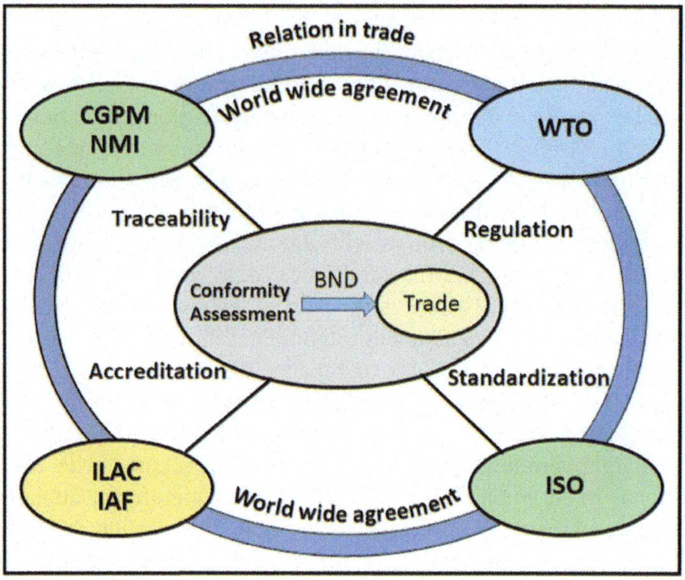

Fig. 18.2 The International trade scenarios

18.1.2.1 International Comparability Through Key Comparison: A Pilot Study

BIPM was established during the meter convention with an objective to work on metrological aspect and measurement standards among all the member states which implies the NMIs. BIPM mainly works for the globalization on the international measurement system and global comparability of measurement with the NMIs of its member states, regional metrology organizations (RMOs) and strategic partners. The APMP is the RMO responsible to promote and support the NMIs under the Asia pacific region. As a result, the individual NMI could build its own capacity as well as demonstrates its metrological capability in the International platform. The respective working group under the consultative committee and technical committee, organizes various types of comparison programme like key comparison, pilot study, supplementary study among NMIs. CSIR-NPL is continuously participating in the comparison studies and showing its own capability in the international forum with greatest impact. For instance, the CSIR-NPL has conducted the comparison studies for the amount of substance emphasis to metrology in chemistry (CCQM). Technical committees under the umbrella of APMP which are listed in [7] Key Comparison Database (KCDB) of BIPM which is given in Table 18.1 [1].

Table 18.1 Participation under BND activity in key comparison studies in the field of inorganic chemistry in CCQM and APMP

S No.	Key comparison studies	Analyte/elements with Matrix
1	CCQM-K143	Copper calibration solution, Organised by NIST (in progress)
2	APMP.QM- P-30	Calibration solutions for As, Pb, Ni, Fe
3	CCQM-K127	Toxic and trace elements in Soil
4	CCQM-P135.1	Trace impurities in high concentrated NaCl solution
5	APMP.QM-P11	Arsenic and arsenobutane content in marine fish (Swordfish)
6	APMP.QM-P10	Cadmium and Lead in Herb
7	CCQM-P97	Cadmium and Lead in Herb
8	APMP. QM-K24	Cadmium in Rice powder
9	APMP.QM P-18	Ca, Cd, Pb & Zn in Herba Eclipta
10	APMP.QM-S5	Cd, As, Fe, Zn in Shrimp sample
11	APMP.QM P-21	Cadmium and arsenic in brown rice

18.1.2.2 ISO & ISO/REMCO Committee

ISO is a non-government international organization having 164 national standard bodies from the government of each country. ISO promotes common standards among all the nations. Implementation of common standards helps to get the quality product, customer satisfaction, hurdles free trade. Regulatory bodies maintain and ensures the quality of products, protect the health of living being and benefits to economy, etc. so as to lead an easier, better and safer life. ISO develops standard document after receiving the request from the stakeholders like industries or consumer group through the national member association. BIS, the national standard body of India is the member body of the ISO with 666 technical committees (TC) and 5 PDC participations. CSIR-NPL, being the NMI is a member of many TCs. Director, CSIR-NPL is the chairman of PDC dealing with ISO/REMCO and many other TCs. This committee works mainly to establish concepts, terms and definitions related to reference materials and to specify the basic characteristics of reference materials as required by their intended use. Further to propose actions on reference materials required to support other ISO activities, prepare guidelines for ISO technical committees when dealing with reference material issues, communicate with other international organizations on reference material matters and to advise the ISO Technical Management Board (TMB) on reference material issues.

Recently, development of ISO/TC/229/WG4 document on liquid suspension of nanomagnetic particles: Characteristics & specification. no. ISO 19807 has been initiated by CSIR-NPL with the involvement of member countries and it was published by BIS in 2019.

The participation and formulation of standards in the committees constituted by ISO is presented in Fig. 18.3.

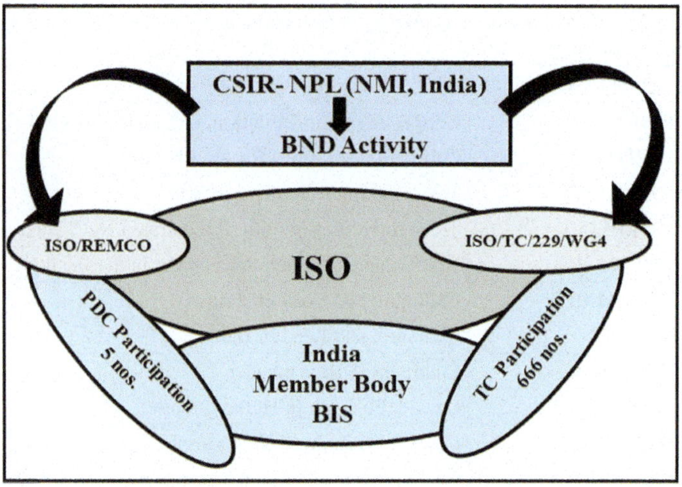

Fig. 18.3 Liaison of BND activities with ISO

Related to reference materials, the ISO/REMCO covers the establishment of terms and definitions, basic characteristics as per intended use, formulation of guidelines, liaison with relevant international bodies like BIPM, ILAC and production of RMs and its applications.

Standards/guidelines published by ISO and ISO/REMCO which are being used routinely for the production/development of BNDs are given in Table 18.2. Under the BND activity, continuous involvement of CSIR-NPL for the formulation of the standards/guidelines [8, 9] has been taken into consideration.

18.1.2.3 APMP Committee/Focus Group for Materials

APMP is a grouping of NMIs from the Asia–Pacific region engaged in improving regional metrological capability through the sharing of expertise and exchange of technical services among Member laboratories. APMP is also an RMO recognized by the International Committee for Weights and Measures (CIPM) for the purpose of worldwide mutual recognition of measurement standards and of calibration and measurement certificates. The framework for international comparability assessment is shown in Fig. 18.4 [10].

APMP's mission is to promote and support a measurement infrastructure in the Asia–Pacific region that facilitates international trade, improves industrial efficiency and competitiveness, ensures equity in the marketplace, and enhances the quality of life and the environment. APMP's vision is to become an inclusive, representative and transparent RMO providing leadership for meeting the measurement challenges of the region.

Table 18.2 Standards/Guidelines commonly referred/followed for development/production of BND [5, 6]

Subject		Name of standard/guidelines	Published by technical committee
Reference material	Selected terms and definitions	ISO Guide 30:2015	ISO/REMCO
	Contents of certificates, labels and accompanying documentation	ISO Guide 31:2015	ISO/REMCO
	Good practice in using reference materials	ISO Guide 33:2015	ISO/REMCO
	General requirements for the competence of reference material producers	ISO 17034:2016	ISO/CASCO in a joint WG with ISO/REMCO
	Guidance for characterization and assessment of homogeneity and stability	ISO Guide 35:2017	ISO/REMCO
Testing & calibration	General requirements for the competence of testing and calibration laboratories	ISO/IEC 17025:2017	ISO

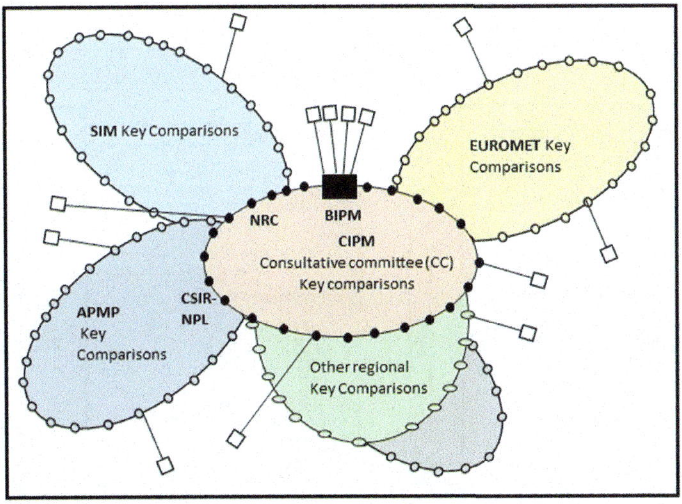

Fig. 18.4 Framework for international key comparisons

At present, under the framework of MoU, APMP operates under the supervision of the executive committee, the managing body, headed by a chairperson, assisted by a secretariat as the coordinating body. APMP has 12 TCs responsible for coordinating technical activities for each field covered by the CIPM's consultative committees. The Developing Economies' Committee (DEC) is responsible for initiating and coordinating work programs to address the needs of APMP member of developing economy countries NMIs (DENs). APMP has also established 5 focus groups (for clean water, climate change and clean air, energy efficiency, food safety, and medical metrology) to promote liaison with and provide measurement support to stakeholders in identified priority fields.

18.1.2.4 The International Database for CRM (COMAR)

The demand for CRM/BND is increasing continuously worldwide to ensure reliability and global comparability of measurement result in every sector including scientific institutes, testing/calibration laboratories, Industries, trade, environmental, health care, etc. In this regard, an international database for CRMs called COMAR has been developed to assist the CRM consumers. The required CRMs can easily be found as per the requirement of the customer, which had been very difficult to locate due to unavailability of universal catalogue in the market in a synchronized way. The office is located at Bundesanstaltfur Materialforsch ungund-prufung (BAM), Berlin Germany. The management and organizational structure of the COMAR is shown in Figs. 18.5 and 18.6, respectively [11].

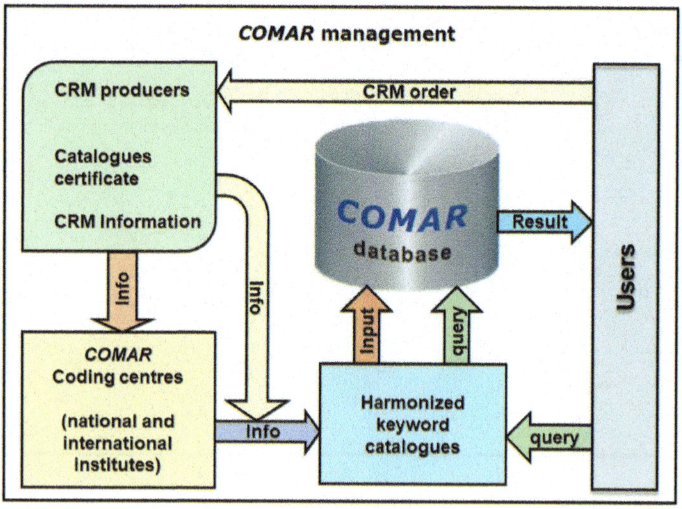

Fig. 18.5 COMAR management

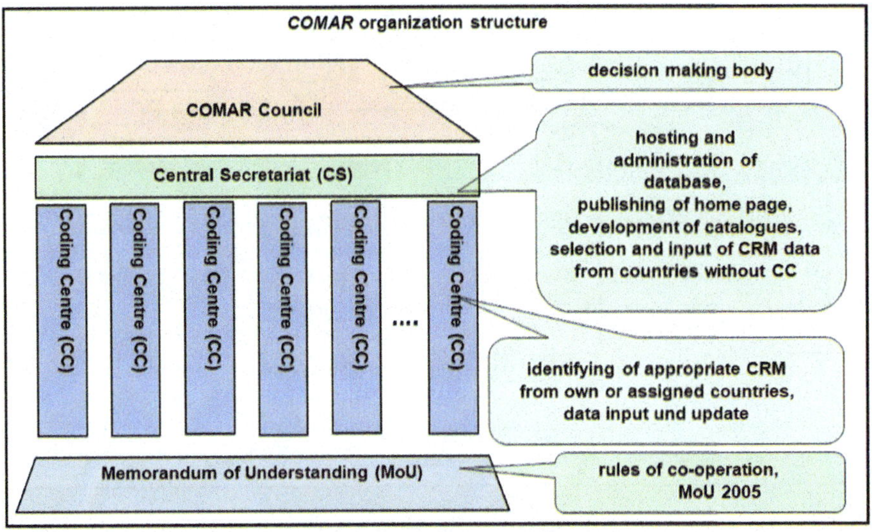

Fig. 18.6 COMAR organization structure

The COMAR central secretariat at BAM database makes varieties of available CRM in a harmonized catalogue with broad scope of single producer. National or International CRM producers have been given access and rights to update in the COMAR coding centres so as to offer an easier platform on the availability of CRMs of world's major CRM producers [4]. Apart from many private CRM producers worldwide, several NMIs/DIs have enrolled their CRMs in the COMAR database. Recently developed BNDs have been uploaded for its global dissemination. Each BND uploaded in the COMAR database has been provided with a unique CRM identification number.

18.1.3 Role of NMI

Each nation has to maintain the measurement standard (traceable to the international standards at BIPM) is named as NMI. The NMIs have to continuously sharpen/strengthen themselves to improve their calibration measurement capabilities (CMC) by sharing through a measurement data by a valid international laboratory comparison programs called key comparison and then disseminate to the consumer. In addition to this, the NMIs are working intensively for the development of varieties of CRMs for the need of their economy as well. In India, the primary and secondary measurements standards are maintained by CSIR-NPL. The main function of CSIR-NPL is realization, maintenance of national standards and dissemination of measurement tractability to industry and national accreditation laboratory

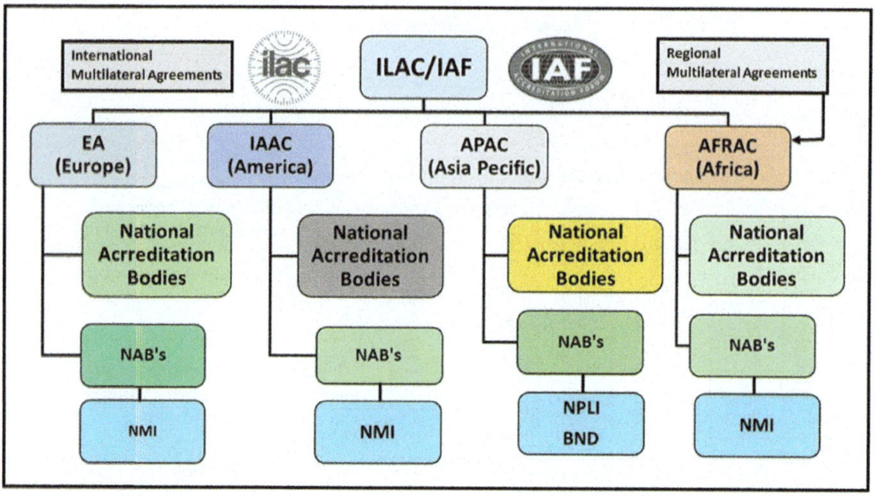

Fig. 18.7 The international recognition framework networking committee and accredited laboratories

(NABL). The reference standards maintained/produced by the NMIs are disseminated through accreditation bodies to various tests and calibration laboratories. Thus, the reference standard (BNDs) produced by CSIR-NPL are competent to support the national/international industries. This unbroken chain of measurement enhances the confidence level in the global market. The national accreditation bodies are connected to the international multilateral agreement (ILAC/IAF) through regional multilateral agreements (Fig. 18.7).

CRMs are only produced by NMIs such as the National Institute of Standards and Technology (NIST) in the United States; the European Commission Joint Research Centre for Directorate, health consumers and reference materials (JRC) (Geel, Belgium); Federal Institute for Materials Research and Testing (BAM) in Germany; National Measurement Institute Australia (NMIA); National Research Council of Canada (NRCC); National Institute of Metrology China (NIMC); National Metrology Institute of Japan (NMIJ); Korea Research Institute for Science and Standards (KRISS); CSIR-NPL, India and many others in the world. The above-mentioned NMIs are producing CRMs in different sectors to cater the need globally as illustrated in Fig. 18.8.

It is not possible for NMIs to develop all the CRM on its own, though they are having all measurement capabilities. For this, there are accredited RMPs in each country which are engaged for the production of CRMs in different sectors in collaboration with NMI of the respective country. Figure 18.9 shows the number of CRMs produced by various economies and the number of RMPs taking part in it. The data is based on BAM 2018. BND, a registered trademark of the CSIR-NPL, is a CRM of CSIR-NPL which is supported by Ministry of Commerce & industries and CSIR. The BNDs are being produced by CSIR-NPL and its associate RMPs in specific areas

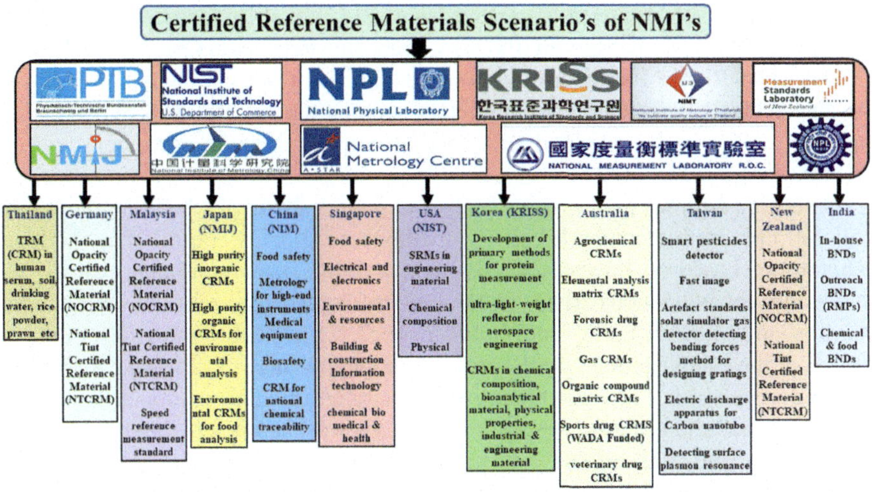

Fig. 18.8 Production of CRMs by various NMIs

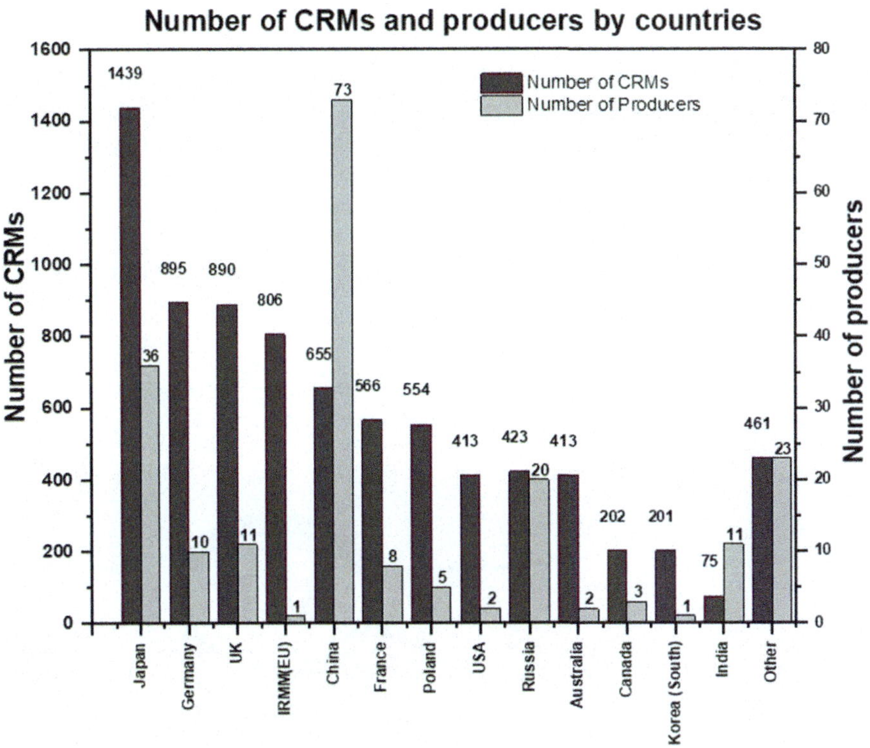

Fig. 18.9 Number of CRMs and producers with various countries (some of the data from BAM, 2018)

such as cement BNDs are produced by National Council for Cement and Building Materials (NCCBM), petroleum BNDs by Hindustan Petroleum Corporation Limited (HPCL) and Bharat Petroleum Corporation Limited (BPCL), chemical BNDs by Aashvi Tech LLP, mechanical parameters such as hardness blocks by Global PT providers according to specific guidelines of BIS which are in agreement with the relevant ISO standard 17034.

BNDs are bedrocks of modern analytical quality assurance because they allow calibration of instruments, validation of methods, and quality control of methods and laboratories based on traceability and comparability of measurement results. The availability of indigenous BNDs in the country is poised to reinforce the national mission on "Make in India" and "self-reliance". CSIR-NPL, owing to its untiring efforts in the establishment, development and maintenance of "Primary Standards" and its derivatives & their dissemination has occupied a pivotal position in the country.

The precise and accurate measurements of various parameters through unbroken chain of traceability, to various government, strategic, public and private sectors have resulted in the overall quality and safety of human beings in the country as well as international trade.

NMI is to ensure the dissemination of measurement traceability to industries and shouldered the responsibility of development of CRMs/BNDs at par with international level. CRMs are widely used for the calibration of measuring apparatus, for the evaluation of measurement procedures and for the internal or external quality control of measurements and laboratories. Currently, India is majorly dependent on few foreign multinational companies and losing a hefty amount of foreign exchange on procuring these CRMs. Moreover, with the ongoing BND programme, CSIR-NPL supports various ministries/organizations/industries to develop indigenous reference standards which will enhance the quality of life in the country and reduce the import and increase the export as shown in Fig. 18.10. The CRMs provide the measurement

Fig. 18.10 Contribution of BND to various fields to support various ministries/organizations

benchmarks, ensure reliability and comparability that help to create confidence and provide traceability for quality assurance. NMIs active interaction with RMOs for pilot studies, round robin test for BNDs directly link to BIPM ensures the authenticity of the measurement results.

18.1.4 Global Market and Impact on Economy

The global analytical standards market of CRM is projected to reach USD 1.73 billion by 2022 from 1.27 billion in 2017, at a compound annual growth rate (CAGR) of 6.3% (Fig. 18.11). The factors driving the growth of this market depend on stringent regulations over pharmaceutical industry, rising public concern related to optimal food safety and quality, increasing public–private investments for various researches and growing use of analytical techniques in proteomics and metabolomics. However, limited public awareness related to effective pollution monitoring strategies and stringent storage requirements for analytical standards are the key factors hindering the growth of the market.

Key industries supplying certified reference materials are specifically, ATCC, Merck, FLUXANA, Paragon Scientific, Starna Scientific, Hellma, Agilent, Phenomenex, Cerilliant, Central Geological Laboratory cover the market globally of billions of USD of reference standards. The main NMIs associated with these companies are from North America (United States, Canada and Mexico), Europe (Germany, UK, France, Italy, Russia and Turkey, etc.), Asia–Pacific (China, Japan, Korea, India, Australia, Indonesia, Thailand, Philippines, Malaysia and Vietnam), and South America (Brazil, etc.) Middle East and Africa (Egypt and GCC Countries). In India, BNDs are used as CRMs by industries, government sectors and academia to

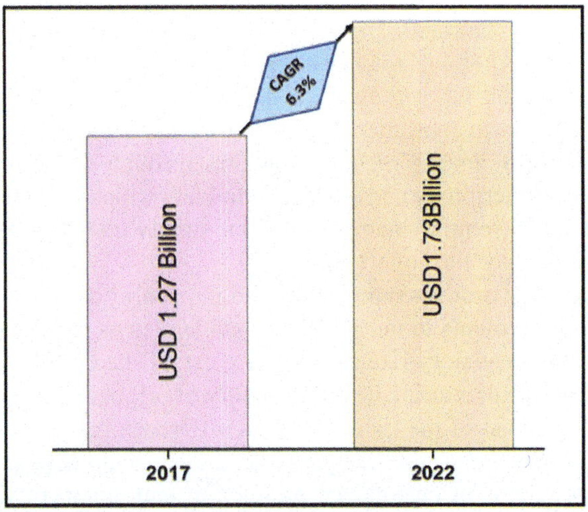

Fig. 18.11 Compound annual growth rates

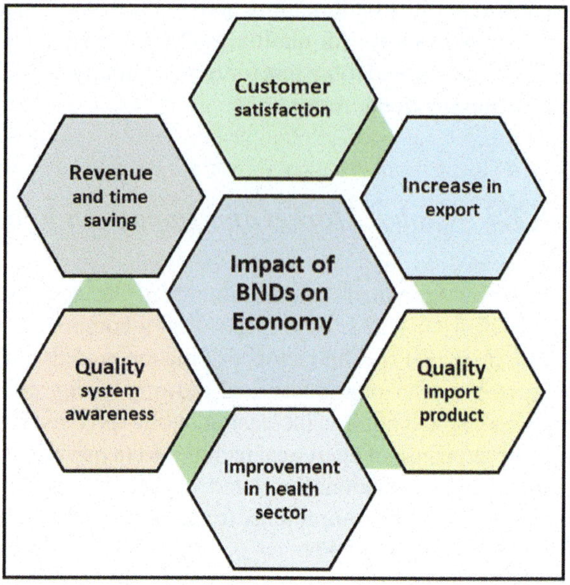

Fig. 18.12 Impact of BND on economy

ensure the highest quality measurements for calibrating laboratory test equipment, checking the trueness of measurements, validation of methods for quality assurance.

As of today India imports CRMs worth more than Rs 10,000 crore. It is CSIR-NPL responsibility to overcome foreign exchange and become self-reliant in CRM production. Therefore, to cater the needs of Indian/global market, CSIR-NPL has started its mission mode program in association with Ministry of Commerce and Industry (MoC) and the line ministry (S&T) for the indigenous production of CRMs. We have signed 11 MoUs with different RMPs in various sectors for producing CRMs/BNDs. In future, BND will provide quality import product, revenue generation, quality system assurance and consumer satisfaction as illustrated in Fig. 18.12.

This national importance program would certainly bring a paradigm shift in socio-economic fabric in almost all the sectors and that would save thousands of crore rupees along with consumer satisfaction. In this context, it needs to take initiative to develop synergy among regulatory bodies through Quality Council of India (QCI), BIS, legal metrology and other stakeholders which will bring enormous impact on quality system, employment generation, supply chain, economy and over all progressive growth of the country.

Any product subjected to the transport between two nations must follow certain requirements to meet with the guidelines as per the international trade regulations. The mandatory requirements are established by the legislative body of the country for manufacturing items in country of origin and imported products. It has been seen that most of the time there is a difference related to mandatory information in the host country and the importing country and because of this manufacturer faces the problems of customized production that enhances the cost of sale. Therefore, the barrier-free trade for all nations may be on high demand. Basically, the base of free trade barriers depends upon four pillars; namely, NMI of the country, organization dealing with standard documentation, accreditation bodies and legislative bodies.

Possession of the measurement capabilities along with new activity of BND (specifically designed for the purpose along with the standard document) will limit import of the products which are not in compliance with the requirement of Indian geographical, agro-climatic conditions, etc. [12]. Further, this initiative of BNDs in India will also save precious foreign exchange and boost economy without compromising the quality at international level.

18.2 Terminology and Definitions of Types of Standards

Basically, the standards are having a reference value that helps to compare the measurand/analyte of the unknown sample for its accurate and precise measurements. The major role of reference standard is to check the quality of the end product, establish its measurement traceability, calibration of the instrument, validation of analytical method/procedure, assignment of property value.

The hierarchy of metrological standards are in accordance to the traceability link. The primary or the national standards are directly linked to the SI units in which the uncertainty is considered to be nil. The derived standards are the secondary standards having more uncertainty with respect to primary standards. As each comparison added the uncertainty of the previously used step, thus the uncertainty of the measurement gradually increases as the traceability chain increases. Thus, the uncertainty is more at the laboratory where the calibration/testing has been performed. Figure 18.13

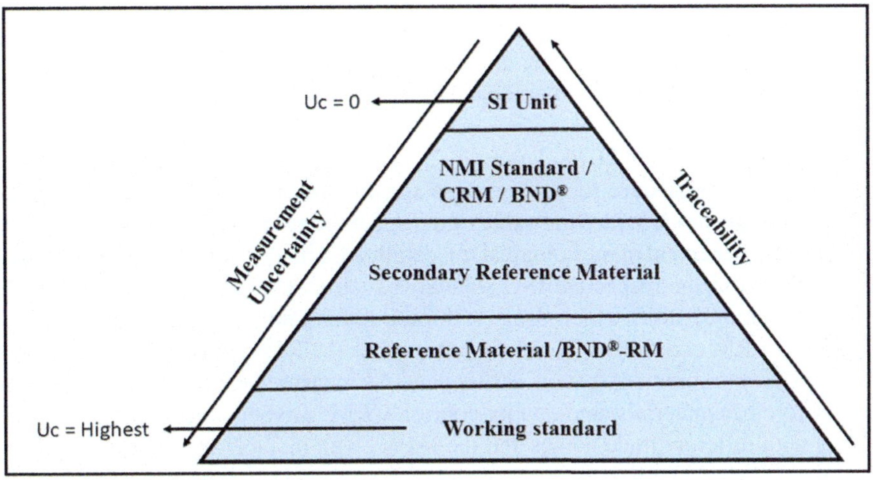

Fig. 18.13 Hierarchy of the reference materials

shows the hierarchy of the reference materials starting from standards prepared by NMI with highest metrological aspect to the standards prepared by the laboratory being prepared for the routine test.

BIPM is actively working for the globalization of the international system of measurement [13–15]. BIPM works with the NMIs and the RMO encourages them for the global comparability of measurements. The International vocabulary of metrology (VIM)—basic and general concepts and associated terms [16] describes different types of reference materials. The role of the ISO/REMCO is also concerned to global harmonization and laid down the standard guidelines ISO 17030:2015 and recommends the terminology and the definition for the reference materials [17–20].

18.2.1 Reference Material (RM)

According to the clause no. 5.13 (6.13) of VIM, RM is the material, sufficiently homogeneous and stable with reference to specified properties, which has been established to be fit for its intended use in measurement or in examination of nominal properties [21–24]. ISO guide 30:2015 describes the definition of RMs in the clause no. 2.1.1 that "material, sufficiently homogeneous and stable with respect to one or more specified properties, which has been established to be fit for its intended use in a measurement process". The measurement process broadly covers both measurement of a quantity and examination of a nominal property.

18.2.2 Certified Reference Material (CRM)

The clause 5.14(6.14) of VIM defines the CRM as the reference material, accompanied by documentation issued by an authoritative body and providing one or more specified property values with associated uncertainties and traceability, using valid procedures [25, 26].

As per ISO guide 30:2015 clause no. 2.1.2, the RM is characterized by a metrologically valid procedure for one or more specified properties, accompanied by an RM certificate that provides the value of the specified property, its associated uncertainty, and a statement of metrological traceability. Each term reflects in the definition is elaborated in the respective ISO standards such as ISO 17034:2016; ISO guide 35; ISO guide 31:2015 which deals with production and certification of RMs, and contents of RM certificates[24, 27, 28]. It implies the RM is applicable for the calibration of a measurement system, assessment of a measurement procedure, assigning values to other materials, and quality control. CRMs are being produced by various NMIs with different trade names and the list is given in Table 18.3.

Table 18.3 The CRMs produced by various NMIs under different trade marks

Name of NMIs	Trade mark	RMO of CIPM MRA
CSIR-NPL, India	Bharatiya Nirdeshak Dravyas (BND)	APMP
NIST, USA	Standard Reference Material (SRM)	SIM
PTB, Germany	Certipur Reference Material (CRM)	EURAMET
IRMM, Belgium	European Reference Material (ERM)	EURAMET
JRC, Belgium	Community Bureau of Reference (BCR)	
NIM, China	China Reference Materials (CRM)	APMP
NIM, Thailand	Thailand Reference Material (TRM)	APMP
NMI/AIST, Japan	Certified Reference Material (NMIJ, CRM)	APMP
LGC, UK	LogiCal, LGC Standard	EURAMET
NMI, Australia	Certified Reference Material (CRM) Australia Certified Reference Material (ACRM)	APMP
KRISS, Korea	Certified Reference Material (CRM)	APMP
NRC, Canada	Canadian Reference Material (CRM)	SIM
HAS, Singapore	Certified Reference Material (CRM)	APMP

18.2.3 Candidate Reference Material

The material, intended to be produced as a RM. A candidate material has yet to be characterized and tested to ensure that it is fit for use in a measurement process. To become an RM, a candidate material needs to be investigated to determine if it is sufficiently homogeneous and stable with respect to one or more specified properties, and is fit for its intended use in the development of measurement and test methods that target those properties. A candidate reference material may be an RM for other properties and a candidate reference material for the target property (ISO guide 30:2015 clause no. 2.1.3).

18.2.4 Matrix Reference Material

According to ISO guide 30:2015, clause no. 2.1.4 [29], reference material is the characteristic of a real sample. Matrix reference materials may be obtained directly from the sources or by spiking the measurand/analyte of interest. The composition of measurand/analyte in the matrix may be in major, minor or trace level concentration. It is primarily used in the analysis of the matrix match sample as the matrix affect the dilution step/interference in the instrument.

18.2.5 Secondary Reference Materials (Secondary Standard)

Secondary reference material is calibrated against a CRM and its traceability is derived from the CRM. Basically, it is used for verification of a secondary reference method and intended for routine test. Usually, it is shown to be commutable (5.5 (6.5) of VIM) [30].

18.2.6 Working Standard

The working standard is derived from a reference standard which may be a CRM or secondary reference material and always calibrated against a reference standard. This working standard is routinely used in the testing laboratories for the calibration of measuring instruments (5.7 (6.7) of VIM).

18.2.7 Bharatiya Nirdeshak Dravya (BND)/Indian Reference Material

BNDs is an Indian reference material (IRM) which is developed/produced by CSIR-NPL and it is trademark of CSIR-NPL. CSIR-NPL is involved in the development of various BNDs, both in-house as well as in joint collaboration with RMPs [31, 32].

18.2.7.1 BND–RM

BND is the RM, which is sufficiently homogeneous and stable with respect to one or more specified properties and has been established to be fit for its intended use in a measurement process. More likely, the property value related to biological parameter in which the traceability and the associated uncertainty value may/may not be established is considered as BND–RM [33].

18.2.7.2 BND-CRM

The BND/IRM is homogeneous, stable materials, certified for one or more properties such as chemical and/or physical parameters. It is traceable to SI unit through an unbroken chain of measurement and contains a certificate. The certified value is associated with a measurement uncertainty value. It is intended to disseminate it throughout the country to provide quality products to the end users through certified parameter in various matrices. BNDs are used for calibration of the instrument,

Table 18.4 Specific requirement of different types of standards

Types of standards	Category	Metrological traceability of certified values	Assignment of the uncertainties of the property values	Level of accuracy provide	Compliance with guidelines (irrespective of version)
Reference material (RM)	Secondary standard	Not mandatory	Not mandatory	Low	ISO 17034
Working standard (WS)	Laboratory grade	Not mandatory	Not mandatory	Lower than SRM	Laboratory—SOP
Secondary reference material (SRM)	Secondary standard	Required	Required	Lower than CRM	ISO 17034
Certified reference material (CRM)	Primary measurement standard	Required	Required	Highest	ISO 17034 & ISO/IEC 17025
Bharatiya Nirdeshak Dravya (BND)	Primary measurement standard	Required	Required	Highest	ISO 17034 & ISO/IEC 17025
Bharatiya Nirdeshak Dravya (BND–RM)	Secondary standard	Not mandatory	Not mandatory	Lower than BND	ISO 17034

validation of analytical method/procedure, assignment of property value of another CRM. Table 18.4 shows the specific requirement of different types of standards.

18.3 Metrological Traceability of BNDs Linking to SI Unit

18.3.1 General Approach for Metrological Traceability

Accuracy and traceability in measurement results are essential to achieve comparability of measurement which helps to overcome non–barrier trade in international market to establish traceability of measurement results, a general principle has been laid down jointly by the international bodies that deal with metrology (BIPM, International Organization of Legal Metrology, OIML), accreditation (ILAC) and standardization (ISO).

According to VIM, 2.41 (6.10) [16, 34–38], "metrological traceability is the property of a measurement result whereby the result can be related to a reference through a documented unbroken chain of calibrations, each contributing to the measurement uncertainty". Metrological traceability, therefore, embodies the concepts of measurement uncertainty and calibrations against a hierarchy of reference standards, i.e. different types of standards/reference materials. The term 'reference' can be a definition of a measurement unit through its practical realization, or a measurement procedure.

Although the metrological traceability can be achieved by the reference value/property value of the RM/CRMs used for the measurement result, but the assigned property value has also to be metrologically traceable. The role of NMI is to establish traceability through CRMs which is shown in Fig. 18.14. The RMP needs to establish, demonstrate, and report the traceability link to SI unit in the RM/CRM certificates for its assigned property value. In this regard, the ISO committee on reference materials (ISO/REMCO), technical report of 16476:2016 [24, 39], investigates, discusses, and specifies in detail about the metrological traceability of quantity values assigned to RM/CRM. The definition of the metrological traceability is given in the VIM along with the traceability of the assigned property value of RM/CRM.

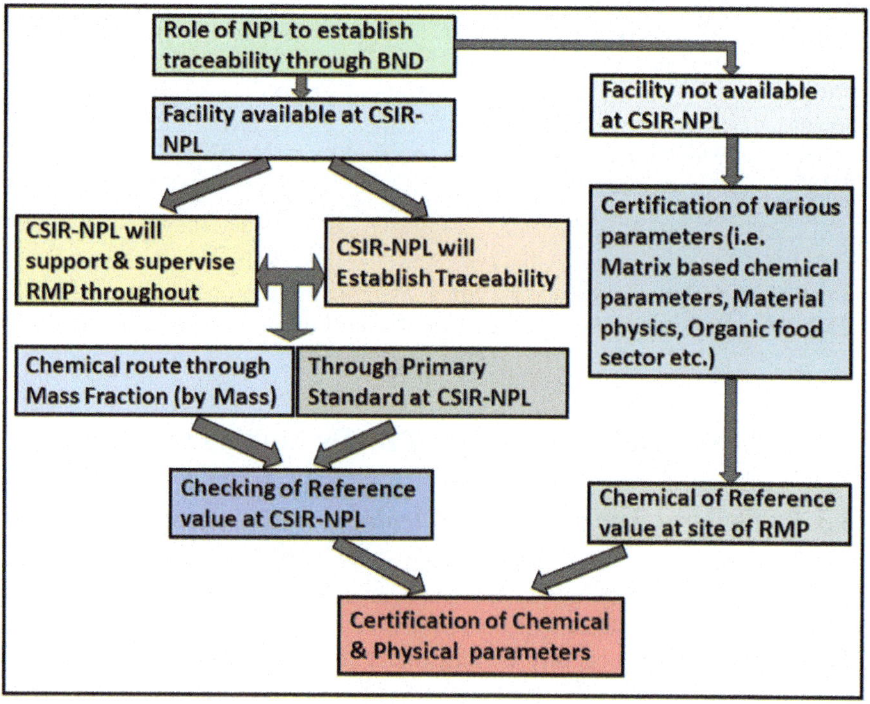

Fig. 18.14 Role of NPL (NMI) to establish traceability through BNDs

The requirements of ISO/IEC 17025:2017 on metrological traceability is broadly given in the following clause no. 6.5.1 is "The laboratory shall establish and maintain metrological traceability of its measurement results by means of a documented unbroken chain of calibrations, each contributing to the measurement uncertainty, linking them to an appropriate reference". The reference represents to measurement unit, procedure and RMs. Using the certified value of a CRM with stated metrological traceability to the SI unit, "the laboratory ensures that measurement results are traceable to the SI Units", as per clause 6.5.2. Thus, it ensures, the metrological traceability established to SI unit through calibration by a competent laboratory [40].

Accordingly, the ISO 17034:2016, clause 7.9 also demonstrates the metrological traceability of the certified values of CRM in compliance with the relevant requirements of ISO/IEC 17025:2017. The RMP provides evidence of the metrological traceability of the certified value to a stated reference. The clause 7.14 of ISO 17034:2016 is discussed briefly about the content of the certificate of a CRM in which it emphasizes to mention the certified/property value with the associated measurement uncertainty and a statement of metrological traceability and how it established [35, 40].

In measurement of chemical parameter, the metrological traceability can be achieved by using (i) a CRM for calibration to SI unit, (ii) a reference measurement method by expert laboratories, and (iii) reference methods by a single expert laboratory [41].

18.3.2 Measurement Pathways to Establish Traceability

The role of NMIs is to maintain the SI unit through calibration, testing and development of RMs. On the basis of the calibration and measurement capability (CMCs), the metrological traceability to international SI units is usually demonstrated and thus exhibits the international comparability and acceptability of the measurement results. The NMIs delivers measurement traceability to the customers and also laydown the quality infrastructure in the country by dissemination of the CRMs. The NMIs shows its capability by implementing the quality system in conformity with ISO 17025 and ISO 17034; participation in key comparison program among other nations [39, 42, 43].

The areas which are of particular interest to CSIR-NPL, covered to support the metrological infrastructure of our country related to manufacturing units for quality product, regulatory purpose and facilitate international trade. It disseminates traceable measurement results to the customer through an unbroken chain of measurements using the reference standards of physical and chemical quantities having highest accuracy and lowest uncertainty. The CMCs with the value of the respective parameter and the range are listed in the Key Comparison Data Base (KCDB) of BIPM [43, 44]. CSIR-NPL disseminates the traceability of the measurement through

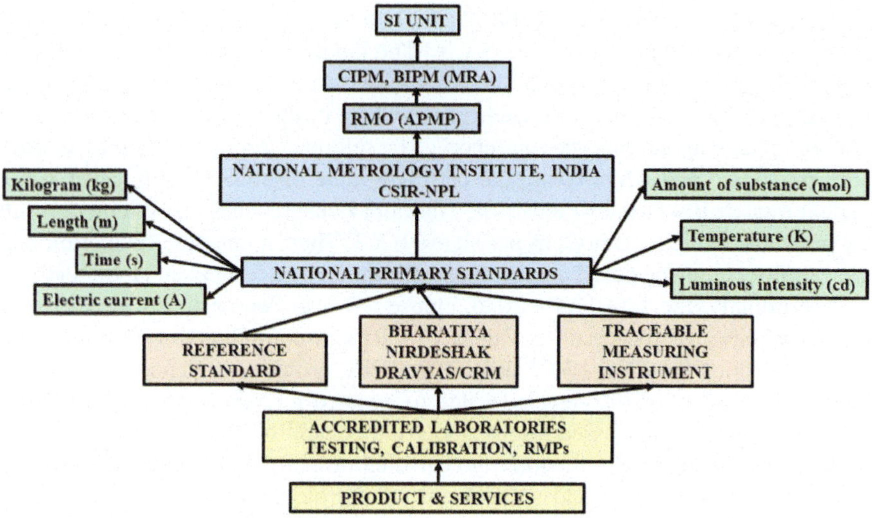

Fig. 18.15 The traceability pathways to SI unit

various pathways which are established based on the comparison with the relevant reference standards, use of BNDs/CRMs, considering the appropriate primary methods.

CSIR-NPL maintains seven SI units which are detailed in the previous chapters of this book. The support given to our country and the international comparability and the mutual recognition shown among the nations are also explained with the results participated in various key comparison programs in the previous chapters. Figure 18.15 shows the traceability pathways followed in CSIR-NPL for the measured quantity.

18.3.3 International Comparisons of the National Standards/BNDs

CSIR-NPL is continuously participating in the key comparison program among NMIs, mainly in supplementary and pilot study programmes in which the measurement traceability is established using national standards and demonstrates its measurement capability [45]. The results of the comparison program show the global acceptability and provide the traceability to the BNDs developed. Participation in any kind of interlaboratory comparison program leads to assess the technical competency and performance of the participating laboratories for testing or calibration purposes of a particular parameter in accordance with the clause 7.7.2 of ISO 17025. The international comparisons are being led by the developed NMIs. In case of physical parameters, the results of the so far participated international comparison program are already discussed in the previous chapters. However, some of the results based on chemical parameters are given below for the ready reference on the comparison program focused on the Inorganic Analysis Working Group (IAWG) conducted on inorganic chemistry by Consultative Committee for Amount of Substance (CCQM) among NMIs/designated institutes (DIs).

Recently CSIR-NPL participated in CCQM-K143 which is on comparison of copper calibration solutions prepared by NMIs/DIs, organized by NIST, USA. The motto of this study is that participant needs to establish traceability link to the International SI unit. In this program, there were twenty-one nos. of NMIs/DIs participated and the individual participant had been prepared the copper calibration solution of 10 g/kg gravimetrically and provided to NIST, USA along with the analytical data. The copper calibration solution has been checked by both NIST, USA and BAM, Germany and CSIR-NPL qualified in this study. The final approval of this study is under process.

PTB, Germany had conducted a pilot study CCQM. P-135.1 on the trace impurities in highly concentrated NaCl solution. The aim of the comparison was to check the competence of the participant laboratories on the determination of impurities present in the salt solution. Figure 18.16 shows the capability of CSIR-NPL on the amount of nitrate impurity determined by ion-chromatography system. The result of the CSIR-NPL is in the equivalence line.

CSIR-NPL is continuously participating in various comparison programs being arranged by the TCs for Quantity of Matter (TCQM) under APMP for BND.

Several focus groups like Clean water focus group; Food safety focus group are working under TCQM, to establish global comparability of measurements with an

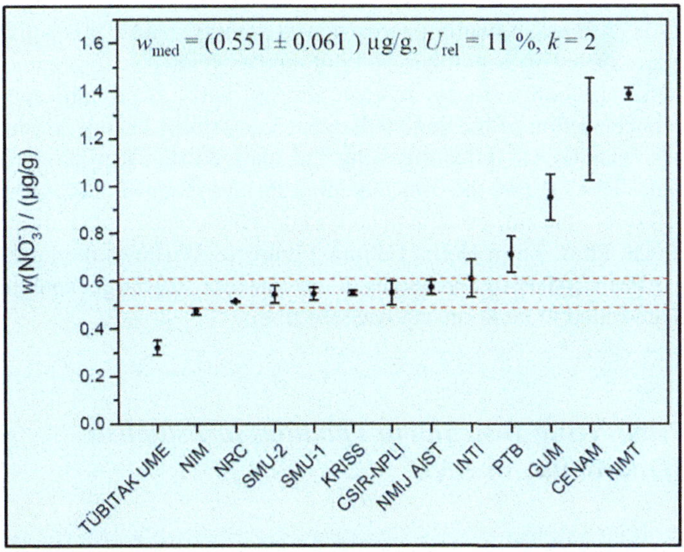

Fig. 18.16 Result of Nitrate content in salt solution under Pilot study CCQM-P135.1

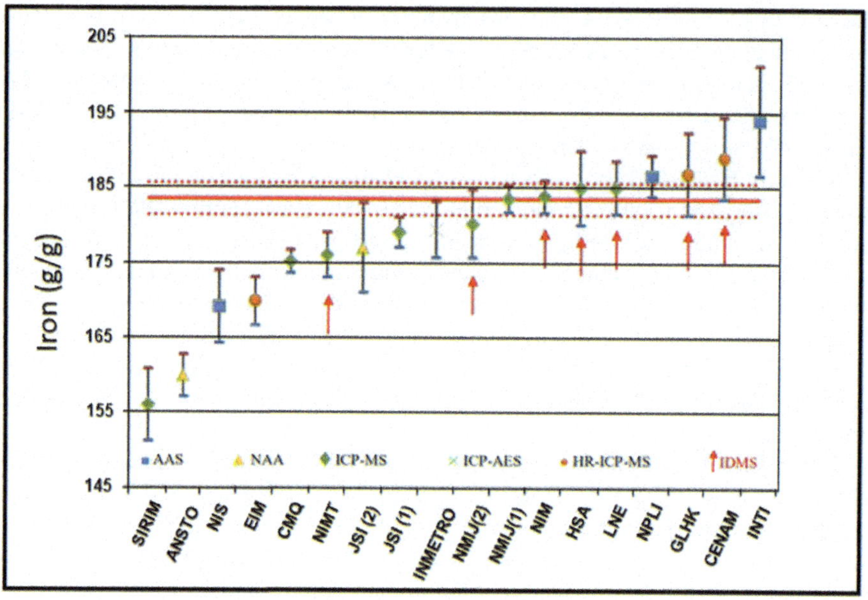

Fig. 18.17 Iron content in shrimp sample under APMP.QM-S5

intention to uphold the measurement infrastructure to promote traceability to SI unit in which CSIR-NPL participate as a member economy.

In Metrology in Chemistry (MiC) area of BND activity several CRMs/BNDs have been developed so far related to aqueous elemental solutions and moreover, this activity is an Internationally peer-reviewed activity. Under this activity, CSIR-NPL has participated in a number of Pilot studies related to Inorganic chemical measurements viz. contaminants in rice, shrimp, herb, swordfish, and soil. The elemental concentration of the analyte had been determined using Atomic Absorption Spectrometer (AAS) after digesting the matrix using a microwave digester system. Figure 18.17 shows the iron content in shrimp sample under APMP.QM-S5. Figure 18.18 shows the zinc content in herb sample under the Pilot study APMP.QM-P18 on Ca, Cd, Pb & Zn in Herba Eclipta. Figure 18.19 shows the lead content in herb under APMP.QM-P10. The results of the content of the target analyte shown in the figures are found to be in the equivalence line.

18.3.4 Measuring Instrument Facilities to Establish Traceability in BND

The metrological traceability is determined as per the level of testing/calibration procedure followed to quantify the measurand. Detail information regarding the

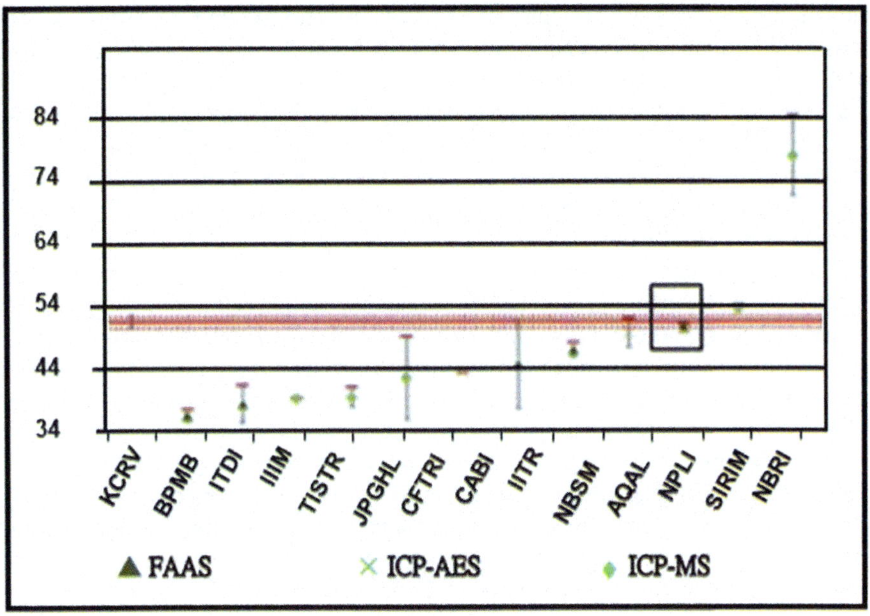

Fig. 18.18 Zinc content in herb under APMP.QM-P18

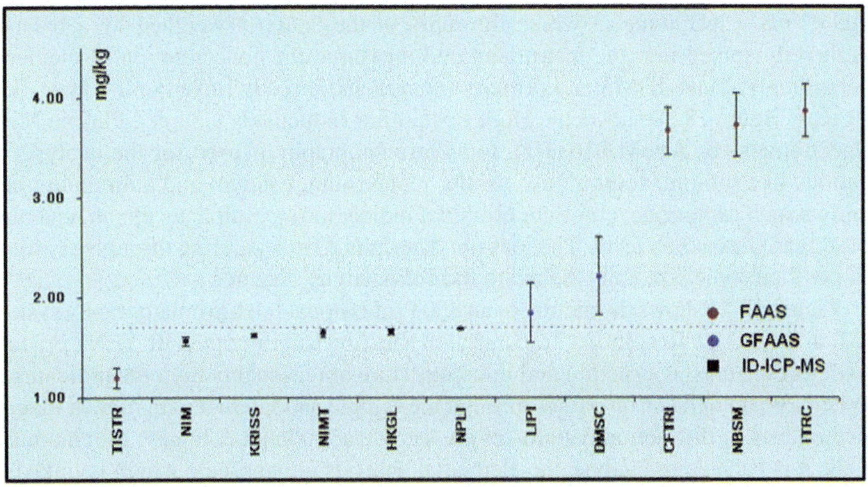

Fig. 18.19 Lead content in herb under APMP.QM-P10

method/procedure opted for analysis is fit for the purpose and the factors significantly affecting the method need to be identified. For the development of inorganic BND/CRM like elemental standard solution, high purity compound; the metrological traceability can also be established by evaluating the purity of the starting material being used for that BND/CRM. Here are some of the instruments which help to develop metrological traceability for the measurement of the chemical/physical parameters either by using a BND/CRM or a valid primary method or by purity evaluation is given below.

Electroanalyser is a technique in which, the purity evaluation of high purity metal is carried out electro gravimetrically by deposition of metals on the electrodes. The hot plate and the stirrer associated with the analyzer facilitate quick deposition of metal oxide. Through oxidation process, the metal ions get deposited as metal oxide on the electrode, which can be weighed out using a calibrated balance and gravimetrically the purity can be determined which is linked to the SI unit.

High resolution inductively coupled plasma mass spectrometry (HR-ICPMS) is a magnetic sector mass spectrometer having high sensitivity with low detection limit and capable of analyzing very trace level elemental concentration. Due to the high resolving power, it is capable to resolve the spectral interference with a wide range of trace level multi-elemental isotopes. The ions reach the detector on the basis of their mass to charge ratio of the isotopes and concentration of elements is determined. The concentration of the unknown sample is determined from the calibration curve obtained by using a BND/CRM. The isotope dilution mass spectrometry (IDMS) is a technique in which the mass of the isotopes weighed by a balance is directly spiked into the instrument and measured the concentration of elements accordingly. Thus, IDMS is a primary method and directly linked to the mass (kg) SI unit. Figure 18.20 shows the High Resolution-Inductively Coupled Plasma Mass Spectrometry of AttoM 016, UK. Ion Chromatography is used for the analysis of cations like lithium, sodium, potassium, magnesium, calcium and ammonium and anions such as fluoride, chloride, bromide, iodide, nitrate, sulphate, etc. in aqueous-based samples at ppb level. The ions are determined by separating through a column as per their ionic size and reached to the conductivity detector.

Figure 18.21 shows the picture of an 850 Professional Ion Chromatography system (Metrohm). High Resolution Transmission Electron Microscope (HR-TEM) is a tool to detect the crystal structure and microstructure at reasonably high magnifications. When a beam of electron passes through the sample and interact with it, then magnified images or diffraction patterns of the sample are obtained. In case of nanomaterials, it is capable to analyse the elemental analysis at nanoscale which is linked to the length SI unit. Figure 18.22 shows the HR-TEM capable to measure the solid sample with thickness less than 50 nm. Magnification calibration HR-TEM BND (BND No. 2006) has been developed by gold lattice imaging as per ISO guidelines and uncertainty has been evaluated by using ASTM data.

Figure 18.23 shows some of the sophisticated instruments that are used for high-end characterization of materials.

Fig. 18.20 High-resolution-inductively coupled plasma mass

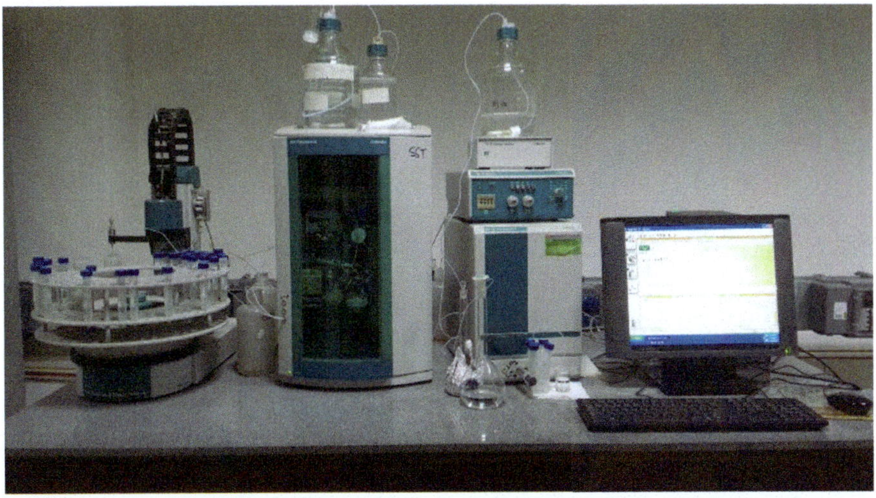

Fig. 18.21 shows the picture of an 850 professional ion chromatography system

18.3.5 Traceability in Various Parameters Compliance with BNDs

The development/production of the BNDs of various parameters is done with an unbroken link from the analyte/measurand of BND to the national standards defined by SI unit. The BNDs related to the physical parameter are directly linked with the

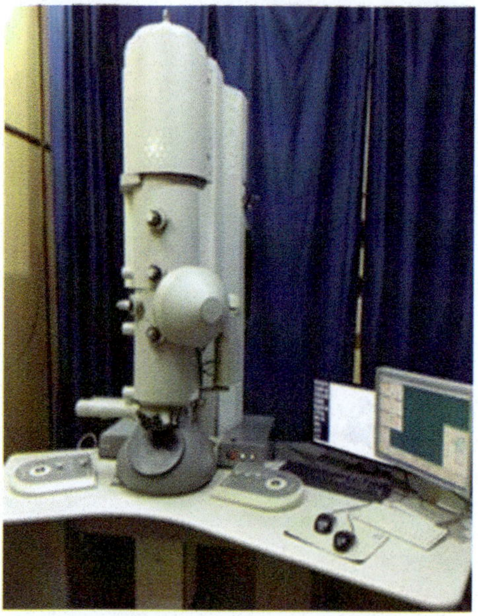

Fig. 18.22 High-resolution transmission electron microscopes (Tecnai G2 F30 STWIN)

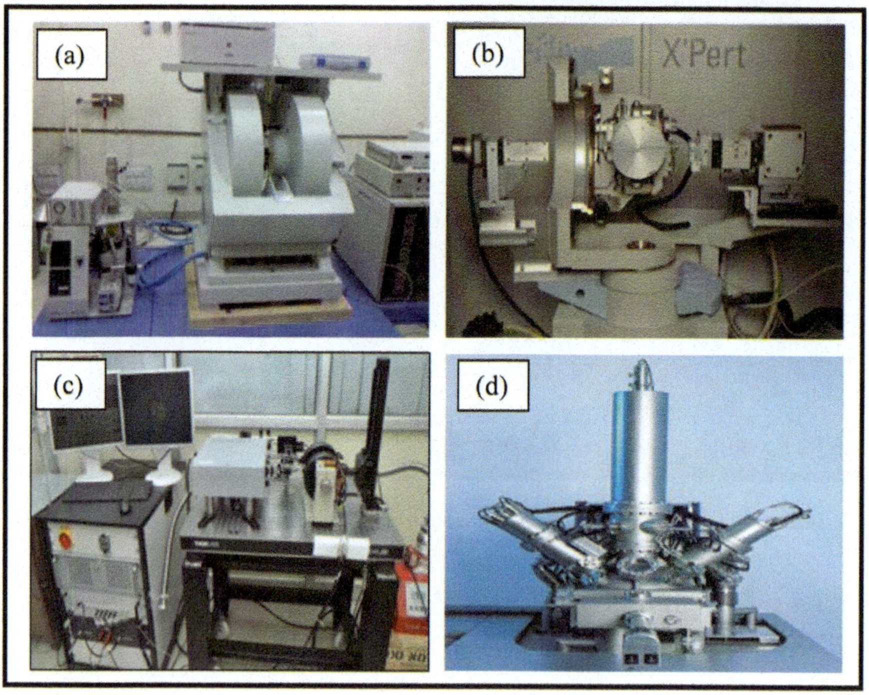

Fig. 18.23 Sophisticated instruments used for high-end characterization of materials **a** Electron paramagnetic resonance (EPR) spectrometer **b** PANalytical X'Pert PRO MRD systems **c** Magneto-Optical-Kerr Effect (NanoMOKE-III), **d** ION TOF's TOF-SIMSS

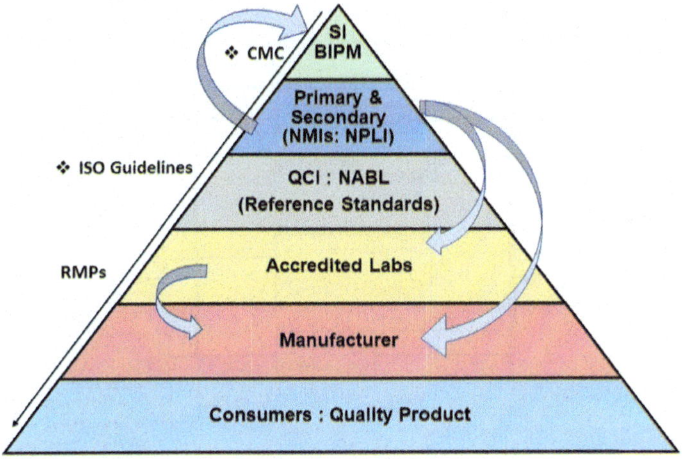

Fig. 18.24 Organizational traceability linkage to SI units

SI unit as all the national standards are related to the physical quantity (Fig. 18.24). The physical BNDs like hardness, viscosity, vacuum and pressure, luminescence, ultrasonic and acoustics, fluid flow, conductivity, length and dimension, magnetic, temperature, etc., are directly linked to the primary standard and then to the SI unit and the uncertainty are evaluated. This way these BNDs are disseminated to the stakeholders.

However, the BNDs of chemical parameters are quite complicated and difficult to establish metrological traceability as the certified values are not directly linked to the SI unit rather it underpinned by the use of CRMs or standard measurement method, etc. In case of matrix-based CRM, to determine the metrological traceability for each parameter is more complicated. According to ISO standard 17034 the metrological traceability of the measurement result may be demonstrated by weighing of the sample, purity of the pure/starting materials, measurement method, and calibration curve. Figure 18.25 reveals the metrological traceability of the BNDs being developed/produced in CSIR-NPL to the SI unit [46]. In the field of chemical parameters, such as inorganic materials including drinking water, high purity material, minerals & ores; petroleum product; building materials; hardness block; toxic elements & pesticide residue in food matrix; law enforcement related to testing in breath alcohol analyzer, etc., the dissemination of CRMs/BNDs underpin the measurement traceability in our country.

The inorganic BNDs developed/produced gravimetrically linking to mass SI unit by evaluating purity of the starting material and considering the primary method is given in Fig. 18.26. The measurement uncertainty increases with the increase in the traceability chain.

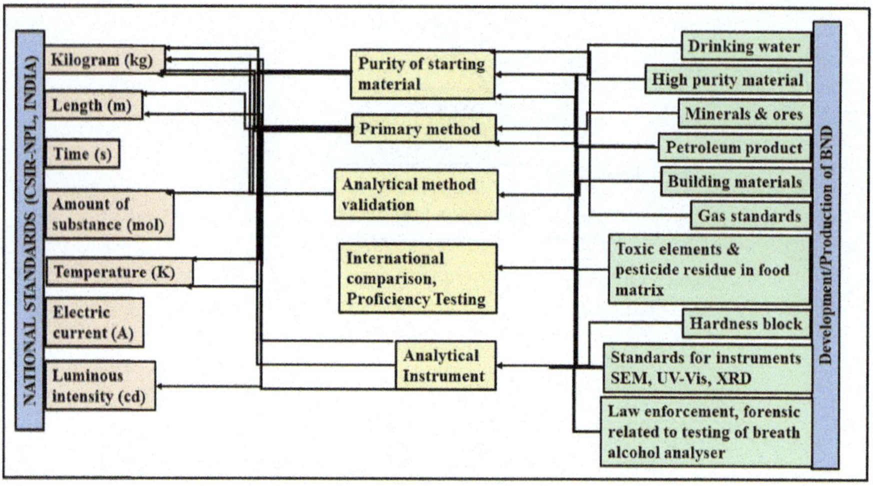

Fig. 18.25 Metrological traceability to the SI unit of BNDs

Fig. 18.26 Traceability chain for inorganic BNDs

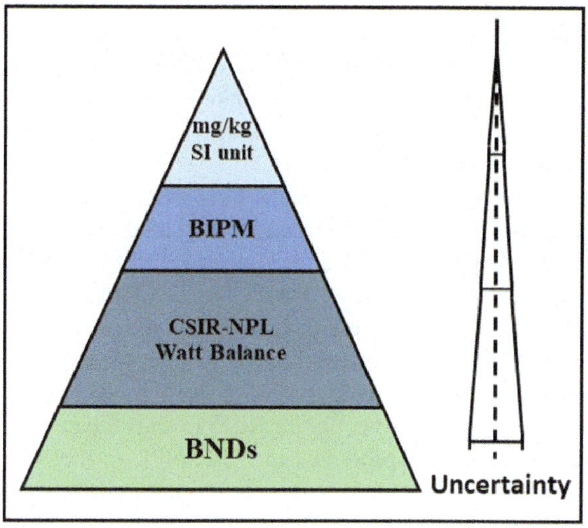

18.4 Use of BNDs in Metrological Perspective

18.4.1 BNDs for India's Need

Accurate and precise measurements are essential to drive the growth engines of Indian science & industry as it removes chaos and prompts innovations, which in turn, would save precious lives, resources and time. Developing India's measurement standards that are internationally accepted and disseminating the measurement capabilities to industry, government, strategic and academia that underpin the India's prosperity and quality of life. CSIR–NPL, owing to its untiring efforts in the establishment, development and maintenance of "Primary Standards" of SI units and its derivatives and their dissemination has occupied a pivotal position in the country.

The rapid advancement in science & technology and globalization of economies poses a strong need for more stringent metrological regulations in trade and commerce. In this endeavor, Legal Metrology Act, 2009 was enforced by Government of India (GOI) on April 1, 2011 throughout the country. One can understand the importance of quality control, failing which has to face the consequences of global trade because of partial compliance of unbroken chain of measurement standards with associated traceability aspects.

In-house BND production for industry and globalization of trade needs comparability of calibration, test outcomes and measurement through traceability to the SI units with some associated uncertainty values. Internal trade, accreditation agreements and markets also desire that trade should not have barriers. Enhancement of the quality of life as well as product quality is vastly assisted by reliable, more accurate and traceable measurements, mostly in extents such as health care, environmental and pollution control, food safety and nutritional content, security, climate change and forensics, etc. Table 18.5 shows use of BND in laboratories as primary measurement standards.

Strengthening the BND programme at CSIR-NPL will improve quality assurance of products and will certainly increase the exports and regulate the inferior imported products. Recently BIS has recommended to use BNDs for all the affiliated laboratories across India and this will strengthen the BND programme which was

Table 18.5 Use of BND in laboratories as a primary measurement standard

Use of BND	Requirement in testing/calibration/RMP laboratories
Material testing	Quality product
Purity evaluation	Establish traceability
Calibration of instruments	Measurement accuracy
Analytical method validation	Accuracy, precision, linearity, repeatability, reproducibility, LOD, LOQ, recovery
Validation property value	Verification of reference value
Quantification of measurand/analyte	Comparison with unknown sample
Interlaboratory comparison studies	Value assignment Comparability testing
Control chart preparation	Stability of instrument, others RM
Homogeneity, stability assessment	RM/CRM production requirement
Assignment of property value of RM/CRM	Establish traceability
Internal quality control	Recovery Data accuracy
Regulatory/Accreditation/Trade	Primary measurement standard

initiated by CSIR-NPL. Keeping this in view the government of India has funded a mega project for the production of BNDs. Similarly, to promote the production of BNDs the Ministry of Commerce and Industry has granted a project for bilateral cooperation for the discussion under the NAFTA (The North American Free Trade Agreement) scheme to visit NIST, USA for the discussion on the reference materials and contemporary reference materials.

18.4.2 BND for Quality Infrastructure

The economy is significantly reliant on exports and stringent control on imported consignment. In order to achieve this goal, the enhancement in quality products as well as establishment of metrological infrastructure is essentially required. The QI of a nation bounds the government, research institutes, NMI and society for harmonized development. Looking at all these aspects, a huge demand of BNDs is being considered for disseminating traceability to all the test laboratories and other organizations in India. There are large numbers of testing and calibration laboratories and academic institutions, small and medium scale industries, health sector, etc. All these organizations required to come under QI. In general, CRMs are desirable as a kind of metrological tool to govern the operation in terms of least requisite accuracy, target compliance and preferred level of confidence. Unrealistic assumptions in measurements have been quite often as a common practice which causes huge amount of error in measurement and hence for sufficiently accepted data of outcomes must be validated through metrologically valid procedures.

Therefore, the accurate and precise measurements through dissemination of BNDs that gets fulfilled through severe conformity assessment which helps in nation's growth through innovation and research in fields of science, sustainable green energy, healthcare, advancement of high-end technologies, pollution-free environment, strong cyber security, improved international trade and formulation and regulation policies that welfare the nation. The proper calibration of all instruments requires the CRMs/BNDs which are having key prominence in creating comparability and accuracy of analytical results.

BNDs can be used for materials testing, calibration of instruments, analytical method validation, validation of property value, traceability of measurement results, checking repeatability & reproducibility of analytical results, control chart preparation, inter-laboratory comparison studies, quality control and quality assurance (QA & QC) of the end product. All these are discussed below.

18.4.2.1 Materials Testing

The testing of a material is important in terms of regulatory compliance and to know the property and composition of the material before utilization in product development. The analysis of ionic and organic species in biological and environmental samples is of prime importance material for regulatory compliance for trade, to prevent pollution and to protect community health.

The testing of the material is being carried out using specified standard test methods based on ASTM, IS, European standard or by validated method, etc. The standard test method specifies the methodology as well as specific requirement. To test the unknown material, the use of reference material is important for accurate analysis, validation of method and also ensuring traceability. BND is also useful to verify test methods and to calibrate measurement systems. For the analysis of composition of unknown material two categories of reference material utilized based on pure substance or matrix match material. The matrix match reference materials have more advantages than pure substance as it identifies interference effects of matrix over analyte during measurement. The classical gravimetric and titrimetric is a lengthy and time-consuming process that is why demand of matrix match reference material is increasing for sophisticated instruments. The major testing of the material requires chemical structure, compositional analysis, microbiological species and several other physical parameters like melting point, viscosity, and optical density. The reference material support measurement for compositional, clinical, physical, engineering properties or some specific tests like taste and odour, octane number, flash point, hardness, etc.

18.4.2.2 Calibration of Instruments

For the testing of unknown materials, calibration of the equipment is important to identify the specific properties. The accuracy of the reference standard should be several times better than the accuracy of the measuring device to be tested. The calibration of equipment is carried out to check the accuracy of the instrument and traceability of the measurement. The calibration of equipment suggests whether the service or repair of the system is required as instruments degraded with time or keeping instrument unused for longer time. The functioning of the instrument is also affected by mechanical or electric shock and by hostile environment. For example, in Powder X-ray diffractometer, the line position (2θ) and intensity play a major role for getting the accurate data of any material by satisfying the Braggs condition $2d\sin\theta = n\,\lambda$, where θ is the angular position of the atom and d is the interatomic distance. The proper alignment and calibration for line positioning, X-ray intensity, lattice spacing and FWHM is achieved using BNDs. Similarly, all analytical equipment's such as SEM, HRTEM, FTIR, UV–Vis., etc. to be calibrated to obtain accurate data using BND/CRM. The routine calibration of the instrument improves accuracy of measuring system which further improves quality of the products. So, the instrument should be calibrated as per instruction given in manual or periodically as per experience or after shifting from one place to another place or after any electrical or mechanical shock or when observations seem to be questionable. There are several experiments that requires calibration certificate, so in such cases, it is always recommended to verify the calibration before commencement of analysis using BNDs/CRMs.

18.4.2.3 Analytical Method Validation

ISO/IEC 17025 specifies the requirement of validated procedures for testing. The method validation is a process which provides information about the suitability of analytical procedure for specific test and method. On the basis of results of method validation, the quality, reliability and consistency of analytical results can be decided. The method validation is being carried out by estimation of bias, i.e. quantification of difference between the measured value and the true value. Method validation can be carried out accurately by using CRM/BND. The estimation of bias should be within the limits of the uncertainty of the certified value. It is important that method should be validated or revalidated before utilizing in routine testing or on changing of conditions like different characteristics of instrument utilized or samples of different matrix. So, on changing condition, methods should be validated for accuracy, sensitivity, specificity, and reproducibility of test methods.

18.4.2.4 Validation of Property Value

The property value of certified reference material accompanied by a certificate, one or more of whose property values are certified by a procedure which establishes its traceability to an accurate realization of the unit in which the property values are expressed, and for which each certified value is accompanied by an uncertainty at a stated level of confidence. The determination of property value is carried out by a procedure traceable to SI unit. For validation of property, the level of that property (concentration), matrix, structure, form (powder, cylinder, block, liquid), stability (shelf-life) and uncertainty should be taken into account. The laboratory generates results in the form of value and uncertainty. The uncertainty denotes reliability of the result. The reliability just cannot be improved by repeating the results. It requires precise and accurate determination by using validated methods. It requires proper measuring device, calibration of system with certified reference material, correct measurement equation, traceability of the measurements and evaluation of uncertainty. The precise and accurate property value can be achieved by method validation, procedure validation standard operating procedure (SOP) validation.

18.4.2.5 Traceability of Measurement Results

In recent years, metrological traceability in measurement concept has gain more attention worldwide. In general terms, it is considered that the use of a CRM or RM for quality control purposes automatically establishes metrological traceability. It is also considered that satisfactory participation in an inter-laboratory comparison, proficiency testing scheme, or external quality assessment scheme automatically provides metrological traceability of the participant's measurement results. The metrologically traceable property value of a measurement result can be related to a reference standard through a documented unbroken chain of calibrations, each contributing to the measurement uncertainty.

The ILAC considers the elements for confirming metrological traceability to be an unbroken metrological traceability chain to an international measurement standard or a national measurement standard, a documented measurement uncertainty, a documented measurement procedure, accredited technical competence, metrological traceability to the SI, and calibration intervals. VIM 2009 also defines traceability as "Property of the measurement result whereby it can be related to a reference through an unbroken chain of calibrations, each contributing to the measurement uncertainty". In a simpler way, three different kinds of traceability can be envisaged (i) traceability to the international system of units. It is universally valid and the values are independent of any validated method and artefact, (ii) traceability to a method (only valid when a specific measurement protocol is strictly followed) and (iii) traceability to an artefact.

18.4.2.6 Repeatability and Reproducibility of Analytical Results

Repeatability and reproducibility are the terms used for measuring precision. The repeatability measures the variation in measurements whereas reproducibility denotes that entire study or experiment can be reproduced. For repeatability carried out by the same person with the same instrument under the same environmental conditions and should be repeatable within a period of time. The repeatability is related to mean of standard deviation. The satisfactory level of repeatability can be determined by repeatability coefficient (r). The repeatability coefficient determines the difference between the repeated test results. It is suggested that if $r < 0.2$ it considered as slight repeatability, if between 0.2 and 0.4 low repeatability, between 0.4 and 0.7 moderate repeatability, if r is between 0.7 and 0.9 high repeatability, if $r > 0.9$ considered as very high repeatability. On the other hand, reproducibility refers to the degree of agreement between the results of experiments done by different person at different places, times and using different instruments. Reproducibility measures the ability to replicate the findings of different persons. The repeatability, reproducibility and quantification of error in a measurement arise from a single or multiple persons that can be done by analysis of variance using one-way ANOVA (analysis of variance) based model using standard deviation across all measurements. But this model does not suggest quantification of sources of error and capacity that means precision to tolerance ratio. However, it can be better explained by using two-way ANOVA-based model.

18.4.2.7 Control Chart Preparation

The control chart is a graphical representation of the measurement studies which suggest changes in process over time. The data recorded is plotted in time order. The central line denotes average, upper line suggest upper control limit whereas lower line suggests lower control limit. The continuous recording and plotting of data suggest stability of performance of the instrument, including calibration, analytical procedure, analyst and influence of environmental conditions. The control charts that are most commonly used are of two types: the Shewhart control chart and Cumulative Sum (CUSUM) control chart. They are graphically presented in Figs. 18.27 and 18.28.

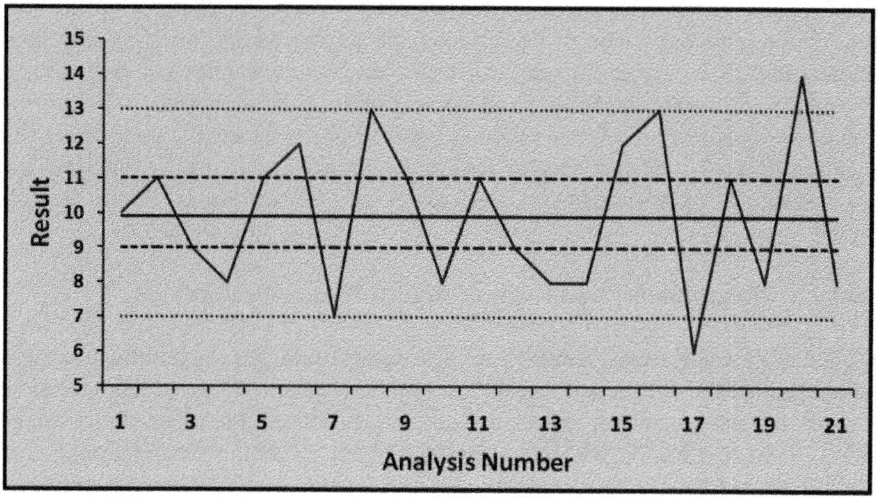

Fig. 18.27 Shewhart control chart (target value is 10; upper and lower warning limits are set at 11 and 9, respectively; action limits are set at 12 and 8)

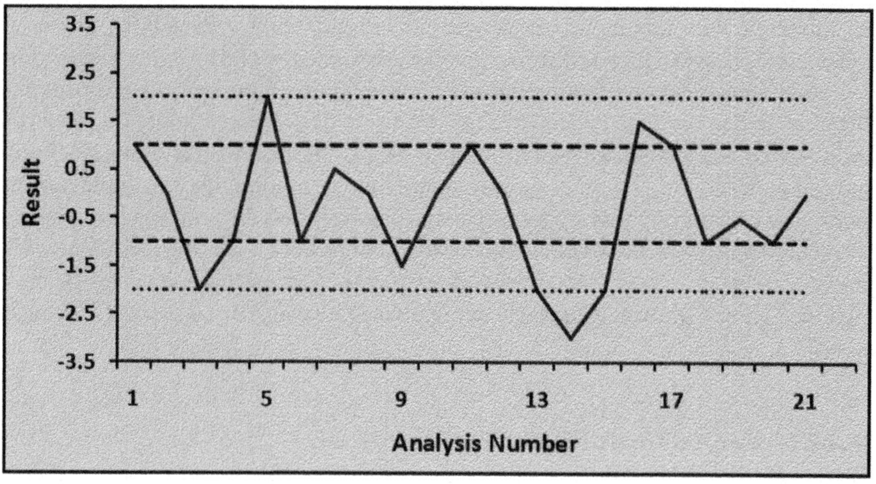

Fig. 18.28 Cumulative Sum (CUSUM) control chart on the same set of data as for the Shewhart Control

18.4.2.8 Inter-Laboratory Comparison Studies

The successful participation in inter-laboratory comparison or proficiency testing (PT) is important for accreditation of laboratory as per ISO 17025. In an interlaboratory studies PT or round-robin test of the samples are carried out by different laboratories and subsequently results are compared. Such studies are carried out for method validation, method standardization or characterization of candidate reference materials. The homogeneity of the materials used for such studies is crucial as variations of results should only reflect reproducibility of the method and not the suitability of the material as reference [47]. Each inter-laboratory study includes the distribution of various units of samples to a group of participants. The basic objectives of these studies are to validate the test method, to determine the characteristics of the product and to check the competency of participating laboratories. The standard requirement of PT/inter laboratory comparisons/round-robin test depends on the objectives of the studies. ISO 5725 defines trueness and precision. The trueness refers to accuracy of the measurement method, and also closeness of agreement between the arithmetic mean of a large number of test results and the true or accepted reference value, whereas precision denotes closeness of agreement between test results. ISO standard 17034 for the production of reference material requires assessment of homogeneity, stability and characterization and value assignment along with evaluation of uncertainty, establishment of the metrological traceability. In the same way ISO 13528:2015 specifies statistical methods for PT providers to design PT schemes and analysis of data. This standard can be applied to demonstrate the measurement results obtained by the laboratories, inspection bodies and individuals meet specified criteria for acceptable performance. ISO/IEC 17043:2010 specifies general requirements for the competence of PT providers about the development and operation of proficiency testing schemes.

18.4.2.9 Quality Control and Quality Assurance (QC & QA) of End Product

QA and QC are the criteria for assurance of quality of product based on set of processes and ensuring product and services fulfil the requirement of customers. Quality assurance is the part of quality management system of any organization which assures to internal management and outside customers about quality of a service or product. It is based on two principles, first is based on "Fit for propose" and "right first time". QA is a way of to minimize the mistakes and defects in manufactured products and service to customer. QA comprises administrative, quality of raw materials, assemblies, components, systematic measurement, comparison with known standard, procedural activities and also monitoring of defined processes [48].

Quality control is also the part of quality management system of any organization which comprises how a process is performed and product developed. It is a documented operational procedure and activities used to fulfil requirement of quality products which should be suitable for the intended purpose and free of defects. QC is a set of procedure which ensures that products manufactured fulfil the requirement of defined criteria as per defined standards and meets the requirement of customers [49]. Physical inspection for product defects like cracks or surface blemishes is one of the major components of quality control before final certification/marketing.

18.5 Conclusion

It is summarized that the NMIs has to be strengthened themselves by improving their measurement capabilities in various parameters. This can be achieved by international collaboration among RMOs and BIPM through CIPM-MRA. The improved measurement capability has to be disseminated to accredited bodies/stakeholders through BNDs for the overall growth of quality system. Similarly, this should reach to the regulatory bodies to examine the quality of import/export products. Networking between the NMIs through bilateral cooperation, participation with international bodies/organization establishes wide recognition. The networking among international organizations accelerates global harmonization in testing/calibration and measurement standards. Further, to overcome the technical barrier in international trade, standardization of measurement with high metrological aspect has to be envisaged. Due to increasing competitiveness in the global trade, reference materials are required for conformity assessment which ultimately promotes international trade. Thus, BNDs produced in India not only fulfil the demand of market but also provide import substitutions.

Acknowledgements The authors are thankful to Director, CSIR-NPL for strong initiative and providing an opportunity to contribute in a book chapter. We would also like to thank the NPL standard division, regulatory bodies (BIS, QCI, NABL), RMPs and sister CSIR labs for their valuable support. All the staff members associated with BND programme: Dr Santosh Singh Golia, Dr Vipin Jain, Dr N Karar, Dr. Asit Patra, Dr E. Selvi, Dr Pallavi Kushwaha, Dr Anuj Krishna, Mr. Dinesh Singh, Dr Jai Tawale, Mr. Praveen Tanwar, Mr. Himanshu Yadav, Ms. Jyoti Pokhriyal, Ms. Dibya Topo, Ms. Geetanjali, Mr. Himanshu Yadav, Mr. Sandeep Singh, Mr. Sumit Mohan are highly appreciated. Further, Ministry of Commerce & Industry under TIES scheme and CSIR (S&T Ministry) for sanctioning a major project for the establishment infrastructure/facilities are gratefully acknowledged.

References

1. R.H. Albert, Metrology in chemistry. Chem Eng News **77**, 2–2 (1999)
2. G.H. White, Metrological traceability in clinical biochemistry. Ann. Clin. Biochem. **48**, 393–409 (2011)
3. S. Trapmann, A. Botha, T.P. Linsinger, S. Mac Curtain, H.J.A. Emons, Q. Assurance, The new International Standard ISO 17034: general requirements for the competence of reference material producers, **22**, 381–387 (2017)

4. A.M.H. van der Veen, Revision of ISO Guide 33: good practice in using reference materials. Accredit Qual Assur **20**, 529–532 (2015)
5. V. Jimenez, R. Companyo, J. Guiteras, Preparation of quality control materials for the determination of sulfonamides in animal feed. Food Addit Contam A **26**, 969–977 (2009)
6. A. Botha, J.P. Hammond, S.J.A. Sauvage, Q. Assurance, Report on the 42nd meeting of ISO/REMCO, **25**, 83–85 (2020)
7. M. Sinojmeri, Final technical report for key comparison CCAUV.A-K2. Metrologia **46**, 09006 (2009)
8. D.K. Singh, A. Sahu, S. Kumar, S. Singh, Critical review on establishment and availability of impurity and degradation product reference standards, challenges faced by the users, recent developments, and trends. Trac-Trend Anal Chem **101**, 85–107 (2018)
9. B. Halamoda-Kenzaoui, U. Holzwarth, G. Roebben, A. Bogni, S. Bremer-Hoffmann, Mapping of the available standards against the regulatory needs for nanomedicines, Wires Nanomed Nanobi, **11**, (2019)
10. C. Thomas, F.S. Bureau International des Poids et Mesures %J Bureau International des Poids et Mesures, The BIPM key comparison database (KCDB): linkage of key comparison results, DOI (2005).
11. T. Steiger, R.J.A. Pradel, Q. Assurance, COMAR: the international database for certified reference materials—an overview, **20**, 47–52 (2015)
12. D.K. Aswal, Quality infrastructure of India and its importance for inclusive national growth. Mapan-J Metrol Soc I **35**, 139–150 (2020)
13. E. Bulska, P. Taylor, Do we need education in metrology in chemistry. Anal Bioanal Chem **377**, 588–589 (2003)
14. P. Armishaw, B. King, R.G. Millar, Achieving traceability in chemical measurement—a metrological approach to proficiency testing. Accredit Qual Assur **8**, 184–190 (2003)
15. V. Barwick, S. Wood, Achieving metrological traceability in chemical and bioanalytical measurement. J Anal Atom Spectrom **25**, 785–799 (2010)
16. P. De Bievre, The 2007 International vocabulary of metrology (VIM), JCGM 200:2008 [ISO/IEC Guide 99]: meeting the need for intercontinentally understood concepts and their associated intercontinentally agreed terms. Clin Biochem **42**, 246–248 (2009)
17. I.G. Tanase, D.E. Anghel, M. Buleandra, Traceability in chemical measurement. Definitions. Reference documents. Elements and principles (for crediting labs). Rev Chim-Bucharest **57**, 4–7 (2006)
18. M. Valcarcel, A. Rios, Traceability in chemical measurements for the end users. Trac-Trend Anal Chem **18**, 570–576 (1999)
19. A. Williams, Traceability and uncertainty—A comparison of their application in chemical and physical measurement (vol 6, pg 73, 2001). Accredit Qual Assur **6**, 283–285 (2001)
20. A. Williams, Traceability and uncertainty—A comparison of their application in chemical and physical measurement. Accredit Qual Assur **6**, 73–75 (2001)
21. P. De Bievre, R. Dybkaer, A. Fajgelj, D.B. Hibbert, Metrological traceability of measurement results in chemistry: Concepts and implementation (IUPAC Technical Report). Pure Appl Chem **83**, 1873–1935 (2011)
22. P. Debievre, R. Kaarls, S.D. Rasberry, W.P. Reed, Metrology and the role of reference materials in validation and calibration for traceability of chemical measurements, Impact of Metrology on Global Trade—1995 Workshop and Symposium, Proceedings, DOI 479-494 (1995)
23. K. Gates, N. Chang, I. Dilek, H.H. Jian, S. Pogue, U. Sreenivasan, The Uncertainty of reference standards-a guide to understanding factors impacting uncertainty uncertainty calculations, and vendor certifications. J Anal Toxicol **33**, 532–539 (2009)
24. T. Saito, A. Botha, Guidance on the contents of accompanying documentation for reference materials (RMs). Accredit Qual Assur **21**, 239–241 (2016)
25. R. Krska, E. Welzig, Certified reference materials and proficiency testing. Anal Bioanal Chem **376**, 316–318 (2003)
26. S.D. Rasberry, Certified reference materials in analytical chemistry - A century of NIST contribution. Accredit Qual Assur **6**, 95–99 (2001)

27. M. Stuart, A. Squirrell, L. Besley, Reference material requirements for laboratories and the role of accreditation bodies. Accredit Qual Assur **9**, 209–215 (2004)
28. M.L. Tang, B.L. Fan, L.Y. Yao, Z.Y. Bao, H. Tian, Z.Z. Ma, Z.Y. An, M. Ma, H.Y. Zhang, Z.B. Fu, Preparation and certification of reference materials (GBW07397, GBW07398, GBW07399 and GBW07400) for Selenium and other trace element mass fractions. Geostand Geoanal Res **44**, 375–384 (2020)
29. M. Thompson, A properly developed consensus from a proficiency test is, for all practical purposes interchangeable with a certified value for a matrix reference material derived from an interlaboratory comparison. Geostand Geoanal Res **42**, 91–96 (2018)
30. A.V. Yashin, General problems of metrology and measurement technique—questions of processing the results of measurements in secondary standards. Meas Tech **49**, 536–539 (2006)
31. A. Botha, J. Hammond, H. Emons, S. Sauvage, Report on the 40th meeting of ISO/REMCO. Accredit Qual Assur **23**, 193–195 (2018)
32. D. Johnson, M. Rettinger, T. Tarbox, B. Marek, S. Pogue, I. Dilek, U. Sreenivasan, Development challenges in the preparation of solution-based phytochemical and vitamin certified reference materials. Anal Bioanal Chem **405**, 4345–4352 (2013)
33. B. Calderon-Jimenez, M.E. Johnson, A.R.M. Bustos, K.E. Murphy, M.R. Winchester, J.R.V. Baudrit, Silver nanoparticles: technological advances, societal impacts, and metrological challenges, Front Chem, **5**, (2017)
34. R. Buccianti, M. Cibien, L. Mari, B.I. Rebaglia, Measurement and calibration: considerations based on the international vocabulary of metrology (Vim, 3(Rd) Ed.) and related standards, Xix Imeko World Congress: Fundamental and Applied Metrology, Proceedings, DOI, 1061–1064 (2009).
35. R. Buccianti, M. Cibien, L. Mari, B.I. Rebaglia, accuracy, trueness, and precision: considerations based on the international vocabulary of metrology (Vim, 3(Rd) Ed.) and related standards, Xix Imeko World Congress: Fundamental and Applied Metrology, Proceedings, DOI, 1065–1069 (2009).
36. G. Ferard, F. Pontet, C. Giroud, M. Dumontet, Introducing the new international vocabulary of metrology. Ann Biol Clin-Paris **67**, 591–594 (2009)
37. L. Mari, Evolution of 30 years of the international vocabulary of metrology (VIM), Metrologia, **52**, (2015)
38. P.J. Potts, Glossary of analytical and metrological terms from the international vocabulary of metrology (2008). Geostand Geoanal Res **36**, 231–246 (2012)
39. W. Hasselbarth, Traceability of measurement and calibration in chemical-analysis. Fresen J Anal Chem **352**, 400–402 (1995)
40. G. Beges, J. Drnovsek, I. Pusnik, J. Bojkovski, Measurement uncertainty, traceability and evaluation of test results in testing laboratories. Meas Sci Technol **13**, 565–572 (2002)
41. M. Freemantle, Metrology in chemistry. Chem Eng News **77**, 29–32 (1999)
42. A. Drolc, M. Cotman, M. Ros, N. Majcen, Measurement traceability and its role in proficiency testing schemes - a case study for wastewater analysis in Slovenia. Accredit Qual Assur **11**, 455–461 (2006)
43. S.L.R. Ellison, A. Williams, Disseminating traceability in chemical measurement: Principles of a new EURACHEM/CITAC guide. Accredit Qual Assur **8**, 483–485 (2003)
44. P. De Bievre, Traceability and comparability of results in food measurements: a CCQM workshop at BIPM on "comparability and traceability in food analysis" . Accred. Qual. Assur. **9**, 521–522 (2004)
45. S. Fakhry, The capability of the National Institute of Standards in the electromagnetic compatibility measurements, 2008 Conference on precision electromagnetic measurements digest, DOI https://doi.org/10.1109/Cpem.2008.4574840 436–437 (2008)
46. O.N. Velichko, Traceability of measurement results at different levels of metrological work. Meas Tech **52**, 1242–1248 (2009)
47. S.L. Ellison, A.J.A. Botha, Q. Assurance, Principles for the assessment of homogeneity and stability in the new ISO Guide 35:2017, **23,** 47–51 (2018)

48. W.J.J.E.m. Shampine, assessment, Quality assurance and quality control in monitoring programs, **26**, 143–151 (1993)
49. C. Langheinrich, M.J.B. Kaltschmitt, Bioenergy, implementation and application of quality assurance systems, **30**, 915-922 (2006)

Chapter 19
Bharatiya Nirdeshak Dravyas (BND®): Indian Certified Reference Materials

Part II: Production and Future Perspectives

R. P. Pant, S. Swarupa Tripathy, D. K. Misra, Vidya Nand Singh, Arvind Gautam, N. Vijayan, G. A. Basheed, K. K. Maurya, S. P. Singh, and Nahar Singh

Abstract Indian Reference Materials BND® represents an inevitable tool for analytical quality and plays an important part in maintaining quality measurements. The reliability of analytical procedures can only be verified with the use of CRMs/BNDs that control and evaluate the quality of analytical results. Rapid industrialization demands reference materials in various sectors like the environment, food & agriculture, petroleum products, physico-chemical materials, etc. To meet the demand for CRM in the market, various countries are producing them with the help of RMPs in their respective countries. In this endeavour, CSIR-NPL is producing BNDs on its own and also in the association of RMPs across the country as per ISO 17034. Usually, these CRMs are sold out worldwide at a high cost and the validity of reference standard is for a limited period and requires to procure again and again which are very tedious and also cause huge recurring loss of national resources. In this aspect, BNDs produced at CSIR-NPL acts as import substitution and makes it economically viable for many academic institutions and industries. Keeping these facts in view, this chapter describes the process for the production of different BNDs, sample handling and storage, homogeneity and stability, uncertainty, assignment of property value, certification, etc. In addition to this, detailed information about certain in-house produced BNDs and in association with RMPs are provided. Few future contemporary on-going BNDs activity such as for thermoelectric metrology, spin current, tensile strength and dynamic viscosity, Newtonian fluids (silicon oils) are also described in this chapter. At the end of this chapter, to support medium, small and large-scale industries future prospects of BNDs are also highlighted. Further, it has been elaborated on how CSIR-NPL is collaborating with accreditation and regulatory bodies to strengthen the quality infrastructure in India.

R. P. Pant (✉) · S. S. Tripathy · D. K. Misra · V. N. Singh · A. Gautam · N. Vijayan ·
G. A. Basheed · K. K. Maurya · S. P. Singh · N. Singh
CSIR-National Physical Laboratory, Dr. K. S. Krishnan Marg, New Delhi 110012, India
e-mail: rppant@nplindia.org

D. K. Aswal (ed.), *Metrology for Inclusive Growth of India*,
https://doi.org/10.1007/978-981-15-8872-3_19

"Whatever the measurement system is, it needs to be consistent, repeatable, and as unbiased as possible."

–Pearl Zhu

19.1 Introduction

Scientific research, technology, innovation and competitiveness require precise, reliable, traceable and accurate measurement results. NMIs in the world are contributing enormously to the developments of the method of analysis, procedures, validation and support to accreditation bodies by providing traceability. In India, CSIR-NPL is also focusing on the contemporary "Metrology" program on the development of BNDs for testing and calibration of instruments. This improves the quality assurance of products which in turn contributes towards the growth of the country's economy. Harmonization among accreditation bodies, regulatory bodies and legal metrology with NMI is an essential part for the quality of life. Therefore, coherence among these bodies with NMIs has to be envisaged for the large-scale production of BNDs. This is the high time when NMIs have to gain expertise for developing varieties of CRMs/BNDs based on the demand of the market. In this regard, CSIR-NPL has taken lead by signing MoUs with various RMPs to fulfil the requirement of the country as well as export. The increased exports will lead to huge employment generation and will play a major role in the country's GDP.

19.2 Process for Production/Development of BNDs/IRM

The production of BND strives to seek support from various Reference Material Producers (RMPs) to cater to the need of the quality products in the market (Fig. 19.1). The demand for BNDs is based on the market-driven and consumer satisfaction. The BND preparation protocol, methods of analysis/characterization of parameters of BND, assignment of certified values associated with uncertainty in measurement, uncertainty budget and statistical methods used for the analysis of data and results are carried out as per the ISO 17034 and ISO 17025.

CSIR-NPL, being NMI of India, is a signatory to CIPM MRA of BIPM and accordingly its calibration, testing and BND activities are assessed for metrological conformity through an international system of peer review by external experts of RMOs as per ISO/IEC 17025: 2017 and/or ISO 17034: 2016 standard(s) (as applicable to respective activity).

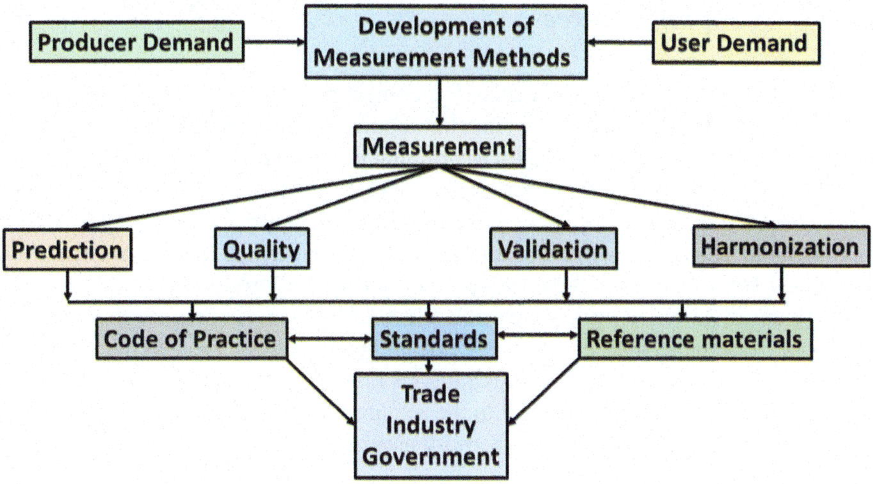

Fig. 19.1 Development of measurement methods as per the demand from producer and user helping in trade, industries and government

19.2.1 Planning for the BND

Planning of BNDs is a prerequisite step to develop BNDs for the intended use. It includes starting from the materials' sources with authentic certificates/verifications, processing, sampling, etc. to finally assigning the properties values as depicted in the schematic given in Fig. 19.2 [1].

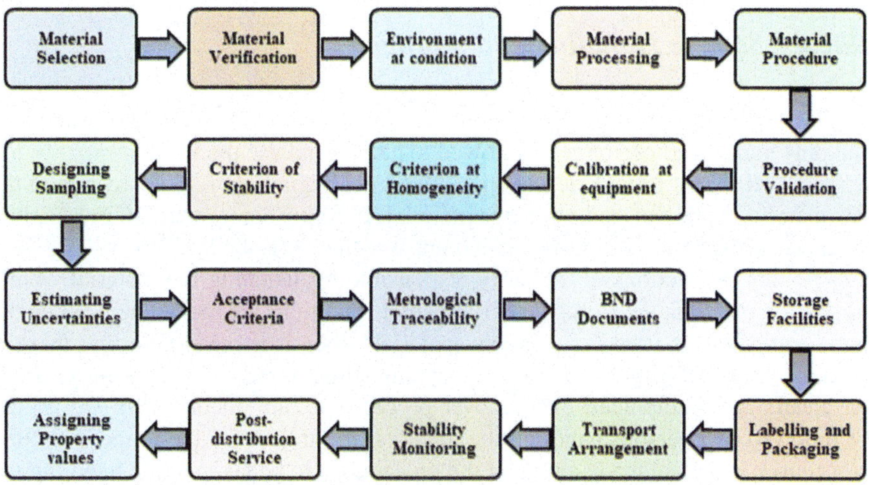

Fig. 19.2 Steps involved in the planning for the BNDs

19.2.2 Sampling Source

Sampling sources are the basis of the source materials of the highest purity with a certificate which can be obtained from reliable sources. If the process or the material is of new variety, proficiency testing for a particular parameter should also be conducted. For example, for making the micron-sized particles, the sands are collected from some river bed or desert and for cement-related BND, the sample can be taken from very reliable producers with certification of source materials. In case of metal, it is purchased from a vendor where the material purity and processes are documented with the certification of source materials [2]. CRM producers procure the materials from various reliable sources which are further purified by the process of crystallization/filtration to remove undesired extraneous matters. The purified material is then further processed to obtain material in the required form for further characterization for purity and homogeneity for conformity assessment.

19.2.3 Sample Processing

The documented procedures to ensure that the material has undergone adequate processing for its intended uses are described in this section. The procedures for material processing addresses the qualitative and/or quantitative analysis for verification of material type and/or identity; synthesis, purification (e.g. distillation, extraction), incubation, and transformation into the final form (e.g. machining, grinding, blending, sieving and riffling, extrusion, melting, etc.) is shown in Fig. 19.3 [3].

19.2.4 Sample Handling and Storage

BNDs developed/produced are handled very carefully by following ISO guidelines and only authorized personnel is allowed to handle and use the BNDs.

The RMPs must follow the specified guidelines beginning from the receipt of the raw material to the finished reference material. If during some stages of production, the material has to go out of the direct control of the RMP, the RMP provides necessary written instructions to the party responsible for handling the material. While storing the material, the storage environmental conditions are maintained. When the same equipment is used for different materials, then it is made sure that there is no cross-contamination or carry-over contamination takes place. Work instructions for cleaning of equipment, change over process, etc. are documented and should be followed. All persons handling the materials are trained on the proper handling procedures. They are made aware of the precautions to be taken while handling the material, as required by clause 6.1.3 of ISO 17034: 2016. It is the responsibility of the RMP to ensure that the packaging and labelling of the reference materials meet

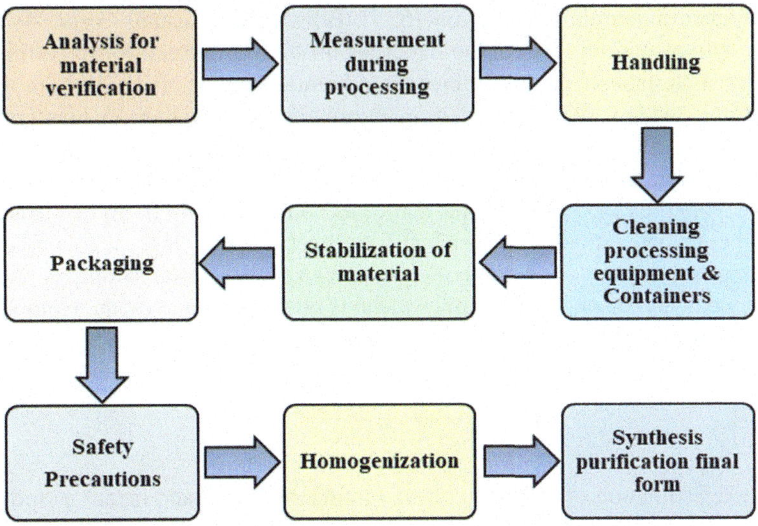

Fig. 19.3 Sample processing steps for BND production

the safety and transport-related regulatory requirements. Reference material must be stored separately from the test materials and other materials in such a way that any adverse effects on their quality/integrity as well as misuse and loss are excluded. If particular storage conditions are specified (e.g. cooling), compliance shall be monitored and documented where applicable, safety measures for occupational health and environmental protection are taken according to the relevant dangerous properties (toxic, flammable, explosive, radioactive, etc.). Access to rooms and facilities where RM is stored as well as withdrawal of RM is regulated and documented [4].

19.2.5 Homogeneity and Stability

The assessment of the homogeneity of candidate BND in its final packaged form is carried out to ensure its fitness for purpose. Whenever BND is developed/produced in multiple batches, the equivalence of the batches is demonstrated along with the homogeneity of each batch evaluated separately, wherever possible. Standard/validated measurement procedures are followed to achieve the precision and selectivity that are fit for the purpose required for the intended measurand. The homogeneity for every property of interest is determined experimentally unless it can be shown, using scientific evidence or previous experience, that particular groups of properties are sufficiently closely associated that measurement of one property in such a group furnishes evidence of homogeneity for other properties in the same group [5].

In order to demonstrate the validity of the certified values and their uncertainties in the analysis of individual units, the homogeneity analysis is of the prime importance for reference materials certification [6, 7]. Homogeneity usually occurs within a few exceptions and inherent in the materials itself. Therefore, prior to certification, reference materials are usually subjected to homogeneity testing. The homogeneity test is employed to verify a random number of units provided they are identical while performing testing of between samples homogeneity. The homogeneity test can be performed by employing ANOVA [8], which has been discussed here briefly.

Here from the Fig. 19.4, samples named as bottle A, bottle B, bottle C and bottle D (for better understanding of reader) have been chosen, and each sample has n number of subsamples. The mean of properties values of each sample is known as sample mean and mean of all samples mean is known as grand mean. Homogeneity test employing ANOVA adopts the following steps:

Step 1. Estimation of the mean of each sample for a set of k samples, i.e.,

$$\overline{X}_1, \overline{X}_2, \overline{X}_3, \ldots, \overline{X}_k$$

Step 2. Estimation of the mean of the sample means (Grand mean) as follows:

$$\overline{\overline{X}} = \frac{\overline{X}_1 + \overline{X}_2 + \overline{X}_3 + \cdots + \overline{X}_k}{\text{No. of samples } k}$$

Step 3. Determination of sum of squares between the samples (SS between):

The sum of squared deviations of the sample means from the grand mean when multiplied by the number of items in the corresponding sample results the sum of squares between the samples (SS_{between}) as

$$\text{SS between} = n_1(\overline{X}_1 - \overline{\overline{X}})^2 + n_2(\overline{X}_2 - \overline{\overline{X}})^2 + \cdots + n_k(\overline{X}_k - \overline{\overline{X}})^2$$

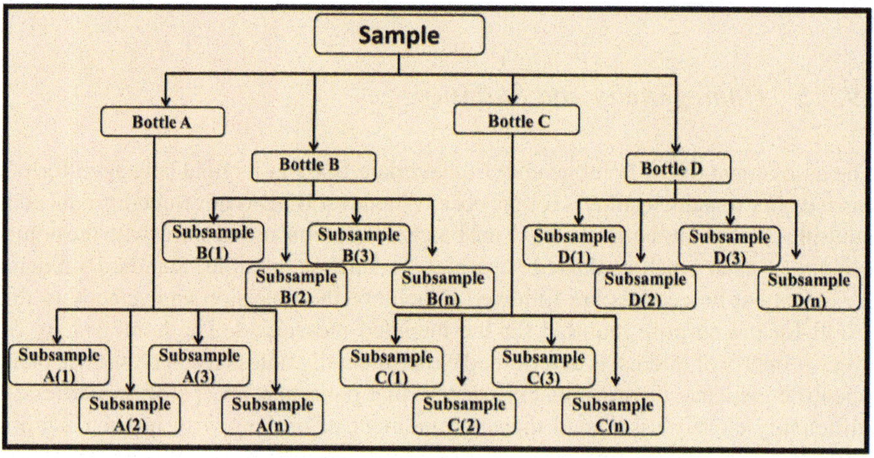

Fig. 19.4 Procedure for homogeneity test

where the deviation of each sample mean from grand mean.

$$n_1(\overline{X}_1 - \overline{\overline{X}})^2, n_2(\overline{X}_2 - \overline{\overline{X}})^2$$

Step 4. Estimation of the mean of the square between the samples: The squared deviation obtained between the samples can be obtained by dividing SS_{between} by the DOF between the samples. Symbolically, it can be written as

$$\text{MS between} = \frac{\text{SS between}}{(k-1)}$$

where k specifies the number of samples and $(k-1)$ is DOF between samples.

Step 5. Determination of sum of squares within the samples: It expresses the variability occurring in measurement of same sample, as each student measures the different volume of same sample. It can be obtained as

$$\text{SS within} = \sum (X_{1i} - \bar{X}_1)^2 + \sum (X_{2i} - \bar{X}_2)^2 + \cdots + \sum (X_{ki} - \bar{X}_k)^2 \quad i = 1, 2, 3, \ldots$$

where $\sum (X_{1i} - \overline{X}_1)^2, \sum (X_{2i} - \overline{X}_2)^2$ are the deviation of each sample value from same sample mean.

Step 6. Estimation of the mean of the square within the samples: This is obtained by standardizing the deviation obtained within the samples which are obtained by dividing SS_{within} by the degrees of freedom within the samples. Symbolically, it can be written as

$$\text{MS within} = \frac{\text{SS within}}{(n-k)}$$

where $(n-k)$ is DOF within samples,

n = total number of items in all the sample

k = number of samples.

Step 7. Determination of the total variance: It is obtained by adding the squares of deviations of individual items and can be expressed as

$$\text{SS for total variance} = \sum (X_{ij} - \overline{\overline{X}})^2 \quad i = 1, 2, 3, \ldots \quad j = 1, 2, 3, \ldots$$

This total variance should be equal to the sum of the result of step 3, and step 5 explained above and thus one can obtain

$$\text{SS for total variance} = \text{SS between} + \text{SS within}.$$

Here the degrees of freedom $(n-1)$ for total variance will be equal to the sum of degrees of freedom for between $(k-1)$ the samples and degrees of freedom within the sample $(n-k)$

$$(n-1) = (k-1) + (n-k)$$

Step 8. Finally, F-ratio can be estimated as

$$F\text{-ratio} = \frac{\text{MS between}}{\text{MS within}}$$

The F-ratio helps identify whether there is an epochal difference between sample means or is just fluctuations due to some human or machine error. To identify it, one can look at F-table [9], and accordingly, depending upon the value of F test, the relevant hypothesis is checked to infer the accurate results.

The variance occurring between the bottles (bottle A, bottle B, bottle C & bottle D and within the bottles (subsample A (1), subsample A (2), subsample A (3) can be described as above. The level of homogeneity can be assessed in terms of uncertainty range within the given [10]. Hence one can find between the bottle uncertainties as formulated below;

$$u_{\text{bb}} = (\text{MS}_{\text{within}}/n)^{1/2}.\left(\frac{2}{\vartheta}\right)^{1/4}$$

ϑ being the DOF of mean square of within samples and n number of replicates performed on each sample. One must not confuse among experimental between-bottle uncertainty and between the bottle uncertainty as experimental uncertainty is a combination of measurement uncertainty (which is evaluated by uncertainty occurring in measurement such as human limits, least count of the machine, etc.) and [11] and can be shown as:

$$u_{\text{exp}}^2 = u_{\text{meas}}^2 + u_{\text{bb}}^2$$

On the other hand, when the between samples variability u_{bb} is significant in the total uncertainty then in such cases, the batch may either be rejected or regenerated, or each sample must be individually certified. In case of within the bottle, uncertainty can be calculated

$$u_{\text{within}} = (\text{MS}_{\text{within}})^{1/2}.\left(\frac{2}{\vartheta}\right)^{1/4}$$

ϑ being the DOF of mean square of within samples. It is worth mentioning that within the bottle homogeneity is also certification for repeatability of the material and one can say if material has repeatability if it is found to be homogeneous with the bottle. Again, the experimental within-unit uncertainty (u_{exp}) is commonly its overestimated value, which is a combination of analytical measurement (u_{meas}) and the material inhomogeneity (u_{within})

$$u_{\text{exp}}^2 = u_{\text{meas}}^2 + u_{\text{within}}^2$$

Stability is among the essential properties of a CRM, and stability assessment is accordingly required for certified reference materials. Stability under long-term storage and also under conditions of transport is both expected to be assessed.

Other reference materials will usually undergo experimental tests of stability at some point prior to the material being distributed for sale. Where reference materials are certified for more than one property, stability is expected to be demonstrated for every certified property. There are two important strategies for CRM stability testing; simple real-time studies and accelerated testing. Real-time studies simply keep units of the material at their planned storage temperature for a suitable period of time and observe the material at intervals. Accelerated studies use a range of more stringent conditions, most commonly increased temperature, to test whether the material is likely to be stable over longer time scales.

Real-time stability studies simply hold a set of RM units at proposed storage temperature and test a proportion of them at regular intervals. The results are usually assessed by inspection and by linear regression to determine whether there is a significant change in measured value over time.

Accelerated stability studies have been in use since mid-1950s. Typically, CRMs are monitored at a range of temperatures and the results are used to predict the rate of change at a proposed, usually low, storage temperature [12].

Often, the prediction uses a well-known degradation model such as an Arrhenius model. The advantage over real-time studies is that results are available sooner and predictions of stability over a much longer period than the stability study can be defended. For some applications, accelerated studies have been described as the only practical approach.

In the absence of a reference method or a higher-order standard, accelerated studies under stress conditions provide the only approach for assessment of stability

Following steps are undertaken to assess the stability of BNDs as per ISO guidelines:

i. by experimentation whenever necessary, the stability of all relevant properties of a BND as per defined storage conditions and pre-treatment criteria, packaging and storage conditions;
ii. by experimentation whenever necessary, the stability of all relevant properties of a BND as per defined conditions for transport, and choose transport conditions to maintain stability during transport;
iii. define the storage conditions and use of BND to maintain stability at the user's premises;
iv. monitor the stability of BND for its long-term storage to detect the change with respect to time;

v. change in certified value due to stability issue exceeding the defined criteria, CSIR-NPL does not allow considering that batch anymore. In case, the stability of BND remains constant during the validity period, validity is extended as per the nature of BND. Validity period is further extended, wherever required, up to constant stability of the BND following the same process;
vi. maintain the stability of the BND, the amount is taken out for use from the BND bottle/container and left-over unused quantity, is discarded.

Experimental assessment of stability of BND is conducted before its release. Unless there is evidence of stability or prior experience of stability from similar BNDs held for an extended period under the same planned storage conditions. Whenever BND is developed/produced in multiple batches, the stability of BND of all the batches is experimentally verified. In case of BND developed/produced through RMP, respective RMP assesses and monitors the stability of the properties of BND. Validity period is further extended, wherever required, up to constant stability of the BND following the same process.

19.2.6 Assignment of Property Value

The certified property values of BNDs are established from the analytical results of regular monitoring of their purity, stability, and z-scores acquired in various national/international PT programmes along with their inter-laboratory comparison study. The use of z-scores is a globally accepted procedure for demonstrating the quality of results [13].

19.2.7 Evaluation of Measurement Uncertainty of Reference Materials

Methods for homogeneity test, measurement of property values, stability testing in terms of measurement uncertainty, the establishment of traceability to standard units of measures, etc. contribute to assign the accurate and precise property values with associated uncertainty. CRM is the whole process of obtaining the property values and their uncertainty, which includes homogeneity testing, stability testing, and characterization. The characterization process only provides information on a few components of uncertainty in the certification model, but it usually dearth with other uncertainty sources such as homogeneity and stability. It is worth mentioning here that the analytical results are only as good as the calibrators used in the analysis are good. Accuracy of quantitative results depends on the robustness of the analytical method; preparation of samples and standards; and the quality, purity and accuracy of the reference standard [14].

Uncertainty is a parameter associated with the result of a measurement that characterizes the dispersion of the values that could reasonably be attributed to the measurand. In general, uncertainty is a measure of the "goodness" of a result. There are three steps to define uncertainty [15] namely type A and Type B uncertainties, Combine Type A and Type B uncertainties and finally the expanded uncertainty.

Primarily, the aim of a measurement is to obtain the true value of the measurand [16]. Many efforts are normally made to optimize the measurement procedure in such a way that the measured value can be close to true value as much possible. However, our measurement results are just an estimate of the true value and the actual true value, which is always unknown to us. Therefore, we cannot reveal exactly how our measured value is close to the true value giving rising an estimate of associated uncertainty. Further, it is important to know about the error which is the difference between the measured value and the true value. The associated uncertainty in measurement of a quantity accounts for the accuracy of the measurement. So, uncertainty can be defined as the dispersion region of the measured value (C_{MEASURED}) in which somewhere true value (C_{TRUE}) lies. The half of that dispersion is known as measurement uncertainty U (Positive always). The same can be better visualized as in the schematic diagram (Fig. 19.5).

The Guide to Uncertainty in Measurement (GUM), published in 1992 by ISO/BIPM has been the origin of determining the uncertainty in measurement. The determination of uncertainty associated with property value can be estimated by using the following steps:

Step 1: Determine Type A and Type B uncertainties

Type A Standard Uncertainty is defined as the standard deviation of the mean of replicate measurements [17].

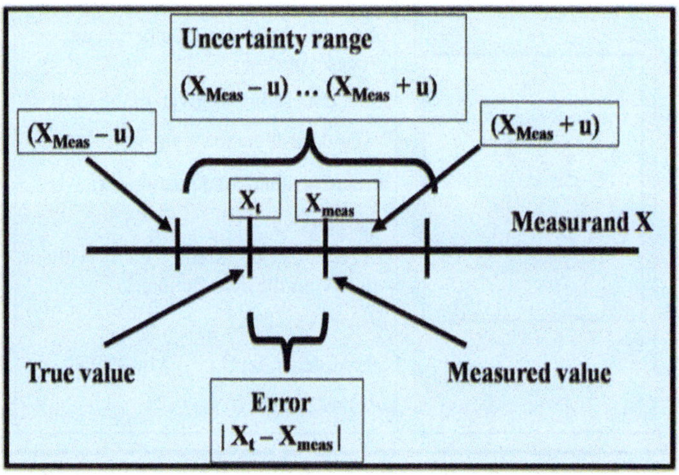

Fig. 19.5 Illustration of interrelations of true value, measured value, error and associated uncertainty

$$U_s = \frac{\sigma}{\sqrt{n}}$$

where σ is standard deviation and n is number of replicates.

Type B Standard Uncertainty is based on judgment using all information available, including measurement data, experience, specifications, and calibration reports. In order to determine this, three common distributions are used which are as follows:

(1) Normal distribution: For a 95% level of confidence, convert to a standard uncertainty by using 1.96 as a multiplier; e.g. uncertainty on a balance certificate (Fig. 19.6).
(2) Rectangular distribution: If the limits do not specify a level of confidence, use a multiplier of $3^{1/2}$; e.g. standard reference material (SRM) uncertainty (Fig. 19.7).
(3) Triangular distribution: For symmetric distribution use the multiplier $6^{1/2}$; e.g. uncertainty associated with volumetric glassware (Fig. 19.8).

Step 2: Combined Type A and Type B uncertainties:

- For only a sum or difference of quantities, where c is a constant and y is a function of the parameters p, q, and r, the combined standard uncertainty $u_c(y)$ can also be originated from the property values estimated from characterization, homogeneity and stabilities as follows:

$$c_{\mathrm{CRM}} = c_{\mathrm{char}} + \delta c_{\mathrm{bb}} + \delta c_{\mathrm{its}} + \delta c_{\mathrm{sts}}$$

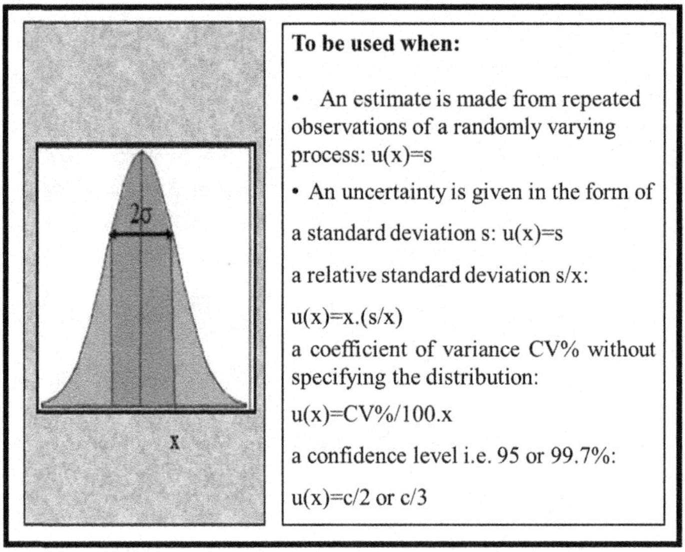

Fig. 19.6 Normal distribution for variance and uncertainty

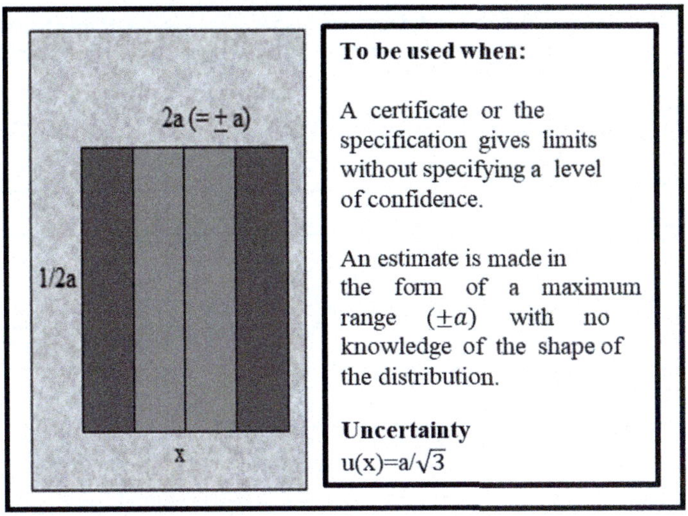

Fig. 19.7 Rectangular distribution for variance and uncertainty

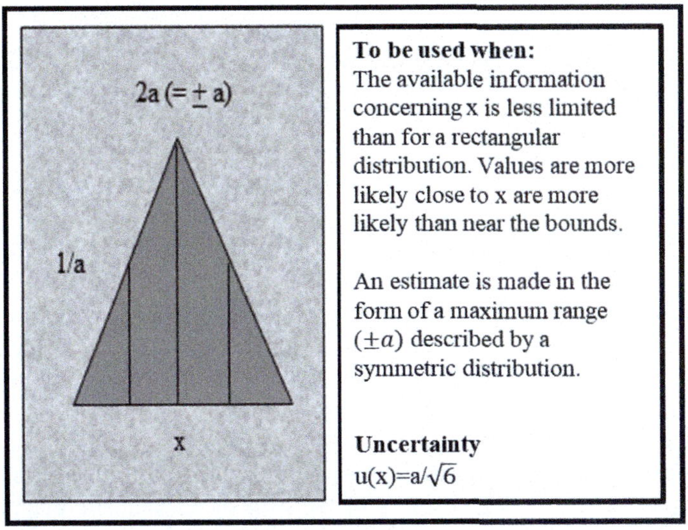

Fig. 19.8 Triangular distribution for variance and uncertainty

where c denotes the property value and c_{char} denotes the property value obtained from the characterization of the batch, δc_{bb} denotes a correction of the property value due to the between-bottle homogeneity study, and δc_{lts} and δc_{sts} present corrections of the property values of the batch due to the long- and short-term stability studies [15, 18].

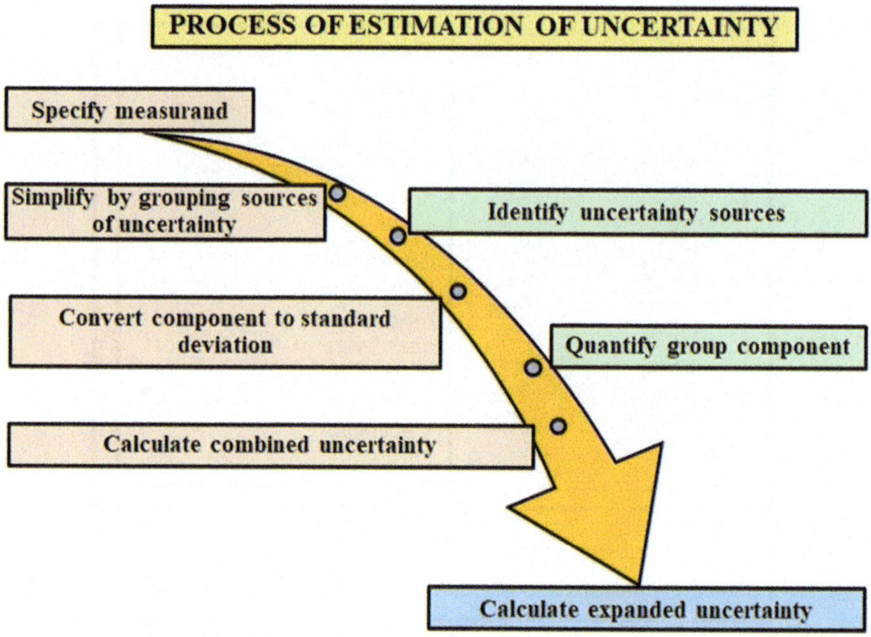

Fig. 19.9 Complete procedure for evaluating uncertainty for certification of reference materials

Thus, uncertainty of a CRM can be expressed as

$$u_{\mathrm{CRM}} = \sqrt{u_{\mathrm{char}}^2 + u_{\mathrm{bb}}^2 + u_{\mathrm{lts}}^2 + u_{\mathrm{sts}}^2)}$$

Step three: expanded uncertainty

It can be estimated using $U = ku_c$, where u_c is the combined standard uncertainty from step two and k is coverage factor. The value of the coverage factor k depends on the desired level of confidence. A coverage factor of 2 ($U = 2u_c$) gives a level of confidence of approximately 95% corresponding to $k = 1.96$ as most vendors use for certifications.

The complete procedure for evaluating uncertainty for certification of reference materials can be better understood with the help of a schematic diagram shown in Fig. 19.9. Figure 19.10 shows the evaluation of measurement uncertainty in the certification process.

19.2.8 Assignment of Property Value

The assignment of property value includes the following procedures (a) details of the experimental designs and statistical techniques used; (b) policies on treatment and

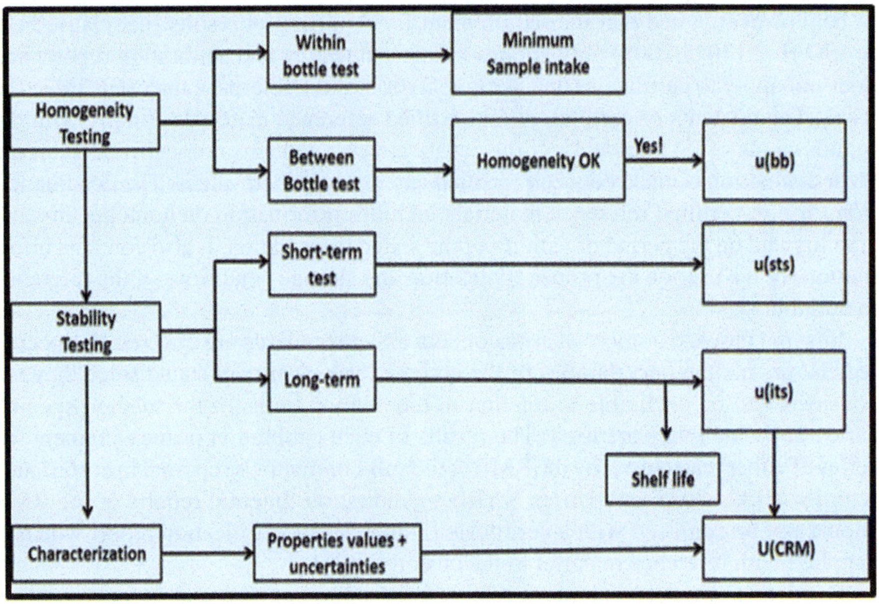

Fig. 19.10 Evaluation of measurement uncertainty in certification process

investigation of anomalous results, including outliers; (c) whether weighing techniques are used for contributions to assigned property values derived from different procedures or laboratories with different measurement uncertainties (d) the approach used to assign uncertainties to the property values (e) any other significant factors that may affect the assignment of property values [19, 20].

The technical information on test methods and equipment, including reported uncertainty information, and of any evidence of BND performance are considered when assigning the property values of interest the following the guide 35 as guidance on valid approaches for value assignment.

The outliers are not rejected solely on the basis of statistical evidence until they have been investigated and, where possible, the reasons for the discrepancies identified. For exclusion of outliers, robust statistical methods are applied where appropriate.

19.2.9 Certification of BNDs

BND developed complying ISO standard 17034 stands for final certification. However, many RMs, particularly older materials and materials not specifically produced as RMs, may not fully comply with ISO guides 31 and 35. RMP perform tests or calibrations for the production of reference materials. Such tests or calibration are performed in accordance with the relevant requirements of ISO/IEC 17025. It has to be, however, noted that the requirement for reporting of results (i.e. clause 5.10 of ISO/IEC 17025:2005) only applies to internal testing and calibration reports and does not apply to certificates and documentation of the reference materials issued to users. The contents of certificates for certified reference materials comply with the requirements of ISO guide 31. If the certificate also contains non-certified values, a clear distinction is made between certified and non-certified values. The documentation for non-certified reference materials includes information on homogeneity and stability and on the period of validity of the stated information. It also contains information for the user on the proper application and storage conditions of the reference material [14].

In some cases, e.g. most pharmacopoeia assay standards are covered by specific legislation and the uncertainties of the assigned values are not stated since they are considered to be negligible in relation to the defined limits of the method-specific assays for which they are used. The results of each calibration or measurement (or series of either) carried out by the RMP or the sub-contractor is reported in accordance with ISO/IEC 17025 and carries NABL accreditation. Internal reports of the RMP should not be confused with a certificate of analysis or certification report which is supplied with reference material to the customer [21].

Certificates or documentation for certified reference material or non-certified reference material contains a unique identification of its production process. This identification takes the form of a reference number, the name of the process or in other suitable information.

19.3 Preparation/Production of BNDs: Large and Medium-Scale Industries

CSIR-NPL is dedicatedly engaged in developing BNDs/CRM in-house and in association with large and medium-scale industries who are working as RMPs in various sectors. The details of BNDs developed are discussed in the subsequent sections.

Reference materials are produced by a metrologically valid procedure following ISO standards as shown in Fig. 19.11. ISO 17034: 2016 is the requirement for the competent of reference material production. ISO guide 35:2017 for characterization and assessment of homogeneity and stability. ISO guide 31:2015 is for certificate and accompanying documentation. The BND is accompanied by a certificate that provides the value of the specified properties with a statement of metrological traceabilities [22].

Measurement capability of the testing laboratories in India, supported by CSIR-NPL is continuously improving its measurement capabilities and the same is being disseminated through BNDs to stakeholders. The RMPs are encouraged by signing MoUs to produce large number of BNDs and disseminate to build up the quality infrastructure. Table 19.1 reveals the RMPs signed MoUs with CSIR-NPL.

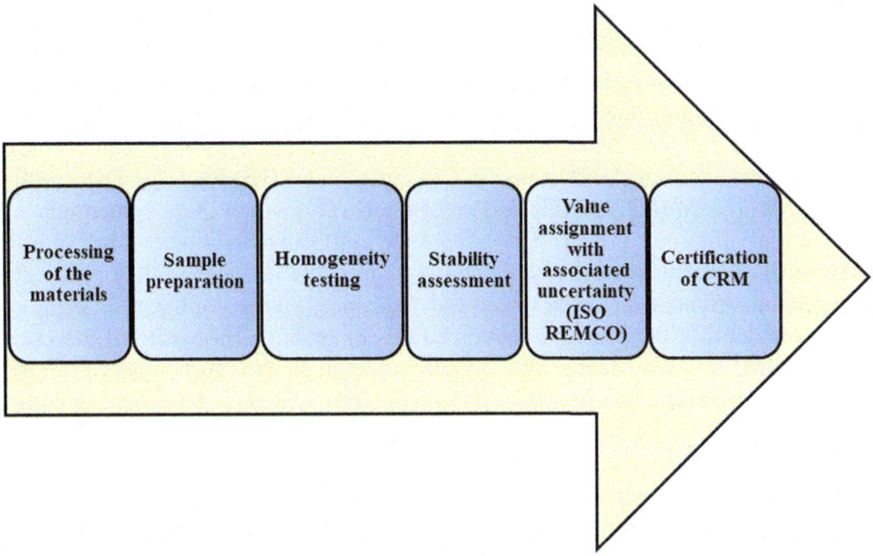

Fig. 19.11 Procedures for the production of reference materials

Table 19.1 MoUs with RMPs in various areas

Name of the RMP	Category	Area
BPCL, QA Laboratory, Mumbai	Physical properties	Petroleum
Aashvi Technology LLP, Ahmedabad	Chemical composition	High purity metals spectro-chemical solution, Pure chemicals, Ion activity
HPCL, Visakhapatnam, Andhra Pradesh	Chemical composition, physical properties	Petroleum products
Global PT, New Delhi	Physical properties	Hardness block
Sums Techno Labs Pvt. Ltd., Bellary, Karnataka	Chemical composition	Inorganic reference material
FARE LABS, Gurugram, Haryana	Chemical composition	Organic compound, Food stuff
CSIR-IITR, Lucknow	Chemical composition	Pesticide residue in food stuff
Jalan Company, Delhi	Chemical composition	Silver metal
MINT Govt., Mumbai	Chemical composition	Gold metal
NCCBM, Ballabgargh	Chemical/Physical	Building materials

19.3.1 Drinking Water and High Purity Materials

The rapid growth of urbanization as well as industrialization has thrown a significant threat to society in terms of contamination of water [23]. The quality of drinking water has become a serious issue due to the toxicity nature of the contaminants. The contaminants of drinking water are mainly organic, inorganic and biological sources. The heavy metal contamination of ground and surface water is mainly due to natural and anthropogenic sources and it increases day by day and becomes a health hazard problem. According to the toxicity nature of the heavy metal contaminants in drinking water, World Health Organization (WHO) has set the permissible limit across the World. Accurate and precise measurement of these contaminants is an essential criterion for the quality assurance in the analytical laboratory. In order to quantify the content of the elemental contaminants in water [24], sophisticated analytical instruments are used which provides many sources of biasness, error and unreliable data. In the previous section of this chapter, the necessity of the use of BNDs/CRMs for the traceable and reliable results in analytical chemistry is already discussed explicitly. More precisely, the use of BNDs/CRMs is a mandatory requirement among others to generate accurate and reproducible data for the testing & calibration laboratory as per ISO/IEC 17025:2017 to satisfy the customer.

As per the metrological point of view, the purity evaluation of the high purity materials is being used starting from the analytical laboratory to the manufacture unit. In an analytical laboratory, high purity materials such as various salt, compounds, primary standards are routinely used in the titrimetric and gravimetric analysis. Thus, impurities present in the high purity materials interfere in the analytical process. According to guide 35, it is stated that in chemical metrology, the metrological traceability may be established by evaluating the purity of the starting material of the CRMs. In the case of development of elemental solution based CRMs, purity evaluation of the starting materials is essential to establish metrological traceability [25].

Aashvi Technology LLP (ATL), Ahmedabad is an Accredited RMP as per ISO 17034: 2016 by QCI-NABL, India has signed Memorandum of Understanding (MoU) with CSIR-NPL. CSIR-NPL, under its metrology in chemistry activity, provides support to the ATL for the development of BNDs on aqueous elemental solutions and high purity materials focusing on the area of drinking water and pure materials. The metrological traceability is established for those BNDs by purity evaluation of the high purity material. The BNDs produced are intended to use as a primary calibration standard for the quantitative determination of elemental content, calibration of instruments and validation of method for the quantification/characterization of the measurand. The mass fraction of the elemental standard solution is certified by the gravimetric value traceable to SI unit and validated by ICP-OES and Atomic Absorption Spectrometer (AAS). The certified value is assigned by Aashvi Technology LLP and further ascertained by CSIR-NPL. Certified value of the concentration of the element along with the associated uncertainty was calculated at 95% confidence level with coverage factor $k = 2$, considering major sources of uncertainty including measurement replication, instrument background, mass taken, buoyance correction, possible heterogeneity and stability factors according to JCGM 100:2008 and ISO guide 35:2017. The schematic diagram of purity evaluation of BND starting material is given in Fig. 19.12.

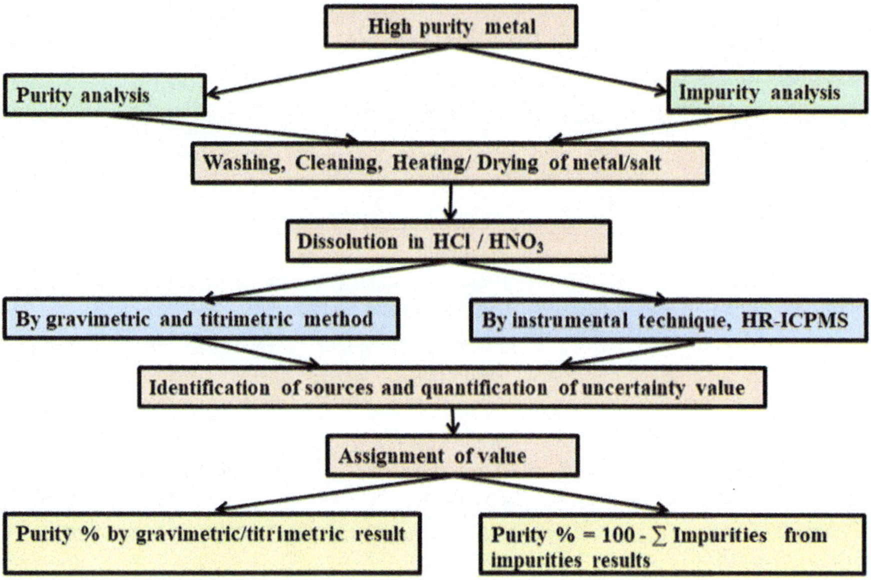

Fig. 19.12 Purity evaluation of starting material used for development of BND

So far, there are 17 BNDs that have been produced by Aashvi Technology LLP on an elemental standard solution of 1000 mg/kg, high purity material, pH standard solutions of pH values 4.0, 7.0, 9.0; and total dissolved solids in collaboration with CSIR-NPL. The list of BNDs developed with name and BND number are given in Table 19.2.

19.3.2 Petroleum Products (BNDs)

CSIR-NPL successfully achieved a significant milestone by releasing 28 petroleum-based BNDs in association with HPCL. Traceability for all vital parameters of petroleum products testing and certification comprising of 12 physical properties, 3 physico-chemical properties and 13 chemical properties including BND for sulphur content measurement at lower concentrations which will be of immense use for BS VI fuels. This unique initiative will be a milestone in ensuring quality assurance of petroleum products in our country. Figure 19.13 shows the protocols described as per ISO guidelines used for the development of petroleum-based BNDs. Table 19.3 shows the list of physical parameters of petroleum BNDs. Figure 19.14 shows honourable Principal Scientific Advisor (PSA) to Govt of India, Dr. R Chidambaram, DG-CSIR Dr. Shekhar C Mande and other dignitaries releasing petroleum BNDs.

Table 19.2 List of BNDs produced by Aashvi Technology LLP on drinking water and high purity compound

S. No.	Name of BNDs	BND No.	Certified property value
1	Copper standard solution	BND 1003	988 ± 6 mg/kg
2	Nickel standard solution	BND 1004	979 ± 5 mg/kg
3	Lead standard solution	BND 1010	981 ± 6 mg/kg
4	Mercury standard solution	BND 1012	936 ± 5 mg/kg
5	Cadmium standard solution	BND 1013	979 ± 6 mg/kg
6	Zinc standard solution	BND 1015	982 ± 6 mg/kg
7	Iron standard solution	BND 1019	983 ± 7 mg/kg
8	Arsenic standard solution	BND 1011	988 ± 5 mg/kg
9	Potassium hydrogen phthalate	BND 1016	100.00 ± 0.22%
10	Potassium dichromate	BND 1017	99.92 ± 0.21%
11	Sodium chloride	BND 1018	99.94 ± 0.27%
12	Sodium carbonate	BND 1023	99.98 ± 0.20%
13	Calcium carbonate	BND 1024	100.00 ± 0.20%
14	pH Buffer solution 4.00	BND 1005	4.01 ± 0.03
15	pH Buffer solution 7.00	BND 1006	7.00 ± 0.03
16	pH Buffer solution 9.00	BND 1007	9.05 ± 0.05
17	Total Dissolved Solids (TDS)	BND 1021	997 ± 7 mg/kg

19.3.2.1 Physical, Chemical and Physico-Chemical Parameter-Based BNDs

The physical properties-based petroleum BNDs have been developed and traceability for all vital parameters of petroleum products has been provided for testing and certification comprising of 12 physical, 13 chemical and 03 physico-chemical properties. Table 19.4 shows the chemical and physico-chemical parameters of petroleum BNDs.

In addition to this, BPCL has signed MOU with NPL to produce BND for making CRM's in the petroleum product segment with SI traceable measurements. This is in synergy with 'Atmanirbhar Bharat' (Self-reliant India) mission of The Honourable

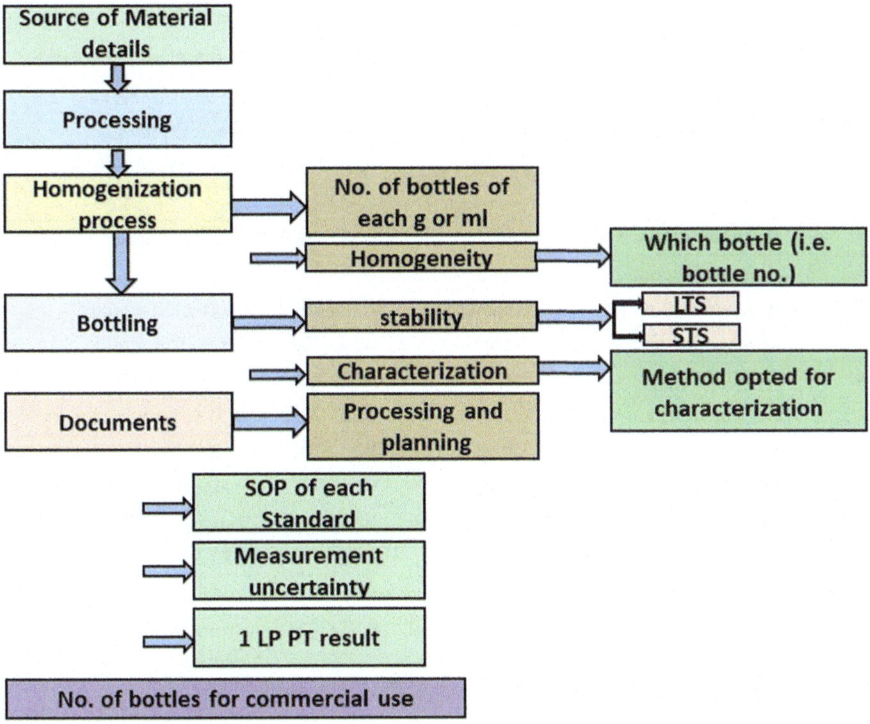

Fig. 19.13 Process of development of petroleum BNDs

Prime Minister of India. This initiative will not only minimize the dependence on overseas reference materials but also reduces the foreign exchange outflow. Apart from serving the domestic customers, BND has immense potential in the overseas market. Some of the BPCL petroleum BNDs are shown in Fig. 19.14b.

19.3.2.2 Dynamic Viscosity BND of Non-newtonian Fluids Using Magneto-Rheometer

BND for rheological properties of non-Newtonian fluids has been developed, which shows shear rate dependence of viscosity. Accurate measurements of these material properties are critical for predicting the behaviour of these fluids in the complex flows they experience during processing and use. Figure 19.15 shows a parallel plate geometry for dynamic viscosity measurements.

Table 19.3 List of physical parameters of few petroleum BNDs

S. No.	Name of BND	Matrix	BND No.	Certified property value
1	Sulphur 2.5 ppm	Diesel	BND 7001	2.5 ± 0.75 mg/kg
2	Sulphur 5 ppm	Diesel	BND 7002	5.0 ± 1.5 mg/kg
3	Sulphur 10 ppm	Diesel	BND 7003	10 ± 1.5 mg/kg
4	Sulphur 25 ppm	Gasoline	BND 7004	25 ± 2.5 mg/kg
5	Sulphur 50 ppm	Diesel	BND 7005	50 ± 2.5 mg/kg
6	Sulphur 50 ppm	Gasoline	BND 7006	50 ± 2.5 mg/kg
7	Sulphur1000 ppm	Kerosene	BND 7007	1000 ± 20 mg/kg
8	Sulphur 2.5 ppm	Gasoline	BND 7026	2.5 ± 0.75 mg/kg
9	Sulphur 5 ppm	Gasoline	BND7027	5.0 ± 1.5 mg/kg
10	Sulphur 10 ppm	Gasoline	BND 7028	10.0 ± 1.5 mg/kg
11	Smoke point	Petroleum oil	BND 7017	23 ± 2 mm
12	Distillation standard	Petroleum oil	BND 7018	Range 34.3 to 193.4 °C
13	Distillation standard	Petroleum oil	BND7019	Range 182.2 to 362.8 °C
14	Total Base Number	Mineral oil	BND 7023	9.22 ± 0.27 mg KOH/g
15	Total Acid Number	Mineral oil	BND7024	1.64 ± 0.12 mg KOH/g
16	FIA-Aromatics, Olefins, Saturates	Petroleum oil	BND7025	Aromatics: 17.0 ± 1.2 (%, v/v) Olefins: 0.9 ± 0.1 (%, v/v) Saturates: 82.1 ± 1.2 (%, v/v)

19.3.3 Building Materials

Increasing industrialization demands for high-quality construction materials. In order to improve the quality of the construction reference standards of building materials are required. For this it may be required a large number of reference standard materials like ordinary portland cement (OPC), white cement, cement composites, concrete, steel-reinforced concrete, polyurethane foam, fly ash bricks, portland pozzolana cement, etc.

Cement is a kind of adhesive for all kinds of building materials used in civil engineering construction. Cements are finely ground powders that, when mixed with water, set to a rigid mass, which is due to a chemical combination of cement compounds and water that yields sub-microscopic material with a high surface area. Due to hydrating properties, these cements, even set and harden under water are often called hydraulic cements.

There are many varieties of portland cement [26]. One of the types is coloured cements that are made by grinding 5–10% of suitable pigments with white or ordinary grey portland cement. Air-entraining cements are made by the addition of grinding of a small amount, about less than one percent, of an organic agent that creates very fine air bubbles in concrete. The air-entraining agent can alternatively be added as

Fig. 19.14 **a** Honourable PSA to Govt of India, R. Chidambaram, DG-CSIR Dr Shekhar C Mande and other dignitaries releasing petroleum BNDs and **b** petroleum BNDs by BPCL

a separate ingredient to the mix when making the concrete. Low-alkali cements are portland cements with a total content of alkalies not above 0.6%. These are used in concrete made with certain types of aggregates that contain a form of silica that reacts with alkalies to cause an expansion that can disrupt a concrete. Water-proof cement is the name given to portland cement to which a water-repellent agent has been added. Hydrophobic cement is obtained by grinding portland cement clinker with a film-forming substance such as oleic acid in order to reduce the rate of deterioration when the cement is stored under unfavourable conditions. Oil-well cements are used for cementing work in the drilling of oil wells where they are subject to high

Table 19.4 Chemical & physico-chemical parameters of petroleum BNDs

S. No	Measurand	Certified value	Expanded uncertainty (Coverage factor $k = 2$)
1	Colour ASTM	4 colour ASTM unit	±1 colour ASTM unit
2	Colour saybolt	24 colour Saybolt	±1 colour Saybolt unit
3	Flash point Abel	40.5 °C	±1.5 °C
4	Kinematic viscosity at 100 °C	10.64 cSt	±0.04 cSt
5	Density (kg/m^3)	799.8 kg/m^3	±3.2 kg/m^3
6	Kinematic viscosity at 40 °C	2.917 cSt	±0.02 cSt
7	Kinematic viscosity at 50 °C	170.2 cSt	±0.6 cSt
8	Kinematic viscosity at −20 °C	3.171 cSt	±0.022 cSt
9	Freezing point	−57.50 °C	±1.5 °C
10	Flash Point PMCC	74.5 °C	±3.5 °C
11	Flash Point COC	212 °C	±15 °C
12	Pour Point	−6.0 °C	±3.0 °C

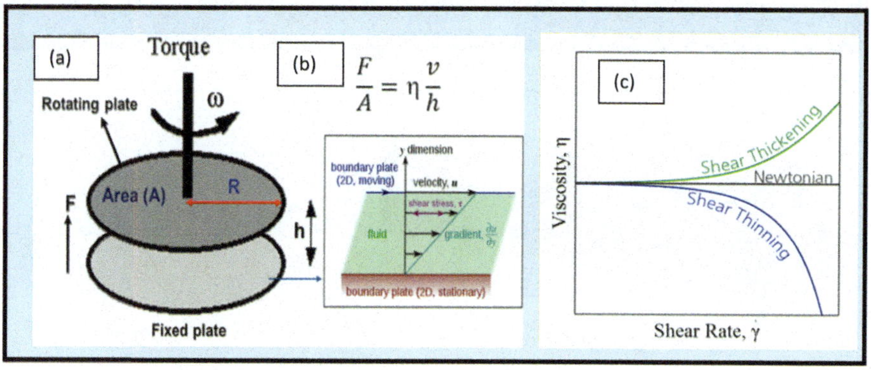

Fig. 19.15 **a**, **b** Parallel plate geometry **c** dynamic viscosity of Newtonian fluid measurements

temperatures and pressures. They usually consist of portland or pozzolanic cement with special organic retarders to prevent the cement from setting too quickly.

The granulated slag made by the rapid chilling of suitable molten slags from blast furnaces forms the basis of another group of constructional cements. A mixture of portland cement and granulated slag, containing up to 65% slag, is known as portland blast-furnace (slag) cement. Properties of these slag cements are broadly similar to those of portland cement, but they have a lower lime content and a higher silica and alumina content. High-alumina cement is a rapid-hardening cement made by fusing at 1500–1600 °C a mixture of bauxite and limestone in a furnace or in a rotary kiln. High-alumina cement attains its ultimate strength within 24 h and has a high resistance to chemical attack.

In India we have been developing a variety of cements like belite calcium sulfoaluminate clinker (BCSA), high MgO clinker for PPC & PSC, composite cements, Portland limestone cement (PLC), multi-component cement, enhanced use of fly ash in PPC, steel slag (LD-slag) as Raw material in clinker making, wollastonite as raw material in clinker making, development of synthetic slag using low-grade limestone, etc. It is imperative to develop synthetic slag using low-grade limestone and other additives to provide the availability of GBF slag in the manufacture of composite cement.

For all such variety of cement, there is stringent need that materials should be sufficiently homogenous and stable with respect to one or more specified properties, which have been established to be fit for its intended use in a measurement process. Blaine fineness method is used for measuring the fineness of Portland cement in accordance with ASTM and BIS standards.

The fineness of cement is measured as specific surface [27]. Specific surface is expressed as the total surface area in square metres of all the cement particles in one kilogram of cement. The higher the specific surface is, the finer cement will be. The principle of air permeability method is in observing the time taken for a fixed quantity of air to flow through a compacted cement bed of specified dimension and porosity (Fig. 19.16). Under standardized conditions the specific surface of cement is proportional to where; t is the time for a given quantity of air to flow through the compacted cement bed. The number and size range of individual pores in the specified bed is determined by the cement particle size distribution which also determined the time for the specified airflow. The method is comparative rather than absolute and therefore a reference sample of the known specific surface is required for calibration of the apparatus.

A bed of cement is prepared in a special permeability cell and has exact porosity $e = 0500$ weight a quantity of cement. The weight of the cement is calculated from

$$m_1 = 0.500 \times \rho \times V \ (\mathrm{g})$$

where ρ is the density of the cement (g cm^{-3}) and V is volume of the cement bed (cm^3).

Specific surface S is expressed as

$$s = \frac{K}{\rho} \times \frac{\sqrt{e^3}}{(1-e)} \times \frac{\sqrt{t}}{\sqrt{0.1\eta}} \ (\mathrm{cm}^2\ \mathrm{g}^{-1})$$

where K is the apparatus constant, e is porosity of the bed, t is the measured time (in second), ρ is density of cement (g cm^{-3}) and η is the viscosity of air at the test temperature.

The apparatus constant is determined by measuring permeability of reference cement of a known specific surface. The procedure have been repeated three times for three samples of cement bed. For each sample the constant K is calculated from the formula:

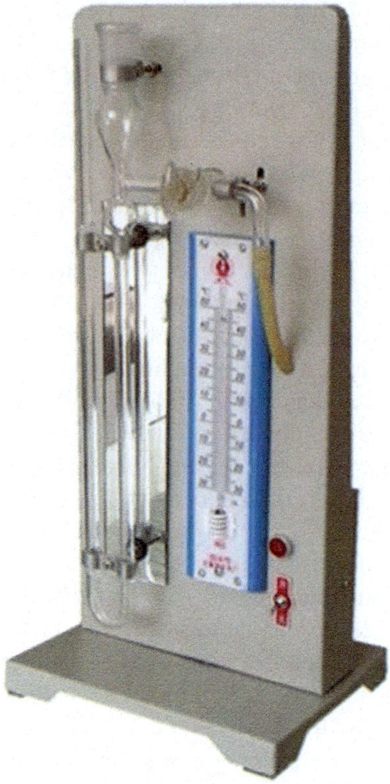

Fig. 19.16 Blain Fineness apparatus for determination of fineness

$$K = S_0 \times \rho_0 \times \frac{(1-e)}{\sqrt{e^3}} \times \frac{\sqrt{0.1\eta_0}}{\sqrt{t_0}}$$

where S_0 is the specific surface of the reference cement ($cm^2\ g^{-1}$), ρ_0 is density of the reference cement ($g\ cm^{-3}$), t_0 is measured time (in second), η_0 is air viscosity at the test temperature. With the specified porosity of $e = 0.500$

$$K = 1414 \times S_0 \times \rho_0 \times \frac{\sqrt{0.1\eta_0}}{\sqrt{t_0}}$$

The average of three values of K is taken for the apparatus.

CSIR-NPL is producing BNDs in collaboration with the National Council of Concrete and Building Materials (NCCBM), Govt. of India in different areas including the building materials; such as ordinary Portland cement, Fly Ash Cement, Portland Pozzolana cement, etc. The BNDs related to building materials are summarized in Table 19.5.

Uncertainty for Blaine's fineness of hydraulic cement [28] is determined by following GUM method. Relative uncertainty of each parameter is determined from

Table 19.5 BNDs related to building materials developed by NPL and NCCBM

BND No.	BND name	Purpose/use	Specific surface area (Blain); m^2/kg
BND 5001A	Ordinary Portland Cement	Calibration of blain air permeability apparatus	274 ± 6.4
BND 5002	Portland Pozzolana Cement	Calibration of blain air permeability apparatus	384 ± 9.3
BND 5003	Portland Slag Cement	Calibration of blain air permeability apparatus	330 ± 9.3
BND 5004	Fly Ash	Calibration of blain air permeability apparatus	333 ± 22.1
BND 5006	Composite Cement	Calibration of blain air permeability apparatus	372 ± 9.3
BND 5007	White Portland cement	Calibration of blain air permeability apparatus	382 ± 8.4
BND 5011A	Ordinary Portland Cement	Calibration of blain air permeability apparatus	419 ± 9.9
BND 5021	Ordinary Portland Cement	Calibration of blain air permeability apparatus	337 ± 12.9

the standard deviation values. Further Combined relative uncertainty and combined standard uncertainty are also determined. Finally expanded uncertainty is derived from the combined relative and combined standard uncertainties based on the effective degree of freedom and probability factor at the confident level of 95%.

NCCBM is exporting cement BND to countries like France, Austria, Oman, Nepal, Bangladesh, Egypt and UAE. Karen L. Scrivener has reported that the projected increase in demand for cement in the coming decades will be in emerging and developing countries [29].

19.3.4 Hardness Block

Hardness testing provides critical information and insights into a material's durability, strength, flexibility and capabilities. It is commonly used to test the methods in many industries to verify heat treatment, structural integrity and quality of components. Hardness testing ensures the materials to be utilized in components in our everyday life and contributes to develop a well-engineered and efficient component [30].

Indentation hardness testing is a convenient means of investigating the mechanical properties of a small volume of materials. Since the technique is based on the impression of indentations in small areas on the surface by pressing a hard indenter of specified geometry into the solid, it is essentially non-destructive in contrast to the methods involving compression, tensile or bending test of bulk samples. Indentation hardness testing has also become increasingly important to industries that deal with micro machines, microelectronics and magnetic recording.

CSIR-NPL in collaboration with Global PT provider Pvt Ltd has developed BNDs related to hardness blocks for measuring the Brinell, Rockwell and Vickers hardness.

The hardness BND blocks are used in automobile, metallurgical testing Lab, steel manufacturing, railway, defence and aerospace industry. In addition to this, it will also help the Indian industries suffering from traceability issues. Calibrated test blocks are an integral part of hardness testing. They ensure accuracy, integrity and traceability of hardness testing processes. They are used to verify instrument performance and provide a means for performing indirect instrument calibrations. These hardness test blocks are used by leading companies in the industry, especially those in the aerospace, medical primary, metals and automotive industries.

Hardness BNDs uses low carbon, oil-hardened steel rods of an appropriate diameter of known chemical compositions and microstructure. The chemical composition was determined by optical emission spectroscopy and its microstructure using a scanning electron microscope.

During the processing of materials, all the samples were cured in such a way to maintain homogeneity conditions. After the processing again, the microstructure was studied. The sample is polished using the abrasive disc. Hardness measurement was carried in between all the processes to ensure homogeneity at five different locations in such a way that the distances between the centers of two adjacent indents were at least three times the diameter of the indent. The thickness of the blocks was maintained as per standard for better accuracy and repeatability and for minimum measurement of uncertainty.

For the Rockwell hardness test, Rockwell hardness tester ZHR4045/4150 was used. It can measure Rockwell hardness using Rockwell methods (ZHR4150) according to ISO 6508 and ASTM E 18 with a pre-load: 10 kg and test loads: 60, 100, 150 kg. The resolution of the instrument was 0.1 digits. It is a rapid and non-destructive technique. The Rockwell test uses a pre-load to cancel out the error. The test is fast, but it lacks the ability to cross-reference to other hardness tests and correlate to material strength as accurately [31, 32].

Figure 19.17a shows the stepwise procedure for measuring the Rockwell hardness. The resulting Rockwell number represents the difference in depth from zero datum position as a result of the application of major load. The entire procedure requires only 5–10 s. It may be mentioned that the use of a minor load greatly increases the accuracy of this type of test because it eliminates the effects of backlash in the measuring system and causes the indenter to break through slight surface roughness.

The result of the proficiency of the Vicker's hardness test by different laboratories is shown in Fig. 19.17b. All the values are quite close and the *Z* score is almost linear.

The certified value of Vicker's hardness for the samples in the batch with stated uncertainty is 735.6 ± 13.9 HV-5. The value is at the expanded uncertainty with a coverage factor $k = 1.96$ at a confidence level of 95%.

The hardness of all the samples was finally measured using Rockwell hardness tester ZHR4045/4150: Rockwell methods (ZHR4150) according to ISO 6508 and ASTM E 18. The pre-load was 10 kg and the test load was 150 kg. For two sets of

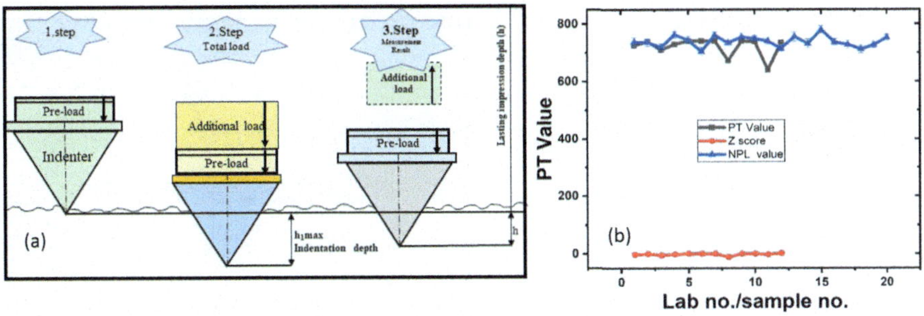

Fig. 19.17 **a** Stepwise procedure for measuring the Rockwell hardness, and **b** results of proficiency of the Vicker's hardness test

Table 19.6 BNDs related to Hardness

S. No.	BND Number	BND Name	Purpose
1	BND 2201	Rockwell hardness (C Scale): 44.33 ± 0.83	Verification of the Rockwell hardness testing machine
2	BND 2202	Rockwell hardness (C Scale): 61.04 ± 0.77	Verification of the Rockwell hardness testing machine
3	BND 2204	Vickers hardness (5 kgf/49.03 N Load): 735.6 ± 13.9	Verification of the Vickers hardness testing machine

samples, the certified value with the expanded uncertainty was 44.33 ± 0.83 (HRC) and 61.04 ± 0.77 (HRC), respectively. Similarly, the values of Vickers hardness were 5 kgf/49.03 N load and 735.6 ± 13.9, respectively. The reported uncertainty is at a coverage factor, $k = 1.96$, which corresponds to a coverage probability of approximately 95% for a normal distribution. Table 19.6 shows BNDs related to hardness [33].

19.3.5 Development of Toxic Elements Induced Basmati Rice Flour BND

The contamination of toxic heavy metals is seen in ground water, surface water, crops and irrigated water of many parts of the world including India. Contamination of toxic metals in the environment enters the human body mainly through the consumption of contaminated water, food and causes many diseases to human beings. Due to the carcinogenic nature of the heavy metals like cadmium, lead, mercury, arsenic causes health hazard for the human being as reported by the International Agency for Research on Cancer. The concentration of the contaminants presents in the soil, water gets transported to the cereals, vegetables and fruits and thus accumulated by the plants. At present, contamination of heavy metals in paddy crops is seen

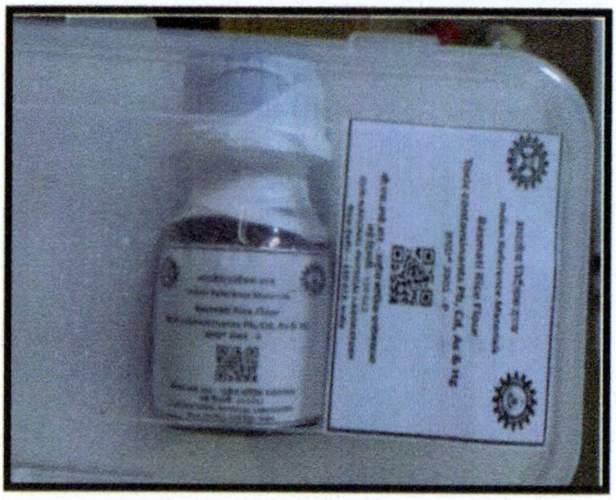

Fig. 19.18 Toxic elements induced rice flour bottle

worldwide for which many countries set regulations limiting the amount of main toxic elements, such as arsenic, lead, cadmium, etc. in rice to protect public health [34].

Precise, accurate and reliable measurement result of the end product is a mandatory requirement for the testing and calibration laboratory. Due to unavailability of matrix-based CRMs in our country, a huge number of CRMs have been continuously importing from the international market. To minimize the difficulties of importing the CRMs with traceability at an affordable cost and prompt supply for the development of toxic elements induced basmati rice BNDs to meet the demand of CRMs to fulfil the requirement for international trade of food related to food safety. To generate the rice grain contaminated with heavy metals and pesticides, Basmati rice of PUSA 1121 was contaminated in a land of 200 square yards area. The spiking of contaminants has been carried out during the ploughing field grain and cultivation of rice and ultimately harvested the contaminated rice. The harvested contaminated rice is processed and followed the homogeneity, stability, bottling, and validation studies. Figure 19.18 shows the picture of the toxic elements induced rice flour bottle.

The process of conducting the proficiency testing program on toxic elements in rice flour to assess the competency of the rice testing laboratories of India. Thereafter, the toxic elements induced rice BND with SI traceability will be disseminated.

19.3.6 Gold for Magnification Calibration of Transmission Electron Microscope

This reference material of evaporated gold is intended for use as a standard for magnification calibration of transmission electron microscope at the high end of the magnification range. The starting material of gold used for evaporation was 99.99% pure. Material is deposited on carbon-coated 300 mesh copper grids by thermal evaporation. The conditions are optimized for uniform deposition of gold on 12 grids mounted on a specific sample holder. The value of the gold (111) lattice plane is found to be 2.36 ± 0.04 Å. The associated uncertainty in the certified value is combined uncertainty with approximately 95% confidence level (coverage factor k = 2), calculated according to the ISO guide 35.

Initially, the substrate TEM grid holder, chamber of thermal vapour deposition instrument and other things are cleaned thoroughly. The holder was then placed in an upside position with carbon-coated grids. Weighted amount of 99.99% pure gold was placed in the Mo boat. High vacuum condition up to the order of 7×10^{-6} torr was achieved. Then current was passed to the molybdenum boat through L.T. Gold evaporated and got deposited on TEM grids at a particular applied voltage.

Lattice spacing was recorded from 15 different regions at constant temperature and humidity from the prepared batch of 12 gold-coated grids randomly. TEM samples were viewed in HRTEM (FEI, Tecnai G2 F30, STWIN operated at 300 keV accelerating voltage). Sample was left inside the microscope for stabilization for half an hour before high-resolution imaging. Sample was adjusted at eccentric height and then fine focus was done at low magnification (e.g. 25000×, 49000×). Further to view fine and clearly resolve lattice images, magnification was increased stepwise till clearly resolved crystal lattice fringes were observed. Once clear lattice fringes become visible, two or three types of lattice fringes were observed. The largest lattice fringes corresponding to (111) plane of gold with d- spacing of 2.36 Å was used for measurements. Figure 19.19 also shows the (a) TEM, (b–e) HRTEM and (f) SAED pattern of gold deposited thin film. Images are taken from various locations in order to check the homogeneity. From the micrograph, it is clear that the film consists of

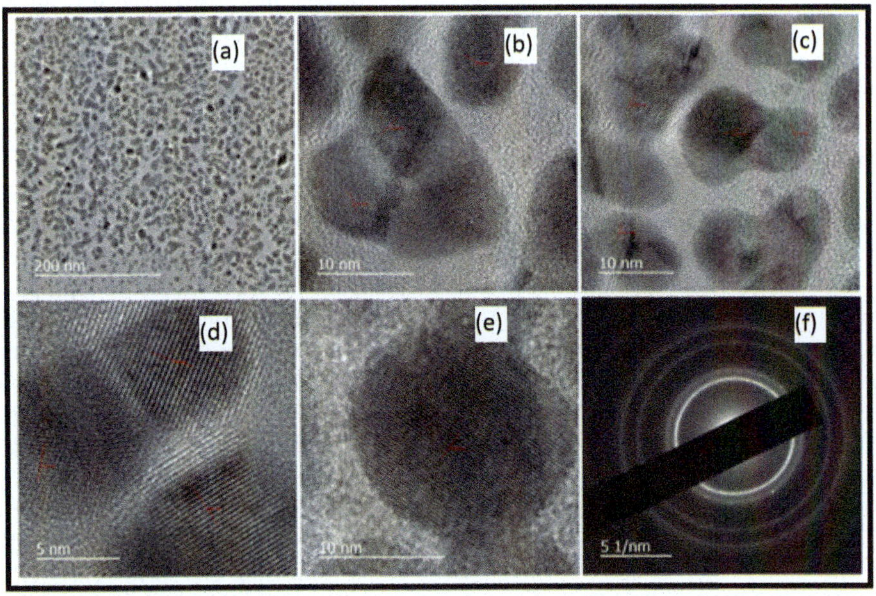

Fig. 19.19 a–f TEM, HRTEM and SAED images of samples from different locations used for measuring the lattice spacing

uniformly dispersed nanocrystals. Average value of lattice spacing was calculated using several measurements.

The instrument was calibrated using standard gold samples provided by Ted Pella (606; Grating replica, parallel, 2160 l/mm, on 3 mm grid). Measurements on test samples were carried out for at least 10 lattices at different locations (throughout the grid) to minimize the error in measurement. The calculated mean value of the measurements for (111) plane is 2.3619 Å and standard deviation value is 0.0130 Å. The major sources of uncertainty were included measurement replication, instrument resolution and possible inhomogeneity. TEM instrument at high magnification was calibrated using oriented gold sample.

The certified value of lattice spacing for the gold (111) by HRTEM is 2.36 ± 0.04 Å with a stated uncertainty $k = 2.16$ (Confidence level 95%). The material is stable in the temperature range of 16–30 °C and relative humidity of 45–60%.

19.3.7 Alpha-Alumina for Powder XRD Standard

XRD is one of the most widely used equipment for the structural analysis of the materials. The use of certified reference materials for the powder X-ray diffractometer (PXRD) having significance in the diminishing presence of systematic errors in 2θ due to absorption, vertical divergence, calibration errors, and sample displacement. However, in order to record accurate data, reference materials are essential and play a major role in advanced research and also in different industries (pharmaceutical, mining, etc.). In India, the total number of Academic Institutions (Universities) is 847 (including Indian Institute of Technology (IITs), Indian Institute of Science Education and Research (IISERs), National Institute of Technology (NITs), state universities), 7500 colleges (Science and Engineering) and more than 2000 mining and pharma industries [35].

All the institutions need BND for calibrating their PXRD. Usually, the data from the instrument without reference material is not authentic [36]. Leading XRD suppliers Rigaku, Bruker, Proto, PANanalytical, Shimadzu are also being benefitted from CSIR-NPL by using the reference material for their customers. A high pure raw material alpha-alumina (α-Al_2O_3) was used for the preparation of internal standard for intensity calibration of powder X-ray diffractometer. The finely prepared powder whose particle size lies in the range of 5–20 μm and is irregular in shape is used for the preparation of BND. The crystallinity of the material was improved by annealing process which is carried out at the temperature range of 1308 °C, in an inert atmosphere of (Ar) using a tubular high-temperature furnace with the accuracy of ±0.5 °C. Annealing process eliminates the defects and improves the crystallinity. The powder X-ray diffraction pattern was recorded in the 2θ range of 10°–150°, with a step size of 0.02°/s and a spinning speed of 30 rpm using 2/3° divergence, 2/3° anti-scattering slit and 0.03° detector slit. The recorded powder XRD pattern shows the material is well crystalline and has high intense diffraction peak of (113) with well-resolved $K\alpha_1$ and $K\alpha_2$.

Thereafter, the material was sent to two different NMI for comparability for the purpose of certification. The NMI, China and NMI, Thailand have participated in round-robin test. The set operating protocol for the measurement procedure has been followed uniformly. The first six intense diffraction peaks (2θ between 10° and 60°) were considered for the determination of intensity with associated uncertainty. The PXRD data (Fig. 19.20) obtained from the participating NMIs were analyzed for the accurate determination of X-ray powder diffraction intensity measurements. The Rietveld analysis has been performed on the XRD data which is depicted in Fig. 19.21. Certified values for relative intensity along with the expanded uncertainty are given in Table 19.7.

Certified Value of Lattice Parameter (Å)

$$a = 4.7584 \pm 0.0038; c = 12.9914 \pm 0.0044$$

In the recent past, CSIR-NPL has released a few BNDs in which alpha-alumina (Table 19.8) and silicon was used for the calibration of powder X-ray diffractometers. The photograph of the released material was given in Fig. 19.22.

In general, for calibration of PXRD, different SRMs are prepared by NIST USA for line position, line shape, etc. as shown in Table 19.9. In this aspect, CSIR-NPL is intending to develop BND for various parameters relating to XRD.

For the calibration of line positioning and intensity as well, Si powder is used for the preparation of BND. It is a chemically stable and robust material with high crystalline in nature. The prepared silicon powder with the particle size of 5–15 μm is achieved and annealed at 1200 °C to reduce the size broadening effects. The recorded

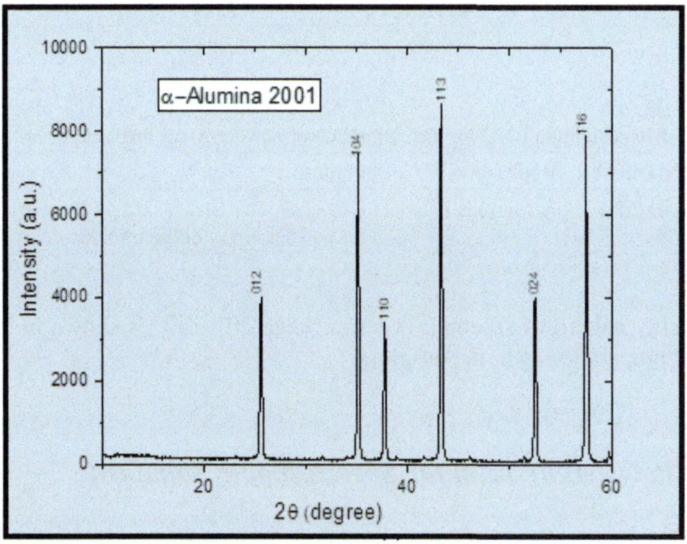

Fig. 19.20 X-ray pattern of alpha-alumina BND

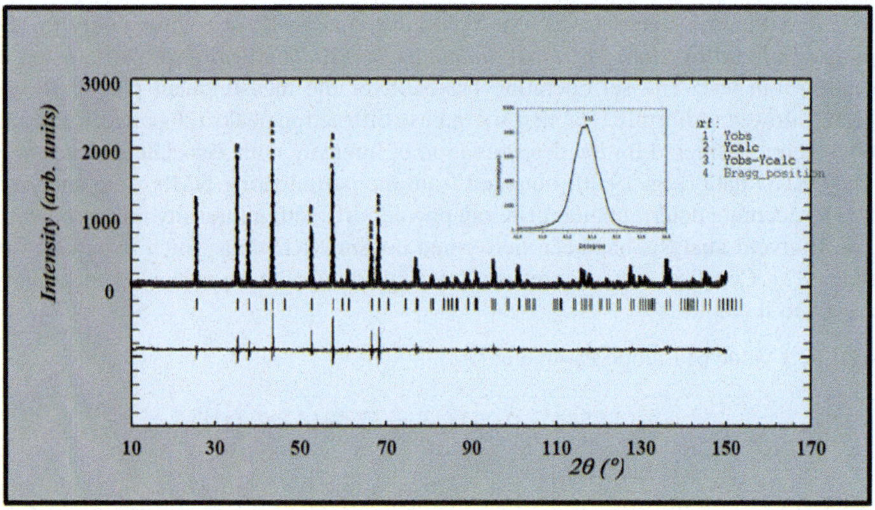

Fig. 19.21 X-ray diffraction of α-alumina for BND® 2001

Table 19.7 Relative intensity along with expanded uncertainty

Reflection indices (h k l)	Relative intensity	Expanded uncertainty ($k = 2$)
012	58.46	±8.25
104	89.99	±7.16
110	39.09	±3.61
113	100	–
024	45.90	±3.43
116	91.08	±9.07

Table 19.8 Alpha alumina BND for the calibration of powder X-ray diffractometer

S. No.	BND number	BND name	Purpose
1	BND 2001	α-Alumina Powder	Use as an internal standard for intensity calibration for powder X-ray diffractometer

powder X-ray pattern of silicon powder in range 20°–140° is shown in Fig. 19.23. This BND preparation is in the progress.

19.3.8 Polystyrene Film for Spectroscopic Standard

Fourier Transform Infrared (FTIR) spectrophotometer is a powerful characterization technique that reveals various fundamental facets of materials in many areas such as

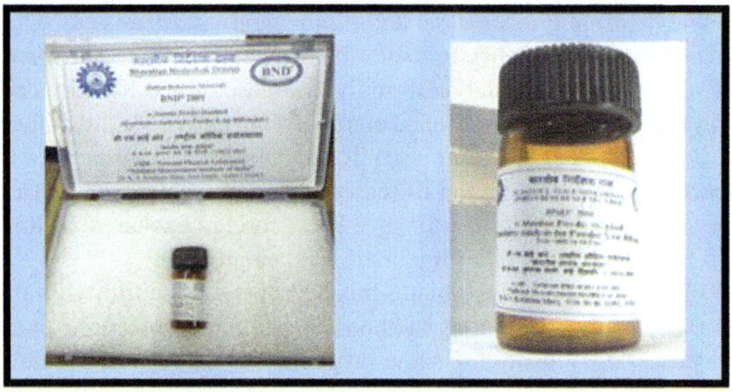

Fig. 19.22 Alpha Alumina BND for the calibration of PXRD

Table 19.9 List of SRMs released by NIST, USA [37]

S. No.	SRM application	NIST SRM No.	Material composition
1	Line position	640e	Silicon (Si)
2	Line shape	660c	Lanthanum hexaboride (LaB_6)
3	Quantitative analysis	676a	Alumina (corundum)
4	Quantitative analysis	1878b	Respirable quartz
5	Line shape	1979	ZnO, 15 nm and 60 nm

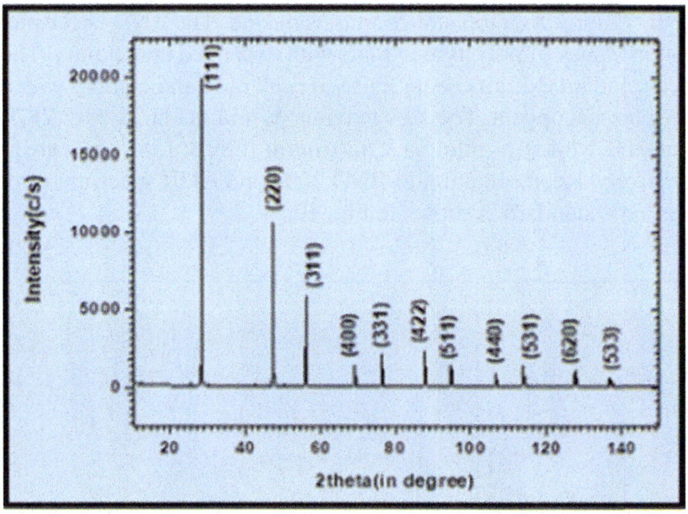

Fig. 19.23 XRD pattern of Si for BND

chemical, analytical, pharmaceutical and biomedical industries, etc. The reference materials can be used to qualify mid-IR - FTIR spectrophotometer for wavelength accuracy. Calibration, testing laboratories, private sector and FTIR making companies so forth is the user of the reference material. Currently in India, importing SRM 1921b from NIST, USA or other imported IR standard for calibrating the wavenumber scale of spectrophotometers on the IR region from 400 to 4000 cm^{-1}. Development of IR reference standard in India will save lot of foreign exchange as importing the SRM1921b from NIST and other imported IR standards [38].

CSIR-NPL has developed the matte-finished polystyrene film with thickness of about 38 ($\pm$2) μm by indigenous method. The film was prepared by indigenous and simple process. By using the same process, it is possible to prepared uniform thickness without using any costly and sophisticated equipment. The films can be used in different applications. The transmission and absorption peaks of polystyrene films were well characterized by FT-IR spectroscopy. It exhibits the clear peaks of standard polystyrene with good resolution. The homogeneity of the films has been checked thoroughly by measuring five different positions of a 25 mm diameter area along with stability.

Using the indigenous matte-finished polystyrene, CSIR-NPL has also developed BND 2004, is primarily for the calibration of wavelength values of FTIR spectrophotometers in the IR spectral region from 4000 to 400 cm^{-1}. The BND 2004 is a matte-finished polystyrene film (38 $\pm$ 2) μm thick with a 25 mm diameter exposed area mounted on a cardboard holder having dimension of 11 cm × 5 cm × 0.15 cm. The center location of the polystyrene film with respect to cardholder is 38 mm from bottom, 72 mm from top and 25 mm from side. The BND is certified using 14 peaks position values of polystyrene films with expanded uncertainty. The spectrum of reference standard exhibits better noise to peak ratio and comply with the criteria of different pharmacopoeia. The Abs maxima & minima at 2849 & 2870 cm^{-1} are greater than 0.60, while Abs minima & maxima at 1589 & 1583 cm^{-1} are greater than 0.16, respectively. Representation of BND 2004 and FTIR spectrum of polystyrene for spectroscopic standard is shown in Fig. 19.24.

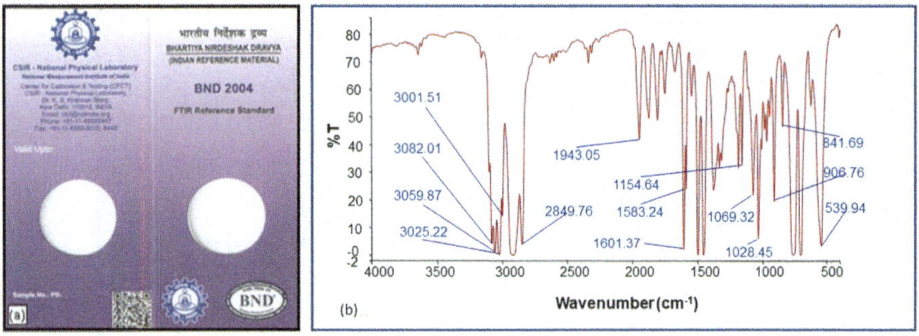

Fig. 19.24 **a** Label and representation of BND 2004 and **b** FTIR spectrum of polystyrene for spectroscopic standard

19.3.9 Development of Graphitized Petroleum Coke BND 5102

Graphitized petroleum coke has high fixed carbon content and very low levels of sulphur, ash and volatile matter which enacts extremely useful as a substitute for graphite in many applications. As a high purity carbon material, it is used for producing sealing and lubricating materials, electricity brushes, compound brakes, etc. It can also be used in plastic and rubber as an additive. A number of industries worldwide are involved in the manufacture/supply of carbon raiser. Contents of different components of available carbon raisers in market range as: carbon 90–98.5%; Ash content 0.8–5.0%; volatile matter 0.5–3.0%; sulphur 0.05–3.0%, particle size 0.5–5.0 mm and moisture <1%. India imports huge amount of carbon raisers from Japan, China, Korea, etc. The cost of the carbon raiser in turn depends upon the purity, i.e., high fixed carbon content and minimum impurities. The percentage of sulphur for this product is specially monitored as the customers of this product require a minimal level of sulphur. The concentration of theses impurities is being routinely analyzed by testing laboratories. However, reliable analytical results are required for compliance with international regulations in almost all areas of analysis. The main purpose of the development of this GPC standard is to provide a reference standard to the Indian industries to evaluate the carbon additive for its quality control and in the evaluation of techniques used in the analysis of graphite/carbon and materials of a similar matrix.

CRM/BND of graphitized petroleum coke has been developed to ascertain the quality of carbon additive and in the evaluation of techniques used in the analysis of graphite/carbon and materials of a similar matrix. A unit of BND 5102 consists of 50 g of graphitized petroleum coke that was ground to pass a 280 μm sieve, homogenized, packaged in an amber glass bottle under an argon atmosphere. It provides certified values of ash, sulphur and heat of combustion and reference values of moisture, volatile matter, and fixed carbon which is reflected in Tables 19.10 and 19.11. Figure 19.25 shows the graphitized petroleum coke BND 5102.

Table 19.10 Certified values for BND 5102 and associated uncertainties

Measurand	Certified value	Expanded uncertainty (±) at $k = 2$
Ash	0.35%	±0.01%
Sulphur (%)	0.31%	±0.03%
Heat of combustion (MJ/kg)	32.66 MJ/kg	±1.25 MJ/kg

Table 19.11 Reference values for BND 5102 and associated uncertainties

Measurand	Reference value	Expanded uncertainty (±) at $k = 2$
Moisture (%)	0.03	±0.01
Volatile matter (%)	0.52	±0.06
Fixed carbon (%)	99.10	±0.08

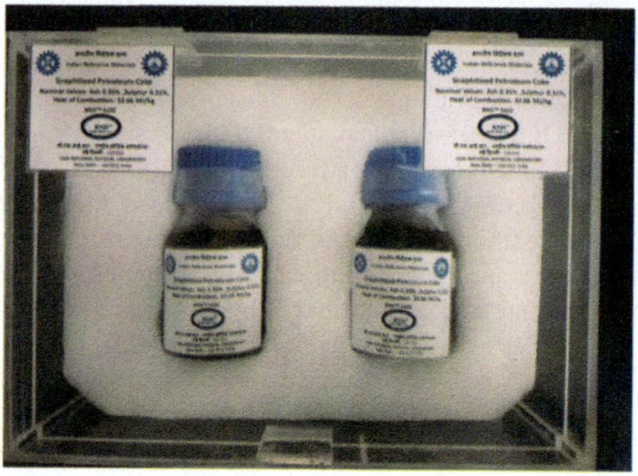

Fig. 19.25 Graphitized petroleum coke BND 5102

19.3.10 Gold BND 4201

India is the second-largest consumer utilizing approximately 849 metrics tons of gold. Therefore, its purity standardization (Ag, Pt, Pd, Rh, Ir, Ru, Os and Cd) is urgently required for consumer satisfaction. CSIR-NPL has signed MoU with Govt. Mint, Mumbai for the production of gold standard. In 2017 India's first home-grown high purity gold reference standard BND-4201 was launched at the India Government Mint (IGM) (Fig. 19.26) in collaboration with NPL as presented in Table 19.12 [39].

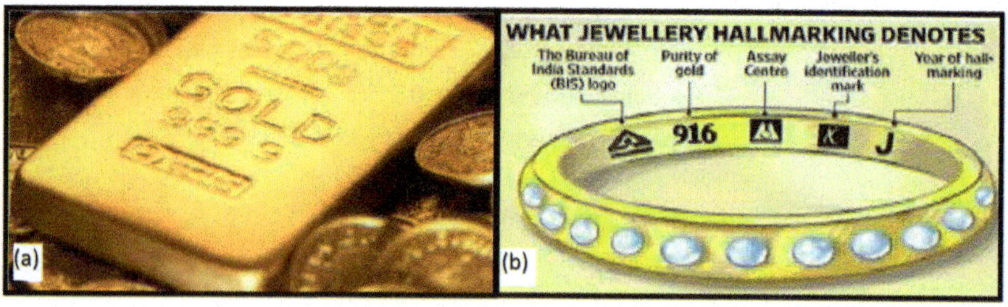

Fig. 19.26 BND-4201, reference material for gold of '9999' fineness (99.99% pure) and BIS hallmarking proposal

Table 19.12 BND 4201 for high purity gold reference standard

S. No.	BND number	BND name	Purpose
1	BND 4201	High purity gold reference standard	4 N Purity of gold

This BND of gold with '9999' fineness (99.99% pure) is beneficial to the consumers and public at large to ensure purity of gold.

BIS in India is standardizing the gold hallmarking process as per government guidelines.

To be very specific, the gold BND provides a quality check for the jewelers by calibrating their equipment. The calibrated equipment for purity gives the actual purity of gold to the customer as per their satisfaction. In addition to this, the gold BND improves the testing of the conventional fire assay method which is time-consuming and not an environmentally friendly as poisonous gases are released in fire assay method [40].

Gold reference standard is indispensable in gold and jewelry hallmarking. This is also useful for collection and purity testing centers to certify the purity of gold deposits under the gold monetization scheme. Govt. of India has decided that from Jan. 2021 onwards only hallmarked gold with purity of 58.33, 75.00 and 91.60% (14, 18, 22 carat) will be available in the consumer market. Further 800 referral laboratories of BIS are looking forward to these purity gold BNDs as mentioned above (Fig. 19.26).

19.3.11 Development of Aqueous Alcohol Standard BND 01.31.01 for Breath Alcohol Analyzer

BND for breath alcohol analyzer significantly contributes to the reliability of routine measurements performed with breath alcohol analyzer [41]. This ensures that the result obtained during calibration as well as verification from breath analysis is accepted as evidence and this improves the accuracy of breath alcohol testing. The impact of this work would directly link with the human health management/enforcement/evidence in medico-legal prosecution for safety, clinical and forensic laboratories. CSIR-NPL has established the above capacity, developed Reference materials (SI traceable standards) and created a calibration facility for accurate and reliable measurements of alcohol by breath analyze. The schematic diagram given below is described the process followed for the certification of aqueous alcohol standard solution. As per the requirement under the Motor Vehicle Act of India, the aqueous alcohol reference materials BND 01.31.01 with 0.031% alcohol content had been released on 31.10.2017. Figure 19.27 shows the certification process of aqueous alcohol reference materials.

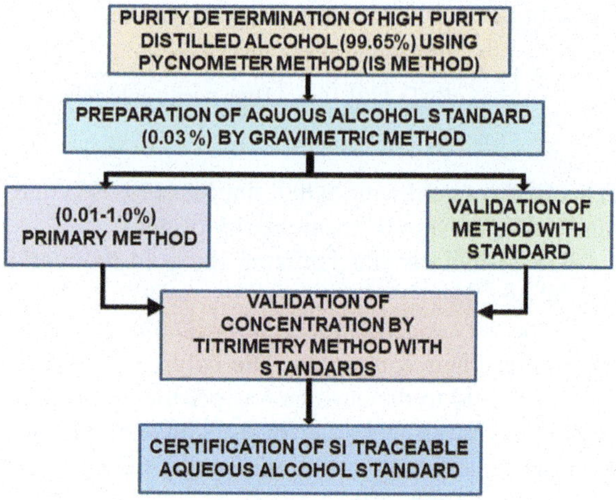

Fig. 19.27 Certification process of aqueous alcohol reference materials

19.3.12 Development of BNDs Related to Gas Mixture

Gas Metrology plays a crucial role in the field of global atmospheric monitoring of Green House Gases (GHGs), National Ambient Air Quality Standards (NAAQS) parameters and vehicular emissions as well. CSIR-NPL maintains the primary Reference Gas Mixtures (PRGMs) and conducts Proficiency Testing (PT) in this field. The laboratory has established facilities for the gravimetric preparation of PRGMs, their analytical validation and impurity analyses.

CSIR-NPL has developed BNDs related to gas mixtures [42]. The list of BNDs related to gas mixtures is shown in Table 19.13. These BNDs of gas mixture were released by Honourable Minister of Science and Technology, Minister of Health and Family Welfare and Earth Sciences, Dr. Harsh Vardhan as shown in Fig. 19.28.

19.3.13 BNDs Related to Synthetic Blood

Personal protective equipment (PPE) is the protective gear to safeguard the health of the workers for minimizing the exposures to the harmful ambience [43]. The potency of PPE depends upon the

stringent testing as per the standard protocols. These tests assure that such PPE would safeguard the health workers or patients and restrict the flow of bacterial/viral transmission. The resistance offered by PPE to synthetic blood is considered as an acid test for their utility in pandemic situations like the worldwide spread of COVID19, which is a viral transmission. The prime requirement of blood penetration tests is synthetic blood with physical properties similar to human blood. NPL-CSIR has taken the lead in developing the standard for synthetics blood, particularly for testing of PPE kits.

Table 19.13 BND related to gas mixture

S. No.	BND number	BND name
1	BND 6001	Nitrogen(N_2) with minor impurities (THC $\leq$ 0.05 mmol/mol, CO $\leq$ 1 mmol/mol, CO_2 $\leq$ 10 mmol/mol, Oxygen $\leq$ 2 mmol/mol, NO_X $\leq$ 0.02 mmol/mol, N_2O $\leq$ 0.02 mol/mol)
2	BND 6002	Synthetic air with minor impurities (THC $\leq$ 0.05 mmol/mol, CO $\leq$ 1 mmol/mol, CO_2 $\leq$ 10 mmol/mol, NO_X $\leq$ 0.02 mmol/mol, N_2O $\leq$ 0.02 mmol/mol)
3	BND 6003	Primary Gas Mixture of CO in nitrogen gas (5 μmol/mol)
4	BND 6004	Primary Gas Mixture of CO in nitrogen gas (50 μmol/mol)
5	BND 6005	Primary Gas Mixture of CO in nitrogen gas
6	BND 6006	Primary Gas Mixture of CO in nitrogen gas (480 μmol/mol)
7	BND 6007	Primary Gas Mixture of CO_2 in nitrogen gas (13% mol/mol)
8	BND 6008	Primary Gas Mixture of CO_2 in nitrogen gas (2.7% mol/mol)
9	BND 6009	Primary Gas Mixture of CO_2 in nitrogen gas (1% mol/mol)
10	BND 6010	Primary Gas Mixture of CO_2 in synthetic air (380 μmol/mol)
11	BND 6011	Primary Gas Mixture of CO_2 in synthetic air (480 μmol/mol)
12	BND 6012	Primary Gas Mixture of CO_2 in synthetic air (800 μmol/mol)
13	BND 6013	Primary Gas Mixture of CH_4 in nitrogen (2 μmol/mol)
14	BND 6014	Primary Gas Mixture of CH_4 in nitrogen (10 μmol/mol)
15	BND 6015	Primary Reference Gas Mixture Standard CH_4 (methane) in nitrogen gas (2.50% mol/mol) in 10 ltr aluminium cylinder filled at 120 bar. (2.5% mol/mol)
16	BND 6016	Primary Reference Gas Mixture Standard NO (nitric oxide) in nitrogen gas (50 micromol/mol) in 10 ltr aluminium cylinder filled at 120 bars—(50 mmol/mol)
17	BND 6017	Primary Reference Gas Mixture Standard NO (nitric oxide) in nitrogen gas (100 micromol/mol) in 10 ltr aluminium cylinder filled at 120 bars—(100 mmol/mol)

The synthetic blood shall meet the following requirements, traceable to national standards.

Surface Tension: (0.042 $\pm$ 0.002) N/m, pH: (7.3 $\pm$ 0.1), Viscosity: (2.7 $\pm$ 0.3) MPa s, Conductivity: (12.0 $\pm$ 1.2) mS/cm.

19.3.14 BNDs for Thermoelectric Metrology

In order to realize the thermoelectric metrology, the precise and accurate measurement of thermoelectric parameters is required. The continued development of new TE materials for high-temperature power conversion applications requires reliable and accurate characterization of the electrical and thermal transport properties [44].

Fig. 19.28 Honourable Minister of Science and Technology, Minister of Health and Family Welfare & Minister of Earth Sciences, Dr. Harsh Vardhan releasing the water and gas mixture BNDs

In general, thermoelectric metrology will empower the Indian industries in power generation based on thermoelectric technology and will also impose a stringent process of quality check of imported TE devices in the country. CSIR-NPL is taking the initiative to develop reference standards for thermoelectric parameters to improve the measurement capabilities to strengthen the medium, large and small (MSME) industries. The Seebeck coefficient and thermal conductivity is an essential physical property in evaluating the potential performance of these TE materials, as it is sensitive to the electronic and crystal structure. However, the deceptive simplicity of measuring the Seebeck coefficient and thermal conductivity has led to the implementation of non-ideal practices that have further complicated the confirmation of reported high ZT materials. To ensure meaningful inter-laboratory comparison of data, thermoelectric measurements must be reliable, accurate, and consistent. The relevant measurement techniques and apparatus designs are required to effectively manage uncertainty, while also providing a reference resource of previous advances in high-temperature thermoelectric metrology [45].

For establishing Seebeck coefficient metrology, the primary requirements involved are (i) the measurement of the voltage and temperature at the same location and at the same time, (ii) the probes must be in very good thermal and electrical contact with the specimen, and (iii) low voltages must be acquired with minimal extraneous contributions. In addition, the implementation must avoid distortion of the temperature-voltage correspondence and must use appropriate temperature differences such that the range satisfies the assumption of linearity in S(T). Further, the thermal conductivity measurement is another important parameter involved in determining the performance of thermoelectric materials [46].

Thus, it would be useful to evaluate various arrangements and methodologies over a broad temperature range and for a variety of materials, and contact interfaces are required to be establishing thermoelectric metrology. CSIR-NPL has taken initiative to develop a reference standard for these parameters and working initially on the development of a reference standard of the Seebeck coefficient.

In this endeavour, the heating-cooling experiment for Seebeck coefficient for the best optimized Bi_2Te_3-based sample was performed in ZEM-3 as presented in Fig. 19.29a. Interestingly, heating and cooling path for Seebeck coefficient retrace the same path indicating the most stable system within the displayed temperature range. Further, we have performed the repeatability test of Seebeck coefficient four times as shown in Fig. 19.29b. One can clearly notice that the value remains almost unchanged throughout the temperature range.

For accurate and precisely reporting the data, the mean value of Seebeck coefficient is estimated at each temperature and is plotted to calculate the expanded uncertainty as shown in Fig. 19.29c. The final mean values of Seebeck coefficient has been fitted with a parametric model given below:

$$m(t) = a_0 + a_1 t + a_2(t - 200)^2 + a_3(t - 200)^3 + a_4(t - 200)^4$$

After polynomial, the values of parameters a_0, a_1, a_2, a_3 and a_4 were estimated as 1206.37556, −12.72871, 0.05458, −9.64891E−5, and 5.97607E−8. It has been

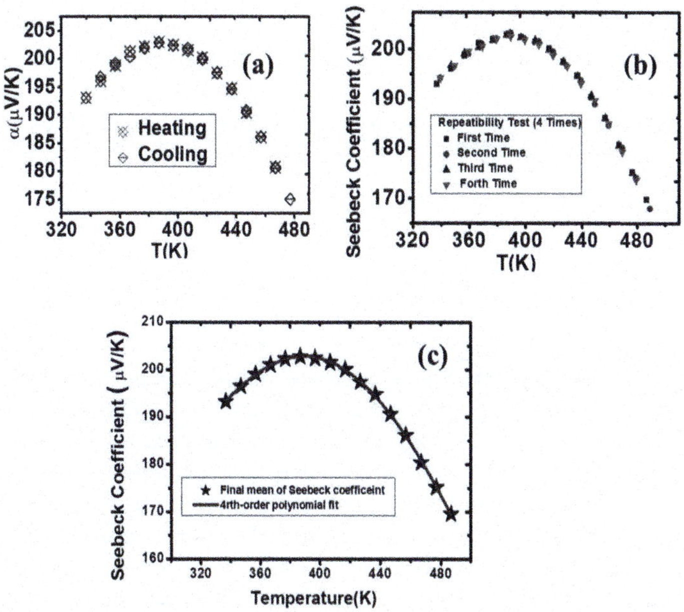

Fig. 19.29 **a** Seebeck coefficient during heating and cooling, **b** repeatability test (four times) of Seebeck coefficient, and **c** fitting of mean value of Seebeck coefficient by 4th order polynomial

Table 19.14 Mean of Seebeck coefficient at all temperatures with expanded uncertainty

S. No.	Temperature (K)	Final mean of Seebeck coefficient (μV/K)	Expanded uncertainty (coverage factor = 2)
1	339.4	193.2128	0.123564
2	349.4	196.4614	0.239707
3	359.4	199.044	0.351613
4	369.4	201.003	0.459954
5	379.4	202.255	0.38444
6	389.4	202.7462	0.179238
7	399.4	202.3964	0.082468
8	409.4	201.5032	0.220678
9	419.4	200.0274	0.117104
10	429.4	197.37	0.128283
11	439.4	194.7004	0.082648
12	449.4	190.567	0.124812
13	459.4	186.0918	0.056333
14	469.4	180.3724	0.21537
15	479.4	175.1816	0.086581
16	489.4	169.5454	0.163733

observed that the fourth-order polynomial gives a better fit for our data. In order to have a confidence level of the measured data and for an accurate value of Seebeck coefficient with associated uncertainty, the final mean of Seebeck coefficient at all temperatures with expanded uncertainty is presented in Table 19.14 and further conformity assessment will be carried out by round-robin test through NMIs of other countries.

19.3.15 BND for Spin Current

Spin Current in magnetic structures has been a great challenge to replace conventional charge current towards a complete quantum world. The simplest way to spin current generation is spin pumping, where the ferromagnetic (FM) layer induces a spin-polarized current, $J_s \alpha\ 1/4\pi\ g_{\uparrow\downarrow}\ (m \times \mathrm{d}m/\mathrm{d}t)$ perpendicular to spin-polarization vector (σ) at resonance condition, and the spin-polarized electrons create a charge current, $J_c\ \alpha\ J_s\ x\ \sigma$ in non-magnetic layer, termed as Inverse Spin Hall effect (ISHE). CSIR-NPL aimed to develop CRM/BND for spin current through ISHE by the use of ferromagnetic resonance (FMR) technique (Fig. 19.30a). $Ni_{81}Fe_{19}$ (Py)/Pt magnetic bilayer thin film is used to detect the spin current where Py (20 nm)/Pt (10 nm) film deposited on MgO (001) substrate by DC magnetron sputtering technique. Resonant microwave absorption (d*P*/d*H* vs. H: Fig. 19.30b) and induced voltage ($V_{\mathrm{ISHE:}}$

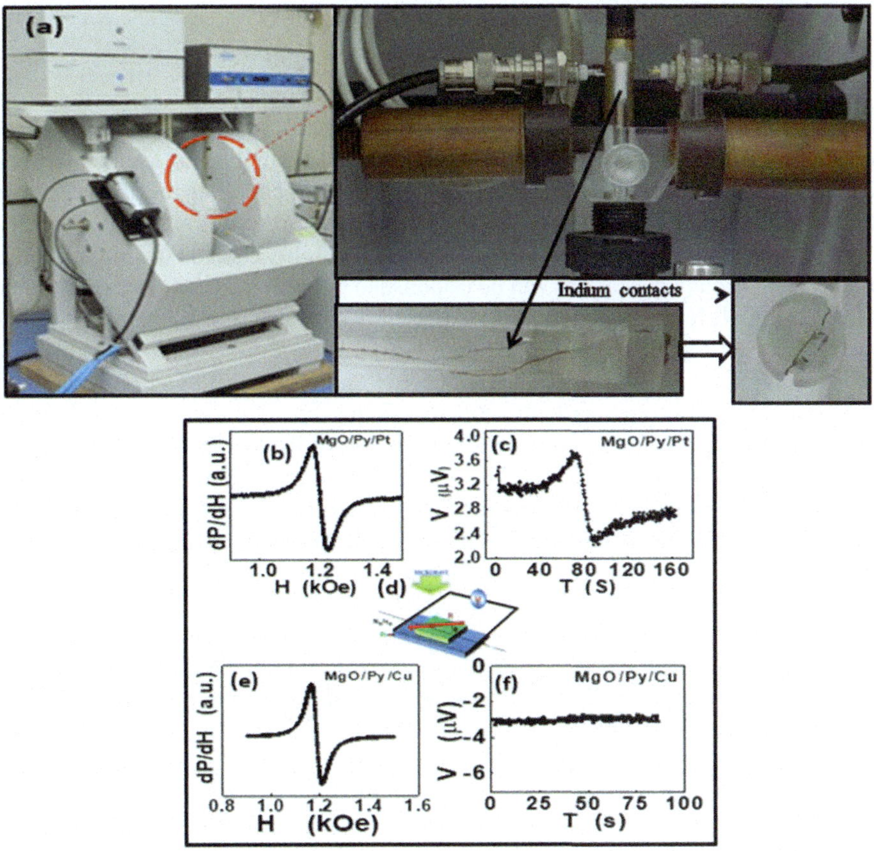

Fig. 19.30 **a** FMR setup for ISHE voltage measurement containing assemble of sample tubes and indium contacts, **b**, **c** voltage spectra recorded for Pt(10 nm)/Py(20 nm)/MgO(001), **d** Schematic illustration of sample (Py/Pt) used in the measurement, **e**, **f** voltage spectra recorded for Cu(10 nm)/Py(20 nm)/MgO(001)

Fig. 19.30c) spectra of Py/Pt thin film were recorded at 9.8 GHz (X-band) frequency by using Electron Paramagnetic Resonance (EPR) Spectrometer. The absorption and voltage spectra recorded at 18 mW of Microwave Power and $\theta = 90°$ (Fig. 19.30d) with maximum induced voltage (of ~3.72 μV) at resonance field (H_{res}). Further, to confirm the origin of pure spin current absorption voltage spectra recorded replacing Pt by Cu, which shows no detectable induced voltage curve (Fig. 19.30e, f). In this direction, CSIR-NPL has taken initiative to develop BND for measuring pure spin current. Further, the trails are going on to enhance the sensitivity of the measurement to detect accurately the induced voltage (V_{ISHE}) for the development BND for spin current.

19.3.16 BNDs for Tensile Strength

The mechanical properties of materials are the most important engineering properties as it is directly related to the safety of structures, machine design, development of transportation systems and the working life of components. Many important mechanical properties of materials including tensile strength, impact strength, creep and fatigue are procedure dependent and therefore test results are traceable to standard test procedures rather than SI base units. Traceability chain in the measurement of these mechanical properties can be established through international round-robin tests among NMIs. In this direction, CSIR-NPL has taken the challenge of developing tensile strength reference standards of three sets of samples with different tensile strength; polypropylene, polycarbonate and polyamide. The traceability is being established through a round-robin test [47].

Polcarbonate (PC) reference sample was prepared by THERMO Haake Minilab II micro compounder. For this, a known amount of PC pellets was taken. It was loaded into an extruder maintained at 225 °C and then heated gel is filled in the Thermo hake Minijet II Injection moulding machine having suitable shape as per ASTM D638 type V and 750 bar pressured was applied. The samples were taken out after the automatic release of pressure by the instrument (Fig. 19.31). The tensile strength measurements for samples were carried out using a calibrated universal testing machine (Instron5967). Samples were produced in a batch and were sent to various accredited laboratories for measuring the tensile strength in order to get a good estimate of the value.

Tensile strength BNDs can be used by various industrial organizations. Some of them are mentioned here: Apollo tyres, SSN motors, Forbes Gokak, Light Alloy Products, Chennai, Timken Engineering, Bangalore, Tessitura Monti India Pvt Ltd, Kolhapur, Gharda Chemicals Ltd, Mumbai, Berger Paints, Kolkata SHAR, Sriharikota, etc.

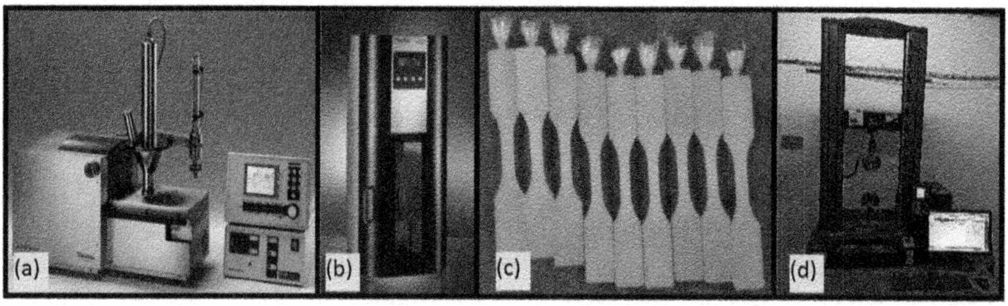

Fig. 19.31 **a** Micro compounder, **b** injection moulding system, **c** test specimen and **d** universal testing machine used for making tensile strength BNDs

19.3.17 Dynamic Viscosity BND® of Newtonian Fluids (Silicon Oils)

Newtonian fluids in particular silicone oils are very vital for almost every industry that deals with liquid/semi-liquids due to their peculiar properties/behaviour. Dynamic viscosity (η) is the main factor to determine the viscous nature of fluids for the assurance of the quality products [48].

Dynamic Viscosity (η)

The dynamic viscosity of a fluid is a measure of the resistance to flow exerted by a fluid, and for a Newtonian fluid it is defined as:

$$\mu = \tau/(\mathrm{d}v/\mathrm{d}n)$$

Where τ is the shear stress, v is the velocity of the fluid in the shear stress direction, and dv/dn is the gradient of v in the direction perpendicular to flow direction.

Dynamic viscosity measuring tool is the concentric (or coaxial) cylinder systems (DIN EN ISO 3219 and DIN 53019) which measure the dynamic viscosity of Newtonian fluids. In this Rheometer, bob and cup geometry is used to compute precisely shear rate and shear stress to calculate dynamic viscosity (Fig. 19.32). The measurements are performed in the temperature.

Range from −20 °C to 100 °C for CC geometry

From CC geometry the shear rate, $\gamma = 2 \times \omega$RC/(RC)2- (Rb) 2; and shear stress, $\tau = M/2\pi$(Rb) 2. These equations are for the shear rate on the bob's surface and shear stress depends on the measured torque, the bob's radius and length. Shear rate (γ) and Shear stress (τ) parameters are used in Newton's law to determine the dynamic viscosity (η) of a Newtonian fluid (Table 19.15).

The traceability to SI units established through mass standard of CSIR-NPL as shown in traceability chart Fig. 19.33.

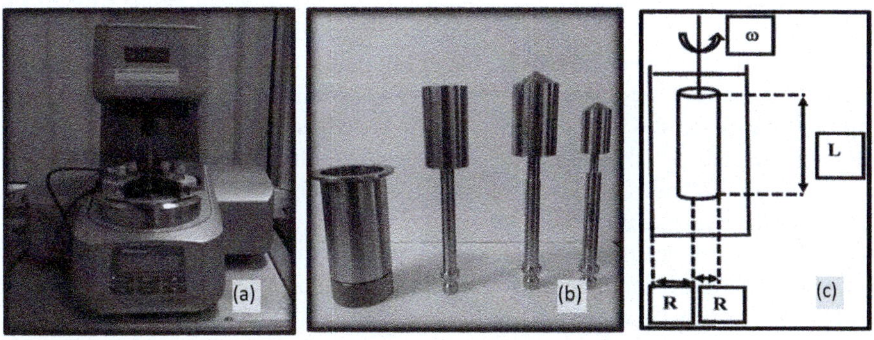

Fig. 19.32 **a** Magneto-rheometer (MCR series) **b**, **c** cup & bobs of concentric cylinder (CC) geometry

Table 19.15 15 Value of dynamic viscosity range and temperature range for a Newtonian fluid

Newtonian fluid	Dynamic viscosity range	Temperature range
Silicon oil	1 Cp–105 Cp	−20 °C to 100 °C

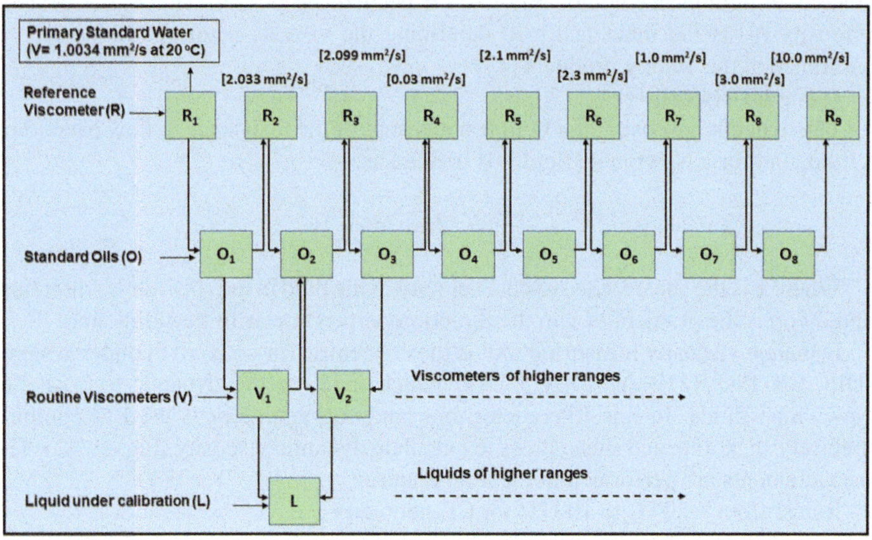

Fig. 19.33 Traceability chart showing traceability to SI units established through mass standard of CSIR-NPL

19.3.18 Development of ISO/TC-229/WG4 Document No. 19807 for Nanometrology

CSIR-NPL has contributed to the preparation of ISO document along with BIS in the area of nanometrology. It is a new direction and one step ahead from India contributing to the ISO document for the development of liquid suspension of nanomagnetic particles. The above-said document was published in 2019. This document will set clear guidelines for the characterization and measurement of nanosuspension of magnetic particles to be used for various applications around the globe.

Nanoparticles have one dimension less than 100 nm in size. The physical properties of materials in nanodimension change drastically and make them technologically important in health, energy harvesting, memory devices, etc. The wide use of nanoparticles required international standard to be defined to give the best products to society. Keeping this in view ISO started to develop guidelines for nanoparticles under ISO/TC229. To keep pace with the world in the nanotechnology field, CSIR-NPL-BIS initiated to prepare ISO/TC 229 document on "Suspension of magnetic nanoparticles—characteristics and measurement" and published under materials specification ISO/TC229-WG4 group.

19.4 BNDs to Strengthen the Quality Infrastructure in India

BND program supports strengthening the quality infrastructure through dissemination of BNDs in various sectors. It is the need of time that various ministries, accreditation bodies in India like NABL, etc., BIS regulatory bodies and all the accredited laboratories [49] should work in harmony to realize the quality products. The interdependent linking through CRMs is shown in Fig. 19.34.

19.4.1 Support to Various Ministries

The economy of the country depends mainly on various resources in nature such as coal, petroleum products, mining, ores, agrotech, etc. In particular, India is rich in Ayurveda based knowledge and Unani for keeping healthy society. By considering this, the Govt. of India has set up different ministries to enrich the R & D for consumer products as per ISO guidelines. Therefore, BNDs of various parameters related to Ayurveda, Unani, and Siddha are required to control the quality and standardization of the products. Similarly, BND related to agricultural, oils, fatty acids and processed food products, etc. have an impact on the economy and therefore CSIR-NPL and M/s Farelabs have jointly signed an MoU for the production of BNDs in these areas. Thus, the BNDs in the area of agricultural and processed food products will also enhance the export and regulate import which certainly supports M/o Commerce & Industry. Further, the petroleum of various concentrations of impurities (like from 2.5 ppm to 1000 ppm of sulphur in diesel) is required by stakeholders. For that CSIR-NPL has signed MoU with Hindustan Petroleum Corporation Limited and Bharat Petroleum

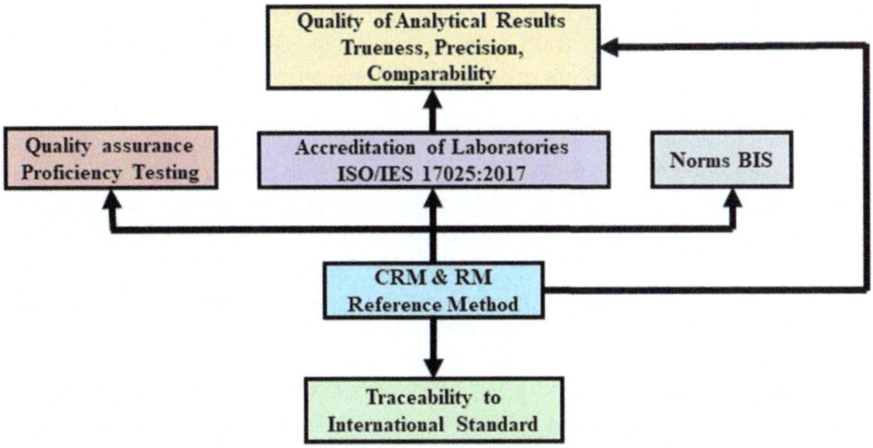

Fig. 19.34 Schematic diagram showing the CRMs linking through NABL, BIS and PT

Corporation Limited and has released 27 petroleum BNDs. Sulphur of 10 ppm in diesel is only permissible to implement Bharat stage VI as per environmental ministry guidelines. Sulphur in diesel is intended to use for the calibration of instruments and evaluation of test method used in the determination of total sulphur in diesel or materials of similar matrix. These BNDs directly support to the ministry of petroleum and natural gases for quality-controlled products. Similarly, for textiles, the purity of yarn is also important for making fabric. Subsequently, an agreement has to be signed with concern departments for the production of required BNDs. BND related to aerosol, water and gas mixture are being produced at CSIR-NPL and supports M/o Environment & Forests (Central Pollution Control Board).

CSIR-NPL in collaboration with M/s SUMS techno is making BNDs related to various minerals, ores, etc., which will help for assuring the quality of mineral-based ores products like Au, Zn, Sn and Cu. Iron and ferro-alloy metals including Cr, Co, Ni, Mn and V and Light metals including Al, Ti, Mg, Li and Be. Similarly, for automobile industry for the ministry of transport, metal hardness BND blocks are developed in joint collaboration with M/s. Global PT Provider. The hardness BND blocks are used to examine the mechanical property of the material which is mainly used for manufacturing automobiles; aviation industry etc. Similarly, the cement and building materials developed in collaboration with NCCBM are helpful to the Ministry of Urban Development for providing housing for all sections of the society.

19.4.2 Services to Accreditation Body (NABL)

CSIR-NPL and NABL having many accredited laboratories across India are closely related to provide testing, validation of methods, calibration of equipment, proficiency testing and traceability to fulfil the requirement of stakeholders in the Government and Private sector. NABL as an accreditation body develops the confidence of stakeholders by providing high-quality value-driven services, fostering APAC/ILAC MRA, empanelling competent assessors, creating awareness among the stakeholders, initiating new programs supporting accreditation activities and pursuing organizational excellence [50]. It also involves a third-party assessment of the technical competence of testing.

NABL in its document (clause 6.6 metrological traceability) related to the general requirement for the competence of testing and calibrating laboratories has mentioned that metrological traceability using CRMs like BNDs is required. There are a number of clauses in ISO/IEC 17025 that cover a wide range of aspects of a laboratory's operations such as validation of methods, measurement traceability, etc. which has already been discussed in the earlier sections.

In order to establish accurate and reliable measurement capabilities, a number of organizations, including ILAC, national accreditation bodies and national measurement institutes, have developed guidance documents and programmes to assist laboratories in compliance with these requirements. For RMPs, the NABL accreditation has an advantage that they can obtain metrological traceability from the NPL and, therefore, can produce and market their CRMs under the brand name of BND, which can reach to the international community through, COMAR database.

19.4.3 Services to National Standard Bodies: BIS

BIS is the national standard body of India which is established under the BIS Act 2016, involved in the formulation of Indian standards through its various technical committees comprising of stakeholders and eminent experts from the subject area. It is formulated under the act of parliament. Particularly for the importance of reference materials in India and global economy, BIS has established a dedicated committee for the work related to standardization in the area of reference materials. This committee acts as a national mirror committee of ISO REMCO and has the right to vote in ballots of ISO REMCO. It will examine, India's stand on various documents emanating from ISO REMCO. Director, CSIR-NPL is the chairman of ISO REMCO committee in India.

For developing country like India, it is the need of the hour to increase the production and use of reference materials. So far 75 BNDs have been released by NPL in specific areas such as cement, petroleum, water, hardness, gold, silver.

BIS has been providing traceable and tangible benefits to the national economy in a number of ways—providing safe reliable quality goods; minimizing health hazards to consumers; promoting exports and import substitute; control over proliferation of varieties, etc. through standardization, certification and testing [51]. BIS has a large no. of accredited and referral laboratories for hallmarking process. All these laboratories are to be supported by CSIR-NPL in terms of measurement traceability by providing BNDs and validation of results. Being the NMI, CSIR-NPL has broad expertise in various physicomechanical, chemical and related measurement techniques is continuously supporting in various committees and developing ISO documents as a member. In nutshell, NMI is tirelessly contributing to the growth of economy of the country.

19.4.4 Conduct Proficiency Testing for the Stakeholders

A large number of measurements are carried out by laboratories all over India. In order to assess the quality of natural and commercial products, food, environment and health, one can use the reference material for examining its quality. Usually, economic losses occur due to incorrect data analysis and wrong interpretation, and this needs to be corrected [52]. To achieve this goal, ISO/IEC 17025 may be used in the routine analysis only after an appropriate validation. Within this validation process, it has to be experimentally demonstrated that the method is statistically correct and evidence for the required trueness has been achieved. The techniques used for the determination of the method's performance include the use of CRMs and

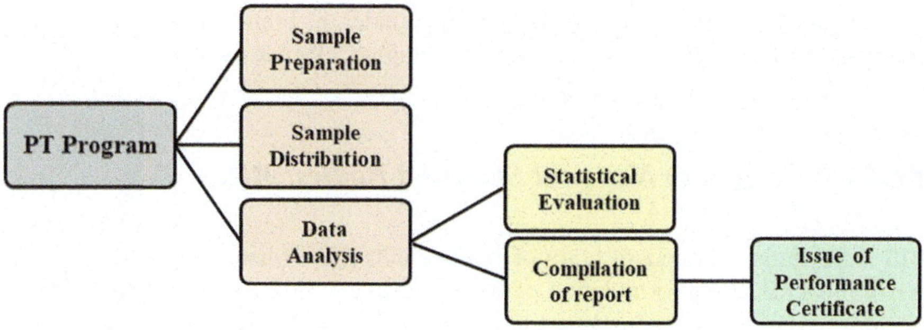

Fig. 19.35 Schematic diagram of the steps followed by CSIR-NPL

inter-laboratory comparisons. According to clause no. 7.7.2 of ISO/IEC 17025:2017, the laboratory involved for testing/calibration needs to show its performance by comparing its result with the other laboratory. To show its capability, the laboratory has to participate in the Proficiency Testing (PT) program.

CSIR-NPL is supporting for the PT time to assess the competency of testing laboratories in various commodities. The certified reference materials (CRMs) of lead [53], chromium and zinc solutions of 1, 2 and 5 mg/L had been prepared gravimetrically at CSIR-NPL and sent as the PT sample to the sixty-seven numbers of participated laboratories. Starting from the protocol preparation to the performance report of the participants had been carried out by CSIR-NPL. The steps followed by the CSIR-NPL to conduct the PT program are given in the schematic diagram like Fig. 19.35. To maintain the confidentiality of the laboratories each laboratory is assigned a code number. Data received from the participating laboratories were statistically analyzed for their quality by calculation of Z-score within the laboratory (ZWi) and between the laboratories (ZBi) in accordance to the NATA guidelines [53], which are in compliance with the requirements of ISO 17043:2010. Figures 19.36 and 19.37 are the Z score of the zinc solution between (ZBi) and within (ZWi) laboratories and most of the laboratories showed satisfactory results. Figure 19.38 shows percent of success w.r.t metal solution of lead, zinc, chromium of 1, 2 and 5 mg/L concentration.

The performance among the sixty-seven numbers of laboratories is shown in Fig. 19.38, for the testing ability of lead, zinc and chromium. In the case of higher concentration of analyte, the analytical competency of the laboratories is very good irrespective of the elements rather than the lower concentration.

19.5 Future Prospects of BNDs

In nutshell the BNDs and its related activities will promote method validation for metrologically valid procedure, calibration, traceability of measurement results, etc. This may enormously help medium, small- and large-scale industries to strengthen

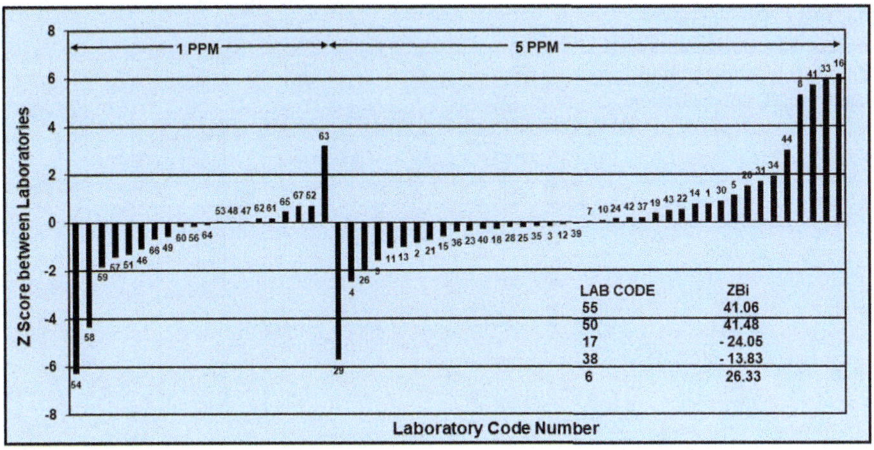

Fig. 19.36 The *Z* score of the zinc solution between (ZBi) laboratories

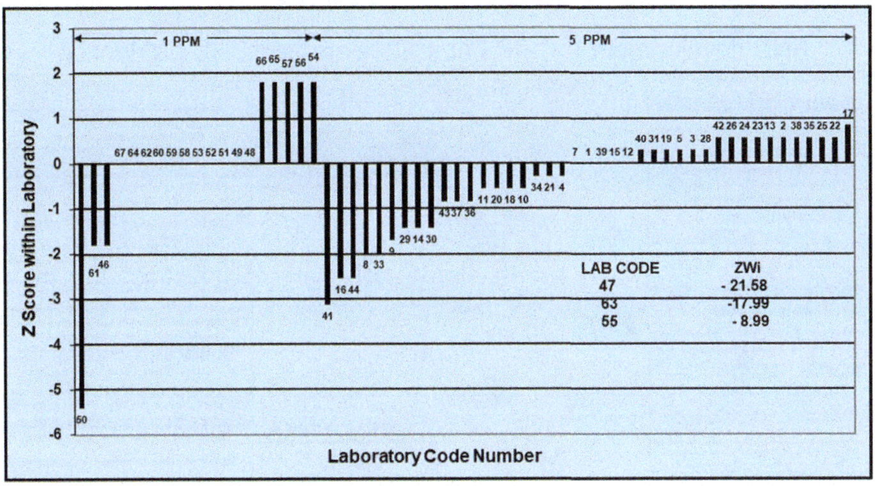

Fig. 19.37 The *Z* score of the zinc solution within (ZWi) laboratories

their capabilities for quality infrastructure along with other sister labs of CSIR (Fig. 19.39). As a result, it may have immense impact on import substitution and export. In this context, India faces huge problems due to poor quality of import and rejection of exports at foreign borders. This causes huge economic losses to the country. The main reason for this is the non-availability of enough BNDs/CRM dealing with quality control of products. As India has nearly four lakhs testing laboratories dealing with the quality control of the product/services. Therefore, the BNDs not only required for calibration, quality control and validation of method rather

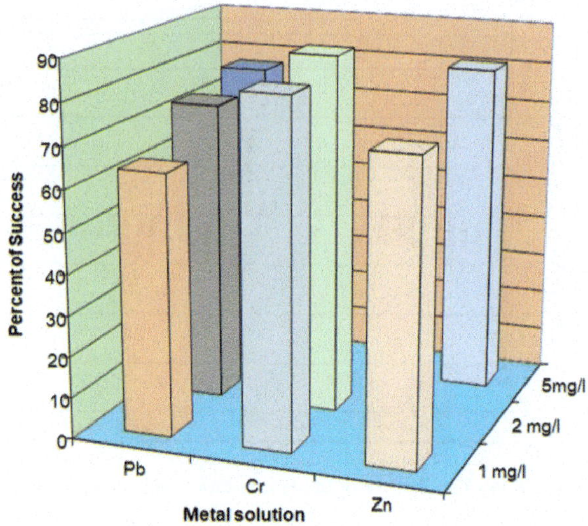

Fig. 19.38 Percent of success w.r.t metal solution of lead, zinc, chromium of 1, 2 and 5 mg/L concentration

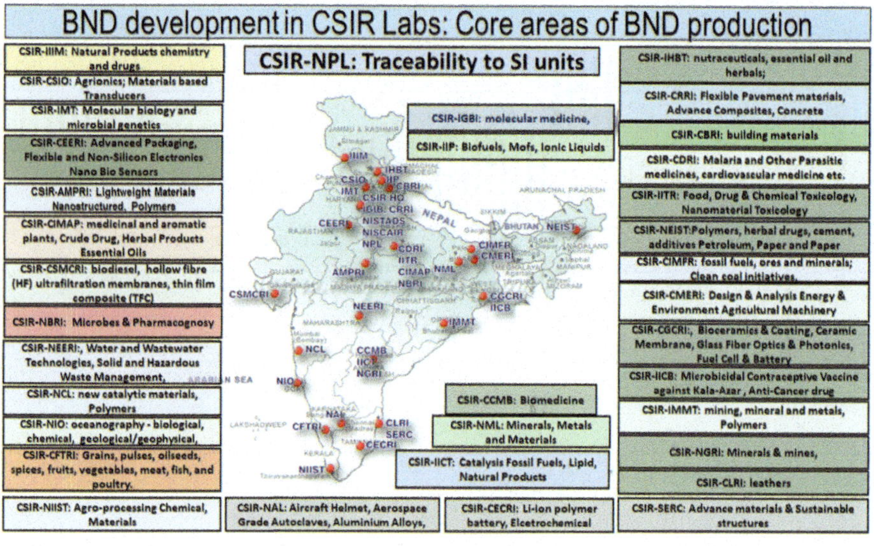

Fig. 19.39 BNDs will be developed in association with RMPs and other sister CSIR laboratories

it affects the export/import and causes hindrance to the economic growth. Therefore, countries need a larger number of BNDs to overcome such problems in all the sectors. In addition to this, a coordinated BND program with various ministries, regulatory bodies, accreditation bodies and stakeholders may improve consumer satisfaction. CSIR-NPL is committed to improve the international competence to produce BNDs at large so as to maintain quality assurance and minimize the import of the CRMs which will make India self-reliance (Atma Nirbhar Bharat). This above-said programme will have an impact not only on the country's quality infrastructure but also will help to generate huge employment in MSME. The country's youth power can be engaged/utilized for enormous productivity. This trend is a step ahead for becoming superpower fulfilling the dream of Dr. APJ Abdul Kalam.

Figures 19.40, 19.41, 19.42, 19.43, 19.44 and 19.45 show the requirement of BNDs related to environment, food, feedstuffs and agriculture, health/safety and industrial products, respectively. CSIR-NPL is committed to improve the quality

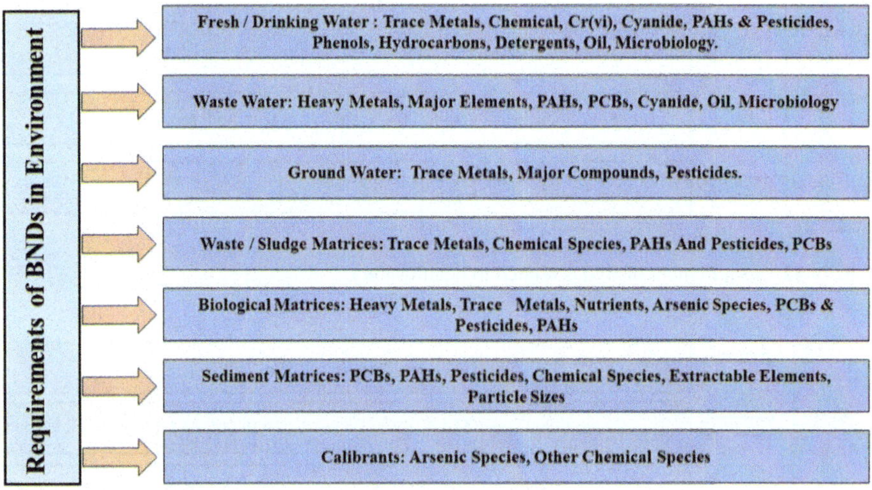

Fig. 19.40 Requirement of BNDs related to environment

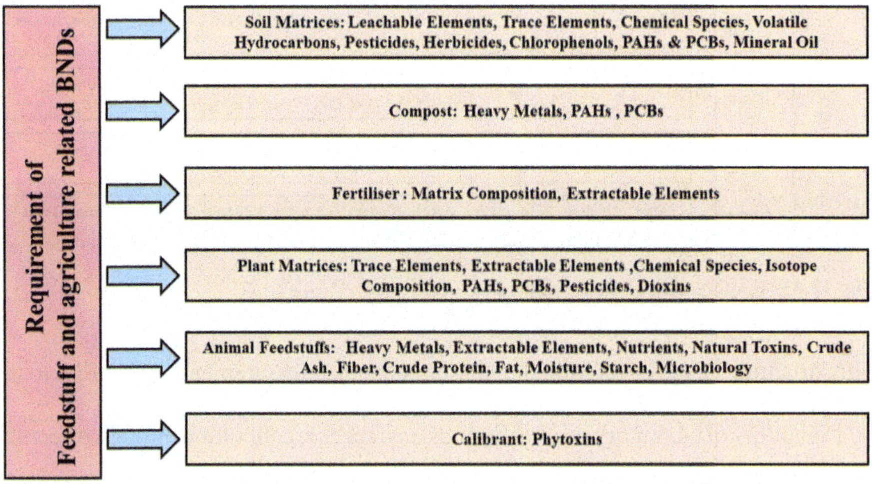

Fig. 19.41 Requirement of BNDs related to feedstuffs and agriculture

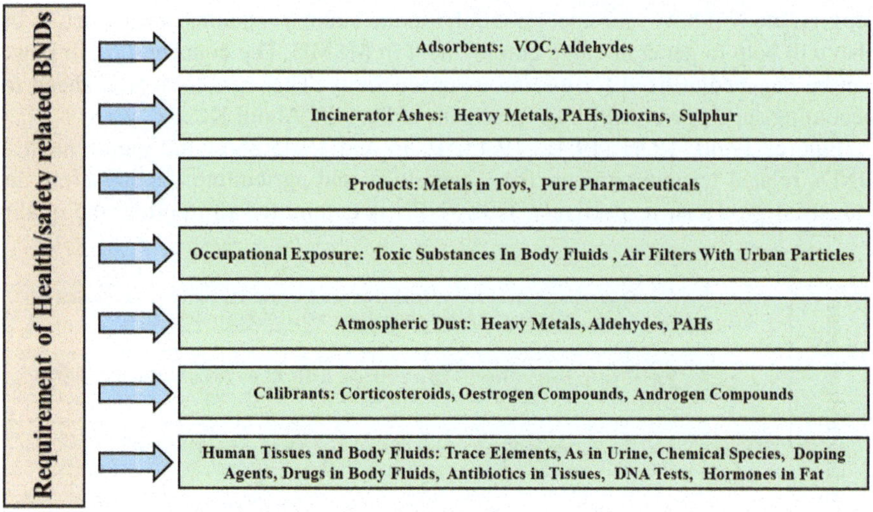

Fig. 19.42 Requirement of BNDs related to health/safety

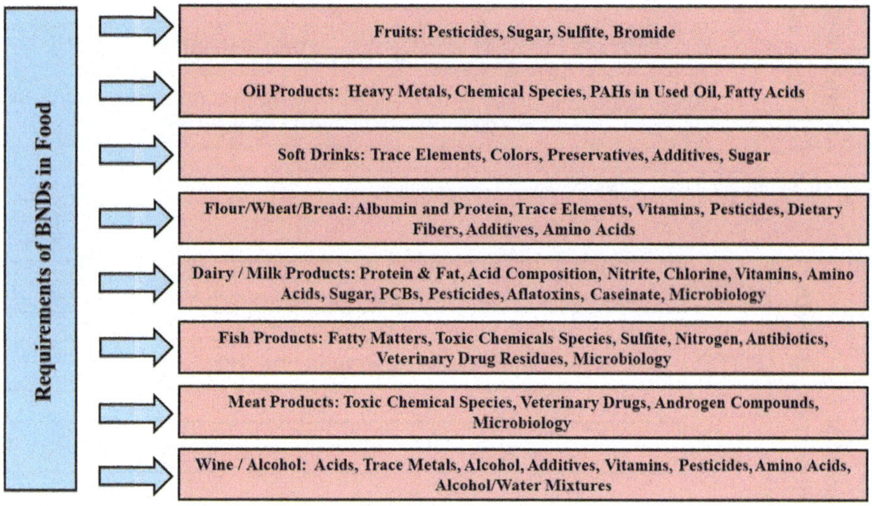

Fig. 19.43 Requirement of BNDs related to food

infrastructure using Aswal Model to disseminate the measurement traceability to the end product.

Reliability of laboratory results depends on traceable calibration through reference standard materials.

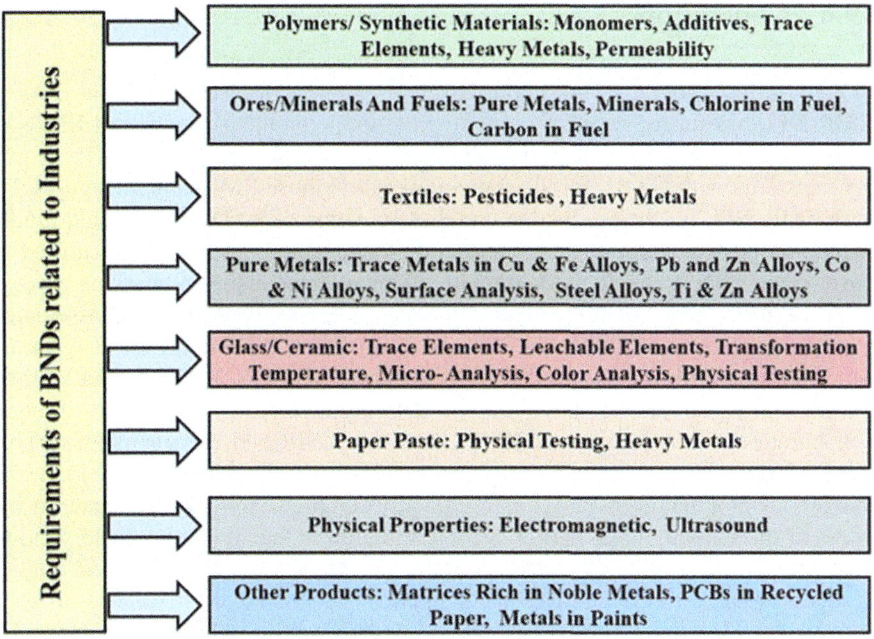

Fig. 19.44 Requirement of BNDs related to food

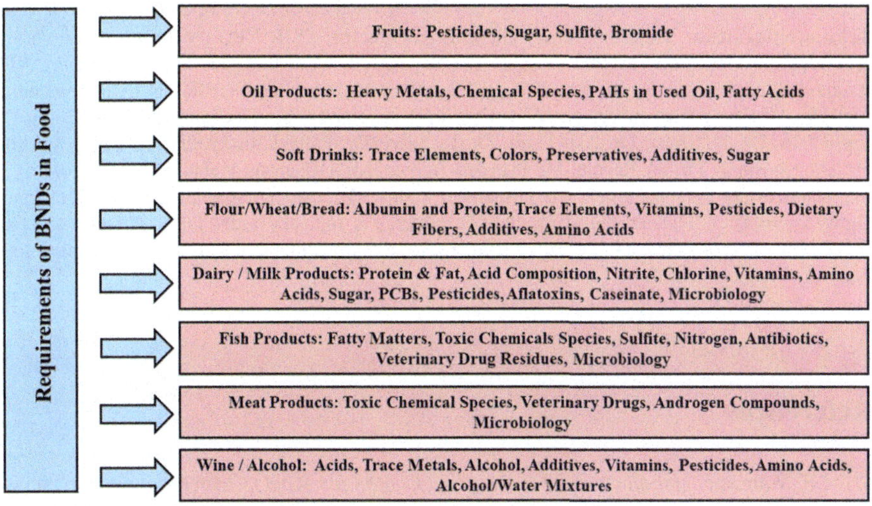

Fig. 19.45 Requirement of BNDs related to food

19.6 Conclusion

Keeping in mind the growing importance of reference materials in global market, CSIR-NPL has started its mission mode programme on the production of BNDs in various sectors. In this context, it is necessary that a large number of RMPs should get registered with NABL for the accreditation and those RMPs should sign MoUs with CSIR-NPL for the production and dissemination of BNDs. In particular, India is the second-largest importer of gold and other precious metals, it is required to develop their BNDs in collaboration with RMPs to assist in the hallmarking process of BIS referral labs. Most of the NMIs have a large number of RMPs associated with the production of CRMs. In the same line of tune, NPL India has also started signing MoUs with various RMPs for the production of BNDs in different sectors. Several other market-oriented BNDs need to be developed in the future through collaboration with other potential RMPs. Recently, CSIR-NPL has got major support from the Ministry of Commerce and Industry along with line ministry (S&T) for the indigenous production of BNDs and gradually commissioning them to cater to the needs of the Indian/global market. Strong initiation in this direction would yield to combat low-quality products and encourage MSME sectors to manufacture quality products at par with the international standards. This national importance of BND activity would certainly bring a paradigm shift in socio-economic fabric of India through quality control assurance for export, import and domestic consumer products in almost all the sectors.

Acknowledgements The authors are thankful to Director, CSIR-NPL for strong initiative and providing an opportunity to contribute in a book chapter. We would also like to thank the NPL standard division, regulatory bodies (BIS, QCI, NABL), RMPs and sister CSIR labs for their valuable support. All the staff members associated with BND programme: Dr. Santosh Singh Golia, Dr. Vipin Jain, Dr. N. Karar, Dr. Asit Patra, Dr. E. Selvi, Dr. Pallavi Kushwaha, Dr. Anuj Krishna, Mr. Dinesh Singh, Dr. Jai Tawale, Mr. Praveen Tanwar, Mr. Himanshu Yadav, Ms. Jyoti Pokhriyal, Ms. Dibya Topo, Ms. Geetanjali, Mr. Himanshu Yadav, Mr. Sandeep Singh, Mr. Sumit Mohan are highly appreciated. Further, the Ministry of Commerce & Industry under TIES scheme and CSIR (S&T Ministry) for sanctioning a major project for the establishment of infrastructure/facilities are gratefully acknowledged.

References

1. T.E. Vollmann, *Manufacturing Planning and Control for Supply Chain Management*, 2005
2. M.H. Ramsey, Sampling as a source of measurement uncertainty: techniques for quantification and comparison with analytical sources **13**, 97–104 (1998)
3. S.J. Lehotay, J.M. Cook, Sampling and sample processing in pesticide residue analysis **63**, 4395–4404 (2015)
4. E.J. Maienthal, D.A. Becker, A survey of current literature on sampling, sample handling, and long term storage for environmental materials, Department of Commerce, National Bureau of Standards, Institute for ...1976

5. M. Ihnat, M. Stoeppler, Preliminary assessment of homogeneity of new candidate agricultural/food reference materials **338**, 455–460 (1990)
6. T.P.J. Linsinger, J. Pauwels, A.M.H. van der Veen, H. Schimmel, A. Lamberty, Homogeneity and stability of reference materials. Accred. Qual. Assur. **6**, 20–25 (2001)
7. A. Kumar, D.K. Misra, A Review on the Statistical Methods and Implementation to Homogeneity Assessment of Certified Reference Materials in Relation to Uncertainty, MAPAN, https://doi.org/10.1007/s12647-020-00383-4 (2020)
8. L. Sthle, S. Wold, Analysis of variance (ANOVA). Chemometr. Intell. Lab. Syst. **6**, 259–272 (1989)
9. J.P. Reiter, Small-sample degrees of freedom for multi-component significance tests with multiple imputation for missing data. Biometrika **94**, 502–508 (2007)
10. R. Willink, Coverage intervals and statistical coverage intervals. Metrologia **41**, L5–L6 (2004)
11. A.M.H. van der Veen, T.P.J. Linsinger, A. Lamberty, J. Pauwels, Uncertainty calculations in the certification of reference materials. Accred. Qual. Assur. **6**, 257–263 (2001)
12. K.C. Waterman, R.C. Adami, Accelerated aging: prediction of chemical stability of pharmaceuticals **293**, 101–125 (2005)
13. N.J.A. Majcen, A need for clearer terminology and guidance in the role of reference materials in method development and validation **8**, 108–112 (2003)
14. A.M. van der Veen, T.P. Linsinger, J. Pauwels, *Uncertainty calculations in the certification of reference materials. Homogeneity study, Measurement Uncertainty in Chemical Analysis*, (Springer, Berlin, 2001), pp. 94–98
15. J. Pauwels, A. Lamberty, H.J.A. Schimmel, The determination of the uncertainty of reference materials certified by laboratory intercomparison **3**, 180–184 (1998)
16. H.-C. Chen, P.-C. Wu, J.-Y. Huang, L.-A. Chen, Uncertainty analysis for measurement of measurand. Measurement **43**, 1250–1254 (2010)
17. A.M. van der Veen, J. Pauwels, *Uncertainty calculations in the certification of reference materials. Principles of analysis of variance, Measurement Uncertainty in Chemical Analysis* (Springer, Berlin, 2000), pp. 88–93
18. C.M. Wang, H.K. Iyer, A generalized confidence interval for a measurand in the presence of type-A and type-B uncertainties. Measurement **39**, 856–863 (2006)
19. W. May, R. Parris, C. Beck, J. Fassett, R. Greenberg, F. Guenther, G. Kramer, S. Wise, T. Gills, J.J.N. Colbert, Definitions of terms and modes used at NIST for value-assignment of reference materials for chemical measurements **260**, 136 (2000)
20. K.E. Sharpless, L.M.J.J.o.A.I. Gill, Value assignment of nutrient concentrations in five standard reference materials and six reference materials **83**, 413–423 (2000)
21. A.M. van der Veen, T.P. Linsinger, H. Schimmel, A. Lamberty, J.J.A. Pauwels, Q. Assurance, Uncertainty calculations in the certification of reference materials 4. Characterisation Certification **6**, 290–294 (2001)
22. M. Nishikawa, Q. Hao, M.J.G.E.R. Morita, Preparation and evaluation of certified reference materials for Asian mineral dust **4**,1037113 (2000)
23. A.K. Agrawal, Certified reference materials of trace elements in water **28**, 373–378 (2005)
24. P. Gupta, C. Kaw, S. Ganju, A. Agrawal, R. Ramchandran, A. Sarkar, K.J. Lal, Preparation of certified reference materials of traces of chromium and arsenic in high purity water and infant milk powder **345**, 278–281 (1993)
25. C. Reimann, M. Birke, P.J.A.G. Filzmoser, Bottled drinking water: water contamination from bottle materials (glass, hard PET, soft PET), the influence of colour and acidification **25**, 1030–1046 (2010)
26. P.E. Stutzman, S. Leigh, Compositional analysis of NIST reference material clinker 8486, in *Proceedings of the International Conference on Cement Microscopy* (International Cement Microscopy Association, 2000), pp. 22–38
27. D.P. Bentz, G. Sant, J.J.J. Weiss, Early-age properties of cement-based materials. I: Influence of cement fineness **20**, 502–508 (2008)
28. C.F. Ferraris, M. Peltz, B.J.N.S.P. Toman, Certification of Standard Reference Material® 46 h: Fineness of cement-Addition of Blaine and Particle Size Distribution **260**, 190 (2018)

29. K.L. Scrivener, V.M. John, E.M. Gartner, Eco-efficient cements: Potential economically viable solutions for a low-CO_2 cement-based materials industry. Cem. Concr. Res. **114**, 2–26 (2018)
30. W.S. Liggett, S.R. Low, D.J. Pitchure, J.J. Song, Technology, capability in Rockwell C scale hardness **105**, 511 (2000)
31. S.R. Low III, *NIST Recommended Practice Guide: Rockwell Hardness Measurement of Metallic Materials* (2001)
32. J.-F. Song, S. Low, D. Pitchure, A. Germak, S. DeSogus, T. Polzin, H. Yang, H. Ishida, G.J.M. Barbato, Establishing a world-wide unified Rockwell hardness scale with metrological traceability **34**, 331 (1997)
33. S. Low, A. Germak, K. Herrmann, *Traceability of industrial Rockwell, Brinell, Vickers and Knoop hardness measurements, IMEKO 2010 TC3, TC5 and TC22 Conferences: Metrology in Modern Context* (National Institute of Metrology (NIMT), 2010), pp. 193–196
34. F.S. Rondan, C.A. Hartwig, D.L. Novo, D.P. Moraes, S.M. Cruz, P.A. Mello, M.F. Mesko, Analysis, Ultra-trace determination of bromine and iodine in rice by ICP-MS after microwave-induced combustion **66**, 199–204 (2018)
35. J.P. Cline, R.B. Von Dreele, R. Winburn, P.W. Stephens, J.J. Filliben, Addressing the amorphous content issue in quantitative phase analysis: the certification of NIST standard reference material 676a, **67**, 357–367 (2011)
36. V.J.C.P. Ramaswamy, Narosa, Structural characterization of catalysis by X-Ray analysis 92–115 (2002)
37. J.K.J.N.S.P. Taylor, *Handbook for SRM users* 260–100 (1993)
38. D. Barnes, G.J.S.E. Dent, Polystyrene films as a performance check for FTIR spectrometers **6**, 8–14 (1994)
39. K. Bansal, *CSIR Technology Showcase-BND 4201: India's Own Gold Standard* (2018)
40. D.S. Rao, V. Balaram, P. Sangurmath, M.J. Satyanarayanan, Major, trace and REE characteristics of the high grade gold geochemical reference material from Middle Reef, Hutti gold mines, Karnataka **85**, 313–321 (2015)
41. K.M. Dubowski, Quality assurance in breath-alcohol analysis **18**, 306–311 (1994)
42. K. Singh, D. Singh, C. Dixit, N. Singh, C. Sharma, S. Sahai, A.K. Jha, Z. Khan, P.K. Gupta, assessment, Chemical characteristics of aerosols and trace gas distribution over North and Central India **184**, 4553–4564 (2012)
43. L. Portnoff, P.A. Jaques, J.L. Furlong, Evaluation, the surface tension of synthetic blood used for ASTM F1670 penetration tests **47**, 1635–1644 (2019)
44. J. Martin, T. Tritt, C.J. Uher, High temperature Seebeck coefficient metrology **108**, 14 (2010)
45. R.N. Rao, S.J. Alvi, HPLC studies on evaluation of purity of pesticide reference standard materials **121** (2004)
46. J. de Boor, E.J. Müller, Data analysis for Seebeck coefficient measurements **84**, 065102 (2013)
47. M.S. Loveday, *Towards a Tensile Reference Material, Harmonisation of Testing Practice for High Temperature Materials* (Springer, Berlin, 1992), pp. 111–153
48. R.P. Chhabra, *Non-Newtonian Fluids: An Introduction, Rheology of Complex Fluids* (Springer, Berlin, 2010), pp. 3–34
49. D.J.M. Aswal, Quality infrastructure of India and its importance for inclusive national growth **1** (2020)
50. A. Kanagasabapathy, P.J. Rao, Laboratory accreditation-procedural guidelines **20**, 186–188 (2005)
51. K.J. Nadvi, Global standards, global governance and the organization of global value chains **8**, 323–343 (2008)
52. C.H. Weiss, Towards the future of stakeholder approaches in evaluation 186–198 (1986)
53. S. Swarupa Tripathy, R.K. Saxena, P.K. Gupta, A. Statistics, Comparison of statistical methods for outlier detection in proficiency testing data on analysis of lead in aqueous solution **2**, 233–242 (2013)

Chapter 20
Human Resources in Metrology for Skill India

Rina Sharma, J. J. Pulikkotil, Nidhi Singh, and D. K. Aswal

Abstract Human resource is fundamental to all strategic and coherent approaches in organizational management. To be a valued asset, human resource management assigns appropriate and skilled people working collectively towards a goal. Therefore, it becomes important for the organization to facilitate the necessary requirements such as infrastructure, education and skills. In this chapter the status of human resource in metrology in India has been reviewed and associated advantages and impediments were analysed. The growths of developed economies are also briefly discussed. The role of human resources in establishing a workforce in metrology, from Indian context, so as to enhance effectiveness of Quality infrastructure and make a socioeconomic impact, is also presented. Considering that human resource development requires both theoretical as well as practical knowledge imparted at laboratory or industry, the need to introduce the essentials of measurements from secondary level education to skill-development in tertiary university levels, as a part of the curriculum were highlighted. Besides, with rapid development in science and technology, it is also important that the industry personnel are updated with the current measurement science and technology by rigorous training. Analysis of gaps, challenges has been made and solutions are proposed for tackling the problem through a revised scheme of education system which includes personnel certification, skill developments and capability enhancement coupled with availability of career progression.

R. Sharma (✉) · J. J. Pulikkotil · N. Singh · D. K. Aswal
CSIR-National Physical Laboratory, Dr. K. S. Krishnan Marg, New Delhi 110012, India
e-mail: rina.npl@nic.in

J. J. Pulikkotil
e-mail: jiji.npl@nic.in

N. Singh
e-mail: singhnidhi.npl@nic.in

D. K. Aswal
e-mail: dnpl@nplindia.org

D. K. Aswal (ed.), *Metrology for Inclusive Growth of India*,
https://doi.org/10.1007/978-981-15-8872-3_20

Any measurement that you make, without any knowledge of the uncertainty, is MEANINGLESS!

–Walter Lewin (1936-84 years)

20.1 Introduction

Metrology is the underpinning science for the growth of the economy. It's all pervading as it finds applications in all areas and affects all of us as consumers, manufacturers, or professionals. For example, the vehicles that touch the road have to undergo several hundreds of tests before being released to the market. These tests are starting from the phase of accepting material to be used, manufacturing of components, assemble and post assembly testing. The smooth experience of driving and safety depends on hard work put into measurements for testing as much as the manufacturing. The same may be extended to all the products that we use. Professionals in different areas rely on measurements for the development and decision making, for example health professionals rely on the medical investigation's reports; engineers depend on inspection report to decide the acceptability as per their design specification and quality control; scientists use measurements for research, for modification or generalisation of theories.

Metrology or accurate measurement is not a new thing in our society; it existed since the beginning of the civilization. The thousands of years old Indian civilization has been known to practice measurements and have skills that could produce architectural wonders and texts describing solar systems and astrology. During the Bronze Age people started using measurements mainly for construction of dwelling units, town planning and exchange of goods. Indus valley civilisation (2600–1900 BCE) [1] (Harappa 3A, 3B, 3C) is well known to have measurement of length and mass [2]. Astronomical measurement and time measurement were also started during Indus valley civilisation and flourished during Vedic period. These measurements were necessitated initially for performance of rituals at correct time and for vedang jyotish. There is enough evidence from excavations of Mohan Jodaro, Harrappa, and DholaVira etc. that our ancestors had measurements skills that were quite advanced in those times. During British period, development of Indian industry was in low key, India was seen by British as source of raw materials and value addition to raw material development of Indian industry for processing and making finished goods was a low priority. Perhaps this was the reason that skill development also had a setback. During First World War, due to supply chain disruptions need to promote Indian industry was felt and testing of materials and products was realised. During that period efforts were made in consultations with Indian brain to create such infrastructure which resulted in creation of Government test house (National test house now) and later establishment of CSIR and NPL.

Worldwide, the need of accurate measurements and harmonisation of units was realized way back in nineteenth century. Harmonization of measurements across the globe was initiated by setting up the metre convention in 1875 as described in Chap. 1. The treaty and mutual recognition arrangements have resulted in realisation of a uniform scale of measurements in last century. In the middle of last century, India also adopted SI units and a uniform measurement scale had been realized and implemented across the nation. Nevertheless, the preponderant significance of precise and accurate measurements in scientific and engineering fields remains at challenge.

Presently India is in the era of liberalised economy and mass production. Mass production opens up an avenue on getting different parts made in bulk in different places and assembling to make final product. The compatibility and interchange ability of parts relies upon the precision measurements that are part of quality control exercise. As India is moving ahead and entering the brigade of developed nations, its export potential needs to be fostered by quality. Quality of local produce is essential for global competitiveness of Indian industry. Appropriate training to comply with the requirements of International mutual recognition arrangements and membership of relevant international organisations are essential for worldwide acceptability. These compliances of then industry produce are supported by Precision measurements, accreditation and standardisation. Rapid growth of India coupled with technology driven industrialisation underlines the need for skill development not only at lower level for employment but also for up gradation of employed workforce to make them capable to survive in the competitive and technologically evolving world. Emerging fields such as automated manufacturing, machine learning, nanotechnology, robotics, genetics, internet of things, quantum computing, energy resources and storage, materials science, biomedical applications etc., are few examples that poise the transformation and modernization at the workplaces. Human skills need to be developed and continuously upgraded to strike a balance with these advances.

> Finding a balance between technology integration and human capital investments will be critical to enhancing productivity.............GCI report 2019 [3]

Further, industry 4.0 in India is expected to have dramatic changes in all human cantered technical, economic, administrative and environment professional fronts. In this view, the HR in metrology will become one of the essential objectives and functioning of national resource management.

Aswal's model discussed in Chap. 1 has brought the metrology in forefront and an essential element for development of society and economic growth and thus requires development of metrology HR for inclusive national growth. The lack of training and skill courses in metrology is a problem that's faced by organizations in developing nations. Developed nations usually have a well-placed metrology education programme to develop Human Resources for the quality infrastructure. As illustrated in Fig. 20.1 Metrology HR is linked to growth of various sectors like academia, industry, and legal metrology, standardisation, accreditation and framing of policies.

Human Resource is the key to success of any organisation, Society or economy perceiving people as asset. Human resources is the set of the people who make up

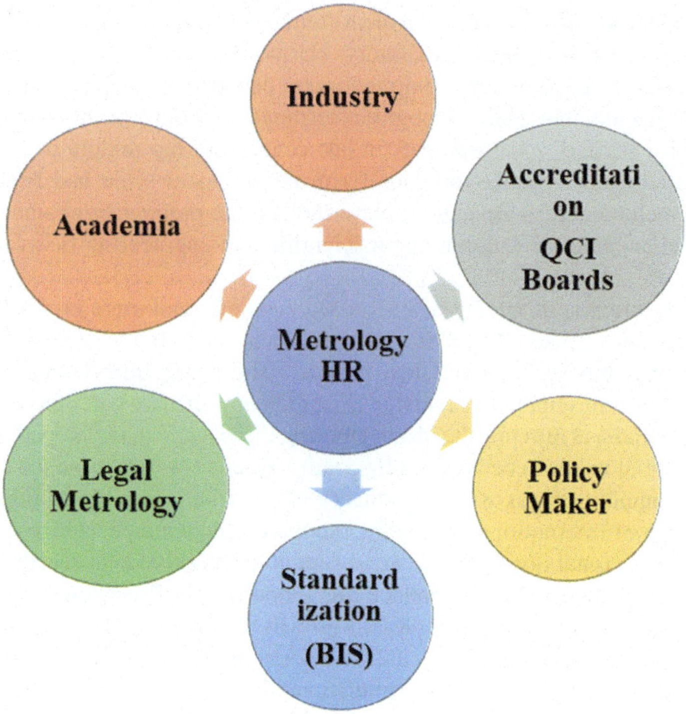

Fig. 20.1 Metrology Human Resource is key to Academia, Industry, Accreditation, Policy Making, Standardisation, and Legal Metrology

the workforce of an organization, business sector, industry, or economy [4]. From the very outset, HR's involvement includes people employment and management, so as to achieve the goals, preserving the organizational and cultural values. Several intriguing challenges are encountered by organization in its endeavour towards having a socio-economic impact. One of such challenge is skill development planning. Skill India mission aims to provide market-relevant skills training to more than 40 crore young people in the country by the year 2022 [5]. The mission intends to create opportunities and space for the development of talents of Indian youth. Further, India has taken a challenge of self-reliance by launching Aatm Nirbhar Bharat mission. These are very much interlinked missions and require an effort into the HR part of the policies for meeting such an ambitious goal in timely manner. It's imperative to analyse the status of HR in various areas, which will be important to meet this goal, analyse the needs and find the solutions that will make a desired impact. Broadly, HR in product quality infrastructure involves calibration technicians, calibration engineers and metrologists. With the advent of sophisticated instruments and methodologies, the work force needs to be continuously updated. HR management includes knowledge and skills updating, establishing networks, fostering intellectual and emotional health and providing motivations. Thus, rather than being a mere established practice at the workplace, HR relies upon policies and strategies enhancing employee capabilities and commitment.

To support the organization in achieving its objectives, it must be ensured that right people are at the right place executing their role at right time. This helps maintain a positive environment in the workflow of the organization. Hence, it becomes quite rudimentary to focus on skill development, optimization of project resources and time. People management and development when confined within the rules and goals of a particular organization, the role of the HR is more or less a constructive mapping of its resources and infrastructure to a socio-economic product. However, in the realms of metrology, HR role is complex as the workflow comprises of several independent but interlinked institutions as represented by the traceability pyramid described in Chap. 1. This chapter examines the national and international scenario of HR in metrology and highlights the need for development of metrology HR for Skill India. Significantly, it includes the highlights and contributions of CSIR-National Physical Laboratory, and also discusses the impediment and challenges to HR in Metrology and various proposals towards mitigating the measures. In view of the need for metrology education, it is recommended that metrology may be introduced in schools and explicit course subject in the university education. A consortium comprising of stakeholders from industry, government, educational institutions, accreditation and standardization bodies and national measurement institutes must be formed to make recommendations and nationwide implementation. Diploma and graduate courses in all disciplines of science, technology and social sciences, should have metrology practiced and the quality infrastructure inculcated. The existing technical institutes such as polytechnics also must be integrated with industries with emphasis on measurement science.

20.2 Demographic Dividend, Educational System and Employability

Demographic dividend is term used in urban industrial states; the term refers to shifts in a population's age structure or increase in workforce and decrease in dependency ratio [6] resulting in the economic growth of the country. This dividend becomes available to a country due to decrease in mortality and fertility rates. As the country goes through a state with better health facility the mortality comes down. The reducing fertility rates result in more investment in child education and health. Participation of women also increases in formal employments. These factors result in a transition period where more and more workforce is available and dependency is lowered, meaning more working hands with lower number of mouths to feed. With passing of years, the development status improves but the lower fertility combined with increased average age results in phasing out of demographic dividend.

Thus various countries get a window of demographic dividend most developed nations have passed their demographic window while India is entering into this window. Shifting age structure can result to demographic dividend window only through a very meticulous planning with investment in education and health system. Planning a policy for education such that, it meets the needs of industry leading to enhanced employability. In absence of such planning there can be negative spill overs like unemployment etc. rather than the dividend. Hence, there is an urgent need for the policymakers to align the developmental policies with this demographic shift

20.2.1 India Demography

India is the second most populated country in the world with $\simeq$1.3 billion counts [7]. The population have doubled during 1975–2010, and it is projected the country would be the most populous by 2024, surpassing China. With its population growth rate deduced as 1.13%, it is anticipated that by 2030 India, its population would become 1.5 billion, and 1.7 billion by 2050 [8]. Analysis also shows that 50% or more of its population age below 25 and more than 65% are of age below 35. This infers that the average age of an Indian will be 29 years (2020), making the country much younger to China (38 years), USA (38 years) Europe (43 years) and Japan (48 years) [9, 10]. Thus, a young human resource and a large working age population for coming 10–20 years, appear to be a promising dividend to make India a self-reliant and a fully developed state. Besides, due to increasing awareness in the society about the benefit of education to female, women who were earlier confined to home will join the work force in larger numbers. India's dependency ratio is expected to fall from 0.7 (in 2010) to 0.4 by 2030. This further suggests that earning population would peak in future.

From the above, it is well understood that India has a demographic advantage. Study of various countries shows that high growth rates were achieved during the window of 'demographic dividend', defined by the United Nations Population Fund (UNFPA) as the growth potential that results from shifts in a population's age structure [11–14]. Japan had its demographic-dividend phase from 1964 to 2004 and an analysis reveals that in five of these years, Japan's growth rate was in double digits where the growth rate was above 8% in two years, and slightly below 6% in one.; growth rate of below 5% in only two of these 10 years [11]. China had reforms of 1972 and saw accelerated growth, the years of demographic dividend of China from 1994 helped sustain growth rate for a very long period. During 1978 to 1994 in an interval of 16 years (post-reform, pre-dividend), China saw eight years of double-digit growth. However, since 1994 China could not cross the 8% growth mark [11]. Analysis the demographic dividend data for two of the four Asian tiger economies also show similar patterns.

India's shift in population structure by more than 60% of its citizen in working is a great advantage in terms of savings and investment rates. However, this demographic dividend poses several challenges [15]. With decreasing fertility rate, the present working class would move into dependent class in a matter of 20–30 years and the average age of the country's population would go higher. This poses a "demographic echo" of aging dependent populace. Further, it also becomes important to enhance scientific temper for a social change. Although at present, the demographics look good statistically, it is less useful unless massive investments are made in health, agriculture and education, and also spread widely to the corners of rural India. Hence, policies need to be framed and/or modified to develop and educate the young minds, nurture their skills, create employment opportunities and create an effective human capital, thereby leading to an overall growth of the Nation. Besides, sector wise discussion for HR policies and programs are also important and necessary. Government of India is making efforts through its various ministries including ministries of Science and Technology, human resource development, skill development, labour and industry to meet the challenge (Fig. 20.2).

In view of the importance of metrology and the need to effectively use the demographic dividend for self-reliance of India, it is required to review the status of HR in India, carry out in depth analysis and resolve the gaps through strategic interventions to convert HR into Human Capital. Even though Metrology is a key driver for the growth of any Nation, the lack of training and skill courses in metrology

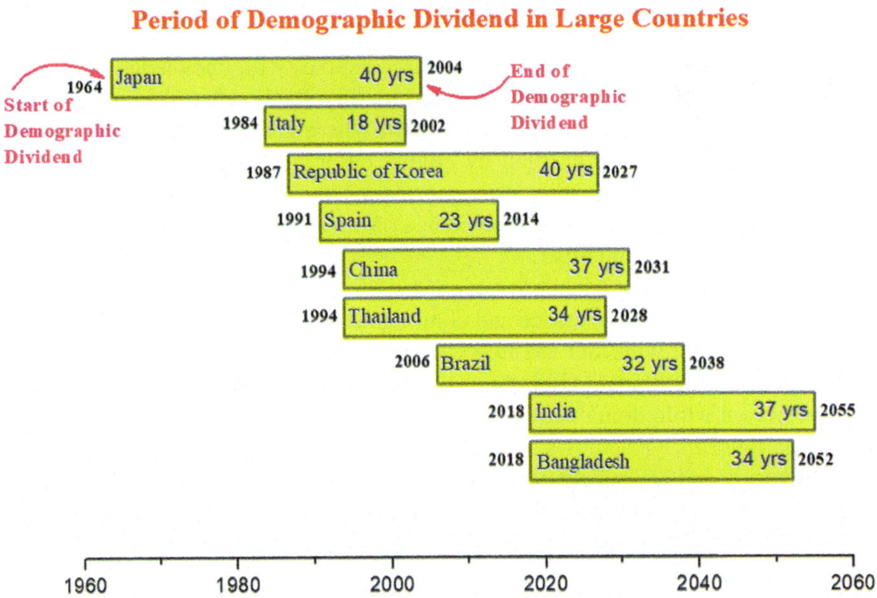

Fig. 20.2 India enters 37-year Demographic dividend window. The demographic dividend (DD) that is increased availability of workforce and decrease in dependency ratio presents unique opportunity for accelerated development. A comparison of DD windows shows among Asian countries Japan has already passed it; Korea and China's window will be passed in next decade. India has this opportunity window till 2055. Planning on education in metrology is one of the factors that can help to reap benefits [10]

is a problem that's faced by organizations. HR development in metrology is necessitated by demand of quality and reliability of products. Broadly, HR in quality infrastructure involves calibration technicians, calibration engineers and metrologists. With the advent of sophisticated instruments and methodologies, the work force needs to be continuously updated with the latest advancements in measurement science and technology. The need to meet the requirements put forwards by the dynamism of conformity assessment standards. For instance, as envisaged by World Economic Forum, Industry 4.0 is set to revolutionize trade and administration by means of digital technology. Emerging fields such as automated manufacturing, machine learning, nanotechnology, robotics, genetics, internet of things, quantum computing, energy resources and storage, materials science, biomedical applications etc., are few examples that poise the transformation and modernization at the workplaces.

As per world economic forum-Global Competitiveness report 2019, India has shown increase in innovation capability and there is need to improve talent base and to strengthen the skills to avoid the negative social spill over's [3]. Further analysis reveals that India has scored low in GCI 2019 on finding skilled manpower. India reports good scores in R&D and innovation however; it has not so impressive overall rank. With industry 4.0 this problem will aggravate further and corrective measures are necessary. **Emerging use of AI will require precision measurement and control in one package.** Interdisciplinary courses with knowledge of ICT and measurements build into core syllabus are now inevitable when we are talking of self-reliant India. We need to create a system where ease of converting scientific innovation to a product is enabled by quality assurance or standardization is not limited to application of quality management systems like ISO 9000, It involves technical competence as well as competence of infrastructure resources as much as quality management standards.

Quality standards applicable in laboratories for example ISO IEC 17025: 2017 requires the technical competence and continual up gradation of skills of personnel, same is the case for with ISO 15189, ISO 17043, and ISO 17034 etc. implementing organizations, i.e., clinical Laboratories, PT Providers, reference material producers etc. Further on while doing calibration of an instrument or testing a product, there are standard norms written by standards bodies like BIS which need to be followed. Manufacturing and measuring these parts without imbibing the culture of metrology is like singing a song in an unknown language. One can of course sing such a song but will never know the meaning as in likely to miss the nuances.

20.2.2 Educational System

The modern system of education in India is due to Lord Macaulay in 1830. With English being adopted as the language of communication, the curriculum largely emphasized on science and mathematics. Subjects like Indian history, literature, philosophy was considered less significant. Further, teaching got confined mainly to the classrooms. This change was a paradigm shift in the educational system, which earlier was based on the gurukul system. In the latter, the education was quite flexible, gurus trained their disciples according to the capability and talents of the student and concentrated upon an overall development; discussion and learning between student and teacher extended from the cosmos to the local political and social issues.

In 1952, several local boards of education, including the Board of High School and Intermediate Education was brought into one umbrella, and called as the Central Board of Secondary Education (CBSE). A second central scheme also exists in India, known as the Indian Certificate of Secondary Education (ICSE). In a matter of time, both CBSE and ICSE have spread not only among all Indian states, but also to countries beyond the international border. Universal and compulsory education for age group of 6-14 was adopted by the government of Republic of India. This was later strengthened with the National Policy on Education (1986) and the Programme of Action (1992), where it was envisaged that quality education may be delivered to all children across the country by twenty-first Century.

As per the constitution of India, school education was the responsibility of the respective state governments and had complete authority on policy making and its implementation. On the other hand, the role of the central government was mostly on setting standards in higher education. However, with the 1976 constitutional amendment, school education policies and programs were to be suggested at the national level, and the states find ways and means to implement it. A Central Advisory Board of Education (CABE), Ministry of Human Resource Development, which was set up in 1935, continue to lead the educational policies and programs of the country. Besides there is National Council for Educational Research and Training (NCERT) and its state counterpart called State Council for Educational Research and Training (SCERT), which plays an additional role in developing educational policies and programs. These units propose the curriculum, strategies, pedagogical schemes and evaluation methodologies.

Figure 20.3 shows the educational mainframe of India. The school system is categorized into four levels: lower primary (from age 6 to 10 years), upper primary (11 and 12 years), high school (13 to 15 years) and higher secondary schooling (17 and 18 years). The curriculum till high school remains more or less very similar across the country, except that the mode of communication. Several of the states adopt the communication to be in their respective regional language, which by itself is an additional subject to the national curriculum. Subject specialization begins at the higher secondary level with options in science, mathematical, social and information sciences. Generally, students were introduced to three languages; English, Hindi and their regional language and/or Sanskrit. Later on, in three language scheme, foreign languages like German, French, Spanish etc. has also been introduced as third language.

During the time of writing this chapter National education policy has been revised. NEP2020 [17] will have revised school years in stages called foundational, preparatory, middle and secondary, in $5 + 3 + 3 + 4$ years of schooling, to be completed during the age 3 to 18 years. Degree courses will be four years. Technical education will be through diploma and degree course.

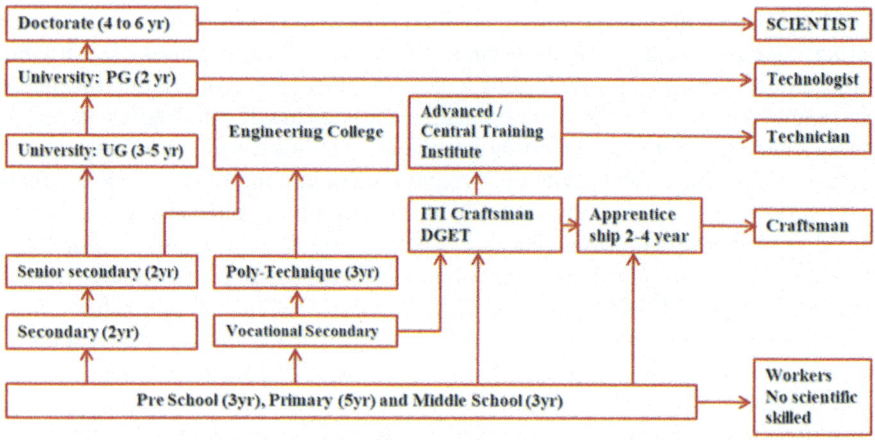

Fig. 20.3 Educational routes in India (Adapted from Ref. [16])

Unlike primary and secondary educations, which are largely controlled by the state, the tertiary level of education is largely set up on the standards laid by the University Grants Commission of India. As of 2020, India has more than 1000 universities including the central, state, and private. There are about 155 Institutes of national importance which include IITs, IIMs, AIIMS, NITs, IIITs and IISERs, and are empowered to examine their own degrees, up to Ph.D. level. In all fronts, the tertiary level of education lies on science, technology and social sciences. Apart from these, there are also several distance learning and open education programs which are monitored by the Distance Education Council. Indira Gandhi National Open University (IGNOU) is the largest Open University in the world by regard to the number of students.

There have been many reforms in the education policies however still lot of scope of improvement exists in the education system. To address the prevailing gaps Indian higher education needs radical reforms so that quality of education as well as world rankings of India Universities can be improved. Efforts for improvements must be primarily on improving the vocational, technical and doctoral education. A strong public-private coordination and professionalization of the courses must be considered at highest priority. In fact, in the early 1990's India did cash on the rising demand in the IT sector and engineering education. However, updating of the IT segments to machine learning, robotics, algorithm developments, artificial intelligence, digitization technology are yet to be implemented. With enormous diversity in population, concerted efforts should also be broadened to include arts and sports education.

The sixth report on All India Survey on Higher Education (AISHE) 2015-2016 finds that the total enrolment in higher education is approximately 34.6 million with 18.6 million boys and 16 million girls, with girls constituting 46.2% of the total enrolment. There is also a surge in the number of foreign students enrolled in higher education in India. The survey finds that the highest share of foreign students come from the neighbouring countries: Nepal (21%), Afghanistan (10%), Bhutan (6%), and from Nigeria and Sudan (5%) each. However, on the contrary, it is also found that there is an increasing trend showing that more Indian students opt for education abroad after their post-graduation from Indian universities. As per the statistics of "The U.S. Council of Graduate Schools" Indian post-graduate students admissions went up by 25% in 2013-14 from its previous year. To limit the brain drain from the country, **Dr A. P. J. Abdul Kalam once quoted "*In decades time, India will need 300 to 500 million employable skilled youth and there's a need to completely change the university education syllabus and secondary school education syllabus*".**

20.2.3 Metrology In Curriculum

To a large extent, quantitative measurements are mathematical in nature, assigning numbers to a physical observable. However, rather than being mere numbers, the measured value actually is a conglomeration of the theory, methods and techniques used to understand the concept (property being measured). A priori, metrology being a science and practice of measurement has several desired factors in anticipation. CSIR-NPL has been advocating on providing exposure to SI units and metrology right from schools. In Indian educational curriculum, **the first fully devoted chapter to measurements is in the class XI physics textbook titled "Units and measurement". Out of seven fundamental physical units, three of them (length, mass and time) are detailed at this level.** However, the recent change in the definition of mass has not been modified in the textbooks, despite CSIR-NPL making recommendations for it. It is therefore proposed that the chapter may be elaborated to include all the seven SI base units and their importance so that students feel motivated at a very young age. Also, the basic definitions as laid by the VIM [18] document of metrology can be well introduced at this level. Statistics, which is an essential tool in measurement science, is included in the mathematical books at the higher secondary levels. Grouping of data into frequency plots, pie-diagrams, use of inferential statistics etc. are known to students by the end of their school career. Nevertheless, there still remains a wider scope to emphasize on the statistical tools to envisage the meaning of the basic laboratory experiments pursued in the school laboratories. Several institutes and universities of India have now implemented metrology as a dedicated credit course at the Bachelor's level. The syllabus includes measurement glossary of terms such as accuracy, precision, resolution, traceability, uncertainty, tolerance and others. The courses also cover information on the various hand-held measuring devices, automated measuring systems, significance of statistical processes relative to measurement theory, understanding of the relationship between quality and metrology, working knowledge of various standards used in everyday practice in the field of electrical, mechanical and electro-mechanical equipment. The contents of the course is expected to benefit students to pursue a career as Laboratory technicians involved with testing and/or measurement, Six Sigma (6 σ) practitioners, Quality control specialists, for engineers and technicians involved with manufacturing facilities and developing critical-to-quality specifications on components and assemblies, etc. Apart from traditional degree courses Indian system offers several Vocational trainings in country which includes vocational secondary or vocational course after 12th. Vocational learning involves essentially hands-on training through which one improves capabilities and experience which is straight way related to occupation in future. It assists scholars to be skilled and thereby, attain better employment openings. The traditional courses offered are sometimes not able to fetch employment. Among the disciplines computers are most sought after course. A package of metrology with electrical/mechanical area can be designed to improve the skills imparted [19].

20.2.4 Employability fall Causes and Concerns

The fall of employability is a long-standing problem in the Indian economy. Employability of university graduates is decreasing, as shown in Table 20.1, it was noted in 2017 that 60% of engineering graduates remain unemployed, while a 2013 study found that 47% of them were unemployed in any skilled occupation. India's overall youth unemployment rate, meanwhile, has remained stuck above 20% in the past few years.

The fall in the employability, in particular concerning that of Industrial Training Institutes (ITI), Industrial Training Centres (ITC) and polytechnic must be urgently considered on priority and necessary steps must be taken to improve the skill sets of students who pass-outs from these segments. In general, the ITI, ITC and polytechnic students constitutes a large portion of the MSMEs. Education and professional degrees in present-day India need drastic restructurings so as to improve employment prospects **plausible solution is to integrate the final course curriculum of these ITIs and ITCs with national and state level laboratories that have advanced and sophisticated instruments and can teach nuances of precision measurements. With multi-disciplinary applications in basic and applied sciences, engineering**

Table 20.1 Employability percentage of various courses

Skill	2014 (%)	2015 (%)	2016 (%)	2017 (%)	2018 (%)	2019 (%)
B.E/B.Tech	51.74	54	52.58	50.69	51.52	57.09
MBA	41.02	43.99	44.56	42.28	39.4	36.44
B.Arts	19.1	29.82	27.11	35.66	37.39	29.3
B.Com	26.99	26.45	20.58	37.98	33.93	30.06
B.Sc	41.66	38.41	35.24	31.76	33.62	47.37
MCA	43.62	45	39.81	31.36	43.85	43.19
ITI	46.92	44	40.9	42.22	29.46	NA
Polytechnic	11.53	10.14	15.89	25.77	32.67	18.05
B.Pharma	54.65	56	40.62	42.3	47.78	36.29

Table source India skill report, 2019 [20]

and technology, metrology require more emphasis as a full-fledged curriculum. The process of measurement, analysis and testing to secure the traceability are very complex and thus require more competent and rigorous specialization. Hence, India may establish a centre or a college dedicated for metrology. **In this perspective, CSIR NPL has already started a diploma programme on "Precision Measurement and Quality Control (PMQC)"which is running under the aegis of AcSIR, India.** The details of this course are given in Sect. 20.5 nevertheless; NPL's intention is to further expand the course, as it has been proven to have 85 to100% job placements to the students who qualify the course. The proposed segments in future would include, machine learning, big-data analytics, automation, certified reference material metrology in chemistry and others.

According to Pravin Sinha [21] the three adopted classification of the Indian labour force are, (i) Rural sector (includes farm labour), (ii) Urban formal sector (includes factory and service industry) and, (iii) Urban informal sector (self-employment and contract/casual wage workers). The rural and informal sectors in Indian market accounted for more than 90% of the employment in 2011. However, at the time of survey, these jobs are not considered by the existing Indian labour laws. On the other hand, World Bank report of 2010 finds that the low-paying and relatively unproductive jobs are dominant in the informal sector jobs [22]. Metrology plays a great role in urban formal sector. There are more than four lakh testing and calibration laboratories. According to association of Indian laboratories, these laboratories face lack of attention by Government. **These laboratories provide essential services; however, their pain was that during COVID 19 situation they faced a problem of neither being recognised under industry norms nor under essential services. This is an example that need of Human resource for such sector is neglected. Their need is the manpower trained in metrology and related quality management aspects. Aswal's model takes a note of this situation when it talks of QI and four helices together**.

20.3 Metrology Training in India and Contribution of NPL India

Regardless of the stream of education, training and knowledge dissemination is integral part of growth and development of any society. In general, school and colleges keep many frontline activities along with traditional course structure in their course curriculum. **However, one of the disconcerting problems associated with education in India is that metrology has not received its due recognition in most of the course designs neither in schools nor in higher education.** In recent past few academic institutes in India have begun to comprehend the significance of metrology and have included basic metrology in their curriculum; however, it is not sufficient to cater the country's requirement. Measurement is an integral part of our daily life as well as many applications like research, automobile, space, pharmaceuticals, satellite etc. Off late, it has been accepted that metrology provides an essential basis for all the engineering and science disciplines along with related areas such as medicine and the law. For example, if the person measuring or sampling the blood for measurement of blood sugar is not trained, the results may not be true or it can be false positive meaning that it falsely exceeds its true value, the specialist recommends medications, or if the measured blood sugar is falsely below its true value, medicament will not be scheduled. In such cases either the patient will consume unneeded drugs and waste money or will face hypoglycaemia leading to risk of injury/death. This scenario is not limited to developing countries only, even in developed countries there are example of measurement errors. One such well-known case of measurement error is NASA's Mars Orbiter where NASA lost a whopping $125 million [23–25] because of software not taking care of unit conversion. English units of measurement were used by Lockheed Martin Engineering Team while the NASA team utilized metric system for a key spacecraft operation, according to an examination finding. Due to this conversion factor not being considered, the propulsion system of the spacecraft was suspected to be overheated which might have triggered an alarm to disable it. That probably stopped the engine from completing its burn. This might have resulted in dipping of Climate Orbiter deeply into the atmosphere [23–25]. That probably stopped the engine from completing its burn, so Climate Orbiter likely ploughed through the atmosphere. This could have been perhaps stopped if software developers were aware of the importance of units.

Further, metrology plays an important role in lowering the technical barriers to trade through the National quality infrastructures comprising of a National Measurement institute, a standard writing body, a body for enforcement of law and a network of accredited testing, calibration and inspection [26] for any Country. In the era of Global trade various economies of the world depend on each other for raw material as well as finished goods. Hence raising quality standards and infrastructure in these trade connected countries becomes responsibility of all. Countries are classified as developed, developing and least developed countries depending on their development index [3, 27, 28]. Despite the importance of metrology and it being the growth engines, the development of Human resource in metrology area that has been talked about by many researchers internationally [19, 29–37] as not being given its due importance. In most developing countries there is no course of specialisation of metrology in formal education. Metrologists acquire knowledge of this trade on the job and by way of short-term courses. As this realisation has come in some countries like Brazil, Lithuania, and Nigeria, steps have been taken for development of HR in these areas [19, 33, 38, 39]. The state of development in country is related to its technology adaption and relates to status of metrology. Developed nations owe, their transformation from developing to developed state, to development of metrology temperament. The developed nations have taken effort to explain even the most complex metrology achievements and the exciting science behind it in simple ways to engage attention of school kids and general public [40]. US and UK have large number of metrology courses. Their ways of implementation may be different but effort on metrology education is seen [33]. The need for development of metrology education is usually recognised by NMIs and other elements of QI

[26], however it usually becomes difficult to convince Governments and stake holder of education system. Thus, these elements resort to international collaboration and training on job. International collaboration in development of HR in metrology has been the way out for global consumerism for a long time. Developed countries help the developing and least developed ones to realise the importance of human resource development in metrology and QI. This helps to maintain the conformity assessment echo system. Not only measurement, the environment/culture also play a crucial role in the measurement or testing results. Development of scientific temperament in society is also as essential as metrology temper. For example, India trades with several countries for more than 200 spices and spice products in whole, powdered, blended and mixed forms which include pepper, cardamom, spice oil etc. [41]. In the last few years, imports of Indian spices have increased many folds in value, however, due to contamination and failure of exporters in maintaining the quality standards of the spices, FSSAI is rejecting lot of imported spices and other spice products. An acquaintance once narrated an anecdote about spices from Indian farmers being rejected due to animal excreta detected in the spice sample. The experts investigated the cause and visited the site and released the technical review of their findings. It was found that the farmers were spreading the spices on the floor for sun drying. As per the prevalent tradition in India an emulsion of cow dung in water is made and a layer of it is applied over the entire flooring. The animal excreta reported in samples was actually traces of cow dung coming from the floor on which spices were sun dried. Thus, the problem was solved with necessary training provided to farmers with an instruction to use appropriate sheets over the floor for sun drying of the spices. Some examples for rejection criterion are; the clove sample does not conform the parameter for Headless clove and Extraneous matter as per standards laid down, Cassia sample fails for the parameter of moisture, coriander samples failed for split fruit and discoloured fruit criterion, Chilli and turmeric exports from the Guntur belt to the Western countries have been rejected due to the presence of excess pesticide residue and toxins in them and their related products etc. [42]. Similar examples exist in various fields, value addition of raw material can be done by grading through testing and further value addition happens by process or manufacturing those results in final products. Compliance to national and international standards is required in such value chain. Small interventions of building skills in quality handling and awareness about testing norms can solve the problems described in Fig. 20.4.

Developed countries have a well structure and experienced quality infrastructure and have trained personnel for the measurements. However, with advancement and the new players in market, they also face this impediment of HR in metrology and short courses become the way of re skilling and up skilling. It has been reported that for many years the formal education in metrology was not available in universities and US military trained the personnel for precision measurement [35, 40]. There are short term courses offered by NIST the NMI of US, its accreditation bodies and few other schools covering various parameters like, mass, temperature, length etc. [40]. In United Kingdom there are metrology courses available from NPL UK in both modes class room courses as well as e-learning courses. NPL UK also cooperates with others to run metrology courses by providing the course material developed

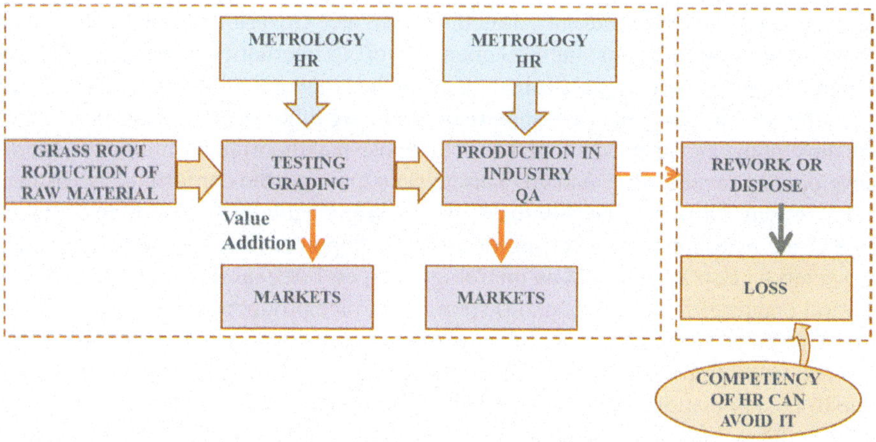

Fig. 20.4 HR competency in measurement and testing is needed for value chain that is value addition by grading, processing, production and delivering quality products

by its learned and experienced metrologists [43, 44]. Educational institutions of US, France, Australia, Germany etc. offer various metrology programs or classes in developed countries [33, 45]. Some of these courses are available internationally for training of metrologists [33]. The data reveals that very few endeavours have been made to introduce specialisation in metrology in higher schools of learning.

Need for this enabling knowledge arises at various levels, e.g. for industrial metrology availability of personnel at shop floor levels to do measurement during production. Similarly, inspectors, calibration or testing technicians, technical managers and quality managers is inevitable. In MSME sectors, one person may play many of these roles. The lack of availability of work force in metrology skills, forces the organisations to take up the available manpower and train them on the job by way of consultancies or training courses available, thereby affecting the productivity of these personnel in initial years. In case of Brazil, they handled the situation differently and set up a master's course of one-year duration [32, 38] which, has been implement with due planning., Brazilian National Institute of Metrology (INMETRO)cooperated with agency for coordination of programs of master's degree and a university to train 60 people in two groups in 1975-76, which were absorbed at NMI and further trained continually at PTB Germany. Brazilian Government created Human resource for metrology programme in 1995. During 1998 to 2002 after detailed mapping of metrology need the first regular course was launched and programme consolidated in 2006. The program is reported to be a great success and of benefit to NMI, accredited labs, R&D and society. Many other studies and review show that the countries have made recommendations and taken steps to mitigate the weakness of metrology education.

As discussed in Chap. 1, CSIR-National Physical Laboratory, India, is National Metrology Institute of India. With the advancement of science and technology and emergence of mass production, need for precise measurement came into the force.

Further, for the development of new materials and new technology accurate measurements are also required. The human resource for NPL was selected with utmost care and interest by the great visionaries like Dr K. S. Krishnan, the first Director, NPL and Dr S. S. Bhatnagar DG, CSIR. Both these leaders are well known for their farsightedness and vision that helped in shaping India's S&T as well as metrology infrastructure. In first two decade of NPL, its service to Industry included advice, testing and specialised services [46, 47]. The facilities at NPL India were established for metrology which in initial years was to support legal metrology and trade. NPL's scientist and officers were par excellence and were groomed by learned scientists who were at the helm of the affairs and involved in policy making and recommending on standardization. These scientists were given international exposures to present their work to learn newly developing methods. While on job, they were trained in areas of emerging methods of metrology and materials in cooperation with other NMIs and universities of leading Nations. Thus, the human resource of NPL was looked after and nurtured with knowledge to meet the ever-increasing demand for modernisation and Industrialisation. These people of NPL trained young minds by way of having them as guest worker. NPL was most obvious choice for young guest workers from different institutions of India, owing to its state of the art facilities, and availability of the brilliant faculty. Later on, this developed into Ph.D. programmes, internship, and training for graduate and post graduate students. Ph.D. Students worked at NPL and registered at various universities.

Around late eighties vocational training programmes of two years duration was organised to provide training in various trades, including three months training in species plant at NPL.

20.3.1 Industries, Testing and Calibration Laboratories

Measurement provides quantitative knowledge about the process, its control, adjustments or modifications. The need for precise measurements has turned out to be essential since the commencement of Industrial revolution. With the advancement in the technology, the requirement of industries keeps changing resulting in a continuous process of learning and training with the changing needs of the industry. Worker from an industry requires theoretical as well as hands on training which will be beneficial for their own research problems. It will be an added advantage if the environment of standard laboratory is provided for such learning. This kind of training and education improves the quality of personnel and workers and promote quality control in industry. In India there are four lac testing laboratories out of which > 5600 are accredited laboratories [26].

A survey was conducted in Industries, testing and calibration Laboratories, to find out their requirement towards type of training courses they need and preference of organization for such training. Following graphs shows the current external training requirement trend and its type (Fig. 20.5).

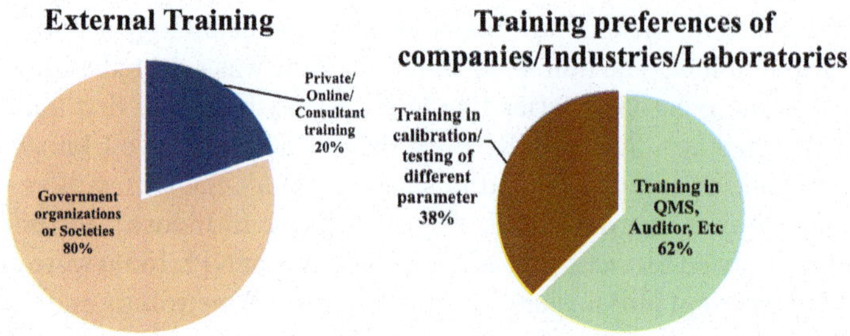

Fig. 20.5 Results of survey on training used by NABL accredited organisations. It indicates that NABL accredited organizations need training in metrology and quality systems. Survey reveals that many freelance and private players are providing training and this indicates necessity of setting standards for such training

Further, Metrology education and training of personnel for repair and maintenance services of scientific and technical equipment should also be planned for the industries so as to make them self-reliant. However, the foremost barrier is to find appropriate education and training programs which is designed based on the needs of industry/laboratory staff who undertake measurement, calibration, testing or verification. By and large many industries and laboratories are ignorant or do not pay attention bout the required set of knowledge or skill to be possessed by a worker in order to be competent for the assigned job. Mostly managers or human resource (HR) persons play a very crucial role in recruiting new worker or assigning job to the existing employees. Students of universities, technical colleges and other institutions of higher learning are the potential managers and professional experts. They are responsible for implementing the quality system in the companies. Mostly the management policies are inclined towards the contentment of HR people instead of the operating process requirements like measuring parameters, tolerances, measurement uncertainty, specifications, etc. Metrology areas are comparatively unfamiliar to several HR specialists as compared to traditional occupations like chemists, physicists, biologists, or the engineering professions. Therefore, special training with overall understanding of Metrology and Quality system is mandatory for managerial positions too. Due to lack of such training, they are not able to do justice with the worker and company. Management team usually has persons with MBA, they choose what they consider the finest training in metrology course for their people. However, this approach does not cater the actual requirements of training of the people carrying out measurement jobs in the institution (Fig. 20.6).

In the absence of structured courses, requirements of the companies in particular of MSME's are fulfilled by hiring workers and providing them on job training. In general turnover of MSME's are quite low, on job trainings are quite taxing on time and money for such companies as it is utilizing hours of pre-existing trained manpower and infrastructure of the company, however not much output is attained from the newly hired person. In view of this, these jobs are underpaid and not very lucrative to the employee/worker. As a result, staff retention in MSME's is a critical

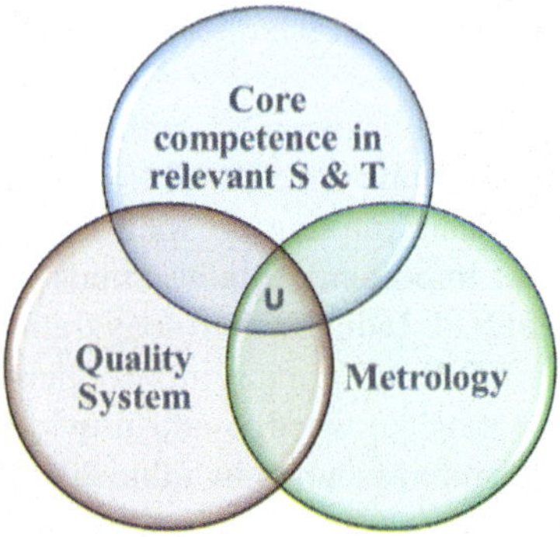

Fig. 20.6 NABL accredited organizations need training in metrology and quality systems. Survey reveals that many freelance and private players are providing training and this indicates necessity of setting standards for such training

issue. After the value addition in their profile, workers tend to move to Big Houses (Large scale Industry) where they are offered handsome pay package along with additional benefits. The MSME's suffers a lot in this process as they have to train again a new manpower with no certainty to retain it. It is a vicious cycle for the small-scale industry sector. Further, there are also some cases, when big houses go for accreditation, they are in desperate need of quality system and metrology trained manpower. They lure trained people from MSME's by giving them higher pay packages. However, as soon as the accreditation process is completed the hired manpower becomes redundant for them and thus is fired.

There can be a situation when, only few comprehend the techniques and structures that are essential to the work in that area, and they go on leave or quit the job. To overcome problems arising from such situations, where only the operational knowledge has to be apprehended prior to the employee's exit, otherwise it would lead to a knowledge void. Most of the time, organizations do not save, record or register the knowledge developed by their people performing measurement functions. The knowledge curve has to be developed again whenever altered or fresh worker embrace the job. Thus, knowledge management is an area which needs to be considered and nurtured in industrial culture. Use of information technology for knowledge management needs to be enhanced. Identification of skill sectors on metrology, statistics, quality managements and accreditation standards, use of ICT in knowledge management need to be taken on priority for skilling and up skilling.

Metrology has been understated and its prominence in the progress of a company/lab has never been understood specifically. It is considered less than the crucial or basic processes of most companies. The significance of metrology is understood in an industry/company typically, (i) when there is a serious problem in production/process which leads to major negative economic impact on the concerned entity which otherwise could have been avoided if good measurements had been warranted.

(ii) When there is a requirement from the client to fulfil the traceability norms and conformance criteria.

Since its Inception, CSIR-NPL is playing significant role in contributing towards the metrological needs of the country. CSIR-NPL is providing calibration, testing, consultancy and education and training to industries, companies, laboratories, SAARC countries and most importantly young minds who are going to be nation builder in future. The role of NPL India in Metrology training will be dealt in detail in the next section. Apart from CSIR-NPL, many more organizations in India are playing important role in disseminating knowledge and providing training in Metrology, quality system and infrastructure as shown in Table 20.2 (Fig. 20.7).

20.3.2 Endeavours of CSIR NPL for Skill Training

From early years CSIR-NPL has been advising Industry towards development of quality infrastructure, precise measurement techniques etc. Gradually, it was felt that imparting training in more formal manner and organizing awareness program is essential for societal benefit and acceptance of MADE IN INDIA products at international forum. In view of this NPL India designed parameter wise classroom/practical courses as per the requirements of the customers from laboratories/industries which continue throughout the year. With the liberalisation and setting up of accreditation infra-structure the new calibration and testing laboratories started coming up and NPL's role in shaping the human resource became more crucial in nineties and early 2000. During this period the new training institute and laboratories depended on human capital of NPL till they had sufficient training of their trainers. These courses were organised in cooperation with MSI, NABL etc. at NPL campus mostly. NABL together with NPL faculty conducted region wise training programmes to help

harmonisation practice in India. NPL's expertise was most sought after in those days.

With changing needs, the courses have also evolved from basic training to specialised training. The recent policies of Government of India have created vast opportunities in India in various sectors. These sectors need skilled manpower. NPL's focus on human resource developments in line with Sustainable Development Goals and Government of India Missions of Skill Development and Make in India, for supporting industries, accreditation bodies, regulators, S&T and other stakeholders. In view of increasing global competition many industries are going for ISO certification, using precision measurements and implementing quality control measures. The industries in India, especially MSME sector need support in terms of up skilling of human resource. NPL, India, being the NMI of India, is uniquely positioned to cater the needs of training in these areas.

Presently NPL is committed to capacity development for strengthening the QI of India to support economic growth. It has variety of programmes designed for different stake holders

Table 20.2 Metrology related training in India [46, 48–61]

Organization	Courses
National Physical Laboratory India	One-year PG Diploma in precision measurements and quality control
National Physical Laboratory India	Short Courses in mechanical, Optical, thermal, material, environment, Electrical, time and biomedical metrology
National Physical Laboratory India	2–6 months training to UG and PG students
Institute for Design of Electrical Measuring Instruments (**IDEMI**)IDEMI, Mumbai	Dimensional Metrology
Fluid Control Research Institute(FCRI), Kerala	Metrology, Pressure, Thermal and Electro technical Measurement and Calibration
Central Manufacturing Technology Institute (CMTI), Bangalore	Metrology and Calibration
Nettur Technical Training Foundation (NTTF) NTTF, Kerala	Certificate Program in Inspection and Metrology
Swayam and National Programme on Technology Enhanced Learning (NPTEL)	Engineering Metrology
Electronic Test and Development Centre(ETDC), (Bangalore, Chennai, Hyderabad, Pune, Goa, Jaipur, Mohali, Solan, Guwahati And Agartala)	Short Courses in Mostly Electrical and Electronics Measurements, Also Mechanical and Other Measurements
Electronics Regional Test Laboratory (ERTL) (N) Delhi (E) Kolkata (W) Mumbai (S) Thiruvananthapuram	Short Courses Testing and Calibration

(continued)

Table 20.2 (continued)

Organization	Courses
Centre for Electronics Test Engineering, (CETE) Pune, Hyderabad, Bangalore, Kolkata, Noida, Kolkata, Delhi	Testing and Calibration Reliability Engineering Programs
Indian Institute of Quality Management	Evaluation of Measurement Uncertainty Testing, Calibration, Quality Management etc.
Standardisation, Test, and Quality Control (STQC) IT Centres at Delhi, Hyderabad, Kolkata, Chennai, Bangalore, Pune, Jaipur, Mohali and Guwahati	Software Quality Engineering
Centre for Reliability (CFR), Chennai	Training and Certification of Industrial Employees in The Field of Reliability Engineering
BIS Training Institute National Institute of training in standardisation NITS, Noida	Quality Systems and Standardisation
Various Boards Like Spice Board	Training in standards and testing
ARAI, Pune	Automotive Testing and Measurements
National Test House	Training in calibration testing
Confederation of Indian Industry CII	Quality system certification, Standards, accreditation basic metrology

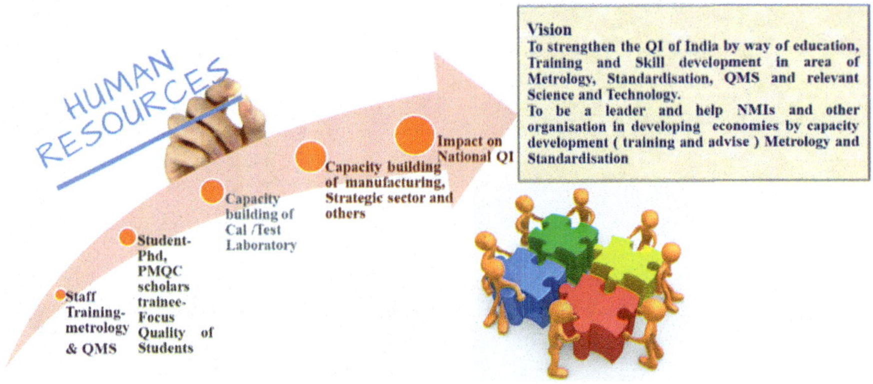

Fig. 20.7 CSIR NPL is strengthening QI by capacity building development

20.3.2.1 Short Term Training for Industries, and Other Sectors

Training programmes consist of theory lectures on various scientific and technical aspects followed by practical demonstration and hands-on training on the related instruments/apparatus/machines. These programs cover Physico-Mechanical Metrology, Electrical, Time and Frequency, Environment and Material's Metrology, radio and environment metrology as shown in Fig. 20.8. These programs can be tailor made according to the exceptional request of the customer or a particular sector. Examples are, a course designed to suit the needs of Air India, or Department of Legal Metrology etc. A glimpse of past decade of training programmes tells that Mass, length, pressure and quality are in great demand as shown in Fig. 20.9. Training of students and faculty from S&T institutes on Material characterisation, Nano science, Photovoltaic was carried out on demand basis. The reason for this may be attributed to the lack of facilities at private institutes of higher learning. Students want to have exposure to these techniques to increase employability (Fig. 20.9).

20.3.2.2 One-Year Full Time PG Diploma Course in Precision Measurement and Quality Control

The recent policies of Government of India have created vast opportunities in India in various sectors. These sectors need skilled manpower. Government of India has launched a "Skill India" programme for supporting industries and other stakeholders. In view of increasing global competition many industries are going for ISO certification, using precision measurements and implementing quality control measures. The industries in India, especially MSME sector need support in terms of providing specific trained manpower. Lack of the trained manpower is not only a major constraint for industries and manufacturing/production sector but also for

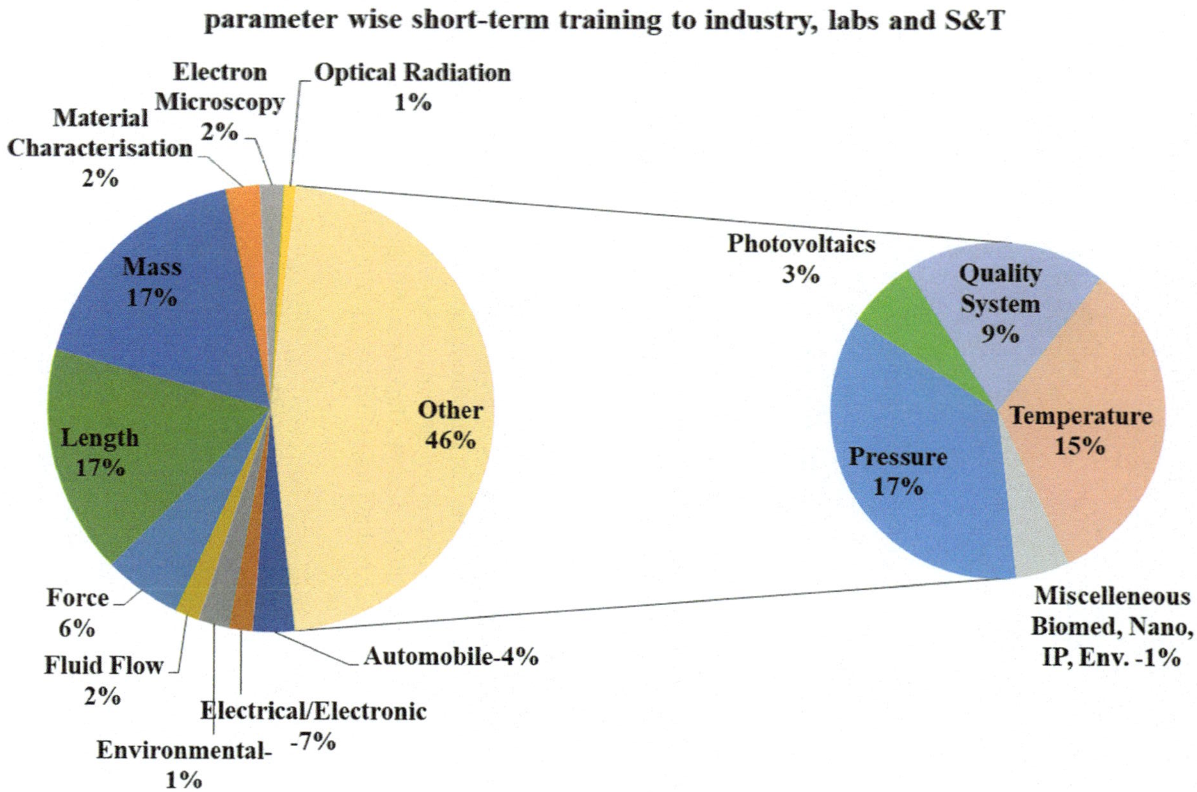

Fig. 20.8 CSIR NPL India's short-term training programmes are organised for various measurement parameters. Length, mass, force, quality system, temperature, electrical and electronics measurements are most sought after by industry for training. Pressure, environmental, materials and photovoltaic and Biomedical have got demand recently

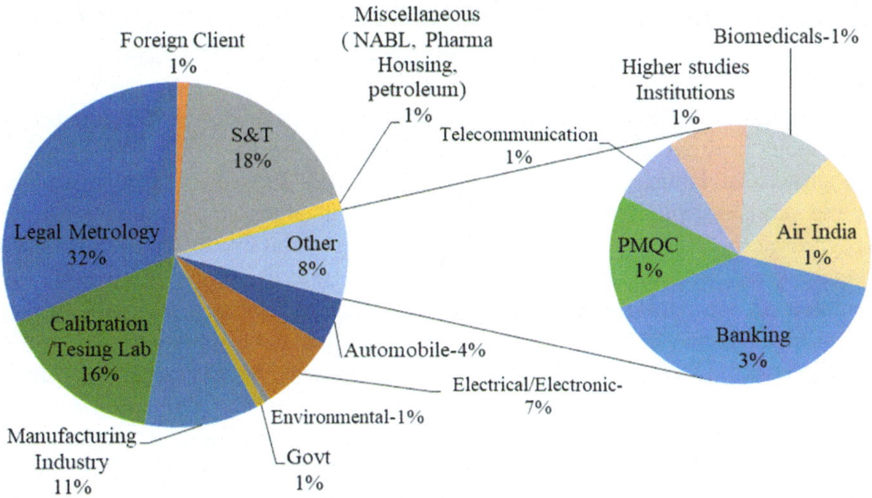

Fig. 20.9 CSIR NPL India's short-term training programmes are availed by various sectors. Legal metrology, S&T and calibration testing labs have been major beneficiaries in last decade

calibration and testing laboratories. NPL, India, being the NMI of India, is uniquely positioned to cater the needs of training in these areas of industrial importance.

Today due to increasing demand of high quality products as well as their global acceptability, Industries and production units are standardizing their products with ISO/IEC conformity assessments following national and international specifications for the product and set up or liaisons with NPL India/accredited Testing and calibration laboratories complying to International Standard; ISO/IEC 17025 (General Requirements for the Competence of Testing and Calibration Laboratories) for traceability needs. Laboratories in medical field conform to ISO 15189: 2012 (Medical laboratories - Requirements for quality and competence). These laboratories seek recognition for their proficiency by participating proficiency testing programmes run as per ISO/IEC 17043: 2010 (Conformity assessment - General requirements for Proficiency testing) and use reference materials from ISO Guide 34: 2009 (General requirements for the competence of reference material producers) complying manufacturers or Bhartiya Nirdeshak Dravya (Trademark of NPL India). These Organisations need human capital enabled with knowledge and skill for measurements and quality control.

Such skilled workforce is a catalyst to enhance productivity, minimise losses and to overcome technical trade barriers and thus facilitates global competence to cater international needs and helps international branding of local produce.

During interactions with Industry, it has emerged that knowledge of precision measurements and quality control is generally available with few personnel like managers or senior engineers who learnt it on the job over the years. However, the people at shop floor level who carry out day to day production and measurement do not have adequate knowledge of the subject and carry out work as per the instructions given. In today's competitive world, in order to maintain the quality of products and reduce the rejection ratio, it is essentially required that the staff actually working at shop floor level must have training and exposure.

NPL, India, being NMI of the country, therefore, came forward to create trained manpower in areas of precision measurements and quality control to accelerate the growth of industry. CSIR, encouraged by Skill India call from honourable Prime minister, Shri Narendra Modi, launched its integrated skill development programme in 2017. Based on CSIR-NPL's strength and expertise in accurate and precise measurements, a well-designed full time one-year certificate course was initiated on precision measurement and control. This course was aimed to meet the needs of National Accreditation Board for Testing and Calibration Laboratories (NABL) accredited testing and calibration laboratories, legal metrology institutes and improvement in efficiency of manpower and industries. Two batches of this course had a good placement. In view of success of this course, it was upgraded to be run at CSIR NPL as 'Post Graduate Diploma in Precision Measurement and Quality Control' under the aegis of Academy of Scientific and Innovative Research (AcSIR) which is an Institute of National Importance under Government of India with Ministry of Human Resource Development (MHRD ID No. U-0713) [62].

Main Features of Course [59]

(a) **Personal Growth** by meticulously designed course modules with a blend of classroom lectures, discussion, tutorials and practical related to measurements, testing and calibrations.
(b) **Transformation** by industrial oriented experimental learning and exposure through hands-on experience in industrial internship.
(c) **Academic Support** to students, world-class infrastructure, free access to SCI journals in library, seminars and lectures,
(d) **Multidisciplinary Skill Development**, faculty with extensive experience in precision measurements.

Placement Opportunities: On successful completion, young minds trained in this discipline have a unique advantage of wide exposure and hands-on training on best equipment/standards for measurements. These professionals have good placement opportunities in quality control(QC), accredited laboratories, manufacturing and production industries, Micro, Small and Medium Enterprises (MSME) and many other related industries, Government, public and private sector. The manpower trained by NPL, India will be attractive to these sectors. They will not only be the most suited candidates for Government or private industries, manufacturing sectors but would also be an asset to Government run legal entities, e.g. National Accreditation Board for Testing and Calibration Laboratories (NABL), Quality control sections, Testing and Calibration Laboratories from Private/public/Government sector, Legal Metrology Department, Standardisation, Testing and Quality Certification (STQC), Quality Council of India (QCI), etc. The demand of skilled manpower for precision measurements in manufacturing sector and quality sectors identified above will be increasing day by day. This way the trained manpower in precision measurements through this certification course would have ample job opportunity.

20.3.2.3 Metrology Skills for Accreditation, Standardization and Legal Metrology

CSIR-NPL has made a huge contribution in creating Human resource for establishment of accreditation infrastructure that exists for testing and calibration laboratories in India today. The visionary MGK Menon called for making an effort on coordination of testing and calibration facilities in India. In 1981 [58]. Department of Science and Technology set up National Coordination of Testing and Calibration Facilities (NCTCF) which started working and made its criteria based on ISO guide 25 for competence of test and calibration facilities. NCTCF recognition was given to various government and private laboratories. NCTCF had two programmes one for calibration and one for testing. Calibration programme of NCTCF was implemented by CSIR-NPL. Assessors for Calibration were initially provided by CSIR NPL and as the volumes grew, further manpower was trained by NPL to run these programmes.

Accreditation refers to third party recognition of capabilities of the organisation being accredited. As described by Aswal [26], accreditation is essential part of quality infrastructure of India. The NMI (NPL), Standard writing body (BIS), Legal Metrology Departments were setup in pre independence era or soon after independence. It may be noted that plans for accreditation of testing and calibration Laboratories were initiated in India by DST as early as 1980 and laboratory recognition activity was started by National coordination of testing and calibration facilities NCTCF and later by National accreditation board for testing and calibration laboratories NABL (1992). The scheme was coordinated by DST and worked from DST and NPL.

With liberalisation in 1991, the need for conformity assessment (CA) infrastructure came to forefront; a committee comprising various stake holders was setup to formulate a system that would enhance international acceptability of CA results. The proposal for such a system was prepared by department of Industries recommending establishment of an entity, with partnership between government and Industry, which was required to be self-sustaining and be away from Government [57, 58]. As a result of the proposal being accepted by government, it was decided to set up Quality Council of India with an aim of having an accreditation structure in India and for proliferation of quality movement in India through a National Quality Campaign. NABL was merged with QCI in year 2017.

QCI provides awareness and training through its various boards. These boards make efforts for spreading awareness in the area of Metrology, organise training, awareness workshops and capacity building. For example, their training programmes cover standards like ISO 17020; ISO 17011, ISO 17065, ISO 15189, and ISO guide 34, ISO 17043 etc. QCI has initiated consolidation of these efforts, by establishing an independent division 'Training and Capacity Building (TCB)' under its ambit. Among QCI Boards, NABL runs awareness and basic training programmes on metrology with faculty from NPL [57]. CSIR NPL has played a role in formative years of NABL to create human resource needed in the areas of its operations. Course on ISO Gum were done by NPL, region wise all over India, in order to provide training and have harmonised practices in India.

Training in area of standardisation is also taken up by BIS, which came into being for this task on 6 Jan 1947. It may be recalled that about the same time CSIR NPL also came into being. Thus, ISI was to formulate standards and testing for compliance was taken up by NPL in addition to its mandate for standards of measurements. With passing of time and due planning, Certification Marks Scheme, under the Indian Standards Institution (Certification Marks) Act, 1952 was started in 1955-56. Legal status was granted to ISI's functioning by BIS Act 1986 [63]. NPL cooperates with BIS and provides expertise necessary for their course's scenario. As a partner in QI, BIS also performs the task of training in standardisation and has established National Institute of Training for Standardization (NITS)' to cater to the demands of the industry for training standardisation, certification and quality management [54], etc. NITS like NPL also conducts programme for developing countries of Asia, Africa, Europe and Latin America every year [54]. NPL's expertise is availed by BIS as and when required. For example, when implementation of ISO GUM was picking up in India, NPL faculty has been nominated to teach these at BIS courses also.

Another important pillar of QI is Legal Metrology which is in purview of department of consumer affairs. For any country, legal metrology is vital for fair trade and to protect consumers. Director, Legal Metrology is a statutory authority with powers and responsibilities prescribed under the Legal Metrology Act, 2009 and has the responsibility for establishment of standards of Legal Metrology and keeping these standards traceable [56]. Legal metrology has Regional Laboratories, HR of these get training on job. Legal metrology officers are recruited with S&T background and trained further at **Indian Institute of Legal Metrology, Ranchi,** [64]. Further, it also liaisons with National Physical Laboratory for training of legal metrology officers in the areas of SI units, length, mass, pressure, temperature, time and frequency etc. The new recruits of legal metrology are sent to NPL for training in areas of mass, length, pressure, and temperature metrology. Time standards' training has also been added to this portfolio.

20.3.2.4 Creating Awareness of SI, Quality Infrastructure and Development of Scientific Temper Among School Students

NPL India believes in imbibing the culture of metrology to Indian citizens from very early age (Fig. 20.10).

With this aim it has been decided to motivate the school students to pursue science of measurements and to have a rendezvous with measurement standards. Visits of School students and summer projects are organised to let them have glimpses of Primary standards-the deities of the measurement world. These activities include most accurate standards of mass, length, time, voltage, current etc. The subject is made simple through use of working models and exhibits including audio-visual shows on the scientific activities of the laboratory. CSIR-JIGYASA programme which is a student - scientist connect programme launched by the government, is one of such programmes [65, 66]. The programme is implemented by the Council of Scientific and Industrial Research (CSIR) during its Platinum Jubilee Celebration

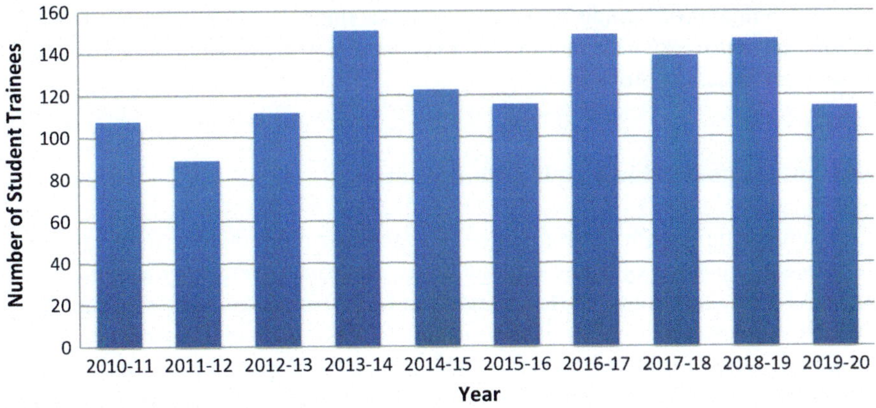

Fig. 20.10 Students from tertiary education system are groomed by NPL India with experience on sophisticated instruments during their training and dissertation. The metrology awareness and the experience are useful for their future. Figure shows number of UG and PG students, trained year wise

Year in collaboration with Kendriya Vidyalaya Sangathan (KVS) at national level for further widening and deepening its Scientific Social Responsibility (SSR). In 2020, CSIR also tied up with Atal Tinkering lab mission and is connecting to students other than KVs through this Mission. The focus of these scheme is on connecting school students and scientists so as to extend student's classroom learning with well-planned research laboratory-based learning. Under these programme students of KVS are connected through three modes- visit to the laboratories, scientists visit to school and summer projects. CSIR NPL is providing the base for metrology training to these students.

20.3.2.5 Programmes for Student Trainee from Educational Institutes UG and PG Level

Concepts of metrology aligned with international vocabulary of terms in metrology and ISO guidelines on uncertainty in measurements should ideally be included in university syllabus but this is not the case in India. To fill up this gap in education system CSIR NPL is taking steps by training of UG and PG students. Science and technology students perusing graduate or post graduate courses in India work at NPL as a trainee for their dissertations. The duration of these programmes is usually between two months to one year. These students get hands on experience with highly sophisticated instruments that are useful for researchers and learn R&D method during the phase of life after which they are going to start professional carrier. Uniqueness of carrying out such dissertation in CSIR NPL is that, besides the development of skills in Physics, chemistry or engineering science and technology, they also get exposure to methods used by metrologists. This exposure is very much required for industries and academia. Hands on experience of measuring instruments enriched with metrology concepts makes these students ready for the job and enhances employability.

20.3.2.6 Higher Education like Ph.D. courses

NPL India is running Ph.D. programme and encouraging students to undertake Ph.D. in the various aspects of Metrology including Material's Metrology as illustrated in Fig. 20.11. Ph.D. scholars started working in NPL from early years of its inception. Initially they used to work with various universities, this had advantage of diversification of work however, and such multi university cooperation had its own share of problems too. With setting up of Academy of Scientific and Industrial Research (AcSIR) [62] in 2011, NPL tied up with AcSIR through the MOU between CSIR and AcSIR. NPL is one of the largest centres of AcSIR for Ph.D. Figure 20.12 shows number of students enrolling for Ph.D. work at CSIR-NPL has been increasing; this may be attributed to increased awareness of NMIs role.

Several students are enrolled in Ph.D. programme who are working in problems related to Metrology. Today more than 300 hundred students are pursuing their doctorate research at NPL. As it has been discussed in the previous section, these students can be an asset for any laboratory/company/industry/academic institution they join. During their Ph.D. program they undergo detailed course work which

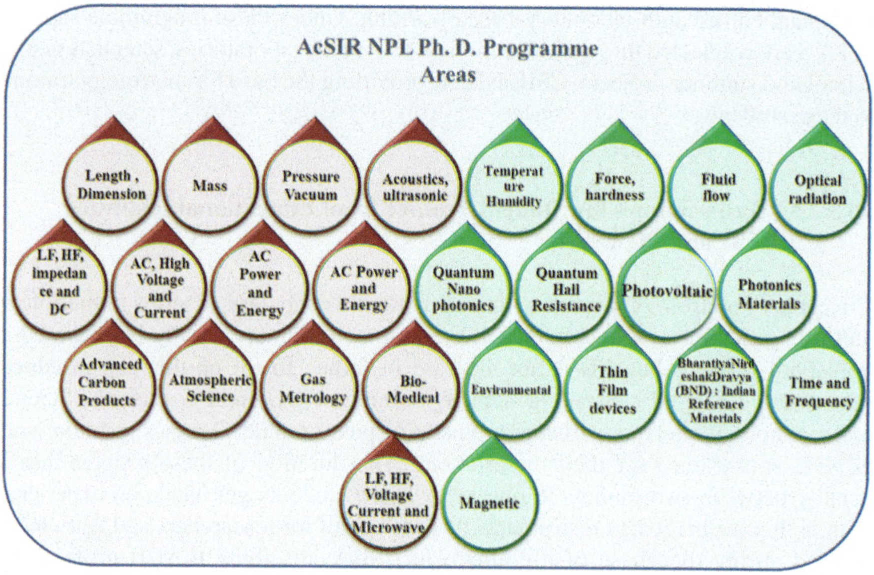

Fig. 20.11 Areas of Ph.D. work at NPL cover science technology and metrology for physico mechanical, optical, electrical and electronics, environment, biomedical, chemical, materials, devices, Bhartiya nirdeshak dravya (reference materials), time and frequency

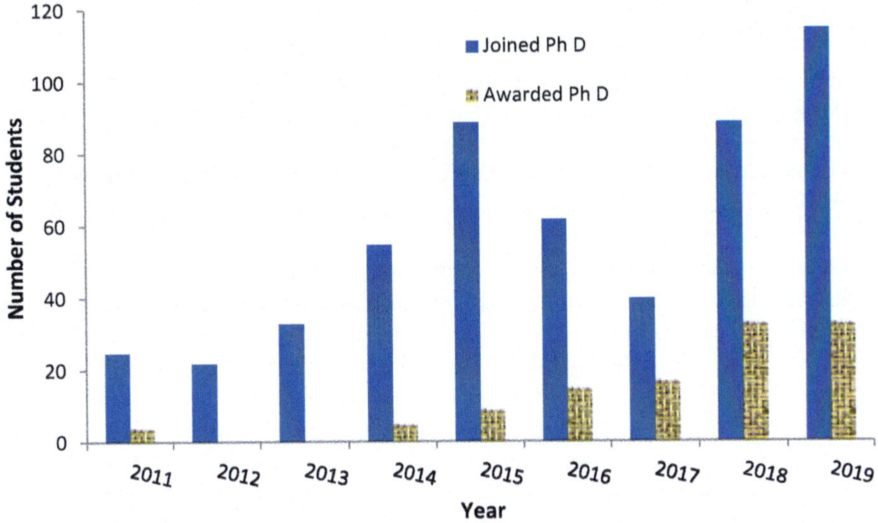

Fig. 20.12 Number of Ph.D. students shown year wise. Working with a national metrology institute has an advantage of strengthening core science and technology research as well as metrology and quality systems skills

includes quality system and basic metrology. Having in-depth knowledge in the subject these students can directly be absorbed at higher positions in academics as well as in Industry. They have an edge over their competitors due to hands-on training and learning extended by CSIR-NPL. Learning of metrology catalyses innovation and help bridge the gap between technology development and product development [67, 68].

20.3.3 NPL's Efforts for Uplifting SAARC and Other Nations

SAARC nations have realised the importance of quality infrastructure and have made efforts in last two decades to develop the QI. Nepal and Sri Lanka have come up with their NMIs offering training course. NMI of Bhutan is coming up whereas some like Afghanistan are still struggling to establish and investing on development of human resource with courses from other NMI and LM institutes. The need for metrology usually arises in the beginning for protecting the interest of consumers. Weight, length, volume, temperature are the initial key areas that are needed by the LDC. Thus, the starting point of quality infrastructure is setting up a body for national measurement and legal metrology. At this stage need for imparting knowledge to managers, policy makers and the measurement personnel arises. Afghanistan is slowly overcoming these hurdles.

In developed countries education efforts for metrology are made and QI infrastructure is strengthened, further the developed economies have been doing capacity building of the developing and underdeveloped economies in terms of providing training and knowledge sharing. As a result developing and under developed nations build quality system, accreditation bodies and legal metrology system in their countries. Further examples from past are the aid from Germany to India in 80's and 90's, from JICA to Thailand and from PTB Germany's initiatives for upgrading quality infrastructure and enhance regional cooperation in the SAARC region. German government supports development of metrology in developing nations through its NMI-Physikalische Technische Bundesanstalt (PTB) under various projects such as SAARC-PTB programme [69] and Metrology-Enabling Developing Economies in Asia (MeDEA) [70] project and various projects. Similarly European commission and many other donors provide support for the cause of metrology development [71, 72]. These are the examples of handholding made available to developing nation and upgrading competence of the national QIs. One may wonder why developed nations care for metrology HR development in developing nations. Let us try to understand this example by considering a situation where Items are exported from a developing or least developed economy as depicted in Fig. 20.13. Now if the economy 'A' with

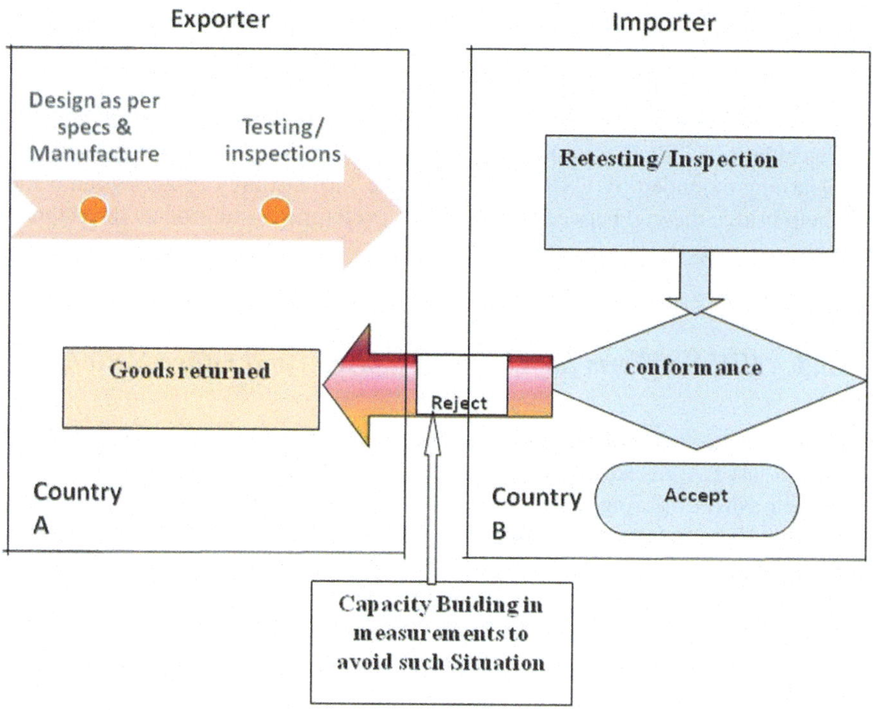

Fig. 20.13 Measurement capacity building is needed to support export –import as it can minimise losses due to retesting and rejection

a weaker QI is the exporter and importer is a quality conscious nation with well laid conformance criteria, the measurements required to decide upon conformance in economy 'A' will not be relied upon by 'B'. On reaching 'B', the receiver arranges retesting of the item and invests time and efforts to find out compliance. If the exported item fulfil the conformance criterion laid down by developed economy then only the product is put in use else both the economies suffer losses. Hence the metrologically advanced nations resort to the development of HR in QI of the nations which are major supplier or potential suppliers for them.

Many donors like UNDP, European commission for trade, German Government (through SAARC PTB Project) have come forward to support the SAARC in its ventures for HR in metrology. SAARC PTB project harnesses the expertise available in the region and catalyses the intra-regional cooperation in metrology.

CSIR NPL realises that the world of metrology and conformity assessment is very much interlinked internationally. Growth of a nation cannot be in isolation from its region and other interregional networks. Young researchers from various nations visit NPL from time to time. The state of metrology in commonwealth countries was of concern in international arena in eighties, a Common wealth India Metrology (CIMET) centre was established at NPL to provide group as well as individual training to participants from commonwealth countries. The centre worked under the advice of Dr A. P. Mitra, then DG CSIR, Dr S. K. Joshi, then Director NPL and Dr K. Chandra from India and other member from Barbados, Australia, Ghana and China. NPL also cooperated with centre for science and technology of the non–aligned and other developing countries (NAM S&T centre) to organise training course in metrology. These programmes continued till 2003. A number of delegates from various countries attended training programmes on basic metrology. Metrology concepts in various parameters like dimension, mass etc. and also training on quality system for ISO 17025 continued till 2003 under CIMET training programmes [47]. In later half of next decade empowerment of human resource with knowledge and skills for metrology was initiated for SAARC. A programme was prepared with funding from German Government and technical expertise from NPL. Under the SAARC-PTB programme, personnel from NMIS of Afghanistan, Nepal, Pakistan, Bangladesh, and Sri Lanka have been imparted knowledge. This support is still continued. It has made impact in the region and NMIs and accreditation structure has come up well in three countries others are also on the road to success.

20.4 Job Opportunities in the Field of Metrology

Scientists, Engineers and Technicians are the promising careers in metrology. Their qualifications differ according to the position and needs of the institute/industry. A job in metrology means understanding accuracy, precision, systematic bias/variances, evaluation of measurement uncertainty and traceability. In fundamental or basic metrology, the target outcome is to establish and perceive new units, devise new measurement methods and/or to upgrade them with better accuracy and lesser uncertainty, build standard references, and to practically help implement the adherence of every industry to the traceability protocols. Near exact measurements are essential in all fields of measurements, especially in trade and commerce. For the latter, operational skill sets associated with instrument and its working principles become very important, and these individuals in metrology pursue their career as engineers and/or technicians.

A career in industrial metrology implies knowledge in specification of tolerances and high emphasis on the inspectional requirements, involved at the very outset of the production processes itself. Such teams need to have expert engineers with multi-level competences, which include understanding of the measurand, design and modelling with advanced technologies and certainly communication skills. Engineers are responsible for making detailed reports of the instruments and hence are required to possess knowledge of the quality control manuals.

A career in legal metrology is about regulation, in relation to the mandatory technical and legal requirement. The focus of this important branch of metrology is to ensure public guarantee in terms of security and accuracy of weights and measures. They include protection of public health, public safety, and the environment.

A plethora of jobs in metrology are those related to calibration technicians. This group of people work with a variety of advanced machinery and simultaneously performs several numerical, analytical and statistical procedures. Calibration technicians in some countries are also known as calibration inspectors or metrology technicians. In common, calibration technicians are anticipated to have good vision, adequate mathematical skills, knowledge of the principles underlying the equipment functionary, good eye-hand coordination, and also good communication skills. They also are required to interpret data and other relevant information in order to achieve the calibration procedures.

20.5 Challenges and Proposed Solution

Having discussed the need for Metrology education in India, it clearly emerges that there is a lack of focus on metrology education in Indian education system. Education system has to cater to the needs of various levels and sectors of employment. Science and Technology workforce coming out of educational institutions is struggling to find suitable jobs, on the other hand businesses also are facing difficulty in getting the needed skills [3]. Global competitiveness report highlights that the industry/Business should have a say in curriculum/education. India is in transition from developing to developed nations, focus on science, technology and Innovation has shown results. India ranked fairly well on research and innovation in global competitiveness. We are shifting from price based economy to innovation and quality based economy. Standards foster innovation [26, 67, 68], compliance to standards is enabled by metrology, which in term is performed/designed by human resource. In view of this, a strategic intervention is necessary to reengineer the courses and accelerate learning for precision measurement. The system has to adapt to changing international standards quickly. A gap between floating of new norms and its adaptation in India will lead to losses due to noncompliance and rejection. The challenge is to plan and implement this intervention quickly as well as effectively.

As mentioned earlier many countries have made recommendations and taken steps to mitigate the weakness of metrology education. In case of Lithuania, they started full time course and made four levels of metrology education with detailed planning of competence evaluation and management. Brazil's Human resource metrology program evolved in many stages from 1995 to 2006 and has become a reference course in the region. Bosnia and Herzegovina implemented metrology education by pooling in capacity of various educational institutes in a consortia approach [34].

The discussion on the education system presented in Fig. 20.2, identifies that there is an impediment in terms of skilling and metrology in our education system. Few of the reasons that one could identify on a cursory glance are:

a) School education does not include theory and practical related to measurements concepts
b) Vocational training does not build up metrology skills
c) Degree programs with outdated syllabus, No exclusive degree programme in Metrology
d) Lack of state-of-art infrastructure for developing measurements skill.
e) Quality of teaching

Although each state has its own freedom to implement the educational policies and strategies, in many cases the curriculum design appears to have failed to consider the local issues.

For India a strategic planning and intervention is required to bridge this gap of metrology knowledge as described in Fig. 20.14. Strategic planning is needed to solve the problem through restructuring the education system, skill development through short term training or full time courses and a certification scheme for personnel. It may be noted that the education falls under the ambit of ministry of human resource. Ph.D. programmes are carried out in areas of thrust as decided by STI policy (Ministry of Science). Ministry of Consumer affairs oversees standardisation, legal metrology and accreditation whereas the National Metrology Institute is under Ministry of Science and technology. Most heavy Industries in public sector which are product of MSME are aspiring to be globally competent. Regulators in various sectors fall under different ministries.

The proposed solutions are discussed in Fig. 20.14.

(i) **Strategic intervention in Education System**

Fit to proper education is job or job creation. Quoting Maimonides, "Give a man a fish and you feed him for a day; teach a man to fish and you feed him for a lifetime" holds quite appropriate in the present case of increasing unemployment problem. On one hand there are sectors like accredited laboratories, QC and QA in manufacturing industry etc. that are facing problem of metrology HR, on other hand there is unemployment. Education system and curriculum needs to be restructured

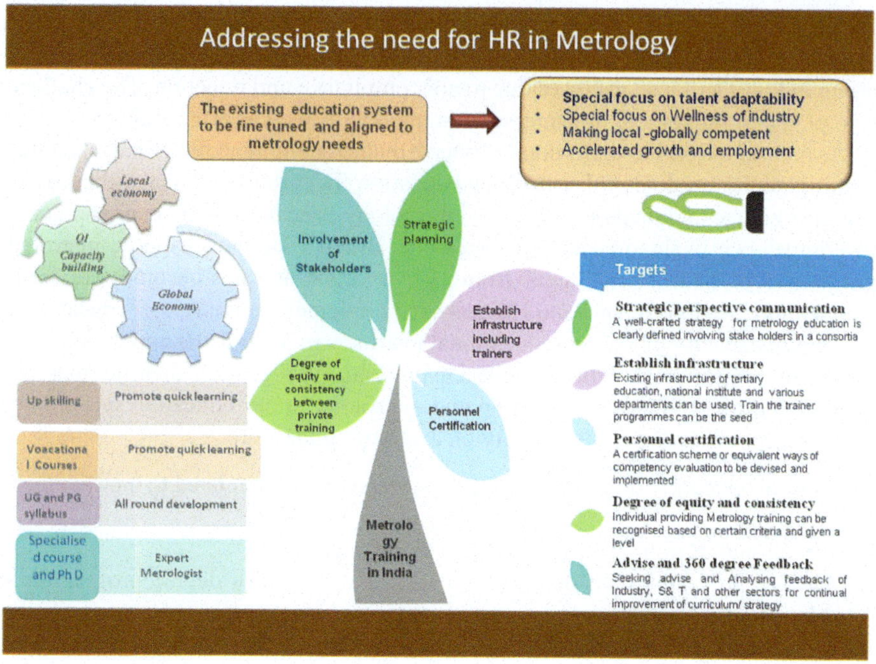

Fig. 20.14 Addressing the need for HR in metrology- requires strategic plan with consultations of stake holder, establishment of infrastructure for industry education. It is also important to achieve quality and equity of freelance trainers and certification of metrology personnel. Various levels of metrology training are needed as described in bottom left. Local economy along with QI can drive India's global economy

to solve these issues. The MSME sector takes most manpower right from school or vocational courses like ITI etc. Hence endeavours should be made in education system right from beginning.

A multipronged approach involving its secondary and tertiary education is suggested. It may be noted that the education falls under the ambit of ministry of human resource. Ph.D. programmes are carried out in areas of thrust as decided by STI policy (Ministry of Science). Ministry of Consumer affairs oversees standardisation, legal metrology and accreditation whereas the National Metrology Institute is under Ministry of Science and technology. Most heavy Industries in public sector which are product of MSME are aspiring to be globally competent. Regulators in various sectors fall under different ministries. A comprehensive policy needs to be designed and in order for it to be effective, authors are of the view that a consortium of various stakeholders and participation of all these stake holders is must. First challenge will be sensitising and synergising all to understand the urgency for development of HR in Metrology. This consortium, once synergised will make recommendations for policies and invite UGC, AICTE, etc. to focus on metrology and other needs of industry, academics and STI.

It is further proposed that the chapter on SI unit in secondary education may be elaborated to include all the seven SI base units and their importance to motivate the students right from young age. Also, the basic definitions as laid by the VIM document of metrology can be well introduced at this level. Scope of statistics, which is an essential tool in measurement science, should be widened to emphasize on the statistical tools to envisage the meaning of the basic laboratory experiments pursued in the school laboratories.

Several institutes and universities of India have now implemented metrology as a credit course of metrology like engineering metrology at the Bachelor's level. Basic terms of metrology, traceability, working knowledge of various standards used in everyday practice in the field of electrical, mechanical and electro-mechanical, biomedical, and chemical measurements should be part of mandatory core syllabus. The contents of the course is expected to benefit students to pursue a career as laboratory technicians involved with testing and/or measurement, Six Sigma (6 σ) practitioners, quality control specialists, for engineers and technicians involved with manufacturing facilities and developing critical-to-quality specifications on components and assemblies, etc. Vocational training should focus on basic concepts and more practical knowledge of measurements in the specific sectors. Besides this, skilling and up skilling should be carried out. The scope of such education is wide and expertise are available in the country but at different places in fragmented manner. Partnership of various tertiary education Institutes as well as vocational training Institutes with NMI and regulators can achieve this ambitious target.

Three Levels of education and skill set may be defined for industrial metrology (Shop floor technician, Inspector and Engineers) in line with recommendations of skill Council. Fourth level will be highly specialised metrologists for scientific metrology and train the trainer schemes. A personnel certification scheme operating as per ISO 17024 criteria is proposed. NMI can play a great role in designing the scheme (Fig. 20.15).

Education policy is to be formed by involving various stake holders from industry, government, educational institutions, accreditation and standardization bodies and national measurement institutes. Though NEP 2020 recognises need to involve industry in technical education, explicit involvement of Industry in making policy itself and syllabus will ensure their needs being identified and taught right from beginning and will enhance the employability. Involvement of NMI, quality infrastructure and regulators will again make the system robust and fulfil the need of skill building suitable to these areas. Nationwide implementation of the policy is to be carried out through education Boards, University Grants Commission, All India council of technical education, Medical education research institute, MHRD Institute and National laboratories with educational institutes for tertiary education and vocational educations. Use of information technology should be made to enhance training in these areas. There can be programmes for trainers, beginners and practitioners. These measures will benefit Industrial metrology and elements of QI namely legal metrology, standardisation, testing and calibration labs, accreditation bodies and new recruiters of NI. However for NMI and regional labs of echelon 1, there is a need for high end programmes like doctorate to prepare HR for scientific metrology. Such

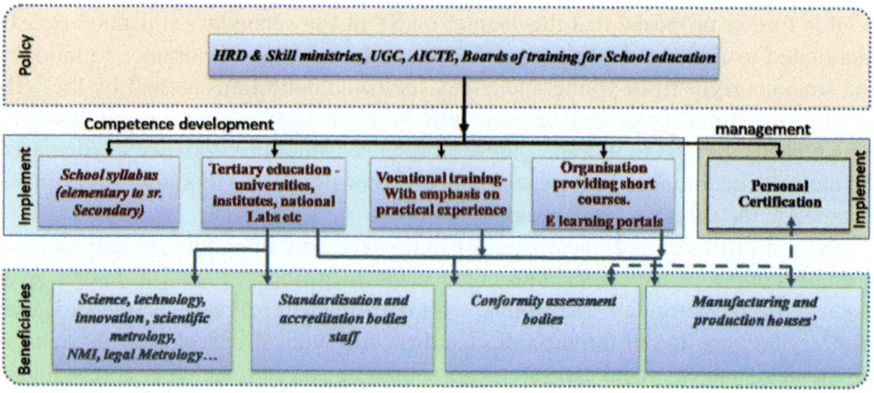

Fig. 20.15 Proposal for policy intervention to bridge the metrology HR in consultation with stakeholder and its nationwide implementation for competence development through educational system and short-term training providers. Certification of metrology personnel will support recognition of capability level

doctorate programmes can be offered by cooperation between NMI, and improvement by feedback from targeted beneficiaries as shown in Fig. 20.15, is expected to give good results.

(ii) **Skilling of workforce of India and availability of career progression Routes**

Skill development has to include metrology training and keep on building it with advancements and adaptation of new technologies in workplace. Individual level expertise is essential to perform a certain job which is utilized by the employer to perform that role. Competency of the team is requirement of the organisation which depends on competence needed by individual for a specific role and complementing the skills of individual to achieve team targets. Skills can be segregated into a hierarchy based on **immediacy, relevancy, and feasibility** [59]. Proficiency, which is needed currently or in forthcoming days, comes under immediacy. The relevancy of ability or aptitude can be assessed based on the ease of transfer of training from a learning atmosphere to the actual work atmosphere. The feasibility of proficiency refers to the potential of the skill to stay relevant in the future. With this hierarchy in place, a plan in terms of the requirement of organization regarding training, hiring, up skilling etc. can be decided more strategically and make the best use of the talented employees already on board.

Career progression is a very important aspect that needs to be taken care of, for existing metrology workforce. Due to lack of structured metrology program, education of metrology is obtained in general on the job. It becomes imperative to provide career progression to experienced individuals who learnt the skills while carrying out day to day calibration or testing activities. Performance improvement can be targeted by skill training which often results in further improvements in the potential of such personnel. If the person of lower qualification shows potential, continued education programmes should be offered which encourages and motivates the worker further. Practical experience gained by the person in parallel to acquiring further education and training on advanced technology can be helpful in taking mid potential/mid performance person to high performance one (Fig. 20.16).

This becomes a very crucial stage where complacency needs to be broken by aligning personnel growth goal to organisation growth targets and strategic intervention of mentoring, training and providing motivation by enlarging their roles. Further impetus for growth can be through measures like accelerated position and role growth and international exposure. Thus, within the organisation the human capital on lowest paddle of career growth can also aspire to reach highest paddle and thus remain motivated. Training of metrology that organisations build in the system will not go waste and a goal oriented HRM can provide the best returns on this investment.

The training of metrology is in line with the missions of *'Self-reliant India', 'Vocal for Local' and 'Skill India missions'*. It is expected to enhance employability of personnel, benefit the industry, help economic growth and improve quality of life.

(iii) **Certification scheme**

Personnel certification program is proposed for the certification of metrology personnel. The programme may be developed to meet requirements of ISO 17024:

The personnel certification will be based on the technical criteria and basic qualifications set up in consultation with stake holders. The focus of metrology education can draw an analogy with computer education focus in nineties. Department of electronics had once come up with open learning course curriculum, with O level, A level curriculums and authorised centres for learning. Thus, having such open education

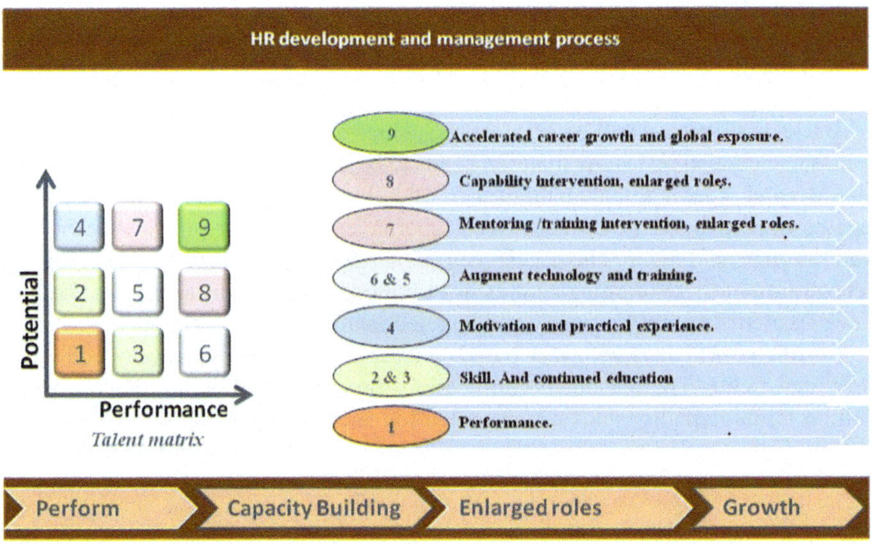

Fig. 20.16 Proposed scheme for continued education and up-skilling to retain the man power and help the trained metrology personnel to reach higher positions

done, a person could appear in exam and get certified. Even working professionals could add up this qualification mode.

Today we have the need for certification in metrology. As discussed, metrology knowledge is required by operators, testing and calibration technicians, engineers and metrologists. In line with these requirement Level O, Level 1, Level2 and level 3 certifications are proposed. **Level O** will be beginner's level targeting operators, basic qualification can be 12th.

Level 1

For Level 1 calibration tasks, testing SOPs, detailed calculation of uncertainty of measurement as per ISO GUM, Inter Laboratory Comparison (ILC) report preparation, knowledge of QI should be included in the curriculum. Basic qualification can be 12th plus 3 years' experience in metrology or ITI plus one-year experience or Diploma

Level 2

For going to Level 2 person must be able to perform error analysis of at least one test or calibration task, have experience of calibration and testing and ILC. The further learning at this stage would be compliance standards, quality management standards, process control tools, validation of new methods etc. Basic qualification can be 12th plus 5 years' experience in metrology and certificate, or ITI plus 7-year experience or Diploma plus three-year experience or Graduation in science/Egg. and passing level 2.

Level 3

In addition to skills of level 2, at this level knowledge of relevant quality measurement standards, development of SOPs, design of metrology experiments, development of new methods and validation, piloting intercomparison is the skills targeted in curriculum

Basic qualification will be Masters of Science and technology.

Course content and study material needs to be developed by NMI, STQC, ARAI, IDEMI, and other national laboratories. A flexible public, private partnership (PPP) can be implemented where accredited laboratories can offer practical learning in various cities and theory classes can be run online by the NMI or national labs. The recognition of these learning centres in PPP mode will have a cascading effect and develop metrology HR in timely and targeted manner.

Certification agency will work based on these criteria and tests developed for each level and fields. The personnel certifications programs will assure that workforce fulfil minimum quality standards specified for each level.

20.6 Conclusion

This chapter portrays the need of HR development in metrology which is not given its due importance in education system. It further presents the national and international scenario of HR in metrology and its relevance to quality infrastructure of Nation. Nationally mostly the metrology is being learnt on the job and through short training programme. The chapter presents a detailed review of present scenario and the courses available nationally and internationally. The development state of a country is related to development of metrology in the nation. Developed economies have a robust quality infrastructure and prepare the human resource for it. From the review presented it can be concluded that India has lots of challenges to face in terms of HR for metrology. CSIR NPL is trying to meet some of the challenges; the efforts of CSIR-NPL for development of human resource in metrology have been presented here. CSIR-NPL is fulfilling the need by a dedicated full-time course/training for industrial metrology and preparing high end skills for scientific metrology.

Review of metrology in curriculum exhibited that there are some initiatives to introduce engineering metrology, launching e-courses like Swayam portal of Indian Government. Further, a large number of organisations and individuals are also offering short term courses in India. Managing equivalence of such programmes and evaluation of competence is a challenge. Significantly, it includes the highlights and contributions of CSIR-National Physical Laboratory, in training HR for metrology by short courses, a full time one year PG diploma and Ph.D. programme under AcSIR [62]. Short courses offered by NPL to industry people, various companies and calibration laboratories leads to utilization of experienced employees in their organization with updated knowledge. In the same line NPL is also helping other developing countries by way of training and advice.

Analysis of gaps, challenges has been made and solutions are proposed for tackling the problem through a scheme for education system, personnel certification, skill developments and capability enhancement coupled with availability of career progression.

References

1. Timeline|Harappa. https://www.harappa.com/content/timeline. Accessed 24 Jul 2020
2. S.K. Srivastava, Measurement units of length, mass and time in India through the ages. Int. J. Phys. Soc. Sci. (2017)
3. K. Schwab, *Insight Report* (2019)
4. Armstrong, Michael and ST Armstrong's Handbook of Human Resource Management Practice. Kogan Page Publishers, 2020
5. Ministry of Skill Development and Entrepreneurship. https://www.msde.gov.in/nationalskillmission.html. Accessed 24 Jul 2020
6. IMF, *Population Shifts—Finance and Development*. March 2016 (2016)
7. United Nation, World population prospects 2019 (2019)
8. Gladstone R (2015) *India Will Be Most Populous Country Sooner Than Thought, U.N. Says*. New York Times

9. Population United States 2019|Statista. https://www.statista.com/statistics/183457/united-states–resident-population/. Accessed 22 Jul 2020
10. Median age over 43 years in the EU—Product—Eurostat. https://ec.europa.eu/eurostat/web/products-eurostat-news/-/DDN-20191105-1. Accessed 22 Jul 2020
11. India enters 37-year period of demographic dividend—The Economic Times. https://economictimes.indiatimes.com/news/economy/indicators/india-enters-37-year-period-of-demographic-dividend/articleshow/70324782.cms?utm_source=contentofinterest&utm_medium=text&utm_campaign=cppst. Accessed 9 Jul 2020
12. W. Joe, A.K. Dash, P. Agrawal, Demographic transition, savings, and economic growth in China and India. IEG Work. Pap. (2015)
13. S. Mehrotra, A. Gandhi, B.K. Sahoo, Estimating India's skill gap on a realistic basis for 2022. Econ. Polit. Wkly. (2013)
14. K. Navaneetham, A. Dharmalingam, A review of age structural transition and demographic dividend in South Asia: opportunities and challenges. J Popul Ageing (2012). https://doi.org/10.1007/s12062-012-9071-y
15. W. Lutz, J.C. Cuaresma, E. Kebede et al., Education rather than age structure brings demographic dividend. Proc. Natl. Acad. Sci. USA **116**, 12798–12803 (2019). https://doi.org/10.1073/pnas.1820362116
16. Vocational Education in India. https://nistads.res.in/all-html/VocationalEducationinIndia.html. Accessed 9 Jul 2020
17. National Education Policy 2020 Ministry of Human Resource Development Government of India
18. BIPM (2012) International vocabulary of metrology-Basic and general concepts and associated terms (VIM) 3rd edition 2008 version with minor corrections Vocabulaire international de métrologie-Concepts fondamentaux et généraux et termes associés (VIM) 3 e édition
19. J. Dobiliene, Importance of education and compe… preview and related info|Mendeley. J. Phys. Conf. Ser. 588(1) 588:1–5 (2015)
20. (No Title). https://www.aicte-india.org/sites/default/files/IndiaSkillReport-2019.pdf. Accessed 9 Jul 2020
21. P. Sinha, Study Combating Youth Unemployment in india (2013). Accessed from http://library.fes.de/pdf-files/iez/09728.pdf
22. World Bank, *India's Employment Challenge—Creating Jobs, Helping Workers*. New Delhi: Oxford University Press, ISBN 978-0-19-806351-3 (2010)
23. When NASA Lost a Spacecraft due to a Metric Math Mistake. https://www.simscale.com/blog/2017/12/nasa-mars-climate-orbiter-metric/. Accessed 9 Jul 2020
24. Why the Mars Probe went off course—IEEE Spectrum. https://spectrum.ieee.org/aerospace/robotic-exploration/why-the-mars-probe-went-off-course. Accessed 9 Jul 2020
25. Mars Climate Orbiter Team Finds Likely Cause of Loss. https://mars.nasa.gov/msp98/news/mco990930.html. Accessed 9 Jul 2020
26. D.K. Aswal, Quality infrastructure of india and its importance for inclusive national growth. Mapan. J. Metrol. Soc. India **35**, 139–150 (2020)
27. Human Development Report 2019: Reader's Guide|Human Development Reports. http://hdr.undp.org/en/content/human-development-report-2019-readers-guide. Accessed 9 Jul 2020
28. Office of the High Representative for the Least Developed Countries, Landlocked Developing Countries and Small Island Developing States|. https://www.un.org/ohrlls/. Accessed 9 Jul 2020
29. C. Paper, S. Leme, Importance of metrology education—B&H Case importance of metrology education—B&H CASE. In: *First regional Metrology Organisation Symposium—RMO2008; 20th international metrology symposium*, Nov 12–15, 2008, Cavtat-Dubrovnik, croatia. pp. 262–265 (2008)
30. A. Meskuotiene, R. Zilinskas, V. Zabolotnas (2008) *Education and Competency Evaluation System* pp. 3–5
31. S. Echeverría-Villagómez, M. Flores-Campos, R. Pantoja-Lessoet al. (2006) Human resource and knowledge management system for competences in metrology using information technologies. 18th IMEKO World Congr. 2006 Metrol. a Sustain. Dev. **2**, 1188–1198

32. G.M. Rocha, R.P. Landim, The brazilian experience in the development of human resources in metrology. 20th IMEKO World Congr. **2**, 760–764 (2012)
33. Ibrahim SM, Bills P, Allport JM (2017) Situation of metrology education in developing countries using Nigeria as a case study. Lit. Inf. Comput. Educ. J. **8**, 2751–2760. https://doi.org/10.20533/licej.2040.2589.2017.0363
34. C. Paper, S. Leme, Importance of Metrology Education—B&H case importance of metrology education—B&H CASE (2014)
35. The Education of a Metrologist. http://asq.org/quality-progress/2001/01/measure-for-measure/the-education-of-a-metrologist.html. Accessed 9 Jul 2020
36. S.M. Ibrahim, BPAJ (2017) Metrology education impediment—what do we need to do? In: proceedings of academic world international conference. Saudi Arabia, Mecca, pp. 6–10
37. D. Newcombe, S. Reynolds, A measurement standard for evaluating metrology positions. NCSLI Meas. **4**, 68–73 (2009). https://doi.org/10.1080/19315775.2009.11721462
38. G.M. Rocha, R.P. Landim, Inmetro's metrology executive master's degree course. Proc. Soc. Behav. Sci. **46**, 4928–4932 (2012). https://doi.org/10.1016/j.sbspro.2012.06.361
39. M. Fonseca, Metrology in vocational education in Brazil
40. National Institute of Standards and Technology|NIST. https://www.nist.gov/. Accessed 9 Jul 2020
41. FSSAI rejected list of imported spices shipments: what went wrong?|Food Safety Mantra Blog. https://www.foodsafetymantra.com/2019/07/fssai-rejected-list-of-imported-spices-shipments-what-went-wrong. Accessed 9 Jul 2020
42. Asian countries import Guntur spices while West rejects|Business Standard News. https://www.business-standard.com/article/economy-policy/asian-countries-import-guntur-spices-while-west-rejects-104050301032_1.html. Accessed 9 Jul 2020
43. NPL Training—measurement and calibration training. https://training.npl.co.uk/. Accessed 9 Jul 2020
44. Education—Cal Lab Magazine. https://www.callabmag.com/education-training/. Accessed 9 Jul 2020
45. NCSL International Training Resources. https://www.ncsli.org/iMIS/TAL/Train.aspx?WebsiteKey=d502eebf-7ea1-4ae1-ac05-e2faa9324627
46. The Human Resource Development Group|National Physical Laboratory. http://www.nplindia.in/human-resource-development-group. Accessed 9 Jul 2020
47. NPL Annual Report|National Physical Laboratory. http://www.nplindia.in/npl-annual-report. Accessed 9 Jul 2020
48. Swayam Central. https://swayam.gov.in/. Accessed 24 Jul 2020
49. Index|CMTI-India. https://cmti-india.net/. Accessed 24 Jul 2020
50. IDEMI, Mumbai (IDEMI), Mumbai—2020 *Admissions, courses, fees, ranking*. https://www.collegedekho.com/colleges/idemi-mumbai. Accessed 24 Jul 2020
51. Fluid Control Research Institute—FCRI India. https://www.fcriindia.com/. Accessed 24 Jul 2020
52. CII. https://www.cii.in/. Accessed 24 Jul 2020
53. (No Title). https://www.nttftrg.com/course/inspection-metrology. Accessed 24 Jul 2020
54. Overview of NITS—Bureau of Indian Standards. https://bis.gov.in/index.php/training-2/overview-of-nits/. Accessed 9 Jul 2020
55. STQC|Standardisation Testing and Quality Certification Directorate. http://stqc.gov.in/. Accessed 24 Jul 2020
56. Weights and Measures|Department of Consumer Affairs|Ministry of Consumer Affairs Food and Public Distribution|Government of India. https://consumeraffairs.nic.in/organisation-and-units/division/legal-metrology/weights-and-measures. Accessed 9 Jul 2020
57. qci india—Google Search. https://qcin.org/
58. Introduction. https://nabl-india.org/about-nabl/introduction/. Accessed 9 Jul 2020
59. NPL India—skill initiative
60. ARAI Home. https://www.araiindia.com/home. Accessed 24 Jul 2020

61. Nptel, online courses and certification, Learn for free. https://nptel.ac.in/about_nptel.html. Accessed 24 Jul 2020
62. AcSIR—Academy of Scientific and Innovative Research|. https://acsir.res.in/. Accessed 9 Jul 2020
63. Home—Bureau of Indian Standards. https://bis.gov.in/. Accessed 9 Jul 2020
64. Indian Institute of Legal Metrology, Ranchi|Department of Consumer Affairs|Ministry of Consumer Affairs Food and Public Distribution|Government of India. https://consumeraffairs.nic.in/weightsmeasures/indian-institute-legal-metrology-ranchi. Accessed 9 Jul 2020
65. Jigyasa|National Portal of India. https://www.india.gov.in/spotlight/jigyasa. Accessed 9 Jul 2020
66. About Jigyasa|CSIR Jigyasa. http://csirjigyasa.niscair.res.in/about_jigyasa. Accessed 9 Jul 2020
67. R.H. Allen, R.D. Sriram, The role of standards in innovation. Technol. Forecast. Soc. Change **64**, 171–181 (2000). https://doi.org/10.1016/s0040-1625(99)00104-3
68. A QI-Reform Toolkit at a Glance
69. Bundesanstalt Braunschweig P-T South Asian Association for Regional Cooperation (SAARC) Strengthening regional integration and cooperation in the field of quality infrastructure in South Asia international cooperation
70. MEDEA Training Program—APLMF. https://www.aplmf.org/medea-training-program.html. Accessed 9 Jul 2020
71. CAPACITY BUILDING—Sim. https://sim-metrologia.org/capacity-building/. Accessed 9 Jul 2020
72. Thailand|Thailand|Countries and Regions|JICA. https://www.jica.go.jp/thailand/english/index.html. Accessed 9 Jul 2020

Chapter 21
CSIR-NPL: Growth Driver for Effective Implementation of National Policies

Nita Dilawar Sharma, Sushil Kumar, T. D. Senguttuvan, Anjali Sharma, Jiji Thomas Pulikkotil, S. Ragam Rao, Deepti Chadha, Nahar Singh, Anuradha Sengar, and D. K. Aswal

Abstract In the previous chapters, the metrological capabilities of CSIR-NPL in various parameters related to physical, chemical, engineering, biomedical, environmental, etc. along with their dissemination to various stakeholders have been explicitly deliberated upon. In this concluding chapter, we discuss how CSIR-NPL has been contributing to the nation by supporting the government in implementing its policies. A brief discussion on the 'Atmanirbhar Bharat' program, which is a paradigm shift towards the indigenous development of product, is presented. The applicability of Aswal model of inclusive growth—metrology is the requisite for coordination

---

N. D. Sharma (✉) · S. Kumar · T. D. Senguttuvan · A. Sharma · J. T. Pulikkotil · S. Ragam Rao · D. Chadha · N. Singh · A. Sengar · D. K. Aswal
CSIR-National Physical Laboratory, Dr. K. S. Krishnan Marg, New Delhi 110012, India
e-mail: ndilawar@nplindia.org

S. Kumar
e-mail: skumar@nplindia.org

T. D. Senguttuvan
e-mail: tdsen@nplindia.org

A. Sharma
e-mail: anjali@nplindia.org

J. T. Pulikkotil
e-mail: jiji@nplindia.org

S. Ragam Rao
e-mail: ragamsr@nplindia.org

D. Chadha
e-mail: deepti@nplindia.org

N. Singh
e-mail: naharsingh@nplindia.org

A. Sengar
e-mail: sengara@nplindia.org

D. K. Aswal
e-mail: dkaswal@nplindia.org

D. K. Aswal (ed.), *Metrology for Inclusive Growth of India*,
https://doi.org/10.1007/978-981-15-8872-3_21

amongst the government, academia, industry, and civil society—for implementing the policies of Atmanirbhar Bharat is discussed. In this perspective, an overview of the contributions made by CSIR-NPL in recent past to various stakeholders (government, industry, and academia) through its services of metrological traceability of SI units, Indian Standard Time (IST) dissemination, development of Bhartiya Nirdeshak Dravyas (BNDs), transfer of technologies, consultancy services, R&D projects and training on metrological programs are elaborated. The efforts of CSIR-NPL in publicizing the role of metrology for the national growth are discussed, which include the celebration of World Metrology Day every year on May 20, inviting students, entrepreneurs, and common citizens to CSIR-NPL for discussions on metrology and its applications, organizing the science festivals like India International Science Festival (IISF-2016), etc. The Chapter is concluded with suggestions on how CSIR-NPL should strengthen its capabilities and discharge its duties to fulfill the dreams of a self-reliant India.

Accurate and precise measurements are essential to drive the growth engines of Indian Science & Industry as it removes chaos and prompts innovations, which in turn, would save precious lives, resources and time. Let us strengthen CSIR-NPL for a better future of the country.

– D.K. Aswal

21.1 Introduction: Make in India and Atmanirbhar Bharat

The Government of India launched the 'Make in India' program on September 25, 2014, with an intention to encourage foreign companies to manufacture their products in the country. This policy was aimed at the growth of the manufacturing sector, the creation of additional jobs in the manufacturing sector, and to enhance the contribution of manufacturing to the GDP of the country [1]. On the other hand, Atmanirbhar Bharat Abhiyan (Self-reliant India Mission) was launched by the Government of India on May 12, 2020, which was aimed at recuperation of the economic slowdown due to COVID-19 pandemic [2]. Under Atmanirbhar Bharat Abhiyan, a stimulus economic package of Rs 20 Lakh Crore (equivalent to 10% of India's GDP) was allotted for different sectors to overcome the damage caused by the pandemic. The approach of Atmanirbhar Bharat resembles that of "glocalization", as discussed in Chap. 1, and focuses on a growth that originates from an internal strength of the country. This has also been well represented using the slogan: vocal for local to make it global. Atmanirbhar Bharat also represents a demand-based economy system, i.e., self-producing and self-consuming, and the surplus for export. In Fig. 21.1, the concepts of Make in India and Atmanirbhar Bharat often referred to as Make in

Fig. 21.1 Schematic representing concepts of Make in India and Atmanirbhar Bharat. In the center, the parameters required for making them a grand success are listed

India 1.0 and Make in India 2.0, respectively, are highlighted along with parameters required to make them a grand success.

Both Make in India and Atmanirbhar Bharat programs are aimed at transforming the country into an international design and manufacturing hub. Both of them envisage products to be manufactured in India, attract foreign investors for production in the country, reduce imports and increase exports, creation of new job opportunities, etc. The similarities and differences between these two programs are analyzed based on the parameters required for their effective implementations, as depicted in Fig. 21.1.

21.1.1 Investment for Manufacturing

Investment for capital and technology is the first and foremost requirement for setting up the manufacturing facilities in the country. In Make in India, foreign companies are encouraged to invest and bring technologies to manufacture their products in the country which can even be exported. For this, the government has created a conducive environment for foreign investors. According to a recent report, Foreign Direct Investment (FDI) inflows in India increased from $35.3 bn in 2014–15 to $56.00 bn during the year 2019–20 [3]. To facilitate FDI programs and to increase such ventures, the country must provide an appropriate environment to make investments profitable.

Under the Atmanirbhar Bharat program, local companies are motivated to manufacture products of international quality with government support and/or foreign investors to fulfil the local demand as well as for export. Under the Atmanirbhar Bharat package, the government not only has changed the definition of MSMEs to their advantages but also provided a big stimulus package, enhanced the scope for private participation in various sectors, and increased FDI in the defence sector, etc.

21.1.2 Innovation

It is well-known that the environmental conditions of India are quite different from those of developed nations [4]. Hence, the technologies that are brought in the country under Make in India program from these countries, need to be adopted to Indian conditions. This necessitates innovative scientific and technological solutions. For instance, cast iron products although commercially acceptable in many countries are not practicable in Indian scenarios as they are prone to rusting in Indian climatic conditions and need to have an innovative protective coating or different composition [5].

In the case of Atmanirbhar Bharat, the indigenous technologies need to be developed from basic concepts to final products through various technology readiness levels (as discussed in Chap. 1). Therefore, MSMEs would require technologies and innovations to develop new products, especially from academia as well as science and technology organizations of the country.

21.1.3 Quality Infrastructure

For both Make in India and Atmanirbhar Bharat programs, a robust national quality infrastructure (metrology, standards, and accreditation: all of them harmonized with respective international counterparts) is essential to ensure that manufactured products are in compliance with regulatory requirements. Atmanirbhar Bharat provides a scope to MSMEs for local production- right from raw materials to the finished product, and to achieve this, they would require the national quality infrastructure at each of the steps [6].

21.1.4 Skill Development

Skilled manpower is mandatory to operate the high-technology industry. Make in India provides employment opportunities for skilled manpower in high-technology manufacturing facilities, whereas nonskilled manpower may not find enough scope for the jobs. In the case of Atmanirbhar Bharat as the products are developed indigenously from scratch, it will build employment opportunities for everyone. In fact, unskilled manpower can be trained to make them capable to operate advanced machineries. To achieve maximum output, the skilled manpower should be trained to understand and implement the requirements of the quality infrastructure.

21.2 Atmanirbhar Bharat from the Perspective of Aswal Model

Atmanirbhar Bharat is a great aspiration of the government to cater to the needs of the citizens through knowledge economy that can grow faster. The vision is to achieve economic growth in quantum jumps and not the conventional incremental increase. The government has made sufficient efforts to create world-class civil infrastructure (i.e., high-quality highways and roads, uninterrupted power supply through maximal generation using renewable sources, providing high-speed internet connectivity across the country, etc.), which are creating an identity of a new and modern India. The government is in the process of developing a new system of governance based on modern technologies to realize the dreams of the twenty-first century. The rules and rituals of the past are expected to be done away with. The vibrant demography and strong demand-and-supply cycle in the economy are the assets, which need to be utilized fully to make the nation self-reliant. Therefore, the efforts made by the government of India for enhancing the economy and making the country self-reliant, stand proved beyond doubt.

The mandate of Atmanirbhar Bharat is to empower each of its citizens (>1.3 billion) to contribute to the knowledge economy of the country. This, in principle, is achievable if Aswal Model for inclusive growth, as discussed in Chap. 1, is applied for the realization of Atmanirbhar Bharat. As depicted in Fig. 21.2, the four helices of the economy, namely the government, the universities/S&T institutions, industrial establishments, and civil society and media, are expected to work together in a cohesive

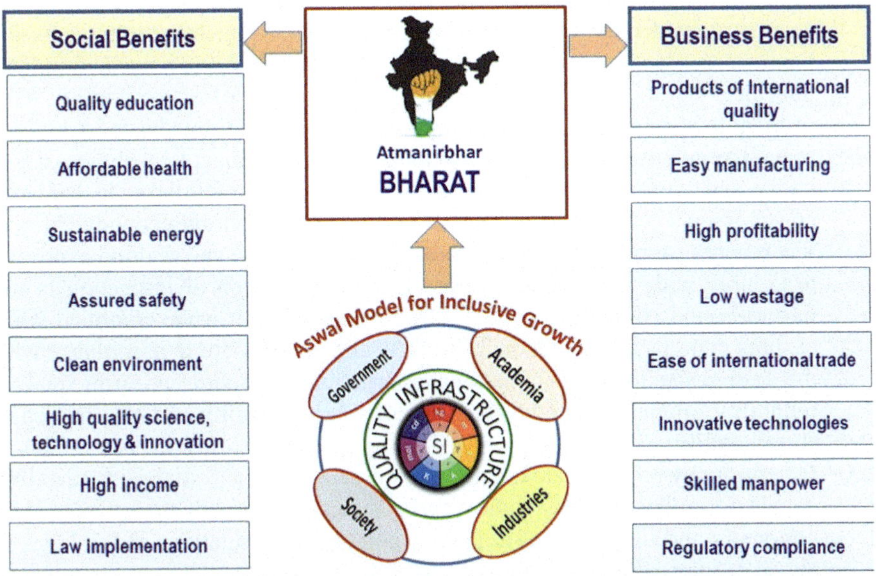

Fig. 21.2 Social and business benefits of Atmanirbhar Bharat implemented using the Aswal Model

manner through the national quality infrastructure, in which all the measurements are metrologically traceable to SI units.

Atmanirbhar Bharat if implemented using the Aswal Model will yield enormous naturally occurring social and business benefits, as listed in Fig. 21.2. Some of the notable examples are illustrated below.

(i) Metrology is the foundation of high-quality science, technology, innovation, and trade. Therefore, providing education on metrology right from schools to graduation/postgraduate levels would create human resource, conscious of national quality infrastructure. Such a high-quality metrology education is necessary for the creation of knowledge economy with high efficiency but without harming the environment.

(ii) To manufacture high-quality, indigenous, and cost-effective goods to fulfill the national needs as well as to compete in international business, meeting the laid down international standards is mandatory. This requires a robust national quality infrastructure for ensuring the conformity assessment, i.e., proven technical and regulatory requirements through well-established calibration, testing, certification, accreditation, and inspection processes. Conformity assessment entails steps taken to evaluate whether the products, processes, systems, or personnel, observe and meet the requirements identified in the concerned and relevant specified standard. These technical and regulatory requirements are strongly associated with stipulated respective standards and provide the end-user with the necessary confidence in the products and services being offered. For this reason, conformity assessment is a critically important aspect of conducting reliable global business and acts as an interface of confidence between the supplier and the customer. The manufacturers or the suppliers need to consider the benefits accrued from observing the requirements for conformity assessment and the customer, on the other hand, needs to assure himself about the business benefits of investing in terms of profitability, ease of manufacturing, ease of marketing including international business aspects as well as quality and regulatory compliance. Without meeting technical, regulatory, and conformity requirements, many export deals fall through, and large consignments get liable to be rejected leading to considerable economic loss. A few recent examples include the failure of basmati rice and seafood exporters to comply with the mandatory testing and certification of inspection in the import-bound countries [7, 8]. Conformity assessment when complied with, ensures manufacturers that their products and services are in accordance with the desired specifications of national/international standards, ensuring quality, reliability, efficiency, safety, effectiveness, interoperability, and environmental sustainability.

(iii) Exceptionally reliable data obtained by ensuring the metrological traceability, enable formulation of regulation and policies that benefit the country both nationally and internationally. The formulation of regulations and policies is aimed at improving the quality of life of the citizens, which keep on evolving with time, and therefore, depend upon trustworthy data obtained through national quality infrastructure.

In view of the above, the role of CSIR-NPL, as NMI of the country, undoubtedly becomes pivotal. In the next section, we analyze how CSIR-NPL has been contributing to the growth of the nation as well as those of neighboring nations.

21.3 CSIR-NPL's Metrology Services Towards National Growth

As shown in Fig. 21.3, the CSIR-NPL through its wide-ranging metrological services, namely primary/national measurement standards, accurate and precise Indian Standard Time (IST), the certified reference materials under the trademark of Bhartiya Nirdeshak Dravyas (BND®), transfer of technologies, consultancy services, sponsored R&D projects, skill development on metrology, etc. have been contributing towards knowledge creation for the overall growth of the nation. The beneficiaries of these metrological services include government ministries, regulatory bodies, public sector undertakings, private industries, MSMEs, strategic sectors, S&T organizations, etc. In addition, CSIR-NPL has been supporting the NMIs of neighboring countries, especially those who belong to the South Asian Association for Regional Cooperation (SAARC) nations.

In the preceding chapters of this book, the contemporary and world-class, accurate, and neoteric services for various measurement parameters have been elaborated. In this section, an attempt is made on the critical analyses of metrological services provided by CSIR-NPL to various sectors, as summarized in Fig. 21.3. The trust and

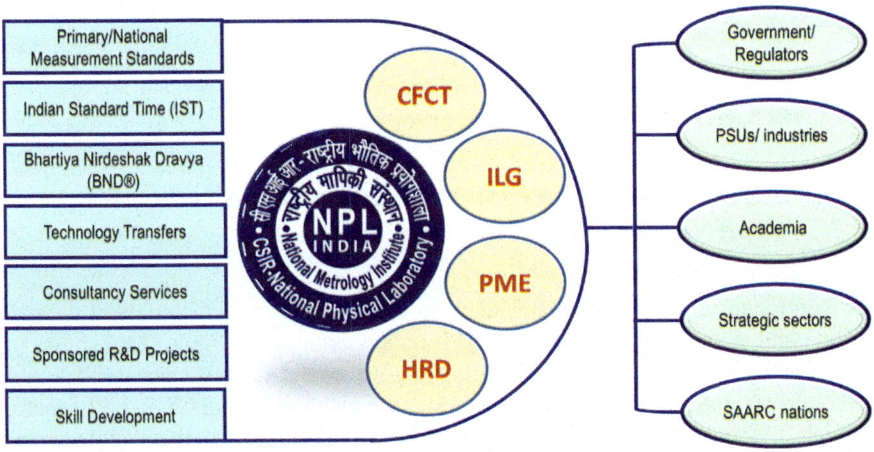

Fig. 21.3 The metrological services provided by CSIR-NPL to various government/regulators, public sector undertaking (PSUs)/industries, and strategic sectors for national growth. In addition, CSIR-NPL supports SAARC nations. Centre for Calibration and Testing (CFCT), Industrial Liaison Group (ILG), Planning, Monitoring and Evaluation (PME), and Human Resource Development (HRD) of CSIR-NPL act as the interface with external agencies

confidence inherent in the dissemination of metrologically traceable measurements by CSIR-NPL are demonstrated as a significant component of the scientific, industrial, economic development of the nation and its role in elevating the status of the civil society.

CSIR-NPL provides various metrological services through its four organs namely, Centre for Calibration and Testing (CFCT), the Industrial Liaison Group (ILG), Planning, Monitoring and Evaluation (PME), and Human Resource Development (HRD). CFCT acts as an interface between CSIR-NPL and patrons requiring the metrological traceability. Currently, CFCT is supporting >3970 patrons from industries, national laboratories, and government organizations from all over the country and abroad. CFCT provides services in a professional manner and most of the desired information is available at its website. PME undertakes the sponsored, grant-in-aid, or collaborative R&D projects from external agencies. ILG acts as an interface for taking up several technical services. These include technology development and transfer, consultancy services in various research areas, joint development of the products, optimization of industrial processes, impact assessment studies, upgradation of technologies, feasibility studies, fabrication of specialized products, modeling and machining, development of specialized device/component/product, design of calibration facilities, development of software for automation, characterization of devices/parts, interference studies, improvements in the design/accuracy, structural studies, specialized calibration and testing, operation, performance checking, troubleshooting, effect assessment studies, instrumentation and automation, development of methodologies, the establishment of the calibration laboratories, performance optimization, etc. Finally, the HRD group deals with the training programs related to metrology and research scholars/students.

21.4 Analyses of Calibration and Testing Services of CSIR-NPL

21.4.1 Metrological Traceability to Government, Industries and SAARC Nations

As mentioned, a total of 3970 agencies have registered at CFCT to avail the calibration and testing services of CSIR-NPL on a regular basis. Figure 21.4 shows the distribution of various agencies belonging to government sectors, central PSUs, state PSUs, private industries, and SAARC nations which benefit from CSIR-NPLs metrological services. As accurate and precise measurements are key to the manufacturing, it is not a surprise that >82% of the CSIR-NPL metrological services are availed by private industries, which include small scale industries and NABL accredited testing and calibration laboratories across the country. The government/PSUs contribute to 17.4%, while SAARC nations to 0.3%. A partial list of the beneficiaries belonging to the different categories mentioned above is given in Table 21.1. The small contribution of SAARC nations as depicted in Fig. 21.4, is quite expected as only the NMIs

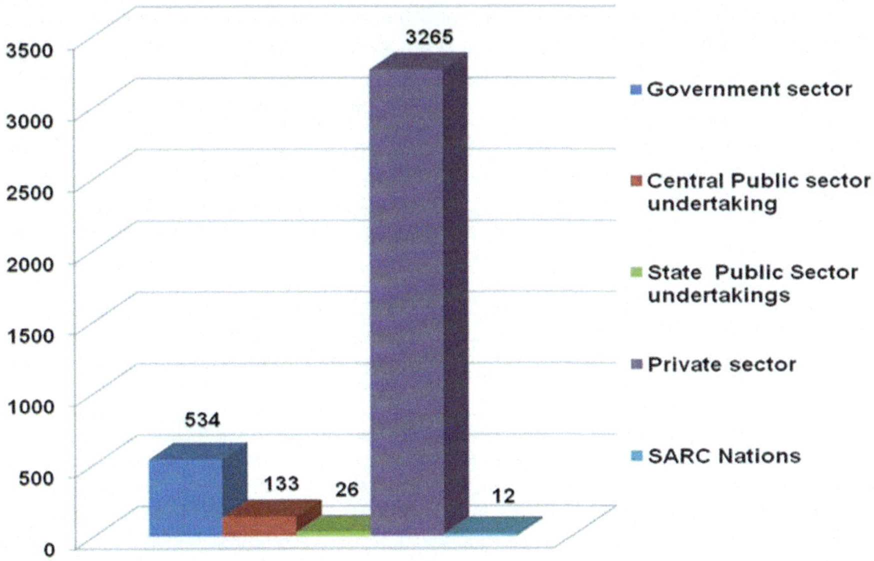

Fig. 21.4 CSIR-NPL provides calibration and testing to several government organizations, private industries and SAARC nations

of these countries approach CSIR-NPL for calibration and traceability. Fairly strong connects of CSIR-NPL with both industries and government is a testimony to the applicability of the Aswal model, that metrology acts as a binding force, which is elaborated further in subsequent sections.

It is worth mentioning here that the validity period for calibration and testing certificates issued by CSIR-NPL is fixed, which depends upon the nature of the parameter and can vary between 1 and 3 years. Therefore, the registered agencies approach CFCT only when the validity of calibration and testing certificates issued to them approach their expiry dates. For financial years 2015–16 to 2019–20, the CFCT data showing (a) number agencies registered for metrological services and (b) number of calibration and testing certificates issued by CSIR-NPL are plotted in Fig. 21.5. It is seen that on an average ~ 1000 agencies register every year for calibration and testing services and correspondingly, the issued average number of certificates is ~2500 per year. This implies that most agencies avail calibration/testing facilities for more than one parameter.

21.4.2 Sector Wise Analyses of Metrological Traceability Data

As discussed in Chap. 1, CSIR-NPL has 236 CMCs at the BIPM for a wide range of parameters, which were elaborated in subsequent Chapters. Here we present

Table 21.1 A partial list of users availing the primary/national calibration facilities of CSIR-NPL

1. Government/semi-government organizations	Air Force; Air India; Bharat Electronics; BHEL; Bhilai Steel Plant; Bureau of Indian Standards; Bharat Dynamics; Central Pollution Control Board; State pollution control boards; Central Power Research Institute; Central Public Works Department; Railway Information System; Central Institute of Mining and Fuel Research; Defense Research and Development Organization; Defense Electronics Applications Laboratory; Delhi Jal Board; Directorate of Border Security Force; Hindustan Aeronautics Limited; Indian Oil; HPCL; ISRO Inertial Systems Unit; Maharashtra State Electricity; Micro, Small and Medium Enterprise Testing Center; NTPC; Nuclear Fuel Complex (DAE); Ordnance Factory; Rail Coach Factory; Steel Authority of India, etc.
2. Private industries	Adani Electricity; Tata Steel; CK Birla Group; GE power systems; ABB India; ACC; AIMIL Ltd.; Alstom India; Ambuja Cement; Binani Cement; Bangur Cement, Birla Tyres; Blue Star; Bureau Veritas; Casio India; Crompton Greaves Limited; Diesel Locomotive Works; Essar Oil Ltd.; Godrej and Boyce Mfg. Co. Ltd; Havells India; Honda Cars; International Zinc Association; J. K. White Cement; JK Lakshmi Cement; Kirloskar Brothers; Larsen and Toubro; Maruti Suzuki; Mysore Paints and Varnish; Philips India; Piramal Healthcare; Ranbaxy; Rapid Metro Rail Gurgaon; Samsung India; Honda Siel; Surya Roshni, Wipro consumer care and lighting; Orient Electric; ITC, Halonix Technologies, Astra lighting; Bharat Forge; Tetronix India; Fluke Technologies; Verka Dairy; etc.
3. SAARC nations	Nepal Bureau of Standards and Metrology (MBSM), Nepal; Bangladesh Standards and Testing Institution (BSTI), Bangladesh; Measurement Units, Standards and Services Department (MUSSD), Sri Lanka; National Physical and Standards Laboratory (NPSL), Pakistan; Bhutan Standards Bureau (BSB), Bhutan; Afghanistan National Standards Authority (ANSA), Afghanistan; Maldives Standards and Metrology Unit (MSMU), Maldives

the CFCT data in terms of metrological services provided to vital sectors, which contribute to the knowledge economy and knowledge society. These are categorized into following twenty sectors: automobile, aviation, biomedical and pharmaceutical, diary, food and beverages, defense, electrical/electronics, energy/power, environment metrology, home appliances, housing/infrastructure, irrigation and agriculture, IT consultancy, and banking, laboratory equipment, petrochemicals and chemicals, railways, S&T and academic, space, testing laboratories, textile, and others. The "other" category includes SAARC nations and those that could not fit into the above-mentioned specialized categories.

Figure 21.6 shows the sector-wise distribution of metrological services of CSIR-NPL for 2015–16 to 2019–20. The testing labs and laboratory equipment are the top

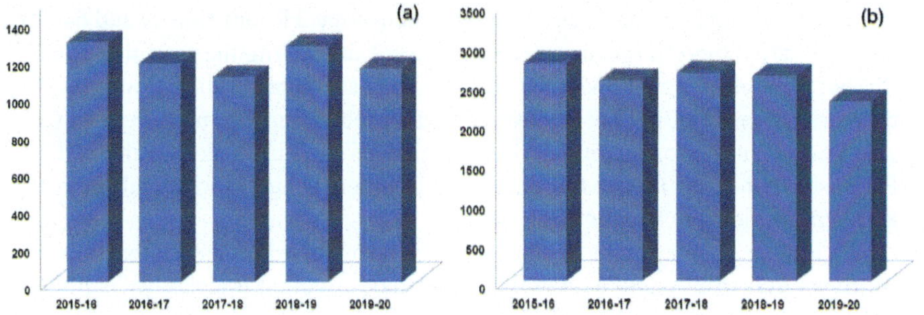

Fig. 21.5 Total numbers of (**a**) agencies registered at CFCT and (**b**) calibration and test certificate issued by CSIR-NPL during FY 2015–16 to 2019–20

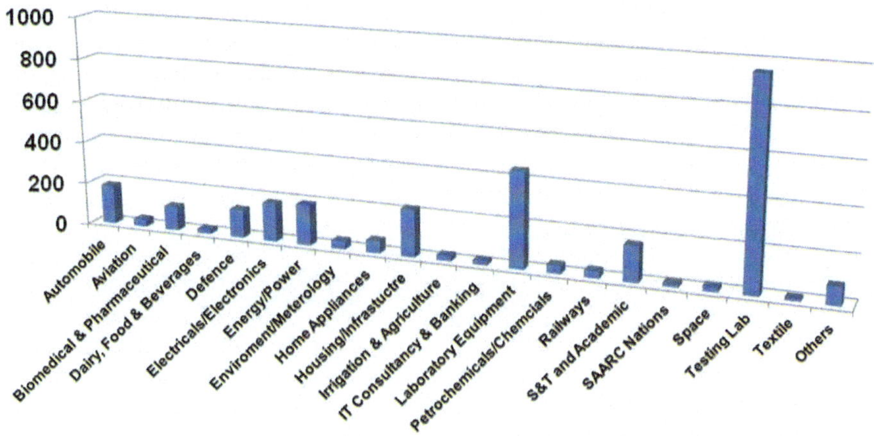

Fig. 21.6 Sector wise distribution of metrological services of CSIR-NPL for 2015–16 to 2019–20

sectors. This is quite expected because the NABL accredited laboratories across the country, need to obtain the mandatory metrological traceability to SI units through CSIR-NPL. As discussed in the next section, most of the government ministries have their own secondary testing and calibration laboratories to ensure the conformity assessments for their regulations as well as for products manufactured in India or imported. Since these secondary laboratories are spread across the country, the state regulators approach these local laboratories for calibration and testing. This also explains why the number of state PSUs that approach CSIR-NPL is relatively lower in number, as observed in Fig. 21.4. These secondary laboratories have their traceability to the SI units through CSIR-NPL.

The other major sectors which obtain measurement traceability to SI units through CSIR-NPL are automobiles, biomedical and pharmaceutical, defense, home appliances, housing/infrastructure electrical and electronics, energy/power, S&T and academic, aviation, railways, and space. The measurement traceability to sectors like diary, food and beverages, environment/ metrology, IT consultancy and banking, textile, etc. are although not significant at present in availing the facilities of CSIR-NPL, but possibilities exist that they may be using secondary calibration and testing laboratories. However, it must be ensured that all sectors utilize measurement values traceable to SI units for quality assurance. As discussed in earlier Chapters, CSIR-NPL is strengthening its capabilities in environmental and biomedical metrology, which will immensely benefit these sectors.

India is a big country and to ensure inclusive growth, it is essential that every part of it grows industrially and economically. As measurement traceability is linked to economic growth, organizations in the various states of India, availing the CSIR-NPL facilities are plotted in Fig. 21.7. New Delhi is the frontrunner, which is expected as most of the government bodies and industries approach CSIR-NPL due to its proximity. Similar is the case with Haryana, being an adjoining state. Otherwise, the ranking in measurement traceability to SI units through CSIR-NPL and the industrial growth appears to be correlated well. According to RBI's recent report, states like Maharashtra, Tamil Nadu, Gujarat, Uttar Pradesh, and Karnataka have a high rate of industrial growth while north-eastern states have the lowest industrial growth. Therefore, to improve the economy of the country, among many other parameters, the respective state governments should be encouraging industries and MSMEs to ensure the measurement traceability for industrial growth through the manufacturing of products of international quality.

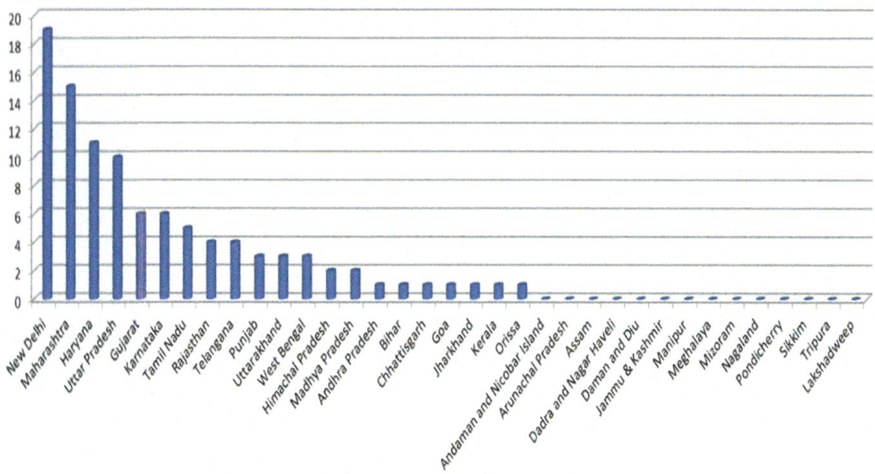

Fig. 21.7 State-wise distribution of metrological services of CSIR-NPL, in percentage, for the period 2015–16 to 2019–20

21.5 Collaboration Between CSIR-NPL and Government Organizations

A critical analysis of the past 5 year's data of CFCT, ILG, PME, and HRD shows that the metrological services of CSIR-NPL have reached the majority of the existing government ministries, as schematically depicted in Fig. 21.8, through regulatory bodies, R&D organizations, and/or public sector undertakings (PSUs) working under them. Major sectors that regularly pursue CSIR-NPLs metrological traceability services include defense, civil aviation, heavy industries, power, steel, oil and natural gas, space, atomic energy, railways, science, and technology, etc. Other sectors seeking the metrological traceability services include dairy, water, railway, health, petrochemicals, environment monitoring, etc. There are many sectors looking for new and contemporary measurement and/or materials metrology services, which have either been initiated or are under the process of the establishment by CSIR-NPL. These sectors include AYUSH, cybersecurity, finance, textiles, etc.

Fig. 21.8 A schematic showing the metrological services provides by CSIR-NPL reaching majority of government ministries through regulatory bodies, R&D organizations, and/or central public sector enterprises (CPSEs)

It is worth mentioning that the Department of Public Enterprises (DPE) formulates policies for Central Public Sector Enterprises (CPSEs), which are mainly related to performance evaluation and improvement, financial delegation and autonomy, and personnel management. DPE, based on the performance and annual profits has listed 10 Maharatna Companies in India namely, Bharat Heavy Electricals Limited (BHEL), Bharat Petroleum Corporation Limited (BPCL), Coal India Limited, GAIL (India) Limited, Hindustan Petroleum Corporation Limited (HPCL), Indian Oil Corporation Limited (IOCL), NTPC Limited, Oil and Natural Gas Corporation Limited (ONGC), Power Grid Corporation of India Limited (PGCIL) and Steel Authority of India Limited (SAIL) [9, 10]. Most of these Maharatna companies and other Navratna companies are beneficiaries of the metrological traceability services offered by CSIR-NPL.

The following sections describe the specific areas where crucial metrological services of CSIR-NPL are being disseminated. Gap analysis of the areas where the expertise and facilities have further potential to grow and/or added is also made and an attempt is made to envisage the future directions.

21.5.1 Defense and Civil Aviation

CSIR-NPL provides regular metrological services to Defense and Civil Aviation sectors. Defense is one of the most important areas as it protects the sovereignty of our borders and very existence as a Nation. To ensure synergy, interoperability, and efficient utilization of resources for defense equipment and applications, maximizing standardization is one of the basic demands. Therefore, all measuring equipment needs to be calibrated to national/primary standards traceable to SI units. Table 21.2 lists the organizations under the defense ministry which are supported by CSIR-NPL for metrological traceability services along with the calibration/testing parameters. It is evident that all three wings of the Defense Ministry, i.e., military, navy, and air force as well as their associated research and scientific organizations are beneficiaries of the national/primary standards of CSIR-NPL. One of the major constituents of defense is the *Defense Research and Development Organization* (DRDO) which is the R&D wing of Defense Ministry with a vision to empower India with cutting-edge defense technologies and a mission to achieve self-reliance in strategic defense technologies and systems while equipping our armed forces with state-of-the-art weapon systems and equipment in accordance with requirements laid down by the three services. It has several laboratories, under different clusters such as Naval Systems and Materials (NS&M), Aeronautical Systems (Aero), Armament and Combat Engineering Systems (ACE), Missiles and Strategic Systems (MSS), Electronics and Communication Systems (ECS), Life Sciences (LS), and Micro Electronic Devices, Computational Systems and Cyber Systems (MED and CoS) [11]. CSIR-NPL provides crucial metrological services for the standardized measurements in these areas as listed in Table 21.2. Bharat Electronics Limited, the electronics wing of DRDO, supplies advanced electronic components to the Indian Armed forces. These include

Table 21.2 List showing the CSIR-NPL's metrological services provided to defence and civil aviation sectors

Ministry	• Organization/PSU/Joint Venture	Traceability/calibration/MoU area
Defense 	• Defense Electronics Application Laboratory (DEAL) • Defense Research and Development Laboratory, (DRDL), • Opto Electronics Factory, Dehradun, Uttarakhand • Mishra Dhatu Nigam Limited (MIDHANI), Hyderabad • Bharat Earth Movers Limited (BEML), Bengaluru • Gas Turbine Research Establishment (GTRE), Bengaluru • Naval Sci. and Tech. Lab, Vishakhapatnam • Military Engineer Services, New Delhi • Directorate of Standardization, New Delhi • Directorate of Border Security Force (Air Wing) • Indian Air Force • Bharat Dynamics Ltd. • Gun Carriage Factory (GCF), Jabalpur, M.P • Indian Ordnance Factories at Ghaziabad, Kanpur, Nagpur	Pressure, vacuum, and ultrasonic metrology
	• Heavy Alloy Penetrator Project (HAPP), Kolkata, W Bengal • AIR Force Station, Palam, Delhi	Mass metrology, optical radiation, time and frequency
	• High Explosives Factory, Pune, Maharashtra	Mass, volume, viscosity, temperature, and humidity
	• Bharat Dynamics Ltd, Hyderabad, Telangana • Defense Bioengineering and Electromedical Laboratory (DEBEL), Bengaluru	Luminance and fluid flow
	• Department of Defense Research and Development: • Defense Research and Development Organization (DRDO)	Electrical and electronics and related parameters
	• DRDO, Ammunition Factories • DRDO Proof and Experiment Establishment Bharat Electronics Ltd. (BEL), Bengaluru • Bharat Electronics Ltd. (BEL), Ghaziabad • Bharat Electronics Limited, Panchkula • Office of Commanding officer, Airforce Station, New Delhi/Goa • Hindustan Aeronautics limited, U.P	Optical radiations, precise timing systems, LF and HF impedance and DC standards, LF and HF voltage, current and microwaves, length and dimension. fluid flow, electrical and electronics, pressure, vacuum, ultrasonics, acoustics and vibrations, magnetic
	• Indian Air Force, Department of Defense Production: Ordnance Factory Board	Electrical and electronics, precise timing systems
	• Hindustan Aeronautics Limited (HAL)	Pressure, vacuum, and ultrasonic length and dimension, acoustic, and vibrations, optical, magnetic

(continued)

Table 21.2 (continued)

	• Directorate General of Quality Assurance, New Delhi	Force and hardness, LF and HF, voltage, current and microwaves
	• Opto Electronics Factory, Dehradun Ordnance Equipment Factory, Kanpur	Acoustics and vibration, optical radiation
	• Ordnance Factories	AC power and energy, fluid flow, length and dimension, pressure
	• Command Met Officer, HQ Training Command Air force, Bengaluru. • Western Air Command, New Delhi-10	Information on earth's ionized media for frequency management; Geophysical and solar activity
	• DRDO	Spark Plasma synthesis of Bi2Te3 alloy-based nano-composites
Civil aviation 	• Air India Limited • Air India Engineering Services	Pressure, fluid flow, optical radiation, magnetics, mass
	• Office of Commanding Officer, Airport Authority of India (AAI), New Delhi	Luminance meters, photometric and radiometric scales
	• Manager-SO&I, IATA, New Delhi • Air India Flight Dispatch, IGI Airport, New Delhi • Air India Flight Dispatch JFK Airport, New York • Sr. Manager - Flight Dispatch Air India	Geophysical and solar activity, HF propagation for daily message

radars, naval systems, C4I systems, defense communication weapon systems, homeland security, electronic warfare, tank electronics, telecom and broadcast systems, electro-optics, professional electronic components, solar photovoltaic systems, etc. The quality control and quality assurance of these depend upon the calibration of all the electronic and electrical parameters from CSIR-NPL.

To ensure the safe flight conditions, both Air Force and Civil Aviation require accurate information on various parameters related to geophysical activities. These include geomagnetic field, solar wind conditions (in last 24 h on any given day and its forecast), solar activity (i.e. sunspot number and value of solar 10.7 cm (2.8 GHz) radio flux and its forecast), and high frequency (HF) propagation conditions and their forecast (normal, fair or bad) at different latitudes, turbulence, thunderstorm, icing, ceiling, visibility, etc. [12]. Information on these crucial parameters is provided by CSIR-NPL to both the Air force and Civil Aviation.

The Ministry of Civil Aviation has three statutory bodies, i.e., Directorate General of Civil Aviation, Airports Authority of India, Bureau of Civil Aviation Security, and Commission of Railway safety. Air India, the national carrier is one of the significant clients to which calibration is provided in the areas of pressure measurement, fluid flow, optical radiation, magnetics, mass metrology, etc. Pressure is a crucial parameter both for maintaining cabin pressures in aircraft as well as for estimation of the height using Altimeters. Also, calibration of luminance meters, photometric and radiometric scales are being given to the *Office of Commanding Officer, Airport Authority of India (AAI)*, New Delhi. In addition, CSIR-NPL also provides radiation dose data to Regional warning centers under the Ministry of Civil Aviation.

21.5.2 Ministry of Consumer Affairs, Food, and Public Distribution

As amply discussed in Chap. 1, the Legal Metrology has a responsibility to ensure that the benefits of metrological traceability of weight and measures to SI units are transferred to all the citizens of the country. Therefore, Legal Metrology has the statutory power to establish and enforce the standards of weights and measures, regulate trade in commerce, and all goods which are either sold or distributed by weight. The legal metrology, therefore, guarantees fair trade in the market by assuring citizens through correct measurements. Legal Metrology has to comply with the requirements of agreement on technical barriers to trade (TBT agreement) of the World Trade Organization (WTO). The Legal Metrology discharges its duty through five regional reference standard laboratories (RRSLs) across the country for weights and measures. These RRSLs obtain metrological traceability from CSIR-NPL, as summarized in Table 21.3, for the parameters of mass, length and dimension, and the temperature. CSIR-NPL provides traceability on mass and length metrology for, respectively, weights (a set of 1 mg to 5 kg weights) and length (meter bar—1 m and Gauge blocks M-112) to each RRSL for maintaining the weights and measures systems in the country. The weights need to be recalibrated every three years, whereas lengths need to be recalibrated every five years. The temperature standards in ranges 44–300 °C are provided using liquid-in-glass thermometers (LIGTs) and in ranges 90–400 °C using platinum resistance thermometers (PRT). Legal Metrology further disseminates the measurement traceability down to state and district level weights and measurement laboratories. Recently Legal Metrology has also started a nationwide program on dissemination of Indian Standard Time, traceable to CSIR-NPL, through its RRSLs, which is expected to help the country in the field of cybersecurity, power sector, telecom sector, digital, aviation, etc.

Table 21.3 The mass, dimension, and temperature reference standards at regional reference standard laboratories (RRSLs) are calibrated at CSIR-NPL

RRSL	Dimension Artefact	Mass	Temperature
RRSL Ahmedabad	Gauge Block Set M-122	1 mg–5 kg	LIGT 275–300 °C/least count 0.05 °C LIGT 44–56 °C/least count 0.02 °C
RRSL, Faridabad	Meter Bar (1 m)	1 mg–5 kg	
RRSL Bhubaneshwar	Meter Bar (1 m)	1 mg–5 kg	LIGT 44–56 °C/least count 0.02 °C
RRSL Guwahati	Gauge Block Set M-122	1 mg–5 kg	
RRSL Bengaluru	Gauge Block Set M-122	1 mg–5 kg	PRT PT-100 with temperature indicator—90–400 °C/Resolution 0.01 °C

The working links between CSIR-NPL and Legal Metrology are strong. CSIR-NPL also provides measurement traceability to RRSLs for many other derived parameters, including pressure, vacuum, fluid flow, force, etc. CSIR-NPL also supports the Indian Institute of Legal Metrology, Ranchi, by providing measurement traceability to its secondary and reference standard weights. CSIR-NPL is also collaborating with Legal Metrology for setting up of following facilities at different RRSLs.

(i) Design, development, fabrication, and establishment of customized indigenous working standards for verification of blood pressure measuring instruments for RRSL, Faridabad
(ii) Establishment of a reference standard for length using laser interferometry for RRSL, Bengaluru.
(iii) Establishment of clinical thermometer testing kit for RRSL, Guwahati.
(iv) Setting up of secondary atomic time scales at all the 5 RRSLs and a disaster recovery primary atomic time scale at Bengaluru.

In addition, CSIR-NPL provides regular training programs for the officers of Legal Metrology in areas of mass, length, temperature, and pressure metrology.

CSIR-NPL also provides traceability to the National Test House (NTH), which is one of the oldest multi-disciplinary industrial testing laboratories in the country. NTH has six regional laboratories located at Kolkata, Mumbai, Chennai, Ghaziabad, Jaipur, and Guwahati. These NTH labs cater to the testing needs of local industries, institutions, and trades. In all these labs, testing and evaluation facilities for raw materials as well as finished products are available for all the branches of engineering disciplines. These NTH labs obtain measurement traceability from CSIR-NPL for various parameters viz. pressure, vacuum, ultrasonic, photometric, radiometric, length and dimension, force and hardness standards, etc.

21.5.3 Departments of Space and Atomic Energy

Department of Space (DoS) and Department of Atomic Energy (DAE) are very important and strategic departments of India. Both the departments have made enormous contributions to international recognition in their respective areas of space technologies and nuclear technologies, which in turn, has contributed to the scientific, strategic, and economic development of India.

DAE is a leader in the development of nuclear power technology and applications of radiation technologies in the fields of agriculture, medicine, industry, and basic research. DAE comprises of five research centers, three industrial organizations, five public sector undertakings, and three service organizations. Bhabha Atomic Research Center (BARC) is the Designated Institute for the ionization radiation, and as per CIPM-MRA, the quality management of BARC is looked after by CSIR-NPL. Table 21.4 shows the services provided to DAE institutions by CSIR-NPL.

DOS has the primary objective of promoting the development and application of space science and technology, to assist in the all-round development of the

Table 21.4 Services provided to DAE institutions by CSIR-NPL

Organization	Traceability/calibration/consultancy area/projects
Nuclear Power Corporation of India Limited (NPCIL)	Mass, temperature, electrical and electronics and related parameters
Electronics Corporation of India Limited (ECIL)	Electrical and electronics and related parameters
Bhabha Atomic Research Center (BARC), Mumbai	• Support as designated institute in ionization radiation • Mass and fluid flow • Development of high-temperature resistant C/C composite tube, development of nuclear grade high-density graphite • Calibration of magnetic fields and magnetic materials for INO project • Development of high density isotropic nuclear grade • Synthesis of high figure-of-merit thermoelectric materials • Development of thermally stable novel nanophosphors for improving color rendering index (CRI) in white-LEDs
Nuclear Fuel Complex, Hyderabad	Force and hardness

nation. Major programs of ISRO include satellite communication, earth observation, launch vehicle, space science, disaster management support, international cooperation, safety, reliability, public relations, economic analysis, human resources development, etc. The CSIR-NPL provides various services to DoS, as listed in Table 21.5. One of the most critical and strategic needs of ISRO is the nano-scale precision in the time synchronization required for the launch and monitoring of rockets and satellites.

Table 21.5 Services provided to DoS institutions by CSIR-NPL

Organization	Traceability/calibration area/MOU
ISRO Telemetry, Tracking and Command Network (ISTRAC), Bengaluru and Lucknow	• Precise timing systems
Vikram Sarabhai Space Center, Thiruvananthapuram	• Mass, pressure, vacuum, and ultrasonic • LF and HF voltage, current, and microwaves
Satish Dhawan Space Centre (SDSC), Sriharikota	• Force and hardness
Semi-Conductor Laboratory, SAS Nagar, Panjab	• LF and HF impedance and DC standards
Space Applications Centre (ISRO), Ahmedabad	• Temperature and humidity, precise timing systems • LF and HF impedance and DC standards • LF and HF voltage, current, and microwaves
ISRO Master Control Facility, Karnataka ISRO-Space Applications Center, Ahmedabad	• Geophysical and solar activity, HF propagation daily message • Indian regional navigation satellite system (IRNSS)

ISRO obtains the strategic time synchronization with nanosecond accuracy, which is traceable to UTC through Indian Standard Time (IST) generated by the national Primary Atomic Clock Ensemble at CSIR-NPL. Atomic clocks in Global Navigation Satellite Systems (GNSS) keep time to within few nanoseconds. All GNSS satellites need to transmit their data signals exactly at the same time, which demands clock synchronization of all satellites to primary reference time, i.e., IST. GNSS signals are monitored constantly and adjusted as needed. In addition, metrological traceability of physicomechanical and electrical parameters is provided to various ISRO centers located at Sriharikota, Ahmedabad, Thiruvanthapuram, and SAS Nagar, Panjab.

21.5.4 Uplifting Science and Technology in the Country

Ministry of Science and Technology has several departments, including DSIR, DST, DBT, and MoES. Various laboratories under these departments work on context-driven research for industrial applications. Several research laboratories under CSIR work in diverse areas including electronics, instrumentation, metallurgy, petroleum, mining, leather, environment, glass, and ceramics, etc. These labs utilize the national measurement standards of CSIR-NPL for the calibration of several equipment required for critical and strategic projects, as summarized in Tables 21.6 and 21.7.

Table 21.6 Metrological traceability services to various research organization under the Ministry of Science and Technology

Organization	Traceability/calibration area/projects
Central Electronics Limited	• Optical radiation parameters
Fie Research Institute	• Force, hardness, torque • Length and dimensional, pressure and • Thermal parameters
CSIR-Central Electrical and Electronics Research Institute	• Acoustics and vibration
CSIR- Central Scientific Instruments Organization	• Mass, volume, viscosity • Fluid flow • LF and HF voltage, current, and microwaves
CSIR-Central Institute of Mining and Fuel Research	• Force and hardness • Acoustics and vibration
CSIR-Central Leather Research Institute	• Fluid flow
CSIR-Indian Institute of Petroleum	• Fluid flow • LF and HF voltage, current, and microwaves
CSIR-Central Mechanical Engineering Research Institute	• Length and dimension • Precise timing systems
CSIR-National Metallurgical Laboratory	• Temperature and humidity • Force and hardness • Acoustics and vibration

Table 21.7 Projects of various research organizations under the Ministry of Science and Technology

Organization	Projects
Department of Science and Technology (DST)	• Investigation of the molecular dynamics at the interface of liquid crystal and immiscible structures • Electronic and magnetic properties of some single-layered cobaltite based compounds • Conversion of waste magnetic energy into useful electricity with Magnetoelectric Composite for wireless sensor nodes • Studies of stabilization of ferromagnetism in MnX(X = Al, Ga) binary alloy thin films by ion beam irradiation • Development of innovative cathode materials and flexible lithium-ion conducting ionic liquid-based gel polymer electrolytes for rechargeable Li-batteries • Advanced single-photon detector and establishment of single-photon detection based quantum standard for QuEST • Development of new interfacial layers for efficient and stable excitonic solar cells • Growth and study of highly conducting delafossite single crystal: device application in metrology • Buried contacts high-efficiency crystalline radial p-n junction Si nano cord solar cell • Advancing the efficiency and production potential of excitonic solar cells • Chemical composition and source apportionment of aerosols using receptor Models at urban sites of the Himalayan Region of India • Designing, development, and calibration of PM2.5 and PM10 high volume sampler • Structure/microstructure magneto-transport correlations in sputtered magnetically doped Bi_2Se_3 topological insulator thin films • Development of interface layer of perovskite solar cells in view of silicon-perovskite tandem solar cell fabrication • Fabrication of efficiency enhanced nanostructured InGaN/GaN • Process development for the deposition of Poly-Si-films on bi-axially textured substrate for solar cell application • Development of efficient, stable, and environmentally friendly metal-halide perovskite photovoltaic cells • Separation, Purification, and Standardization of Glucose for Glucose meter and other clinical applications • Study of atmospheric brown clouds over the Indo-Gangetic Plain and their impact on monsoon and agro-ecosystem • First principles investigation of electronic, thermal, magnetic properties of solids • Design and fabrication of perovskite-Si tandem solar cells and flexible perovskite solar cells • MEMS sensor array for environmental monitoring • Fabrication of cadmium-free large area kesterite (Cu_2ZnSnS_4) solar cells • Exploring the role of weak interactions in biomimetic host-guest complexes for efficient solar energy conversion

(continued)

Table 21.7 (continued)

Organization	Projects
	• Fabrication and characterization of perovskite-based electronics devices: different types of low voltage field-effect transistors and resistive switching memory • Investigation of the effect of misoriented/ patterned substrate on the properties of Nitrides structures grown by PAMBE technique • Semiconducting Thiophene based electronic materials for organic solar cells • Development of Carbon-based force sensors for artificial intelligence and biomedical applications • Novel graphene supported metal anode catalysts and PWA based electrolyte for direct ethylene glycol fuel cells • Design of Rare-earth substitution at a metal-organic framework to highly efficient luminescent security feature for anti-counterfeiting application • Search for quasi-two-dimensional electron gas in new oxide hetero-interfaces: Theory and Experiments • Development of stable and luminescent hybrid organic-inorganic nanocomposites of Pb-free/Pb-based perovskite quantum dots for LED applications • Organic molecular tracers and isotopic compositions of atmospheric aerosols at varying altitudes of Himalayan Region • Designing of graphene quantum dots for photocatalytic organic transformations • Design and fabrication of high-efficiency silicon solar cells: Passivated Emitter and Rear Cell (PERC) • A novel strategy of AlGaN/GaN heterostructure based pollution monitoring system: Nitrogen dioxide gas sensing under UV illumination • Design and development of novel autonomous self-healing graphene-carbon nanotubes hybrid-based polymer nanocomposites • Investigation of Sub-auroral ionosphere region for different space weather conditions and its global impact for effective communication and navigation application • Organic intercalated 2D Layered transition metal sulfide dichalcogenides (TMSDCs) hybrid nanostructures for thermoelectric application • Investigation on the unclonable security functions generated from plasmonic nanostructures for anti-counterfeiting application • Indigenous Development of PEDOT: PSS type materials employing novel methods for cost-effective and flexible excitonic solar cell applications (clean energy) • Introducing magnetoelectric coupling in rare-earth orthoferrite and barium thin films and in multi-layers through magnetic anisotropy • Development of cost-effective water purification set up on carbon-based nanocomposite for rural areas of India • Force spectroscopy study of modified polydimethyl siloxane surface using atomic force microscope • Synthesis of cost-effective environmentally friendly luminescent materials for solid-state lighting devices • Simple and efficient frequency offset locking scheme for lasers in atomic physics experiments

(continued)

Table 21.7 (continued)

Organization	Projects
	• Development of non-invasive blood glucose monitor using electrical bioimpedance spectroscopy • Development of lightweight high-performance carbon fiber fabric-carbon nanofibers hybrid polymer nanocomposites • Studies on physical and optical properties of atmospheric aerosols and clouds using in-house developed multiwavelength-dual polarization lidar • Engineering chemical structure to improve device efficiency: novel organic polymers/macromolecules and their nanocomposites for photovoltaic application
Ministry of Earth Sciences (MoES)	• An Integrated study of air pollutant sources in the Delhi NCR (ASAP-Delhi) • Megacity Delhi atmospheric emission quantification assessment and impacts (Delhi Flux)
Indian Council of Medical Research (ICMR)	• Study of the impact of exposure to ultraviolet radiation and aerosol exposure on ocular health in India (Phase-II)

Central Electronics Limited, originated from CSIR-NPL, is a Govt. of India Enterprise under Department of Scientific and Industrial Research (DSIR), Ministry of Science and Technology. It works with an objective to commercially exploit the indigenous technologies developed by national laboratories and R&D institutions in the country. The lux meter and other optical radiation measurements for CEL are carried out by CSIR-NPL. Fie Research Institute is also a NABL accredited calibration and testing laboratory for parameters like force, hardness, dimension, torque, pressure, thermal, etc. The metrological traceability for these parameters is obtained from CSIR-NPL.

21.5.5 Power, Steel, and Heavy Industries

The Power, Steel, and Heavy Industries form the backbone of any growing economy for technical advancement and especially for self-reliant services and products. Modern industrialization relies heavily on a strong steel base as well as the power to drive the plants to produce heavy machinery for the industrialization and progress. India is the world's sixth-largest energy consumer, accounting for 3.4% of global energy consumption. India is also the sixth-largest in terms of power generation. CSIR-NPL plays a significant role with a myriad of parameters providing metrological traceability through calibration to various state and center owned organizations related to power, steel, and heavy industries as briefly summarized in Table 21.8.

In the heavy industry sector, BHEL is involved in the design, engineering, manufacturing, construction, testing, commissioning, and servicing of a wide range of products, systems, and services for the core sectors of the economy, viz. power,

Table 21.8 Traceability services rendered to power, steel and heavy industries

Ministry	Organization	Traceability/calibration area
Steel 	National Mineral Development Corporation Limited (NMDC)	Pressure, vacuum
	Steel Authority of India Limited (SAIL)	• Acoustic and vibration • Pressure, vacuum
Power 	Damodar Valley Corporation (DVC)	• AC high voltage and AC high current • Magnetics, magnetic parameters
	Central Power Research Institute (CPRI), Bangalore, Karnataka; National Hydroelectric Power Corporation Limited (NHPC); Power Grid Corporation of India Limited (POWERGRID); Rural Electrification Corporation Limited	• Pressure, vacuum, and ultrasonic • AC Power and energy • Fluid flow • Magnetics, magnetic parameters • Acoustics and vibrations • AC high voltage and AC high current • LF and HF voltage, current, and microwaves
	National Thermal Power Corporation NTPC	• AC high voltage and AC high current • Acoustics and vibrations
	State Electricity Boards • Ajmer Vidyut Vitran Nigam Ltd • Gujarat Electricity Board • Kerala State Electricity Board Ltd • Maharashtra State Electricity Board	• Magnetic parameters • Electrical and related parameters
	BEE (Bureau of Energy Efficiency)	• Creation of testing and calibration facility for LED and LED-based lighting at NPL India as per national/international standards
Heavy Industries and Public Enterprises 	Bharat Heavy Electricals Limited (BHEL)	• Electrical and electronics and related parameters • Pressure, vacuum and ultrasonic, optical radiation • Acoustics and vibration • LF and HF impedance and DC standards

(continued)

Table 21.8 (continued)

Ministry	Organization	Traceability/calibration area
	Cement Corporation of India Limited (CCI)	• LF and HF impedance and DC standards • Mass, temperature, and humidity • Acoustic and Vibration
	Automotive Research Association of India (ARAI)	• Mass, temperature, and humidity • Optical radiation
	Fluid Control Research Institute (FCRI)	• Mass, temperature and humidity • Optical radiation • Pressure, vacuum and ultrasonic

transmission, renewable energy, water, oil and gas, aerospace and defense. CSIR-NPL provides significant calibration services to BHEL for parameters like electrical, pressure, acoustic parameters, etc.

The Light Emitting Diodes (LED) industry in India is also a major sector that requires calibration/testing of a large variety of products for consumer utilization. To cater to this contemporary need of lighting industries, CSIR-NPL in association with the Bureau of Energy Efficiency (Ministry of Power) is establishing a national testing and calibration facility for LEDs as per national/international standards. CSIR-NPL has also provided consultancy to Tata Steel, Jamshedpur for the development of value-added products from coal tar, as well as for characterization of different kinds of coal tar.

21.5.6 Certifying/Testing of Environmental Monitoring Equipment

Under the Ministry of Environment, Forest and Climate Change (MoEF&CC), the Central Pollution Control Board (CBCB) is a statutory body with the primary responsibility of advising the Central Government on any matter concerning prevention and control of water and air pollution and improvement of the quality of air. The state pollution control boards carry out similar tasks in their home states. The pollution can be of air, water, or even noise. Currently, India has no scheme for the certification of ambient air as well as industrial emission monitoring equipment. Consequently, such monitoring equipment is sent abroad for calibration, which is not only very expensive but also may not get calibrated as per the Indian environmental conditions, which are significantly different from the certifying countries. Therefore, such monitoring equipment is susceptible to erroneous readings. To circumvent this

Table 21.9 Services provided by CSIR-NPL in the field of air pollution monitoring

Statutory body	Traceability/calibration/testing area
Central Pollution Control Board (CPCB)	• CSIR-NPL designated as national verification agency for certifying instruments and equipment for monitoring emissions and ambient air • Acoustics and vibration • Fluid flow • BND related to aerosol, water, and gas mixtures • Delineation of airshed for air quality management in Delhi-NCR
Delhi Pollution Control Board	• Acoustics and vibration • Fluid flow • Environment metrology
Chhattisgarh State Pollution Control Board	• Improvement of quality of air pollution monitoring equipment • Quality checks of the data and instruments working at air quality monitoring stations in Raipur city
Ministry of Environment, Forest and Climate Change (MoEF)	• Establishment of type testing calibration and certification facility for online continuous emission monitoring system (OCEMS) and continuous ambient air quality monitoring system (CAAQMS) • Carbonaceous aerosols emissions, source apportionment, and climate effects

situation, MoEF&CC and CBCB have entrusted CSIR-NPL to setup the national environment standard laboratory and the Indian certification scheme for air quality monitoring instruments. This is in anticipation of rising demand by States—against the backdrop of the National Clean Air Campaign—for low-cost air quality monitoring instruments that can monitor levels of nitrous oxides, ozone, and particulate matter. CSIR-NPL is also producing standard gas mixtures for measurement reference. Various services provided by CSIR-NPL in the field of air pollution monitoring are summarized in Table 21.9. Apart from air pollution, noise is another pollution of major concern, especially in metros and big cities. To tackle the noise pollution, CSIR-NPL provides acoustics and vibration measurements to CPCB and SPCB, e.g., Delhi, etc. The traceability in fluid flow is also being rendered by CSIR-NPL to SPCB for calibration of gas and water flowmeters.

21.5.7 Ministry of Petroleum and Natural Gas

The Petroleum and Natural Gas Regulatory Board Act, 2006, states that "it aims to protect the interest of consumers by fostering fair trade and competition amongst the entities; register entities, lay down the technical standards and specifications including safety standards in activities relating to petroleum, petroleum products, and natural gas, including the construction and operation of pipeline and infrastructure projects related to downstream petroleum and natural gas sector, etc.". The implementation of these regulatory requirements goes a long way in ensuring the safe

Table 21.10 Services provided by CSIR-NPL to the petroleum and natural gas industries

Organization	Traceability/calibration/BND Area
Engineers India Limited (EIL)	• Optical radiation • Mass, density
GAIL (India) Limited	• Pressure, vacuum and ultrasonic • Temperature and humidity
Hindustan Petroleum Corporation Limited (HPCL) Indian Oil Corporation Limited (IOCL) Oil and Natural Gas Corporation Limited (ONGC) Bharat Petroleum Corporation Limited (BPCL)	• Mass, volume, density • Temperature and humidity • Pressure, vacuum, and ultrasonic
HPCL and BPCL	BND related to petroleum, petroleum products, and natural gas are being produced at HPCL/BPCL in association with CSIR-NPL

functioning, minimization of loss, and efficient execution of product distribution and standardization.

In this context, CSIR-NPL not only provides metrological traceability to a myriad of equipment in use at the many PSUs across India, but it has also embarked on an ambitious plan aiming towards self-reliant India, in the area of production of Standard Reference Materials, trademarked as "Bhartiya Nirdeshak Dravyas" as described in Chaps. 18 and 19. CSIR-NPL is producing BNDs related to petroleum, petroleum products, and natural gas in association with HPCL/BPCL [13, 14]. This will help in the quality assurance of the products. The BNDs relate to quantization of sulfur impurities in diesel, gasoline, and petrol as well as for density and viscosity evaluation in petrol samples. As summarized in Table 21.10, the major oil and gas giants, i.e., ONGC, IOCL, GAIL, HPCL, and BPCL obtain measurement traceability from CSIR-NPL in the areas of pressure, mass, density, viscosity, temperature, humidity, optical radiation, etc.

21.5.8 Ministry of Electronics and Information Technology

Various organizations under the Ministry of Electronics and Information Technology (MEITY), namely Standardization, Testing and Quality Certification (STQC), Electronics Regional Testing Laboratories (ERTL), and Electronic Test and Development Centers (ETDC) require traceability for various physico-mechanical and electrical parameters from CSIR-NPL, which are listed in Table 21.11. The STQC, ERTL, and ETDC are engaged in supporting industry and trade as well as in protecting consumer interest in the electronic and information technology sector by providing customer-centric, accredited quality assurance services as per international standards for global acceptance. There are a number of regional testing and development centers working

Table 21.11 Services provided by CSIR-NPL to various research organization of the Ministry of Electronics and Information Technology

Organizations/departments	Traceability/calibration/consultancy area
Standardization, Testing, and Quality Certification (STQC), Electronics Regional Testing Laboratories (ERTLs) at Delhi, Kolkata, Mumbai, and Thiruvananthapuram Electronic Test and Development Centers (ETDCs) at 10 locations across India	• Mass, volume, density, and viscosity • Pressure, vacuum • Temperature and humidity • LF and HF impedance and DC standards • LF and HF voltage, current, and microwaves • Optical radiation • Force and hardness, • Acoustics and vibration, • AC high voltage and AC high current • AC power and energy standards • Magnetic parameters
National Informatics Centre Services Inc. (NICSI)	• Implementation of IST service using NPL controlled remote oscillator system for national knowledge Network at National Informatics Centre

under the aegis of STQC, ERTL and ETDC, spread across the length and breadth of the country. These regional centers in turn further disseminate these testing and calibration services to the local industries and organizations and contribute significantly to the GDP of the nation. CSIR-NPL India, in collaboration with MEITY, has established the state-of-the-art nano-metrology facilities, which are aimed at supporting nano-technology development through very low level electrical current measurements, nanoscale dimension measurements, etc.

21.5.9 Increasing Reliability of Dairy Products

CSIR-NPL is contributing to the milk industry by way of calibration of Butyrometers and Lactometers [15]. Butyrometer is a measuring instrument used to measure fat content in milk or milk products in general. Fat content is determined by Gerber's method and it helps in the detection of adulteration like watering and skimming of milk. Lactometer is a special type of hydrometer. It is an instrument that is used to check for the purity of milk by measuring its density. It is constructed and graduated such that lactometer reading is related to the specific gravity of milk based on the ratio of the milk to water weight of a unit volume at a specific temperature. These instruments are calibrated at CSIR-NPL for correct density measurements of milk.

Table 21.12 Services provided by CSIR-NPL to railways

Organization	Traceability/calibration area
Diesel Locomotive Works, CMT	Ultrasonic measurements
	Force and hardness
Chief Workshop Manager's Office, Mumbai	Acoustics and vibration
Research Designs and Standards Organization (RDSO), Manaknagar, Lucknow	Lux meter/optical radiation
	Ultrasonic
Rail Coach Factory	Electrical and electronics and related parameters
Railways, N.E.	Ultrasonic measurements,
	Mass calibration
RITES Ltd.	Acoustic and vibration
Delhi Metro Rail Corporation	
Bengaluru Metro Rail Corporation Limited	

21.5.10 Safety for Railways

Indian Railways (IR) is the fourth largest railway network in the world and the largest one in Asia used for public facilities and transportation which is managed under the umbrella of the Ministry of Railways. Passenger safety is of prime importance for the rail journey. The detection of critical inherent cracks in the railhead, quite apparently, is a major challenge for the railway industry. For the defect-free production of rails, each rail is tested ultrasonically by various ultrasonic transducers connected in different orientations as per requirements. If any defect(s) are found the rail is rejected. At present, CSIR-NPL also provides metrological traceability of various parameters, e.g., acoustics, vibration, electrical, electronics, force, hardness, optical, etc. to the major rail manufacturing industries and suppliers of the Indian railway, as listed in Table 21.12. CSIR-NPL, in collaboration with CPCB is also working on the measurement of noise levels in the railway compartments for passenger safety.

21.5.11 Micro, Small and Medium Enterprises

The MSME sector has established itself as a highly vibrant and dynamic sector of the Indian economy over the last few decades. MSMEs not only play a crucial role in providing large employment opportunities at comparatively lower capital cost as compared to large industries but also help in industrialization of rural and backward areas, thereby, reducing regional imbalances, assuring more equitable distribution of national income and wealth [16]. MSMEs are complementary to large industries as

Table 21.13 Services provided by CSIR-NPL to MSMEs

Organization	Traceability/calibration/grant-in-aid project
MSME Kolkata MSME Mumbai	• Photometric and radiometric parameters
Institute for Design of Electrical Measuring Instruments, IDEMI, Mumbai	• Mass, volume • Fluid flow • Pressure, vacuum and ultrasonic • Temperature and humidity • Acoustic and vibrations • Multi-ferroics and magnetics • Magnetic parameters • Josephson Junction • LF and HF voltage, current, and microwaves
MSME Testing Centre New Delhi	• Pressure, vacuum, and ultrasonic • Mass, optical radiation • Length and dimension • Force and hardness
MSME (Ministry of Micro, Small and Medium Enterprises)	• Development of carbon fiber composite limbs for recurve bow

ancillary units and this sector contributes enormously to the socio-economic development of the country. India's MSME base is the largest in the world after China and provides a wide range of services and manufactures >6000 products—ranging from traditional to hi-tech items. To strengthen MSMEs for the production of high-quality products of international quality, CSIR-NPL supports them by providing the metrological traceability for a large number of parameters, as listed in Table 21.13. CSIR-NPL also collaborates with MSMEs for the development of various products, e.g., indigenous development of archery recurve bow limb based on carbon fiber in association with Technology Development Centre, Meerut, ultraviolet disinfection casket for coronavirus in association with M/s Motras, production of structural tiles using waste plastics, production of PM2.5 samplers, etc.

21.5.12 Health Through Confidence in Critical Equipment/Measurements

The human happiness and wellbeing depend upon their state of health. The health of the citizens in any country also makes an important contribution to economic progress, as healthy populations live longer, are more productive, and save more. This has been amply proven in the recent Covid-19 pandemic which has stalled the world's greatest economies. Hence, the Ministry of Health is undoubtedly an important factor, along with other government departments, organizations, civil society groups, and communities themselves.

The Central Drug Standardization and Control Organization (CDSCO) is the national regulatory body for Indian pharmaceuticals and medical devices. The various pharmaceutical companies engaged in the manufacturing of critical and life-saving drugs require an exact ratio of the constituents by weight. CSIR-NPL provides metrological traceability for mass to a number of pharma companies in India. In addition, CSIR-NPL supports the calibration of clinical thermometers, blood pressure measuring machines, Magnetic resonance imaging (MRI), etc. A detailed discussion on the contribution of CSIR-NPL in the health sector is described in Chap. 15.

21.5.13 Preservation of Original Constitution of India

In the largest democracy of the world, parliamentary affairs are extremely vital for ensuring the functioning of democratic processes and principles. These are documented in the iconic "Constitution of India" which is required to be safeguarded in its original document state. The responsibility of safeguarding the original calligraphic copies of the Constitution of India, kept at Parliament Library, is proudly, the privilege of CSIR-NPL. A discussion on this has been made in Chap. 1.

21.5.14 Sound and Noise Measurements: Housing and Urban Affairs

It is well known that noise pollution affects people's health and quality of life. Construction and roadwork activities, cars, trains, airplanes, and other forms of transportation are some of the worst contributors to noise pollution [17]. Prolonged high levels of noise can cause hearing loss and stress-related illnesses and even cardiovascular diseases. To mitigate its ill-effects, mapping, and segregating the sources of noise are a basic requirement. CSIR-NPL plays a strategic role in creating social awareness apart from giving crucial noise mapping services to Central public works department, Delhi Development Authority, Roads organizations such as Border Roads Organization, etc. This is done by way of calibration of electro-acoustic equipment such as microphones, accelerometers, sound calibrators, sound level meters, vibration meters, testing of sound absorptive and insulating materials, as well as industrial noise and vibration measurements and control. Sound/noise insulation is provided through acoustically insulating panel designs, as well as through technical advisory consultancy in architectural acoustics, noise absorptive barriers, etc.

21.5.15 Ministry of Commerce and Industry

The QCI under the Department of Promotion of Industry and Internal Trade in the Ministry of Commerce and Industry is an important constituent of the quality infrastructure with a primary role to establish and operate the national accreditation structure for conformity assessment bodies; providing accreditation in the field of education, health, and quality promotion. The NABL is an important wing of QCI, primarily for the accreditation system accepted across the globe by providing high quality, value-driven services, fostering APAC/ILAC MRA, empanelling competent assessors, creating awareness among the stakeholders, initiating new programs supporting accreditation activities, and pursuing organizational excellence [18]. CSIR-NPL's contribution to NABL is immense in more ways than one, and include expert assessors for almost all parameters who are routinely a part of accreditation activities for private and public laboratories across the length and breadth of the country. Innumerable calibration/testing is also regularly carried out for NABL accredited laboratories along with participation in the accreditation process, awareness programs, and formulation of policies and initiation of new programs. Ministry of Commerce has supported CSIR-NPL in the development of BNDs, as discussed in Chaps. 18 and 19. In association with National Council for Cement and Building Materials (NCCB), Ballabgarh, an Institute under Commerce Ministry, CSIR-NPL has not only produced and marketed a number of cement BNDs, but many of them have also been exported.

21.5.16 Ministry of Earth Sciences

CSIR-NPL is one of the important scientific partners of the Ministry of Earth Sciences and National Centre for Antarctic and Ocean Research Laboratory (NCAOR) and has been participating in the Indian scientific expeditions to Antarctica right from the first expedition in the year 1981 to date. CSIR-NPL also provided national leadership to four Indian scientific expeditions viz. 8th, 10th, 14th and 31st Indian scientific expeditions to Antarctica. The 8th and 31st Indian scientific expeditions to Antarctica were marked as important milestones for the Indian Antarctic mission when the two permanent Indian scientific bases, i.e., Maitri and Bharati were established by India and the country joined the elite club of countries having more than one scientific base in Antarctica. CSIR-NPL has also established a state-of-the-art Indian Polar Space Physics Laboratory (IPSPL) at Maitri for continuous and real-time monitoring of high latitude ionosphere. A detailed discussion on this is presented in Chaps. 13 and 14.

21.5.17 Polymer Engineering

Central Institute of Plastics, Engineering and Technology (CIPET) operates under the Ministry of Chemicals and Fertilizers. It is an apex Institute of International repute in the field of polymer science and technology to ensure sustainable growth in plastic engineering. It is engaged in the fields of design, tooling, plastics processing, testing, and quality assurance and inspection services to the plastics industries through a quality management system and does dedicated research on the development of new polymeric materials. The requirements of CIPET for metrological traceability in the parameters of mass and density, force and hardness, pressure, vacuum, and ultrasonic metrology, etc. are met from CSIR-NPL.

21.5.18 Contributions to Finance Sector

Among the regulatory bodies under the Ministry of Finance is the Reserve Bank of India (RBI), which regulates the issue of banknotes and keeping of reserves with a view to securing monetary stability in India and generally to operate the currency and credit system of the country to its advantage; to have a modern monetary policy framework to meet the challenge of an increasingly complex economy, to maintain price stability while keeping in mind the objective of growth. The "Security Printing and Minting Corporation of India Limited", SPMCIL, a PSU working directly under the ministry, is engaged in the manufacture/ production of currency, security paper, non-judicial stamp papers, postal stamps, and stationery, travel documents viz. passport, security certificates, cheques, bonds, warrants, special certificates with security features, security inks, Circulation, and Commemorative Coins, Medallions, Refining of Gold and Silver, and Assay of precious metals [19]. CSIR-NPL, in coordination with SPMCIL, has launched India's first 9999 Pure Gold BND. In addition, CSIR-NPL is also closely working with SPMCIL for indigenous development of color shift intaglio ink (CSII). Details of these are presented in Chap. 17.

21.5.19 New and Renewable Energy

The Ministry of New and Renewable Energy (MNRE) is the nodal Ministry of the Government of India for all matters relating to new and renewable energy including solar photovoltaic technologies and systems. MNRE is the regulatory body in policy framework and implementation of various programs and activities through central electricity authority, state nodal agencies (SNAs), academic institutions, national laboratories, public sector undertakings, and state and central government departments. To support the Ministry, there are five institutions consisting of three autonomous bodies, i.e., National Institute of Solar Energy (NISE), National

Table 21.14 Projects of CSIR-NPL with MNREs

Organization	Grant-in-aid project
Ministry of New and Renewable Energy (MNRE)	• National primary standard facility for cell calibration • To conduct regular workshops and hand-on-training on solar photovoltaic for bridging the gap between industry and academia

Institute of Wind Energy (NIWE) and National Institute of Bio-Energy (NIBE) and two public sector undertakings, i.e., Indian Renewable Energy Development Agency (IREDA) and Solar Energy Corporation of India (SECI). CSIR-NPL in association with MNRE worked on R&D projects on solar photovoltaics and systems. Currently, CSIR-NPL is establishing "National primary standard facility for solar cell calibration", which will be used for the dissemination of measurement traceability to PV industry and academic institutes in the country. Projects undertaken by CSIR-NPL with MNRE are listed in Table 21.14.

21.5.20 Jute Recycling

The Ministry of Textiles, Government of India has constituted the National Jute Board (NJB) in the exercise of the powers conferred by Sub-Sect. (1) of Sect. 3 of the National Jute Board Act, 2008 (No. 12 of 2009). CSIR-NPL in association with National Jute Board is working on development of activated carbon from waste jute stick biomass. The technological solution provided by CSIR-NPL for conversion of jute sticks biomass into activated carbon will be a game-changer in the whole chain of jute business involving industries, farmers and people of the area because they will be benefited from the additional income from waste jute sticks. It will also create an ecofriendly environment with an emphasis on the conversion of ecology into the pristine status and achieve an ever-elusive pollution-free environment.

21.5.21 Providing Confidence to the Flow of Jal Shakti

Jal or water is the basis for all life. For regulatory purposes, all water flow meters in use need to be calibrated. The fluid flow facility of CSIR-NPL is utilized by several state water boards to calibrate/test these water flow meters, including Delhi Jal Board, New Delhi Municipal Corporation, Uttar Pradesh Jal Nigam, Uttarakhand Jal Sansthan, and Tamilnadu Water Board, etc. Water meter testing is carried out up to DN50 size for threaded type domestic water meters. The bulk type water meters of sizes DN50-DN200 are tested in the primary water flow calibration facility for accuracy. The traceability to these boards is further disseminated to the consumer through various manufacturers and NABL accredited laboratories.

21.5.22 Skill Development Initiatives

The Skill Development Ministry is responsible for co-ordination of all skill development efforts across the country, removal of disconnect between demand and supply of skilled manpower, building the vocational and technical training framework, skill up-gradation, building of new skills, and innovative thinking not only for existing jobs but also jobs that are to be created. CSIR-NPL along with other CSIR labs has recently emerged as a skill development center in addition to its dominion over metrology for the Nation. Diploma courses, area-specific training courses, seminars as well as workshops are regularly being organized for building skilled man-power, up-gradation of skills for industrial job requirements, etc. The details of the skill initiatives at CSIR-NPL are described in Chap. 20.

21.6 Collaboration Between CSIR-NPL and Industries

As shown in Fig. 21.5, private industries are major beneficiaries of the calibration and testing facilities of CSIR-NPL. Many of the small industries take the metrological traceability from the NABL accredited laboratories. However, most of the multinational companies manufacturing their products in India, as well as many large industries obtain measurement traceability to SI units from foreign NMIs. One of the reasons for this could be the limited CMCs of CSIR-NPL at BIPM, as discussed in Chap. 1. Therefore, CSIR-NPL is working on enhancing its CMCs.

Apart from the measurement traceability, during the past few years, CSIR-NPL has started supporting private industries through the transfer of technologies, consultancies, and/or collaborations. A partial list of the notable collaborations established with various private industries is listed in Table 21.15. The technology transfers are mostly for the societal applications, e.g., PM2.5 sampler, recycling of the waste plastics, utilization of jute, UVC based disinfectant casket, etc. One of the major successes of CSIR-NPL has been to rope in the private industries for the development of BNDs in the areas of water, food, hardness, petroleum, etc. Such collaborations are very important for becoming self-reliant in the area of reference materials, which are undoubtedly critical for quality control and quality assurance. CSIR-NPL also has a collaboration with M/s IFR for time signal (IST) broadcast using long-wave transmitters.

Table 21.15 gives a glimpse of the technologies propagated by CSIR-NPL for societal benefits. Many in India are still not aware of the fact that during each general election in India, nearly 40 million people wear a CSIR-NPL mark on their fingers, known as the ‘indelible Ink’ [20]. This Indelible ink is a time-tested gift of CSIR-NPL

Table 21.15 A list of collaborations established with various industries by CSIR-NPL

Private industry	Technology
M/s Mysore paint and Varnish Limited (MPVL), Mysore, Karnataka	Development of indelible ink marker pen (IIMP)
M/s Max Super Specialty Hospital, Delhi, India	Infrared spectroscopic study for tumor diagnosis
M/s Kataline Infra Products Pvt Ltd,	Long afterglow phosphors
M/s Tata Steel Ltd.	Feasibility study on development of value-added carbon products from coal tar
M/s Shayna Ecounified India Pvt. Ltd., Delhi	Recycling of waste plastic bags into tiles for structure designing
M/s Joy Trading Company, New Delhi	Low-cost Peltier based refrigerator
M/s Cell Propulsion Pvt Ltd., Karnataka	Development of carbon foam as a heat sink for high energy Lithium ions batteries for electrical vehicles
M/s Environmental Solutions, Noida	High-volume $PM_{2.5}$ impactor sampler
M/s Addin Infra Private Limited, Rajkot	Recycling of waste plastic bags into tiles for structure designing
M/s ATOS Instruments Marketing Services, Bangalore	Ferroelectric loop tracer
M/s IFR Information Dissemination Pvt Ltd	Dissemination of IST by long-wave transmitters
M/s Rajasthan Electronics and Instruments Limited (REIL), Jaipur	Solid-state Peltier refrigerator
M/s Aashvi Technology, Ahmedabad	Production of water Bhartiya Nirdeshak Dravyas
M/s Sreeni Labs Pvt. Ltd, Hyderabad	Production of Poly (3,4-ethylenedioxythiophene)-poly(styrenesulfonate) (PEDOT:PSS)
M/s Vyzag Bio Energy Fuel Private Limited, Madhavadhara, Visakhapatnam	Recycling of plastic waste into useful tiles
M/s Reliable Diesel Engineers (P) Ltd, Faridabad	Noise absorptive barrier for metro/railway/ highway/airport/ noise abetment
M/s Global PT Provider (P) Limited, Okhla Industrial Area, New Delhi	Production of hardness Bhartiya Nirdeshak Dravyas
M/s Physics Instruments Co., Chennai, Tamil Nadu	• Improved variable frequency ultrasonic interferometer for velocity and attenuation measurement in liquid • 3 T auto range pulsed/static field gaussmeter
M/s Jalan and company	Production of silver Bhartiya Nirdeshak Dravyas
M/s FARE Labs Private Limited, Gurugram, Haryana	Production of food Bhartiya Nirdeshak Dravyas
M/s NAS Industries, Thane	Recycling of plastic waste into useful tiles
M/s Motras, Manesar	Ultraviolet C based disinfectant casket

to the spirit of democracy in India. Today the know-how for the same has been given to M/s Mysore Paints, which is the sole organization authorized to manufacture it. It is also exported to Sri Lanka, Indonesia, Turkey, and other democracies. Likewise, the technology for converting waste plastic into useful tiles for construction purposes, low-cost Peltier based refrigerator which works on solar power and has been designed for rural areas, indigenous BNDs which save precious foreign exchange, disinfectant casket to mitigate the vicious spread of Covid-19, PM2.5 samplers for monitoring pollution levels, noise mitigation barriers to decrease noise pollution, etc. are only a few of the highlights of CSIR-NPL's recent contributions to the civil society for benefit of the masses of this country [21, 22]. A complete list of the technologies developed at CSIR-NPL for transfer to start-ups/industries is available on the CSIR-NPL website [23].

21.7 CSIR-NPL's Connect with Society

CSIR-NPL, being the NMI of the country, has the responsibility to disseminate the knowledge created at the international level in the domain of metrology and its applications to common citizens of the country. In addition, the initiatives of the GoI also need to be communicated to the masses. The Ministry of Science and Technology (MoST) is resolute to cultivate scientific temper amongst the young minds, who in turn, would be the country's most trustworthy assets for the implementation of the various GoI missions. Scientific temper is a way of thinking and action that uses scientific methods and involves the application of logic for problem-solving. Scientific temper includes questioning, hypothesizing, observing physical reality, testing, analyzing, and communicating, and fairness, equality, and democracy are an integral part of it. CSIR-NPL executes this mandate of MoST, as described in the following sections, as it also allows CSIR-NPL to have direct interactions with masses to disseminate the basics and benefits of metrology to them.

21.7.1 Communicating the World Metrology to Masses

Each year BIPM celebrates World Metrology Day on May 20 to inform the world about the contributions the measurements make to our daily life [24]. Each year, BIPM releases a poster on this day with a particular theme, as summarized in Table 21.16 for the last five years. CSIR-NPL also commemorates the World Metrology Day by arranging a day-long seminar on the suggested theme by BIPM each year and releasing the theme poster. The ideas behind the themes are communicated to common people through FM radio, television interviews, newspapers, and social media. In addition, the need to implement the changes suggested by BIPM, e.g., the revisions of SI units, are communicated to the concerned government departments, as described in Chaps. 2 and 3.

Table 21.16 Themes of last five World Metrology Days

Year	Theme of world metrology day	Brief description
2020	Measurements for global trade	To create awareness amongst masses that metrology ensures that products fulfill the quality expectations and meet standards and regulations. Metrology plays a key role in facilitating fair global trade. The message that metrology is behind the scientific discovery and innovation, manufacturing, trade, which in turn improves the quality of life and protect the environment
2019	The International System of Units—Fundamentally better	To create awareness that on 16 November 2018, the CGPM has agreed to the implementation of new revised SI units based on the fundamental constants of and will come into force worldwide on 20 May 2019
2018	Constant evolution of the International System of Units	To create awareness that the CIPM has in principle agreed upon the new definitions of SI units, in which each SI unit is linked to quantum laws of physics, and will be presented in the next CGPM meeting for approval. The new SI units have the advantage for further improvements to meet the future needs of science and technology
2017	Measurements for transport	To create awareness that transport plays a key role in our daily life and the measurements are responsible for advancement in transport. A range of measurements also control the quality of food, clothes, goods, etc. as well as the quality of raw materials they are made from, and that too without any environmental impact
2016	Measurements in a dynamic world	To create awareness that measurement science is changing with a rapid pace and it has an impact on the world around us

21.7.2 Indian International Science Festival

To promote the scientific temper amongst young minds and the masses, the MoST and Ministry of Earth Science (MoES) in association with Vijnana Bharti (VIBHA) conceptualized the India International Science Festival (IISF). IISF is a uniquely intellectualized festival that showcases science and technology with the same freshness and flair that takes place in arts or music festivals. VIBHA is a non-profit organization and started from a movement on "Swadeshi Science", which in a larger picture can be epitomized as Atmanirbhar Bharat. Swadeshi science represents development of indigenous science and technology to cater to the national needs with the appropriate linkage of traditional, modern, and spiritual sciences. VIBHA is engaged in the popularization of science to the masses and a collective application of ancient science and modern technology. The first IISF-2015 was organized at the Indian Institute of Technology, Delhi during 4-8 December 2015, which was a mega success. CSIR-NPL organized the second India International Science Festival (IISF-2016) during December 7-11, 2016 under the theme of "Science for the Masses" [25]. The Hon'ble Home Minister of India, Shri Rajnath Singh inaugurated IISF-2016 in the benign presence of Hon'ble Union Minister for Science and Technology and Earth

Fig. 21.9 The inauguration of IISF-2016 by Hon'ble Home Minister of India, Shri Rajnath Singh

Sciences Dr. Harsh Vardhan and other dignitaries namely Dr. K. Vijay Raghavan, Dr. Girish Sahani, Dr. V. P. Bhatkar, Dr. M. N. Rajeevan, and Dr. D. K. Aswal, as shown in Fig. 21.9. In IISF-2016, India's quest for self-reliance in science and technology was demonstrated through mega exposition of outstanding achievements in metrology, space, defense, atomic energy, indigenous products by grassroots innovators, new scientific findings of young researchers, industry-academia interactions, etc. IISF-2016 had ~10,000 participants along with additional thousands of visitors (common public) every day. The participants included young school students from rural India, college students, young researchers, innovators, farmers, politicians, renowned scientists, ministers, diplomats from abroad as well as the common public. Shri Rajnath Singh emphasized that he would rather not like India to emerge as a Super Power, but aim to be a "guru" or teacher for the world. He proposed an addition to the popular slogan of 'Jai Jawan, Jai Kisan, Jai Vigyan', the term 'Jan Vigyan' that is science for the masses.

IISF-2016 was well organized and had numerous dedicated programs, which are briefly outlined below.

21.7.2.1 Pre-event Activities

A National level competition was one of the themes used as the curtain raiser event to propagate the activities of IISF 2016 to the masses. In this program regional problems were taken up, and for which the scientific solutions were invited from the public and researchers. For this, several CSIR, DST, and DBT labs across the country had an open day for the public. In addition, the burning issues of society in different parts of the country were shared through photos/videos and write-ups in the web portal of IISF-2016.

21.7.2.2 Events for Students

(a) **Science Village**: The Science Village was a unique experimental concept of CSIR-NPL which was introduced for the first time in the country with an intention to reach the unreached and was well coupled with the 'Sansad Adarsh Gram Yojna' of GoI (see Fig. 21.10a). The Science Village program was sculpted in such a way that 1769 students were selected and nominated by the Honorable Members of Parliament (Lok Sabha and Rajya Sabha) from their respective adopted villages in their constituencies, under the aegis of Pradhan Mantri Sansad Aadarsh Gram Yojana and therefore the Science Village demonstrated a mass representation from the rural part of India. The concept of Science Village emanated from the fact that in India there exists a large gap in terms of scientific awareness between the urban and rural population, in spite of India hosting a majority of its citizens in its rural countryside. The untapped potential of students belonging to the villages and small towns needed to be utilized and brought at par with that of the students from the metropolitan cities. This was needed at a time when India is reclaiming its past scientific glory. But the task to execute it required microscopic planning and was undeniably mammoth in nature. The Science Village was distributed under 8 different houses named after renowned scientists produced by our country, namely, K. S. Krishnan, A. P. J. Abdul Kalam, Aryabhatta, J. C. Bose, S. Chandrasekhar, Srinivasa Ramanujan, H. J. Bhabha, and C. V. Raman. In a 5 day long program these students were exposed to hands-on science experiments in physics and chemistry, visit mega science expo, visit to INSPIRE projects, interaction with scientists, know your capital (Delhi Visit), visit CSIR-NPL laboratories, and introduced to SI units, international science film festival, night sky watching laser shows, and popular talks from speakers like Dr. Suresh Prabhu, Hon'ble Union Minister for Railways, Shri. Anil Madhav Dave, Hon'ble Union Minister of State for Environment, Forest and Climate Change and Dr. G. Satheesh Reddy, Scientific Advisor to

Fig. 21.10 **a** Hon'ble Minister of S&T &ES Dr Harsh Vardhan inaugurating Science Village, **b** a glimpse of DST-INSPIRE national camp, and **c** 550 school students dressed up as Albert Einstein during IISF 2016

the Defense Minister. The exposure received by the students motivated many of the participates to take science as a career option.

(b) **DST-INSPIRE National Camp**: The innovative science models and presentations created by students were exhibited in this camp, as a part of 'Innovation in Science Pursuit for Inspired Research (INSPIRE)', a program sponsored and managed by the DST for attracting young talent from schools and encouraging the pursuit of science. Around 560 selected awardees from all over the country displayed their projects in this event (see Fig. 21.10b). These participants were selected to display their achievements/innovations after two levels of screening. The first screening was done at the District level, where 10,000 students were shortlisted out of 140,000 contestants. The second level screening was done at the State level, where these 560 projects were selected for IISF-2016 and were evaluated by a high-level Jury for National Award. This program encourages young innovators to take up science as a career that reinforces the human resource for the national R&D base.

(c) **Dressing students as lookalikes of Einstein**: With an objective to develop an awareness among students about the phenomenal contributions of Albert Einstein to modern-day science, 550 school students dressed to look like the great Einstein (see Fig. 21.10c). This gathering of lookalikes of Albert Einstein was aimed to break the earlier world record of Berkeley's Black Pine Circle School (USA).

21.7.2.3 Young Scientists' Conclave

The Young Scientists' Conclave (YSC) comprised of scientists/researchers/science innovators/professionals below the age of 45 years, which later emerged as one of the most fascinating events at IISF2016 due to the innovative deliberations under various national flagship programs of GoI, such as Swachh Bharat, Swastha Bharat, Namami Gange, Smart City, Digital India, Make in India, Green Energy, Sustainable Development and cleaner production, Materials for sustainable development, and Indigenous Science and Technology. 21 plenary lectures on themes of national importance, Agriculture, Health, Water, Information technology, Space and defense, and Energy, were delivered to ignite the young minds by eminent scientific leaders. To echo the call of Hon'ble Prime Minister to the youth to become the job giver rather than a job seeker, workshops on patent filing, copyright, proposal drafting, and ethics were organized. The youngsters, who have established successful start-ups, were given platforms to present their journey and exhibit their ideas in the pavilion. YSC achieved the objective of scouting, augmenting, and propagating the national needs of the young researchers and stimulated them to play a constructive role in transforming the nation into Sashakt Bharat.

21.7.2.4 Industry-Academia Interactions

Industry-academia interactions were one of the major events during IISF-2016 and had ~950 participants (45% from industries, 40% from academia, and remaining 15% from media, laboratories, and other organizations) from all over India. The main emphasis was placed on theme-based technical discussions, including technologies available with national laboratories (e.g. CSIR, DRDO, DST, DBT, MoES, ISRO, ICAR, ICMR) for commercialization; industry-oriented curriculum development, entrepreneurs for executing laboratory ideas to implementable technologies in small industries; metrology for time and frequency, electrical, and electronics metrology; mechanical and optical metrology; environment pollution monitoring Initiatives; and agriculture and industries.

21.7.2.5 International Science Film Festival

In the extravagant journey of bridging the gap between scientific temperament and communication, extraordinary science films from various nations were screened during IISF-2016, as listed in Table 21.17. These films demonstrated the concept of science for sustainability, advancement, prevention, growth, and prosperity. Besides the screening of these science films, it also focused on interactive sessions with eminent scientists, filmmakers, and communicators of India. The international science film festival was inaugurated by Mr. Ofi Akunis, Hon. Minister of Science, Technology and Space, Israel (see Fig. 21.11). The science film festival set a benchmark of green thoughts and ideas and inculcated a practical approach in the minds of people, both with scientific backgrounds as well as non-scientific backgrounds, making the idea of "Science for masses" a reality with necessity. The film festival kept the viewers engrossed, physically, emotionally, and practically, and was successful in decoding the essence of science for masses and science for the future. The program had aimed at understanding the value and need of simplification of complexity in different domains of science for its acceptance in layman's life.

National NGOs' Conclave

IISF-2016 also organized the National non-governmental organizations (NGOs) conclave. 71 NGOs from all over the country participated and discussed various key points related to specialized field works done by them in the areas of environment, agriculture, women empowerment, child development, and education, renewable energy, health, rural development, and science and technology. NGOs vowed to make India a prosperous and healthy nation by spreading the fruits of science into the villages (Fig. 21.12).

Table 21.17 The list of International films screened during IISF-2016

Country	Science films
Canada	• The connected universe
France	• Soil for food security and climate • A future without water • Perfect project towards the autonomous car • CNES-ISRO space research collaboration
Germany	• Quantum physics: tap-proof by randomness • Light gets on your nerves • Clouds: more shades in the greenhouse
Israel	• Israel space agency • Israel inside: how a small nation makes a difference
Norway	• Bio-economy: an ocean of opportunities • Arctic mission • Drifting with the ice-life on the arctic expedition • iProcess • Harvesting and cultivating seaweed • Thin ice studies in the arctic-N-ICE2015
Russia	• Secret of nature: wild Russia-Siberia • Look up-roscosmos
Sweden	• Innovative Swedish eye-tracking technology company looks to the future • Sweden leading the world of innovation • Swedish design can help millions
UK	• Changing climates: the impact inspiring change • The farms of the future • Changing lives in the Indian Himalayas
USA	• Weapons of mass destruction • Gravity • Racing extinction

21.7.2.6 Mega Science, Technology and Industrial Expo

IISF 2016 Mega Science, Technology, and Industrial Expo were the main attractions for the visitors, which was aimed at showcasing the Indian contribution in the world of science, both ancient and modern, and to emphasize the inter-dependence of science-technology-innovation and society. In the mega-expo, the "working models" of outstanding technologies developed by various R&D Labs of ISRO, DRDO, CSIR, DST, DBT, MOES, ICAR, ICMR, Ministry of Urban Development, Ministry of IB, Ministry of Shipping, DAE, IITs for Unnat Bharat Abhiyan (Rural Development initiatives), and many other ministries/organizations like ONGC, IOCL, GAIL, etc., were displayed. In a special Vibrant India pavilion, 25 State Councils for Science and Technology, showcased the scientific programs of their respective states. AICTE had put up a special Pavilion on the technical education for the students. The expo was arranged theme-wise, i.e., agriculture; water; environment and energy; space and defense; health, medical, and biotechnology; unnat Bharath Abhiyaan, etc. In

Fig. 21.11 The international science film festival was inaugurated by Mr. Ofi Akunis. The (L-R): Dr. D. K. Aswal, Director, NPL, Mr. Peretz Vazan, Director General, Ministry of Science and Technology, Israel, Mr. Ofi Akunis, Hon. Minister of Science, Technology and Space, Israel, Dr. Harsh Vardhan, Hon'ble Minister of Science and Technology and Earth Sciences, GoI, H. E. Mr. Daniel Carmon, Ambassador of Israel to India, Prof. Ashutosh Sharma, Secretary, Department of Science and Technology, Govt. of India, Shri A. Jayakumar General Secretary, VIBHA

Fig. 21.12 Glimpses of Mega Science, Technology and Industrial Expo at IISF-2016, which was visited by more than 300,000 people

addition, a start-up pavilion with sessions for the youth to share their journey and products was also arranged.

To sum-up, through IISF 2016, CSIR-NPL along with VIBHA could demonstrate to the masses that India is on a path to self-reliance. On the other hand, IISF-2016 also brought policy-makers, ministers, parliamentarians, scientists, young innovators, students, and common citizens under one roof to realize the task of nation-building through science and technology.

21.7.3 Interaction with Society

CSIR-NPL also interacts with society through various programs. Every year, CSIR-NPL invites several thousand school/college students and teachers to visit its facilities on an open day, which is usually celebrated one day before or after the foundation day of CSIR, i.e., 26 September. Visitors are explained about the mandate of CSIR-NPL, SI, and derived units and their applications in all walks of life. Various advanced experiments are explained to them. In addition, students can visit CSIR-NPL anytime during the year, under the program known as 'Jigyasa'. Also, start-ups, entrepreneurs, or any common person can approach CSIR-NPL anytime through ILG, PME, or HRD for discussion on any scientific subject, joint development of devices or to solve any social problem. Many of the collaborations listed in Table 21.13 are the outcome of such discussions.

21.8 Concluding Remarks

There is no doubt that India's March towards a developed nation with a high economy and quality of life for each of its citizens can be achieved only when the manufacturing is promoted in the country. Recent government policies on the Atmanirbhar Bharat program to support MSMEs are to enhance the manufacturing of the country. In this chapter, we have elaborated that in order to make Atmanirbhar Bharat program a success, the coordination among the government, industries, academia, and civil society needs to be extraordinarily strong. For this, as suggested by Aswal model of inclusive growth, the metrology program of the country should be strengthened equal to that of developed economies. CSIR-NPL, being the NMI of the country, has been discharging its duties diligently over the past 75 years. CSIR-NPL has established 236 calibration and measurement capabilities (CMCs) at the key-comparison database of BIPM and has been participating in the Asia Pacific Metrology Program as its founder member. CSIR-NPL has been supporting Legal Metrology, NABL, and BIS in strengthening the quality-infrastructure of India. CSIR-NPL has generated new capabilities in the areas of IST with an uncertainty of <3 ns, development of indigenous certified reference materials under the trademark of BND®, new initiation on environmental monitoring, optical metrology, biomedical metrology, etc.

CSIR-NPL has been supporting the government, industries, academia, and society. However, there is enough scope for improvement, and some of the suggestions are as follows.

(i) The number of CMCs at KCDB, BIPM should be enhanced by initiating new metrological programs. In addition, all seven SI units based on fundamental constants of nature should be realized at CSIR-NPL. These activities would require high funding, a dedicated team of scientists, and collaborations with other NMIs on advanced fields of metrology.

(ii) Industry 4.0 is still evolving and the success will depend upon how soon its metrology gets developed. CSIR-NPL along with other national labs should initiate a program on the development of metrology for Industry 4.0. As discussed in Chap. 1, the development of digital SI units is an essential requirement for Industry 4.0. In addition, the operations in Industry 4.0 are expected to be accurate, fast, safe, and flexible, which in turn, would require touchless calibration. Consequently, an unprecedented requirement for digital metrology is evolving for the digital transformation of metrological services, and remote calibrations wherever possible to keep pace with the time-efficient error identification to minimize production delays [26, 27]. If the calibration/reference laboratories and the client laboratories can be connected by electronic means for data transport, instead of instrument transport, then the overall calibration process is possible on-line and in real-time. This would directly manifest into a saving of both time and money. This can be achieved through establishing a procedure to produce, transfer and control calibration data which will compare the values of the reference standards to the values of the artefacts without endangering accuracy and traceability necessary for calibration. This concept is referred to as touchless calibration. However, the new conformity assessment infrastructure would be feasible only with smart transmitter/sensors, high-speed data transfer, cybersecurity and reliability, and integrity of data processing, calculations and operations, and even robotics. Remote calibration technology currently appears feasible with the use of smart sensors, which can either be used remotely or transported after calibration to the test site and used for remote calibration. A remote calibration technology for disseminating pressure standard and calibration has been developed [28] which is one of the ways CSIR-NPL can disseminate SI units to Industry 4.0 in the foreseeable future. In the remote calibration, a transfer standard can be transported from a calibration laboratory to a client's site where a calibration item is used. Using the transfer standard and the IoT, the calibration item can be calibrated on-site. For such remote calibrations, integrated transfer standards can also be newly developed in various other parameters for SI units, both basic and derived. In addition, since digital metrology is expected to be heavily reliant on data transfers, cloud computing, cybersecurity, simulations, and even artificial intelligence, etc., the metrological aspect of data transfer and IoT, e.g., validated statistical procedures, infrastructure for digital calibration certificates, metrology in the analysis of large quantities of data, synchronized communication systems, and simulations are a few of the futuristic needs.

(iii) Currently, the dissemination of metrological traceability from CSIR-NPL to stakeholders also takes place through Legal Metrology and NABL. This needs to be further strengthened. All the calibration and testing laboratories falling under the ambit of central and state government/ministries should ensure that they obtain the metrological traceability to SI units from CSIR-NPL either directly or through Legal Metrology and NABL. This is a massive job, but once implemented, will reap huge benefits to the country by providing the much-needed uniform and consistent test results for quality control and quality assurance.

(iv) India to a major extent lacks in appropriate national agencies that can provide Type approval or certificate of conformity to indigenously developed products and to ensure the compliance of imported equipment for the local environmental conditions. Type-approval implies that a product has met a minimum requirement of technical, regulatory, and safety standards. Although BIS provides ISI marks for daily consumer products and Legal metrology issues, for the Type approval of certain medical devices, there are no agencies especially for high-end sophisticated analytical equipment, biomedical equipment, etc. Recently, the Ministry of Environment and Climate Change has notified CSIR-NPL as an institute for the Type-approval of environmental monitoring equipment. CSIR-NPL in coordination with ministries can ensure that the country creates the Type-approval for all the products made in India.

Acknowledgements The authors are grateful to all the employees of CSIR-NPL for working together as a team to realize various national obligations elaborated in the Chapter.

References

1. Make in India program, Department for promotion of Industry and Internal trade. https://dipp.gov.in/programmes-and-schemes/industrial-promotion/investment-promotion. Accessed 24 July 2020
2. Finance Minister announced Government Reforms and Enablers across Seven Sectors under Aatma Nirbhar Bharat Abhiyaan (2020). https://pib.gov.in/PressReleasePage.aspx?PRID=1624661. Accessed 24 July 2020
3. The Data Book state of India-2020 Edition (July 2020) p. 166
4. The Little Green Data Book 2017, World Bank group, p. 104
5. Rob Francis, The Australian corrosion association (2019). https://membership.corrosion.com.au/blog/rust-free-iron-from-ancient-india/. Accessed 24 July 2020
6. D.K. Aswal, Quality-infrastructure of India and its importance for inclusive national growth. MAPAN-J. Metrol. Soc. India **35**, 139–150 (2020)
7. Explained: Why Basmati export has come to a standstill in Punjab (2019). https://indianexpress.com/article/explained/explained-why-basmati-export-has-come-to-a-standstill-in-punjab-6164891/. Accessed 24 July 2020

8. India wants EU to warn seafood exporters before blacklisting (2018). https://economictimes.indiatimes.com/news/economy/foreign-trade/india-wants-eu-to-warn-seafood-exporters-before-blacklisting/articleshow/62706818.cms. Accessed 24 July 2020
9. Public sector undertakings in India (2020). https://en.wikipedia.org/wiki/Public_sector_undertakings_in_India. Accessed 24 July 2020
10. List of Maharatna and Navratna Companies in India (2020). https://www.jagranjosh.com/general-knowledge/list-of-maharatna-and-navratna-companies-in-india-1467721897-1. Accessed 24 July 2020
11. Technology Cluster DRDO (2020). https://www.drdo.gov.in/about-drdo. Accessed 24 July 2020
12. Aviation Weather Hazard Safety, US National whether Safety (2020). https://www.weather.gov/zme/safety. Accessed 24 July 2020
13. List of BND (2020). http://www.CSIR-NPLndia.in/sites/default/files/Available-Bhartiya-Nirdeshak-Dravya-list-13-02-20.pdf. Accessed 24 July 2020
14. List of BND (2020). http://www.CSIR-NPLndia.in/sites/default/files/BND-ongoing-projects-list-13-02-20.pdf. Accessed 24 July 2020
15. Instrumentation in the dairy industry, Rui Oliveira, Petia Georgieva, Sebastião Feyo de Azevedo, Instrumentation and Process Control—Instrumentation, in Encyclopedia of dairy sciences ed by H. Roginski, J.W. Fuquay, P.F. Fox (Academic Press UK, 2002), pp. 1392–1401
16. Ministry of Micro, Small and Medium Enterprises (MSME) (2020). https://msme.gov.in/about-us/about-us-ministry. Accessed 24 July 2020
17. How does noise pollution affect people (2017). https://sciencing.com/noise-pollution-affect-people-4563995.html. Accessed 24 July 2020
18. National Board of Accreditation. https://nabl-india.org/about-nabl/vision-and-mission/. Accessed 24 July 2020
19. Profile: Security printing and minting corporation of India Limited (2020). https://www.spmcil.com/Interface/AboutUsspmcil.aspx?menue=1. Accessed 24 July 2020
20. National Physical Laboratory of India (2020). https://www.wikiwand.com/en/National_Physical_Laboratory_of_India. Accessed 24 July 2020
21. Low-cost and maintenance-free Thermoelectric Cooler/Refrigerator (2019). http://www.nplindia.in/iglpdf/thermo-electric-cooler-1.pdf. Accessed 24 July 2020
22. Recycling of Plastic Wastes into Tiles for Structure Designing for Societal Usage (2020). http://www.nplindia.in/iglpdf/Recycling-of-Plastic-Waste.pdf. Accessed 24 July 2020
23. Entrepreneur's Corner: Technologies/Know-hows ready for Transfer (2020). http://www.nplindia.in/industrial-liaison-group. Accessed 24 July 2020
24. World Metrology day (2020). http://www.worldmetrologyday.org/. Accessed 24 July 2020
25. Shilpa Pandey, Md Firoze Quamar, India International Science Festival 2016. Curr. Sci. **113**, 842 (2017)
26. Metrology Is Making Industry 4.0 A Reality. https://metrology.news/metrology-is-making-industry-4-0-a-reality/. Accessed August 05, 2020
27. Sasho Andonov and Marija Cundeva-Blajer, Calibration for Industry 4.0 Metrology: Touchless calibration, J. Phys.: Conf. Ser. 1065, 072019 (2018)
28. Development of remote calibration system for pressure standard, Tokihiko Kobata Momoko Kojima Hiroaki Kajikawa. Measurement **45**(10), 2482–2485 (2012)

The manufacturer's authorised representative in the EU is Springer Nature Customer Service Centre GmbH, Europaplatz 3, 69115 Heidelberg, Germany. If you have any concerns regarding our products, please contact ProductSafety@springernature.com

Printed and bound by CPI Group (UK) Ltd, Croydon, CR0 4YY

30/09/2026

02225257-0001